T0136045

SPACE SCIENCE SERIES

Tom Gehrels, General Editor

Planets, Stars and Nebulae, Studied with Photopolarimetry
Tom Gehrels, editor, 1974, 1133 pages

Jupiter
Tom Gehrels, editor, 1976, 1254 pages

Planetary Satellites
Joseph A. Burns, editor, 1977, 598 pages

Protostars and Planets
Tom Gehrels, editor, 1978, 756 pages

Asteroids
Tom Gehrels, editor, 1979, 1181 pages

Comets
Laurel L. Wilkening, editor, 1982, 766 pages

Satellites of Jupiter
David Morrison, editor, 1982, 972 pages

Venus
D.M. Hunten, L. Colin, T.M. Donahue and V.I. Moroz, editors,
1983, 1143 pages

Saturn
Tom Gehrels and Mildred S. Matthews, editors, 1984, 968 pages

Planetary Rings
Richard Greenberg and André Brahic, editors, 1984, 784 pages

Protostars and Planets II
David C. Black and Mildred S. Matthews, editors, 1985, 1293 pages

Satellites
Joseph A. Burns and Mildred S. Matthews, editors, 1986, 1021 pages

The Galaxy and the Solar System
Roman Smoluchowski, John N. Bahcall and Mildred S. Matthews, editors,
1986, 485 pages

Meteorites and the Early Solar System
John F. Kerridge and Mildred S. Matthews, editors, 1988, 1269 pages

Mercury
Faith Vilas, Clark R. Chapman and Mildred S. Matthews, editors,
1988, 794 pages

Origin and Evolution of Planetary and Satellite Atmospheres
S.K. Atreya, J.B. Pollack and M.S. Matthews, editors, 1989, 881 pages

Asteroids II
Richard P. Binzel, Tom Gehrels and Mildred S. Matthews, editors,
1989, 1258 pages

Uranus
Jay T. Bergstralh, Ellis D. Miner and Mildred S. Matthews, editors,
1991, 1076 pages

The Sun in Time
C.P. Sonett, M.S. Giampapa and M.S. Matthews, editors, 1991, 996 pages

Solar Interior and Atmosphere
A.N. Cox, W.C. Livingston and M.S. Matthews, editors, 1991, in press

Mars
H.H. Kieffer, B.M. Jakosky, C.W. Snyder and M.S. Matthews, editors,
1992, in press

THE SUN IN TIME

THE SUN IN TIME

C.P. Sonett

M.S. Giampapa

M.S. Matthews

Editors

With 83 collaborating authors

THE UNIVERSITY OF ARIZONA
TUCSON

About the cover:
The Sun's hot outer atmosphere, or corona, color coded to distinguish levels of brightness, reaches outward for millions of miles. A coronagraph, one of Skylab's eight telescopes, masked the Sun's disk, creating artificial eclipses. It permitted 81 months of corona observation compared to less than 80 hours from all natural eclipses since use of photography began in 1839.

The University of Arizona Press

Copyright © 1991
The Arizona Board of Regents
All Rights Reserved

This book was set in 10/12 Times Roman.
⊚ This book is printed on acid-free, archival-quality paper.
Manufactured in the United States of America

95 94 93 92 91   6 5 4 3 2 1

Library of Congress Cataloging-in-Publication Data

The sun in time / Charles P. Sonett, Mark S. Giampapa, Mildred S.
   Matthews, editors : with 89 collaborating authors.
      p.   cm. — (Space science series)
   Includes bibliographical references and index.
      1. Sun.   2. Earth.   3. Geophysics.   4. Astrophysics.   I. Sonett,
   Charles Philip.   II. Giampapa, M. S.   III. Matthews, Mildred
   Shapley.   IV. Series.
   QB521.S863   1991
   523.7—dc20   91-29614
                                                                CIP
ISBN 0-8165-1297-3

British Library Cataloguing-in-Publication data are available

CONTENTS

COLLABORATING AUTHORS

Anderson, R. Y., *543*
Baliunas, S., *809*
Barry, D. C., *633*
Basri, G., *682*
Beardsley, B., *59*
Becker, R. H., *389*
Beer, J., *343*
Berger, A., *498*
Bertout, C., *682*
Bochsler, P., *98*
Bonino, G., *562*
Brown, T. M., *59*
Cabrit, S., *682*
Caffee, M. W., *413*
Castagnoli, G. C., *562*
Cisowski, S. M., *761*
Coyne, G. V., *3*
Damon, P., *360*
Dearborn, D. S. P., *159*
Delache, Ph., *59*
de la Zerda Lerner, A., *317*
Feigelson, E. D., *658*
Foukal, P. V., *11*
François, L. M., *463*
Fröhlich, C., *11*
Gaffey, M. J., *710*
Geiss, J., *98*
Gérard, J.-C., *463*
Ghil, M., *511*
Giampapa, M. S., *658*
Goswami, J. N., *413, 426*
Grinspoon, D. H., *447*
Herbert, F., *710*
Hickey, J. R., *11*

Hohenberg, C. M., *413*
Hood, L. L., *761*
Hudson, H. S., *11*
Hunten, D. M., *463*
Jokipii, J. R., *205*
Kasting, J. F., *447*
Kerridge, J. F., *389*
Kocharov, G. E., *288*
Kuhn, J. R., *59*
Laclare, F., *59*
Lal, D., *221*
Leister, N. V., *59*
Levy, E. H., *589*
Lingenfelter, R. E., *221*
Marti, K., *260*
Morfill, G. E., *30*
Nichols, R. H., Jr., *413*
O'Brien, K., *317*
Olinger, C. T., *413*
Pedroni, A., *413*
Pellas, P., *740*
Pepin, R. O., *389*
Provenzale, A., *562*
Radick, R., *787*
Raisbeck, G. M., *343*
Ramaty, R., *232*
Reedy, R. C., *260*
Ribes, E., *59*
Ruzmaikin, A. A., *589*
Ruzmaikina, T. V., *589*
Saar, S. H., *848*
Scheingraber, H., *30*
Shea, M. A., *317*
Signer, P., *389, 413*

Simnett, G. M., *232*
Smart, D. F., *317*
Smith, E. J., *175*
Soderblom, D. M., *832*
Sonett, C. P., *30, 360, 710*
Spruit, H. C., *118*
Stauffer, J. F., *832*
Swindle, T. D., *413*

Vrba, F. J., *658*
Walter, F. M., *633*
Wieler, R., *389, 413*
Willson, R. C., *11*
Wolfsberg, K., *288*
Wood, J. A., *740*
Yiou, F., *343*

PREFACE

Our goal in this unique volume in the Space Science Series is to achieve an improved understanding of our nearest star, the Sun. In doing so, we have adopted a highly interdisciplinary approach that utilizes the data and analytical tools from a broad range of fields of active research. The result is encompassed in 36 chapters that, in total, trace the past evolution of the Sun and provide a review of our current understanding of both its structure and its role in the origin and evolution of the solar system.

After its luminous output and gravitational field, the magnetic field-related activity of the Sun exerts the next most significant influence on the environment of the solar system. Moreover, magnetic fields determine the angular momentum evolution of the Sun (and other late-type stars) and modulate its radiative and particle output. Sunspots are among the most vivid manifestations of the hydromagnetic solar substrate. The observational record of the sunspot index and the cycles that the record reveals clearly provide clues to the nature of the solar dynamo while simultaneously baffling scientists over 400 years after the first telescopic observations of sunspots.

Much material—anecdotal, historical, and physical—is already available; we are fortunate to provide in this volume a new slant on the role of Galileo and his contemporaries on this subject. Though the continuous sunspot record is only about 280 years old, it is a daily record and consequently some 100,000 data are available. This very long record makes it possible to try the Grassberger-Procaccia dimensional algorithm on a long natural record.

This book covers the final establishment of the absolute irradiance variability of the Sun. The question of variability in solar diameter as well as the theory is also reviewed.

The road to understanding the evolution of the Sun is usually inferred from the HR diagram which, in turn, is an expression of the theory of stellar structure corroborated by observation of star types. A major break in this completely observational approach to astronomy began with the study of the geochemistry of meteorites, spearheaded significantly by H. Urey in the 1950s. With some exceptions, meteoritics is a study of the early solar system of which the Sun is only a part. Curiously our consequent understanding of the early Sun is followed by an intellectually bare menu for most of the next 4.55 Gyr up until 1 Myr before the present when the Earth record becomes decipherable. By some standards, the more recent record is plagued by a mix of climatic and putative extraterrestrial forcings; on the other hand, it constitutes a record of unparalleled diversity and hidden complexities.

The production of isotopic species on and about the Earth, the Moon, and meteorites is an important source of information on fluctuations in the flux of cosmic rays. In effect this becomes a monitor of the history of intensity variations of the interplanetary magnetic field as it modulates the intensity of the galactic cosmic-ray flux. Part II ends with a contemporaneous review of the solar neutrino problem. The division of the isotope record into the two parts; energetic particles (Part II) and isotopes (Part III) might be viewed as somewhat arbitrary. The distinction is admittedly fuzzy; it is made on the basis that the energetic particle section tends to some extent emphasize the role of physics, whereas the isotope section emphasizes the record in geologic and meteoritic matter.

Part IV is devoted to the terrestrial record which, as expected, has a strong climatic flavor. This is obviously a subject with a vast and partially speculative, even anecdotal, literature. The selection for this book was based on a combination of the judgement of the editors, querying the advisory committee, and the contributing authors. Because of the extensive literature, we took the point of view that, insofar as possible within the limits of this book, a global perspective should be taken.

That the ancient terrestrial record is sparse is indicative of the somewhat strange circumstance that information or solar history is most dense during primordial time and the Recent. It is a well-known fact that the Earth record tapers into ignorance as the Hadean is approached. Unfortunately the banded iron record is still so conjectural with respect to a possible solar role that it could not participate on a basis which was sufficiently rigorous.

The early Sun is treated in Part V beginning with a review of the hydrodynamics of the young Sun together with the evolution of its magnetic field. The reason, of course, is the significant role that the Sun's field has in establishing the interplanetary magnetic field which in turn, is the primary modulator of isotope production. Along with this review is the possible relation of T Tauri-like objects and solar evolution.

The T Tauri stars are the young precursors to Sun-like stars. If the Sun followed a normal path of evolution, then these pre-main-sequence objects should yield insights on the nature of the early Sun and the origin of the solar system. Chapters in Part IV and V discuss the evidence for a "T Tauri Sun." Evidence from primordial meteorite thermal effects of strong early fields and plasma flow are reviewed while sampling of the solar nebula and/or its sequel is considered by review of evidence of relict magnetic fields in meteorites.

In Part VI we have assembled reviews of the properties of solar-like stars. The investigation of these solar counterparts places the study of the Sun in a stellar context. No longer can the Sun be considered especially unique in the Galaxy of stars. For example, the pioneering efforts of O. Wilson at the

Mount Wilson Observatory during the 1960s disclosed the occurrence of cycles similar to the solar cycle in late-type stars. This work and its continuation reveal the full range of cycle behavior through the observation of many solar-type stars at various phases in their cycles. In this way, the probable trends of future solar activity are indicated. Indeed, evidence has recently emerged that Sun-like stars undergo periods of prolonged quiescence analogous to the Maunder-minimum. This promising avenue of astrophysical research will provide a new window into the evolution of the Sun by the study of the correlated bolometric output and magnetic field variability of such stars.

In making topical selections we have been materially aided by an organizing committee of experts whose opinions we are grateful for. The task of organizing the book and the prefatory conference could not have been successfully concluded without the dedication of M. Magisos and M. Guerrieri.

Charles P. Sonett
Mark Giampapa
Mildred S. Matthews
Tucson, Arizona, April 1991

PART I
The Sun

SUNSPOTS: THE HISTORICAL BACKGROUND

GEORGE V. COYNE, S.J.
Vatican Observatory

*After a brief review of the earliest observations of sunspots, this chapter con-
centrates on the controversies of the 16th and 17th centuries concerning the
nature of sunspots and evaluates these controversies in light of the traditions
inherited from Aristotle and against the background of the conflict between
heliocentric and geocentric cosmologies. The role played by the early study of
sunspots in the development of the scientific method is also discussed.*

I. THE EARLIEST OBSERVATIONS OF SUNSPOTS

Theophrastus of Athens (ca. 370–290 B.C.), a pupil of Aristotle, provides the earliest known reference to a sunspot (Sarton 1947). It is in China,
however, that we have the earliest records that preserve with some continuity
the naked-eye observations of sunspots. Some sunspot observations are also
recorded in annals of Japan and Korea. In contrast to these numerous records
on sunspots in the Orient, there are very few recordings in the West throughout the first fifteen centuries of this era. Some have attributed this contrast to
the overwhelming influence in the West of the Aristotelian teaching that the
Sun was a perfect body without blemish. Although the influence of Aristotle
was very prominent in the West, it is obviously difficult to establish in this
case a causal connection.

A brief review of the early pre-telescopic observations of sunspots is
given by Bray and Loughhead (1964). *A Revised Catalogue of Far Eastern
Observations of Sunspots (165 B.C. to A.D. 1918)* has been published by Yau
and Stephenson (1988). It contains 235 entries; however, it is estimated that
this number represents as little as one-thousandth of the number of spots that

could have been seen with the naked eye during that interval of time. Eddy et al. (1989) have discussed the reasons for this paucity of recorded citings. What we have in the records was clearly not the result of any systematic continuous patrol for sunspots in the countries of the Far East. In discussing the scientific value of these early naked-eye observations, Eddy et al. (1989) conclude that careful analysis of these records can still give an indication of the general level of solar activity and thus they have a value in the study of secular variations on the Sun.

II. THE FIRST TELESCOPIC OBSERVATIONS OF SUNSPOTS

The first telescopic observation of a sunspot was probably made in the year 1611, although Galileo made a claim to have observed spots at Padua in 1610. It is difficult to establish historically the priority of observation among the gentlemen mentioned below; at any rate, such a curiosity of history is of little importance for this chapter. It is rather the contests over the interpretation of the spots, the preconceptions that influenced those interpretations, and the effect those contests had upon the cosmological thinking of the 17th century and upon the development in general of the scientific method that is of interest to us. In order to set the scene for these more interesting considerations, a brief history is given of those gentlemen who were most closely associated with the first telescopic observations of sunspots.

Johann Goldschmid (1587–1616), a Dutchman whose name in the literature is Fabricius, was the son of a pastor in the city of Osteel. His father was a keen observer of the skies and a friend of Brahe and Kepler. From his observations of spots he inferred that the Sun must rotate, but it appears that he did not appreciate the importance of this and did not further pursue his investigations (Fabricius 1611). Christopher Scheiner (1575–1650), born in Swabia, Germany, became a Jesuit priest and a professor of mathematics at the Jesuits' Roman College (Collegio Romano). He first recorded his sunspot observations in three letters which he wrote to a certain Mark Welser, a wealthy merchant of Augsburg and a friend of Galileo. Scheiner's Jesuit superiors, cautious of the consequences of any scientific investigations which might threaten an orthodox interpretation of Scripture and yet not of a mind to squelch such investigations completely, required Scheiner to publish under a pseudonym, which he did, calling himself Apelles. Welser, of course, communicated these letters to Galileo and so there began a most interesting exchange over a period of three decades between Galileo and Scheiner concerning both priority of discovery and the interpretation of sunspots. The exchange climaxed with two great works: The *Rosa Ursina* of Scheiner in 1630 and the discussions on the Third Day of Galileo's *Dialogo* published in 1632. This exchange will serve as the focal point of our discussion in the following section regarding the importance of the interpretation of sunspots for the cosmology of the 17th century and the development of the scientific method. We might mention that in support of his claim Galileo provided a daily sequence of solar drawings for June and part of July 1612, and that these were used in modern times (Herr 1980) for a determination of solar rotation. To complete the list of "discoverers" of sunspots there was Thomas Harriot (1560–1621) in England, a tutor in mathematics to Sir Walter Raleigh, and Francesco Sizzi (died 1618), a Florentine patrician who was executed for treason at the court of the French king, Louis XIII (Drake 1970). Although Sizzi's attempt to refute the telescopic observations of sunspots by Galileo is considered trifling, he may have played an important role in Galileo's coming to the knowledge of the annual variation in the paths of the sunspots across the face of the Sun (Drake 1970). We shall soon see the importance of this.

III. THE CONTROVERSY OVER THE PHYSICAL INTERPRETATION OF SUNSPOTS: THE FACTS

From the very beginning of the attempts to understand the sunspots, inherited preconceptions influenced all parties. It was in the struggle to overcome these prejudices that significant contributions were made, not only or even principally to solar physics, but rather to the practice itself of doing science and to the understanding of the universe, at least that small part of it which was known at that time, namely, the planetary system. In this section, the historical facts are established, as accurately as possible and Sec. IV proceeds to discuss the significance of what happened for cosmology and for the development of science.

The principal characters in this history are Scheiner and Galileo, although many others play significant roles. In the first of his three letters to Welser, dated 12 November 1611, Scheiner (1612a) affirmed that he had seen sunspots seven months earlier but that he had not paid attention to them until he saw them again in October. He struggled to explain them by eye defects, blemishes in his telescope, or atmospheric disturbances. He rules out all of these possibilities and argues for the fact that they are bodies orbiting the sun, excluding the possibility that they are on the Sun with the following argument: ". . . if they were on the Sun their motion would imply that the Sun rotates, and we should see the spots return in the same order and in the same position they had among themselves and with respect to the Sun. So far they have failed to reappear although other spots have followed the first ones across the solar disk. This is a clear argument that they are not on the Sun. . ." (Scheiner 1612a).

Had Scheiner observed more assiduously and for a longer interval at that time he might have discovered the rotation of the Sun and, as we shall see, this would have been a significant discovery in terms of the heliocentrism-geocentrism controversy. Later on, having observed the spots for a period of two months, he failed to detect any periodic return of the spots. He failed, of course, to take into account that the shape and size of the spots could change considerably. At this time, he was undoubtedly prejudiced by the Aristotelian notion of the immutability of the heavenly objects and he could not put a "blemish" on the Sun. Galileo responded in his first letter criticizing Scheiner for not admitting that the spots could undergo genuine alteration. In a second letter to Welser, Galileo propounded for the first time his own theory on the sunspots. Briefly, he affirmed that the spots were contiguous to the Sun, that their properties were analogous to those of clouds, and that they were carried around by the rotation of the Sun. He gave an elaborate mathematical treatment of foreshortening to show the spots had to be contiguous to the solar surface (Galileo 1613). Scheiner likewise attempted to give a mathematical treatment to his interpretation of sunspots in his *Accuratior Disquisitio* (Scheiner 1612*b*). At this time he realized that his interpretation would crumble if he were not able to predict and observe a spot on the Sun due to the transit of Venus. He failed to do so and thus remarked: "It is still doubtful whether the spots are on the Sun or away from it . . . . But this seems certain: the common teaching of the astronomers about the hardness and constitution of the heavens can no longer be maintained . . . . It is fitting, therefore, . . . to start thinking of some other cosmic system." (Scheiner 1612*b*).

This statement belies, at least in the case of Scheiner, any of those who think that the Aristotelian notions inherited by the gentlemen of this century were irreformable. The same statement can be made, for instance, of Bellarmine, one of the principal protagonists in the Church's dealings with Galileo (Baldini and Coyne 1984). To have preconceptions and to be open-minded are not necessarily mutually exclusive in the case of all human beings. In 1630 Scheiner published his collected observations of sunspots in the volume *Rosa Ursina* (Scheiner 1630). The title of the work comes from the escutcheon of the Orsini family, to whom the book was dedicated, which consisted of a rose (*Rosa*) and a bear (*Orso*). The collection of sunspot drawings presented there represents an important contribution as they have helped to trace the sunspot cycle back to the time of the first telescopic observations. Unfortunately, *Rosa Ursina* also has a polemical tone, because Scheiner is replying to what he considers to be an accusation of plagiarism leveled at him, he thinks, by Galileo in the *Il Saggiatore* (Galileo 1623).

It has been generally accepted among historians that Galileo brought all his troubles upon himself by vigorously attacking the Jesuit Scheiner in the *Il Saggiatore* and that, as a result, the Jesuits mounted a campaign against him, which finally resulted in the condemnation of 1632. Drake (1970) has made a strong case for the fact that Galileo was not attacking Scheiner in *Il Saggiatore* and that Scheiner was wrong in thinking that he was. Furthermore, there is no historical evidence which supports a thesis of a concerted attack by the Jesuits.

The admonition to Galileo in 1616 by Bellarmine on behalf of Pope

Urban VIII (Baldini and Coyne 1984) was that he was not to teach Copernicanism as if there were *physical* proofs for it, but only that it might offer a mathematically or philosophically simple way of presenting the observations. Nonetheless, Galileo in the *Dialogo* offers two physical proofs for the motion of the Earth: one based on the tides and the other on the apparent paths of the sunspots. In his letters on sunspots (Galileo 1613), he described the paths of the sunspots as parallel to the ecliptic. It is only in 1632 that Galileo mentions an inclination to the ecliptic of the Sun's axis of rotation. The argument for heliocentrism is critically dependent upon recognizing that inclination. Galileo claimed that he knew of the annual variation in the paths of the sunspots before 1614. The question arises as to the long silence of Galileo, from 1614 until 1632, with respect to the annual variation in the sunspot paths (and, therefore, the evidence for the tilt of the Sun's axis of rotation with respect to the ecliptic). Because this tilt, as we shall see, is critical for heliocentric arguments from sunspots, why did Galileo not speak of it in those twenty years? The thesis held for a long time is that Galileo actually learned of the tilt only from Scheiner's *Rosa Ursina* published in 1630 and that he plagiarized from Scheiner. This is, of course, a serious charge and one which would have deserved, so the story goes, the wrath of the Jesuits, who would have intrigued to the bitter end to see Galileo condemned. Such is the scenario presented by many.

A significant piece of evidence was uncovered by Stillman Drake in 1970 and this has allowed a very different reconstruction of the events, a reconstruction which proves to be very interesting from the point of view of the heliocentric-geocentric polemics and of the development of the scientific method. Let us begin with Galileo's earliest position with respect to the tilt of the Sun's axis of rotation. In his third letter to Welser in 1612, Galileo stated that the straight lines described by the spots were a necessary proof that the Sun's axis was perpendicular to the ecliptic. Scheiner was actually more cautious and did not venture to state that the motion of the spots was parallel to the ecliptic. The problem was that Galileo's observations extended only over the period May to July, precisely one of the periods when the Earth is crossing the intersection of the planes of the Sun's equator and the ecliptic. However, in the *Dialogo* Galileo claims that the annual variation in the paths of the spots had been known to him in Salviati's lifetime. Salviati died in March, 1614. If we are to avoid attributing an outright lie to Galileo, we must accept that between 1612 and 1614 he learned of the annual variation in the paths of the sunspots and, therefore, of the tilt of the Sun's axis of rotation. Drake (1970) finds evidence that this is precisely what happened.

In 1613 Francesco Sizzi, whom we have mentioned above, wrote a letter to a certain Father Horatio Morandi of Rome in which he reports, although in a somewhat cryptic manner, the annual variation of the paths of the sunspots (Sizzi 1613). Within three months of receiving it, Morandi sent this letter to Galileo. A copy of the letter exists with corrections to it in Galileo's own hand. There can be no doubt that he studied it attentively. The letter reports the following:

> . . . the angles made by the spots at the equinoxes with the imaginary perpendicular line in the Sun and parallel to our view differ from those made at the solstices—which in turn differ between themselves since the angle which at one solstice will be considered as in one of the four quadrants of the solar surface will (at the other solstice) be in the opposite quadrant . . . (Sizzi 1613).

Although it is cryptic, the hints supplied by Sizzi's letter were more than adequate to direct the attention of an interested astronomer to the existence of systematic variations in the paths of the sunspots at various seasons of the year. Let us assume, therefore, that this was the source of Galileo's knowledge of the annual variation in the paths of the sunspots and try to reconstruct the course of events.

When Galileo went to Rome in 1616 to defend Copernicanism he was not yet prohibited from offering *physical* proofs. So he offered his argument from the tides, but at this time and in this context he says nothing of the sunspots. Why not, since he knew of the annual variation of the spots since 1613? The only logical explanation is that he did not make the connection between the sunspot variations and the motion of the Earth. He potentially had with the sunspots an independent argument than the one from the tides, but at this time he did not realize it. The argument first appears in the *Dialogo* where it occupies ten pages. It depends entirely upon the tilt of the Sun's axis of rotation with respect to the ecliptic. In Galileo's published writings there is no mention of this tilt until the *Dialogo* of 1632. Scheiner, however, had already two years before in 1630 published his *Rosa Ursina* in which the annual variation in the paths of the sunspots was correctly described. Did Galileo take the material for his sunspot argument in the *Dialogo* from Scheiner's *Rosa Ursina*? Drake (1970) gives the following convincing arguments that this could not have been the case. Galileo's argument is extremely vague concerning the specific direction of the tilt or else it implies a tilt and a seasonal variation in the sunspot paths that do not fit the observations of Scheiner reported in the *Rosa Ursina*. Scheiner correctly states that, with respect to the ecliptic, the rectilinear motion of the spots occurs in summer and winter and the inclined motion in March and September. Galileo would surely not have given his opponent the edge by being so vague or by proposing a theory that was contradicted by his opponents observations. Conclusion: Galileo had not yet seen the *Rosa Ursina*.

There is also external evidence to support the thesis that Galileo's arguments in the *Dialogo* were not dependent upon Scheiner's *Rosa Ursina*. When the *Rosa Ursina* was published in 1630 the *Dialogo* was already half printed. Now while it was physically possible that Galileo then inserted fully ten pages on the argument from the sunspots, it is extremely unlikely. The printer of the book would never have accepted this material without previous censorship both in Florence and Rome. The book was already notoriously controversial and had suffered many delays and revisions due to the censors. In particular, the sunspot argument was integrated into the closing statement of the book and this closing statement had been through extremely careful scrutiny by the censors.

The only reasonable thesis to explain the appearance of the sunspot argument in the *Dialogo* is that of Drake (1970). Galileo first heard of the annual variation in the paths of the sunspots from the letter of Sizzi sent to him by Morandi in 1613. We, therefore, accept his statement in the *Dialogo* that he knew of the effect before 1614. He did not, however, at that time realize the significance of it as an argument for the motion of the Earth. It was only in 1629 when composing the *Dialogo* that he came to that realization and thus wrote the Third Day of the *Dialogo* as a "thought-experiment." A fifteen year period elapsed, therefore, between his knowledge of the effect and his understanding of it. While the logic of the "thought-experiment" may have been quite correct, the results of it did not correspond at all to the observations, because Galileo did not know the details of the observations. Sizzi's report was too cryptic and he had not yet seen Scheiner's. As we shall now see the connection between the annual variation of the sunspot paths and the Earth's movement about the Sun is not an obvious one. But it is a very weighty one and probably represents the strongest "physical proof" that Galileo had. Had his censors recognized this, he may have had even a harder time than he did.

IV. THE ARGUMENT FROM SUNSPOTS FOR HELIOCENTRISM

If we accept the following four statements as true, then we have a weighty argument for a heliocentric universe: (1) the spots are on the Sun; (2) a given spot moves across the surface of the Sun with a 29 day period; (3) there is an annual (seasonal) variation in the tilt of the path of the spots across the face of the Sun; (4) there is no obvious diurnal variation in the spots. What is required to explain these facts, respectively, in a geostatic and a heliocentric scenario? In each scenario the Sun rotates on its axis in 29 days. In heliocentrism we furthermore require the diurnal rotation of the Earth and the annual orbit of the Earth about the Sun; these motions and the respective tilts of the axes of rotation of the Sun and the Earth are all that is required and the explanation is particularly simple. In the geostatic case, we require that the Sun's axis of rotation precess systematically in one year as the Sun traces out the ecliptic and that this same axis of the Sun's rotation be locked to the Earth's orientation over the course of a day. It is the simultaneous linkage of facts (3) and (4) which makes the geostatic model far more difficult to understand. One and the same axis must perform an annual precession while being locked on a daily basis to the orientation of a separate body. While this cannot be excluded by purely abstract geometrical and kinematical considerations, it is a much less simple and economical explanation than heliocentrism and Galileo capitalized on it in the Third Day of his *Dialogo*. Galileo argues against a geostatic model because of its complexity and the incongruity of the motions. In arguing for simplicity over complexity, he made a significant contribution to the scientific method. What did he mean by incongruous? Galileo had studied in detail freely floating spheres. For him to have a literal motion of the Sun's axis so that it kept the same face to Earth over a day (to explain the lack of diurnal variation in the spots) and yet slowly rotated about the center of the Sun in a year (to explain the annual variation in the paths of the spots) was incongruous. In terrestrial spheres, he was not able to generate and conserve a double rotation of different periods in the same axis of a freely floating sphere. It seemed to him incongruous that it should occur in a heavenly sphere.

The argument was far from constituting a physical proof, however, because the situation is clearly one of relative motion in which the kinematics must ultimately be equivalent. In the case of a moving Earth, Copernicus required multiple motions for the Earth that were just as preposterous as those ridiculed by Galileo for a moving Sun. Simplicity and economy are in the eye of the beholder and can cut both ways. Yet, by showing a case where the moving Earth did indeed make the solar phenomena more easily explained, Galileo helped make the heliocentric arrangement persuasive. Thus the sunspots, while not providing the physical proof that he sought, nevertheless played an important role in helping his readers think seriously about the possibly physical reality of the Copernican system.

V. CONCLUSION

The attempts to interpret the earliest telescopic observations of sunspots contributed significantly to the resolution of the conflict between geocentric and heliocentric cosmologies and to the development of the scientific method. In the course of understanding the sunspots, the preconceptions of both Aristotelianism and the Church's literal and fundamentalistic view of the Scriptures were found wanting. Even before the discovery of the aberration of light and of parallax, the explanation of sunspots provided a weighty argument for heliocentrism.

Acknowledgment. I am most grateful to O. Gingerich for his critical remarks which have contributed significantly to modifications in my original presentation of this chapter.

SOLAR IRRADIANCE VARIABILITY FROM MODERN MEASUREMENTS

C. FRÖHLICH
Physikalisch-Meterologisches Observatorium Davos

P.V. FOUKAL
Cambridge Research and Instrumentation

J.R. HICKEY
Eppley Laboratory

H.S. HUDSON
University of California, San Diego

and

R.C. WILLSON
Jet Propulsion Laboratory

Direct measurements from satellites of the solar "constant"—the total irradiance at mean Sun-Earth distance—during more than ten years show variations over time scales from minutes to years and decades. At high frequencies solar oscillations contribute to the variance. The most important influences are related to solar activity: during the passage of active regions on the solar disk (sunspots and faculae) changes of a few 0.1% lasting for several days are observed. The effect of spots can be well reproduced by the projected sunspot index, whereas the influence of faculae have to be modeled from proxy data like the Ca-K plage index or the He I index. Long-term trends have also been detected which are connected to the 11-yr solar activity cycle showing a difference

*between the maximum and minimum of nearly 0.1% with the irradiance being
higher during solar maximum. This modulation can be modeled by proxy data
as the He I index or the 10.7-cm flux indicating the importance of the magnetic
network. Moreover, it can also be explained by observed temperature variations
with latitude over the solar cycle. Whether the latter is physically linked to the
explanation by a varying bright network is not yet clear.*

The irradiance from the Sun at the mean Sun-Earth distance, integrated
over the energetically important wavelength range (hence total irradiance) is
called the solar constant. Observations of the solar constant have a long history, starting with the measurements by the Smithsonian Institution from
mountain stations at the turn of the century and including data from the beginning of the space era. Many of the early measurements were inconclusive,
mainly because of lack of sufficient radiometric precision, but also due to
influences of the Earth's atmosphere upon observations from the ground and
airplanes (for reviews see, e.g., Fröhlich 1977,1987a; Hoyt 1979; Angione
1981). The first clear evidence of solar constant variability appeared at the
beginning of the 1980s with the data from radiometers on the Solar Maximum
Mission and on NIMBUS 7, proving that the Sun is indeed a variable star.
Interest in total solar irradiance variability on all time scales has since then
greatly increased. Not only are the atmospheric physicists and climatologists
concerned, because of possible effects on the Earth's energy balance, but also
the solar physicists. The existence of global changes based on the solar output
had been doubted for a long time and their observation obviously leads to
new ways to understand the Sun.

I. MODERN MEASUREMENTS OF TOTAL SOLAR IRRADIANCE

Modern measurements from satellites have been made by the Earth Radiation Budget Experiment (ERB) of the NIMBUS 7 satellite in a near-polar
orbit since 16 November 1978 (Hickey et al. 1988,1989), by an active cavity
radiometer on the Solar Maximum Mission Satellite in a 27° orbit since 14
February 1980 (Willson 1984) and more recently by a series of radiometers
within the ERBE experiment on NASA ERBS in a 57° orbit and the NOAA
9 and 10 satellites in near-polar orbits starting in October 1984, January 1985
and October 1986, respectively (Mecherikunnel et al. 1988). In all these experiments, the solar irradiance is measured by electrically calibrated cavity
radiometers.

Radiometry and Operation

In the ERB experiment the sensor channel 10C is one of a family of
thermopile-based radiometers commonly referred to as the H-F type (Hickey
et al. 1977). The Solar Maximum Mission radiometer is called ACRIM (Active Cavity Radiometer Irradiance Monitor) and was designed by Willson

(1979) based upon developments in the late sixties. The term "active cavity" radiometer refers to its method of operation, as described in detail below. The radiometers of the ERBE experiment are similar to the active cavity radiometer of Willson and have been described by Lee et al. (1987).

The radiometers are based on the measurement of heat flux by using an electrically calibrated heat-flux transducer. The radiation is absorbed in a cavity which ensures a high absorptivity over the spectral range of interest for solar radiometry. ACRIM uses as cavity a 30-deg cone illuminated on the inside, whereas the H-F design has a 60-deg inverted cone. Schematic drawings of both detectors are shown in Fig. 1. The heat flux transducer consists of a thermal impedance and thermometers to sense the temperature difference across it. In the case of ACRIM, these are resistance thermometers made out of nickel wire wound around the top and bottom of the thermal resistor; in the case of the H-F radiometer, the transducer consists a thermopile made of constantan wire with alternating parts covered by electroplated copper, forming a copper-constantan thermopile. Heat developed in the cavity is conducted to the heat sink of the instrument and the resulting temperature difference across the thermal impedance is sensed. The sensitivity of the heat flux transducer is calibrated by shading the cavity and measuring the temperature difference while dissipating a known amount of electrical power in a heater element mounted inside the cavity. It is advantageous to determine the electrical power that is needed to produce the same temperature difference as was observed with the cavity irradiated, as in this case the heat losses are the same during radiative and electrical heating, even if nonlinear effects are involved. The absolute accuracy of these radiometers depends on the accurate knowledge of the area of the precision aperture and of the amount of all possible deviations from ideal behavior, such as the nonequivalence of electrical and radiative heating, the absorptivity of the cavity, the influence of stray-light and diffraction, and the linearity and calibration of the electrical power measurement. The present state-of-the-art accuracy is of the order of 0.2%; the precision, however, is much better.

ACRIM Radiometer on the Solar Maximum Mission.   During operation of ACRIM, an electronic circuit maintains the temperature signal constant by proportionally controlling the power fed to the cavity heater. The servo operation is independent of the mode, i.e., whether the cavity is shaded or irradiated. The substituted radiative power is then equal to the difference in electrical power as measured during the shaded and irradiated periods, respectively. This mode of operation is called the active mode, hence the designation ACR, the active cavity radiometer. The heater power is calculated from the square of the voltage across the cavity heater. The absolute accuracy of the ACRIM instrument is of the order of 0.2%, and the resolution of the measuring system of ACRIM is 1 in 5000 of the individual reading. The

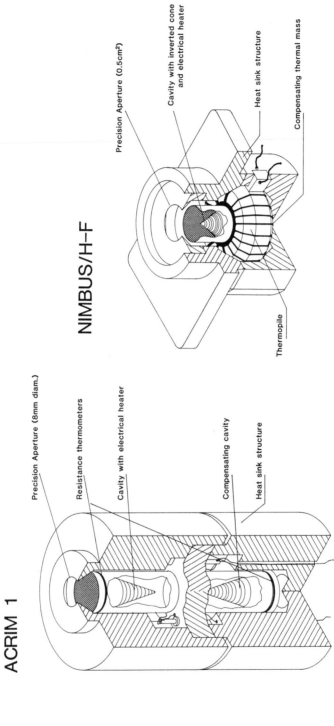

Fig. 1. Schematic drawing of the sensors of the ACRIM (left) and H-F (right) detectors.

precision, a measure of the short-term stability or resolution of the radiometer itself, is much better than the resolution of the electrical measuring system.

The ACRIM experiment contains three identical radiometers A, B and C. All are operated continuously: A is always exposed to the Sun, B from time to time and C very rarely. From the ratios A/B, A/C and B/C the degradation due to exposure-dependent influences can be determined. The basic timing of the measurements is 131.072 s with a 50% duty cycle of solar observations. Individual readings of heater voltages are taken every 1.024 s and the mean of the last 32 of a cycle are used for further evaluation. The Solar Maximum Mission (SMM) is a three-axis stabilized spacecraft, pointed very accurately at the Sun until December 1980 and after April 1984. In December 1980 the stabilization system failed and the spacecraft was put into a spinning mode with its axis roughly pointing to the Sun. During this period ACRIM had to be operated differently: the shutter opened at sunrise of the orbit, closed after sunset and the few readings with the Sun in the field of view were used for irradiance determinations with the night part of the orbit as reference phase calibrated by the off-Sun (deep space) values during the sunlit part. During spring 1984, SMM was successfully repaired during a Challenger shuttle mission and has worked well up to the writing of this review; high-precision data are available from before the failure and after repair. These data consist of continuous measurements during about 55 min of each 96-min orbit, yielding up to ~ 400 averaged shutter values per day. During the spinning mode far fewer readings per day are available, but the noise in the data is still lower than that from, e.g., NIMBUS 7/ERB with one measurement point per orbit.

H-F Radiometer on NIMBUS 7.   The H-F is operated in the passive mode: it is electrically calibrated once every 12 days by heating the cavity electrically while the sensor views space (Hickey et al. 1989). As the NIMBUS satellite is Earth observing and no solar pointed platform is available, measurements with the solar sensors are performed during the passage of the Sun through their field of view when crossing the southern terminator of the polar orbit. The measurements consist in monitoring the thermopile output every second for a 3-min period of solar acquisition every 104-min orbit. The radiometric accuracy of the H-F instrument is of the order of 0.5%. The resolution of the individual measurements is 1 in 1800 for a typical value.

During the early mission years, the solar measurements were performed on a 3-day-on/1-day-off schedule. The data taken during the thermal transient period (the first measuring day of the 4-day period) were considered less precise and were filtered out of most of the early results. Continuous operation is available from October 1982 to February 1983 and after November 1984. Moreover, during the periods of April through June 1986 and late April through late August 1987, special operations were required due to the mission constraints. On a normal day 14 orbital values are available, which are the mean of some 150 individual 1-s readings.

ERBE Solar Monitors. The measurement with this ACR consists of a reference phase and an observational phase of 32 s each. As for the ERB, the Sun drifts through the field of view of 13.6 deg during the observation. The shutter of the radiometer is normally closed, and opened only during the 32 s of solar observations. The data from the last 4 s of each phase are used to evaluate the irradiance. The radiometric accuracy is stated to be 0.2% (Lee et al. 1987). Due to operational constraints (e.g., thermal environment, offset angle) this accuracy is probably not achieved in orbit.

On all three satellites solar irradiance observations are taken approximately once biweekly during two adjacent orbits.

First Evaluations and Corrections

Corrections, taking into account the instrument performance and the position of the spacecraft relative to the Sun, have to be applied in the first evaluation of the raw data. The instrument-related corrections include temperature effects of the electric measuring system and the precision aperture. In the case of the ACRIM and ERBE radiometers, the temperature of the shutter yields the thermal reference radiation for the closed values; the correction is of the order of 600 ppm. The reference of the H-F instrument is the deep-space radiation during the calibration by electrical substitution. The navigational information from the spacecraft are used to reduce the data to 1 AU (annual variation of $\pm$ 3.46% and orbital modulation of 85 ppm peak to peak) and to correct for the relativistic Doppler shift of the energy within the spectrum due to the radial velocity $[1 + (v/c)^2]$ with v being the satellite velocity towards the Sun); this latter factor approaches $\pm$ 50 ppm. In the early data products of the ERB experiment, the distance corrections were based on an approximate formula for the Sun-Earth distance and have a limited accuracy. For the ERB results presented in this review, a more accurate algorithm is used. The Doppler correction is only applied to the ACRIM data, as the readings are taken during the whole sunlit part of the orbit with the Sun-spacecraft velocity varying from plus to minus 8 km s^{-1} (Hudson and Willson 1981). For the ERB and ERBE radiometer, the Doppler correction is neglected, yielding a bias of up to 50 ppm. Furthermore, for all instruments, pointing data are used to correct for the angular distance between the instrument's axis and the direction to the center of the Sun assuming a cosine dependence.

Data Sets

Figure 2 shows the data of the three experiments from 1978 to 1988. The most obvious features are the differences in absolute values and the general behavior of a downward trend from 1978 to 1986, followed by an increase common to all three observations. In order to evaluate the differences among the experiments, Table I compares the means of the values from 23 days in the period October 1984 through March 1986 where data from all experiments are available. The observed absolute differences are within the stated uncertainties of the different experiments. Moreover, the ACRIM data during the accurate pointing (before December 1980 and after April 1984) exhibit the lowest standard deviation, indicating some dominating noise of nonsolar origin in the two other data sets and during the period 1981–1984 of ACRIM.

II. INTERPRETATION OF VARIATIONS ON TIME SCALES UP TO MONTHS

The modern data on total solar irradiance reveal a broad spectrum of variations (see, e.g., Hudson 1988a). This section describes the rapid variations and the explanations that have been put forward to explain them. Two overviews of the data rapidly communicate their nature: the time series themselves (Fig. 2) and estimates of the spectral power density of the solar variability (Fig. 3) representing the behavior of solar irradiance variability in time and frequency domain, respectively.

Rapid Variations

ACRIM sampled up to a Nyquist frequency of 3.815 mHz, corresponding to the 131-s period of its shutter mechanism. This review will not dwell on the "rapid variations," defined here as the band between a one-day period (frequency 11.57 μHz) and this ACRIM shutter-cycle limit. Briefly, however, this range of the spectrum contains an approximately $1/f$ continuum power spectrum, plus resonances due to solar global oscillations. These include both the p-mode or 5-min oscillations (see, e.g., Woodard and Hudson 1983), and possibly the signatures of deeper-seated g-mode oscillations (see, e.g., Fröh-

TABLE I
Comparison of ERB, ACRIM and ERBE[a]

Experiment	Mean (W/m2)	Stand.Dev. (W/m2)	Diff (W/m2)	Percent
ERB	1370.16	0.30		
			3.17	0.23
ACRIM	1366.99	0.14		
			1.97	0.14
ERBE[b]	1365.02	0.44		

[a]The values are means over 23 simultaneous days in the period from October 1984 through March 1986 (data from Mecherikunnel et al. 1988).
[a]Mean of ERBS and NOAA 9.

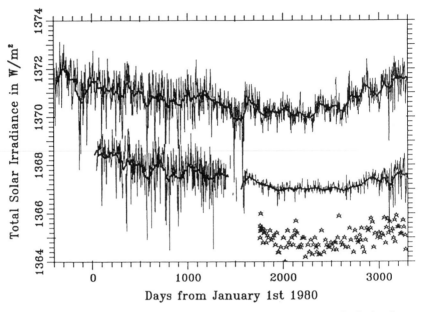

Fig. 2. Time series of ERB (top), ACRIM (middle) and ERBE (bottom) total solar irradiance
measurements. For ERB and ACRIM the individual points represent daily mean values, the
lines 81-day running means. The ERBE data points are individual measurements taken at about
a bi-weekly rate.

lich 1987b). The 1/f broadband spectrum has a *rms* fractional amplitude of
tens of ppm over the frequency range 0.1 to 1.0 mHz (Hudson and Woodard
1983).

Variations of Days to Months: Models and Explanations

Observations.   The time scales associated with solar magnetic active
regions reveal extremely interesting variations of the total solar irradiance.
The exact range of such time scales, however, is not precisely definable, as
the eruption of magnetic flux on the Sun exhibits correlations on the longest
time scales yet studied (a few hundred yr) and on the largest spatial scales
that the surface area of the Sun permits. For the purpose of this review, we
take time scales of a few solar rotations (about 0.1 μHz) down to two days
(the 5.8-μHz Nyquist frequency of daily sampling). The dominant feature in
the time series for this range of periods is the "dip," a negative excursion of
a few days' length and a depth ranging up to a few 0.1% of the normal total
irradiance. As explained below, these dips result from large sunspot groups
rotating past the central meridian. Between the dips the total irradiance fluc-
tuates on a wide range of time scales, as demonstrated by the broadband
character of the power spectrum.

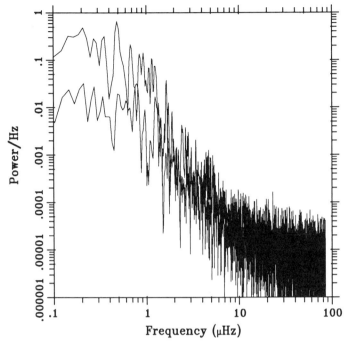

Fig. 3. Broadband power spectrum of the ACRIM time series for 1980 and 1985. The major difference between 1980 (upper) and 1985 (lower) is at low frequencies and is due to the different levels of activity.

Dips and Spots.   Prominent dips appeared during the first few months of data from the ACRIM instrument on board SMM; Willson et al. (1981) described them in terms of the Projected Sunspot Index (PSI) function similar to the models of sunspot darkness noted earlier by Foukal and Vernazza (1979) and Hoyt (1979). Hudson and Willson (1982) defined this projected-area index as

$$PSI = \alpha \cdot \Sigma \{[(3 \cdot \mu + 2)/2] \cdot \mu \cdot a\} \tag{1}$$

where μ is the central angle of the sunspot group and a its area. The factor α takes into account the umbra/penumbra area ratio and the effective temperature of the sunspot photospheres, in the simplest possible way. This index, calculated from the synoptic data for sunspot areas and with textbook calculation of the value of α (i.e., no adjustment of parameters whatsoever), removed about half of the variance of the total ACRIM time series for the initial period of good data in the active year 1980. This success was quite surprising in view of the grossly simplifying assumptions involved in the construction of PSI. Figure 4 shows an example of a clear dip of the new solar cycle centered on 1 July 1988 as observed by ACRIM, the corresponding PSI function derived from spot observations and a series of pictures of the spots moving across the solar disk.

For the period 1980 to 1988, the PSI function is shown in Fig. 5 together with the irradiance data from ACRIM and ERB corrected for spots by this function $(S + \mathrm{PSI})$. The success of the simple PSI explanation of sunspot effects immediately implied that a bright ring surrounding the sunspot did not directly re-radiate the blocked convective flux presumed to be welling up continuously through the solar interior. Clearly, any empirical knowledge of the disposition of this missing flux would greatly help in understanding the physics of the solar convection zone; this strong relationship offered the possibility.

Modeling.   The success of the simple matching of ACRIM total irradiance and the PSI model of sunspot-projected area led to a flurry of model-building, in which various authors attempted to improve on the fit (see, e.g., Hoyt and Eddy 1982; Oster et al. 1982). There are several reasons for this kind of model-building activity (Hudson 1988b):

1. The models may help to identify the cause of a particular variation;
2. The models may lead to the determination of some key physical parameters;
3. The model may provide a proxy for the estimation of solar irradiance variations in the absence of data; and finally
4. The models' reduction of variance may unmask more subtle effects due to different physical mechanisms.

The attempts to improve on the naive fit of the PSI index have generally failed, owing perhaps to the additional noise in the correlation introduced by random and systematic errors in the synoptic data (Schatten et al. 1985).

Facular Effects.   Willson (1982) called attention to the effects of faculae in modulating the irradiance and having the potential to compensate the sunspot deficit. More direct photometric approaches have yielded some insight into the relationship between the irradiance excesses due to faculae of an active region and the sunspot deficits. Given the almost simultaneous existence of dark spots and bright plage in active regions, this prompted the question as to whether the energy excess and deficit actually balanced. If so, this would imply some mechanism for local storage via the subphotospheric magnetic field structure. Indeed, at the simplest level both the Ca proxy measure of facular brightness (Hirayama et al. 1984; Chapman et al. 1986) and direct facular photometry (Chapman 1984; Lawrence et al. 1985) show a rough balance (see Chapman [1987] for a complete review).

However, the observational situation is not clear cut, because the divi-

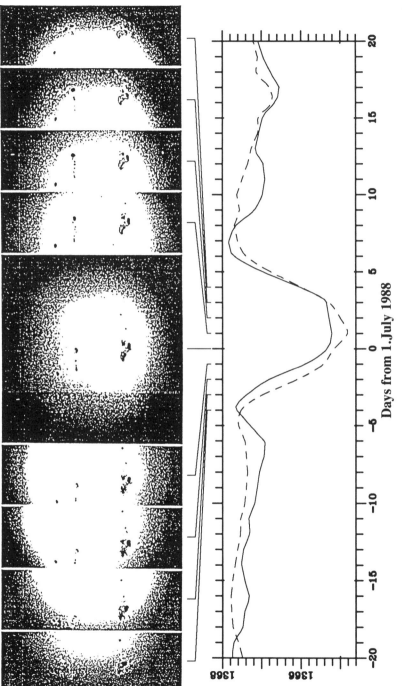

Fig. 4. Influence of sunspots on solar irradiance during the dip in July 1988 illustrated by a series of pictures of the passage of the spot groups over the solar disk compared to the observed solar irradiance (solid line) and the downward plotted PSI function (dashed line) (pictures by San Fernando Observatory, California State University, Northridge, CA, U.S.A.).

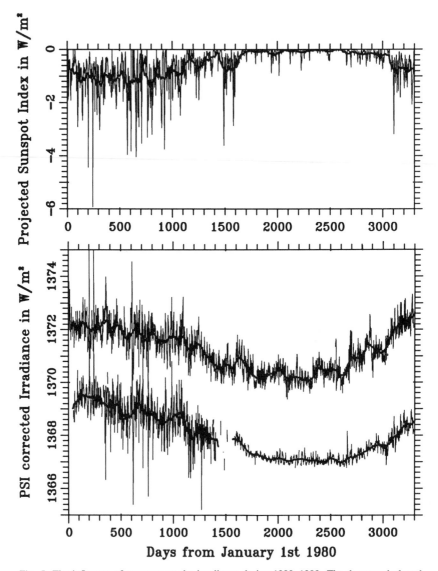

Fig. 5. The influence of sunspots on the irradiance during 1980–1988. The downward plotted PSI function (top panel) and the PSI corrected irradiance (S-PSI) of ACRIM and ERB with the solid lines indicating a 81-day running mean (bottom panel).

sion of facular elements into active regions is somewhat ambiguous; see the discussion in Sec. III concerning the possible relationship between the active network and the solar-cycle dependence of total irradiance. Furthermore, a given active region has a complicated structure both in space and time, with sunspots appearing and disappearing in nests of activity that may persist for many rotations. Finally, the idea of detailed balancing has received little theoretical support, general opinion holding that thermal effects in the solar interior far outweigh any possible magnetic effects.

Multivariate Analysis. Multivariate spectral analysis is a powerful tool for the investigation of multiple influences on a time series, as, for example, the quasi-independent spot and facular contributions discussed above. Fröhlich and Pap (1989) have applied this technique to the ACRIM data, both in 1980 and after the repair mission in 1984/85. They find for the former period that more than 90% of the variance of the ACRIM data can be explained by the effects of sunspots and magnetic elements, the latter including both active-region and network facular elements (represented by the equivalent width of the He I 1083-nm line). The analysis reveals the presence of power spectral peaks near 27 and 9 days, the latter suggestive of heretofore unknown processes within the solar interior.

Physical Theories of Sunspot and Facular Effects. Why are sunspots dark and faculae bright? We have a simple diffusion theory of thermal energy transport in the solar convection zone that offers at least partial answers for both questions (Spruit 1977; Foukal and Fowler 1984; Chiang and Foukal 1985; see Spruit [1988] for a review). This theory estimates a diffusion coefficient from the mixing-length theory of convection, that yields a large value for the thermal conductivity. This eddy conductivity grows rapidly with depth, to the extent that a localized temperature disturbance should spread rapidly around the surface of the Sun.

The thermal diffusion theory readily explains the absence of a bright ring around sunspots, postulated to exist for the purpose of compensating the sunspot deficit of radiant luminosity. This ring, if localized in space and simultaneous with the spot deficit, would compensate for the sunspot darkness and result in a reduced dip in total irradiance. This conflicts with the ACRIM observations (Hudson and Willson 1981). Furthermore, direct high-precision surface photometry shows no sign of bright structures near sunspots, beyond the brightening expected from faculae (Fowler et al. 1983).

III. INTERPRETATION OF LONGER-TERM VARIATIONS

Comparison of the ERB and ACRIM radiometry has revealed common signals on time scales exceeding the two weeks expected from rotation of active regions across the photospheric disk, and extending to the longest time scales observed understandable in terms of the solar activity cycle. After smoothing the time series (solid lines in Figs. 2 and 4) both data sets show common variations on time scales of 4 to 9 mon, as well as a general slow downturn, which has since reversed during 1986, after passage of the solar minimum.

Several ideas and models have been put forward to explain these variations. One approach is to account for the variations purely in terms of the effects that magnetic flux tubes seem to have on the radiation and convection in the relatively shallow photospheric layers that emit most of the Sun's luminosity. Others have invoked deep-seated changes in solar convection, perhaps involving even variations in the nuclear-burning core (see, e.g., Gough 1989).

A relatively straightforward approach to both the 4 to 9 mon and 11-yr variations has been put forward by Foukal and Lean (1988) and Willson and Hudson (1989). In the Foukal and Lean study, it was shown that the residual irradiance variations remaining, after a correction for sunspot blocking is made to the smoothed ACRIM or ERB data, correlate well with indices of facular area such as provided by the Ca-K plage index and the He I index. This is not surprising since facular area variations were previously shown to account for day-to-day variations in these residuals (Hudson and Willson 1981; Foukal and Lean 1986).

To test this model, the regression relations found were then used, together with daily facular index data, to reconstruct the irradiance residuals obtained from ACRIM and ERB data in 1981 to 1984. As can be seen in Fig. 6, both the 4 to 9 mon variations and the slow downtrend between 1981 and 1984 are remarkably well simulated when the 10.7-cm flux or the He I index is used. The reconstruction based on the Ca-K plage index is about equally successful in simulating the 6 to 9 mon variations, but fails to reproduce the slow downtrend (Foukal and Lean 1986).

The He I index and the 10.7-cm flux represent contributions from all bright magnetic elements on the disk, including the network, while the Ca-K plage index takes into account only the largest faculae in active regions. Thus, one may conclude that the 6 to 9 mon variations are caused by the tendency of major complexes of activity to persist for about this number of solar rotations (Gaizauskas et al. 1983). This time scale in persistence of solar activity episodes has been documented before in studies of the He I index time series (Harvey 1984) and may be related to the persistence of certain active longitudes on the solar surface, noted already by Dodson and Hedeman (1970). No well-accepted explanation of this 6 to 9 mon activity time scale (or of the active longitudes) exists as yet, although the recent ideas of Wolff (1984; Wolff and Hickey 1987) in terms of Rossby waves and g-mode oscillations and their possible interference are interesting, and deserve more attention. Wolff and Hickey have also proposed that such oscillations may modu-

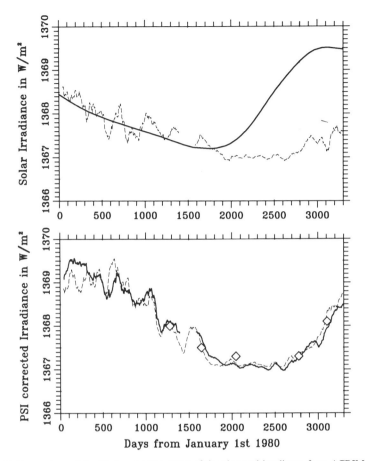

Fig. 6. Comparison of the 81-day running mean of the observed irradiance from ACRIM with different models. Top panel: irradiance (*S*, broken line) and the model (solid line) of Schatten (1988). Bottom Panel: the residuals (*S* + PSI, broken line) with the irradiance residual calculated from the 10.7-cm flux (solid line), and the irradiance data deduced from d*T* and facular contribution (from Kuhn 1989) (diamonds).

late heat flow to the photosphere and thus contribute to irradiance variation directly.

The finding that simulation of the downtrend between 1981 and 1984 requires use of a global index, as mentioned above, indicates that this downtrend is associated with some kind of a slow change in the solar atmosphere not reflected in the Ca-K plage index, but well modeled by the He I index or the 10.7-cm flux. The simplest explanation is a slow decrease in the emission from the bright magnetic network outside active regions. This identification cannot be considered proven until the changes in area (and possibly also intensity) of the network are measured in white light over a solar cycle. However, no other explanation is able to account so convincingly (i.e., using a unique relation between total irradiance and He I index or 10.7-cm flux) for the relative amplitude of both the 4 to 9 mon variations discussed above, and the slower trend over years. Livingston et al. (1988) have since shown that the correlation between total irradiance residuals and He I index pointed out by Foukal and Lean (1988) for the period 1981 to 1984 holds through at least the beginning of 1986. Figure 6 suggests that the whole solar cycle can possibly be modeled by these indices.

Figure 7 shows a reconstruction of the irradiance for the last three solar cycles, using the 10.7-cm solar microwave flux as a global index for faculae, instead of the He I index, which is not available before 1975. The most interesting point in Fig. 7 is the model's estimate that cycle 21 produced a much larger irradiance variation than cycle 19, which was the largest in the 130-yr history of reliable sunspot records.

Schatten (1988) has put forward a model in which the same photospheric structures (sunspots, active region faculae and magnetic network) are used to reproduce the slow downtrend of the ERB and ACRIM data until 1986. His handling of the facular and network terms is somewhat different from that of Foukal and Lean (1988), as he computes their behavior from measurements of the 11-yr evolution of the spherical harmonic structure of the Sun's field together with photometric contrast of faculae taken from the literature. Schatten's model is able to reproduce the variation in total irradiance reasonably well (Fig. 6), but the phase within the cycle seems to be off. Schatten's model provides some additional support for the view that the 11-yr irradiance variation can be explained entirely in terms of photospheric magnetic activity, although it relies on highly uncertain data on facular and network contrast. It further suggests that concentration of enhancements in the facular network at lower latitudes as the cycle progresses may be significant in determining the network's effect in the 11-yr irradiance variation. This latitudinal migration of facular influence is claimed to account for the significant phase advance of the peak irradiance, which contradicts, however, the observations for the onset of the cycle 22.

More recently, Kuhn et al. (1988) have reported observations of the limb brightness which they interpret to indicate a substantial contribution to the total irradiance variation of the solar cycle. The observations are broadband, two-color photometric measurements of the brightness distribution in a narrow annulus 20 arcsec wide, just inside the solar limb. The solar limb flux observed as a function of latitude can be divided into a "facular" and "temperature" part based on the assumption that the "temperature" part is constant over the 4-mon observing summer period and that the "facular" part shows up as intermittent bright regions. The component of excess brightness moves toward the equator between 1983 and 1985, and then reappears at relatively high latitudes again in 1987. The corresponding equivalent blackbody tem-

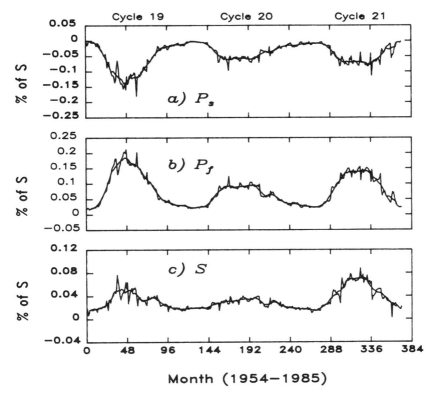

Fig. 7. Behavior of the total solar irradiance between 1954 and 1985 reconstructed with the ACRIM 10.7-cm flux correlation. The heavy line represents a 12-mon running mean (figure from Lean and Foukal 1988).

perature excess dT (determined from the color ratio) for the years 1983, 1984, 1985, 1987 and 1988 is shown in Fig. 8. The excess brightness responsible for dT is due to features which are not resolved by this observation, and it could be due to the bright network in and outside the active regions. The decreasing contribution of dT (including the facular component) can account for the total irradiance decrease between 1983 and 1987, and its increase in 1988 (Kuhn 1989) as shown Fig. 6.

Only the zero-order term of the polynomial expansion of the latitudinal variation of the dT signal is responsible for the irradiance variation. However, from Fig. 8 it is obvious that higher-order terms are present. In a separate paper Kuhn (1988) compared the expansion coefficients needed to explain the latitudinal variation of dT with the corresponding ones of splitting data of medium-order p-modes from Duval et al. (1986), Brown and Marrow (1987) and Jefferies et al. (1988) and developed a model in which the underlying physical effects are parameterized by a local effective sound speed where dc^2/c^2 is directly related to the surface dT/T observed. He is able with his model

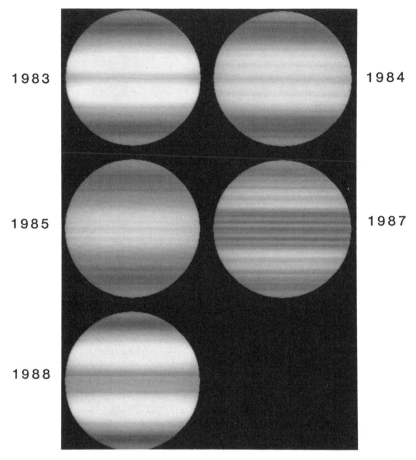

Fig. 8. Solar temperature distributions during the summers 1983, 1984, 1985, 1987 and 1988. The range from light to dark corresponds to a fractional intensity variation of about 2×10^{-3} ($dT \approx 3$ K) (Kuhn et al. 1988; Kuhn 1989). Photo by courtesy of J. Kuhn.

to explain consistently the p-mode splitting coefficients (2nd and 4th order) from the corresponding coefficient of the dT. Moreover, the calculated centroid frequency shifts (zero-order "splitting") correspond well to the reported frequency shifts of low-order p-modes reported by Woodard and Noyes (1985), Gelly et al. (1988) and Palle et al. (1988). These independent results strongly support the model that the observed dT is due to a global effect manifested by an asphericity in c^2 and T_{eff} in a shallow layer below the surface that varies with the solar cycle. How this explanation of the solar cycle modulation of the irradiance by observed dT is physically linked to the explanation by a varying bright network as indicated by, e.g., the He I index is not yet clear, but we note that the two interpretations are not necessarily inconsistent with each other.

IV. SUMMARY

The past decade has seen the introduction of radiometers with sufficient precision and sampling to measure a variety of small variations of the total solar irradiance. These variations are interesting to solar physicists in several ways. In addition, the presence of a distinct 11-yr modulation of total irradiance suggests that longer-term variations may be significant for the Earth's climate.

The large variance imposed on the total solar irradiance by active-region effects, and in particular by sunspots, demands that careful modeling efforts be made. From the frequency-domain point of view, the solar variability in this band consists of the sunspot deficits and facular excesses modulated by the solar rotation (nominally at ~ 0.43 μHz), superposed on a continuum caused by the hierarchy of granulation (including mesogranulation and supergranulation, as well as possible smaller features). To characterize the active-region effects well, and to search for the granulation and other signals in the total irradiance, will require several new developments:

1. Acquisition of higher-quality photometric data, with which the PSI and analogous facular terms can be estimated with higher accuracy;
2. Careful analysis of the existing time series of total irradiance measurements;
3. Acquisition of spectral irradiance measurements of comparable stability to those of ACRIM on these time scales, in order to study the spectral redistribution of the variations formed in different regions of the solar atmosphere.

Even more important than these new developments, however, is the maintenance and extension of the precise total-irradiance data base. Overlapping measurements are especially desirable, so that the longest-term variations can be reliably determined. For this purpose we urge that future meteorological satellites carry solar irradiance instrumentation. This instrumentation does not require much weight, power or telemetry, and the precise long-term monitoring of total solar radiative input is a crucial element for the quantitative understanding of global environmental problems, such as the greenhouse effect.

SUNSPOT NUMBER VARIATIONS: STOCHASTIC OR CHAOTIC

GREGOR E. MORFILL, HERBERT SCHEINGRABER,
WOLFGANG VOGES
Max-Planck-Institut für Physik und Astrophysik

and

CHARLES P. SONETT
University of Arizona

In this chapter we analyze the sunspot record over time scales of weeks or months, which gives information about the magnetic field transport in the solar convection zone and, possibly, also information about mechanisms responsible for the production and annihilation of sunspots. The analysis we use is new; it has not yet been employed in the study of this phenomenon. We employ the standard Fourier or power spectral analysis and supplement this with a correlation integral analysis. The latter allows us in principle to determine the "correlation dimension" and to investigate the minimum number of Fourier modes sufficient to describe the data record. (In practice, we use the slope of the correlation integral over the full range of scales for a comparison of models with the observed data.) This in turn should provide us with important clues about the nature and origin of the fluctuations in the sunspot record on short (weekly) time scales. Problems exist due to the superposition of stochastic processes and the 11-yr (dynamo associated) sunspot cycle. The corresponding signatures are analyzed by reference to synthetic data. Regarding the observed sunspot data, it can be shown, rather surprisingly, that a heuristic model (which has as its only major assumption a stochastic, but 11-yr modulated, occurrence of spots in time) does not fit the observations well. This implies that the assumption about a stochastic (i.e., independent) occurrence of sunspots cannot be correct. There is some "memory" in the sunspot cycle that correlates the observed short-term fluctuations over a characteristic time period. This time period has been found to be ~ 50 days. In conjunction with numerical experiments on Boussinesq convection, this suggests an association with giant convection cells as the main driving process for the "memory" as well as the number of Fourier modes. The observed phenomenon that occasionally new sunspots appear in the same sites where others have just dissipated, is presumably related to this association.

I. INTRODUCTION

This chapter addresses an investigation of the short-period dynamical properties of the solar sunspot number record; i.e., short-term fluctuations in the high-frequency record of sunspots—the daily record, suitably averaged as the case may be. The daily record permits investigation of the dynamically interesting monthly time scale which, in turn, provides information on the physics of the convective zone and of dissipation mechanisms of sunspot fields. The study reported here is cast in a somewhat unusual mode. The generation of a suite of models to be compared to computation of power spectra and correlation integrals of the sunspot number record makes contact with Bayesian statistics in that specific *a priori* models are the basis of comparison. However, the mathematical basis underlying a full Bayesian calculation of likelihood is not presently available and the comparison is qualitative and nonunique.

Sunspots are conspicuous small dark regions, in which the luminosity is diminished with respect to the general solar surface. They thus appear dark although the absolute luminosity remains high. Early naked-eye observations were carried out in the Orient (see the accounts by Wittman and Xu [1987], Xu and Jiang [1982], Zhang [1983] and Clark and Stephenson [1978]. Modern observations began with the telescope; more or less uninterrupted records are available since 1610, the year of Galileo's telescopic observations (see Waldmeier 1961; NOAA 1988; chapter by Coyne).

Hale's (1908) discovery of strong magnetic fields associated with sunspots leads to our knowledge that spots are the optical expression of a much more involved magnetohydrodynamic phenomenon associated with the Sun's global magnetic field. Sunspots tend to occur in bipolar pairs with a leading (preceeding or *p*) spot and a following (*f*) spot. The polarity of the magnetic field connecting the spots depends on the polarity of the global field. Thus the polarity of the bipolar pairs follows the polarity reversal of the Sun's global field every 11 yr. [One should note of course that the 11-yr period is not exactly 11 yr, and only referred to as such for brevity. The recent NOAA (1988) compilation yields 11.1 yr. Apparently the discovery of the 11-yr periodicity is really due to Schwabe (1843) with Wolf responsible for the modern sunspot index.]

The spatial distribution of sunspots also reflects the 11-yr solar activity cycle. The butterfly diagram (Maunder 1922), which gives the distribution, in superimposed rotations, of spots over the solar surface, clearly shows the migration towards the equator with increasing time during the cycle. The breakout of new high latitude spots tends to take place prior to the complete loss of near-equatorial spots of the previous cycle.

The magnetic field in sunspots is estimated to be of order of 1 to 10 kgauss whereas the global-average solar field is only of the order of 1 gauss. The current viewpoint is that sunspots are vortices formed during convective upwelling of matter. This matter transports the initially toroidal magnetic field through the convection zone towards the surface, local buoyancy effects being responsible. Differential rotation and Coriolis forces are also vital factors in forcing spot formation (see, e.g., Babcock 1961; Leighton 1964,1969).

Over long time periods (e.g., the 11-yr cycle and possibly longer periods), sunspots appear to reflect the amplitude (though not the sign) of dynamo activity. Over medium scales in time, they provide a visible signature of physical processes such as transport effects in the convection zone. They are less useful in the study of processes characterized by short time scales (e.g., solar oscillations, granulation) because their own lifetime is too long and they integrate out these rapid variations. It is the medium range which is the subject of this chapter.

To investigate the dynamical structure in this period range, we use both the power spectral density (PSD) and the slope of the correlation integral (Grassberger and Procaccia 1983a,b). This parameter can help decide whether a dynamical system is regularly periodic, deterministic chaotic, or stochastic. Subsequently, when we refer to "deterministic chaos" we shall simply use the term "chaos" in its modern usage. Although magnetohydrodynamic (MHD) systems inherently contain an infinite number of degrees of freedom, in most problems they can be described adequately by a finite number of modes. This obviously leads to far simpler descriptions of the system than the full treatment. This in turn allows for a more thorough parameter study and perhaps a better physical and analytical understanding.

In Sec. II, we briefly describe the sunspot index, its derivation and the quality of the long-term record; this section also summarizes published work relating to sunspot index periodicities. As we shall deal with short-term variability on the order of days to months, we only briefly touch on longer periods in the data record, with some reference to possible physical origins. As the central aim of this chapter is the analysis of the temporal structure of the sunspot activity, on time scales of days to weeks using the slopes of the correlation integral, the analytic approach is sketched in Sec. III.

The principal features of the wide-band sunspot index record and the correlation integral analysis of the data are reviewed in Sec. IV. The aim of this is the comparison with synthetic models of sunspot generation discussed in the ensuing Sec. V. Finally the comparison is made of these models with the sunspot index time series and applied in the last section to an analysis of the physics of the convective zone.

Because a new analysis technique is used here for a complicated system, which contains random noise as well as modulation, we demonstrate the properties of these effects using some simple synthetic data sets. In the appendix to this chapter, we compare modulated random noise and a modulated chaotic signal and investigate edge and projection effects on the slopes of the correlation integral.

II. THE SUNSPOT INDEX AND ITS FREQUENCY SPECTRUM

The sunspot record is one of the oldest uninterrupted scientific records and is readily available. Hence it has become the cornerstone for much of that part of geophysical time-series analysis aimed at recognition of solar-related variability. Observationally, the canonical parameter used to indicate sunspot frequency and activity is the sunspot index N, an empirical number invented by Wolf; it is issued once per day, currently from Brussels, determined from a combination of N_S, the number of sunspots, G the number of groups, and a weighting factor W, i.e.,

$$N = W(N_S + 10\,G) \qquad (1)$$

where W is a figure of merit associated with individual observers and instruments (see, e.g., Gibson 1973).

What is especially appealing about the sunspot index is its universal quality as an indicator of global solar activity, in spite of the personal term W in Eq. (1). This universality was demonstrated by the proportionality of Wolf's numbers to those published by Greenwich (see Norton's comment in Schove [1983]) and by others. Because sunspots are regions of lessened surface solar temperature, an association with total radiance is expected (Spiegel and Weiss 1980; Lean and Foukal 1988). The ACRIM radiometer reveals that radiance does vary with sunspot number or index (Willson, et al. 1986; Willson and Hudson 1988; see also the chapter by Fröhlich et al.).

The dominant period in the sunspot record which has occupied most attention since its discovery by Wolf is the 11-yr cycle. Additional properties of potential importance contained in the sunspot index sequence are the asymmetry in the magnitude of the index in alternating cycles, the variability in solar cycle period expressed in the index, and the several other variables reported over many years including rise and fall times that differ and are related to the maximum cycle amplitude. These properties are tabulated by Wilson (1988).

Lomb and Anderson (1980) list key studies of sunspot index periods reported by various workers. Their analysis using the prewhitened periodogram (see Fig. 1) exhibits 14 spectral features of which only 3 appear to be linearly independent and the remainder due to amplitude modulation. Their choice of independent line frequencies are 0.0111 yr^{-1}, 0.0181 yr^{-1} and

SUNSPOT INDEX AUTOCORRELATION

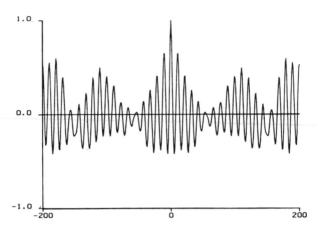

Fig. 1. Autocorrelation $\rho(t)$ vs lag parameter t (in yr) of the sunspot index. The dominant 11-yr solar cycle (high frequency) is modulated by the 90-yr Gleissberg period. The important asymmetry of $\rho(t)$ about $\rho(0)$ is an effect outside the scope of this chapter.

0.0909 yr^{-1}, corresponding approximately to the Gleissberg period (Gleissberg 1966), its second harmonic, and lastly the 11-yr period. Interestingly, Feynman and Fougere (1984) confirm this spectral feature at a period of 88 yr using maximum entropy method (MEM). Lomb and Anderson also discuss the application of autoregressive modeling to the sunspot index. Barnes et al. (1980) generated an autoregressive-moving average model using narrowband gaussian noise; that model predicts the major features of the sunspot index sequence. Cohen and Lintz (1974) also compute the power spectrum of the sunspot index using both MEM and the periodogram. They identify major features at periods of 95.8 yr and at 11.0, 9.8 and 8.3 yr but these are probably dependent somewhat upon MEM order.

Inspection of the sunspot index time record also suggests that the ~ 90-yr period modulates the basic 22-yr Hale period. Some evidence for this exists in the spectrum; of the 13 spectral lines appearing in the MEM spectrum all but those at ~ 90 and ~ 45 yr are clustered into two groups ~ 11 and 22 yr. A model calculation suggests amplitude modulation of the 22-yr period by the 90-yr Gleissberg period with

$$F(t) = \{[1 + \delta\cos(\omega_m t)][\Delta + \cos(\omega_c t)]\}^2 \tag{2}$$

where $2\pi/\omega_c$ is the 22-yr (carrier) period, $2\pi/\omega_m$ an amplitude modulation (Gleissberg) period, Δ is a small offset conjectured to be a fossil core field, and δ the amplitude modulation index (Sonett 1982).

Expansion of Eq. (2) discloses 13 spectral lines, a quintet ~ 11 yr, another ~ 22 yr, the ~ 90 yr Gleissberg period, its 45 yr second harmonic, and finally a small zero frequency offset. An alternative method of computation based on Eq. (2) which yields relative line probabilities and approximate intensities is reported by Bretthorst (1988). Periods too long to be detected in the sunspot record are inferred from the radiocarbon record (see the chapter by Damon and Sonett). The 11- and 22-yr cyclic activity, as well as possible prolonged minima in activity have been discussed in terms of a simplified (and specialized) solar dynamo model by Zel'dovich and Ruzmaikin (1980).

From various indicators of solar activity, a number of intermediate periodicities have been reported. Rieger et al. (1984) find possible evidence for a 154-day periodicity in the occurrence of flares; Bai (1987) reports a dominant periodicity in the comprehensive flare index during solar cycle 19 at 51 but not at 154 days. Because flare production appears to be correlated with the total magnetic energy of an active region (Mayfield and Lawrence 1985) and this in turn with sunspot size (Allen 1976), these measurements are relevant for processes generating and transporting magnetic fields. Recently, Lean and Brueckner (1989) have investigated the possible occurrence of intermediate periodicities in the sunspot blocking function, the 10.7-cm radio flux, the Zürich sunspot number and the CAII K plage index. They report a significant signal at periods of 155 days and near 323 days. The latter periodicity was also reported earlier by Delache et al. (1985); however, it is prominent only in cycle 21. Currie (1974) reports a full spectrum of 1 to 11-yr periods using MEM.

III. THE CORRELATION INTEGRAL AND THE CORRELATION DIMENSION

To assess whether the short-period (high-frequency) sunspot record is chaotic or contains merely random noise components, we resort to the phase space dynamical representation to provide a measure of the structure of the putative "attractor." (An "attractor" is that subspace in phase space that is occupied by the system in the limit $t \rightarrow \infty$.) The technique employed is the correlation integral algorithm (Grassberger and Procaccia 1983a,b) to determine the (possibly fractal) dimension of the system. Higher fractal dimension implies increasingly developed turbulence in analogy to the (Fourier) mode representation of MHD flows. Periodic or multiple periodic processes have integer dimensionality with the dimension corresponding to the number of modes (for references, see Morfill et al. 1989).

Low-order fractal dimensionality is of special interest, because for systems displaying such low order, one may possibly succeed in describing them in mathematically simple terms—by just a few ($n = 3$ or more) coupled nonlinear ordinary differential equations. Such a description, which must satisfy observational constraints, is potentially a very powerful tool for a deeper understanding of the underlying nonlinear dynamics.

The attractor is loosely defined as the subspace in phase space finally occupied by the system during its normal evolution starting from a set of different initial conditions. A given limited data record will obviously not fill all the attractor; there are not enough points to do this. However, it should be large enough to be representative of the structure of the attractor and of the corresponding dynamics. Hence a second and third data set of similar quality should produce similar results, unless the system undergoes significant evolutionary or topological changes, other than the normal evolution (e.g., external influences).

The sunspot data base is too short for a meaningful analysis of the long-period (yearly) record, a problem recognized earlier by Zel'dovich and Ruzmaikin (1980). From our experience, reasonable results are obtained for a data record of ≥ 1000 points, with a temporal resolution of $\sim 1/10$ of the correlation time (or of the period range of interest). Attempts to interpret significantly shorter data sets with respect to their underlying processes are fraught with difficulties. Current analysis techniques yield either meaningless results, or at best, inspired guesses. The daily sunspot record contains a sufficiently large number of data points; it is uninterrupted and equally spaced. The data provides a direct measure of solar activity, and finally, it permits investigation of the dynamically interesting monthly time scale related to the physics of the convection zone and dissipation mechanisms of sunspot fields.

We now briefly review the correlation algorithm as developed by Takens (1981) and by Grassberger and Procaccia (1983a). The data are represented as shifted time series

$$
\begin{aligned}
&x(t_1), \ldots , x(t_N) \\
&x(t_1 + \Delta t), \ldots , x(t_N + \Delta t) \\
&\quad \cdot \cdot \cdot \\
&\quad \cdot \cdot \cdot \\
&\quad \cdot \cdot \cdot \\
&x(t_1 + (d - 1)\Delta t), \ldots , x(t_N + (d - 1)\Delta t)
\end{aligned}
\tag{3}
$$

where $x_i = \{x(t_i), x(t_i + \Delta t), \ldots x(t_i + (d-1)\Delta t)\}$ is a d-dimensional vector, defining a point in a d-dimensional space. In this way, the attractor is reconstructed from a scalar time series of data from a single variable (i.e., the sunspot number). The real attractor, which gave rise to the particular data sequence, is completely embedded in the artificial d-dimensional phase space, if $d \to \infty$. The procedure is then to determine $|x_i - x_j|$ for $d = 1$, and to calculate the qth-order correlation integral $C_{d=1}^{(q)}(r)$. Next, the dimension of the artificial phase space is incremented, and $C_{d=2}^{(q)}(r)$ is calculated and so on until the slope of the correlation integral no longer changes. According to the embedding theorem (Takens 1981) $d \geq 2n + 1$ is adequate, where n is the actual, but unknown, real dimension of the attractor. The attractor is then embedded in the d-dimensional artificial phase space. In principle, this can be done for any multipoint correlation q (Grassberger 1983,1988; Pawelzik and Schuster 1987; Atmanspacher et al. 1988) although the two-point correlation suffices for many purposes.

The embedding theorem assumes an infinite number of data points of infinite precision, free of noise, and an arbitrarily small sampling interval (Takens 1981). Successful reconstruction of attractors is restricted by data precision that sets fundamental limits on the dimension of the attractor, which can be reconstructed independently of the length of the data set. The slopes of the correlation integrals converge with increasing d towards the qth order dimension $D^{(q)}$

$$D^{(q)} = \lim_{d \to \infty} \lim_{r \to 0} \frac{\log_2 C_d^{(q)}(r)}{\log_2 r} \qquad (4)$$

Different orders q of the correlations signify local substructure in phase space, i.e., a different scaling of regions of different point densities. The dimension $D^{(q)}$ is in general a real number, though not integer. For $q = 2$, we obtain the two-point correlation or "correlation dimension" $D^{(2)}$. The dimension $D^{(1)}$ is termed "information" dimension. Generally, for the different order dimensions, the inequality $D^{(q)} \leq D^{(q-1)}$ holds.

In practice, the limits given in Eq. 4 are unattainable but fortunately not essential. Also for practical applications, very high dimensions n are not tractable. We have found that an embedding dimension, $d \approx 20$ provides reasonable results (significant convergence of the slopes) in most cases of low dimensional chaos. The unavoidable random noise present in the data shows up in the correlation integral $C_d^{(q)}(r)$ as a gradient increasing with increasing embedding dimension. These random processes are superimposed on any chaotic processes and occupy regions of small r. Hence plotting the slope of $C_d^{(q)}(r)$ for given d as a function of r helps in disentangling these two components, provided the random part is not too large. Otherwise, it may mask an underlying chaotic signature. In principle, the analysis can be carried further to obtain the Kolmogorov entropy and the spectrum of scaling indices. We shall not describe this here; the reader is referred to the original literature (Grassberger and Procaccia 1983b; Halsey et al. 1986; Pawelzik and Schuster 1987).

While the effect of random noise superimposed on a modulated signal has been studied already using these analysis techniques (see, e.g., Morfill et al. 1989), no study of a modulated chaotic signal has yet been performed. In the appendix, we investigate this modulation, as well as other effects, and determine the associated signature in the algorithms used.

IV. ANALYSIS OF THE SUNSPOT DATA

The principal features of the daily sunspot record shown in Fig. 2a depict the measurements of the daily Sun activity index for the time period 1878–1945 binned in units of 12 days. The overall time period comprises 6 solar cycles (number 12–17); binning in units of 12 days gives a total of 2048 data points. The length of the record allows the 27-day signature of the solar rotation period up to the 11-yr cycle to be detected.

A. The Power Spectral Density

Figure 2b shows the power spectrum of the record thus averaged with the 11-yr peak. The 22-yr periodicity is also visible, although somewhat masked by a broad noise level including a spectral feature extending from $\sim 5 \times 10^{-9}$ Hz upwards to the Nyquist frequency. This broad feature is not constant in amplitude but drops by about an order of magnitude from $\sim 2 \times 10^{-8}$ Hz to 2×10^{-7} Hz exhibiting an f^{-1} spectrum. There is an additional indication that the power decreases again at frequencies below $\sim 2 \times 10^{-8}$ Hz until this component merges with the 11-yr signature. The 27-day solar rotation period (4.3×10^{-7} Hz) observed in the spectrum is associated with long-lived sunspots or sunspot groups that persist for more than one rotation period.

B. Correlation Integral

Figure 2c shows the slopes of the two-point correlation integrals, computed for embedding dimensions of the artificial phase space between 1 and 20, and as a function of the correlation scale r. It contains a wealth of information which we now interpret. The slope generally increases with increasing embedding dimension. The small knee seen in the correlation integral in Fig. 2c, more pronounced at higher embedding dimensions and large scale $\log_2 r \sim 11.5$, corresponds to a slope of ~ 1. This is taken to be the signature of the near sinusoidal variation of sunspot activity, which, as a singly periodic signal, should yield a dimension of unity (see the appendix). That this signature is found at large correlation scales is understandable because, for large values of r, the key relation is between maxima and minima of the signal. For smaller r, the 11-yr variation is masked by short-term fluctuations.

An approximate plateau is present at intermediate values of the correlation scale r with a local slope of the correlation integral of 1.5 to 2.0. This may possibly be the signature of the alternation in the peak sunspot number in successive solar cycles. Thus in this regime of correlation scales, the period of 22 yr is picked up. That this should occur at lower r is understandable, if we compare the amplitude of the peak-to-peak variations with the 11-yr modulation. This amplitude gives a measure of the correlation scale for this second harmonic.

With smaller correlation scale, slopes generally increase but one pos-

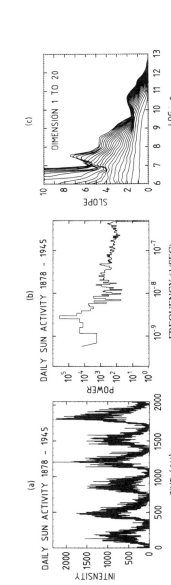

Fig. 2. (a) Daily sunspot index binned over 12 days during the period from 1878 to 1945. (b) Power spectrum of the time series (a). (c) Slopes of the correlation integrals of the time series (a) for embedding dimensions 1 to 20 (reading from bottom to top).

sible flattening occurs between 4 and 6, as indicated by the knee seen in the curves with artificial phase space dimension $\gtrsim 13$. For small correlation scales r, the correlation integral slopes increase, consistent with the existence of small-amplitude random noise. Theoretically, slopes should approach the artificial phase space dimension as $r \rightarrow 0$. In practice, finite data sets, signal discretization, etc. prevent this (see, e.g., Smith 1988). According to an argument of Ruelle (Ruelle 1990), one always ends up with $D^{(2)} \leq 2 \log_{10} N$ using the Grassberger-Procaccia algorithm. In our case, the limit for the correlation dimension is $D^{(2)} \leq 6.6$. If the algorithm yields slopes near the limiting value one can no longer make quantitative statements.

V. MODELS

In this section, we carry out the idealizations of the sunspot number index which we then match to the power spectrum and correlations computed in the last section. This is a key to our understanding of the correlations.

A. Model with 11-yr Substrate and Poisson Noise

If the long-term sunspot level were determined by the basic solar dynamo activity, and the mean sunspot level is merely a manifestation of the mean dynamo magnetic field, then short-term variations could just be imposed by superimposition and have a distinct and different physical origin or be purely statistical. To test this, the sunspot data of Fig. 2a is first averaged and then the smoothed curves shown in Fig. 3a are computed. These are $\sin^2(t)$ fits for the leading and trailing edges of the data, taking into account the nonzero levels at different sunspot minima. This smoothed data set was subjected to the same power spectral and correlation integral analysis using the same time binning as was done to the real original sunspot data. The results are shown in Figs. 3b and c, where Fig. 3c shows only the slope of the correlation integral for the artificial phase space dimension of 20.

The power spectral analysis repeats the 11-yr and the 22-yr peaks as well as an f^{-2} intermediate frequency tail. This tail could be due to the intrinsic digitization of the data, and will not be considered further. Comparison with Fig. 2b shows that only the power in the dominant 11-yr peak is well matched. In contrast, the correlation integral shows little evidence of the doubly periodic property of the smoothed data set. The slope barely rises to 1 for large correlation scale and the doubly periodic signature, which sets in at $\log_2 r \geq 9$, does not rise to 2. The reason for this is due both to the phase space projection and the small number of cycles in the data set. (See the appendix for a discussion of projection effects.)

Next, the uncorrelated fluctuations (amplitude proportional to the *rms* sunspot intensity) are superimposed upon the substrate. Figures 4a, b and c show the model for Poisson noise (mean amplitude 0.15 relative to the smoothed background signal of Fig. 2a), and Figs. 5a, b and c for an inter-

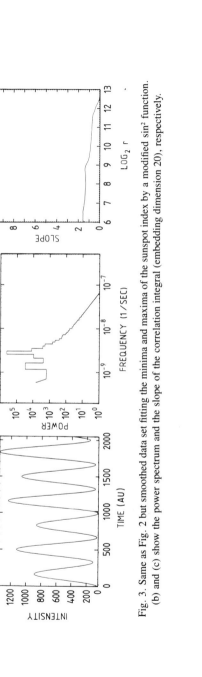

Fig. 3. Same as Fig. 2 but smoothed data set fitting the minima and maxima of the sunspot index by a modified sin^2 function. (b) and (c) show the power spectrum and the slope of the correlation integral (embedding dimension 20), respectively.

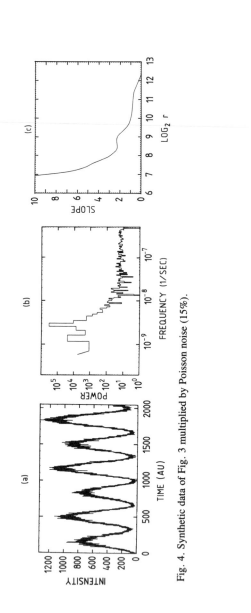

Fig. 4. Synthetic data of Fig. 3 multiplied by Poisson noise (15%).

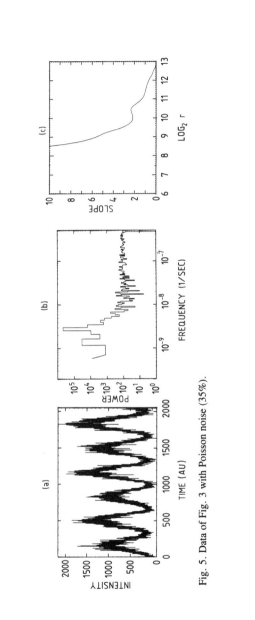

Fig. 5. Data of Fig. 3 with Poisson noise (35%).

mediate noise level of 0.35. The noise level of 0.15 amplitude in Fig. 4 was chosen so that the increase in the slope of the correlation integral at small correlation scales r coincided with that of the original data. Correspondingly, the noise level of 0.35 in Figs. 5 was chosen so that the power at high frequencies (10^{-7} Hz) was of similar magnitude as that of the original data. Figure 5b yields power at high frequencies at a level consistent with the sunspot data although the noise spectrum is flat, as expected for white noise or purely stochastic fluctuations.

The correlational integral (Figs. 4c and 5c) shows the steps leading to slopes 1 and 2 and the increase in the slope with decreasing r associated with noise. The random signal has two effects on the plateau. The first is to increase its level (in the noise-free case the slope in the doubly periodic domain remained below 1.5 due to projection effects, whereas in Figs. 4c and 5c it exceeds 2). The second effect is to extend it to larger values for r; the random fluctuations increase the dynamic range of the signal and hence the correlation scale r is stretched. This is clearly seen in Fig. 5c, where the upper extent of $\log_2 r$ has increased to 12.85 compared to 12.55 in the noise-free case (Fig. 3c). Also the limit, where the doubly periodic signature raises the plateau level, has changed from $\log_2 r \sim 9$ to ~ 10.7. Increasing the noise level further eventually swamps the second step, and it is no longer recognizable.

Comparison of power spectra and correlation integrals of these models with the original data (Figs. 2a, b, c) shows little correspondence. It is clear that we can match either the high-frequency power (as was done in the "high-noise" model), or the correlation-scale location of the onset of the noise signature in the correlation integral plots (as was done in the "low-noise" model). However, we cannot do both. In addition, the shapes of the spectral continua are inconsistent: the original data exhibit an f^{-1} dependence, whereas white noise is frequency independent. In the low-noise model, the power at high frequencies ($\geq 10^{-8}$ Hz) is about a factor of 10 too small (Fig. 4b). While the synthetic data in the high-noise model agrees with the original data for correlation scales $\log_2 r \geq 10.5$, below this it is dominated by the signature of the stochastic signal, and the slope of the correlation integral increases sharply with decreasing values of r for $\log_2 r \leq 9.5$ (Fig. 5c). By contrast, the original data has a second plateau for slopes between 4 and 5.

B. Model with Spatial and Temporal Sunspot Dependence

A more realistic but still simple heuristic model assumes that the occurrence of sunspots in longitude on the Sun is stochastic, with sunspots (or groups) rigidly co-rotating with the Sun; and the formation of sunspots themselves is spontaneous and stochastic in time apart from the observed long-term (11-yr) modulation. From the measured mean sunspot index variation (Fig. 3a), we define a temporal probability or modulation curve for sunspot formation using a mean sunspot (or sunspot group) production rate $\dot{N}$. Using random numbers yields a sequence of formation times for sunspots (or groups) which corresponds closely to the observations of Fig. 2a. The important assumption in this model is the stochastically modulated occurrence of sunspots (groups). Next, using random numbers, we determine those events to be regarded as single sunspots and those to be regarded as sunspot groups. A mean relative number of groups and $f_s \equiv 1 - f_g$ of single sunspots is assumed. For most cases, we use $f_g = 0.1$. We then determine the growth time t_g, the full strength life time t_f, the decay time t_d and the total life time $t_L \equiv t_f + t_d$ of the sunspot or sunspot group.

For individual sunspots, t_g and t_d are effectively zero, whereas groups of sunspots have growth/decay times of 1 to 3 weeks. Our model assumes a growth time of $t_g = 7(1 + \tau_g)$ days, where τ_g is determined using normally distributed random numbers with half width 0.5. For the sunspot group decay time, we again use random numbers, such that $t_d = 7(1 + 3\tau_d)$ days. Here $0 \leq \tau_d \leq 1$ is obtained from random numbers using a white noise probability distribution. The total spot lifetime t_L is then given as the sum of the period while the sunspot group is at full strength, t_f, and the decay time is t_d. We use $t_f = (1.5 + 0.5\tau_f)\tau_0$ for individual spots and $t_f = (1.5 + \tau_f)\tau_0$ for groups. In both cases, τ_f is determined individually using random numbers and a gaussian probability distribution of half width 0.5, and $\tau_0 = 27$ days is the solar rotation period, which for this model we assume to be latitude independent in the $\pm 30°$ latitude band of sunspots. Finally, we need to know which sunspots are visible from Earth. To compute this, we specify the sunspot longitude on the Sun using random numbers and assuming a constant distribution in longitude. The sunspots are then assumed to co-rotate rigidly with the solar rotation period τ_0.

This heuristic sunspot model is based on available observational data (see, e.g., Bray and Loughhead 1964). It gives the number of individual sunspots N_S, and sunspot groups G in the Earth-facing hemisphere at any given moment in time. The results are shown in Figs. 6a, b and c. Panel 6a shows the computed synthetic data sets for three different sequences of random numbers, panel 6b the associated power spectra and panel 6c the correlation integrals for embedding space dimension 20.

The power spectrum (panel 6b) shows reasonable agreement with the observations (Fig. 2b) with both the 11-yr feature and the peak at the solar rotation frequency, although the 22-yr period is comparatively weak. An overall lack of power in the heuristic model exists at low frequency. Some of this may be due to the asymmetry in the cycles which is not modeled. Alternatively, this could be due to the fact that there are only three 22-yr cycles in the data set. The plateau at intermediate frequencies occurs at about the expected power level, indicating that the choice $\dot{N} = 0.3$ spots/day is reasonable. The power level appears to drop off more sharply towards higher frequencies (at around 10^{-7} Hz) in the heuristic model than in the original data. The addition of more short-lived sunspots could possibly rectify this.

Based on the power spectra alone, it might be concluded that the heuris-

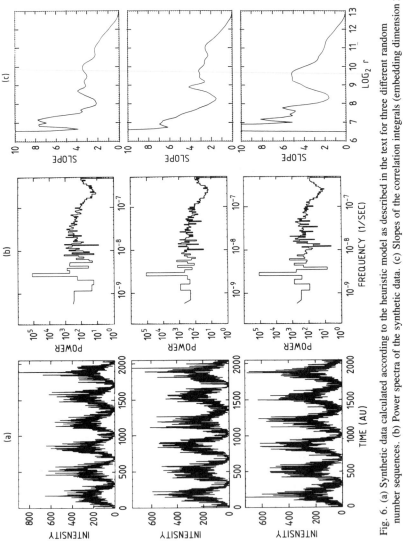

Fig. 6. (a) Synthetic data calculated according to the heuristic model as described in the text for three different random number sequences. (b) Power spectra of the synthetic data. (c) Slopes of the correlation integrals (embedding dimension 20).

tic model (with "fine tuning") provides a good description of the observed sunspot variations. If so, then the two principal assumptions (i.e., random distribution of sunspots in longitude and their stochastic occurrence within an 11-yr modulation envelope) are reasonable. But the power spectrum itself cannot provide information on correlations and ordering and the structure of the attractor that defines the data set. Figure 6c shows the latter for embedding space dimension 20. The curves for three sets of random numbers shown in the panels are superficially similar: they all exhibit the classical signature of random noise at small correlation lengths r, i.e., slope rising to large values, a minimum in slope at intermediate r, and a more or less well defined plateau at correlation scales above this minimum. Although the levels vary between 3 and 5, the gradual drop to zero slope at the largest correlation scales is similar for the three examples.

Figure 6c should be compared with the corresponding curve of the real data set in Fig. 2. Note the change in the range of the correlation scale r due to different normalizations of the absolute intensity values I. On a (natural) logarithmic scale, this can be adjusted by shifting the curves horizontally by $\log_2 (I_1 / I_2)$. Outside the range $5 \leq \log_2 r \leq 9$ in Fig. 6c, corresponding to $7 \leq \log_2 r \leq 11$ in Fig. 2c, agreement is reasonable. This corresponds to the stochastic signature at small correlation lengths and the decrease in the slope to zero as the maximum correlation scale r is reached.

At intermediate values of r the agreement is poor. The plateau with slope ~ 2 at $\log_2 r \sim 10$ (corresponding to $\log_2 r \sim 8$ in Fig. 6c) is not well defined in the heuristic model and, in particular, the knee at $\log_2 r \sim 8$ (corresponding to $\log_2 r \sim 6$ in Fig. 6c) is absent. The slope at these comparable scales are ~ 4 to 6 in the original data, and ~ 2 in the heuristic data. The lack of agreement at intermediate values of $\log_2 r$ is compelling evidence that the heuristic model is lacking an essential physical element in its description. Hence we conclude that although the model adequately describes some solar sunspot features, from the standpoint of the correlation integral it is severely deficient.

The model assumption of a stochastic longitudinal distribution of sunspots is of minor importance. It contributes to the power spectral feature at the solar rotation frequency but not to the correlation. It is the assumption that sunspots occur stochastically within an 11-yr modulation envelope which is at fault, corresponding to the observational evidence for a strong variability in the sunspot index with solar cycle.

C. Model Based upon Lorenz Equations

In construction of a model encompassing an attractor, we assume, as before, that the observed long-term means in the sunspot numbers reflect the dynamo activity, and thus provides the substrate for the shorter-scale processes, which are modulated accordingly. We assume that shorter-scale transport of magnetic flux tubes through the convection zone may be described by a suitable Lorenz model including some stochastic noise. The latter is at an insignificant level compared with the general scale of fluctuations; nevertheless it is reasonable to include it, if only to simulate variations in seeing conditions, observational bias, weather, discretization of data etc.

The Lorenz equations represent a well-known chaotic system and describe the physics of a gravity and pressure-confined viscous fluid heated from below and cooled at the surface. Such a system can be either static, large-scale convective and/or chaotic, as shown by Lorenz (1963). We see a close analogy here to the solar convection zone. The system is heated from below by the energy released in nuclear fusion reactions, and cooled at the surface by radiation according to the Stefan-Boltzmann law. The associated strong temperature dependence (T^4) ensures that the mean surface temperature T_s does not fluctuate too greatly.

The Lorenz equations are

$$
\begin{aligned}
\dot{X} &= -\sigma\dot{X} + Y \\
\dot{Y} &= -XZ + rX - Y \\
\dot{Z} &= XY - bZ.
\end{aligned}
\tag{5}
$$

They can be derived from the hydrodynamic equations using the method of mode reductions. The variables X, Y, Z are the intensity of the convective motion, the temperature difference between the upward and downward moving fluid, and the departure from linearity of the vertical temperature profile, respectively. The parameters σ, b and r represent the Prandtl number ν/κ, where ν is the kinematic viscosity and κ the thermal conductivity, the quantity $b = 4/(1 + a^2)$, where a is the ratio of horizontal/vertical scale of the cells, and $r = R/R_c$, the normalized Reynolds number which is proportional to ΔT.

For our model $\sigma = 10$, $b = 8/3$ and $r = 28$, and the correlation time $\tau = 54$ days. With these values the solution of the Lorenz system is chaotic with a correlation dimension ~ 2.06 (Grassberger and Procaccia 1983a). The Lorenz sequence is then multiplied by the (approximately $\sin^2 \omega t$) average solar cycle period and the random noise is assumed to be Poisson as before. The sequence was scaled by a constant factor to obtain the same power level as for the real data (equivalent to choosing $\dot{N}$).

In Figs. 7a, b, and c, we show three synthetic data sets based on the modulated Lorenz sequence described above. Different regions of a chaotic solution of the Lorenz system are used for the simulations. Panel (a) shows the synthetic data set, intensity plotted against time. In panel (b), we show the power spectra. Comparison with Fig. 2b shows an almost exact correspondence: the 22-yr peak is somewhat less pronounced in the real data, but the dip at $\sim 10^{-8}$ Hz, the f^{-1} dependence of the broad region above 3×10^{-8} Hz and the general power level compared with the 11-yr peak are

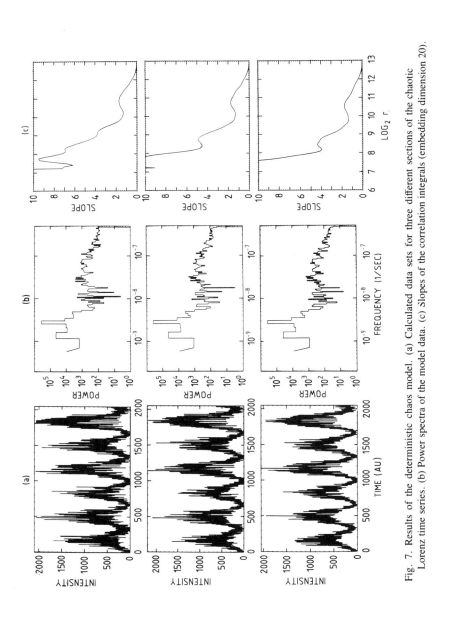

Fig. 7. Results of the deterministic chaos model. (a) Calculated data sets for three different sections of the chaotic Lorenz time series. (b) Power spectra of the model data. (c) Slopes of the correlation integrals (embedding dimension 20).

practically exact fits. The only shortcoming of this model is the lack of the solar rotation peak. We know the origin of this feature, and as we did not include it in the model, its absence is hardly surprising.

Next we compare the results of the correlation analysis, panel (c), with that for the sunspot data (Fig. 2c), and note again that the comparison is close. From large correlation scales to the small scales, the same features are formed in the synthetic data set as in the real data: the gradual increase in the slope of the correlation integral to a value of ~ 2, the dip, the increase to form a shoulder at around 4 to 5 and the rise to large values. The one shortcoming of this particular model sequence seems to be the stochastic component, which appears too large. In these examples, the stochastic signal was modulated with the square root of the mean signal strength, simulating Poisson noise. Other, equally likely scenarios can be envisaged, e.g., a constant noise level. This will shift the steep rise towards lower r, but should leave the signature at larger r values essentially unchanged.

VI. PHYSICAL INTERPRETATION AND CONCLUSION

Solar magnetic field production is canonically based upon magnetic dynamo theory (see, e.g., Parker 1955b,1979a; Steenbeck and Krause 1969; Stix 1972,1979). Linear dynamo theory yields periodic solutions for the mean magnetic field. The role of differential rotation, helicity, diamagnetism, buoyancy, etc. as important driving mechanisms for such self-oscillating dynamo systems is reviewed in, e.g., Parker (1979a) and Zel'dovich and Ruzmaikin (1980). The total magnetic flux in a sunspot is approximately proportional to its area; thus the sunspot index provides a rough measure of the total enhanced magnetic flux (or energy) at the surface of the Sun. This implies that sunspots contain, in principle, information about processes responsible for magnetic field production, annihilation and transport.

We briefly summarize the requirements for the generation of a mean field in a turbulent MHD fluid. These are (a) a "seed" field amplified by the dynamo process, and (b) turbulence with nonzero mean helicity. Mirror symmetric turbulence can either transport or destroy the mean field; it cannot amplify it. The averaged induction equation (averaged over scales much larger than the coherence length of the turbulence) is

$$\frac{\partial \mathbf{B}}{\partial t} = \nabla \times \mathbf{E} \tag{6}$$

where $\mathbf{E}$ is regarded as local. Equation (6) may be rewritten as

$$\frac{\partial \mathbf{B}}{\partial t} = \nabla \times \alpha \mathbf{B} + \nu_t \Delta \mathbf{B} \tag{7}$$

where the second term implies dissipation with a turbulent viscosity v_t, and α is a pseudoscalar function which is directly proportional to the mean helicity (velocity × vortex spin) where

$$\alpha = -\frac{\tau}{3}\mathbf{v}(\nabla \times \mathbf{v}) \tag{8}$$

where τ is the correlation time for the turbulence and is related to v_t via $v_t = \tau v^2 / 3$. Field generation takes place with a characteristic growth rate $\gamma \sim k\alpha$, where k is the wave number. A field with helicity $|\delta\mathbf{B}|\nabla \times \delta\mathbf{B}$ in one sense will grow, whereas the opposite helicity will decay. For a body rotating with angular velocity Ω, we have $\alpha \approx \Omega/k$, so that the growth rate is $\gamma \approx \Omega$.

The back reaction of the generated magnetic field on the fluid yields two effects. First, the increase in magnetic energy density adds a pressure that will affect the flow characteristics somewhat. Second, it will weaken the mode interactions and delay the transfer of energy to small scales. This effect reduces the dissipation of turbulent energy. Viewed in another way, the magnetic energy and helicity spectra will shift to smaller wave numbers (larger scales), an effect known as inverse cascades (see, e.g., Pouquet et al. 1976). These large structures have a longer lifetime than they would in Kolmogorov-type fully developed turbulence.

We have investigated three models: a stochastic noise model, a heuristic model based upon observed sunspot properties and a chaotic model involving the Lorenz equations. Sunspot activity is not well described by the stochastic model, modulated by the long-term dynamo variation. The results of the analysis are consistent with the existence of some "memory" with a nonperiodic correlation time of $\sim$ 50 days. This memory cannot be attributed to the possible dominance of individual long-lived sunspot groups; the heuristic model contains such groups and yields acceptable power spectra including their intermediate frequency signature.

At least one of the two assumptions made in the heuristic model is wrong. These were a stochastic occurrence of spots and spot groups in time, modulated only by the smoothed average function taken from the data itself (Fig. 3a) and a stochastic occurrence in longitude. It was argued that the latter translates mainly into the signature of the 27-day peak in the power spectrum. As the heuristic model reproduces this feature quite well, one may conclude that it is the first assumption (the stochastic occurrence in time) which is responsible for the discrepancy between this model and the observations.

The deterministic chaos model which we described in the third alternative and for which we generated synthetic data records for analysis and comparison, rather surprisingly fits the data best. The coherence time required for this model has to be in the region $\sim$ 50 days. Significantly different coherence times fail to reproduce the measurements in the correlation integral analyses. We cannot claim, on the basis of the comparisons—even with a whole range of test runs for the various models (many more than the samples we show here)—that the short-term variations of the sunspot index must be the consequence of a chaotic attractor of the Lorenz type, which affects spot production on short time scales ($\sim$ 50 days). However, our first two models are not satisfactory; there exists "memory" or "correlation" in the sunspot index with a characteristic time scale of $\sim$ 50 days, and this memory is not a manifestation of the mean lifetime of sunspots or sunspot groups. The Lorenz system, chosen for its original application to large-scale convective systems appears to have physical relevance to sunspot dynamics, if the memory is related to transport in the convective layer of the Sun.

From the point of view of introducing a memory into sunspot generation on time scales of the order 50 days or so, the following possibilities exist:

1. Effects on the seed field, e.g., by compressions, solar oscillations (g-modes?), distortion.
2. Effects on the mean helicity of turbulent eddies, e.g., differential rotation, Coriolis forces.
3. Processes affecting the field growth rate in the dynamo, e.g., length scales of turbulent eddies.
4. Changes in the structure of the turbulence, e.g., due to the back reaction of the magnetic field and helicity, and associated transport effects.

A. Transport in the Convection Zone

In spite of the success of the mixing length theory (Böhm-Vitense 1985) in describing stellar convection, it is nevertheless expected that the convective zone should be turbulent on all allowed scales rather than purely convective at one dominant scale $2\pi / k = \Delta R$, the extent of the convection zone. The success of the mixing length theory argues somewhat in favor of a dominant mode analysis. Recent work by Urata (1986) supports this conjecture. He examined three-dimensional Boussinesq convection in a box with insulated vertical walls heated from below. The conservation equations were written in Fourier expansions in x and y, with a finite cartesian mesh employed in the z direction, parallel to the gravitational field and the thermal gradient. For a Prandtl number $\sigma = 3$, the numerical experiment settled down into a limit cycle, whereas for a $\sigma = 1$, the system was irregular. Further analysis showed that the correlation dimension of this system was 3.3 $\pm$ 0.06, i.e., it exhibits low-order chaotic behavior.

At first sight such a result is surprising. The conservation laws are, of course, nonlinear—the $\mathbf{v} \cdot (\nabla \mathbf{v})$ term alone guarantees that—but they are partial differential equations, whereas the numerical results would suggest that the system may equally well be described with a set of four ordinary-coupled nonlinear differential equations. Thus a system with inherently infinite degrees of freedom (partial differential equations) is reduced to one which is adequately described by $n = 4$ ordinary coupled nonlinear differential equations. (Note from the embedding theorem that the analysis requires a Fourier expansion up to $2n + 1 = 9$ modes. Urata used 12 modes and hence the results do not suffer from embedding problems.)

The physical explanation of this result would appear to be that at Prandtl number $\sigma = 3$, the system develops large convection cells, whereas at Prandtl $\sigma = 1$, it becomes chaotic, but is still dominated by a low number of modes. Interestingly, the system did not develop into a Lorenz type.

B. Structures and Time Scales

Observations of the solar photosphere show a number of characteristic structures, and associated time and length scales. These are:

a. Granules having scales typically of the order 10^8 cm, gas velocities typically 10^5 cm s^{-1}, and characteristic time scales of $\sim 10^3$ s.

b. Supergranulation with dimensions typically of 1.5×10^9 cm, and gas velocities $\sim 5 \times 10^4$ cm s^{-1} and associated time scale of 3×10^4 s.

c. Giant convection cells (although evidence for these is still somewhat controversial) with characteristic length scales of $\sim 10^{10}$ cm and gas velocity $\sim 5 \times 10^3$ cm s^{-1}. The time scale is then 2×10^6 s, or ~ 20 days. The time scale for complete turnover is perhaps a factor of 4 longer.

The correlation time, or memory in the sunspot data suggests an association with giant convection cells based on these characteristic scales. This association is also physically plausible. Giant convection cells presumably are the most likely candidates to scoop up the magnetic field from the base of the convection zone and bring it to the surface, enhancing the field along the way into the kgauss range.

The numerical experiments are very useful in the sense that they give strong hints about the self-organization of unstable fluids. It should not surprise us, therefore, if the Sun also evolved in the simplest (= lowest number of modes?) way to get rid of the energy produced in its interior. This could imply, as we have assumed in our third model, that the short-term fluctuations in the sunspot index are governed by the self-organization of the convection zone. The excellent agreement with the chaotic model using a Lorenz attractor may still be fortuitous, although the possible behavior of the convection zone which it implies is entirely plausible.

In summary, we have taken a new look at one of the oldest uninterrupted scientific records, the sunspot index, and analyzed it with modern techniques based on using the correlation integral method. We have found that the sunspot index appears to be chaotic on short time scales, with a correlation period of ~ 50 days. Of several possible physical origins for this behavior, it seems plausible that transport through the solar convection zone is responsible. If so, sunspots may be used as tracers of the physics of solar convection.

APPENDIX
THE INFLUENCE OF NOISE AND MODULATION

A real chaotic system will exhibit stochastic fluctuations superimposed on a chaotic signature. Stochastic effects may arise from real random variations in flow patterns due to integrating the system spatially over resolved but uncorrelated fluctuating regions. In practice, this only works if either the power in the random signal is sufficiently low, or if separated by different time (length) scales. These general considerations apply also to the situation where regular signals are extracted from a noisy data set—the difference occurs in the meaning of sufficient in both instances.

A. One-Dimensional Sine Wave Model

For simplicity, we begin by computation of $C^{(2)}$ for a simple sine wave. The slope of the correlation integral in this primitive example yields a correlation dimension $D^{(2)} = 1$. A periodic signal $x(t)$ is represented by a closed curve in phase space; however, if the time resolution of the data coincides with the signal period, the phase space orbit will contract to a fixed point. The left panel of Fig. A1 is the phase space representation of the signal $x(t)$ plotted vs $x(t + l\tau)$ for a fixed ratio of period t_p and with time resolution τ = 0.0038 t_p. In the left panels of Fig. A1 panel (a) represents $l = 3$ (fine-scale resolution), for the center (b) $l = 10$, while for panel (c), $l = 33$.

The corresponding right panels (reading from the top down) give the slope of $\log_2 C_{10}^{(2)}(r)$ plotted as a function of $\log_2 r$. The expected dimension value of unity is found in all cases; however, the upper graph shows that the projection is not optimal as witnessed by the step in the slope of $\log_2 C_{10}^{(2)}$ at large $\log_2 r$. The lower graph (the one with the best projection) has a more pronounced hump than the center panel.

B. The Influence of Noise

When there is random noise present, the data points suffer random scattering about the original attractor. As noise is uncorrelated, this component tends to occupy the whole phase space. The slope of $\log_2 C_d^{(2)}$ of this random component increases with increasing dimension d of the embedding artificial phase space as already mentioned earlier. When the noise component is small, we expect this effect to occur at small r, but depending on the noise level, it might still blanket the underlying regular or chaotic signal.

Correlation integrals of $x(t)$ with superimposed noise are shown in Fig. A2 where the slope of $\log_2 C_{10}^{(2)}(r)$ is shown as a function of $\log_2 r$ with fractional noise levels of 0%, 6% and 20% relative to the amplitude of $x(t)$. The slope of C increases rapidly with decreasing r. Although not shown explicitly, it approaches a maximum dimension of $d = 10$ as $r \to 0$. At large r the slope of $C \sim 1$ as expected for the sine wave alone. For the signal-to-noise ratio ≥ 5, $\log_2 C_{10}^{(2)}(r) \to 1$ can no longer be attained as the superposition of

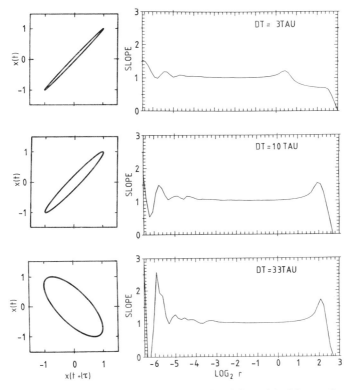

Fig. A1. Phase space representation of varying resolution (left panels) of the one-dimensional
sine wave and corresponding correlation integral slopes (right panels). Correlation integral
slope is unity in all cases except for edge effects.

noise and of the peak leaves too small a constant range in r. Processes may
occur on small time scales relative to modulation by an uncorrelated process
occurring on a much larger time scale (e.g., cosmic ray transport in the heliosphere, X-ray emission from accreting and rotating neutron stars, convective accretion disks and sunspot activity modulated by long-term dynamo
effects).

C. A Model Lorenz System

As examples, Fig. A3 illustrates the effect of modulation of a data set
on its correlation integral. From top to bottom, the left panel shows a synthetic data set of 2048 points taken from a chaotic system (Lorenz 1963), the
same set multiplied by a sine wave, a data set containing white noise and
finally the latter multiplied by a sine wave. The right panel shows the corresponding correlation integral plotted as a function of scaling parameter r for
an artificial phase space dimension of $d = 20$.

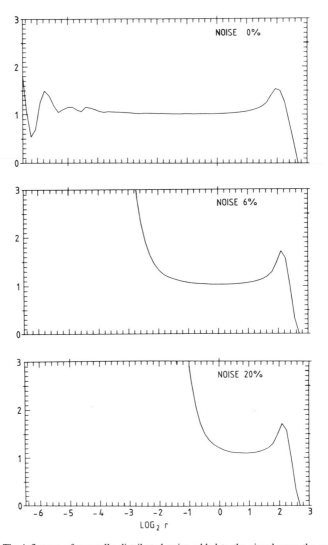

Fig. A2. The influence of normally distributed noise added to the signal upon the correlation integral. With increasing noise (top to bottom) the slope at small r increases and the scaling range (constant slope) decreases. The delay time $\Delta t = 10\tau$.

The Lorenz attractor has a correlation dimension ~ 2.06 (Grassberger and Procaccia 1983b). Within the statistical uncertainties, the slope of $\log_2 C_d^{(2)}(r)$ (right upper panel) gives the right result, except for the bump at large r, which is an edge effect. (The height of this hump depends on the correlation time of the Lorenz sequence.) The stochastic system (second panel) shows a rapid increase in slope towards smaller r, which, as noted earlier, is the signature of purely random data sets (see Fig. A2). The modu-

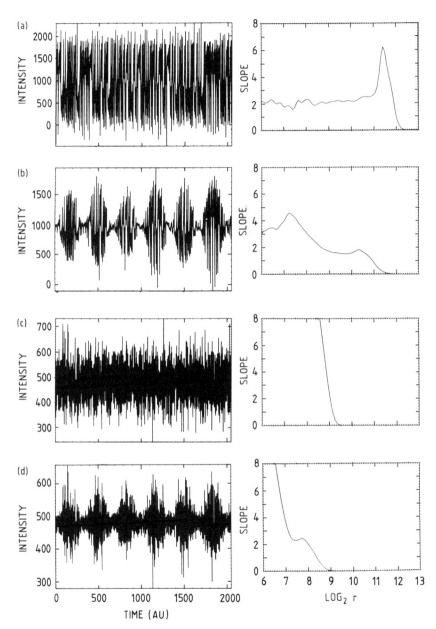

Fig. A3. The effect of strong modulation of a time series on the correlation integral. Example (a) is calculated from a system of Lorenz equations; (b) system of (a) multiplied by a sine wave. The slope is lower and more variable and the edge effects are modified; (c) normally distributed random noise; (d) case (c) multiplied by a sine wave.

lated chaotic time series shows a completely different behavior due to the multiplicative periodic modulation. (An additive periodic modulation would simply increase the slope of the correlation integral by 1.) Panel b of Fig. A3 shows that for large values of r the slope quickly increases to a value around 2, then slowly drops to ~ 1.5 at intermediate r, and finally peaks between 4 and 5 at small values of r.

The correlation integral of the modulated random noise also shows an increase of the slope to ~ 2 at large r, but then it climbs steeply with decreasing r towards the embedding dimension d. In our particular example, the intermediate region with decreasing slope merges with this steep climb, so that only a small step can be seen in panel d. A similar phenomenon appears in Fig. A2. Naively, one could be led to believe that this system had a plateau giving a slope ~ 2.4 and was, therefore, chaotic. However, a careful analysis of the shape of the whole curve, the relation of the maximum value of r to the strength of the stochastic component and consideration of all embedding dimensions (together with inter-comparisons of synthetic data sets) can in general reveal the true nature of the plateau if it is a modulated stochastic system. Note, however, that in contrast to our experience, Norris and Matilsky (1989) have argued that chaotic systems cannot be differentiated from modulated noise based on the correlation integral technique.

THE VARIABILITY OF THE SOLAR DIAMETER

E. RIBES
Observatoire de Paris

B. BEARDSLEY
University of Arizona

T. M. BROWN
National Center for Atmospheric Research

Ph. DeLACHE, F. LACLARE
Observatoire de la Côte d'Azur

J. R. KUHN
Michigan State University

and

N. V. LEISTER
Universidade de Sao Paulo

We discuss here possible variations of the solar radius, which are of interest both because of their astrophysical significance, and because of their potential relation to the terrestrial climate. Observations of the solar radius span more than three centuries, and display considerable variation. We argue that most of this dispersion arises because the various observation methods do not directly measure a true radius, e.g., the radius of a surface of given optical depth. Rather, they measure properties of the Sun's limb darkening funtion, and are affected by many sources of degradation of the solar image. The exact forms of these dependencies are difficult to calculate, especially for visual observations; those uncertainties make interpretation of such observations difficult. For these reasons, we discuss mainly long series of visual observations made by a single observer, or recent photoelectric observations. We find evidence for periodicities in the apparent radius that occur in both modern and historical records. The magnitude of the observed variations is quite different in visual and photoelectric observations, suggesting that the process responsible for the periodicities is either one that modifies the solar limb darkening function, or one that causes systematic variations in image blurring by the Earth's atmosphere. A connection between solar magnetic activity and apparent radius seems likely, with evidence for such a relation dating back as far as the Maunder minimum. Interesting questions that cannot yet be answered concerning the apparent radius variations include whether they have any connection with a deep-seated unknown phenomenon or with the recently observed time-varying asphericity of the Sun, or with the quasi-biennial oscillation seen in the Earth's stratospheric circulation.

I. INTRODUCTION

The solar radius has been measured with an accuracy of a few parts in 10^{-4}. Since the radius can be measured so precisely, it is natural to ask whether this precision can be used to tell us something useful about the processes at work within the Sun. The actual value of the radius turns out to be relatively uninteresting. Due to uncertainties in other solar parameters and (much more importantly) in the theory of stellar structure, solar models can be constructed that match any radius within a fairly wide range. If, however, the solar radius were observed to change with time, the constraints on the models would be much more important.

Traditional theory of stellar structure concerns itself with stars that are spherically symmetrical, nonrotating and devoid of magnetic fields. Within the limits of this theory, one expects a star with the Sun's mass, composition and age to have a radius that increases by at most a few 10^{-11} of its own value per year. Any measurable change in radius would therefore be gigantic by the standards of stellar structure theory, and would imply the action of processes that are not treated in the standard theory. Many such processes exist: rotation, convection and magnetic effects, for example. The difficulty is that there are no adequate theories to treat these processes. Mixing-length theory is almost certainly inadequate to treat the dynamics of the convection zone in the presence of important time scales associated with rotation and the magnetic cycle. Even numerical calculations that are, in some sense, more realistic (Glatzmaier 1985a; Gilman and Miller 1986) seem to do poorly: for example, they predict quite different variations of the Sun's angular velocity with depth and latitude from those inferred from recent helioseismological evidence (Libbrecht 1988). Thus, while there are theoretical reasons to expect variations of the solar radius on time scales connected with the solar cycle (see the chapter by Spruit for a review of the mechanisms), the amplitudes of such proposed variations cannot be estimated even approximately.

The real motivation for observing the solar radius is the suspicion that it might be variable, and the role of the results is to guide theory, not to verify it. As early as 1771, Lalande reported the existence of a possible trend in the apparent semi-diameter of about 0.5 arcsec per century, and concluded that the trend would probably result from an improvement of the telescopes and optics. More recently, impetus for much of the current work on the solar radius came from a study by Eddy and Boornazian (1979) of meridian transit data from the Royal Greenwich Observatory. Using records dating back as far as 1836, they found evidence for a secular decrease in the apparent solar semi-diameter approaching 1 arcsec per century. This work prompted many other studies of the same and related data sets (Shapiro 1980; Parkinson et al. 1980; Dunham et al. 1980; Gilliland 1981), leading to the conclusion that a large secular trend was almost certainly not present during the last two centuries or so. These investigations did, however, leave open the possibility (and in some cases, to be described below, actually suggested) that the apparent solar radius might vary by as much as a few tenths of an arcsecond (roughly 1 part in 10^{-4}), on some time scale longer than a day. Much of the work to be discussed in the body of this chapter is intended to determine, either by taking new measurements or by re-examining old data, whether such radius variations actually occur.

Before this question can be answered, one must clarify what is meant by the solar radius. Particularly confusing, in this context, are assumptions that the Sun's disk is circular, sharp-edged or uniformly bright. If these assumptions were true, then all methods of estimating the limb position would give the same result. Unfortunately, at the precision we need, none of these assumptions holds: the Sun's limb is intrinsically fuzzy, atmospheric seeing makes it more so, and there is now good evidence (Hill and Stebbins 1975; Beardsley 1987; Kuhn 1989; Beardsley et al. 1990) for latitude- and time-dependent temperature variations that cause significant variability even in the shape of the limb-darkening function. As a result, different ways of defining the limb position yield different results. Worse, the difference between the radius measured in different ways is not likely to be constant in time, but will change, depending on the methods' relative sensitivity to image circularity, blurring or the shape of the limb-darkening function. As we will see below, reasonable variations in these parameters may account for apparent radius variations of several tenths of an arcsecond. For this reason, it is important to distinguish between the Sun's *radius* and its *apparent radius:* by radius, we mean the solar size given in terms of some theoretical construct such as density, gravitational potential or optical depth at some specified wavelength; by apparent radius, we mean the result of a measurement using some specified technique. In fact, one usually measures not the apparent Sun's radius but rather its apparent diameter. We shall therefore describe observational results in terms of the apparent *semi-diameter;* this expression will always refer to an observed quantity, and is exactly equivalent to apparent radius R.

It is important to remember that variability in apparent semi-diameter may indicate a variable radius, or it may result from quite unrelated phenomena on the Sun, in the Earth's atmosphere, or in the instrument or to the observer.

II. METHODS FOR MEASURING THE SOLAR DIAMETER

A recent review by Wittman and Debarbat (1990) has given a very comprehensive description of the various methods that are in use for measuring the apparent solar semi-diameter R; there are basically two approaches for measuring it. One consists in directly measuring the angle (or the distance on an image) between two opposite limbs. This necessitates the observation of two simultaneous contacts and proves to be somewhat difficult. The other type of methods consists in timing the duration between successive contacts of opposite limbs with fiducial lines on the sky. The case of a vertical meridian line yields the meridian values of the horizontal transit time, whereas a horizontal line such as an almucantar defined by a mercury bath (i.e., a horizontal mirror) and a constant prism angle corresponds to astrolabe transit times. The determination of the radius necessitates a universal definition, and should take into account the blurring of the solar limb by the Earth's atmosphere. Such a definition has been available since the pioneer work on the FFTD (Finite Fourier Transform Definition) by Wittman (1973) and Hill et al. (1975a).

However, the limb is generally defined by a visual contact between an image and a wire, or an edge, or a reversed image of the same limb; this means a personal appreciation of which picture corresponds exactly to the contact. As a result, there is a bias from one observer to the other, which is the easiest part of the "personal equation." Indeed, it can be calibrated from one observer to another if one deals with a large number of observations obtained by a set of observers working at the same time. The order of magnitude of the bias can be as high as 0.5 arcsec. It is thus essential that the analysis takes properly into account all the relevant information, which means that the name of the observer should be recorded together with the observed value. This remark explains why published analyses come more often to significant conclusions when they deal with homogeneous time series.

It is not as easy to deal with that part of the personal equation which results in more than a simple offset. For example, how sensitive is the retina + brain system to the limb-darkening function? Active regions like plages and filaments may change the limb's appearance over some domain in latitude, close to the contact region. Or again, what is the extent of the domain over which information is really retained by the brain? And to what extent do these effects falsify the measurements? There are no clear answers to these questions, which is why there is an urgent need to record and analyze two-dimensional images which modern techniques permit. There is, however, an advantage in image processing by the human brain which has sometimes been overlooked by those who do not practice visual observations themselves: the selection of good images out of a series of poor ones. This is well known as an explanation for the extraordinary resolution obtained by visual binary observers. It is also true for solar radius observations, provided that they are carried out on systems allowing enough time to perform the actual measurements, say of the order of 20 s at least.

The observations which are discussed in this chapter are visual and photoelectric. The visual determinations of the contact are subject to systematic errors and contain the signature of the seeing process. Photoelectric methods attempt to decouple the Earth atmosphere effect from the solar signal, thus allowing a more objective analysis of the apparent variability. Although these methods have not yet been in use for a full 11-yr cycle, their overlap with modern visual time series may help to interpret the historical visual observations. In Sec. III, we discuss visual observations, both modern and historical, and in Sec. IV, photoelectric observations.

III. VISUAL OBSERVATIONS

A. Meridian Timings

1. Method.   The type of observations available since the application of the lens to astronomical research are essentially based on timings of the transit of the solar image. A long debate has been continuing, to decide whether or not meridian visual timings are sufficiently accurate to be used for the purpose of radius determination. As we have already pointed out, meridian observations are subject to defects such as the variability of the focal length, the reference system and personal bias which are difficult to control. Additional problems exist in long time series. One is the discontinuity in measurements due to a replacement or a change of the telescope. This has been clearly shown in the vertical apparent semi-diameter series obtained at Greenwich (Parkinson et al. 1980), where changes in mean telescope properties over long periods, like changes in observers, can induce fictitious trends. Another source of error affects the timing of the transit duration. In the seventeenth century, observers watched the Sun while listening to the ticking of a pendulum. The exact transit time was determined by interpolation between two successive ticks. This severely limited the accuracy of the semi-diameter measurement. In addition, the low quality of seventeenth century optics tended to increase the size of the solar image. Therefore, the comparison of independent observations has been questioned (O'Dell and Van Helden 1988a).

Only long time series obtained by a single observer using the same instrument are exempt from the above mentioned difficulties. It is fortunate that some such apparent radius time series are in fact available. The period spanned by La Hire's observations (1683–1718) covers a large part of the so-called Maunder minimum (1645–1705), when little surface activity was present at the solar surface. This dearth of sunspot activity at that time was coincident with a cold period in Europe and Atlantic region, the Little Ice Age (Leroy-Ladurie 1967). A unique record of temperatures and cloudiness over Paris was made by Morin, over the same period (1665–1713). Similar homogeneous time series have been obtained over the last century: meridian timings have been performed at the Monte Mario and Neuchatel Observatories with the aim of studying a possible sunspot effect. These time series are of long duration, generally greater than (or of the order of) one solar cycle, so that one can search for shorter-term variability of the apparent semi-diameter.

B. Micrometer Measurements

1. Method.   The principle of this method consists in measuring the linear size of the solar image in the focal plane of a telescope (or lens). This is made possible by the use of an eyepiece micrometer, first invented by Gascoygne in 1640, and considerably improved by Auzout (1729). In the seventeenth century, a serious problem was to perform a rapid measurement at the time of the Sun's passing through the zenith, as the telescope was not fixed in the meridian plane before 1682. Only very skilfull astronomers were able to achieve accurate measurements of apparent semi-diameters (Chapman 1987). The micrometer method was extensively used by Picard (1662–1682) to measure the Sun and planet semi-diameters. An example of his skill is shown by the observations of Jupiter, at various positions with respect to the Earth and Sun (O'Dell and Van Helden 1988*a*; Ribes et al. 1988). Picard reported an apparent solar radius larger than the present one by 5 arcsec. As pointed out by O'Dell and Van Helden (1988*b*), the nighttime observations of planets were more accurate than the solar disk's measurements, because of ground convection during the day. So, the apparent solar radius to be compared to modern observations should be reduced by ∼ 2 arcsec. It is possible that the solar radius might have been larger at that time.

Although it is of interest from an historical point of view to discuss the reliability of micrometer measurements, the number of solar observations performed during the seventeenth and eighteenth centuries are not numerous enough to study periodicities. Therefore we focus the discussion on later results based on the micrometer associated with a heliometer built by Savery in 1743 and improved by Bouguer and Dollond (Lalande 1771). This instrument is a refracting telescope whose objective lens has been split into two halves. Each half objective forms a solar image, and the two halves are moved until the two images are in contact (Danjon and Couderc 1979). The apparent semi-diameter is then proportional to the number of screw divisions. The problems associated with this method are: (1) the calibration of the micrometer screw, which is usually done at night using stars of known separation (this implies a correction for daytime observations); and (2) the small size of the solar image. Systematic and accurate determinations of the apparent solar radius have been performed at Göttingen, by Schur and Ambronn, from 1844 to 1888 (Ambronn 1905).

One should stress that transit observations exhibit a frequently observed annual oscillation, in the sense that the apparent radius looks smaller in winter, with a difference of ± 0.3 arcsec. This effect has been explained by Wittman (1980) as the result of neglecting the refraction effect upon the declination in the drift-time formula. In the discussion of the results below, all observations have been corrected for this bias.

2. *Historical Results.* The Maunder minimum (1645–1705) deserves special attention as it is a period characterized by the scarcity of sunspots (Maunder 1894) and coincident with a cold climate on the Earth (Leroy-Ladurie 1967). Meridian timings were performed almost daily during most of the period of low sunspot activity (La Hire 1683). They show clearly a modulation of the apparent horizontal radius, over a period of almost four solar cycles, which corresponds to the period spanned by the Paris observers (Ribes et al. 1989a). The effect is clearly seen in the apparent radius power spectrum, where much of the power is concentrated around the period 9 to 12 yr (Fig. 1). This suggests that the solar "oscillator" was still functioning, even though there was little sunspot activity. The evidence for aurorae reported during the second half of the seventeenth century, though reduced (Link 1977; Schröder 1988), supports the conclusion that the magnetic cycle was not interrupted. A further confirmation of the existence of a solar cycle is given by the modulation of the ^{14}C and dendrochronology (Kocharov 1986; see the chapters by Damon and Sonett and by Jokipii).

The amplitude of the 11-yr modulation of the apparent radius (1 arcsec) was larger during the Maunder minimum than the value of 0.2 arcsec reported on the modern series. In both cases, however, the apparent radius is larger when the Sun is less active. The contention of a bulge associated with the dearth of sunspots is also supported by the lowering of the rotation of sunspots during the Maunder minimum (Ribes et al. 1987). There are many other contemporaneous observations of the apparent solar radius during the seventeenth century, and on the average, the semi-diameter values reported during the Maunder minimum (Wittman et al. 1981) are significantly larger than the present estimates, lending support to the conclusion that the apparent horizontal radius had increased during the period of reduced magnetic activity. As individual historical timing was severely affected by the ear and eye approach used for measuring the solar edge contacts, sparse measurements cannot serve as a check. This is the case for observations such as those of Mouton (Wittman et al. 1981). On the other hand, if micrometer measurements performed by Picard may be considered reliable, they clearly would show an horizontal bulge of 1.5 arcsec on the apparent semi-diameter, in spite of the correction associated with daytime observations. Such a bulge coincides with

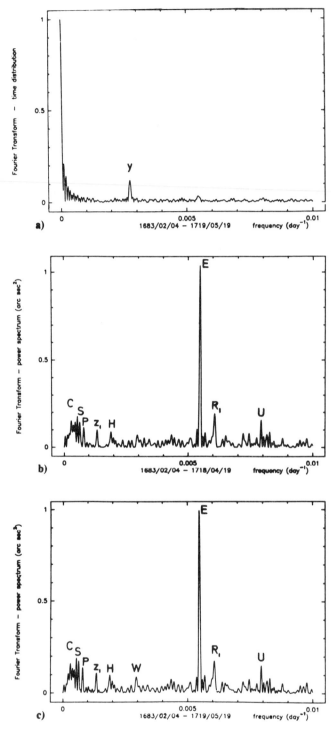

Fig. 1. (a) Fourier transform of the time distribution of the meridian timings performed by La Hire from 1683 to 1718. (b) Amplitude of the Fourier transform vs the inverse of the period (in days), for the same period. (c) Idem as (b) but for the period 1683 to 1719.

a period of complete magnetic calm (1666–1673). Finally, there was the solar eclipse observation timed by Halley and collaborators in May 1715. The border of totality was determined, thus providing a solar radius estimate of 959.63 ±0.5 arcsec. Assuming that this is the best estimate available, it provides a way of calibrating the timings of La Hire and indicates a correction of 3 arcsec which would reduce the largest apparent semi-diameters measured in 1683 to 962.5 arcsec. It is immediately evident that the apparent increase of the horizontal semi-diameter associated with the weak magnetic activity remains a real phenomenon. What the eclipse results imply is that the apparent radius of the Sun was not much different in 1715 from what it is now. This is not surprising as the surface activity had resumed from 1712 onwards.

The question then arises as to whether or not the whole envelope had expanded during the Maunder minimum or only part of it. As mentioned previously, the vertical semi-diameter is much more difficult to measure with accuracy, and is very sensitive to any seeing effect. There is some indication, however, that the change could have been local and not the result of a global pulsation. A Fourier analysis of the data reveals a strong semi-annual periodicity, at the 15 σ level. Such a periodicity occurs as a result of the elliptical shape of the Sun, because the horizontal radius obtained from meridian timings approximates the true equatorial radius only twice a year, at the time of solstices. The amplitude of the phenomenon was 2 orders of magnitude larger than variations in recent determinations of the solar diameter (Ribes et al. 1989a).

Variable asphericity, possibly enhanced by the sunspot activity, seems to have occurred during the Maunder minimum. The amplitude of the apparent bulge was so large that one cannot exclude an image blurring effect due to the particular atmospheric conditions prevailing during the Little Ice Age (Legrand and Legoff 1987). Some other periodicities were present, which can also be found in modern Fourier spectra. The significance of these periodicities is not well established. However, the large amplitude of the signals (compared to the modern ones) is an indication of some magnetic anomaly with a possible amplifying effect in the Earth's atmosphere.

During the nineteenth century, observers focused their interest on the size of the solar envelope as well as the solar disk shape (see Poor 1908 for a review). The variability of the apparent radius (both vertical and horizontal) has been thoroughly explored, and an attempt has been made to relate this variability with solar activity. That an active Sun looks smaller was claimed already by Secchi (1872). From July 1871 to July 1872, 187 observations of the apparent solar radius were obtained at the Monte-Mario Observatory (Secchi 1872). The solar image was projected on a screen and to increase the accuracy in the timing, each daily radius was taken as the average of twenty contacts of the projected solar image with wires. The accuracy of the daily measurements could attain 0.16 arcsec on the semi-diameter. The observations were all made by the same observer eliminating systematic variations in the observer bias as much as possible. Some interesting comments were made, chiefly that the author (Secchi) noticed that there were some variations larger than the noise, from month to month. A study was made of these apparent semi-diameters according to their heliographic latitudes, as the apparent horizontal semi-diameter spans heliographic regions varying from $\pm 26°$, in latitude over the year. The surprising result was that the apparent radii were smaller at latitudes where solar activity was located, the maximum radius occurring near $\pm 20°$. The difference in semi-diameter (between the equator and $\pm 22°$) could reach as high as 0.78 arcsec, corresponding to a change of 10^{-4} in the solar shape. This result, independently corroborated in 1884 by Hilfiker at the Neuchatel Observatory, has not been universally accepted (Poor 1908; Wittman 1980). However, whenever caution was taken to reduce fictitious trends (i.e., no change in the technique nor in the observer), two dominant periodicities seem to prevail: a semi-annual periodicity and an 11-yr periodicity, both of which can be interpreted as signatures of sunspot activity. The available heliometer measurements (see Poor 1908) show a fluctuation in the solar shape corresponding in period with the sunspot cycle, though the amplitude found (0.1 arcsec at most) is smaller than the value deduced from the transit timings. A possible interpretation of the cycle dependence of the shape of the Sun is given in Sec. IV.C.

Another long series of meridian timings was made, early in this century, at Monte Mario Observatory (Gialanella 1943; Cimino 1946). To reduce the scatter of measurements due to observer bias, the timing of the projected solar image was recorded simultaneously by four observers, over the period from 1910 to 1945. A 22-yr periodicity was detected, with an amplitude of 0.1 arcsec on the apparent semi-diameter. Again, a negative correlation between the sunspot cycle and the apparent semi-diameter was ascertained and corroborated by a re-analysis of the Greenwich timings, the U.S. Naval Observatory timings and eclipse data (Gilliland 1981). Finally, a recurrent 6 to 8-yr signal appears in the Monte-Mario data, which bears some resemblance to the 2 main oscillations found today in the active phase of the solar cycle (Laclare 1987). Such a periodicity is found in the Fourier analysis of the 45-yr record of Ottawa radio flux measured in the 10.7-cm wavelength (Ribes et al. 1989a).

B. Astrolabe

More recently, astrometrists have measured the apparent solar radius by means of an astrolabe, to determine the Earth's orbit parameters. Although the measurements are still visual, the instrument is well adapted for metrology measurements.

1. Method.   The instrument is a Danjon-type astrolabe (Danjon and Couderc 1979) which has been adapted for solar observations. The single equilateral prism has been replaced by a series of vitro ceramic prisms allowing observations at various zenith distances, and hence better coverage of the apparent orbit of the Sun (Laclare et al. 1980). The procedure consists of timing the point at which a directly received image coincides with one that is reflected off a level mercury surface, when the upper (or inner) edge of the Sun crosses the almucantar fixed by the prism angle, as shown in Fig. 2. The observer keeps the two images in contact during the crossing of the solar edge by means of a micrometric screw, which compensates the change of the zenith distance and almucantar. An astrolabe, with its several prisms, can be used for more than twenty independent measurements per day.

2. Data Reduction Procedure. The method is based upon the comparison between the observed timings of the Sun at the zenith distance defined by the prism. The general method for determining the orbit parameters of the Earth were simplified by Laclare as they concern solar radius measurements. The geocentric zenith distance (corrected for parallax) at the time of observations, is determined by resolving a spherical triangle. Once the local atmospheric conditions (pressure, temperature and hygrometry) are measured, the atmospheric refraction can be calculated. The observed radius is half the difference between the zenith distances of the upper and inner solar edges, and should be reduced to one astronomical unit distance: The estimate of the contact of the two images (direct and reflected one), is specific to each observer, and explains the systematic difference among observers.

3. Advantages of Astrolabe Observations. Astrolabe observations are not affected by errors in atmospheric refraction. An error in R_i affecting the apparent zenith distance Z_{c_i} cancels out, as the radius is half the difference between the zenith distances of the upper and inner solar edges. The only

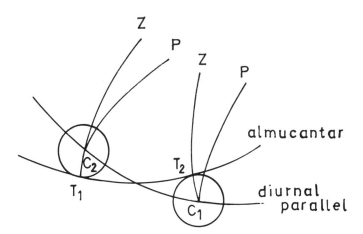

Fig. 2. Astrolabe method. The semi-diameter is derived from the timings of the two contacts of the upper and inner edges of the solar image.

(small) error would be caused by a change in the atmospheric properties between the two timings ($i = 1, 2$). This important advantage is specific to
astrolabe observations. An advantage of the astrolabe over meridian timings
concerns the stability of the almucantar from which the timings of the solar-
image contacts are made. The reference system (the horizon defined by the
mercury plane and the zenith distance defined by the prism angle) is stable
with time. The Earth's rotation provides the most reliable motor. Furthermore, the drift of the focal length during the transit is well under control in
the case of the astrolabe observations.

4. Limitations of Visual Observations.   Personal biases remain one of
the main unknowns. They are extremely difficult to estimate, as we do not
know which part of the limb profile the eye is sensitive to. Moreover, there
is no reason personal bias should remain constant with time for a given observer. Although CCD and other photoelectric devices offer a more objective
determination of the solar limb, their stability also is never certain. There
remains the most critical problem, namely image blurring due to atmospheric
turbulence. The limb profile as well as the position of the inflection point
might be affected by the properties not only of the low atmosphere of the
Earth, but of its middle and upper reaches as well. The limb profile might
also be dependent upon the local environment of the observing station. It is
likely that no proper modeling is possible as these effects are not sufficiently
well understood. This problem will be addressed further below (see Sec.
IV.A.4).

5. Astrolabe Results.   The CERGA time series began in 1976 and measurements were regular from 1978 onwards. The Cerga astrolabe allows a
large number of observations each day. Depending on the number of available
reflector prisms, as many as 22 independent measurements have been made
per day over the last 5 yr at the Cerga observatory. The uncertainty of a given
apparent radius measurement is of the order of ± 0.3 arcsec and the standard
deviation of daily averages is ± 0.15 arcsec over 1 yr. The observer bias in
visually estimating the effective contact between the two images, however,
can go as high as 0.5 arcsec. To work with a homogeneous time series of
apparent radius data, measurements only made by a single observer have been
selected. The series consists of 3000 independent measurements distributed
over all seasons and covers solar cycle 21, with some observational gaps in
the winter time. The apparent semi-diameters (including all heliographic latitudes) are shown in Fig. 3. Most radii correspond to heliographic latitudes
above 45°, referred hereafter as CERGA-100. A number of periodicities are
present above the 5σ level, in the Fourier analysis of the data (Delache et al.
1985; Laclare 1987; Delache 1988), with one of the most prominent signals
occurring near 1000 days.

The Brazilian time series made with an astrolabe began in 1978 and

C.E.R.G.A. MONTHLY AVERAGE

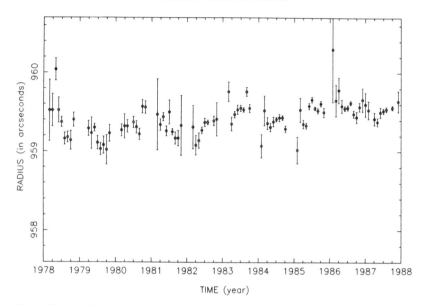

Fig. 3. Visual radius time series (daily means) obtained by Laclare with the CERGA astrolabe. The capital letters (E and W) refer to the onset of the stratospheric QBO, the easterlies and the westerlies, at 10 mbar pressure.

became regular from 1982 onwards (Leiter 1989*a*). The number of observations is substantially less, due to the small number of prisms used at Sao Paulo. For comparison, the CERGA time series contains more than 2000 observations compared to 800 for the Sao Paulo time series, for the period 1982 to 1988. Therefore the error bars corresponding to the monthly mean semi-diameters are close to the error on each individual measurement, i.e., 0.39 arcsec. Moreover, the observations were made by two observers (N. V. Leister and W. Monteiro). Although there is a remarkable coherence between the two observers (the systematic difference being less than 0.1 arcsec on each observation), the combination of the observations adds some additional noise. The semi-diameters measured at Sao Paulo correspond to heliographic latitudes closer to the Sun's equator, in contrast with the CERGA time series. Results are shown in Fig. 4. The Fourier analysis of this time series (Leister 1989*a,b*) indicated: two signals: a 1000-day periodicity and some power around 240 days. We shall mainly concentrate on the former one which lies above the 3 σ level. The amplitude of the signal is ±0.25 arcsec, so the signal is similar, in amplitude and phase, to that of CERGA astrolabe series.

The main difference between the two time series does not lie in the technique but rather in the geographic location of the observatories: 43°5 N, for the CERGA Observatory (France) and 23°5 S for the Abraho de Moraes

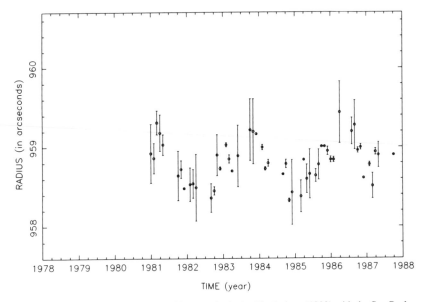

Fig. 4. Visual radius time series (monthly means) obtained by Leister (1989) with the Sao Paulo astrolabe.

Observatory (Brazil). A second difference could be the heliographic latitudes of the apparent radius measured at the two sites.

It should be emphasized that visual meridian timings were obtained at the Belgrade Observatory during the same period (1978–1987) and show the same 1000-day periodicity (Ribes et al. 1988) in spite of differences (techniques, noise, sampling, etc.). Although no theoretical mechanism is known for explaining a 1000-day modulation in visual time series, the similarity in the observed apparent semi-diameter points to a solar cause.

IV. PHOTOELECTRIC METHODS

A. Photoelectric Meridian Transit Measurements

1. Method.   Since 1981, measurements of the solar radius have been performed almost daily at the High Altitude Observatory in Boulder, Colorado. HAO's Solar Diameter Monitor (henceforth SDM) and some of its data analysis algorithms have been described in detail by Brown et al. (1982) and by Brown (1982,1988*b*). Briefly, the instrument is a dedicated, fully automated photoelectric transit telescope. Each day it provides measurements of the horizontal (E-W in the sky) and vertical (N-S in the sky) radii, as well as indices of the solar brightness distribution near the limbs, and values for a number of instrumental and atmospheric parameters. The horizontal semi-diameter measurement is obtained by timing the duration of the Sun's meridian transit, while the vertical semi-diameter is obtained by measuring the size of the solar image, relative to a standard invar spacer in the SDM's image plane. The vertical measurement therefore may be subject to some systematic errors (for example, slow changes in image scale) that have no effect on the horizontal measurement. However, since early 1986, the solar diameter effort at HAO has been reduced, resulting in a significant drop in the number of measurements, and some lessening of the data quality.

Because of the fuzzy nature of the solar limb, the effects of atmospheric seeing and scattering, and the poorly understood response of the human eye, solar radius values have always suffered from some ambiguity as to what exactly is being measured. The Solar Diameter Monitor (SDM) has several features that are intended to minimize this ambiguity. Most importantly, the SDM measures the solar limb position in several different ways, each with different (and calculable) sensitivity to changes in the limb darkening function. These different limb positions are combined in such a way that one may obtain not only an edge position, but also accurate measurements of the slope and curvature of the limb-darkening function in the immediate vicinity of the limb. Since all practical definitions of the solar limb position have some sensitivity to limb-darkening, this additional information is necessary to distinguish between solar radius variations and changes in the shape of the limb-darkening function. Another useful feature of the SDM is that the technique used to define the solar limb (the FFTD (Fast Fourier Transform Definition) and variants thereof [Hill et al. 1975a; Brown 1982]), is quite insensitive to atmospheric seeing and scattering. Moreover, the SDM measures these atmospheric parameters at the time of each measurement, so that corrections can be made for residual effects. Finally, the SDM horizontal semi-diameter measurement depends only on timing and on the Earth's rotation, and is therefore independent of most systematic instrumental effects. This is not true of the SDM vertical semi-diameter measurement, which depends on the telescope's image scale, and on accurate measurements of distances in the image plane.

2. HAO Results.   The precision of the SDM's measurements may be estimated from their internal scatter on single days, or from the day-to-day scatter over a short time interval. The method yields a typical error of 0.14 arcsec for daily radius measurements (either horizontal or vertical); the day-to-day scatter over a few months is about 0.23 arcsec. In both cases, there is good evidence that the noise arises principally (perhaps entirely) from atmosphere-induced image motion. The discrepancy between the two estimates probably reflects the relative importance of high- and low-frequency components of the motion, the latter being inadequately sampled during the SDM's daily 2-min observation periods.

The raw semi-diameter values are corrected for gain and dark-current patterns in the SDM's linear photodetector arrays, for atmospheric seeing and scattering, and (for the vertical radius) for differing atmospheric refraction between the top and bottom of the solar image. The most important source of instrumental error is probably the correction for gain and dark-current changes in the SDM's two Reticon detectors. Gain and dark current are measured on a daily basis, by taking images at the center of the solar disk and well off the solar limb. The gain and dark current are found to change slowly, but the daily values are often contaminated (by solar activity near disk center, for instance). For this reason, the corrections for these variations are based on measurements on specific days, typically six months apart, that are known to be free of observational problems. Corrections to the semi-diameter from this source are typically of order 0.0125 arcsec, while the change between calibrations may be about 0.01 arcsec. The measurement of the slope of the limb-darkening function depends on differences between apparent radii measured with different limb definitions. It is therefore a differential measurement that is largely independent of image motion. Two separate measures of internal scatter suggest that slope measurements in the horizontal direction have daily errors of about 1%, while the errors in the vertical direction are about 10%. Particularly in the case of the horizontal measurements, the day-to-day variations are much larger than the internal errors, and are surely of solar origin.

   3. Limb-Darkening-Function Variability.   Observations with the SDM began in August 1981, and have been analyzed through December 1987; a thorough discussion of the results and the method of analysis may be found in Brown (1982). The SDM observations of the slope of the limb-darkening function show clear indications of solar variations. A parameter α, which is proportional to the limb-darkening function slope dI/dr evaluated at the solar limb, is derived from the horizontal radius measurements.
   Figure 5 shows daily variation of α and Fig. 6, corresponding monthly averages. It is evident that, at least at the low solar latitudes sampled by the horizontal measurement, α is strongly variable from day to day, with extreme values differing by more than 20%. Two facts point to solar activity as the source of most of this variation. First, the variations were substantially larger (by about a factor of 2) in 1981–82, when solar activity was high, than in 1985–86, when activity was near a minimum. Second, the temporal power spectrum of α shows most of its power at frequencies corresponding to even multiples of the Sun's rotation frequency. This suggests that the signal comes from localized regions on the Sun that affect the α measurement twice per month, as rotation sweeps the regions past the east and west solar limbs. Since the horizontal semi-diameter is an equatorial semi-diameter only twice per year, the presence of such features at both limbs implies that these regions must be at least moderately extended in latitude, and must extend across the

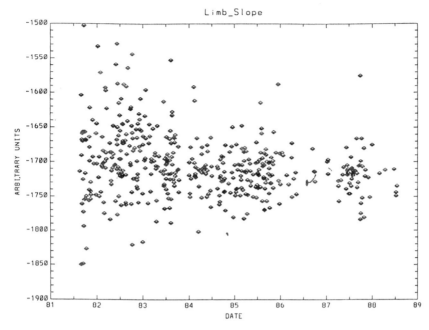

Fig. 5. Solar Diameter Monitor (SDM) observations of the slope of the limb (daily values) obtained by Brown (1989) with the High Altitude Observatory (HAO) meridian telescope.

equator. The slower variations of α, as seen in the 30-day averages, may be a different phenomenon, and requires further study.

Given a quantitative means for defining the position of the solar limb, one may express these changes in α as equivalent changes in the apparent radius. For a typical SDM limb definition, the observed daily variations in α correspond to $\pm \approx 0.11$ arcsec changes in the apparent radius. The best of the five horizontal limb definitions has about half of this sensitivity, the worst about twice as much. This result suggests that apparent radius measurements that do not account for this effect are almost certain to be variable at the level of a few tenths of an arcsec.

4. Seeing Parameter. The slope of the limb-darkening function at the inflection point is measured at the time of observation (Fig. 7). This parameter should, in principle, characterize the state of the Earth's atmosphere as the limb profile gets shallower for a poor seeing. The situation is complicated by the fact that the limb darkening might change over the solar cycle, independently from the atmospheric blurring (see the slower variation of α in Fig. 6). A strong correlation of the seeing parameter with the solar activity index, characterized by the Ottawa 10.7-cm radio-flux emission, clearly indicates that solar activity is leaking into the seeing parameter, leading to an overcor-

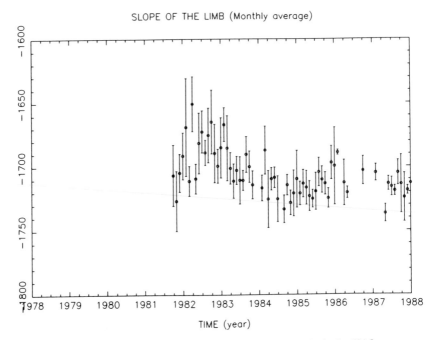

Fig. 6. SDM observations of the slope of the limb (30-day average) obtained at HAO.

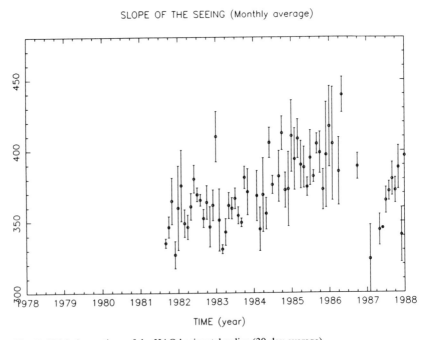

Fig. 7. SDM observations of the HAO horizontal radius (30-day average).

rection of the HAO radii. A procedure tending to decorrelate the seeing parameter and the solar parameters (slope of the limb darkening and HAO radii) has been applied. Moreover, a strong annual periodicity reflecting the seasonal atmospheric conditions have been removed from the HAO data.

5. Horizontal and Vertical Semi-Diameter Variability. The horizontal and vertical semi-diameter results (averaged over 30-day intervals, i.e., roughly 1 solar rotation) are illustrated in Figs. 8 and 9. It is immediately evident that the horizontal solar semi-diameter as measured by the SDM is less variable than indicated by modern astrolabe observations. The HAO horizontal radius does not exhibit a 30-day average differing from the mean value by more than 0.20 arcsec. The vertical semi-diameter is slightly more variable, as one might expect, because of possible systematic errors. To assess the significance of such variations (solar, or arising from the various instabilities to which the vertical semi-diameter measurement is susceptible), an intercorrelation with other time series is necessary (see Sec. V). The result is that the SDM data provide no compelling evidence for any real semi-diameter variation whatever, within the limits set by atmospheric noise (in the case of the horizontal radius) or instrumental drift (in the case of the vertical radius). For the horizontal semi-diameter, these limits are roughly 4×10^{-4} for time scales of 1 day, or 5×10^{-5} for time scales of a year. It should be stressed

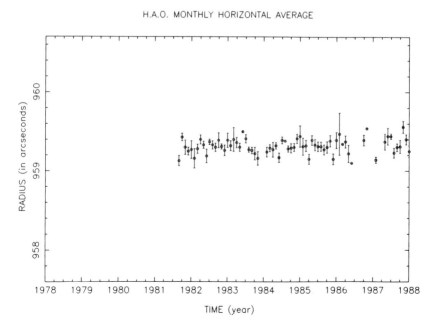

Fig. 8. SDM observations of the HAO vertical radius (30-day average).

H.A.O. MONTHLY VERTICAL AVERAGE

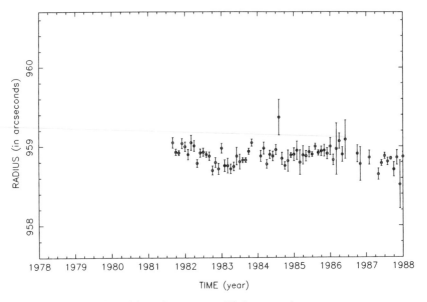

Fig. 9. SDM observations of the seeing parameter (30-day average).

that the latter value is ten times smaller than the variance present in the visual astrolabe data.

B. Photoelectric (SCLERA) Measurements

1. Method.   The SCLERA (an acronym for the Santa Catalina Laboratory for Experimental Relativity by Astrometry) photoelectric measurements continuously record up to 700,000 limb profiles in a 12-hr observing day. These limb profiles are located at several predetermined solar latitudes, typically the poles, equator, and $\pm 45°$. With the use of the FFTD, a daily mean value of what is known as a differential radius measurement δr can be derived from these limb profiles for each solar latitude. The quantity δr has been defined as

$$\delta r(\alpha_i, \alpha_j, \theta, t) = \rho(a_i, \theta, t) - \rho(\alpha_j, \theta, t). \qquad (1)$$

The quantity a is the amplitude over which a sinusoidal scan of the profile across the detectors has been made. The heliographic latitude of the observation is θ and the time is t. ρ is the edge location that has been obtained from the FFTD using a scan amplitude $a(i)$ or $a(j)$ with $a(j) \gtrsim 2\, a(i)$. Since the edge location is very sensitive to both a and the shape of the limb-darkening function, δr can effectively detect small slope changes in the limb-darkening function for fixed values of $a(i)$ and $a(j)$. For this chapter, we have selected data for $a(i) = 8$ and $a(j) = 24$ arcsec, but other edge locations using different values of a are also available. Each different edge location, obtained by changing a parameter, yields valuable information about the shape of the limb-darkening function (Hill et al. 1975a).

The quantity δr is closely related to the slope parameter α used at HAO. With the latter instrument (SDM), the edge location ρ has been calculated using 5 values of a equal to 6.1, 8.7, 12.3, 17.3 and 24.5 arcsec. A quadratic fit for ρ as a function of a was then obtained. The slope was then evaluated at $a = 12.5$ arcsec and has been defined as the quantity α. Since the quadratic coefficient has been determined to be very small, a crude approximation for the relationship between δr and α can be approximated by $\delta r \simeq -16\,\alpha$.

In a simple model of the limb-darkening function such as found in Brown (1988b), where the solar intensity gradient is assumed to be constant from the center to the edge of the Sun, α and δr would both be zero when there is no limb darkening. Equivalently, the solar intensity profile would resemble a step function with a sharp transition in intensity occurring at the edge of the Sun. In this model, it would be expected that the solar luminosity and apparent solar radius would be maximum with this type of limb-darkening function. As the limb darkening (or intensity gradient) increases, α becomes more negative and δr becomes more positive. At such times, the total solar luminosity and apparent semi-diameter would decrease if the effective temperature at disk center remained constant. Therefore, one might expect that the values of δr (or $-\alpha$) to be anti-correlated with radius or luminosity measurements. Although both HAO and SCLERA techniques are relatively unaffected by changes in atmospheric turbulence, scattering, differential refraction and instrumental aberrations, and are unaffected by changes in atmospheric transparency, there are some important differences. An advantage of the HAO data set is good data continuity between 1981 and 1987; however, the SCLERA data set extends back to 1973 and has been analyzed through 1988. Also, meridian observations at HAO cannot exclude an influence of active regions at lower heliographic latitudes. SCLERA observations are performed at heliographic latitudes relatively free from active regions, although some contamination still remains. However, because of the large number of observations and different latitudes available for analysis, a significant amount of activity associated with the remaining transient phenomenon has been removed from the results. This may make HAO observations correlate with parameters that change with the solar cycle. SCLERA results should not correlate as well with the solar cycle and may show other underlying phenomena associated with the intrinsic temperature gradient of the Sun. For the SCLERA observations, the dispersion in the daily averages is larger than the 1σ error associated with a daily mean. Because of the FFTD, most of the day-to-day variation in the daily averages of δr is believed to be solar in origin and not atmospheric or instrumental.

2. SCLERA Results.   Differential radius measurements have been obtained from the period 1973 to 1988 (Beardsley et al. 1989,1990). Combining these observations (from 1973 onwards) with other limb profiles (Rogerscn 1959; Gaustad and Rogerson 1961), differential radius measurements can be traced back to 1957. The basic conclusion is that the slope of the limb-darkening function at the equator has been changing over several times scales. The dominant variation appears to be consistent with the 22-yr Hale cycle (Fig. 10). A downward trend in δr consistent with a longer-term variation may also exist from 1957 to the present time if the results from balloon experiments in the late 1950s are included. However, the balloon observations were made photographically and at slightly different wavelengths, so it is difficult at this time to know if the δr obtained from different experiments can be directly compared.

SCLERA observations show that the solar temperature gradient had been increasing between 1973 and 1981 and remained approximately constant after 1983. A negative or inverse correlation was also found between satellite total-irradiance data (ACRIM and Nimbus 7) and δr between approximately 1979 and 1987. The SCLERA observations are not regularly distributed over the solar cycle, so it is difficult to extract any significant shorter period variations. In particular, the significance of the 1000-day modulation present in the vi-

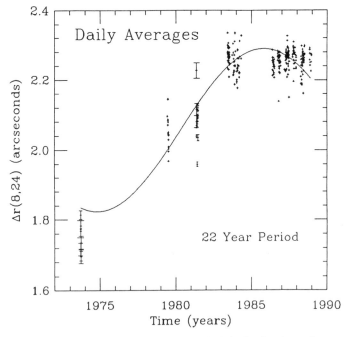

Fig. 10. SCLERA observations of the slope of the limb-darkening function at the equator. A 22-yr trend is present in the data.

sual time series cannot be assessed. However, they may be subject to some correlation analysis with other time series (Sec. VI).

C. Solar Shape

Both the Sun's radius and shape help to determine the solar luminosity. Furthermore, since most radius measurements are obtained from a few positions along the limb, we should also consider how the Sun's geometrical figure may affect measurements of its radius. For example, even a constant solar oblateness can produce a biannual radius variation in meridian transit measurements. The changing orientation of the Sun's oblate figure with respect to the terrestrial E-W direction produces a corresponding modulation in timing data. Modern observations seem to suggest that the solar oblateness can contribute to the variable radius signal for at most 0.04 arcsec. On the other hand, any other asphericity in, for example, the limb brightness or limb-darkening function may similarly affect radius measurements. The magnitude of the shape contribution to radius data depends closely on how sensitive the photometric semi-diameter measurement is to the limb-brightness profile. Thus, conclusions about small radius or luminosity variations rest, in part, on an understanding of the solar asphericity. Modern limb measurements may provide some insight into decoupling the contributions of the shape, radius and limb-brightness profile to the observations. In addition, we can expect this problem to be improved by nonphotometric solar radius and shape observations as helioseismology is beginning to provide clues to changing global solar structure.

1. Method.   Modern shape and radius measurements rely on the accurate identification of fiducial points in the photosphere to define the solar figure. An instrument, originally built by Dicke and Goldenberg (1967), the Solar Distortion Telescope (SDT), effectively defines the limb position by integrating the light that extends beyond a circular occulting disk. In 1980, Dicke, Kuhn and Libbrecht rebuilt this instrument to obtain measurements of possible long-term or solar-cycle variations in the limb shape and brightness. The SDT measures differential flux variations at different position angles around the solar limb. By changing the size of the image, which is centered on a circular occulting disk, the flux from only the outer 5 to 25 arcsec of the image is measured. The shape of the limb-darkening function is determined by incrementally changing the amount of limb exposure beyond the occulting disk. The fluxes from opposite sides of the disk (corresponding to photospheric regions separated 180° in position angle) are combined and the full circle of limb data is divided into 256 angular bins. These flux measurements are obtained simultaneously in two broadband colors at wavelengths near 525 and 800 nm (cf. Kuhn et al. 1985).

This technique determines the limb figure from the occulting disk geometry and limb brightness. The measurement is largely insensitive to small angular-scale seeing effects but is sensitive to differential atmospheric absorption and refraction. Sky transparency variations preclude making absolute radius observations with the SDT. In addition, since the instrument measures brightness, it is sensitive to bright and dark faculae and sunspots near the limb. Nevertheless, shape and brightness components of the data are easily distinguished by comparing measurements with different annular widths of exposed limb. This separation is made by expressing the intensity measurements at each angle around the limb in terms of the equivalent local limb displacement that would yield such a brightness excess. This is possible because the radial gradient of the limb-darkening function is known at the edge of the occulting disk. If the flux excess is due to a displacement of the limb (for example, a real projected-shape variation due to the Sun or differential refraction), then this signal will be independent of the amount of exposed limb. On the other hand, if the excess is due to a variation in the brightness of the photosphere then, in general, the derived signal may depend on the amount of exposed limb.

2. Results. The slope of the limb-darkening function does not change much within 5 and 20 arcsec of the limb. This implies that the shape of the limb-darkening function is invariant, just the scale caused by a temperature change. The effective limb displacement that produces a brightness excess equal to the flux due to a localized limb brightening is, in general, only weakly dependent on the amount of exposed limb in the SDT measurements. This assertion has been questioned by the HAO and SCLERA observers who detect changes in the slope of the limb-darkening function on time scales of a few hours, minutes or days (Hill et al. 1975b). Assuming, however, that the apparent temperature variation can be expressed in terms of an apparent brightness change, we can discuss various brightness contributions in terms of their equivalent limb displacement (in arcsec). We find that the limb displacement that produces a brightness excess corresponding to a large facular patch, or active region, can be > 0.2 arcsec. The mean equivalent limb displacement due to faculae, for example in 1983, was about 0.1 arcsec (Kuhn et al. 1985). For comparison, differential atmospheric refraction produces a difference between the horizontal and vertical apparent solar radii in excess of 0.1 arcsec when the Sun is more than 30° from the zenith. Sunspots are not easily seen near the limb because of the Wilson depression; but they may produce occasional effective limb-brightness variations as large as 0.1 arcsec.

One conclusion from these observations is that photospheric activity (sunspots and faculae) can produce apparent limb-displacement signals of order 0.1 arcsec. The spatial resolution of the SDT observations allows the separation of such brightness signals from limb-shape variations. We do not know how historical observations, which were obtained using a visual definition of the limb, were sensitive; therefore we have no model of the sensitivity of these observations to the limb brightness and it is difficult to relate them directly to the SDT results. On the other hand, the historical measurements do rely on "unspecified" brightness estimates to define fiducial limb points, so it is reasonable to assume that the modern Solar Distortion Telescope determinations of the brightness contribution to the limb position are at least representative of the historical measurements. In this case, we could expect time and latitude variable limb-brightness contributions to the radius measurements, of the order of about 0.1 arcsec.

Having set the scale for the brightness contributions to the radius measurements, we may also use the modern observations to discuss the corresponding time variation. Figure 11 shows the total mean effective limb displacement as measured by the SDT between 1983 and 1988. Several features are obvious. The apparent solar shape is prolate with a solar cycle-dependent bump at the active latitudes that moves toward the equator during the solar cycle. These mean data consist of a smooth brightness contribution that is well described as a thermal photospheric-flux excess plus a short-term variable facular flux contribution. Figure 12 plots the thermal and mean facular effective shape contributions between 1983 and 1988. One should stress that the prolate shape described by Kuhn et al. (1988) refers to a solar shape definition defined by an apparent temperature distribution rather than the

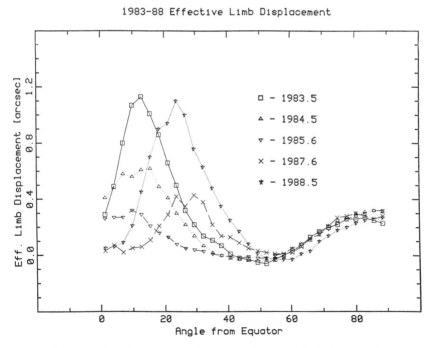

Fig. 11. Mean effective limb displacement as measured by the Solar Distortion Telescope between 1983 and 1988.

E. RIBES ET AL.

Limb Temperature 83-88

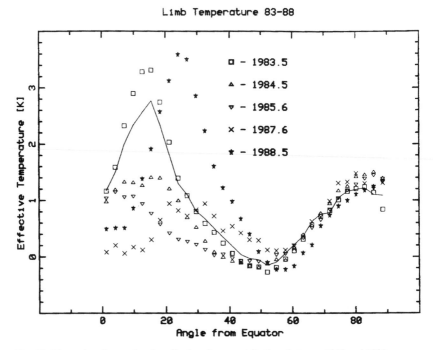

Fig. 12. Thermal and mean facular effective shape contributions between 1983 and 1988.

nearly spherical or oblate shape associated with the mass distribution (Dicke and Goldenberg 1967; Dicke et al. 1984,1985).

The modern shape observations do not extend over a complete solar cycle; it may be dangerous to extrapolate these data, but several qualitative conclusions are appropriate. Foremost is the observation that the scale of the limb-brightness contribution to the semi-diameter measurements is about 0.1 arcsec and it is time dependent over a range of time scales from days (the rotation of faculae across the limb) to years (the solar-cycle evolution of the facular and smooth background-brightness variations). The periodic variation in the projected angle between the Earth's and the Sun's axes of rotation, and the 11-yr evolution in the latitudinal brightness distribution of the Sun will also modulate these radius contributions in, for example, meridian transit observations, thus producing peaks and sidelobes in radius-data power spectra.

V. ECLIPSES: SOLAR AND PLANET

A. Method

Observations of the Mercury transit across the Sun (Parkinson et al. 1980) provide a very accurate method of detecting long-term changes in the solar radius. The geometry of the orbits of Earth and Mercury are such that transits occur only in May and November. As reported by Parkinson et al. (1980), the internal contacts, at which Mercury appears completely inside the limb of the Sun, are more definite than the external ones. With a maximum transit duration of 8 hr in May and 6 hr in November, the standard deviation of each transit observation is in the range of 0.5 to 1.0 arcsec. There are ~14 events in a century; 2000 observations spread over 300 yr were collected by Morrison and Ward (1975). The Fourier analysis of the apparent semi-diameters deduced from the Mercury transits clearly shows an 80-yr periodicity. Moreover, a linear regression analysis of the diameters deduced from Mercury transits produces a secular trend of −0.14 ± 0.08 arcsec per century. An analysis of a smaller set of Mercury transit observations made by Shapiro (1980) yielded a similar decrease.

Solar eclipses have been observed over the last 250 yr and can be used for detecting long-term changes in the solar radius. Timings of the duration of the same total eclipse were made by independent observers, and the dispersion of the resulting solar semi-diameters is an indication of the accuracy of the method. The solar diameter is obtained with an accuracy of ±0.4 arcsec. The time sampling, however, prohibits the detection of shorter-term periodicities, such as the 11-yr cycle.

B. Results

If the diameter deduced from solar eclipses is included with the Mercury transit data in the linear regression analysis, the secular trend in the semi-diameter reduces to −0.08 arcsec per century (Parkinson et al. 1980; Parkinson 1983). This effect is 1 order of magnitude smaller than the rate claimed by Eddy and Boornazian (1979). Two conclusions can be drawn. The first concerns the evidence of an 80-yr periodicity, which also appears in the sunspot series, and the second, the secular trend. According to Parkinson et al. (1980), the secular trend is < 0.01 arcsec. A larger estimate has been made by Gilliland (1981). However, there is no large effect of the amplitude announced by Eddy and Boornazian (1979).

At this stage, one should state that a genuine change of luminosity and radius would imply some exchange between the various energy reservoirs. Along this line of thought, it is worth mentioning that the apparent radius variability is accompanied by fluctuations in the rotation rate. Spectroscopic observations of the rotation of the plasma velocity and magnetic tracers show variations. Although the whole process is not well understood, active cycles seem to be characterized by a larger equatorial velocity, while a reduced rotation rate seems to be associated with low-amplitude solar cycles.

VI. INTERCORRELATIONS OF THE TIME SERIES

In a recent review (Delache 1988), an attempt has been made to assess the value of recent visual apparent solar-radius measurements on the following grounds: if there is anything intrinsic to the Sun in series it is likely of solar-radius measurements that part of the signal should be the signature of other types of solar variability such as sunspot activity or irradiance affecting the observation, as we have mentioned previously. A more optimistic view would be that one may expect real radius variations to be related to variations in activity, irradiance, etc. In either case, any significant correlation between apparent-radius time series (denoted hereafter as $R_\odot$) and time series related to solar activity would be an assessment of the reliability of the visual radius determinations. We shall consider some causes for the apparent changes in the measured solar radius, whatever their origin (solar and/or atmospheric).

$S_\odot$ defines the set of those global solar quantities, which includes the solar irradiance monitored by SMM/ACRIM, the 10830-Å helium infrared line and the 10.7-cm radio flux monitored in Ottawa. The result of such a study has been presented at length by Delache (1988). To summarize, let us say that visual transit-time measurements show significant correlations with $S_\odot$ series, and among them, the astrolabe (which permits up to 20 measurements per day) stands out from the other visual methods, as measured by the significance of the correlation coefficient. This is already some indication that the observed variability in the visual apparent radius is, partly at least, of solar origin.

Other quantities have been made available since the time of the previous review: the HAO horizontal and vertical radius time series, the SCLERA time series relative to three different heliographic latitudes and the Brazilian astrolabe observations (monthly means). A correlation has been performed between these various time series. As a large dispersion (larger than the noise) is present in the various time series, one may suspect that the statistical error associated with the daily or monthly measurements contains some pertinent solar information. A test can be made at this stage by calculating also the correlation coefficient between the statistical error associated with each photoelectric radius time series and the solar activity. The seeing HAO parameter (see its definition in Sec. IV.A) should be kept as witness of the atmospheric effects.

The coefficients of intercorrelation are calculated in averaging the measurements over successive time intervals of 30 days and 120 days, respectively. Tables I and II show the correlation coefficients corresponding to monthly and 3-month average with a confidence level higher than 90%. The boldface values correspond to correlation with a confidence level higher than 98%. It is clear that the errors attributed by each observer to his observations are strongly anti-correlated with solar activity. On the other hand, the seeing parameter, which should account for the Earth's atmospheric effects, is also affected by solar activity. The negative correlation with the $S_\odot$ series is interpreted to mean that an active Sun looks smaller. This leads to the conclusion that a comparative study (or Fourier analysis) of the time series will not lead to the same results whether the data are or are not weighted by their errors.

TABLE I
Coefficients of Linear Correlation: 30-day Averages of the Sigmas*

Data	σSeq	σS45	σSpo	σBra	σHeq	σHpo	Limb	Seeing	ACRIM	Ottowa
σS-eq[a]	**1.00**	**0.92**	**0.92**	—	—	—	**0.68**	—	**0.54**	**0.68**
σS-45°[a]	**0.92**	**1.00**	**0.99**	—	—	—	—	**0.77**	—	**0.66**
σS-po[a]	**0.92**	**0.99**	**1.00**	—	**0.59**	—	—	**0.84**	—	**0.64**
σBraz	—	—	—	**1.00**	0.36	0.33	—	**0.49**	—	—
σH-eq[b]	—	—	**0.59**	0.36	**1.00**	**0.54**	—	**0.48**	−0.22	−0.24
σH-po[c]	—	—	—	0.33	**0.54**	**1.00**	—	0.33	—	—
Limb[d]	**0.68**	—	—	—	—	—	**1.00**	—	**0.60**	**0.52**
Seeing[e]	—	**0.77**	**0.84**	**0.49**	**0.48**	0.33	—	**1.00**	**−0.62**	**−0.56**
ACRIM[f]	**0.54**	—	—	—	−0.22	—	**−0.60**	**−0.62**	**1.00**	**0.60**
Ottawa[g]	**0.68**	**0.66**	**0.64**	—	−0.24	—	**0.52**	**−0.56**	**0.60**	**1.00**

*Boldfaced entries correspond to correlations with a confidence level > 98%.
[a] Statistical errors of the SCLERA time series at three latitudes (45°, equatorial, polar), denoted σSCL-eq, σSCL-45°, σSCL-po (S-eq, S-45°, S-po).
[b] Errors on the horizontal HAO (transit timing) determination, σHAO-eq (H-eq).
[c] Errors on the vertical HAO determination, σHAO-po (H-po).
[d] Limb parameter of HAO (slope of the limb function).
[e] Seeing parameter is a measurement of the seeing quality on the day when the observation is recorded at HAO.
[f] Solar irradiance monitored with ACRIM.
[g] Ottawa 10.7-cm flux, as a $S_\odot$ series.

Because of this, two cross-correlations have been calculated for time intervals of 30 days: in one case (Table III), observations within each time series were given the same weight (sigmas are taken to be constant), while in the other case (Table IV) the sigmas are those given by the observers.

A. Discussion

The stronger correlations are between the SCLERA-45° radius and the CERGA-po (Table IV). This high correlation results from the lack of contamination of the measurements by the magnetic activity of the equatorial zones. Over 30-day averages, visual and photoelectric radius observations become significantly anticorrelated with the solar activity, except for the HAO radii. It would be nice if the seeing did not influence the HAO radius determination, but unfortunately this is not the case (Table III). There appears to be an anti-correlation (coefficients −0.56 and −0.47, significant at 98%) between the seeing and the solar activity as measured by the 10.7-cm flux and by the irradiance data. We cannot accept the idea that it is a real effect (the Earth's atmosphere being more, or maybe less, turbulent when the Sun is active). On the contrary, we suggest that the term "seeing" encompasses part of those effects that we have been discussing previously, namely the departure of the limb line from a perfect circle due to solar active regions. If this is the case, then a wholly satisfactory use of this seeing signal should ultimately succeed in extracting that false part of the "apparent radius" signal and leave a semi-

TABLE II

Coefficients of Linear Correlation: 120-day Averages of the Sigmas[a]

Data	σSeq	σS45	σSpo	σBra	σHeq	σHpo	Limb	Seeing	ACRIM	Ottawa
σS-eq	**1.00**	**0.98**	**0.98**	—	—	—	**0.82**	—	0.64	—
σS-45	**0.98**	**1.00**	**1.00**	—	—	—	—	**0.87**	—	**0.90**
σS-po	**0.98**	**1.00**	**1.00**	—	—	—	—	—	—	**0.90**
σBraz.	—	—	—	**1.00**	—	—	**0.54**	—	—	—
σH-eq	—	—	—	—	**1.00**	**0.74**	—	**0.61**	—	−0.41
σH-po	—	—	—	—	**0.74**	**1.00**	—	0.42	—	—
Limb	**0.82**	—	—	**0.54**	—	—	**1.00**	—	**0.72**	**0.59**
Seeing	—	**0.87**	—	—	**0.61**	0.42	—	**1.00**	−**0.71**	−**0.70**
ACRIM	0.64	—	—	—	—	—	**0.72**	−**0.71**	**1.00**	**0.60**
Ottawa	—	**0.90**	**0.90**	—	−**0.41**	—	**0.59**	−**0.70**	**0.60**	**1.00**

[a] See Table I footnotes for explanations. Boldfaced entries correspond to correlations with a confidence level > 98%.

diameter free of solar activity and of terrestrial effects. The fact that there subsists some correlation between seeing and both HAO radii is an indication that this extraction may not yet be complete. The high correlation between the seeing measured at HAO and other independent radius measurements is a further indication that the slope of the limb-darkening function at the inflection point responds to both atmospheric effects and solar activity.

Conclusions could now be drawn on the slope of the limb α: this HAO parameter is negative and should therefore be positively correlated with $S_{\odot}$ variability. This is indeed the case (see Table III): an active Sun reduces the curvature of the limb, the slope being less negative. The degree of correlation with activity, however, is smaller than that of the SCLERA radii. We interpret this as a consequence of the previous remark, namely that the seeing parameter has taken with it some information pertaining to the Sun itself. This could lead to an overcorrection of the HAO radius. According to Brown (1988), the magnitude of the correction (with an upper limit of 0.005 arcsec on the radius) is not detectable. Most of the variance in the seeing signal is an annual variation resulting from the change of the solar zenith angle at noon. So the correlation between the HAO seeing parameter with all apparent-radius time series might be an indication that all observations have annual problems of their own.

A correlation exists between the HAO limb parameter α and other radius time series'. α is negatively correlated with the CERGA time series, as expected. If the change of the limb is caused by active regions, it is natural that only the radius corresponding to equatorial zones be affected. The negative correlation found between the HAO limb parameter and visual radius data show that the effect of active regions on α may be to make the slope parameter appear closer to zero than would be obtained from quiet-limb profiles. However, to use α by itself as a proxy for radius measurements may not be appro-

TABLE III

Coefficients of Linear Correlation: 30-day Averages of Unweighted Data with σ's Taken as Constant*

Data	S-eq	S-45	S-po	C-eq	C-po	Braz	H-eq	H-po	Limb	Seeing	ACRIM	Ottawa
S-eq[a]	1.00	0.97	0.97	0.68	0.60	—	-0.41	-0.52	—	—	-0.74	-0.71
S-45°[a]	0.97	1.00	0.99	—	—	—	—	—	—	-0.76	-0.62	-0.52
S-po[a]	0.97	0.99	1.00	—	—	—	—	—	—	—	-0.68	-0.53
C-eq[b]	0.68	—	—	1.00	0.64	—	—	—	-0.29	0.35	-0.31	-0.56
C-po[c]	0.60	—	—	0.64	1.00	—	—	—	-0.23	0.36	-0.31	-0.44
Braz[d]	—	—	—	—	—	1.00	—	—	—	—	—	—
H-eq[e]	-0.41	—	—	—	—	—	1.00	—	—	—	—	—
H-po[f]	-0.52	—	—	—	—	—	—	1.00	—	—	—	—
Limb[g]	—	—	—	-0.29	-0.23	—	—	—	1.00	—	0.35	0.52
Seeing	—	-0.76	—	0.35	0.36	—	—	—	—	1.00	-0.47	-0.56
Acr.[h]	-0.74	-0.62	-0.68	-0.31	-0.31	—	—	—	0.35	-0.47	1.00	0.72
Ottawa[h]	-0.71	-0.52	-0.53	-0.56	-0.44	—	—	—	0.52	-0.56	0.72	1.00

* Boldfaced entries correspond to correlation with a confidence level > 98%.

[a] SCLERA time-series at three heliographic latitudes (S-eq, S-45°, S-po).

[b] CERGA astrolabe data restricted to latitudes from −45 to 45 degrees, denoted hereafter as equatorial (C-eq).

[c] Complement to the equatorial, i.e., rather the polar latitudes (C-po).

[d] Brazilian astrolabe time series (monthly means, all latitudes, though predominantly equatorial).

[e] Horizontal HAO (transit timing) determination, denoted HAO-equ (H-eq).

[f] Vertical HAO determinations, HAO-po (H-po).

[g] Limb parameter (slope of the HAO limb-darkening function) (Limb).

[h] Irradiance measured with ACRIM and the Ottawa 10.7-cm flux, termed $S_\odot$ series.

TABLE IV
Coefficients of Linear Correlation: 30-day Averages of the Data Weighted with their σ's

Data	S-eq	S-45	S-po	C-eq	C-po	Braz	H-eq	H-po	Limb	Seeing	ACRIM	Ottawa
S-eq	1.00	0.83	0.87	0.59	—	—	—	-0.59	—	—	—	-0.37
S-45	0.83	1.00	0.92	—	0.77	—	—	—	—	—	—	—
S-po	0.87	0.92	1.00	—	0.58	—	—	—	—	—	—	—
C-eq	0.59	—	—	1.00	0.64	—	—	—	-0.28	0.35	-0.37	-0.56
C-po	—	0.77	0.58	0.64	1.00	—	—	—	-0.23	0.36	-0.44	-0.44
Braz	—	—	—	0.43	—	1.00	—	-0.34	—	—	—	—
H-eq	—	—	—	—	—	—	1.00	—	—	—	—	—
H-po	-0.59	—	—	—	—	—	—	1.00	—	—	—	0.25
Limb	—	—	—	-0.28	-0.23	—	—	—	1.00	—	0.31	0.52
Seeing	—	—	—	0.35	0.36	—	—	—	—	1.00	-0.45	-0.56
ACRIM	—	—	—	-0.37	-0.44	—	—	—	0.31	-0.45	1.00	0.70
Ottawa	-0.37	—	—	-0.56	-0.44	—	—	0.25	0.52	-0.56	0.70	1.00

[a] See Table III footnote for explanations. Boldfaced entries correspond to correlation with a confidence level >98%.

priate because the intrinsic limb-darkening function may also be variable, and furthermore, have a different period than the solar cycle as shown by SCLERA observations. The correlation between the HAO horizontal radius and the limb function is insignificant, showing that, indeed, the limb profile is correctly taken into account by the FFTD process. On the other hand, if the seeing contains some solar signal, the HAO horizontal semi-diameter could have been overcorrected. Recall that the HAO radii are obtained for 5 different values of scan amplitude a, and the true radius is extrapolated for a "zero" scan amplitude. Correction due to changes in the limb-darkening function is expressed in terms of the slope parameter evaluated at 12.5 arcsec, according to the following relation:

$$D_{obs} = D_{true} - K \alpha\, a \qquad (2)$$

where D_{obs} is the observed radius using scan amplitude a. α is the slope of the limb-darkening function at 12.5 arcsec and K is obtained from a quadratic fit to the data. On the one hand, K is probably not constant as assumed by Brown (1988b), and a small error in K could easily account for an overcorrection of the HAO data. On the other hand, the absence of correlation between the HAO limb parameter and the HAO radii indicates that K has probably been calculated correctly.

We shall examine the possibility that the HAO radii have been overcorrected by the seeing parameter. Slight fluctuations in seeing will cause very minimal changes in the edge location for scan amplitudes larger than the seeing. However, for bad seeing ($\geqq$ 6 arcsec), the smaller scan amplitude a corresponding to 6 arcsec becomes very sensitive to seeing fluctuations. Some intrinsic solar variability could be removed with the seeing parameter correction K. As the slope of the limb-darkening function is sensitive to the same type of limb-darkening functional changes, one would anticipate a correlation between α and the HAO radii. This does not seem to be the case, as shown in Tables III and IV. The residual HAO radii (after correction of the seeing and the changes in the shape of the limb-darkening function) do not show much correlation with other radius time series. The correlation coefficient between HAO-eq and CERGA-eq is 0.31 at the confidence level of 82%. One should stress, however, that the variance in the HAO residual radii is of the same order of magnitude as the variance in the CERGA data when limited to the period 1982–87. The two main oscillations occurring before 1982 contribute most to the variability in the visual time series.

These conclusions conform with the discussions presented in Delache (1988), where it was suggested that short-term variations (up to about 300 days) could well be due to leakage of $S_{\odot}$ variability into visual-radius determinations. For longer characteristic times, it was hoped that the correlation coefficients between $R_{\odot}$ and $S_{\odot}$ time series would increase with the increase of the elementary time intervals over which daily data are averaged. This is

TABLE V
Unweighted Data
Coefficients of Linear Correlation: 120-day Averages of the Data with σ's taken as Constant

Data	S-eq	S-45	S-po	C-eq	C-po	Braz	H-eq	H-po	Limb	Seeing	ACRIM	Ottawa
S-eq	1.00	0.96	0.96	0.75	0.67	—	—	-0.54	-0.89	—	-0.71	—
S-45	0.96	1.00	0.99	—	—	—	—	—	—	—	—	-0.82
S-po	0.96	0.99	1.00	—	—	—	—	—	—	—	—	-0.82
C-eq	0.75	—	—	1.00	0.79	—	—	—	-0.42	0.69	-0.71	-0.76
C-po	0.67	—	—	0.79	1.00	—	—	—	-0.39	0.50	-0.56	-0.56
Braz	—	—	—	—	—	1.00	—	—	—	—	—	—
H-eq	—	—	—	—	—	—	1.00	—	—	—	—	—
H-po	-0.54	—	—	—	—	—	—	1.00	—	—	—	—
Limb	-0.89	—	—	-0.42	-0.39	—	—	—	1.00	—	0.59	0.59
Seeing	—	—	—	0.69	0.50	—	—	—	—	1.00	-0.60	-0.70
AC-RIM	-0.71	—	—	-0.71	-0.56	—	—	—	0.59	-0.60	1.00	0.92
Ottawa	—	-0.82	-0.82	-0.76	-0.56	—	—	—	0.59	-0.70	0.92	1.00

[a] See Table III footnotes for explanations. Boldfaced entries correspond to correlation with a confidence level >98%.

the case, as shown in Table V, the same as Table III but for 120-day averaging. For longer time scales, the comparison is not possible, as boxes larger than 120 days are not numerous enough and the samplings of the various time series are irregular.

We now examine the 1000-day periodicity present in the visual apparent-radius time series. No conclusion can be drawn from the SCLERA photoelectric observations, because they are unevenly distributed. Concerning the HAO semi-diameters, the period of observations spanned by the SDM do not cover the period 1978 to 1981 where the oscillation is the most enhanced. On the other hand, one cannot discard the possibility that the 1000-day periodicity present in the visual $R_\odot$ time series could be produced by some atmospheric effects. If this were the case, the periodicity would also be apparent in the seeing parameter measured at HAO. However, the main variance of the seeing parameter is due to the annual signal mentioned above.

Next, we explore the possibility that the apparent changes detected from the $R_\odot$ time series come from a limb shift due to solar activity. An increase of the effective surface temperature as measured by Kuhn (1988) would cause a bump at the latitudes where the solar activity is located. In other words, an active Sun should look larger, at least over the range of latitudes affected by the sunspot activity. To test this idea, we partition the various time series (CERGA, Brazil, and SCLERA) according to latitude ranges (between 0 and 30°, 30° to 60°, and 60° to 90°). In the case of the SCLERA measurements, we use the unweighted data as well, in order to preserve the integrity of the solar signal. Table VI shows the correlation coefficients for two levels of significance (98% and 90%). The only correlation (0.65 at the rather low confidence level of 90%) is found for the sigma SCLERA with the limb temperatures. The absence of a correlation between CERGA apparent radius and limb temperature is not too surprising as the CERGA measurements contain few radii in the activity belts where the effect should be large. This is not the case for the Brazil observations. On the other hand, the Brazil astrolabe does

TABLE VI
Matrix of Correlation Coefficient for Latitude Bins = 30°[a]

	CERGA	Brazil	SCLERA	σ-Scl	KuhnT_1	$K(T_1 + T_2)$	σ-Kuhn
CERGA	**1.00**	**−0.52**	**0.85**	—	—	—	—
Brazil	**−0.52**	**1.00**	**−0.83**	—	—	—	—
SCLERA	**0.85**	**−0.83**	**1.00**	—	—	—	0.72
σ-Scl	—	—	—	**1.00**	—	—	—
KuhnT_1	—	—	—	—	**1.00**	**0.78**	—
$K(T_1 + T_2)$	—	—	—	—	**0.78**	**1.00**	**0.65**
σ-Kuhn	—	—	0.72	—	**0.65**	—	**1.00**

[a] Boldfaced entries correspond to correlation with a confidence level > 98%. T1 is the equivalent smooth background temperature; T2 is the equivalent facular temperature.

not contain many observations during the boreal summer, for which limb temperatures are measured. At present, the available data do not allow us to draw any firm conclusion in favor of the assumption that the apparent radius is governed by the surface temperature of the limb. That the solar shape can be modified by the solar activity is certainly a reality, as shown by the asphericity present in the helioseismic data as well as the apparent radius. However, an active Sun looking smaller is common to all visual and SCLERA time series and cannot be currently explained by a photospheric limb shift only.

VII. CONCLUSIONS

The first conclusion that can be drawn from the comparison between the modern visual and photoelectric time series is the fair agreement of the mean solar apparent radius corresponding to the common period of observations (from August 1981 to end of 1987). The measured mean radius is consistent with the errors (959.44 ± 0.08 arcsec) for the CERGA data and (959.321 ± 0.024 arcsec) for the HAO data.

The apparent radius is not constant in time. There are a number of repeatable signals which cannot be attributed only to the technique used nor to observer biases. Some short-term variations in the measured radius are obviously caused by the change of the limb due to active regions. As a result, the solar limb shape changes through the solar cycle. A semi-annual periodicity in the Fourier spectra of Belgrade horizontal observations can be explained along this line of thought, although the signal is marginally significant (Ribes et al. 1988). The historical data, however, do show a strong seasonal (semi-annual) periodicity which may be a hint of a change in the solar limb shape, due to the effect of the solar activity on the limb definition. This was particularly pronounced at the time of the Maunder minimum.

Special attention should be paid to the HAO photoelectric measurements. The intercorrelation analysis shows that the HAO seeing parameter contains some solar information. Moreover, the slope of the limb-darkening function is negatively correlated with SCLERA and visual series, showing the influence of the activity on the slope of the limb-darkening function and the determinations of the apparent radius. On the other hand, the SCLERA differential radii are less contaminated by the sunspots and faculae than the HAO radius observations, and show a better consistency with the visual observations. Therefore, it is likely that some of the variability in the visual apparent radius do not simply reflect local changes of the limb profile due to sunspots and faculae. On the one hand, the solar bright features are correlated with the surface temperature, leading to an increase of the apparent photometric radius, as shown by Kuhn et al. (1988). On the other hand, this effect may reduce the curvature of the limb (the slope is less negative) leading to an apparent radius smaller than the apparent radius of a quiet Sun. The latter result, specific to visual observations, is not present in the photoelectric data. The relation between the radius and the limb-darkening function is apparently not simple.

Part of the variance at least may reflect the change of the temperature gradient at the photospheric level. This effect can be invoked to account for the variability of the granulation as well. The question arises as to whether or not the phenomenon is local or global. There are arguments and counterarguments in favor of a real change of the radius, in terms of a change in the solar mass distribution. That the variability of the polar apparent radius seems to be in phase with that of the apparent horizontal radius suggests a global phenomenon (see the positive correlation between SCLERA-eq and SCLERA-po as well as CERGA-eq and CERGA-po in the tables). The negative correlation between HAO-eq and HAO-po has probably a different explanation (the vertical radius is much more sensitive to atmospheric effects).

We should mention the observation of p-mode frequency changes over a solar-cycle time scale (Woodard and Noyes 1985). Using the ACRIM satellite solar irradiance data, these workers argued that the low degree ($1 < 2$) p-mode frequencies decreased by about 0.4 nHz between 1980 and 1984. Subsequent observations by Fossat et al. (1987) using velocity data provided consistent evidence for p-mode frequency variability. Additional measurements by Gelly et al. (1988) and Palle et al. (1986) tend to support this conclusion. Although there are conflicting reports (Palle et al. 1988) concerning the evidence for this variability, we consider, for the moment, that the variability is a solar phenomenon.

The implications of the helioseismic observations for solar radius studies are complicated by the fact that solar structural changes during the activity cycle may also affect the seismic data. If we believed the frequency shifts were due entirely to a real radius change, then the Sun must have swollen by almost 10^2 km between the sunspot maximum (near 1980) and 1984. This is an enormous radius change. For comparison, the difference in polar and equatorial solar radii due to the rapid rotation of the solar core is an order of magnitude smaller. If only the solar convection zone participates in this expansion, it would require more energy than the total solar luminosity integrated over an activity cycle.

There are additional arguments for interpreting the small frequency shifts as not due to a changing solar radius. If the surface brightness were to remain constant, the solar luminosity would be maximum at solar minimum, though the ACRIM satellite observations have shown that the solar irradiance tends to a minimum value at the sunspot minimum (Willson and Hudson 1988). If the irradiance is a measure of the luminosity, then the satellite observations imply a comparatively larger fractional luminosity variation of 10^{-3} over the solar cycle. Thus, it is clear that we must consider changes in the thermal structure in order to account consistently for the helioseismic and irradiance observations. For a broad range of possible perturbations, numerical work also suggests that, unless the Sun changes beneath the convection zone, the fractional radius variations are typically 2 orders of magnitude smaller than the corresponding fractional luminosity change (cf. Dearborn and Blake 1980a).

It has been shown that a time-dependent photospheric brightness change (Kuhn et al. 1988) can account for the ACRIM irradiance variability without implying a real radius change. The arguments are based on the limb-brightness observations described in Sec. IV.C. The limb observations imply a variable aspherical thermal structure, or an asphericity in the sound-speed distribution. Since the solar p-modes are acoustic waves, the observed eigenfrequencies are sensitive to this asphericity, with a time-dependent mean sound speed or temperature variation. The direct result of the asphericity is to cause modes of the same spherical harmonic degree l, but different spherical order m, to have slightly different frequencies. This frequency splitting can be calculated (Kuhn 1988) and related to helioseismic frequency observations (Jefferies et al. 1988). A broad consistency between the helioseismic data (Beardsley 1987; Kuhn 1989) and the photospheric-brightness variations lends additional support to the conclusion that solar structural (i.e., thermal stratification) changes during the solar cycle are associated with helioseismic variations. If the long-term variability of the SCLERA-eq radius (or the 1000-day variability very evident in the visual observations) are related to the trend present in the irradiance data (chapter by Fröhlich et al.), one can invoke a long-term trend in the convective processes that transport energy to the surface. Such a trend could explain some of the radius variability recorded over the past three centuries (11-yr, 80-yr and long minima). We cannot rule out the possibility of solar radius changes over many-year time scales, although the recent photoelectric and visual radius observations are best described by brightness variations near the limb, which are consistent with the ACRIM satellite photometry and recent helioseismic observations.

There remains the possibility that the Earth's atmosphere is also responsible for some long-term variations, such as those found by Laclare (1983). Djurovic and Paquet (1988) have clearly established the presence of short-term periodicities common to solar activity and terrestrial atmosphere properties, such as its kinetic momentum and rotation (Feyssel and Nitschelm 1985). An intriguing correlation between the stratospheric-circulation reversal with the 1000-day oscillation of the apparent radius has been noted by Ribes et al. 1988.The extrema of the 1000-day variation of the radius seem to be associated with the onset of the stratospheric Quasi-biennial Oscillation (QBO) (Naujokat 1986). An association of this kind could also have a terrestrial origin. Since the QBO's are located within the tropics, it would be surprising that they affect stations at different latitudes at the same time. That the 1000-day periodicity is roughly in phase for astrometric stations such as CERGA 43°5 N, Belgrade 43°5 N and Sao Paulo 23° S cannot easily be explained by an atmospheric effect. If the correlation between the Earth's temperatures and the solar cycle through the QBO (Labitzke 1987; Labitzke and van Loon, 1988) is real, this tends to support a solar origin.

Although solar activity undoubtedly modifies the limb darkening function, it is still unclear whether the phenomenon can account for all radius variability. The variability of the polar semi-diameters in phase with the equatorial semi-diameters is puzzling. Very little is known about the network variability at high latitudes, or the variability of the various convective scales (granule, mesogranule, giant cell, azimuthal roll). No doubt a better understanding of the convective and magnetic processes as well as the rotation will help our understanding of the internal structure of the Sun and its evolution.

Acknowledgments.   We are indebted to R. C. Willson for having generously sent the solar constant data. The data was provided through research carried out by the Jet Propulsion Laboratory, California Institute of Technology, under a contract with the National Aeronautics and Space Administration. We are also grateful to P. Morel (Worls Meteorology, Genova) and A. Wittman, (University of Gottingen, Germany) for interesting discussions.

LONG TIME VARIATIONS IN SOLAR WIND PROPERTIES: POSSIBLE CAUSES VERSUS OBSERVATIONS

J. GEISS AND P. BOCHSLER

University of Bern

Three decades of observation have given us an overview over averages and variations of the physical properties and the chemical composition in the solar wind. Significant time variability in the elemental composition is observed, and also systematic differences between average solar wind and solar composition have been established: helium is generally depleted by a factor of two relative to hydrogen, and in the low-speed solar wind, the elements with ionization potentials below 10 eV are overabundant by a factor of ~ 3. Reasonable theoretical interpretations have been developed for the anomalies in solar wind abundances: (a) an ion-atom separation process in or directly above the upper chromosphere causes a preferential supply of elements with low first ionization potential to the corona; and (b) the acceleration of the solar wind in the corona causes further fractionation that can lead to significant short-time variations in elemental composition. The theoretical models that have been established to interpret elemental abundances and their variations indicate that with the exception of $^3He/^4He$ isotopic ratios should remain relatively unaffected. The few direct observations of isotopic abundances in the solar wind seem to confirm this prediction. In lunar surface material, hydrogen and the noble gases are dominantly of solar origin. Thus, noble-gas isotopes trapped in this material provide us with a unique record of the solar wind flux and composition in past epochs, covering the main-sequence life of the Sun. The isotopic abundances of solar wind noble gases collected at the lunar surface during different epochs show little variability, in agreement with expectation. There is some evidence for a modest secular increase of ~ 10% in the $^3He/^4He$ ratio. Such a change could be explained in two ways: (1) the solar wind acceleration dynamics may have been different in the past, and this could well cause a change in the relative proportion of ^3He and ^4He, considering their relatively large mass ratio; or (2) slow secular mixing of the Sun could have brought up material from the ^3He

bulge in the Sun, thus causing an increase of the ^{3}He abundance in the outer convective zone of the Sun. Changes in solar wind dynamics as the cause for the large variations found in the isotopic composition of trapped nitrogen are improbable. It appears that in addition to solar nitrogen there is one or even more than one nonsolar nitrogen component trapped in lunar surface material causing the large variability in the isotopic composition of this element. The most puzzling finding is that in lunar surface material there are 2 noble gas components which are distinguished by their isotopic composition. The helium and neon isotopic composition in the light component *is identical with the measured present-day solar wind composition. The isotopically* heavier component *could either represent a solar wind gas modified by strong loss effects like diffusion out of grains or it is an energetic or suprathermal particle population of solar origin. Both explanations encounter severe difficulties, but they imply in any case that particle fluxes were higher in the past than they are observed today.*

The solar wind is an important source of information on processes occurring above the solar surface and inside the Sun. Furthermore, for some aspects of the composition in the outer convective zone of the Sun, solar wind and solar energetic particles are presently the only sources of information. Solar isotopic abundances have been estimated from solar wind and solar energetic particle data. Some differences have emerged, as in the case of the neon isotopes, which are not yet explained.

The record of implanted solar wind and the more energetic particles left in the lunar soil, can provide information on the composition, and on absolute fluxes from the time when these soils were formed a few Gyr ago. Hence, in the lunar data, we have clues regarding the solar wind, the solar activity and the composition of the young and middle-aged main-sequence Sun. To interpret these clues correctly, a thorough understanding of properties and systematics of the present-day solar wind and the more energetic solar particles is required.

I. CONTEMPORARY SOLAR WIND

Three decades of observation have taught us that the solar wind is a highly variable phenomenon. All properties including flux, velocity and composition vary significantly. Concerning the bulk properties and the general dynamics of the solar wind, we refer to the chapter by Smith. For the purpose of comparison with the lunar record, two main types of solar wind should be distinguished: the fast, low-density streams coming out of the coronal holes, i.e., the open field-line regions at the solar surface, and the slow solar wind that typically has a higher density. It has been pointed out (Steinitz and Eyni 1980; Schwenn 1983a) that (at least in the ecliptic plane) the different types of solar wind carry a rather constant momentum flux, i.e., particle flux and velocity are inversely correlated.

Shortly after the experimental verification of the continuous outflow of plasma from the Sun into interplanetary space, it was recognized that the solar wind not only consists of protons and electrons but that there is a highly variable contribution of $^4\text{He}^{++}$ (cf. Neugebauer 1981). It became also clear that the average $^4\text{He}^{++}/\text{H}^+$ ratio of $\sim$0.04 does not reflect the solar surface abundance ratio, but that there must be varying fractionation processes in the solar wind feeding zone near the Sun's surface and in the acceleration region in the inner corona. Subsequently, it became evident that there is a close correlation between the helium/hydrogen ratio in the solar wind and the magnetic structure and flow properties in the source region (Hirshberg et al. 1970; Borrini et al. 1981). Not only does the helium abundance exhibit strong variations, but the heavy ion content also varies significantly (Bame et al. 1975; Bochsler et al. 1986; Gloeckler et al. 1989).

Because typical persistence times in elemental abundance ratios and in $^3\text{He}/^4\text{He}$ are of the order of 10 to 30 hr (cf., Coplan et al. 1984; Bochsler 1984), much of the compositional variability averages over days. This is evident from the comparison of the $^3\text{He}/^4\text{He}$ and He/Ne data obtained by the foil collection technique from the Apollo Solar Wind Composition experiment (Geiss et al. 1970a,1972) with the data obtained by the *in situ* mass spectrometry from ISEE-3 (Coplan et al. 1984; Bochsler et al. 1986). Examples are

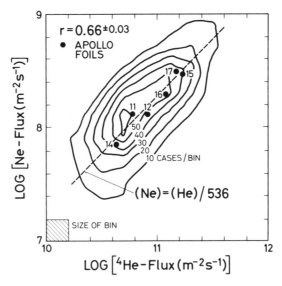

Fig. 1. Correlation of elemental neon and helium fluxes in the solar wind. The contour lines represent 950 spectra taken by the ISEE-3 mass spectrometer (Bochsler et al. 1986). The data from the Apollo 11–17 foil collection experiments are shown as dots (Geiss et al. 1970a,1972); *r* is the correlation coefficient for the ISEE-3 data. The two experiments give virtually the same mean value for the He/Ne ratio ($\sim$540). The variations around this mean value are significant for both experiments. They are smaller in the case of Apollo because of the integrative nature of the solar wind collection experiment.

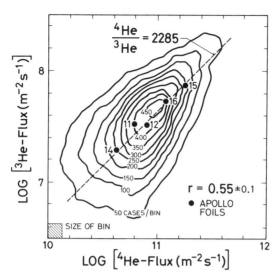

Fig. 2. Correlation of ^{3}He and ^{4}He fluxes in the solar wind. The contour lines represent 27,000 spectra taken by the ISEE-3 mass spectrometer (Coplan et al. 1984; Bochsler 1984). The data from the Apollo 11–16 foil collection experiments are shown as dots (Geiss et al. 1970a,1972); r is the correlation coefficient for the ISEE-3 data. Within experimental errors, the two experiments give the same mean value for the ^{4}He/^{3}He ratio. The variations around this mean value are significant for both experiments. They are smaller in the case of Apollo because of the integrative nature of the experiment.

presented in Figs. 1 and 2. The six Apollo He/Ne ratios representing integration times between 77 min and 42 hr deviate not more than 20% from the average, the five Apollo ^{4}He/^{3}He ratios by only $\pm 15\%$ from the average. The much broader distribution of the abundance ratios obtained by the ISEE-3 mass spectrometer is partly due to the statistical error but also reflects short-time variations (<1 day) in solar wind composition.

However, even for long-time averages there remains a systematic difference in the elemental composition of the slow solar wind and the fast streams (Ogilvie and Wilkerson 1969; Bame et al. 1977; Neugebauer 1981; Mitchell et al. 1983; Galvin et al. 1984; Ipavich et al. 1986; Gloeckler and Geiss 1989; Gloeckler et al. 1989). As the occurrence and the extension of coronal holes at low solar latitude is strongly coupled to the solar cycle, some changes of solar wind composition (near the ecliptic plane) with the phase of the solar cycle are to be expected. So far, such solar-cycle-related changes have been clearly established only for the He/H ratio where they amount to a factor of about 1.5 (Formisano and Moreno 1971; cf., Neugebauer 1981; Ogilvie et al. 1989).

In Table I, we give average elemental and isotopic abundances in the contemporary solar wind. As not all data correspond to the same observational conditions, some of these averages may be somewhat biased towards

J. GEISS AND P. BOCHSLER

TABLE I
Elemental Abundances in the Solar Wind, SEP-Derived Solar Corona and
Solar System[a]

	Average Solar Wind			SEP-Derived Corona (7)		Solar System (8)
H	1900	± 400	(1)	—		1170
He	75	± 20	(2)	—		114
C	0.43	± 0.02	(3)	0.414	± 0.049	0.42
N	0.15	± 0.06	(3)	0.123	± 0.011	0.13
O	≡ 1			≡ 1		≡ 1
Ne	0.17	± 0.02	(2,4)	0.138	± 0.017	0.14
Si	0.19	± 0.04	(5)	0.176	± 0.011	0.042
Ar	0.0040	± 0.0010	(4)	0.0042	± 0.0007	0.0042
Fe	0.19	$\begin{array}{c} + 0.10 \\ - 0.07 \end{array}$	(6)	0.224	± 0.031	0.038

[a]References. 1: Bame et al. 1975; 2: Bochsler et al. 1986; 3: Gloeckler et al. 1986; 4: Geiss et al. 1972; 5: Bochsler 1989; 6: Schmid et al. 1988; 7: Breneman and Stone 1985; 8: Anders and Grevesse 1989.

and some against the fast-stream composition. However, this effect is not of great significance for the discussion of the ancient solar wind particles that are found at the lunar surface. The most conspicuous result in Table I is the systematic overabundance of elements with low first ionization potential (FIP) in solar flare particle populations (Hovestadt et al. 1973a; Meyer 1981,1985; Breneman and Stone 1985) and in the solar wind (Geiss and Bochsler 1985,1986; Bochsler and Geiss 1989) relative to the measured or inferred abundance in the outer convective zone (Anders and Grevesse 1989).

Average elemental abundances in the solar wind and in flares normalized to the abundances in the outer convective zone and normalized to oxygen are plotted in Fig. 3 vs FIP. An overabundance of the elements with FIP < 10 or 11 eV by a factor of about 3 is evident. This observation is most readily explained by an atom-ion separation process (Cassé and Goret 1978; Meyer 1981; Veck and Parkinson 1981). Comparison of atomic data with the observed abundances implies that this process takes place at relatively low density and at temperatures around 10^4 K (Geiss 1982; Geiss and Bochsler 1985), conditions that occur in the upper chromosphere. Several separation mechanisms have been proposed. It seems to us that leakage of neutral species out of fine magnetic structures is the most promising approach (Bochsler and Geiss 1989; von Steiger and Geiss 1989). The element fractionation resulting from this mechanism in the case of a 10-km wide filament is also shown in Fig. 3. The theoretical model reproduces the observed abundances satisfactorily, even in some detail. In the context of our discussion, isotopic effects are of particular importance. Extensive model calculations confirm that isotopic fractionation in the atom-ion separation process is very small (von Steiger and Geiss 1989). This is expected, because (a) atomic levels and rates are virtually identical for isotopes, and (b) the differences in atomic mass are quite small compared to the mass differences of the investigated elements which do not show a significant mass dependency in their abundances (cf. Fig. 3).

Recently, Gloeckler et al. (1989) have found that the FIP effect is much weaker in the fast solar wind streams coming out of coronal holes than it is in the average solar wind. This observation could be explained if systematically coarser magnetic structures existed in coronal hole areas.

A question not settled definitely is the level of contamination of the corona by meteoritic and cometary matter (cf. Geiss and Bochsler 1986; Lemaire 1989). A glance at Fig. 3 shows that the overabundant elements are just those that would be expected from the addition to solar matter of refractory material as it exists in meteoritic or cometary grains. The estimated dust influx (Grün et al. 1985) is much too small to produce a substantial effect. However, other sources, such as mini-comets (cf. Fernandez and Ip 1989) entering near solar space, cannot yet be completely ruled out as significant contributors. To be sure, there are several strong arguments against assuming that the abundance pattern shown in Fig. 3 is mainly due to contamination of the corona: (a) the near-absence of the FIP dependence of elemental abundances in fast streams does not support the contamination hypothesis; (b) if contamination was the main source of the low FIP elements, it would be surprising to have identical heavy-element abundance patterns for solar wind and flare particles; (c) the observed charge states of elements in the solar wind are consistent with a regular expansion of the solar gas through the 1 to 2 $\times$ 10^6 degree zone of the corona. In the case of evaporation of solid particles, we would expect at least a tail of low charges for elements like Si or Fe which

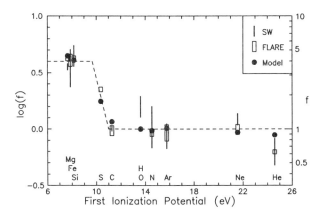

Fig. 3. Observed average abundances of elements in the contemporary solar wind and solar flare particle (SEP) populations. For reference see Table I. The dots represent the results of von Steiger and Geiss (1989) for a model in which element fractionation results from a competition of ionization and leakage out of fine magnetic structures at the solar surface.

is not observed. However, adequate charge state investigations are still scarce. Specific studies of charge states of solar wind and suprathermal ions, as well as of compositional variations in the solar wind with time and with solar latitude, would be desirable in order to establish a safe upper limit for the level of contamination by foreign material in the corona.

The atom-ion separation process could be partly responsible also for the low He/H ratio in the solar wind (Geiss 1982; von Steiger and Geiss 1989). In particular, a mechanism that depends on the differences between the ionization time of elements would discriminate against He because of its exceptionally long ionization time under solar surface conditions (Geiss and Bochsler 1985).

Further fractionations occur in the corona. The varying competition between the forces caused by Coulomb drag, separation E-field, thermal diffusion, momentum transfer from waves and gravity produce at times substantial variations in the composition of the solar wind, and by inference also a certain compositional stratification in the corona (Hirshberg et al. 1970; Hundhausen 1972; Borrini et al. 1981; Bürgi and Geiss 1986). Probably, most of the transient variations cancel when averages are taken but stratification must lead to a somewhat enhanced return to the Sun of elements heavier than hydrogen, and a complementary depletion in the solar wind.

Mass effects are not dominant in the variations of solar wind elemental composition that are caused in the corona. Thus, systematic effects on isotopic ratios ought to be small, and this is consistent with the absence of variations in the neon isotope ratios of the contemporary solar wind (Geiss et al. 1972). An exception is $^3He/^4He$ for which significant changes are observed (Geiss et al. 1972; Coplan et al. 1984) and expected from theory (Bürgi and Geiss 1986).

In summary, the observations available so far and results of theoretical modeling give us confidence to make some predictions about possible secular changes in solar wind composition. There is not much room for long-time changes of solar wind isotopic compositions, if fluxes were higher in the past: changes of $\sim 30\%$ in $^3He/^4He$ and of a few percent in other isotopic ratios is all we could expect. Only if fluxes were much lower than they are today, a "solar breeze" might have prevailed causing a depletion of heavier isotopes. However, as we shall see in Sec. IV, the solar wind flux in the distant past was higher rather than lower, and thus, it was even more hydrodynamical than today, making separations even more difficult. The record in meteorites and on the Moon seems to be consistent with this prediction (cf. Sec. V).

II. THE MOON AS A MONITOR OF EARLY
SOLAR IRRADIATION

The Moon provides us with the most direct information on solar corpuscular radiation in ancient times. The currently preferred view is that Moon and Earth were formed jointly and separated during the early epoch in the history of the planetary system. The presently observed deceleration of the rotation of the Earth implies that the Moon-Earth distance was once much smaller. It is, however, impossible to give a quantitative description of the evolution of our two-body system, as during its life-time the exchange rate of angular momentum has probably changed appreciably.

Early in the Moon's history, global melting led to the formation of the anorthositic crust (cf. Warren 1985). Thus, by the time the Moon began to form its first lasting regolith, the Sun had already settled on the main sequence. Later, further losses of collected gaseous solar atoms occurred during the phase of mare basalt extrusion which ended ∼ 3 Gyr ago. Afterwards, losses of regolithic material were mainly caused by impact-induced melting or deep burying. Thus, the solar record on the Moon is restricted to the main-sequence Sun, but for this epoch in the Sun's history it is the most readable and clean record we have—in particular, if we investigate material such as the mineral ilmenite of mare basalts that manifestly did not exist until the Moon was some 500 Myr old. Another factor limiting the access of solar wind particles to the early lunar surface could have been an ancient global lunar magnetic field for which evidence has been found (Runcorn 1983).

Volatile elements are extremely rare in lunar samples. As a consequence, many noble gas isotopes of solar origin found at the lunar surface can be unambiguously identified and distinguished from noble gas components of other origin (cosmic-ray products or products of radioactive decay like ^{40}Ar). Also a portion of nitrogen and carbon is of solar origin, but for these elements it is more difficult to identify the admixtures of other components.

Thus for N and C, the addition of volatiles other than those from the solar wind or solar flares has to be taken into account. Various mechanisms and evidence for the occurrence of such volatiles have been discussed (cf. Geiss and Bochsler 1981; chapter by Kerridge et al.) such as (1) outgassing of volatiles from the lunar interior, and (2) extralunar contribution of volatiles, e.g., by frequent cometary impacts, or from repeated passages of the solar system through interstellar clouds.

It has been proposed that the early Moon might largely have been confined inside the Earth's magnetosphere (cf. Fig. 4). At that time, the magnetospheric fluxes were possibly stronger and more extended due to a stronger geomagnetic field and a more intense ultraviolet bombardment of the atmosphere. Thus, on the one hand, the terrestrial magnetosphere might have inhibited direct solar wind irradiation, but on the other, it might have been a source of intense particle bombardment with ions partially originating from the Earth's atmosphere. (See also Reedy and Marti chapter.)

Which would be the most sensitive tracers for such a process? Judging from the composition of the early terrestrial atmosphere, the prime candidates are C, N, O and ^{40}Ar. ^{40}Ar is indeed found in trapped gases in the lunar regolith. However, it is generally ascribed to retrapped ^{40}Ar emanating from

TRANSFER OF ^{14}N FROM EARTH TO MOON?

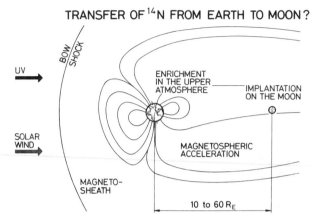

Fig. 4. Tidal forces in the Earth-Moon system transfer angular momentum from Earth's rotation to the Moon's orbital motion (i.e., in the past the Moon was closer to the Earth and the Earth rotated faster implying also a stronger terrestrial magnetic field). The absolute time scale of these changes is not yet known. However, there ought to have been an early period during which the free solar wind could not reach the Moon, and only gradually the fraction of time during which the wind had free access to the lunar surface increased to its present value. On the other hand, contamination of the Moon by ions from the Earth's atmosphere was probably more intense in the past when the Moon was closer to the Earth and the terrestrial magnetic field was stronger. Considering relative elemental abundances in the Sun, on the Moon and in the Earth's atmosphere, we suggest that nitrogen is one of the most promising elements for which a terrestrial origin on the Moon could be established. In this nitrogen, ^{14}N would probably be enriched.

the lunar interior following the decay of ^{40}K, although a terrestrial origin has also been considered. Its high abundance in virtually all lunar minerals excludes oxygen as a practical tracer. Thus, there remain C and N, for which a terrestrial component cannot be readily excluded, and perhaps a fraction of ^{40}Ar.

If nitrogen had been transferred in this way from the Earth to the Moon, we expect that it was isotopically fractionated with ^{14}N enriched over ^{15}N due to the difference in scale height of isotopes in the high atmosphere (cf., Geiss et al. 1975; Lind et al. 1979).

For studying the history of lunar bombardment by solar wind or solar flare particles, methods are needed to estimate the time of their implantation. The ^{235}U/^{136}Xe method (Eugster et al. 1983b) is the only method we know of giving absolute times, although with rather large errors. Relative indicators of antiquity of exposure can be calibrated against the ^{235}U/^{136}Xe age scale. In bulk samples with high-maturity indices (Fe/FeO, glass content, fine-grain size), solar gases were generally implanted earlier than in immature soils (cf., chapter by Kerridge et al.). Parentless ^{40}Ar (Heymann et al. 1970), ^{129}Xe and ^{244}Pu-fission-Xe (Marti et al. 1973; Drozd et al. 1975) in the regolith are due to outgassing of the Moon and subsequent retrapping (Manka and Michel

1971). Because of the relatively short lifetimes of their progenitors, these isotopes have been successfully used as empirical indicators for the time of surface exposure of a sample (cf., Eugster 1985).

Of particular interest are samples that were exposed for a certain time interval and shielded from the solar wind afterwards, such as the grains in soil breccias which were typically exposed in the distant past, up to a few Gyr ago (Stöffler et al. 1975; Geiss and Bochsler 1981; Eugster 1985; Becker and Pepin 1989).

III. EVIDENCE FROM ASTRONOMICAL OBSERVATIONS CONCERNING THE TEEN-AGE SUN

As outlined in the preceding section, the usefulness of the lunar record for studies of the solar wind and flare particle history is limited to the main-sequence life of the Sun. Thus, in this section we give a short summary of results for solar-type stars of an age of 1 to 2 Gyr, in order to have an idea of the expected properties of the Sun in that age interval that for brevity we call here the "teen age of the Sun."

There is general agreement that the teen-age Sun was rotating at a faster rate and exhibited a stronger activity than today. The surface rotational velocity was probably higher than today, by about a factor of 2 (chapter by Stauffer and Soderblom). Regarding the differential rotation (both radial and latitudinal), the difference between the teen-age and the present Sun may have been as large or even larger, because a time lag in angular momentum adjustment throughout the Sun is plausible as long as its surface velocity is markedly decreasing. Also, the solar activity was higher, by a factor of 2 to 4, depending on the index considered (chapter by Walter and Barry). The strongest change is indicated for the magnetic flux, with a decrease by a factor of ~ 5 between the teen-age and the present Sun. This decrease appears to be mainly due to the filling factor and not to the magnetic field strength (chapter by Saar).

Newkirk (1980) has discussed consequences of changes in magnetic properties of the Sun for the earlier solar wind. According to his parametric study, the Alfvén point would have been farther out by about a factor 3 from where it is now, and thus there would have been an increased angular momentum loss from the Sun.

As shown in Sec. IV, the lunar evidence points to a substantially higher flux of solar wind ions or suprathermal particles or both in the past. Two effects could have contributed to this. First, due to the stronger differential rotation, the solar corona could have been more efficiently heated. Enhanced heating below the critical (sonic) point ought to result in an enhanced solar wind flux. Second, the higher filling factor (chapter by Saar) could have led to more mass being supplied to the corona, which could have enhanced the solar wind flux even further. Concerning suprathermal particles, amplification may have resulted from further causes. One should not overlook the likelihood of a stronger coronal and interplanetary shock activity. Because of the higher solar rotation rate, different solar wind domains would more often have overtaken one another and interacted in the inner part of the solar system, thereby contributing to more frequent and possibly more intense particle acceleration events. One could even question whether during the solar teen-age anything like an approximately stationary solar wind existed, as we observe it today at 1 AU. Once better models for the contemporary slow solar wind with its structured coronal flow tubes have been developed, it will be easier to speculate about the behavior of the early solar wind and to compare these models with the observational evidence.

IV. SOLAR WIND VELOCITY AND FLUX IN THE PAST

Solar wind irradiation over several Gyr has left two kinds of signatures in lunar surface material: (a) solar wind atoms trapped in the lattices of lunar grains; and (b) solid state alterations. The latter are directly demonstrated by the track technique (cf., Crozaz 1980) or in high-energy electron microscope pictures such as the one reproduced in Fig. 5. The amorphous layer of a few tens of nanometers around silicate crystals are produced by the enormous energy dosage deposited by the solar wind ions. The thickness of the layer is inferred to be a measure of the average solar wind speed at the time of the grains' exposure. By comparing grains exposed at different times, Borg et al. (1980) have concluded that the average solar wind velocity has increased with time.

Using average measured xenon contents in regolith material along with estimates of average regolith thicknesses, Geiss (1973) found that the total inventory of solar Xe on the Moon is a factor of 2 or 3 higher than would be expected from solar wind bombardment over 3 or 4 Gyr with present-day intensity. Comparing time integrals of galactic cosmic-ray (GCR) flux (derived from spallation products) and solar wind (SW) flux, an SW/GCR ratio resulted that again was a factor of 2 to 3 higher than would be expected from the present-day ratio. Later, Clayton and Thiemens (1980) supported these findings by using implanted nitrogen. A factor of 2 or 3 does not seem very much in these estimates. However, the finding was really surprising because one would have expected to find a significant deficit in xenon in view of observed or inferred gas-loss effects: a substantial fraction of the older soils is highly altered (agglutinized) and has certainly lost a portion of its gas content. Moreover, through larger impacts, regolith material has been evaporated, deeply buried or ejected from the Moon. These considerations further increase the difference between observed and expected solar gas inventory on the Moon and lead to the suggestion that the average solar wind flux during the main-sequence epoch of the Sun was significantly larger than the contemporary flux (Geiss 1973; Clayton and Thiemens 1980).

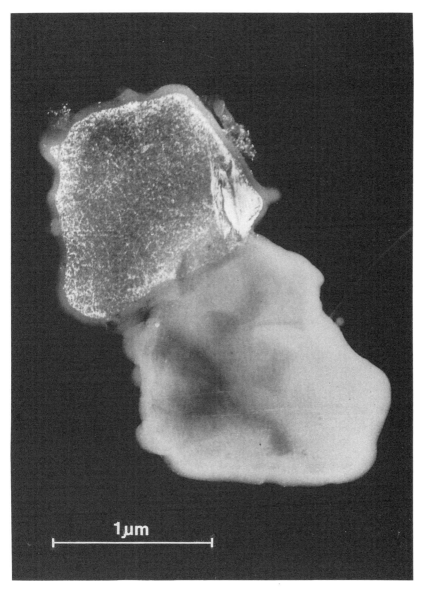

Fig. 5. Amorphous layer produced by solar wind particle irradiation around a silicate grain of an Apollo 11 soil sample (M. Maurette, personal communication, 1989). The micrograph is obtained with a 1 MeV electron microscope (Dran et al. 1970). The crystalline nature of the inner parts of the grain and the amorphous nature of the outer ~40-nm layer were verified by X-ray diffraction.

This view has gained considerable support from more recent observations showing two isotopically rather distinct noble gas components in lunar soils and breccias. These components have been distinguished by refined noble gas extraction techniques, the stepwise hydrolysis extraction employed at Minneapolis (Frick et al. 1988; Becker and Pepin 1989) and the on-line etching technique developed at Zurich (Wieler et al. 1985,1986). The latter technique gives very clean depth profiles in the individual grains of the implanted gases. Examples are shown in Fig. 6 and in Fig. 4 of the Kerridge et al. chapter. The stepwise gas-release patterns shown in these figures imply that the isotopically lighter component resides closest to the grain surfaces, whereas the heavy component is more deeply embedded. In the ilmenite of soils, recognized to be the mineral with the best gas retentivity long ago (Eberhardt et al. 1970), the isotopic compositions of He and Ne of the light component are virtually identical with the composition in the contemporary solar wind (see Figs. 2 and 6 and also Fig. 4 in the chapter by Kerridge et al.). The abundance ratio of the heavy component to the light component increases by a factor of ~2 (Wieler et al. 1983) with antiquity, i.e., the time elapsed since the grains of the sample were exposed at the surface (cf. chapter by Kerridge et al.).

   Of the two lines of interpretation for the wide-spread existence of the two components, in one view (Becker and Pepin 1989), the heavy component is made of solar wind particles implanted at some time in the past and isotopically changed by diffusive losses. We could hardly expect the fraction of the

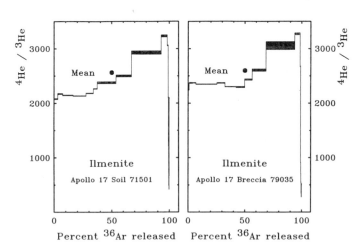

Fig. 6. Helium isotopes in lunar grains (125 to 175 μm); results obtained by the on-line etching technique for ilmenite samples in a soil (71501) and a breccia (79035). $^4He/^3He$ is plotted as a function of the cumulative amount of argon released. The plateau on the left corresponds to the light component, the rise towards the right is due to the heavy component which is inferred to be located deeper in the grains (Benkert et al. 1988; Benkert 1989).

gas that was lost to be uniform. However, isotopic enrichments might come out more uniform, because it is a fact of diffusion theory in a homogenous medium that after an initial change, the abundance ratio of two components implanted below the grain surface remain nearly constant over a wide range of loss factors (Geiss and Bochsler 1981). Employing the mirror-source method, one can show that, approximately, the resulting fractionation factor H between two types of implanted particles depends only on the ratio of their diffusion constants D_i and the ratio of their depths of implantation d_i:

$$H = \left(\frac{D_1}{D_2}\right)^{1/2} \frac{d_2}{d_1}. \tag{1}$$

Equation (1) is valid for $d_i < \sqrt{D_i t} < r$, where t is the duration of diffusion and r is the grain radius. If we assume that for isotopes the diffusion constant D_i is inversely proportional to the square root of the mass and that the depth of implantation d_i is proportional to the energy (Schiøtt 1970) and considering that in the solar wind energy is proportional to mass, we obtain

$$H = (m_2/m_1)^{5/4}. \tag{2}$$

After initially increasing towards this value, H remains constant until the inward diffusing particles begin to fill up the grain center, whereafter volume diffusion sets in and H increases further. The observed relation between isotopic ratios of the heavy and the light component is more like (Benkert et al. 1988)

$$H = (m_2/m_1)^2. \tag{3}$$

To be sure, there are many details in the occurrence and distribution of the noble gas isotopes that are not so easily interpreted by simple diffusion considerations. Moreover, not only the isotopic ratios but also the elemental abundances have to come out right. Equation (1) would also be applicable for elemental abundances, but the relative magnitudes of the diffusion constants D_i are more difficult to assess. Nevertheless, the observed values of H (Eq. 3) for isotopes are not far enough away from the near-invariant value predicted by the simple diffusion estimate (Eq. 2) that with the present evidence we could totally dismiss the diffusion explanation.

In the alternate model (Wieler et al. 1986), the two components represent solar particles of different energy; the isotopically light component is the solar wind, the heavy component represents implanted suprathermal particles. So far, suprathermal particle observations in the 10 to 10^3 keV energy range are rather scarce, and the energy coverage is often incomplete. In Fig. 7, we have assembled some data of observed proton fluxes in the range from

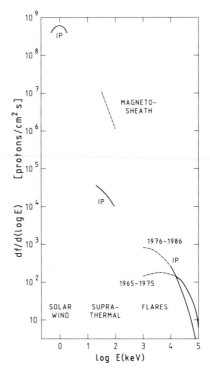

Fig. 7. Differential fluxes of protons in interplanetary space at 1 AU. The scale on the ordinate, df/dlog$E$, allows us to assess directly the relative contributions of the different energy domains. All curves marked by IP are for interplanetary fluxes at ~1 AU. In the domain of flare particles ($E \gtrsim 10^3$ keV), estimated averages for two 11-yr periods are given. Data in the suprathermal domain are scarce. The curve gives a rough estimate made from ISEE-3 data obtained in 1978–79. In the solar wind domain, the well-established long-time average of the contemporary solar wind proton flux is distributed over an energy range that roughly corresponds to the variations in solar wind speed and temperature. We have also included suprathermal fluxes observed by the CHEM-AMPTE satellite in the magnetosheath, the region behind the Earth's bow shock that contains heated and accelerated solar wind ions. For the sources of data, see the text.

solar wind energies to 10^5 keV. The two curves in the 10^4 keV region are based on spacecraft measurements during the periods 1965–1975 (Reedy 1977) and 1976–1986 (Goswami et al. 1988). In this region, the data are well represented by an exponential rigidity spectrum (chapter by Reedy and Marti):

$$f(R)dR = \frac{F}{R_0} e^{-R/R_0} dR \qquad (4)$$

where R_0 is the average rigidity and F the integrated flux. The present-day particle fluxes in this energy region that are due to flares are by a few orders of magnitude too low to account for the abundances of the heavy component in the trapped lunar gas (Wieler et al. 1986).

Average fluxes at 10 to 10^3 keV are very uncertain, but they are undoubtedly higher than fluxes with $E > 10^3$ keV. From raw data of the low-energy proton experiment on ISEE-3 (Sanderson et al. 1985), kindly provided by K.-P Wenzel (personal communication, 1989), and using the spectral analyses given by Wenzel et al. (1985), we have made a rough estimate of the proton flux in the energy range 30 to 56 keV for the 2-yr period 1978–79. Finally, the long-time average for the present-day solar wind of 3×10^8 proton cm^{-2} s^{-1} is also shown. Recent data obtained with the CHEM instrument onboard the AMPTE-CCE satellite indicate that the suprathermal flux behind the bow shock of the Earth is substantially higher than in free interplanetary space at 1 AU. An estimate by Gloeckler (personal communication, 1989) of the flux in the magnetosheath is included in Fig. 7. The figure is suggestive in several ways:

1. Between 10 and 10^2 keV there are probably one order of magnitude more particles than there are above 10^3 keV.
2. The suprathermal particles (10 to 10^2 keV) are better candidates for the isotopically heavy noble gas component than flare particles with $E \gtrsim 10^3$ keV.
3. Still, present-day average fluxes of particles with energies above, e.g., 10 times the typical solar wind energy of 0.8 keV amu^{-1} appear to be a few orders of magnitude smaller than the solar wind flux. Thus, the second explanation for the two lunar noble gas components would require suprathermal fluxes in the past that were several orders of magnitude higher than they are today. Not only breccias but also more recently exposed soils like 71501 contain large amounts of the heavy component (cf. Fig. 6). This indicates that the average suprathermal flux even during the last 1 Gyr would have been much higher than we estimate from the contemporary spacecraft data.
4. The acceleration mechanism near coronal and interplanetary shocks, which presumably creates the suprathermal population, would on average have had to favor heavier over lighter particles.

The diffusion model would require that the average flux of the solar wind during the last 3 to 4 Gyr was an order of magnitude higher than it is today. The explanation of two different primary particle populations would require that during much of the Sun's history suprathermal fluxes were a few orders of magnitude higher than they are estimated from spacecraft data.

In Sec. III, we have discussed the likelihood of enhanced fluxes in these two energy domains during the teen-age epoch of the Sun. Whereas an increase by an order of magnitude for the solar wind may be plausible, the increase of several orders of magnitude required for the suprathermal particle explanation is severe. The activity parameters observed in solar-type stars of

1 to 2 Gyr of age are increased only by the rather modest factors of 2 to 5, and thus these would have to be combined and amplified enormously when it comes to shock acceleration of particles.

In the magnetosheath behind the Earth's shock front, Gloeckler (personal communication) has observed average suprathermal particle fluxes that are about 2 orders of magnitude higher than those estimated for the interplanetary space (Fig. 7). These particles could contribute even more to the lunar inventory during earlier times when the Moon was closer to the Earth. At the present time, the Moon spends only a fraction of the time in those parts of the magnetosheath where suprathermal fluxes of the magnitude measured by Gloeckler (personal communication, 1989) are encountered. Still, Fig. 7 suggests that even today the Moon may intercept more suprathermal particles of solar origin in the magnetosheath than in the open interplanetary space.

As the isotopically heavier component has also been seen in meteorites (Wieler et al. 1987; Pedroni 1989), particles accelerated in the near-Earth environment do not seem to be the sole explanation for the occurrence of this component.

V. SOLAR WIND COMPOSITION IN THE PAST

If the solar wind flux in the past was higher than it is today, densities in the source region were probably also higher than today. High densities hamper all separation processes, and thus we would expect average solar wind composition and solar composition to have been even closer in the past than they are today. Thus, as a general rule, secular changes of the solar wind composition would have been limited in magnitude to the present-day differences between solar wind and solar abundances (cf. Sec. I). In our view, the lunar record on solar wind abundances during the main-sequence epoch of the Sun is not at variance with this expectation. Isotopic ratios of noble gases trapped at different times in the outermost surface layers of selected minerals from lunar soils, especially ilmenite (Eberhardt et al. 1972; Benkert et al. 1988; Benkert 1989), are very similar to the corresponding noble gas component in meteoritic minerals (metal [Hintenberger et al. 1965]; enstatite [Marti 1969; Signer et al. 1977; Pedroni 1989]) and to the directly measured ratios in the contemporary solar wind (Geiss et al. 1970a,1972; Coplan et al. 1984). Also the abundances of Ne, Ar and Kr in lunar samples of different antiquity show only modest trends, if any (Hintenberger et al. 1971; Eberhardt et al. 1972; Becker and Pepin 1989). Among the noble gas abundance ratios, there are two for which a measurable secular change is indicated: the $^3He/^4He$ ratio and the abundance of Xe relative to Ne, Ar and Kr.

It was suggested long ago that there exists a secular increase in the $^3He/^4He$ of the trapped solar wind gas which cannot be solely caused by preferential diffusion loss of ^3He (cf. Eberhardt et al. 1972; Geiss 1973). The recently achieved, cleaner separation of the helium component closest to the surface of ilmenite grains again indicates a difference of ~10% between a breccia and a soil sample (Benkert et al. 1988; Benkert 1989; cf. Fig. 6 above). We consider that all these findings do not yet establish a change but rather an upper limit for any change in the average solar wind ^{3}He/^{4}He ratio over the last few to several Gyr. A change of ~10% in this ratio would not be surprising as there are 2 mechanisms which could readily cause it:

1. The ^{3}He/^{4}He ratio in the solar wind is observed to change significantly (cf. Fig. 2), and the rather different dynamical behavior of the two helium isotopes in the corona (cf., Bürgi and Geiss 1986) suggests that the average ^{3}He/^{4}He ratio in the contemporary solar wind is somewhat higher than the solar ratio. If fluxes were higher in the past, the difference in dynamical behavior would have a smaller fractionation effect.

2. As a result of incomplete hydrogen burning, there must be a ^{3}He bulge at intermediate depths, around $M_r/M_\odot = 0.5$, of the present Sun with a maximum ^{3}He/^{4}He ratio (assuming standard conditions) of 1 to 2% (Bochsler and Geiss 1973). This is a factor of ~30 higher than the solar surface value (nature has been kind to us in this case), and therefore even very slow or partial turbulent diffusion down to intermediate solar depth would significantly increase ^{3}He/^{4}He at the solar surface and become visible in the solar wind. Thus, the upper limit of the long-time changes of ^{3}He/^{4}He in the solar wind places severe limits on models of the Sun that take into account secular compositional changes caused by turbulent diffusion (Schatzman and Maeder 1981; Lebreton and Maeder 1987; Bochsler et al. 1990) or mass loss (Guzik et al. 1987).

Whereas changes as a function of antiquity of the relative abundances of trapped solar wind He, Ne, Ar and Kr are small, the relative abundance of solar wind Xe is found to be systematically higher in samples with the earliest time of surface exposure, i.e., the soil breccias (Hintenberger et al. 1971; Eberhardt et al. 1972). In a forthcoming paper by von Steiger and Geiss, this finding will be discussed in the light of the atomic properties of xenon, such as mass, ionization potentials and rates.

A pronounced secular change of ~30% in the ^{15}N/^{14}N of nitrogen trapped at the lunar surface has been found (Kerridge 1975) and confirmed by comparative measurements in breccias and soils and by employing the stepwise heating techniques (Becker and Clayton 1977; Thiemens and Clayton 1980). Many causes for this secular change have been discussed (cf. the chapter by Kerridge et al., and references therein). Some of the evidence (i.e., the close correlation between the nitrogen and the noble gas abundances) suggests that trapped nitrogen is mainly solar in origin, whereas other observations suggest that a large fraction of the trapped lunar nitrogen is of nonlunar origin and causes the isotopic change. There are two reasons for favoring the second view: (1) a change of 30% of ^{15}N/^{14}N in the solar wind is very difficult to explain, either by changes in solar wind source dynamics, by

nuclear reactions, or by contamination of the solar surface (cf. Geiss and
Bochsler 1982); on the other hand, (2) there are several nonsolar sources that
could explain the nitrogen isotopic anomalies on the Moon without much
affecting the noble gas record (cf. Geiss and Bochsler 1982; also see Sec. II
and Fig. 4). Whereas the 5 noble gases show very similar trends, trapped
nitrogen behaves completely differently in important respects, i.e., its abundance is an order of magnitude too high relative to the noble gases (cf. Fig.
7 in the chapter by Kerridge et al.), and the variation in isotopic abundances
in stepwise heating experiments has the opposite trend as compared to the
noble gases.

VI. CONCLUSIONS

Measurements of the composition in the contemporary solar wind have
led to tentative identifications of causes for compositional variations and have
allowed us to estimate the magnitude of systematic differences between solar
wind and solar composition. Experiments with improved ion identification
capability and time resolution on forthcoming space missions (Ulysses and
others) should give results enabling us to confirm the existence of the fractionation processes in the chromosphere and corona that the present studies
suggest. When the underlying mechanisms are understood, more accurate
estimates of systematic differences between solar composition and solar wind
composition could be derived.

The data on the composition of solar wind particles trapped in lunar
surface material of different antiquity show that secular variations in the elemental and isotopic abundances of the noble gases are limited in magnitude;
we consider them to be compatible with the present view on the fractionation
processes in the chromosphere and the corona.

Spacecraft data on composition and average interplanetary fluxes in the
suprathermal energy range (10 to 10^2 keV amu^{-1}) in interplanetary space are
still scarce. More systematic studies in this area are urgently needed for an
interpretation of the most puzzling and intriguing result of the studies on solar
particles trapped at the lunar surface: the abundant occurrence of two noble
gas components with different elemental and isotopic composition. At the
same time, the distribution on the Moon (and in meteorites) of these components as a function of antiquity should be further studied. One aspect to be
addressed again is the macroscopic and microscopic redistribution of noble
gas atoms trapped at the lunar surface. Is it after all possible that particles
stemming from an early phase of the Sun with an extremely intense solar
wind constitute a major fraction of the present inventory in the fine-grained
lunar soils and breccias? So far, the observed distribution of the two components of trapped noble gases, e.g., their regular occurrence in ilmenites, argues strongly against this.

Considering the presently available evidence on the absolute abundances of trapped solar gases on the Moon and the occurrence of the two components, we conclude that either the solar wind flux and/or the suprathermal flux were significantly higher in the past than they are today, even well after the Sun had settled on the main sequence.

Twenty years of studies on lunar samples have demonstrated that we have on the Moon significant and intriguing information about the history of the Sun. The data obtained have allowed conclusions to be made on solar mixing, on past solar activity, and on the solar wind flux and composition during the main-sequence life of the Sun. Some of the still tentative conclusions could be drawn more firmly by further investigations with refined techniques of the material collected by the Apollo astronauts or sampled by the Luna spacecraft. A definite reading and full exploitation of the lunar record on the history of the Sun requires further, specifically collected samples. In the planning for future lunar landings, a sampling program specifically aimed at studying the history of the Sun should be among the primary goals.

Acknowledgments.   The authors are indebted to G. Gloeckler, P. Signer and R. Wieler for many interesting discussions and constructive comments. They thank G. Troxler for manuscript preparation. This work was supported by the Swiss National Science Foundation.

THEORY OF LUMINOSITY AND RADIUS VARIATIONS

H.C. SPRUIT
Max Planck Institut für Astrophysik

Theoretical calculations of luminosity and radius variations are reviewed. As sources for these variations, two kinds of effects are considered: (i) temporary conversion of thermal energy inside the convection zone into a different energy form (such as magnetic fields); and (ii) the effect of surface magnetic fields on the energy emissivity of the solar surface. Because strength, extent and location of the subsurface fields are uncertain, emphasis is on upper limits, derived from general thermodynamic considerations, and from the buoyancy of strong fields. With either of these limits, fields located at the base of the convection zone do not produce luminosity or radius changes of observational interest. Fields closer to the surface, but still embedded entirely inside the convection zone can in principle have measurable effects, mainly through their influence on the efficiency of convective energy transport. If the toroidal fields from which active regions erupt are located at a depth $> 10,000$ km, their effect on the solar radius is likely to be $< \Delta R/R = 3 \times 10^{-6}$ during a solar cycle and their effect on the luminosity $<10^{-4}$. The vertical fields seen at the surface itself have effects on the luminosity that compare well with observations. The dominant effect in this case is a modulation of the surface emissivity (reduction in spots, enhancement in small-scale fields). The radius changes associated with this effect are negligible.

I. INTRODUCTION

There are a number of known causes of variation of the solar luminosity and radius in time. The simplest variation is just that due to stellar evolution, the secular effect caused by the burning of hydrogen into helium in the core. This is qualitatively and quantitatively well understood, but does not produce observable changes on a historical time scale. The next obvious cause of variations are processes in the convection zone. The formation and decay of convective cells at the surface produce small random variations in luminosity. Also, the convection zone excites p-mode oscillations that produce tiny effects on the solar irradiance and radius at 5-min periods. This is discussed in the chapter by Fröhlich. Then, the appearance of magnetic fields at the surface is a known cause of small variations, because sunspots are darker and small-scale magnetic fields brighter than the average solar surface. Generalizing this, we may ask what the effect of magnetic fields could be, both those visible at the surface and those lingering somewhere inside the convection zone, on radius and luminosity. It is this question that we address primarily in this chapter. The theory used can be adapted easily to a class of other processes, namely those where thermal energy is converted (temporarily) into some other known form, such as kinetic energy of differential rotation, for example. Although this makes the arguments used fairly general, the conclusions derived do have a limited range of applicability and do not include more exotic causes of variation.

Magnetic activity can influence the solar luminosity in a number of ways. For the purpose of this review, I group them into two classes:

a. *Subsurface effects.* The periodic creation and destruction of magnetic fields within the convection zone implies temporary sinks and sources in the thermal energy of the Sun. Depending on how close to the surface these take place, some of this modulation is transmitted to the surface and appears as temporary variations in luminosity and radius.

b. *Surface effects.* Magnetic fields contribute visibly to the solar output by their energetic activity in the atmosphere: by dissipating magnetic energy while erupting from below the surface and by channeling energy across the photosphere in mechanical form. More importantly, they do this at the photospheric level itself by modifying the amount of radiation that is emitted at the surface. An example of such a modification is the reduction of the heat flux in sunspots.

Analytical estimates and detailed numerical models have been made by several authors to study the effects of magnetic activity. Most of these effects can be illustrated with a simple analytically tractable model of energy transport in the convection zone. This model was introduced by Spruit (1982a); it is elaborated further in this chapter. It lacks the accuracy of a full stellar evolution calculation but has the advantage of showing the dependence on properties of the stellar model more explicitly than numerical results do. It also provides a convenient framework for discussing the numerical results.

In any calculation, a model is needed for the transport of energy in the convection zone. In essentially all work, the mixing-length model is used to relate the heat flux to the instantaneous entropy gradient. We argue that this is sufficient to obtain results that are accurate to an observationally relevant level. The theory described here applies equally well to other stars with convective envelopes (except in cases where the envelope is significantly non-adiabatically stratified throughout, as in supergiants). The effect of large spots on luminosity and colors of main-sequence stars has been calculated (Spruit and Weiss 1986).

II. EFFECTS OF MAGNETIC FIELDS BELOW THE SURFACE

In this section we discuss changes due to fields buried below the solar surface. They are distinct from the effects of fields at the surface itself, which are discussed in Sec. III.

Luminosity and radius variations due to magnetic fields in the convection zone have been considered by several authors (Endal and Twigg 1982, and references therein; Gilliland 1982; Sofia and Chan 1982; Däppen 1983; Endal et al. 1985; Gilliland 1988; Spruit 1988). The periodic creation and destruction of magnetic fields during the solar cycle implies periodic sources and sinks of thermal energy in the convection zone. Such changes (β-perturbations) cause local changes in the energy flux, a part of which appears at the surface as a luminosity variation, after modulation by the convection zone. In addition, magnetic fields interfere with the efficiency of convective energy transport, also causing changes (α-perturbations) in the thermal structure of the convection zone. The time dependence of these is different: α-perturbations peak at the maximum of the field strength; β-perturbations during the buildup and decay of the fields.

Radius changes due to magnetic fields have two sources. When in hydrostatic equilibrium, thermal changes imply changes in the solar radius. We call such changes *thermal* radius changes. In addition, magnetic fields have a *direct* effect on the stellar radius. The presence of a magnetic field increases the stiffness of the effective equation of state of the gas so that creation of a magnetic field requires a slight expansion.

Especially for changes taking place in the lower convection zone, the associated luminosity and radius effects turn out to be very small (see, e.g., Endal et al. 1985). They are significantly less than the surface effects discussed in Sec. III. The main reason for this is the enormous thermal inertia of the deeper layers of the convection zone, which effectively absorbs energy sources and sinks on the time scales of observational interest. The thermal inertia is a very steep function of depth in the convection zone, however, and if the presence of strong fields closer to the surface can be made plausible, they could in principle result in observable effects. This holds especially for α-perturbations (Gilliland 1988).

In Sec. II.A through Sec. II.F, we outline the physics of thermal adjustments in a convective envelope, in the mixing-length approximation. Model problems for α- and β-perturbations showing the dependence of the adjustments on position and time are then given in Secs. II.G through II.I.

A problem in calculating the actual amplitude of the luminosity and radius effects is of course that it is hard to be sure about the strength of the field and at what depth it can plausibly exist. In Sec. II.J, we use the maximum possible magnetic field, setting the magnetic pressure equal to the external gas pressure. This is sufficient to rule out magnetic effects in the lower part of the convection zone as a significant source of radius and luminosity variations. In Sec. II.K, we use a (less general) argument to derive upper limits, assuming that the buoyancy of magnetic fields is what actually limits the field strengths.

Although the calculations are done with magnetic fields in mind, the developments are more general and can also be applied to other processes that act as sources or sinks of thermal energy. One can think of differential rotation or other hydrodynamic processes, for example.

A. Energy Transport in the Convection Zone

The main purpose of this subsection is to show that time-dependent energy transport in the convection zone, in the mixing length approximation, can be calculated from a single transport equation, with only minimal reference to the other equations defining the structure of the star. This allows one to calculate with little effort all *thermal* effects associated with magnetic fields, such as the surface flux variation. In order to obtain radius variations as well, a modest amount of additional knowledge about the structure of the star is needed.

Let $\mathbf{F}$ be the energy flux (convection plus radiation); then the first law of thermodynamics can be written as

$$\rho T \frac{dS}{dt} = -\mathrm{div}\mathbf{F} + G \qquad (1)$$

where ρ is the density, T the temperature, S the entropy per unit mass and d/dt the total (co-moving) derivative. G includes the sources and sinks of heat (but not the conduction or advection of heat, which is described by $\mathbf{F}$). The sources include viscous dissipation and the energy released in the destruction of magnetic fields, while the generation of steady flows and magnetic fields act as sinks of thermal energy. By using the entropy as thermal variable the gravitational energy does not appear in G, allowing a clean separation between purely thermal adjustments and the effects of expansions and contractions of the star. This way the thermal effects can be calculated without having to take the gravitational energy into account explicitly.

In the mixing-length approximation, $\mathbf{F}$ is written as a function of ρ, T and the local entropy gradient ∇S. The dependence on the gradient is nonlinear, but for the modest perturbations in the heat flux found in most cases, it can be accurately approximated by a linear dependence (cf. Spruit 1977,1982a,b):

$$\mathbf{F} = -\kappa\rho T \nabla S \qquad (2)$$

where κ is a turbulent diffusivity, given by a typical convective velocity times a convective length scale. In mixing-length models of main-sequence stars, the convective velocity and the length scale conspire to make κ nearly independent of depth (for the Sun within a factor of 3). The model of turbulence underlying mixing-length convection implies that κ is roughly the same for energy transport in vertical and horizontal directions (though in stellar structure calculations, only the vertical transport coefficient plays a role). Thus, it is justified to approximate κ by a single scalar constant. For a typical solar model, it would be of the order $\kappa = 10^{13}$ cm^{2}s^{-1}. Thus Eq. (1) becomes

$$\rho T \frac{dS}{dt} = \kappa \nabla \cdot (\rho T \nabla S) + G. \tag{3}$$

As the next step, we introduce the *quasi-hydrostatic* approximation:

$$\frac{dP}{dr} = -g\rho \tag{4}$$

where g is the local acceleration of gravity. This approximation is also used in most stellar evolution calculations. It means that only processes that are slow compared with the hydrodynamic adjustment time (about an hour for the Sun as a whole) are taken into account. Possible effects of stellar oscillations on the mean luminosity and radius, for example, cannot be calculated within this approximation. We consider one-dimensional (spherically symmetric) solutions to the diffusion Eq. (3) (for two-dimensional calculations of the sunspot blocking effect, see Spruit [1982b], Foukal et al. [1983] and Chiang and Foukal [1984]).

In the spherically symmetric case, there are no large-scale circulations transporting heat, so that the problem can be treated as a pure diffusion problem. Perturbations involving large-scale flows have not been studied so far (see, however, the related work by Glatzmaier [1985a,b]). One may wonder whether the large-scale motions set up by the temperature perturbation could lead to larger effects at the surface than the turbulent diffusion of heat discussed here (for an early discussion of the role of circulation in the sunspot blocking problem see Sweet [1955]). The heat carried by large-scale circulations, however, is limited by the turbulent viscosity due to convection. Assuming this to be of the same order as the turbulent diffusivity κ, one finds that circulations can only carry a flux of the same order as the diffusive flux (though the spatial dependence can be rather different), so that for order of magnitude estimates it is sufficient to consider only diffusion.

The problem can be further simplified if we restrict application to relatively thin convection zones (depth small compared with stellar radius), such that g is constant. This is a fair approximation for the Sun. Due to the hydrostatic approximation, the gas pressure is then a Lagrangian coordinate, i.e., it is a function of the mass coordinate only:

$$P = - \int \rho g \, dr \approx \frac{GM}{4\pi R^4}(M - M_r). \tag{5}$$

As depth coordinate we use

$$\mu \equiv \ln(P/P_0) \tag{6}$$

where P_0 is an arbitrary reference pressure. Equation (3) then can be written as (Spruit 1982a)

$$H^2 \frac{dS}{dt} = \kappa \frac{\partial^2 S}{\partial \mu^2} + \kappa(1 - \nabla) \frac{\partial S}{\partial \mu} + \frac{H^2}{\rho T} G \tag{7}$$

where $H = -dr/d\mu$ is the pressure scale height, and $\nabla = \partial \ln T/\partial \mu$ is the logarithmic temperature gradient. Because our coordinate is a Lagrangian one, this equation is valid *including* the expansions or contractions of the convection zone that by hydrostatic equilibrium accompany changes in the thermal structure.

As most of the luminosity changes to be discussed are small, very few, if any, of the effects to be discussed depend on the nonlinearity of the equations of stellar structure. Thus we may assume the source G to be weak, and linearize Eq. (7). This yields (an index $_1$ indicates the perturbation):

$$H^2 \frac{dS_1}{dt} = \kappa \frac{\partial^2 S_1}{\partial \mu^2} + \kappa(1 + \nabla_a - 2\nabla) \frac{\partial S_1}{\partial \mu} + \frac{H^2}{T\rho} G \tag{8}$$

(Spruit 1982a[a]), where $\nabla_a = 1 - 1/\gamma$ is the adiabatic gradient, which has been assumed to be unaffected by the perturbation. The only perturbed quantity in Eq. (8) is the entropy; the equation can therefore be solved, given an equilibrium model, without reference to equations determining the other perturbed quantities. Because the pressure is a Lagrangian variable in our approximation ($P_1 = 0$), the entropy perturbation can also be written in terms of the relative temperature perturbation:

$$S_1 = c_p \frac{T_1}{T}. \tag{9}$$

---

[a]There are a number of errors in the equations in this paper: in Eq. (9), $-2\delta - \nabla_a$ should read $1 - 2\delta - \nabla_a$; in (27) the division by c_p should be deleted, in (35) ζ_b should be ζ_b^2.

The solution of Eq. (8) can be described by a superposition of modes, each decaying exponentially in time with its own time scale and spatial structure. Given a sufficiently simple model for the basic state, these time scales can be calculated analytically.

B. Time Scales

The basic time scales to be expected in the solutions of Eq. (8) can be estimated by going back to the original mixing-length diffusion Eq. (3). With $\nabla = \partial/\partial z$, and in the absence of sources we get

$$\frac{dS}{dt} = \kappa \frac{\partial^2 S}{\partial z^2} + \frac{\kappa}{H} \frac{\partial S}{\partial z} \qquad (10)$$

where $H = [\partial \ln(\rho T)/\partial z]^{-1}$ is the pressure scale height. If the first term on the right-hand side dominates, the equation is a standard diffusion equation. Its solution evolves on a time scale

$$\tau_d = d^2/\kappa \qquad (11)$$

where d is the depth scale over which the perturbation varies. This is the *diffusive time scale,* the time it takes the turbulence to transfer information over a distance d. If the second term dominates, write $S = c_v(\ln P - \gamma \ln \rho)$, then using Eq. (2) the equation becomes

$$\frac{d}{dt}(\ln P - \gamma \ln \rho) = -\frac{F}{\rho T c_v H}. \qquad (12)$$

With $\rho T c_v = U$, the internal (thermal) energy of the gas per unit volume, this yields a time scale

$$\tau_t \approx U H/F. \qquad (13)$$

This time scale apparently depends on depth, and is the time it would take the solar flux to deplete the thermal energy contained in one scale height at that depth. This is called the *thermal time scale.*[b] Due to the steep increase of P, H with depth, $U H$ at depth z is roughly equal to the thermal energy of the envelope down to z. So $\tau_t(z)$ can be interpreted as the thermal time scale of the mass of the envelope above z.

For solar parameters, the diffusive time scale is of the order of one year for d equal to the depth of the convection zone. The thermal time scale is a

---

[b]In Spruit (1982a) and subsequently also by other authors, this time scale was called the Kelvin Helmholtz time scale of the convection zone. This is confusing because the Kelvin Helmholtz time is defined as the thermal time scale of a star as a whole.

very steep function of depth, due to the rapidly increasing temperature and density. It is roughly equal to the solar cycle time scale of 10 yr at a depth of 16,000 km and it is about 10^5 yr at the base of the convection zone. When a perturbation is applied, for example, a temporary energy sink at some depth, it will in general evolve on both thermal and diffusive time scales. For depths below the superadiabatic surface layers the two time scales are quite different. The perturbation will then first adjust on the diffusive time scale, followed by a much longer adjustment that brings the stratification back to thermal balance. A unique time scale for thermal adjustments does not exist; both time scales play an essential role in a convective zone of a star. The reason for the existence of two kinds of time scale is the presence of two energy transport processes. At the stellar surface, the luminosity is connected with the surface temperature through radiation. In the convection zone, a much more effective process operates, with corresponding short time scales. In radiative stars, energy transport can still be described by an equation very similar to Eq. (3), and time scales defined similarly. In this case, the time scales are of the same order of magnitude because energy is transported only by radiation (see also Gough 1981b).

The difference between the thermal and the diffusive time scales is large only below a certain depth. Equating the diffusive time scale at depth z, z^2/κ, to the thermal time scale (Eq. 13), one finds that this depth is a few pressure scales heights; it is equal to the depth of the nonadiabatic part of the convection zone. In the calculation of radius and luminosity variations, this layer plays the role of a boundary layer (Sec. II.E).

C. A Pseudopolytropic Model

Transport Eq. (8) can in principle be solved for perturbations around an arbitrary equilibrium stratification, but for analytical estimates the model has to be fairly simple. The stratification of density and temperature in an equilibrium model of the solar convection zone can be rather accurately described by a polytrope, that is a linear variation of temperature with depth. A good fit is given by

$$H = H_0 \zeta \qquad (14a)$$

$$P = P_0 \zeta^{n+1} \qquad (14b)$$

$$\rho = P_0/(gH_0)\zeta^n \qquad (14c)$$

where

$$\zeta = 1 + z/((n + 1)H_0) \qquad (14d)$$

where n is the polytropic index. A good fit for the solar convection zone is $n \approx 2$, $H_0 \approx 1.5 \times 10^7$cm, $P_0 \approx 4 \times 10^5$ erg cm^{-3}. The photosphere is at $z = 0$ ($\zeta = 1$). This model defines the temperature, density and pressure stratification but not the superadiabaticity. The logarithmic temperature gradient is fixed by the value of n:

$$\nabla = \frac{d \ln T}{d \ln P} = 1/(n + 1). \tag{15}$$

In order to describe the transport of energy in the equilibrium model, we must include the superadiabaticity, which we define here by

$$\delta = \nabla - \nabla_a. \tag{16}$$

From the mixing-length expression for the convective energy flux in the equilibrium state, $F_{conv} = \rho c_p T v_{conv} (l/H)\delta = F_\odot$, $l \approx H$, $lv_{conv} \approx \kappa \approx$ const, we find that the superadiabaticity δ must depend on depth as

$$\delta = \delta_0 \zeta^{-n} \tag{17a}$$

with

$$\delta_0 = F_\odot H_o/(T_o \kappa \rho_o c_p). \tag{17b}$$

This can be reconciled with the given value (Eq. 15) of ∇ if we assume that the adiabatic gradient varies through the star as

$$\nabla_a = 1/(n + 1) - \delta(\zeta). \tag{18}$$

Although the actual variation of ∇_a differs from this, the difference is rather modest because δ is small except in the uppermost layers. As in actual models c_p varies somewhat near the surface, there is some arbitrariness in the value of δ_0. With $\delta_0 \approx 0.25$, a quite acceptable fit to the entropy distribution in a standard mixing-length model of the solar convection zone is obtained, except very close to the surface.

The model defined in this way differs from the usual polytropic model, in which both ∇ and ∇_a are constants. By exploiting the fact that ∇_a varies, depending on properties of the equation of state that are not directly relevant for the problem at hand, we have produced a polytropic model that has the same simple stratification of P and T but with a much more realistic entropy gradient. All of the developments below are based on this model.

D. Thermal Modes of the Pseudopolytropic Model

Let us consider first regions without sources or sinks $(G = 0)$. The perturbation Eq. (8) then has solutions of the form

$$S_1 = c_p y(\zeta)\exp(-at). \qquad (19)$$

where $y(\zeta)$ satisfies

$$\zeta^2 \frac{\partial^2 y}{\partial \zeta^2} + (n + 1)\zeta \frac{\partial y}{\partial \zeta}(1 - \delta_o \zeta^{-n}) + \frac{(n + 1)^2 a H_0^2}{\kappa}\zeta^2 y = 0. \qquad (20)$$

The corresponding heat flux perturbation is found by taking the Lagrangian perturbation of Eq. (2). This yields for $n = 2$:

$$F_1/F_\odot = -y + \frac{\zeta^3}{3\delta_o} \frac{\partial y}{\partial \zeta}. \qquad (21)$$

Note that with Eq. (9), y is the amplitude of the (relative) Lagrangian temperature perturbation T_1/T. The general solution of Eq. (20) is a superposition of a number of modes, each with its own decay rate a. The initial conditions determine the relative amplitude of these modes. The higher modes, with the faster decay times, play a role only at short times after the perturbation is switched on. For long time scales, only the slowest decaying mode plays a role.

Equation (20) has a singular behavior near the surface. This is because the coefficient of the second derivative vanishes for $\zeta \to 0$, and because the coefficient of the first derivative varies rapidly near the surface. The equation can be solved by boundary-layer procedure (see, e.g., Bender and Orszag 1978). Near the surface, this procedure is equivalent to neglecting the last term in Eq. (20). The solution of the resulting equation is matched smoothly to the solution in the bulk of the convection zone, where the term involving δ_0 is negligible (resulting in a Bessel equation). Details are given in Spruit (1982a). From now on we use $n = 2$. To find solutions of Eq. (20), we apply the boundary conditions

$$F_1/F = 0 \quad (\zeta = \zeta_b); \qquad F_1/F = 4y \quad (\zeta = 1). \qquad (22)$$

The top boundary condition assumes that radiation emitted at the surface is close to a blackbody spectrum. Also, it assumes that the boundary condition is to be applied at a fixed mass. This does not take into account that the mass at which optical depth 2/3 is reached actually depends somewhat on temperature. (Corrections to these assumptions will be made in Sec. II.E). The decay time of the *fundamental* mode (in which the perturbation $y(z)$ has no nodes) is then:

$$\tau_0 = \frac{\gamma}{\gamma - 1}\frac{H_0 P_0}{qF_\odot}\zeta_b^4 = \frac{\gamma}{\gamma - 1}\frac{H_b P_b}{qF_\odot} \approx 1.56\frac{H_b P_b}{F_\odot} \tag{23}$$

where ζ_b is the value of ζ at the base of the layer considered, H_b, P_b are the scale height and pressure there, and q is a numerical factor of order unity; for the parameters given $q = 1.6$. With the thermal energy per unit volume given by $U = P/(\gamma - 1)$ we have $\tau_0 \approx H_b U_b/F_\odot$. Thus τ_0 is essentially the *thermal* time scale of the mass above z_b (cf. Sec. II.B).

For $\zeta_b \gg 1$, that is, for a layer whose depth is much larger than the superadiabatic boundary layer near the surface, the temperature and heat flux perturbations in the fundamental mode are:

$$\frac{T_1}{T} = A\left\{\frac{5}{4}\exp\left[\frac{3}{2}\delta_0(1 - \zeta^{-2})\right] - 1\right\} \tag{24}$$

$$\frac{F_1}{F} = A[1 - (\zeta/\zeta_b)^4] \tag{25}$$

where A is the amplitude of the perturbation (cf. Fig. 1). The entropy perturbation is constant with depth below the boundary layer (ζ^{-2} term). The entropy perturbation at the surface is reduced with respect to the interior by a factor $(5\exp(3/2\delta_0) - 4)^{-1}$, which is of order unity. The first *harmonic* mode (in which S_1 has one node) has a time scale:

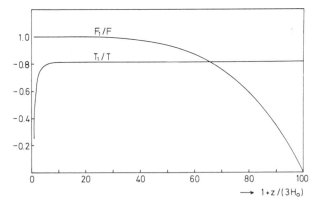

Fig. 1. Flux and temperature perturbation (temperature and flux minus their values for $t \to \infty$) in the fundamental mode of thermal decay in the convection zone, for a lower boundary at 45,000 km ($1 + z/3H_0 = 100$). This mode decays on the thermal time scale of the layer.

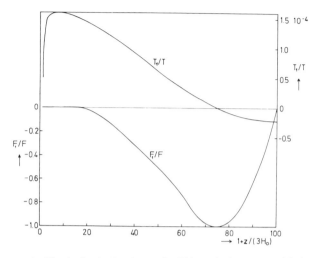

Fig. 2. Same as for Fig. 1, for the first harmonic. This mode decays at a diffusive time scale (∼ convective turnover time at the base of the layer).

$$\tau_1 = 9j_{21}^{-2}\zeta_b^2 H_0^2/\kappa = 0.34 H_b^2/\kappa \qquad (26)$$

where j_{21} is the first zero of the Bessel function J_2. This is apparently the diffusive time scale over one scale height. Because the mixing length is of the order H, τ is also the convective turnover time scale at the base of the layer considered. The higher harmonics (with larger numbers of nodes) have shorter time scales, corresponding to their smaller-length scales.

The depth dependence of these modes is shown in Figs. 1 and 2. Notice the thin boundary layer near the surface. (The depth of the convection zone has been reduced to 45,000 km in these figures for clarity.) The relative temperature perturbation drops by a factor of ∼ 3.3 across this layer, for all (except the highest) modes. This has to do with the fact that convection becomes inefficient close to the surface. If one is not interested in the temperature profile in the boundary layer, its effect may be absorbed into the upper boundary condition (Eq. 22) by changing the coefficient 4 into a smaller value, and simultaneously removing the term proportional to δ_0 from Eq. (20). This is discussed further in the next section.

We conclude that the thermal response of a convective envelope is not fundamentally different from that of a radiative one; what is unusual is the large difference in time scale between the fundamental and the higher modes of thermal adjustment.

E. The Surface Boundary Layer

In the preceding section we showed that thermal readjustments in the envelope have a spatial dependence that can be separated into an interior part and a surface boundary layer. It turns out that artifacts due to the plane-parallel approximation made in the previous section can be removed when dealing with the surface boundary layer. Because the luminosity effects seen at the surface are significantly modulated by the boundary layer, it is useful to look at its response to luminosity and radius variations in more detail.

We choose a depth z_b such that $H_0 \ll z_b \ll D$, and call this the base of the boundary layer (the precise value of z_b does not matter as long as this inequality is satisfied).

Our aim here is not to resolve the boundary layer itself, but to find its effect on the perturbations seen at the surface. We can then incorporate these effects by replacing the radiative boundary condition $F_1/F = 4T_1/T$ by an effective new condition applied directly to the perturbations in the envelope. If the time scales considered are much longer than the thermal time scale of the boundary layer (on the order of hours, down to a depth of 1000 km), the boundary layer is in thermal equilibrium, i.e., the heat flux F is constant with depth. The solution of the heat flow problem in the boundary layer gives the surface temperature T_s in terms of the heat flux F through the layer, the radius R of the star and the entropy S_b at the base of the layer:

$$T_s = T_s(R,F,S_b). \qquad (27)$$

Once this relation is known, the radiative boundary condition $F = \sigma T_s^4$ determines a relation between R,F,S_b that can be used as boundary condition for the deeper layers. For small perturbation Eq. (27) yields:

$$T_{s1} = \left.\frac{\partial T_s}{\partial R}\right|_{F,S_b} R_1 + \left.\frac{\partial T_s}{\partial F}\right|_{R,S_b} F_1 + \left.\frac{\partial T_s}{\partial S_b}\right|_{R,F} S_{b1}. \qquad (28)$$

To calculate the first term one needs the dependence of the boundary layer on surface gravity g. An estimate based on the mixing-length model for convection (a calculation that will not be repeated here) yields $(\partial \ln T_s/\partial \ln g)_{F,S_b} \approx 0.3$. This number is determined mostly by the dependence of opacity at the surface on temperature and density. With $g \sim R^{-2}$ this yields:

$$\partial T_s/\partial R \approx -0.6T_s/R. \qquad (29)$$

This was the only term that depends on the stellar radius. The last two terms in Eq. (28) can be calculated from the pseudo-polytropic model, as before. To calculate the second term in Eq. (28) we need the response of the surface layers to a small change in heat flux, at constant radius and base entropy. This is found by solving Eq. (21) with $y \to 0$ for $\zeta \to \infty$. The solution is

$$\frac{T_1}{T} = \frac{F_1}{F}\left[\exp\left(-\frac{2}{3}\delta_0\zeta^{-2}\right) - 1\right]. \qquad (30)$$

Evaluated at $\zeta = 1$ corresponding to $\tau = 2/3$:

$$\frac{\partial T_s}{\partial F} = \frac{T_s}{F}\left[\exp\left(-\frac{2}{3}\delta_0\right) - 1\right]. \qquad (31)$$

The last term in Eq. (28) can also be evaluated by constructing an appropriate solution of Eq. (21); we find $\partial \ln T_s/\partial S_b = 1/c_p$, hence the last term is just

$$\frac{1}{T_s}\frac{\partial T_s}{\partial S_b}S_{b1} = \left(\frac{T_1}{T}\right)_0 \qquad (32)$$

which is the temperature perturbation at the top of the envelope, as found from solving the heat-flow problem in the convection zone, excluding the boundary layer. The radiative boundary condition to be applied at the surface is

$$\frac{T_{s1}}{T_s} = \frac{1}{4}\frac{F_1}{F}. \qquad (33)$$

With Eq. (28) this yields

$$\frac{F_1}{F} = -0.6\eta\frac{R_1}{R} + \eta\left(\frac{T_1}{T}\right)_0 \qquad (34)$$

where

$$\eta = \left[\frac{5}{4} - \exp\left(-\frac{2}{3}\delta_0\right)\right]^{-1}. \qquad (35)$$

With $\delta_0 = 0.25$ as before:

$$\eta \approx 1.8. \qquad (36)$$

This expression gives the surface flux change if changes in radius and temperature below the superadiabatic layer are given. It is the effective boundary condition to be applied to the interior perturbations in the bulk of the convection zone. With $L = 4\pi R^2 F$, Eq. (34) can be written in terms of the luminosity variations:

$$\frac{L_1}{L} = (2 - 0.6\eta)\frac{R_1}{R} + \eta\left(\frac{T_1}{T}\right)_0. \qquad (37)$$

This relation can now serve as a boundary condition for the heat-flow problem in the bulk (adiabatic part) of the convection zone. The values of the coefficients in this expression are only approximate and intended for the estimates below.

The relations between the changes in surface flux, luminosity and radius derived, combined with the pseudopolytropic model developed above for the thermal adjustments of the convection zone, allow us to calculate the effect of various perturbations on the thermal structure of the convection zone. The model is somewhat hybrid in character: the response of the surface boundary layer is calculated in a way that includes effects of the spherical geometry, in particular the effect of changes in radius on the luminosity. Below the surface boundary layer, the model assumes a plane-parallel approximation, so that the mass displacements accompanying thermal adjustments can be taken into account easily and implicitly. Although the accuracy of this approximation is finite for the solar convection zone, I believe the combination of the two procedures reproduces all the main features of the thermal adjustments taking place.

F. Radius Changes

The radius change in general has two distinct components (cf. Endal et al. 1985). First, the magnetized fluid, at a given temperature and in pressure equilibrium with its surroundings, has a lower density. The production of such a field therefore produces a direct expansion on a hydrodynamic time scale. Let us call this the *direct* effect. Second, a certain amount of energy is needed for the creation of the field. This energy comes ultimately (via convection and differential rotation) from the thermal energy at some place in the star. Hence it is accompanied by a perturbation in the energy flux through the star, and a change in the temperatures. By hydrostatic equilibrium, the perturbed thermal profile produces a change in radius. We call this second effect the *thermal* part of the radius change. (See the chapter by Ribes et al.)

The direct effect was considered separately from the thermal effect in early papers (Thomas 1979; Dearborn and Blake 1980a). This was criticized by Gough (1981b), who argued that the two effects tend to cancel, so that the direct effect may strongly overestimate the radius change. It turns out that this cancellation is only very partial. Because the thermal perturbation due to the magnetic field diffuses through the convection zone, the radius effect associated with it has a time dependence that is in general quite different from that of the direct effect, and the parts of the convection zone involved also differ. Even if the thermal effect is confined to the same volume as the magnetic field (as it is initially following a sudden change in the magnetic field), there is a net effect because the magnetic fluid has a different equation of state (cf. Endal et al. 1985; see also below). Although it is true that the two effects are of opposite sign, the net effect in this case is still of the same order of magnitude as the direct effect. In practice it is more convenient to treat the two effects separately at first, as we shall do presently. Gough also argues that the expansion effect of a magnetic field is much smaller if the field is in the form of isolated tubes. The argument is that if a tube is held suspended in the convection zone, the outside is in the same hydrostatic equilibrium as in the absence of the tube. As the radius follows directly from hydrostatic equilibrium with the given temperature stratification, the effect of the tube is the same as that of adding (adiabatically) to the Sun an amount of mass equal to the mass displaced by the tube on account of its lower density. Because the density in the convection zone is low compared with the solar average, Gough argues that the effect would be very small. The flux tube is buoyant however, and while rising through the convection zone exerts a stress on its surroundings through viscous or turbulent drag. A field divided into many small tubes can thereby transmit its buoyancy effectively to the fluid, so that it behaves like a fluid of lower density, like water containing air bubbles. This shows, however, that the effect is transient, lasting as long as it takes the tubes to rise. Because tubes with a strong field rise fast, the strength of the field enters in. This is discussed further in Sec. II.K.

Let us consider first the thermal part of the radius change, which we denote here by an index $_t$. It follows from the temperature perturbation through the condition of hydrostatic equilibrium. In the plane-parallel approximation used above, it is given by

$$R_{1t} = \int \frac{T_1}{T} dz. \tag{38}$$

For a calculation of the direct radius effect and the associated luminosity effect, details must be known about the energetics of the field buildup process. If the field is produced in temperature equilibrium with its surroundings, for example, the result is not the same as when adiabatic changes are assumed. In principle, it is even possible to envisage a field that causes no density change at all, if it is produced in gas that is somewhat cooler than average for the level considered. These details are unknown; one can only put some global limits on the range of possibilities (see Secs. II.J. and II.K). Thus we are forced to generalize somewhat. Let us assume that no special cases occur, so that the change in gas density due to the magnetic field is of the order

$$\frac{\Delta\rho}{\rho} \approx -\frac{B^2}{8\pi P} \tag{39}$$

while

$$\Delta P = -\frac{B^2}{8\pi}.$$ (40)

In the plane-parallel approximation, Eq. (40) is exact for spherically symmetric changes in density, because the total pressure $P + B^2/8\pi$ depends only on the mass coordinate in this case (cf. Sec. II.A). The density change given by Eq. (39) is then exact for isothermal changes. Assume that the magnetic field extends over a narrow depth range $d \ll H$. The change in thickness of this layer due to the field is then $-d\Delta\rho/\rho$. In the plane-parallel approximation this is also the direct radius change R_{1d}:

$$R_{1d} \approx \frac{B^2 d}{8\pi P}.$$ (41)

It must be remembered that this number depends on the way in which the field is produced. If the fields are created in density equilibrium ($\Delta\rho = 0$), we have $R_{1d} = 0$, and the only radius change is through the thermal effect. The energetics involved in the field production is considered further in Sec. II.G. With Eq. (37) we can calculate the luminosity change if only a direct radius effect is present (i.e., the energy required to produce the field is ignored). Even in this case the response will vary in time, because of the slight change in the upper boundary condition produced by the change in acceleration of gravity at the surface (cf. the discussion in Sec. II.E). Consider time scales long enough that the surface layers are in thermal equilibrium, but short enough that the deeper layers respond adiabatically. This includes most time scales of observational interest, as the upper layers respond on a time scale of days at most. Then $(T_1/T)_0 = 0$, and with Eq. (37) we get the relation between luminosity and radius change for such a perturbation:

$$\frac{L_{1d}}{L} = (2 - 0.6\eta)\frac{R_{1d}}{R} \approx 1.1\frac{R_{1d}}{R}.$$ (42)

This shows that, if the direct radius effect dominates, the luminosity and radius changes are of the same sign and order of magnitude.

G. Response to a Sudden Perturbation

In this section, we calculate the response of the envelope to the sudden and local appearance of a magnetic field at some depth, with an associated local sink of thermal energy. This sudden perturbation is the subject of most numerical studies. Perturbations varying on the time scale of the cycle behave differently in some respects, because the cycle period is considerably longer than the diffusive time scale of the convection zone. They are treated separately in Sec. II.H.

Perturbations in the energy transport equation can be of two kinds: (1) an additional source or sink of energy; (2) a change in the energy transport coefficient. These are called β- and α-perturbations, respectively, in the terminology of Endal et al. (1985). A magnetic field has both kinds of effect. The buildup of the field requires energy, that ultimately (through convection and differential rotation) comes from the thermal energy content of the star (and, by the virial theorem, from the gravitational energy). Secondly, the field changes the transport coefficient by interfering with convection.

The two effects differ in their time dependence. The sinks or sources have their maximum when the rate of change of the magnetic energy is maximal, the effect on the transport coefficient when the magnetic energy itself is maximal. The strength of the effects, for a given magnetic field strength, is also rather different. In the following, we concentrate on the source/sink effect; this is the more important one in the deeper parts of the convection zone. The effect on the transport coefficient is discussed afterwards in Sec. II.I.

Assume that at $t = 0$ a magnetic field is suddenly produced over a narrow range $d \ll H$ at depth z_b (where $z = R_\odot - r$). The process is to be fast, but still slow enough that it occurs in hydrostatic equilibrium. The magnetic field is generated by fluid motions which otherwise would have dissipated into heat. Ultimately the field therefore acts as a thermal energy sink. There is some freedom to specify where in the convection zone the thermal energy is taken out, as this depends on details of the field generation process. We assume here that the energy is taken out of the volume containing the magnetic field itself. This differs from the perturbations studied by Endal et al., which imply a net input of energy (from an unspecified source) at some depth. The results in this section differ a little from those of Endal et al., probably because of this difference.

Because the perturbation takes place suddenly, the total energy of the star is conserved in the production process. If Δe_g, Δe_t and Δe_m denote the changes in gravitational, internal and magnetic energy, their sum vanishes:

$$\Delta e_g + \Delta e_t + \Delta e_m = 0. \tag{43}$$

Let the mass in the volume containing the magnetic field be m. The gravitational energy change can be calculated easily in the plane-parallel approximation (constant gravity). In this case, the pressure at each depth is a function only of the mass above that depth. If an individual layer changes its thickness, this happens at constant pressure. Also, it only lifts the overlying layers uniformly, without affecting the layers below. In the plane-parallel approximation, the gravitational energy change is therefore just the work done by the mass against the external pressure:

$$\Delta e_g = mP\Delta(1/\rho) = -\frac{m}{\rho}\frac{\Delta\rho}{\rho}P. \tag{44}$$

For an ideal gas the internal energy change is

$$\Delta e_t = m\Delta U = \frac{m}{\rho}\frac{P}{\gamma-1}\frac{\Delta T}{T} \tag{45}$$

and the magnetic energy change is

$$\Delta e_m = \frac{m}{\rho}\frac{B^2}{8\pi}. \tag{46}$$

Thus

$$-\frac{\Delta\rho}{\rho}P + \frac{P}{\gamma-1}\frac{\Delta T}{T} + \frac{B^2}{8\pi} = 0. \tag{47}$$

Still in the plane-parallel approximation, the external pressure does not change during the expansion caused by the field, so that

$$\Delta P = -\frac{B^2}{8\pi}. \tag{48}$$

With the ideal gas law and Eq. (47) this yields

$$\frac{\Delta T}{T} = -\frac{B^2}{4\pi P}\frac{\gamma-1}{\gamma} \tag{49}$$

$$\frac{\Delta\rho}{\rho} = \frac{B^2}{8\pi P}\frac{\gamma-2}{\gamma}. \tag{50}$$

The last equation demonstrates the well known fact that on expansion perpendicular to the field lines a magnetic field behaves as a fluid component with $\gamma = 2$. For $\gamma = 5/3$, the value we are assuming throughout, Eq. (50) shows that the magnetic field still causes a net expansion, in spite of the cooling of the gas implied Eq. (49).

Equations (49 and 50) represent the initial state of the perturbation. Because of the lower temperature in the field, it starts acting as a thermal energy sink on its surroundings. This thermal perturbation then starts evolving on the various time scales discussed in Sec. II.C. The result is a reasonably complicated time dependence, during which the surface luminosity perturbation can change sign one or two times. Instead of a detailed discussion of these effects, I consider the perturbation here mainly at two times: immediately after the initial state, and on time scales $\tau_d(D) \ll t \ll \tau_t(D)$, that is after the adjustment on the diffusive time scale has taken place, but short compared with the thermal time scale of the convection zone (the depth of the convection zone is D). Initially, the only effect is that of an adiabatic expansion due to the field. If the field extends over a depth d, the radius change is approximately (in the plane-parallel approximation: exactly, see the discussion above)

$$R_{1d} = -d\Delta\rho/\rho = \frac{e_m}{P}\frac{2-\gamma}{\gamma}, \qquad (t \ll \tau_d) \qquad (51)$$

where

$$e_m = \frac{B^2}{8\pi}d \qquad (52)$$

is the magnetic energy of the layer per cm² of horizontal surface. As the thermal perturbation has not yet spread, the convection zone has responded adiabatically to this change, and Eq. (42) applies. The luminosity change at this stage is therefore positive, and of the same order of magnitude as the radius change.

Next consider a time scale $\tau_d \ll t \ll \tau_t$, for example, the cycle time scale of 10 yr. The initially cool magnetic field has extracted an amount of heat ΔQ from the convection zone given by

$$\Delta Q = \rho c_p\Delta Td = Pd\frac{\gamma}{\gamma-1}\frac{\Delta T}{T} = -e_m. \qquad (53)$$

This amount of heat diffuses through the convection zone until the stratification is again adiabatic (except for the surface boundary layer), but at a slightly different entropy. From then on, the further development is on the thermal time scale of the envelope (cf. Secs. II.B and II.C). Again using the Lagrangian perturbations of Sec. II.A., the entropy difference between this state and the initial state is denoted by S_1, and it satisfies (using $S_1 = c_pT_1/T$, and $c_p = \gamma/(\gamma-1)P/\rho T$ for an ideal gas):

$$\Delta Q = \int_0^D \rho TS_1dz = \frac{T_1}{T}\frac{\gamma}{\gamma-1}\int_0^D Pdz. \qquad (54)$$

The integral can be evaluated with the pseudopolytropic model of Sec. II.C. With Eq. (53) this yields

$$\frac{T_1}{T} = -\frac{4}{3}\frac{\gamma - 1}{\gamma}\frac{e_m}{P_D H_D}, \qquad (t \gg \tau_d(D)) \tag{55}$$

where P_D, H_D are the pressure and scale height at the base (D) of the convection zone. This temperature change has an associated radius change; with Eq. (38):

$$R_{1t} = D\frac{T_1}{T}, \qquad (t \gg \tau_d(D)). \tag{56}$$

In addition, there is the direct radius change due to the expansion of the magnetic layer. The layer has the same temperature as its surroundings, which is higher than initially, and is now thicker. Instead of Eq. (51), the value for isothermal fields (Eq. 39) applies. It is larger by a factor of 5 (for $\gamma = 5/3$). The total radius change is the sum of the two:

$$R_1 = R_{1d} + R_{1t} = \frac{e_m}{P_b}\left(1 - \frac{P_b}{P_D}\frac{D}{H_D}\frac{4}{3}\frac{\gamma - 1}{\gamma}\right), \qquad (t \gg \tau_d(D)) \tag{57}$$

where P_b is the pressure at the depth (z_b) of the magnetic layer. As the pressure increases steeply with depth, the direct radius change dominates unless the layer is close to the base, and is positive:

$$R_1 \approx \frac{e_m}{P_b}, \qquad (z_b \ll D). \tag{58}$$

The associated luminosity effect is, with Eq. (42) and $\gamma = 5/3$:

$$L_1/L \approx 1.1\frac{e_m}{P_b R}. \tag{59}$$

This agrees with the findings of other authors (Endal et al. 1985; Gilliland 1982). On the other hand, if the magnetic layer is close to the base, the thermal effect can be somewhat larger than the direct effect, so that the radius change can be negative:

$$R_1 \approx -2\frac{e_m}{P_D}, \qquad (z_b = D). \tag{60}$$

This last result should not be taken too literally, however, because we have ignored the diffusion of the thermal perturbation into the radiative layers below the convection zone, which affects somewhat the value of T_1/T derived above.

We conclude that both in the initial and the late phases, the luminosity and radius changes are due mostly to the direct radius effect, first discussed by Thomas (1979). It does not follow, however, that this also holds for intermediate times. If the magnetic layer is fairly close to the surface, for example, a signal of its low temperature reaches the surface by diffusion before diluting itself over the entire convection zone. This yields a transient episode of low surface temperatures, which lasts for a time of the order $\tau_d(z_b)$, the time it takes the perturbation to diffuse to the surface. The surface luminosity then has the time dependence sketched in Fig. 3. The luminosity perturbation changes sign twice, at times depending on the depth of the layer. To see this quantitatively, consider the perturbation at a time of the order $\tau_d(z_b) = z_b^2/\kappa$. The cool region then has spread to the surface and to a depth of order z_b below z_b. By the same calculation as that leading to Eq. (55), the temperature perturbation in this region is found to be of the order

$$\frac{T_1}{T} \approx -\frac{e_m}{H_b P_b}. \tag{61}$$

This produces a surface flux change of the same order of magnitude and sign, that competes with the luminosity increase due to the direct radius increase Eq. (59). The direct effect is therefore smaller than Eq. (61) by a factor H_b/R. Thus there is a transient period of reduced luminosity if the disturbance occurs not too deep below the surface.

I believe that this is responsible for the changes of sign of the luminosity perturbation seen by Endal et al. (Their explanation in terms of the factor γ $-$ 4/3 that appears in stellar stability does not look convincing. The gravitational energy change is a global quantity that does not sense only the local value of γ. In any case, the effect just discussed is strong and should also be clearly visible in numerical simulations.)

H. Effect of Fields Created and Destroyed on the Cycle Time Scale

This case is perhaps of more practical interest than the case of an impulsive disturbance. Since the solar cycle is long compared with the turbulent

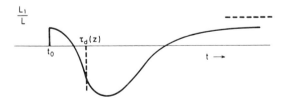

Fig. 3. Surface luminosity variation in time following the sudden conversion of thermal into magnetic energy at a depth z_b (schematic). The initial and final luminosity increase is caused by the expansion of the envelope due to the magnetic pressure. The intermediate reduction is due to the cooling of the magnetic layer, which spreads to the surface on a time scale $\tau_d(z_b)$.

diffusion time across the convection zone, simplifications can be made that allow a complete analysis.

Assume again that the magnetic field is created and destroyed in a thin layer at depth z_b. As before, let the total magnetic energy, per cm^2 of horizontal surface be $e_m = B^2 d/8\pi$, where d is the thickness of the layer ($d \ll H$). Then the source term appearing in Eq. (8) is

$$G = \frac{de_m}{dt}\delta(z - z_b) \tag{62}$$

where δ represents the Dirac delta function. The solution of Eq. (8) with this source is found by solving the equation separately in the layers above and below z_b, where no sources are present, and matching these at z_b with the condition

$$F(z_b + \epsilon) - F(z_b - \epsilon) = \frac{de_m}{dt}. \tag{63}$$

Since we are considering a sufficiently long time scale, the solution involves only the fundamental (thermal) mode (cf. Secs. II.B and II.C). Instead of solving this problem explicitly, the results of interest are found easily from physical considerations.

In the fundamental mode the relative temperature perturbation T_1/T is constant with depth except in the superadiabatic boundary layer. As the temperature must be continuous across z_b, this constant is the same in the layers above and below z_b. The amount of thermal energy spent in creating the field at each instant is just e_m. This amount must be equal to the change in energy content of the convection zone represented by the value of T_1/T. T_1/T follows from this as in Sec. II.G; the result is given by Eq. (55). The heat flux perturbation can also be found; it is sketched in Fig. 4.

The most remarkable property of this result is that the heat flux disappearing into the sink comes almost entirely from below z_b, so that the disturbance produces only a very weak signal at the surface. This behavior has an analogy in the heat flux blocking problem of sunspots (Sec. III.B). A larger signal can reach the surface if the sink varies on time scales on the order of the diffusive time scale or shorter. This was illustrated in Sec. II.G.

I. Effect of Magnetic Fields Through Changes in Convective Efficiency

By interfering with convective motions, a magnetic field can locally increase the entropy gradient required for transporting a given amount of energy. The luminosity and radius changes produced by such α-perturbations have been studied by several authors (Spiegel and Weiss 1980; Dearborn and Blake 1980a; Gilliland 1982; Endal et al. 1985). The details of the interaction

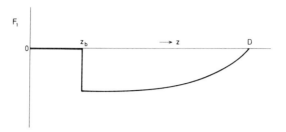

Fig. 4. Heat flux perturbation through the convection zone for a thermal sink located at z_b and varying on a time scale long compared with the diffusive time scale of the convection zone. The energy transport in the convection zone adjusts itself such that the sink is supplied almost entirely from the layers below z_b.

between fields and convection are uncertain but the simple model for this interaction used by these authors is probably adequate for order of magnitude estimates.

In this model the magnetic field energy density is compared with the kinetic energy density in convective turbulence in the absence of a magnetic field. When the two are equal, the magnetic field has the so-called equipartition strength. From the mixing-length expression for the convective velocity, it is seen that this happens when the product $\beta\delta$ is of order unity, where $\delta = \nabla - \nabla_a$ as before, and $\beta = 8\pi P/B^2$ is the plasma β. At this strength, the field is strong enough to start reducing the degrees of freedom of the convective flow, so that a larger entropy gradient is required to transport the same energy flux. A simple model to include this effect is to change the mixing-length expression for the convective energy flux into

$$F_{\text{conv}} = \kappa\rho c_p \frac{T}{H}(\delta - q/\beta) \qquad (64)$$

where q is a numerical factor of order unity.

Suppose this effect is switched on suddenly ($q = 0$ before $t = 0$, $q = 1$ thereafter), in a layer of depth d at depth z_b. Initially, this reduces the heat flux through the layer, so that the temperature above it starts dropping, and below it starts increasing. The sequence of events is sketched in Fig. 5.

Initially the temperature distribution is unchanged, and the heat flux is reduced near z_b. The initial response of the layers above and below z_b is governed by adjustments taking place on the turbulent diffusion time scales of these layers. Because of the separation in time scales, we can again consider times long compared with these diffusion time scales (longer than a year or so). On time scales short compared with the thermal time scale of the convection zone below z_b, the temperature below z_b does not change. Assuming that the reduction of the transport efficiency is sufficiently large to reduce the heat flux initially by a significant factor (which may in fact require conditions that are not entirely plausible for the solar convection zone), the temperature in the layers above z_b starts dropping because of the mismatch of the heat flux at z_b and that at $z = 0$. This happens on the thermal time scale of these layers. Unless z_b is close to the convection zone depth D, this time scale is much shorter than that of the layers below z_b. On time scales $\tau_t(z_b) \ll t \ll \tau_t(D)$, the layers above z_b have returned to thermal equilibrium, while the layers below are still slowly heating up. The surface heat flux is somewhat reduced during this time. The magnitude of this reduction can be calculated from the increase in the temperature drop across the layer d due to the magnetic field. From Eq. (64), this temperature change is of the order

$$\frac{T_1}{T} \approx -\frac{q}{\beta}\frac{d}{H}. \tag{65}$$

The surface flux perturbation is of the same order of magnitude. This is the luminosity effect derived by Spiegel and Weiss. It follows from the above that it takes a time $\tau_t(z_b)$ for this effect to reach its full amplitude. This time equals the cycle time of 5 yr if $z_b = 16{,}000$ km (see Sec. II.B). The Spiegel-Weiss estimate is therefore valid only if the field is located at a depth z_b *less* than about 16,000 km. For $t \gg \tau_d(z_b)$, the relative temperature change is independent of depth down to z_b (see also Fig. 5), and is given by

$$\frac{T_1}{T} \approx -\frac{q}{\beta}\frac{d}{H_b}(1 - e^{-t/\tau_t(z_b)})e^{-t/T_t(D)}. \tag{66}$$

For the time scales of interest, $t \ll \tau_t(D)$, this yields a surface flux change of

$$\frac{F_{s1}}{F} \approx \frac{T_1}{T} \approx -\frac{q}{\beta}\frac{d}{H_b}(1 - e^{-t/\tau_t(z_b)}). \tag{67}$$

With the definition of β, $e_m = B^2d/8\pi$ as before, and with $q \approx 1$ Eq. (67) can be written as

$$\frac{F_{s1}}{F} \approx -\frac{e_m}{P_bH_b}(1 - e^{-t/\tau_t(z_b)}). \qquad (\tau_d(D) \ll t \ll \tau_t(D)). \tag{68}$$

The sign of this flux change is opposite to that of the luminosity change associated with the expansion due to the field given by Eq. (59). Its magnitude can be larger or smaller. On the time scale on which the field is built up, $t = \tau_c/2 \approx 5$ yr, the α-effect (Eq. 68) is of the same order (but of opposite sign) as the direct effect when

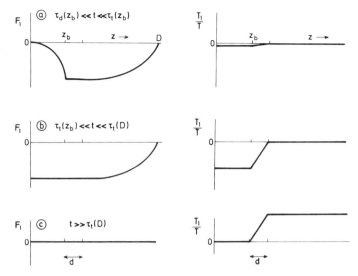

Fig. 5. Effect of a reduction in energy transport efficiency between z_b and $z_b + d$, at three characteristic instants (schematic): (a) after the adjustments on the diffusive time scale have taken place but before the layers above z_b have returned to thermal equilibrium; (b) after the upper layers have returned to equilibrium but before the deeper layers return to equilibrium (the surface flux is reduced during this interval); (c) on very long time scales ($\approx 10^5$ yr), such that the entire convection zone is back in equilibrium.

$$\tau_c/\tau_t(z_b) \approx H_b/R. \tag{69}$$

If the magnetic layer is at a relative shallow depth such that $\tau_t < \tau_c R/H$, the α-effect is larger; below this depth, the α-effect rapidly becomes insignificant compared with the direct effect (see also Figs. 6 and 7). With expression (23) for the thermal time scale, one finds that this critical depth is about 50,000 km.

The radius change can be calculated from the temperature change (Eq. 67), along with the hydrostatic equilibrium condition (Eq. 38). For the time scales of interest ($t \ll \tau_t(D)$), the relative temperature perturbation is independent of depth (except for the superadiabatic surface layer) down to the magnetic layer, and vanishes below it (Fig. 5). Hence

$$R_1 = z_b T_1/T. \tag{70}$$

The surface flux follows from T_1/T with relation (34), which includes the effect of the superadiabatic boundary layer. This yields

$$W = \frac{R_1}{R} \frac{F_s}{F_{sl}} \approx 0.6 \frac{z_b}{R}. \tag{71}$$

Because this is small compared to unity, the radius change does not contribute to the luminosity change (cf. Sec. II.F). Thus the relative luminosity changes are identical to the surface flux changes in the above equations:

$$\frac{L_1}{L} \approx -\frac{e_m}{P_b H_b}(1 - e^{-t/\tau_r(z_b)}). \tag{72}$$

The result (Eq. 71) can be compared with the results of Endal et al. (1985, Fig. 2). After conversion from depth to a mass scale one finds agreement to within 50%.

J. Thermodynamic Limit on Luminosity Variations

Up to this point, we have only discussed the relative magnitudes of various kinds of luminosity and radius effects. To obtain real numbers, we must assign a value to the variations in the magnetic energy stored in the convection zone, and the depth where the fields are located. There are considerable uncertainties in both these parameters. In the following, we give upper limits on the luminosity effects, using thermodynamic arguments. We assume that the fields are in a horizontal layer somewhere in the convection zone, but that its depth is essentially unknown. For the strength of the fields, we use a number of constraints of increasing tightness, but decreasing level of generality.

The most general constraints follow from thermodynamic considerations. In the periodic creation and destruction of fields during the cycle, the *creation* phase requires conversion of *thermal* into *nonthermal* energy. There are limits on the rate and efficiency with which this can be done. In the absence of as yet unknown sources of energy (little black holes somewhere in the Sun?), a plausible maximum to the energy available for production of fields is the solar luminosity. Another limit follows from the equilibrium of pressure between the field and its surroundings. This limits the field strength to

$$B_{max} = (8\pi P)^{1/2}. \tag{73}$$

Suppose that the fields are built up over a depth range of 1 scale height (as this is a significant fraction, $\approx 1/3$, of the depth from the solar surface, this is sufficiently general for our purposes). If the length of the solar cycle is τ_c, this takes a time of the order $\tau_c/2$. Then the limit (Eq. 73) corresponds to a limit on the magnetic energy per cm² of horizontal surface:

$$e_m < H B^2_{max}/8\pi = H_b P_b \tag{74}$$

where the index $_b$ indicates the value at the depth z_b of the magnetic layer. For the luminosity variation due to β-perturbations, *i.e.*, those produced jointly by the direct radius increase and the thermal sink effect, Eq. (59) is valid. Denote these changes by an index $_\beta$. Then Eq. (74) yields

$$\frac{L_{1\beta}}{L} < 0.8\frac{H_b}{R}. \tag{75}$$

As discussed above, we must have $F_1(z_b) < F_\odot$. In Sec. II.H, we showed that fields located at depth z_b and varying on the cycle time scale τ_c produce a flux perturbation below z_b given by $F_1 = de_m/dt$. Thus e_m is limited by

$$de_m/dt \approx 2e_m/\tau_c < F_\odot. \tag{76}$$

With Eq. (59):

$$L_{1\beta}/L < 0.8\frac{F_\odot\tau_c/2}{P_bR} = 0.6\frac{H_b}{R}\frac{\tau_c}{\tau_t(z_b)} \tag{77}$$

where we have used expression (23) for the thermal time scale. We find that the surface flux perturbation is therefore limited by

$$\frac{L_{1\beta}}{L} < \frac{H_b}{R}\min\left(0.8, 0.6\frac{\tau_c}{\tau_{tb}}\right). \tag{78}$$

The corresponding radius changes are (with Eqs. 58 and 59)

$$\frac{R_{1\beta}}{R} = 1.1\frac{L_{1B}}{L}. \tag{79}$$

These limits are shown in Fig. 6. The luminosity change has a maximum of 3×10^{-3} at a depth where $\tau_t \approx \tau_c$; this is at about 16,000 km. If $z_b = D$ is the base of the convection zone, where $\tau_t = 2 \times 10^5$ yr, we have with $\tau_c = 10$ yr

$$L_{1\beta}/L < 3 \times 10^{-6}, \quad (z_b = D). \tag{80}$$

This excludes the possibility that energy storage in magnetic fields at the base of the convection zone is responsible for the observed luminosity variations. The observed level of variation, around 3×10^{-3}, is barely compatible with upper limit (Eq. 78) even if the energy is stored near the optimal depth of 16,000 km.

The limits are a bit different for the effect of fields on the convective transport efficiency. Denote this effect by an index $_\alpha$ (for α-perturbations). With Eq. (72), the limits (Eqs. 74 and 76) yield

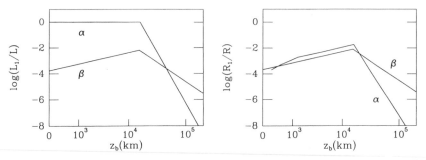

Fig. 6. Upper limit on luminosity and radius variations on the time scale of the solar cycle, using thermodynamic limits on the field strength: (β) including the expansion due to magnetic pressure and the thermal energy sinks due to field production (β-perturbations); (α) limit for the effect of magnetic fields on convective efficiency (α-perturbations). Breaks are near the depth where the thermal time scale equals half the solar cycle.

$$\frac{L_{1\alpha}}{L} < 1, \quad (\tau_t < \tau_c) \tag{81a}$$

$$\frac{L_{1\alpha}}{L} < \frac{\tau_c^2}{2\tau_t} \frac{F_\odot}{H_b P_b} = 0.8 \frac{\tau_c^2}{\tau_t^2} \quad (\tau_t > \tau_c) \tag{81b}$$

(using Eq. (23) for τ_t). The radius changes follow from Eq. (71):

$$\frac{R_{1\alpha}}{R} = 0.8 \frac{z_b}{R} \frac{L_{1\alpha}}{L}. \tag{82}$$

The thermodynamic limits are compatible with strong luminosity variations if the fields are located sufficiently close below the surface; they allow variations $> 3 \times 10^{-3}$ if $z_b < 40,000$ km. The interference of magnetic fields with convective transport efficiency is the most effective way of producing variations at these depths.

In reality, these limits are likely to overestimate the luminosity and radius effects, because they still imply rather high field strengths. As strong fields are unstable and buoyant on short time scales, there are further limits on the field strength (though these are of a less fundamental nature). They are discussed in the next section.

K. Buoyancy and Equipartition Limits

For strong fields in temperature equilibrium, the density reduction (Eq. 50) makes the field buoyant, so that it starts rising through the convection zone. If the rise is too fast, the field cannot be kept in the convection zone. The question is what is "too fast." Setting the rise time equal to the cycle time gives Parker's (1975) limit of about 100 G for fields created at the base of the convection zone. This is quite restrictive; for observational reasons one would like field strengths of at least 1000 G (Zwaan 1978), that are 100 times more buoyant. It is conceivable, however, that magnetic fields can stay within the convection zone until their buoyant rise speed becomes comparable with the typical convective velocity. The magnetic field could be systematically trapped in convective downflows, for example. A related effect is the topological pumping mechanism (Drobyshevski and Yuferev 1974; Arter 1983). This effect can keep fields up to the equipartition value from rising. The speed of rise of fields in temperature equilibrium is of the order of the Alfvén speed. (The speed is less if the field is in the form of tubes that are thin compared with a scale height [cf. Schüssler 1979; Moreno Insertis 1983,1986]. Such fields have a small filling factor, however, so that they also contribute less to the luminosity effect.) Thus a more conservative, but still plausible limit is obtained by equating the Alfvén and typical convective speeds. This limit is the so-called equipartition field B_{eq}:

$$\frac{B_{eq}^2}{8\pi P} = \frac{1}{4}\delta \tag{83}$$

which is about 10,000 G at the base of the convection zone. Assuming again that the field extends over 1 scale height, the limit on e_m is

$$e_m < \frac{B_{eq}^2 H_b}{8\pi} = \frac{1}{4}P_b H_b \delta. \tag{84}$$

The limits on the luminosity and radius changes for β-perturbations are then (cf. Eqs. 58 and 59):

$$\frac{L_{1\beta}}{L} < 0.27\frac{H_b}{R}\delta_b \tag{85a}$$

$$\frac{R_{1\beta}}{R} < 0.25\frac{H_b}{R}\delta_b. \tag{85b}$$

The limits for α-perturbations are (Eq. 68):

$$\frac{L_{1\alpha}}{L} < 0.25\delta_b(1 - e^{-\tau_c/2\tau_t}) \tag{86a}$$

$$\frac{R_{1\alpha}}{R} < 0.2\frac{z_b}{R}\delta_b(1 - e^{-\tau_c/2\tau_t}) \tag{86b}$$

where $\delta(z)$ is given by Eq. (17). These limits are shown in Fig. 7.

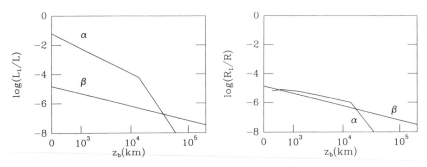

Fig. 7. As Fig. 6, but for magnetic fields limited by the equipartition value.

If the equipartition limit is accepted, Fig. 7 shows that only the effect of magnetic fields on the convective efficiency has a chance of producing an observable luminosity change. This has been found before by Gilliland (1988). Even then, a luminosity change exceeding 3×10^{-3} is obtained only for fields located not deeper than 1600 km, i.e., inside the superadiabatic boundary layer. This is so close to the surface that observations arguably provide more reliable indications of the field strength and amount of flux in this layer. If this is accepted, the field in this layer has a higher strength but much lower filling factor than assumed in the buoyancy limit discussed above (for which a filling factor of unity was used). But more importantly, this field is vertical rather than horizontal. The luminosity effects of such a field differ significantly from that of a quasi uniform horizontal field of equipartition strength. This is the subject of the next section.

As it will turn out that the radius changes produced by the effects discussed in the next section are negligible, the most likely source of radius changes are still the α-perturbations discussed above. They depend critically on the assumed depth of the horizontal fields producing them. For example, if the horizontal fields from which active regions erupt are located at a depth greater than 10,000 km, Eq. (86a) predicts that they will cause a change during the solar cycle of $< 3 \times 10^{-6} \, R_\odot$. Present upper limits on radius variations (Brown 1988a; chapter by Ribes et al.) are still much above this value.

III. SURFACE EFFECTS

The way in which atmospheric magnetic fields influence the luminosity is rather different from the subsurface effects discussed above. As we shall show, the surface effects on the solar luminosity can be *larger* than the subsurface effects. This is because, by modifying the effective emissivity of the photosphere, the surface fields have an influence that stays as long as the magnetic structure exists, even if nothing changes in it. The change in energy flux integrated over their life can be far larger than their own energy content. For this reason, the limits found in the previous sections do not apply to the surface magnetic fields, and we have to consider their effects separately. To illustrate this point, consider the case of a weak average field consisting of many small (subarcsec) tubes of kG field strength. Most of the magnetic flux at the surface at any given time is in the form of such fields. Due to the way in which surfaces of constant gas pressure are distorted by the magnetic field, both the effective radiating area and the temperature of this area can be larger than in the absence of magnetic fields (Spruit 1976,1977; Deinzer et al. 1984; Knölker et al. 1988). This effect, seen in the ACRIM and Nimbus data, was predicted theoretically. To quote Spruit (1977, abstract): "Through the lateral influx, small tubes such as are found in the quiet network act as little 'leaks' in the solar surface through which an excess heat flux escapes from the convection zone." The effect was calculated to yield an effective emissivity up to 100% above the normal solar value, for the smallest tubes. [To get the average excess emissivity of the area, this number has to be multiplied by the filling factor of the kG fields in the area.]

Consider, for example, an area of active network (the remnant of an active region). This can live for a time $\tau \approx 60$ d or more, and the magnetic flux in it is in the form of the very small elements that are predicted to be the most efficient radiators. Assume that the excess emissivity ϕ is 40%. The excess energy emitted per unit of magnetic flux during the 60 d period, E_e is

$$E_e = \phi \tau F_\odot / B_s \qquad (87)$$

where B_s is the surface field strength, $B_s \approx 10^3$ G. Compare this with the energy content of the flux tube. Assuming that the tube extends all the way from the surface to the bottom of the convection zone, and that it has an average field strength $\bar{B}$ (which need not be the same as the surface field strength), this energy E_B, again per unit of magnetic flux, is

$$E_B = D\bar{B}/8\pi \qquad (88)$$

where D is the depth of the convection zone. Assuming $\bar{B} = 10^4$ G, the maximum of the equipartition strength in the convection zone, we find that the ratio of the two energies is

$$E_e/E_B = \phi \frac{8\pi\tau F_\odot}{DB_s\bar{B}} \approx 15 \qquad (89)$$

i.e., the excess surface emissivity may produce effects that exceed the energy content of the field by substantial factors, even though only a very small volume of the magnetic field is involved in the process. Considering that the luminosity effect of subsurface magnetic fields additionally suffers an amplitude reduction between the level of origin of the field and the solar surface (see Sec. II), this demonstrates the importance of surface emissivity effects.

The excess emission from small tubes is one example of a magnetic influence on the surface emissivity. Another is the reduced emission from sunspots. There are more possibilities; they are reviewed in Sec. III.B. Finally, additional energy is emitted from the atmosphere due to the energy dissipated during the eruption of magnetic fields.

Luminosity and radius effects due to magnetic fields near the surface have also been considered by Dearborn and Blake (1982). They model the effect of the fields by average changes in the pressure and density of the gas. This is equivalent to a change in the equation of state. This influences the structure of the superadiabatic boundary layer. The changes are of the same order as the convective efficiency changes discussed in Sec. II.I. The effects seen by Dearborn and Blake are therefore of the same order as found in Sec. II.I for fields just below the surface. Dearborn and Blake attribute these effects to the direct expansion due to magnetic pressure. As shown in Sec. II.F, however, this effect is only of the order H_0/R. Also their averaging does not do proper justice to the influence of vertical fields of small filling factor, as argued in Sec. II.I. Dearborn and Blake also consider the effect of blocking by spots under the assumption that all the blocked flux is re-radiated elsewhere at the surface (however, compare with Sec. III.B below).

A. Changes of the Effective Surface Emissivity

There are several ways in which the surface fields can produce changes in the amount of emitted radiation. Effects that change the emission from the photosphere are:

1. Reduction of the energy flux in sunspots;
2. Increased emission from small magnetic elements (flux tubes); [In addition there are processes that are not related directly to the photospheric radiation, but whose net effect also corresponds to an increased energy output.]
3. Mechanical energy transport in the form of magnetic waves excited in the convection zone followed by dissipation in the atmosphere. The waves can be of various kinds; but Alfvén waves and transverse tube waves (also called kink waves) have the best chances of reaching higher layers of the atmosphere (Hollweg 1984; Spruit 1984,1988; Ulmschneider and Stein 1982, and references therein). An average energy flux into the upper atmosphere by such waves of 10^5 erg cm^{-2} s^{-1} is quite possible (Spruit 1984). Some authors obtain energy fluxes up to 10^7 ($\approx 10^{-4} F_\odot$) in active regions (Hollweg 1981; Gordon and Hollweg 1983; Hollweg and Stirling 1984; Ionson 1984).
4. Magnetic energy buildup in the atmosphere by (slow) footpoint motions. The (quasi-static) field in the atmosphere can take up stresses produced by changes taking place in its photospheric boundary conditions, and thereby store energy. This energy is released again in magnetic reconnection processes, the subject of study by many authors (Gold 1964; Parker 1974;

Rosner et al. 1978; Ionson 1984; Heyvaerts and Priest 1983; Parker 1983a,b). The rate of energy transport into the atmosphere by this process is rather uncertain, because it depends critically on properties of the reconnection process that are still insufficiently known. It could be as large as the energy input by waves or larger. As stressed by Parker (1974), this process has the advantage that it explains in a rather natural, although still qualitative way, the observed connection between coronal emission and the evolution of active regions.

Although considerable uncertainties exist in the theoretical treatment of the last two processes, their importance for the solar luminosity can be assessed to some extent by observations. This is because to the extent that the dissipation of the energy takes place in the upper atmosphere, it shows up mostly in the UV and EUV. Separate observations of the UV and EUV emission can be used to calculate the effect of mechanical dissipation in the upper atmosphere (middle chromosphere and upwards) on the luminosity. Unfortunately, it is not clear whether this takes into account all the mechanical energy dumped in the atmosphere by magnetic fields, because excess emission from the lower chromosphere and the temperature minimum region does not have a very characteristic spectral signature. Energy transported by waves may actually be dumped preferentially in these regions, as is the case with purely acoustic waves, for example (cf. Herbold et al. 1985). Comparison with theoretical radiative transfer models (Vernazza et al. 1981; Anderson and Athay 1989a,b) could in principle answer this question, but as yet the required accuracy (10^{-3} or better) cannot be reached.

B. Sunspot Blocking

Sunspots are dark and emit only about 25% as much as the ordinary photosphere. Because the magnetic field of a spot appears to be well ordered and stable, and such a field is known to interfere strongly with the efficiency of convective energy transport (see, e.g., Proctor and Weiss 1982), this darkness is traditionally understood as a natural consequence of the spot's magnetic field (Biermann 1941). The blocking of convective transport by the field must lead to a temperature increase below the spot. This has raised the familiar bright-ring problem: why is this temperature increase not visible as a ring of enhanced brightness around the spot (see, for example, the references in Fowler et al. 1983). It was already realized by Sweet (1955), however, that such a bright ring, though it must be present, would probably have a very low contrast because the very large heat capacity of the gas in the convection zone would make the temperature enhancement below the spot quite small. This was demonstrated explicitly in numerical results by Spruit (1977). Although there are definitely some unsolved problems concerning the existence and structure of sunspots (see, e.g., Parker 1979b), the bright ring problem is not one of them. Especially after the calculations by Spruit (1982a,b) and Foukal et al. (1983), the theory of thermal perturbations in the convection zone due to sunspots can be considered to be in a rather satisfactory state. A central result from this theory is that most of the energy flux blocked by the spot is stored in the entire convection zone, on its thermal time scale (10^5 yr). Because of this very large time scale, the missing flux of sunspots is truly missing on all time scales of observational interest; on such time scales, spots produce a real reduction in the emissivity of the solar surface. Instead of being re-emitted elsewhere at the surface, the blocked flux is (mostly) spent in heating up the convection zone. The observation that the reduced flux from spots roughly compensates the excess emission in facular areas (for a detailed discussion of these observations see Foukal and Lean [1988]) has stimulated speculations that there exists a physical but as yet unknown connection between the spot deficit and the facular excess (Oster et al. 1982; Schatten et al. 1985; Chapman et al. 1984; Schatten and Mayr 1985). Lacking a specific idea or model for such a physical connection, it is hard to discuss this proposal. Published ideas (Schatten and Mayr 1985) cannot be considered to have reached the status of a reasonably well-defined model. In any case, the mechanisms outlined in this section produce effects by sunspots and small-scale fields that are of opposite sign and roughly equal amplitude, although it would indeed be a somewhat curious coincidence if the observations were to show that the two cancel to great accuracy. That such is not the case, however, at least on time scales of a few years and less, was recently shown by Foukal and Lean (1988).

The sunspot blocking calculations were made by Spruit (1977,1982a,b) and Foukal et al. (1983); Chiang and Foukal (1984) use a mixing-length description for two-dimensional convective energy transport, which leads to a transport equation of the form of Eq. (3). It treats the energy transport as a turbulent diffusion process, and does not take into account the systematic flows that can also be set up by the thermal perturbation. Let us define the spot blocking problem as follows. We start with a time-independent spherically symmetric temperature field that is the solution of the transport equation. At $t = 0$, this solution is perturbed by reducing the turbulent diffusivity in a sunspot: a volume extending from the surface down to a depth d, and with horizontal extent R. To be determined is the evolution of the temperature perturbation in time and space in the convection zone, and in particular, the evolution of the total flux emitted at the surface. In Spruit (1982a,b) it was shown that the main results can be found analytically; the model used in this calculation is described in Sec. II.A above. This calculation concentrates on time scales large compared with a diffusive time scale but short compared with the thermal time scale of the convection zone. Detailed time-dependent numerical calculations were done by Foukal et al. (1983) and Chiang and Foukal (1984). These calculations show the same effect, storage of most of the blocked heat flux in the convection zone. Because, for numerical reasons, the depth of the layer computed by Foukal et al. was much less than the actual depth of the convection zone, the thermal time scale in their calculations is also much smaller than the 10^5 yr of the convection zone (cf. Eq. 20). This produces only a qualitative difference, however; it reduces the time scale on which the later evolution stages take place.

   As in the case of subsurface magnetic fields, more than one time scale plays a role in the evolution of thermal perturbations due to spots. With the aid of the description in terms of the thermal modes in Sec. II.E, we can draw the following qualitative picture of the evolution. After the blocking is switched on at $t = 0$, the gas below the spot starts heating up. The first sign of this heating appears at the surface on the diffusive time scale $\tau_d = d^2/\kappa$ (d is the depth of the blocking effect due to the spot). As a result, a bright ring appears around the spot which spreads horizontally. After a time $\tau_D = D^2/\kappa$ the perturbation has reached the base of the convection zone and the bright ring has reached its full amplitude. After this stage, the evolution is determined entirely by the fundamental mode, and takes the form of a slow heating of the entire convection zone. Because of the large heat capacity of the convection zone, this takes a long time, $\tau_t(D) \approx 10^5$ yr.

   It may seem strange that the entire convection zone takes part in a perturbation caused by a very superficial effect. The reason lies in the large efficiency of convective energy transport. This communicates temperature fluctuations (or rather, entropy fluctuations) across the convection zone on a turbulent diffusion time scale τ_D. Because of this, it is impossible to change the entropy locally in the convection zone except on short time scales. On longer time scales, the energy associated with the change is spread over the entire convection zone. The large heat capacity then ensures that the deviation of the temperature distribution from the original state is quite small. As the heat flux perturbation at the surface is related directly to this temperature change (through the Stefan-Boltzmann law), this implies that the heat flux from the non-spot surface of the Sun is also unchanged. The luminosity change, for times much smaller than the thermal time scale τ_t, is therefore just determined by the reduced emission from the spot areas.

   From the above, it is clear that this picture applies only as long as the blocking takes place in a region where the efficiency of convective transport is large. Near the surface, inside the superadiabatic region, the convective efficiency is not very large. Thus, if the base of the blocking region lies *within* the superadiabatic layer, a more or less significant bright ring does indeed develop, and not all of the blocked heat flux disappears into the deep convection zone. This effect has been calculated by Spruit (1982a). It is defined by α, the *fraction of the blocked heat flux that reappears at the surface* in the form of a bright ring. This number is in general a function of time. If the fraction of the solar surface covered by spots is f (the spots are assumed to be completely black for simplicity), the luminosity can be expressed in terms of α as

$$L = L_\odot[1 - f(1 - \alpha)].$$ (90)

With the analytic model presented in Sec. II.C, one finds

$$\alpha \approx 0.5[1 + d/(3H_0)]^{-2}, \qquad (\tau_d \ll t \ll \tau_t).$$ (91)

The value $3H_0$ is a measure of the depth of the superadiabatic layer. This shows that the bright ring effect rapidly vanishes as the spot becomes deeper than the superadiabatic layer. The actual depth to which a sunspot blocks the heat flux is not well known; obviously it must be larger than the observed Wilson depression (600 km) of spots. The observed field strength of 3000 G could substantially inhibit convection down to several thousand km depth, however. Even for a depth of 1000 km the predicted value of α is only 0.05. Thus it seems fairly safe to assume that on time scales of observational interest, practically none of the blocked flux reappears at the surface. Conversely, if the value of α can be determined from observations, it can be used to estimate the depth of the blocking in spots (Fowler et al. 1983).

C. Excess Emission from Small-Scale Magnetic Fields

A sufficiently small magnetic flux tube, oriented vertically at the solar surface, can emit an *excess* of radiation caused entirely by two-dimensional effects of radiative energy transport. This can be the case even if the *convective* energy transport along the tube is negligible (Spruit 1976,1977; Knölker et al. 1988; Grossmann-Doerth et al. 1989). We review here briefly the mechanisms responsible for this effect.

A flux tube of 1000 G, with an observed life time of at least 5 min, must be in sideways pressure equilibrium with its surroundings (it takes the pressure gradients only 10 s to equalize the pressure across a tube of radius 100 km). Due to the pressure exerted by the magnetic field, the gas pressure inside the tube is less than in the surroundings, at the same depth. If the tube is sufficiently thin, horizontal exchange of radiation across the tube will keep it at the same temperature as its surroundings, at each depth. Due to the lower pressure inside, the opacity is less, so that optical depth unity is reached at a deeper level (the equivalent of the Wilson depression in spots). At $\tau = 1$, the temperature is therefore higher inside the tube than outside, and the tube emits more radiation. Because the temperature inside such a very thin tube is determined entirely by exchange of photons with the surroundings, the effect works even if there is no convective transport of energy along the tube at all. This picture applies to very thin tubes, with radii of the order of the mean free path of a photon at $\tau = 1$ (50 km) or less. For somewhat larger tubes, the exchange of heat with the surroundings becomes less effective. Assuming that the convective energy transport along the tube is in fact reduced to roughly the same degree as in spots, the central parts of such a tube will now be cooler than the surroundings. From a certain size (radius around 200 km), the tube will no longer appear bright when seen at disk center. Even such a tube, however, can contribute a net excess emission, because it can radiate more effectively than the normal photosphere in directions close to the horizontal. This is illustrated schematically in Fig. 8.

When seen near the solar limb, the emission from the tube is dominated by the brightness of its walls. This brightness can be greater than that of the surrounding photosphere for two reasons:

1. A hot wall effect, if the wall temperature is higher than photospheric;
2. A bright wall effect, due to a different center-to-limb variation.

The photosphere is darker near the limb because optical depth unity along the line of sight is reached at a higher level in the atmosphere where the temperature is lower. The tube walls, on the other hand, are seen nearly face-on, and have their maximum brightness when seen near the limb. In other words, the tube emission shows the opposite effect, a limb brightening. As the photospheric limb darkening is a strong effect, this difference can contribute significantly to the excess emission from tubes.

Quantitative models of radiative energy transport in tubes show that the *hot wall effect* is actually a *bright wall effect* because tube walls are not very hot. Their typical temperature is around the photospheric temperature of 6400 K (Spruit 1976; Knölker et al. 1988; Grossmann-Doerth et al. 1989). This is not really surprising; because of the high opacity below the photosphere, a significant energy flux from the walls can be sustained only by a strong horizontal temperature gradient. The temperature of a tube wall, if it contributes significantly to the emission, can be nowhere near the convection zone temperature at the same geometric level.

Just like spots must be surrounded by a bright ring (if only one of unobservably low amplitude), thin tubes emitting excess radiation must be surrounded by a dark ring. In analogy with the case of spots (Sec. III.B), we can define a fraction α' of the excess emission that is compensated by the dark ring. In the diffusion approximation for convective energy transport near

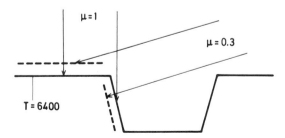

Fig. 8. Bright wall effect. The wall temperature as seen at $\mu = 1$ is assumed to be the same as the photospheric temperature at $\mu = 1$. At $\mu < 1$, a lower temperature is seen in the normal photosphere, but the observed wall temperature is higher.

the tube, this quantity would again be given by Eq. (91). The two-dimensional energy transport effects that produce the excess emission take place *within* the superadiabatic surface layers of the convection zone, roughly at the Wilson depression level ($d \approx H_0$). Hence we should expect that a significant part of the excess emission is indeed compensated by a dark ring. The numerical models show that this is the case (Spruit 1977). Nevertheless, a significant net effect remains. These early results suffer from significant uncertainties due to the difficulties of modeling the turbulent energy transport near the tube; this situation should improve considerably as two-dimensional radiative hydrodynamic simulations become available (Grossmann-Doerth et al. 1989).

Models based on two-dimensional energy transport in the surface layers of tubes of strong field in which convection is inhibited thus yield the following predictions. Very small tubes are brighter than the photosphere at all disk positions, and produce a positive excess emission. As the tube size is increased, their cores become darker, until at a specific size (around 200 km radius) they have no net contrast at disk center. At this size, they still produce a net excess emission, because of the flatter center-to-limb variation of their emission. At a somewhat larger size, the excess emission also changes sign and the tubes start behaving like pores and spots. A quantitative comparison of some of these predictions with observations was made by Spruit and Zwaan (1981).

Can these models be used to predict quantitatively the effect of faculae and the network on the solar luminosity? Unfortunately, this is not yet the case; the situation for small-scale fields is as difficult as the sunspot case is easy. There are two reasons for this. First, the calculations of energy transport in and around small tubes are complicated, and significant improvements are still needed before quantitatively reliable results can be obtained. Second, the emission from small tubes depends very sensitively on their size. To calculate a luminosity effect, one must know the size distribution of the tubes, for which no dependable theory is available (though attempts have already been made [cf. Knobloch and Rosner 1981]). Observationally, this is a hard problem because of limited spatial resolution (see, however, Spruit and Zwaan 1981).

D. Effects of Surface Fields on Radius

The radius effects due to magnetic fields near the surface turn out to be extremely small. As in the case of subsurface fields, magnetic fields in the form of vertical flux tubes near the surface have a direct mechanical effect on the radius due to the mass they are displacing, as well as a thermal effect, the hydrostatic response to the thermal perturbation due to the tubes. Consider first the thermal effect. Related to the temperature increase below a blocking spot, there is a radius increase of the order (Spruit 1982a)

$$R_{1r} = H_0 f \frac{3dH_0}{(d + 3H_0)^2}. \tag{92}$$

In this expression f is the effective area covered by spots, that is, if the contrast in surface flux of a spot is $\phi_{spot} = F_{spot}/F_\odot - 1$, and the fraction of the solar surface covered is f_A:

$$f = -\phi_{spot} f_A. \tag{93}$$

The radius change (Eq. 92) also applies to the effect of excess emission from small-scale fields, if ϕ_{spot} and f_A in Eq. (93) are replaced by the corresponding flux contrast and filling factor, and d is set to some value of the order of the Wilson depression in the small tubes. For spots, we get a radius increase; for the small tubes a decrease. Because Eq. (92) has a definite maximum at $d = 3H_0$, the absolute value of the radius change for both cases satisfies

$$|R_{1r}| < 0.25 H_0 |f|. \tag{94}$$

For $|f| \approx 10^{-3}$, we get $R_{1r}/R_\odot < 5 \times 10^{-8}$, a negligible effect. The direct effect due to the lower density inside the tubes can also be calculated; it is even smaller. This is because the buoyancy of vertical flux tubes does not act directly on their surroundings, but via the magnetic tension at the level where the field turns horizontal. This is different from the case of a horizontal layer of field (see Sec. II), which lifts the mass above it due to its buoyancy, without affecting much the mass below it. The radius effect of vertical surface magnetic fields is therefore completely negligible.

CONCLUSIONS

The effects of surface magnetic fields on luminosity is qualitatively different from that of fields buried below the surface. The strongest effect of subsurface fields on the luminosity is through the effect on the convective energy transport (α-perturbations), of fields located close below the surface. Deeper fields have a smaller effect, and influence the luminosity mainly through the radius increase associated with the magnetic pressure (β-perturbations). This radius increase is also larger than the other effects that cause radius changes. Thermodynamic constraints are derived for the effects, as a function of the depth of the magnetic fields. These upper limits are below the observed luminosity variation if the fields are stored deeper than 40,000 km but allow significantly larger than observed amplitudes for storage above this level. It is argued that a more plausible upper limit is obtained by assuming a field in equipartition with convection. With this limit, significant luminosity variations are possible only by α-perturbations, and the fields involved must be $\leq$ 2000 km below the surface. If the fields are located deeper than 10,000 km, the luminosity effect of such perturbations is $< 10^{-4}$, and the radius change $< 3 \times 10^{-6}$.

The effects of surface magnetic fields (i.e., fields that cross the surface more or less vertically) are different because they have a direct effect on the rate at which radiation is emitted from the photosphere. In this way, they can during their life time mediate a change in the emitted energy that can be much larger than their magnetic energy content. These effects result from the inhibition of convection (in the case of sunspots) and from peculiarities in the two-dimensional transport of energy in small magnetic elements. Detailed analyses of the thermal perturbations due to spots and small-scale fields show that on time scales of observational interest ($< 10^5$ yr), the heat flux missing from spots is stored in the convection zone. Similarly, a significant fraction of the excess emission from small-scale fields is supplied by the entire convection zone. Although the two effects are of opposite sign, there is no physical connection between them, and no reason why they should cancel exactly. The luminosity changes predicted in this way from spot blocking and from small-scale emission agree roughly with the observations. It is stressed that for a more accurate theoretical determination of the excess emission, it is necessary to determine the distribution of sizes in the small-scale field to a fair accuracy. The effect of surface fields on radius is predicted to be below the 10^{-7} level.

Acknowledgment.   I thank B.R. Durney for his detailed comments on an earlier version of the text, which greatly helped in improving it.

STANDARD SOLAR MODELS

DAVID S. P. DEARBORN
Lawrence Livermore National Laboratory

Solar/stellar comparisons provide improved understanding of stellar astron-omy, cosmology and various problems in physics. These and the physical pro-cesses included in stellar evolution calculations are presented, along with those of nonstandard phenomena. The bases of physical models are discussed in con-junction with observational data by which solar models and their uncertainties may be judged.

I. INTRODUCTION

While the Sun is a very special object to us, it is nevertheless just an average star, similar to a hundred billion others in the Galaxy. Its close prox-imity provides us with a detailed look at some properties of stars only hinted at in the observations currently possible such as, for example, magnetic field structure and surface velocity fields. In turn, with the aid of stellar evolution theory, observations of stars provide us with an ensemble of stellar ages which can give us a view of the past and future of the Sun.

After celestial mechanics, the theory of stellar evolution is perhaps the most quantitatively successful in astronomy. It reproduces the observed char-acteristics of stars to an accuracy limited only by the physical model. Using a minimal set of assumptions, stellar evolution theory links an extraordinary variety of types and classes of stars as milestones in an evolutionary sequence (Iben 1967), while seemingly able to ignore physical phenomena that would appear to affect significantly the character of stars such as rotation, magnetic fields and element diffusion or separation.

Among the successes of the theory of stellar evolution is the distribution of stars in color magnitude diagrams of clusters (Eggen and Sandage 1964). As stars at the upper end of the main sequence evolve, their luminosity increases in a manner consistent with the predictions of stellar and solar models. This luminosity increase is a direct consequence of hydrogen-to-helium conversion. During later evolution off the main sequence, the products of this nuclear fusion are transferred to the surface, causing the composition changes observed in red giants (Dearborn and Eggleton 1976). The Sun's mass is known with extraordinary precision, and its composition is relatively well known. The application of these parameters to solar models reproduces the observed solar luminosity at the correct age; this is a strong indication that the theory is not grossly wrong. Further support for the accuracy of the solar structure as calculated by stellar evolution theory comes from the detection of surface waves (Leighton 1961) coupled with "seismic" interpretations (Ulrich 1970; Iben and Mahaffy 1976; Rhodes et al. 1977; Deubner et al. 1979; Duvall et al. 1988). The spectrum of observed p-mode frequencies can be used to determine the sound travel time through the solar envelope (Ulrich and Rhodes 1977).

While the preponderance of the evidence available demonstrates that the stellar evolution theory is substantially correct, it is perhaps most widely known for one of its failures. Neutrino observations from the chlorine experiment (Davis et al. 1968; Cleveland et al. 1984; Bahcall et al. 1985; Rowley et al. 1985; Rosen 1991) provide a probe of the Sun's deep interior that is unavailable for any other star (excepting briefly for SN1987a). The deficiency of observed neutrinos from the number expected in solar models has been a long-standing problem. Because the chlorine experiment is primarily sensitive to the high-energy neutrinos produced in the rarest branch of the proton-proton (p-p) chain, the expected count rate is extremely sensitive to the Sun's central temperature (see the chapter by Wolfsberg and Kocharov).

Classical solutions of the neutrino deficiency are based on a reduction of the central temperature by $\sim 8\%$ from the value in the latest standard solar model (Bahcall and Ulrich 1988). While many ways of doing this have been proposed, they involve either phenomena which cannot be rigorously calculated or conditions which are not observed. Some of these proposed solutions remain possible, but none of them provides a compelling solution for the low observed emission rate of solar neutrinos, namely 2 SNU (where 1 SNU = 10^{-36} captures per chlorine atom per second). Other solutions for the dilemma of the neutrino deficiency involve properties of the neutrinos themselves. If electron neutrinos are transformed into other unobservable species by decay or matter-induced oscillations (Wolfenstein 1978, 1979; Mikheyev and Smirnov 1986; Bethe 1986; Rosen 1991), the expected flux is reduced. Conversely, solar models can be used to constrain the neutrino properties (mass differences and mixing angles) necessary for a reduction of the flux to the observed values. However, the accuracy of the required constraints is limited by the uncertainties in the solar model (Dearborn and Fuller 1989).

In spite of the neutrino problem, stellar evolution has been sufficiently successful for its results to be used in other branches of astronomy. From the lifetimes calculated for the oldest clusters, 16 ± 3 Gyr (Vandenberg 1983; Janes and Demarque 1983), stellar evolution places strong limits on the age of the universe. The use of CCD photometry (Stetson and Harris 1988), reduces the error from the photometry to ~ 1 Gyr, but does not address uncertainties resulting from physical approximations used in the calculations. Age determinations using radioactive decay are also sensitive to calculations of the rate at which elements are produced in stars (11 ± 1.6 Gyr; Fowler 1987). Calculations of the production and annihilation of light elements in stars (Rood et al. 1976; Dearborn et al. 1986a,b; Deliyannis et al. 1989) are used to correct the observed abundances of those elements to their primordial values. From this, limits are set on the number of neutrino species and the baryon density of the universe (Wagoner et al. 1967; Yang et al. 1984). Understanding the destruction of lithium has become particularly important to constrain inhomogeneous cosmologies (Applegate and Hogan 1985; Alcock et al. 1987; Applegate et al. 1987).

In the next section, we examine the observed properties of the Sun which are used to define an acceptable solar model. In Sec. III, we discuss the physical basis of modeling commonly used in stellar evolution calculations, and in the next section, the nonstandard phenomena that may be important in some stars. Section V briefly discusses the basic behavior of the stellar models and the subject is summarized in Sec. VI.

II. OBSERVATIONS DEFINING THE SOLAR MODEL

The principal requirement for a successful solar model is its ability to match the luminosity and radius of the Sun using the known mass, age and composition. The primary parameter which defines a star's spectral type and subsequent evolution is the mass. For the Sun, the product, $GM = 1.322712438 \times 10^{26}$ cm^{3}s^{-2}, has been determined to a high degree of precision from celestial mechanics. The Sun's mass is found using a laboratory determination of the gravitational constant G; the current value of the mass is 1.9891×10^{33} g. The accuracy of this mass determination rules it out as a significant source of model uncertainty. Additionally, the present rate of mass loss, 10^{-14} M$_\odot$yr^{-1}, via the solar wind is too low to have an impact on the Sun's mass over its lifetime. Pre-main-sequence stars are observed to have much stronger winds, but these winds appear to decrease substantially as stars approach the main sequence. Still an unobservable wind of 10^{-9} M$_\odot$yr^{-1} could have a significant effect if it continued over a long period after the beginning of the main-sequence phase (see the chapter by Bertout et al.).

Recently, solar models have been proposed (Guzik et al. 1987) in which an epoch of extensive mass loss occurs during early main-sequence evolution. These models exhibit the correct luminosity and radius at the Sun's age, but the outer envelope where lithium could survive was ejected (the surfaces of these models contain no lithium). This is in conflict with the observation that shows lithium exists in the solar photosphere. For consistency, an early epoch of extensive mass loss requires a mechanism to resynthesize the lithium, not just in the Sun, but in other stars. Observations of clusters (Boesgaard 1987; Boesgaard et al. 1988; Hobbs and Pilachowski 1988) indicate that lithium is present in low-mass stars, and that it exhibits a systematic decrease with temperature and age. If these stars lose more than a few percent of their mass subsequent to arriving on the main sequence, this surface lithium would have to be resynthesized. Until such a mechanism can be demonstrated, the assumption that the Sun's mass has been effectively constant since arriving on the main sequence is most attractive.

The age of the Sun is inferred from model ages of meteorites and lunar rocks using the plausible assumption that planetary matter did not predate the formation of the Sun. The consistency of ages determined from many meteorites provides some confidence that the meteoritic age of 4.55 ± 0.1 Gyr (Wasserburg et al. 1977, 1980) is also the Sun's age.

The outer 2% of the mass or 28% of the radius of the Sun is convective in solar models. The radius of the Sun, and the depth of this convective region is a function of the mixing length used in the calculations. There is little to constrain the value of the mixing length so it is simply tuned to match the Sun's radius. The calculated depth is then compared to interpretations of the observed 5-min surface oscillations. Measurements of p (pressure) modes on the solar surface have been used to determine the sound speed through the envelope by a means called asymptotic inversion (Christensen-Dalsgaard et al. 1985a, 1988). Accuracies of 1% have been obtained between 0.4 and 0.9 solar radii (the outer 20% of the Sun by mass). A discontinuous change in the sound speed is found at approximately 0.7 solar radii. This probably indicates the location of the bottom of the convection zone where the temperature gradient shifts from one determined by opacities to an adiabatic temperature gradient.

An interesting constraint on the behavior of stars could be provided from the time history of the Sun's radius (Ribes et al. 1987). Unfortunately, long-term historical measurements are subject to systematic errors that are difficult to evaluate (see the chapter by Ribes et al.). Recent measurements (Brown 1988b) can only set upper limits on radius changes. Luminosity fluctuations like those observed in other solar-type stars (Giampapa 1984) should be accompanied by radius variations (Sofia et al. 1979), but the magnitude of these variations is quite small (Dearborn and Newman 1978; Dearborn and Blake 1980a,b; Gilliland 1980). Nevertheless, long-term radius variations would provide valuable clues to the stability of the convective region.

The metals (elements other than hydrogen and helium) contribute to the Sun's structure through their effect on the opacity. Even in the center of the Sun, these elements contribute nearly 40% of the opacity (Turck-Chieze et al. 1988), and their abundance with respect to hydrogen is highly uncertain. The abundances of most elements have been determined to an accuracy ranging from 5 to 15% (Anders and Ebihara 1982), though there are important exceptions. The ratio of mass fraction of heavy elements to hydrogen has changed from 0.0228 to 0.0276 between the Bahcall et al. (1982) solar model characterized by 5.8 SNU, and the more recent Bahcall and Ulrich (1988) model with 7.9 SNU. Much of the change in metallicity is due to a dramatic increase in the accepted neon abundance. The increase in this element by itself can account for most of the rise in the predicted neutrino rate. In Table I, the relative number densities of the most abundant species are listed from Anders and Ebihara (1982), Grevesse (1984) and Aller (1986).

Using the hydrogen and helium values of Anders and Ebihara (1982) together with the uncertainties they assign, the hydrogen mass fraction is 0.74 ± 0.04, helium is 0.24 ± 0.03 and the metallicity is 0.02 ± 0.004. Most of the uncertainty in the metallicity is in the CNO isotopes and the noble gases whose abundances are determined from astronomical measurements of the Sun, other stars and nebulae. The relative abundances of Mg, Si, Fe and most of the other elements are determined with high precision from meteoritic measurements and normalized to the abundances of these elements which are determined with lower precision from observations of the Sun (the uncertainty in the abundances of these elements results primarily from the normalization to the solar values).

Once a metallicity has been chosen, the opacity can be calculated for various hydrogen/helium ratios. Modeling the Sun then involves determining the initial value of the hydrogen/helium ratio necessary to obtain the observed solar luminosity of 3.85×10^{33} ergs^{-1}, following from the solar constant of 1368 W M^{-2} incident on the Earth (Wilson 1984; see the chapter by Frölich et al.). Increasing the metallicity generally increases the opacity, requiring an increase in the helium to satisfy the luminosity-age constraints. Determina-

TABLE I
Relative Number Densities

Element	AE[a]	Grevesse	Aller	Error[a]
C	.2879	.2976	.2817	30
N	.0590	.0594	.0589	40
O	.4782	.4938	.5009	20
Ne	.0895	.0607	.0691	30
Mg	.0255	.0232	.0257	4
Si	.0238	.0217	.0269	4
S	.0123	.0098	.0105	13
Ar	.0025	.0023	.0023	20
Fe	.0214	.0285	.0240	3

[a]From Anders and Ebihara (1982).

tions of the solar helium abundance from direct observations are poor because helium has no observable photospheric lines in the Sun. The high ionization potential of the noble gases in general raise questions of fractionation due to differential acceleration causing concerns for abundance determinations made by measuring the composition of the solar wind (see the chapter by Geiss and Boschler). Observations of metal-poor systems have been extrapolated to an initial helium mass fraction for the universe of 0.23 to 0.24 (Pagel et al. 1986). The Sun should have a value above that, and most models require an initial helium mass fraction of 0.27 or 0.28 (Turck-Chieze et al. 1988).

Additional constraints on the solar helium abundance, equation of state, and internal rotation rates can be obtained from helioseismology. At a given frequency, eigenmodes of lower degree penetrate more deeply into the Sun. This allows observations of the spectrum of oscillations to be used to determine the sound speed through the Sun's outer envelope (20% by mass). The composition in this region has not been significantly perturbed by nuclear processing, so the helium present should represent the initial value. The sound speed depends on the equation of state, the mean molecular weight and the temperature. In radiative regions, the temperature depends on the opacity. The accuracy with which the sound speed can be used to determine the helium abundance is limited by knowledge of these other factors.

Guenther et al. (1989) claim that the mass fraction of helium is in the range 0.27 to 0.29 for the uncertainties in the solar model. Similar calculations by Christensen-Dalsgaard et al. (1985b) found values from 0.23 to 0.27. Therefore, while observations of the p-modes in the envelope constrain the helium abundance more than surface spectroscopy, they are model dependent and of questionable sensitivity (Guenther and Sarajedini 1988). Unfortunately, the p-modes (5-min oscillations) are most sensitive to conditions in the outer envelope. Improved precision in the observations may allow deeper probing in the future. But, at present, the inversion fails in the core (inner 40% by radius or 80% by mass) where information relating to solar evolution and the solar neutrino problem is to be found (Christensen-Dalsgaard 1985b).

Gravity g-modes are predicted to penetrate deeper into the radiative interior of the Sun. Oscillations with high l-values are, however, damped through the convective region, so their amplitudes are low. Nevertheless, detection has been made (Hill and Caudell 1979; Caudell and Hill 1980; Delache and Scherrer 1983), and modes tentatively identified (Severny et al. 1984). These observations have been interpreted as a composition gradient, and suggest diffusive mixing in the solar core (Berthomieu et al. 1984), but the identification and interpretation of these modes is complex and currently controversial (Christensen-Dalsgaard et al. 1985b; Wentzel 1987).

III. PHYSICS OF SOLAR MODELS

Calculating a stellar or solar model involves the determination of pressure, temperature, luminosity and composition as a function of mass or radius through the star (or Sun). This must be consistent with four differential equations of structure and an equation of state (Schwarzschild 1958; Cox and Giuli 1968). Composition changes with time caused by nuclear reactions in the core result in a continuously evolving structure, the calculation of which adds another differential equation to the set to be solved. Finally, boundary conditions are placed on the luminosity and mass in the center, and on the temperature and pressure at the surface (Table II).

A. Hydrostatic Equilibrium

The first differential equation (see Table II) describes the balance of forces at any point in the star and equates the pressure gradient to the gravity. This equation of hydrostatic equilibrium allows radius changes, but requires the kinetic energy involved in any bulk motion (expansion or contraction) to be small compared to the gravitational potential of the star. This is justified by the global stability of our Sun over historic times, as well as observations in other main-sequence stars of solar type. The second equation (Table II) constrains the integral of the density over volume to be equal to the mass.

B. Thermal Equilibrium and Energy Sources

Thermal equilibrium describes a balance at each point between the energy production (nuclear fusion) or gravitational contraction, and changes in luminosity. Mechanisms such as neutrino production, which cause en-

TABLE II
The Equations of Stellar Evolution

Hydrostatic Equilibrium:	$\dfrac{dP}{dr} = -\rho \dfrac{GM_r}{r^2}$
	$\dfrac{dM_r}{dr} = 4\pi r^2 \rho$
Thermal Equilibrium:	$\dfrac{dL}{dr} = 4\pi\, r^2 \rho\, (\varepsilon_{nuc} + \varepsilon_{th} - \varepsilon_\nu)$
Energy Transport:	$\dfrac{dT}{dr} = \nabla \dfrac{T}{P} \dfrac{dP}{dr}$
	$\nabla = \dfrac{3}{16\pi ac} \dfrac{K_r L_r P}{G\, M_r T^4}$ radiative
	$= \left(\dfrac{\partial \ln T}{\partial \ln P}\right)_{adiabatic}$ convective
Boundary Conditions:	
Interior	$L = 0$ and $R = 0$ at $M = 0$
Exterior	$L = 4\pi R^2 \sigma T^4$ and $P = \dfrac{GM\tau}{R^2 K_r}$ at $M = M_{total}$
Composition Change:	
P = pressure	K_r = Rosseland mean opacity
ρ = density	τ = optical depth
T = Temperature	M_r = the mass at radius r

ergy loss directly from any point in the star, may be treated as a negative energy production term. Over most of a star's main sequence lifetime, the energy produced by nuclear reactions approximately equals the surface luminosity. Small inequalities between the nuclear energy production rate, the neutrino energy loss rate, and the luminosity result in structure changes. The magnitude of the structure changes is determined from the increase or decrease in the potential energy of the system.

This exchange of thermal energy for gravitational potential is very significant during the post-main-sequence phase when a star expands to become a red giant. During this stage, the nuclear energy production rate increases, but the extra energy production is trapped (unable to be transported to the surface due to high opacities) causing it to expand, and thereby increasing the potential energy of the star. If enough energy is trapped, the surface luminosity may actually decrease.

For the Sun in hydrostatic equilibrium, the virial theorem shows the kinetic energy to be equal to one-half of the potential energy (Schwarzschild 1958, p. 32). This leads to a lifetime for the Sun known as the Kelvin time scale, i.e., $T_k = GM^2/2RL = 15$ Myr where T_k is the Kelvin time, M is the mass of the Sun, R is the Sun's radius and L is the Sun's luminosity. While the astronomers were initially satisfied with this age for the Sun, overwhelming geologic evidence in the 19th century demonstrated a fundamental inconsistency with the age of the Earth. Resolution of this difficulty was made by the discovery of radioactivity, an entirely new source of energy. Recognition that the Earth (and Sun) were much older than 15 Myr led Eddington (1920) to speculate on some new energy source for the Sun and other stars, including the total annihilation of matter, or the conversion of hydrogen to helium. A decade later, Bethe (1939) enumerated the basic set of nuclear reactions that convert hydrogen into helium.

The evidence for nuclear fusion as the solar energy source is compelling. The most direct observation supporting nuclear fusion is the observed neutrino flux. While the high-energy neutrino flux is lower than expected, it is still sufficient to account for most of the Sun's energy as a result of fusion. Nuclear-processed material is also observed on the surface of red giants (Wallerstein 1988, and references therein) including freshly synthesized elements such as short-lived ^{99}Tc (Wallerstein and Dominy 1988). This nuclear ash is transported from the interior by deeply penetrating convective regions. Finally, the luminosity increase seen in low-mass stars near the main-sequence turn-off of both open and globular clusters is a direct consequence of the change in the pressure per gram as material in the core is converted from hydrogen to helium.

The energy produced by fusion of hydrogen to helium has been measured in the laboratory, and the luminosity of the Sun gives a rate at which helium is produced in the solar core. The known age of the Sun then sets a limit on any additional energy sources or sinks (energy loss mechanisms) that can be operating. Additional energy loss shortens the life of the Sun, and when the loss is too high, it becomes impossible to produce an acceptable solar model. Conversely, additional energy sources slow the apparent rate of evolution of the Sun and an alternate source would make it impossible to evolve to the observed luminosity at the Sun's age (for an acceptable initial helium abundance). Also, such solar models would produce too few ^{8}B neutrinos. In Sec. IV, we discuss such energy loss mechanisms.

Most of the energy produced in the Sun originates from the proton-proton (p-p) chain which proceeds by the fusion of hydrogen with itself, or the fusion of the direct products of that hydrogen fusion. The p-p chain will operate in a star devoid of heavier elements. The CNO cycle uses isotopes of carbon, nitrogen and oxygen to convert the hydrogen into helium. It operates satisfactorily in the Sun to modify these isotopes (leading to the composition changes observed in the red giants), but is not the principal energy source there as it is in higher-mass stars. The nuclear reaction rates that are important for the stars are measured by a number of groups, and a regularly updated list has been published (Fowler et al. 1967, 1975).

During the contraction of the pre-main-sequence Sun, the material becomes more tightly bound gravitationally, and the central temperature increases. Eventually, the temperature is sufficient for hydrogen fusion and contraction ceases. The fusion of hydrogen is limited by the slowest reaction, which for the Sun is $p(p, \nu\, e^+)d$. The cross section of this rate is too small to be measured directly and is based on theory and the neutron lifetime. Improvements in the measured neutron lifetime lead to a 27% increase in this rate between the two publications of Fowler et al. (1967, 1975). This increase allows the solar luminosity to be satisfied at a slightly reduced temperature, and it reduces production of the highest-energy neutrinos from the production of ^{8}B. It does not, however, result in a significantly shorter lifetime for the Sun, because this is set by the luminosity which depends more upon mass than upon the nuclear energy generation rates. Finally, while fusion produces energy at the expense of mass, energy conservation is usually maintained separately from the conservation of mass. As only $\sim 5 \times 10^{-4}$ solar masses are converted into energy over the lifetime of the Sun, this is an acceptable assumption.

C. Convective Energy Transport

The energy produced in a star's interior must be transported to the surface to replace the energy radiated away. Available mechanisms are radiation, convection and conduction. In the absence of WIMPs (weakly interacting massive particles) or other exotic particles, only radiative and convective energy transport are believed to be significant in the Sun (electron conduction is important in stars with degenerate cores, but not in the Sun). In the Sun, convective transport is active in the outer regions where the opacity becomes too high for radiative transport to be efficient. The convective region contains

~ 2% of the Sun's mass, and 28% of its radius. Because convection does not extend through the inner 98% (by mass), the rate of evolution, or the neutrino flux is relatively insensitive to its treatment.

Convective energy transport is generally treated by a parameterization called mixing-length theory, and the boundaries set by a stability criterion (Cox and Giuli 1968). However, the lack of rigor in mixing-length theory, especially in places where time dependence is involved (like the helium flash or the production of lithium in red giants), is a major concern. Nevertheless, its application to solar models is probably adequate for most purposes. Opacities are high through the bulk of the convection zone, leading to the adiabatic transport of energy. Only near the surface, where the density and opacity drop, can significant amounts of energy again be transported by radiation. This region contains about 10^{-6} $M_\odot$ and is called the superadiabatic region. It is in this region where the mixing length determines the convective efficiency and sets the adiabat for the bulk of the convection zone.

As mentioned above, the adiabat of the convective region is adjusted via the mixing length to produce a model with the observed radius, and this choice affects the depth of the convective region. This depth is usually too shallow to agree with that interpreted from the p-modes. The discrepancy can be resolved if the opacity near the base of the convective region has been underestimated by 20 to 30% (Ulrich 1986; Korzennik and Ulrich 1989). Quite independently, Stringfellow et al. (1987) have shown that a 30% increase in the opacity in the same region results in the lithium depletion observed in the Hyades cluster (which is otherwise underestimated). Most (90%) of the opacity in this region is due to the heavy-element contribution; and opacity uncertainties of 20 to 30% in this region do not seem excessive.

Increasing the depth of the convective region by invoking an opacity increase at its base is not a unique solution. Korzennik and Ulrich (1989) also found that opacity reductions of 50 to 70% in the core could resolve the discrepancy, and a reduction would resolve the low flux of neutrinos observed in the chlorine experiment. While an actual opacity reduction of 50% seems impossible for the conditions in the solar core, an enhanced energy transport equivalent to such reductions is obtained in WIMP models (Press and Spergel 1985; Faulkner and Gilliland 1985; Gilliland and Däppen 1988; Faulkner and Swenson 1988).

Because the superadiabatic region determines the adiabat of the convective region, the convection zone is sensitive to changes in its structure. The small amount of mass involved results in a short ($<$ 1 yr) thermal time scale. It is at least possible for the structure of the region to change on such time scales, resulting in expansions or contractions of the convection zone, and the exchange of gravitational energy with the thermal energy there. In the Sun, this would lead to short time scale luminosity and accompanying small radius fluctuations (Ulrich 1975; Dearborn and Newman 1978; Sofia et al. 1979; Dearborn and Blake 1980a; Gilliland 1980).

Structure changes in the superadiabatic region might occur if the material located there is only exchanged slowly with the bulk of the convection zone, and there were variations in the local convective efficiency. Stochastic fluctuations are found in the convective energy transport of some laboratory flows (Busse and Whitehead 1974). While it is possible that such fluctuations also occur in the Sun, convective theory is inadequate to prove that they exist.

A mechanism that could drive structure changes in the superadiabatic region is changes in the magnetic field (Thomas 1979; Dearborn and Blake 1980b,1982; Tandon and Das 1982; Spruit 1988). The emergence of a magnetic field changes the energy content (pressure) of this surface region, as well as its ability to transport energy. Recent comparisons of the surface magnetic activity of the Sun and the solar constant find a positive correlation (Foukal and Lean 1988). A decrease in the magnetic activity has in the last solar cycle coincided with a decreasing solar constant. This suggests that the presence of the magnetic field leads to a higher density through the superadiabatic region, and the remainder of the convection zone contracts slightly to maintain hydrostatic equilibrium. A better treatment of convection and its interaction with magnetic fields might allow a more rigorous evaluation of how the superadiabatic region is perturbed by the solar cycle, as well as how this region evolves with the Sun.

D. Radiative Energy Transport

Energy transport through the interior (98% of the Sun's mass) is radiative. As the mean free path of a photon (typically a fraction of 1 cm) is short compared to the distance over which the temperature changes, diffusion is an accurate approximation throughout the interior, and the efficiency of energy transport here is determined by the opacity. Lower opacities can transport the Sun's luminosity with a shallower temperature gradient, resulting in a lower central temperature. This is significant for issues like the neutrino rates which are highly sensitive to the central temperature.

Since the early 1960s, the most common source of the opacities has been the Los Alamos Scientific Laboratory (Cox 1965; Huebner et al. 1977; Magee et al. 1984). These opacities were a great improvement over those available earlier, and have served well for investigating many problems of stellar evolution. Unfortunately, the accuracy of the physical approximations used to calculate the opacities is difficult to determine, leading to indeterminate error in the results of stellar evolution calculations. Opacity also depends on the material composition, but uncertainty in abundances can be formally evaluated, and the effect on opacity can be determined by making calculations within the probable range of compositions.

While it is difficult to evaluate the accuracy of the approximations used in calculating astrophysical opacities, it is possible to compare the results of different computer codes. Bahcall et al. (1982) compared the results of Livermore (HOPE code of B. Rozsnyai) to those of Los Alamos. They demonstrated general agreement to within 10% for individual elements. While this suggests a limit to the uncertainty, it is difficult to assess the true implication of this comparison. The HOPE code and the Los Alamos codes use different physical models, but they were not developed completely independently.

Calculating opacities requires the solution of complex problems in both plasma and atomic physics. It was, and continues to be, necessary to make approximations that may compromise the accuracy of the calculations. These concerns have led the astrophysics community to support an independent effort to calculate astrophysical opacities (Mihalas 1988). At the same time, a Livermore group has begun a project to develop a code with more detailed atomic modeling than is present in the Los Alamos Astrophysical Opacity Library (Iglesias et al. 1987).

The Rosseland mean of the opacity (appropriate for use in the diffusion approximation) smooths over many of the details of the atomic physics, and approximations that lead to quite different spectra may give roughly the same opacity. Nevertheless, an increase in the number of photoionization edges and spectral lines reduces the number of available "windows" in the spectrum and raises the Rosseland mean opacity. The number and strength of lines that contribute to the spectrum depend on the occupation numbers calculated for each state. In a dense LTE plasma, this may differ from the Boltzman distribution due to continuum lowering (Hummer and Mihalas 1988). Uncertainty in the occupation numbers and the resulting line strengths are sources of potential error. Line widths can also affect the opacity significantly. In the region of a dense line spectrum, a precise treatment of line widths becomes very important because of line overlaps (Iglesias et al. 1987). In sparse line regions, the bound-bound transitions will be less important, but the wings of very strong lines can still contribute to the opacity. The proposed improvements in the physics seem most likely to increase the opacity.

E. Equation of State

In addition to the four differential equations (and associated physics), stellar structure calculations require an equation of state to relate the pressure, temperature, density and composition, as well as provide thermodynamic quantities. Stellar interiors span a range of temperature and density over which one must consider radiation pressure, Fermi pressure and Coulomb pressure, as well as normal gas pressure. In the Sun, however, the pressure behaves nearly like a perfect gas. The density is high enough that some plasma effects begin to be important, and precise solar modeling requires their inclusion.

One such effect is pressure ionization. Simple application of the Saha equation leads to the erroneous result of a substantial population of neutral hydrogen present in the Sun's center. This is usually corrected with lowering the continuum using a Debye-Hückle approximation. The energy necessary to ionize a species is reduced by an amount equal to the Coulomb energy in the gas, so for consistency, a small Coulomb term must be included in the pressure (Zel'dovich and Raizer 1966). The Debye-Hückle approximation for cutting off the partition function is somewhat artificial, and a less *ad hoc* procedure involves the calculation of an equation of state which includes scattering states. The population in any particular atomic state can then be calculated using a renormalized activity expansion of the grand canonical ensemble (Rogers 1981, 1983, 1986). Ulrich (1982) used an equation of state based on this technique, and found a lower pressure by an amount equal to the Coulomb pressure. This results in an initial helium mass fraction that is 0.01 to 0.02 lower than models using the same opacity but a different equation of state.

IV. NONSTANDARD PHYSICS

In addition to the phenomena discussed in Sec. III, there are important characteristics that are not usually included in solar or stellar models. These are mass loss, magnetic fields, rotation and nonconvective mixing, though mass loss is not currently a significant influence on the global structure of the Sun (however, when coupled with the magnetic field, it may be responsible for angular momentum loss from the solar surface—see the chapter by Stauffer and Soderblom).

A. Magnetic Fields

The short time scale of 11 yr for reversal of the polarity of the solar magnetic field supports the model of a dynamo source (Parker 1955*a,b;* Gilman 1985). The average value of the solar magnetic field is ~ 1 Gauss, but the field is far from homogeneous and is distributed over a small fraction of the Sun's surface (< 1%) in flux regions of relatively high intensity (kilogauss). Below the surface, the structure of the Sun's magnetic field is known only through inference. In the convection zone where flux tubes are dynamically buoyant (Parker 1975), the magnetic field is probably as chaotic as it is on the surface. The field is no doubt better organized near the base of the convection zone and may reach an amplitude of 10^5 Gauss (Parker 1985a,b; Moss 1987). A toroidal field is generated from a poloidal seed field by differential rotation with an intensity in the range 0.5 Gauss to 5000 Gauss (Mestel and Weiss 1987). It is then buoyed into the convection zone on a thermal time scale which leads, in turn, to additional mixing (Hubbard and Dearborn 1981).

Mixing by this mechanism cannot be rigorously calculated due to uncertainty in the depth from which the magnetic field arises, and how well the material is locked to the field. If the emerging magnetic field brings up a small amount of lithium-depleted material each cycle, it would lead to an exponential decrease in the surface lithium. Mixing 5×10^{-8} $M_\odot$ of such material into the convective region each cycle would result in a decrease of lithium abundance by a factor of 100 in 1 Gyr. The magnetic field occupies about 10^{-3} of the solar surface at solar cycle maximum, and a crude integral indicates that 10^{-7} of the mass of the convection zone is contained by the magnetic field (assuming that the flux is constant with depth in the convection zone, and that the material is frozen to the field). This is close to the amount of mass that must be mixed each cycle to cause the observed solar lithium depletion. Parker (1985b) and Moss (1987) have turned this problem around, and used the mixing driven by magnetic fields to limit the strength of the field below the convective region to 10^5 Gauss.

While the strength of the field in the solar interior is very uncertain, a field strong enough to affect the structure of the Sun (or other stars), should be too buoyant to remain in the interior over the Sun's (star's) lifetime. There is, however, little direct evidence excluding such fields (see the chapter by Levy et al.). If strong magnetic fields existed in massive main-sequence stars, the suppression of convection would lead to slightly shorter lifetimes, but the consequences would be difficult to observe. A similar lack of observability might be expected from WIMP models applied to the upper main sequence (Hubbard and Dearborn 1980). The expected effects of strong fields on lower-mass stars like the Sun are even more subtle.

Even fields as weak as those at the surface of the Sun can affect energy transport. In photospheric regions, a kilogauss field is strong enough to in-hibit local convection leading to the generation of sunspots in the case of larger flux tubes; such regions must affect the energy transport through the superadiabatic regions, and potentially result in luminosity variations, as dis-cussed in Sec.III.C.

B. Nonconvective Mixing

Nonconvective mixing can be driven by anything that perturbs a star's potential away from spherical symmetry. Rotation is universal in stars, and necessarily leads to such mixing (Fricke and Kippenhahn 1972; Tassoul and Tassoul 1984; Pinsonneault et al. 1989, and references therein). The low ^{12}C/^{13}C ratios observed in some red giants suggests that a slow mixing occurs in some stars, but the depth of mixing is limited by molecular weight barriers (Dearborn and Eggleton 1977; Endal and Sofia 1981). Because such mixing is inhibited in regions where the composition has changed significantly, it does not supply fresh hydrogen to the core and perturb the evolution of stars. It may, however, play a role in the surface depletion of lithium. Observations of the Hyades, and other open clusters show a systematic depletion of lithium with temperature. If this results from nonconvective mixing driven by rota-tion, it requires the development of a relation between the angular momentum and spectral type.

C. Exotic Particles

The Sun is the only main-sequence star for which a direct probe of the interior via helioseismology and neutrino observations can currently be made. Because the Davis experiment is primarily sensitive to the high-energy boron neutrinos produced in the rarest branch of the p-p chain, the expected count rate is extremely sensitive to the central temperature, i.e., $\sim T^{18}$. In the absence of any neutrino decay or transformation, this provides an exceedingly precise thermometer of the solar core.

Many means of reducing the central temperature have been proposed, but they involve either phenomena which cannot be rigorously calculated or conditions that have not been observed. These solutions include WIMPs, exotic particles (Takahashi and Boyd 1988), composition changes in the core (Sienkiewicz et al. 1988; Dearborn et al. 1987) and various other hypotheses.

V. BASIC BEHAVIOR OF SOLAR MODELS AND THEIR TIME SCALES

Given the diversity of the physics necessary to describe the solar model, and the plausible phenomena that are not included in such calculations, it is perhaps amazing that models of the Sun (and stars) succeed as well as they do. A star like the Sun begins with a pre-main-sequence collapse phase in which the energy that is radiated results from gravitational contraction. The time scale of this phase is limited to the Kelvin time (Sec.III.B) by the available gravitational potential energy.

The main-sequence phase commences when the central temperature of a star becomes sufficient for the fusion of hydrogen and ends when it has converted all of the hydrogen to helium in the inner 10% of its mass (Schönberg and Chandrasekhar 1942). The luminosity and surface temperature evolve in response to the slow composition change in the core. The time required for a star like the Sun to convert its core hydrogen into helium is $T = 0.1MEX/L = 8$ Gyr, where T is the main-sequence lifetime of the Sun, M is solar mass, X is the mass fraction of hydrogen, E is the energy obtained by converting 1 g of H to He and L is the luminosity of the Sun. A detailed calculation results in an age of ~ 9 Gyr.

As discussed in Sec.III.C, luminosity variations around the mean luminosity may occur due to changes in the structure of the convective region. Such fluctuations could occur on time scales as short as months and have magnitudes of $dL = GMM_c/2R \times dR/dt$ where dL is the brightness change, M is the Sun's mass, M_c is the mass in the convection zone and R is the solar radius; alternately $dL = 1.5 \, dR/dt$ with dL solar luminosities and dR/dt measured in seconds per century. Unfortunately, the lack of a fundamental understanding of convection prohibits us from a genuine predictive capability of luminosity and radius changes due to magnetic fields (or other phenomena).

Some indicators of such variations have been suggested, but this remains an area for further investigation.

Post-main-sequence evolution involve both adjustments in thermal time scale like the expansion to the red-giant branch, and in shorter time scale nuclear evolution as the Sun (star) ascends the giant branch. Eventually, after all available nuclear fuel is expended, the Sun will undergo a cooling phase as a white dwarf star. Discussion of these phases can be found in Iben (1967).

A final time scale that can be mentioned is the dynamic time scale, or the sound travel time through the Sun. While this is pertinent for interpreting p-mode observations, it is not relevant for the structural changes that we have discussed here. Large structural changes on dynamic time scales (hours) can occur only with an expenditure of energy comparable to the binding energy of the Sun, i.e., 10^{48} erg. Fortunately for those of us who live in the solar neighborhood, there are no ready sources of this amount of energy, and the Sun is expected to remain with us for a considerable time to come.

VI. SUMMARY

Stellar evolution is very successful in reproducing many observable quantities of both the Sun and other stars. Most of these, like the luminosity function, are quantities depending on the relative time spent at various phases of evolution. The Sun provides a calibration of stellar evolution theory with a precision unavailable in other stars, and from independent measurements giving an absolute value for the age. It also provides a detailed example of behavior patterns that are only hinted at in observations of other stars. In turn, stellar evolution theory allows us to identify T Tauri stars (naked or classical) as the pre-main-sequence phase of solar evolution, and shows that the Sun's future involves becoming a red giant and then a white dwarf. In addition, short time scale luminosity variations observed in other stars give us clues to classes of behavior that may be occurring in our own Sun.

THE SUN AND INTERPLANETARY MAGNETIC FIELD

EDWARD J. SMITH
Jet Propulsion Laboratory

The interplanetary magnetic field (IMF) serves as a link between the Sun, the response of the Earth to solar activity and variations in galactic cosmic radiation. The IMF originates as a solar-coronal magnetic field that is transported into space by the solar wind. The close connection between solar magnetic fields and the origin and structure of the solar wind is described. The solar wind forms the heliosphere, a cavity containing the magnetized solar plasma from which the interstellar plasma and field are excluded. The entry of galactic cosmic rays into the heliosphere and their strong interaction with the IMF are discussed, this topic being of primary importance to the production and temporal variations of radiogenic elements. The profound influence of the IMF on geomagnetic activity and the aurora is discussed within the context of merging or reconnection with the planetary field. The physical connection is thus established between solar magnetic fields, magnetic storms and aurora. The state of the solar wind and IMF during the Maunder minimum is considered and an explanation for the (relative) absence of sunspots and aurora is proposed. The mechanism is an interruption of the oscillatory solar dynamo, a consequent reduction in the heating of the corona, a cessation of the supersonic solar wind and a weakening or absence of southward-directed magnetic fields in the vicinity of the Earth.

This review of the interplanetary magnetic field (IMF) has four major elements. The first is the origin of the IMF in the solar corona and the influence of coronal structure and events on the properties of the field. The second section concerns the heliosphere and the effect of the IMF on galactic cosmic rays, in particular, their modulation by the solar activity cycle. The third addresses issues relating to the interaction of the IMF with the terrestrial magnetosphere and with the production of magnetic storms and aurorae. In the final section, the Maunder minimum is considered; here we speculate about the possible character of the IMF and the solar wind during this unusual epoch. Knowledge of the IMF can make a vital contribution to the theme of this book by serving as a link between solar magnetic fields, cosmic ray behavior and the response of the Earth to solar activity. Of necessity, the discussion is very general and provides only an overview of each of the major topics and their relationships. Details of the principal physics involved is available in the figures, many of which are accompanied by extended captions. Relevant references are also provided for those wishing to pursue these diverse topics in greater depth.

I. THE SUN AND THE INTERPLANETARY MAGNETIC FIELD

The interplanetary magnetic field (IMF) brings to mind a picture like Fig. 1. The view is looking down on the solar equator or ecliptic. The Sun is

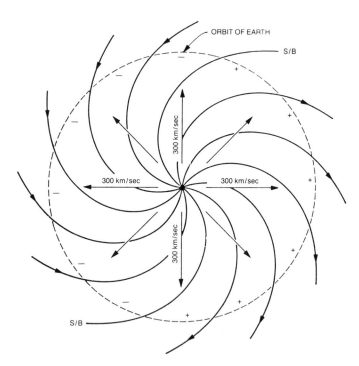

Fig. 1. The interplanetary magnetic field viewed in the solar equatorial plane. This is an early view of the spiraled IMF. The fields and radial solar-wind streamlines originate at the Sun (the center of the figure). A more typical solar-wind speed based on years of observation is 400 km s^{-1}. The origin of the term "magnetic sector" is evident from the pluses and minuses along the circle which represents a radius of 1 AU (figure adapted from Parker 1963a)

at the center of the radial lines that represent streamlines of the solar wind and the familiar Archimedian spiral ($r \approx \phi$) representing lines of force of the IMF. This representation is rather simplified because of the neglect of two other important properties of the field. The field is irregular, not smooth, with large fluctuations superposed on the average field directions shown here. Typically, these fluctuations are approximately equal in magnitude to the average field strength or $\Delta B/B \approx 1$. The other simplification which has been introduced is to ignore some of the most important large-scale structure, by and large associated with the underlying solar-wind structure, i.e., alternating fast and slow solar-wind streams. One large-scale feature which has been added is the polarity of the IMF which is indicated by $+/-$ signs lying along the dashed circle representing a distance of 1 AU (the orbit of Earth) and by the arrowheads which are outward from "noon" to "six" o'clock and inward from "six" to "noon." This property is the so-called sector structure, there being two sectors in this instance separated by a sector boundary (S/B in the figure). There is no doubt about the marked tendency of the field to point in or out along the spiral direction. Figure 2 is a histogram of the longitudinal or azimuthal angle in a coordinate system aligned with the Parker spiral. As is evident, the field tends to favor orientations corresponding to 0 or 180°.

The origin of the IMF is well understood (Parker 1963a). It is a solar-coronal field that is transported or convected into space by the solar wind. The field is "frozen" into the collisionless solar wind at the source region. As is well known, the solar wind originates as a hydrodynamic expansion of the hot coronal plasma acting against the constraint of solar gravity. Figure 3 is a representation of the familiar de Laval nozzle analogy. As a result of the throat, whose role in the case of the solar wind is played by the gravitational field, the random motion of the hot gases in the combustion chamber is converted into a directed motion at a supersonic speed.

The hydrodynamic expansion model, on which this analogy is based, is not the whole story, however, because it neglects the effect on the flow of magnetic fields at the source. It has been known from the earliest space measurements that the solar-wind energy density, extrapolated back to the corona, is typically less than the extrapolated energy density associated with the magnetic field (Davis 1966), and thus the magnetic field could obstruct the flow unless it was oriented more or less radially. It is now recognized that, although the solar wind is observed at all times and (as far as we know) in all locations remote from the Sun, it actually originates from restricted regions on the Sun and subsequently expands to fill all of surrounding space.

The magnetic field observations, especially the polarity measurements, have been very helpful in identifying the solar-wind source regions. Basically, it has been found that the solar wind originates on open field lines, such as those originating at high latitudes in the accompanying schematic (Fig. 4), i.e., those which have one end rooted in the corona with the other being carried off by the solar wind. The equatorial region near the Sun is occupied

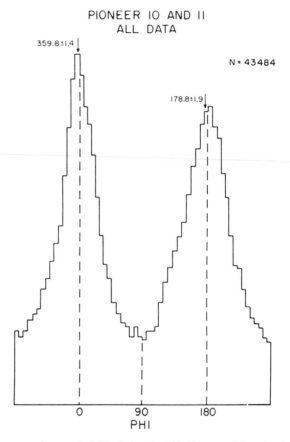

Fig. 2. Measurements of magnetic field spiral angle. This histogram is based on large numbers of observations by two Pioneer spacecraft between 1 and 8 AU. Since the spiral angle between the field and the radial direction changes over this distance range from 45° to approximately 90°, the observations were transformed into a coordinate system with one axis along the Parker spiral which then corresponds to 0° and 180°. There is a remarkable agreement between theory and observation (figure from Thomas and Smith 1980).

by closed field lines, i.e., those which begin and end in the corona or on the solar surface. In the equatorial region far from the Sun, are open field lines which are oppositely directed. By Maxwell's equations, they must be separated by a thin current sheet.

The equatorial topology corresponds to a coronal streamer which is rather commonly observed in coronagraph images as a bright "bottle-shaped" structure made visible by the presence of a relatively high density of electrons trapped on the closed field lines. By contrast, the open field lines are populated by much less dense plasma as a result of the solar wind having drained these regions. The open field regions which are a source of the solar wind

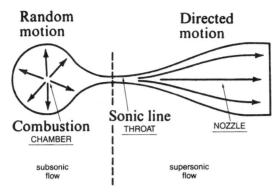

Fig. 3. Supersonic hydrodynamic expansion of a gas in a de Laval nozzle. The chances are that many readers have even less knowledge of supersonic flow through nozzles than of the supersonic expansion of the solar wind. However, the figure is shown for two reasons. It is a dramatic reminder that the solar wind is not simply evaporation from the corona. Second, the analogy was important historically in convincing theorists that the origin of the solar wind did not require special or unusual conditions (figure adapted from Brandt 1970).

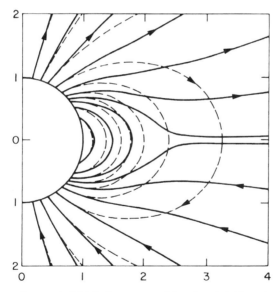

Fig. 4. Solar-coronal magnetic fields in the presence of the solar wind. Two sets of field lines are represented. The (initial) dashed lines correspond to the solar dipole magnetic field. The solid lines represent a solution of the hydrodynamic equations with the solar wind, including currents in the corona. The main features of note are the "opening" of field lines at high latitudes and the formation of an equatorial current sheet or neutral sheet (figure from Pneuman and Kopp 1971).

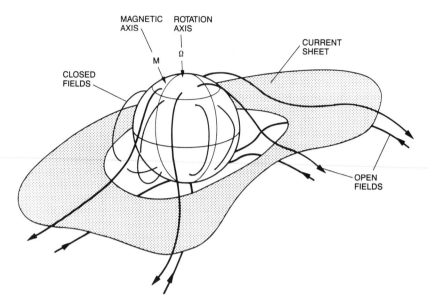

Fig. 5. The current sheet near the Sun in three dimensions. Closed fields, beginning and ending in the corona, are shown at low latitudes. The open fields at higher latitudes pass over and under the current sheet that separates oppositely directed fields. The open fields are beginning to spiral as a result of the solar rotation about axis Ω. The current sheet, whose normal is M, which constitutes, in effect, a magnetic equator is tilted relative to the Sun's equator (figure from Smith et al. 1978).

appear as "coronal holes" in coronagrams or in X-ray images such as the outstanding examples obtained by Skylab (Zirker 1977). Coronal holes can extend from the polar cap to well past the equator. The bright areas surrounding the dark holes are hot gases trapped on closed field lines.

The three-dimensional picture of the interplanetary field which has emerged from these considerations is shown diagrammatically in Fig. 5. The rotation axis is identified by Ω. Closed magnetic loops are shown crossing the equatorial region and open field lines from the polar caps extend out into space. The polarity of the field is the same as that of the polar caps in which they originate, in this example, outward in the north and inward in the south. The thin current sheet separating oppositely directed fields is shown to divide the solar wind into two distinct hemispheres. The current sheet represents a solar magnetic equator and is commonly inclined relative to the solar equator as indicated by the equivalent tilt between the magnetic symmetry axis (M) and the solar rotation axis (Ω). This inclination typically results from the asymmetric location of the polar coronal holes with respect to the rotational axis. As the Sun rotates, an observer or spacecraft will be located below the current sheet for a time and record inward (negative) fields followed by observations above the current sheet dominated by outward (positive) fields.

Thus, the current sheet corresponds physically to the sector boundary noted in Fig. 1.

The magnetic polarity measurements are useful in determining which hemisphere solar-wind streams come from. The upper panel in Fig. 6 displays solar-wind velocity data over a solar rotation consisting of two distinct high-speed streams. (In this representation, the velocity is plotted as a function of Carrington solar longitude so that time progresses right to left in the diagram.) The lower panel shows the corresponding coronagram in which the main features to be seen are the white "equatorial" steamer belt (equivalent in a sense to the magnetic equator of the Sun) and the polar coronal holes extending equatorward from the north and south polar regions. The polarity of the coronal holes have been determined by magnetographs and are, as

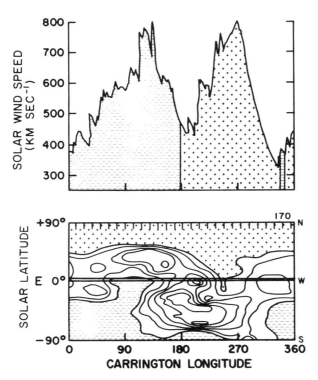

Fig. 6. An example of the relation between polar coronal holes, fast solar wind and magnetic sectors. The upper panel shows the solar-wind speed observed near the solar equator as a function of solar longitude. Two fast streams are present having opposite magnetic polarities (− from 0° to 180°, + from 180° to 360°). The lower panel contains equal-brightness contours for the corona that have the general form of a tilted disk (streamer belt). A coronal hole with negative magnetic polarity extends up from the south pole near 90° longitude and is identified with the fast stream in the figure above it. A positive polar coronal hole extends down to the equator near 240° and is correlated with the other solar-wind stream (figure from Hundhausen 1977).

shown, positive in the north and negative in the south. These polarities coincide with those measured in conjunction with the high-speed streams in the
upper panel so that the leading (right-hand) stream originated in the north
hemisphere and the following (left-hand) stream arose in the south.

These magnetized high-speed streams are responsible for imposing a
large-scale structure on the solar wind. The streams in Fig. 6 were long lasting, i.e., they endured with only modest changes for many solar rotations,
and, hence, co-rotated with the Sun. As the Sun rotates, the fast streams
sweep out into space and overtake the slower-moving solar wind preceding
them (Fig. 7). As a consequence, there is a pile-up along a spiral shaped
ridge (analogous to the effect of a snow plow) which is a region of high
pressure created by the compression of the plasma and magnetic field (Davis
1966; Siscoe 1972; Belcher and Davis 1971; Hundhausen 1977). Such features are called co-rotating interaction regions. Generally, there are 2, sometimes 4, of these high pressure interaction regions which spiral around the
Sun as they extend into space. Beyond 1 AU, the large-amplitude waves
which form the co-rotating interaction region boundaries, steepen into a pair
of shocks and gradually widen so that at large distances (beyond 10 AU) they
begin to merge to form a single region (Razdan et al. 1965; Colburn and
Sonett 1965; Smith and Wolfe 1977; Smith 1985).

The situation described so far is characteristic of the minimum in solar
activity when the solar wind is dominated by co-rotating structures. During
solar maximum, however, the large-scale structure is altered as a result of the
dominance of transient solar events over co-rotating solar features. The major
change is the appearance of relatively large numbers of coronal transients or
mass ejections (CMEs) emitted sporadically from essentially all regions of
the Sun (Howard et al. 1986). An example of a CME is shown in the Skylab
coronagraph image (Fig. 8). Large blobs of plasma (plasmoids) containing
what appear to be loop-like magnetic fields have been found to be emitted
several times a day on average. A time sequence of coronagraph pictures
reveals that they accelerate away from the corona eventually to pass out of
the field of view at speeds of a few hundred km s^{-1}. The structural features
evident in the figure, e.g., the two bright loops, are the subject of on-going
research directed toward identifying whether they are shocks, a contact surface or other forms which arise in plasmas. CMEs are well correlated with
other solar events such as flares and eruptive prominences. They have been
observed farther from the Sun by spacecraft, principally HELIOS, and the
characteristic properties of the fields and particles associated with CMEs have
been investigated (Schwenn 1983b).

Not all of the solar wind at solar maximum originates from CMEs. There
is another low-velocity component that is continually present, presumably
flow from polar coronal holes, which are gradually decreasing in area, or
from other short-lived coronal holes which are typically present for less than
one solar rotation. The faster-moving CMEs force their way through this pre-

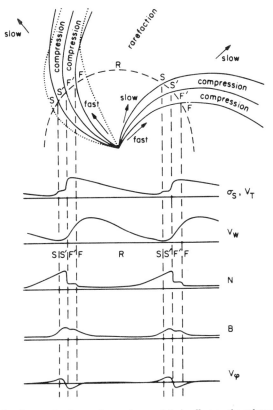

Fig. 7. Schematic of corotating interaction regions and their effect on the solar wind. The upper-half figure shows the encounter of two fast streams (the long arrows representing the solar-wind velocity) with the preceding slow solar wind. The solid curves represent the spiraled interaction or compression fronts which develop. The dotted curves are two representative magnetic field lines that are more strongly spiraled in the slow wind and less strongly spiraled in the fast wind. At increasing radial distance, they are overtaken by the leading and trailing edges of the compression region. The dashed circle designated R shows the relative motion of an observer (spacecraft) as the complex co-rotates (counterclockwise) with the Sun. Four subregions are identified as S (slow wind), S' (slow wind accelerated by the oncoming fast wind), F' (fast wind decelerated by its interaction with the slow wind) and F (fast wind). Two compression regions are shown separated by a rarefaction region in which the solar wind is expanding because its velocity is decreasing. The vertical dashed lines leading to the lower-half figure connect to profiles of various solar-wind parameters influenced by the interaction. The solar-wind speed and azimuthal velocity component are V_W and V_ϕ and the density is N. The magnetic field magnitude is B, and σ_s represents the standard deviation in the superposed magnetic field fluctuations, that is correlated with the solar-wind temperature or, in this instance, thermal speed V_T (figure from Belcher and Davis 1971).

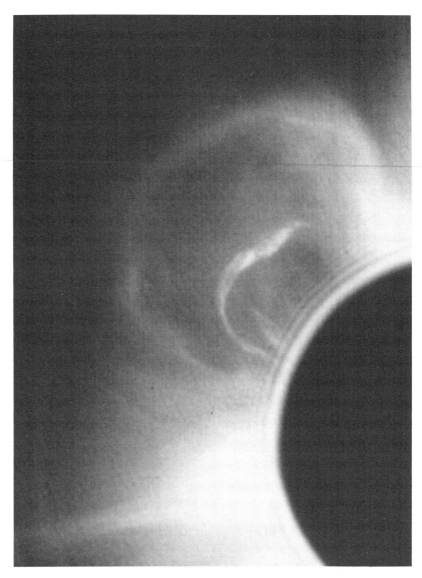

Fig. 8. Coronal mass ejection. This coronagraph image shows the eruption of a major portion of the corona as it is in progress. The structure visible, consisting of bright outer and inner loops separated by a dark void, is a typical feature.

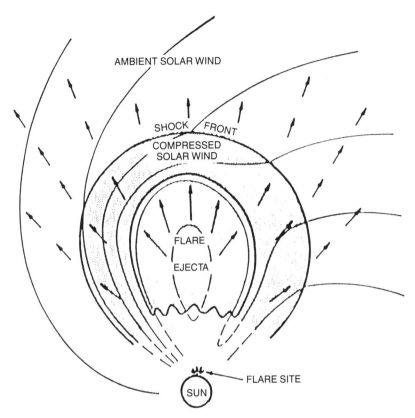

Fig. 9. Schematic of the interaction between flare ejecta/CME and the ambient solar wind. The solar-wind velocity vectors of differing length (speed) diverge slightly within the ejecta (plasmoid). The fast plasma compresses the preceding ambient plasma and a shock wave forms at the leading edge. The main features of the interaction are very different from those diagramed in Fig. 7 (figure from Hundhausen 1972).

existing solar wind, giving rise to a characteristic structure such as shown schematically in Fig. 9. Typically, the plasmoid, as indicated, is accompanied by a collisionless bow shock which develops near the Sun and behind which the preceding solar wind is diverted to flow around the ejecta. The limited extent of the plasmoid in solid angle and radial distance leads to a structure that is very different from that associated with co-rotating flows (cf. Fig. 7).

II. THE INTERPLANETARY MAGNETIC FIELD AND THE HELIOSPHERE

As the solar wind flows into the outer solar system beyond 1 AU, it pushes the local interstellar plasma and magnetic field out of its way eventually to form the heliosphere (Fig. 10; Parker 1963; Axford 1972; Lee 1988;

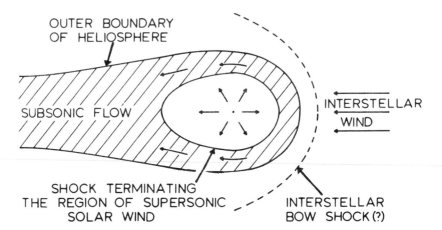

OUTER BOUNDARY
OF HELIOSPHERE

SUBSONIC FLOW

INTERSTELLAR
WIND

SHOCK TERMINATING
THE REGION OF SUPERSONIC   INTERSTELLAR
SOLAR WIND   BOW SHOCK(?)

Fig. 10. The heliosphere. The central dot represents the Sun from which the solar-wind vectors are seen to be diverging radially. Several thin surfaces resulting from the effect of the interstellar medium on the solar wind are shown. The heliosphere, containing solar plasma and fields, is separated from the interstellar plasma and fields by the asymmetric boundary that extends off to the left. The shaded region inside the boundary contains solar wind that has been decelerated at the egg-shaped "terminal" shock and is flowing tailward as indicated by the arrows above and below the shock. A possible bow shock outside the heliosphere is also shown which could exist if the interstellar medium is flowing past at high speeds (figure from Axford 1972).

Holzer 1989). The outward pressure associated with the solar wind and the inward pressure exerted by the interstellar medium are expected to equalize along a surface called the heliopause. The result is the formation of a cavity containing solar plasma and magnetic fields from which interstellar plasma and fields are excluded. As shown in the diagram, a shock is expected to form inside the heliosphere (Jokipii and Lee 1974) where the solar wind undergoes an abrupt transition to subsonic flow so that it can be diverted to move parallel to the heliopause and exit through an open tail ultimately to merge with the interstellar medium. Neither the inner shock nor the heliopause have yet been detected by Pioneer or Voyager so that these boundaries must lie beyond 50 AU and should be considered reasonable but unconfirmed predictions.

Although the general effect of the solar wind is to exclude interstellar material, several constituents are able to cross the heliopause and enter the heliosphere. These constituents are neutral gases (principally H and He), dust and galactic cosmic rays. The latter are relativistic ions and electrons whose large gyroradii and individual particle behavior (absence of collective interactions) enable them to cross the heliopause. Once inside the heliosphere, the cosmic rays sense the electric field associated with the outward flow of the solar wind and the IMF about which they begin to spiral. In order to reach the inner heliosphere, the cosmic rays must overcome the dual effects of the outward convection and ever present irregularities in the IMF which act as scattering centers and impose a random walk on the particle motion (Parker 1963a). The latter effect is customarily described as diffusion and the combined effect is known as convection-diffusion. It results in a gradient in the cosmic ray intensity with the intensity being significantly greater outside the heliosphere than inside it at 1 AU (Van Allen 1980; McDonald et al 1981; Webber and Lockwood 1981; McKibben et al. 1982; Venkatesan et al. 1984). The primary distribution function or energy spectra of the galactic cosmic rays are also altered rather drastically to cause a decrease in the lower-energy particles.

In addition to convection and diffusion, the cosmic rays are subject to guiding center drifts which result from the large-scale gradients associated with the IMF (Jokipii et al. 1977). A simplified schematic of these drifts is shown in Fig. 11 in which the cosmic rays enter from above the polar caps and drift toward the equator. The diagram includes the current sheet discussed above which is represented as a wavy surface oscillating above and below the solar equator. An important feature of the large-scale drifts is that they lead to the current sheet at which point they undergo a rapid motion along it which leads them out of the heliosphere. This effect is associated with the abrupt reversal in the direction of the field above and below the current sheet which causes a corresponding reversal in the sense of gyration of the cosmic rays. These drifts depend on the polarity of the IMF (in this case outward in the north) and during solar cycles in which the polar cap fields are reversed (inward in the north and outward in the south); the direction of the motion is also reversed with the particles drifting sunward along the current sheet and exiting above the poles. The complete description of the particle motion requires a combination of all three effects, convection, diffusion and drift, which are obviously not mutually exclusive.

In order to study the details and relative importance of convection, diffusion and drift, advantage is taken of the major perturbation associated with solar-cycle modulation, the well-established anticorrelation of cosmic-ray intensity with solar activity. Measurements of the solar-wind speed and the power in the magnetic field fluctuations have not yet revealed a significant dependence on solar cycle and do not appear to be primarily responsible for modulation as had been originally supposed. Instead, the effect of large-scale structure on modulation has become apparent. Modulation is an implicit aspect of gradient drifts associated principally with the changing inclination or extent in solar latitude of the heliospheric current sheet. Basically, when the current sheet is highly inclined near the Sun, the peaks and troughs evident in Fig. 11 extend to higher latitudes. The drifting cosmic rays then encounter the current sheet sooner and are expelled more readily from the heliosphere. It has been known for some time that the inclination of the current sheet undergoes a characteristic variation with the solar cycle (Svalgaard and Wilcox 1974). The current sheet has a low inclination at sunspot minimum (the origin or zero value on the time axis) and a high inclination, essentially a

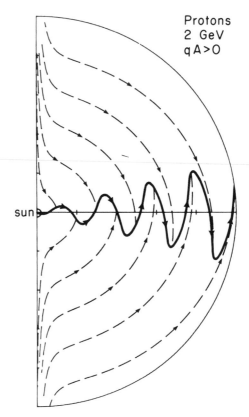

Fig. 11. Drift trajectories of cosmic rays in the large-scale heliospheric magnetic field. In this meridional cross section, the Sun is at the left-center and the north pole is upward. The solid line is the heliospheric current sheet which develops peaks and troughs at the maximum and minimum latitudes reached as the Sun rotates about its axis that is inclined by 20° to the magnetic axis. The dashed lines show the trajectories of 2 GeV protons which enter over the Sun's magnetic poles (which are assumed to be positive/outward in the north and negative in the south as indicated by $qA>0$) and drift equatorward in the magnetic field (gradient and curvature drift). Eventually, the drifting particles reach the current sheet and move rapidly outward along it (because of the abrupt field reversal). The trajectories represent the average motion on which would be superposed a random walk and outward convection (figure from Jokipii and Thomas 1981).

north-south orientation, at solar maximum. According to the gradient drift theory outlined above, this behavior should result in a change in the cosmic-ray intensity corresponding to the observed modulation. Confirmation of this prediction is shown in Fig. 12 which compares the cosmic-ray count rate at Earth (Deep River) with the maximum latitudinal extent of the current sheet near the Sun as inferred from source surface calculations (see the chapter by Jokipii). The latter involve the extrapolation of the observed photospheric fields to 2.5 solar radii and the imposition of a boundary condition that the coronal field become radial at that altitude corresponding to the outflow of the solar wind (Schatten et al. 1969; Altschuler et al. 1969; Hoeksema and Scherrer 1982). The field-strength contours over this spherical surface include the current sheet (or neutral sheet at which $B = B_r = 0$). There is evidently a good correlation between the two parameters which includes short-term increases and decreases in the cosmic-ray intensity, presumably in association with less extensive flapping or oscillatory motion of the heliospheric current sheet.

An alternative explanation involving large-scale structure has been proposed to account for modulation. It involves the effect of CMEs whose distribution in the heliosphere is shown schematically in Fig. 13. The proposal is that the enhanced magnetic fields associated with the plasmoids oppose the entry of cosmic rays into the inner heliosphere. A further extension of this hypothesis is that at large distances, the individual plasmoids will merge to

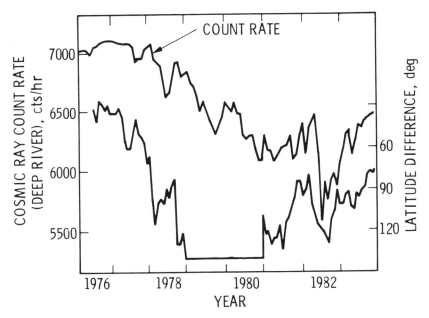

Fig. 12. Cosmic-ray count rate and the maximum latitudes reached by the current sheet. Monthly averages of the cosmic-ray intensity (in counts per hr) measured by the Deep River neutron monitor are shown from 1976 to 1983. The decrease to a minimum beginning in 1977, followed by a recovery, is typical of the modulation associated with solar activity which was a maximum in 1979–80. The lower curve is the difference in the maximum latitudes of the heliospheric current sheet in the north and south solar hemispheres. The latitude difference is derived from maps of the current sheet along a source surface located at 2.5 solar radii to which the observed photospheric magnetic fields are extrapolated. A close correspondence between changes in the current sheet and changes in the cosmic-ray intensity is evident (figure from Smith and Thomas 1986).

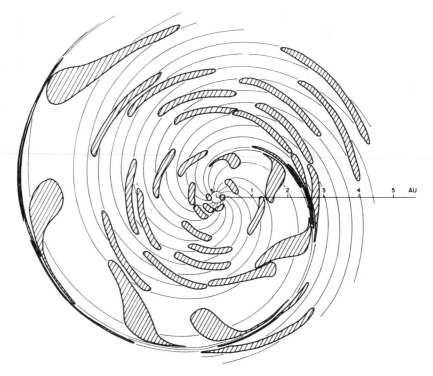

Fig. 13. Distribution of CMEs at sunspot maximum. This figure is similar to Fig. 1 with the Sun at the center and the spiral "arms" of the heliosphere magnetic field. The shaded "blobs" are schematic representations of CMEs emitted in succession by the rotating Sun. As a result, the CMEs cover a range of distances and longitudes. It was proposed in this model that the many plasmoids tend to keep cosmic rays out of the heliosphere and thus lead to modulation. A further elaboration of the basic model involves the merging of CMEs at large distances ($\geqslant$ 10 AU) to form "barriers" that close off the inner heliosphere (figure from Newkirk et al. 1981).

form ring-like magnetic barriers that tend to seal off the inner heliosphere (Burlaga et al 1984). This model has been promoted principally by Voyager investigators who have demonstrated a correlation between plasmoids and step-like decreases in the cosmic rays. Obviously, this effect and the effect of the heliospheric current sheet are not mutually exclusive and both may be operative.

III. THE INTERPLANETARY MAGNETIC FIELD AND THE EARTH

The following discussion will emphasize the influence of the IMF on aurorae and magnetic substorms. The physical phenomenon central to understanding of substorms and aurorae is the merging or reconnection of oppositely directed magnetic fields. The motion of the plasma and magnetic fields in the vicinity of the reconnection region (identified with the so-called "x-type neutral point" because the magnetic fields appear to cross where $B = 0$) is shown in Fig. 14. The surface between the fields with opposing polarity also corresponds to another thin current sheet as indicated. Southward fields (left-hand side) are to be identified with the IMF while northward fields (right-hand side) are to be considered the geomagnetic field so that the plane of the diagram contains the solar-wind velocity vector (the arrow approaching from the left) and the Earth's magnetic dipole or, in other words, the magnetic meridian plane near local noon. Magnetospheric plasma approaches the neutral point from the right and the combination of solar and magnetospheric plasma exits in opposite directions along the current sheet and transports reconnected field lines which have one end in the solar wind and the other end in the magnetosphere. This mechanism provides one of the few means capable of transferring collisionless plasma from one set of fields to another or, in this case, from the solar wind into the magnetosphere.

The effect of reconnection in creating the "open" magnetosphere is shown in Fig. 15 which shows magnetic lines of force at various instants of time as denoted by the sequence of numbers attached to them. Briefly, after reconnection at the noon neutral point, the fields lines are swept downstream to form the long geomagnetic tail. At large distances down the tail, the fields are again antiparallel above and below the magnetotail current sheet and are able to reconnect at a second neutral point located in the midnight meridian. Fields lines outside this point such as number 8 are reconnected interplanetary fields which rejoin the solar wind. Field lines inside the neutral point, such as those labeled 8′, return to the magnetosphere. This topology is representative of a steady state and produces the convection or circulation of plasma and magnetic fields inside the magnetosphere from the tail to the nose.

From the standpoint of this model, an important aspect of the solar wind is the absence of a north-south component of the IMF when averaged over long periods of time (intervals of hours to days). The requirement that $B_z = 0$ on average is a direct consequence of the radial steady flow of the solar wind and is implicit in Parker's theory. Figure 16, which is a histogram of the latitude angle of the IMF, shows that the average and most probable value of $\delta = 0$. However, the ever-present field fluctuations result in significant deflections of the field over short intervals both northward and southward. Consequently, there are intervals during which the IMF is oriented nearly southward and fulfills the conditions described above.

It is basically the turning southward of the IMF that is responsible for magnetic substorms and aurorae (the velocity also enters, principally as the product $B_z V$, i.e., the electric field in the vicinity of the neutral point). The physics of the neutral point, electric field and merging is discussed in detail in, e.g., Vasyliunas (1975), Parker (1979a) and Priest (1984). The correlation of southward interplanetary fields with geomagnetic substorms has been demonstrated in numerous studies (see, e.g., Arnoldy 1971; Tsurutani and Meng

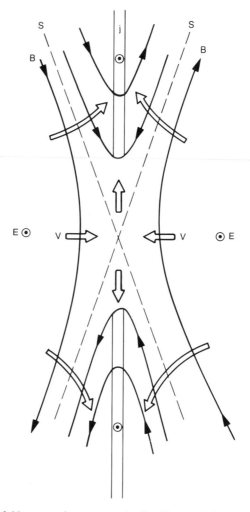

Fig. 14. Magnetic field reconnection at a neutral point. The two dashed lines pass through the x-type neutral point (where $B = 0$). They separate the fields (solid lines) into regions having four different topologies. Oppositely directed fields are swept into the neutral point from the left and right by the converging plasma flows (open arrows). The plasma diverges from the x-point and convects the reconnected field lines upward and downward. The regions across which the field directions reverse contain a current sheet (j, which is outwardly directed). An electric field ($E = -V \times B$) is present throughout the entire region. When applied to the Earth's magnetosphere, the solar wind is approaching the neutral point from the left with B representing a southward interplanetary field. The plasma approaching from the right is magnetospheric plasma with the geomagnetic field being northward. The left ends of the reconnected field lines (above and below) are in the solar wind whereas their right ends pass through the magnetosphere and reach the Earth's surface. They are called "open" field lines to distinguish them from the "closed" field lines (on the right) which begin and end at the surface. This representation is an adaptation of a figure in Dungey (1961).

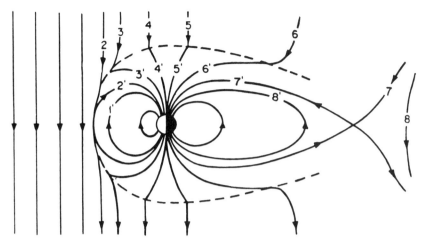

Fig. 15. Transport of field lines when the magnetosphere is "open." The IMF is seen approaching from the left in the solar wind. It merges with lines of force of the Earth's magnetic field (one of which, labelled 1', is being convected toward the magnetopause by magnetospheric plasma). A geomagnetic field line (2') is shown attached to the interplanetary field (2) according to the details provided in Fig. 14. As the IMF is then swept downstream by the solar wind, the "dipole" field lines to which they are connected (3' - 6') are also carried downstream to form a magnetic tail. Near the center of the tail, fields become oppositely directed, a second neutral point occurs and the fields are again reconnected to leave a dipole field line (8') convecting earthward and an interplanetary field (8) anti-sunward. The corresponding plasma motion involves a circulation of magnetospheric plasma with vortices in the dawn and dusk hemispheres (figure from Levy et al. 1963).

1972; Hones 1984a). A schematic of the resulting substorm (Fig. 17) shows the development of a second x-type neutral point inside the steady-state neutral point introduced in Fig. 15. Reconnection at this inner neutral point results in the formation of a plasmoid which moves rapidly tailward (perhaps with the formation of additional neutral points as the result of the tearing instability as shown) and the return of reconnected fields to the magnetosphere.

It is these latter fields containing what was previously magnetotail plasma that are responsible for substorms. The development of field-aligned currents, the result of stresses exerted on the reconnected fields, which flow down to and through the polar ionosphere, give rise to the large magnetic perturbations that characterize substorms. The plasma populating these fields is energized as the field lines shorten and their magnitude increases. The precipitation of these particles then gives rise to the aurorae.

This sequence of events, combined with the tendency for the magnetic field strength to be larger at solar maximum and for CMEs and other transients to give rise to stronger southward fields, results in correlations of the

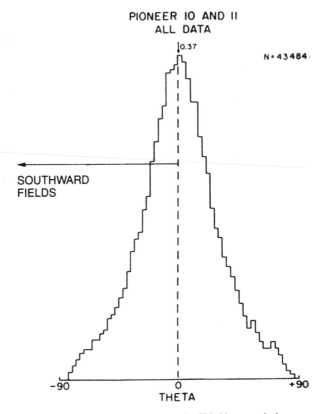

Fig. 16. Probability distribution of the IMF polar angle. This histogram is the counterpart of that shown in Fig. 2. Although there is no average component in the north-south direction (the average of B_8 over one or more days is zero), in agreement with Parker's theory, transient southward components occur regularly, some of them persisting for several hours. These transient episodes of southward fields are the primary source of magnetic storms and substorms (figure after Thomas and Smith 1980).

kind shown in Fig. 18. The upper curve represents the average number of days with aurorae made by a single observer between 1883 and 1932. The middle curve is the *aa* index which is essentially a measure of the rate of occurrence of substorms. Aurorae and substorms are well correlated with solar activity as represented by the sunspot number in the bottom curve.

IV. THE MAUNDER MINIMUM AND THE INTERPLANETARY MAGNETIC FIELD

The Maunder minimum is characterized by the absence, or near absence, of sunspots and aurorae (Maunder 1890,1922; Spörer 1889; Eddy 1980). A "block diagram" (Fig. 19) demonstrates how sunspots, as an input, can result

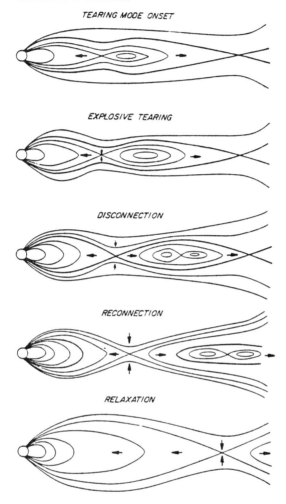

TAIL SUBSTORM DYNAMICS

TEARING MODE ONSET

EXPLOSIVE TEARING

DISCONNECTION

RECONNECTION

RELAXATION

Fig. 17. Dynamics of the geomagnetic tail during a substorm. The several panels show successive stages in the development of a substorm, a characteristic magnetic disturbance accompanying aurorae, according to one of the popular models. The top panel shows the neutral point in the distant tail discussed in Fig. 15 and the development of a second neutral point nearer the Earth. This neutral point could result from an instability of the distended oppositely directed fields or be the response to enhanced magnetic merging at the day-side magnetopause in conjunction with a southward turning of the IMF. The succeeding panels show the development of a plasmoid, bounded by the two neutral points, which is expelled tailward. Reconnected fields inside the inner neutral point return to the magnetosphere transporting plasma that is simultaneously energized and ultimately precipitated into the Earth's atmosphere at high latitudes to cause the aurora (figure from Hones 1984b).

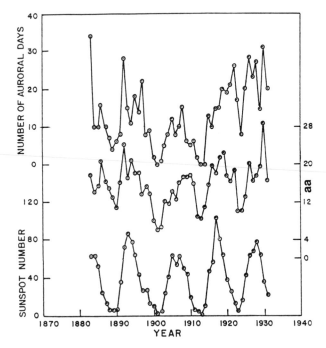

Fig. 18. Correlation between aurora, magnetic activity and sunspots over 4 solar cycles. The parameter at the top is the number of days with aurora observed each year at a single ground station. The middle panel is the *aa* index, a standard observatory measure of magnetic disturbance. Annual average of sunspot numbers from 1883 to 1931 are shown in the bottom panel. The auroral observations are the work of a single, dedicated observer (figure from Silverman and Blanchard 1983).

in auroral output. The links in this rather lengthy chain are based on our current understanding of the relation between the many physical processes involved. It makes use of many of the concepts and observations described above. Sunspots are generally thought to be manifestations of toroidal magnetic fields lying below the photosphere which are brought to the surface by buoyant forces. The toroidal fields are intimately related to the solar poloidal field by virtue of a connection that is a fundamental part of magnetic dynamo theory. The poloidal fields are large scale and extend up into the corona. They are thought to be an essential accompaniment to the heating of the corona to high enough temperatures to produce the solar wind. The other elements in the sequence then follow along the lines of the above discussion. The solar wind brings southward magnetic fields to the magnetopause, where they merge with the geomagnetic field, etc. to cause aurorae.

    The physical mechanisms in this sequence which have not yet been discussed involve the relation between toroidal and poloidal fields and coronal heating. To complete this story, dynamo theory can be described in general

CONNECTION BETWEEN SUNSPOTS AND
AURORAE

SUNSPOTS
↓
BUOYANT TOROIDAL FIELD
↓
GLOBAL POLOIDAL FIELD
↓
CORONAL HEATING
↓
SOLAR WIND
↓
INTERPLANETARY FIELD
↓
SOUTHWARD FIELD COMPONENT
↓
MERGING OF INTERPLANETARY AND
GEOMAGNETIC FIELDS
↓
FIELD RECONNECTION IN THE MAGNETOTAIL
↓
SUBSTORMS
↓
AURORAE

Fig. 19. The physical connection between sunspots and aurora. This block diagram shows the large number of links in the chain connecting the two phenomena. The physical processes referred to are all discussed in the text.

terms using Fig. 20. Panel (a) consists of a poloidal field which threads the interior as well as exiting above the surface of the Sun. Panel (b) shows how differential rotation (the ω effect) deforms the poloidal field component below the surface to form a toroidal component as in panel (c). Panel (d) demonstrates the effect of Coriolis forces on the submerged toroidal fields which, together with buoyancy, twists and lifts them simultaneously to form poloidal loops (equivalent to a current parallel to the toroidal field, the α effect). The final step is thought to involve magnetic merging of the individual loops to yield the large-scale poloidal field as shown in panel (a) (with the notable exception that the polarity of the field has been reversed). The sequence, which represents the so-called α-ω dynamo, is obviously cyclic and can be repeated, e.g., with an 11-yr period. Although many of the details are uncertain and no adequate quantitative model is available, this type of model can reproduce, at least qualitatively, the major properties of both solar and planetary magnetic fields.

Irrespective of details, a major feature of the models and observations is the interchange between poloidal and toroidal fields. The buildup and decay of the two is 90° out of phase so that the dipole field is dominant when the toroidal field is weakest (solar minimum) whereas the dipole field is weak when the toroidal field is dominant (solar maximum). Another way of stating

E. SMITH

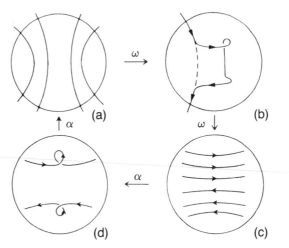

Fig. 20. Relation between poloidal and toroidal magnetic fields: the α - ω dynamo. The four panels show the interchange between global poloidal and toroidal magnetic fields that is a basic feature of hydromagnetic dynamos. The sequence is based on a specific, popular dynamo model that depends on a physical effects characterized by the parameters α and ω. Panel (a) shows the stage of the cycle when the poloidal field is dominant (normally, solar minimum). In panel (b), the sense of the poloidal field is assumed to be southward and the deformation of the field by differential rotation (the ω effect) is shown. Panel (c) shows the phase when continual stretching of the toroidal field and the rotation of the Sun have resulted in a predominantly toroidal field (normally, at sunspot maximum). The α effect shown, in panel (d), involves currents flowing parallel to the toroidal magnetic fields and producing a helicity as indicated. The final stage is merging of the meridional field loops to re-form the poloidal field. Note that the poloidal field now has the opposite polarity to that with which it started (compare panels b and d) (figure from Stevenson 1983*b*).

this relation is that the toroidal field derives from the poloidal field of the previous half cycle and vice versa.

It is widely accepted that coronal heating is magnetic and follows directly from the topology and evolution of the solar fields. Three main categories of magnetic heating have been proposed (Kuperus et al. 1981). The corona may be heated by hydromagnetic waves generated at or below the photosphere and propagating along field lines to high altitudes. An alternative proposal (not necessarily inconsistent or incompatible with the wave hypothesis) is that the heating involves current sheets that form between adjacent but oppositely directed magnetic fields. The heating might involve waves in the current sheet or the development of electric fields parallel to the magnetic field as a consequence of anomalous electrical resistivity (a collisionless plasma physical effect associated with basic instabilities). Finally, it has been suggested that the heating results from magnetic reconnection, which converts stored magnetic energy into accelerated particles, perhaps of the flux tubes that emerge spontaneously all over the Sun, observed as bright points in solar X-ray images, and then decay or otherwise disappear.

With this information and Fig. 20 as background, the question may be considered of what the solar wind was like during the Maunder minimum. We offer the following suggestion which is diagramed in Fig. 21. The left-hand side of this diagram corresponds to observations and the right-hand side to implications (which we consider plausible but not necessarily conclusive). In the interest of simplifying the description, we adopt the expedience of speaking as though no aurorae, etc. existed. In actuality, a few sunspots, aurorae, etc. were observed so the conditions we infer must be treated as intermittent and our model considered to represent extreme conditions.

The starting point in the inferences is taken to be the (relative) absence of aurorae. This fact implies the absence of magnetic reconnection. At this point, there are several obvious possibilities: (1) The supersonic solar wind may have been present but unmagnetized, or (2) a magnetized solar wind

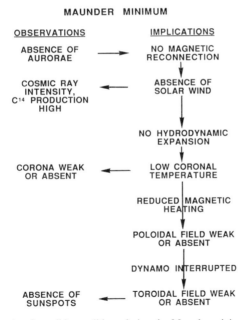

Fig. 21. A schematic of possible conditions during the Maunder minimum. The left column consists of the basic observations to be explained ("absence" of aurora, radiogenic production of carbon, absence of sunspots, possible weak or absent corona). The right column consists of a number of plausible implications drawn from the observations and inferences regarding the nature of the solar wind during the Maunder minimum. The sequence is speculative but may be testable.

may have been present but much less likely to give rise to southward fields, presumably because the solar-wind flow is much smoother and more regular. In light of Fig. 21, we argue that an unmagnetized solar wind is unlikely if the magnetic field is indispensable to coronal heating. The second alternative has been proposed before (Suess 1979) and has received support, based on new evidence (see the chapter by Jokipii). However, we proceed to develop a further hypothesis, namely, that the solar wind was absent (or intermittent) during the Maunder minimum.

The cessation of the solar wind (at least as we know it) could occur if the temperature of the corona were to be reduced. It has been calculated that, for a polytrope corona obeying an equation of state of the form $\rho \sim \rho^\alpha$ with α = 1.1, the solar wind would not expand supersonically for a reduction in coronal temperature by a factor of 3, i.e., for a temperature $< 0.7 \times 10^6$ K, at 1.43 solar radii (Hundhausen 1972).

A low temperature, perhaps substantially lower than this limit, would be possible if the heating were to be reduced significantly. Reduced heating would presumably follow from the absence (or weakening) of the poloidal/coronal magnetic field. The discussion of dynamo theory above stressed that the poloidal field is derived from the preceding toroidal field. If the latter were absent, as implied by the disappearance of sunspots, it could mean that the solar dynamo was stalled (a term introduced recently by Parker).

Although the foregoing scenario is physically plausible, it lacks a quantitative basis. However, it may be possible to explore what the coronal expansion would be like for a cooler, less-extended corona than the one with which we are familiar. Other solutions to the problem of the extension of the corona into space were investigated in the early history of the solar wind. Chapman (1957) investigated the consequences of an isothermal corona in hydrostatic equilibrium. Chamberlain (1960) emphasized the branch of the hydrodynamic solution which does not pass through a sonic transition so that the expansion is subsonic both near the Sun and far from it, i.e., a solar "breeze." These solutions were evaluated on the basis of the present corona and the supersonic solar wind was favored even before direct measurements confirmed the major outlines of the theory. A re-examination of the different possible solutions for a corona appropriate to the Maunder minimum (a significantly reduced heat flux) might be very revealing.

Additional implications which are consistent with this scenario are indicated in the left column of Fig. 21. The absence of the solar wind would result in an increase in the cosmic-ray intensity. This inference follows from the above discussion of the heliosphere and the modulation of galactic cosmic rays which would allow their penetration into the inner heliosphere in the absence of the magnetized solar wind. Based on the production of cosmogenic ^{14}C, the cosmic-ray intensity is known to have been high during the Maunder minimum (Stuiver and Quay 1981).

A drastic reduction in coronal heating would presumably also lead to a less-prominent, less-structured corona. This appearance of the corona during the Maunder minimum has been investigated and it is reported to be "pale" and without observed structure (Eddy 1980). However, observational records prior to the Maunder minimum do not mention coronal structure, so some doubt is cast on the significance of the observations at Maunder minimum. It can be said, at a minimum, that such observations as exist are not inconsistent with a corona that is much diminished compared to the present.

In conclusion, we make no exaggerated claims for our speculations regarding the solar wind at Maunder minimum. However, our hypothesis is based on plausible inferences drawn from our present understanding of the physics involved (even though it is severely limited with respect to details in some instances). In our view, what is needed at this stage of research is a number of competing hypotheses based on physical models of various kinds. The advantage of a model based on physics (as distinct from simple correlations of various phenomena which may or may not be physically related) is that it should lead to implications (predictions) that are testable. That, after all, is what the scientific method is all about.

Acknowledgment. This review was originally begun with the collaboration of A. Hundhausen. His helpful comments during the early phase of the work are greatly appreciated. Thanks are due to J. Feynman for useful discussions and for making Fig. 18 available. The research reported here was carried out by the Jet Propulsion Laboratory, California Institute of Technology, under a contract with the National Aeronautics and Space Administration.

PART II
Energetic Particles

VARIATIONS OF THE COSMIC-RAY FLUX WITH TIME

J. R. JOKIPII
University of Arizona

The physical foundations of the modulation of the galactic cosmic ray flux by the Sun are reviewed and related to heliospheric structure and dynamics. The basic physical effects—diffusion, convection, adiabatic cooling and drifts—are evaluated and shown to be all important. The results of numerical models are briefly presented and compared with observations. Present-day modulation is shown to reflect the combination of diffusion-convection effects and drift effects. The Maunder minimum is conjectured to be a period when a quiet Sun and smooth solar wind resulted in the dominance of drift effects.

I. INTRODUCTION

In the context of the study of long-term variations to in the Sun, the solar modulation of galactic cosmic rays plays a major role. This is because a proxy indicator, such as ^{14}C, which is used to infer solar behavior in the past, depends on the results of cosmic-ray interactions with the Earth and its atmosphere. Therefore, in order to relate observed variations to proxy indicators of solar variations, we must understand the physical mechanisms with which the Sun changes (or modulates) the galactic cosmic-ray flux striking the Earth. This chapter addresses the current understanding of the modulation of galactic cosmic rays by the Sun.

Figure 1 illustrates the observed quiet-time energy spectrum of cosmic-ray protons and helium at Earth. The steep dashed line is an estimate of the time-averaged solar cosmic-ray flux. Solar cosmic rays are present only sporadically, and particularly at high energies, disappear quickly after solar flares. As the amount of ^{14}C depends on the integrated flux over long periods

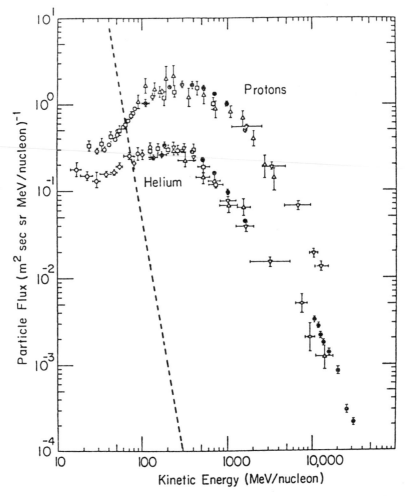

Fig. 1. Spectrum of cosmic ray protons and helium nuclei obtained during sunspot minimum (Meyer 1989). The dashed line is the approximate intensity of solar cosmic rays averaged over a sunspot cycle (courtesy of J. King).

of time, and depends on energies greater than a few hundred MeV per nucleon, it is appropriate in this chapter to concentrate on the solar modulation of galactic cosmic rays. In addition, it should be noted that the theory of the production and acceleration of *solar* cosmic rays is not yet at a state where definite predictions of the relationship of the flux to conditions of the Sun can be made. Not indicated in Fig. 1 is the unmodulated galactic flux which is present outside of the heliosphere. This flux is unknown at present because we have not yet flown a spacecraft outside of the heliosphere nor have we been able to develop theories that are sufficiently accurate to make possible accurate extrapolation from observations from within the heliosphere.

Figure 2 illustrates one major effect of the solar wind on galactic cosmic rays. Shown is the counting rate of the neutron monitor at Climax, which responds essentially to the flux of ≈ 1.6 GeV protons just outside the Earth's magnetosphere. This record of the counting rate starts in 1953, and the last three sunspot minima are shown. The 11-yr solar sunspot-cycle related variation is clearly evident. Deep minima in the cosmic-ray intensity occur at the maxima in the sunspot count. Also apparent is a possible 22-yr variation, in that the shapes of alternate maxima are flat or sharply peaked. As we will see below, such a 22-yr variation is a natural result of the present theory. From such observations, independent of specific models, we may readily infer that longer-time variations observed in various proxy indicators of the cosmic-ray flux could be produced by physical processes similar to those which cause the sunspot-cycle related variations.

However, we must first consider the possibility that the variations observed are the result of variations in the *interstellar* flux rather than variations in the Sun. Interstellar variations can be of two kinds: the Earth could pass through effectively static cosmic-ray variations in its motion relative to the nearby interstellar medium; or dynamical cosmic-ray variations associated,

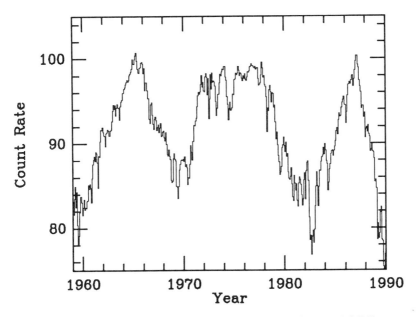

Fig. 2. The counting rate of the Climax neutron monitor (corresponding to ≈ 1.6 GeV protons) for the last three sunspot cycles (figure courtesy of R. Pyle).

for example, with shock waves in the interstellar gas could envelop the solar system. It may be shown that it is not likely that cosmic ray intensity variations observed at Earth on time scales of up to of the order of 10^5 yr can be produced by the solar system passing through quasi-static variations. For such variations to exist long enough for the motion of the solar system to bring the Earth through them, the transport of galactic cosmic rays would have to be much less rapid than is currently thought to be possible. The near-isotropy of the cosmic rays implies *diffusive* transport in the galactic magnetic field. Consider a fluctuation in the cosmic-ray density of characteristic scale L (both normal to and parallel to the local magnetic field), which would have a lifetime τ of order L^2/κ, where κ is the cosmic ray diffusion coefficient. If the solar system is moving at a speed V_E, it will take a time L/V_E to cross this fluctuation. Therefore, we require

$$\frac{L^2}{\kappa} \gg \frac{L}{V_E}. \tag{1}$$

Setting $V_E \approx 10$ km s^{-1} and $\kappa = 1/3\lambda c$, where c is the speed of light, the diffusion mean free path λ must be at least $>$ several cosmic ray gyro-radii in the interstellar magnetic field of a few μ-gauss. We find that $L \gg 3 \times 10^{17}$ cm, which would be traversed in a time $\tau \approx 10^5$ yr or more.

In view of this, the only way that short time scale (i.e.<a few $\times 10^5$ yrs) variations could be observed in the interstellar flux would be for *dynamical* processes to propagate past the Earth. This would include for example strong shock waves from a nearby supernova (see, e.g., Sonett et al. 1987). Such fluctuations will tend to occur in isolated time periods, so that, in general, recurring fluctuations on time scales $< 10^5$ yr which one sees in many cosmic-ray records probably are not the result of plausible interstellar processes.

We are left, then, with attempting to interpret the observed cosmic-ray variations in terms of physical processes associated with changes in the heliosphere. Figure 3 is a schematic illustration of a plausible heliospheric structure generated by the flow of the solar wind and its interaction with the interstellar gas. The heliospheric cavity is the result of a radial outflow of solar wind from the Sun. The wind flows supersonically out to a spheroidal "termination" shock, thought to be at some 60 to 160 AU from the Sun, where the supersonic flow becomes subsonic. The shock is followed by a gradual deflection of the flow to a direction downwind from the solar system. (Downwind here refers to the direction of motion of the interstellar plasma which confines the solar wind.) The separatrix between the solar plasma and the interstellar plasma is a contact surface which will slowly erode as the gas flows downwind of the solar system because of nonideal processes such as interdiffusion, etc. Much of this structure is speculative and spacecraft have not explored beyond $\approx$50 AU or at heliographic latitudes greater than 30°.

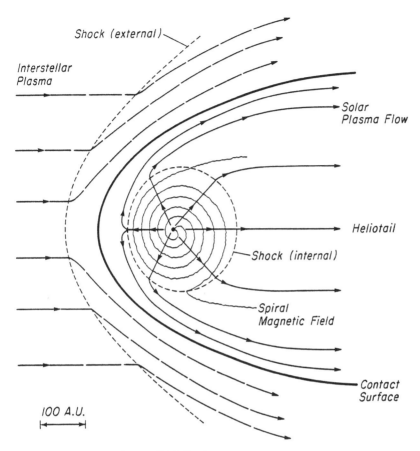

Fig. 3. Schematic sketch of a plausible heliosphere.

However, the picture is based on sound fluid dynamics, and uncertainties should not affect the general nature of the cosmic-ray modulation models discussed below.

II. THE TRANSPORT OF COSMIC-RAY PARTICLES IN THE HELIOSPHERE

The first problem before us then is to obtain a transport equation that governs the motion of fast charged particles in a magnetic field, which is convected outward with the solar wind. This magnetic field has an average given by the Archimedean spiral of Parker. Superposed on this average field are magnetic irregularities that scatter the cosmic-ray particles and cause a random walk. It is important to note that collisions of cosmic rays with the ambient plasma are completely negligible, so that the magnetic field determines the motion. This transport problem has been studied intensively over the past few decades, with the equation that is now used first devised by Parker (1965). In this approximation, particle transport is basically the combination of 4 major effects. These are:

1. *Convection:* The solar wind convects the magnetic field and cosmic rays outward at the solar wind velocity U. (The solar wind is hydromagnetic and the magnetic field is frozen in.)
2. *Diffusion:* A random walk or diffusion caused by the scattering of the cosmic rays by the irregularities in the magnetic field. This scattering produces the observed directional near-isotropy of the cosmic-ray flux. The associated diffusion tensor is denoted κ_{ij}, which can, in principle, be calculated in terms of the irregularity spectrum. In practice, it is an assumed parameter, adjusted to fit observations.
3. *Drifts:* The large-scale variations of the average magnetic field causes coherent guiding-center drifts of the energetic cosmic-ray particles which is in addition to the random walk. These are the usual guiding center drifts, which one studies in basic plasma physics problems and are well understood. The drift velocity is given in terms of the average magnetic field B_0 and the particle momentum p, speed w and charge q by $V_d = (pcw/3q)\,\nabla \times (B_0 / B_0^2)$.
4. *Adiabatic Energy Change:* In addition to the outward convection with the wind, its radial divergence or expansion $\nabla \cdot U$ causes the cosmic rays to lose energy to the wind continuously as they propagate in the solar system. It may be shown to be essentially the same as the energy change caused by the drift motions against the $V \times B$ electric field in the wind. This energy loss is a significant factor in the modulation process. In addition, where the fluid is compressed, such as at the termination shock, the particles gain energy from the flow.

These effects were first combined by Parker (1965) to obtain the generally accepted transport equation for the quasi-isotropic distribution function $f(r,p,t)$ of cosmic rays of momentum p at position r and time t. See also reviews by Jokipii (1971), Völk (1975) and Toptygin (1985).

$$\frac{\partial f}{\partial t} = \frac{\partial}{\partial x_i}\left(\kappa_{ij}\frac{\partial f}{\partial x_j}\right) \quad (diffusion)$$

$$-U_i\frac{\partial f}{\partial x_i} \quad (convection)$$

$$-V_{di}\frac{\partial f}{\partial x_i} \quad (guiding - center\ drift) \quad (2)$$

$$+\frac{1}{3}\frac{\partial U_i}{\partial x_i}\left(\frac{\partial f}{\partial \ln p}\right) \quad (energy\ change)$$

$$+Q(x_i,t,p) \quad (source).$$

The labels next to each of the various terms indicate the associated physical effect. This transport equation is remarkably general, and has been used in nearly all discussions of cosmic ray transport and acceleration over the past two decades. It appears that it is a good approximation whenever there is enough scattering by the magnetic irregularities to keep the distribution function nearly isotropic. In addition, it requires that the particles have random speeds substantially larger than the background fluid convection speed U. In particular, the velocity need not be a continuous function of position. Shocks can be discussed within the framework of this equation and all that happens is that the divergence of the velocity of the flow velocity in this part of the equation becomes a delta function.

III. SOLAR MODULATION OF GALACTIC COSMIC RAYS.

The essential problem in the solar modulation of galactic cosmic rays is to solve the differential equation (Eq. 2) by analytical or numerical methods, subject to the boundary condition that the intensity at the boundary of the heliosphere is the galactic cosmic-ray spectrum, which is assumed to be constant in time. It may be shown that the characteristic time scale for these various transport processes to bring the cosmic-ray distribution into equilibrium throughout the heliosphere is of the order of a year. Hence, the study of cosmic-ray variations on time scales of the 11-yr sunspot cycle or longer, requires only that we solve the differential equation in the static limit (neglecting the term in $\partial f/\partial t$ and then predict the intensity at various times by looking at the solution for the different magnetic-field configurations or different values of the parameters expected at different times. We have recently carried out some simulations which retained the full time dependence, and verified that it is not important in the sunspot or solar magnetic cycle variations. Some observed effects (such a differential time dependence for particles having different energies) require use of the full time-dependent equation, but this is more difficult and modulation studies have not utilized time dependence in any detail. A number of further, less well-justified approximations have been utilized over the years.

The random walk or diffusion of the cosmic rays through the magnetic field as they are convected and cooled in the expanding solar wind effectively is all that was contained in the old view of modulation which was current some ten years ago. In this picture we have the solar wind flowing outward to a boundary at some distance D from the Sun. At this point the solar wind ceases, the convection ceases, and the intensity of cosmic rays is equal to its time-invariant galactic value. The cosmic rays tend to diffuse or random walk into the inner solar system from this point, but their progress is impeded by the outward convection of the magnetic field and plasma. In addition, they are cooled in the expanding flow. This competition between the inward diffusion and the outward convection, the so called *diffusion convection/ adiabatic cooling* model, gives rise to a depressed intensity. Clearly as one increases the convection velocity V, as might happen during periods of higher solar activity, we will lower the intensity at a given point in the solar system. Similarly if we decrease the diffusion coefficient κ, corresponding to more scattering, we would also decrease the intensity in the inner solar system. Hence if one is willing to vary V or κ in an appropriate way by using this procedure, one can easily get the 11-yr solar cycle variation. However, there is no clear observational evidence for such systematic variations in V or κ which are correlated with the sunspot cycle. In addition, because the random walk does not depend on the sign of the interplanetary magnetic field, any 22-yr variation obtained from this model would have to be *ad hoc*.

An approximation to the above simplified model neglecting drifts has gained some currency, presumably because of its exceedingly simple form. It turns out that if one neglects the drift term in Eq. (2), assumes spherical symmetry, and then considers only high-energy cosmic rays, for which the dimensionless quantity rV/κ is small, a very simple analytic solution can be obtained (Gleeson and Axford 1968). The form of this solution corresponds exactly to that obtained for charged particles influenced by a force field with a potential energy given as a function of heliocentric radius r by $\Phi(r) \propto \int_r^p (V_w/\kappa)dr$. Note that this cannot be a real electrostatic potential because it affects positively and negatively charged particles in the same way. Attempts to fit the data yield values of $\Phi \approx 300$ MeV. Because of the use of an effective potential energy, this approximation is called the "force-field" solution. Although such fits are made, it is not clear what meaning to attach to Φ, because the model is so restrictive. For example, observations show a strong dependence of the cosmic ray intensity on heliographic latitude which is assumed to be absent in the force-field model. A model which includes more physics is clearly needed. We will find, below, that such a model leads to a picture in which an electric potential plays a role and which might explain why the force field model yields results that approximately fit the observations.

In order to proceed to a more correct model, we must first specify the heliospheric configuration more carefully. The large-scale structure of the magnetic field has been clarified considerably by observations carried out on the Pioneer 10 spacecraft (Smith and Thomas 1986), and the inferred relationship to observed coronal structure. However, observations are available only for regions within some 30° of the heliographic equator. It is found that during the years around each solar sunspot minimum, the field is generally organized into two hemispheres separated by a thin current sheet across which the field reverses direction. In each hemisphere the field is a classical Parker Archimedean spiral, with the sense of the field being outward in one hemisphere and inward in the other. Observations indicate that the waviness, or inclination, of the current sheet is a minimum at sunspot minimum and increases as a function of time away from sunspot minimum. At sunspot maximum the sense of the field changes sign in each hemisphere, and its structure is not simple and probably not well described in terms of a simple current sheet. The over-all magnetic field direction therefore alternates with each 11-yr sunspot cycle, so that during the 1975 sunspot minimum, the northern field was directed outward from the Sun, but in 1965 the northern field pointed inward.

In the rest of the present discussion, it will be generally assumed that the overall magnetic structure is given by this model, recognizing that it may not be a particularly good representation in the few years around maximum sunspot activity. In addition, recent work suggests that the magnetic field near the solar rotation axis, at large heliospheric radii may differ significantly from this (Jokipii and Kota 1989), but it is sufficient for present purposes to ignore this additional complication.

With the heliospheric structure established, we next examine Eq. (2) to determine whether any of the transport effects are small enough to be neglected. We can estimate the drift velocity magnitude, the gradient of the cosmic rays, and the diffusion coefficient. The results are summarized in Table I.

One finds that all four terms, convection, diffusion, energy change and drifts, are comparable in magnitude. They are all complicated and one really must do a full numerical solution to gain even qualitative insight into the solutions. The simple picture discussed above which neglects the drifts is clearly not justified.

As an example of the complication introduced by drifts, Fig. 4 shows typical drift streamlines for positively charged cosmic rays in the 1975 heliospheric magnetic field configuration. It is readily apparent that the particles drift in over the poles and are ejected very rapidly along the current sheet that separates the regions of opposite sign of the magnetic field. Negatively charged particles such as cosmic ray electrons would drift in the opposite

Table I
Size of Transport Effects for $\sim$ GeV Protons

Use the values

$V_w = 400$ km s^{-1}

$|V_D| \sim 3 \times 10^8$ cm s^{-1}

$|\nabla_n| \approx 5\%/\text{AU} = 3 \times 10^{-15}$ cm^{-1}

$\kappa \approx 10^{22}$ cm^2 s^{-1}

which yield the contributions to $\dfrac{1}{n}\dfrac{dn}{dt}$

Diffusion $\sim 3 \times 10^{-7}$ s^{-1}
Convection $\sim 3 \times 10^{-7}$ s^{-1}
Energy Change $\sim 2 \times 10^{-6}$ s^{-1}
Drift 10^{-6} s^{-1}

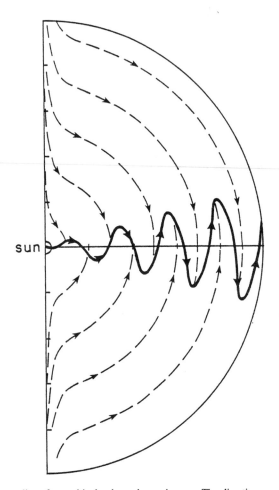

Fig. 4. Drift streamlines for positively charged cosmic rays. The directions correspond to the
1975 sunspot minimum (when the polar solar magnetic field was directed outward from the
Sun) and would be reversed in alternate minima. The heavy wavy line is the interplanetary
current sheet, along which particles drift extremely rapidly. Similarly, the arrows would be
reversed for negatively charged particles.

direction. The heliospheric magnetic field structure alternates with each succeeding sunspot minimum, so the direction of the arrows (for a given sign of
the charge) would reverse for the minima in 1985 and 1965. In current models, this particle drift turns out to be probably the most significant effect on
the particle transport in the standard model. There is no simple way to avoid
the fact that because the drifts are coherent over large distances in the heliosphere, they can be more important than the random walk and convection
terms. For the standard spiral we find that nearly all of the particles we see
in the inner solar system during the 1975 sunspot minimum came from a very

small region near the poles of the boundary. Conversely, the ones we see in 1985 have come more or less along the current sheet.

Recent numerical work has confirmed the importance of the drift terms. It has been shown that one could in fact produce the basic solar-cycle variation of the cosmic rays entirely in terms of the drifts using the observed temporal variation of the heliospheric magnetic field configuration, with *no change* in any of the other parameters. Of course, these other changes do occur, and the actual cosmic ray variation must be viewed as the combination of drift and diffusion-convection-cooling effects. In this picture, the overall, large-scale structure of the cosmic ray distribution is determined primarily by the drift motions except during the few years around sunspot maximum. Smaller-scale variations, fluctuations, forbush decreases, on the other hand, are determined by local waves and other disturbances that affect the local diffusion and other transport properties. These processes peak near sunspot maximum, when they dominate the modulation.

IV. SAMPLE MODEL SIMULATIONS USING THE FULL TRANSPORT EQUATION

The full differential equation containing all of these effects can be solved with two- or three-dimensional, time-independent heliospheric models, although proper inclusion of the termination shock appears to require a time-dependent code. The effects of helicity and cosmic-ray viscosity have not yet been shown to be significant and will not be discussed further here. A number of computer codes to accomplish this have been developed over the last several years. It is clear that we can quite naturally reproduce many of the phenomena observed in the cosmic ray flux, including many 22-yr cyclic effects (see, e.g., Kota and Jokipii 1983; Jokipii 1986; Potgieter and Moraal 1985).

Examples of agreements of the model with predictions are given in Figs. 5 and 6. In Fig. 5 is reproduced the energy spectrum given by Jokipii and Davila (1981), superimposed on the observed spectrum during the 1965 sunspot minimum. The dashed line is the assumed interstellar spectrum. Clearly, the models can reproduce the observed spectra.

Figure 6 shows a sample 22-yr solar-magnetic-cycle variation computed by Kota and Jokipii (1983). In this model, the only parameter which varies during the solar cycle is the tilt, or latitude excursion of the current sheet. The wind velocity and cosmic-ray transport coefficients are otherwise left unchanged. The part of the cycle when the northern polar field is directed outward from the Sun is labelled positive and the other half is negative. The times near sunspot maximum, when the field is disordered, and the theory is not accurate, are not plotted. One can readily see the rapid decrease of the intensity away from solar minimum, as the tilt angle increases for the negative phase, and the much weaker variation in the positive phase, that is insensitive to model parameters. This characteristic behavior of the model, which

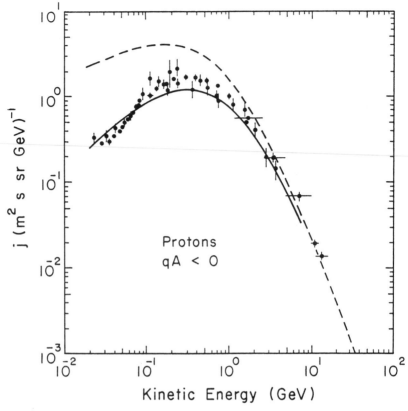

Fig. 5. Computed energy spectrum of protons compared with observations for a typical heliospheric model. The dashed line is the (assumed) interstellar spectrum at the heliospheric boundary.

is caused by the change in direction of the drifts, may be identified naturally
with the alternating sharply peaked and flat-topped cycles observed (see Fig.
1). Although only three cycles have been observed, the fact that this is a
strong prediction supports the model.

Another intriguing observation which has a natural explanation in terms
of drifts, and which would otherwise have to be explained in an *ad hoc* manner, is the correlation between the geomagnetic aa index and the observed
cosmic ray intensity at Earth. The aa index is a measure of activity in the
solar wind at Earth. The correlation was carefully studied by Shea and Smart
(1981). They found that in the several years around the 1965 solar minimum
there was an excellent negative correlation between the cosmic-ray intensity
and the aa index, as expected. However, in 1975, the next sunspot minimum,
there was a very poor and essentially insignificant correlation between the
two. This is easily explained in terms of the diagram in Fig. 4, in which the

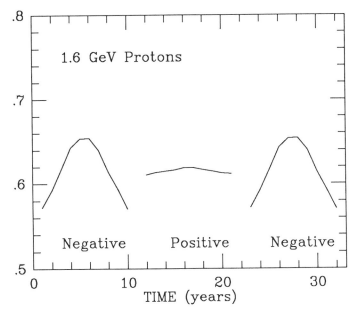

Fig. 6. Illustration of the time variation of 1.6 GeV protons during a hypothetical 22-yr solar magnetic cycle, in which only the tilt of the current sheet varies. qA positive corresponds to a period in which the polar magnetic field is outward (1975 minimum) (figure from Kota and Jokipii 1983).

sense of the drifts is indicated. Cosmic rays observed in the equatorial regions of the inner heliosphere in 1975 entered over the poles of the heliosphere, and therefore are not sensitive to fluctuations in the interplanetary medium observed at Earth (which are carried outward in the equatorial regions). On the other hand, in 1965, when the drifts are in the opposite direction, the cosmic rays come in along the equatorial regions and would be sensitive to solar wind activity in the equatorial regions. It is perhaps noteworthy that the energy loss of the cosmic rays as they drift in the electric field of the solar wind is ~ 200 to 300 MeV, and this may explain the success of the force-field model discussed above (Sec. III).

A detailed analysis, using data obtained from a variety of spacecraft during the last (1985) sunspot minimum to study both spatial and temporal variations, has verified many of the basic predictions of the theory (Cummings et al. 1987; McDonald and Lal 1986; Venkatesan et al. 1987). Of particular importance was the observation of the predicted change in sign of the latitudinal gradient in cosmic rays away from the solar equatorial plane from that observed at the previous sunspot minimum.

A summary of comparisons between theory and observations as of 1987 was presented by McKibben (1988). At the present time, it seems that the only really significant discrepancy is that the time variation of cosmic ray electrons seems to be significantly different from model predictions (see, e.g., Garcia-Munoz et al. 1986). However, it has very recently been recognized that the presence of a significant fraction of positrons (which would behave more like protons) will change this conclusion. Preliminary studies suggest that the model is in quite reasonable agreement with the data if proper allowance is made for positrons (Moraal et al. 1991). We may conclude from these analyses that there is a plausible, comprehensive model of the effects of the solar wind on galactic cosmic rays, which is based on sound physical principles. Detailed numerical simulations of the fundamental equations plus consideration of the basic physics involved has led to a picture in which both drifts and diffusion play important roles. The model accounts well for a variety of phenomena observed over the past two sunspot cycles.

In a sense then we may say that the net or total modulation of galactic cosmic rays at the present time can be regarded effectively as a combination of drift effects and diffusion-convection effects. The diffusion-convection effects can naturally produce only an 11-yr periodicity, which is the dominant variation observed today. The drifts also play a significant role in this variation, but in addition they predict significant 22-yr variations. These 22-yr periodicities are for the most part observed, indicating that these effects are also present.

As we go back further in time where we have no direct interplanetary measurements, but only have measurements which effectively give us the cosmic ray flux at the orbit of the Earth, we must interpret any variations observed in terms of the physical effects discussed above. As the model seems to indicate that two effects are significant, isolating them in the historical record can be difficult.

V. THE MAUNDER MINIMUM

As one example of the application of these ideas to pre-historical observations of the cosmic-ray flux, we should discuss the Maunder minimum. [14]C data for the period 1600 to 1730 A.D. was presented by Kocharov (1987) and is reproduced in Fig. 7. It is quite clear in looking at this record that, prior to 1640, a period of presumably normal solar activity, there is a clearly discernable approximately 11-yr periodic variation. We also know from other data that after the Maunder minimum there is a clearly observable 11-yr cyclic dependence. However, these data clearly suggest that, during the Maunder minimum (taken here to be from 1660 to 1720), the variation actually had a larger period. A power spectral analysis of these data by Sonett (personal communication) lends some support to this in that there is an indication of two periods in the data, one of which is close to the 11-yr sunspot cycle; the other period is consistent with a substantially larger period, perhaps twice as long.

In the light of the discussion in the previous section, it is attractive to

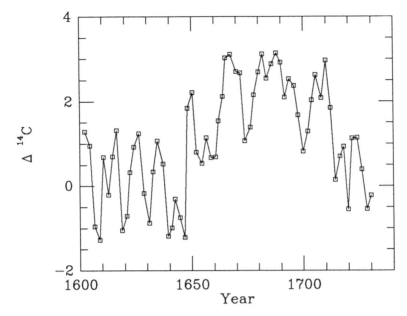

Fig. 7. The measured variation of $\Delta^{14}C$ with time just before and during the Maunder minimum (figure from Kocharov 1987).

try interpreting this variation of the intensity in terms of the drift component of the modulation during this period of the Maunder minimum, which would naturally have the 22-yr period suggested by the data. The consequences of this idea has been explored by the author in a series of time-dependent computer simulations which modeled a 44-yr period. The results are summarized in Fig. 8. Shown are two separate time histories of 1 GeV protons in the inner solar system. The dashed line shows the intensity for a normal model, in which all of the variation comes from varying the tilt of the current sheet and changing the sign of the magnetic field. Clearly apparent is the 11-yr variation with the alternating flat and sharp profiles discussed above. This behavior is consistent with the neutron monitor data for the past few sunspot cycles.

The solid line is a 44-yr period for a model in which the current sheet was assumed to remain flat at all times, but there was a change in sign of the field every 11 yr, as at present. Also, it is thought that the solar wind was extremely quiet during the Maunder minimum (Suess 1979), so the diffusion coefficients were changed to be those appropriate to a quiet wind ($\kappa_{\parallel}$ increased by a factor of 2 and $\kappa_{\perp}$ decreased by a factor of 3). It is clear that the 11-yr variation has been suppressed and the 22-yr variation is all that remains. Although this is based on a very idealized model that oversimplifies a complex situation, my conclusion is that this may provide a reasonable start to an improved understanding of the behavior of cosmic rays during the Maunder minimum.

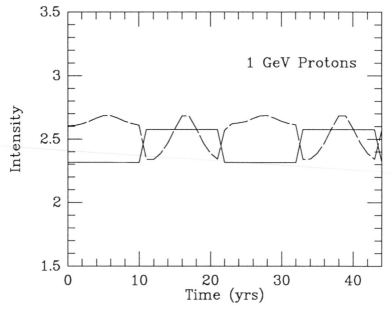

Fig. 8. Illustration of the time history of 1 GeV protons in two simulation of the solar modulation. The dashed line corresponds to the "standard" model in which the tilt of the current sheet varies to give a maximum every 11 yr (see also Fig. 6). The solid line shows the effect of suppressing the tilt of the current sheet but retaining the change in sign of the magnetic field every 11 yr, giving a maximum every 22 yr. In addition, the diffusion coefficients were changed to reflect less scattering in a presumably quieter solar wind.

VI. CONCLUSIONS

The study of the modulation of galactic cosmic rays by the solar wind has progressed to the point where the basic physical ideas appear to be in place. Most of the observations obtained in the past few decades, both from Earth and from spacecraft, seem to be reasonably well explained by a plausible model in which diffusion-convection effects and drift effects play significant roles.

Application of these ideas to ^{14}C data obtained during the Maunder minimum suggests that the expected smoother, quieter solar wind at that time would result in cosmic-ray variations that are dominated by a 22-yr variation. This is consistent with ^{14}C data obtained during the Maunder minimum.

Acknowledgments. I am grateful to C. P. Sonett for helpful discussions. This work was supported in part by grants from the National Science Foundation and the National Aeronautics and Space Administration.

HISTORY OF THE SUN DURING THE PAST 4.5 GYR AS REVEALED BY STUDIES OF ENERGETIC SOLAR PARTICLES RECORDED IN EXTRATERRESTRIAL AND TERRESTRIAL SAMPLES

D. LAL and R. E. LINGENFELTER
University of California, San Diego

Recent observations of solar particles and radiation have greatly improved our understanding of a variety of physical processes operating in both the near-surface region and the interior of the Sun. These observations include studies of solar irradiance, solar wind, and a great variety of solar energetic particles and radiation: heavier nuclei, protons, neutrons, gamma rays, X rays and neutrinos. Observations and studies of the effects of the interactions of energetic solar particles in terrestrial and extraterrestrial material have also enabled us to deduce important information on the prehistory of such solar particles all the way back to 4.5 Gyr. Here we give a brief overview of what we have learned and can learn about the history of energetic particles emitted by the Sun.

I. INTRODUCTION

Considerable progress has been made in studying the prehistory of energetic particles from the Sun. Much of this has resulted from extensive observations of the records of solar wind, energetic solar particles and galactic cosmic rays in extraterrestrial samples. The solar wind record in the most ancient samples, irradiated within 1 Gyr of the formation of the Sun, has been studied for elemental and isotopic composition, and compared with contemporary observations of the solar wind (see chapters by Kerridge et al. and by Geiss and Boschler). The galactic cosmic ray and solar energetic particles tag the early solar system materials; the nuclides produced by them are indicative of the "geometry" and "time" of their exposure to these radiations. The record can be interpreted in a meaningful manner as a result of our increased understanding of the production rates of isotopes by both galactic cosmic rays and solar energetic particles. A major contribution in recent years has come from simulation studies (Michel et al. 1989) using artifically accelerated particles in target geometries that mimic space exposure. The number of accurately determined cross sections has also grown steadily.

The solar wind is an important source of information on processes occurring above the solar surface as well as in the interior. The evolution of solar surface isotopic abundances has been deduced from studies of implanted solar wind observed in particular grains in gas-rich meteorites. These meteorites were exposed to solar wind before compaction in regolith materials on the lunar surface, and in targets exposed to solar wind on the lunar surface. A detailed discussion of the work carried out to date, and the implications of the results, are given by Geiss and Boschler and Kerridge et al. (see their chapters).

The gas-rich meteorites also contain information on the fluxes and energy spectra of solar flare-accelerated particles. This record is decipherable from studies of nuclear tracks in silicate crystals and from studies of stable noble gas isotopes produced by spallation reactions in the mineral grains. The work to date on the studies of nuclear tracks is summarized by Caffee et al. (1988) and Goswami (see his chapter), and that of solar-particle-produced noble gases by Caffee et al. (his chapter) and by Olinger et al. (1989). Comparisons of these studies with observations have been made to help understand the history of the solar nebula, important stages in the formation of planetesimals, and in the recent irradiation history of meteorites. In favorable cases, it may be possible to deduce the temporal variations in the fluxes of both the solar wind and the solar energetic particles.

The degree of confidence with which a record can be interpreted, however, is often a matter of concern. When one is dealing with a record of 4.5 Gyr duration, it can be argued that several uncertainities may creep in, e.g., the estimation of the time of exposure of grains to the radiation, partial annealing of the materials, overprint from previous exposures, etc. As our information on the nature of energetic particles emitted by the Sun improves, from contemporary observations of the Sun and Sun-like stars at different stages of evolution, we will be able to interpret better the paleorecord. Also we now have a better understanding of the evolutionary history of the meteorites, in particular on the nebular collapse time scales and temperatures (see Kerridge 1988). Thus, it should be possible in the coming years to improve considerably on the information extractable from meteoritic materials.

It is now possible to measure quantitatively the isotopic composition of several elements in solar energetic particles and galactic cosmic rays, which permits a direct comparison with records in extraterrestrial samples. The studies of isotopic compositions in the contemporary particles have been going on for more than two decades, but only recently have the data become good enough to make meaningful comparisons with other abundance determinations. Similarly, there have been significant improvements in the studies of the concentration of cosmogenic nuclides in terrestrial and extraterrestrial samples. In meteorites one would also expect to see recent solar particle irradiation effects, but these are confined to the outer few centimeters which are generally ablated during atmospheric entry. The availability of accelerator mass spectrometry has led to a tremendous increase in the number of high-precision measurements of radionuclides produced by solar energetic particles in lunar samples and by galactic cosmic rays in terrestrial samples. Long time series are available for the first time of ^{10}Be in ice cores (see the chapter by Beer et al.).

This terrestrial record is, as we shall see, a function of the cosmic ray intensity, the climatic effects on the removal of these nuclides from the atmosphere, and their storage in the upper layers of the Earth. The cosmic ray intensity incident on the Earth changes appreciably due to changes in the solar activity and thus if the ^{14}C and ^{10}Be records can be deconvolved to determine the changes in the cosmic ray intensity, we will have a useful proxy record for changes in solar activity during the past 10^4 to 10^5 yr (see chapters by Damon and Sonett and by Beer and Raisbeck).

Below we will first summarize the important results obtained to date for the various solar particles from studies of extraterrestrial sources and then discuss the terrestrial record of ^{14}C and ^{10}Be nuclides.

II. SOLAR WIND COMPOSITION

A wealth of data has accumulated on the secular changes in the elemental and isotopic compositions of solar wind. These data on He, Ne and Ar are summarized in the chapter by Kerridge et al. and also discussed in detail for He and Ne in the chapter by Geiss and Bochsler.

The magnitude of secular changes during the past 3 Gyr are well outside the errors of the measurements and there seems to be a general agreement between results from different investigations. The changes in the isotope ratios are generally quite large, $\sim \pm 20$ to 50%, except for the ^{20}Ne/^{22}Ne ratio, that is only $+3\%$. The observed changes are summarized in Table I. These changes may be primarily due to changes in the solar wind composition; or the solar wind acceleration process may have undergone a significant change with time with the degree of fractionation of the species alone having changed. Alternatively, material implanted in meteorites may represent a mixture of solar wind and another component with the proportion of these components changing with time. These issues have been discussed by several authors (cf. chapter by Kerridge et al.).

None of the observed changes can yet be understood unambiguously.

TABLE I
Changes in Isotope Ratios

Ratio	Present/Past (~ 3 Gyr) (% change)
$^4He/^{36}AR$	-50
$^4He/^3He$	-20
$^{15}N/^{14}N$	50
$^{20}Ne/^{22}Ne$	3
$^{132}Xe/^{36}Ar$	50

Geiss and Bochsler in their chapter emphasize the importance of understanding interplanetary fluxes of suprathermal ions in the energy range between 10^4 and 10^5 eV amu^{-1}. They also ask whether the major fraction of implanted ions in fine-grained lunar soils and breccia did not in fact derive from the early T Tauri phase of the Sun. All observations to date are consistent with the deduction that either the solar wind flux or the suprathermal ion flux (or both) was substantially higher in the past, even after the Sun had settled on the main sequence (see chapters by Bertout et al. and by Geiss and Bochsler).

III. SOLAR ENERGETIC PARTICLES

Long-term average fluxes and energy spectra of particles of $E > 10$ MeV, determined over different half lives from 5700 yr to 5 Myr and recent values for solar cycles 19, 20 and 21 (1954–1986) are summarized in the chapter by Reedy and Marti. There appear to be possible variations in the flux and the spectrum of the long-term emission which may have been harder than that observed during the last three solar cycles (Lal 1988a). This, if established by further observations, could be due either to a more general change in solar activity, or just to re-acceleration in the interplanetary medium in a few large events that can dominate the average, due to the statistics of a few events.

On longer time scales, the extraterresterial samples provide information only on events which happened in approximately the first Gyr period after the collapse of the solar nebular disk. A number of observations made in gas-rich meteorites have been interpreted as evidence for a more active Sun, presumably during the T Tauri phase.

IV. T TAURI PHASE OF THE SUN

There is some evidence (chapter by Caffee et al.; Olinger et al. 1989) for excess ^{21}Ne in grains in Murchison, Murray, Cold Bokkeveld and Kapoeta materials that have been irradiated with energetic solar particles. The excess

²¹Ne is interpreted as due to an early T Tauri phase of solar-flare energetic-particle emission. Existence of this excess is supported by Padia and Rao (1989) in their observations of Kapoeta and Fayetteville materials. However, this interpretation has been questioned by Weiler et al. (1989), as ²¹Ne can be produced in a few 100 Myr of galactic cosmic ray irradiation, although such irradiation time scales are difficult to accomodate within the framework of our understanding of the early evolution of the asteroidal regolith (Olinger et al. 1989). These exciting observations that must be understood so far are practically the only way we have to study the possible existence of a T Tauri stage of the Sun.

Complementary information comes from current observations (and models) of pre-main-sequence stars (see the chapter by Walter and Barry). These observations led to the discovery of the T Tauri phase in early stars and suggest that the Sun might have undergone this phase soon after its formation 4.6 Gyr ago. The so-called T Tauri wind is now considered to be the mechanism for clearing out the solar nebula during the formation of the planetary complex. Although no direct observations exist on energetic particles emitted during the T Tauri/pre-main-sequence phase, there exist data on X ray and ultraviolet emission ($\geq$ 10 eV) from pre-main-sequence stars (Vaiana 1981; Stern 1983; Feigelson and Kriss 1989). The X-ray observations could possibly be used as a proxy for the energetic particles, as X-ray emission can be indicative of large, nonthermal energy dissipation.

The far-ultraviolet and X-ray fluxes are very large, $\sim 10^7$ photon cm^{-2} s^{-1} during a 10 Myr epoch and $\sim 10^6$ photons cm^{-2} s^{-1} during a 10 to 100 Myr epoch (personal communication, G. Gahm). If such X-ray and far-ultraviolet data are indeed indicative of the energetic particles, we might expect the early Sun to be a profuse emitter of energetic particles, in addition to large fluxes of solar wind and suprathermal particles. Based on the extensive work done to date (Simon et al. 1985; chapter by Walter and Barry), one might expect (Gahm 1989) typical solar energetic particle fluxes of $\sim 4 \times 10^4$ particles cm^{-2} s^{-1} for 5% of the time during the first 10 Myr and $\sim 4 \times 10^3$ particle cm^{-2} s^{-1} for 1% of the time during the subsequent 100 Myr, for protons of > 10 MeV kinetic energy.

If so, since some solar system material must have seen these fluxes, the gas-rich meteorites and the most primitive meteoric materials should contain their records. We have no consensus at present, however, on the existence of T Tauri irradiated materials, based on analyses of meteorite samples studied so far. Future work will require studies of records from that epoch that have not undergone appreciable alterations.

V. ¹⁴C, ¹⁰Be AND SOLAR TERRESTRIAL RELATIONSHIPS

As mentioned earlier, extensive time series are available for ¹⁴C and ¹⁰Be in terrestrial samples (see the chapters by Damon and Sonett and by Beer and

Raisbeck). The terrestrial records of cosmogenic nuclides contain important information on temporal changes in the mean cosmic-ray flux incident on the Earth. Studies of changes in the temporal variations in the cosmic-ray flux, based on long-lived cosmogenic nuclides in meteorites and lunar samples, have led to the conclusion that the long-term averaged galactic cosmic-ray fluxes, over periods of 10^4 to 10^7 yr have remained essentially constant within 30%. The short-lived nuclides, ^{60}Co, ^{22}Na and ^{54}Mn do show the effects of solar modulation of the galactic cosmic-ray flux; their activities in meteorites also depend on the date of their fall. The interstellar flux of cosmic radiation would also be expected to be constant since the mean cosmic-ray age in the Galaxy has been deduced to be of the order of 15 Myr (Simpson and Garcia-Munoz 1988), which is very long compared to the mean time between supernovae events (50–100 yr). Thus the principal causes of secular variation in the solar system should be related to processes within the solar system, principally due to solar wind modulation within the heliosphere. The nature of modulation-driven variations of the cosmic-ray flux are discussed in the chapter by Jokipii. The cosmic-ray flux on the Earth would also depend on variations in the geomagnetic dipole field of the Earth (see, e.g., Lal and Peters 1967; Ramaty and Lingenfelter 1971; chapter by O'Brien et al.).

In interpreting the paleorecord of a cosmic-ray nuclide for changes in solar activity, one has also to consider the geophysical and geochemical processes involved in mixing and transfer of the nuclide from the atmosphere. To begin with, the nuclide production rates are latitude and altitude dependent, and their removal from the atmosphere is not uniform (Lal and Peters 1967). From considerations of its removal from the atmosphere, ^{14}C offers the simplest and most plausible model. The specific activity of ^{14}C (i.e., ^{14}C/ ^{12}C ratios) in the atmosphere is known to be very uniform and its transfer and mixing within the dynamic carbon cycle is fairly well understood. Further, the time-dependent record is available in the form of ^{14}C/^{12}C ratios, which filters off much of the secular variations in the exchange parameters. The nuclide ^{10}Be, on the other hand, presents severe problems; its fall-out is expected to depend strongly on meteorological factors: stratosphere-troposphere exchange, meridional mixing within the troposphere, and the rate of wet precipitation (which is the principal mechanism) of ^{10}Be from the atmosphere. Furthermore, ^{10}Be is scavenged from the atmosphere along with aerosols and here one studies the ratios ^{10}Be/water or ^{10}Be/soil, as the case may be. These ratios are climate dependent. In fact, it would be appropriate to say that ^{10}Be is an excellent nuclide for studying climatic changes in the past.

The nuclide ^{14}C, although ideal from many considerations, is not free from problems. It also responds to climatic changes directly as well as indirectly. The air-sea exchange is climate dependent, and climatic changes induce changes in the carbon-cycle parameters, in the exchange parameters as well as in the sizes of the reservoirs.

VI. THE ¹⁰BE PALEORECORD

Beer et al. in their chapter present a brief summary of the studies to date on ¹⁰Be in polar ice cores. They conclude that the ¹⁰Be record contains a faithful record of solar activity as judged by the correlations between the ¹⁴C and ¹⁰Be records. However, in the case of ¹⁰Be, it can be shown that the expected climatic changes associated with solar-activity changes would also lead to changes in ¹⁰Be concentrations in ice which would mimic the expected flux changes due to variations in solar activity (see Lal 1987). Furthermore, Lal has shown that the magnitude of changes in ¹⁰Be concentrations is much larger than estimated on the basis of expected changes in the cosmic-ray flux with changes in the solar activity. Thus, even though the ¹⁰Be time series shows a prominent secular variation which shows a general agreement with changes observed in the ¹⁴C record in tree rings, it does not necessarily follow that the variation is in large part due to changes in the cosmic-ray flux. This reservation is further reinforced by the fact that secular variations in ¹⁰Be parallel those in ¹⁸O/¹⁶O ratios; this is primarily indicative of mean changes in the global temperatures.

VII. THE ¹⁴C RECORD

A very detailed high-precision record of ¹⁴C is now available in the form of ¹⁴C/¹²C ratios in tree rings (Stuiver et al. 1986b). The record has been analyzed for principal components by several authors, including the pioneering analyses by Stuiver and Suess (for a review, see the chapter by Damon and Sonett).

The principal component, which should be attributed largely to variations in solar activity, is the 208-year periodicity, known as the Suess Wiggles. There is, however, much more to be exracted from the record: (a) possible variations in the cosmic-ray flux caused by changes in the geomagnetic field; and (b) other periodicities, caused directly by changes in cosmic-ray flux and/or climatic changes.

In the interpretation of the ¹⁴C record, one must explicitly model the variations in the dynamic carbon reservoirs, because the atmosphere comprises only 2% of the total carbon, and the ¹⁴C/¹²C ratios are fairly well equilibrated within the reservoirs; the extreme difference is only about 25%. Unfortunately, at the present stage of our understanding of the large-scale circulation in the oceans in the Holocene, it is not possible to quantitatively assess the role played by climate in modulating the atmospheric ¹⁴C/¹²C ratios. Some of these aspects have been considered by Lal et al. (1990) and Damon and Sonett (see their chapter).

In conclusion, studying ¹⁴C/¹²C ratios allows a more quantitative examination of the secular changes in solar activity, but ¹⁴C/¹²C ratios must also be examined carefully with a view to isolate direct climatic-effect-induced changes in the atmosphere.

VIII. SUMMARY

A great deal of proxy information is contained in the extraterrestrial and terrestrial samples on variations in the intensity of a variety of energetic particles in space. Clever attempts have been made to unravel the information contained in these records. However, there are two limitations: (1) the records are not continuous in time, and (2) the interpretation of the record is not unique. In the case of the extraterrestrial samples, solar data can be recovered for the first 0.5 Gyr or so after the formation of the solar system, and for the recent past, < 10 Myr before the present. The records for the intermediate time period are either not available or are too complex to interpret. In the case of the terrestrial "solar" records, the records tapped so far go back to $\sim 10^5$ yr (ice sheets), but the most useful records of solar activity are found in tree rings, which go back only to about 10^4 yr.

In interpreting the records, the situation is generally not straightforward. Some time-dependent transfer functions are usually encountered between individual processes and the corresponding records. Usually, the record is a sensitive function of the transfer function so that this record can be used to study either the solar process or the transfer function. We will qualify this general remark for the case of a few processes discussed above.

For example, we now have unequivocal evidence of secular variations in the composition of solar wind during the past 4.5 Gyr, but it cannot be asserted that this implies that the composition of the outer layers of the Sun has changed. One cannot rule out a change in the degree of fractionation in the solar-wind acceleration process since the early Sun was formed. According to the Standard Solar Model (chapter by Dearborn), solar luminosity has steadily increased from about 70% of its present value. This is not universally accepted, and there are some speculative models which predict at least the current luminosity in the past and others which predict appreciable luminosity changes over many time scales (Gilliland 1989).

Information on solar energetic particles is confined to the past < 10 Myr or during the very early period. The short-term data do not show any marked variations in either the time-averaged fluences or hardness of the spectra. Even if they did, it would be difficult to connect their observations directly to changes in a variable Sun.

The meteoritic evidence for the presence of a T Tauri phase during the early Sun is still being debated, as discussed above. Here the difficulties in interpretation seem to arise from uncertainties in the exposure history of the samples. It should be borne in mind that the meteoritic record may be conspicuously devoid of records of the main T Tauri phase because any solids irradiated during that period may have been blown out of the solar system.

However, solids irradiated during the declining phase of T Tauri may have been preserved. Thus any records of high fluxes in the early solar system should be considered as lower limits. The terrestrial record of ^{14}C and ^{10}Be seems to have a great deal of information on the solar activity, as well as influences of climatic changes. But the latter must be unambiguously separated from the former. The process is difficult and the progress to date is slow.

Thus, in a most rigorous sense, we cannot cite an example of a clear-cut solar record, but most of the records studied so far are the best that are available and future efforts will certainly be made to improve on the reading of these records.

IX. FUTURE PERSPECTIVE

More work is certainly needed to improve on the interpretation of "solar" records. The important advances in the paleontological studies of solar particles and radiation will come forth only when (1) our present state of understanding of solar processes improves, and (2) when we understand more clearly the solar-terrestrial relationships, specifically the cause-effect interplay between the solar irradiance and the character of the principal biogeochemical cycles.

Paleontology of the solar particles from the terrestrial/extraterrestrial record obviously involves interpretation of the record in terms of certain models. It is an inverse problem requiring explicit models for the history of both the energetic particles and the detectors. This is a bootstrap procedure. Clearly the more we learn about the general processes of energetic-particle acceleration and propagation at the present, the better we will be able to interpret the record of their interactions in material irradiated in the past. The more we know from contemporary observations, the better we know what to look for in the record.

An example of this comes from detailed studies (Hua and Lingenfelter 1987a,b; Ramaty and Murphy 1987) of the recent observations from the Solar Maximum Mission (SMM) of solar-flare gamma rays and neutrons made by nuclear interaction of accelerated particles in the solar atmosphere (see, e.g., Chupp 1984). These studies, reviewed in the chapter by Ramaty and Simnett, have provided new information on the acceleration, interaction and escape of energetic flare particles, which have direct bearing on the particle irradiation of solar system material. They provide new measures of the elemental and isotopic abundances in the solar chromosphere and photosphere, that can be compared with coronal and solar wind abundances to study possible temporal variation and fractionation, as discussed above.

Of particular importance is the fact that these studies show that in impulsive flares both the spectral shape and the ratio of the total number of accelerated particles (trapped in magnetic loops and interacting in the solar atmosphere to produce the neutrons and gamma rays) to the number of accelerated particles observed in the interplanetary medium vary greatly with the size of the flares (Hua and Lingenfelter 1987b). In small impulsive flares nearly all of the accelerated particles are trapped in the solar atmosphere and only a very small fraction ($> 10^{-3}$) escape into the interplanetary medium, while in the largest flare nearly all of the accelerated particles escape and they undergo further acceleration in the process, so that the energetic particles in the interplanetary medium have much harder spectra. Thus the accelerated-particle interactions in the solar atmosphere may be dominated by small flares with relatively soft spectra while the particle interactions in outer solar system material are dominated by the largest flares (Lingenfelter and Hudson 1980) with relatively hard spectra.

We already have at hand very high-quality data on the secular variations in the isotopic compositions of helium, nitrogen, neon and some other elements. In the case of Ne, data are available for both solar wind and solar energetic particles (Padia and Rao 1989; Weiler et al. 1989; Olinger et al. 1989). These solar particles involve quite different acceleration processes. Nevertheless it seems worthwhile to study the character of temporal changes in recent flares and attempt to determine if any limits can be placed on maximum fractionation effects in solar winds, guided by solar wind acceleration models (Geiss and Bochsler 1990).

The present evidence for the T Tauri phase of the Sun is shaky; nevertheless, meteorites seem to be the only probes to determine whether this phase existed after solids had formed in the solar system. It seems plausible that solids irradiated during the declining T Tauri phase are preserved in meteorites. Data are now rapidly accumulating on X-ray and ultraviolet emission (≥ 10 eV) from solar-type stars (Feigelson 1982; Gahm 1986, 1989). Further advances in this field will only come when we better understand both the broad evolutionary history of the T Tauri phase and the formation of protoplanetesimals and planetary accretion processes. Advances in these two quite different directions are necessary in order to read unambiguously the meteoritic evidence.

There are also other directions in which future work could grow. Additional new information of both the past and present Sun has come from the recent determination of the ^{3}He/H abundance in the solar photosphere from SMM observations of solar-flare gamma rays. Measurements of the time dependence of 2.223 MeV gamma-ray line flux from neutron capture on hydrogen can also provide a direct means of determining the ^{3}He abundance in the photosphere, because the (n,p) reaction on ^{3}He, which has a cross section 1.6×10^4 times that of H(n,γ), can compete effectively for the capture of neutrons. Thus, the time dependence of the 2.223 MeV gamma rays from capture on hydrogen is quite sensitive to the presence of ^{3}He, if its abundance exceeds 10^{-5} that of H. An analysis (Hua and Lingenfelter 1987a) of the SMM measurements of the time dependence of the 2.2 MeV line emission from the flare of 3 June 1982 implies a ^{3}He/H ratio of $(2.3 \pm 1.2) \times 10^{-5}$ at the 90% confidence level.

This new value of the ^{3}He/H ratio in the solar photosphere can be compared with that of $(3.4 \pm 1.7) \times 10^{-5}$, estimated (Geiss 1982) for the outer convective zone of the Sun from measurements of ^{3}He/^{4}He in meteorites and from estimates of the protosolar D/H and ^{4}He/H ratios. The previous upper bound was high enough that it could have allowed for a significant contribution of ^{3}He mixed into the photosphere by turbulent diffusion (Schatzman and Maeder 1981) from the solar interior where it can be made by deuterium burning. The present value, however, is close enough to that expected (Yang et al. 1984) solely from primordial nucleosynthesis to suggest that such turbulent diffusion does not make an important contribution to the photospheric ^{3}He abundance. Taking a photospheric ^{4}He/H ratio of 0.07 to 0.08 (see, e.g., Geiss 1982), the present determination also gives a ^{3}He/^{4}He ratio of $(3.1 \pm 1.6) \times 10^{-4}$. This is marginally lower than either the solar wind values (Geiss and Reeves 1972; Ogilvie et al. 1980) of $(4.3 \pm 0.3) \times 10^{-4}$ and $(4.7 \pm 1.2) \times 10^{-4}$, or the coronal prominence value (Hall 1975) of $(4 \pm 2) \times 10^{-4}$, suggesting a possible modification of the solar wind and coronal ^{3}He abundance by some processes of ion fractionation.

The studies of Castagnoli et al. (1984) seems to provide evidence for the presence of 11-yr periodicity in the ^{10}Be record in ice, and of solar activity-like cycles in the thermoluminescence data in some lake sediments. These variations are clearly not directly due to changes in the cosmic-ray flux, but a consequence of the solar activity variation-induced climatic change. These changes must be studied in detail with a view to learn how climate affects the principal biogeochemical cycles.

Thus, even though we have such records that may enable us to understand the long history of solar activity, their interpretations at the present time are still clouded with considerable uncertainties.

Acknowledgments. This work was supported by grants from the National Science Foundation. We thank J. Goswami and R. C. Reedy for helpful criticisms.

ACCELERATED PARTICLES IN SOLAR FLARES

R. RAMATY

Goddard Space Flight Center

and

G. M. SIMNETT

University of Birmingham

There is considerable evidence that the Sun has been a source of energetic particles over much of its life. It is of current interest to understand the ways in which this might be achieved in the present Sun. We first review the modern evidence from space-borne instruments for energetic solar-particle emission. We then relate this to the fluences at the Sun by detailed examination of the gamma-ray, hard X-ray and radio emissions produced at the time of energetic solar flares. Possible acceleration mechanisms are reviewed, together with interaction models of both relativistic and nonrelativistic ions and electrons in the solar atmosphere. The correlation between the particle populations escaping from the Sun and those actually accelerated in the solar atmosphere is dependent on the trapping, propagation and escape from the corona. Different acceleration phases and their relationship to other transient phenomena such as coronal mass ejections are discussed. We also compare the current average rate of accelerated ion production in solar flares with the ion production rates at earlier phases in the evolution of the Sun suggested by flare activity on T Tauri stars and isotopic anomalies in meteorites.

Evidence for particle acceleration at the Sun comes from the direct detection of the accelerated particles in space and in the Earth's atmosphere, and from the observation of a variety of neutral radiations, in particular gamma rays, X rays, radio emission and neutrons, which are unambiguous signatures of accelerated particle interactions. In this review we discuss the current state of the observations of charged particles; what their signatures are in the associated flares; how they may be accelerated and how they interact in the solar atmosphere; and what the observations tell us about coronal trapping, chemical abundance and mass ejections. We also evaluate the current average rate of accelerated ion production in solar flares and compare it with ion production rates at earlier phases in the evolution of the Sun. There is evidence from meteorites (Caffee et al. 1987) for intense activity above ~ 10 MeV in the early Sun. They have argued that the Sun, in its early stage of evolution, was as active as the T Tauri class of stars, which are characterized by strong stellar winds. The meteorite evidence points to proton fluxes > 10 MeV several orders of magnitude higher than current levels, which is consistent with the ratio of the optical flare luminosity on T Tauri stars to the present average solar-flare optical luminosity. Thus, it would appear that the present Sun is quite benign compared to its early life.

I. REVIEW OF THE OBSERVATIONS

During active times, the Sun is capable of accelerating ions occasionally to energies of 15 GeV/nucleon, and electrons to energies up to ~ 100 MeV. It is difficult to detect, in space, particles above these energies, so it remains to be established if these energies are close to the real limit of the solar accelerator. The first solar particles to be detected, in the 1940s, were highly relativistic ions observed with groundbased detectors. The generation of relativistic charged particles is spasmodic and is loosely correlated with the sunspot cycle. The large energetic events are almost invariably associated with solar flares; it is widely believed that any apparent exceptions are associated with flares on the backside of the Sun. The production of subrelativistic particles, ions up to ~20 MeV/nucleon and electrons of up to ~50 keV, is quite different. Such particles are readily produced outside of flares, and at lower energies still, the Sun is a quasi-continuous source of nonthermal particles. Traveling interplanetary shocks accelerate ions readily to 1 MeV. In this section we review the observations.

A. Particle Observations in Space

For several decades, observations from space have recorded in detail the emission of charged particles from the Sun (see, e.g., McDonald et al. 1974). Both the intensity fluctuations with time of ions and electrons, and their energy spectra have been studied. The chemical and isotopic abundances of the ions have also been investigated. The most energetic particles are seen following large solar flares; however, the size of the Hα flare is only loosely correlated with particle emission. Also, there are times when the accelerated particles appear to be temporarily trapped at the Sun, in which case the subsequent event seen in space has a quite different time history from the direct, impulsive event.

Direct observations of an impulsive particle event from well-connected flares from the visible western hemisphere of the Sun show that the differential energy spectrum of the protons in the energy range from about 20 to 80 MeV follows a power law (Van Hollebecke et al. 1975). But over a broader energy range, in many cases, the observed spectrum deviates from a power law and can be better fit with a Bessel function or an exponential in rigidity (McGuire and von Rosenvinge 1984). The space measurements of solar protons do not extend above $\sim$ 1 GeV, but a few events/solar cycle produce a response at low-latitude groundbased neutron monitors, indicating that the spectrum occasionally extends to beyond 10 GeV. In addition to protons, α particles and heavier nuclei are also observed. The relative abundance of these particles varies from flare to flare. The most dramatic variation is in the ^{3}He abundance that may vary by at least 3 orders of magnitude (Ramaty et al. 1980; Kocharov and Kocharov 1984). Enhancements in the abundance of heavy ions are also seen during ^{3}He-rich flares (see, e.g., Reames 1990).

The Sun generates intense MHD shocks which are usually accompanied by protons up to around 10 MeV. However, their role in particle acceleration is still not completely clear as Van Ness et al. (1984) showed that around half of the shocks are not accelerating 10 MeV protons at 1 AU. But there is no doubt that at times such shocks are an important source of protons in the MeV region, and that they may be responsible for much of the quasi-continuous solar emission at these energies. Some, but not all, shocks are associated with Hα flares.

Electron emission has been studied over 8 orders of magnitude in energy; a summary of space observations is shown in Fig. 1 (Evenson et al. 1984; Cane et al. 1986; Moses et al. 1989). The majority of electrons observed below 10 MeV are solar, although there are small fluxes of relativistic electrons from Jupiter. As with the ions, solar flares generate impulsive events, and the fastest events have a rise time to maximum intensity of $<$ 1 hr; proton events are somewhat slower. Occasionally the spectrum extends up to 100 MeV. The energy spectrum usually is a double power law, with a break at around 100 to 200 keV. Typical spectral exponents are -1.3 below 200 keV and -3 in the relativistic region. For impulsive flares (as defined from soft X-ray observations; Pallavicini et al. 1977), the relativistic electron spectrum flattens around $\sim$ 1 MeV (Moses et al. 1989).

Other impulsive electron events are those associated with type III radio bursts (see Sec. I.D). These travel without scattering along the Archimedean spiral lines and typically have energies of a few tens of keV (Lin 1985). They not only have fast rise times, but also fast decays, and are consistent with an impulsive acceleration at the base of open, coronal magnetic field lines. An unexplained accompaniment to these events are the ^{3}He enhancements

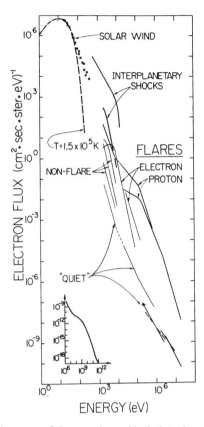

ENERGY (eV)

Fig. 1. A summary of the spectra of electrons observed in the interplanetary medium (figure after Lin 1985).

(Reames et al. 1985); they are almost always associated with impulsive, scatter-free low-energy electron events. On the other hand, not all of the low-energy electron events are accompanied by ^{3}He enhancements, but this effect could simply be due to the lower sensitivity of the detectors to ^{3}He fluxes rather than to low-energy electron fluxes. Electron events are associated with the MHD shocks referred to above; however, these do not appear to accelerate electrons beyond around 20 keV, and only exceptionally to this energy.

There are also long-duration events that occur just a few times over the 11-yr solar cycle. Fig. 2 illustrates an example of this type of event which occurred in 1979. The flux of relativistic electrons ($\gamma = 2$ to 10) increased over a period of several days to 3.5 orders of magnitude above the quiet-time level, and then gradually decayed over the next month. The primary source of the particles is unquestionably the Sun, but what fraction of the electrons represent continuous solar emission, and what fraction are merely diffusing in the inner solar system, is an open question.

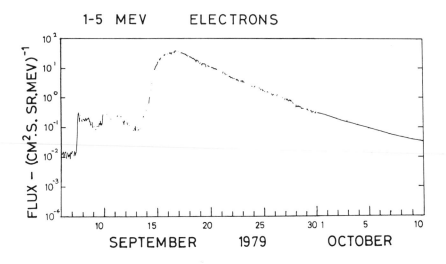

Fig. 2. The long-duration relativistic electron event in September-October 1979, which lasted over one solar rotation (figure after Simnett 1986*a*).

B. Gamma-Ray and Neutron Observations

Gamma-ray lines from the Sun were first observed from two flares in 1972 with a spectrometer on the seventh Orbiting Solar Observatory (OSO-7; Chupp et al. 1973). Several other observations followed, but it was not until 1980 that routine observations of gamma-ray lines and continuum from solar flares became possible with the much more sensitive gamma-ray spectrometer (GRS) on the Solar Maximum Mission (SMM; see e.g., Chupp 1984). Important additional observations were also carried out with a spectrometer on the Japanese satellite Hinotori (Yoshimori et al. 1983). Neutrons escaping from the Sun were observed with the SMM/GRS detector, as well as with groundbased neutron monitors (see Chupp 1988*b*). Protons resulting from the decay of neutrons in interplanetary space were observed with a charged-particle detector on ISEE-3 (Evenson et al. 1983). The gamma-ray and neutron emissions have provided new insights into the physics of particle acceleration and transport (Sec. II.A), and afforded new methods for determining elemental and isotopic abundances in the solar atmosphere (Sec. III). The gamma-ray observations can also be used to calculate the irradiation rate of the Sun due to flare-accelerated ions (Sec. II.B).

Gamma rays and neutrons result from nuclear reactions of flare-accelerated protons and nuclei with ambient gas in the solar atmosphere (see Ramaty and Murphy 1987). The principal mechanisms for the production of gamma-ray lines are nuclear de-excitation, neutron capture and positron annihilation. The nuclear de-excitation spectrum consists of narrow lines, resulting from protons and α particles interacting with the ambient gas, and broad lines, resulting from the interaction of accelerated carbon and heavier nuclei with ambient hydrogen and helium. Neutron capture on ambient hydrogen in the photosphere produces a strong line at 2.223 MeV, and neutrons escaping from the Sun can be detected at Earth or in interplanetary space. Nuclear reactions also produce positrons, which annihilate to produce a line at 0.511 MeV. The narrow nuclear de-excitation lines, and the 2.223 and the 0.511 MeV lines are superimposed on a continuum composed of the broad nuclear de-excitation lines and bremsstrahlung from primary electrons. Broadband gamma-ray emission extending to high energies ($\sim$ 100 MeV) results from the decay of pions, which are also produced in the nuclear reactions. Neutral pions decay directly into gamma rays, while charged pions decay via muons into secondary positrons and electrons which produce gamma rays by annihilation in flight and bremsstrahlung (Murphy et al. 1987).

These observations have demonstrated that protons, heavier nuclei and relativistic electrons are accelerated in many solar flares, not just in the most energetic ones. Moreover, the acceleration of these particles is very impulsive. There are flares for which the gamma-ray line and hard X-ray emissions are coincident within the temporal resolution of the SMM/GRS detector ($\sim$ 2 s). The hard X-ray emission, most probably nonrelativistic electron bremsstrahlung, is the hallmark of impulsive energy release in flares (see, e.g., Dennis 1985; see also Sec. I.C). The temporal coincidence of the X rays and gamma rays implies that ion and relativistic electron acceleration is also intimately related to the impulsive energy release. The gamma-ray observations allow the accurate calculation of the total energy content in protons of energies greater than several MeV. This energy amounts to at least several percent of the total flare energy (Ramaty 1986), showing that ion acceleration is an important ingredient in the overall flare process.

Gamma-ray emitting flares are observed from sites located predominantly near the limb of the Sun (see, e.g., Rieger 1989). This effect was observed for flares detected at energies > 0.3 MeV, but it is at energies > 10 MeV that the effect is particularly pronounced (Fig. 3). Since in both of these cases the bulk of the emission is bremsstrahlung from primary electrons, these results imply that the radiating electrons are anisotropic. As demonstrated recently (Miller and Ramaty 1989), the anisotropy could result from the mirroring of the charged particles in the convergent chromospheric magnetic fields.

The solar gamma-ray emission exhibits a complex spectrum, with nuclear lines observed at the energies expected from de-excitations of the most abundant constituents of the solar atmosphere. An example, observed (Forrest 1983) with the SMM/GRS detector, is shown in Fig. 4. The lines of ^{16}O at 6.129 MeV, ^{12}C at 4.438 MeV and ^{20}Ne at 1.634 MeV are particularly well

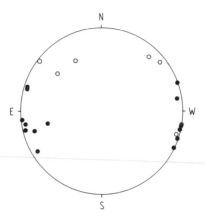

Fig. 3. The location on the solar disk of flares observed in > 10MeV gamma rays (figure after Rieger 1989).

defined. The ^{56}Fe, ^{24}Mg and ^{28}Si lines at 0.847, 1.369 and 1.778 MeV, as well as the neutron capture and positron annihilation lines at 2.223 and 0.511 MeV, are also quite evident. The feature just below the positron annihilation line results from reactions between accelerated α particles and ambient He nuclei leading to excited states of ^{7}Li and ^{7}Be. The smooth curve is the best fitting theoretical spectrum obtained from a model calculation using compiled and directly measured line-emission cross sections. Abundances in the ambient medium were derived (Murphy et al. 1985) from this spectrum (see Sec. III).

The 2.223 MeV line resulting from neutron capture by protons in the photosphere is a probe of the photospheric ^{3}He abundance (Wang and Ramaty 1974). This is because nonradiative capture on ^{3}He competes with capture on H, allowing the determination of the ^{3}He abundance from observations of the time-dependent 2.223 MeV line flux (Hua and Lingenfelter 1987*b*). While the 2.223 MeV line is produced predominantly by neutrons of energies < 100 MeV, at higher energies GeV neutrons are observed at Earth with groundbased neutron monitors (Debrunner et al. 1983). These high-energy neutrons are produced by GeV protons, and hence provide information on the acceleration of the highest energy solar flare particles. Independent information on the GeV protons is provided by the observations of gamma rays from pion decay (Forrest et al. 1985). The pion and neutron observations indicate that protons can be accelerated to hundreds of MeV in < 10 s and to several GeV in < 1 min. But the combined analysis of the neutron and pion data indicates that the proton spectrum cannot continue as an unbroken power law to energies much in excess of a GeV (Murphy et al. 1987). This high-energy turnover probably reflects the limits of the accelerator.

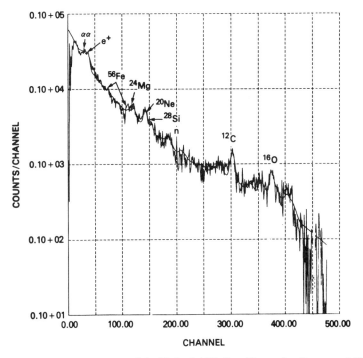

Fig. 4. The gamma-ray spectrum of the 27 April 1981 flare (figure after Ramaty and Murphy 1987).

C. Hard X-ray Observations

Hard X-ray observations made in 1959 (Peterson and Winkler 1959) first indicated how important this wavelength interval would be in the detailed study, analysis and interpretation of solar flares. At current sensitivity limits, it is only from flares that X-rays above 20 keV are seen, so that at other times the Sun may be regarded as a quiet, non-X-ray-emitting star. Solar hard X-ray observations have been made over a significant fraction of the last three decades, generally without spatial resolution. Around the 1980 solar maximum, observations were made with $\leq 10''$ spatial resolution, giving us for the first time detailed information on the structure of flares at nonthermal energies.

As might be expected, the incidence of hard X-ray flares is well correlated with the sunspot cycle. However, a more subtle periodicity of ~ 152 to 158 days has been reported by Rieger et al. (1984) in the gamma-ray data and this has been confirmed by hard X-ray (Dennis 1985), microwave (Bogart and Bai 1985) and H α (Ichimoto et al. 1985) observations. The size spectrum of X-ray events above ~ 30 keV has been measured with the Hard X-Ray Burst Spectrometer (HXRBS) on SMM over the 5-yr period February 1980 to February 1985, covering the decline from maximum of the last solar cycle. This has the form

$$n(P) = 100 \, P^{-1.8} \text{ flares day}^{-1}(\text{counts s}^{-1})^{-1} \qquad (1)$$

where $n(P)$ is the differential rate per day of flares detected at a peak counting rate P measured in counts s^{-1} (Dennis 1985). If we integrate this above $P = 1000$ counts s^{-1}, we find that the average rate of occurrence of such flares is 0.55 day^{-1}; since this covers the entire period from solar maximum to minimum, this rate is approximately the long-term average rate of occurrence.

The energy spectrum of solar X-ray bursts varies not only from event to event but also throughout an event (see Fig. 5). For the impulsive bursts, the spectrum tends to be soft at the start of the burst, harden around the time of maximum intensity, and then becomes soft again during the decay. The differential number spectrum can conveniently be expressed as a power law in photon energy. At the time of peak intensity the spectral index may vary from flare to flare, from -3 to steeper than -7. In flares with other signatures of energetic particle production, such as gamma-ray and strong microwave emission, the X-ray spectrum is typically between -3 and -4. The first observation of a hard X-ray burst with a high-energy-resolution Ge spectrometer was made by Lin et al. (1981). They recorded a spectral index at the peak of the event above 30 keV of around -3, but in the decay phase the thermal spectrum in the 13 to 30 keV region was equivalent to E^{-11}. Thus it would appear that the X-ray emission resulting from the presence of energetic charged particles (electrons) has a spectral index around -3 to -4, while the steeper spectra are more likely due to thermal bremsstrahlung from heated flare plasma.

For the long-lived, slowly varying flares, the energy spectrum starts off soft and gradually hardens with time. Such flares have source locations high in the corona ($> 4 \times 10^4$ km) and are almost certainly associated with energetic particle events and/or coronal mass ejections.

Turning to the spatially resolved measurements of hard X-rays, SMM data showed that at the onset of the impulsive phase, the hard X-ray emission is from the footpoints of a magnetic loop (Hoyng et al. 1981; Duijveman et al. 1982; Simnett 1983). One advantage of spatially resolved observations over a wide spectral range is that emissions from different parts of the flare region can be separated unambiguously and the different magnetic structures participating in the flare development can be identified. One of the surprising results from these analyses is that the flare does not just involve a loop, or an arcade of loops, but that a whole hierarchical structure of magnetic loops is involved, often stretching to an adjacent active region. Figure 6 shows an example of this from a flare on 5 November 1980 (Dennis 1985). A contour plot from the Hard X-ray Imaging Spectrometer (HXIS) on SMM shows the

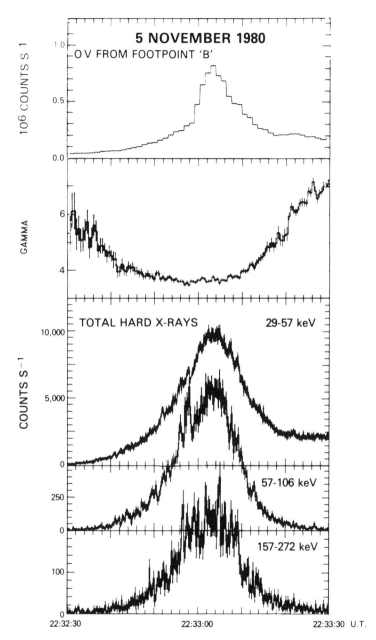

Fig. 5. Intensity-time profiles for the hard X-ray burst covering the onset of the 5 November 1980 flare observed with the Hard X-Ray Burst Spectromoter. Also shown in the evolution of the spectral index γ of the X-ray burst, and the intensity-time profile of the ultraviolet O V emission from close to one of the flare footpoints shown in Fig. 3a (figure after Dennis 1985).

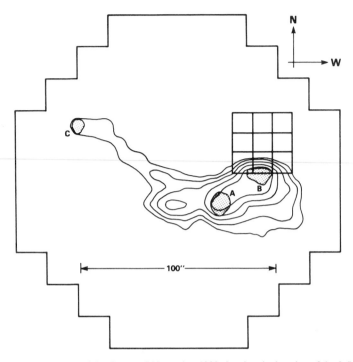

Fig. 6. The X-ray image of the flare on 5 November 1980 showing the location of the 3.5 to 8.0 keV emission (solid contours) and the 16 to 30 keV emission (shaded). The 3 × 3 array of squares is the area covered by the ultraviolet O V emission shown in Fig. 3b (figure after Dennis 1985).

extent of the 3.5 to 8.0 keV emission while the shaded regions labeled A,B and C are the sites of the 16 to 30 keV X rays. Figure 5 shows the hard X-ray intensity-time profile from the same flare in several energy bands, together with the evolution of the spectral index γ. In the upper panel of Fig. 5 is the intensity-time history of the OV line from the Ultraviolet Spectrometer and Polarimeter on SMM; the field of view is given by the 3 × 3 area of pixels to the north of the bright X-ray footpoint B in Fig. 6. Although a detailed discussion of this event is beyond the scope of this chapter, briefly, the event started in a compact magnetic loop with footpoints at A and B; particles accelerated in the primary energy release were guided by a much larger loop linking the loop A-B to point C; hot plasma generated in the loop A-B expanded upwards into the larger loop where it was detected through the thermal X-ray emission generated by plasma at temperatures between 10 and 20 × 10^6K. The ultraviolet observations, plus others at different wavelengths, help us to understand details of the evolution of the event, especially with regard to the interaction of flare-produced energetic particles.

One very important diagnostic which places important boundary conditions on the generation of energetic particles is the observation of high-temperature plasma upflows from the chromosphere at the same time as, or in some events slightly before, the onset of the hard X-ray burst (Antonucci et al. 1982,1984). The X-ray line spectrometers can identify the velocity of the upflows via Doppler line shifts, while the imaging instruments can track the ablation of the chromospheric plasma into low coronal loops. Occasionally, in events where the density in the corona becomes significantly enhanced while the energy release and particle acceleration is still occurring, even the hard X-ray source can shift to the corona. Figure 7 illustrates a good example of this phenomenon from an event on the solar limb; the HXIS images shown at 22 to 30 keV are in the chromosphere at the onset of the flare, and as the coronal loop fills with dense ablated plasma the energy can no longer penetrate to the chromosphere so that the hard X rays are generated in the corona. It appears that not all events have a phase where the hard X-ray emission is in the chromosphere, which suggests that during the very early phase of some flares the ablation of chromospheric material into the corona may take place at relatively low temperatures. The Hinotori spacecraft recorded images of hard X-ray flares from February 1981 to September 1982. However, there were relatively few events where the flares exhibited hard X-ray emission from the loop footpoints; instead, a significant fraction of the events observed by Hinotori were gradually varying hard X-ray events from the corona.

D. Radio Observations

The Sun is a highly variable radio source from the microwave to decametric part of the spectrum. The first comprehensive attempt to define and interpret these emissions was made by Wild et al. (1963) who recognized that the radio data gave important information on energetic-particle production. In particular, the radio observations address the energetic electrons, because it is these that are responsible, in the main, for radio emission.

The Type III radio bursts, which are interpreted as plasma oscillations excited by beams of electrons traveling out from the Sun along open magnetic field lines, are the most common radio transient. From the frequency drift rate of the burst, the velocity of the exciting electrons may be inferred, since the plasma frequency is proportional to the square root of the ambient electron density. For most bursts the electron exciter moves with a velocity 0.3 to 0.7 c (where c is the velocity of light), corresponding to electron energies in the 30 to 200 keV region. These electrons normally propagate without appreciable scattering into the interplanetary medium (see Sec. I.A); however, they are constrained to follow the magnetic field lines and a small fraction of them propagate on closed coronal field lines, whereby they return to the Sun. The beam of electrons needs to develop a suitable velocity distribution before the plasma oscillations are excited; therefore it is quite likely that

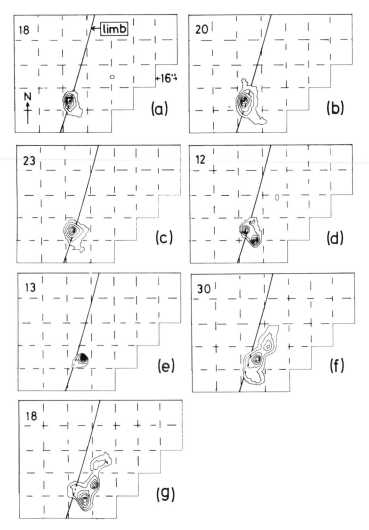

Fig. 7. 22 to 30 keV X-ray contour maps of the evolution of the limb flare on 18 November 1980. The corona is to the right. The time periods go sequentially from frame (a) to (g). Note how the hard X-ray source suddenly shifts into the corona after frame (c) (figure after Simnett and Strong 1984).

many of the beams emitted onto closed coronal field lines are never observed. Ion beams are rarely associated with Type III bursts.

Type II radio bursts were originally thought to be the signature of rela-tivistic charged-particle production. They are slowly negative-frequency-drifting bursts which are also attributed to plasma radiation and they typically start several minutes after the impulsive phase of a flare. The exciting agent has a typical speed in the 500 to 1000 km s^{-1} region, which is generally in excess of the Alfvén velocity in the corona. It is, of course, much slower than the velocity of nonthermal particles. The most likely explanation is a collisionless MHD shock wave moving out through the corona. Many events have fast-frequency-drifting bursts to both higher and lower frequencies—herringbone structure—that are presumably short bursts of shock-accelerated electrons moving away from the shock front (cf. Type III bursts). Unlike the Type III bursts, which may start at decimetric frequencies, the Type II bursts usually originate at frequencies below ~ 150 MHz, indicating that they originate in the corona. The notion that the Type II burst was responsible for energetic-particle acceleration stemmed from the unfortunate tendency for a large solar flare to produce such a plethora of emissions that everything was correlated with everything else. It was therefore understandable that when considering a large flare in isolation, researchers seized on the concept of a strong, outwardly propagating shock as the accelerating agent for the energetic particles.

We now know that this is only partially true. Dodge (1975) first showed that most Type II bursts were only associated with small flares and Wright (1980) quantified this analysis; ~40% of Hα flares associated with Type II bursts were subflares, and ~40% were class 1. Neither flare category is customarily noted for energetic-particle production, although it should be pointed out that our understanding of solar-particle acceleration to modest MeV (proton) levels is rather poor. Now that we have the recent results (see Sec. I.B) of the timing of gamma-ray emission relative to the impulsive phase, it is clear that in gamma-ray flares relativistic ions are present within seconds of the onset of the hard X rays, well before any Type II emission. This then removes the need to interpret the Type II exciter as the relativistic-particle accelerator in the impulsive phase of solar flares. The existence of two classes of flare-acceleration phenomena is discussed in Sec. II, and Type II radio emission may still be a signature of shock acceleration responsible for the second class. Indeed, from a careful analysis of the broader topic of coronal mass ejections, Type II bursts and charged-particle events, it is the events which propagate well into the interplanetary medium that are likely to be accompanied by energetic particles.

The impulsive microwave emission is the most compelling radio evidence for the presence of energetic, even relativistic, electrons in flares. Several decades ago, it was realized that the intensity-time structure of the impulsive hard X-ray and microwave bursts were very similar. The microwaves are usually attributed to gyrosynchrotron radiation from mildly relativistic electrons trapped in a magnetic field. Since hard X-ray bursts require similar populations of electrons, it is not too surprising that the time structure is similar.

The time resolution of the two data sets is now good enough that we know that the microwaves are delayed slightly from their X-ray counterparts (Cornell et al. 1984; Costa et al. 1984). At this time, a satisfactory explanation for this is still being sought. One very simple solution is to allow the electron acceleration time to increase with increasing energy. Then one would expect the 30-keV X rays to precede the microwaves, as the latter probably require electron energies in excess of 100 keV. Another possibility is that the electrons are produced in the chromosophere where the X-ray production is relatively efficient but the microwave emission is rather inefficient. Then the bulk of the associated microwaves must wait until some of the electrons have propagated to the top of the coronal loop, thus producing a slight delay. Yet another explanation (Cornell et al. 1984) calls for the X rays to be produced by electrons propagating along the coronal loop with close to zero-degree pitch angle, while the microwaves are delayed by the time taken for the remainder of the trapped electron distribution to accumulate at the loop top. Spatially resolved observations are helping to resolve this issue. Kundu (1985) has reviewed measurements made with the Very Large Array (VLA) in New Mexico and his conclusions are that the nonthermal microwave emission comes generally from the flare-loop footpoints while there is delayed, but thermal, emission from the loop top (cf. ablation of hot chromospheric plasma).

Following a large flare there is generally a microwave continuum generated, which is accepted to be gyrosynchrotron radiation from electrons energized in the flare and trapped in the corona. Such bursts are referred to as stationary Type IV, and they may extend from microwave to decametric wavelengths. A Type IV burst typically starts at the highest frequency and gradually extends to metric wavelengths on a time scale comparable to the drift of a Type II burst. Thus, they are consistent with an electron-source population slowly moving into a region of lower magnetic field, and possibly losing energy at the same time. Type IV events are always flare associated, although subsequently they may persist for days.

II. PARTICLE ACCELERATION AND INTERACTION MODELS

It is now quite certain that a significant fraction of the total energy of a solar flare appears in the form of accelerated particles. This energy, which ultimately must be derived from subphotospheric mechanical energy (e.g., differential rotation and convection), is most probably stored in magnetic fields. Rapid reconnection of magnetic fields, probably in the corona, is thought to result in the acceleration of charged particles, but the nature of the dominant mechanism or mechanisms is not well understood. Closely tied to the problem of acceleration are the problems of transport of particles at the Sun and energy deposition in the solar atmosphere, and the escape of the particles to interplanetary space. We discuss these problems below.

A. Acceleration and Transport in Solar-Flare Loops

We first provide arguments for the acceleration and transport in solar-flare loops, emphasizing ions above several MeV and relativistic electrons.

A comparison of the number of particles that escape from the flare with the number that remain and interact at the Sun can be made using gamma-ray line and interplanetary charged-particle observations. Because the line emission is emitted essentially isotropically, independent of the angular distribution of the interacting particles, the number of interacting protons of energies greater than several MeV (the line production energy threshold) can be calculated with an accuracy of better than a factor of 2. This calculation is much more reliable than calculations of the number of electrons at the Sun obtained from either gamma-ray continuum observations or radio observations, since these are very strongly influenced by the unknown angular distribution of the electrons, and, in the case of the radio data, by complicated absorption effects.

The comparison of the number of interacting and escaping protons (the latter derived from the interplanetary observations to an accuracy of about an order of magnitude) shows that for gamma-ray flares with impulsive hard X-ray time profiles, the interacting particles are generally more numerous than the escaping ones (see, e.g., Ramaty and Murphy 1987). This implies that the protons are efficiently trapped and forced to interact at the Sun. Because of the well-known existence of magnetic loops in the solar atmosphere (Sec. I.C), it is reasonable to assume that the particles are accelerated in loops, and subsequently produce gamma rays by interacting with gas in the same loops.

In addition to this trapping argument, the observation of > 10 MeV gamma-ray continuum from flares on the solar disk (Rieger 1989) provides another argument for the production of the gamma rays in loops; it is almost certain that this continuum emission is bremsstrahlung from ultrarelativistic electrons whose radiation pattern is highly collimated. Since it is much more likely that these electrons are accelerated in the corona rather than in the photosphere (see below), there must exist a mechanism which reflects the electrons and allows them to interact on their way up in the solar atmosphere. Multiple bounces between magnetic mirror points in convergent flux tubes is a viable mechanism (Miller and Ramaty 1989). The preferential detection of gamma-ray flares from sites close to the solar limb (Sec. I.B) can also be explained by relativistic electron mirroring in solar-flare loops (Miller and Ramaty 1989).

On the other hand, there are flares for which there are more escaping than interacting protons (Cliver et al. 1989). Here the escaping particles are probably accelerated on open field lines with ready access to interplanetary space. The existence of two classes of solar-flare particles has been suggested by Cane et al. (1986) and Bai (1986). Indeed, a variety of observations suggest the existence of two distinct populations of accelerated particles, belonging to two classes of acceleration phenomena (see Reames 1990). Particles of the first class have much larger electron-to-proton and $^3\text{He}/^4\text{He}$ ratios and are much more abundant in heavy elements than those of the second class.

First-class particle events are well correlated with impulsive solar-flare phenomena, such as Type III radio emission, impulsive X-ray emission and non-relativistic electron emission. Second-class events are correlated with coronal mass ejections and Type II and IV radio emissions. It has been suggested that the particles in first-class events are accelerated from hot-flare plasma, while those of second-class events are accelerated from ambient coronal gas (Lin 1987). This suggestion was based on the higher ionization state of Fe found in first-class events relative to that in second-class events (Luhn et al. 1987). Further support for this suggestion is provided by the correlation of the Mg/He, Si/He and Fe/He ratios in first-class events with the temperature of the hot-flare plasma, as measured by soft X-ray emission. Gamma-ray production in solar flares is mostly due to accelerated particles belonging to the first class. However, in some solar flares (e.g., 3 June 1982) evidence was found for gamma-ray production by particles belonging to the second class (see Murphy et al. 1987).

Any successful solar-flare acceleration theory must be capable of accounting for all of the above phenomena. Among the best-studied particle acceleration mechanisms in solar flares are stochastic acceleration and shock acceleration (see, e.g., Forman et al. 1986). Stochastic acceleration in particular has been applied extensively to gamma-ray and neutron production (Ramaty and Murphy 1987). It has been shown that resonant interactions of ions with Alfvén waves and electrons with whistlers can accelerate the particles stochastically (Miller and Ramaty 1987). On the other hand, shock acceleration (see, e.g., Ellison and Ramaty 1985) could be responsible for the acceleration of the particles which escape to interplanetary space. Both stochastic and shock acceleration require pitch-angle scattering by MHD turbulence, which could be produced in the corona during the primary-flare energy release and could exist in the ionized corona, but is expected to be damped quickly by collisions between ions and neutral atoms in the chromosphere and below. This is an important argument favoring acceleration in the corona. Many aspects of solar-flare acceleration remain unexplained. An important problem is that of injection or pre-acceleration. It is thought that the spectacular ^{3}He enrichments (Sec. I.A) are due to resonant heating by ion cyclotron waves (Fisk 1978).

B. The Current Rate of Accelerated Ion Interactions at the Sun

We proceed now to calculate the ion irradiation of the Sun, i.e., the number of accelerated ions incident on the solar atmosphere per unit time. We carry out the calculation for solar cycle 21 which peaked around 1980. We start from the size distribution of gamma-ray flares measured in 0.3 to 1.0 MeV continuum emission (Vestrand et al. 1987). To determine the ion irradiation, we need the flare size distribution measured in nuclear line emission, but this distribution has not yet been determined from the SMM data. However, because of the good correlation between the 0.3 to 1.0 MeV continuum emission and the 4 to 7 MeV nuclear line emission (Forrest 1983) and the availability of a reliable scaling between these two emissions (Murphy et al. 1990), we do not expect much error from the use of the 0.3 to 1.0 MeV size distribution. To minimize the effects of the anisotropic angular distribution of the electrons, we use the size distribution observed from flares near the solar limb. For flares at heliocentric longitudes 60° to 90°, observed from March 1980 to February 1986 (i.e., approximately one half of a solar cycle), this size distribution can be approximated by $dn/dF_{0.3-1.0} \simeq 8.5\, F_{0.3-1.0}^{-1.1}$, where $F_{0.3-1.0}$ is the observed 0.3 to 1.0 MeV continuum fluence at Earth measured in photons cm^{-2} per flare (Dermer 1987). This expression is valid for $F_{0.3-1.0}^{-1.1}$, $\lesssim 6500$ photons cm^{-2}, and vanishes at higher fluences. The total number of flares per unit $F_{0.3-1.0}$ is given by the above expression multiplied by a factor of 12, where a factor of 6 takes into account the whole solar surface and another factor of 2 corrects for half of the solar cycle for which there is no data. We thus have that over the whole solar surface and a complete solar cycle

$$\frac{dn}{dF_{0.3-1.0}} \simeq 100\, F_{0.3-1.0}^{-1.1} \text{ flares } [\text{unit } F_{0.3-1.0}]^{-1},$$

$$F_{0.3-1.0} \lesssim 6500 \text{ photon cm}^{-2}. \tag{2}$$

The bulk of the gamma rays observed from solar flares are produced in thick-target interactions, i.e., by ions which are trapped and forced to interact at the Sun (Sec. II.A). The relationship between the observed nuclear line emission and the number of ions which interact at the Sun has been accurately calculated (Murphy and Ramaty 1984) for a variety of accelerated ion energy spectra. It has been shown (Ramaty and Murphy 1987) that a variety of gamma-ray and neutron observations can be interpreted in terms of ion spectra derived from a stochastic acceleration model (Sec. II.A). These spectra are given by a Bessel function characterized by the parameter αT (α = acceleration efficiency; T = trapping time), such that larger values of αT correspond to harder spectra (Ramaty 1979). It has been shown that for solar flares αT is typically in the range of 0.015 to 0.04 (Hua and Lingenfelter 1987a).

The ratio between the 0.3 to 1.0 MeV continuum emission and the total nuclear line emission as a function of the heliocentric longitude was studied recently by Murphy et al. (1990). For flares near the limb, this ratio is about 4.5. From studies of nuclear de-excitation line production, we obtain a ratio of ~ 6 between the total and the 4 to 7 MeV nuclear line emission. We need the latter because in the thick-target calculations of Murphy and Ramaty (1984) it was the 4 to 7 MeV photon production that was explicitly related to the number of protons of energies > 30 MeV incident on the solar atmosphere: $F_{4-7} \simeq 1.1 \times 10^{-31} N$ (> 30), for $\alpha T = 0.025$ (a typical value). Thus, $N(> 30) \simeq 3.4 \times 10^{29} F_{0.3-1.0}$.

Using this relationship and the size distribution given above, we can calculate the total number of protons of energies > 30 MeV incident on the Sun per solar cycle by evaluating the integral

$$Q(> E) = \int_{F_{min}}^{6500 \text{ photons } cm^{-2}} \frac{dn}{dF_{0.3-1.0}} N(> 30) \, dF_{0.3-1.0}. \tag{3}$$

By taking the lower limit $F_{min} = 0$, we derive the desired result, $Q(> 30$ MeV) $\simeq 1.0 \times 10^{35}$ protons per solar cycle. Using the Bessel-function spectrum with $\alpha T = 0.025$, this result yields $Q(> 10 \text{ MeV}) \simeq 1.0 \times 10^{36}$ protons per solar cycle, or 3×10^{27} protons s^{-1}.

In addition to irradiation by protons accelerated in solar flares, the Sun may also be irradiated by protons accelerated in the solar atmosphere during nonflaring periods. However, since currently there are only upper limits on the quiet-time solar gamma-ray emission, we can only place an upper limit on such irradiation. The most compelling current upper limit is that on the flux of the 2.223 MeV neutron-capture line, < 10^{-3} photons cm^{-2} s^{-1} (G. H. Share, personal communication, 1989). Using the calculations of Murphy and Ramaty (1984) with $\alpha T = 0.025$, this limit implies an upper limit < 10^{29} > 10 MeV protons s^{-1}. This upper limit is a factor of 30 larger than the finite irradiation rate deduced from the flare observations.

The above calculation, of course, gives only the irradiation rate at the Sun. However, from the results of Cliver et al. (1989), we estimate that, on the average, the rate of interplanetary proton production by solar flares is about equal to the irradiation rate at the Sun. Furthermore, interplanetary acceleration, particularly in the outer solar system, could serve as an additional source of protons.

C. Deposition of Energy in the Solar Atmosphere

The flare radiations across a wide part of the electromagnetic spectrum are caused by the deposition of energy. It is thought that this energy is derived from accelerated particles. In this section, we discuss the processes by which such particles, particularly those at low energies, interact with the solar atmosphere to produce what we recognize as a flare. The traditional wavelengths for flare observations are almost certainly just witnessing the aftermath of the main event, and as such are looking at totally secondary phenomena.

The nature of energetic-particle spectra, as we have seen, is such that the bulk of the energy content resides in the low-energy regime where the spectral index flattens. For protons (ions) this is in the 0.1 to 1 MeV range, whereas for electrons it is below about 100 keV. Therefore, although some of the least-ambiguous signatures come from the highest-energy particles, it is the low-energy part of the spectrum which must power the flare. This of course has been recognized for many years and, starting with Brown (1971), serious attempts have been made to interpret the hard X-ray burst as nonthermal bremsstrahlung from beams of accelerated electrons interacting with a thick target, the chromosphere. Of the two candidates outlined above for carrying the flare energy, the electrons are easily the more visible. Thus it became, almost by default, reasonably well established that the bulk of the energy which powered the impulsive-phase emissions came from beams of ≤ 100 keV electrons (Lin and Hudson 1976) accelerated in the corona. At that time, it was still believed that relativistic-particle (ion) acceleration occurred later in the flare; that pre-heating of the flare plasma was negligible; and that gamma-ray emission occurred after the onset of the impulsive phase. Also chromospheric ablation, and in particular its timing relative to the hard X-ray burst, was not soundly established. Coupled to this is the relative ease with which electrons betray their presence in the solar atmosphere and the invisible nature of low-energy ions.

Following the 1980 solar maximum, and the analysis of data from SMM and P78–1, plus the firmly established results from SKYLAB, it began to be realized that the nonthermal electron hypothesis as the means of powering solar flares was perhaps not so well established after all. The gamma-ray results showed that relativistic ions were present in the impulsive phase. The energy budget for flares began to be oversubscribed, as it was calculated (see, e.g., Antonucci et al. 1984) that on the thick-target electron-beam hypothesis there was already enough energy in electrons >25 keV to power the whole flare. Then the timing results (Sec. I.C) on chromospheric ablation with respect to the hard X-ray burst showed that electrons could not drive the ablation without first producing hard X rays (Simnett 1986b). Previously, it had been assumed that both were caused by electrons of energies up to several hundred keV. The relationship between the X-ray emissions and the plasma upflows has been studied in detail by Karpen et al. (1986) who question that there is any relationship between the electrons producing the X rays and the energy driving the ablation. In fact, Simnett (1986b) has argued that the ablation is driven by energy deposition by low-energy (~ 1 MeV) protons, while Simnett and Haines (1990) have discussed a mechanism whereby such protons will, under certain circumstances, produce energetic electrons in the chromosphere to generate the impulsive hard X rays. One of the attractions of this concept is that the energetic particles powering the flare emissions are ions, rather than electrons. With a mean ion energy of 1 MeV, the number of charged particles required to deposit a given flare energy is much smaller than the number that would be required if the deposition were due to electrons with a mean energy of only a few tens of keV. A reduction in the number of energy-carrying particles greatly alleviates the flare energy-transport problem and the large electric-current problem implied by the electron beam models. As we have seen in Sec. I.B, gamma-ray emission is frequently observed from flares, although the sensitivity of the instruments is not yet high enough to ascertain whether all flares accelerate ions.

In summary, the interaction models needed to account for the impulsive-phase emissions involve the deposition of nonthermal charged particles into a thick target, the chromosphere. Electron beams have been used successfully to account for many aspects of solar-flare emission, including the X-ray burst. There are difficulties in accounting for some of the other characteristic flare signatures, especially when detailed comparisons are made between different data with high time, spectral and spatial resolution. The proton beam model has some attractive features, but it would be premature to suggest that it has been developed in sufficient depth to be a replacement for the electron beam model. What is clear, however, is that the Sun has been a prolific source of nonthermal particles (electrons, ions or both) whenever there has been evidence of sunspot activity.

D. Trapping and Escape of Charged Particles

As we have seen above, the energetic particles accelerated impulsively in flare loops have difficulty in leaving the Sun. Moreover, different particle species do not behave identically. Most energetic-particle observations are made from near the Earth and flares from which particles are expected to be seen most readily are located at the root of the interplanetary magnetic field lines linking the Sun to the Earth. Thus typically, a flare from ~55 W solar longitude should be the best connected. In fact, in the absence of significant interplanetary diffusion (cf. scatter-free electron events), we should, unless particles diffuse in the corona, only observe particle events from well-connected flares. The actual situation is far removed from this suggestion, and we have good evidence that we receive particles not only from east-limb flares, but also from flares behind the visible solar disk. But the position is more complicated still, as it has been known for many years (see, e.g., Holt and Ramaty 1969) that only a small fraction of accelerated particles leave the Sun. This may be calculated from a comparison of the released energetic electrons with those required to produce the hard X-ray and microwave bursts. An even better test (Sec. II.A) is to compare the escaping energetic protons with the gamma-ray emission, since the gamma rays are not subject to absorption and scattering as the X rays and microwaves, and, moreover, their production mechanisms and radiation patterns are much better understood. The conclusion from analysis of both the electron and proton populations is that the majority appear to interact catastrophically with the dense chromosphere; they do not appear to have easy access to the interplanetary medium.

Simnett (1971) first pointed out the long delay (≥ 30 min) between the impulsive phase and the arrival of the relativistic electrons at the Earth even for "well-connected" events. The expected minimum delay for a relativistic electron traveling along the Archimedean spiral field line is ~11 min. This point has been addressed by Cliver et al. (1982) with respect to the onsets of 2 GeV protons, 1 MeV electrons and 100 keV electrons from solar events energetic enough to produce a ground-level neutron monitor response. They find, systematically, that the relativistic electrons follow the GeV protons by at least 5 min, and that these protons in turn are delayed with respect to the 100 keV electrons. These effects are probably due to delays in the acceleration, or rigidity dependent escape from the Sun. On the other hand, Kane et al. (1985) find that in one flare relativistic electrons escaping from the Sun were accelerated simultaneously with the electrons which produce the hard X rays and gamma-ray continuum at the Sun.

Within a few days of a large solar proton event, the proton flux becomes almost uniform (within a factor of 2) in space at all longitudes. This was established in the 1960s when the Pioneer spacecraft was deployed at ~1 AU around the Sun. The observations demonstrated that there were two components to the flare emissions; the first decays rapidly and has an intensity that is strongly dependent on the flare longitude with respect to the observer, and the second is emitted over a wide range of solar longitude and slowly decays. The latter appears to contain intensity features that co-rotate with the Sun. This is most easily achieved if the particles are stored in the corona and slowly released over a period of days from coronal regions where the field is more open. During the period of trapping, some azimuthal diffusion takes place in the corona, so that eventually particles are released from all around the Sun. Diffusion in the interplanetary medium cannot account for the observations.

Energetic particles from flares that are far removed from 55 W longitude do not normally have a direct, impulsive component because the particles must first propagate around the Sun to ~55 W. If the flare is energetic enough that high-energy neutrons are generated, these are not hampered by the magnetic field and neutron-decay protons may be injected within minutes onto local magnetic field lines where they will be detected directly. Such events have been observed; Evenson et al. (1983) recorded a good example from a flare at 72 E longitude where the neutron-decay protons (Sec. I.B) had virtually disappeared before the delayed flare protons started to arrive some 12 hr later.

The most convincing example of coronal particle trapping comes from the very long duration events, such as that shown in Fig. 2. We know from observations of scatter-free electron events, and ^{3}He-rich events, that interplanetary diffusion across the field lines is negligible compared with transport along the field lines. Therefore, interplanetary diffusion is not responsible for maintaining the intensity at a high level for so long, if we assume an impulsive flare injection at the beginning of the event. The regular 13-month observation of relativistic Jovian electrons at the Earth also indicates how well the electrons are tied to the interplanetary magnetic field lines. The long-duration events nearly always can be associated with a major parent flare; exceptions are when the flare is hidden behind the Sun, and it must be attributed to a known active region. Therefore the likely, if not the only, explanation is that the particles are trapped in the corona. During the period of trapping, some azimuthal diffusion takes place in the corona so that eventually particles are released from virtually all around the Sun. This interpretation is supported from observations of long-duration Type IV radio events, that are attributed to energetic coronal electrons.

The duration of the trapping, and hence, from collisional energy-loss considerations, and the altitude, are not known for certain. Protons are easier to trap than electrons. Around solar minimum, long-lived, co-rotating streams of energetic (10 MeV) protons are observed over several solar rotations. How frequently the source is replenished is not known, but it is most unlikely that the particles are being accelerated continuously at the base of the field lines along which they propagate.

E. Coronal Mass Ejections

To complete the discussion of energetic-particle acceleration and release, we shall briefly discuss coronal mass ejections. These represent the ejection from the Sun of large amounts (10^{16} to 10^{17} g) of low-velocity plasma which may be detected near the Sun via observations of Thompson-scattered photospheric light off the free electrons. Such ejected clouds carry with them magnetic field and, if they intersect the Earth, we may detect not only the magnetic field enhancement but also a significant increase in particle density.

Coronal mass ejections (CME) are often observed together with metric type II radio bursts (Sheeley et al. 1984). More usually such events have fast CME speeds (>400 km s^{-1}) and are associated with shocks. However, such associations may be fortuitous since close to half of all high-speed CMEs have no metric Type II burst. Although traveling interplanetary shocks generally have a CME associated with them, the converse is not true. Therefore any associations in this area may also be fortuitous. We conclude that there is nothing inherent in the physical conditions around the CME that is accelerating charged particles.

III. CHEMICAL AND ISOTOPIC ABUNDANCES

The gamma-ray line observations from solar flares have provided two new techniques for determining abundances in the solar atmosphere. The first technique (Murphy et al. 1985; Ramaty et al. 1990) is based on nuclear de-excitation line emission, which provides information on elemental abundances of both the ambient gas in the interaction region and the accelerated particles that interact with the ambient gas to produce the observed gamma rays. As discussed above, the interaction region is probably in the chromospheric and upper photospheric parts of a flare loop. The second technique (Wang and Ramaty 1974; Hua and Lingenfelter 1987b) is based on the time profile and the absolute intensity of the 2.223 MeV neutron-capture line, which can determine the abundance of ^{3}He in the photosphere.

The nuclear de-excitation line studies are based on the spectrum shown in Fig. 4, which was observed with the SMM/GRS detector. The comparison of this spectrum with a calculated gamma-ray spectrum using photospheric abundances for both the ambient gas and the accelerated particles revealed a significant deficiency of the calculated intensities of the lines of C and O relative to those of Ne, Mg, Si and Fe (Forrest 1983). Subsequent analysis (Murphy et al. 1985), in which the observed spectrum was compared with a variety of theoretical spectra obtained by varying the elemental abundances of the ambient gas, showed that the data can be fit with C/Mg and O/Mg abundance ratios which are lower by a factor of about 3 than the corresponding photospheric values and with Si/Mg and Fe/Mg ratios similar to those of the photosphere.

The suppression of the C and O abundances relative to those of Mg, Si and Fe is also evident from the comparison of the solar energetic-particle abundances observed from large solar flares with photospheric abundances. It has been shown previously (see, e.g., Breneman and Stone 1985) that these solar-flare abundances represent coronal abundances and that they differ significantly from photospheric abundances, in that the abundances of elements with high first-ionization potential are reduced. The fractionation between the photosphere and corona could be caused by charge-dependent mass transport (Vauclair and Meyer 1985). Since the photosphere is collisionally ionized at a relatively low temperature, the transport could be less efficient for elements with high ionization potential, leading to suppressed abundances for such elements. The suppression of the chromospheric flare-loop C and O abundances could have a similar origin.

Analysis of the gamma-ray data also showed that the Ne abundance is quite high, the Ne/O ratio exceeding by about a factor of 3 the large flare solar-particle value, Ne/O $\simeq$ 0.15, which is also the coronal value. Thus it appears that relative to the corona the flare-loop material is either enhanced in Ne or suppressed in O. The origin of this difference is not understood; but the high Ne/O ratio obtained from the gamma-ray data is supported by a recent study (Reames et al. 1988) of the solar-particle abundance of Ne in ^{3}He-rich flares. These workers have shown that the average solar particle Ne/O in these flares, 0.42 ± 0.03, is higher than Ne/O in large solar-particle events, and that this enhancement is probably not due to preferential acceleration but represents the ambient flare-plasma abundance.

That observations of the 2.223 MeV line from solar flares could provide information on the photospheric ^{3}He abundance was first pointed out by Wang and Ramaty (1974) after observations of this line from the 4 and 7 August 1972 flares (Chupp et al. 1973). More recent observations and analysis have used this technique to derive an upper limit, ^{3}He/H $< 3.5 \times 10^{-5}$ (Hua and Lingenfelter 1987b). This limit is sufficiently low to be consistent with the ^{3}He abundance expected solely from primordial nucleosynthesis, suggesting that the contribution of turbulent diffusion from the solar interior, where ^{3}He is made from deuterium burning, is not very important.

IV. FLARE ACTIVITY ON THE ANCIENT SUN

It is of course not possible to extrapolate reliably from today's level of flare activity to that which might have existed during the early life of the Sun. However, there are at least three avenues open to us to try to understand the evolutionary properties of flares. The first is to look at other stars which are not as evolved as the Sun. The second is to recognize that the source of the flare activity almost certainly is the differential rotation of the present Sun, together with the primordial magnetic field. And the third is to look at evidence for isotopic anomalies in meteorites and the Moon which could have resulted from large accelerated-particle fluxes of solar or interplanetary origin.

If we are to attempt to investigate flare activity on other stars, we must infer every conclusion we come to from an analysis of the spectrum and its time variability, since we have no spatial information. Also, the Sun is the only star which is monitored continuously; therefore inferences about conditions on other stars are made from spasmodic observations, generally with restricted wavelength coverage. Despite these obvious handicaps, considerable information has been obtained that is pertinent to stellar flaring. Underhill and Fahey (1984) deduced the presence of a large bipolar magnetic region on the surface of ρ Leo, a B1 Ib star, from spectroscopic observations with the International Ultraviolet Explorer. They detected large and discrete absorption features in the ultraviolet spectrum that were consistent with the release of packets of gas, interpreted as magnetic clouds, from an altitude deduced to be $\sim 2\,R_*$. These clouds are analogous to the magnetic clouds, or coronal mass ejections, seen in the solar wind. The spasmodic nature of the release, together with its inferred large size, was plausibly interpreted in terms of sudden release from a large bipolar magnetic region.

A useful class of stars to study are the T Tauri stars. These stars are inferred to be in an early evolutionary state, and are probably of a similar mass to the Sun. They are 1 to 2 orders of magnitude more luminous than the Sun and are probably < 10 Myr old. Therefore their observed properties may well give us a good indication as to how the Sun might have appeared in its early life. Worden et al. (1981) have studied the variability of T Tauri stars in the optical band and observed many flare-like events. It should be remembered that for a flare to be visible on a young hot star above the normal continuum, it must represent a substantial energy release, orders of magnitude above that estimated for a large solar flare. Worden et al. estimate an average optical flare energy in the T Tauri phase (lasting for about 6 Myr) of 1.7×10^{31} erg s^{-1}, which exceeds the current average solar flare optical luminosity by about a factor of 5×10^5.

Further evidence for violent events on T Tauri stars comes from recent X-ray observations. Feigelson and DeCampli (1981), have observed soft X-ray events from the Einstein Observatory with luminosities $\sim 10^{30}$ erg s^{-1} from stars in Taurus and $\sim 10^{31}$ erg s^{-1} from stars in the Orion cloud. They point out that the X-ray emission cannot be generated from the stellar envelope as a whole, and is likely to come from a thin hot region in the outer layer of the star. Some T Tauri stars exhibit rapid variability, suggesting that energetic flares are occurring and that highly nonthermal particle fluxes are present.

Additional estimates of the energy involved in large flares from young stars come from soft X-ray studies. On the Sun, a reasonably good estimate of the total energy release is obtained from the energy in the soft X-ray emission, this being perhaps within an order of magnitude of the total. Stern et al. (1983) observed a large soft X-ray flare in the Hyades which had a peak luminosity in soft X rays $>10^{31}$ erg s^{-1}, which is several thousand times brighter than the most intense solar flares currently seen. They infer that the flare, from the G dwarf HD27130, covered a significant fraction of the stellar surface; this is believed to be a common feature of young, rapidly rotating stars. Their analysis showed that there was no need to postulate strong, kilogauss magnetic fields on the star. The large energy release was believed to result from a large volume, rather than a strong field. Montmerle et al. (1983a) from Einstein observations, have identified a flare from a star in the ρ Ophiuchi dark cloud that had an X-ray luminosity $\sim 10^{32}$ erg s^{-1} and a total X-ray energy release $>8 \times 10^{35}$ erg. This would be $\sim 10^5$ greater than a class 3 solar flare. The sources in ρ Oph are most likely pre-main sequence, or very young, objects. Other observations of this region in the microwave region of the spectrum (Feigelson and Montmerle 1985) have also seen rapid variability in the emitted flux, and the general conclusion is that young stars exhibit very high levels of magnetically induced surface activity (see the chapter by Feigelson et al.). The strong microwave fluxes suggest that not only does the emitting volume need to be large (a few times the stellar radius) but that the energy of the electrons needs to be in the MeV region, somewhat higher than that frequently assumed for electron energies responsible for large solar microwave bursts.

It may not only be in the very early life of a star that the flare levels are very high. Landini et al. (1986) reported some EXOSAT observations of a strong X-ray flare in an otherwise normal solar-type star, π^1 UMa, which is spectral type G0 V. This star has a very high rotation rate and a quiescent X-ray luminosity some 2 orders of magnitude greater than the typical quiet-Sun value. This star is a member of the Ursa Major cluster and is a factor of ~ 2 younger than the Hyades and ~ 15 younger than the Sun. Landini et al. argue that the energy release in the flare they observed was at least an order of magnitude larger than that emitted in the largest solar flare.

Turning to the isotopic anomalies, we first examine the suggestion of Worden et al. (1981) that the > 10 MeV proton-irradiation rates on T Tauri stars are 6×10^{34} to 6×10^{35} s^{-1}. We believe that these are excessive since they exceed the current flare irradiation rate calculated in Sec. II.B by factors of 2×10^{7} to 2×10^{8}, while the ratio of the optical luminosities is $< 10^{6}$. Next, the suggestion of Heymann and Dziczkaniec (1976) that the ^{26}Mg anomalies observed in meteorites could be of spallogenic origin produced shortly after the formation of the solar system also seems untenable since it requires an irradiation rate in > 10 MeV protons of $\sim 10^{6}$ cm^{-2} yr^{-1}, which exceeds the current solar irradiation rate by at least a factor of 3×10^{8}. Another anomaly is that of a secular increase of ^{15}N/^{14}N in lunar regolith, which is thought to be due to deposition by the solar wind. Kerridge et al. (1977) have shown that the ^{15}N cannot be produced by proton irradiation at the Sun. In agreement with this result, we find that the minimum steady state ^{15}N production of $\simeq 5 \times 10^{5}$ s^{-1} at the surface of the Sun implies a > 10 MeV proton irradiation rate of $\sim 10^{34}$ s^{-1} over the entire lifetime of the Sun, and this exceeds the current irradiation rate from flares by 3×10^{6} and the upper limit from quiet-time observations by 10^{5}.

On the other hand, the suggestion of Caffee et al. (1987) that the enrichments of ^{21}Ne and ^{38}Ar found in meteorite grains could be due to spallation is quite acceptable. These workers find that the required > 10 MeV proton irradiations are 10^{16} to 10^{18} cm^{-2}. Using an irradiation period of 10^{5} yr and an irradiation site located at 3 AU (as suggested by Caffee et al.), these values require a > 10 MeV proton production of 8×10^{31} to 8×10^{33} s^{-1} at the Sun. These production rates exceed the current solar irradiation rate (Sec. II.B) by factors of 3×10^{4} to 3×10^{6}, which, when compared with the ratio of optical luminosities of 5×10^{5}, do not seem excessive. Furthermore, additional irradiation, due to interplanetary particle acceleration could have an important effect at a distance of 3 AU from the Sun.

The picture that emerges is that there is considerable evidence, of which the above discussion is merely representative, that young, rapidly rotating stars are extremely active, and that they are emitters not only of strong X-ray fluxes, but of large fluxes of energetic particles as well. The Sun is still rotating fast enough to produce the active regions responsible for present-day flare activity. Isaak (1981) has postulated, from an analysis and interpretation of the rotational splitting of the nonradial p-modes of the 5-min global oscillation, that the Sun may have a strong magnetic field in the rapidly rotating core. The internal rotation rate may be some 2 to 9 times the observed surface rotation rate. Isaak suggests that in the early life of the Sun, the magnetic energy might have been a significant fraction of the total energy of the star. With time, the magnetic flux will decrease both from ohmic dissipation and through solar-wind loss. Thus we have good reason to believe that in its early life, the Sun might not have been too different from the young active stars in our Galaxy which are now beginning to be observed by modern instruments. There is strong evidence (Pallavicini 1981) that in dwarf stars the coronal X-ray activity is well correlated with rotation rate. Many years ago, Schatzman (1962) showed that if gas emitted from the star is kept rotating by magnetic torques out to large distances, then it carries off far more angular momentum per unit mass than it had when it was at the stellar surface. This is clearly an effective way of braking the star's rotation. Examination of the solar wind shows that both magnetic braking and magnetic flux loss are continuing. Thus, even when the Sun was half its present age and presumably rotating much faster than now, flare activity should have been substantially above current levels. We speculate that the young Sun was (a) rapidly rotating, (b) had a stronger magnetic field than at present, and (c) had a very much more violent flare pattern than today. But, except for the ^{21}Ne anomaly, enhanced proton irradiations from the ancient Sun could not have been responsible for the observed isotopic anomalies.

V. SUMMARY

We have reviewed the variety of accelerated particle phenomena in solar flares, including particle observations in interplanetary space, and gamma-ray, neutron, hard X-ray and radio emissions. We have emphasized that a significant or perhaps even large fraction of the total solar-flare energy is in accelerated particles. However, it is still not clear whether the bulk of this energy is in protons or electrons. The energy content in protons of energies greater than several MeV can be accurately calculated using gamma-ray observations, but estimates of the energy content in lower-energy protons and electrons are very uncertain. We have pointed out that the majority of the particles that produce the impulsive flare phenomena remain trapped at the Sun, probably due to confinement in the flaring magnetic loops. Clear evidence for loops structure exist from a variety of other observations, including X-ray imaging data. We have discussed some of the better-understood acceleration mechanisms, but have emphasized that much remains to be learned on particle acceleration in solar flares. Using gamma-ray observations, we have calculated the current rate of irradiation of the Sun by MeV protons accelerated in solar flares. Assuming that, on the average, the number of escaping and interacting protons are approximately equal, we have compared the current number of escaping protons with irradiation rates of meteorites implied by observations of isotopic anomalies. We find that except for the ^{21}Ne anomaly, enhanced proton irradiations from the ancient Sun could not have been responsible for the observed isotopic anomalies.

SOLAR-COSMIC-RAY FLUXES DURING THE LAST TEN MILLION YEARS

ROBERT C. REEDY
Los Alamos National Laboratory

and

KURT MARTI
University of California, San Diego

The fluxes of energetic (E $\geq$ 10 MeV) solar particles in the vicinity of the Earth in the past can be determined from nuclides that they produced in the top centimeter of lunar rocks. Activity-vs-depth profiles of short-lived radioactivities measured in the top centimeter of lunar rocks agree with profiles calculated with directly measured solar-proton fluxes since about 1965 and were used with indirect observations to get solar-proton fluxes back to 1956. Lunar-rock profiles for long-lived radionuclides (ranging from 5730-yr ^{14}C to 3.7 Myr ^{53}Mn) have been used to infer solar-proton fluxes averaged over several time periods in the past. New results are reported for solar-proton-produced ^{81}Kr measured in lunar rock 68815. Activities of 76,000-yr ^{59}Ni can be used to get fluxes of solar α particles averaged over the last $\sim 10^5$ yr. The average solar-proton fluxes in the past are not greatly different from those observed during the last three 11-yr solar cycles. We discuss the work that needs to be done to determine more and better fluxes of energetic particles from the Sun in the past.

I. INTRODUCTION

Large flares on the Sun release huge amounts of energy and can produce many high-energy particles, such as X rays, γ rays, electrons, neutrons and ions (see, e.g., Chupp 1988a). A fraction of these energetic radiations escape into interplanetary space where they can be detected; they have only been directly observed during the last few decades. Solar γ-ray lines were first observed in August 1972 (cf. Chupp 1988a), and energetic ($E \gtrsim 50$ MeV) solar neutrons were first measured in June 1980 (Chupp 1988b). Energetic electrons from the Sun have only been detected since about 1965 (cf. Simnett 1974). There are very few records, either direct or indirect, of solar γ rays, neutrons or electrons much before these dates. Our best records of the radiations produced as a result of large solar flares in the past are for the energetic ions that reached the vicinity of the Earth. Energetic ions from the Sun have been directly measured in space since about 1958, and indirect Earth-based observations with ionization chambers and neutron monitors extend back to 1942 (Pomerantz and Duggal 1974; Shea and Smart 1990). These energetic particles from the Sun will be called solar cosmic rays (SCR) in this chapter, although they also are often called solar energetic particles or solar-flare-associated particles. This chapter discusses these energetic particles from the Sun, their interactions in the Moon, and their record during the past few Myr. It includes much material on the SCR record from a previous review (Reedy 1980) and extends it up to the end of 1989. Other reviews that discuss the SCR record in the past include Reedy et al. (1983) and Lal (1988a).

The record considered here is that of the nuclides made in lunar samples by energetic solar protons and α particles. The record for heavy nuclei ($Z \gtrsim 20$) in the SCR have been presented elsewhere (see, e.g., Walker 1975; Reedy et al. 1983) with Crozaz (1980) Goswami et al. (1980) and the chapter by Goswami discussing the record of solar energetic heavy nuclei in the past. We only consider those nuclides made by SCR particles over the last ~ 10 Myr. Possible nuclide production by energetic SCR particles in the early solar system, ~ 4.5 Gyr ago (Caffee et al. 1988), is discussed in the chapter by Caffee et al. We only present and discuss the record of solar protons and α particles near the Earth. We cannot infer whether these particles were accelerated at or near the Sun or in interplanetary space, although modern observations indicate that significant interplanetary acceleration is fairly rare and requires special conditions (see, e.g., Pomerantz and Duggal 1974; Shea and Smart 1990). In contrast to atmospheric-produced cosmogenic radionuclides such as ^{14}C and ^{10}Be in terrestrial surface samples, which are differential records covering fairly short time periods, the SCR records from lunar samples are integral ones, the time periods usually being determined by the radionuclides' half-lives or the sample's exposure history.

The properties of SCR particles in the vicinity of the Earth is briefly reviewed and followed by a discussion of how SCR and galactic cosmic-ray (GCR) particles interact with matter, with emphasis on the production of nuclides in the Moon by nuclear reactions. The nuclides made by cosmic-ray particles are often referred to as "cosmogenic," and those made by energetic (spallation) reactions are also called "spallogenic" nuclides. Cosmic-ray-produced nuclides are usually used to study the history of the irradiated object, such as determining the length of time that a lunar rock has been on or near the surface of the Moon (e.g., see Reedy et al. 1983; Lal 1972). In the case of SCR-produced nuclides, we also can use the measurements to study the incident particles as well as the target's history. In this case, we need to know the target's history well enough to correct the record so as to get the nature of the incident SCR particles. For example, as will be discussed below, we need to know that the lunar sample has a simple history (i.e., deeply buried so as to be almost completely shielded from cosmic-ray particles then only having one exposure geometry on the Moon's surface) while the SCR-produced nuclides were being made, and we have to determine the rate that material was being eroded from its surface while it was on the Moon.

As the threshold energies for most nuclear reactions of interest are $\gtrsim 10$ MeV, we cannot say anything about solar protons and α particles at lower energies. While nuclide production by the high-energy (~ 0.1 to 10 GeV) GCR particles dominates in most lunar samples, certain nuclides made by the low-energy SCR particles are observable in the very tops of lunar samples, within ≤ 1 cm of the surface. It is this record of SCR-produced nuclides in lunar samples returned by the Apollo astronauts that is presented here. Some new results for SCR-produced Kr isotopes are presented and interpreted as an example and as a relevant recent development. The fluxes of solar protons over the last few decades and averaged over time periods into the past ranging from $\sim 10^4$ yr to almost 10 Myr ago as determined by others are reviewed. The long-term averages for the fluxes of solar protons are compared with the modern record. The nuclides made by SCR particles in the Moon have not been completely measured nor the measurements fully unfolded, and the tasks that need to be done in the study of the past SCR record are discussed.

II. COSMIC-RAY PARTICLES AND THEIR INTERACTIONS

The properties and recent observations of SCR particles are discussed in more detail by Shea and Smart (1990), who also summarize proton fluxes in the major SCR-particle events observed from 1956 to 1986 in addition to a discussion of a few earlier solar-particle events. In brief, SCR are energetic electrons and ionized nuclei emitted from the Sun mainly during large solar flares. The nuclei in the SCR are almost entirely protons, with only a few percent of α particles and a trace of heavier nuclei (McGuire et al. 1986), although the composition of SCR particles varies dramatically with time and energy. The composition of the nuclear component of the GCR ($\sim 87\%$ H, $\sim 12\%$ He and $\sim 1\%$ heavier nuclei) is discussed by Simpson (1983). The major differences between SCR particles and the other particles seen near the Earth are energy and flux. The GCR particles have low fluxes of much higher energy ($E \sim 0.1$ to 10 GeV) particles. The solar wind has very high fluxes of quite low energy ($E \sim 1$ keV) particles. Only the trapped radiation belts around the Earth have particles with energies similar to those in the SCR (~ 10 to 100 MeV), but these trapped particles do not reach the Moon, the target of interest here for recording the past histories of interplanetary particles.

The distribution of SCR particles as a function of energy is one with the flux decreasing rapidly with increasing energy. Relatively few SCR particles have energies $\gtrsim 100$ MeV/nucleon, although many solar-particle events have some particles with energies above 1 GeV (cf. Shea and Smart 1990). SCR particles near the Earth can be anisotropic, especially early in an event when the first particles arrive from the Sun. This anisotropy could be a problem when studying cosmogenic nuclides made by SCR particles in only a few events, but should not be a problem for the longer time periods ($\gtrsim 10^4$ yr) considered below because the effects of such anisotropies over thousands of events should average out.

At any time during an event or averaged over an event, the flux F of SCR particles for relatively narrow energy ranges behaves roughly as a power-law in kinetic energy E

$$dF/dE = \text{const. } E^{-\gamma} \qquad (1)$$

where γ is typically between 2 and 4 (Van Hollebeke et al. 1975). For broad energy ranges, a better fit is obtained using an exponential rigidity shape (McGuire and von Rosenvinge 1984). Lal (1972) compared these two spectral shapes. The differential flux per unit rigidity is (see Reedy and Arnold 1972)

$$dF/dR = \text{const. } \exp(-R/R_0). \qquad (2)$$

Rigidity R (sometimes denoted P), is the momentum of a particle per unit charge, pc/ze, and for SCR particles is typically given in units of megavolts, MV. To get rigidity in units of MV from energy in MeV, one uses

$$R^2 = (E^2/z^2) + (2mE/z^2) \qquad (3)$$

where z is the charge of the particle and m is its mass in MeV (e.g., 1 and 938.256 MeV for a proton). The spectral parameter R_0 for solar protons in the energy range of ~ 5 to 200 MeV usually ranges from 20 to 150 MV for event-integrated spectra (see, e.g., Goswami et al. 1988). An exponential in rigidity with $R_0 \approx 100$ MV has been found in many of the works discussed below to be a reasonably good fit to the spectra of solar protons averaged over $\sim 10^4$ to 10^7 yr.

The spectral shape of GCR particles in the vicinity of the Earth is quite different from that for the SCR, being very roughly described by

$$dF/dE = \text{const. } (1000 + E)^{-2.65} \qquad (4)$$

where the kinetic energy E is in MeV (Reedy and Arnold 1972). More exact descriptions of GCR-particle spectra in interstellar and interplanetary space can be found in the chapters by O'Brien et al. and by Jokipii, and Lal (1988b). At energies$\gtrsim$ 10 GeV, this shape approaches a power-law in energy, $E^{-2.65}$, and GCR particles with energies up to $\sim 10^{20}$ eV have been observed (Simpson 1983). Below $\sim$ 1 GeV/nucleon, the flux of GCR particles as a function of energy is fairly flat or decreases with decreasing energy. The mean energy of the GCR particles is a few GeV, whereas that for the SCR is much lower, of the order of tens of MeV.

The intensities of the SCR and GCR particles also are different. The omnidirectional integral flux of GCR particles is low, only $\approx$2 to 5 particles cm^{-2} s^{-1} (this range of fluxes being that from the time of maximum solar activity to that for a typical solar minimum). The fluxes of SCR particles near the Earth at 1 AU at any time range from essentially zero at most times up to $\sim 10^6$ protons cm^{-2} s^{-1} for a few hours around the peak of the biggest events, such as those of 14 July 1959 or 4 August 1972. The long-term average omnidirectional flux of solar protons, as presented below, is $\approx$ 100 protons cm^{-2} s^{-1} for energies above 10 MeV.

Because of the large differences in the energies of SCR and GCR particles, each tends to interact with matter in different ways. In a thick target like the Moon, most solar protons are stopped by ionization energy losses in the top centimeter (a few g cm^{-2} of lunar rock or soil). The lowest-energy protons are most rapidly stopped. For example, it takes only 0.17 g cm^{-2} (less than a mm) of lunar rock to stop a 10-MeV proton. Higher-energy protons are slowed to lower energies (e.g., a 40-MeV proton is slowed to about 38 MeV in 0.17 g cm^{-2}). The spectrum of solar protons with a spectral shape of $R_0 = 100$ MV is shown in Fig. 1 along with the proton fluxes calculated at various lunar depths assuming a 2π isotropic flux incident on a slab and considering only ionization energy losses (Reedy and Arnold 1972). Notice how rapidly the flux of lower-energy ($E \lesssim 30$ MeV) protons decreases with increasing depth while the flux of higher-energy ($E \gtrsim 100$ MeV) protons varies much more slowly. Solar α particles are stopped even more rapidly (Lanzerotti et al. 1973). Thus only a small fraction of solar protons and solar α particles incident on the Moon induce nuclear reactions, and these reactions are mainly in the top cm ($\approx$2 to 3 g cm^{-2}) of the lunar surface. Few secondary particles, especially neutrons, are emitted in SCR-induced reactions (Armstrong and Alsmiller 1971a,b), so only primary SCR protons and α particles need to be considered in calculating the production of nuclides in the Moon.

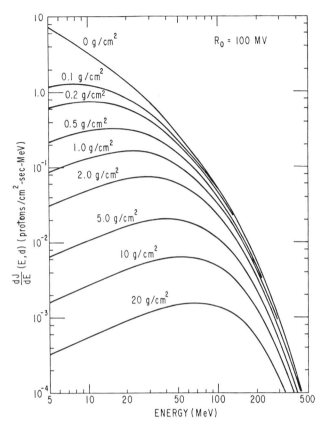

Fig. 1. The flux-vs-energy distribution of solar protons as a function of depth in the Moon. The curve for 0 g cm^{-2} is the flux of protons incident on a slab with a spectral shape of $R_o = 100$ MV and an omnidirectional flux above 10 MeV of 100 protons cm^{-2} s^{-1}. The fluxes in the Moon were calculated using only ionization energy losses (Reedy and Arnold 1972). In lunar soil or rock, a depth of 1 cm is $\approx$ 2 g cm^{-2} and $\approx$ 3 g cm^{-2}, respectively. This figure shows that few solar protons penetrate much more than a centimeter into the Moon.

In contrast, the high-energy particles in the GCR can travel much farther into the Moon and induce a wide range of nuclear reactions and produce many neutrons to depths of meters (Reedy and Arnold 1972). These secondary GCR-produced neutrons are penetrating and usually produce most cosmogenic nuclides in lunar samples. Several models (see e.g., Reedy and Arnold 1972; Yokoyama et al. 1972) have been developed that describe fairly well the production of radionuclides in the Moon by GCR particles. The measured and calculated production rates of cosmogenic noble gases were discussed by Hohenberg et al. (1978) and Regnier et al. (1979). Because most primary and secondary GCR particles are so penetrating, the rates at which they produce nuclides varies slowly with depth in the Moon, especially in the top tens of cm (Reedy and Arnold 1972).

III. LUNAR COSMOGENIC NUCLIDES AND
THEIR PRODUCTION RATES

The production rate for a cosmogenic nuclide at depth d in the Moon $P(d)$ can be calculated as the sum over target elements i, the sum over different types of particles j, and the integral over energy E

$$P(d) = \sum_i N_i \sum_j \int \sigma_{ij}(E)\, F_j(E,\, d)\, dE \qquad (5)$$

where N_i is the abundance of the ith element, $\sigma_{ij}(E)$ is the cross section as a function of energy for making the nuclide from element i with particle j, and $F_j(E,d)$ is the flux of particle j at energy E and depth d (cf. Reedy and Arnold 1972). For SCR particles, only the primary particles are considered, as discussed above and shown in Fig. 1. For GCR particles, both the primary and secondary particles need to be considered. For most GCR-produced cosmogenic nuclides, production is mainly induced by the secondary neutrons (cf. Reedy and Arnold 1972; Reedy 1987a).

The Moon is an evolved planetary body and the lunar rocks and soils have a range of compositions. However, there are very few elements near the lunar surface that are heavier in atomic mass than iron, so most cosmogenic nuclides studied in lunar samples have atomic masses less than about 56. The most abundant elements in a typical lunar rock, in approximate order of decreasing abundance by weight, are oxygen, silicon, aluminum, iron, calcium, magnesium and titanium. Other elements seldom constitute more than one percent by weight in lunar samples. The abundances of the major target elements for each cosmogenic nuclide need to be known for each sample studied, especially if small samples are analyzed and the lunar rock or soil being studied is not chemically homogeneous.

Other forms of extraterrestrial matter that are exposed to SCR particles can have compositions different from those seen in the Moon. However, while SCR particles produce cosmogenic nuclides in the surficial layers of meteorites (Reedy 1987b; Michel et al. 1982), ablation of the meteorite's surface while passing through the Earth's atmosphere removes most of the surface layers that contain the SCR record. There are only a few meteorites in which a clear SCR record has been observed in the bulk of the meteorite, e.g., Salem (Evans et al. 1987) and ALHA77005 (Nishiizumi et al. 1986), although there are cases where SCR production in pieces from the meteorite's surface has been observed (see, e.g., Lal and Marti 1977). The flux of SCR particles varies considerably with distance from the Sun, so one needs to know the orbital history for a meteorite or other forms of extraterrestrial matter (such as cosmic dust or the spherules found in deep-sea sediments or glaciers) in which SCR production of nuclides can be observed. The SCR fluxes at 1 AU for a given event also vary with position along the Earth's orbital path; this appears to have been the case for ^{56}Co in the Salem meteorite (Evans et al. 1987; Goswami et al. 1988), which only a few months before it hit the Earth saw a much higher flux of solar protons than was observed at the Earth. Thus the quantitative studies of SCR particles with cosmogenic nuclides rely only on documented lunar samples.

For a cosmogenic nuclide to be readily detectable in a lunar sample, it usually is produced from the more abundant elements or can be detected in low concentrations, such as the minor isotopes of the noble-gas elements helium, neon, argon, krypton or xenon. Radionuclides with half-lives less than ~ 100 Myr are not present in extraterrestrial matter unless made fairly recently by some nuclear process and thus are ideal particles for being identified as having been made by cosmic rays (Reedy et al. 1983). The noble gases are normally only present inside lunar rocks at extremely low abundance levels, and cosmic rays made these elements in different isotopic ratios than any other noble-gas component likely to be present in a sample. For example, terrestrial neon has a ^{20}Ne/^{21}Ne ratio of 338 but these isotopes are made with a production ratio of ≈ 1 by cosmic rays. The radionuclides used to study SCR particles are given in Table I in order of increasing half-life along with their half-lives and the major target elements from which they are made in the Moon. Except for ^{56}Co, these same radionuclides are also readily made by GCR particles. (GCR particles make little ^{56}Co because there are relatively few of the low-energy protons that readily make ^{56}Co and very low abundances of targets heavier than iron from which ^{56}Co can be made by neutrons.) There are also a few cosmogenic nuclides, such as 269-yr ^{39}Ar, that are readily made by GCR particles but only are made in low concentrations by SCR particles. Cosmogenic ^{81}Kr is made from target elements that are heavier than krypton. Some stable isotopes, such as ^{83}Kr, can be detected as having been produced by SCR particles.

The production rates for these nuclides in the Moon by GCR particles are usually determined from measurements below the depths reached by SCR particles. Theoretical GCR production profiles (see, e.g., Reedy and Arnold 1972; Yokoyama et al. 1972) are used to extrapolate measured GCR production rates to the surface. Lunar measurements of several nuclides not made by SCR particles have shown that the theoretical profiles are fairly good, especially as GCR production rates vary slowly with depth. More details on predicting the production rates and depth-vs-concentration profiles of cosmogenic nuclides can be found in Reedy (1987a) along with a discussion of the cross sections needed for GCR-produced nuclides. In most cases presented below, adequate corrections for GCR production can be made to the measured concentrations of cosmogenic nuclides to obtain a clear signal of the SCR production.

The production rates of cosmogenic nuclides by solar protons and α

particles are calculated using Eq. (5), SCR fluxes like those in Fig. 1, and
excitation functions, $\sigma_{ij}(E)$, cross sections as a function of energy. Examples
of excitation functions for the production of several radionuclides from iron
by protons are shown in Fig. 2. These excitation functions are mainly based
on measurements by Michel et al. (1979) and Michel and Brinkmann (1980)
for ^{56}Co and ^{54}Mn, the sources listed in Reedy and Arnold (1972) for ^{55}Fe,
and measurements by Gensho et al. (1979) and K.Nishiizumi and M. Furu-
kawa (personal communication, 1988) for ^{53}Mn. The excitation functions in
Fig. 2 have peaks at various energies, with ^{56}Co being made mainly at low
proton energies ($\sim$5 to 15 MeV) by the ^{56}Fe(p,n)^{56}Co reaction (in which a
proton reacts with a ^{56}Fe nucleus to produce a neutron and a residual ^{56}Co
nucleus), while ^{54}Mn is made mainly by protons with energies of $\sim$30 to 60
MeV. As the fluxes of protons with different energies vary relative to each

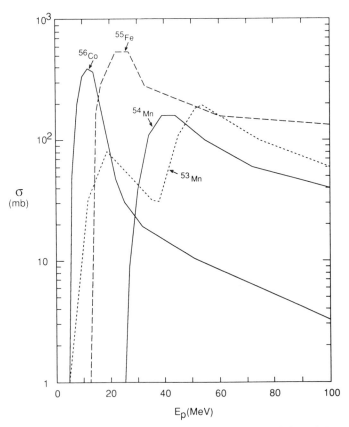

Fig. 2. Cross sections (in millibarns, mb) as a function of energy (in millions of electron volts,
MeV) for the proton production of several radionuclides from iron. See text for sources of
cross sections.

other with depth (see Fig. 1), the production profiles for nuclides like ^{56}Co and ^{54}Mn are different (Reedy and Arnold 1972). The depth-vs-concentration profile for ^{56}Co (see Fig. 3) decreases the most rapidly with depth for a solar-proton-produced nuclide. Only ^{59}Ni, made by solar α particles, has a steeper profile (Lanzerotti et al. 1973). In contrast, the production profile for a high-energy product like ^{10}Be is flat as most of the production is by more penetrating protons with energies of $\geq$ 50 MeV (Nishiizumi et al. 1987; Reedy and Arnold 1972).

IV. SCR-PRODUCED NUCLIDES IN LUNAR SAMPLES

The activities of ^{56}Co in Fig. 3 were measured in layers that were successively removed from lunar rock 12002 (Finkel et al. 1971). A schematic diagram of the samples' locations in this rock is shown in Fig. 4. The first two layers were 1 mm thick, and the following layers were from depths of 2 to 4 mm, 4 to 9 mm and 9 to 20 mm. There were also several deeper samples

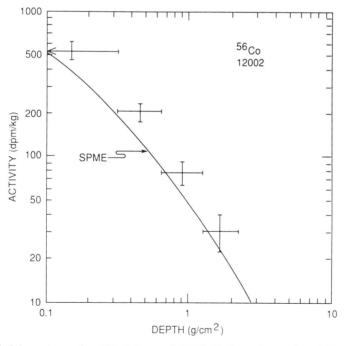

Fig. 3. Solar-proton-produced ^{56}Co in lunar rock 12002. The four points are the activities of ^{56}Co measured by Finkel et al. (1971) as a function of depth in the rock. The curve is the activity calculated for 78-day ^{56}Co using the proton fluxes directly measured by the Solar-Proton Monitor experiment of Bostrom et al. (1967–1973) and the Fe(p,xn)^{56}Co cross sections in Fig. 2. This calculated curve is very similar to that of Reedy (1977).

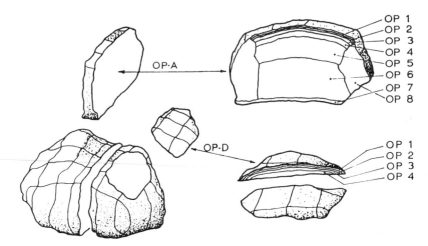

Fig. 4. Schematic of lunar rock 12002 showing where the samples analyzed by Finkel et al. (1971) were obtained. The slab OP-D provided a large area for samples within 1 cm of the top. Deeper samples came from the slice OP-A.

to get the GCR production terms (Finkel et al. 1971). The production-rate curve in Fig. 3 was calculated with the ^{56}Co cross sections of Fig. 2 and solar-proton fluxes as a function of energy and depth that were calculated from the measured proton fluxes near the Earth for several ^{56}Co half-lives prior to Apollo 12 (see Reedy 1977 for details). Prior to the Apollo 12 landing on 12 November 1969, two fairly large SCR events occurred on 2 November 1969 and 12 April 1969 (Bostrom et al. 1967–1973; King 1974; Feynman et al. 1990), and protons from these two events made most of the ^{56}Co observed in rock 12002. The fluxes determined for ^{56}Co in lunar rock 12002 from these three sources (Bostrom et al. 1967–1973; King 1974; Feynman et al. 1990) agreed well, except that above 60 MeV the flux from King (1974) was much greater than that of Bostrom et al. (1967–1973). (This difference above 60 MeV mainly affects production rates deeper than ≈3 g cm^{-2}.) In the calculations here, as in Reedy (1977), the fluxes of Bostrom et al. (1967–1973) were adopted. The production curve is for the time of the Apollo 12 mission and was calculated assuming that the reported fluxes of solar protons were isotropic (see Reedy 1977 for details). The agreement is similar to the agreements among the fluxes by various satellite experiments and shows that lunar samples are good recorders for solar protons. Activities of 35-day ^{37}Ar in the tops of lunar samples from each of the six Apollo missions that returned samples from the Moon correlated well with the solar-proton fluxes observed for several months prior to each mission (Fireman 1980).

One of the more serious problems in determining past fluxes of solar protons from lunar cosmogenic nuclides has been the lack of cross sections for the important nuclear reactions, especially at energies near the reaction thresholds. Different sets of cross sections also appear to be the main cause of differences in rates and profiles calculated for solar-proton nuclides in lunar samples as calculated by various groups (Reedy and Arnold 1972; Yokoyama et al. 1972; Tanaka et al. 1972; Michel and Brinkmann 1980). Thus the sources of cross sections used to unfold the measurements for SCR-produced nuclides in lunar samples should be noted. For ^{59}Ni made by solar α particles (Lanzerotti et al. 1973), the cross sections measured by Wahlen (1969) were used. The more recent measurements by Yanagita et al. (1978) for this reaction agree well with those of Wahlen (1969); they also extend the cross-section measurements to higher α-particle energies.

For the solar-proton results presented below, the following sources were used for production cross sections: Reedy and Arnold (1972) for ^{22}Na, ^{55}Fe and ^{26}Al; Tamers and Delibrias (1961) for ^{14}C; Tuniz et al. (1984) for ^{10}Be; and Gensho et al. (1979) for ^{53}Mn. However, there are no measured cross sections for the production of ^{10}Be below 150 MeV, so the interpretations of SCR production of ^{10}Be are limited (Nishiizumi et al. 1987). Similarly, no good cross sections at low proton energies exist for ^{36}Cl (Nishiizumi et al. 1988). For ^{41}Ca, only a few preliminary cross sections above 40 MeV have been reported (Fink et al. 1987). It should be noted that the ^{14}C production cross sections of Tamers and Delibrias (1961) were revised slightly using better monitor cross sections by Audouze et al. (1967) but now are in doubt as recent preliminary measurements for the ^{16}O(p,3p)^{14}C cross section at $\approx$150 MeV (Fireman and Beukens 1989) are about three times higher than the earlier published ones. Additional cross sections for the main reactions making most of these solar-proton-produced radionuclides will probably be measured during the next few years.

Earlier results based on ^{81}Kr (Reedy 1980; Yaniv et al. 1980) used the cross sections of Regnier et al. (1979). As will be seen below, we present and use new measurements for solar-proton-produced Kr isotopes in a lunar rock; we have revised the cross sections used to unfold the lunar measurements. These new cross sections for ^{81}Kr (B. Lavielle, personal communication, 1989) are shown in Fig. 5 and are based on work by Regnier et al. (1982), Lavielle and Regnier (1984) and Lavielle (1987). However, measured cross sections at proton energies near the reaction thresholds are still missing, especially for strontium, and these low-energy cross sections are the most important ones in unfolding the measured concentrations of SCR-produced nuclides.

Several techniques have been used to measure the radionuclides listed in Table I or other cosmogenic nuclides. Many radionuclides, including most of the shorter-lived ones ($T_{1/2} \lesssim 10^4$ yr), are measured by counting the radiations emitted when a radioactive nucleus decays. In some cases, usually when γ rays are emitted during the decay, the sample can be counted nondestructively, as has been done for ^{26}Al and ^{22}Na (see, e.g., Bhandari et al. 1976; Fruchter et al. 1982). Usually the subsample is dissolved and each element

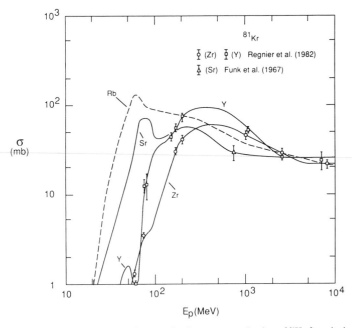

Fig. 5. Cross sections as a function of energy for the proton production of [81]Kr from its important target elements (rubidium, strontium, yttrium and zirconium) as provided by B. Lavielle (personal communication, 1989). These cross sections are based on measurements by Funk et al. (1967) and Regnier et al. (1982) and on cross-section systematics similar to those of Regnier et al. (1982) and Lavielle (1987).

of interest is chemically separated and counted in a chemically pure form (see, e.g., Finkel et al. 1971; Kohl et al. 1978). The concentrations of 3.7-Myr [53]Mn is usually measured by an activation technique in which a few decays per minute of [53]Mn are converted in a high flux of thermal neutrons to high activities of 312-day [54]Mn (cf. e.g., Finkel et al. 1971).

Since about 1980, the longer-lived radionuclides are usually measured by accelerating the isotopes of an element to high energies and using nuclear techniques to identify and count the atoms of the radioactive isotopes together with other methods to get the stable isotopes. This method is called accelerator mass spectrometry and is many orders of magnitude more sensitive for long-lived radionuclides than counting the decay radiations (cf. Elmore and Phillips 1987). In lunar samples, accelerator mass spectrometry has been used for [10]Be (Nishiizumi et al. 1987; Klein et al. 1988), [26]Al (Klein et al. 1988) and for [36]Cl (Nishiizumi et al. 1988). Accelerator mass spectrometry also is routinely used for [14]C (see, e.g., Fireman and Beukens 1989) and 17-Myr [129]I (Elmore and Phillips 1987) and should soon be regularly used for 0.1-Myr [41]Ca (see, e.g., Fink et al. 1987; Klein et al. 1989) and 76,000-yr [59]Ni.

TABLE I
Radionuclides Often Used to Study Energetic Solar Particles

Radionuclide	Half-life[a] (yr)	Major Targets
^{37}Ar	0.096	Ca, K
^{56}Co	0.213	Fe
^{54}Mn	0.855	Fe
^{22}Na	2.60	Mg, Al, Si
^{55}Fe	2.7	Fe
^{3}H	12.3	O, Mg, Si, Fe
^{14}C	5730.	O
^{59}Ni	7.6×10^4	Fe[b]
^{41}Ca	1.0×10^5	Ti, Ca, K
^{81}Kr	2.1×10^5	Sr, Zr, Y
^{36}Cl	3.0×10^5	Ca, K
^{26}Al	7.1×10^5	Si, Al, Mg
^{10}Be	1.5×10^6	O
^{53}Mn	3.7×10^6	Fe

[a] Browne et al. (1986), except for ^{10}Be (Hofmann et al. 1987), ^{26}Al (Norris et al. 1983) and ^{59}Ni (Nishiizumi et al. 1981).
[b] Mainly by α-particle reactions.

Stable and long-lived (e.g., 0.21-Myr ^{81}Kr) noble-gas isotopes are measured by static mass spectrometry. Each sample is melted, the evolved gases are purified, and all noble-gas isotopes are measured in a static mass spectrometer by jump-scanning the magnetic field (see, e.g., Lightner and Marti 1974). To obtain the cosmogenic component, other sources of stable isotopes (e.g., solar wind or, for Kr and Xe, fission components) need to be identified and subtracted from the measured mixture of components. This can be achieved by using each component's distinctive signatures of isotopic ratios. The ratios of stable GCR-produced Kr isotopes and the radioisotope ^{81}Kr permit calculations of the time interval of exposure to cosmic radiation in cases of a simple one-step exposure (Marti 1967, 1982). In the next section, we concentrate on the Kr isotopes observed in lunar rock 68815 as an example and present new results for 0.21-Myr ^{81}Kr and stable ^{83}Kr made by solar protons.

V. KRYPTON-81 IN LUNAR ROCKS

Among the many lunar samples returned by the Apollo 16 astronauts was the top from a meter-high boulder near South Ray Crater that was given the number 68815. Its orientation on the lunar surface was determined from documentation photographs taken by the astronauts before and after sampling. The length of time that this rock had been exposed to cosmic rays was determined from GCR-produced ^{81}Kr and other krypton isotopes to have been 2.0 Myr (Drozd et al. 1974; Yaniv and Marti 1981; Yaniv et al. 1980; this chapter). This [81]Kr-Kr age is in agreement with those for other rocks taken around the South Ray Crater and is attributed to the formation of the crater 2 Myr ago. The results presented below for 0.7-Myr [26]Al and 3.7-Myr [53]Mn by Kohl et al. (1978) include analyses performed on samples from 68815, and details of the sampling done on this rock can be found in Kohl et al. (1978).

Krypton was measured in pieces of the surface layer and three subsurface chips of lunar rock 68815. The krypton isotopes were measured by static mass spectrometry with peak switching techniques (Lightner and Marti 1974). Given in Table II are the measured [81]Kr concentrations, the cosmogenic [83]Kr/[81]Kr ratios, and the inferred exposure times (T_e) based on the abundances of cosmogenic [81]Kr and [83]Kr and the production ratios of Marti and Lugmair (1971). Large amounts of solar-wind-implanted [83]Kr made it impossible to determine the cosmogenic [83]Kr content of the surface sample. Also given in Table II are the cosmogenic Kr measurements for a sample deep in this rock by Drozd et al. (1974). The reason for the poor agreement between the [81]Kr concentration for our deepest sample and for that of Drozd et al. (1974) is not known but could possibly be due to inhomogeneity of the target elements. The [81]Kr concentration-vs-depth data in Table II are displayed in Fig. 6. Included in Fig. 6 are [81]Kr concentrations observed in rocks 67015 and 67095 (Hohenberg et al. 1978; Regnier et al. 1979). Both rocks were recovered on the rim of North Ray Crater and were exposed in a fixed geometry. The sample depths are known, and the compositions of the target elements for cosmogenic Kr are similar to those in rock 68815 (Hohenberg et al. 1978; Regnier et al. 1979). These other [81]Kr concentrations fit well into the profile for rock 68815. The rise of [81]Kr concentrations near the surface reflects the SCR component, while [81]Kr at larger depths is dominated by the GCR component.

Figure 6 also shows the depth-vs-concentration profiles calculated for [81]Kr in the top of lunar rock 68815 by both GCR and solar protons. These production rates were calculated with the following abundances (in ppm by weight): rubidium (8.8), strontium (180 ± 20), yttrium (64.4), and zircon-

TABLE II
Cosmogenic [81]Kr and [83]Kr in Lunar Rock 68815.

68815 Subsample	Depth (mm)	[81]Kr $(10^{-12}$ cm^{3}STP g$^{-1})$	[83]Kr/[81]Kr	T_e (Myr)
234 (1A + 1C)[a]	0 − 0.5	0.33 ± 0.06	—	—
234	8 − 11	0.19 ± 0.03	9.7 ± 1.8	1.8 ± 0.4
234	12 − 15	0.16 ± 0.02	11.9 ± 1.3	2.2 ± 0.3
207	~18	0.140 ± 0.012	11.5 ± 1.0	2.2 ± 0.2
113[b]	~64	~0.23	10.0 ± 0.4	2.04 ± 0.09

[a]Part of subsample analyzed by Kohl et al. (1978).
[b]Drozd et al. (1974), subsample number and depth from Regnier et al. (1979).

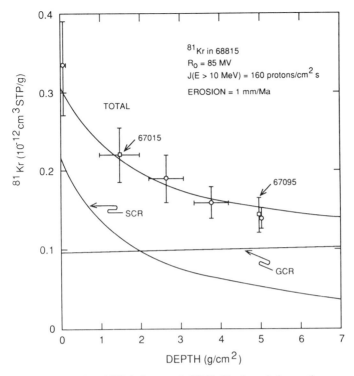

Fig. 6. Solar-proton-produced [81]Kr in lunar rock 68815. The four circles are the measured [81]Kr concentrations plotted as a function of the depth of the sample, using a density of 2.8 g cm^{-3} to convert depths in cm to g cm^{-2}. Also shown are [81]Kr measurements in two other lunar rocks with similar chemistries (Hohenberg et al. 1978). The curves are the calculated [81]Kr production rates for GCR particles and for solar protons with the indicated spectral shape, flux and erosion rate.

ium (220) (Regnier et al. 1979; H. Palme, personal communication, 1982). As there was not a good [81]Kr measurement for a sample well shielded from solar protons, we do not have a measured concentration for GCR-produced [81]Kr in this rock. Thus we treated the GCR production rate as an adjustable parameter along with the parameters for the solar protons. The erosion rate used here for 68815, 1 mm Myr^{-1}, is that determined by Walker and Yuhas (1973) with tracks, and the use of other erosion rates did not improve the fit. The match shown in Fig. 6 is not the only one possible, but the parameters for other matches were less reasonable, e.g., GCR rates that were too low. Even the GCR rate use in Fig. 6 is on the low end of the expected range.

The match used for Fig. 6 corresponds to integral proton fluxes above ≈60 MeV that are less than those previously reported by Yaniv et al. (1980) for [81]Kr in lunar rock 12002. Yaniv et al. (1980) reported a fairly hard spectral shape for the match to their [81]Kr data for 12002, but such a hard spectral shape is not consistent with our measurements for ^{81}Kr in 68815 (as it would require an unreasonably low GCR rate). These fluxes are also ~50 % higher than those of Nishiizumi et al. (1987) that fit the ~1-Myr radionuclides (^{26}Al, ^{10}Be and ^{53}Mn) in this rock. However, these estimates of ^{81}Kr-based proton fluxes are still uncertain until we have reliable production cross sections at low proton energies, as mentioned above. Additional ^{81}Kr measurements in surface and deep samples of lunar rocks with known target element abundances are also required to separate clearly the SCR and GCR components. The most important target element is strontium, from which ~0.66 of the solar-proton-produced ^{81}Kr was calculated to have been made, and then zirconium is usually the next most important. Even ~ 10 % or more of cosmogenic krypton can come from rubidium and yttrium, so these elements can be important. Unfortunately, the abundances for these target elements for cosmogenic krypton are often poorly known or variable in lunar rocks.

VI. SCR FLUXES FROM OTHER LUNAR-SAMPLE MEASUREMENTS

The fluxes of SCR particles averaged over various times in the past are summarized below. There are no quantitative results for solar α particles, but the ability to use 76,000-yr ^{59}Ni to study the solar α-particle fluxes averaged over the last ~ 10^5 yr is discussed below. For solar-proton fluxes, results or possibilities for seven time periods other than that covered by ^{81}Kr are presented below. These solar-proton fluxes are given in Table III, along with directly measured fluxes of solar protons for the last two 11-yr solar cycles. Note that, for several radionuclides, the lack of cross sections to interpret the lunar measurements has prevented us from getting quantitative solar-proton fluxes.

It should be noted that for some radionuclides, such as ^{10}Be, the lowest energy that a proton needs to induce a nuclear reaction making the desired nuclide can be fairly high. These reaction thresholds range from 5 MeV for ^{56}Co from iron to ≈60 MeV for ^{81}Kr from zirconium. Thus, for some of the radionuclides in Table III, there are blanks at the lowest energies. Few thresholds are much below 10 MeV, so no solar-proton fluxes below 10 MeV are given. Most of the half-lives of these radionuclides are fairly well measured, usually to within about ± 10%. Except for ^{81}Kr, the half-life does not affect the interpretation if the rock's exposure age is long enough that the radionuclide is in equilibrium with its production rate. For ^{81}Kr, the half-life is needed to convert the measured atoms into decay rates for comparisons with calculated production rates. In Table I, the most recently evaluated half-lives are given for these SCR-produced radionuclides.

In Table III, the average fluxes of solar protons measured directly in space during the last two 11-yr solar cycles, numbers 20 (1965–1975) and 21 (1976–1986), are given. Solar cycles are based on sunspot numbers and date

TABLE III
Solar-Proton Integral Fluxes Averaged Over Various Time Periods

Time Period[a]	Data Source	Flux References	R_0[b] (MV)	Integral Fluxes[c]			
				E>10	E>30	E>60	E>100
1976–1986	IMP-8[d]	Goswami et al. 1988	40	63	5	0.6	~0.2[e]
1965–1975	SPME[d]	Reedy 1977	90	92	30	8	—[e]
1954–1964	^{22}Na, ^{55}Fe	Reedy 1977	100	378	136	59	26
~ 10^4 yr	^{14}C	Boeckl 1972	100	—[f]	~72[g]	~26[g]	~9[g]
~ 2×10^5 yr	^{41}Ca	Klein et al. 1989	—	—[f]	—[h]	—[h]	—[h]
~ 3×10^5 yr	^{81}Kr	this work	~85	—[f]	—[f]	—[h]	—[h]
~ 5×10^5 yr	^{36}Cl	Nishiizumi et al. 1988	—	—[f]	—[h]	—[h]	—[h]
~ 10^6 yr	^{26}Al	Kohl et al. 1978	100	70	25	9	3
~ 2×10^6 yr	^{10}Be	Nishiizumi et al. 1987	≥70	—[f]	~35[h]	~8[h]	~2[h]
~ 5×10^6 yr	^{53}Mn	Kohl et al. 1978	100	70	25	9	3

[a] Time periods are last three 11-yr solar cycles or mean-lives of the radionuclides.
[b] Spectral shape in rigidity, usually between 10 and 30 MeV.
[c] Fluxes are omnidirectional in protons cm^{-2} s^{-1}, and the four energies are in MeV.
[d] Direct proton measurements by experiments on several satellites.
[e] Not measured (1965–1973) or sometimes not reported (1973–1986).
[f] Energy is below main reaction thresholds.
[g] Needs additional cross sections to check those used to unfold the measurements (see text).
[h] Few or no cross sections available for unfolding the lunar-radioactivity measurements.

back over the last two centuries for which there are good sunspot observations. Each solar cycle starts when the smoothed distribution of sunspot numbers reaches a minimum. These two sets of solar-proton fluxes are based on compilations by Reedy (1977) and Goswami et al. (1988) for proton fluxes during individual events. The fluences of protons measured during each solar cycle were divided by the number of seconds in 11 yr to get the averages given in Table III. Somewhat different values for the proton fluxes during the last two solar cycles and more details on the proton events during this time period are given by Feynman et al. (1990) and Shea and Smart (1990). These solar-proton fluxes measured during the last two solar cycles are our only direct measurements for energetic particles from the Sun. During this time period, only the proton fluxes during August 1972 were much higher in intensity than those for other SCR events. (From preliminary data, several solar-particle events during 1989 appear to be very large, possibly like those prior to 1962; M. A. Shea, personal communication, 1989.) Below, lunar radioactivities used to convert Earth-based observations during solar cycle 19 (1954–1964) into solar-proton fluxes are discussed.

Solar α-Particle Fluxes from ^{59}Ni

As α particles are a small fraction (~2%, McGuire et al. 1986) of SCR particles, special circumstances are needed to see the production effects of solar α particles above those of solar protons. The one case where solar-proton production is unimportant is for ^{59}Ni in lunar samples (Lanzerotti et al. 1973). In this case, the lunar abundances of nickel and cobalt, from which GCR or solar-proton production of ^{59}Ni will occur, are very low relative to lunar iron abundances. The next best case for solar α-particle production is for 71-day ^{58}Co. In several lunar samples, 272-day ^{57}Co has been measured, and this radionuclide is made roughly equally by solar protons and solar α particles (Lanzerotti et al. 1973).

Lanzerotti et al. (1973) presented fluxes of solar α particles measured from 1967 to 1969 from 1 to ≈ 12 MeV/nucleon. The average α-particle flux over this time period and energy range was fit with a function having the shape $E^{-1.9}$ and an omnidirectional flux above 2.5 MeV/nucleon of 8 α particles cm^{-2} s^{-1}. The lunar production profile of ^{59}Ni was calculated with this incident flux. The cross sections for the main reaction, ^{56}Fe(α,n)^{59}Ni, are well measured (see discussion above). As the production profile is very steep, the activity of ^{59}Ni in the top few millimeters of a lunar rock depends on the erosion rate for the rock's surface. Lunar rocks typically have erosion rates that range from a fraction to several mm Myr^{-1} (Walker 1975; Burnett and Woolum 1977). Only a few measurements have been reported for ^{59}Ni in the tops of lunar samples. These ^{59}Ni measurements are qualitatively (within a factor of 4, according to Lanzerotti et al. [1973]) in agreement with the calculated ones. To get the average solar α-particle flux over the last ~10^5 yr, good depth-vs-activity profiles of ^{59}Ni need to be measured and erosion rates for that rock's surface independently determined. As the radiations emitted during the decay of ^{59}Ni are very hard to count, this work will probably be done when ^{59}Ni can be measured well by accelerator mass spectrometry.

Solar-Proton Fluxes During Solar-Cycle 19 (1954–1964)

Prior to about 1960, there were only indirect observations of the fluxes of solar protons using neutron monitors and the effects that these SCR particles have on the absorption of extraterrestrial radio noise near the polar caps (see, e.g., Webber et al. 1963; McDonald 1963; Shea and Smart 1990). Reedy (1977) compiled these indirectly determined solar-proton fluxes during solar-cycle 19. The lunar radioactivities measured for 2.6-yr ^{22}Na and 2.7-yr ^{55}Fe were much higher than those calculated with the proton fluxes determined for solar-cycle 19 (1954–1964) using these indirect observations (cf. Finkel et al. 1971). Reedy (1977) used these lunar radioactivities and the relative distribution of solar protons (based mainly on the compilation of Webber et al. 1963) to determine absolute proton fluxes during this period. Reedy's (1977) fluxes determined for solar-cycle 19 from the lunar ^{22}Na and ^{55}Fe measurements are given in Table III. As these results depend on the relative distribution of the proton fluxes during solar-cycle 19, they would change if this distribution is changed. However, the change in the fluxes will most likely be ≤50 % as these radionuclides have half-lives that are about half of the period (1956–1961) during which most of the solar protons during solar-cycle 19 were emitted. The old indirect measurements during this time period should be re-examined to get a better determination of the relative distribution of proton fluxes, especially for lower proton energies (down to 10 MeV), although this re-evaluation for some of the older solar-particle events in solar-cycle 19 may be difficult (cf. Shea and Smart 1990).

Gaps in Time Windows into the Past

Cosmogenic nuclides made *in situ* are integral records of cosmic radiations, and most of the time periods studied are determined by the half-lives of the SCR-produced radionuclides. For a constant production rate, half of the equilibrium (saturation) activity of a radionuclide is made over one mean-life of the radionuclide (a mean-life being the half-life divided by 0.693, the natural logarithm of 2). There are several time regions for which there are no suitable radionuclides with which to determine solar-proton fluxes. As can be seen in Table I, the first big time gap is from 12 to 5730 yr. Although 12.3-yr ^{3}H (tritium) is listed in Table I, the lunar measurements are spotty (see, e.g., Fireman 1980) and often poorly reproducible. The large fluxes of solar protons during 1954–1964 also make it hard to determine solar-proton fluxes prior to 1954 from the lunar ^{3}H measurements. There is also a fairly large uncertainty in the GCR production rates of ^{3}H (Reedy 1977), and some lunar samples have anomalously high ^{3}H contents released at low temperatures (Fireman 1980) that further complicates data interpretation. This ^{3}H released at low temperatures appears to be contamination and not due to tritium in the solar wind (Fireman 1980). The firmest conclusion from the lunar ^{3}H measurements is that the solar-proton fluxes prior to 1954 were not usually high, probably having a flux above 10 MeV of very roughly 150 protons cm^{-2} s^{-1} (Reedy 1977).

Several other radionuclides with half-lives between 12 and 5730 yr were listed in Reedy (1980), but most (e.g., 680-yr ^{91}Nb) are unsuitable for solar-proton studies, mainly because of the lack of target elements. Michel and Brinkmann (1980) considered the production of 47-yr ^{44}Ti by solar-proton reactions with titanium. The calculated SCR production rates of ^{44}Ti are low, ($\sim$ 0.5 atoms min^{-1} kg^{-1}) for a lunar rock with 1% by weight of titanium. Thus, improved ^{44}Ti measurement techniques are needed as presently ^{44}Ti is not easily measured, and a lunar rock with high titanium content would be preferable (Michel and Brinkmann 1980). If ^{44}Ti measurements are to be done on Apollo samples, they should be done soon as the ^{44}Ti is fairly rapidly decaying.

It would be very nice to determine the SCR-particle flux during the period of the Maunder Minimum, which was from about 1645 to 1715 when there were many indications (e.g., the lack of sunspots) that the Sun's activity was low (Eddy 1976a). However, there are no suitable radionuclides with the appropriate half-life, except 269-yr ^{39}Ar. However, ^{39}Ar is made in very low yields by SCR particles but is readily produced by GCR-secondaries by the ^{40}Ca(n,2p) ^{39}Ar and ^{39}K(n,p)^{39}Ar reactions. In iron meteorites, ^{39}Ar is made from iron by GCR particles with energies $\gtrsim$100 MeV. Forman and Schaeffer (1980) note that it might be possible to see the effects of the Maunder Minimum in GCR-produced ^{39}Ar in meteorites that fell during the 18th century. They also point out that observed ^{39}Ar activities in recently fallen meteorites are consistent with a strong enhancement of GCR production during the Maunder Minimum. From 269-yr ^{39}Ar to 5730-yr ^{14}C and from ^{14}C to 76,000-yr ^{59}Ni, there are also no good candidate radionuclides with which to study SCR-particle fluxes in the past.

Solar-Proton Production of 5730-yr ^{14}C

Solar protons readily make ^{14}C by the ^{16}O(p,3p)^{14}C reaction (Reedy and Arnold 1972). An activity-vs-depth profile was measured by Boeckl (1972), and the proton fluxes given in table III were determined from this profile using the cross sections of Tamers and Delibrias (1961) discussed above. Begemann et al. (1972) measured several activities of ^{14}C for different depths in lunar rock 12053 that are consistent with the profile of Boeckl (1972) except for their surface sample. As noted by Fireman (1980), there is some evidence for solar-wind-implanted ^{14}C on lunar samples. The higher ^{14}C content observed in the top sample of Begemann et al. (1972) compared to the corresponding sample of Boeckl (1972) may be due to such solar-wind-implanted ^{14}C.

While it would be good to measure another ^{14}C activity-vs-depth profile in a lunar rock, the major need now is to be certain of the cross sections for proton production of ^{14}C from oxygen. As mentioned above, a preliminary cross section for this reaction at $\approx$150 MeV (Fireman and Beukens 1989) was ~3 times that used to unfold the lunar ^{14}C measurements. If the proton fluxes listed for ^{14}C in Table III are reduced by a factor of 3, then they would be in good agreement with those determined over the last few Myr. Solar-wind ^{14}C on the very surface of lunar samples means that the topmost sample may be compromised in its use in getting solar-proton fluxes, so in any new measurements of lunar ^{14}C the next sample from the surface should not be too far from the very top.

Solar-Proton Fluxes During the Last Few Hundred Thousand Years

Besides 0.21-Myr ^{81}Kr discussed above, two other solar-proton-produced radionuclides have half-lives in this range, 0.1-Myr ^{41}Ca and 0.30-Myr ^{36}Cl. Some preliminary measurements have recently been made using accelerator mass spectrometry for each of these radionuclides in the tops of lunar samples, but unfortunately there are very few or no good cross sections at low-proton energies with which to unfold the measurements.

Klein et al. (1989) have observed an excess of ^{41}Ca in the top of lunar rock 74275 due to solar-proton production. The major target element for producing ^{41}Ca in many rocks is titanium, so some cross sections for these reactions should be measured to confirm and extend those reported by Fink et al. (1987). The production of ^{41}Ca from calcium also must be known, since it could be important.

Nishiizumi et al. (1988) measured a clear solar-proton signal for ^{36}Cl in the top of a lunar core. There could have been some movement of lunar soil (gardening) in the top of this core, which would make it difficult to unfold the measurements in terms of solar-proton fluxes. A ^{36}Cl activity-vs-depth profile should be measured in a hard lunar rock with fairly high contents of the major target element, calcium. As discussed in Nishiizumi et al. (1988), proton cross sections need to be measured near the threshold energies before measurements for solar-proton-produced ^{36}Cl in lunar samples can be unfolded.

Solar-Proton Fluxes During the Last Few Million Years

Of all the time periods in the past, that back to ~1 Myr ago has had the most attention. One reason is that there are three radionuclides with half-lives in this time range: 0.7-Myr ^{26}Al, 1.5-Myr ^{10}Be, and 3.7-Myr ^{53}Mn. Both ^{26}Al and ^{53}Mn are readily made by solar protons and are fairly easily measured. Another set of time windows are lunar rocks that were excavated from depth during this period. The time that a rock has been on the lunar surface can be determined (cf. Burnett and Woolum 1977) from studies of the tracks made by heavy cosmic-ray nuclei or of GCR-produced nuclei (e.g., using the ^{81}Kr-Kr dating method; Marti and Lugmair 1971). The surface age for a lunar rock needs to be known in order to interpret the cosmogenic nuclides measured in it. Care must be taken to avoid lunar rocks with complex histories, especially ones with a relatively recent excavation from fairly shallow depths (cf. Bhandari 1981). Many rocks have simple exposure histories and have been on the surface for short enough periods that the activities of some of these long-lived radionuclides, especially ^{53}Mn, are not in equilibrium with their production rates.

Another complication for studies of SCR-produced nuclides with half-lives $\gtrsim 10^5$ yr is that the surfaces of lunar rocks are eroded by micrometeoroid impacts at a rate of ~1 mm Myr^{-1}, and SCR production rates can change appreciably over 1 mm. Lunar rocks can have a fairly wide range of erosion rates, with hard igneous ones eroding slowly while friable breccias can erode at much higher rates. For example, the erosion rates used by Kohl et al. (1978) for their three rocks were 0.5, 1.3 and 2.2 mm Myr^{-1}. Thus, the rate that a lunar rock has been eroding needs to be independently determined, such as with tracks of heavy cosmic-ray nuclei (Walker 1975; Burnett and Woolum 1977).

The solar-proton fluxes given in Table III for ^{26}Al and ^{53}Mn are those of Kohl et al. (1978) in lunar rocks 12002, 14321 and 68815. An independent and more thorough interpretation of the measurements of Kohl et al. (1978) for rock 68815 was performed by Russ and Emerson (1980), who also got the same flux. However, as both sets of authors noted, these fluxes depend on the erosion rate used for a rock. One can fit a measured solar-proton-produced profile for a nuclide with a number of combinations of flux, spectral shape and erosion rate. For example, the ^{26}Al and ^{53}Mn measurements of Kohl et al. (1978) for rock 68815 could also be matched with a higher integral flux above 10 MeV, a softer spectrum, and a higher erosion rate (cf., e.g., Nishiizumi et al. 1987). Kohl et al. (1978) also were able to match the ^{26}Al and ^{53}Mn profiles in rocks 12002 and 14321, both of which had long surface ages, with the same solar-proton fluxes and spectral shape as they obtained for 68815. Thus, the time period covered by ^{53}Mn in Table III is given as ~5 Myr, the mean-life of ^{53}Mn. As rocks 12002 and 14321 probably were on the surface for at least 10 Myr, the measurements for ^{53}Mn imply that the solar-proton fluxes from ~5 to 10 Myr ago were not greatly different than those over the last 2 Myr. Another indication of the constancy of the solar-proton flux over the last few Myr is the work of Bhandari et al. (1976) and Bhandari (1981), who nondestructively measured ^{26}Al in six lunar rocks with surface ages ranging from about 0.5 Myr to 5 Myr and saw less than about $\pm$ 20% (the experimental error in the ^{26}Al measurements) variation in the average solar-proton fluxes for this time period.

While there is little indication of significant variations of the fluxes of solar protons over the period of ~0.5 to 10 Myr, there is considerable dispute about what were the solar-proton fluxes and spectral shape. Bhandari (1981) argues that the fluxes of solar protons over the last few Myr were higher and the spectrum harder (a higher value for R_o, 125 MV, than the 100 MV value obtained by Kohl et al. 1978) than those corresponding to the fluxes given in Table III. Bhandari (1981) prefers integral fluxes above 10 and 60 MeV of 125 and 25 protons cm^{-2} s^{-1}, respectively. This flux above 60 MeV is $\approx$3 times that of Kohl et al. (1978). Another independent determination of the solar-proton fluxes was made by Rancitelli et al. (1972), who fit their ^{26}Al measurements with the proton spectral shape of $E^{-3.1}$ and an integral flux above 10 MeV of 60 protons cm^{-2} s^{-1}, which corresponds to an integral flux above 60 MeV of only $\approx$1.4 protons cm^{-2} s^{-1}. Langevin et al. (1982) found that any solar-proton spectral shape with $R_o \geq$ 120 MV was "not compatible with experimental profiles" for ^{26}Al and ^{53}Mn in seven lunar cores and could generally explain the lunar-core measurements with their gardening models and the fluxes listed in Table III for these two radionuclides.

Other more recent measurements of SCR-produced ^{26}Al in lunar rocks are in fairly good agreement with those of Kohl et al. (1978). Two groups have measured activity-vs-depth profiles of ^{26}Al in lunar rock 74275. Fruchter et al. (1982) nondestructively counted ^{26}Al in $\approx$1-cm-thick horizontal sections of this rock and fit their measurements with the same proton fluxes as determined by Kohl et al. (1978). Klein et al. (1988) used accelerator mass spectrometry to measure ^{26}Al and ^{10}Be in 0.5- to 1-mm-thick slices of this rock. Their ^{26}Al measurements agreed with those of Fruchter et al. (1982). Klein et al. (1988) noted that their surface of rock 74275 only saw space with a solid angle of 1.8π and that the rock had a complex exposure, having been excavated from a depth of $\sim$40 g cm^{-2} about 2.8 Myr ago. They found that the solar-proton fluxes for ^{26}Al in Table III did not fit their measurements very well, although they have several ways of removing the discrepancy, like lowering the assumed density of the rock or decreasing their adopted aluminum content for the rock.

Recently, there have been detailed measurements of activity-vs-depth profiles of ^{10}Be in the top of lunar rocks, by Nishiizumi et al. (1987) in rock 68815 and by Klein et al. (1988) in 74275. Both saw no evidence of ^{10}Be in excess of that produced by GCR particles in the tops of their rocks, and they both set limits of $\leq$1 disintegration min^{-1} kg^{-1} (dpm kg^{-1}) for solar-proton-produced ^{10}Be. The ^{10}Be production rate calculated at the surface of a lunar rock using the solar-proton fluxes given in Table III for ^{26}Al and ^{53}Mn is 2 atoms min^{-1} kg^{-1}), which at equilibrium is an activity of 2 dpm kg^{-1}, and this rate is slowly varying with depth and not very sensitive to erosion rate since ^{10}Be is mainly made by high-energy ($E \sim 100$ MeV) protons.

Nishiizumi et al. (1987) cite two possible explanations for the low observed solar-proton-produced ^{10}Be: (1) incorrect cross sections (noting that there are no experimental cross sections for ^{10}Be from oxygen below 135 MeV); or (2) the solar-proton fluxes over the last few Myr were higher than 70 protons cm^{-2} s^{-1} above 10 MeV and softer ($R_o < 100$ MV) than those adopted for the other two long-lived radionuclides, and the erosion rate for rock 68815 was higher than 1.3 mm Myr^{-1}. For example, a flux above 10 MeV of 150 protons cm^{-2} s^{-1}, a spectral shape of $R_o = 70$ MV, and an erosion rate of 3 mm Myr^{-1} would fit all of the measurements for 68815. The fluxes listed in Table III for ^{10}Be correspond to this latter fit to all three radionuclides in 68815. The ^{81}Kr results discussed above, therefore, are not in serious conflict with the ^{26}Al, ^{53}Mn and ^{10}Be results. Note that for fluxes above 30 MeV, this fit and those for ^{26}Al and ^{53}Mn are fairly similar. Because ^{10}Be is not made by protons with energies below 34 MeV, no flux is given in Table III for $E > 10$ MeV. However, cross sections should be measured for several energies below 100 MeV before one can say anything definite about what the ^{10}Be measurements mean in terms of solar-proton fluxes in the past. The lack of solar-proton-produced ^{10}Be in lunar rocks is inconsistent with the high flux and hard spectral shape advocated by Bhandari et al. (1976) and Bhandari (1981). Using the fluxes and spectral shape of Bhandari (1981) and the ^{10}Be cross sections estimated by Tuniz et al. (1984), the calculated equilibrium activity of ^{10}Be in the top of a lunar rock would be 6 dpm kg^{-1}. After ^{10}Be production cross sections are measured at lower proton energies, the combination of the high-energy product ^{10}Be with the low-energy products ^{26}Al or ^{53}Mn should help to get more definitive solar-proton fluxes over the last few Myr (cf. Nishiizumi et al. 1987).

Another test for the average flux and spectral shape of solar protons over the last few Myr would be the activities measured for ^{26}Al and ^{53}Mn as a function of depth in lunar cores. While avoiding the problem with lunar rocks of erosion, lunar cores usually have more serious problems (cf., e.g., Langevin et al. 1982). For example, a few cores probably had lunar soil material (often that near the surface) physically lost between the time just before the astronauts took the core on the Moon's surface until the core was opened in Houston, and core material could have been vertically mixed during this period. Another complication with core samples on time scales of $\gtrsim 10^5$ yr is gardening, the gradual but continuous vertical and horizontal mixing of the lunar soil. Vertical mixing alone would not necessarily be a serious problem as the integral production over depth of these nuclides would be preserved. However many cores appear to have had material horizontally removed from or added to them. As mentioned above, Langevin et al. (1982) used the lunar core measurements to infer that the average spectral shape of solar protons over the last ~ 1 to 5 Myr was one with $R_o \lesssim 120$ MV. The lunar core data may also be able to constrain the softness of the spectrum over this time period and to eliminate some combinations of fluxes and spectral shapes.

VII. SUMMARY AND CONCLUSIONS

The fluxes of solar protons averaged over various time periods are summarized in Table III. Most of these fluxes have been available for at least a decade, and there is little new in the table. However, as discussed below, most of the blanks in Table III for solar-proton fluxes in the past should be filled fairly soon. There also should be some new fluxes for the contemporary SCR, such as those in solar-cycle 22 (which started in 1986 and appears as if it will have much solar activity, including large SCR events). However, there is much that can be concluded from what has been presented above. There is also much that needs to be done to use fully the information that has been preserved in lunar samples about past fluxes of SCR particles.

One of the first conclusions from the studies of solar-proton-produced radionuclides in lunar samples was that the integral fluxes of solar protons above 10 to 100 MeV averaged over the last few Myr were not very different (factors of several) from the contemporary fluxes. The similarity of the activity-vs-depth profiles of 2.6-yr ^{22}Na and 0.71-Myr ^{26}Al, which are made by similar proton-induced reactions, qualitatively showed this conclusion (see, e.g., Shedlovsky et al. 1970). Table III is more quantitative and shows that the ranges of solar-proton fluxes during the last three solar cycles are greater than those determined for average fluxes during various periods over the past $\sim$ 10 Myr. The sparsity of numbers in this table illustrates the need for more measurements of the contemporary SCR as well as the ancient SCR.

One of the more interesting (but tentative) conclusions from early work involving solar-proton fluxes in the past was that the fluxes determined from ^{14}C measurements by Boeckl (1972) and those reported using ^{81}Kr (Yaniv et al. 1980) were at least a factor of 2 higher than those for the last $\sim$ 1 Myr. These results suggested that the Sun's output of energetic particles may have varied over long periods of time in the past. As noted then (see, e.g., Reedy 1980), these results needed to be confirmed. If the preliminary cross section by Fireman and Beukens (1989) for ^{14}C for oxygen is correct, then the solar-proton fluxes over $\sim$ 10^4 yr are similar to those over the last $\sim$ 1 Myr. Similarly, the new results reported here for krypton suggest that the old ^{81}Kr-derived fluxes were also too high. While cross sections need to be measured for both of these radionuclides and additional lunar measurements should be done, it appears that the average fluxes and spectral shapes of solar protons have not differed greatly (i.e., within factors of several) over time periods of $\sim$ 10^4 to 10^7 yr into the past. Solar-proton fluxes determined from 0.1-Myr ^{41}Ca and 0.3-Myr ^{36}Cl will help to fill in this picture for the temporal nature of solar-proton fluxes during the last 1 Myr.

One question that has been asked but has not been well answered is whether there are any differences in the properties of SCR observed since 1956 and those averaged over the last $\sim$ 10^4 to 10^7 yr. Lal (1988a) argues that the mean spectral shape of recent events is softer than the average spectral shape over the last 1 Myr. However, Lal (1988a) mainly considered the mean for the spectral shapes for all recent events, and this may not be representative because a single SCR event can dominate a cycle, as the August 1972 events dominated solar-cycle 20. Lal (1988a) also used the groundbased data for solar-cycle 19, and the fluxes in table III for this cycle as corrected using lunar measurements are much harder ($R_0 \approx$ 100 MV). One problem in this comparison is that we really do not know the properties of modern SCR events well, especially for the rare, very large events for which we have few statistics (cf., e.g., Goswami et al. 1988). There are also indications that the long-term spectral shape is softer than the value in Table III for ^{26}Al and ^{53}Mn (R_0 = 100 MV). As discussed above, the ^{10}Be measurements suggest that the solar-proton spectrum averaged over the last $\approx$2 Myr could have been softer than one with an R_0 of 100 MV. We conclude that with more and better data for both modern and ancient SCR events, we should be able to determine how representative are the modern energetic solar particles compared to those over the last $\sim$ 10 Myr.

The proton fluxes given in Table III for solar-cycle 19 (1954–1964) could possibly be improved by re-evaluating the groundbased data for the relative distributions of solar protons for several energies above 10 MeV during the major SCR events in that cycle. This particular solar cycle is of great interest because it had several events with very large fluences ($\gtrsim 10^{10}$ protons cm^{-2}) of SCR particles, while the 4 August 1972 event was the only such very large event in the subsequent two solar cycles. These relative distributions for the solar protons could then be used with the lunar ^{22}Na and ^{55}Fe measurements, as discussed above, to get absolute fluxes during solar-cycle 19. Improved fluxes for events from 1956 to 1963 would provide additional data for statistical studies of the proton fluxes in SCR events. Such improved statistics would allow better predictions for the probabilities of proton fluxes in future SCR events, especially for the highest fluxes.

Another interesting but more speculative subject is the occurrence of huge SCR events in the past. During the last three solar cycles, the largest events had peak fluxes above 10 MeV of $\sim 10^6$ protons cm^{-2} s^{-1} and event-integrated fluences above 10 MeV of $\sim 10^{10}$ protons cm^{-2}. Events of this size are very serious radiation hazards to people and electronics in space (Rust 1982; Letaw et al. 1987). Much larger events could have affected life on Earth, either directly by their radiation or indirectly (by depleting the ozone and allowing high fluxes of ultraviolet radiation to reach the Earth's surface— see the chapter by Hunten et al.). Several authors (see, e.g., Lingenfelter and Hudson 1980; Lal 1988a) have observed that the probability distribution of having an event of a certain size per unit time vs its size is roughly a straight line on a log-log plot up to integral fluences of $\sim 10^9$ to 10^{10} protons cm^{-2} but that this simple trend cannot be extended to much higher fluences. As noted by these authors, the results from lunar cosmogenic nuclides show that the probability of huge ($> 10^{11}$ protons cm^{-2}) events drops considerably below this trend line for more typical ($< 10^{10}$ protons cm^{-2}) events, at least for SCR events during the last ~ 10 Myr.

We note that, since the reviews of Reedy (1980) and Reedy et al. (1983), considerable progress has been made on the study of SCR particles in the past using lunar cosmogenic nuclides. The use of accelerator mass spectrometry to measure quickly and very accurately long-lived radionuclides in small samples has very recently added three isotopes, 0.1-Myr ^{41}Ca, 0.30-Myr ^{36}Cl and 1.5-Myr ^{10}Be, to the list of those used to study ancient SCR particles. The big need now for these three radionuclides, and for ^{14}C and ^{81}Kr, is to measure cross sections for their production by low-energy protons. If we had such cross sections now, many of the blanks for solar-proton fluxes in Table III would have numbers in them. Proton cross sections that should be measured then include:

1. ^{14}C from oxygen (best if $E_p \approx 40$ to 50 MeV) to confirm those of Tamers and Delibrias (1961), and from magnesium, aluminum and silicon from $\approx$ 50 to 60 MeV to ~ 100 MeV;
2. ^{81}Kr near and above the reaction thresholds with strontium ($\approx$60 MeV) and zirconium ($\approx$80 MeV);

3. ^{10}Be from $\approx$40 MeV to $\sim$ 150 MeV for oxygen and with $E_p \gtrsim$ 50 to 60 MeV from magnesium, aluminum and silicon;

4. ^{36}Cl from calcium for $E_p \gtrsim$ 45 MeV and from potassium starting at 20 to 25 MeV;

5. ^{41}Ca from titanium for $E_p \gtrsim$ 35 MeV and from calcium for several low proton energies.

Several groups of investigators are now planning to measure many of these cross sections, and soon some of the blanks in Table III may be filled.

We also conclude that additional measurements of SCR-produced nuclides in lunar samples, especially rocks, are needed for ^{81}Kr, ^{36}Cl, ^{41}Ca, ^{59}Ni and ^{14}C. These activity-vs-depth profiles need to be measured in the tops of hard lunar rocks with known densities (or with sample depths well determined in g cm^{-2}). It would be worthwhile if some profiles for the other long-lived radionuclides, ^{26}Al, ^{10}Be and ^{53}Mn, are also determined in these samples. The rocks used for such studies should be well documented to have been horizontal on the Moon's surface with a known, simple exposure history.

Acknowledgments. We have benefited in countless ways from discussions and collaborations with J. R. Arnold, J. Nishiizumi, D. Lal, B. Lavielle, H. Palme, A. Yaniv and many other colleagues doing research involving SCR-produced nuclides. We thank N. Bhandari, G. F. Herzog, and an anonymous reviewer for valuable comments and suggestions on earlier drafts of this chapter. We dedicate this chapter to our late colleague Ed Fineman, who was involved in many aspects of the work presented here. Most of this research was supported by NASA. The work at Los Alamos was done under the auspices of US DOE.

Note added in proof: In several places in this chapter, it was noted that the cross section for the production of ^{14}C from oxygen as reported by Fireman and Beukens (1989) near 150 MeV was about 3 times higher than that of Tamers and Delibrias (1961) at this energy. The cross sections of Tamers and Delibrias (1961) were used to unfold the lunar ^{14}C measurements. According to J. Sisterson (personal communication, 6 August 1990), the number of protons hitting Fireman's target was 2.26 times higher than the value used by Fireman and Beukens (1989) to get their cross section. Thus their cross section should be lowered by this factor, bringing it into much better agreement with the cross section reported near this energy by Tamers and Delibrias (1961).

Another set of unpublished cross sections for the ^{16}O(p,3p) ^{14}C reaction, measured by H. Roman for several energies below 63 MeV and reported in the McMaster University Accelerator Laboratory 1988 Annual Report, are also in good agreement (much less than a factor of 2 difference) with those of Tamers and Delibrias (1961). These agreements with the cross sections of Tamers and Delibrias (1961) imply that the high solar-proton fluxes averaged over the last 10,000 yr (the mean-life of ^{14}C) as reported in Table III are probably not in serious error because of the cross sections used to unfold the lunar ^{14}C measurements.

SOLAR NEUTRINOS AND THE HISTORY OF THE SUN

KURT WOLFSBERG
Los Alamos National Laboratory

and

GRANT E. KOCHAROV
A. F. Ioffe Physical-Technical Institute

Neutrinos may be the only particles created in the thermonuclear processes of the Sun that escape directly. Therefore, their measurement provides important tests of solar models and of fundamental properties of neutrinos. There is a serious discrepancy between the results of the Homestake and Kamiokande experiments that measure the high-energy neutrino flux at the present time and the values predicted by the Standard Solar Model. Interpretations of this famous puzzle fall into two classes: (1) problems in cross sections or velocity distributions for the high-energy, neutrino-producing reactions or in the understanding of the fundamental properties of neutrinos; and (2) adjustments to models of the Sun and its evolution involving energy transport, new types of particles, various types of mixing, or more burning of ³He. The gallium detectors, now coming on-line, will measure lower-energy neutrinos to address some of these solutions. New detectors have been proposed to measure other features of the current neutrino fluxes. Geochemical experiments can measure long-lived nuclides in the Earth's crust that were produced by solar neutrinos and can offer information on the neutrino fluence over millions of years. The molybdenum-technetium experiment, also currently underway, is one such experiment.

I. NEUTRINO GENERATION IN THE SUN

In the 1930s, the hypothesis was presented (see, e.g., Atkinson and Houtermans 1929; von Weiszäcker 1937; Gamow 1938; Bethe and Critchfield

288

1938; Bethe 1939) that energy generation in stars resembling the Sun is due to reactions involving conversion of hydrogen to helium. Although mankind has learned to achieve thermonuclear reactions in the laboratory and in explosions, the origin of solar energy still remains a hypothesis. The hypothesis is elegant and well founded but, nevertheless, requires a direct test. Neutrinos provide a way of testing the theory of nuclear generation in the nearest star, the Sun, directly and quantitatively. They are the only particles produced by the reactions that can penetrate from the center of the Sun into space. Thus neutrinos offer a unique probe for studying the solar interior. Experiments to measure solar neutrinos are crucial tests of theories of stellar evolution.

It is universally accepted that fusion of hydrogen to helium in Sun-like stars should proceed via the proton-proton (p-p) or carbon-nitrogen-oxygen (CNO) cycles (Table I). In both cycles the result is the same: four protons fuse to form a ^{4}He nucleus. The reaction also involves the formation of 2 neutrinos, positrons, which annihilate with electrons, and gamma rays. The reaction liberates 26.7 MeV of energy, the difference of the rest masses of the

TABLE I
The p-p and CNO Cycles in the Sun[a]

Reaction	Termination (%)	Energy (MeV)	Type of Neutrino
p-p Reaction Chains			
$p + p \rightarrow {}^2\text{H} + e^+ + \nu_e$	99.75	$\leq$0.420	p-p
or			
$p + e^- + p \rightarrow {}^2\text{H} + \nu_e$	0.25	1.442	pep
${}^2\text{H} + p \rightarrow {}^3\text{He} + \gamma$	100		
${}^3\text{He} + {}^3\text{He} \rightarrow {}^4\text{He} + p + p$	85		
or			
${}^3\text{He} + {}^4\text{He} \rightarrow {}^7\text{Be} + \gamma$	15		
${}^7\text{Be} + e^- \rightarrow {}^7\text{Li} + \nu_e$	15	0.861 (90%)	${}^7\text{Be}$
		0.383 (10%)	
${}^7\text{Li} + p \rightarrow {}^4\text{He} + {}^4\text{He}$	15		
or			
${}^7\text{Be} + p \rightarrow {}^8\text{B} + \gamma$	0.02		
${}^8\text{B} \rightarrow {}^8\text{Be}^* + e^+ + \nu_e$	0.02	<15	${}^8\text{B}$
${}^8\text{Be}^* \rightarrow {}^4\text{He} + {}^4\text{He}$	0.02		
or			
${}^3\text{He} + p \rightarrow {}^4\text{He} + e^+ + \nu_e$	0.00002	$\leq$18.77	hep
CNO Reaction Chains			
${}^{12}\text{C} + p \rightarrow {}^{13}\text{N} + \gamma$			
${}^{13}\text{N} \rightarrow {}^{13}\text{C} + e^+ + \nu_e$		$\leq$1.99	${}^{13}\text{N}$
${}^{13}\text{C} + p \rightarrow {}^{14}\text{N} + \gamma$			
${}^{14}\text{N} + p \rightarrow {}^{15}\text{O} + \gamma$			
${}^{15}\text{O} \rightarrow {}^{15}\text{N} + e^+ + \nu_e$		$\leq$1.732	${}^{15}\text{O}$
${}^{15}\text{N} + p \rightarrow {}^{12}\text{C} + {}^4\text{He}$			

[a]Table from Bahcall 1989.

4 protons and the ^{4}He nucleus. All of the kinetic energy and gamma rays produced are absorbed in the solar material, the thermal energy ultimately leaking from the solar surface as photons after millions of years. Only the neutrinos escape from the Sun directly according to the standard model. Even for these particles, we shall consider the possibility of flavor changes in the Sun.

The Standard Solar Model (SSM) is a theoretical model that is constructed from the most refined physical calculations and data available at the time a given model is constructed. Detailed descriptions are given by Dearborn in the chapter on Standard Solar Models. The principal assumptions used in forming the SSM are the following:

1. Hydrostatic equilibrium. The Sun is assumed to be in hydrostatic equilibrium. The radiative and particle pressures are exactly balanced by gravity.
2. Energy transport by photons or convection. In the deep interior, which is most important for the solar-neutrino problem, energy transport is primarily by photon diffusion. Radiative opacity is crucial. Other things being equal, the larger the opacity coefficient the higher the temperature of the core and, therefore, also the more abundant the high-energy neutrino fluxes. That is why the opacity problem is the subject of a number of recent detailed studies (see Bahcall and Ulrich 1988).
3. Energy generation by nuclear reactions. The primary energy source for the radiated photons and neutrinos is nuclear fusion.
4. Elemental abundance changes by nuclear reactions are monotonic. The primordial solar interior was chemically homogeneous.
5. The initial solar composition was uniform and equal to that presently observed in the photosphere.

Constructing a solar model is constrained by the observed luminosity (in photons) $L_\odot$ and the outer radius $R_\odot$ at an elapsed time of 4.6 Gyr. Table II

TABLE II
Neutrino Fluxes[a]

Source	Flux (10^{10} cm^{-2} s^{-1})
p-p	6.0 (1 ± 0.02)
pep	0.014 (1 ± 0.05)
hep	8×10^{-7} (1 ± ?)
^{7}Be	0.47 (1 ± 0.015)
^{8}B	5.87×10^{-4} (1 ± 0.37)
^{13}N	0.06 (1 ± 0.50)
^{15}O	0.05(1 ± 0.58)
^{17}F	5.2×10^{-4}(1 ± 0.46)

[a] Table from Bahcall and Ulrich 1988.

gives the best current estimates for each of the neutrino fluxes with the associated theoretical uncertainties (Bahcall and Ulrich 1988). The neutrinos in
the β^+ decay of ^{17}F, generated in the $^{16}O(p,\gamma)^{17}F$ reaction, provide, in principle, a possibility for measuring the initial oxygen abundance of the Sun
(Bahcall et al. 1982).

Figure 1 shows the information that is contained in the individual neutrino fluxes. While the total solar-neutrino flux is only weakly dependent on
the actual physical condition in the solar core, the fluxes of the individual
neutrino groups are strongly dependent on the temperature T. According to
Bahcall (1989), the ^{7}Be neutrinos will vary with T^8, the ^{8}B neutrinos with
T^{18}, and hep neutrinos approximately with T^{-n}, where n is between 3 and 6.
The p-p flux is much less sensitive to temperature, varying with $T^{-1.2}$. As a
result, the ^{8}B-neutrino generation rate drops dramatically in moving outward
from the center of the Sun. The dominant source of these neutrinos is a small
volume. Now imagine several telescopes on the Earth, each designed for
viewing a particular type of neutrino. The ^{8}B-neutrino telescope will produce
data close to the center of the Sun, within a radius of 1 to 4 $\times$ 10^4 km. We
may also observe temperature changes with time, answering the question of
whether the solar interior "breathes." We may probe regions more distant
from the center with the ^{7}Be, p-p and hep telescopes. In contrast to optical
telescopes which see the enormous solar disk, the neutrino telescopes would
see only the single, burning interior regions. It is difficult even now to imagine the wealth of information that could be obtained from these idealized
neutrino, optical, X-ray and γ-ray telescopes and by instruments registering
corpuscular radiation.

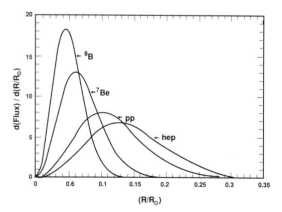

Fig. 1. Neutrino production as a function of radius in the Standard Solar Model (Bahcall et al.
1988). The fraction of neutrinos that originate in each fraction of the solar radius is [dFlux/
d($R/R_\odot$)][d($R/R_\odot$)].

II. THE SOLAR-NEUTRINO PUZZLE

A. The Chlorine-Argon Experiment

The remarkably low interaction probability of the neutrino, on the one hand, provides unique tools for probing the interior of the Sun; on the other hand, it makes neutrino detection on the Earth extremely difficult. Pontecorvo (1946) first proposed the reaction $^{37}Cl(\nu, e^-)^{37}Ar$ as a method for detecting solar neutrinos. This concept was brought into realization by Davis and his collaborators (1968), and the experiment is still in operation. The experiment, which is located in the Homestake Mine in South Dakota, employs a tank containing 615 tons of perchloroethylene. The liquid had been purified to remove argon and other impurities that would produce unwanted backgrounds. Several tens of atoms of ^{37}Ar should form during each run of 2 to 3 months according to the SSM and the neutrino cross section. At the end of each collection period, the ^{37}Ar is purged from the tank, purified and counted in a low-level gas proportional counter. The ^{37}Ar decays with a half-life of 35 days by electron capture back to ^{37}Cl with the emission of X-rays and/or auger electrons with energies of 2.8 keV and 280 eV. Detecting the decay of the few atoms of ^{37}Ar was a formidable problem that was solved. A typical background now corresponds to one decay per month.

This experiment was the only operating solar-neutrino experiment for two decades. Figure 2 shows the experimental data obtained for the period 1970.3 to 1988.3. It is seen that the values after 1986 are higher than the average from the earlier years. Table III contains the experimental averages for the pre- and post-1986 periods. Capture rates are expressed in terms of solar-neutrino units (SNU), 10^{-36} captures per second per target atom. Table IV gives the capture rates in ^{37}Cl from the various neutrino groups predicted by the standard model as calculated by Bahcall and Ulrich (1988); their com-

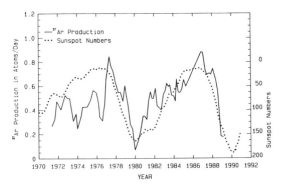

Fig. 2. ^{37}Ar production rates in the Homestake perchloroethylene solar-neutrino detector. Individual points are a running average of 5 experiments. The variation in sunspot number is plotted on a reverse scale (Courtesy of R. Davis, Jr., 1989).

TABLE III
Summary of ^{37}Ar Production Rates[a]

Period of Observation	1970.3 – 1988.3	1986.3 – 1988.3
Avg. ^{37}Ar production rate[b]	0.518 ± 0.036	0.87 ± 0.13
Cosmic ray background[b]	0.08 ± 0.03	0.08 ± 0.03
^{37}Ar above known backgrounds[b]	0.438 ± 0.047	0.79 ± 0.13
^{37}Ar production rate in SNU	2.33 ± 0.25	4.2 ± 0.7

[a] Table from Davis et al. 1988.
[b] In atoms/day/615 tons C_2Cl_4.

bined capture rate is 7.9 ± 2.6 (3σ) SNU. According to another standard-model calculation by Turck-Chièze et al. (1988), the combined capture rate in ^{37}Cl is 5.8 ± 1.3 (1σ) SNU.

Note that, of the total 8-SNU capture rate in Table IV, 6 SNU are contributed by ^{8}B neutrinos. The difference between the average experimental value of 2.33 SNU and the predicted value of 7.9 SNU is known as "the solar-neutrino puzzle." Possible explanations for the discrepancy between the observations and standard theory are presented in Sec. III. We note here that the experimental ^{37}Ar production rate in the Homestake detector is the first experimental confirmation of calculations that predict that a star must be more massive than the Sun for the CNO cycle to dominate. If the CNO cycle were the dominant source of energy production, the capture rate due to ^{13}N and ^{15}O neutrinos would be 29 SNU, 10 times higher than the observed value. This assumes that electron neutrinos are not changed by neutrino oscillations.

B. The Kamiokande-II Experiment

The Kamiokande-II 2140-ton water Cerenkov detector is located at 2700 mwe (meters water equivalent) in a zinc mine in the Japanese Alps (Hirata et

TABLE IV
Predicted Capture Rates for a ^{37}Cl Detector[a]

Neutrino Source	Capture Rate (SNU)
p-p	0.0
pep	0.2
hep	0.03
^{7}Be	1.1
^{8}B	6.1
^{13}N	0.1
^{15}O	0.3
^{17}F	0.004
Total	7.9 ± 2.6 (3σ)

[a] Table from Bahcall and Ulrich 1988.

al. 1987,1988). Cerenkov radiation is emitted by electrons that are scattered in the forward direction by neutrinos. The neutrino-electron scattering experiment can provide new information that cannot be obtained from the radiochemical experiment. It gives unique information about the neutrino-energy spectrum. It can determine the direction of the neutrino. Lastly, it records the exact time of each event. Results from this detector are very important because until now all experimental solar-neutrino data, which are in conflict with theory, have come from the chlorine detector discussed in the previous section (Sec. II.A).

Figure 3 gives results obtained during the period from January 1987 to May 1988. There is an indication of forward peaking of the recoil electrons along the direction of the Earth-Sun axis. According to these data, the flux of ^{8}B neutrinos is 0.46 ± 0.13 (statistical) ± 0.08 (systematic) $\times$ SSM (Totsuka 1988; Nakahata 1990; Hirata et al. 1988), in substantial agreement with the Homestake experiment.

C. Recent Results

Bazilevskaya et al. (1982) and Davis et al. (1988) pointed out an apparent anti-correlation of the neutrino capture rate in the Homestake detector with the sunspot cycle. Indeed, early observations during the onset of the maximum activity in solar cycle 21 in 1989 indicate continuation of the trend. A significant decrease in the ^{37}Ar production rate was observed during 1989 at the Homestake detector that follows the same pattern with sunspot number as 11 years before (Davis, personal communication). There is a suggestion

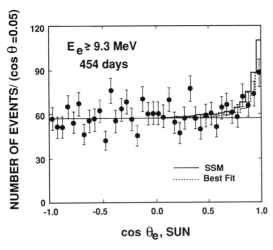

Fig. 3. The Kamiokande-II detector results obtained during 1987 and 1988 (Mann 1988; Nakahata 1990).

of a downward trend in the solar-neutrino signal at the Kamiokande-II detector in the time period from January 1987 to July 1989 (Kim 1989).

D. The Puzzle

Therefore, on the basis of the results of two completely different experiments, we can conclude that the ^8B neutrino flux is considerably less at the present time than the SSM predicts. It was noted above that there is an important indication of an increase in the capture rate in the chlorine experiment during 1987 and 1988 and a decrease in 1989. Does this mean that the neutrino fluxes vary with time? We will have to wait for new results for the final conclusion. Measurements that will be made with both detectors over the next several years in the early 1990s can help settle the question of time dependence. We cannot exclude the possibility that there will be time variation in the chlorine experiment and not in Kamiokande II if time variations are connected with the lower-energy neutrino flux. In the Kamiokande experiment, only the high-energy portion of the ^8B neutrinos is measured, whereas neutrinos with lower energies, including ^7Be neutrinos as well as ^8B neutrinos, interact in the chlorine detector. The possibilities of nonstationary neutrino-producing reactions have been considered (Kocharov 1980,1984, and references therein).

III. EXPLANATIONS FOR THE DISCREPANCY

There are two types of "solutions" to the solar-neutrino puzzle:

1. The present-day astronomical theory of stellar evolution and structure predicts the temperature distribution in the Sun's core. Attempts are made to reduce the predicted capture rates in the experiments by effects that reduce the high-energy production rates in the Sun or by invoking mechanisms for neutrino losses during propagation from their source to the detector. We discuss effects from propagation of uncertainties in cross sections in Sec. III.A and energy distributions of the interacting nuclides in Sec. III.B. In Sec. III.C we cover the physics of the neutrino itself, first the question of oscillations in vacuum or in matter, and then the effect of a possible magnetic moment.

2. The current theory of stellar evolution is wrong in temperature predictions, the present central temperature being lower than the calculated value. One looks for possible reasons to account for a lower temperature. There have been several proposals explaining how the solar interior could have cooled. Possible problems in calculating energy transport from the center of the Sun are discussed in Sec. III.D, and the recent concept of WIMPs (weakly interacting massive particles) in Sec. III.E. The postulates of a different primary nuclear fuel, at least some of the time, and of solar mixing are reviewed in Sec. III.F.

A. Nuclear Physics

The explanations in this section are connected with uncertainties of extrapolating the experimentally measured excitation functions of the various reactions in the p-p cycle to the low energies in the Sun. Consideration shows that the solar-neutrino problem cannot be due to known uncertainties in these functions; this assumes that there are no unknown differences between the real cross sections and those used in the calculations (Kocharov 1977,1980; Bahcall 1989).

Bahcall (1989) considers the possibility that the ^{8}B production cross section, ^{7}Be$(p,\gamma)^8$B, is zero. This assumption would have no significant consequences for stellar evolution, but the consequences for solar-neutrino investigations would be unfortunate. The new experiments that will use ^{2}H, ^{11}B, ^{40}Ar or ^{98}Mo for detectors or neutrino-electron scattering at energies above 5 MeV would yield only upper limits or very reduced rates that are only due to the rare hep neutrinos from the ^{3}He$(p,e^+)^4$He reaction (Table I). Only the experiments designed to detect p-p neutrinos would have signals comparable to those expected from the standard model. Nevertheless, the predicted capture rate of 1.8 SNU for the chlorine experiment would be in agreement with observation.

B. Plasma Physics

Kocharov (1973) and Clayton (1974) proposed a solution to the discrepancy involving the velocity distribution of pairs of interacting nuclei in the solar interior. A small deviation of the distribution from Maxwellian can lead to a drastic decrease in the ^{8}B neutrino flux. The reactions that produce high-energy neutrinos take place at energies between 10 and 20 kT (where k is Boltzman's constant and T is the temperature), whereas those in the main p-p branch take place at 2 to 4 kT. Thus, if the relative number of particles in the high-energy tail of the distributions is less than Maxwellian, the high-energy (^{8}B, ^{7}Be, ^{13}N and ^{17}F) neutrino production would be severely depressed while the p-p neutrino flux, which plays a decisive part in solar-energy production, is practically unchanged. Also, if the p-p reaction rate is increased because of transfer of particles to lower energies in the 2- to 4-kT region, the central temperature of the Sun would be less than predicted by the SSM. This means a significant decrease in high-energy neutrino fluxes.

The physical explanation for any distortion in the particle distribution is not established. Clayton (1974) proposed that many-particle interactions can deplete the high-energy tail in a dense plasma. Kocharov (1977,1980) and Vasiliev and Kocharov (1977) considered a distortion of the distribution due to an interaction between the particles and the plasma turbulence, the energy density of which would be $\sim 10^{-3}$ of the thermal energy. The predicted capture rate in the chlorine detector would be ~ 0.2 SNU, primarily from hep

neutrinos; the capture rate in gallium (see below) would be ~70 SNU from
p-p neutrinos.

C. Elementary Particle Physics

Oscillations and Solar Neutrinos.   Electrons, muons and tauons are
leptons (weakly interacting fermions, i.e., elementary particles with spin 1/2
that obey Fermi-Dirac statistics) of different flavors. Each of these massive
leptons has an associated neutrino of the same flavor. Long before the first
solar-neutrino experiment, Pontecorvo (1958) hypothesized that the electron
neutrino v_e oscillates to its muon or tau counterpart, v_μ or v_τ, in the vacuum
between the Sun and Earth (Gibrov and Pontecorvo 1969). If the solar-
neutrino puzzle is really connected with vacuum neutrino oscillations, the
difference in the neutrino masses squared should be $\Delta m^2 \geq 10^{-12}$ eV² (Pontecorvo and Bilenkij 1987). Thus, the flux of v_e (averaged over the v_e-
producing region of the Sun, the distance to the Earth, and the neutrino energy) may be as low as $1/N$ of the average value expected in the absence of
oscillations, where N is the number of neutrino flavors. It is seen that substantial mixing and three lepton flavors present an appealing explanation for
the factor of ~3 discrepancy of the solar-neutrino puzzle (Pontecorvo and
Bilenkij 1987).

Another theory of oscillations is believed by many to be the most likely
explanation for the puzzle. It was shown recently that matter effects may
substantially affect neutrino oscillations—the MSW effect (Wolfenstein 1978;
Mikheyev and Smirnov 1985,1986,1987). In this theory, there is an enhancement by the medium on the propagation phase of the neutrinos. In matter, the
phase of v_e varies relative to v_μ and v_τ because elastic scattering of v_e by
electrons is larger. Electron neutrinos interact with electrons through both
charged and neutral currents, whereas v_μ and v_τ interact only by neutral currents. This is why neutrino oscillations are different in matter than in vacuum.
Because of the variation in the density of the Sun from center outwards, it is
possible that, instead of a true oscillation, the original v_e will adiabatically
transform to a state v_2 that is primarily v_μ or v_τ. Because of its attractiveness
and because of the importance placed on this new theory, many investigators
have presented treatments of the subject (see, e.g., Bethe 1986; Rosen and
Gelb 1986; Haxton 1987). The depressed v_e flux observed in the chlorine
experiment can be explained (see Bahcall et al. 1988, and references therein)
if the parameters roughly satisfy

$$\Delta m^2 \leq 10^{-4} \text{ eV}^2$$
$$(\sin^2 2\theta)\Delta m^2 \geq 3 \times 10^{-8} \text{ eV}^2$$
$$\sin^2 2\theta < 0.8 \tag{1}$$

where $\tan \theta$ is the relative amplitude of v_1 and v_2 in the v_e wave function. For
a typical amount of mixing expected theoretically [$\sin \theta \sim 0.2$ and $m(v_2)$

$\gg m(\nu_1)$], this corresponds to a value of $m(\nu_2)$ between 5×10^{-4} and 10^{-2} eV. This range appears to be consistent with Grand Unified Theories (GUTs).

Matter-enhanced oscillations may also play an important role in the passage of neutrinos through the Earth. This phenomenon would give rise to a day-night effect that would be seen in real-time detectors. There is also a proposal to modify the chlorine detector to look for the effect.

A very important peculiarity of the neutrino-oscillation phenomenon is a distortion of the original solar-neutrino spectrum because the transformation of ν_e to ν_μ or ν_τ by this mechanism is energy dependent. If we can measure the neutrino spectrum, we can offer a solution to the oscillation question. The importance of the MSW theory goes beyond the solar-neutrino problem to the fundamental physics of elementary particles.

Neutrino Magnetic Moment.   If the neutrino has as large a magnetic moment as 10^{-11} or 10^{-10} μ_B (Bohr magnetons), there could be a decrease in the detectable solar-neutrino flux. Some of the familiar left-handed neutrinos, in passing through magnetic fields of a few kG, could become sterile (right handed) due to precession of the magnetic moment. If this hypothesis is true, one should observe neutrino-flux variations that are correlated with the 11-yr periodicity in the variation of solar-magnetic activity (Veselov et al. 1987, and references therein). The magnetic field in the solar convective zone may change with sunspot number. The ^{8}B neutrinos pass through varying paths through the convective zone depending on the season of the year because of the slight inclination of the Earth's orbit relative to the equator of the Sun. Further observations in both the chlorine experiment and the Kamiokande-II detector during solar-cycle 22 maximum will help test this interesting idea.

D. Opacity Calculations and the Low-Z Model

The temperature gradient of the solar interior depends on the mechanism of outward energy transport. Ignoring other conditions, the calculated central temperature of the Sun will increase with increasing absorption probability of the radiation produced in the *p-p* cycle in the thermonuclear furnace or in its immediate vicinity. The predominant contribution to gamma-ray absorption comes from elements heavier than helium. If the abundance ratio of heavy elements to hydrogen in the interior of the Sun is the same as on the surface, about half of the opacity of the interior is contributed by heavy elements. Bahcall and Ulrich (1971) constructed models that had 10 times lower abundances of heavy elements in the core. This reduced the predicted neutrino capture rate to 1.5 SNU, in agreement with the observations of the chlorine experiment and the Kamiokande detector. Bahcall and Ulrich (1988) concluded that low-Z models lead to disagreement between observed and calculated *P*-mode oscillation frequencies. Bahcall (1989) states that this discrepancy makes low-Z models less attractive but that it cannot be used to reject the hypothesis completely.

Another way to reduce the opacity is to have iron precipitated out of the gas phase (see Bahcall 1989, and references therein). Iron-like elements contribute up to 25% of the opacity of the solar interior. Dearborn et al. (1987), using an iron-free radiative opacity, calculated that the capture rate for the chlorine experiment would be 4.0 SNU. This value would be in agreement with the recent results of the chlorine and Kamiokande detectors.

E. Weakly Interacting Massive Particles

A very interesting theory was proposed by Press and Spergel (1985) and Faulkner and Gilliland (1985) that would solve not only the solar-neutrino puzzle but also the problem of the "missing mass" in the universe. It is postulated that weakly interacting massive particles (WIMPs) were produced in the early universe in abundances sufficient to explain the missing galactic mass. If they are accreted in appreciable quantity in the Sun, they can provide another channel for transporting energy out of the solar interior, thereby smoothing out the temperature gradient and reducing the ^8B neutrino flux. Thus WIMPs have an effect on the high-energy neutrino fluxes similar to a decrease in opacity.

WIMPs should have certain necessary properties (see Bahcall 1989, and references therein) to be a solution to the puzzle. In order to confine the WIMPs to the region where ^8B neutrinos are produced, their mass must be > 2 GeV. So that they can reach out of the region of ^8B-neutrino production, the mass should be <10 GeV. An optimum scattering cross section on protons, $\sim 5 \times 10^{-36}$ cm^2, for transporting energy would allow WIMPs to interact about 40 times per orbit.

If this cross section is realistic, a particle abundance of WIMPs relative to protons of $\sim 10^{-11}$ is sufficient to reduce the ^8B-neutrino flux. Gilliland et al. (1986) showed that the flux can be reduced by a factor of 4 relative to the standard model. The observed decrease would be highest for the high-energy neutrino detectors and less (only $\sim 25\%$ for the gallium detector) for the low-energy neutrino detectors.

F. The ^3He Hypothesis and Mixing

We have already noted that the discrepancy between the experiments and theory can be explained with a solar interior that is cooler than predicted by the SSM. To provide for the observed luminosity at a lower temperature, Kocharov and Starbunov (1969,1970a,b) postulated a solar fuel capable of producing energy at a lower temperature. The hypothesis uses the $^3He(^3He,2p)^4He$ reaction, which liberates ~ 13 MeV. This idea was subsequently developed by a number of authors (Abraham and Iben 1970; Ulrich 1971a,b; Kocharov and Starbunov 1971; Mihalas 1972).

Figure 4 and Table V show the main characteristic fluxes of solar models for various assumed abundances of ^3He (Kocharov and Starbunov 1969,1970a,b,1971). The ^3He abundance in the solar core must be more than

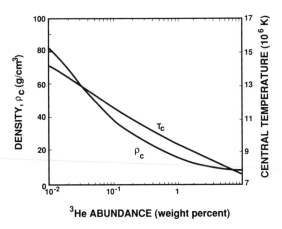

³He ABUNDANCE (weight percent)

Fig. 4. Central density and temperature vs ^{3}He abundance (figure from Kocharov and Starbunov 1969,1970a).

3×10^{-5}. This lower limit is less than that in the abundance of ^{3}He in the solar wind (10^{-4}). It is much less than that accumulated by reactions of the p-p cycle in regions far from the center; at a distance of 0.28 $R_\odot$ from the center, the SSM gives an abundance of 3.2×10^{-3} (Bahcall and Ulrich 1988).

The main difficulty with the hypothesis lies in its not assuming a relatively high ^{3}He abundance in the solar interior at the present time. If the luminosity is to be accounted for by the ^{3}He(^{3}He,2p)^{4}He reaction throughout the Sun's evolution, the ^{3}He content of the primordial Sun must have been a few percent; although not impossible, this does not seem plausible.

Even if the primordial Sun did not contain any ^{3}He, it would be possible to explain the observed low high-energy neutrino fluxes within the hypothesis of Kocharov and Starbunov by invoking Fowler's (1972) suggestion of sudden mixing. In this scenario, hydrogen burning occurs first, and the gradual accumulation of ^{3}He takes place, especially in regions far from the center of the Sun. Then ^{3}He moves to the central region due to mixing, and ^{3}He becomes the main source of energy. After the ^{3}He burning, the temperature of the core increases, and energy is generated again by hydrogen burning with new generation of ^{3}He. This efficient cycle of producing new fuel, burning it and following the main evolutionary path would repeat.

During the last 20 yr, several different mixing models have been proposed. The first such model was considered by Ezer and Cameron (1968), who noted that the expected flux of ^{8}B neutrinos could be reduced if the central temperature is lowered by the introduction of fresh hydrogen into the solar core. A number of papers followed investigating different effects of composition mixing on solar evolution and on the neutrino fluxes (see Sienkiewicz et al. 1989, and references therein; Bahcall 1989, and references

TABLE V

Characteristics of the Solar Models[a]

³He Abundance (%)	T_c (10⁶K)	ρ_c (g cm⁻³)	p-p	pep	hep	⁸B	³⁷Cl capture rate (SNU)
			(10¹⁰ cm⁻²s⁻¹)				
10	7.7	9.5	6.5×10^{-2}	2×10^{-5}	10^{-6}	low	3.4×10^{-4}
5	8.2	10	9.7×10^{-2}	3.5×10^{-5}	10^{-6}	low	6.0×10^{-4}
1	9.3	17	3.5×10^{-1}	2×10^{-4}	1.5×10^{-6}	low	3.0×10^{-3}
0.5	10.0	20.5	5.8×10^{-1}	4.2×10^{-4}	1.7×10^{-6}	$\sim 10^{-7}$	8.4×10^{-3}
0.1	11.6	39	1	2×10^{-3}	2×10^{-6}	$\sim 10^{-6}$	4.8×10^{-2}
0.01	14.0	81.7	1.5	1×10^{-2}	2×10^{-6}	5×10^{-5}	0.85
7.7×10^{-4}	15.6	148	6.0	1.4×10^{-2}	7.6×10^{-7}	5.8×10^{-4}	7.9

[a]Last line of the table from SSM (Bahcall and Ulrich 1988) assumes the ³He abundance at the solar center. Table from Kocharov and Starbunov 1971.

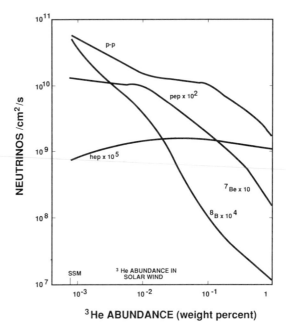

^{3}He ABUNDANCE (weight percent)

Fig. 5. Dependence of solar-neutrino fluxes on the ^{3}He abundance (figure modified from Kocharov and Starbunov 1971).

therein). For example, gravity-mode instabilities cause such mixing in the "Solar Spoon" model of Dilke and Gough (1972). The mixing, expansion of the core, and cooling could occur every 200 Myr., followed by a decrease of solar luminosity for about 3 Myr. Several authors cite the similar spacing of glacial epochs as evidence for such solar variability.

Sienkiewicz et al. (1989) have calculated evolutionary sequences of the Sun assuming models of continuous or episodic mixing; these results are considered in the following treatment. The variation of the expected neutrino fluxes at the Earth with the abundance of ^{3}He are shown in Figs. 5 and 6. The fluxes of all groups, with the exception of the hep group, decrease drastically with increasing ^{3}He. The hep neutrino flux increases from the SSM value of 7.6×10^3 cm^{-2} s^{-1} (Bahcall and Ulrich 1988) to twice that at a ^{3}He abundance of $\sim 0.01\%$. This is important for the ^{2}H, ^{40}Ar (see below) and perhaps Kamiokande-II detectors. The hep neutrinos will be detected by measuring electron-recoil energies above 13 MeV, providing a test of the ^{3}He hypothesis.

For a postulated massive mixing event, Sienkiewicz et al. (1989) showed that the flux of p-p neutrinos is strongly reduced for 10 Myr following such mixing; the ^{8}B-neutrino flux, for >100 Myr. The flux of hep neutrinos is significantly increased for 0.1 Myr. Figure 7 shows the time dependence of these fluxes for a model in which 60% of the Sun was mixed 60 Myr ago. If such a mixing scenario is the solution to the solar-neutrino puzzle, the p-p

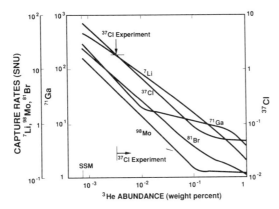

Fig. 6. Calculated neutrino capture rates in solar-neutrino detectors as a function of the ³He abundance. The limits imposed by the results from the chlorine detector are shown.

and ⁸B neutrino fluxes will be depressed today with an enhancement of the hep neutrino flux. The hep neutrinos will produce more energetic recoil electrons than are possible for the ⁸B neutrinos in the ²H and liquid argon detectors.

Radiochemical detectors, such as the chlorine detector, can measure variations that occur over time scales governed by the half-life of the daughter nucleus produced and the duration of the experiment. Real-time detectors such as Kamiokande-II have very short resolving times; thus, these types of detectors will provide information on variability up to many tens of years. Suitable geochemical detectors, the information from which will be governed by the much longer half-life of the daughter nucleus, can integrate past neutrino fluences. Comparison of the results from the contemporary-flux detectors with those from the past will answer questions about long-term variations of neutrino fluxes.

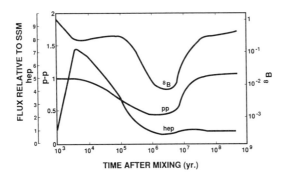

Fig. 7. Time dependence of the p-p, ⁸B and hep neutrino fluxes for a model in which 60% of the Sun was mixed 60 Myr ago (from Table 3 of Sienkiewicz et al. 1990). The various flux values are in units of the SSM flux at a solar-model age of 4.6 Gyr.

IV. MEASUREMENTS OF THE CURRENT
SOLAR-NEUTRINO FLUX

A number of experiments have been proposed to address the puzzle presented by the Homestake and Kamiokande experiments, to measure other energy components of the solar-neutrino spectrum and to measure the past luminosity of the Sun. We touch briefly on some of the experiments to measure the contemporary neutrino flux that are in progress or being planned. There are a number of conference proceedings and surveys (see, e.g., Friedlander 1978; Cherry et al. 1985; Friedlander and Weneser 1987; Bahcall et al. 1988; Bahcall 1989) that deal with the subject in detail.

A. The Gallium Solar-Neutrino Experiments

The $^{71}\text{Ga}(\nu,e^-)^{71}\text{Ge}$ reaction has a threshold of 0.233 MeV and is, therefore, sensitive to the fundamental p-p neutrino flux. The predicted capture rate for this detector is 132^{+20}_{-17} (3σ) SNU (Bahcall and Ulrich 1988) and 125 $\pm$ 5 (1σ) SNU according to Turck-Chièze et al. (1988). The p-p neutrinos produce 54% of the total reactions within the framework of the SSM, and the ^7Be and ^8B neutrinos produce 26% and 11% of the captures, respectively. Two gallium solar-neutrino experiments are in progress.

The GALLEX experiment, involving collaboration of Western-European and American scientists, is located in an underground laboratory in the Gran Sasso Tunnel in Italy. The detector will be loaded with 30 tons of gallium as gallium chloride in aqueous solution, and germanium will be removed by gas sweeping (see Hampel 1985, and references therein).

The SAGE (Soviet-American Gallium Experiment) is a collaboration of Soviet and American investigators. The detector, located in a laboratory underneath a high mountain in the Baksan Valley in the Caucasus region of the Soviet Union, consists of 60 tons of gallium metal. Germanium is removed by sweeping it from liquid gallium metal (see Barabonov et al. 1985, and references therein).

The SAGE collaboration began experiments with 30 tons of gallium in early 1989 (Abazov et al. 1989; Gavrin 1990), and the GALLEX group hopes to have the full amount of gallium by late 1989 and start the first run in January 1990 (Kirsten 1990). The initial chemical extraction of germanium from the gallium will be different for the two experiments, but the final chemical procedures will be similar. The nuclide ^{71}Ge decays by electron capture with a half-life of 11.43 days. The emitted Auger electrons and X rays result in an energy spectrum with a 1.2-keV L peak and a 10.4-keV K peak. A counting rate of about 1 event per day is expected. In both experiments, the ^{71}Ge will be counted in low-level counting systems based on miniaturized proportional counters.

The gallium experiments will provide the first experimental observations of the fundamental p-p neutrino flux from the Sun and may answer basic

Plate 1. Aerial view of Red Mountain, Colorado. The Henderson molybdenite ore body, containing about 0.5% molybdenum, is located at a depth of 1130 to 1500 meters. Thousands of tons of the ore must be processed to measure the technetium isotopes produced by solar-neutrino interactions. (Reproduced by permission of Economic Geology.)

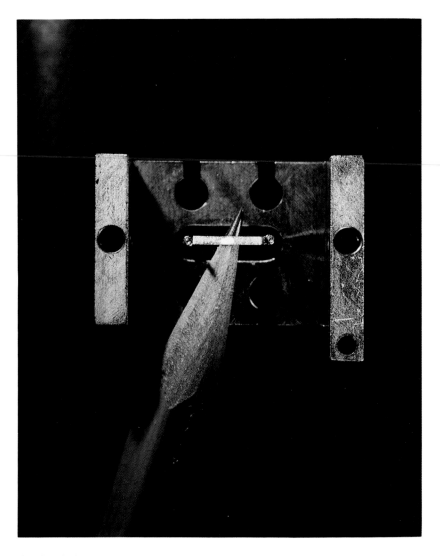

Plate 2. Initial metallurgical purification by the AMAX Corporation and final chemical processing at the Los Alamos National Laboratory produces a massless sample that contains the technetium isotopes on a rhenium filament. The sample is analyzed by negative thermal-ionization mass spectrometry.

questions about this assumed principal reaction. The experiments may indicate which classes of solutions, such as MSW oscillations or the ^{3}He hypothesis, are incorrect. Comparison of the two experiments is important for confidence in the results, assessing possible systematic errors and checking unusual events.

B. The Heavy-Water Detector

The SNO (Sudbury Neutrino Observatory) collaboration will build and operate a deep detector using 1000 tons of D_2O (Chen 1985; Sinclair et al. 1986; Earle et al. 1988). A laboratory will be constructed at a depth of 6800 feet in a mine near Sudbury, Ontario, Canada. The proposal is to surround the D_2O vessel with 4 m of normal H_2O. The specific details of construction and the method for detection are still being planned by Canadian and US collaborators. An early proposal called for 2000 50-cm-diameter photomultiplier tubes to measure the paths of relativistic electrons by way of the Cerenkov light produced in the vessels. The Sudbury detector will have significant advantages over the Kamiokande detector because reactions in D_2O allow for the measurement of additional properties of the neutrino spectrum. The ν_e flux, energy and direction would be measured by the charged current reaction:

$$\nu_e + d \rightarrow p + p + e^-. \tag{2}$$

The almost monoenergetic electron has an energy ($E_\nu = 1.44$) MeV. Secondly, the detector will measure all neutrinos scattering on electrons by the neutral-current process:

$$\nu_x + e^- \rightarrow \nu_x + e^-. \tag{3}$$

Because the cross section for ν_e reactions by this mode is 6 to 7 times greater than the ν_μ or ν_τ reactions, the electron scattering, measured in both vessels, will provide an independent measurement of the ν_e flux.

The total neutrino flux is determined by a neutral-current reaction that disintegrates the deuteron:

$$\nu_x + d \rightarrow p + n. \tag{4}$$

The cross section for this third reaction is not dependent on neutrino type. Measurement of this reaction could, in conjunction with the others, be an indication of the extent of vacuum or MSW oscillations. The neutrinos from this reaction might be measured either directly with neutron detectors or by their capture in a distinctive dopant nuclide added to the tank.

In addition to the three neutrino reactions, there are anti-neutrino reaction signatures in both the H_2O and D_2O. These are important for measuring anti-neutrino sources such as supernovae. It is the hope of the collaboration

that the SNO detector will have a sufficiently low background to allow measurements of neutrino spectra above 4 or 5 MeV and to determine the hep
neutrino flux.

C. The Indium Detector

The proposed indium detector (Raghavan 1976) is a real-time experiment in which the ^{115}In(ν_e,e^-) reaction is measured. Neutrino capture produces the 613-keV, 3.3-μsec state of ^{115}Sn. The excited state decays through
a 497-keV state; the 160-keV gamma ray having a conversion efficiency of
~50% with the emission of a 90-keV electron. The conceptual experiment
measures time-coincident pulses in a large, shielded amount of indium. The
electron energy is related to the neutrino energy. One expects a rate of about
1 event per day in 3.5 tons of natural indium. There are possibilities for
performing the measurement in large liquid scintillator tanks, with plastic
scintillators, or with many superconducting indium semiconductor detectors
(Booth 1987; de Bellefon et al. 1985). The effective neutrino threshold for
the reaction is 0.119 MeV. Although 73% of the 639-SNU capture rate is due
to p-p neutrinos (Bahcall and Ulrich 1988), high backgrounds would make
the low-energy measurements difficult. Because of the neutrino-energy resolution, this detector has great appeal for measuring the ^{7}Be neutrinos.

The ^{8}B-neutrino flux results from proton capture on ^{7}Be in the Sun, and
the monoenergetic ^{7}Be neutrino flux results from electron capture on ^{7}Be.
Mann (1988) has made the unique suggestion that the ratio of the two fluxes,
with the ^{7}Be flux measured in an indium detector, can lead to the central
temperature of the Sun, which would be a precise test of the Standard Solar
Model. This is because the two reactions in the Sun have different, but well
understood, dependencies on the temperature and the comparison is independent of the amount of ^{7}Be.

D. The Bromine-Krypton Detector

A contemporary bromine-krypton experiment (Hurst et al. 1984,1985)
would be similar to the Homestake chlorine-argon experiment, with the extraction of krypton from a liquid bromine-containing target such as CH_2Br_2.
The proposed measurement technique for measuring as few as 1000 atoms of
2.1×10^5-yr ^{81}Kr involves laser resonance ionization and mass spectrometry
with a counting technique known as "Maxwell's Demon." There was much
initial interest in this proposed detector because it was thought to be sensitive
primarily to ^{7}Be neutrinos. Recent calculations have indicated that the capture
reaction will have a large contribution (55%) from ^{8}B neutrinos (Bahcall and
Ulrich 1988).

E. The Iodine-Xenon Detector

Haxton (1988) has proposed yet a third radiochemical detector in which
a noble gas is removed from a liquid containing a halogen, in this case,

iodine. Extraction of 36.4-day ^{127}Xe and counting would be similar to that of the chlorine experiment. Several organic or inorganic iodine-containing liquids are candidates for the neutrino target. The ^{127}I(v_e,e^-)^{127}Xe reaction has a threshold of 664 keV, and Haxton estimates a neutrino capture rate 10 times that of chlorine. Some implications of the large capture rate are reduction in statistical fluctuations among runs, the possibility of a good calibration with an intense neutrino source and the prospect of exploiting the eccentricity of the Earth's orbit to demonstrate the solar origin of the observations.

F. Other Contemporary-Time Detectors

The proposed ^{11}B detector (Raghavan et al. 1986) would sample ^{8}B electron neutrinos via the charged current ^{11}B(v_e,e^-)^{11}C reaction. Concurrently, it would also measure neutrinos, regardless of flavor, via neutrino-electron scattering or neutral-current excitation, ^{11}B(v,v')^{11}B*. The gamma de-excitation of the excited ^{11}B* would be measured. The attraction of the experiment is that the neutral current measurement could provide evidence for neutrino oscillations.

Similar neutral current excitations are possible in the ^{40}Ar detector (Raghavan 1986; Bahcall et al. 1986) being planned for the Gran Sasso laboratory. The detector will also have a unique signature for hep neutrinos.

V. GEOCHEMICAL EXPERIMENTS

A. What Can be Learned?

What can information stored in geologic records offer to these intriguing multidisciplinary endeavors? If the Standard Solar Model is even approximately correct, electron neutrinos have been irradiating the Earth throughout its history. Unless they are transformed into other particles, electron neutrinos have been inducing reactions in the Earth's crust similar to the reactions discussed in previous sections. In observations of long-term changes that can be attributed to solar neutrinos, the methodologies of isotope geochemistry can contribute unique information concerning the history of the solar-neutrino flux striking the Earth. The motivation for seeking information on long-term variation in the neutrino fluxes was discussed in Sec. III. After discussing features and requirements common to all geochemical detectors, we will discuss the thallium-lead, molybdenum-technetium, bromine-krypton and potassium-argon experiments in detail.

For low-to-moderate energies, the neutrino reactions are simple inverse beta decays to ground state or low-lying excited states of the daughter nucleus, which simply increase the nuclear charge by one unit through the conversion of a neutron to a proton. For more energetic neutrinos, highly excited states of the daughter nucleus can be produced to allow neutron emission; the daughter nucleus may be one mass unit lighter than the parent. Haxton and

Johnson (1988) have calculated that neutrinos produced by galactic collapses may be energetic enough to induce significant reactions of the latter type.

The first requirement, then, for a geochemical experiment is a suitable parent-daughter pair of nuclides. Because neutrino capture rates are of the order of 10^{-35} captures per target atom per second (2×10^6 captures in 1000 kg or 6×10^{27} atoms of a medium-mass element per Myr), a number of very stringent conditions must be satisfied. The daughter nuclide should be sufficiently long-lived to integrate over the long time period of interest. An ultrasensitive detection method is essential because only 10^6 to 10^8 daughter atoms will be assayed, usually in the presence of many orders of magnitude more of another isotope of the same element. The detection of the product atoms would be very difficult in the presence of any stable nuclides of the same element. (For example, a 1 part-per-trillion abundance of a different and stable isotope of the product element in 1000 kg of the target in the above case represents 6×10^{18} atoms that must be discriminated against in the measurement of a few 10^6 atoms.) Thus, it is desirable to choose an element that has no stable isotopes. Geologic conditions must be adequate so that background effects for other modes of production of the same nuclide from natural radioactivities and cosmic radiation are tolerable. This means that the ore deposit must be at sufficient depth, uranium and thorium concentrations must be low and concentrations of nearby elements that can be transformed into the nuclides of interest must also be low. The target element and the product nuclide must have remained geologically immobile; if the ore body has become enriched or depleted in either, then one cannot determine production rates over geologic time periods. The age of the deposit must be known relative to the product nuclide. It is necessary to know or infer the cross section for neutrino capture. Calculations involving Gamow-Teller strength distributions are model dependent; Gamow-Teller transitions are the category of transitions in which the electron and neutrino are emitted with mutually parallel spins in a relative triplet state (Marimier and Sheldon 1969). The distributions are being measured by forward-angle (p,n) reactions (Rapaport et al. 1985). One must have separation methods for isolating the submicrogram quantities of the element of interest from the ore material. Chemical separation factors of $>10^{20}$ are required. Thus, extreme care must be taken to avoid extraneous contamination, for example from fallout, and to prove that this has not occurred.

B. Background Requirements

Neutrino interactions with matter, either through charged-current or neutral-current interactions, are rare events. Therefore, one must evaluate and minimize background reactions from other sources. In this section, we discuss background problems for measurements of both the contemporary flux in experiments discussed in prior sections and for those of geochemical experiments because many of the problems are similar. However, in the case of experiments to measure the present-day flux, measures can be taken to mitigate the problems (1) by choosing the location and depth of the detector; (2) with the addition of tailored shielding and anti-coincidence detectors; and (3) by the use of the purest materials in all parts of the detectors. For a given geochemical neutrino experiment, only a limited number of deep geologic deposits are available, and the experimenter can only assess their suitability for the experiment.

A requirement for experiments designed to measure extraterrestrial neutrinos is sufficient depth to overcome large contributions from secondary reactions from cosmic-ray muons. Fast muons interact with rock and detector materials by nuclear cascades or spallation reactions to produce neutrons, protons and pions. These secondary particles can interact to produce background products or signals. Fireman et al. (1985) have measured the cosmic-ray-produced ^{37}Ar in a ^{39}K detector as a function of depth in the Homestake mine. The response agrees well with muon measurements and theory. The response decreases by more than 4 orders of magnitude at a depth of 4000 mwe (meters water equivalent). The contribution from stopped muons is much lower at depths below 1800 mwe. In any radiochemical experiment the product produced by a (ν_e, e^-) reaction is also produced by a (p,n) reaction on the same nuclide. In addition, (p,xn yp), (n, xn yp), (n,α) and similar reactions must be considered for nuclides in the detector. The activities from short-lived spallation products can increase backgrounds in real-time detectors. Burial depths of detectors range from about 1800 mwe (IMB) to 2400 mwe (Kamiokande), 4200 mwe (Baksan and Homestake) and 6200 mwe (Sudbury). The present depths of the ore deposits for the geochemical experiments at Red Mountain (Henderson Mine) and Allchar are about 3500 and 550 (expected) mwe, respectively. However, because of surface erosion, these deposits were buried more deeply during a substantial period of the geologic past while the product of interest was accumulating.

Naturally occurring radionuclides in the rock or the detector can be troublesome in contemporary real-time and radiochemical experiments as well as in geochemical experiments. Alpha particles can produce the nuclide of interest directly by (α,p), (α,n) and (α,γ) reactions in impurities; however, this only occurs within the energetic portion of the range of the alpha particle. Neutrons and protons produced by (α,n) and (α,p) reactions on any element may induce secondary reactions to make interfering amounts of the nuclide of interest. In real-time detectors, high-energy gamma rays resulting from neutron capture can produce interfering signals. Ewan et al. (1987) discuss in detail the very stringent requirements for shielding the SNO detector and controlling the contamination in all parts of the detector. In radiochemical detectors, particularly the geochemical detectors, careful attention must be paid to reactions from the secondary neutrons and protons on all components of the system, major and minor. Rowley et al. (1980) discuss these processes for the bromine and thallium experiments, and Cowan and Haxton (1982a) deal with them for the molybdenum experiment.

If the product of interest is a fission product ([81]Kr, [97,98]Tc), one must estimate the contribution from fission (spontaneous fission of [238]U if the neutron flux is low). Fortunately, these krypton and technetium isotopes are shielded nuclides and the contributions from this source will be low.

C. The Thallium—Lead Experiment.

Freedman et al. (1976) proposed a geochemical experiment to measure the long-term neutrino fluence using the [205]Tl(ν_e, e^-)[205]Pb detector. The half-life of [205]Pb is 14 Myr. Capture to the 2.3-keV isomer has a neutrino-energy threshold of 43 keV. The reaction therefore would be sensitive to low-energy neutrinos. However, Bahcall and Ulrich (1988) estimate that contributions from [7]Be and [8]B neutrinos are comparable to those from p-p neutrinos. Problems with the experiment include the availability of a site with suitable geochemical conditions, the estimation of the neutrino-capture cross section of [205]Tl and the measurement of [205]Pb in a purified lead sample. Progress in resolving these difficulties is addressed in Freedman (1978), Henning et al. (1985), Morinaga (1986) and Nolte and Pavicevic (1988).

Theoretical calculations of the $(1/2)^+ \rightarrow (1/2)^-$ transition for neutrino capture are uncertain. Freedman (1988) discusses a planned resolution in which the decay of bare [205]Tl[81+] ions to [205]Pb[81+] is measured in a heavy-ion storage ring to yield the necessary nuclear matrix elements. The transition of interest is the only energetically allowed channel.

Thallium crystalline minerals occur only rarely. The proposed geologic setting for the experiment is the lorandite (TlAsS$_2$) deposit in Allchar, Yugoslavia. The age of the deposit is about 5 Myr, and the mineral can be found presently at depths exceeding 150 m. Uranium, lead and thorium concentrations are 39, 790 and 9.2 ppb, respectively. Background reactions from mercury and bismuth impurities must also be considered. A few tens to a few hundreds of tons of ore are required depending on the measurement technique.

Chemical separation of lead from the mineral will result in samples containing macro amounts of stable lead isotopes in which small numbers of [205]Pb atoms must be measured. From the standard solar model, [205]Pb/[205]Tl and [205]Pb/[206]Pb ratios of about 3×10^{-20} and 3×10^{-14} are expected. The degree of chemical purification depends on the measurement method, which may be laser-resonance fluorescence, low-energy mass spectrometry or accelerator mass spectrometry.

D. The Molybdenum-Technetium Experiment

The molybdenum-technetium solar-neutrino experiment, conceived by Cowan and Haxton (1982a,b)—one of the authors (KW) is a participant—is the only geochemical experiment under way. In the near future The Los Alamos team will measure technetium isotopes in deeply buried molybdenum ore produced by the following reactions:

$$^{98}\text{Mo} + \nu_e \rightarrow {}^{98}\text{Tc} + e^-$$
$$^{97}\text{Mo} + \nu_e \rightarrow {}^{97}\text{Tc} + e^-$$
$$^{98}\text{Mo} + \nu_e \rightarrow {}^{97}\text{Tc} + n + e^-. \tag{5}$$

The first reaction has an effective threshold of 1.74 MeV which means that it is effectively sensitive to only ^{8}B neutrinos from the Sun. The second reaction has a lower threshold and can be induced by ^{7}Be neutrinos as well. The third reaction is sensitive to higher-energy neutrinos from the Sun (^{8}B and hep) and to high-energy neutrinos from galactic collapses. Haxton and Johnson (1988) have calculated that, for a rate of 0.09 supernovae per year, 40% of the ^{97}Tc signal may due to galactic stellar collapses. The half-lives of ^{98}Tc and ^{97}Tc are 4.2 Myr and 2.6 Myr, respectively; the abundances of ^{98}Mo and ^{97}Mo are 24% and 9.6%. One expects about 10^8 atoms ^{98}Tc in about 50 tons of the mineral molybdenite, MoS_2, at equilibrium. The only deep molybdenite deposit that is mined commercially is the Henderson ore body located at a depth of 1130 to 1500 m in Red Mountain, Colorado (see Color Plate 1). The AMAX Corporation mines this ore, which is about 0.5% in molybdenite. Fortunately the uranium concentration in the molybdenite mineral is only 1.3 ppm, making background reactions tolerable for the production of ^{98}Tc and perhaps for ^{97}Tc. The problems are (1) to chemically separate technetium from thousands of tons of ore, and (2) to measure 10^6 to 10^8 atoms of ^{98}Tc and ^{97}Tc, using mass spectrometry. The nuclide ^{99}Tc, produced by background-neutron capture on ^{98}Mo and by spontaneous fission of ^{238}U should have an abundance 6 orders of magnitude greater than ^{98}Tc. The absolute concentration of ^{99}Tc in molybdenite can be measured on a laboratory-size sample spiked with ^{97}Tc tracer during dissolution. The known concentration of ^{99}Tc then serves as an internal-yield monitor for determining ^{98}Tc and ^{97}Tc from their ratios relative to ^{99}Tc starting with multi-ton quantities of molybdenite.

This experiment (Wolfsberg et al. 1985) begins with much help from the AMAX Corporation in the initial chemical separations. This would not be possible within the resources available to the experimental team. In their milling operation, AMAX produces a concentrate that is >90% MoS_2 by flotation. The concentrate, which is shipped to the AMAX Ft. Madison, Iowa, plant, is roasted in excess air at temperatures as high as 740°C to produce MoO_3. The roaster exhaust gas is principally SO_2, which is known to carry halides and volatile compounds of arsenic, selenium and rhenium. It has been shown also to carry the rhenium homolog technetium. Prior to conversion of the SO_2 to H_2SO_4, the SO_2 is scrubbed with water to remove the volatile impurities. Indeed, rhenium and technetium are found in the acid-scrub waste stream and not in the sulfuric acid. The AMAX Corporation granted permission to the Los Alamos team to oxidize the acid scrub stream with NaOCl, to ensure that the technetium would be present as TcO_4^-, and to intercept the stream with large commercial anion-exchange resin columns to absorb and concentrate technetium. They performed five 12-hr separations on the 20-gal min^{-1} waste stream. This step should have achieved >90% recovery of technetium from each 40-ton lot of molybdenite on 300 liters of resin.

The rest of the chemistry is performed at Los Alamos. The resin is first eluted with an NaOH-NaCl solution, still at the several-hundred-gallon scale for one 70-liter resin column, representing the technetium in 10 tons of molybdenite, to remove co-absorbed molybdenum. Next the resin is ashed at 440°C to reduce the sample to the 100-gram scale. Ashing conditions are sufficiently reducing to maintain the technetium in the ash, but most of the rhenium is fortunately volatilized. The ash is dissolved in HNO_3 and HF. The technetium is purified by a lengthy series of solvent-extraction, precipitation and ion exchange steps. The mass is finally reduced sufficiently to absorb the entire sample on 1 ml of anion-exchange resin.

Final purification (Rokop et al. 1990), in preparation for mass spectrometry, is carried out under clean-room conditions with ultra-pure reagents. The technetium is stripped from the resin with 8 M HNO_3 which is evaporated. Technetium is then absorbed on a 50-μl anion-exchange resin column from which it is again eluted with HNO_3. Although the sample is essentially mass free now, there are still poisons present that prevent efficient ionization of technetium for the mass spectrometric measurement. Much effort has gone into achieving a satisfactory interface between chemistry and mass spectrometry. Purification is continued with micro-distillation of technetium from reverse aqua regia, collecting the product as a droplet on a cold teflon surface. Very pure technetium is extremely volatile from this distillate, even at room temperature; extreme care must be taken to avoid loss of the precious sample for the final concentration. Hydroiodic acid is used to reduce the technetium to a less volatile state while the distillate is evaporated. The dried sample is then dissolved in 1 M NH_4OH.

Mass spectrometric isotopic measurements of technetium are difficult. Thermal ionization requires low-volatility compounds and relatively low-ionization potentials. In most forms, technetium is very volatile, and its ionization potential, 7.3 eV, is too high to achieve efficient ionization by standard positive thermal-ionization processes. An additional complication is that the most ubiquitous isobaric impurity is molybdenum, which has similar ionization characteristics and atomic masses. To overcome these difficulties, a negative thermal-ionization technique was developed (see Color Plate 2). Pertechnate, TcO_4^-, ions are produced with La_2O_3 ion enhancers with high efficiency, ~5%, and no isobaric molybdenum beams. Molybdenum is produced as the MoO_3^- ion which is far removed from the technetium spectrum.

Isobaric impurities contributed by the ionization technique are measured at 6 $\times$ 10^5 atom equivalents of ^{98}Tc, ^{97}Tc and ^{99}Tc. Measurement limits on pure samples of ^{98}Tc and ^{97}Tc are of the order of 10^6 atoms (Rokop et al. 1990).

By the end of 1989, the investigators had performed essentially all of the chemical purification of technetium from the roasting of ~10 tons of molybdenite, measured the ^{99}Tc in this sample and were attempting to make the first measurements of ^{98}Tc.

E. Other Proposed Geochemical Detectors

The ^{81}Br$(\nu_e,e^-)^{81}$Kr reaction (see above) has also been proposed for use to integrate the neutrino flux over a geologic time interval (Scott 1976; Friedlander 1978; Rowley et al. 1980; Hampel 1980). Interest in this detector has also declined because of the sensitivity of the reaction to higher-energy neutrinos and also because of the difficulty in finding an underground deposit to meet requirements for the experiment. Kuzminov et al. (1988) have recently proposed the Volgograd bishofite deposit or East Siberian deep brines to accomplish this geochemical experiment.

Haxton and Cowan (1980), examining the requirements for a geochemical neutrino detector, proposed the ^{41}K$(\nu_e,e^-)^{41}$Ca reaction. The product ^{41}Ca has a half life of 1.03×10^5 yr, and the effective threshold for the reaction is 2.36 MeV. Anticipated difficulties for this experiment are the preconcentration of ^{41}Ca from ^{40}Ca prior to a quantitative measurement and potential contamination from other sources of ^{41}Ca. The authors later deemed that the technetium experiment held more promise.

The material in this chapter covers the state of solar neutrino knowledge and theory through 1989. The situation is expected to change dramatically in the early and mid 1990s. The reader who goes on to survey the literature of the 1990s is promised exciting reading.

Note added in proof: The SAGE experiment (Sec. IV.A) had analyzed five extractions of germanium from 30 tons of gallium as of the summer of 1990. The standard solar model predicts that ~18 events due to solar neutrinos should have been detected for these five runs. Events selected as ^{71}Ge-decay candidates, based on energy and pulse shape, are analyzed by a maximum-likelihood procedure. The assumption is made that the events are due to either ^{71}Ge decay or background. The ^{71}Ge event rate should decrease exponentially with an 11.43-day half-life, and the background is assumed to be constant with time. The best fit to the candidate events results in a zero signal with 68% and 95% confidence limits of 70 and 135 SNU, respectively, for the capture rate (Abazov et al. 1990; Elliot 1990). We anticipate that this dramatic result, if corroborated by improved statistics and by the GALLEX experiment, will lead to a host of comments, interpretations and solutions based partly on material presented in this chapter and perhaps on new concepts yet to be developed.

PART III
Isotopes

THE PRODUCTION OF COSMOGENIC ISOTOPES IN THE EARTH'S ATMOSPHERE AND THEIR INVENTORIES

KERAN O'BRIEN, ALBERTO De La ZERDA LERNER
U. S. Department of Energy

M. A. SHEA and D. F. SMART
United States Air Force Geophysics Laboratory

Production rates of cosmogenic isotopes in the Earth's atmosphere and their dependence on solar modulation and geomagnetic field intensity have been calculated. Spallation cross sections were also obtained using the Silberberg-Tsao equations and solar modulation effects were calculated using the force-field model. The current geomagnetic field is treated in detail, and past magnetic fields are modeled based on the archeomagnetic record. Radiocarbon and radioberyllium inventories so obtained are in good agreement with current values. The neutrino-emitting radioactivity of the Earth's atmosphere is shown to add a negligible contribution to the flux from the Sun.

Lal and Peters (1962) have determined cosmogenic isotope production rates in the atmosphere based on extensive studies of nucleons and nucleon reactions in the Earth's atmosphere. The result of their efforts is a valuable and widely cited reference. In this chapter, both production rates and inventories are determined using a different approach. We have carried out detailed calculations of the physical processes involved in cosmogenesis and the buildup of tellurian inventories. In this approach, we take into account the detailed structure of the geomagnetic field and the interaction of galactic cosmic rays with it and with the solar wind, the subsequent hadronic cascade in the atmosphere, and the accumulation and decay of cosmogenic radioactivity on the planet. As a consequence, our results can be extended into conditions not experimentally accessible, such as the remote past.

There have been a number of such cascade calculations both before and since Lal and Peter's seminal work (Benioff 1956; de la Zerda Lerner and O'Brien 1987; Hess et al. 1961; Light et al. 1973; Lingenfelter 1963; Merker 1970; Newkirk 1963; O'Brien 1979), stimulated chiefly, although not entirely, by the need to understand the balance between the production and the quantity of radiocarbon in its reservoir, and the time dependence of the production rate. Small differences between the results presented in this chapter and those given earlier by O'Brien (1979) and by de la Zerda Lerner and O'Brien (1987) are due to an improved representation of the geomagnetic field and its behavior in the distant past.

I. THE PROPAGATION OF GALACTIC COSMIC RAYS THROUGH THE HELIOSPHERE AND INTO THE EARTH'S ATMOSPHERE

Solar Modulation

The primary cosmic-ray flux in the solar system at the Earth's orbit is a mixture of energetic protons and α particles, as well as a small admixture of heavier nuclei, electrons and photons which are omitted in this treatment. The flux and its composition vary over the 11-yr solar activity cycle; thus to calculate cosmic-ray phenomena as a function of time, the cosmic-ray flux must be obtained as a function of time and energy. The solar system is filled with an expanding fully ionized plasma, the solar wind which is the outer part of the Sun's corona, streaming outwards, containing frozen-in irregular magnetic fields. Cosmic rays undergo many scatterings from these irregularities and undergo a random walk in the solar wind. The cosmic-ray population outside the heliosphere diffuses inward and during this diffusive process undergoes deceleration by adiabatic cooling produced by the expanding solar wind. It has been shown theoretically (Gleeson and Axford 1967) that the effect on the galactic cosmic-ray spectrum in passing through the interplanetary medium is approximately the same as would be produced by a heliocentric electrical potential with a magnitude at the Earth's orbit equal to the energy lost by the cosmic rays in interacting with the solar wind. As solar activity waxes and wanes with the 11-yr solar cycle, the energy lost by cosmic rays in penetrating the interplanetary medium to the Earth's orbit rises and falls with it. An increase in the heliocentric potential causes a decline in the cosmic-ray intensity, while a decrease results in a rise. This approximation at cosmic-ray energies above about 50 MeV appears to be a good one and has been used by Ehmert (1959), Freier and Waddington (1965, 1966) and Cleghorn et al. (1971). Energies below 50 MeV contribute a negligible amount to atmosphere cosmic-ray intensities.

Cosmic-ray proton and α-particle spectra in the neighborhood of the Earth's orbit are determined from the heliocentric potential model equations (O'Brien and Burke 1973). These are obtained by applying Liouville's theorem to the unmodulated energy spectra "at infinity," i.e., beyond the heliopause, and by calculating the energy spectra at the Earth's orbit in terms of the energy lost per unit charge in penetrating the interplanetary medium and are given by

$$N(T) = N_o(E) [P (E) /P (T)]^2$$
$$T = E + ZU \qquad\qquad (1)$$
$$P(x) = (1/c)\sqrt{x^2 + 2 \, Amc^2 x}$$

where m is the nuclear mass in MeV, c is the speed of light, N_o is the unmodulated galactic spectrum of atomic weight A and atomic number Z per (cm² s MeV sr) having an energy E in MeV, and U is the modulating potential in MV. For $N_o(E)$, the unmodulated proton and α-particle spectra of Garcia-Munoz et al. (1975) were used for energies below 10 GeV. For energies above 10 GeV, the spectra of Pal (1967) and O'Brien (1972) were used (Fig. 1). The latter spectra are not derived as a functions of modulation: they are insensitive to modulation because of their energy, as a glance at Eq. (1) will indicate, and are used as given in the original reference for $N_o(E)$ to obtain $N(T)$. (Cosmic-ray modulation is reviewed in the chapter by Jokipii.)

The Deep River Neutron Monitor

The dependence of the isotope production rates on solar modulation was obtained by calculating the hourly counting rate of the Deep River cosmic-ray neutron monitor as a function of U (O'Brien and Burke 1973). We have

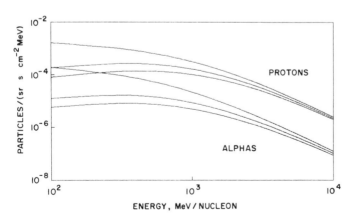

Fig. 1. Cosmic-ray α and proton spectra incident on the Earth's atmosphere for modulation potentials of (top to bottom) 0, 600 and 1200 MV.

also calculated the Deep River neutron monitor counting rate corresponding to the unmodulated Garcia-Munoz et al. spectra impinging on the Earth's atmosphere and obtained the counting rates vs U (Table I). The historical record of the measured Deep River counting rates can then be used to obtain U as a function of time (as we have determined the dependence of U on the Deep River monitor counting rates), and from U one can obtain the cosmic-ray energy spectra at the Earth's orbit (Eq. 1). The May 1965 solar minimum resulted in 2.147×10^6 counts hr^{-1} (Steljes 1973). The 1958 solar maximum corresponded to about 1.7×10^6 counts hr^{-1}.

The Geomagnetic Field

The magnetic rigidity (i.e., momentum per unit charge) of cosmic-ray particles is defined as $r = pc/q$, where p and q are the momentum and charge of the particle and c is the velocity of light. The magnetic field of the Earth deflects the incoming cosmic rays depending on their rigidity and angle of incidence. For each angle of incidence, there is a critical rigidity below which the incoming particle cannot reach the Earth's atmosphere.

The equation of motion of a charge particle in a magnetic field does not have a solution in a closed form. Reasonably accurate geomagnetic cutoff rigidities (the cut-off being the critical rigidity) can be obtained by the numerical integration of cosmic-ray trajectories using mathematical models of the geomagnetic field. However, the calculation of these values for a large number of locations and directions involves a formidable amount of computing time. The precision of these calculations is limited only by the accuracy of the geomagnetic field used.

TABLE I

Calculated Deep River Neutron Monitor Counting Rate as a Function of Solar Modulation Expressed in Terms of the Modulating Potential U

U (MV)	Counting Rate $\times$ 10^6 (per hr)
0	2.5468
100	2.4137
200	2.3032
300	2.2091
400	2.1277
500	2.0562
600	1.9928
700	1.9359
800	1.8845
900	1.8378
1000	1.7950
1100	1.7557
1200	1.7192

There are long-term secular variations in the cosmic-ray cutoff rigidities directly reflecting the long-term secular changes in the geomagnetic field. Over an approximate 10-yr period, or epoch, these changes are sufficient to be experimentally observed, and, for very precise analyses, these secular variations should be considered.

Because of the complexity of cutoff rigidity calculations, a number of approximations are generally employed. For many purposes it is sufficient to know the vertical cutoff rigidity as a function of location over the Earth's surface. We used the world grids of vertical cutoffs for Epochs 1955, 1965, 1975 and 1980 (Shea and Smart 1975, 1983; Shea et al. 1968) and determined the nonvertical cutoffs with the aid of the computer code ANGRI (Bland and Cioni 1968). ANGRI's results depend on accurate values for the vertical cutoff. The cutoffs determined by ANGRI as a function of zenith and azimuth were used as high-pass filters, to cut off cosmic-ray α and proton fluxes with rigidities below the critical rigidity.

Archeomagnetic data indicate that the Earth's magnetic field has varied from about 0.5 to 1.5 times its current value over the last 12,000 yr (Bucha 1969, 1970; Cox 1969; Creer 1988; Kitazawa 1970; see also the chapter by Damon and Sonett). There is evidence from excavations in the Lake Mungo area of southeastern Australia that between 25,000 and 35,000 yr ago the geomagnetic field may have reached 3 times its current value during a geomagnetic excursion (Barbetti 1976).

To stimulate the effect of a change in the Earth's magnetic field in past epochs by some factor M, the vertical cutoff rigidity was multiplied by M before calculating the cosmic-ray fluxes in the Earth's atmosphere and integrating over the globe. The justification for this approach is that a charged particle describing a fixed curved path in a magnetic field (above the Earth's atmosphere) must have a rigidity proportional to the magnetic field strength.

Atmospheric Cosmic-Ray Propagation

The behavior of cosmic rays in the Earth's atmosphere is governed by the stationary form of the Boltzmann equation. The Boltzmann equation is an integrodifferential equation describing the behavior of a dilute assemblage of particles. It was derived by Ludwig Boltzmann in 1872 to study the properties of gases. It applies equally to the description of the behavior of radiation, which for the purposes of this chapter will comprise the primary and secondary particles of cosmic radiation: nuclei, leptons, mesons, baryons and energetic photons.

Boltzmann's equation is a continuity equation in phase space which is made up of the three space coordinates of euclidean geometry and the kinetic energy, and the direction of motion of the particles:

$$B_i\phi(\mathbf{x},E,\mathbf{\Omega},t) = \Sigma_j Q_{ij} \qquad (2)$$

where

$$B_i = \mathbf{\Omega} \cdot \mathrm{grad} + \sigma_i + d_i - (\partial / \partial E)S_i \qquad (3)$$

$$Q_{ij} = \int_{4\pi} d\mathbf{\Omega}' \int_E^{E_{max}} dE_B \, \sigma_{ij} \, (E_B \to E, \mathbf{\Omega}' \to \mathbf{\Omega}) \, \phi \, (\mathbf{x}, E_B, \mathbf{\Omega}', t) \qquad (4)$$

and

$$d_i = [\sqrt{(1 - \beta_i}] / (T_i c \beta_i). \qquad (5)$$

Here, B_i is the Boltzmann operator, σ_i is the absorption cross section for particles of type i; d_i is the decay probability per unit flight path of unstable particles (such as muons or pions) of type i; S_i is the stopping power for charged particles of type i (assumed to be zero for uncharged particles); ϕ $(\mathbf{x}, E, \mathbf{\Omega}, t)$ is the particle flux at location $\mathbf{x}$, energy E, direction $\mathbf{\Omega}$ and time t; Q_{ij} is the "scattering-down" integral, the production rate of particles of type i with a direction $\mathbf{\Omega}$, an energy E at a location $\mathbf{x}$, by collisions with nuclei or decay of j type particles having a direction $\mathbf{\Omega}'$ at a higher energy E_B; σ_{ij} is the doubly differential inclusive cross section for the production of type-i particles with energy E and a direction $\mathbf{\Omega}$ from nuclear collisions or decay of type j particles with an energy E_B and a direction $\mathbf{\Omega}'$; β_i is the speed of a particle of type i with respect to the speed of light ($= v_i / c$); T_i is the mean life of a radioactive particle of type i in the rest frame, and c is the speed of light in a vacuum.

As the incoming particles that interact with the Earth's atmosphere have energies largely over 100 MeV, the cross sections used were geometric (equal to the size of the nucleus) and constant with energy (Barashenkov et al. 1968). The stopping powers were taken from O'Brien (1972). The production spectra were derived from a combination of experimental and theoretical models (O'Brien 1971, 1975a).

The Boltzmann equation was solved for ϕ $(\mathbf{x}, E, \mathbf{\Omega}, t)$ in a form based on the work of Passow (1962). The code incorporating this solution transports neutrons, protons and pions. While they are interacting with the solar wind and the Earth's geomagnetic field, α particles are treated as two protons and two neutrons bound into a single neclecus. However, the protons and the neutrons in the incident primary cosmic-ray α particle flux are treated as unbound after entering the Earth's atmosphere (the so-called superposition approximation). In dilute media such as the atmosphere, kaons, which are produced in nuclear interactions, generally decay before undergoing a nuclear collision. For this reason, kaon production was taken into account (to conserve energy), but kaon transport was ignored. The code yields results that are in good agreement with cosmic-ray pion, proton, neutron and muon spectra (O'Brien 1971, 1975a,b,).

For this calculation, the atmosphere was modeled as a spherical shell with an inner radius of 6.371×10^8 cm, and a radial structure corresponding to the standard atmosphere (Kallmann-Bijl et al. 1961).

As the Earth's atmosphere is over a thousand g per cm^2 or 13 collision lengths deep, the equivalent in mass of 10 m of water, the atmosphere absorbs almost all the primary and secondary cosmic rays. Hence, atmospheric production of a cosmogenic isotope is essentially the total production of the isotope, with the exception of a few isotopes such as ^{36}Cl and ^{41}Ca resulting from neutron capture by nuclei in the Earth's crust with large neutron capture cross sections.

The Computational Method

The computer code described above will yield cosmic-ray spectra in one location in the Earth's atmosphere at a given altitude, at a given latitude and longitude (which determine the local geomagnetic field), for a given level of solar modulation. In order to obtain production rates both as a function of latitude and globally, points were chosen along the axes of altitude, latitude and longitude corresponding to the nodes prescribed by the Gauss-Legendre numerical integration technique; calculations were performed for each of these points and numerically integrated (integrations along the longitudinal axis had, of course, to be weighted with the cosine of the latitude to account for the spherical surface of the Earth). Sixteen points along the altitude axis, 17 points along the latitude axis and 11 points along the longitude axis, for a total of 2992, were found to be sufficient. These calculations were performed for 5 equally spaced solar modulating potentials from 0 to 1200 MV for the 1955 and 1980 epochs. Only 0 and 1200 MV modulating potentials were used in the 1965 epoch calculations. The three 1° by 1° data sets for each of the three epochs were used for these calculations. The 1980 epoch was used for estimates of cosmogenic production rates in the distant past.

Cosmogenic Isotope Calculations

The most convenient approach to determine cosmogenic isotope production is first to calculate the concentration of high-energy inelastic collisions (called "stars" because of their appearance in a photographic plate) in the air and then to determine the spallation yield per star for each particular isotope. There is some sacrifice of accuracy in using this method of about 10% (Yasyulenis et al. 1974). A more accurate approach would be to calculate the production rates directly, multiplying each hadronic species by cross section for a particular isotope, as the cross sections are energy dependent and as the hadronic spectra change slightly in shape with altitude, solar modulation, latitude and longitude. At the top of the atmosphere, of course, the proton spectrum and the primary α-particle spectrum are discontinuous in energy at each zenith angle and azimuth, owing to the effect of the cutoff field.

However, the computational advantage gained by our approximation is enormous and well worth the small sacrifice in accuracy. New yields or revised yields can be easily applied to the same body of computed data. The volume of the computation (which is nonetheless of some considerable size) is very much reduced and the applicability of the results is considerably broadened. This procedure is only applicable to spallation products. The solution to the Boltzmann equation used in these calculations is only applicable to hadrons with energies above 100 MeV (O'Brien 1971). This low-energy limit has little or no effect on charged cosmic-ray hadron transport because of particle stopping which limits the number of these particles at low energies. Neutrons, however, are uncharged and their energies extend all the way down to thermal energies.

In order to account for isotope production below 100 MeV, the low-energy neutron spectrum has been assumed to have the same shape as the spectrum reported by Hess et al. (1961). This necessitates the assumption that there is no change in the shape of the neutron spectrum below 100 MeV with altitude, modulation and changes in the geomagnetic field. This assumption is a reasonable one. All atmospheric neutrons are secondaries resulting from collisions between the primary cosmic rays and the constituents of the atmosphere. Hence, they reflect the interaction of the primary spectra with production mechanisms and atmospheric absorption cross sections. The shape of the energy spectra of protons and neutrons above 100 MeV is essentially independent of these factors, so that the production of low-energy neutrons from high-energy collisions and the resulting spectra, when they are far from any boundary, will follow suit. At the top of the atmosphere, some diffusion hardening will take place that will not be reflected in this treatment; a small overestimate will result in the flux calculated by the method used in this chapter (O'Brien 1971) within a few g cm^{-2} of the top of a few percent. The expected invariance of low-energy neutron spectral shape with altitude is clearly evident in the measurements of Hess et al. (1961) and Armstrong et al. (1973), justifying the adequacy of this assumption.

Scope of the Transport Calculations

Radiocarbon and star productions were calculated as a function of altitude in the Earth's atmosphere, latitude and longitude, and were integrated over the total atmosphere for solar activity corresponding to conditions ranging from a Deep River neutron monitor hourly counting rate ranging from 2.5468×10^6 (no modulation) to 1.7192×10^6 (strongly attenuated). The Earth's magnetic field was varied from zero to 5 times the current magnetic field strength ($0 \le M \le 5$). The production rate in the stratosphere alone was also obtained for the full range of modulation conditions, but only for the current geomagnetic field strength ($M = 1$).

II. DESCRIPTION OF THE RESULTS

By using five sets of proton and α-particle spectra, such as those of Fig. 1, from 0 MV (unmodulated) to 1200 MV (strongly modulated), star densities and radiocarbon production rates were calculated for ($0 \leq M \leq 5$). The star density values obtained as a function of altitude and geographic latitude for $M = 1$ are shown in Figs. 2 a,b,c,d,e for modulating potentials of 0, 300, 600, 900 and 1200 MV. The corresponding values for radiocarbon production are show in Figs. 3 a,b,c,d,e. These calculations were all performed using the data from epoch 1980. Radiocarbon production using epochs 1955, 1965 and 1980 are compared in Fig. 4 for $U = 0$ as a function of geographic latitude. It can be seen that epochs 1965 and 1980 are essentially equivalent for the purpose of cosmogenic isotope production, the 1955 production being somewhat lower. Essentially identical results are obtained for star production. The values of star density and radiocarbon production integrated over the whole Earth are summarized in Tables II and III.

Spallation Yields

The yield, per star, of some cosmogenic isotope x is defined to be

$$
y_x = \Sigma_i \left\{ \left[\int_a^\infty \sigma_x(E)\phi_i(E)dE \right] / \left[\int_a^\infty \sigma_{\text{nonel}}(E)\phi_i(E) \, dE \right] \right\}
$$

$$(i = p,n, \pi^\pm)$$

$$(6)$$

where ϕ_i represents a particle in the cosmic-ray flux (proton, neutron or pion), $\sigma_x(E)$ is the cross section for the production of x as a function of energy, $\sigma_{\text{nonel}}(E)$ is the total nonelastic cross section, and a is the lower-energy limit of the calculation (100 MeV).

The yields y_x depend primarily on the slope of the neutron and proton spectra. These in turn depend on the primary nucleon spectrum and show relatively little change of shape with changing conditions of altitude, latitude or modulation. The pion spectrum, which does change drastically with altitude, contributes less than 5% to the total yield; hence it is assumed that the spallation yield y_x does not vary with time or location in the atmosphere. To calculate y_x, hadron spectra are chosen from conditions corresponding to the mean global production rate: at 45° latitude, at an altitude of 10 km, $U = 457$ MV and $M = 1$. Yasyulenis et al. (1974) indicate that the error in assuming y_x constant everywhere is on the order of 10%.

Spallation cross sections were calculated using the formalism of Silberberg and Tsao (1973a,b). The composition of the atmosphere was taken from Kallman-Bijl (1961). The results of this calculation are shown in Table IV.

Experimental Yields

Lal and Peters (1962) have published a number of spallation yields, some of which are based on experimental data. In addition, Young et al.

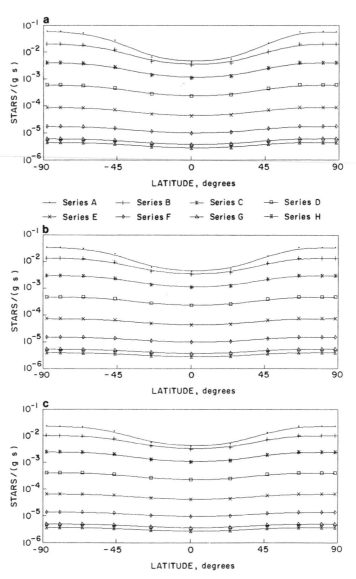

Fig. 2. (a) Star production rates for $U = 0$ and $M = 1$. (b) Star production rates for $U = 300$ and $M = 1$. (c) Star production rates for $U = 600$ and $M = 1$. (d) Star production rates for $U = 900$ and $M = 1$. (e) Star production rates for $U = 1200$ and $M = 1$. **A** corresponds to 24,000 m altitude, **B** to 14,000 m, **C** to 9000 m, **D** to 5500 m, **E** to 3000 m, **F** to 1200 m, **G** to 270 m and **H** to sea level.

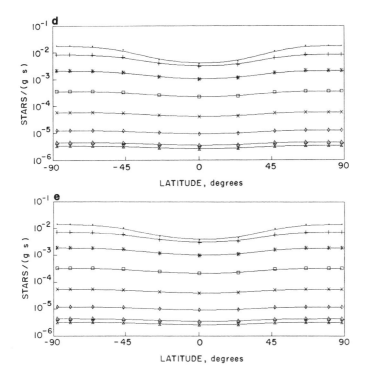

(1970) have published isotope production rates in argon that were measured at an altitude of about 20 km near 50° geomagnetic latitude (with a vertical cutoff of about 2.5 GV). Combining these data with the calculated production rates allows the inference of the corresponding spallation yields. Experimental yields from both sources are exhibited in Table V. No values for the total experimental error is given in either paper, but the results given in Table V are consistent with apportioning about 30% error both to calculation and to measurement.

III. ISOTOPIC INVENTORIES

Beryllium-7

Aircraft and balloon measurements of [7]Be in the stratosphere carried out under the direction of the Environmental Measurements Laboratory (EML) from 1970 to 1974 (Feely et al. 1988) are shown in Fig. 5. The time-dependent [7]Be inventory was calculated by using the Silberberg-Tsao spallation yield of Table IV along with the stratospheric star density totals of Table II to construct an array of [7]Be inventories vs heliocentric modulating potential. Deep River neutron monitor data (Steljes 1973; Wilson and Bercovitch

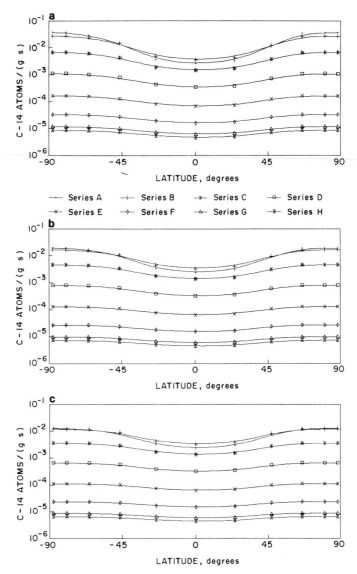

Fig. 3. (a) ^{14}C production rates for $U = 0$ and $M = 1$. (b) ^{14}C production rates for $U = 300$ and $M = 1$. (c) ^{14}C production rates for $U = 600$ and $M = 1$. (d) ^{14}C production rates for $U = 900$ and $M = 1$. (e) ^{14}C production rates for $U = 1200$ and $M = 1$. **A** corresponds to 24,000 m altitude, **B** to 14,000 m, **C** to 9000 m, **D** to 5500 m, **E** to 3000 m, **F** to 1200 m, **G** to 270 m and **H** to sea level.

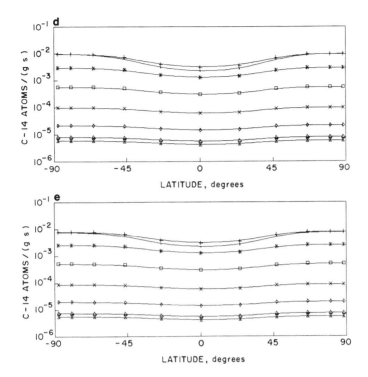

1975) were then used to obtain the modulation corresponding to each measurement period and hence the ^{7}Be production rate. To pass from isotope production to radioactivity, a half-residence time of 10 months was assumed (Krey and Krajewski 1970) and the equilibrium of decay and production was assumed. Thus,

$$\lambda_i I(t) = 138 \, y_{Be7} \, S_s(t) \, [\lambda_d / (\lambda_R + \lambda_d)]$$
$$\lambda_i = \ln 2 / H_i \quad i = R, d \tag{7}$$

where $\lambda_d I(t)$ is the time-dependent inventory in MCi, the constant 138 converts from radioberyllium atoms in units of $cm_e^{-2}s^{-1}$ to MCi (where the subscript e signifies that results are averaged, after integration, over the whole Earth); y_{Be7} is the ^{7}Be yield per star from Table IV; $S_s(t)$ is the time-dependent worldwide production rate for the stratosphere (Table II); λ_i is the removal rate due to radioactive decay or transfer out of the stratosphere; H_i is the corresponding half-life; R refers to stratospheric residence; d refers to radioactive decay. Overall agreement is seen to be very good. The time dependence is correctly given.

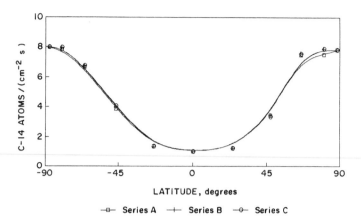

Fig. 4. ^{14}C production rates vs latitude for three Epochs with $U = 0$. **A** corresponds to 1955, **B** to 1965 and **C** to 1980.

TABLE II
Star Production Rate ($cm_e^{-2}s^{-1}$)

| Modulating Potential | Multiple of Current Geomagnetic Field Strength | | | | | |
| | Total Atmosphere | | | | | Stratosphere |
MV	$M = 0$	0.5	1.0	1.5	5	$M = 1$
0	8.38	4.60	3.00	2.73	1.43	2.29
300	5.21	3.43	2.30	2.12	1.14	1.67
600	3.82	2.81	1.92	1.80	0.99	1.33
900	3.02	2.41	1.68	1.59	0.89	1.03
1200	2.50	2.12	1.51	1.44	0.82	0.97

TABLE III
Radiocarbon Production Rate ($cm_e^{-2}s^{-1}$)

| Modulating Potential | Multiple of Current Geomagnetic Field Strength | | | | | |
| | Total Atmosphere | | | | | Stratosphere |
MV	$M = 0$	0.5	1.0	1.5	5	$M = 1$
0	7.99	4.25	2.79	2.54	1.34	1.83
300	4.77	3.14	2.12	1.97	1.07	1.32
600	3.48	2.59	1.78	1.67	0.93	1.05
900	2.76	2.23	1.57	1.48	0.84	0.90
1200	2.30	1.97	1.42	1.35	0.78	0.77

TABLE IV
Calculated Spallation Yields

Isotope	Half-life	Yield per Star
^{10}Be	1.50×10^6 yr	1.37×10^{-2}
^{26}Al	7.16×10^5 yr	2.22×10^{-5}
^{36}Cl	3.08×10^5 yr	4.74×10^{-4}
^{32}Si	280 yr	3.55×10^{-5}
^{22}Na	2.62 yr	1.99×10^{-5}
^{35}S	87.9 days	3.17×10^{-4}
^{7}Be	53.6 days	3.04×10^{-2}
^{33}P	24.4 days	1.79×10^{-4}
^{32}P	14.3 days	1.86×10^{-4}
^{28}Mg	21.2 hr	9.30×10^{-6}
^{24}Na	15.0 hr	3.83×10^{-5}
^{38}S	2.87 hr	7.75×10^{-6}
^{31}Si	2.62 hr	8.68×10^{-6}
^{18}F	1.83 hr	1.67×10^{-5}
^{39}Cl	55.7 min	5.11×10^{-4}
^{38}Cl	37.3 min	3.19×10^{-4}
^{34m}Cl	32.0 min	2.12×10^{-5}
^{29}Al	6.56 min	5.46×10^{-5}
^{37}S	5.04 min	2.16×10^{-5}
^{24}Ne	3.4 min	2.84×10^{-6}
^{30}P	2.55 min	2.29×10^{-5}
^{28}Al	2.3 min	8.08×10^{-5}

TABLE V
Calculated and Measured Spallation Yields per Star

Isotope	Calculated Yield	Measured Yield	Percent Deviation[c]	Coefficient of Variation of Counting Statistics
^{10}Be	0.014	0.025[a]	-44	
^{3}H		0.14		
^{35}S	3.2×10^{-4}	8.6×10^{-4} [a]	-63	
^{7}Be	0.030	0.045[a]	-33	
^{33}P	1.8×10^{-4}	3.8×10^{-4} [a]	-53	
^{32}P	1.9×10^{-4}	4.6×10^{-4} [a]	-59	
^{28}Mg	9.3×10^{-6}	1.2×10^{-5} [b]	-22	15
^{24}Na	3.8×10^{-5}	4.4×10^{-5} [b]	-13	18
^{38}S	7.8×10^{-6}	2.9×10^{-5} [b]	-73	20
^{18}F	1.7×10^{-5}	8.1×10^{-6} [b]	110	36
^{39}Cl	5.1×10^{-4}	1.1×10^{-3} [b]	-54	5
^{38}Cl	3.2×10^{-4}	7.0×10^{-4} [b]	-54	10
^{34m}Cl	2.1×10^{-5}	7.7×10^{-5} [b]	-73	13

[a] Lal and Peters (1972).
[b] Inferred from Young et al. (1970).
[c] Percent deviation = [(calculated − measured)/measured] × 100.

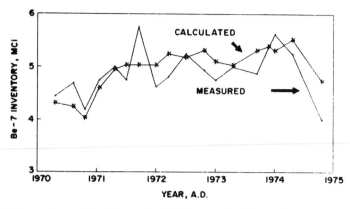

Fig. 5. Measured and calculated stratospheric ^{7}Be inventories during 4.5 yr.

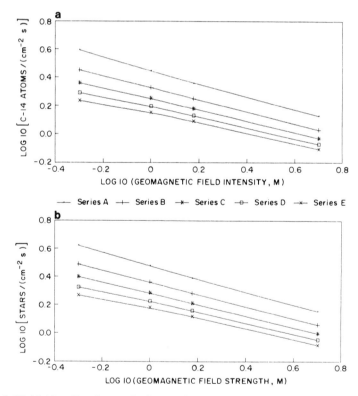

Fig. 6. Worldwide radiocarbon production as a function of M, the geomagnetic field strength in units of the current value. **A** corresponds to 0 MV, **B** to 300, **C** to 600, **D** to 900 and **E** to 1200.

Radiocarbon

Long-Term Variations. Elsasser et al. (1956) showed that measured radiocarbon production is related to the Earth's geomagnetic field strength, by $Q \propto M^{-0.52}$, where Q is the production rate and M is the geomagnetic dipole moment in units of the current value, i.e., with today's field strength taken as unity. From the log-log plot (Fig. 6) of the data from Table III, it can be seen that this power-law formula describes, approximately, the radiocarbon dependence for all conditions of modulation for $0 \leq M \leq 5$. Not unexpectedly, as Fig. 7 shows, the Elsasser et al. formula applies equally well to spallation-produced isotopes. The bulk of radiocarbon production is via thermal and resonance capture, but there is a small contribution from spallation.

In order to account for the effect on the total accumulated quantity of radiocarbon of the previously described geomagnetic field variations, we converted the archeomagnetic data points of Creer (1988) to "relative rate of production points" by first renormalizing the virtual dipole moment so that the value at the present time is unity, and then raising the resulting values to the -0.44 power (a slightly lower value than the Elsasser et al. result, but it fitted our data somewhat better over the range of our calculations). When the result is analyzed by the maximum entropy method, a peak at 1.1×10^{-4} yr^{-1} (0.055 when a 500-yr width is used), or 9090-yr period is prominent (Fig. 8). In Fig. 9, we display the production data fitted with the sinusoidal curve

$$K(t) = 1 + 0.116 \sin[(2\pi) \, 1.1 \times 10^{-4} \, t] \qquad (8)$$
$$= 1 + 0.116 \sin(6.912 \times 10^{-4} \, t)$$

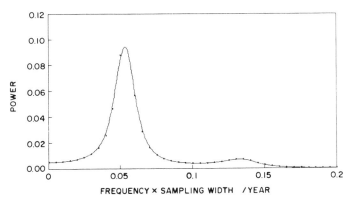

Fig. 7. The power spectrum of the cosmogenic isotope production rate over the last 12,000 yr. The sampling interval is 500 yr.

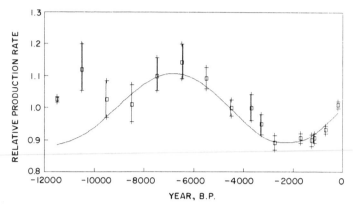

Fig. 8. The relative production rate over the past 12,000 yr fitted with a sine curve with a period of 9090 yr. Data are based on Creer (1988).

where $K(t)$ is the relative production rate in $cm_e^{-2}s^{-1}$. All but the last two points, earlier than 10,000 yr BP, seem to fit the curve quite adequately. The time dependence of the inventory is given by

$$dI / dt = - \lambda I + Q(t) \qquad (9)$$

where λ is the decay rate of the isotope and $Q(t) = Q_o K(t)$ is the absolute production rate of a specific isotope, and Q_o is the production rate at $t = 0$, i.e, at present. Substituting $Q(t) = Q_o(1 + a \sin bt)$ in the equation above and solving yields

$$I = Q_o \{1 / \lambda + [\lambda / (\lambda^2 + b^2)] (\lambda \sin bt - b \cos bt)\}. \qquad (10)$$

Expressing the inventory in terms of the current production rate, $M = 1$, $t = 0$ yields

$$\lambda I / Q_o = 1 - ab\lambda / (\lambda^2 + b^2) \qquad (11)$$

or perhaps more conveniently, using values of a and b obtained from Eq. (8), with the half-life

$$\lambda I / Q_o = 1 - 116H / (H^2 + 10^6). \qquad (12)$$

This solution implies that the relative production rate we have fitted to a sine curve extends back in that form to $t = -\infty$. There is no physical reason to suppose this. However, short-lived isotopes are unaffected by the shape of the curve in ancient times, and as long as the current field is close to the average value of the field, as it is over the past 12,000 yr, the effect of the exponential term in the integral of I is to filter out all variations from the mean for half-lives greater than a few thousand years. The maximum deviation of $\lambda I \, / \, Q_o$ from unity occurs for half-lives of a thousand years and amounts to little less than 6%.

Short-Term Variations.   Stuiver and Quay (1980) have suggested that the geomagnetic *Aa* index might reflect some of the properties of the solar wind and might serve as a measure of cosmogenic isotope formation.

The Aa index is derived from two observatories situated at approximately antipodal locations in England and Australia. Its development in the early 1970s was motivated by the availability of records from two old observatories (Greenwich and Melbourne) and has resulted in an index which spans from 1868 to the present. Using the relationship between neutron counting rates and U in Table I, the annual neutron monitor data compiled by Rao (1972) can be utilized to yield mean annual values of U. We have plotted these values vs the corresponding aa indices, taken from Stuiver and Quay (1980) for the years between 1937 and 1967. These data are shown in Fig. 10. They are the identical data underlying Stuiver's and Quay's (1980) Fig. 3. They can be fitted by

$$U = 23.4 \, \text{Aa} \qquad\qquad (13)$$

with a standard deviation of 6%.

Stuiver and Quay (1980) have compiled annual Aa indices for the period from 1868 to 1967. They have also synthesized Aa indices averaged over

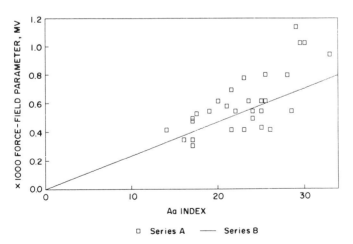

Series A, Data; B, Least Squares Fit

Fig. 9. U vs annual Aa indices for the period from 1937 to 1978. **A** corresponds to the data and **B** is the best fit in the chi-squared sense.

K. O'BRIEN ET AL.

each solar cycle from observations of the radiocarbon record from 1005 to 1860. Using the relationship between the Aa index and the modulating potential above, and the neutron monitor data and Table I, allow us to determine the behavior of the modulating potential over the last millennium, which we show in Fig. 11. An enlarged portion of Fig. 11 (Fig. 12) shows the Wolf, Sporer and Maunder minima.

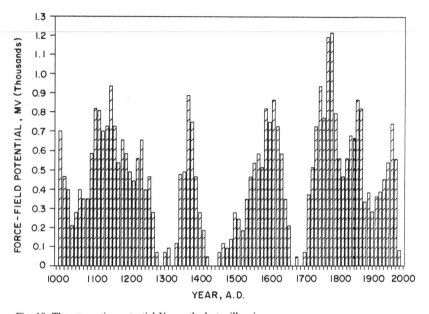

Fig. 10. The attenuating potential U over the last millennium.

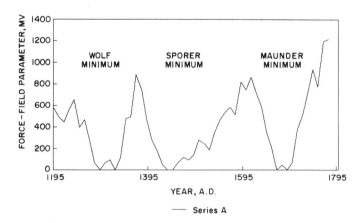

Fig. 11. The Wolf, Sporer and Maunder minima in terms of their attenuating potentials.

Combining the modulating data of Fig. 11 with the fit to the relative production data $K(t)$ yields the rates of radiocarbon and star production over the last millennium, integrated over the Earth's surface. We show these results in Figs. 13 and 14. The three peaks corresponding to the 3 solar minima are quite prominent. Millennial means obtained from these data are summarized in Table VI.

The Radiocarbon Inventory.   Because we do not have either geomagnetic or solar modulation data extending back for several times the radiocarbon half-life, it is convenient to assume that short-term solar and long-term geomagnetic effects are separable. We take Q_0 to be the radiocarbon production rate corresponding to 457 MV, the mean modulating potential over the last millennium, which we obtained by interpolating among the values in Table III. This value is inserted into the formula for the inventory we derived above, yielding

$$\lambda I = 1.924[1 - 116H / (H^2 + 10^6)] = 1.89 \text{ cm}_e^{-2}\text{s}^{-1} \qquad (14)$$

for the current radiocarbon reservoir.

This is in excellent agreement with direct estimates of the radiocarbon reservoir based on analyses of the specific activity of ^{14}C on the Earth's surface. Damon et al. (1978) give 1.98 $\text{cm}_e^{-2}\text{s}^{-1}$; Grey (1972) gives $1.8^{+0.18}_{-0.04}$,

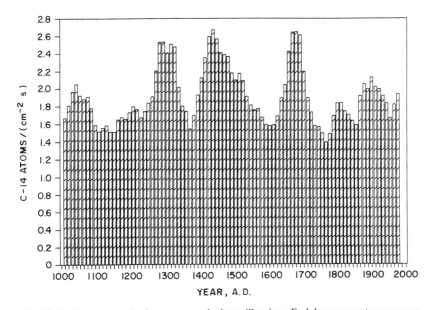

Fig. 12. Radiocarbon production rates over the last millennium. Each bar represents an average over 1 solar cycle.

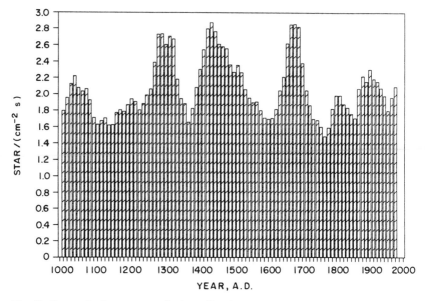

Fig. 13. Star production rates over the last millennium. Each bar represents an average over 1 solar cycle.

and Seuss (1965) gives 1.76. All these results lie within a range of 12% and are within the error bounds on Grey's value. Lingenfelter and Ramaty (1970) have calculated a radiocarbon inventory of 2.2 $cm_e^{-2}s^{-1}$ using experimental neutron flux and density measurements in the literature.

Two calculations of radiocarbon production, those of Merker (1970) and Light et al. (1973), are compared with the results of this study in Table VII. Both are higher than our results by 16 and 20%, respectively.

Beryllium-10

The [10]Be production rate corresponding to the mean modulating potential over the last millennium is

$$Q_o = y_{Be10}S_o \tag{15}$$

TABLE VI
Mean Values Associated with Cosmogenic Isotope Production over the Last 1000 yr

Modulating Potential U	457 MV
Star Production	1.93 $cm_e^{-2}s^{-1}$
Radiocarbon Production	1.79 $cm_e^{-2}s^{-1}$

TABLE VII
Comparison of Radiocarbon Production Calculations
$(cm_e^{-2} s^{-1})$

Year	Light et al. (1973)	Merker (1970)	This Study
1964	2.42	2.15	1.93
1965	2.58		2.02
1966	2.39		1.90
1967	2.10		1.85
1968	2.03		1.72
1969	1.93	1.86	1.69
1970	1.90		1.72

where S_o is the mean star production rate (from Table II) corresponding to 457 MV, and y_{Be10} is the yield (from Table IV). The ^{10}Be inventory is then

$$\lambda I = Q_o[1 - 116H / (H^2 + 10^6)] = 0.0285 \, cm_e^{-2} s^{-1} \tag{16}$$

which agrees well with Ku (1979) who measured $0.023 \pm 0.008 \, cm_e^{-2} s^{-1}$.

Krypton-81 and Krypton-85

^{81}Kr is produced by peripheral spallation reactions from ^{86}Kr, ^{84}Kr, ^{83}Kr and ^{82}Kr, as well as neutron capture by ^{80}Kr. ^{85}Kr is produced by peripheral spallation reactions from ^{86}Kr and neutron capture from ^{84}Kr.

The half-life of ^{81}Kr is 2.13×10^5 yr, and its inventory is given by

$$\lambda I = 4.06 \times 10^7(\lambda I / Q_o) = 4.06 \times 10^7 \, cm_e^{-2} s^{-1}$$
$$= 0.031 \, d / m \, l^{-1} \text{ of krypton.} \tag{17}$$

This compares with the measured values of $0.10 \pm 0.01 \, d / m \, l^{-1}$ (Oeschger et al. 1970) and $0.076 \pm 0.004 \, d / m \, l^{-1}$ (Kuziminoff and Pomansky 1983). The cause of the roughly factor-of-three disagreement is not known. Kuzminoff and Pomansky suggest that other reactions at energies below 100 MeV and the photonuclear reaction $^{82}Kr(\gamma,n)^{81}Kr$ contribute $0.03 \, d / m \, l^{-1}$. This, if it were so, would reduce the discrepancy to 25%.

The half-life of ^{85}Kr is 10.76 yr, and its inventory is given by

$$\lambda I = 1.05 \times 10^{-7} (\lambda I / Q_o) = 1.05 \times 10^{-7} \, cm_e^{-2} s^{-1}$$
$$= 14.5 \, Ci. \tag{18}$$

Diethorn and Stocko (1972) estimate an inventory of 10 Ci. This is clearly negligible with respect to the fission-produced inventory of 0.1 MCi (UN-SCEAR 1972).

Cosmogenic Isotope Inventories

On the basis of the results obtained in the previous section, we can compile an average inventory of cosmogenic isotopes in the environment. This is shown in Table VIII, accounting both for geomagnetic and solar effects. Spallation yields were the Silberberg-Tsao yields of Table IV, except for tritium, which was taken from the experimental yields of Table V, radiocarbon and the isotopes of krypton, which are partly produced by both spallation and neutron capture (O'Brien 1983).

IV. NEUTRINOS FROM COSMOGENIC ISOTOPES

Measurements (Rowley et al. 1985) of the flux of neutrinos emitted as by-products of hydrogen fusion reactions in the Sun's core bear on questions ranging from stellar evolution theories to the question of the neutrino mass (Bahcall 1978, 1985; see also the chapter by Wolfsberg and Kocharov). Attention has been devoted to time variations in the neutrino flux data and some investigators have found an inverse correlation with the 11-yr sunspot cycle (Bazilevskaya et al. 1982; Davis 1987, 1988; Raychaudhuri 1984), although there is still some debate about the statistical significance of this correlation (Bahcall et al. 1987). According to standard theory, the neutrino luminosity of the Sun should increase slowly, 5% in 1 Gyr, ruling out any variation with time scale less than 1 Myr, and therefore alternative sources of neutrinos have been invoked to explain these temporal variations. Galactic cosmic rays have been considered a likely candidate, as their intensity at the Earth's orbit varies inversely with the activity of the Sun, as discussed earlier.

The techniques of Sec. III were applied to 40 positron emitters (which are therefore neutrino emitters) produced by spallation of atmospheric nitrogen, oxygen and argon, using the spallation cross sections of Rudstam (1966). The most abundant neutrino emitters produced are ^{13}N, ^{11}C and ^{15}O, with a yield of about 1% each. The results of the calculation are shown in Table IX. The mean neutrino flux resulting from cosmogenic radioactivity is calculated for any point inside of the thin atmospheric shell which makes up the atmosphere. The results can be seen to be on the order of 0.1 $cm^{-2}s^{-1}$. These neutrinos have energies between 1 and 10 MeV, and they contribute on the order of 10^{-9} SNU (a "solar neutrino unit" or SNU = 10^{-36} captures per target atom per second) to the capture rate in the Davis chlorine detector. This contribution to the neutrino capture rate is some 9 orders of magnitude smaller than the event rate actually observed in the chlorine solar neutrino detector.

Thus, the apparent negative correlation between neutrino capture rates and the solar activity cycle cannot be accounted for by the cosmic-ray activation of the atmosphere. This finding, together with the results of Gaisser and Stanev (1985), which showed that the contribution from atmospheric

TABLE VIII
Cosmogenic Isotope Inventories

Isotope	Half-life	Inventory ($cm_e^{-2}s^{-1}$)	Inventory (MCi)
^{10}Be	15×10^6 yr	2.85×10^{-2}	3.9
^{26}Al	7.16×10^5 yr	4.61×10^{-5}	0.0064
^{36}Cl	3.08×10^5 yr	9.83×10^{-4}	0.14
^{81}Kr	2.13×10^5 yr	4.06×10^{-7}	5.6×10^{-5}
^{14}C	5730 yr	1.89	261
^{32}Si	280 yr	7.50×10^{-5}	0.010
^{3}H	12.3 yr	0.290	40
^{85}Kr	10.8 yr	1.05×10^{-7}	1.4×10^{-5}
^{22}Na	2.62 yr	4.13×10^{-5}	0.0057
^{35}S	87.9 days	6.59×10^{-4}	0.091
^{7}Be	53.6 days	6.31×10^{-2}	8.7
^{33}P	24.4 days	3.37×10^{-4}	0.051
^{32}P	14.3 days	3.86×10^{-4}	0.053
^{28}Mg	2.12 hr	1.93×10^{-5}	0.0027
^{24}Na	15.0 hr	7.95×10^{-5}	0.011
^{38}S	2.87 hr	1.61×10^{-5}	0.0022
^{31}S	2.62 hr	1.81×10^{-4}	0.025
^{18}F	1.83 hr	3.46×10^{-5}	0.0048
^{39}Cl	55.7 min	1.06×10^{-3}	0.15
^{38}Cl	37.3 min	6.62×10^{-4}	0.091
^{34m}Cl	32.0 min	4.41×10^{-5}	0.0061
^{29}Al	6.56 min	1.13×10^{-5}	0.0016
^{37}S	5.04 min	4.49×10^{-5}	0.0062
^{24}Ne	3.4 min	5.89×10^{-6}	8.1×10^{-4}
^{30}P	2.55 min	4.76×10^{-5}	0.066
^{28}Al	2.3 min	1.68×10^{-4}	0.023
Total			314

TABLE IX
Neutrino Flux in the Earth Produced by Cosmogenic Radioactivity in the Atmosphere

Solar Activity Level	Neutron Monitor Counting Rate (counts hr^{-1})	Star Production Rate (cm_e^{-2} s^{-1})	Total Positron Activity (MCi)	Neutrino Flux (cm^{-2} s^{-1})
Minimum	2.15×10^6	2.20	14.9	0.108
Mean	2.05×10^6	2.08	14.1	0.103
Maximum	1.77×10^6	1.59	10.8	0.078

secondary particle decay in the hadronic cascade is also insignificant, imply that, if such variations are real, their cause is most likely to be derived from the Sun itself.

Acknowledgments. The authors wish to thank M. Forman for pointing out the significance of the Aa indices as a device for treating solar modulation, and R. Davis, Jr. for suggesting the calculation of the neutrino flux from cosmogenic isotopes.

TIME VARIATIONS OF ^{10}BE AND SOLAR ACTIVITY

J. BEER

Institute for Aquatic Sciences and Water Pollution Control

G. M. RAISBECK and F. YIOU

Centre de Spectrometrie Nucleaire et de Spectrometrie de Masse

The main goal of this chapter is to demonstrate that ^{10}Be is a useful tool to reconstruct the history of solar activity. A comparison of neutron flux at the Earth's surface with the number of sunspots exhibits a clear negative correlation which is due to solar wind interaction with the galactic cosmic-ray flux. Since cosmogenic ^{10}Be production is proportional to the n-flux in the atmosphere, it also shows the inverse correlation with solar activity. Other sources of ^{10}Be variations in geologic reservoirs are changes of the geomagnetic dipole field and transport, and deposition processes within the atmosphere. Polar ice cores record the atmospheric fallout over the last ca 100,000 yr. Detailed ^{10}Be studies in these cores reveal the expected negative correlation and phase lag with sunspots. Periods of low solar activity, such as the Maunder minimum, are clearly reflected by higher ^{10}Be concentrations. In order to reduce the climatic signal introduced by atmospheric transport, ^{10}Be records from Greenland and Antarctica were combined and compared with ^{14}C tree-ring data. The generally good agreement of these comparisons strongly indicates that the main source of the short-term ^{10}Be variations is solar modulation of the cosmic rays and that several Maunder-minimum-type periods occurred during the last few millenia.

Magnetic activity is a fundamental feature of the Sun, and probably of many other stars. An understanding of the source and temporal variability of the activity level is therefore obviously of fundamental importance in understanding the structure and evolution of the Sun. This is particularly true if, as recent work suggests, the solar irradiance level itself is related to the activity level (chapter by Froehlich et al.).

Since direct observations of solar activity are available only for a very recent period of the Sun's lifetime, it is important to look for proxy indicators that can extend this record further into the past. In the last few years it has been shown that the cosmogenic (cosmic-ray-produced) isotopes ^{14}C in tree rings and ^{10}Be in ice cores can serve as proxy indicators. In this chapter, we discuss the principles for using ^{10}Be in this way, and summarize the currently available data. We also directly compare ^{10}Be records from Arctic and Antarctic ice cores, and combine them to produce composite records. It is hoped that such composite records will tend to minimize local and regional meteorological noise in the ^{10}Be record, thus enhancing the global signal due to the solar activity-induced production variations. Finally, we compare the ^{10}Be records with other indicators of solar activity, namely sunspots and ^{14}C in dated tree rings.

I. PRODUCTION OF ^{10}BE

We use the term galactic cosmic rays here to describe the flux of charged particles (~90% protons, ~10% alpha particles, ~1% heavier particles) moving more or less isotropically through interstellar space. The differential flux dF/dE (particles m^{-2} s^{-1} sr^{-1} MeV^{-1}) for kinetic energies $E > 1000$ MeV can be approximated by the equation (see the chapters by Jokipii and by O'Brien et al.)

$$\frac{dF}{dE} = A \ (m + E)^{-\gamma} \qquad (1)$$

with A, m and γ being constants (Fig. 1). Those particles that penetrate into the atmosphere of the Earth interact with nuclei of N, O, and Ar producing a large variety of secondary particles (protons, neutrons, mesons) and radioisotopes. The two long-lived radioisotopes with the highest mean global production rates are ^{14}C (half-life = 5730 yr, production rate: ~2.5 atoms cm^{-2} s^{-1}) and ^{10}Be (half-life = 1.5 Myr, production rate: ~3.5 × 10^{-2} atoms cm^{-2} s^{-1}) (Monaghan 1987). Since ^{14}C is discussed elsewhere (see the chapter by Damon and Sonett) we concentrate here on ^{10}Be.

The ^{10}Be production rate is altitude as well as latitude dependent. The altitude dependence is caused by the propagation of cosmic-ray particles through the atmosphere. In the first 100 to 200 g cm^{-2} (16 to 12 km) the production rate increases due to the effects of secondary particles. Then going deeper into the atmosphere, more and more particles are stopped and the production rate decreases. Since protons are subject to Coulomb interactions and lose their energy rather rapidly, most high-energy nuclear reactions (spallation) are caused by secondary neutrons.

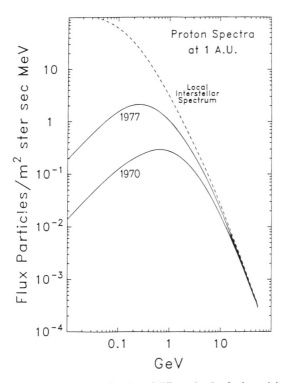

Fig. 1. Cosmic-ray proton spectra as a function of different levels of solar activity ranging from 1970 (activity maximum) to 1977 (activity minimum). An estimate of the unmodulated spectrum is shown as the dotted line (figure adapted from Lal 1988).

The latitude dependence of the production rate is caused by the geomagnetic dipole field which prevents charged particles below a certain rigidity (momentum per charge) from penetrating into the atmosphere. The shielding effect is largest at the equator, decreasing toward the magnetic poles according to

$$R\,(\lambda) = R_{o}\,\cos^{4}\lambda \tag{2}$$

with R_{o} = magnetic rigidity at the equator and λ = geomagnetic latitude, i.e., nuclei of rigidity $R < R(\lambda)$ are reflected back into interplanetary space and cannot reach the atmosphere (see the chapter by O'Brien et al.).

II. TRANSPORT AND DEPOSITION OF ^{10}BE

After production, ^{10}Be attaches quickly to aerosols (solid or liquid particles) and follows the motion of the air masses in which it was formed. Since the production rate is higher in the upper atmosphere, it turns out that about

⅔ of the ^{10}Be is produced in the stratosphere (above 10 to 15 km) and ⅓ in the troposphere. From the study of radioactive fallout of nuclear bomb tests, it is known that the mean residence time of aerosol-bound radioisotopes in the troposphere is of the order of days to weeks whereas the same isotopes need about 1 to 2 yr to settle from the stratosphere down to the surface of the Earth. Similar values have been found for cosmogenic ^{10}Be (Raisbeck et al. 1981a). The main reason for the relatively long stratospheric residence time is the tropopause, which separates the two layers. However, changes in the atmospheric circulation patterns during spring at mid-latitudes lead to injection of stratospheric air into the troposphere.

Radioisotopes are mainly removed from the lower troposphere by wet precipitation (rain and snow). The ^{10}Be fallout at most locations (but perhaps not at very high latitudes; Raisbeck and Yiou 1985) is thus approximately proportional to the amount of precipitation. This means that, even in the case of a constant mean global production rate of ^{10}Be, transport and deposition processes within the atmosphere may cause variations of local fallout, especially when we consider time scales of ≤ yr. A more detailed discussion of cosmogenic production processes and the distribution of radioisotopes in the atmosphere is given by Lal and Peters (1967; see also chapters by O'Brien et al. and by Lal and Lingenfelter).

Most of the cosmogenic ^{10}Be becomes stored in geologic archives such as ice sheets or lacrustine and marine sediments. The question then arises: what can we learn by measuring a detailed record of the ^{10}Be concentration in such an archive? To answer this question we simply have to ask what are the parameters determining the ^{10}Be flux.

As shown in Fig. 2, there are several potential causes influencing the ^{10}Be flux into an archive. Let us discuss first variations of the production rate. The production rate is proportional to the cosmic-ray flux interacting with the atmosphere. This may change due to several factors:

(1) Changes in the primary cosmic-ray flux. Based on radioisotope concentrations in extraterrestrial material (meteorites, the Moon) it is estimated that the mean galactic cosmic-ray flux has been constant to about 30% over the last several Myr (Reedy et al. 1983; chapter by Reedy and Marti). However, short-term variations during this time due to cosmic events (e.g., supernova explosion) cannot be excluded.

(2) Solar activity. From the surface of the Sun, a plasma is streaming out into the solar system. This plasma, called the solar wind, carries frozen in magnetic fields which deviate the charged particles of the cosmic rays. Since particles with low momentum are affected the most, the solar wind causes a depletion at the low-energy end of the cosmic-ray spectrum (Fig. 1). Energetic particles from the Sun (often referred to as solar cosmic rays), which are roughly in phase with the solar activity, also interact with the Earth's atmosphere. However, their energy is too low to contribute significantly to the average production of ^{10}Be. Although the details of the modulation pro-

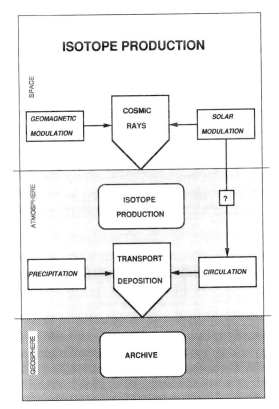

Fig. 2. Schematic representation of production of cosmic-ray-produced nuclides in the atmosphere.

cess between solar wind and cosmic-ray particles are not yet understood (see the chapter by Jokipii), experimentally there is a clear anticorrelation between solar activity and the cosmic-ray flux reaching the Earth. Typical time scales of solar-activity variations are in the range of hours to several 100 yr. Longer-term time changes cannot be excluded.

(3) Secular changes of the geomagnetic dipole field. As discussed above, the geomagnetic field intensity sets a lower limit (cut-off) to the momentum of the cosmic-ray particles which are able to penetrate into the Earth's atmosphere. For small changes in the geomagnetic field, the relationship between the production rate Q and the geomagnetic dipole moment M can be described approximately by the following equation

$$\frac{Q_1}{Q_0} = \sqrt{\frac{M_0}{M_1}} \qquad (3)$$

where the index 0 and 1 correspond to values of the field at two different times. For an almost-zero dipole field, as may occur during a magnetic reversal, the mean global production rate is enhanced by a factor of $\sim$2.5 (O'Brien 1979).

Even if the production rate is constant the transport and the removal processes within the atmosphere may lead to regional variations of the ^{10}Be deposition to the Earth's surface. Although the mean residence time of ^{10}Be in the stratosphere ($\sim$1.5 yr) is long enough to result in significant mixing, the short tropospheric residence time results in fallout patterns which are strongly determined by the regional meteorological conditions (wind directions, precipitation). This has been clearly demonstrated by the radioactive cloud emitted by the burning Chernobyl reactor. Significant variations of the regional ^{137}Cs fallout have been observed depending on the meteorological conditions (rain) during passage of the radioactive cloud (Clark and Smith 1988). However, such meteorological effects are short-term effects which should become less important as one goes to longer time scales ($\geq$ 1 yr). Nevertheless, long-term features of wind and precipitation patterns will give rise to corresponding inhomogenities in ^{10}Be deposition.

It is obvious from the above discussion that ^{10}Be variations measured in geologic archives contain a great deal of information over widely different time scales. The problem, of course, is how can we distinguish between the different causes; the answer to this question is not straightforward. Different approaches can be used. For example: (i) comparison of ^{10}Be records with known records of solar activity, paleomagnetism and climate; (ii) comparison of ^{10}Be with other radioisotopes having a different geochemical behavior (^{14}C); (iii) intercomparisons of ^{10}Be records from different locations. Before discussing these approaches, some information on the properties of the different potential archives is needed.

III. POTENTIAL ARCHIVES

In order to record the atmospheric fallout in an ideal way, an archive should fulfill several requirements:

1. Stable stratigraphy. The fallout should be recorded in a stratigraphically undisturbed way;
2. Preservation. The deposited material should be preserved. No mixing and no chemical and biological processes should occur;
3. Time information. The sedimentation rate should be known in order to establish the precise relationship between depth and time;
4. The time resolution should be high and the time span covered by the archive should be long;
5. Location. Comparable archives should be available at many places over the globe.

In reality, of course, these partly contradictory requirements are never all fulfilled. In the following, we discuss briefly the three main archives of ¹⁰Be.

A. Deep Sea Sediments

Deep sea sediments provide the longest ¹⁰Be records. Due to low sedimentation rates (typically 0.1 to 10 cm per 1000 yr) they usually cover the last 1 to 100 Myr. However, time resolution is limited mainly by two effects:

a. Oceanic residence time. The oceanic residence time of ¹⁰Be can be expressed as $T = M/Q$, where M is the inventory of ¹⁰Be in the ocean and Q the rate of input. T has been estimated to be 400 to 4000 yr (Kusakabe et al. 1987). It probably depends on the chemistry and the biological productivity within the ocean.

b. Bioturbation. Small organisms penetrating into the top 5 to 10 cm of the sediment mix up the material, usually limiting the time resolution to >1000 yr. Based on the data available so far, it seems that the ¹⁰Be concentration in deep-sea sediments reflects primarily changes of the ocean system. The only production effects that may be detectable are those due to magnetic reversals, with production increases of a factor of 2.5 (Raisbeck et al. 1985) or changes over long periods ($>10^4$ yr).

B. Lacustrine sediments

Lakes are geographically quite homogeneously distributed and the recovery of lacustrine sediments is rather simple. Time resolutions of about 1 yr are possible in favorable cases (laminated sediments) and time spans of 10 to 10^5 yr are quite common. Accurate dating is, however, often a problem because the sedimentation rate depends on different and highly variable parameters such as precipitation rate, drainage area, and chemical and biological conditions. The two most serious problems are first, the drainage area which is often several times as large as the actual surface of the lake and may be variable with time. Due to its strong affinity to particles, "old" ¹⁰Be may be brought into the lake due to erosion during periods of strong precipitation or changing vegetation. Another source of ¹⁰Be is that brought in on dust during windy and dry-weather conditions. A second problem is that after deposition on the surface of the lake, and before incorporation into the sediment, ¹⁰Be is subject to complex transport processes (focusing) within the lake. These also depend on different parameters and may lead to additional variations. Based on the few ¹⁰Be data available so far, it seems that the ¹⁰Be concentration in lacustrine sediments is strongly determined by the processes of sediment formation, so that atmospheric production variations are difficult to detect.

C. Polar Ice

For high-resolution studies, the archive that comes closest to fulfilling the ideal requirements appears to be polar ice. Polar snow directly samples the atmospheric fallout and keeps it frozen. Mixing processes due to wind shift of snow are rather small as long as the ice sheet under consideration is flat. The accumulation rates are high (2 to 50 g water cm^{-2} yr^{-1}) and the time span covered from presently available cores ranges from a few 10^3 to several 10^5 yr. Young ice ($<10^3$ yr) from high accumulation rate (>10 cm yr^{-1}) areas can be dated by measuring and counting the seasonal variations of δ ^{18}O and δ D, or H_2O_2, δ ^{18}O and δ D expressing the deviation in per mil of the isotopic ratio ^{18}O/^{16}O and D/H, respectively, from a standard value. These deviations are related to the temperature at the time of precipitation and reveal a clear seasonal variation. H_2O_2 is produced by a photochemical reaction in the atmosphere leading to a strong production maximum during summer. Absolute dates during the last few centuries can be found in the form of acidity peaks caused by historically dated volcanic eruptions.

However, there are also some properties of ice which are far from ideal. Ice is not stable. Due to the high pressure of the overlying layers, ice starts slowly flowing from the central to the coastal part which results in a thinning of the annual layer thickness with increasing depth. As a consequence dating of ice near bedrock becomes a real problem (Hammer 1980). Another limiting factor regarding ice is that it is restricted for obvious reasons to the polar regions. High-altitude glaciers are less favorable due to strong wind erosion and shorter lifetimes.

This discussion of the potential archives clearly indicates that the archive best suited for detecting solar modulation effects is polar ice, especially as far as short-term variations (11-yr cycle, Maunder-minimum-type periods) are concerned. Therefore, it is not surprising that practically all published ^{10}Be data dealing with solar modulation effects to date have been obtained from ice cores. We discuss these data in the following sections.

IV. POLAR ICE DATA: CURRENT STATUS

The most prominent feature observable on the Sun are sunspots, which wax and wane periodically. A continuous record of decreasing quality, of sunspot numbers has been constructed back to $\sim$1600 A.D. Beside the very strong 11-yr or Schwabe cycle, there are periods when the sunspot maxima numbers are relatively low (1900,1800) or almost zero (Maunder minimum). In the following, we discuss how well these features can be traced into the past by means of ^{10}Be.

A. 11-Yr or Schwabe Cycle

The whole idea of using radioisotopes to reconstruct the history of solar activity is based on the solar modulation of the galactic cosmic-ray flux. A crucial test is therefore to compare the cosmic-ray flux, as recorded by ionization chambers or neutron monitors, with the sunspot record (Fig. 3). The

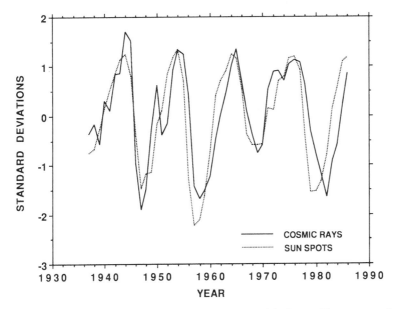

Fig. 3. Comparison of groundlevel cosmic-ray secondary-particle fluxes with sunspot number over the past 150 yr. Sunspot number has been plotted on an inverse scale. The amplitudes are given in units of standard deviations (see text).

cosmic-ray flux data (Flückiger, personal communication) are compiled from the ionization chamber of Cheltenham (1937–1953) and the neutron monitor of Climax (1950–1985). In order to correct for the different cut-off rigidities of the two neutron counters, the mean value $\bar{x}$ and the standard deviation (σ) for the corresponding periods were used to calculate the normalized values according to

$$x' = \frac{x - \bar{x}}{\sigma_x} . \tag{4}$$

This procedure is justified to some extent by the good overlap of the two data sets. The same normalization procedure was applied to the sunspot data that are plotted in Fig. 3 on an inverse scale for easier comparison. All data points represent annual mean values. The good anticorrelation between the two curves is expressed by the high correlation factor of -0.8.

Since the production rate of ¹⁰Be and the other cosmogenic isotopes is proportional to the cosmic-ray flux, we expect similar variations in the abundance of these isotopes in appropriate archives.

Such a study has begun on a 300 m ice core from station Dye 3, Greenland, using annual samples (Beer et al. 1990). The core was dated by count-

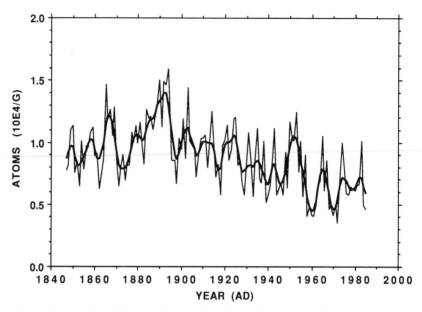

Fig. 4. Concentration of [10]Be in an ice core from Dye 3, Greenland as a function of time. The smooth curve is a 9-point running mean.

ing the seasonal variations of H_2O_2 and the dating was checked by absolute time marks (maximum of nuclear bomb fallout 1963, Tambora eruption 1815, Laki eruption 1783). The measured [10]Be concentrations, in 10^4 atoms g^{-1}, are shown in Fig. 4. The [10]Be curve exhibits a periodic variation superimposed on which there is some noise possibly due to atmospheric transport and deposition processes. In order to remove this noise, a low-pass filter was applied (thick curve). For the filtered data the cross-correlation function with the corresponding sunspot numbers was calculated and is displayed in Fig. 5. The periodic behavior of the correlation function confirms the presence of the Schwabe cycle. In the mean, the [10]Be concentration lags behind the sunspots by ~2 yr, which agrees well with the sum of the modulation delay (0.5 yr) and the atmospheric residence time of [10]Be (~1.5 yr). In Fig. 6, the measured [10]Be concentration is plotted together with the inverse sunspot record shifted by 2 yr for easier comparison. The [10]Be concentration appears to reflect not only the Schwabe cycle, but also the long-term trend of the amplitude of the sunspot numbers, with a minimum around 1900. Unfortunately there is at present no sufficiently high-resolution [10]Be record from an Antarctic ice core to compare with the above work. In fact, over much of the Antarctic Plateau precipitation rates are so low (< 5 g cm^{-2} yr^{-1}) that identification and retention of annual layers are difficult or impossible; so it would be hard to demonstrate an 11-yr cycle. For the South Pole (precipitation rate

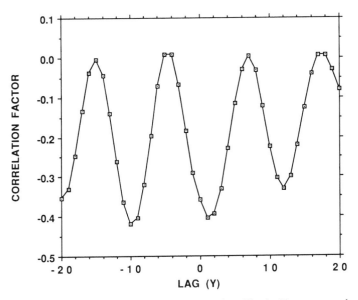

Fig. 5. Cross correlation of smoothed ¹⁰Be concentration from Fig. 4 with sunspot number.

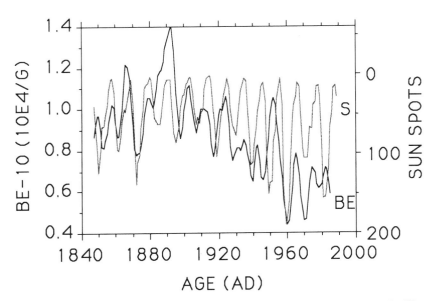

Fig. 6. Comparison of smoothed ¹⁰Be concentration from Fig. 4 with sunspot number. The ¹⁰Be data have been shifted by 2 yr for easier comparison (see text).

$\sim$ 8 g cm^{-2} yr^{-1}), annual layers have been detected and there are plans to try to construct a high-resolution ^{10}Be record at this location.

B. Maunder-Minimum-type Variations

The Maunder minimum period (1645 to 1715 A.D.) is an especially interesting period. Direct observations that are relatively reliable (chapter by Rabin) reveal exceptionally few sunspots; ^{14}C measurements on dendrochronologically dated tree rings show a clear peak; and at this time the climate in Europe reached the climax of the so-called Little Ice Age. Similar periods named Spoerer (1420–1540 A.D.) and Wolf Minima (1280–1350 A.D.) are observed earlier in the ^{14}C record (Stuiver and Quay 1980), but are not so well documented by direct observations.

Comparison Between Greenland and Antarctica.   In Fig. 7, we show two ^{10}Be records covering the time period of the three minima mentioned above. One record is from Milcent, Central Greenland, with a time resolution of 3 to 7 yr (Beer et al. 1983), and the other record is from the South Pole with a time resolution of 5 to 10 yr (Raisbeck and Yiou, in preparation).

The Milcent core is dated by a high-resolution δ^{18}O profile and claimed to be precise to $\sim$ 1 yr (Hammer et al. 1978). The South Pole core is dated by identification of peaks representing known volcanic eruptions (Kirschner 1988). Assuming that these eruptions have been correctly identified, the dating of these times should also be correct to $\sim$1 yr. However, between these peaks, the dating may be uncertain by as much as several tens of years. As seen most clearly for the smoothed curves, both records show the same general behavior, with enhanced concentrations during periods of low solar activity, as expected. It is interesting to note that in the South Pole record, the Maunder minimum is smaller and the Spoerer minimum higher than in the Milcent record. In order to combine the data, both sets were resampled with a fixed time interval of 3 yr and normalized to a mean overall concentration of unity. The two were then combined and averaged in the following way: for periods where both records exist, they were averaged and in periods where one record was absent, the value of the other record was retained. In this composite curve, the three periods of enhanced production show up quite clearly even in the unsmoothed data. This confirms our expectations that combining records from different localities should help remove meteorological noise.

In order to extend the ^{10}Be record back further in time, we show in Fig. 8 two ^{10}Be records from Camp Century, Greenland (Beer et al. 1988) and Dome C, Antarctica (Raisbeck and Yiou 1988). In this case, comparison of the two curves is considerably more delicate, because neither has a solid time scale and several data points are missing. Therefore we only compare the period 2500 B.C. to 1000 A.D. Once again, we find a very good degree of similarity between the two curves, although the agreement is obviously not

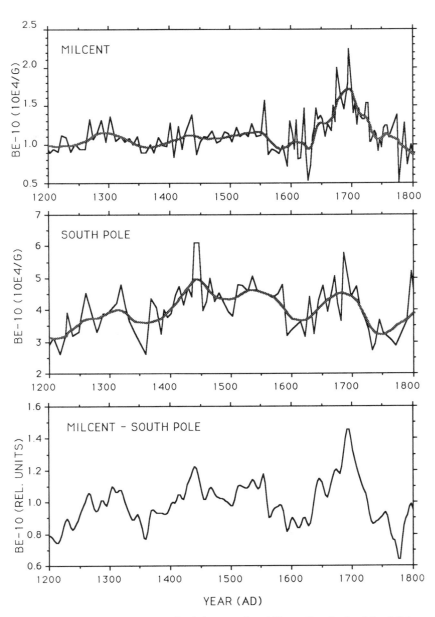

Fig. 7. Comparison of ¹⁰Be concentration in ice cores from Milcent, Greenland and South Pole, Antarctica. Composite curve has been constructed as described in text.

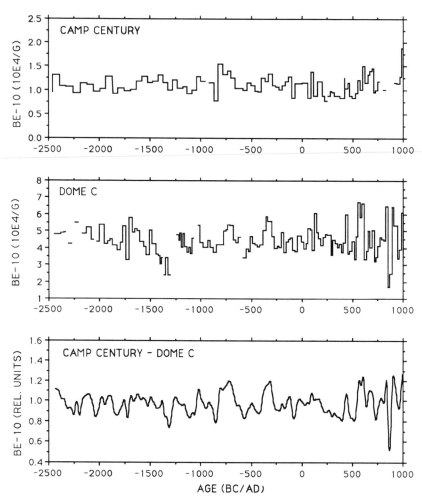

Fig. 8. Comparison of [10]Be concentration in ice cores from Camp Century, Greenland and Dome C, Antarctica. Composite curve has been constructed as described in the text.

perfect. In order to produce a composite record, we have again resampled the two original records, this time with an interval of 5 yr, and normalized to a mean concentration of unity. This composite curve is also shown in Fig. 8.

Comparison of [10]Be and [14]C.   In order to compare [10]Be and ^{14}C, one has to take into account the different geochemical behavior of these two isotopes. As mentioned before, [10]Be becomes attached to aerosols and is removed from the atmosphere by precipitation. [14]C, on the other hand, oxidizes to CO_2 and is exchanging between the reservoirs of the atmosphere, ocean and biosphere. Due to the relatively large sizes of these reservoirs, production variations are attenuated by a factor depending on the frequency. For 11 yr the attenuation factor is about 100 which makes the detection of the 11-yr signal difficult. For Maunder-minimum-type variations, the attenuation is still about an order of magnitude. To account for these fundamental differences, Beer et al. (1983,1988*a*) chose the following procedure to compare the ¹⁰Be record with the ¹⁴C record. Postulating that the measured ¹⁰Be concentration in the ice core directly reflects the mean global-production variations (neglecting climactic and local effects), they calculated the expected ¹⁴C variations using the box diffusion model of Oeschger and Siegenthaler (1975). The results showed general agreement between the calculated and the measured ¹⁴C curve, although the Maunder minimum turned out to be too high and the Spoerer minimum too low. (We plan in the future to apply this technique to the combined Arctic and Antarctic records developed here.) For the moment, however, we use a simpler, though more qualitative procedure (Raisbeck and Yiou 1988), which is to smooth the ¹⁰Be data with a binomial filter to give approximately the type of time resolution seen in the ¹⁴C tree-ring record. These comparisons are shown in Fig. 9 (1200 to 1800 A.D.) and Fig. 10

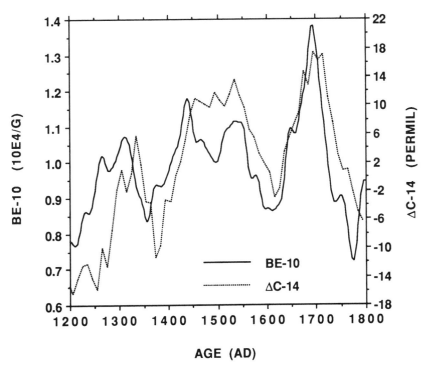

AGE (AD)

Fig. 9. Comparison of smoothed ¹⁰Be composite concentration curve from Milcent and South Pole with tree-ring Δ¹⁴C record. Δ¹⁴C expresses the relative deviation in per mil of the atmospheric ¹⁴C concentration from the one of 1850 AD.

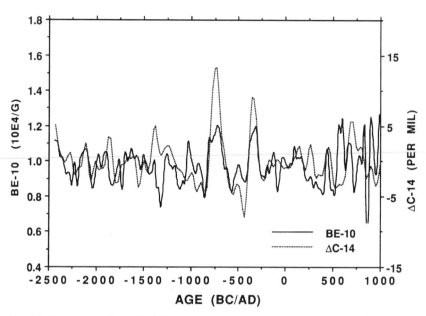

Fig. 10. Comparison of smoothed ^{10}Be composite concentration curve from Camp Century and Dome C with tree-ring ^{14}C record.

(2500 B.C. to 1000 A.D.). In Fig. 9, the general agreement between these two types of records, that involve systems of very different geochemical behavior, strongly supports the hypothesis that the observed variations are indeed production variations rather than climatic effects. In Fig. 10 the agreement is not so good, probably due to the uncertainties in the ice core dating. However, there are at least two peaks at 750 B.C. and 300 B.C. that stand out as clearly as the Maunder minimum in Figure 9 and are very likely due to solar modulation.

C. The Pre-Holocene Record

While the ^{10}Be ice-core records are obviously an important complement to the ^{14}C tree-ring records as a proxy solar-activity indicator, particularly for short-period variations such as the 11-yr cycle, they also offer perhaps the unique potential of extending such a record much further back in time. For the moment there exist no continuous high-resolution ^{10}Be data for pre-Holocene ice. However comparisons between Holocene (last 10,000 yr) and pre-Holocene ice has shown a factor of ~2 variation in ^{10}Be concentrations (Raisbeck et al. 1981a; Beer et al. 1984). This has generally been interpreted as due to large-scale (perhaps global) change in the precipitation rates for these two very different climatic periods. It is thus clear that it will be difficult to extract solar activity induced changes during periods of strong climatic change. Nevertheless, Raisbeck et al. (1987) have observed two strong enhancements of ¹⁰Be in the Vostok ice core at ~ 35,000 and 60,000 yr ago, which did not appear to be associated with climate change. The former was also found in the Dome C ice core, and there are plans to look for it in the Byrd and Camp Century cores. While primary cosmic-ray flux increases and geomagnetic field changes are other possible causes (Raisbeck et al. 1987; Sonett et al. 1987), solar modulation remains a strong contender for the source of the ¹⁰Be peaks. Since these enhancements last approximately an order of magnitude longer than the Maunder-minimum-type events, a solar origin would obviously imply solar variability changes on yet another time scale. Longer records of solar variability would of course also allow more significant tests of the degree of cyclicity in the solar-activity variability.

CONCLUSIONS

The use of ¹⁰Be to reconstruct the history of solar activity is still in the development and testing stage. This chapter shows that ¹⁰Be measurements in polar ice are very promising in this respect. The real value in the future will be in applications where no other method presently exists (for example: study of 11-yr cycle during Maunder minimum and variations in the pre-Holocene period).

The problem of disentangling the different potential causes can be partly overcome by using additional records of ¹⁴C, δ¹⁸O and ¹⁰Be from different latitudes. There is much hope that by a general improvement of our knowledge about atmospheric circulation and transport processes, a better separation of the different causes may become possible. Trace constituents with known source functions may help one day to correct for climatic-induced variations. We have discussed the serious restriction caused by changing climate which probably limits the application of ¹⁰Be in polar ice to times of relatively constant climatic conditions. Even this, however, opens up the possibility of extending the solar activity record in a very significant way, and may greatly improve our understanding of the Sun, the Earth and their mutual relationship.

Acknowledgments. The authors would like to thank their colleagues from Bern, Zurich, Buffalo, Grenoble, Saclay and Orsay who participated in the various phases of ice core collections, sampling and analysis which led to the original data on which this chapter is based. We also thank T. Jull and T. Montmerle for carefully reviewing this chapter.

SOLAR AND TERRESTRIAL COMPONENTS OF THE ATMOSPHERIC ^{14}C VARIATION SPECTRUM

PAUL E. DAMON and CHARLES P. SONETT
University of Arizona

This chapter reviews the variability of atmospheric ^{14}C concentration. The work is motivated by the role the Sun plays in causing this variability either directly by forcing changes in the rate of production of ^{14}C, or indirectly by possible bolometric effects on climate. Medium and short-term variations of periods 208, 88 and 11 yr are in large part produced by solar wind modulation of cosmic-ray production of ^{14}C. A bolometric component of this variability is inferred from correlations between climatic variability and the Δ^{14}C record. A 2300-yr period in radiocarbon modulates the 208-yr period. The source of this period is enigmatic. The ^{14}C record provides an important extension in time of solar variability.

I. INTRODUCTION

The ultimate source of terrestrial radiocarbon is traceable to the cosmic-ray flux incident upon the atmosphere from which, by spallations, an atmospheric neutron sea is generated. Radioactive ^{14}C is produced terrestrially primarily by the specific nuclear reaction

$$^{14}N(n,p) \rightarrow \,^{14}C \tag{1}$$

where ^{14}N is atmospheric. ^{14}C decays by

$$^{14}C \rightarrow \,^{14}N + \nu^- + \beta^- \tag{2}$$

where ν^- is the antineutrino and β^- the electron. The half life of radiocarbon is $T_{1/2} = 5730$ yr (Lederer, et al., 1967). Thus it is neutrons that participate in the $N(n,p)$ reaction yielding ^{14}C (see the chapter by O'Brien; Lingenfelter and Ramaty 1970; O'Brien 1979). If the cosmic-ray flux were constant, the atmospheric ^{14}C inventory would be in secular equilibrium and radiocarbon would be an absolute archaeological-geological clock, but it would then be of lesser interest geophysically.

Variability of the atmospheric radiocarbon concentration ($^{14}C/^{12}C$ is conceptually divisible into three parts (Table I): (1) variations in the global rate of radiocarbon production; (2) variations in the rate of exchange between geochemical reservoirs; and (3) variations in the total CO_2 in the atmosphere, biosphere and hydrosphere. In the first case, solar wind modulation of galactic cosmic rays (Castagnoli and Lal 1980; Jokipii 1971; see also the chapter by Jokipii) and changes in the solar cosmic-ray flux (Lingenfelter and Ramaty 1970; Damon et al. 1990) give rise to short- and medium-term variations (10 to 10^2 yr), but the long-term change (10^4 yr) is commonly attributed to changes in the strength of the terrestrial magnetic field. In the second case, variations in the rate of exchange between geochemical reservoirs can be forced by changes in climate. In this case, variability in solar irradiance cannot be arbitrarily ruled out as the cause of climate change. Lastly, changes in the total ambient CO_2 have been produced by combustion of fossil fuels and

TABLE I
Possible Causes of Radiocarbon Fluctuations[a]

I. Variations in the global rate of radiocarbon production
    1. Variations in the cosmic ray flux throughout the solar system
        a. Cosmic ray bursts from supernovae and other stellar phenomena
        b. Interstellar modulation of the cosmic ray flux
    2. Modulation of the cosmic ray flux by solar activity
    3. Modulation of the cosmic ray flux by changes in the geomagnetic field
    4. Production by antimatter meteorite collisions with the Earth
    5. Production by nuclear weapon testing and nuclear technology

II. Variations in the rate of exchange between geochemical reservoirs and CO_2 reservoir inventory
    1. Control of CO_2 solubility and dissolution and residence time by temperature variations
    2. Effect of sea-level variations on ocean circulation and capacity
    3. Assimilation of CO_2 by the terrestrial biosphere
    4. Dependence of CO_2 assimilation by marine biosphere

III. Variations in the total CO_2 in atmosphere, biosphere and hydrosphere
    1. Changes in input rate of CO_2 by lithospheric degassing, e.g., vulcanism
    2. Combustion of fossil fuels from industrial and domestic activity
    3. Changes in long-term storage in the sedimentary reservoir

[a]Table adapted from Damon et al (1978).

may conceivably be produced by other processes such as increased weathering and sedimentation or dissolution of oceanic carbonates. Lal (1985,p.231) has argued that "an appreciable part of the slow variation in atmospheric ^{14}C/ ^{12}C ratio could in fact be due to carbon cycle changes rather than changes in the geomagnetic dipole field."

The time variability of the radiocarbon record is a central problem of geophysics, geochemistry and possibly solar physics. The record of variations contains numerous periods ranging from ~ 2300 yr downwards to the 11-yr period of solar activity. Houtermans (1971) first reported ~ 2300 and 200 yr periods, while later Suess (1980a) reported previously unpublished calculations of Kruse of the power spectrum of the La Jolla Δ^{14}C record. [Δ^{14}C is the convention for isotope (δ^{13}C)-corrected delta radiocarbon. The lower case (δ) is also used for uncorrected radiocarbon.] Suess' report showed spectral lines at 2400, 930, 498, 308, 202 and lesser years; later Neftel et al. (1981) studied the 208-yr line exhaustively. A possibly important 150-year period was first uncovered by de Jong and Mook (1980; see also Sonett 1984).

As the basic aim of this chapter is directed towards the possible role of the Sun in forcing the radiocarbon variability, it is necessary to discuss both terrestrial (climatic) and solar forcing in order to provide a basis for separating the two. This requires discussion of the evidence for variability of ^{14}C and model forcing during the last fourteen millennia. The solution to this problem is potentially important for solar physics but is not completely understood; much of what is said is uncertain or even speculative. The gravest difficulty as with much of solar-terrestrial physics is the lack of an underlying conceptual structure for aspects of solar forcing that are inferred from inspection of data. The primary tools used in the analysis of the radiocarbon record are time sequences, spectra both in the form of Fourier transforms and power spectral density, and finally covariances (or correlations).

Radiocarbon dating was established for archeological and Quaternary research with publication of *Radiocarbon Dating* (Libby 1952). It was then commonly assumed that the radiocarbon inventory of atmospheric radiocarbon was invariant, leading to the radiocarbon time scale being an absolute new time standard for archeological research. But shortly thereafter, Suess (1955) produced the first evidence for anthropogenic changes in atmospheric ^{14}C activity caused by combustion of radiocarbon-free fossil fuels. This was followed within a few years by evidence that ^{14}C variations had also occurred in medieval times from natural causes (de Vries 1958,1959).

As the majority of radiocarbon records are based on wood, it serves most often as the reference time base. [It should be mentioned that ^{14}C also serves as an important tracer of total CO_2 in the global ocean and other ^{14}C reservoirs, but except for specific reference, the subject lies outside the scope of this chapter.] All Δ^{14}C values discussed in this chapter are from measurements on dendrochronologically dated wood with the parameter Δ^{14}C being the age and fractionation-corrected activity of the wood at any time in the past relative to mid-19th century wood with $\delta^{13}C = -25$ per mil. The deviations from the mid-19th century value are per mil values that represent the fluctuation of the ^{14}C/C ratio of the atmosphere or the ^{14}C activity.

de Vries correlated the atmospheric ^{14}C fluctuations during the 16th to 18th century with climate change, specifically the waxing and waning of montane glaciation during the Little Ice Age. The medieval data of de Vries, fractionation-corrected by Lerman et al. (1970), are shown in Fig. 1 along with the data of Stuiver and Quay (1980,1981) for the last millennia and an underlying graph from analysis of the residuals around a sixth-order logarithmic trend curve based on a compendium of 1200 medium-precision data from an NSF Workshop (Klein et al. 1980). The de Vries data are in remarkable agreement with later work.

Stuiver (1961,1965) was the first to show a convincing relationship between the de Vries effect and solar activity. He showed that the observed relationship between the sunspot index R, and ^{14}C production during the 11-yr solar cycle could be used successfully to proxy model the 17th through the 18th century ^{14}C fluctuations, including the fluctuation labeled e in Fig. 1, which has become known as the Dalton solar minimum. Eddy (1976a,b) pointed out that the ^{14}C maximum (d in Fig. 1) was correlated with the extremely low sunspot activity during the well-known Maunder period of low sunspot activity ($\sim$1640 to 1720 AD). Using their radiocarbon record, Stuiver and Quay (1980) showed that two other radiocarbon maxima, b and c in Fig. 1, corresponded in time to the Wolf and Spörer sunspot index minima. Although the relation between the Maunder minimum and the low in the sunspot index is broadly accepted today, Kocharov (1987) has recently argued that the ^{14}C record during the Maunder minimum is more complex than hitherto recognized in that solar modulation continues at zero sunspot number.

The ^{14}C minimum at a (Fig. 1) is associated with high solar activity during the Medieval Warm Epoch (Schove 1955). The marked ^{14}C decrease during the late 19th and 20th centuries is the Suess effect, resulting from dilution of the atmospheric concentration of radiocarbon from the combustion of fossil fuels.

II. THE SECULAR VARIATION IN THE Δ^{14}C RECORD

The most prominent component of the Δ^{14}C time sequence over the past 10,000 yr is the variation for which a model sinusoid has an offset $= 32\%o$, half amplitude $= 51\%o$, period $= 11,300$ yr and phase lag $= 2.29$ radians (Damon and Linick 1986), but there is no *a priori* reason to suspect that this variation is periodic. The variation is equally well modeled by a polynomial. Indeed, the recent sequence of combined Becker et al. (1986) data yields an additional (local) decrease, so far unexplained, during the first $\sim$ 1000 yr (6000 to 7200 BP). [BP means "before present" used alternatively to AD and BC. It is actually measured relative to an AD 1850 standard age-corrected to

364

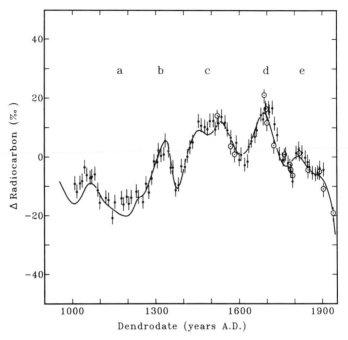

Fig. 1. Δ¹⁴C time sequence (solid line) for the last millennium from residuals of 6ᵗʰ-order polynomial fit to NSF Workshop data (Klein et al. 1980). Error bars are 1σ. Open circles are original data of de Vries (1958,1959) fractionation corrected (Lerman et al. 1970). Dots are high-precision data of Stuiver and Quay (1980). Letters mark times of (a) Medieval Warm, (b) Wolf, (c) Spörer, (d) Maunder and (e) Dalton epochs.

AD 1950 as "present."] See Fig. 2. The intense Milankovitch periods with periods of 23,700, 22,400 and 19,200 yr (see the chapters by Berger and Ghil) lie outside the possible bandwidth of this feature (except for a weak period at ~ 2400 yr).

   The major radiocarbon variation is generally attributed to a change in the terrestrial dipole moment and is qualitatively supported by paleomagnetic data. Figure 2 shows the ~ 11,000-yr sinusoid model curve upon which is superimposed data for the Earth's dipole moment assembled by McElhinny and Senanyake (1982; see the chapter by O'Brien). Dipole moment data from individual sites are inherently noisy. This is not only due to measurement uncertainties, but because of nondipole components and, particularly, their westward drift, together with regional components of the nondipole field and differences due to the magnetic properties of the Earth's crust at the sample locations. The Δ¹⁴C fluctuation in Fig. 2 between D and C is less well established, as the underlying tree-ring record has not yet been fully certified as chronologically absolute for times earlier than 6731 BC (8681 BP), and den-

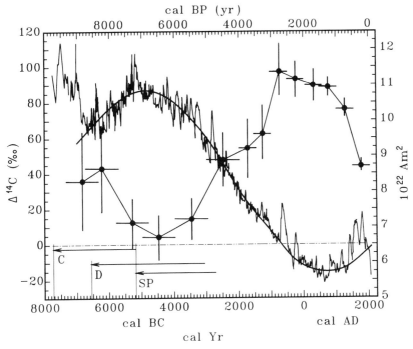

Fig. 2. Δ¹⁴C time sequence for the past 9 millennia from compendium of Seattle (Stuiver and Pearson 1986; Pearson and Stuiver 1986; Pearson et al. 1986) to 7150 BP (before present) and Seattle (Stuiver et al. 1986a), La Jolla (Linick et al. 1985), Tucson (Linick et al. 1986) and Heidelberg (Kroner et al. 1986) for 7150 BP to 9700 BP radiocarbon data. The underlying absolute tree-ring chronology exists to 6554 BC (Ferguson and Graybill 1983). Dipole moment values are from McElhinny and Senanayake (1982). ¹⁴C dating of dipole ages to 6500 BP are from Clark (1975). Earlier dipole dates are based on Stuiver et al. 1986a). Arrow marked SP indicates extent of Seattle and Belfast data, D based on an independent tree-ring chronology not "wiggle matched" and C combined data of all laboratories. Parameters of underlying sine wave are offset = 32 per mil, half amplitude = 51 per mil, period = 11,300 yr, and phase lag = 2.29 radians (131 deg).

drochronologically calibrated values in Fig. 2 extend back only to 6554 BC (8504 BP).

Figure 2 also displays a more or less continuous record of variations or fluctuations, whose analysis has been a central issue in radiocarbon analysis. The longer-term variations are shown explicitly in the low-pass filtered (~ 200 yr cutoff) superimposed La Jolla (Suess) and Belfast (Pearson) data in Fig. 3 which, in spite of the differing laboratories, counting techniques and tree material (Bristlecone pine for La Jolla and Irish peat bog oak for Belfast), convincingly exhibits the 208-year variation. Here we shall use the relatively exact value of 208-yr as a working value, for we think it is established with reasonably high confidence.

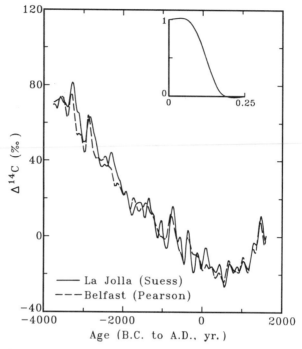

Fig. 3. La Jolla and Belfast Δ¹⁴C sequences truncated to common length and filtered but undeci-
mated, using transfer function shown in inset. Common 208-yr period from laboratories using
different counting techniques and differing wood samples is evidence of the reality of this
period (figure from Sonett 1985).

III. THE ~ 2300 YR PERIOD

Aside from the aforementioned long (secular?) variation, the strongest
feature in the $\Delta^{14}C$ record is the long period of ~ 2300 yr. This component
of the $\Delta^{14}C$ spectrum, in addition to the 208 yr period, was first reported by
Houtermans (1971). Its source is enigmatic but probably not attributable to
the geomagnetic dipole field, for no periodic geomagnetic dipole field change
of the required *amplitude* has been detected. Various proposals have been
made for a connection between the ~ 2300-yr ($\delta^{18}O$ and $\Delta^{14}C$) cycle and the
~ 2300-yr cycle in the directional components and westward drift of the
nondipole part of the Earth's magnetic field (Creer 1988), i.e., storage and
release of water as ice and melt might affect the Earth's magnetic dynamo.
However, to our knowledge none of these proposals is based upon actual
evidence or upon computational geophysical models.

A ~ 2400-yr periodicity has also been reported for the directional components of the paleomagnetic field obtained from late Quaternary Minnesota
lake cores (Lund and Banerjee 1985). Lund and Banerjee also suggested that

harmonics are present at 1200 yr, 800 yr and 600 yr and observe that "the fundamental wavelength of 2400 yr is very noteworthy because that is the estimated time for the nondipole field to rotate 360 degrees due to westward drift (Lund and Banerjee 1985,p.821). Creer has extended this analysis to both North American and United Kingdom data from late Quaternary cores and finds a dominant period in the inclination record of 2200 to 2400 yr associated with westward drifting sources, giving a drift rate of about 0.15 deg. per year. However, he concludes that declination variations are characterized by longer periods of 3000 to 4000 yr which differ by about 30% for the two regions (Creer 1988,p.390).

There is no compelling reason to suppose that the relatively small variation in direction of the dipole would cause a long-term variation in the cosmic ray flux at the top of the atmosphere. Thus there are only weak grounds for supposing that the 2300-yr radiocarbon period might be associated with the Earth's field. But candidate sources for this variation still admit of both terrestrial and extraterrestrial forcing sources. A quasi-periodicity of about 2400 yr is also found in the $\delta^{18}O$ record in ice cores and foraminifera from ocean cores (Pestiaux et al. 1987,1988). Glaciation shows this period as does the Middle Europe oak dendroclimatic record and Dansgaard et al. (1984) report the most prominent period in their Camp Century $\delta^{18}O$ core to be a line at 2550 yr. A terrestrial origin can account for the radiocarbon line if it has the role of a tracer of globally varying CO_2 inventory and exchange rates. An additional factor to be considered in conjunction with this problem is the possibility of an association between the 2300-yr period and the ubiquitous period at 208 yr. All of these reports strengthen the supposition that the ca. 2300-yr period is a climatic feature.

IV. THE OVERALL RADIOCARBON SPECTRUM

Radiocarbon has a well-defined spectrum whose striking clarity has only recently become exposed, primarily as a result of the intensive efforts over the past three decades of a small and dedicated group of workers toiling so to speak "in the wilderness." The spectrum shown in Fig. 4 is the discrete Fourier transform (DFT) of ^{14}C data concatenated from the La Jolla, Belfast, Seattle, and La Jolla/Becker sequences (SBSP sequence) and which extends backwards in time to 7199 BC. The combined sequence was detrended by least squares using a fifth-order polynomial reference. The insert shows the periodogram for frequencies in the range of $0.03 \leq f \leq 0.01$ yr^{-1} or in period $33.3 \geq \tau \geq 20$ yr. Major line features are present at 2272, 909, 649, 207, 149 and 88 yr in the large panel and 37.2, 26.3, 20.8 and 20.1 yr in the inset. These high frequencies are discussed in Sec. VIII.

The four longest periods shown in Fig. 4 suggest relatively sharp (high Q) sources while the 149 and 88 yr features are broader, in turn suggesting variable or ill-defined forcing periods. The 20.1- and 20.8-yr lines suggest

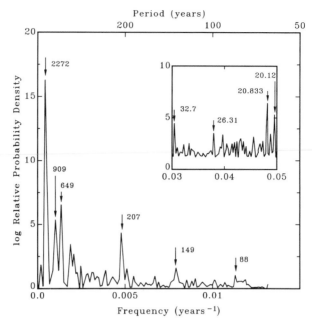

Fig. 4. Amplitude of discrete Fourier transform (DFT) of concatenation of La Jolla-Becker-Stuiver-Pearson ^{14}C sequences. Detrended using least squares fit to 5th-order polynomial, data is interpolated using cubic spline. Major features are demarked by period in years. Some periods of lesser amplitude are unmarked. Insert is from high-frequency segment periodogram of (BSP) Becker-Stuiver-Pearson sequence extended to 20 yr to emphasize higher frequencies. Major high-frequency lines are present at 20.12, 20.833, 26.31 and 32.7 yr. These high-frequency lines may be aliased contributions.

association with the Hale period. The issue of the exact period for the solar activity cycle (or sunspot index period) although never the center of great controversy, does range from 10 yr upwards to ~ 11.5 yr. Bretthorst's (1988) value is nearly 10 yr, while most other values are nearer to 11.5 yr. It is important to note that these spectra are subject to certain errors, both in amplitude and line frequency assignment discussed later in this section. The 20.1 and 20.8 yr lines are also subject to severe spectral folding, being near to the mean Nyquist period.

Although minor (but still sometimes important) differences exist between the SBSP sequence and the sequence composed of data from Stuiver, Becker and Pearson (in Stuiver and Kra 1986), henceforth referred to as the BSP sequence, both appear nearly the same as shown in the dual periodograms of Fig. 5. The reader's attention is directed to the minor differences between the SBSP upper panel and the lower panel showing the BSP periodogram. As would be expected, on the basis of a larger data base, the SBSP spectrum appears marginally less noisy.

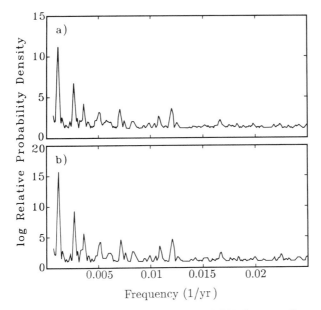

Fig. 5. Periodograms of ¹⁴C from SBSP (upper panel) and BSP (lower panel) sequences, de-trended using least squares fit to 5ᵗʰ-order polynomial. Only very minor differences appear showing the basic robustness of these data towards their spectra.

The spectra obtained from (a) Bayesian modeling (Bretthorst 1988), (b) maximum entropy method (MEM) (see, e.g., Burg 1975), and (c) the DFT (discrete Fourier transform) are varied. The Bayesian estimator is a recent development where prior models (here sinusoids) are fitted to the data. For details the reader is referred to Bretthorst's thesis and Bretthorst (1988). We make extensive use of it primarily for calculating line positions (Table II).

The assessment of MEM includes the possibility of artifact line splitting and the difficulty of making an optimum choice of autoregressive (AR) order. Aikaike's criterion (1969a,b) appears too conservative and most of the resolution advantage of MEM is lost by adhering to it, but "pushing" the AR order can lead to instabilities in the computed spectrum. Finally, a satisfactory theory of line amplitude (except perhaps in the asymptotic limit) is not available.

The DFT has the advantage that a true Fourier transform can be obtained but there are uncertain contributions from (1) window side lobes, (2) computational sampling on line shoulders or minima rather than at peaks, and signal line shape. Because of its relative freedom from side lobe interference the Bretthorst algorithm has an intrinsic advantage.

Differences between the MEM and Bretthorst line position estimates are generally greatest for the longer periods. The reason for the differences is

TABLE II
¹⁴C Bayesian Line Source Assignment

Frequency (yr⁻¹)	σ	Period
2.1066×10^{-4}	$\pm 1.4370 \times 10^{-6}$	4747
4.1925×10^{-4}	$\pm 1.1240 \times 10^{-6}$	2385
1.0468×10^{-3}	$\pm 4.1000 \times 10^{-7}$	955
1.3637×10^{-3}	$\pm 4.7800 \times 10^{-7}$	733
1.3674×10^{-3}	$\pm 1.3830 \times 10^{-6}$	512
2.2702×10^{-3}	$\pm 3.0720 \times 10^{-6}$	440
2.8359×10^{-3}	$\pm 1.3050 \times 10^{-6}$	353
4.8158×10^{-3}	$\pm 1.2530 \times 10^{-6}$	208
6.6442×10^{-3}	$\pm 1.2960 \times 10^{-6}$	151

uncertain. They lie outside the 2σ limits of the Bretthorst estimates; tentatively they are assigned to intrinsic instabilities of the MEM algorithm. The persistence and solar origin of the 208-yr line in the spectrum of radiocarbon has been repeatedly stressed by Suess (1980a). This 208-yr line has been shown to extend, with strong amplitude modulation, over the entire 8500 yr La Jolla $\Delta^{14}C$ record by Sonett (1984). As noted earlier its reality has also been noted previously by close matching of filtered records of the La Jolla and Belfast sequences as shown in Fig. 3 (Sonett 1985).

There seems little doubt that fundamental periods (ca. 2300, 208 and 88 yr) exist in the radiocarbon spectrum. Exact values vary with computational algorithm to some extent. The 150- and 88-yr lines clearly are of lower Q than the longer-period features. As they encompass more cycles in the record, it seems likely that their greater width signifies unstable or variable periods. Alternatively, this means that the Gleissberg may vary from 88 yr downward in period to about 81 or 82 yr. A similar result for the line at 150 yr is inferred. This line was the strongest feature noted by de Jong and Mook (1980) in their high-precision data for the 33rd through 34th century BC.

Stuiver and Braziunas (1989) reported the spectrum of combined Seattle and Pearson data using MEM (maximum entropy) claimed to be at AR order 20 but possibly higher. They noted a set of spectral line harmonics with the fundamental estimated at a frequency $f \sim 2.4 \times 10^{-3}$ yr⁻¹ (420 yr period) and harmonics, most notably the second with a period of ~ 210 yr and the third, a period of ~ 140 yr. The comparison to the DFT suggest too low a resolution for resolving the lines they reported.

The MEM (maximum entropy) power spectrum (AR = 120) of the Stuiver-Pearson data is shown in Fig. 6 with the line at ~ 2240 yr the most prominent. Figure 6 discloses numerous other line features including the period at 208 yr and the line at 504 yr endowed with much more power than the 427-yr period. Assuming a fundamental at 520 yr, there are weak harmonics at $\sim 260, 175, 131$ and 105 yr. This figure also shows harmonics of the

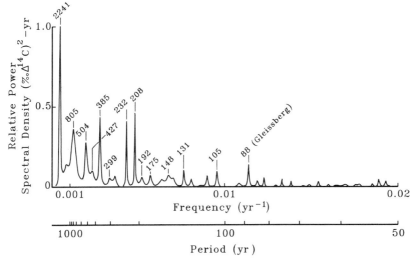

Fig. 6. Maximum entropy (MEM) power spectrum of combined Belfast and Seattle data, de-trended via least squares sine wave fit and cubic tension spline interpolated to 20-yr intervals. Autoregressive order 120.

~ 410-yr period at 208 yr and about 148, 105 and 84 yr. The second and third harmonics of the 2230 yr fundamental appear at ~ 1160 yr and ~ 805 yr. Until the differences in spectra are resolved, it seem premature to view the multi-hundred-year region of the spectrum as composed of a fundamental at 420 yr followed by a string of harmonics. The Bayesian (Bretthorst) estimator confirms the major features of the MEM and DFT spectral analyses (Table II). A more detailed discussion of the radiocarbon spectrum is given by Sonett and Finney (1990).

From their classification of the strongest 9 lines in the combined La Jolla and Belfast Bretthorst periodogram, Finney (1988) and Sonett and Finney (1989) proposed a series of harmonics and side bands based on fitting to within 2σ frequency limits, but this fit is nonunique. They find that all 9 lines can be fitted within the constraint where only 3 lines are nonuniquely fundamental.

Nonlinear forcing in geophysics is not a rare phenomenon. In the circumstances surrounding the generation of the atmospheric inventory of radiocarbon, not only is the incident cosmic-ray flux inconstant, but additional processes of terrestrial origin can influence the eventual inventory. Simple nonlinear (finite amplitude) forcing can lead to the generation of a harmonic spectrum, while a mixture of two or more periods in a nonlinear device leads to amplitude modulation (AM). Frequency modulation (FM) is a more complex process and involves periodic or aperiodic changes in source-forcing frequency, by and large a problem outside the scope of this chapter, although

FM cannot easily be ruled out as an adjunct to some of the spectral features of radiocarbon. The ability to detect and understand modulation can lead to a significant possibility for identifying forcing mechanisms and separating those of terrestrial from solar origin.

Sonett (1984) showed how the nonlinear dependence of the radiocarbon production rate upon changes in the Earth's dipole moment (Elsasser et al. 1956) could provide the essential basis for amplitude modulation. The corresponding statistical analysis, using a combination of the La Jolla and NSF Workshop data provided marginal evidence for modulation. The 2300-yr feature has also been argued (Damon and Linick 1986; Damon 1988; Damon et al. 1990) to modulate strongly the 208-yr period. The evidence for modulation is strengthened by analysis of the more complete Becker-Stuiver-Pearson (BSP) time sequence. But direct detection of symmetric side bands about the 208-yr period (viewed here as a "carrier" wave) is negative, due possibly to a low signal/noise (S/N) ratio or interference between signal and window side lobes. Neither MEM nor the DFT discloses symmetric side bands; even the higher resolution of the Bretthorst algorithm fails this test. A major failing shared with all attempts to explain the full spectrum is the absence of aforementioned visible symmetric side bands. In short, the appearance often is akin to single side-banding as found in communication practice but we know presently of no natural mechanism for so filtering radiocarbon sequences.

As an alternate approach to the problem of detection of modulation, a narrow-banded version of the BSP sequence infers the presence of modulation of the 208-yr period by the 2300-yr period. This is seen in Fig. 7 from the BSP sequence bandwidth limited to a range of $237 < \tau < 190$ yr (where τ is the period) about a central period of 208 yr showing intense modulation of approximately 100%. The procedure for narrow-banding is to select a narrow range of frequencies from the Fourier transform and invert into the time domain. The bandwidth is just sufficient to accommodate the two side bands generated by amplitude modulation (AM) by a 2300-yr period. We cannot with certainty rule out FM (frequency modulation) but as a working hypothesis AM modulation suggests two forcing functions, an extraterrestrial 208-yr (carrier) period and a terrestrial 2300-yr (modulation) period. Figure 7 suggests very strong ($\sim 100\%$) modulation. If both forcing functions were commonly terrestrial or extraterrestrial, the likelihood of frequency modulation would be increased.

VI. THE MAUNDER MINIMUM

This historically important minimum in the sunspot index ~ 1640 to 1720 AD is of special interest to the problem of the radiocarbon variations because it is suggested by the earliest direct continuous telescopic record of

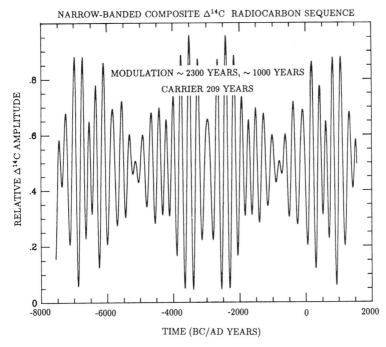

NARROW-BANDED COMPOSITE Δ¹⁴C RADIOCARBON SEQUENCE

MODULATION ~ 2300 YEARS, ~ 1000 YEARS

CARRIER 209 YEARS

RELATIVE Δ¹⁴C AMPLITUDE

TIME (BC/AD YEARS)

Fig. 7. Δ¹⁴C BSP time sequence narrowbanded to $237 \leq \tau \geq 190$ where τ is the period in yr. Two major periods are 208 yr ("carrier") and ~ 2400-yr "modulation." Amplitude modulation of approximately 100% is inferred.

the sunspot index record (mainly based on the resurgence of interest spawned by Eddy); but there is little direct historical evidence for earlier Maunder-like solar minima. (Oriental records are intermittent and irregular.) The most compelling evidence for earlier sunspot index minima comes from the proxy construction of the record for the last millennium using radiocarbon measurements of Pacific Northwest wood (Stuiver and Quay 1980). The correspondence between the Maunder minimum and the earlier minima proxy-disclosed by their data and the radiocarbon spectrum is uncertain. The spacing between the minima during the last 1000 yr averages 208 yr but with significant variability as does the length of the minima.

The three strong and three weaker increases in the radiocarbon inventory of the millenial record of Fig. 1, which corresponds in time to the Stuiver and Quay record, infer corresponding decreases in the interplanetary modulation index, a causative agent for the measured Δ¹⁴C variation. The 208- and 88-yr lines in the spectrum can be brought into approximate coincidence with this record, thus opening up the possibility that these lines are the spectral representation of the Maunder and Gleissberg minima. But, if variability is a key requirement, as suggested by the record reported by Stuiver and Quay, the weaker and broader line at 150 yr should be taken into consideration. The association of the major solar minima with the 208- and 150-yr periods in the spectrum provides a basis for believing that these variable features of the Sun have extended over the past 10 millennia represented by the BSP record, and furthermore if the major minima are primarily associated with the stronger 208-yr line, then these major minima are amplitude (AM) modulated with a period of $\sim$ 2300 yr.

Kocharov (1986) has recently reported that during the Maunder minimum the 11-yr period of the radiocarbon inventory (difficult to observe but not necessarily impossible as discussed later) changed from 11 to 22 yr with a return to 11 yr at the end of the Maunder period. This hypothesized FM is consistent with diffusion in association with the 11-yr period and the interplanetary electric field polarity in association with the 22-yr period (see the chapter by Jokipii). The Kocharov result is sufficiently unusual that substantive corroboration is planned.

VII. THE GLEISSBERG PERIOD

This period in the sunspot index and aurorae first reported by Gleissberg (1944) has been confirmed by many workers. From the sunspot index, Cohen and Lintz (1974) estimated the Gleissberg period to be 89.6 yr from the 12-month smoothed sunspot numbers and also suggested the existence of 2 $\times$ 89.6 = 179.2-yr period which is so far not detected in the radiocarbon record. The uncertainty of this period is emphasized by the short length of the record and is a risky supposition as the sunspot time sequence is limited to 282 yr. Siscoe (1980) and Attolini et al. (1988a) observe periodicities of 88 and 131 yr in the historical auroral record. The 131-yr period also shows up in the $\Delta^{14}C$ power spectrum (Fig. 4). Feynman and Fougere (1985) obtain a Gleissberg period of 88.4±0.7 yr from the AD 450 to 1450 auroral record. Castagnoli and Bonino (1988), based on the time profile of the relative thermoluminescence in a recent sea sediment core, observe periods of about 132, 84, 12.1 and 10.8 yr. A similar period is evident in the MEM (AR 120) spectrum (Fig. 4) at 88.2 yr. Sonett (1984) using MEM and Bretthorst (1988) using a Bayesian estimator obtain a strong period at about 90 yr in the spectrum of the yearly averaged sunspot index.

The appearance of the Gleissberg period in the radiocarbon spectrum and the $\delta^{18}O$ discussed in Secs. IX and X is a significant observation. The Bretthorst algorithm applied to the BSP sequence yields a value of 83 yr and the periodograms of Fig. 5 indicate a wide line with peak intensity at the high end $\sim$ 88 yr, but the caveats about the periodogram given earlier (Sec. IV) should be remembered. Stuiver and Braziunas (1989) also report this line. Based on the auroral evidence alone the Gleissberg period is clearly a hydromagnetic phenomenon associated with the activity cycle of the Sun. Its appearance also in the radiocarbon record provides a strong argument for linking ¹⁴C variability with the Sun and is perhaps the best indicator so far of the extraterrestrial origin of this periodicity in the ¹⁴C spectrum.

VIII. THE SOLAR CYCLE

As a consequence of the atmosphere's role as a low-pass filter (Houtermans et al. 1973; Siegenthaler et al. 1980; Lazear et al. 1980) and the observed variation in neutron flux and subsequent Δ¹⁴C production (Lingenfelter and Ramaty 1970), the 11-yr cycle is expected to exhibit $<$ 4 per mil peak in the radiocarbon record for the most active solar cycles (Damon et al. 1983,1990). For example, using the Lingenfelter and Ramaty (1970) production function and the box diffusion model parameterized according to Damon and Sternberg (1990) and assuming only changes in ¹⁴C production, the attenuation factors for the 11, 88, 210, 424, 2300, and 11,000 yr cycles are 54, 19, 16, 11, 6, and 2.6, respectively.

In spite of intense atmospheric attenuation and damping of spectral lines, the Bretthorst algorithm does identify the lines mentioned in Sec. IV present in the neighborhood of 20 yr (20.12 and 20.83 yr) which we tentatively identify with the Hale period. The additional lines appearing at 26.31 and 32.7 yr in Fig. 5 are so close to an expected Nyquist folding axis that they give unreliable true frequency estimates. The irregular sampling period of most radiocarbon records increases the likelihood that alias may corrupt the spectrum at high frequency. (See Bracewell 1978 for treatment of Nyquist frequency limits for sampling at multiple frequencies.)

The 11-yr signal is further obscured by production of ¹⁴C by solar flares and which tends to be of opposed phase with respect to "steady"-state ¹⁴C production (Lingenfelter and Ramaty 1970; Damon et al. 1990). Accordingly, the total variation during an 11-yr solar cycle may be as much as 10 per mil peak to trough, but no more than 4 per mil of the peak-to-trough fluctuation is the result of modulation of galactic cosmic rays by the solar wind. The remainder is ascribed to proton events accompanying solar flares. The very large peak-to-trough variations during the 11-yr cycle reported by Baxter and Walton (1971) and by Baxter et al. (1973) have not been verified by other workers (Damon et al. 1973a,b; Stuiver and Quay 1981; Burchuladze et al. 1980).

Recently, peak-to-trough fluctuations of 20 per mil during the 11-yr cycle have been reported by Fan et al. (1986) and by Kocharov (1986). Both groups measured Δ¹⁴C in trees growing under extreme climatic conditions near the northern forest boundary. In addition, Dai and Fan (1986) reported a latitudinal gradient following the 1964 peak of atmospheric bomb ¹⁴C. This gradient of atmospheric ¹⁴C from 68°N to 27°N persisted for at least three years after 1964. Neither of the two groups made an attempt to distinguish between variations in ^{14}C resulting from modulation of galactic cosmic rays and ^{14}C produced by protons associated with solar flares.

IX. MULTI-HUNDRED-YEAR RADIOCARBON PERIODS AND THE δ^{18}O RECORD

Early publications of Dansgaard's group discuss periodicities in the δ^{18}O record for the time interval from 1200 to 2000 AD (Johnsen et al. 1970; Dansgaard et al. 1971). Their chronology was based on the assumed average accumulation rate of ice. Their power spectrum for this time interval shows prominent periods at 78 and 181 yr. They associated the 78-yr period with the Gleissberg sunspot periodicity. In a longer section of the core going back to about 10,000 yr, they obtained a period of 350 yr, again using the ice accumulation time scale (Dansgaard et al. 1971), and at about 35,000 BP ice accumulation rate years, they found a persistent oscillation with a period of ~ 2000 yr.

The assumption that these periods correspond to certain Δ^{14}C periods can be tested via a linear time-scale correction (based upon the Gleissberg period from other sources) and then multiplying by the factor 88/78 = 1.13 (using the Gleissberg period estimate of 88 yr) to obtain corrected periods of 205, 396 and 2262 yr, very close to the Δ^{14}C spectral periods (Fig. 6) of 88.2, 208 and 2241 yr (MEM AR 120), with the exception of the 396 period, which falls between the strong peak at 358 yr and the weak peak at 427 yr. The contraction of the uncorrected periods is probably due to missing intervals as pointed out by Dansgaard et al. (1984,p.292): "Such holes might be caused by essential cease (sic) of Greenland accumulation, or by ablation (due to wind erosion, as observed in some areas close to Vostok, East Antarctica)."

X. COMMENTS ON CLIMATE AND THE δ^{18}O RECORD

Subsequent to the early work, ice strata for the past 8500 yr has been dated by stratigraphic methods to an accuracy that is considered to be better than 2% (Hammer et al. 1978), although the earlier ice core chronology is less accurate due to poorly known strain and accumulation rates. Figure 8 shows the δ^{18}O data (A) of Dansgaard et al. (1984) for the Camp Century, Greenland, ice core back to an age little more than 20,000 yr BP. Also shown is a band-pass filtered version (B) (2200 to 2800 BP) of the raw data (A). According to Dansgaard et al. (1984), the MEM spectrum of the Camp Century ice core for the last 8500 yr shows that, apart from the longer-term trend, a 2550-yr oscillation dominates the spectrum containing 17% of the spectral power (Dansgaard et al. 1984,p.293). (Note that MEM amplitude estimates can be uncertain.) They also correlate the ~ 550-yr oscillation with a similar oscillation of the extent of glaciation (C). The ~ 2550-yr oscillation contin-

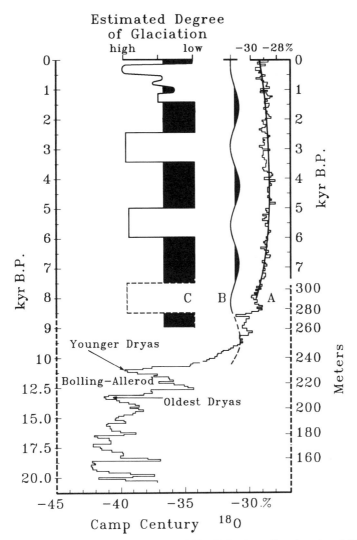

Fig. 8. δ¹⁸O Camp Century ice core data. (A) δ¹⁸O; (B) bandpass filtered version of (A); (C) estimated world-wide glaciation from Denton and Karlen (1973) (figure adapted from Dansgaard et al. 1984).

ues during the last ice age prior to 10,000 BP, and the δ¹⁸O oscillation of the glacial ice undergoes an oscillation from Oldest Dryas to Younger Dryas comparable with the oscillation in Δ¹⁴C. The δ¹⁸O oscillations during the Wisconsin (Würm) glacial are an order of magnitude greater than during the Holocene (post-10,000 BP). Damon et al. (1990) have used a cubic spline as a low-pass filter to remove the 11,300 yr trend curve of the Δ¹⁴C data (Fig. 9). The resulting curve describes a quasi-sinusoid ~ 2300-yr period and peak-

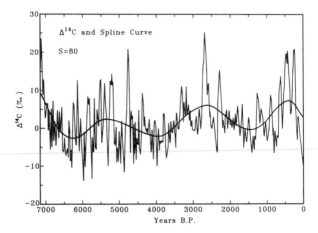

Fig. 9. After removal of the secular trend, a cubic spline with stiffness 80 used as a low-pass filter, takes the form of a quasi-sinusoid with wavelength ~ 2300 yr and peak-peak amplitude of 7 per mil. the spline curve follows the amplitude of the Suess "wiggles" (de Vries effect) which are being modulated by the ~ 2300-yr period and the geomagnetic field (figure after Damon et al. 1990).

to-peak amplitude of 7 per mil although this is an order of magnitude less than the $\Delta^{14}C$ variation shown in Fig. 10 for the Oldest through Younger Dryas. Fisher (1982) has demonstrated a significant cross correlation between $\delta^{18}O$ and $\Delta^{14}C$ in ice cores from both Camp Century and Devon island for a ~ 2600-yr cycle.

The Middle Europe oak dendroclimatology (Schmidt and Bruhle 1988) indicates that the Little Ice Age, AD 1500 to 1800 (Lamb 1977,p.104), was preceded by the Hallstattzeit cold epoch, 750 to 400 BC, and an earlier cold epoch between 3200 and 2800 BC (Schmidt and Bruhle 1988). These climatic epochs are between 2200 and 2300 yr apart, and each coincides with $\Delta^{14}C$ fluctuations that drastically affect the calibration of the radiocarbon cycle, so much that Schmidt and Gruhle (1988,p.179) point out that "Der ^{14}C-Physiker nennt diese Zeitspanne das 'Hallstattdesaster.' "

Pestiaux et al. (1987,1988) analyzed foraminifera from four Indian Ocean cores with high sedimentation rates. Their chronology consisted of 11 radiocarbon dates corrected for the age of surface water and identified isotopic transitions dated by uranium-series disequilibrium dating methods extending the chronology back to transition 5/6 at 27,000 yr. Spectral power was resolved into quasi-periodicities within 3 energy bands: $10,200 \pm 1200$ yr, 4600 ± 300 yr and 2300 ± 200 yr. The correspondence of these 3 periods with the 11,300 yr model sinusoid, the ~ 2300 yr, and the putative 4600 yr $\Delta^{14}C$ period, strengthens the argument that radiocarbon oscillations are related to climate, either as a response to solar luminosity or terrestrially induced changes in ocean-atmosphere carbon reservoir parameters.

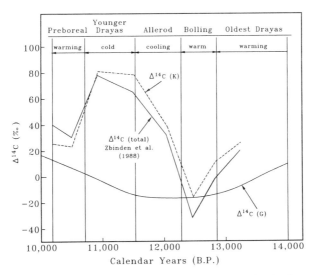

Fig. 10. Δ¹⁴C obtained by AM *S* measurement of ¹⁴C in terrestrial microfossils, mainly birch fruit, in sediment cores from Swiss lakes that have also been palynologically studied (Zbinden et al. 1990). Δ¹⁴C (G) is obtained by extrapolation of the 11,300-yr period sinusoid in Fig. 2. The fluctuation, Δ¹⁴ (K), includes half of a ~ 2300-yr cycle.

During the glacial climatic episode, the temperature difference between equator and pole(s) is greatest, producing much lower values of $\delta^{18}O$. The climate is also much less stable. The ~ 2300 yr $\delta^{18}O$ cycle has a peak-to-peak amplitude of only 0.3 per mil during the Holocene (postglacial), whereas it attains a peak-to-peak amplitude ≤7 per mil during the glacial, which is up to a factor of 23 greater amplitude (Fig. 8). The 208-yr oscillations have a varying amplitude during the glacial episode which is as high as 5 per mil, whereas after 8000 BP, the ~ 208 yr $\delta^{18}O$ oscillations do not exceed 1 per mil peak-to-peak. During the transition from full glacial to postglacial from about 10,000 BP to 8,000 BP, the 208-yr oscillations gradually dampen. In addition, 10^4-yr $\delta^{18}O$ oscillations are observable in the longer $\delta^{18}O$ record (not shown in Fig. 8); they obtain a peak-to-peak amplitude of about 3 per mil during the ice age.

The largest $\delta^{18}O$ fluctuation involves the ~ 2300-yr cycle. With reference to changes in atmospheric CO_2, Oeschger et al. (1984,p.104) propose that CO_2 changes were a consequence of a change in the oceanic turnover rates, as low turnover (ice age), leaving enough time for a complete nutrient consumption involving low pCO_2, and a faster turnover (Holocene), leading to incomplete nutrient consumption with a correspondingly higher CO_2. Broecker et al. (1985,p.25) stress "two modes of ocean-atmosphere-biosphere-cryosphere operation. Each oscillation observed in the ice cores involves a jump from one mode of operation to the other." We emphasize the continuity of the oscillations, with the jump being a rapid transition from a higher amplitude to a lower amplitude, *without* distinct change in the period of the oscillation; such a change in oscillation period would be caused by increasing the eddy diffusion constant K and the transfer of larger volumes of water and CO_2 without a corresponding change in the transit time of the ocean cycle. During the glacial episode, K changes drastically during the 2300-yr cycle but only slightly during the Holocene.

The above discussion leaves open the possibility that, in addition to geomagnetic variability, a climatic component exists in the long-term $\Delta^{14}C$ trend (Lal 1985,1990). There is a long-term $\delta^{18}O$ trend of about 0.75 per mil, but it is out of phase with the $\Delta^{14}C$ trend (see Figs. 2 and 10) and appears to be related to the climatic optimum prior to 4000 BP and the following climatic deterioration. It seems that, considering the phase and amplitude of the secular change in $\delta^{18}O$, climate change is unlikely to contribute a major part of the long-term secular variation of $\Delta^{14}C$ during the Holocene; variation of the geomagnetic field seems to be an adequate forcing function for the long-term trend.

XI. VARIATIONS: EXTRAHELIOSPHERIC OR MAGNETOSPHERIC?

Since radiocarbon production in the atmosphere is a consequence of the incident flux of cosmic rays, whether the interstellar cosmic-ray flux arriving at the heliosphere has significantly varied over the past 10 to 20 millennia would be reflected in the eventual atmospheric inventory. Gamma rays, capable of causing spallation (Konstantinov and Kocharov 1965; Lingenfelter and Ramaty 1970; Kocharov 1986) have been studied by workers at A. F. Joffee Physical Technical Institute in Leningrad who show that a pulse with the fast rise anticipated for the gamma-ray burst from Tycho Brahe's, Kepler's and Cassiopeia supernovae are below the threshold of detection, setting an upper limit on the total energy contained in the gamma-ray component at 10^{49} erg. Castagnoli et al (1988) report a thermoluminescence signal in a Tyrennian sea core which correlated with the dates of several supernovae early in this millenium. However, neither of these provide a measure of radiocarbon production in the atmosphere.

Sonett et al. (1987) ascribe two high concentrations of ^{10}Be in Antarctica, found at $\sim$ 33,000 and $\sim$ 60,000 yr BP (which show no correlation to the $\delta^{18}O$ climatic record), to a possible cosmic-ray enhancement from interstellar shock waves passing over the heliosphere. Whether the enhancements alternatively are due to excursions of the Earth's dipole field (Raisbeck et al. 1987) is uncertain. Recent unpublished work by Kocharov infers that the Antarctic source is so close to the geomagnetic south pole that the geomagnetic cosmic ray cutoff is so low as to make considerations of cutoff unimportant.

An excursion of the geomagnetic field (aborted reversal) (Merrill and McElhinny 1983) is reported at Lake Mungo, Australia, and at Laschamp, France, 33,000 yr BP. Barbetti and McElhinny (1976) estimate the duration of the Lake Mungo excursion at 2500 to 3000 yr which is similar to the ^{10}Be events. Less secure evidence suggests it is recorded in the Gulf of Mexico, Canada and Denmark and possibly in Lake Biwa, Japan. However Creer (personal communication) disputes the reality of 33,000 and 60,000 yr excursions. There are no documented cases of magnetic field reversals younger than the Bruhnes-Matuyama around 700,000 yr ago (Cox 1969); on this kind of time scale the isotope record provides only an ambiguous look at either geomagnetic or extrasolar system cosmic-ray variations or perhaps some combination.

XII. FURTHER DISCUSSION ON THE SECULAR TREND IN RADIOCARBON

The de Vries effect (Suess "wiggles") (Fig. 1) has the greatest amplitude when the terrestrial dipole moment $M(t)$ is small, which supports the view that the radiocarbon trend is the result of changes in $M(t)$. In addition, Damon (1988) has observed that the $\sim 50\%$ enhancement of the Suess wiggles is in accord with theoretical expectation for the observed low values of $M(t)$ (Lingenfelter and Ramaty 1970, Fig.3,p.525; Lal 1988,Table V, p.225). Also, Stuiver and Braziunas (1988, Figs. 4a–4j,p.260–264) have estimated ^{14}C production rates during the last 9600 yr, assuming that the de Vries effect (Suess wiggles) is the result of modulation of the galactic cosmic-ray flux by the solar wind (Stuiver and Quay 1980).

Their estimated production rate changes during Maunder, and Spörer-like variations seem to include active and quiet episodes of 1150-yr duration on the average back to 7090 BP. These episodes are preceded by a long active period from 7090 to 9490 BP of ~ 2300-yr duration. (This period may only coincidentally be the same as the pronounced period discussed earlier.) But if we associate this long active period with a prevailing low value of $M(t)$, then the midpoint of the $M(t)$ low at 8290 BP precedes the ^{14}C minimum by 1440 yr, leading to a close correspondence in modeling Δ^{14}C, using the box diffusion model, between both $M(t)$ minimum and Δ^{14}C maximum and their inverses. This lends further credence to the extrapolation of the sinusoidal trend curve in Fig. 2.

XIII. THE ROLE OF THE SUN

That the Sun exhibits variable luminosity has been long suspected but was apparently too small to be detected on the surface of the Earth. Recent satellite measurements of total irradiance I (Willson et al. 1986; Willson and Hudson 1988) plotted vs the Wolf sunspot index (R_x) display a linear relationship where $S(W/m^2) = 1366.63 + 0.0089\,R_x$ (Fig. 11; also see the chapter by Fröhlich). For the most extreme solar cycle with $R_z = 200$, the increase in total irradiance would be 1.3 per mil. However, the increase is not distributed equally over all wavelengths. For example, observations by Heath and Thekaekara (1977) indicate that the solar flux at 1750 Å is about a factor of 2.5 greater at solar maximum than at solar minimum, and at 3000 Å the effect is about 18%. The impact of these measurements on the long-term periodicity problem is unclear.

The well-known 208-yr Suess wiggles have been generally attributed to solar modulation of $\Delta^{14}C$ production (Suess 1968; Damon 1968; Stuiver and Quay 1980; Sonett 1984; see Damon et al. [1978] for review of earlier literature). Most recently, solar modulation has been discussed by Lal (1988b) as a function of the modulation parameter ψ. Lal concludes that the lowest effective ψ value that could be observed during periods of low solar activity would be ~ 100 Mv and that, based on the sunspot record, a typical solar cycle could well be characterized by $\psi = 300$ Mv (solar minimum) and 900 Mv (solar maximum); the typical average would lie at about 450 Mv (Lal 1988,p.230). Thus, applying the attenuation factor of 54 (see Sec. VIII) for the 11-yr cycle and 16 for the 208-yr cycle, we would expect, referring to Lal's table (1988,Table III,p.223), an 11-yr cycle of about 5 per mil peak to peak, which is about twice the variation predicted from Lingenfelter and Ramaty's (1970) relationship between ^{14}C production and sunspot number. It is also greater than observed (Damon et al. 1990), suggesting that the attenuation factor of 70 calculated by Siegenthaler et al. (1980,Fig.7,p.189) is closer

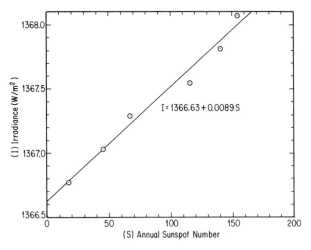

Fig. 11. Total solar irradiance S from the NASA Solar Maximum Mission (Willson et al. 1986) vs annual sunspot number R_x (figure after Damon 1988).

to correct (3.9 per mil peak-to-peak) or that Lal (1988) has overestimated the average variation in the modulation parameter ψ for a typical solar cycle. For a Maunder minimum-like change, we would expect ψ to have been 100 Mv followed by the average value of 450 Mv, yielding a peak-to-peak value of 13 per mil with the attenuation factor of 16 and the same for Siegenthaler et al. (1980).

The observed radiocarbon variation, $\Delta^{14}C$, during the Maunder minimum is about 20 per mil. As we will see, this is reasonable because there must be a climatic component to the de Vries effect. The semi-empirical relation between the radiocarbon production rate Q and the incident cosmic-ray flux can be expressed by the linear approximation (Damon et al. 1983)

$$Q = 2.43 - 0.002645\,R_x \qquad (3)$$

where Q is the production rate in $s^{-1}cm^{-2}$. Referring to Fig. 1, we attribute the Dalton minimum (i.e., sunspot minimum, $\Delta^{14}C$ maximum) to the Gleissberg period. $\Delta^{14}C$ maximizes at AD 1820. The Dalton sunspot minimum occurs between AD 1795 and 1825, with its mid-point at 1810. The time lag between sunspot minimum and the corresponding $\Delta^{14}C$ maximum is 10 yr, and the peak-to-peak amplitude is about 6 per mil, in reasonable agreement with theory (Damon et al 1983). Damon and Linick (1986,p.268) suggested that the Spörer $\Delta^{14}C$ maximum is split into two peaks separated by a valley due to a frequency closely approximating the Gleissberg frequency.

The expected increase in $\Delta^{14}C$ can be calculated from Eq. (1) for the 25-yr period centered around AD 1690 relative to the 25-yr period centered around AD 1781 following the Maunder minimum. The average R_x for the period centered around AD 1690 is 2, and the average R_z centered around AD 1781 is 72. The reference time for calculation of ¹⁴C is mid-19th century. The average R_z for the period centered around AD 1850 is 59. From Eq. (3), the change in production rate is 8.1%, and, dividing by the attenuation factor of 16, the predicted ¹⁴C increase is only 5 per mil compared to 13 per mil calculated by using the modulation parameter ψ rather than the sunspot number R_x. This leaves from 7 per mil to 15 per mil of the variation unexplained by solar wind modulation of the galactic cosmic-ray flux and consequent ¹⁴C production rate.

In light of the previous discussion, the most probable cause of the discrepancy is the dependence of ¹⁴C on climate because of the observed similarity in ¹⁸O and ¹⁴C periodicities for Maunder-like variation. Stuiver reconciled the low predicted $\Delta^{14}C$ values for the Maunder minimum by assuming that solar wind modulation continued at zero sunspot number as a result of a residual solar wind that may originate from the Sun's polar region. The sunspot number R_x is a proxy indicator of solar wind modulation, whereas the modulation parameter ψ is a more direct measure of solar wind modulation, and the lower values of ^{14}C calculated by Eq. (3) are probably the results of the inadequacy of this relationship. However, ψ leaves 35% of the Maunder minimum $\Delta^{14}C$ increase to be explained.

Siegenthaler et al. (1980) studied the carbon cycle-induced ^{14}C variation that may have been significant in the transition from glacial to post-glacial conditions. We have used their estimates of climatic parameters during glacial and post-glacial times to evaluate the effect of climate on ^{14}C. The results are given in Table III Changes affecting the rapidly communicating (years to a few decades) ambient reservoirs tend to cancel (Damon 1970) even for conditions drastically different from Recent conditions that prevailed during the last glacial period (Wisconsin or Würm). These reservoirs have similar ratios $R = {}^{14}C/{}^{12}C$. However, changes affecting ocean circulation involving the transfer of ^{14}C to the deep ocean (where R_{ocean} is only 84% of the atmospheric ratio R_{atm}), are critical. The other significant ambient reservoirs differ by only a few percent from the atmospheric value. In the Oeschger-Siegenthaler (1975) box diffusion model, changes in ocean circulation are modeled by the eddy diffusion coefficient K. They estimate that during full glacial conditions K would be only 50% of the Recent postglacial more stable environmental state. This would result in a $\Delta^{14}C$ increase of $+104$ per mil, which is close to the change observed in the Swiss lake estimates of ^{14}C during the transition from the warm Bølling climatic conditions to the full glacial Younger Dryas climatic episode (Fig. 10).

XIV. GLOBAL WARMING: GREENHOUSE EFFECT OR SOLAR HEATING?

Slightly over two decades ago, Damon argued that atmospheric temperature variations of $\pm 0.8°C$ during the Christian era are consistent with a solar activity-solar modulation theory as suggested by Suess (1968). Higher-than-average sunspot number correlated with higher-than-average annual temperature due to increase in solar irradiance and with a lower Q (^{14}C production) due to solar modulation of interplanetary magnetic fields (Damon 1968). Unfortunately, the available data indicating variation in total solar irradiance, using groundbase and radiosonde balloon measurements, were not conclusive in 1968 nor were they at the time of the symposium on *The Solar Output and Its Variation* (White 1977). However, the results of the Solar Maximum-Solar Minimum Mission showing a rise and fall of total solar irradiance during the solar cycle are now widely accepted (see Fig. 11). The $\Delta^{14}C$ and $\delta^{18}O$ spectra are complex but at least for the last millennium (Fig. 1), 3 periods dominate the record. These are the ~ 88 yr (Gleissberg), 208 yr (de Vries) and the 2400 yr (Hallstadttzeit). It is of interest to ascertain, as a working hypothesis, the ability of these periods to explain both a direct relation between, on the one hand, solar activity and global temperature, and also an inverse relation between $\Delta^{14}C$ and solar activity.

TABLE III
Effect of Glacial Epoch (17,000 yr BP) Climate Change on Δ^{14}C Relative to AD 1850

Climate Parameter	Causal Factor	Δ^{14}C‰
Ocean volume: $\Delta V_O = -3\%$	increased ice and thermal contraction	-7
Ocean Area: $\Delta A_O = -12\%$	as above, $K_{am} = -12\%$	$+6$
Wind velocity: $\Delta u = +20\%$	increased $(T_{eq} - T_{po})$ $K_{am} = +50\%$	-20
Atmospheric $CO_2 = \Delta N_a = -33\%$	increased oceanic biological productivity	$+25$
Terrestrial biosphere: $\Delta N_b = -50\%$	decreased productivity	-4
Sum of all factors affecting surficial reservoirs:		$\Sigma = 0$
Eddy diffusion rate: $K = -50\%$	decreased ocean circulation	$+104$

[a]Data from Siegenthaler et al. (1980).

The Hallstadtzeit period in climate contributed to the Little Ice Age; on a periodic basis its next maximum should occur around AD 3700. As its period is long its effect during this century ought to be small but should nevertheless contribute slightly to the global warming trend that has generally persisted during the 20th century, notwithstanding the 0.2 °C net cooling within the northern hemisphere from the early 1940s through the 1970s (Mitchell 1972). Cooling of this magnitude did not occur in the southern hemisphere (Damon and Kunen 1976) nor persist in the northern hemisphere (Hansen and Lebedeff 1988; Jones et al. 1986).

Our hypothesis suggests that the 208 yr de Vries period should have a greater effect on climate variability in the near future. But first, it is of interest to project it into the past, retrospectively. If the mid-point of the Maunder minimum at AD 1690 (Eddy 1976a) is identified with the 208-yr period (keeping in mind the 19-yr lag time between maximum production and maximum Δ^{14}C), then radiocarbon maxima should have occurred at AD 1293, 1501 and 1709 and minima at AD 1189, 1397, 1605 and 1813. These predicted maxima do occur on average every 208 yr despite the effect of less intense periodicities (see Fig. 1). The Suess effect obscures these maxima and minima during the late 19th and 20th centuries. Sunspot maxima and minima should have occurred 19 yr earlier than the Δ^{14}C extrema. The last relatively weak de Vries minimum occurred between AD 1878 and 1913. Its midpoint is ~ AD 1896 or 206 yr after the approximate midpoint of the Maunder minimum. Solar activity has been generally increasing since AD 1913 and accordingly should reach a maximum during the first half of the next century if the periodicities observed during the past millenium extend into the next century. If the Dalton minimum (which is of short duration) is identified with the Gleissberg period, it should have reached minima at AD 1898 and 1986 and will not climax until $\sim$ 2030. Both periodicities would then contribute to solar warming until the mid-21st century. Thus the greenhouse effect would be obscured by increasing solar activity and the two effects combined may exacerbate global warming. Such a concatenation suggests the approach to a maximum similar to the Medieval Warm Epoch but with the addition of greenhouse warming. The Vikings took advantage of the milder climate and ventured westward to Vinland (North America) while the Maoris colonized New Zealand during the Medieval Warm Epoch, but at the same time the southwest and southern Great Plains in the U.S. suffered severe droughts.

XV. DISCUSSION ON SOLAR VARIABILITY

That the Sun is a star of variable luminosity has been an uncertain issue until the recent and current measurements by the ACRIM satellite radiometer experiment (Willson et al. 1986; Willson and Hudson 1988; also see the chapter by Frölich) which has shown a small variability with a period which is likely to be 11 yr, i.e., coordinated with the solar activity period. Other than this result and inferences from solar-like stars (see the chapter by Baliunas) there is no direct evidence that the Sun is variable on other time scales. Yet the isotopic data infer climatic and cosmic-ray flux variations of longer period. To be sure, those that are assignable to climate cannot be directly attributed to the Sun, although the underlying forcing of climate is probably not all due to Milankovich cyclicity and therefore some is likely still solar. Unfortunately, a secure conceptual edifice cannot be built alone upon arguments of likeliness.

Those geophysical parameters such as the aurorae and, insofar as we can say, $\Delta^{14}C$ features which appear at $\sim$ 88 yr Gleissberg period, more tightly infer an electrodynamic cycle which must have its origin in the solar atmosphere. Other periods which appear in the radiocarbon record also face the problem of partitioning between climate and electrodynamics. As noted, the climatic component may be caused by solar bolometric variability. The 208-yr period appears in both the tree-ring record from Campito Mt. Bristlecone pines in addition to the radiocarbon record (Sonett and Suess 1984). It seems to us unlikely that both electrodynamic and bolometric phenomena are required to produce the observed modulations and their correlations.

The gravest difficulty with the idea that the Sun is endowed with long-period bolometric and hydromagnetic variability rests upon the lack of a conceptual framework. All the known periods that we tend to associate with the Sun, save for the Hale cycle and convection, e.g., seismic, supergranulation, rotation, etc. are less at most than the 27-day rotation period (Gough 1988). Seismic transit times are certainly restricted to several hours although arguments can be made for longer hydromagnetic periods. The mean Alfvén speed over the full solar radius for 83 yr is ~ 0.05 cm s^{-1} corresponding, for uniform density $\rho = 1$ g cm^{-3}, to an internal magnetic field intensity ~ 0.2 gauss. Although these values are not too useful because the solar interior is radially very inhomogeneous, it is nevertheless possible to envision hydro-magnetic modes within the radiative zone which very approximately satisfy transit times consistent with the radiocarbon observations.

However, it is more difficult to reconcile such "eigenmodes" with the sources necessary to excite them if one holds to the view that the core is isothermal and hydrodynamically "benign." Among the few solar physicists who do take seriously the possibility of long-period modes, Dilke and Gough (1972) and Gough (1988) propose episodic breakdown in the solar core, but typical time scales are in the hundreds of Myr. The onset takes place subsequent to the buildup of ³He some hundreds of Myr after formation. (This needs to be tested in the sense of luminosity vs the existence of liquid water 3.7 to 3.9 Gyr ago inferred from the sedimentary Isua formation (see the chapter by Kasting and Grinspoon). Even if such "thermohaline" convection existed, time scales would have to be revised downwards drastically to match the periods discussed herein.

XVI. SUMMARY

Tree rings provide an ideal precisely dated archive of past variations of atmospheric ¹⁴C expressed as Δ¹⁴C (per mil) relative to a mid-19th century standard age-corrected to AD 1950 as zero BP. The three major causes of Δ¹⁴C changes are fluctuations in solar activity, the terrestrial dipole magnetic moment and climate. Δ¹⁴C fluctuations consist of a long-term secular trend modulated by fundamental periods T of 2300 yr, 208 yr, 88 yr and 11 yr. The 11-yr period is very weak and masked by out-of-phase solar flare-produced ¹⁴C. The 2300 yr period, in contrast to the shorter periods, is not modulated by the geomagnetic field but strong amplitude modulates the 208-yr period. It correlates with δ¹⁸O variations in the Camp Century ice core and δ¹⁸O ocean sediment foraminifera as well as Hallstadttzeit climate events, and thus appears to be related to climate-forced variations in reservoir parameters. The higher frequencies are related to production rate changes correlated with solar activity such as the Gleissberg 88-yr period, Schwabe 11-yr period and Maunder-Spörer solar minima $\sim$ 208-yr period. A possible solar forcing of climate change has been suggested by the correlation of high solar activity with the Medieval Warm Epoch (AD 1100–1250) and generally low solar activity with the Little Ice Age (AD 1400–1800). Further support is derived from the occurrence of periodicities similar to the Δ¹⁴C periodicities in the δ¹⁸O spectrum of the high altitude-latitude Camp Century, Greenland ice core and the spectrum of tree-ring widths of high altitude Bristlecone pine. Increasing solar activity during the 20th century leaves open the possibility that we may be entering a warm epoch that is masking the onset of greenhouse warming and may exacerbate anthropogenic warming during the 21st century. However, the relationship between total solar irradiance S and solar activity R_z observed during solar cycle 21 is insufficient in itself without amplification to produce a significant change in global temperature. Such a change seems to require that S be a function of T as well as R_z.

Climatically induced changes in carbon reservoir parameters in addition to changes in the ^{14}C production rate may contribute to the inverse relationship between Δ^{14}C and solar activity. The most sensitive parameter involves exchange between the mixed layer of the ocean and the deep sea. However, Holocene ($\leq$ 10,000 yr) temperature changes are not sufficiently great or of the right phase to contribute significantly to the long-term Δ^{14}C secular trend. It seems quite certain that long-term Δ^{14}C is primarily the result of changes in the Earth's dipole moment.

We now know that the Sun is a star of variable luminosity like many other stars (at least on an 11-yr time scale). Precise observations are limited to solar cycle 21. Yet, climatic and cosmic-ray flux variations of longer period may be inferred from isotopic data. It seems likely, although not definitively proven, that the 88-yr and 208-yr cycles involve changes in solar luminosity. The 2300-yr cycle is more enigmatic although the underlying force could still be solar.

Acknowledgment.   PED received support from a grant from the National Science Foundation.

LONG-TERM CHANGES IN COMPOSITION OF SOLAR
PARTICLES IMPLANTED IN EXTRATERRESTRIAL MATERIALS

J. F. KERRIDGE
University of California, Los Angeles

P. SIGNER, R. WIELER
Eidgenössische Technische Hochschule, Zürich

R. H. BECKER and R. O. PEPIN
University of Minnesota

Analysis of lunar surface samples for elements implanted therein by solar corpuscular radiation reveals evidence for the following compositional changes over a time period between 1.5 and 3 Gyr: 50% decreases in the ratios $^4He/$ ^{36}Ar and $Xe/^{36}Ar$; a 20% increase in the ratio $^3He/^4He$; a 3% increase in the ratio $^{20}Ne/^{22}Ne$; and a 50% increase in the ratio $^{15}N/^{14}N$. The causes of these changes are not resolved at this time but may include (a) a change in acceleration conditions of the solar wind, (b) a change in flux of solar energetic particles relative to that of the solar wind, and (c) a change in composition of the solar convective zone. There is good evidence for a long-term decrease in the solar-wind flux.

In this chapter we explore the way in which surface samples from certain solar-system objects can yield information about long-term changes in the composition of solar corpuscular radiation, by analysis of the elements implanted in such surface-exposed samples at different times. Only objects lacking an atmosphere and an intrinsic magnetic field can sample solar charged particles directly; thus the Moon and the asteroids, in the form of meteorites, provide the samples used in this study. (The possible effects of significant magnetic fields early in the history of the Moon and asteroids are not well understood at this time. The Moon now has only vestigial fields of high order [see, e.g. Sonett and Mihalov 1972]; some meteorites of asteroidal origin contain evidence of substantial paleofields [see the chapters by Hood et al. and Levy et al.].) These objects are also subjected to meteorite and micrometeorite bombardment, resulting in comminution of their surface rocks into a relatively fine-grained layer of soil termed a regolith. The heat and/or pressure generated by meteoroid impact also weld together previously fragmented material into a second generation of rocks, termed breccias, as well as glass-bonded composite particles termed agglutinates. Impacts also garden the regolith so that an individual soil grain does not spend all its life at the same location but can be transported over great lateral distances and can move stochastically throughout the depth of the regolith. The average depth of the lunar regolith is generally believed to be a few meters; asteroidal regolith depths are uncertain.

Because we are concerned with variations in composition of the implanted radiation as a function of time, we need a means of determining the time of implantation, referred to as "antiquity." This chapter therefore begins with a discussion of possible measures of antiquity, none of which have proven to be completely satisfactory at present. We then summarize all those observations of apparent secular compositional change that have been reported by one or more laboratories. Currently, five different compositional parameters meet that criterion. We then assess those five observations for what they might be telling us about long-term changes in the physics and/or chemistry of the Sun. We shall argue that for none of the five is an unequivocal interpretation available at this time and therefore conclude with an outline of possible future lines of inquiry.

Despite these uncertainties, we emphasise that the evidence is strongly suggestive of real long-term changes in either reservoir composition or mode of acceleration of solar particles. This is a fertile area in which much important work remains to be done.

I. THE MEASUREMENT OF ANTIQUITY

As noted above, an individual soil grain can move vertically within the regolith, giving it a finite probability of multiple exposures to solar radiation on the very surface of its planetary object. Consequently, even for a single grain, the concept of antiquity is not necessarily straightforward and for the multigrain soil or breccia samples employed in studies to date, this problem is vastly compounded. It follows that for a secularly varying solar outflow sampled by a parcel of regolith, any radiation-related parameter measured for that parcel represents some kind of average over the range of compositional variations sampled; furthermore, the antiquity derived for that average composition is itself some average of the antiquities of the individual grains. Note also that grains do not necessarily keep their individuality, but may become incorporated into agglutinates or breccias. For surface soil samples, in general, not even the time when the constituent grains were finally brought together can be determined, but for regolith breccias or subsurface core samples, a lower limit to their antiquity may be derivable if the compaction age or the cosmic-ray exposure age of the overburden, respectively, is known.

In fact, depth within the regolith can itself be thought of as a rough measure of antiquity in that older samples will in general be buried deeper than younger ones. However, the stochastic nature of the impact-driven excavation/deposition cycle, resulting in marked nonlinearity in accumulation rate and frequent reversals in stratigraphic sequence, restricts the quantitative antiquity information available from core data to the lower limits mentioned above for certain deep strata. Simple depth within a core is at best a qualitative indicator of antiquity.

A useful measure of antiquity, though one that is indirect and somewhat difficult to calibrate in absolute terms, can be derived from the exposure age of the regolith sample to galactic cosmic rays (*GCR*). Such radiation induces nuclear reactions in the constituent elements of the sample, resulting in production of certain cosmogenic nuclides in amounts proportional to the time that the sample spends within ~2 m of the surface. If those nuclides have low natural abundances in the regolith, the increase due to nuclear reactions in the soil can be measured and converted, via known production rates, into duration of residence within the top few meters of the regolith. That measure of duration is not equivalent to antiquity, in the sense defined above, but is statistically related to it in that the longer a grain has spent within the dynamically stirred upper part of the regolith, the longer ago, on average, it first received exposure to the solar radiation. Several nuclides have been used as monitors of cosmic-ray exposure, most commonly ^{21}Ne, which is straightforward to measure though it suffers from a relative ease of loss from the sample by thermal diffusion. A more robust cosmic-ray indicator is ^{15}N (Becker and Clayton 1977; Clayton and Thiemens 1980), which has the additional advantage that its production rate is essentially independent of composition for most rocks, being produced mainly from ^{16}O. However, distinguishing cosmogenic from trapped ^{15}N depends upon the generally higher release temperature of the former during analysis, and this does not yield a clean separation.

A more direct measure of antiquity would be one based on accumulation of a radiogenic daughter nuclide during exposure to the solar radiation. In principle, the radioactive parent could be of either solar or lunar origin but in practice only lunar species are effective. There are three candidates for such a nuclide: ^{40}Ar from decay of ^{40}K, ^{129}Xe from ^{129}I, and $^{131-136}$Xe from ^{244}Pu. These parent nuclides have quite different half-lives of 1.28 Gyr, 17 Myr and 82 Myr, respectively. Obviously, the last two radionuclides are now extinct, so that their daughter nuclides yield information on an early time scale, beginning with the formation of the Moon or the solar system. However, ^{40}K is still live and can therefore potentially date events relative to the present. Evidence for all these daughter nuclides, unsupported by their radioactive parents, can be found in many regolith samples but interpretation in terms of antiquity is hampered by uncertainties concerning their modes of incorporation into the regolith. Two possibilities exist. First, the gaseous daughter nuclide may leak out from the lunar interior into the transient lunar atmosphere, where it is photoionized by solar ultraviolet and then accelerated by the solar-wind electric field into the lunar surface (Heymann and Yaniv 1970; Manka and Michel 1971). Such implantation would parallel solar particle irradiation and could yield a useful measure of antiquity provided (a) the ratio daughter/ solar being implanted is known as a function of time, and (b) the daughter and solar nuclides are more or less uniformly distributed among the grains in the sample with neither concentrated in a subpopulation of the sample. Second, the daughter nuclide may be directly adsorbed onto regolith grains while percolating out from the lunar interior (Drozd et al. 1976). In this case, accumulation of the daughter would be decoupled from exposure to solar radiation. Although the adsorptive capacity of the regolith is apparently not sufficient to supply the observed quantities of parentless gas (Bernatowicz et al. 1982), there is some evidence for decoupling of the parentless and the solar components (Swindle et al. 1986). This issue is currently unresolved. Several studies have employed parentless ^{40}Ar as a chronometer of antiquity; examples include Eugster et al. (1980) and McKay et al. (1986).

A method to date the time at which a sample was exposed to galactic cosmic rays has been developed by Eugster and coworkers (Eugster et al. 1979, 1983a). This is accomplished by determining the amount of ^{136}Xe produced by neutron-induced fission of ^{235}U, the neutrons themselves resulting from cosmic-ray interactions with lunar material. This method yields meaningful measures of antiquity only for samples that have experienced simple *GCR* exposure under well-defined shielding conditions. With the aid of a few such samples, spanning a range of antiquity, the method therefore has the potential to calibrate other measures of antiquity (Eugster et al. 1983a).

Until such samples are run, a reliable measure of antiquity is lacking. In what follows, we choose to rely mainly on the ratio ^{40}Ar/^{36}Ar in trapped Ar, corrected for *in situ* decay of ^{40}K. This ratio is independent of the duration of surface exposure and depends only on the time when the exposure took place, i.e., the antiquity.

II. COMPOSITIONAL ANALYSIS OF IMPLANTED SOLAR RADIATION

Eight elements are sufficiently depleted in the Moon so that the contribution due to solar radiation dominates the analysis of regolith samples. They are: hydrogen, helium, carbon, nitrogen, neon, argon, krypton and xenon.

TABLE I
Relative Retentivities[a]

	He	N	Ne	Ar	Kr	Xe
Ilmenite	20 – 40	1270	25–50	100	190	340
Pyroxene/Olivine	1 – 3	1160	20–40	100	240	450
Plagioclase	0.3– 1	2540	5–10	100	220	365

[a]Retentivities of major lunar-regolith minerals for solar-wind-implanted noble gases and N. These are normalized for each mineral to Ar = 100. Data are from Signer et al. (1977) and Frick et al. (1988).

The efficiencies with which they are retained by bulk lunar regolith vary markedly: H and He are highly saturated, with retention of the noble gases increasing systematically with increasing atomic weight, while C and N are apparently very efficiently retained (DesMarais et al. 1974). Note that the four lightest elements are each characterized by only two stable isotopes, severely limiting our ability to distinguish nucleogenetically distinct components in those cases.

Most of the data obtained to date have been for bulk regolith samples, but we focus here mainly on the results of some recent analyses of well-characterised mineral separates. Though relatively few in number, these data have several advantages over bulk-sample data. First, pure samples of different minerals reveal very different retentivities for some of the implanted elements (Table I). Consequently, retention in a bulk sample consisting of several different minerals is an ill-defined quantity. Second, for the light noble gases, i.e., Ne, Ar and, to a less extent, He, the most retentive mineral fractions, most notably ilmenite, yield elemental and, especially, isotopic ratios that approach the values measured directly by the Solar Wind Composition Experiment on the lunar surface (Geiss et al. 1972) and those inferred for the Sun itself (see, e.g., Cameron 1982; Anders and Grevesse 1989) (Table II). (Note, however, that solar tabulations rely in part on solar-wind measurements.) Third, although individual mineral grains in a regolith contain lower doses of implanted solar radiation than the generally more abundant composite particles, such as agglutinates and microbreccias, it is evident that they have experienced a simpler exposure history, making their record of that exposure easier to decipher. Also, mineral grains have generally a shorter lifetime in the regolith than composite particles. Thus, they are less likely to be saturated with implanted gases. Fourth, the different grains making up a mineral separate should all have closely similar chemical and physical properties, thus simplifying interpretation of the progressive release of implanted species at increasing extraction temperatures. Finally, by analyzing grains within a fairly narrow size range, generally between 100 and 200 μm, drastic variations in exposure history stemming from differences in grain size can be minimized.

TABLE II

Comparison of Elemental and Isotopic Composition of He, Ne and Ar Derived
from Lunar-Regolith Ilmenite with Analogous Data Estimated for the Sun and
Measured in the Solar Wind by the Apollo Foil Experiment

	$^4He/^{36}Ar$	$^4He/^3He$	$^{20}Ne/^{36}Ar$	$^{20}Ne/^{22}Ne$
Sun [Cameron 1982]	20,300	5625[a]	25.9[b]	8.2[c]
[Anders & Grevesse 1989]	31,450	7050[a]	37	13.7[d]
Solar Wind [Geiss et al. 1972]	25,650	2350	45	13.7
Ilmenite[d] [Frick et al. 1988]	14,500	2520	46	13.5
[Benkert et al. 1988]	—	2170	—	13.7

[a] Prior to D burning in the Sun.
[b] Cameron's Ne data reflect his choice of the so-called "planetary" Ne component ($^{20}Ne/^{22}Ne$ = 8.2) as the value representative of the Sun.
[c] From Geiss et al. (1972).
[d] From 71501, a recently irradiated soil.

From Table I it is clear that ilmenite is the recording medium with the least distortion. This holds true for cosmogenic and radiogenic noble gases as well as for the trapped gases considered here, with the proviso that He does show some depletion. The following discussion therefore rests heavily on data obtained for ilmenite separates, though some results for other minerals and for bulk samples will be included. In particular, separates from two lunar regolith samples having quite different antiquities have been studied. Soil 71501 has an antiquity probably not exceeding 100 Myr, whereas breccia 79035 was probably exposed to solar corpuscular radiation some 1.5 to 3 Gyr ago (Clayton and Thiemens 1980; Wieler et al. 1983).

Before describing the apparent secular changes in composition of the solar radiation recorded by the lunar surface, it is worth noting that the overall abundance of well-retained elements of solar origin (e.g., N and Xe), integrated over the depth of the regolith, exceeds by a factor of a few what could be delivered by the present solar-wind flux operating over 4 Gyr (Geiss 1973; Clayton and Thiemens 1980). More precise calculation of this excess is precluded by uncertainties in several factors, such as gas-retention efficiency, regolith depth and gas contents in the deep regolith (see, e.g., Pepin 1980), but it is difficult to avoid the conclusion that the mean flux over the past few Gyr was significantly greater than it is today (see also the chapter by Geiss and Bochsler).

Establishing a variation in the abundance of an element requires normalizing its abundance to that of some other element that should be well retained and of constant input proportion. The two normalizing elements most commonly used are N and Ar. Nitrogen has the advantage that it is the best retained of the solar-implanted elements, but it suffers from an uncertainty surrounding possible nonsolar N components, as discussed in Sec. III

TABLE III
Inferred Compositional Changes in Solar Gas Implanted in Lunar Regolith Samples

Parameter	Ancient Value	Recent Value	Magnitude of Change
$^4He/^{36}Ar$	1.1×10^4	5.1×10^3	50% decr.[a]
$^4He/^3He$	3.3×10^3	2.6×10^3	20% decr.
$^{20}Ne/^{22}Ne$	13.3	13.7	3% incr.
$^{132}Xe/^{36}Ar$	1.3×10^{-4}	5.8×10^{-5}	50% decr.
$^{15}N/^{14}N$	2.9×10^{-3}	4.4×10^{-3}	50% incr.

[a] Sense of change from 1.5–3 Gyr ago to present.

under "Nitrogen" below. Argon is apparently less well retained than N (see below) but the isotope ^{36}Ar is unambiguously solar and is therefore used here for normalization.

Analyses of lunar regolith samples reveal long-term changes in five measured quantities: these are summarized in Table III. Note that these variations do not *necessarily* have a solar origin, as discussed below. In the following section, we describe the results on which these conclusions have been based.

III. EXPERIMENTAL METHODS AND OBSERVATIONAL RESULTS

Helium and Neon

The principal data sets for these elements have been obtained using two different approaches to the analysis of mineral separates. Because the two sets of results are not wholly in agreement, leading in turn to some differences in interpretation, it is necessary to go into some detail about the approaches used and the manner in which different data sets have been interpreted.

In one analytical approach, the sample is heated to progressively higher temperatures, initially (from 300 to 800°C) in an atmosphere of pure O_2 (oxidation), and then in vacuum up to about 1500°C (pyrolysis) (Frick et al. 1988; Becker and Pepin 1989). In the other, the sample is etched (by an acid such as HF) in a series of steps, if possible until it is totally dissolved (Wieler et al. 1986; Benkert 1989). The two approaches are illustrated schematically in Fig. 1. In both cases, the gases released in each step are analyzed mass spectrometrically. Sequential etching is a better approach than simple pyrolysis would be for probing the depth distribution of an implanted gas, because of the danger of thermal redistribution of gas during the latter. Whether this remains a problem with the stepwise oxidation/pyrolysis approach is not clear; progressive oxidation of a mineral such as ilmenite can be thought of as a kind of chemical etching, albeit at elevated temperatures.

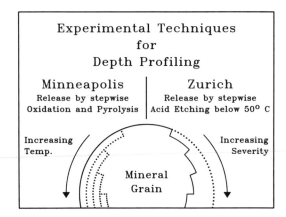

Fig. 1. Schematic representation of the two techniques used for revealing the depth distribution
of trapped noble-gas components in mineral separates from regolith samples.

Oxidation/pyrolysis yields two gas components during stepwise release.
One component, designated *LT*, is released at relatively low temperatures and
is therefore believed to be surficially sited. Its composition is calculated by
integrating over all steps up to a temperature that varies from element to
element but is around 700°C (Frick et al. 1988). The second component,
designated *HT*, is presumably more deeply sited within mineral grains, since
it is released at temperatures above about 700°C (Frick et al. 1988; Becker
and Pepin 1989). Elemental ratios for noble gases in the two components are
illustrated in Fig. 2, which shows that the LT component, except for Xe in
the ancient regolith breccia 79035, is characterized by an elemental composition identical, within uncertainties, to solar abundances (Cameron 1982).
By contrast, the *HT* component is heavily fractionated, with the heavier elements enriched relative to their solar proportions.

The two components are also clearly resolved in the Ne-isotopic data
(Fig. 3); note the sharp decrease in $^{20}Ne/^{22}Ne$ after about half of the Ne has
been released, for both the recent and the ancient sample, with the latter
giving values that are systematically a few percent below the former.

Calculations based on Fick's Law diffusion indicate that the observed
elemental and isotopic differences between the *LT* and *HT* components in
ilmenite could be accounted for by production of the latter from the former
by thermal diffusion (Frick et al. 1988). Such diffusion would cause some of
the initially surficial *LT* gas to be lost from grain surfaces, with the rest,
enriched in heavy elements and isotopes, diffusing into grain interiors to form
the *HT* component. Repeated cycles of implantation and diffusion could occur
to build up the concentrations of heavier noble gases in the *HT* component,
with a final incorporation of surface-implanted but undiffused gas producing
the observed *LT* component. Whether the diffusion coefficients needed to gen-

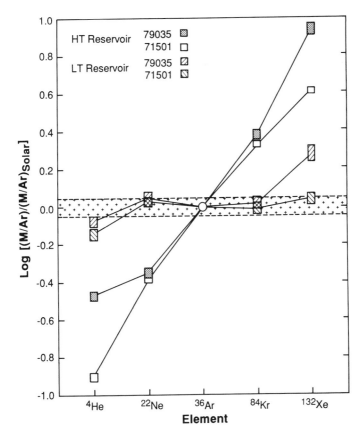

Fig. 2 Logarithms of abundance ratios of species M relative to ^{36}Ar (M/Ar) in the *HT* and *LT* reservoirs of ilmenites from "recent" soil 71501 and "ancient" breccia 79035, normalized to solar ratios (Cameron 1982). Solar values ± 10% are indicated by stippled band (figure from Becker and Pepin 1989).

erate the *HT* elemental distribution would have the appropriate values under lunar-regolith conditions is not known, but the isotopic fractionations observed by Frick et al. (1988) for Ne are attainable by diffusion. Achievement of the observed elemental composition of the *HT* component was found to require loss of 70 to 95% of the implanted solar-wind Xe, and correspondingly larger losses of the lighter noble gases (assuming that observed Xe/N ratios result from diffusion rather than additional sources of nitrogen [see below]. Otherwise, smaller Xe losses are allowed).

   The gases released by stepwise etching, characterized by the severity of the etch needed to release them, can also be viewed as comprising two isotopically distinct components. These resemble isotopically the components identified by stepwise oxidation: a readily released component termed *SW*, and a more resistant, apparently more deeply sited, component termed *SEP*

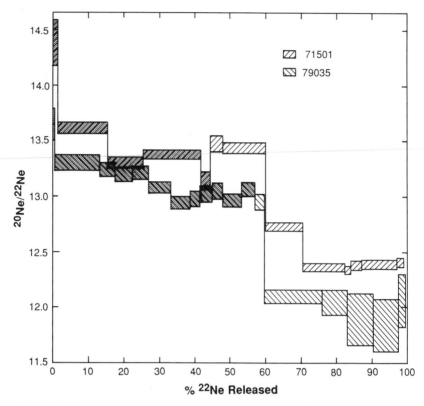

Fig. 3. $^{20}Ne/^{22}Ne$ ratio in each step as a function of the percent of total ^{22}Ne released by oxidation/ pyrolysis from 71501 and 79035 ilmenites. Shaded boxes indicate steps done by oxidation. Note the sharp division between LT and HT components at $\sim 60\%$ ^{22}Ne release, and that this division is not identical to that between oxidation and pyrolysis.

(Wieler et al. 1986; Benkert et al. 1988). These components are shown in Fig. 4, which is a three-isotope plot for Ne. In such a plot, mixtures of two components fall on the straight line joining the end-member compositions. With progressively more severe etching, the data in Fig. 4 follow first the straight line from SW to SEP. (This same trend *could* be generated by mass fractionation of either SW or SEP by a process such as diffusion.) The data then turn towards the third component, GCR, the cosmic-ray-produced spallogenic Ne mentioned earlier. Plots very similar to Fig. 4 have been obtained for different mineral separates as well as for different lunar samples (Etique et al. 1981; Wieler et al. 1983,1986; Nautiyal et al. 1986) and even for samples of gas-rich meteorites (Wieler et al. 1989a; Benkert 1989; Pedroni 1989), although the isotopic composition of Ne for the initially released gas is generally below the 13.7 found for SW in the ilmenites. The frequent oc-

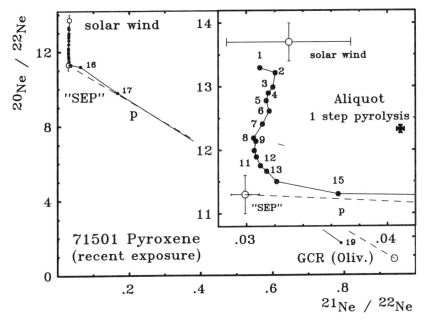

Fig. 4. Three-isotope plot for Ne released by stepwise etching of pyroxene from soil 71501. Note how the trend of the data resolve the three Ne-isotopic components; *SW, SEP* and spallation Ne, termed *GCR* (Oliv) in this plot (figure from Benkert 1989).

currence of Ne with a $^{20}Ne/^{22}Ne$ ratio close to 11.3, i.e., *SEP* in Fig. 4, is taken as an indication that it is a genuine and fairly widespread component.

The siting of the *SEP* component at a nominal depth of up to 30 μm (Etique 1981) and its different, possibly fractionated, isotopic composition relative to that of solar wind have led to the suggestion that it is implanted at energies above that of the solar wind (Wieler et al. 1986). This was originally proposed by Black (1972) on the basis of stepwise-heating analysis of gas-rich meteorites and lunar regolith samples. He concluded that a Ne-isotopic component, Ne-C, released at high temperatures, represents implanted solar-flare Ne. Additional support for such a view comes from the fact that *SEP* and Ne-C have $^{20}Ne/^{22}Ne$ ratios similar to that of contemporary solar flares, 11.3, 10.6 and 9.2 $^{+1.9}_{-1.8}$ (Mewaldt et al. 1984), respectively. However, using the isotopic composition of implanted and retained Ne measured in total-extraction experiments, on the one hand, and that of the *SW* and *SEP* components as deduced from stepwise-etching experiments, on the other, Wieler et al. (1986) calculated for various minerals that the amount of *SEP* Ne comprises up to 50% of the retained solar Ne. This contribution exceeds that of the average solar-flare flux by 2 to 4 orders of magnitude, even if allowance is made for various factors distorting the ratio *SEP/SW* for Ne in lunar-soil grains. Part of this discrepancy may be explained by a possible systematic difference in effective exposure times for a grain to solar-wind and solar-flare particles, respectively, resulting from possible coating of mineral grains by a sub-μm-sized dust layer (Housley 1980; Wieler et al. 1986). Still, the *SEP* flux required in order to supply the observed Ne contributions seems to be much higher than that of the actual solar flares (see also the chapter by Geiss and Bochsler). Note that *SEP* gases are detected in mineral grains to a nominal depth of $\sim$ 30 μm (Etique 1981), corresponding to implantation at energies ranging from just above those of solar-wind particles up to several MeV/nucleon, i.e., to just about the low end of the spectrum for solar-flare particles. A more appropriate nomenclature for what the Zurich group calls the *SEP* component may be "suprathermal ions." However, until their nature is resolved and a possible hiatus in nature and isotopic composition between suprathermal and solar-flare ions is established, we continue to apply the term *SEP* to the more deeply sited trapped noble-gas component in regolith samples.

The etching analyses that point to the existence of isotopically distinct *SW* and *SEP* components in the Ne also indicate the presence of two distinct isotopic compositions for both He and Ar (Benkert et al. 1988; Benkert 1989). For all three elements, the isotopic compositions of the pure *SW* and *SEP* components are characterized by the property that the fractional difference in isotopic ratio between the two components is approximately twice as large as the fractional mass difference between the respective isotopes. This proportionality must be explained by any model that attempts to relate the two components.

In the following, we compare the results obtained by the two analytical techniques. The stepwise release of the Ne isotopes and the Ne/Ar ratio is illustrated in Fig. 5 for the two different techniques applied to both ancient and recently exposed samples. Essentially identical ilmenite separates were used for the two approaches (aliquots in the case of 71501). The isotopic patterns of the Ne released by both techniques are quite similar (Fig. 5a,b). However, the systematic difference in ^{20}Ne/^{22}Ne ratio between ancient and recent samples observed in the oxidation/pyrolysis data (see also Fig. 3) is not revealed in the etch data. Also, there are real differences between the two approaches in the patterns for the Ne/Ar elemental ratios (Fig. 5c,d). The nature and cause of these differences are controversial at this time. The isotopic and elemental compositions of the different components postulated on the basis of the two different experimental approaches are given in Table IV and are shown schematically, for the Ne isotopes, in Fig. 6.

To what degree that differences between *SW* and *LT* and between *SEP* and *HT* in Table IV and Fig. 6 are significant or just due to the different analytical approaches used, is not clear. Nevertheless, both approaches have revealed two distinct populations of implanted gas in lunar ilmenite grains. There is also agreement that the *SW* and *LT* components both represent im-

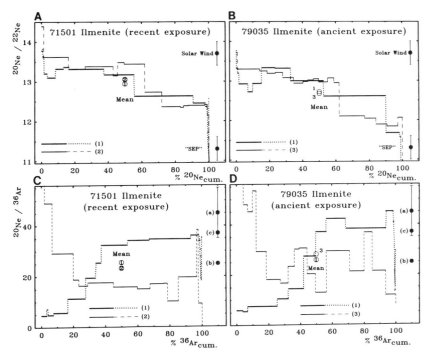

Fig. 5. (a) ^{20}Ne/^{22}Ne ratios during stepwise release from 71501 ilmenite. Progress of gas release is monitored by cumulative yield of ^{20}Ne. The point labeled "Solar Wind" is from Geiss et al. (1972); the point labeled "*SEP*" is defined by Fig. 4. Stepwise-etching data, curve (1), are from Benkert (1989); stepwise-oxidation/pyrolysis data, curve (2) are from Frick et al. (1988). Note the broad similarity between the two different approaches. (b) Analogous plot to (a) for ilmenite from 79035. Terminology as for (a), except that stepwise-oxidation/pyrolysis data, curve (3), are from Becker and Pepin (1989). (c) Ne/Ar ratios as function of cumulative ^{36}Ar release during stepwise analysis of 71501 ilmenite, corresponding to the Ne-isotopic data in (a). Points labeled as follows: (a) solar wind, from Geiss et al (1972); (b) solar abundance, from Cameron (1982); (c) solar abundance, from Anders and Grevesse (1989). Otherwise terminology is as in plot (a). Note the dramatic difference between the two approaches, with oxidation/pyrolysis yielding the solar ratio early in the release and subsequently decreasing, whereas stepwise etching starts with low values and then rises close to the solar value. (d) Analogous plot to (c) for ilmenite from 79035. Terminology as in (b) and (c). Systematic differences between the two analytical approaches are apparent here as in plot (c).

planted solar wind that is isotopically not severely fractionated. Controversy centers, however, on the elemental ratios of the *LT/SW* component and the nature and interrelationship of the *HT* and *SEP* components. Whereas integration of the low-temperature oxidation steps yields an *LT* component with solar elemental abundances, in the early release from stepwise etching, He and Ne are severely depleted with respect to Ar. Concerning the nature of the *SEP* and *HT* components, including the question of how they are related to each other, the etching experiments lead to postulation that the *SEP* compo-

TABLE IV
Data for Stepwise Release from Ilmenite Separates[a]

	^{4}He/^{36}Ar	^{4}He/^{3}He	^{22}Ne/^{36}Ar	^{20}Ne/^{22}Ne	^{36}Ar/^{38}Ar
Recent					
LT	14500	2516	3.39	13.46	5.492
SW	N.D.[b]	2170	N.D.[b]	13.70	5.48
HT	2520	3380	1.30	12.74	5.17[d]
SEP	*[c]	4340	*[c]	11.3	4.91
Ancient					
LT	16900	2790	3.53	13.12	5.445
SW	N.D.[b]	2170	N.D.[b]	13.70	5.48
HT	6820	3724	1.42	12.00	5.004[d]
SEP	*[c]	4340	*[c]	11.3	4.91

[a] LT and HT data derived by combustion (Becker and Pepin 1989); SW and SEP data obtained by on-line stepwise etching (Benkert et al. 1988 and unpublished data from Zurich). Note systematic differences between ancient and recent samples for LT/HT data, not apparent in SW/SEP data; see text.
[b] N.D. = not determined due to apparent fractionation on the lunar surface.
[c] * = detailed calculations not yet completed.
[d] Not corrected for spallation.

nent results from directly implanted solar particles with higher energies than the solar wind (see, e.g., Wieler et al. 1986). In this view, the HT component represents SEP material elementally fractionated at the elevated temperatures during the preceding stages of the extraction procedure itself. In the competing view, the fractionation of the HT component is interpreted as resulting from diffusion of initially LT (= SW) material from the surface into the interior of grains while in the lunar regolith (Frick et al. 1988). In this chapter, we present the conflicting views without favoring one over the other. In fact, it is likely that the truth lies between the two extremes, i.e., that deeply sited solar gas consists of both kinds of material, so that the key question is probably their relative proportions rather than whether they exist or not. (For further discussion of these observations, see the chapter by Geiss and Bochsler.)

Within the context of the two-component models described above, we are now in a position to assess the evidence for long-term changes in ^{4}He/^{36}Ar, ^{3}He/^{4}He and ^{20}Ne/^{22}Ne in trapped gas in the lunar regolith. The ease with which He is diffusively lost from mineral grains reduces the certainty with which changes in either its abundance or isotopic composition can be identified. However, the decrease of ~ 50% in ^{4}He/^{36}Ar and increase of ~ 20% in ^{3}He/^{4}He in the trapped gas appear to be real (Eberhardt et al. 1972; Kerridge 1980; Becker and Pepin 1989). The suggestion that ^{20}Ne/^{22}Ne in lunar trapped gas has increased by a few percent over time (Pepin 1980) has been recently confirmed by analysis of ilmenite separates (Becker and Pepin 1989). Note that the data from stepwise etching, while yielding the same

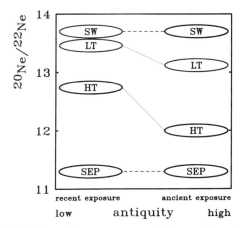

Fig. 6. Summary of the Ne-isotopic components revealed by stepwise oxidation/pyrolysis (*LT*, *HT*) and by stepwise etching (*SW*,*SEP*) of ilmenites, and how they have apparently evolved through time. The abscissa represents antiquity, calibrated according to our current best estimate. "Ancient" data are derived from 79035, "recent" data from 71501.

^{20}Ne/^{22}Ne ratio for the initial release from both recent and ancient ilmenites, show the same systematic shift in *mean* values found in the oxidation approach (Benkert et al. 1988; Benkert 1989).

The two different two-component models have led, inevitably, to two different interpretations of these changes in He and Ne. According to the *SW/SEP* view, the changes are due to a long-term decrease in the relative proportion of the *SEP* to *SW* component (Wieler et al. 1986). If the *SEP* component can indeed be identified as solar particles with energies between ~ 10 and 1000 keV/nucleon and having a flux exceeding that of present-day solar flares by several orders of magnitude, this implies that some 1.5 to 3.0 Gyr ago the ratio of the *SEP* flux to that of the solar wind was even about a factor of 2 larger than its present (unknown) value. Although there is no other evidence for such a change, it is not inherently implausible and is probably not inconsistent with the lunar-rock track record, though if a moderately enhanced flux of intermediate-energy particles implied a much larger solar-flare flux (i.e., if the rigidity of the solar energetic particle flux was low), there may be a conflict.

If, on the other hand, the *HT* component is generated from the *LT* component by diffusive fractionation on the lunar surface (see, e.g., Frick et al. 1988), the differences in the *LT* composition between ancient and recent ilmenites (Becker and Pepin 1989) suggest a secular change in elemental and isotopic composition of the solar wind itself, specifically a 50% decrease in ^{4}He/^{36}Ar, a 20% increase in ^{3}He/^{4}He and a 3% increase in ^{20}Ne/^{22}Ne. The consequences of these different explanations for issues of solar evolution are explored in Sec. IV.

TABLE V
Relative Abundances of Ar, Kr and Xe in the *LT* Component in Ilmenites from
a Relatively Recent Soil and an Ancient Regolith Breccia, Compared with
Estimated Solar Abundances

	^{36}Ar	^{84}Kr	^{132}Xe
71501 Ilmenite (Recent)	100	0.026	0.0019
79035 Ilmenite (Ancient)	100	0.028	0.0033[a]
Sun (Cameron 1982)	100	0.026	0.0017
(Anders and Grevesse 1989)	100	0.030	0.0015

[a] Note that the ancient sample has a Xe/Ar ratio significantly different from both the recent value and that of the Sun.

Xenon

Early analyses of some regolith breccias revealed somewhat higher Xe concentrations than found in soil samples of comparable maturity (Hintenberger et al. 1974). Subsequently, this difference was shown to be a function of antiquity (Kerridge 1980), rather than an intrinsic difference between loose soils and breccias, the breccias being in most cases more ancient than the soils. Recent analyses of ilmenite separates have confirmed the reality of this observation (Becker and Pepin 1989). Table V gives the Ar, Kr and Xe abundances, normalized to Ar, for two ilmenites of different antiquity, compared with recent estimates for those quantities in the Sun (Cameron 1982; Anders and Grevesse 1989). The data for the older ilmenite reveal a clear excess of Xe, evident in Fig. 2, in contrast to the consistent value for Kr/Ar (Becker and Pepin 1989). The magnitude of this change in Xe/Ar in ilmenites agrees well with that found earlier for bulk samples. A possible explanation of this effect is discussed below.

Nitrogen

Before assessing the N-isotopic record, it is necessary to discuss the abundance pattern of N in the lunar regolith, because that issue is important to interpretations of isotopic variations. The ratio ^{36}Ar/^{14}N in regolith samples is at least a factor of 13 less than the solar value estimated by Cameron (1982); see Fig. 7. Use of other estimates of solar abundances yields a similar conclusion. Three possible explanations may be considered. First, perhaps the solar Ar/N ratio has been severely overestimated. Given the large factor involved, this seems implausible. Also, direct measurements of Ar/N in solar flares (Breneman and Stone 1985) and the solar wind (Bochsler 1987) are in good agreement with the solar abundance estimates. The second possibility is that, assuming that ^{36}Ar is predominantly implanted solar-wind gas, an additional, nonsolar-wind source could be supplying N to the lunar surface

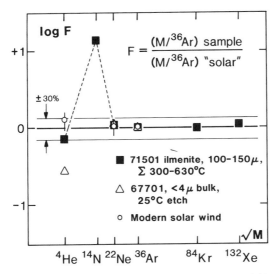

Fig. 7. Abundances of the noble gases and N in two samples believed to yield good estimates of the composition of gas recently implanted in the lunar surface. The abundances are normalized to the solar abundances of Cameron (1982) and are compared, for the light noble gases, with results from the Solar Wind Composition Experiment (Geiss et al. 1972). Note that N is enriched by more than an order of magnitude relative to the noble gases (figure from Frick et al. 1988).

(Geiss and Bochsler 1982; Signer et al. 1986). However, the magnitude of the effect requires that at least 90 to 95% of regolith N is of nonsolar-wind origin, which is difficult to reconcile with the close correlation observed between N content and measures of solar-wind exposure. Furthermore, the close correlation of C and N in the lunar regolith yields a C/N ratio only a factor of 2 below the solar value. Thus, the hypothetical nonsolar-wind N component must have a C content within a factor of 2 of the solar abundance, which seriously limits its possible source, given the very different chemical properties of C and N. Additionally, the secular change in isotopic composition of regolith N, shown in Table III and discussed below, requires the $^{15}N/^{14}N$ ratio of the putative nonsolar-wind component to have changed by at least 50%, severely constraining plausible sources for that N.

The third explanation is that Ar is lost from the regolith much more readily than N (Frick et al. 1988). The closely parallel release of Ar and N during stepwise release in the laboratory may argue against this possibility. However, in the absence of a plausible alternative, the possibility of substantial diffusive loss of Ar and the other noble gases under lunar conditions of temperature and time, even from the retentive ilmenite, cannot be excluded at this time (see, e.g., Frick et al. 1988). Clearly, this issue is not resolved.

The secular increase in $^{15}N/^{14}N$ in lunar regolith N has been abundantly

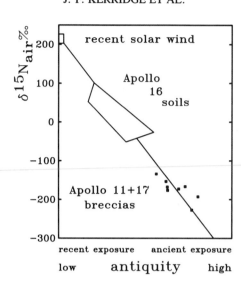

Fig. 8. Summary of dependence of $^{15}N/^{14}N$ on antiquity. The $^{15}N/^{14}N$ ratios are expressed as deviations in parts per thousand from the terrestrial-air value of 3.67×10^{-3}. The measure of antiquity represents a rather loose synthesis of the various lines of evidence available, but this is not reliably calibrated, particularly for the Apollo 16 samples, which have been positioned by bringing their internal trend line into coincidence with that for the other data. The timespan represented by the abscissa is believed to correspond to between 1.5 and 3 Gyr. Recent solar-wind datum from Becker (1989), Apollo 16 data from Kerridge (1980) and Apollo 11 and 17 from Clayton and Thiemens (1980).

confirmed and its reality is not in doubt (see, e.g., Clayton and Thiemens 1980; Kerridge 1975,1980,1989). It is illustrated in Fig. 8, which is based on a mixture of bulk-soil, mineral-separate and stepwise-release data. Discussions of the possible explanations for this effect are given by Kerridge et al. (1977), Kerridge (1980,1989) and Geiss and Bochsler (1982). Their conclusions are summarized in the next section.

IV. INTERPRETATION OF THE OBSERVED CHANGES IN SOLAR RADIATION

As will have become apparent in the foregoing discussion, none of the five compositional variations tabulated above currently has a robust explanation. Consequently, interpretation of those variations in terms of possible solar phenomena is constrained by uncertainties in the proximal causes of the observed effects, e.g., the nature of the two components in He and Ne. In relating the observations to the "Sun in Time," our discussion is therefore markedly model dependent. See also the chapter by Geiss and Bochsler for further discussion of several of the phenomena described here.

Helium and Neon

Model I

If the *LT* and *HT* components are both regarded as originating from solar-wind material, the systematic differences in $^4He/^{36}Ar$, $^3He/^4He$ and $^{20}Ne/^{22}Ne$ between ancient and recent samples must be due to a secular change in those ratios in the solar wind itself. In principle, this change could reflect either a change in the composition of the solar-wind reservoir or a change in fractionation during acceleration from that reservoir into the wind. A long-term increase in $^3He/^4He$ in the convective zone due to leakage from deeper levels in the Sun seems plausible (see, e.g., Bochsler and Geiss 1973; Schatzman and Maeder 1981; MacElroy and Manuel 1986), but suitable changes in $^4He/^{36}Ar$ and $^{20}Ne/^{22}Ne$ are more problematical, though they may be indicative of a "nonstandard" solar structure and/or intrasolar diffusion (MacElroy and Manuel 1986).

The mechanism responsible for acceleration into the solar wind of elements originating in the solar convective zone is not understood in all its details. Two regimes may be identified: transition of mostly neutral gas into the corona, which can lead to ionization and ion-neutral separation, followed by acceleration of coronal material into the solar wind (see, e.g., Geiss 1982). Coupling of minor ions to the proton flux during that acceleration apparently involves Coulomb collisions and can be described by a parameter termed the proton drag factor (*PDF*) (Geiss et al. 1970*b*; Geiss 1973). ^4He is characterized by a value for *PDF* that is anomalously small compared with those of most other species considered here; see Table VI. What this means is that, during acceleration by Coulomb drag as the corona expands into the solar wind, ^4He will be unusually sensitive to changes in acceleration conditions. If the solar-wind proton flux were to have decreased for some reason, ^4He could have become underrepresented in the wind relative to, for example,

TABLE VI
Proton Drag Factors of Some Minor Ions in the Solar Wind

Species	*PDF*[a]	Species	*PDF*[a]
$^3He^{2+}$	1.33	$^{22}Ne^{8+}$	1.83
$^4He^{2+}$	0.80	$^{36}Ar^{9+}$	1.31
$^{14}N^{6+}$	1.71	$^{84}Kr^{14+}$	1.28
$^{15}N^{6+}$	1.57	$^{132}Xe^{15+}$	0.91
$^{20}Ne^{8+}$	2.06		

[a]These factors are a measure of the ease with which each species would be accelerated into the solar wind by Coulomb drag. Their values have been calculated following Geiss et al. (1970) and Geiss (1973).

Ar. Thus, the observed secular decrease in $^4He/^{36}Ar$ might reflect a decrease with time in the intensity of the wind, i.e., the solar wind was apparently stronger in the past, as suggested above from other data (Geiss 1973; Clayton and Thiemens 1980).

However, there is a major problem with this interpretation of the He data. The value of *PDF* for ^3He is closely similar to that for ^{36}Ar so that $^3He/$ ^{36}Ar should not have changed significantly, whereas since the $^3He/^4He$ ratio has only increased by 20% while $^4He/^{36}Ar$ has decreased by a factor of 2, an independent change of about 40% in ^3He must be postulated (Becker and Pepin 1989). (For further discussion of the long-term change in the $^3He/^4He$ ratio, see the chapter by Geiss and Bochsler.) In addition, although such an interpretation may also explain the Xe data (see below), there is no obvious way in which it can accommodate the change in Ne isotopes, because the *PDF* values for those nuclides are closely similar to each other (Table VI).

Model II

If the *SW* and *SEP* components are interpreted as due to implantation of solar wind and solar energetic particles of intermediate energy, respectively, the elemental and isotopic patterns described earlier lead to the following implications for the evolution of the solar radiation. First, for neither *SW* nor *SEP* have the isotopic compositions of Ne and Ar changed significantly over 1.5 to 3 Gyr. This would mean that there is no evidence for a long-term change in those values for the solar convective zone. For He, the etching data are compatible with a secular increase of the $^3He/^4He$ ratio in the solar wind by ~ 10% over the same period. These observations have the further implication that the large short-term variations observed for $^3He/^4He$ in both solar wind and solar flares, and presumably therefore also in *SEP*, are essentially averaged out over the time scale represented by irradiation of a single sample, possibly over ~ 25 Myr.

The apparent long-term change in the proportion of *SEP* to *SW*, with the former declining by ~ 50% over 1.5 to 3 Gyr, clearly implies a change in the acceleration conditions for solar charged particles, but without more detailed knowledge of the relevant acceleration mechanisms, those implications cannot be explored further at this time. The systematic, apparently mass-dependent isotopic differences between *SW* and *SEP* also presumably carry information about the nature of the acceleration mechanism(s), but here too the implications are not yet clear. (For further discussion of fractionation in the solar charged-particle radiation see Fisk [1978] and Mullan [1983].) It is interesting that the elemental ratios for He, Ne and Ar in the *SEP* component appear to be similar to those in the solar wind, as measured by the Apollo Solar Wind Composition Experiment. If this turns out to be the case, it means that the mechanism responsible for the isotopic fractionation between *SW* and *SEP* does not fractionate simply on the basis of mass dependence (see, e.g.,

Mullan 1983). With further study, this could provide a valuable clue to the acceleration mechanism(s) of the solar radiation.

Xenon

Like ^{4}He, all isotopes of Xe are characterized by unusually low *PDF* values; see Table VI. Consequently, a decrease in solar-wind flux below the level needed for complete acceleration of Xe into the wind could lead to a long-term decrease in the Xe/^{36}Ar ratio in the wind, analogous to that postulated above for ^{4}He/^{36}Ar. However, in addition to the problem of accounting for the magnitude of the observed change in ^{3}He/^{4}He (see above), this interpretation runs into difficulty in explaining the actual Xe/Ar ratios observed. The Xe/Ar ratio measured for recently exposed samples is very close to the solar ratio, estimated on the basis of nuclear systematics (see, e.g., Cameron 1982), whereas the value measured in more ancient samples, that should be closer to the true value, is a markedly worse match; see Table V and Fig. 2 (Becker and Pepin 1989). The shortcomings of this simple explanation may be suggesting that acceleration of the solar wind involves a more complex mechanism than simply Coulomb drag. No other explanation has been advanced for the change in Xe/Ar ratio.

Nitrogen

Many explanations have been advanced for the secular increase in ^{15}N/^{14}N: these have been assessed in detail elsewhere (Kerridge et al. 1977; Kerridge 1980, 1989; Geiss and Bochsler 1982), with the conclusion that none are consistent with the full range of observational data. In brief, this conclusion is based on the following arguments. Mechanisms capable of changing isotopic compositions fall into one or another of three basic categories: mass-dependent fractionation; nuclear transformations; and mixing of nucleosynthetically distinct components. In the case of lunar-regolith N, all mass-fractionation mechanisms may be ruled out because of the lack of comparable fractionations in elements that are closely associated with N on the lunar surface and which are therefore believed to share a common origin with N. This argument is particularly strong for C, which correlates very closely with N, but also applies to the noble gases. Note that the apparent secular change in ^{20}Ne/^{22}Ne is much smaller than that in ^{15}N/^{14}N, reflects a decrease, rather than an increase, in the heavier isotope, and fails to correlate with the change in ^{15}N/^{14}N for the relatively few samples for which appropriate data are available.

Explanations based on production of ^{15}N by nuclear reactions in the lunar surface yield an effect of opposite sign to that observed. Hypothetical nuclear production of ^{15}N, or destruction of ^{14}N, in the Sun leads to predictions of either far more intense solar activity than is observed, either directly at this time or indirectly via nuclear-reaction effects in lunar rocks, or concomitant production of boron in the Sun far in excess of that observed spectroscopically.

Finally, explanations based on admixture of a second, i.e., nonsolar-wind, component need to invoke a reservoir of material either heavily enriched or heavily depleted in ^{15}N. The former cannot be ruled out on the basis of lunar-regolith data but are sufficiently rare in the Galaxy to make explanations based on admixture of such a component quite implausible. On the other hand, the latter type of material is common in the Galaxy (hence the far higher abundance of ^{14}N compared with ^{15}N) but the presence of significant proportions of such a component in the lunar regolith is difficult to reconcile with N-abundance systematics (Kerridge 1989).

In a nutshell, all explanations advanced so far either have low plausibility or are in actual conflict with observation. Although most recent attempts at explanation have invoked processes taking place on the lunar surface, the weight of evidence suggests a phenomenon taking place in the source region of the solar wind, despite lack of a plausible mechanism. (For a different view, see the chapter by Geiss and Bochsler.)

Summary

In Table VII, we combine a summary of the interpretations discussed above with the observations given in Table III. The foregoing discussion sug-

TABLE VII
Summary of Inferred Compositional Changes in Solar Gas Implanted in Lunar Regolith Samples and Possible Explanations

Parameter	Magnitude of Change	Model Interpretation	Reference[a]
$^4He/^{36}Ar$	50% decr.[b]	I. Change in SWC[c,d]	(1, 2)
		II. Decr. in SEP/SW	(3)
$^4He/^3He$	20% decr.	I. Change in SWC[d]	(1, 2, 4)
		II. Decr. in SEP/SW[e]	(3)
$^{15}N/^{14}N$	50% incr.	I. Change in SWC	(4)
		II. Non-SW	
		component	(5)
$^{20}Ne/^{22}Ne$	3% incr.	I. Change in SWC	(2)
		II. Decr. in SEP/SW	(3)
$^{132}Xe/^{36}Ar$	50% decr.	I. Change in SWC[d]	(2, 4)

[a] Representative references; many other relevant publications exist; (1) Eberhardt et al. (1972); (2) Becker and Pepin (1989); (3) Benkert, et al. (1988); (4) Kerridge (1980); (5) Geiss and Bochsler (1982).
[b] Sense of change from ~3 Gyr ago to present.
[c] SWC = solar wind composition.
[d] Possibly caused by change in acceleration conditions brought about by decrease in solar-wind intensity.
[e] Also possible 10% change in SWC (Benkert 1989).

gests that implanted solar species provide information concerning the following changes in solar corpuscular radiation during the past 3 or so Gyr: (1) an apparent decrease in absolute flux of the solar wind; (2) possible changes in the acceleration conditions of the solar wind and/or in the relative flux of solar energetic particles *vis-à-vis* the solar wind; and (3) a possible change in isotopic composition of the solar-wind reservoir. We conclude by considering some ways in which these issues might be clarified in the future.

V. FUTURE DIRECTIONS

A quantitative interpretation of the effects described above must eventually be consistent with the time scale over which the phenomena have taken place. As noted earlier, we currently lack a reliable measure of antiquity, so that development of such a measure has a high priority. Probably the most straightforward route to such information is via the compaction age of regolith breccias, which would in at least some cases supply useful limits on the antiquity of irradiation of the constituent grains. This goal has been long sought in both lunar and meteorite studies, with little success so far, but several possible approaches would probably repay further effort, most notably those based on either fission tracks or Xe isotopes from ^{244}Pu, or ^{136}Xe from induced fission of ^{235}U.

At the other end of the time scale, spacecraft measurements of ^{15}N/^{14}N in the contemporary solar wind will help to constrain theories for the long-term change in that ratio, and similar measurements for the flux and isotopic composition of He and Ne in solar radiation in the range 10 keV/nucleon to 1 MeV/nucleon will demonstrate whether such particles can be responsible for the *SEP* component.

A spacecraft that will probably generate data of value to these issues is Galileo as it should, among other measurements, determine the isotopic composition of N and Ne in the atmosphere of Jupiter. If the primitive solar system was homogeneous for those isotopes, a qualification that should be clarified by the general run of Galileo isotopic data, the Jovian values can become our best estimates of the average solar-system values, replacing estimates based on meteoritic data that are subject to unknown fractionations and mixtures of different nucleogenetic components. If the Sun formed initially from average solar-system material, the Galileo data would then establish the zero-age end point for any secular changes observed in the solar corpuscular radiation, assuming that the observations pertain directly to the reservoir and are not perturbed by unaccounted-for fractionations, or nuclear processes in the early Sun.

It should be possible to resolve the apparent experimental discrepancies between the etching and oxidation approaches, if not the differences in their interpretation, by means of hybrid experiments. Thus, partial etching followed by stepwise oxidation/pyrolysis, or the reverse, should reveal whether the *HT* component is formed by diffusion on the Moon or in the laboratory. Also, extending the stepwise etching experiments to include the heavy noble gases and N would be most informative about systematic differences between the two approaches.

Finally, our understanding of all five apparent long-term effects will be greatly aided by further theoretical modeling of such issues as the evolution of solar structure, fractionation during propagation of both solar wind and solar energetic particles, and evolution of solar activity.

Acknowledgments.   Financial support from the National Aeronautics and Space Administration for JFK, RHB and ROP and from the Swiss National Science Foundation for PS and RW are gratefully acknowledged.

DO METEORITES CONTAIN IRRADIATION RECORDS FROM EXPOSURE TO AN ENHANCED-ACTIVITY SUN?

M. W. CAFFEE
Lawrence Livermore National Laboratory

C. M. HOHENBERG, R. H. NICHOLS JR., C. T. OLINGER
McDonnell Center for the Space Sciences

R. WIELER, A. PEDRONI, P. SIGNER
Institut für Kristallographie und Petrographie

T. D. SWINDLE
University of Arizona
and
J. N. GOSWAMI
Physical Research Laboratory

Astronomical observations of young solar-mass stars indicate that, while en route to the main sequence, they go through a period of increased activity, the so-called "T Tauri" phase. There is no apparent reason to conclude that the Sun is any different, so it is likely that the primitive Sun may have been characterized by increased solar-flare activity associated with such a phase. Even before these instabilities were observed, some researchers proposed a local (solar-system) origin for a number of isotopic effects observed in meteorites. However, these models were not widely accepted at that time because they were not consistent with the observed isotopic structure of other elements sensitive to the required energetic particle fluence. Discovery of solar-flare tracks in some of the grains of gas-rich meteorites indicates that ancient irradiation records can be preserved and the scarcity of such grains suggests that only about 1% of

[413]

solar-system material has been involved in pre-compaction irradiation, thus removing some of the early objections. Recently, an active early Sun has been proposed to explain the quantity of spallogenic ^{21}Ne observed in meteoritic grains containing solar-flare tracks (reflecting pre-compaction irradiation). This hypothesis has been challenged by other investigators who would explain such effects by conventional galactic cosmic-ray irradiation of near-surface material on the meteorite's parent body. We review the evidence both for and against the active early solar origin of these effects in gas-rich meteorites.

There is some astronomical evidence (see, for example, chapters by Walter and Barry, Feigelson et al. and Bertout et al.) that suggests that many young stars (e.g., T Tauri) are more active than their older counterparts. In fact, such activity may be a typical part of a solar-type star's early history. There is no reason to believe that the Sun is atypical, but neither is there any firm evidence that it went through such an active phase. Studies of meteorites, some of which contain material that has remained nearly unprocessed for more than 4.5 Gyr may be the best hope for obtaining such firm evidence. Over the years, an active early Sun has been invoked to explain a number of isotopic and elemental effects in meteorites and planetary atmospheres. However, there is not a single case where an effect has been demonstrated to require an active early Sun. An effect that has generated considerable attention recently has been the observation of particle irradiation effects in meteorites that require either an active early Sun or rather long exposure histories.

We focus here on "pre-compaction" irradiation effects, effects recorded before the final consolidation (compaction) of the meteorite parent body, and the evidence that these pre-compaction irradiation effects do (or do not) require an active early Sun. However, we also review several other effects attributed to such early activity. For some of these effects, specific explanations involving the Sun have failed, or other explanations now seem more likely. For others, the observations remained unexplained and may be providing as yet poorly understood clues to the Sun's early behavior.

I. ISOTOPIC EFFECTS ATTRIBUTED TO AN ACTIVE EARLY SUN

Fowler et al. (1962) proposed that the Li, Be and B abundances in the solar system might be explained by an irradiation of 3.5×10^{21} neutron cm^{-2}. They also suggested that such an irradiation might produce ^{129}I (16 Myr half-life), whose presence in meteorites had been demonstrated by Jeffery and Reynolds (1961). To produce the neutrons, they proposed an irradiation of meter-sized icy planetesimals. This model generated numerous searches for isotopic variations that might be associated with incomplete mixing of irradiated and unirradiated material. The lack of observed variations led to an upper limit of the variation in mixing fractions of about 1% (see review by Reynolds 1967).

Later, with the discovery of excesses of the stable nuclides ^{22}Ne and ^{16}O and the discovery of the presence of 0.7 Myr half-life ^{26}Al, several models were proposed that might have produced one or more of these excesses within the solar system by particle irradiation (see, e.g., Herzog 1972; Audouze et al. 1976; Heymann and Dziczkaniec 1976; Clayton et al. 1977; Lee 1978) or ultraviolet irradiation (Thiemens and Heidenreich 1983). Typically, the particle irradiation models required enhanced fluxes of low-energy (10 MeV or less) protons. However, to produce only the observed effects, rather specific mass distributions or energy spectra were often required. The presence of ^{22}Ne anomalies is now generally attributed to the incorporation of unmixed interstellar phases (Anders 1988), while the discovery of ^{26}Al in the interstellar medium (Mahoney et al. 1984) has lessened the need for a local source, so most of these models, too, have fallen by the wayside. For ^{16}O, incorporation of unmixed interstellar phases is a satisfactory explanation, but the possibility of local production, particularly by ultraviolet irradiation, cannot be ruled out.

More recently, Kaiser and Wasserburg (1983) invoked an active early Sun to explain their data on Pd and Ag isotopes in iron meteorites. They found that excesses of ^{107}Ag correlated with the abundance of ^{108}Pd, and plausibly attributed the excess ^{107}Ag to the decay of short-lived (3 Myr half-life) ^{107}Pd, just as ^{129}Xe and ^{26}Mg excesses had been attributed to the decay of ^{129}I and ^{26}Al. Perhaps more importantly, they found that in one group of iron meteorites (IVB), the correlation line on a plot of ^{107}Ag/^{109}Ag vs ^{108}Pd/^{109}Ag had an ordinate intercept near zero rather than at the solar value of ^{107}Ag/^{109}Ag, suggesting that the ^{107}Pd and ^{109}Ag were synthesized together, and not subsequently mixed with "normal" solar system material. As these meteorites are believed to be cores of differentiated (i.e., molten) bodies, Kaiser and Wasserburg argued that the lack of mixing could only be explained if production was after differentiation. Wasserburg (1985) has also pointed out that most extinct radionuclides were apparently present with abundances relative to nearby stable isotopes (e.g., ^{129}I/^{127}I, ^{26}Al/^{27}Al, ^{107}Pd/^{108}Pd) of 10^{-4}, despite the fact that their respective half-lives vary by a factor of 25. Having about 10^{-4} of the solar-system material synthesized near the time of condensation of solid matter in the solar system would be a simple way to explain the remarkable constancy of that number. Although no specific model has yet been proposed, intense irradiation of a small fraction of solar system material seems to be consistent with observations of substantial spallation effects in a small fraction of meteoritic grains.

If the Sun went through an active early phase, an enhanced output in the extreme ultraviolet (less than 1000 Å wavelength) could provide prodigious amounts of energy to young planetary atmospheres, leading to rapid escape of primordial hydrogen atmospheres. It has been proposed that hydrodynamic effects might help explain the apparent elemental and isotopic fractionation of heavier atmospheric species as they accompany the escaping hydrogen, entrapped in the outward hydrodynamic flow. Sekiya et al. (1980) suggested that this mechanism could explain the abundances of the noble gases in the terrestrial atmosphere. Hunten et al. (1987), Zahnle and Kasting (1986) and Sasaki and Nakazawa (1988) have developed promising models to explain various noble-gas isotopic features in the atmospheres of Earth, Venus and Mars. Although a complete treatment of all relevant features has not been achieved, the difficulties with these models usually lie in simultaneously satisfying all the observed isotopic features (Zahnle 1988). If the timing of nebular dissipation, as well as the timing and degree of planetary degassing are right, the size of the enhancement in the solar extreme ultraviolet estimated from astronomical observations (Zahnle and Walker 1982) seems adequate to drive the necessary processes.

II. METEORITES AS DIRECT IRRADIATION MONITORS

Meteorites have been exposed to the energetic particle environment in the solar system at various stages of their evolution. The energetic particle environment encompasses species originating from the Sun (solar flares, often the source of solar cosmic rays, SCR) as well as from outside our solar system (galactic cosmic rays) (cf. Caffee et al. 1988). Can we use the irradiation effects observed in some meteorites to gauge the activity of the Sun 4.5 Gyr ago? To investigate such activity we need samples meeting the following criteria:

1. Identifiable constituents of meteorites must have been exposed to energetic particles from the Sun some 4.5 Gyr ago;
2. This material remained unaltered ever since;
3. This material remained buried deeply enough (at least a few meters) within the meteorite parent body to shield it from further exposure to energetic particles.

Meteoritic material of some sort can certainly meet these criteria as it is clear that, while most material has never been exposed to energetic particles except during the recent (normally short) cosmic-ray exposure age, the gas-rich meteorites (cf. Goswami et al. 1984, and references therein and the chapter by Goswami) clearly contain a record of pre-compaction exposure to energetic particles.

The term "gas-rich" is derived from the occurrence of solar-wind-implanted noble gases in some material within some meteorites (Gerling and Levskii 1956; Suess et al. 1964; Signer 1964; Eberhardt et al. 1965; Wänke 1965). The gas-rich meteorites are breccias, i.e., they consist of broken rock fragments, devoid of solar wind gases, cemented together by a fine-grained gas-rich matrix (Wlotzka 1963; Fredriksson and Keil 1963). Matrix and solar-gas-free inclusions are usually easily distinguished by their dark and light colors, respectively. After the discovery of solar-wind-implanted noble gases came the discovery of solar-flare VH (very heavy, mostly Fe group) ion tracks in mineral grains from gas-rich meteorites (Lal and Rajan 1969; Pellas et al. 1969). The short range of solar-flare VH particles (and clearly the solar wind as well) requires that those grains containing such effects must have been exposed to the Sun with essentially no shielding, demonstrating pre-compaction exposure.

In some instances fragments exist that could not have originated on the parent body of the meteorite, but, rather, are constituents from other solar system objects (Wilkening 1973,1977; Pedroni et al. 1988; cf. Bunch and Rajan 1988). The achondrite Kapoeta, for example, contains carbonaceous clasts that were evidently incorporated in the parent body regolith as a result of bombardment of the regolith by foreign material. Today it is clear that the gas-rich meteorites formed in a parent body regolith produced by bombardment processes, in a setting where impact comminution, solar wind implantation and exposure to solar and galactic cosmic rays were concurrent processes. For this reason, gas-rich meteorites are frequently compared to the lunar regolith. It should be emphasized that gas-rich meteorites are not exclusive to any particular class of meteorite. They are found among both achondrites and chondrites, and in the latter, they comprise both the ordinary and carbonaceous chondrites, demonstrating that records of near-surface processes are preserved on all types of stony meteorites.

Material from the gas-rich meteorites has certainly been exposed to energetic particles from the Sun at some time in their evolution prior to compaction into the final object, but this exposure need not have occurred 4.5 Gyr ago. As discussed later, the timing of the regolith development on gas-rich meteorite parent bodies and the duration of the active regolith, defining the pre-compaction period, are of critical importance. If for instance the regoliths of these parent bodies developed 2 Gyr ago, then we could safely conclude that these meteorites contain little information about the state of an ancient Sun 4.5 Gyr ago. On the other hand, if the regoliths themselves are ancient, then it is plausible that they do contain information about the activity of the ancient Sun.

III. OBSERVATIONS

In gas-rich meteorites only a small but variable fraction of the grains contains solar-flare VH tracks. Using the presence of these tracks as a criterion for exposure to energetic particles during the pre-compaction era, Caffee et al. (1983,1987) found that samples consisting of grains containing tracks have considerably more ^{21}Ne produced from energetic-particle spallation reactions than samples of track-free grains. We operationally define this "excess spallogenic ^{21}Ne" as that quantity of ^{21}Ne in excess of that which could have been produced by galactic cosmic-ray (GCR) irradiation during the time the meteorite spent as a small object, called the cosmic-ray exposure age (typically a few Myr). This episode of the cosmic-ray exposure began at disruption of a larger host body, in which the meteoritic material is assumed to be totally shielded; it is terminated by Earth impact.

The initial studies by Caffee et al. (1983) were done on the carbonaceous chondrite Murchison and the howardite Kapoeta. They were subsequently extended to the carbonaceous chondrites Murray and Cold Bokkeveld as well as the ordinary chondrites Fayetteville and Weston (Caffee et al. 1986,1987). The magnitude of the excess spallogenic ^{21}Ne in some of the track-rich grains corresponds to minimum GCR exposure times in a regolith of hundreds of Myr. This result was somewhat surprising since the conventional wisdom on meteorite regoliths is that, in comparison to the lunar regolith, meteorite regoliths are immature, i.e., the upper portions were exposed to irradiation effects, micrometeorite bombardment, etc. for a short period of time (cf. Goswami et al. 1984). Evidence of this comes from the flat distribution of grain sizes in gas-rich meteorites in comparison to the lunar soil, the lack of impact glasses, the scarcity of micrometeorite craters (Brownlee and Rajan 1973; Goswami et al. 1976), and the paucity of grains containing solar-flare tracks (Goswami et al. 1976). Furthermore, in most meteorites, spallogenic stable isotopes (for instance ^{21}Ne) are dominated by those produced most recently during the meteorite's transit to Earth. Those constituents having detectable pre-compaction exposures are the exception, whereas in the lunar regolith, virtually all grains contain large quantities of pre-compaction spallogenic isotopes (Caffee et al. 1986).

Several models of regolith growth on asteroidal bodies (Anders 1975; Anders 1978; Housen et al. 1979) proposed that regoliths on asteroidal bodies accumulated on a rapid time scale, implying that exposure ages should be of the order of 1 Myr (although Langevin and Maurette [1980] predict longer ages under some circumstances). It has also been observed that only those grains containing solar-flare VH tracks contain any appreciable excess spallogenic ^{21}Ne (Caffee et al. 1983). More recent data confirm the association between excess spallogenic ^{21}Ne and solar-flare VH tracks (see Hohenberg et al. 1990, Fig. 3). The amount of excess spallogenic ^{21}Ne observed in these grain sets and the association of solar-flare tracks with excess spallogenic ^{21}Ne led Caffee et al. (1983) to hypothesize that the excess spallogenic ^{21}Ne came from interactions with solar-flare protons rather than GCR protons. In the current energetic particle environment, GCR protons, and their secondaries, produce far more spallation products than SCR protons. Therefore, an additional requirement of the solar-flare bombardment model is that the proton flux of the ancient Sun be several orders of magnitude higher than is observed in the contemporary Sun (see, e.g., the chapter by Ramaty and Simnett). Since this was proposed, a lively debate has developed, accompanied by more studies of various components of these gas-rich objects.

Wieler et al. (1989a) and Pedroni et al. (1988) studied inclusions and matrix material from Fayetteville and Kapoeta. They determined spallogenic

^{21}Ne and solar-wind-implanted ^{36}Ar in bulk samples, both from matrix and inclusions, rather than concentrating on individual grains (Fig. 1). As in the single-grain studies, the dark matrix material (showing solar exposure effects), as well as several inclusions, contain more ^{21}Ne than can be accounted for by recent exposure to GCR. An upper limit for the GCR contribution (during the conventional exposure age) is given for each of the two meteorites by the component with the lowest spallation content (near the ordinate). One observes a reasonable linear correlation, in the matrix samples, between excess spallogenic ^{21}Ne and solar-wind-implanted noble gases, interpreted as due to a concurrent exposure of grains to solar wind and GCR in a well-mixed regolith. The line labeled "lunar mare" points toward bulk samples from the mare regolith. If one scales the spallogenic neon for differences in target-element chemistry and the higher flux of solar particles at 1 AU (compared to 2 to 3 AU for meteorites), the slope of the lunar line is within a factor of 2 equal to those of the best-fit lines through the points of the meteoritic matrix samples. Because the spallogenic ^{21}Ne produced in the lunar regolith is unquestionably produced by GCR, it is tempting to conclude that the

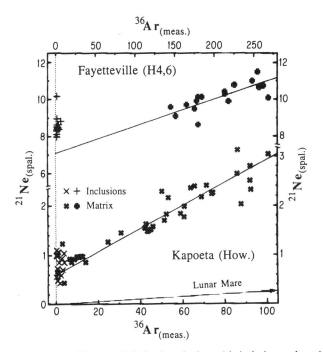

Fig. 1. Spallogenic ^{21}Ne vs ^{36}Ar (essentially implanted solar gas) in inclusions and matrix samples of the gas-rich meteorites Fayetteville and Kapoeta. Note the change in the scales on both axes; they are such that the slopes of the best-fit lines through the matrix points are directly comparable. The lower line, representing the lunar regolith, is drawn between the lunar mare measurements and the origin as all lunar regolith exposure is pre-compaction.

excess spallogenic ^{21}Ne in Fayetteville and Kapoeta must also be from GCR irradiation. Pedroni et al. (1988) further observe that carbonaceous and other "exotic" clasts in Kapoeta contain excess spallogenic ^{21}Ne. It is likely that these aliens acquired this excess as small bodies in space during GCR irradiation of up to several 10 Myr before their incorporation in the Kapoeta parent body regolith.

Taking a different approach to the problem, Rajan and Lugmair (1988) observed excess ^{150}Sm in samples of Kapoeta and Fayetteville. In this case the excess comes from neutron capture on ^{149}Sm. These secondary neutrons, in turn, are the result of high-energy proton interactions. Because current solar-flare protons have much lower energies than GCR protons and cannot produce a large secondary cascade, a plausible explanation of these excesses is exposure of this material in parent body regoliths to GCR. Rajan and Lugmair calculated an exposure duration of at least 20 Myr similar to that observed by Wieler et al. (1989a).

Padia and Rao (1989) have also studied ^{21}Ne in Fayetteville and Kapoeta, in mineral separates freed of solar wind contributions by chemically etching away the surfaces of the grains. Ideally, this leaves three components: solar-flare-implanted neon, GCR spallation products and neon produced by spallation from solar cosmic-ray (SCR) protons. For the pyroxenes they studied, the isotopic structure of the GCR and solar spallation components are too similar to be resolved, so an amount of GCR spallation expected during the conventional exposure age (determined from the light material, which is free of solar effects) is assumed. They conclude that there must be a considerable quantity of spallogenic ^{21}Ne produced by solar-flare protons, requiring an irradiation by a more active Sun. If the exposure of those grains occurred over a 10^5-yr period, then it would require an energetic proton flux 10^3 times higher than that of the contemporary Sun. This is consistent with the interpretation of the St. Louis group, based on studies of individual grains. The Zurich group (Pedroni et al. 1988; Wieler et al. 1989a) argues that the data of Padia and Rao (1989) are also consistent with an extended conventional regolith irradiation. So, although the data sets of Padia and Rao (1989) and those of Wieler et al. (1989a) are compatible, their respective interpretations are in direct conflict.

The St. Louis group has recently extended its studies of Murchison and has also measured the spallogenic ^{21}Ne in Murray and Cold Bokkeveld. However, unlike the previous experiments, Hohenberg et al. (1990) measured the spallogenic ^{21}Ne from single grains rather than in bulk material, mineral separates or subsets of selected grains, as was done in the earlier work. This process eliminates any averaging of exposure ages inherent in the analysis of bulk material or sets of grains. Figure 2 shows the single-grain results for Murchison and Murray, with the longest regolith exposure ages being about 150 Myr for each. Because this age is based on peak production rates for ^{21}Ne, it is a strict lower limit for the duration of regolith activity of the parent

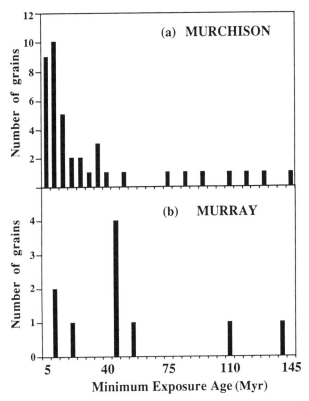

Fig. 2. Model 2π exposure ages for (a) irradiated Murchison olivines and (b) irradiated Murray olivines. Maximum ^{21}Ne production rates are assumed for the entire exposure, so these are minimum GCR exposure ages (figure from Hohenberg et al. 1990).

body. Using the lunar regolith as a model, a more realistic age of 300 Myr, is estimated for the duration of regolith activity on the parent body. As before, there is a good correlation between the presence of excess spallogenic ^{21}Ne and the presence of solar flare VH ion tracks (Fig. 3), although the relative magnitude of the two effects differs (Hohenberg et al. 1990).

IV. DISCUSSION

In the light of these recent measurements, can we reconcile the results of the various groups and is there yet any firm evidence for an active early Sun? For Fayetteville, the noble-gas results obtained by the Zurich group and the Sm isotopic results of Rajan and Lugmair (1988) can be explained by an exposure of the parent-body regolith to GCR for about 20 Myr. Similarly long parent-body exposures had been inferred already by Lorin and Pellas (1979) for the chondrite Djermaia. A re-evaluation of noble-gas data of the chondrite

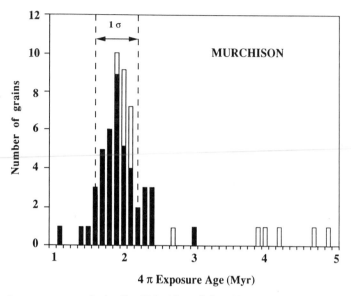

Fig. 3. 4π exposure ages of unirradiated Murchison olivines (closed boxes) and irradiated Murchison olivines with 2π ages < 10 Myr (open boxes). Note that there is little overlap between the irradiated and unirradiated sets (figure from Hohenberg et al. 1990).

Weston and St. Mesmin (Schultz et al. 1972; Schultz and Signer 1977) leads to similarly extended parent body exposures of these objects (cf. Wieler et al. 1989*b*). Also, Schultz and Signer (1977) found clasts in St. Mesmin that are only 1.4 Gyr old. Taken together, these studies indicate that regolith activity on the ordinary chondrite parent body was by no means limited to the early solar system. The radiometric clocks of clasts from St. Mesmin must have been reset by shock melting, associated with collisional bombardment of this parent body. Clasts as young as 3.5 Gyr are found in Kapoeta, suggesting that regolith activity lasted at least 1 Gyr, perhaps even more than 2 Gyr (Huneke et al. 1977). Some constituents of such parent bodies could have been exposed to GCR for long periods, perhaps even the 600 to 700 Myr required by one Kapoeta grain studied by the St. Louis group (Olinger et al. 1988). The most straightforward conclusion is that the exposure of the ordinary chondrite and howardite parent bodies to galactic cosmic rays spanned a period lasting tens of Myr. This suggests that models predicting shorter regolith exposures for these parent bodies need revision. Langevin and Maurette (1980) produced a model that predicted 10-Myr exposures for regoliths on "strong" asteroids, but this result was not emphasized at the time, due in part to the lack of experimental evidence. Thus for the ordinary chondrites and howardites, it seems unnecessary to invoke exposure to an enhanced activity early Sun (for a dissenting view on Fayetteville, the reader is referred to Padia and Rao [1989]).

The situation for carbonaceous chondrites is quite different. By all measures, the regoliths of the carbonaceous chondrite parent bodies, such as Murchison, are quite immature. Although there is evidence for regolith activity in that solar-flare tracks are observed in a few percent of the mineral grains, carbonaceous chondrites are minimally altered samples of early solar system material: they are volatile rich; many of the mineral phases are not in chemical equilibrium with one other; and they show no evidence of either thermal processing or extended processing in a regolith. Yet there is evidence of intense irradiation of individual grains. The critical issue thus is, "When did the exposure of these grains, now incorporated within carbonaceous chondrites, occur?" The available evidence indicates that the compaction of carbonaceous chondrite material into rigid objects, which would end the differential exposure of individual grains, occurred very early (cf. Caffee and Macdougall 1988; Macdougall and Lugmair 1989). For the parent body of the CM carbonaceous chondrites some evidence comes from fission track studies of individual grains (Macdougall and Kothari 1976). This technique yields a compaction time of 4.4 Gyr for Murchison and 4.5 Gyr for Murray, referring to the time that actinide-free olivines came into intimate contact with matrix material. For the CI parent body, the evidence comes from the study of initial Sr in authigenic phases from Orgueil (Macdougall et al. 1984; Macdougall and Lugmair 1989). These results indicate that the formation of the CI parent-body regolith probably occurred within 10 Myr of 4.56 Gyr, and no later than 100 Myr. However, the fission-track technique is open to interpretational ambiguity (cf. Caffee and Macdougall 1988) and the Sr isotopic constraint refers strictly to CI meteorites, not the CM's. At this time, it has not been possible to measure spallogenic ^{21}Ne in individual solar-flare irradiated grains from Orgueil or any other CI because of their rarity and the small size of such grains. The need for a reliable chronometer for dating the compaction of these objects is apparent. Without such constraints, the evidence for an active early Sun on the grounds of the required exposure duration remains largely circumstantial.

An additional constraint on this problem is the association of excess spallation neon with the presence of solar-flare tracks. This observation provides a strong argument that the two effects are closely coupled, although there is not a corresponding association between track density and precompaction spallation neon concentration (Hohenberg et al. 1990). The association between solar-flare tracks and excess cosmic-ray-produced-spallation neon seems surprising since these two groups of energetic particles have energies that differ by orders of magnitude. This difference in energy means their penetration depth in solid material is different (100 μm for solar-flare tracks vs a few cm for SCR spallation and a few m for GCR spallation). In a static regolith, only those grains on top would contain solar-flare tracks, whereas any grain in the upper several m could contain spallation neon. In this case, these two irradiation effects should not be correlated.

Regoliths are not static, so let us suppose that some mixing occurs, as is the case on the lunar surface. Caffee (1986) analyzed grains from a lunar soil and found that grains on the lunar surface did not display the same irradiation effects as those in gas-rich meteorites. In particular, lunar grains were found that did not have solar-flare tracks but did have excess spallation neon. There are at least two possible explanations for this observation. The first is that the lunar regolith is a poor analog for meteorite regoliths, perhaps because lunar soil samples come from the GCR active zone. The second is that the particle environment in the early solar system was very different (i.e., there may have been an early active Sun producing those effects observed in Murchison). Regarding the first possibility, Wieler (1989a) has proposed that this association could be explained if the regolith is a mixture of two components: one very mature (containing both solar flares and GCR neon) and one immature (containing neither tracks nor spallation neon), having been admixed rather freshly from a depth below the reach of both GCR and solar wind. In other words, the process of regolith formation on meteorite parent bodies would be one that is dominated by the cratering processes that excavate material from depths considerably below the penetration depth of galactic cosmic rays rather than a continual gardening of the upper layers of the regolith.

Alternatively, an SCR origin for the observed pre-compaction spallation effects does not require a two-component regolith because the range of solar-flare protons is only a few cm. Mixing in the uppermost few cm is quite efficient, so associations between solar-flare tracks and spallation Ne are more easily understood. If the model GCR regolith exposure age of the CM parent body is 300 Myr, and if, like the CI parent body, compaction of the CM parent body occurred within 100 Myr after solar system formation at 4.56 Gyr, then it is unlikely that pre-compaction spallation in these objects can be due to galactic cosmic rays. The alternative is that the excess spallogenic ^{21}Ne in these grains comes from enhanced solar-flare activity in the early Sun. If so, a minimum fluence of 4×10^{17} proton cm^{-2} from solar flares would be required for a grain irradiated at 1 g cm^{-2} average shielding (for an energy spectrum similar to contemporary solar flares). If solar-flare activity were enhanced by a factor of 100 over that of the present Sun, pre-compaction exposure times of only a few Myr would be required. As the production of spallation neon by solar-flare protons (and their secondaries) falls off rapidly with depth, with little production in the lower portions of a well-mixed surface zone of the parent body regolith, this is clearly a lower limit for the proton budget.

A further constraint is the observation that the solar-flare tracks density observed in gas-rich meteorite grains is rather low in comparison to solar-flare track densities in lunar grains (cf. Goswami et al. 1976). If we accept an enhanced solar-flare proton flux in the very early solar system, we would expect a corresponding increase in those heavy ions responsible for the production of tracks in meteorite grains. Such an increase in track densities is not observed. On the basis of this evidence, we conclude that the enhanced solar particle flux hypothesis may also require the *ad hoc* assumption that in the early solar system the SCR H/Fe ratio differed from the contemporary ratio, either because of a difference at its source or because of partial shielding by residual nebular gas.

VI. SUMMARY

Did our Sun go through a phase of increased activity? Observations of other young solar-mass stars, discussed in this book (see chapters by Bertout et al., Walter and Barry; Radick and Herbert et al.), provide strong circumstantial evidence for the affirmative. Is there any evidence in the meteorite record that the Sun was still in a period of enhanced activity during the formation of meteorite parent bodies? At this time this question is still open.

We have presented the case for gas-rich meteorites. It is very evident that before questions about the ancient Sun can be investigated by using the irradiation records in some gas-rich meteorites, a few more pieces of the puzzle need to be found. It seems that GCR irradiation can account for the excess spallogenic ^{21}Ne observed in many ordinary chondrites and differentiated achondrites. This does not seem to be the case for the carbonaceous chondrites. If we accept the regolith formation ages of the CM parent body as given by the fission-track method and supported by a more rigorous and short compaction time for the CI's based upon Sr isotopes, then we are forced to invoke enhanced particle fluxes in the early solar system to explain the pre-compaction irradiation effects in Murray and Murchison. However, before the issue can be settled with any certainty, we will need both a deeper understanding of meteorite parent body regolith evolution with more rigid constraints on the times of compaction and measurements of pre-compaction spallation effects in meteorites where the pre-compaction ages may be better constrained, such as the CI meteorites. Both of these are difficult tasks, but results could tell us whether meteorites retain the signature of an enhanced active Sun.

Acknowledgments   Reviews were provided by D. Woolum, J. Kerridge, and P. Pellas. Their comments were appreciated and we believe the manuscript is improved owing to their efforts. This work was supported in part by grants from the National Aeronautics and Space Administration and by the Swiss National Science Foundation. Portions of this work were supported under the auspices of the U.S. Department of Energy by Lawrence Livermore National Laboratory.

SOLAR FLARE HEAVY-ION TRACKS IN
EXTRATERRESTRIAL OBJECTS

J. N. GOSWAMI
Physical Research Laboratory

Lunar samples, meteorites and interplanetary dust particles contain fossil tracks produced by energetic solar flare heavy ions. Solar flare track records in certain primitive meteorites provide us with information on the activity of the Sun more than 4.4 Gyr ago. Similar records in lunar samples allow us to probe the level of solar activity at different epochs during the last 1 Gyr. Solar flare track records in interplanetary dust particles serve as a potential source for information about solar activity over the past thousand years. Studies of lunar samples and meteorites have established the similarity in long-term averaged spectral shape of solar flare iron-group nuclei at widely separated epochs. It has also been shown that the preferential enrichment of heavy ions at low energies, seen in contemporary solar flares, was also present in ancient times and continued beyond the iron group. No definitive evidence for marked variations in solar activity can be found in the lunar sample records. However, recent investigations of primitive meteorites that combine solar flare track and noble gas studies suggest the possibility that the Sun had gone through an active early (T Tauri) phase.

I. INTRODUCTION

The passage of energetic charged particles in dielectric solids leads to ionization-induced damage in the structure of the solids along the paths of the charged particles. If the ionization loss rate exceeds a certain critical value, characteristic of each dielectric solid, the structural damage trails are stable under normal environmental conditions, and can be seen directly by transmission electron microscopy or can be enlarged by suitable chemical etching and seen by conventional optical microscopes. The latent or chemically etched damage trails produced by the passage of energetic charged particles in solids are called nuclear tracks. The registration thresholds for different dielectric solids differ widely. While a sensitive plastic like cellulose nitrate can record tracks produced by low-energy α particles, silicate minerals, which are the main constituents of extraterrestrial samples like lunar rocks, meteorites, and interplanetary dust particles (IDP), record tracks only of energetic particles with atomic number $Z > 20$, i.e., primarily of the iron group and heavier ions. As the ionization loss-rate of charged particles in any medium reaches a maximum near the end of their range, nuclear tracks are produced only near the stopping point of the ions irrespective of their initial energy. The track length is thus a function of the atomic number of the track-producing particle. For example, energetic iron-group ions register tracks of ~ 10 to 20 μm towards the end of their range in silicate grains, while the tracks produced by trans-iron-group ions are >20 μm. For an exhaustive discussion on the principles of nuclear tracks in solids and their application in various fields of research, the reader is referred to the book *Nuclear Tracks in Solid* by Fleischer et al. (1975). In this chapter, the discussion focuses on fossil track records produced by energetic solar flare heavy ions in extraterrestrial objects and what they can tell us about the activity of the ancient Sun. Complementary information derived from studies of solar flare proton-produced interaction records in extraterrestrial objects are discussed in the chapters by Caffee et al. and by Reedy and Marti. An overview of all the related observations is also found in the chapter by Lal and Lingenfelter.

II. NUCLEAR TRACK RECORDS IN EXTRATERRESTRIAL OBJECTS

The principal sources of nuclear tracks in extraterrestrial objects are (i) energetic cosmic ray iron group and heavier ions and (ii) fission fragments from the spontaneous fission of uranium ^{238}U and trans-uranium elements, such as now-extinct ^{244}Pu (half-life ~ 83 Myr). Additional contributions come from cosmic ray proton and secondary neutron induced fission and energetic proton induced spallation recoils of heavy ions. Suitable selection of samples and adoption of appropriate observational criteria allow one to identify and analyze tracks produced by each of the above sources. For example, the range of spallation recoils in silicate materials is extremely small, less than 1 μm, and the short spallation recoil tracks can easily be distinguished from the fission or cosmic ray heavy-ion tracks. The length of fission fragment tracks in silicates, on the other hand, is similar to those produced by cosmic ray iron-group nuclei. However, a sample having an extremely small uranium concentration, but with a long cosmic ray exposure will primarily have tracks produced by cosmic ray heavy ions. One can also find samples where the situation is the opposite. When we consider tracks produced by cosmic ray heavy ions only, we have contributions from both solar and galactic cosmic rays. These two populations of tracks can be distinguished based on features that are characteristic of the solar and galactic cosmic rays. The low-energy (1 to 100 MeV/n) solar flare heavy ions, with ranges of < 1 mm in material of silicate composition, are characterized by steeply falling energy spectra and their time-averaged fluxes (over a solar cycle) at low energies, in the inner solar system, are orders of magnitude higher than the corresponding galactic cosmic ray fluxes. Thus, any extraterrestrial object directly exposed to solar flare radiation will exhibit a sharp negative gradient in the number density of tracks, as a function of shielding depth, within the topmost few hundred μm of its surface. This is a direct manifestation of the steeply falling energy spectra of the low-energy solar flare heavy ions. The identification of an observed track record in any extraterrestrial object with solar flare irradiation is therefore made on the basis of either (1) a track density gradient observed over a typical shielding-depth range of 10 to 100 μm, or (2) high track densities that are more than an order of magnitude higher than that expected on the basis of galactic cosmic ray irradiation alone. Examples of solar flare track irradiation records in lunar samples and meteorites are shown in Fig. 1.

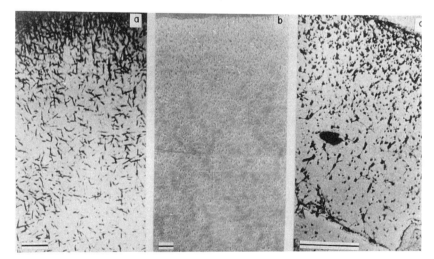

Fig. 1. Solar flare heavy-ion tracks in individual grains from the gas-rich meteorites Kapoeta (a), Cold Bokkeveld (b), and lunar sample 12037 (c). The etched tracks seen in (a) and (b) are observed by optical and scanning electron microscope, respectively; (c) is a transmission electron micrograph of carbon replica of etched tracks. Scale bar is 10 μm.

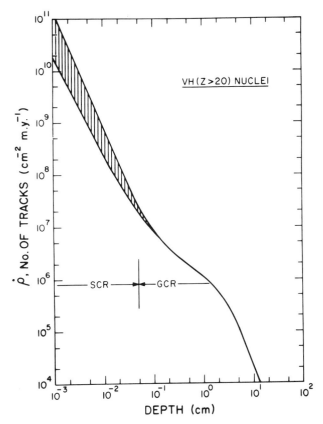

Fig. 2. Time-averaged production rate of tracks produced by solar flare and galactic cosmic ray iron-group particles in lunar samples.

Different techniques are used to observe these track records (see figure legend). We also show in Fig. 2 the typical production rate of cosmic ray heavy-ion tracks in extraterrestrial objects exposed at 1 AU. It is obvious from the figure that there are uncertainties in the production rate estimates of solar flare heavy-ion tracks. This is discussed later (Sec. III. C.2).

Deciphering the solar flare track records found in extraterrestrial objects, in terms of ancient solar activity, demands an adequate knowledge of the cosmic ray exposure histories and particularly solar flare irradiation histories of the objects. In addition, it is also necessary to take into account the environmental effects that can alter the track record. Unfortunately, our understanding of the exposure histories of the objects is not complete enough to derive quantitative information like the exact epoch and duration of solar flare irradiation of a given sample. There are, however, a few notable exceptions. Regarding environmental effects, the most important are the thermal annealing and shock-induced deformation of latent tracks (see Fleischer et al. [1975] for a detailed discussion on this topic), and it is necessary to consider these effects for a proper interpretation of the track record. It is also possible at times to choose specific samples where these effects are minimal (see Sec. III.B). It is worthwhile at this point to give a brief summary of the nature and extent of solar flare heavy-ion track records found in the different types of extraterrestrial objects. The nature and complexity of the fossil records stored in extraterrestrial objects have been reviewed earlier by Walker (1980). (See the chapters by Reedy and Marti and by Ramaty and Simnett.)

A. Meteorites

The solar flare heavy-ion track records in meteorites are generally lost due to the ablation of the outermost layers of the meteorites during their atmospheric transit. However, there is a special class of meteorites where one can find interior components that contain distinct signatures of solar flare irradiation records (see Fig. 1). These meteorites are termed gas rich as they also contain high concentrations of noble gases like He, Ne etc., which have conclusively been shown to be of solar wind origin (Gerling and Leviski 1956; Wänke 1965; Eberhardt et al. 1965). The first observations of solar flare heavy-ion tracks in individual grains of such meteorites was made by Lal and Rajan (1969) and Pellas et al. (1969), prior to the arrival of the lunar samples. The most important aspect of the solar flare track irradiation records in meteorites is the fact that this irradiation predates the time of compaction of the meteorites, i.e., the time when individual components presently found in the meteorites were cemented together by some events. The time of compaction of most of the gas-rich meteorites is believed to be very early in the evolutionary history of the solar system. However, only in the case of a particular subclass of gas-rich meteorites (the carbonaceous chondrites), does there exist experimental data that point towards a compaction time of >4.4 Gyr ago (Macdougall and Kothari 1976; Caffee and Macdougall 1988). Thus, the solar flare records in meteorites provide a unique opportunity for studying the activity of the early Sun. For an elaborate discussion on this topic, readers are referred to reviews by Goswami et al. (1984) and Caffee et al. (1988) (see also chapter by Caffee et al.).

B. Lunar Samples

The lunar samples returned by the Apollo and Luna missions contain abundant records of solar flare tracks. However, these records are difficult to decipher as most of the lunar samples did not have a simple one-stage exposure to solar flare radiation. The lunar samples, in general, experienced complex cosmic ray exposure histories that include multiple exposures to low-energy solar flare particles, while residing on the lunar surface, interspaced by shielded exposure much beyond the range (~1 mm) of the solar energetic particles. Such complex exposure of lunar samples is a direct result of the continuous meteoritic bombardment of the lunar surface, which leads to excavation of shielded material to the lunar surface as well as blanketing of pre-existing material exposed at the surface (Gault et al. 1974). There are, however, certain exceptions, namely rocks and boulders that were excavated in recent times and had a continuous exposure on the lunar surface before being picked up by the astronauts. An additional parameter that affects the solar flare records in exposed lunar rocks is the erosion of the rock surfaces by micrometeoritic impacts. The average erosion rate of lunar rock surfaces is estimated to be 0.3 to 2.0 mm Myr^{-1}. This restricts the "effective exposure" of the lunar rocks to the short-range solar flare particles to about a few Myr. The mixing or gardening of the uppermost lunar soil layers by micrometeorite impacts also does not allow the lunar soil grains to be exposed on the topmost lunar surface for more than 10^4 to 10^5 yr at a stretch (Bibring et al. 1975; Poupeau et al. 1975). They may, however, have repeated exposures on the lunar surface. While the lunar rocks and soils exposed on the lunar surface provide information on solar flare activity in the recent past, the soil samples from different layers of the deep drill cores, which sampled soils up to $\sim$2.5 m beneath the lunar surface, and certain types of lunar breccias provide us with information on solar flare activity at more distant epochs up to a few Gyr. Unfortunately, it is not possible to establish the exact durations or the epochs of solar flare irradiation for these samples. Instead, we only have certain model-dependent relative time scales. The lunar samples, therefore, provide us with a record of solar flare activity, averaged over typical time scales of 10^4 to 10^6 yr, at different epochs over the last 1 Gyr.

C. Interplanetary Dust Particles

The interplanetary dust particles (IDPs), whose direct collection in the stratosphere was initiated in the seventies, were considered for a long time to be a potential carrier of solar flare irradiation records and particularly solar flare heavy-ion tracks. However, it was only in 1984 that the first solar flare track records were observed in these samples (Bradley et al. 1984; see Fig. 3). With a probable lifetime of $\sim$$10^3$ to 10^4 yr in interplanetary space (Dohnanyi 1978), the IDPs contain records of solar flare activity for the most recent time, compared to the records stored in the lunar and meteorite samples. A quantitative analysis of the solar flare track records in IDPs has, however, to contend with several important problems like the orbits of the IDPs in interplanetary space, their exact exposure geometry (including shielding) in space, and finally the possibility of partial annealing of tracks due to atmospheric heating.

III. ANCIENT SOLAR FLARE ACTIVITY

The solar flare track records in extraterrestrial objects are cumulative records that give us information on the time-averaged solar flare activity at

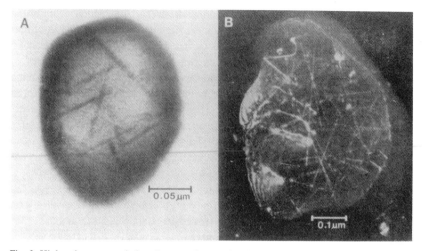

Fig. 3. High-voltage transmission electron micrographs of solar flare tracks in interplanetary dust particles. (a) is a bright-field image and (b) is a dark-field image (figure courtesy of Bradley et al. 1984 from *Science* 226, 1432, 1984).

different epochs in the past. Since the time scales involved are generally 10^4 to 10^6 yr, only the intense flare events will dominate the record; the contributions as well as characteristics of the weak solar flares are suppressed by the stronger events. In this section we summarize the meaningful information on past solar activity in terms of energy spectra, composition and intensity of solar flare heavy ions that have been obtained from studies of solar flare track records in extraterrestrial objects.

A. Energy Spectra of Ancient Solar Flare Heavy Nuclei

The track-density profiles as a function of shielding depth, seen in extraterrestrial objects exposed to solar flare irradiation (e.g., see Fig. 1), reflect the intensity and spectral shape of the ancient solar flare heavy ions averaged over the exposure durations of the objects. If we assume a power-law representation for the solar flare heavy nuclei spectra, $dN \propto E^{-\gamma} dE$, the track production rate as a function of shielding depth in a sample can be written as $\rho(x) \propto X^{-\eta}$, where the exponent η for the track density profile is related to the exponent γ in the power-law spectra through the relation:

$$\eta = (1 / \beta)(\gamma + \beta - 1) \tag{1}$$

where β is the exponent in the range-energy relation of interest and can be approximated as $R \propto E^{\beta}$, R being the range of the track-producing ion.

Equation (1) follows from the fact that tracks are produced by energetic heavy ions only towards the end of their range, irrespective of their initial energies, and as such to a first approximation the track density $\rho(X)$ at a shielding depth X can be written as

$$\rho(X) \propto \left(\frac{dN}{dR}\right)_X = \left[\left(\frac{dN}{dE}\right)\cdot\left(\frac{dE}{dR}\right)\right]_X \qquad (2)$$

Interested readers are referred to papers by Fleischer et al. (1967a). Lal (1972) and Bhattacharya et al. (1973) for more exact forms of these relations. It is obvious from Eq. (1) that the slope (η) of the measured track-density profile in a sample can be directly interpreted in terms of time-averaged spectral shape of the solar flare heavy ions incident on the sample. The early work of Lal and Rajan (1969), Wilkening (1971) and Poupeau and Berdot (1972) showed that the track records seen in irradiated grains of gas-rich meteorites yield a value of 1.0 to 1.5 for the exponent γ which is much different from the value of $\sim$3 for contemporary flares. Initial results obtained from analysis of returned lunar samples during the early seventies also yielded values ranging from 1.0 to 2.5 for the exponent γ. While it is tempting to interpret these results as indications for a harder energy spectra (smaller value of γ) for ancient solar flare heavy nuclei compared to contemporary flares, one must remember that Eq. (1) holds good only when the samples are directly exposed to solar flare radiations and have a simple one-stage exposure history. However, as noted in Sec. II, this is rarely the case. In fact, one can have a shallower track density profile (small value of η) leading to an underestimate of the value of γ, if the analyzed samples were shielded by a few tens of μm of dust layer during their exposure to solar flares (which is quite common for lunar soil samples [Poupeau et al. 1975]), or if they had undergone multiple exposures or had their surface layers removed by micrometeorite-induced erosion effect. More detailed studies of meteorite grains by Rajan (1974), Goswami et al. (1976, 1980) and Macdougall and Phinney (1977), and of specially selected lunar samples (Hutcheon et al. 1974; Blanford et al. 1975; Morrison and Zinner 1977; see also Zinner 1980) showed that one can find a range of slopes for the track-density profiles in both lunar and meteorite samples, some of which are consistent with a value of 3 for the power-law index γ. Thus it seems appropriate at present to consider the track records in both lunar and meteorite samples to be consistent with an exponent in the power-law spectrum for ancient solar flare heavy ions that is similar to the value for contemporary flares. A representative sample of track-density profiles measured in a set of selected lunar samples and meteorites are shown in Fig. 4 along with the track-density profile measured in a portion of the camera glass of the Surveyor spacecraft which was exposed for a 2.6 yr duration during 1967 to 1969 on the lunar surface, before being retrieved by the

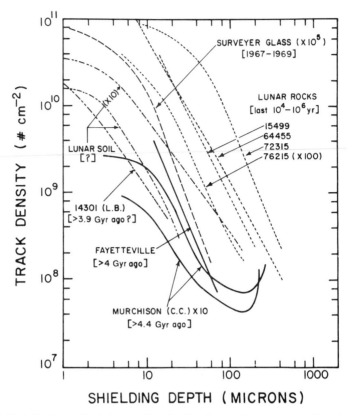

SHIELDING DEPTH (MICRONS)

Fig. 4. Track-density profiles in lunar rocks and soils and meteoritic grains exposed to solar flares at different epochs in the past (L.B. = lunar breccia; C.C. = carbonaceous chondrite). The measured track-density profile in the Surveyor camera glass is also shown for comparison.

Apollo-12 astronauts. The most probable epochs when the solar flare irradiation of these samples took place are also indicated. The similarity in the track-density profiles in samples exposed to both contemporary and ancient solar flares, when one chooses suitable samples for study, is clearly evident. A more important observation in this context is the failure so far to find evidence for a track-density profile that will imply a steeper slope for the solar flare heavy-ion energy spectra (i.e., $\gamma > 3$). The only report for an inferred value of $\gamma > 3$ comes from observation of track gradient in samples from the meteorite Allende (Kashkarov et al. 1977). However, unless there is some additional evidence for a steeper spectrum, the presently available lunar and meteorite data indicate a close similarity for the long-term (10^4 to 10^6 yr) average spectral shape for the solar flare heavy ions at widely separated epochs when the lunar samples and the meteorites received their solar flare irradiations. The similarity of the ancient solar flare heavy-ion spectral shapes with those for contemporary flares, as evident from the Surveyor data, suggest that the spectral shape for the solar flare heavy ions remained similar over the last several Gyr. The spectral shape for the solar flare heavy ions is determined by electromagnetic processes operating at the flare site as well as during their propagation from the flare site to their point of observation (1 to 3 AU) in interplanetary space. From the time-averaged records, as seen in lunar and meteorite samples, it is not possible to decouple these processes. Nonetheless, the similarity between the contemporary and long-term averaged spectral shapes for solar flare heavy ions indicates that the electromagnetic processes involved in flare acceleration and propagation, at least for the large flares that dominate the long-term records, have not changed significantly over the last 4 Gyr.

B. Composition of Ancient Solar Flare Heavy Nuclei

The long duration exposure (10^4 to 10^6 yr) of the lunar and meteorite samples to solar flare particles makes them ideal detectors for the study of the iron and trans-iron group of nuclei by employing the nuclear track technique. The low abundances, particularly of the trans-iron group of nuclei, make their study extremely difficult in contemporary space-borne experiments, most of which are characterized by small collecting power. In fact, the existence of the trans-iron group of nuclei in cosmic rays was first established from studies of nuclear tracks in meteorites (Fleischer et al. 1967b). The identification of the atomic number of the track-forming ions is made primarily on the basis of track length measurements (Fleischer et al. 1967b, 1975; see also Fig. 5), although other approaches like measurement of track diameter and response to controlled annealing (Goswami and Lal 1975; Dyrtyge et al. 1978) have also been attempted. Energetic ions from accelerators are generally used for calibration (Price et al. 1973). An unavoidable problem that needs to be explicitly considered in all these approaches is the possible environmental effects on the fossil tracks, and particularly possible thermal annealing of tracks over the long durations of exposure and storage in space. Comparison of lengths of fossil and "fresh" tracks, the latter being produced by energetic ions from an accelerator, generally enables one to tackle this problem. Even then, it is generally not possible to have a fine charge resolution from the fossil track length data. The studies of ancient solar flare heavy nuclei composition, therefore, refer to the abundance ratio of two groups of elements, the iron group ($Z>20$) also termed as the very heavy (VH) group of nuclei and the trans-iron group ($Z\geq30$) or the very, very heavy (VVH) group of nuclei.

Bhandari et al. (1973a) reported the first systematic measurement for solar flare heavy nuclei compositions from a study of lunar samples. They found that the VVH/VH abundance ratio in flare particles increases rapidly at low energies (<20 MeV/n) and at 6 to 10 MeV/n energy interval, the estimated abundance ratio is $\geq 7 \times 10^{-3}$, much higher than the photospheric

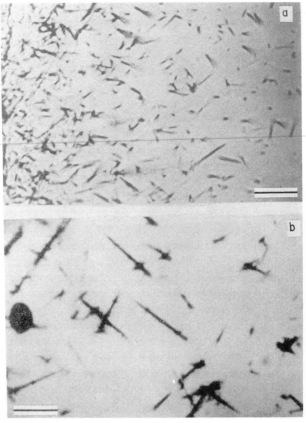

Fig. 5. Photomicrograph of tracks due to solar flare iron group (VH) and trans-iron-group (VVH) particles in grains from the carbonaceous chondrite Murchison. The VVH nuclei tracks (b) are characterized by their longer track length. Scale bar is 10 μm.

abundance ratio of $\sim 10^{-3}$ for these two groups of elements (Ross and Aller 1976). At energies >20 MeV/n, the abundance ratio is essentially constant and matches the photospheric value. Goswami and Lal (1975) extended this work to study the systematics of the VVH enhancement pattern down to energies of ~ 1 MeV/n. They confirmed the enhancement in the VVH/VH abundance ratio in solar flares up to the lowest energy studied. However, they also found that the enhancement factor did not rise monotonically with decrease in energy.

Solar flare heavy ion composition was in fact first determined by Lal and Rajan (1969) from a study of solar flare irradiated grains from gas-rich meteorites. However, their results suffer from the lack of proper calibration studies and cannot be considered as accurate. Additional studies of solar flare heavy-ion composition using meteorite samples were made later by Goswami et al. (1979,1980). The long storage time of solar flare tracks in meteorites, which can exceed 4 Gyr in certain cases, could make environmental effects on the track records quite important. Fortunately, some of the gas-rich meteorites, particularly the carbonaceous chondrites, have never experienced any severe thermal or shock metamorphism during their entire evolutionary history that could have adversely affected the fossil track records in these meteorites. Goswami et al. (1979,1980) have chosen a set of such meteorites for their work and have also calibrated the analyzed samples for possible environmental effects by comparing the characteristics of fossil and fresh ion tracks in the analyzed samples. The results of this work are shown in Fig. 6 where the data from earlier work on lunar samples are also included. It is evident that the enhancement in the solar flare VVH/VH abundance ratio at low energies is seen in data for both lunar samples and meteorites that have

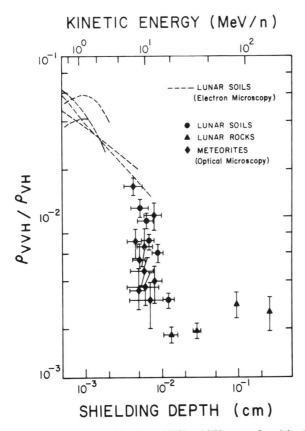

Fig. 6. Ratios of measured track densities due to VVH and VH group of particles (ρ_{VVH}/ρ_{VH}) in different lunar samples and meteorites plotted as a function of shielding depths. The equivalent kinetic energies are also given.

sampled ancient solar flares at widely separated epochs, from the recent past to more than 4 Gyr ago. For the meteorite samples the VVH/VH abundance ratio in the 5 to 15 MeV/n energy interval ranges from (3 to 12) $\times 10^{-3}$. The spread of data does not allow us to draw any definite conclusion regarding the energy dependence in the enhancement pattern, although the lunar data alone suggest a progressive enhancement with decrease in energy. At higher energies (>50 MeV/n), the results primarily reflect the VVH/VH abundance ratio in the galactic cosmic rays, which matches the solar system abundance ratio of $\sim 1.3 \times 10^{-3}$ for these two groups of elements (Anders and Ebihara 1982).

Enhancement of low-energy heavy nuclei in contemporary solar flares was first reported by Price et al. (1971) based on their analysis of the solar flare track records in the Surveyor camera glass exposed to solar flare particles on the lunar surface for a 2.6-yr period. This result was confirmed by Hovestadt et al. (1973b) based on data from the satellite-borne experiment. Further studies using both rocket and satellite-borne detectors showed that there is a progressive enrichment of solar flare heavy ions with increasing atomic number. A relation between first ionization potential and the enrichment factor was also suggested. While some of the experimental data indicated a systematic enhancement of heavy ions with decreasing energy, this was not substantiated by event-integrated data obtained from satellite experiments. Summaries of progressive development in this area can be found in McDonald et al. (1974), Crawford et al. (1975), Mewaldt (1980) and McGuire et al. (1986). Most of the contemporary observations are for elements up to iron. The only study of the trans-iron group of ions in contemporary solar flares was made by Shirk (1974), who found that the enhancement of heavy ions in solar flare continues beyond the iron group, a result reported earlier by Bhandari et al. (1973a) from studies of lunar samples. The enhancement of heavy ions seen in contemporary solar flares is indeed a long-term phenomenon and must have occurred even during the early history of the Sun. The preferential enhancement of low-energy heavy ions in solar flares may represent a compositional bias in the source region itself (Bertsch et al. 1974) and/or may result from preferential acceleration mechanisms operating in the flare region. The initial suggestions for one- or two-stage acceleration models cannot explain well the observations (see Shirk 1974). Models proposed to explain preferential enhancement of ^{3}He (Fisk 1978; Kocharov and Kocharov 1978) predict enhancement of heavier ions if their charge states in the source region satisfy the criterion for resonance heating necessary for preferential injection of these ions into the accelerating region. However, it is doubtful if these models can be extended to explain heavy-ion enhancement in large flares, that generally do not show ^{3}He enrichment, but effectively dominate the long-term averaged solar flare records seen in lunar and meteorite samples. A combination of electromagnetic and plasma processes operating at the flare site, prior to and during the acceleration phase, seems to be responsible for the heavy-ion enhancement seen in contemporary and ancient solar flares.

C. Solar-Flare Intensity in the Past

The interpretation of the observed solar flare heavy-ion records in extraterrestrial objects in terms of past solar flare intensity is possible only if the exposure conditions and the epoch and the durations of solar flare exposures of the analyzed samples are precisely known. Unfortunately, as stated earlier, these parameters cannot be determined accurately in most cases. Attempts to infer past solar flare intensity from heavy-ion track records in lunar samples, meteorites and IDPs are based primarily on the choice of appropriate samples whose solar flare exposure conditions are better known and whose exposure durations to solar flare particles are obtained from model-based estimates or by utilizing independent chronometers. Combined studies of solar flare heavy-ion tracks and solar flare proton-produced spallation effects have also turned out to be a useful approach in this regard (Caffee et al. 1983,1987). A brief discussion of the results obtained so far on the variations in ancient solar flare intensity from studies of nuclear tracks in different types of extraterrestrial objects is presented below.

1. Results from Studies of Interplanetary Dust Particles. The solar flare track records in IDPs (see Fig. 3) contain information on the time-averaged solar flare intensity during the recent epoch ($\sim 10^4$ yr) characterizing the average interplanetary lifetime of these particles (Dohnanyi 1978). Unfortunately, a quantitative estimate of solar flare intensity from IDP data is not yet possible as both the lifetime and orbit of the IDPs in interplanetary space are not well constrained by the available data. Additional uncertainty is also introduced by the possibility of partial annealing of tracks due to atmospheric heating. Nonetheless, the track densities of 10^{10} to 10^{11} cm^{-2} observed in the IDPs (Bradley et al. 1984) are consistent with an exposure age of $\sim 10^4$ yr for these particles, given the uncertainties both in the production rate of tracks in μm-sized particles at 1 AU and the orbits of the IDPs in interplanetary space. While cometary origin for IDPs is most likely, there could be contribution from the asteroidal belt as well. In fact, Sanford (1986) has proposed a possible distinction between these two sources, based on the expected differences in the distribution of solar flare track densities in IDPs resulting from the differences in their orbital evolution. Further work on solar flare track records in IDPs and estimation of their solar wind and solar flare exposure durations from measurements of solar wind noble gases and solar flare proton-produced spallogenic products (e.g., the noble gases in them) should allow us to have more quantitative information in this regard. Mass-spectrometric studies have already led to identification of solar wind noble gases in IDPs analyzed in groups (Rajan et al. 1977; Hudson et al. 1981) and recent improvement in experimental capabilities have made it possible to carry out such measurements on single IDPs (Nier and Schlutter 1990). At present, the best estimate of solar flare intensity during the recent epoch ($\sim 10^4$ yr) comes from radionuclide studies of lunar samples (see the chapter by Reedy and Marti).

2. Results from Studies of Lunar Samples.   Solar flare heavy-ion track records in lunar samples have primarily yielded information on the level of solar activity during the past 10^4 to 10^6 yr, the time scale representing the "effective" solar-flare exposure durations of lunar rocks. Attempts to obtain information on solar flare activity during the past 1 Gyr, and during even earlier epochs, from studies of lunar drill core soil samples, lunar breccias and feldspathic grains derived from anorthositic rocks from the lunar highlands, have not yet been successful. Comprehensive discussion of the results obtained from solar flare track studies of lunar samples can be found in a series of articles that appeared in the proceedings of the conference on *The Ancient Sun* (Crozaz 1980; Zinner 1980; Zook 1980). A brief summary is presented below.

An important requirement for a quantitative interpretation of the solar flare heavy-ion track records in lunar samples, in terms of solar flare intensity, is a knowledge of the time-averaged production rate of solar flare tracks at 1 AU. Initially, the data obtained from the analysis of the Surveyor glass, exposed for a 2.6 yr duration on the lunar surface, were considered as a standard. Several problems associated with this data set led to a concerted effort by several groups to determine the time-averaged solar flare track production rate from track records in specially selected lunar rock samples. The solar flare exposure durations of these samples were ascertained from data on spallogenic noble gases (e.g., ^{81}Kr; Blanford et al. 1975) and radionuclides (e.g., ^{26}Al; Morrison and Zinner 1977) or by indirect normalization to galactic cosmic ray heavy-ion tracks in the same sample (Hutcheon et al. 1974). Unfortunately, the results obtained from these efforts showed that there are significant uncertainties in estimating the time-averaged solar flare heavy-ion energy spectrum (see Fig. 7) and hence uncertainty in the track production rate as well (see Fig. 2). Thus the feasibility of obtaining information on small-scale fluctuations in solar activity from the solar flare track records in lunar samples had to be ruled out. Nonetheless, the track data obtained from these and other specially selected lunar rock samples suggest a similarity in solar flare heavy-ion fluxes (within a factor of 2) averaged over the last 10^5 to 2×10^6 yr, the time span covered by these samples. Indirect evidence for a constancy in solar flare heavy-ion fluxes over the last 10^4 to 10^6 yr also comes from the close correlation between solar flare track and microcrater records in exposed lunar rock surfaces (Morrison and Zinner 1977). Zook (1980), who has made a correlation study of solar flare-induced effects (e.g., production of tracks, radionuclides, etc.) and other surface correlated processes (e.g., microcraters) in lunar samples, had argued for a large enhance-

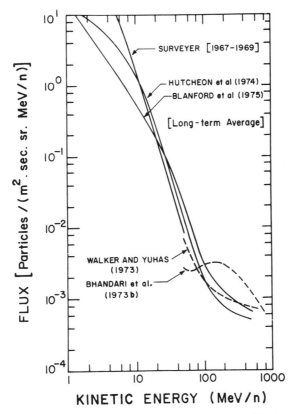

Fig. 7. Long-term averaged energy spectra of solar flare iron-group particles based on analysis of specially selected lunar rocks. Also shown are the Surveyor spectrum and deduced spectra for galactic cosmic ray iron-group particles at higher energies.

ment in solar flare intensity at $\sim 10^4$ yr before present. Recent data on ^{14}C activities in lunar rocks are, however, not consistent with this suggestion (chapter by Reedy and Marti).

The lunar drill core soil samples do contain records of solar activity over the last 1 Gyr. Lunar soil grains (feldspars) derived from anorthositic lunar highland rocks, which were presumably formed during the very early history of the Moon, may also contain records of flare activity from the early Sun. Poupeau et al. (1973), in fact, claimed evidence for an active early Sun based on their analysis of solar flare track records in anorthositic feldspars from the Luna 16 drill core soils. However, more detailed work of Crozaz et al. (1974) did not support this claim. Crozaz (1980), who has considered various aspects of track records in lunar drill core soil samples, has rightly remarked that "there is no compelling reason to invoke past changes in the activity of the Sun to explain the existing data." It must, however, be stressed that our only chance of detecting large variation in solar flare activity over durations of 1000 to 10,000 yr in different epochs during the last 1 Gyr still rests on a careful analysis of the lunar drill core soil samples, which have stored records of solar activity over this enormous time span. Another set of lunar samples, the highland breccias like 14301 and 14318, may contain records of solar flare activity from distant epochs (>3.9 Gyr before present). These samples are analogous to the gas-rich meteorites and very careful studies (Bernatowicz et al. 1979; Swindle et al. 1985) suggest their compaction ages to be >3.9 Gyr. Detail studies of solar flare records in these samples will complement the results obtained from gas-rich meteorites. (For nuclide production by solar cosmic rays over the past 10 Myr, see chapter by Reedy and Marti).

3. Results from Studies of Gas-rich Meteorites.   Solar flare irradiation of individual components of the gas-rich meteorites predates their compaction ages and is believed to have taken place very early in the evolutionary history of the solar system (see Sec. II.B). The solar flare track records in the different types of gas-rich meteorites are significantly different. For example, the maximum track densities seen in solar flare irradiated grains from the carbonaceous chondrites are almost 2 orders of magnitude lower than the corresponding track densities observed in grains from gas-rich chondrites and achondrites (Rajan 1974; Price et al. 1975; Goswami and Lal 1979; Goswami et al. 1976,1984). Although it is tempting to attribute such differences to variations in the activity of the early Sun, the present consensus is that this difference is a manifestation of the widely differing pre-compaction solar flare exposure conditions/durations of the individual components of the different types of gas-rich meteorites.

A systematic study to look for meteoritic evidence for an active early Sun was taken up in the 1980s when it was clearly established that flare activities in young Sun-like stars could be extremely high, about 2 to 4 orders of magnitude higher than that for the contemporary Sun (Worden et al.1981; Feigelson and DeCampli 1981; Feigelson 1982; see also the chapter by Feigelson et al.). As studies of solar flare track records alone cannot provide an unambiguous answer in this regard, a novel approach in which one combines studies of both solar flare heavy-ion tracks and solar flare proton-produced spallation noble gases was pursued for this investigation. Individual grains from gas-rich meteorites having solar flare irradiation records were identified from track studies and these were analyzed in groups or individually by extremely sensitive conventional as well as laser-probe mass spectrometers. Grains without solar flare heavy-ion tracks from each meteorite were also analyzed for comparison. The results obtained from these studies (Caffee et al. 1983, 1987; Hohenberg et al. 1990) clearly demonstrated that grains with solar flare track records also contain excess spallogenic noble gases. In the case of the gas-rich carbonaceous chondrites, Murchison, Murray and Cold-Bokkeveld, and the gas-rich basaltic achondrite Kapoeta, the excess spallogenic noble gases in the solar flare irradiated grains were found to be more than 1 to 2 orders of magnitude higher compared to that measured in the nonirradiated grains of these meteorites. The one to one correspondence between solar flare track records and excess spallogenic noble gases was taken by these authors to indicate that the energetic particles responsible for producing the excess spallation noble gases are of solar origin. However, if one considers contemporary intensity of solar flare protons, extremely long solar flare exposure durations in excess of 10 to 100 Myr are required to explain the excess noble gas contents. Such long solar flare exposure durations are extremely unlikely given our present understanding of the evolution of small objects during the early history of the solar system. These observations led Caffee et al. (1987) to conclude that the excess noble gases in the solar flare irradiated grains were produced by enhanced solar energetic particle emission, 100 to 1000 times the contemporary level from an active early (T Tauri) Sun. Similar results have also been obtained by Padia and Rao (1989) based on noble gas studies of gas-rich meteorites. A differing viewpoint has, however, been offered by Wieler et al. (1989a) who suggested that the excess noble gases in grains with solar flare tracks could be due to an extended exposure of these grains to galactic cosmic rays coupled with a short duration solar flare exposure. A recent study by Hohenberg et al. (1990) has considered all the various possibilities and proposed that the results obtained from studies of primitive carbonaceous chondrites are more consistent with the hypothesis of an active early Sun (see the chapter by Caffee et al. for the present status in this area). Obviously, it is important to pursue this problem further by including additional tracers (e.g., spallogenic ^{15}N) that may allow us to establish whether we are seeing an effect due to low-energy solar particles or high-energy galactic cosmic rays (Murty 1990).

IV. SUMMARY AND CONCLUSION

Solar flare heavy-ion track records in extraterrestrial objects (lunar samples, meteorites and IDPs) contain a wealth of information on the past solar activity going back to the earliest stages of the evolution of the Sun. Unfortunately, the records are complex to decipher as one has to decouple the history of solar activity from the history of the objects themselves. Further, the solar flare records seen in extraterrestrial objects are dominated by large solar flares and it is not possible to study solar processes that are associated with weak solar flares. In spite of these problems and limitations, several important conclusions can be drawn about past solar activity from the solar flare track records in extraterrestrial objects. These include:

1. Evidence for flare activity in the Sun over its entire evolutionary history, going back to more than 4.4 Gyr ago and suggestion for an active early (T Tauri) Sun when the flare activity might have been 100 to 1000 times higher than that for contemporary Sun.

2. Near constancy in the time-averaged solar flare heavy ion intensity over the last 10^4 to 10^6 yr. Of course, the data cannot rule out variations on small scales (less than a factor of 2) or over short durations ($<10^4$ yr).

3. Similarity in the spectral shape of solar flare heavy ions at widely separated epochs implying similar production, acceleration and propagation characteristics for solar energetic particles over billions of years.

4. Evidence for enrichment of trans-iron group of nuclei in ancient solar flares that complements the contemporary observations of heavy ion enhancement in Sun up to the iron group.

The results summarized above point towards a broad similarity between the ancient and contemporary solar flare records except for a possible enhancement in flare activity during the very early evolutionary history of the Sun. There are, however, significant gaps in the ancient solar flare record that need to be filled in by careful investigations. Two important areas where more work is needed are: (i) short-term (10^3 to 10^4 yr) variations in solar flare activity during different epochs over the last 1 Gyr, and (ii) activity of the early Sun. Understanding the solar flare record in interplanetary dust particles is another challenge at hand. The most ideal samples for the study of possible short-term variations in ancient solar flare activity are the lunar soils from the deep drill cores. Changes in spectral shape and composition of solar flare heavy ions are the two important aspects that should be emphasized in these studies. A multi-prong approach is needed for a better understanding of the early solar activity and particularly to resolve the question of whether the Sun has passed through a T Tauri phase. While analysis of individual grains of gas-rich meteorites by nuclear track and laser-probe mass spectrometry seems to be the most promising approach for this study, it is essential to include additional tracers like spallogenic ^{15}N, in addition to noble gases, to distinguish between the solar cosmic ray and galactic cosmic ray alternatives. Ion microprobe studies of spallogenic lithium and boron isotopes in solar flare-irradiated grains may also provide additional clues in this regard. Complementary studies of solar flare heavy-ion tracks and solar flare proton-produced spallogenic nuclides, in specially selected extraterrestrial samples will hopefully consolidate and further our present understanding of the past solar activity.

PART IV
The Sun and Climate

THE FAINT YOUNG SUN PROBLEM

JAMES F. KASTING
Pennsylvania State University

and

DAVID H. GRINSPOON
University of Arizona

The Sun was almost certainly much less bright during the early stages of its main-sequence lifetime. In the absence of some compensating factor, the Earth's mean surface temperature would therefore have been below the freezing point of water prior to ~ 2 Gyr ago. Geologic evidence for liquid water as early as 3.8 Gyr ago implies that the Earth was never this cold. This discrepancy can be resolved if the greenhouse effect of the early atmosphere was much larger than today. The most likely cause of an enhanced greenhouse effect is an increase in atmospheric carbon dioxide (CO_2) concentrations by a factor of ~1000 or more compared to today. Such an increase could have resulted from feedbacks inherent in the carbonate-silicate geochemical cycle that controls the atmospheric CO_2 level over long time scales. Indeed, if the continents were originally much smaller, the rate of silicate weathering would have been slow, and the CO_2 partial pressure could have been as high as 10 bar. In this case, the early Earth could have been quite warm (~85°C), despite the low solar luminosity. A major unresolved question is the extent to which the carbonate-silicate cycle, and thus the Earth's CO_2 level and climate, is affected by the presence of life. If the biota strongly enhance the rate of silicate weathering, they may play an important role in regulating the Earth's surface temperature.

I. INTRODUCTION

According to well-accepted models of stellar evolution, the solar constant is not really constant but has been increasing steadily throughout the main-sequence lifetime of the Sun (Newman and Rood 1977; Gough 1981c; Gilliland 1989; chapter by Dearborn). The increase in luminosity is a consequence of the conversion of hydrogen into helium: this change increases the mean atomic weight and density of the Sun, which in turn produces higher core temperatures, a corresponding increase in the rate of fusion reactions, and hence an increase in luminosity. Gough (1981c) estimated that the luminosity of the young Sun was approximately 30% less than the present value. According to Gough, the rate of increase of luminosity with time can be represented as

$$L(t) = [1 + 0.4(1 - t/t_o)]^{-1}L_o \qquad (1)$$

where L_o is the present solar luminosity and t_o (= 4.6 Gyr) is the present age of the Sun.

It should be emphasized that the predicted increase in solar luminosity is a very robust conclusion of stellar-evolution theory (Newman and Rood 1977; Gilliland 1989). Minor revisions in the standard solar model which may be required to resolve the solar-neutrino problem are unlikely to overturn this prediction (*ibid.;* see the chapter by Wolfburg and Kocharov). Radical revisions of the standard model, such as the main-sequence mass-loss hypothesis proposed by Willson et al. (1987), could reverse this conclusion and result in a bright young Sun; however, such models are considered unlikely and would, in any case, create more problems than they would solve (Sec. V).

A change in solar luminosity over time would have affected the Earth's radiation balance and, thus, its climate. If the Earth is considered to radiate like a blackbody, its effective radiating temperature T_e can be determined by balancing the solar radiation absorbed by the planet with the infrared radiation emitted to space. A simple geometry calculation yields

$$T_e = [S(1 - A)/4\sigma]^{1/4} \qquad (2)$$

where σ is the Stefan-Boltzmann constant, S is the solar constant (~ 1360 W m^{-2}) and A is the planetary albedo (≈ 0.3) (see, e.g., Chamberlain and Hunten 1987). The relevant albedo here is the Bond albedo, which is the percentage of the total incident solar radiation reflected back into space. Figure 1 shows T_e as a function of planetary albedo for three values of S, representing the present solar luminosity and the approximate luminosity 3.0 and 4.6 Gyr ago. The present effective radiating temperature of the Earth is ~ 255 K. The combination of Eqs. (1) and (2) results in an increase of T_e by about 20 deg over geologic time if the albedo of the Earth is assumed to remain constant.

The Earth's mean surface temperature T_s is higher than T_e by about 33 K because of the greenhouse effect of the Earth's atmosphere. The greenhouse effect is caused by the difference in atmospheric opacity in the visible and

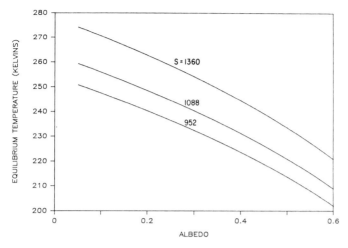

Fig. 1. Effective radiating temperature of the Earth T_e as a function of planetary albedo A for three different values of the solar constant S. The present solar constant is 1360 W m^{-2}. The two lower values represent decreases in solar luminosity by 20 and 30% (figure from Grinspoon 1988).

infrared regions of the electromagnetic spectrum. The atmosphere is relatively transparent to incoming solar radiation, but absorbs a large fraction of the outgoing infrared radiation. Most of the absorption is attributable to vibration-rotation bands of H_2O and CO_2 and to the pure rotation band of H_2O.

Sagan and Mullen (1972) first pointed out the implications of this change in solar luminosity for the Earth's climatic evolution. Using a simple model of the greenhouse effect, they showed that lower solar luminosity would have resulted in T_s below the freezing point of water for roughly the first 2 Gyr of the Earth's history if the atmospheric composition and planetary albedo had remained constant. A more recent calculation with a one-dimensional, radiative-convective, climate model (Kasting 1989a) shows the magnitude of the problem (Fig. 2). The solid curve in Fig. 2 shows the change in solar luminosity with time (Eq. 1); the lower dashed curve shows the effective radiating temperature for the case of a constant planetary albedo of 0.3 (Eq. 2); the upper dashed curve shows the surface temperature calculated by the climate model, assuming a fixed distribution of relative humidity. Details of the climate model can be found in Kasting and Ackerman (1986). The shaded area between the dashed curves represents the greenhouse effect. The greenhouse effect increases as the Earth warms because a warmer atmosphere contains more water vapor.

The surface-temperature curve shown in Fig. 2 cannot, however, be correct. As Sagan and Mullen pointed out, the presence of pillow lavas, mud cracks and ripple marks in 3.2 Gyr-old rocks strongly suggests that liquid

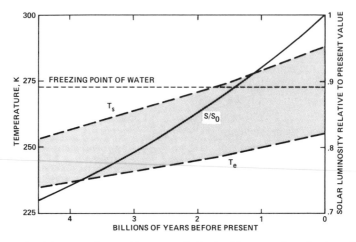

Fig. 2. The faint-young-Sun problem, as calculated with a one-dimensional, radiative-convective, climate model. The solid curve is Gough's solar-luminosity parameterization (Eq. 1); the dashed curves represent the effective radiating temperature T_e and the surface temperature T_s. The shaded area shows the magnitude of the greenhouse effect (figure from Kasting 1989a).

water was already present on the Earth's surface by this time. Indeed, sedimentary rocks (which must have formed in liquid water) are now known to have been deposited as early as 3.8 Gyr ago (Holland 1984). The discrepancy between the predictions of simple climate models and the actual climate record has since come to be known as the "faint-young-Sun" paradox.

Sagan and Mullen realized that the paradox could be resolved if the Earth's albedo were significantly lower in the past or if the greenhouse effect of its atmosphere were larger. They concluded that a large change in the Earth's albedo was unlikely: any decrease in cloudiness that might result from lower surface temperatures would likely be compensated by an increase in snow and ice cover. This argument might not hold if the Earth's surface was mostly water covered, and various investigators have continued to suggest that changes in cloud cover are the answer to the problem (Henderson-Sellers 1979; Rossow et al. 1982). While this hypothesis cannot be ruled out, it makes sense only if the early Earth was cold. From appearances, however, the climate of the early Archean Earth was actually warmer than today, based on the absence of glaciation prior to about 2.7 Gyr ago (Frakes 1979; Kasting 1989a, and references therein). (Indeed, oxygen isotopes from cherts [Knauth and Epstein 1976] and chert-phosphate pairs [Karhu and Epstein 1986] imply warm surface temperatures throughout the Precambrian; however, they may reflect temperatures during diagenesis rather than the temperature of the oceans.) Cloud/climate feedbacks are certainly possible, but they are probably not the solution to the faint-young-Sun problem.

The solution favored by Sagan and Mullen was an increase in the greenhouse effect caused by the presence of ~ 10 ppm (parts per million) of atmospheric ammonia (NH_3). Detailed climate-model calculations by Kuhn and Atreya (1979) confirmed that 10 to 100 ppm of ammonia could have kept the Earth's surface temperature above freezing. However, their own photochemical calculations and those of Kasting (1982) showed that the lifetime of NH_3 against photochemical decomposition to N_2 and H_2 is extremely short. Thus, it is doubtful that even these relatively low NH_3 concentrations could have been maintained on the primitive Earth. The question of ammonia stability may need to be reopened in light of new developments involving sulfur photochemistry. This is discussed further in Sec. V.

The solution that most authors have recently settled on, however, is an increase in atmospheric carbon dioxide concentrations. The idea that the early Earth was warmed by enhanced levels of CO_2 was first suggested by Hart (1978). (Actually, Hart also assumed the presence of various reduced gases, including NH_3, in his model.) Since Hart's paper was published, several independent studies employing radiative-convective climate models have shown that the CO_2 concentrations proposed in Hart's model (a few tenths of a bar early in the Earth's history) could have provided sufficient warming to keep the oceans from freezing (Owen et al. 1979; Kuhn and Kasting 1983; Kasting et al. 1984; Kiehl and Dickinson 1987; Kasting 1989a). The Owen et al. calculations have been shown to be in error (Kiehl and Dickinson 1987), but the results of all the other models agree quite closely, despite differences in CO_2 absorption coefficients and other model assumptions. There is little doubt that CO_2 partial pressures of a few tenths of a bar or higher could have kept the early Earth warm. The real question is whether such high CO_2 concentrations are reasonable. We argue here that high CO_2 levels in the primitive atmosphere are not only reasonable but almost unavoidable, unless other factors (e.g., changes in cloud cover, the presence of NH_3) helped to keep the Earth warm. To understand this argument, it is necessary to consider the process that controls atmospheric CO_2 concentrations over long time scales, namely, the CO_2 geochemical cycle (also referred to as the carbonate-silicate cycle). The next section describes how this cycle operates.

II. THE CO_2 GEOCHEMICAL CYCLE

The Earth's total surface reservoir of carbon is about 10^{23} g, enough to produce a carbon dioxide partial pressure of ~ 60 bar, were all of it present as gaseous CO_2 (Ronov and Yaroshevsky 1967; Holland 1978). Most of this carbon is contained in carbonate rocks on the continents. A much smaller amount (4×10^{19} g) is present in the ocean as carbonate ($CO_3^=$) and bicarbonate (HCO_3^-) ions, and a still smaller (but growing) amount (7×10^{17} g) is present in the atmosphere. Over time scales longer than ~ 1000 yr, the atmosphere and the ocean remain in approximate equilibrium with each other and can be treated as a single reservoir. CO_2 is removed from the atmosphere/ocean reservoir primarily by the weathering of silicate rocks on the continents followed by the deposition of carbonate sediments on the sea floor. About 20% of atmospheric CO_2 is removed by photosynthesis followed by burial of organic carbon; we address this complication in Sec. III. If one represents silicate rocks in general by the mineral wollastonite ($CaSiO_3$), the CO_2 loss process can be described by the reactions

$$CaSiO_3 + 2\ CO_2 + H_2O \rightarrow Ca^{++} + 2\ HCO_3^- + SiO_2 \qquad (3)$$

$$Ca^{++} + 2\ HCO_3^- \rightarrow CaCO_3 + CO_2 + H_2O \qquad (4)$$

$$\text{Net: } CaSiO_3 + CO_2 \rightarrow CaCO_3 + SiO_2. \qquad (5)$$

Reaction (3) represents the silicate weathering process; reaction (4) represents the formation of calcium carbonate. The overall reaction (except for the uni-directional arrow) is the same as the *Urey equilibrium* that was thought by Urey (1952) to control directly the CO_2 partial pressure on the Earth. Urey's equilibrium model is no longer accepted, but the general importance to atmospheric pCO_2 of interconversions between carbonate rocks and silicates is now universally recognized.

CO₂ is returned to the atmosphere/ocean system when old sea floor is subducted and carbonate sediments are subjected to higher temperatures and pressures. Under these conditions, reaction (5) goes in the opposite direction: calcium silicate is reformed, and gaseous CO_2 is released. Much of this CO_2 escapes to the Earth's surface through volcanoes. This reverse process is termed carbonate metamorphism or, equivalently, silicate reconstitution. On the young Earth, the rate of carbonate metamorphism could have been augmented by faster rates of tectonic cycling and by impact processing of carbonate-rich sediments.

From the perspective of the faint-young-Sun problem, the most interesting feature of the carbonate-silicate cycle is that the rates of the weathering reactions [(3) and other analogous reactions] are strongly dependent on temperature. This dependence arises in two ways. First, the reaction rates themselves increase with temperature. Second, and more importantly, weathering reactions require liquid water, and rates of precipitation and runoff increase with temperature, according to atmospheric general-circulation models (see, e.g., Manabe and Wetherald 1980). (The reason is that increased temperature causes increased evaporation, which must be balanced by an increase in precipitation.) Various parameterizations of the relationship between surface temperature and weathering rates have been proposed (Walker et al. 1981; Berner et al. 1983; Volk 1987); these differ in detail but all suggest a strong positive correlation.

The temperature dependence of the silicate weathering rate leads to an overall negative feedback between atmospheric CO_2 levels and surface temperature, as first pointed out by Walker et al. (1981). If the surface temperature were to decrease for some reason (such as, for example, a decrease in solar luminosity), the weathering rate would also decrease, and carbon dioxide would begin to accumulate in the atmosphere. The increase in CO_2 would cause an increase in the greenhouse effect, which would tend to counteract the original temperature decrease. Exactly the reverse would happen if the climate became warmer: the weathering rate would increase, pCO_2 would decrease, and the greenhouse effect would become smaller. An implicit assumption here is that the rate of CO_2 production from carbonate metamorphism is unaffected by changes in surface temperature. This assumption is reasonable, because the large size of the carbonate rock reservoir guarantees that carbonate metamorphism would continue even if carbonate deposition were to cease.

The consequences of this negative feedback mechanism for the faint-young-Sun problem are straightforward. The climatic effects of reduced solar luminosity should have been offset by an increase in atmospheric carbon-dioxide concentrations. The equilibrium surface temperature must have remained above freezing because without liquid water, silicate weathering would come to a virtual halt, and carbon dioxide would simply accumulate in the atmosphere. The modern rate of CO_2 release from volcanoes would create a 1-bar CO_2 atmosphere in only 20 Myr if carbonates were not forming (Walker et al. 1981), so the response time of the system is quite fast from a geologic perspective. Low atmospheric CO_2 levels would be possible in the past only if some other mechanism was keeping the surface warm, so that silicate weathering could proceed, or if CO_2 was being efficiently removed by other processes (see Sec. V). Barring very unusual circumstances, it appears as if the faint-young-Sun problem is effectively solved.

III. EFFECTS OF THE BIOTA

The CO_2 geochemical cycle is presently modulated by the biota. One obvious example is found in the oceans. Calcium carbonate formation (reaction 4 above) can be largely attributed to the secretion of shells by plankton and other marine organisms. This particular mode of biological intervention has no obvious effect on atmospheric CO_2 levels. If the biota were not catalyzing carbonate formation, oceanic calcium and bicarbonate concentrations would simply increase until $CaCO_3$ began precipitating abiotically. No change in atmospheric pCO_2 is required. The factor that ultimately controls the rate of carbonate deposition is the availability of calcium ions, not bicarbonate.

The biota do affect pCO_2 in other important ways, however. Land plants enhance silicate weathering rates by pumping up the carbon dioxide partial pressure in soils by a factor of 10 to 40 over the atmospheric value (Lovelock and Whitfield 1982). Photosynthesis on the land and in the oceans creates organic carbon, some of which is buried in sediments. Both of these processes tend to draw down atmospheric CO_2 levels. Thus, the Earth today is probably cooler than it would be in the absence of life. This has led some authors to suggest that the Earth's climate is controlled by the biota (Margulis and Lovelock 1974; Lovelock 1979,1988). Lovelock calls his theory the "Gaia Hypothesis," in reference to the ancient Greek goddess of mother Earth. According to Lovelock, the Earth would have long since become uninhabitable were it not for the "homeostatic" modulation of climate by organisms.

Because the issue of biological control of the Earth's climate is of great interest, it is worth examining in more detail. The effect of land plants on atmospheric CO_2 levels can be studied using existing numerical models of the CO_2 geochemical cycle. According to Berner et al. (1983) (henceforth BLAG), the dependence of the silicate weathering rate f_w on surface temperature T is given by

$$f_w = 1 + 0.087(T - T_o) + 0.0019(T - T_o)^2 \qquad (6)$$

where T_o is the present mean surface temperature (288 K). The weathering rate factor f_w is equal to 1 for the modern Earth, because the carbonate-silicate cycle is assumed to be in balance. No direct dependence of the weathering rate on CO_2 partial pressure is assumed in this model. The CO_2 greenhouse effect is parameterized in the BLAG model as

$$T - T_o = 2.88 \ln(P/P_o) \qquad (7)$$

where P indicates atmospheric pCO_2, and P_o represents the present atmospheric CO_2 partial pressure.

If this parameterization of the carbonate-silicate cycle were complete, land plants would have no effect whatsoever on atmospheric carbon-dioxide concentrations because the weathering rate would be independent of CO_2 levels. Laboratory studies indicate, however, that the weathering rate of silicate minerals varies approximately as $pCO_2^{0.3}$ for CO_2 partial pressures of 2 to 20 bar and temperatures of 100 to 200°C (Lagache 1965,1976; Walker et al. 1981). This dependence is not included in the BLAG model because the authors realized that soil pCO_2 is generally independent of atmospheric pCO_2 on the modern Earth. If one wishes to examine the effect of land plants on the CO_2 cycle, however, the effect of CO_2 partial pressure on the weathering rate must be explicitly included.

To pursue this question quantitatively, let us assume that the high-temperature, high-pressure laboratory data of Lagache can be extrapolated to conditions on the Earth's surface, so that the weathering rate varies as $pCO_2^{0.3}$. (This assumption is difficult to test, because the rates of silicate weathering reactions are extremely slow at low CO_2 partial pressures. We will see, however, that this is only one of many uncertainties involved in quantitatively estimating silicate weathering rates.) To maximize the climatic effect, let us further assume that removing land plants from the system would reduce soil pCO_2 everywhere by a factor of 40. Equation (3) can then be rewritten as

$$f_w = [1 + 0.087(T - T_o) + 0.0019(T - T_o)^2] [P_s/(40 P_o)]^{0.3} \quad (8)$$

where P_s is the CO_2 partial pressure in the soil. On the modern Earth, $P_s = 40 P_o$ so we obtain $f_w = 1$, as before. On a vegetation-free Earth, P_s would presumably be equal to the atmospheric value P_o. Thus, if land plants were removed, and if the surface temperature remained constant, f_w would be reduced by a factor of $40^{-0.3} \approx 1/3$.

The carbon cycle is only balanced, however, when f_w is equal to 1. Hence, in the absence of land plants, atmospheric pCO_2 and surface temperature would have to increase to bring the silicate weathering rate back to its present value. The new equilibrium values for CO_2 partial pressure and surface temperature can be obtained by substituting Eq. (7) into Eq. (8) and solving for P/P_o. The result is: $P/P_o \approx 9$, and $T - T_o \approx 6.3$ K. In other words, even if land plants are assumed to pump up soil CO_2 by a full factor of 40 (which is probably an overestimate), the effect of eliminating them would be to increase the Earth's surface temperature by only about 6 deg. These figures are approximate because Eqs. (6) and (7) are both somewhat uncertain. Nevertheless, this calculation implies that the cooling caused by present-day land plants is relatively modest.

The net cooling effect of the biota should be larger than this because of the influence of the organic carbon cycle. Today, about 20% of the carbon in sediments is organic carbon rather than carbonate (Holland 1978). Since the isotopic composition of carbonate carbon has generally remained close to its present value, this same sedimentary ratio of organic carbon to carbonate has apparently been maintained throughout most of the Earth's history (Schidlowski et al. 1975). Thus, if life were suddenly eliminated and only the carbonate-silicate cycle remained in operation, f_w would have to increase to 1.25 to compensate for the lack of organic carbon burial. According to Eqs. (7) and (8), this would raise surface temperature by another 1.4 deg, bringing the total increase to approximately 8 K. Even so, a lifeless modern Earth would apparently be no warmer than the real Earth was during the Cretaceous, when the dinosaurs and many different varieties of plant life flourished. Thus, Lovelock's Gaia Hypothesis is probably only partly correct. The biota evidently do modulate the Earth's climate, but the Earth would probably still be habitable even if it were not inhabited.

One caveat should be added to the above discussion. Our analysis of the influence of land plants on silicate weathering presumes that the only important effect of vegetation is to enhance soil CO_2 levels. Recently, Schwartzman and Volk (1989) have argued that this is not a safe assumption. The biota may also accelerate chemical weathering by stabilizing soil (so that the silicate minerals stay in contact with carbonated water), by generating organic acids, and by contributing to physical weathering through microfracturing of mineral grains. Instead of enhancing weathering rates by a factor of 3, as Eq. (8) predicts, they argue that the biota may accelerate weathering by a factor of up to 1000. If this were true, the CO_2 partial pressure on a lifeless Earth might be as high as a few tenths of a bar, and the surface temperature might be up to 60 K warmer than observed. One cannot use Eq. (7) for this analysis, because the greenhouse calculations on which this parameterization is based do not extend to high enough CO_2 levels. If Schwartzman and Volk are correct, an uninhabited Earth might be only marginally habitable; 75°C is too warm for all organisms other than a handful of thermophilic bacteria. We feel that they have overestimated the biotic effect on weathering; however, we acknowledge that the interaction of life with its environment is complex, and view this as an interesting topic for further research.

IV. A WARM EARLY EARTH?

As mentioned in Sec. I, the geologic record of the early Archean Earth (3.5 to 2.8 Gyr ago) contains no evidence of continental glaciation. Taken at face value, this implies either an absence of polar continents during this time or an absence of polar ice. This argument is not iron clad, since the rock record becomes increasingly sparse as one goes back further and further in time; however, let us assume for the moment that it is correct. Since 700 Myr is a long time for (presumably) drifting continents to stay out of the polar regions, the most likely interpretation is that the early Archean Earth was warm and had no polar caps. In a sense, this should not be considered unusual, since the Earth appears to have been ice-free during some 80 to 90% of its history (Frakes 1979).

Since the Sun was less luminous at this time, however, this implies that the greenhouse effect of the early atmosphere was substantially greater than would have been necessary simply to keep the oceans from freezing. If carbon dioxide was responsible for keeping the Earth warm, its concentration must have been bolstered by something other than the negative feedback described in Sec. II. Either the CO_2 production rate from carbonate metamorphism was higher, or the CO_2 loss rate from silicate weathering was lower than one would estimate from Eqs. (7) and (8). (Equation (8) is more appropriate here than is Eq. (4) because vascular land plants did not appear until the late Silurian Period, around 480 Myr ago.)

Fortunately, either one of these possible explanations makes good physical sense. A younger, hotter Earth should have been more active tectonically, so rates of sea-floor creation and destruction would presumably have been faster. Thus, carbonate sediments should have been recycled more rapidly. The continents may also have been substantially smaller at this time (Walker 1985, and references therein) and this would have affected the carbonate-silicate cycle in two ways. First, a reduction in the area of the continents would have reduced the amount of carbonate rock that could be stored on them, leaving more CO_2 to be partitioned between the atmosphere, ocean and sea-floor sediments; and second, a smaller land surface would have provided a smaller exposed area on which silicate weathering could occur. If the continents were half as large as today, f_w would have to have doubled to compensate. If the rate of carbonate metamorphism was also twice as high, f_w would have had to increase by another factor of 2. Thus, the carbonate-silicate cycle could have done more than simply damp the cooling caused by the faint young Sun; other plausible changes in the cycle could well have produced a dense CO_2 atmosphere and warm Archean Earth.

Walker (1985) carried this argument one step further and developed a model of the carbonate-silicate cycle on an entirely ocean-covered early Earth. Such conditions are not likely to have existed during the Archean period, but might possibly have applied during the Hadean Era, prior to 3.8 Gyr ago. Walker's conclusions were as follows. If the Earth's full CO_2 inventory of 60 bar were present at the surface (Sec. II), as much as 10 bar of CO_2 may have resided in the atmosphere. This dense CO_2 atmosphere should have persisted until the continents began to form and weathering of silicate rocks on land became an important loss process for CO_2. Following up on Walker's suggestion, Kasting and Ackerman (1986) performed radiative-convective climate-model calculations for dense CO_2 atmospheres, assuming a 30% decrease in solar luminosity. Their results are summarized in Fig. 3. A 10-bar CO_2 atmosphere on the early Earth would have produced a mean surface temperature of approximately 85°C, according to their model. This is warm compared to today, but not so warm as to preclude either the existence of liquid water, which is stable all the way up to its critical temperature (374°C) (*ibid.*), or the origin of life. Indeed, some theories of early biological evolution favor a warm early Earth (Pace et al. 1986; Lake 1988). As the above discussion indicates, a warm early Earth is perfectly consistent with a faint early Sun.

V. OTHER CLIMATIC FACTORS ON THE EARLY EARTH

Two additional mechanisms have been suggested recently for keeping the early Earth warm. The first is astrophysical in nature. Willson et al. (1987) and Guzik et al. (1987) have suggested that the Sun continued to lose appreciable amounts of mass during its first billion years on the main sequence as a consequence of a vigorous, pulsation-driven, solar wind. If true, this would imply that the young Sun was more massive and, hence, more luminous than the present Sun. The possibility of significant mass loss has

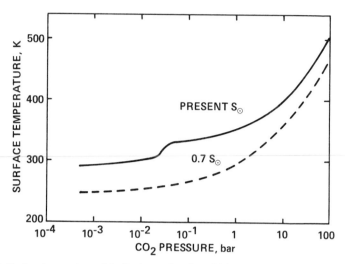

Fig. 3. Surface temperature of the Earth as a function of CO_2 partial pressure for present (solid curve) and 30% reduced solar luminosity (dashed curve). Calculations were performed with a radiative-convective climate model (Kasting and Ackerman 1986). The assumed background atmosphere was 0.8 bar of N_2, plus 0.2 bar of O_2 for the present luminosity case.

not been ruled out by astronomical observations, even though it is totally at odds with standard theories of stellar evolution. It is not an attractive hypothesis from the standpoint of the Earth's climate. A bright young Sun would be much more difficult to deal with than a faint one. According to the Guzik et al. model, the young Sun could have been as much as 15 times more luminous than today (see Gilliland [1989] for a more complete discussion). The increase in the solar flux at the Earth's orbit would have been even greater, since the radius of the orbit would have decreased in response to the stronger gravitational force of the Sun. Such intense solar heating would almost certainly have vaporized the oceans and driven off all the Earth's water into space. A recent theoretical study by Kasting (1988) suggests that a 40% increase in solar luminosity relative to today's value would completely vaporize the oceans, creating a runaway-greenhouse atmosphere (Fig. 4b). Indeed, the stratosphere could become wet, causing the Earth to lose its water by photodissociation and hydrogen escape, if solar luminosity was to increase by as little as 10% over the present value (Fig. 4a). The actual solar fluxes at which these effects would occur could be higher than these values if increased cloud cover counteracted the rise in surface temperature. Nevertheless, the calculations imply that the critical solar flux required to destabilize the Earth's oceans is not much greater than the present flux. If the Sun was brighter than this for some period in its early history, the Earth's present water inventory must have been derived entirely from subsequent outgassing or from come-

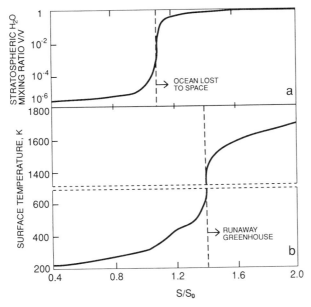

Fig. 4. Response of the Earth's atmosphere to changes in solar flux, according to the climate model of Kasting (1988): (a) stratospheric H_2O mixing ratio; (b) surface temperature. S_o is the present solar flux at the Earth's orbit.

tary impacts. While such a scenario is possible, it becomes less and less plausible as the duration of the enhanced luminosity phase is increased.

A second mechanism for warming the early Earth is to stabilize ammonia against photolysis by shielding it with sulfur gases (Kasting 1990b). The idea of ammonia as a greenhouse gas in the primitive atmosphere is not new; this was Sagan and Mullen's original suggestion (Sec. I). What is new is the hypothesis that sulfur gases may have played a critical role in ammonia photochemistry. Both SO_2 and H_2S are released from volcanoes today, and both gases absorb strongly in the critical 200 to 225 nm wavelength region where most ammonia photolysis would have occurred (Fig. 5). Elemental sulfur vapor, consisting of S_8 rings and other sulfur molecules of different lengths, should also have absorbed radiation at these wavelengths (Kasting et al. 1989). Sulfur vapor would have been created from SO_2 and H_2S by photochemical reactions. If these sulfur gases provided an effective ultraviolet shield, the photochemical lifetime of ammonia could have been greatly extended, and atmospheric ammonia concentrations could have been correspondingly enhanced. This possibility deserves investigation with a photochemical model.

We have now identified a variety of mechanisms for warming the early Earth. Any, or all, of these processes could have helped to solve the faint-

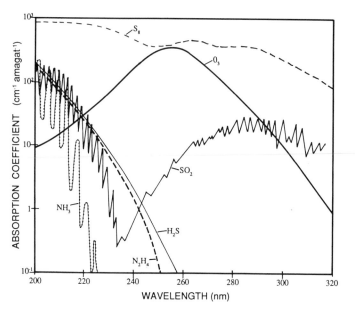

Fig. 5. Absorption cross sections for various gases in the near-ultraviolet region of the spectrum. The S_8 curve is that of sulfur dissolved in ethyl alcohol. The N_2H_4 curve has been extrapolated beyond the existing measurements (figure from Kasting 1989*b*).

young-Sun problem. Before shelving the problem altogether, however, let us mention two factors that might have tended to cool the early climate. The first is chemical in nature. Proponents of dense, CO_2 primitive atmospheres (see, e.g., Walker 1985; Kasting and Ackerman 1986) have assumed that such an atmosphere is stable against photolytic conversion to CO and O. This assumption is grounded on the observation that Venus and Mars have stable CO_2 atmospheres, together with photochemical modeling studies (see, e.g., Kasting et al. 1984) that support the idea of CO_2 stability. (The question of why the Martian atmosphere is stable was resolved almost 20 yr ago when it was learned that the recombination of CO and O could be catalyzed by the by-products of water vapor photolysis [McElroy and Donahue 1972].)

However, previous investigators have not considered the effect of impacts on early atmospheric composition. Based on the lunar cratering record, the primitive Earth should have been bombarded regularly and sometimes spectacularly up until approximately 3.8 Gyr ago (Shoemaker 1982). Models based on this cratering record suggest that the impact flux during this epoch was enhanced over the present-day flux by many orders of magnitude (Hartmann 1981; Maher and Stevenson 1988). If these impactors contained elemental iron, some of this iron should have reacted with atmospheric CO_2 according to the reaction

$$Fe + CO_2 \rightarrow FeO + CO. \tag{9}$$

The production rate of CO is uncertain, but it could have been quite large during the first few 100 Myr of the Earth's history when the impact flux was high. A recently completed study (Kasting 1990a) indicates that a significant fraction of the CO_2 in the Earth's atmosphere could have been converted into CO by this mechanism prior to $\sim$ 4 Gyr ago. The precise amount depends on the impact rate, on the percentage of iron in the impactors, and on the density of the primitive atmosphere. As CO does not absorb well in the infrared, this conversion could have reduced the magnitude of the greenhouse effect and thereby cooled the Earth's surface. Some fraction of the atmosphere's carbon would doubtless have remained as CO_2, so the greenhouse effect may still have been quite large. However, the question of whether the early Earth was hot or cold becomes even more confused.

Another way in which impacts may have affected the Earth's climate is by lofting dust into the stratosphere. The large impact event which ended the Cretaceous Period 65 Myr ago may have raised enough fine-grained debris to create an optically thick global dust cloud (Toon et al. 1982). This cloud could have lasted long enough to create a profound climatic disturbance, which may have been a major contributor to the mass extinctions that occurred at the Cretaceous-Tertiary (K-T) boundary. Objects of the size inferred for the K-T impactor are presently thought to strike the Earth about once or twice every 10^8 yr. Severe impact-induced climatic perturbations are therefore extremely rare. The frequency of such events on the primitive Earth would have been much greater. Grinspoon and Sagan (1987,1991) have modeled the effect of these more frequent impact events on the early climate. They found that an optically thick dust cloud could have been continuously maintained if the early impact flux exceeded the present flux by a factor of 10^6 to 10^7. Although model uncertainties are large, such a flux might have persisted for 150 to 250 Myr after the close of accretion.

The climatic effects of this dust are uncertain. Toon et al. (1982) found that, for the case of heavy dust loading, the lower atmosphere becomes isothermal, so that $T_s = T_e$. In this case, the atmospheric greenhouse effect is effectively shut down, since essentially all of the incident solar radiation is absorbed in the stratosphere. The Earth's surface might therefore have become quite cold. At the same time, though, the dust would have helped to trap any heat released at the surface by the impactors, along with any residual heat left over from accretion and core formation. Thus, the earliest post-accretion period may still have been warm (Grinspoon and Sagan 1987,1991). This epoch of continuous dust shrouding was presumably followed by a long period of intermittent dust loading, the frequency of which diminished as the intensity of the bombardment subsided (*ibid.*). The Earth's mean surface temperature should have gradually stabilized at some value determined by the various other climatic factors discussed earlier.

VI. CONCLUSIONS

In summary, the faint-young-Sun problem was most likely solved by an increase in atmospheric carbon dioxide concentration in the Earth's atmosphere brought about by the CO_2 geochemical cycle. Because the loss process for atmospheric CO_2 requires liquid water, and because the Earth is continually resupplying atmospheric CO_2 by carbonate metamorphism, the surface temperature should never have fallen below the point at which the oceans would freeze. Indeed, the early Earth may have been quite warm if carbonate metamorphism was faster and if the continents were originally smaller, so that silicate weathering was inhibited.

Climatic conditions during the first 0.5 Gyr of the Earth's history remain problematic. The geologic record indicates that the early Archean Earth was warm, so we might expect that the period preceding the Archean was warm also. Carbon dioxide, and possibly ammonia, could have overcompensated for the decrease in solar luminosity. On the other hand, the effect of impacts in converting CO_2 into CO and in creating stratospheric dust could have outweighed these other factors and kept the Earth cool. Further investigation may help to shed light on this question but is unlikely to resolve it, given the magnitude of the uncertainties.

The subsequent evolution of the Earth's climate was determined by the interplay between a slowly brightening Sun and a CO_2 geochemical cycle that was slowly winding down. The negative feedback involving silicate weathering ensured that the mean surface temperature remained within reasonable bounds. The feedback process was modified to some extent by the biota, which should have reduced atmospheric CO_2 levels and cooled the Earth's climate. The cooling was probably small (< 10 deg), but could have been larger if the weathering process was strongly affected by the presence of life. The task of elucidating the interaction of life with its environment should occupy many future generations of the Earth's scientists.

Acknowledgments. The authors thank C. Sagan for his comments on the manuscript. This work was supported by a grant to J. K. from NASA's Exobiology Program and from the National Science Foundation.

THE ATMOSPHERE'S RESPONSE TO SOLAR IRRADIATION

D. M. HUNTEN
University of Arizona

J.-C. GÉRARD
University of Liège
and
L. M. FRANÇOIS
University of Michigan

The variability of solar ultraviolet and corpuscular emissions is described, along with their known effects on the atmosphere. The discussion excludes variations in the total output and their possible climatic effects. Subtle variations are seen at and above 30 km, growing to factor-of-2 changes above 200 km. All claims of correlation of the solar cycle with weather and climate must be viewed with great suspicion, if only because any plausible driving forces are exceedingly small. The primitive atmosphere was deficient in the oxygen and ozone that keep the varying ultraviolet away from the surface, and there are some possibilities for effects that might have survived in the geologic record. Even further back, the abundances and isotopic patterns of surviving noble gases suggest an epoch during which a greatly enhanced ultraviolet flux drove a hydrodynamic escape flow of hydrogen.

I. INTRODUCTION

This chapter discusses the effects, known and postulated, of solar variations on the atmosphere. Solar variability is large at very short wavelengths, and has a correspondingly large effect at altitudes above 100 km. Detectable variations persist to the region above 200 nm, but not as far as 300 nm. These wavelengths penetrate as deep as 30 km, and small variations in temperature and ozone amount are found on the time scale of the solar cycle. The mass of the affected regions is around 1/10 of that of the whole atmosphere. There is no established mechanism by which known solar variations can affect surface weather and climate. Although many such effects have been claimed, none of them has stood up to the examination of unbiased critics.

Sections I and II describe the basics of the solar ultraviolet output, how its various wavelength bands and corpuscular output vary, and how it interacts with the atmosphere. Section III discusses the known responses of the middle and upper atmosphere, and refers to the many empirical studies made attempting to establish correlations of solar activity with meteorological variables. With this background, in Secs. IV and V, we turn to the historical record and discuss the possibilities for using it as a guide to the past behavior of the Sun. Curiously, one basic feature that would be expected seems to be absent from the record: the monotonic increase of some 30% in the solar output is not reflected in estimates of past temperatures. It is necessary to offset the expected behavior by postulating an enhanced greenhouse effect. Even further in the past, the noble-gas composition, both elemental and isotopic, may have been controlled by a greatly enhanced ultraviolet output for the first 10 or 100 Myr. These ideas are, however, too new for a consensus to have developed on whether or not they are unique.

A. General Characteristics of the Solar Radiation

The first four columns of Table I summarize the part of the solar spectrum that is known to exhibit substantial variations, and a typical magnitude for these variations. The soft X-ray flux, normally rather small, can be enhanced by orders of magnitude during solar flares. The other regions tend to correlate with various measures of solar activity, such as sunspot numbers and 10.7 cm radio flux. As the fourth column indicates, the size of the variation becomes smaller and smaller at the longer wavelengths; it is undetecta-

TABLE I
Principal Effects of Solar Radiations on the Middle and Upper Atmosphere

Wavelength (nm)	Name[a]	Energy[b]	Variability[c]	Effect	Height range (km)
1–10	soft Xrays		Sporadic	Ionize all	70–100
10–100	XUV	2 ppm	2×	Ionize N_2, O, O_2	100–300
100–120	EUV	6 ppm	30%	Ionize NO	80–100
120–200	VUV	150 ppm	10%	Dissociate O_2	40–130
200–240	UV	0.12%	5%	Dissociate O_2, O_3	20–40
240–300	UV	1.0%	<1%	Dissociate O_3	20–40

[a] Abbreviations in the second column are: X-ray ultraviolet; extreme ultraviolet; vacuum ultraviolet; ultraviolet.
[b] Energies are in fractions of the solar constant, 1.4×10^6 erg cm^{-2} s^{-1}.
[c] Variability is discussed in Sec. II.

ble at wavelengths longward of 300 nm (excluding the very small changes of total flux established by satellite radiometric measurements).

In addition to its electromagnetic radiation, the Sun emits a solar wind which interacts with the geomagnetic field to form the magnetosphere. Inputs to the atmosphere include precipitation of electrons and protons (aurora), and fairly large ionospheric currents which cause joule heating. All these phenomena are concentrated at high geomagnetic latitudes, but the heating effects can spread equatorward by convection and conduction. The most comprehensive observational and modeling study is associated with the Dynamics Explorer mission and is summarized in Sec. III.A.

B. Mean Atmospheric Structure: Penetration Depths

Figures 1 to 3 illustrate the typical structure (pressure, temperature) of the atmosphere and typical $1/e$ penetration depths for various wavelengths. Figure 1 includes the names of the various spheres, whose boundaries are defined by maxima and minima in the temperature profile. At 100 km, the density is 10^{-6} of its surface value; the next factor of 10^{-6} is only attained at 400 km, reflecting the much higher temperature of the thermosphere. This temperature is strongly dependent on solar activity; it can vary by a factor as large as 2. The major sources of heat and of variability are solar ionizing photons (wavelengths $<$ 100 nm) and magnetospheric processes whose energy source is traceable to the solar wind (Sec. III).

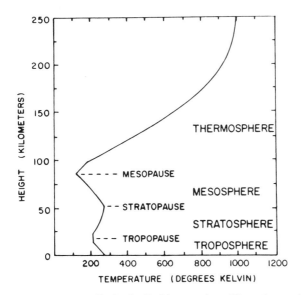

Fig. 1. Average temperature profile for the Earth's atmosphere. The various regions and their boundaries are defined, as shown, by the maxima and minima in the temperature (figure from Chamberlain and Hunten 1987).

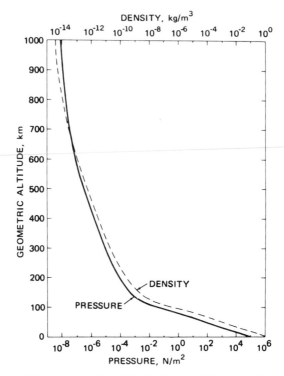

Fig. 2. Pressure and density as functions of height, from the U.S. Standard Atmosphere (1976).

C. Radiation-Atmosphere Interactions: Physical and Chemical Processes

The last two columns of Table I summarize the effects of the various wavelength bands and the height ranges most affected (see also Fig. 3). The principal part of the ionosphere is produced by X-ray ultraviolet (XUV), which is very strongly absorbed; a byproduct of the ionization and recombination is the heating of the thermosphere. Although, as the table shows, the energy involved is small, it is absorbed in a very tenuous medium which can only lose heat by conduction downwards. The first and third ranges are interesting because they penetrate to lower altitudes, even though the energy available for ionization is much smaller. Dissociation of O_2 is so strong that above 120 km, atoms are the principal form of oxygen; however, there is enough vertical mixing and diffusion to keep the ratio of O_2 to N_2 near 0.1 throughout the lower thermosphere. Oxygen atoms continue to be produced as low as 30 km, and most of them combine with O_2 to form ozone, which attains a peak mixing ratio of around 10^{-5} near 30 km. Ultraviolet absorption liberates the atoms, but most of them recombine into ozone very quickly; the result is a

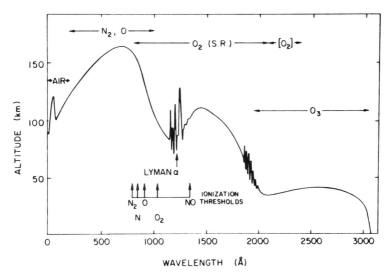

Fig. 3. Effective penetration depths for different wavelengths, that is, the level of $1/e$ absorption for vertical incidence. (Chamberlain and Hunten [1987] based on earlier versions by H. Friedman and L. Herzberg.)

major heating of the middle atmosphere, with the peak at 50 km as illustrated in Fig. 1. Temperature variations in this region are discussed in Sec. III. Photons just around 300 nm, which reach the surface, can still dissociate ozone with production of electronically excited $O(^1D)$, which is metastable and highly reactive and drives a large fraction of urban pollution chemistry.

II. VARIABILITY OF THE SOLAR SOURCE

A. Present Day

This section gives more detail on the variations summarized in Table I. Three solar components affecting differently the structure and composition of the atmosphere may be distinguished:

1. The total irradiance or "solar constant" which exhibits small variation about a mean value of 1367 W m^{-2} (see the chapter by Fröhlich et al.);
2. The ultraviolet, XUV, and X-ray regions which represent a small fraction of the total energy input of the Sun but show substantial temporal variations;
3. The energetic particle component (solar wind and occasional proton events) which is strongly modulated by solar activity and disturbances. Its complex interaction with the Earth's magnetic field controls the energy and geographic characteristics of the particle precipitation into the atmosphere.

Radiative Component. The total amount of solar energy reaching the Earth's atmosphere has probably changed considerably during the planet's history. Fundamentally, two sources of variations of the solar energy input to the terrestrial atmosphere may be considered. First, variations of the incident energy and its distribution with latitude are induced by changes in the orbital and rotational parameters of the Earth's motion (Berger 1979). According to the astronomical ("Milankovitch") theory of paleoclimates, the insolation at a given latitude is a function of the total intrinsic solar irradiance, the semi-major axis and the eccentricity of the Earth's orbit around the Sun, the obliquity of the planet and the longitude of perihelion (see the chapter by Berger). The combination of these quasi-periodic variations generates a complex frequency spectrum with major components at 100,000, 23,000 and 19,000 yr (chapter by Berger).

Superimposed on these changes of purely mechanical origin, intrinsic variations of the Sun's luminosity on a wide spectrum of time scales are known or suspected to exist. A long-term increase of the solar luminosity by roughly 6% per Gyr (10^9 yr) is predicted by theories of the evolution of stellar interiors (see the chapter by Dearborn). It is a consequence of the progressive transformations of hydrogen into helium by thermonuclear processes during the Sun's evolution on the main sequence. This variation of the total energy interacting with the Earth's atmosphere and climatic system is in contrast with the relative stability of the global climate, the apparent absence of global glaciation before 2.3 Gyr ago and the continuity of life records since it appeared about 4 Gyr ago. Various mechanisms have been proposed to compensate the past reduced solar irradiance. The most widely accepted one involves larger amounts of CO_2 in the ancient atmosphere and subsequent enhanced greenhouse efficiency (see the chapter by Kasting and Grinspoon).

On a much shorter time scale, recent satellite observations suggest that the total solar irradiance is modulated by the 11-yr solar cycle. Part of the energy is blocked by dark sunspots and subsequently released in faculae. It appears however that the screening effect of spots, which is greatest at solar maximum, is overcompensated by the energy storage and release on a yearly time scale. Recent measurements made with the ACRIM 1 experiment on board the Solar Maximum and the ERB experiment on the Nimbus-7 satellites show a positive correlation between the solar cycle activity as measured by the sunspot index and the total solar irradiance. A peak to peak variation of about 1 W m^{-2} (out of 1367) is observed between solar maximum and minimum (see the chapter by Fröhlich et al.; Willson and Hudson 1988; Foukal and Lean 1988). Somewhat larger fluctuations, up to 0.2 %, occur on time scales of days to weeks (Willson et al. 1986). Since only a little over half a solar cycle was continuously observed with the same space instrument, it is impossible to extrapolate this behavior into the distant past, although it is reasonable to expect the overcompensation to be a stable feature of the solar energy balance.

Long-term modulation of the solar output lacks direct observational support; it is necessary to use proxy data or solar-activity indicators to infer secular or longer-period changes of the solar constant. Sunspot index time series consistently indicate the presence of a 76–80 yr "Gleissberg cycle" modulating the amplitude of the 11-yr cycle (Sonett 1982; Berry 1987). The presence of this cycle has also been claimed in various sets of proxy data: auroral activity (Feynman and Fougere 1984); isotopic composition of ice cores (Johnsen et al. 1970); tree growth (Svenonius and Olausson 1979); solar radius variations (Gilliland 1981); and modern and ancient sedimentary rocks (Sonett and Williams 1985; Williams and Sonett 1985); and sea-surface temperature (Gérard 1990). Gilliland (1981) reported a 76-yr cycle in the solar radius, inferred from 258 yr of transits of Mercury, solar eclipse reports and meridian transit measurements. He concluded that the solar disk size is negatively correlated with the climatic solar cycle. Similarly, Ribes et al. (1987; see also their chapter) presented evidence that the diameter of the Sun was greater in the 17th century (the period of the Maunder minimum), and decreased when solar activity increased in the early 18th century. As discussed in Sec. V, the work of Williams and Sonett (see the chapter by Herbert et al.) has been re-interpreted in terms of a tidal mechanism. The significance of the other claimed periodicities and correlations has generally not been clearly established by quantitative statistical tests.

Ultraviolet radiation.   The solar electromagnetic radiation is the primary source of energy for the terrestrial environment. The ultraviolet radiation shortward of 320 nm represents only about 2% of the total solar irradiance. About 0.01 % of the incident flux is absorbed in the thermosphere above 80 km and 0.2% above the stratopause near 50 km. However, this spectral range is of primary importance since it interacts directly with atmospheric constituents through photochemical processes and controls the thermal structure of the atmosphere above the tropopause.

As discussed in Sec. I, all solar radiation at wavelengths below 300 nm is totally absorbed above the tropopause and therefore does not directly influence the present Earth's climate. In contrast, the stratospheric composition is directly controlled by absorption and dissociation of O_2 in the 175 to 240 nm range (Schumann-Runge bands and Herzberg continuum) leading to the formation of ozone. The 205 to 295 nm spectral region which is predominantly absorbed by O_3 controls the photodissociation and the production of heat in the stratosphere. Consequently, variability of the solar irradiance between 175 and 300 nm is especially important in establishing possible connections between solar activity and climate variability through stratosphere-troposphere-surface couplings. Radiation below $\sim$ 175 nm is absorbed by O_2 in the mesosphere and thermosphere and controls the exospheric temperature. Although the solar ultraviolet radiation very strongly influences the temperature, pressure and fractional composition of the thermosphere, its influence on the troposphere is even more remote and any such influence would imply mechanisms presently uncertain or unknown. The short-term ultraviolet irradiance variations are ascribed to the evolution and rotation on the solar disk of plage regions with enhanced ultraviolet brightness. The periods of short-term variations are typically 27 or 13 days, the later period being not just a harmonic of the former. More generally, the amplitude of the ultraviolet rotation-induced modulation varies from one rotation to another (Donnelly et al. 1983) and depends on the stability of the active regions during the 27-day period.

Photons with wavelengths below 300 nm, in addition to their photochemical effects, form the main energy source for the middle and upper atmosphere. Their variability induces changes in the thermal structure, composition and dynamics. Two quasi-periodic components have been identified in the time variation of the ultraviolet solar radiation. A mean 27-day periodicity is associated with the solar rotation and an 11-yr period is also observed in phase with the solar activity indicators. The 27-day solar rotation variability has been extensively studied and these measurements are not hindered by instrumental stability problems which plague long-term variability investigations. Therefore, studies of the short term variability and the underlying physical processes are presently on firmer grounds than for the 11-yr cycle.

The 27-day variability at wavelength above Lyman α has been reviewed by Heath (1980), Simon (1981), Simon and Brasseur (1983), Lean (1987) and Simon et al. (1988). Most recent sources of information were gathered by the Solar Backscattered Ultra Violet (SBUV) experiment on the Nimbus-7 satellite (Heath and Schlesinger 1984,1986; Donnelly 1988) and the solar ultraviolet spectrometer on board the Solar Mesosphere Explorer (SME) (Rottman and London 1984; Rottman 1985; Simon et al. 1988). Both sets of data exhibit the same type of variation with similar wavelength dependence: a maximum amplitude on the order of 30% at Lyman α declining to about 5% at 205 nm and $< 1\%$ at 300 nm. Fourier analysis of the dependence on solar cycle of the amplitude of the 27-day variability in SME data shows a short-term variability at both Lyman α and 205 nm that decreases drastically from solar maximum to solar minimum conditions. This is a consequence of the absence of active regions on the solar disk at low solar activity.

The XUV variability due to solar rotation was measured with the OSO-3, Atmosphere Explorer (AE: Hinteregger et al. 1977) and Aeros satellites (Schmidtke 1981,1984; Schmidtke et al. 1977). The variability in the XUV is characterized by a larger amplitude of the emission lines formed in the corona than lines originating from the lower chromosphere.

The wavelength dependence of the solar ultraviolet irradiance over a solar cycle is still debated. A very good instrumental stability and periodic recalibrations are needed for long-period satellite observations. Rocket observations carried out with similar instrumentation at different phases of the solar cycle do not necessarily guarantee total homogeneity of the measurement series. A general view of the measurements of the wavelength dependence of the amplitude of the 11-yr cycle obtained over the last 15 yr is illustrated in Fig. 4. Below 102.5 nm, the Atmosphere Explorer data base is the most widely used source of information for aeronomical modeling of the neutral and ionized upper atmosphere. The curve labeled AE-E refers to the corrected reference solar maximum and minimum spectra (21 REFW and F79050N, respectively) (Torr and Torr 1985). The variability in the wavelength region of the Schumann-Runge continuum deduced from the AE data by Torr et al. (1980) is also indicated. The LASP (University of Colorado) rocket observations were made between 1972 and 1977 (minimum) by Rottman (1981) and in 1979–1980 (maximum) by Mount et al. (1980) and Mount and Rottman (1981) using similar instruments and calibration techniques. The SBUV curve was deduced from an empirical relationship based on the temporal variation of the ratio between core and wing irradiances of the Mg II doublet at 280 nm (Heath and Schlesinger 1986), and is thus not a direct measurement. This method was used since aging problems of the instrument prevented the long-term observations from being used directly for solar-cycle studies. The amplitude of the variability deduced from these observations is somewhat larger than that derived from the SME mission, but there are indications that results from these two data sets are converging. The ultraviolet solar spec-

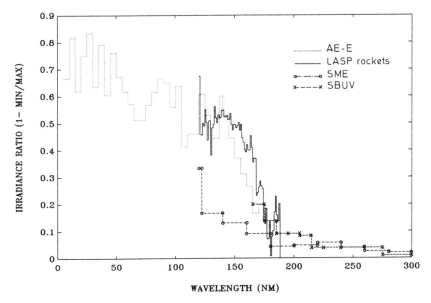

Fig. 4. Wavelength dependence of the solar-cycle variability of the ultraviolet solar irradiance determined from recent rocket rocket and satellite experiments (Simon et al. 1988).

trometer on board the SME satellite obtained solar spectra from 1982 to 1988 which exhibit a maximum/minimum ratio not exceeding 5% at 200 nm. The general behavior of the ultraviolet solar spectrum during a solar cycle is a decreasing amplitude of variability from the X rays to the visible. However, the exact quantitative wavelength dependence of this solar-cycle variation is still debated in spite of major instrumental and calibration improvements.

Lyman α observations indicate a variation by a factor of 2.5 according to the AE-E observations and 1.8 according to the LASP rockets. The OSO-5 (Weeks 1967) and SME (Rottman 1987; Barth et al. 1987) data are consistent with the latter value. At wavelengths above 120 nm, the various data sets do not provide a unique estimate of the magnitude of the solar-cycle variability, although the general trend is a global decrease towards longer wavelengths with an amplitude < 1% at 300 nm. Simultaneous observations between SBUV and SME in 1982 indicate a different rate of decrease at 205 nm (Lean 1987). However, both experiments yield a variability between 200 and 300 nm much smaller than previously reported by Heath and Thekaekara (1977) which reached 35% at 250 nm. Virtually nothing is known concerning possible differences between individual solar cycles and the secular variability of the solar ultraviolet output.

Energetic Particles.   A third important component of the Sun-terrestrial atmosphere interaction is corpuscular bombardment resulting from solar wind and other charged-particle fluxes. Various contributions may be distinguished: the energetic electrons reaching the high-latitude thermosphere after interacting with the geomagnetic field and acceleration, the high-energy solar protons whose flux is considerably enhanced during periods following large solar flares, and galactic cosmic rays (GCR) which originate from outside the heliosphere but whose access to the Earth is partly controlled by solar activity (see the chapter by Jokipii). As discussed in Sec. III, these particles reach the upper stratosphere and lead to catalytic destruction of ozone.

The flux of galactic and solar cosmic rays is modulated by solar activity and exhibits an 11-yr period. The galactic cosmic-ray component is mostly composed of protons (83%) and α particles ($\sim$12%). The geomagnetic field favors access to the high-latitude regions with open field lines. It is modulated by solar activity as a result of the increased interplanetary density which deflects the GCR flux away from the Earth at solar maximum and causes a variation in phase opposition with solar activity. The peak of the ionization rate is located near the tropopause and increases by about 30% from solar maximum to solar minimum conditions (Thorne 1980). A parameterization of the GCR dependence on solar activity and latitude was given by Heaps (1978). This source of NO_2 is small compared to the N_2O oxidation by $O(^1D)$ in the stratosphere. Although it may be the major *in situ* source of odd nitrogen in the lower stratosphere, its contribution is swamped by the NO_x transported from elsewhere.

Intense fluxes of energetic protons (10 to 10^4 MeV) penetrate the polar cap regions as a result of strong flare activity on the Sun. Ionization produced by these solar proton events is considerably more variable than the GCR component. As a consequence of the more modest energies, they deposit most of their energy between 100 and 20 km. Although an event rarely lasts more than a few hours or a few days, large numbers of ions and NO_x molecules are produced. The subsequent ozone depletion predicted by the photochemical theory was observed after major solar proton events in the mesosphere (Weeks et al. 1972; Solomon et al. 1983) and in the stratosphere (Crutzen et al. 1975; Heath et al. 1977; Fabian et al. 1979; Reagan et al. 1981). In August 1972, an abrupt decrease in the ozone concentration was observed by the Nimbus-4 satellite above 4 mbar. This depletion persisted for several weeks above the polar cap while the mid-latitude time-response was obscured by horizontal transport processes.

From a study of the comparison of the strength of the various middle-atmosphere NO_x sources during the period between 1954 and 1979, Jackman et al. (1980) concluded that, on a yearly basis and for latitudes above 50°, the GCR and N_2O oxidation sources are of comparable and relatively constant magnitude. The solar-proton-event source contribution is highly variable, negligible in some years and dominant in other years such as 1972. On a global geographic scale however the GCR and solar-proton-event sources are at least an order of magnitude smaller than the $N_2O + O(^1D)$ source.

Relativistic electron precipitation events were discussed by Thorne (1977, 1980) as a possible variable source of ionization and odd nitrogen production at high latitudes. However, this source is limited to altitudes above ~ 80 km and is not likely to affect directly the ozone in the mesosphere.

Finally, auroral electrons in the keV energy range are a dominant source of ionization at high latitudes. The effect is essentially confined to the thermosphere and only occasionally do they penetrate below the homopause. As discussed in Sec. III, they are efficient sources of NO_x.

The input of auroral electrons into the Earth's atmosphere is highly sporadic. However, groundbased and satellite observations indicate that the total energy flux deposited in the thermosphere by auroral particles is modulated by the 11-yr solar cycle. On the basis of particle energy measurements on board the Atmosphere Explorer—C satellite, Spiro et al. (1982) related the total energy flux and average energy to the Kp index. A statistical study of the auroral electron precipitations measured with the DMSP F2 and F4 and STP P78–1 satellites from 1977 to 1979 was made by Hardy et al. (1987). They found a complex structure depending on magnetic local time, geomagnetic latitude and Kp. Again, a clear positive correlation was found between the total precipitated energy flux and the Kp magnetic index, whereas the average electron energy variation was shown to depend on the region of the auroral zones. The Kp index is not directly connected to the 11-yr solar cycle but is known to be positively correlated with the solar activity as measured by other indicators.

Among other phenomena affected by changes in the characteristics of the auroral precipitation is the electrical conductivity in the auroral zone, which is strongly dependent on the level of geomagnetic (and thus solar) activity (Hardy et al. 1987; Rees et al. 1988).

Another important consequence of the solar-induced modulation of the flux of auroral electrons is the 11-yr dependence of ionization and dissociation in the lower thermosphere. As indicated before, these processes control the production and concentration of nitric oxide (and atomic nitrogen) in the thermosphere and mesosphere. The variability and the dependence of the thermospheric NO density at high latitudes was observed with the OGO-4 (Rusch and Barth 1975; Gérard and Barth 1977), AE (Cravens and Stewart 1978; Cravens et al. 1985) and SME satellites. Subsequent downward turbulent transport into the stratosphere can lead to ozone depletions, especially at high polar winter latitudes, where the absence of NO photodissociation can lead to significant chemical lifetimes and the possibility that NO_x can reach the stratosphere (Garcia et al. 1984; Jackman et al. 1980). This effect appears to be present in SME data (Rusch and Clancy 1989).

B. The Pre-Main-Sequence Phase

During the early life of the Sun, the ultraviolet flux was much higher than today. It has been clear for some time that T Tauri stars are in this class, and more recently a new class, the naked T Tauri stars, has been identified (see the chapters by Walter and Barry and by Bertout et al.). The wavelength regions of particular interest for atmospheric evolution are the XUV and the soft X rays, which are absorbed at the top of the atmosphere and can, if strong enough, drive hydrodynamic escape flows as discussed in Sec. V below. Two independent studies of the variation were published almost simultaneously, both based on observations of T Tauri stars by the International Ultraviolet Explorer (IUE). The results of Zahnle and Walker (1982) are illustrated in Fig. 5; they found that the data could be represented by a flux decreasing as t^{-s} with time t, where s is between 0.5 and 1, depending on the wavelength. Similar results were obtained by Canuto et al. (1982). The enhancements are extremely large, around 150 and 20 at ages of 10 and 100 Myr. Canuto et al. (1983b) made a series of photochemical calculations, based on their estimated fluxes as a function of wavelength; the results are summarized in Sec. IV.C.

Unfortunately, there is not enough information on the relation of this time scale to that of planetary formation and early evolution. Walter and Barry in their chapter suggest that planets are unlikely to be present before 100 Myr; this is true according to current ideas, but these ideas may yet change. In addition, Prinn and Fegley (1989) have pointed out that residual

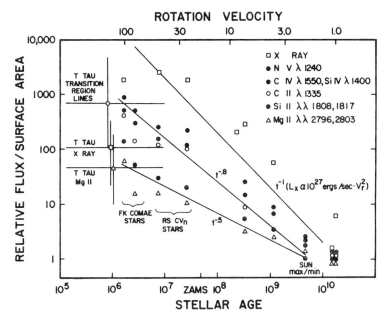

Fig. 5. The XUV flux at the Earth as a function of time, according to Zahnle and Walker (1982).

dust or even gas in the ecliptic plane might prevent any ultraviolet from reaching the planets. One counter-argument is that the same effect would prevent us from seeing the fluxes that we do, but perhaps this is because the direction to the Earth from another system seldom coincides with its putative ecliptic.

Walter and Barry (see their chapter) discuss the considerable body of knowledge of young stars acquired since the surveys quoted above. In particular, the power-law dependence on time does not seem to be well supported, and an exponential with a characteristic time of nearly 1 Gyr is preferred (see their Fig. 5). Although, as they state, the corresponding fluence is large, the long time scale causes difficulties (Pepin 1991). A large flux for a short time (~ 100 Myr) is more efficient from the planetary viewpoint than the same fluence lasting 10 times as long. The latter case requires an enormous initial supply of hydrogen to carry the other gases away. Pepin does stress one important point: plots such as Walter and Barry's Fig. 5 refer to the flux at the surface of the star, not out in the planetary system. In the early phases, the star is still shrinking, and the astronomically interesting fluxes must be multiplied by the square of the radius. The result is substantially more like the power laws indicated in Fig. 5.

From the point of view of understanding the behavior of the early Sun, the theory of hydrodynamic escape is unfortunately too flexible: many scenarios can be accommodated because there are many adjustable parameters. This situation can be expected to improve as we gain more understanding.

III. RESPONSE OF THE PRESENT ATMOSPHERE

As indicated by Fig. 3, solar radiations below 300 nm do not reach the surface, and those below 100 nm barely penetrate past 100 km. With perhaps a few rare exceptions, the effects of particle precipitation and other magnetospheric phenomena are confined to heights of 40 km, and mostly much higher. These same inputs are the ones that show substantial variations, which certainly do affect the middle and upper atmosphere. Small, but significant, effects occur in the lower stratosphere. The behavior of the troposphere is dominated by the total absorbed solar energy, and the changes that are so prominent at higher altitudes are at best very subtle and perhaps nonexistent. The three parts of this section discuss the upper, middle and lower atmosphere.

A. Thermosphere and Exosphere

As indicated in Fig. 1, the thermosphere is the region of positive temperature gradient above 90 km and extending to a height near 250 km. The upper boundary should logically be called the *thermopause,* but the term *exobase* is generally used instead. The exosphere is actually defined as the region in which collisions are negligible and particles can be regarded as executing ballistic orbits, and its base does approximately coincide with the thermopause. The most important heat source is absorption of solar XUV (Table I), which creates the ionosphere; most of the energy winds up as heat, which must be conducted down to the mesopause where it can be radiated. The exosphere is approximately isothermal because it lies above the level where most of the energy is deposited, and because it has a short thermal time constant; both these facts stem from the low density. It is convenient to describe the whole region by the exospheric temperature, which has variations on diurnal and longer time scales. Although the thermospheric gradient also changes, to first order the temperature follows a universal curve until it approaches the exospheric temperature, and levels off there. The typical behaviors discussed here are adapted from the U.S. Standard Atmosphere Supplements (1976), which contains extensive discussions of atmospheric modeling as well as a large number of tables. A much bigger and more recent data base is incorporated in the "MSIS" (mass spectrometer and incoherent scatter) model (Hedin 1983,1987), but the basic results are similar.

The diurnal variation shows a day/night ratio of 1.28 over the equator, with the peak occurring at about 2 p.m. The nighttime behavior can be understood approximately in terms of the heat draining down by conduction from the top of the thermosphere to its base, the mesopause, where the heat is radiated. On this simple picture, one would expect little or no change in the rest of the profile, in reasonable accord with observation. However, the phase of the variation is not well understood, although it has been the subject of many papers.

The effect of solar rotation can be expressed in terms of the $F_{10.7}$ index, which is based on the radio flux at 10.7-cm wavelength, in units of 10^{-22} W m^{-2} Hz^{-1}. The multiplying factor to give temperature is 1.8 deg per unit of flux.

The effect of the sunspot cycle is enormous: the daytime exospheric temperature is observed to vary from around 700 K at minimum to 1200 K (occasionally 1500 K) at maximum. Similarly, the peak electron density varies by a factor around 2, a matter of great interest to users of short-wave radio communication. Temperature changes are obtained from the 10.7-cm index with a multiplier of 3.60. This value, twice that for 27-day variations, presumably represents not only changes in XUV, but also a contribution from auroral heating, that is, energy traceable to the solar wind. This energy is manifested partly in particle precipitation and partly in ionospheric currents that flow mainly at and above 110 km. The input is in the auroral regions but the resulting winds carry much of the heat to considerably lower latitudes. These processes have been studied in detail by the Dynamics Explorer (DE) satellites, which also found major inputs of momentum from the magnetosphere, coupled to the ionosphere through the magnetic field (see reviews by Killeen 1987; Killeen and Roble 1988). These observations have been enriched by a comprehensive theoretical program (Killeen and Roble 1988).

Both ultraviolet and particle precipitation have chemical effects, the most interesting of which for the present purpose is the production of N, NO and NO_2, collectively called NO_x. In the homosphere, discussed below, a single efficiency per ion pair can be defined, but above 90 km the presence of atomic oxygen in substantial amounts makes this impossible. The primary ions N_2^+, O_2^+ and O^+ all convert to NO^+, which has the lowest ionization energy of all major atmospheric ions. The following reactions contribute to NO_x production in the thermosphere but not in the homosphere:

$$O^+ + N_2 \rightarrow NO^+ + N$$
$$N_2^+ + O \rightarrow NO^+ + N \text{ (ion − atom interchange)} \qquad \text{(R1)}$$
$$N_2^+ + e \rightarrow N + N$$
$$NO^+ + e \rightarrow N + O \text{ (dissociative recombination).}$$

Consequently, in the thermosphere each N_2^+ ion formed is capable of producing 2 atomic nitrogen atoms and the total production of NO_x/ion pair increases with altitude from 1.3 to about 3.3.

B. Mesosphere and Stratosphere

This region extends from the tropopause to the mesopause, at approximately 90 km, and is distinguished by being approximately in local radiative equilibrium except for the heat flowing in from the thermosphere. There is also a small amount of ionization by Lyman α, which acts on the rare constituent NO, and by soft X rays. The radiative heating is by absorption of planetary radiation by CO_2 (15 μm) and O_3 (6.3 μm) and the absorption of solar ultraviolet by O_3. The CO_2 band is the principal radiator. Because the O_3/CO_2 ratio decreases upwards, so does the ratio of heating to cooling, and the result is the observed negative temperature gradient in the mesosphere, which is arrested only when the thermospheric heat becomes appreciable.

In the stratosphere, the ultraviolet that is absorbed by ozone begins to be more and more attenuated, enough to more than offset the increasing amounts of ozone. We thus observe a temperature maximum around 50 km, which defines the stratopause, and a steady decline to lower altitudes. The stratopause is not a true boundary: the stratosphere and mesosphere share the same dominant physics, and are sometimes lumped together as the "middle atmosphere."

The only term in the heat balance that depends on solar activity is the ultraviolet flux; one thus expects only a small temperature change, which has been well observed in both the stratosphere and the mesosphere and is discussed below. There can, however, be substantial transient effects on the ionization and, to a lesser degree, the chemistry. Solar flares can greatly increase the soft X-ray flux, giving short-wave radio fadeouts. The protons from large solar proton events can sometimes penetrate the entire mesosphere but more often are confined to high latitudes.

Changes in the ultraviolet flux have only a modest direct effect on ozone amounts, because both the production (by photolysis of O_2) and the destruction are affected in the same way. The net effect depends on wavelength, but, as shown in more detail below, ozone increases with the ultraviolet when the flux at 205 nm is taken as the indicator.

Energetic solar protons or relativistic electrons that penetrate into the middle atmosphere produce considerable amounts of NO_x, which catalytically destroys ozone. Secondary reactions, including those shown above in (R1), convert most of the primary ions into NO^+ and then, by recombination, into N atoms. Consideration of the relative magnitude of dissociation and ionization cross sections leads to the conclusion that approximately 1.3 N atoms are formed for each ion pair created below 100 km (Porter et al. 1976; Rusch et al. 1981). The subsequent reaction of N atoms with O_2 or O_3 forms NO_x, most of which must slowly migrate downwards until it reaches the troposphere and is rained out. Any such enhancement of NO_x increases the destruction of ozone at high altitudes, where the effect of a single large proton event can be seen (Heath et al. 1977). It has also been suggested that the cumulative effect of particle events can explain an inverse correlation of ozone amounts with solar activity (Ruderman and Chamberlain 1975); however, the magnitude of the modeled effect is considerably too small.

Observational tests.   To summarize the above discussion, the main effects of solar variability in the middle atmosphere are expected to be small changes in temperature and ozone amount due to variation in the ultraviolet flux, and NO_x increases due to particle events. Satellite data sets give excellent information on ultraviolet, temperature, and ozone profiles, and have been studied extensively. Better correlations have been found when the ultraviolet is represented directly by satellite measurements at 205 nm than by the 10.7-cm index. The following text begins with a paraphrase of the introductory material of Hood and Cantrell (1988), which can be consulted for references to work earlier than that cited here.

It has been difficult to obtain adequate data at the time scale of the solar cycle because of instrumental drifts and because the typical lifetime of an instrument (or its parent satellite) seldom exceeds 5 yr. Attention in recent years has therefore focused on shorter time scales, especially the solar rotation period. It is also helpful to concentrate on low-latitude data, where the amplitudes of planetary waves are weak. In the stratosphere, the ozone response is caused primarily by changes in production from O_2, and has a maximum value of 0.5% for 1% change in the ultraviolet at 205 nm. Higher up, in the mesosphere, the Lyman α line produces a small anticorrelation of -0.15% for 1% change at 122 nm (Keating et al. 1987). This may be a catalytic effect from increased production of odd hydrogen.

The response of the temperature has been studied by Hood (1986, 1987b) and Keating et al. (1987) between 30 and 0.2 mbar (24 to 60 km) and by Clancy and Rusch (1989) up to 90 km. All the studies are in excellent agreement, and Figs. 6 and 7 indicate the behaviors of ozone and temperature, respectively, for a 1% change of the 205-nm flux. For example, the upper panel of Fig. 6 shows the 0.5% response already quoted, at around 2 mbar, and the lower panel shows that it leads the supposed driving function by a day or two. From Fig. 7, it can be seen that the very small temperature response lags the ultraviolet by 4 to 14 days.

These figures include theoretical curves from Brasseur et al. (1987), and it is obvious that the data are reproduced to some degree but with serious deviations. It is suspected that these deviations are caused by dynamical effects, which have been modeled in a preliminary way by Callis et al. (1985) but are not understood well enough to be included in the aeronomical model. The model includes the direct effect of ultraviolet on oxygen and ozone, the indirect effect through catalytic destruction, and the effect of temperature. The destruction rate of ozone increases with temperature; from numbers given by Brasseur et al. (1987), the sensitivity is between -1.1 and -1.8% K^{-1}. The disagreements seen in Fig. 6 are greatly reduced if the observed temperatures and phase lags are used instead of computed ones. It can thus be concluded that the observed effects are reasonably well understood in principle, but not very well in detail.

A study by Angell and Korshover (1973,1976) seemed to establish a correlation of ozone column with solar activity, with a peak-to-peak variation approaching 10% at 70° latitude and perhaps 4% at 47°. Attempts to account

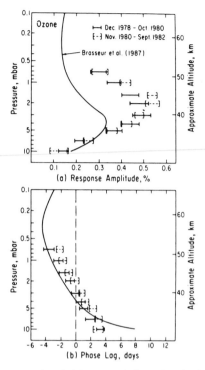

Fig. 6. Response of ozone at various heights to a 1% change in solar flux at 205 nm, expressed as an amplitude (a) and phase (b) (figure from Hood and Cantrell 1988). The lines represent a theoretical calculation.

for this consistently predicted much smaller changes and different phase lags (see, e.g., Ruderman and Chamberlain 1975; Garcia et al. 1984). According to Hood and Douglass (1989), models similar to those illustrated in Figs. 6 and 7 predict no more than 1% change in the column due to the sunspot cycle, although the mixing ratio in the upper stratosphere could change by more than 2%. This story is by no means over, but present indications favor at most a very small effect.

Understanding is almost totally lacking in the case of a correlation of polar stratospheric temperatures and solar activity, established by Labitzke (1987) and further discussed by Labitzke and Van Loon (1988) and by Kerr (1988). The correlation is present only in one phase of the stratospheric quasi-biennial oscillation, which is a more or less periodic reversal of winds in the lower equatorial stratosphere, with an average period of 27 months. If attention is confined to times when the wind is from the west, the temperature at 30 mbar over the north pole in midwinter varies in phase with the 10.7-cm index; typical values are $-60°C$ at maximum and $-75°$ at minimum. At

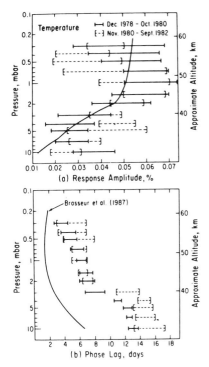

Fig. 7. Responses as in Fig. 6, but for measured temperatures instead of ozone (figure from Hood and Cantrell 1988).

ground level, discussed in the next section, small, but still significant, correlations seem to appear.

C. Troposphere: Climate

Only wavelengths above 300 nm penetrate to the troposphere and surface. As discussed in Sec. II.A and in the chapter by Fröhlich et al., this part of the solar spectrum is very nearly invariant over the period for which sufficiently accurate data exist. A peak-to-peak variation of 1 part in 1400 in the total solar flux over most of a single cycle has been observed. As shown in previous sections, observable effects of solar activity become more and more feeble at lower altitudes, generally tracking the smaller and smaller variations in those parts of the ultraviolet spectrum that reach each level and the greater mass that must be influenced. The troposphere contains 90% of the total mass of the atmosphere and is subject to a nearly constant driving energy.

In the century or so since the discovery of the sunspot cycle, there have been innumerable attempts to find correlations with various meteorological phenomena, as well as physiological, social, economic and other variables.

Given the number of conceivable variables, it is not surprising that some of them should have shown apparent correlations. What tends to be forgotten is that any tests of significance should take into account the number of tests that failed; in other words, they should be applied to the entire initial data set, not just the fraction that is found to correlate. It is necessary to have a prior hypothesis before any statistical test can be applied, and the only possible one is that every variable to be tested will show a correlation. A thorough review of these matters is given by Pittock (1978,1979,1983); the second of these is a short version of the first, and the third is an update. Useful collections are Bandeen and Maran (1975) and McCormac and Seliga (1979).

Nevertheless, in any search for cause-and-effect mechanisms, it is useful to be in possession of a driving function whose period, at least, is known. A similar opportunity is offered by a discrete occurrence such as a solar flare or proton event. As discussed above, these ideas have been put to good use in studies of the middle and upper atmosphere. The present authors concur with Pittock in the opinion that there are no significant established correlations of tropospheric weather phenomena with the solar cycle or with discrete events. A possible exception is the effect discovered by Labitzke (1987) and introduced in the previous section. None of this rules out detectable changes in radioactive isotopes due to modulation of the cosmic-ray input, such as those discussed for ^{14}C and ^{10}Be in the chapters by Jokipii and by Damon and Sonett. Such tracers are a direct reflection of events occurring in the stratosphere.

If the troposphere, driven by a large fraction of the total solar output, is to be significantly influenced by the tiny changes of the solar ultraviolet, some subtle but very strong mechanism of amplification ("trigger mechanism") must exist. Since the atmosphere has no resemblance to a loaded gun, such ideas deserve a good deal of suspicion. Many of them seem to be based on a conviction that there is a significant correlation to be explained, rather than starting with a plausible physical effect. A sample of them may be found in McCormac and Seliga (1979). Several chapters in that book consider transfer of magnetospheric effects by electric fields, perhaps including effects on thunderstorms. Another hypothesis (Hines 1974) suggests that "solar-induced modifications of the middle and upper atmosphere may alter the transmissivity of the stratosphere to upwardly propagating atmospheric waves" (Callis et al. 1985). These authors tested the mechanism by numerical modeling and failed to find a significant effect on the troposphere.

The previous section introduces the mysterious effect discovered by Labitzke (1987), in which temperatures in the polar winter are jointly influenced by the solar cycle and the quasi-biennial oscillation. The corresponding tropospheric aspect is discussed by Van Loon and Labitzke (1988). Tinsley (1988) has considered another influence, on the latitude of storm tracks over the North Atlantic. As discussed by Kerr (1988), this field is too new and too active for any kind of mature assessment.

Climatic Effects. In the chapter by Rabin, the evidence is reviewed for the extended absence of sunspot activity during the 17th century known as the "Maunder minimum", and an earlier event called the "Spörer minimum" (for an earlier summary see Eddy 1988). Both periods seem to coincide with periods of reduced global temperatures, the more recent of which is often called the Little Ice Age. Eddy has suggested that the connection is causal, but the required reduction of solar input would have to be much greater than the tiny amplitude detected on the time scale of a solar cycle. If the suggested association is more than a coincidence, additional amplifying factors must have been operating.

IV. RESPONSE OF THE ANCIENT ATMOSPHERE

Based upon standard solar models (see the chapter by Dearborn), the response of the Earth's atmosphere to solar irradiation can be expected to have been much different in the past than it is today. The reasons for this difference are that, (1) the solar output has varied throughout the Earth's history, with a lower luminosity, a higher ultraviolet flux and possibly a more intense solar wind in the remote past of the planet and, (2) major changes in the atmospheric chemical composition and thermal structure have occurred, as photosynthetic organisms have gradually released oxygen into the atmosphere. During the earlier stages, the solar ultraviolet that is now stopped by the ozone layer may have reached the surface, so that there is a real possibility of effects in the geologic record that do not occur today (see also the chapter by Kasting and Grinspoon).

The history of our atmosphere may be divided into four stages. The composition of the earliest atmosphere, during or soon after the accretion of the planet, is not well known. The abundance of molecular hydrogen (Walker 1982; Hunten et al. 1987) may have been high at that time, as a consequence of possible retention of a remnant of solar nebular gas, very active degassing from the interior, and important photodissociation of atmospheric water vapor by enhanced ultraviolet flux. This large ultraviolet flux would have provided the energy to drive large hydrodynamic escape fluxes from the atmosphere (see Sec. V). Thus, during this first stage, the atmospheric composition may be thought of as reducing.

The second stage, beginning about 3.5 Gyr ago, corresponds to the epoch when the abundance of this early hydrogen had decreased, but when photosynthetic organisms had not yet released oxygen. The atmosphere during this period is generally referred to as prebiologic. Its composition has been studied extensively with one-dimensional photochemical models (Kasting et al. 1979; Kasting and Walker 1981; Levine et al. 1982). A small amount of atmospheric oxygen was present due to the photodissociation of water vapor followed by escape of hydrogen to space.

The third stage began after the appearance of photosynthesis releasing oxygen, perhaps 2 Gyr ago. During this period, the oxygen abundance at ground level progressively increased from very low to $\sim$ 1 to 10% of the modern value. Simultaneously, an ozone layer developed in the stratosphere. This progressive development has profoundly modified the temperature structure and the radiation field in the atmosphere, because radiation, composition and climate are intimately coupled.

During the last stage, the atmospheric composition and climate became essentially modern, with the oxygen being somewhat less than or close to the present value, with a well-developed biosphere and surface biogeochemical processes similar to those prevailing today.

This progressive modification of the atmospheric composition has altered the response to the solar ultraviolet flux, raising the level to which this radiation penetrated as the oxygen amount increased. Sporadic perturbations of the atmosphere may have also occurred, such as increases of NO_x and climate changes induced by extreme solar-proton events or galactic cosmic-ray enhancements. Nevertheless, before analyzing these atmospheric responses to irradiation or particle fluxes, it is useful to describe in further detail the long-term history of the atmosphere, considering both the major and the minor constituents and their effects on the climate and the radiation field.

A. Major Atmospheric Constituents Through Time

The history of the most abundant atmospheric species, molecular nitrogen, is not known. One scenario has been proposed by Hart (1978), based on a geochemical calculation for the whole Earth's history. In this scenario, the amount of atmospheric N_2 is equal to $\sim$10 to 30% of the present value during approximately the first 2 Gyr. It increases rapidly afterwards to reach a value comparable to the modern atmospheric abundance. This behavior reflects the assumed degassing history, and more importantly the existence of large quantities of atmospheric NH_3 which dissolved into the Archean ocean. However, such a large amount of atmospheric NH_3 is precluded by its very short lifetime against photolysis, as shown by photochemical calculations (Kuhn and Atreya 1979; Kasting 1982). As a result, Hart's scenario must be thought of as very speculative and too much reducing during the Archean. Most recent calculations devoted to the chemistry and climate of the paleoatmosphere assume the abundance of N_2 to have been equal to its modern value. This assumption may be justified, to some extent, in view of the lack of reactivity of nitrogen and because degassing is commonly though to have occurred early in the Earth's history (Walker 1977,1982; Stevenson 1983a). Although molecular nitrogen is inactive both chemically and climatically, its abundance may nevertheless be of some significance for the evolution of atmospheric composition and climate, due to its important contribution to total pressure. Indeed, on one side, N_2 can be involved as a third body in some chemical reactions (such as ozone formation) and on the other side, the climatic consequences of line broadening by N_2 pressure of the gases active in the infrared are not negligible (Levine and Boughner 1979; François and Gérard 1988).

Only broad limits may be imposed from biological and geological considerations on the history of oxygen, the second constituent in importance of our atmosphere today (Walker 1977; Holland 1984; Kasting 1987). In the early atmosphere, before the origin of photosynthesis, molecular oxygen was formed through ultraviolet photolysis of stratospheric water vapor followed by escape of hydrogen to space. Photolysis of water vapor and carbon dioxide produces O and H atoms and OH and CO, as indicated in the first two lines of the following schemes (R2). The third lines show production of O_2 molecules, and the fourth recycles the CO_2. The net effects of the two schemes are equivalent. The main source of oxygen atoms to run the process is CO_2, but the rapid oxidation of CO by OH recycles them rapidly, and the net source of molecular oxygen is the water vapor. Other reactions, not shown here, convert most of the hydrogen atoms to molecules, and both forms escape rapidly. There is also some recycling back to water vapor.

$$
\begin{aligned}
&H_2O + h\nu \rightarrow OH + H \quad (2\times) \\
&CO_2 + h\nu \rightarrow CO + O \\
&OH + O \rightarrow O_2 + H \\
&CO + OH \rightarrow CO_2 + H \\
\hline
&2H_2O + 3h\nu \rightarrow O_2 + 4H
\end{aligned}
\qquad \text{(R2)}
$$

$$
\begin{aligned}
&H_2O + h\nu \rightarrow OH + H \quad (2\times) \\
&CO_2 + h\nu \rightarrow CO + O \quad (2\times)
\end{aligned}
\qquad \text{(R3)}
$$

$$
\begin{aligned}
&O + O + M \rightarrow O_2 + M \\
&CO + OH \rightarrow CO_2 + H \quad (2\times) \\
\hline
&2H_2O + 4h\nu \rightarrow O_2 + 4H
\end{aligned}
\qquad \text{(R4)}
$$

The first cycle is dominant between 2 and 30 km, while the second prevails in the higher stratosphere and mesosphere, as well as near the ground (Kasting et al. 1979). The net rates of these two cycles are governed by that of hydrogen atom escape, rather than by the rate of water or carbon dioxide photolysis. The escape flux is governed by the balance between release and removal of hydrogen and other reduced compounds to and from the atmosphere (Kasting et al. 1984; Kasting 1987). Prebiological photochemical models taking into account all these reactions (as well as many others) show that the mixing ratio of O_2 was very low at ground level (between 10^{-18} and 10^{-12}) at that time, while it progressively increased to a peak value of $\sim 10^{-5}$ at an altitude of $\sim$50 km (Kasting et al. 1979; Canuto et al. 1983a; Kasting 1985).

When the rate of O_2 release by photosynthesis became larger than the rate of supply of reduced gases to the atmosphere (possibly more than 3.5 Gyr ago), the abundance of hydrogen and other reduced gases should have rapidly decreased, so that O_2 accumulated in the atmosphere. At that time, the abundance of atmospheric O_2 was maintained at lower values than today due to rapid reaction with dissolved ferrous iron in the deep ocean (Holland 1984; François and Gérard 1986; François 1987). The surface layer of the ocean where photosynthetic organisms were living was approximately in equilibrium with the atmosphere, but subsidence of this surface water exhausted it from its oxygen content as soon as it came into contact with the iron-rich deep water. Photochemical studies show that the mixing ratio of O_2 became independent of altitude when its ground-level abundance became larger than $\sim 10^{-4}$ of the present atmospheric level (PAL) (Kasting and Donahue 1980).

The control of atmospheric oxygen by deep oceanic iron stopped between $\sim$2.3 and $\sim$1.7 Gyr ago, when the deposition of extensive banded iron formations had progressively been halted (Walker et al. 1983). Geochemical processes became essentially modern. At this time, the abundance of atmospheric oxygen was between 10^{-2} and 10^{-1} PAL and it progressively increased to the present level, during the end of the Proterozoic and possibly the beginning of the Phanerozoic. As at present, the control of oxygen was provided by the limitation of the rate of organic matter burial by aerobic respiration of organisms living in marine sediments (Walker 1977; Kasting 1987).

The concentration of carbon dioxide is thought to have responded to the long-term evolution of solar luminosity, in order to maintain a surface climate favorable to the development and the persistence of life, and compatible with the presence of liquid water at the Earth's surface since at least 3.8 Gyr ago, the deposition time of the oldest known rocks (Isua, West Greenland), in which chemical and detrital sediments have been found (Ashwal 1985). A feedback process linking the CO_2 loss rate from the ocean-atmosphere system to the climate is provided by the dependence on temperature of the global weathering rate of silicate rocks (Walker et al. 1981; Marshall et al. 1988). Thus, according to the classical theory of solar evolution, the abundance of atmospheric CO_2 should have progressively decreased throughout the Earth's history, as the solar luminosity increased.

Finally, it is useful to mention that in view of the presumed rapid early degassing, there should have always been important quantities of liquid water at the surface, so that an ocean of size comparable to that of the modern oceans has existed throughout the Earth's history. This means that the abundance of atmospheric water vapor was always governed by the climatic conditions, the lower layers being close to saturation. Usually, in paleoclimatic modeling, a modern atmospheric relative humidity profile is adopted.

B. Minor Constituents, Temperature Structure and Radiation Field

The main contributors to the absorption of solar radiation in the Earth's ancient atmosphere were O_2, O_3, H_2O, CO_2 and to a lesser degree NO, NO_2, CH_4 and H_2O_2. Below $\sim$210 nm, the absorption is due mainly to O_2, CO_2 and H_2O, whereas between 210 and 300 nm the principal absorber is ozone (Hartley continuum and bands). Ozone has also some weaker bands in the near ultraviolet (Huggins bands) and in the visible (Chappuis bands), where the solar radiation flux is larger. For that reason, ozone plays an important climatological role today by providing a sufficient heating rate to warm the stratosphere. Nitrogen dioxide also absorbs visible radiation and, if its concentration was higher, its climatological role could be comparable to that of ozone. Finally, H_2O and CO_2 absorb in the infrared part of the solar spectrum, where they (mostly H_2O) also have an important climatic contribution. Atmospheric absorption of infrared terrestrial radiation ($\lambda \geq 4$ μm) is due mainly to H_2O, CO_2, O_3, CH_4 and N_2O. The infrared transmission of these gases is considered in the calculation of the thermal structures of the present and ancient atmospheres.

The smaller amount of oxygen in the ancient atmosphere resulted in a lower abundance and a different altitude distribution of ozone. Further, these changes in O_2 and O_3 modified the radiation field and the temperature structure. Calculations of the ozone distribution in the prebiological atmosphere show that its density was 10^6 to 10^9 cm^{-3} above $\sim$10 km, while it decreased to very low values below (Kasting et al. 1979; Kasting 1985; Canuto et al. 1983a). By comparison, the present peak is 4×10^{12} cm^{-3} at 25 km. The integrated column density of ozone was in the range of 10^{12} to 10^{15} cm^{-2}, depending on the assumed CO_2 amount and the adopted solar ultraviolet flux (Canuto et al. 1983a). The effects of the larger ultraviolet fluxes on the early atmosphere are discussed below.

The ozone distributions for larger oxygen amounts (10^{-5} to 1 PAL) are presented in Fig. 8. A similar variation of the ozone profile with the decrease of the atmospheric oxygen level has been obtained by several investigators (Kasting and Donahue 1980; Levine et al. 1979,1981). As the oxygen amount is reduced from the present level to 10^{-2} PAL, the altitude of the ozone peak is lowered from $\sim$25 to $\sim$15 km, but the peak density remains essentially unchanged. Consequently, the total ozone column remains close to that at 1 PAL of O_2. This behavior is observed because the region where atomic oxygen is produced is moved to lower altitudes, as a result of the deeper penetration of solar ultraviolet radiation photodissociating O_2. A further decrease of the oxygen level ($\leq 10^{-3}$ PAL) reduces the altitude of the ozone peak by only

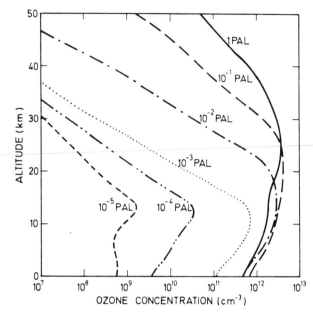

Fig. 8. Vertical profiles of the ozone concentration in the ancient atmosphere for O_2 levels varying from 10^{-5} to 1 PAL (figure from François and Gérard 1988).

2 or 3 km, because ultraviolet radiation with $\lambda \leq 200$ nm is absorbed by abundant tropospheric water vapor independently of the O_2 level. However, the oxygen decrease is now accompanied by a rapid reduction of the peak ozone concentration and thus by an important decrease of its total column density.

The reduction of the oxygen level from 1 to 10^{-5} PAL also results in somewhat larger abundances of the OH and HO_2 radicals (Kasting et al. 1985; François and Gérard 1988). The slight increase in OH is important, because it should result in a lower amount of atmospheric methane (the reaction between CH_4 and OH initiates the CH_4 oxidation chain) if the biological production of this gas is held constant (Kasting et al.1985). Consequently, a higher concentration of methane in the ancient atmosphere would have required a much higher biological production rate of this gas at that time. Similarly, the concentration of N_2O should also be smaller at low O_2 levels if its biological production is unchanged (Kasting and Donahue 1980; Kasting et al. 1985). In these conditions, the densities of NO and NO_2 would be much reduced when compared to the present time (François and Gérard 1988).

Some calculated temperature profiles at reduced oxygen levels are shown in Fig. 9. As the O_2 level is increased from 10^{-5} to 1 PAL, the stratosphere is progressively warmed from very low temperatures as a result of enhanced ozone heating. A thermal maximum appears when the oxygen level is larger

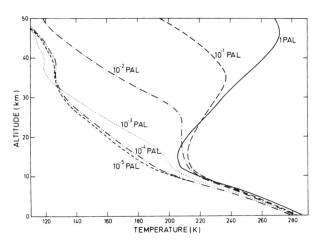

Fig. 9. Vertical temperature profiles in the ancient atmosphere for O_2 levels varying from 10^{-5} to 1 PAL (figure from François and Gérard 1988).

than $\sim 10^2$ PAL. The tropospheric thermal structure shows little change, as the O_2 amount is increased. However, the variation of the surface temperature is relatively important, since a decrease of 5° (Kasting 1987) to 7°K (François and Gérard 1988) of the mean global temperature is predicted when all oxygen and ozone are removed from the modern atmosphere. Similar climatic calculations have also been performed for higher CO_2 amounts.

Figure 10 shows the penetration of ultraviolet radiation in the Earth's atmosphere calculated for O_2 levels varying between 10^{-5} and 1 PAL. The solar radiation with wavelength in the range 200 to 300 nm, which is absorbed in the higher stratosphere today, penetrates progressively deeper into

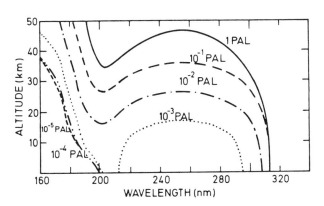

Fig. 10. Altitude of unit vertical optical depth in the ultraviolet. The results are shown for atmospheres containing from 10^{-5} to 1 PAL of O_2 (figure from François and Gérard 1988).

the atmosphere as O_2 is removed and finally can reach the surface when the O_2 level becomes lower than 10^{-4} PAL. The radiation below ~ 200 nm never reaches the surface, since it is absorbed by H_2O and CO_2. The absorption by these two species becomes dominant with respect to that of O_2, when the O_2 level is lower than 10^{-4} PAL. In such calculations, it is important to perform fully coupled climatic and photochemical calculations. Indeed, the low temperatures in the stratosphere reduce the scale height in this region, which results in a much smaller total density at high altitude and, hence, in a further penetration of solar ultraviolet (François and Gérard 1988). We can conclude that in view of the composition and thermal structure of the ancient atmosphere, the atmospheric effects of solar ultraviolet should have occurred at much lower altitude when oxygen was less abundant in the past.

C. Response to Radiative Perturbations

In this section, the response of the ancient atmosphere to perturbations of the solar radiation flux is analyzed. The atmospheric response to the long-term increase of the solar luminosity has been studied by many authors. This topic is analyzed in some detail in the chapter by Kasting and Grinspoon and only a short summary will be given here. In the absence of an additional atmospheric greenhouse gas, the Earth would freeze over if the solar luminosity was decreased by 30%, since even without taking into account the ice-albedo feedback, the mean global temperature would be reduced by more than 1 K for every percent decrease of the solar luminosity. As already mentioned, this faint-young-Sun problem is generally solved by the assumption that CO_2 was more abundant in the past. For the Earth to escape freezing when the solar luminosity was 30% lower than at present, the amount of atmospheric CO_2 must exceed the present value by a factor of several hundred (Kasting 1985,1989a). Other factors may nevertheless have helped to prevent the Earth from freezing, such as a reduction in cloudiness under cold climatic conditions (Rossow et al. 1982), a reduction of the global planetary albedo associated with a smaller land-sea ratio (Cogley and Henderson-Sellers 1984) or a change in the rate of meridional heat transfer (Endal and Schatten 1982). Further, a recent solar model (Wilson et al. 1987; Gilliland 1989) suggests that the Sun possibly lost an important amount of mass in its early history. According to this model, the solar luminosity would have been higher than for a classical constant-mass model and the faint-young-Sun paradox might even be transformed into a bright-young-Sun one (chapter by Kasting and Grinspoon).

Another important problem is the response of the ancient atmosphere to the enhanced solar ultraviolet flux during the T Tauri phase and the early stages spent by the Sun on the main sequence. This question has been studied by Canuto et al. (1982,1983a,1983b) with a complete photochemical model of the early atmosphere. Using data acquired for three bright T Tauri stars,

Canuto et al. (1983*b*) calculated the wavelength dependence of the enhancement of the solar ultraviolet flux during this phase with respect to the present time. Using this composite T Tauri ultraviolet flux, they showed that the prebiological ground-level concentration of O_2 was increased by a factor of $\sim 10^2$ with respect to a case with a modern ultraviolet irradiation. In the stratosphere the O_2 increase would have been of similar magnitude. The total ozone column varies by 1 to 3 orders of magnitude, depending on the CO_2 and H_2 mixing ratios. Another result obtained by Canuto et al. (1983*b*) and possibly relevant to the origin of life is that the ultraviolet enhancement increases the CO/CO_2 ratio in the stratosphere. Indeed, above $\sim$30 km, this ratio can become close to unity under strong ultraviolet irradiation. The formation of organic molecules would be much easier under these more reducing conditions. Photochemical calculations describing the state of the early atmosphere have been performed by Canuto et al. (1983*a*) for uniform enhancement (which does not correspond to the real situation) of the ultraviolet flux by factors of 10, 20, 100 and 300 with respect to present conditions. Their results are comparable to that obtained for the T Tauri phase, except that the perturbations are of somewhat smaller magnitude. Thus, since the largest O_3 column calculated for the T Tauri phase ($\sim 4 \times 10^{16}$ cm^{-3}) is much too low to afford a substantial ultraviolet screen, we can conclude that the larger amount of O_3 in the early atmosphere cannot have helped to protect the primitive life against solar ultraviolet radiation.

Finally we address the response of the ancient atmosphere to the ultraviolet variability associated with the sunspot cycle. As already noted, the atmospheric effects of solar ultraviolet irradiation should occur at lower altitude as the oxygen level is reduced. Thus, the composition and temperature changes over a solar cycle which are presently observed in the higher stratosphere and mesosphere should have taken place in the past in the lower stratosphere or troposphere. Since the ultraviolet variability over a solar cycle is not well known, two extreme cases were considered, one where the ultraviolet fluctuations are restricted to $\lambda \leq 200$ nm and another where the variability extends up to 300 nm. Figure 11 shows the profiles of the ozone concentration changes over a solar cycle for the second hypothesis on the ultraviolet variability and for oxygen levels varying from 10^{-3} to 1 PAL. As the oxygen level is reduced, the altitude where the ozone variation is maximum is lowered. At 10^{-3} PAL, this altitude is only $\sim$12 km but the peak amplitude is reduced with respect to that at higher O_2 levels due primarily to the slightly lower abundance of ozone at 10^{-3} PAL (see Fig. 8). The surface temperature change ΔT_s is very small (0.01 to 0.02 K) for all the oxygen levels considered, except when the amplitude of the variability is increased (as a sensitivity test) by an important factor (Gérard and François 1987). However, for the second hypothesis on the ultraviolet variability, the values of ΔT_s may become substantial, peaking at ~ 0.5 K at 3×10^{-3} PAL, changing sign between

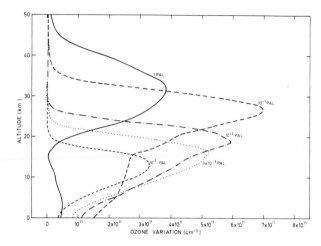

Fig. 11. Vertical structure of the ozone concentration change between solar maximum and minimum, as a function of the O_2 level. The solar ultraviolet variability is assumed to extend up to 300 nm.

10^{-2} and 10^{-1} PAL and reaching approximately -0.1 K under present O_2 levels. The exact values of ΔT_s are, however, somewhat dependent on the CO_2 amount and the average surface temperature. Despite the numerous uncertainties involved, these calculations nevertheless give a quantitative illustration of the specific response of the ancient atmosphere to ultraviolet irradiation.

D. Energetic Particles and Extinctions

The history of life development on Earth has been punctuated by periods of greatly accelerated extinction affecting both terrestrial and marine animals. A widely accepted, but still controversial, explanation was suggested for the Cretaceous-Tertiary event by Alvarez et al. (1980): an impact of an asteroidal body, followed by a global dust cloud which suppressed photosynthesis and created catastrophic changes in atmospheric and climatic conditions. It has also been suggested that major mass extinctions are not independent events but rather are triggered by a single cause recurring at regular intervals. A period of 26.2 ± 1 Myr was derived by Sepkoski and Raup (1986) from the marine mass-extinction time series. A more recent, modified version of this scenario was proposed by Hut et al. (1987) who suggested periodic comet showers, each one consisting of several cometary impacts over a 1 to 2 Myr period. The reality of any such periodicity has been challenged on several grounds, along with the validity of various suggested explanations (e.g., see the news item by Kerr [1985]).

The problem of minor extinctions is equally complex and less documented. These events seem to occur sporadically and there are suggestions of close correlations between climatic conditions and geomagnetic field reversals. These are based on studies of fossil radiolaria extinctions in deep-sea cores (Hays 1971; Mann 1972), oxygen isotopic ratio and planktonic foraminifera (Wollin et al. 1971) and carbon/nitrogen ratio of lake organic sediments in Kashmir (Krishnamurthy et al. 1986). The nature of the causal relationship between the two phenomena has been a subject of extensive speculation reminiscent of similar notions about extraterrestrial perturbations of atmospheric composition and climate.

Among these proposed mechanisms, some involve the stronger interaction between solar energetic particles and the atmosphere. Reid et al. (1976) suggested that, during polarity reversals, the decrease of the dipole moment of the geomagnetic field allows energetic solar protons and galactic cosmic rays to access a larger fraction of the planet. This would make it possible for strong solar-proton events to inject considerable energy fluxes into the middle atmosphere. As described above, the interaction between these particles and atmospheric constituents generates enhanced concentrations of NO_x and subsequent depletions of stratospheric ozone. In addition to photochemical consequences, climatic perturbations are expected as a consequence of the increase of the concentration of nitrogen dioxide and the resulting absorption of visible solar radiation (Reid et al. 1978). Indeed, NO_2 plays a minor role in the thermal budget of the atmosphere in normal conditions but, during periods of significant NO_2 enhancement, may become significant. This possibility was examined with a coupled photochemical-radiative-convective model by Hauglustaine and Gérard (1989). They found that, in the case of a steady solar-proton flux similar to the August 1972 event, substantial perturbations of the stratospheric composition are predicted, including an ozone column depletion by 55%, a doubling of the NO_2 amount and a resulting surface temperature drop by 6°C.

An interesting consequence of the interaction of solar cosmic rays with the middle or lower atmosphere is the predicted increase of the HNO_3 rainout due to the enhanced concentrations of NO_x. The accumulation of major solar proton events during periods of solar maximum activity and the above described scenario is a possible explanation of the correlation found in the Antarctic between solar activity indicators (R_z and aa) and the nitrate content of the snow (Zeller and Parker 1981; Laird et al. 1982, Dreschhoff and Zeller 1989). The largest peak in annual South Pole NO_3^- deposition obtained by Laird et al. (1982) over the 51-yr period sampled closely follows the 1972 event as may be expected from the unusual magnitude of this event. The mechanism proposed by these authors is oxidation of NO_2 by OH radicals in water vapor, forming NO_3^- hydrates and large water cluster ions and aerosols. However, this interpretation of the origin of NO_3^- in ice cores was challenged by Risbo et al. (1981) and Weertman and Peel (1981). Recent observations of the stratospheric composition in the Antarctic ozone hole indicate that nitrate compounds play a major role in nucleation and formation of polar stratospheric clouds.

V. ATMOSPHERIC AND GEOLOGIC EVIDENCE FOR SOLAR VARIATIONS

It is clear from the previous sections that the major identifiable effects of solar variability occur in regions of the atmosphere far from the surface, and the closer to the surface we look, the smaller the effects we see. Thus, the chances are slim of finding traces of any of these effects in the geologic record. Two exceptions stand out, both pertaining to rather early times: the elemental and isotopic ratios of noble gases, and rather strong evidence that the mean temperature has always been similar to what we find today. Curiously, the latter observation flies in the face of what we firmly believe about the early Sun, that its luminosity was only about 70% of the present value. To compensate, we must make models of early enhanced greenhouse effects on both Earth and Mars (Sec. IV; chapter by Kasting and Grinspoon). Possible escape from a primitive atmosphere is discussed in Sec. V.A and B, followed by a discussion of possible relics from the period before the establishment of a substantial ozone layer.

A. Evidence from Noble Gases

It has been clear since just before 1950 that the atmosphere is deficient in noble gases by a factor that varies smoothly with mass between 10^{-10} for neon and 10^{-6} for xenon (Brown 1949; Suess 1949). A decade later it was discovered that there is an enormous fractionation, $> 1\%$ per mass number, of the xenon isotopes when they are compared against a meteoritic standard (Krummenacher et al. 1962). Only in the last few years has there been a viable model to account for these effects (Hunten et al. 1987; Zahnle and Kasting 1986; Sasaki and Nakazawa 1988). This model postulates a very large hydrodynamic escape flux of hydrogen in the first 100 Myr or so, which carries away heavier atoms by gas drag with a substantial mass selectivity. Thus, the observed compositions can be considered as evidence for the presence of this escape flux, and for the energy source required to drive it. Details may be found in several recent reviews (Hunten et al. 1988,1989; Hunten 1990) and a comprehensive paper by Pepin (1991).

Hydrodynamic escape can only exist if a large quantity of XUV energy is deposited in a region corresponding to the ionosphere in a static atmosphere; otherwise, the rapid flow induces an adiabatic cooling that throttles the flow (Watson et al. 1981; Sekiya et al. 1981). The required flux is around 100 times what is emitted by the present Sun, and is reasonably consonant with what is observed from pre-main-sequence stars (Sec. II.C; chapter by Walter and Barry). Since it seems highly probable that there was a phase of hydrodynamic escape, it is equally probable that the early Sun did produce an enhanced XUV output.

B. Light Gas Escape

In addition to the massive hydrodynamic escape discussed above, there continues to be a much more modest flux of hydrogen and helium. A recent review, from a historical viewpoint, is given by Hunten (1990). The hydrogen escape rate is equivalent to a few meters of water over the age of the solar system, and is regulated by the low humidity of the stratosphere. The actual escape from the exosphere is by a combination of two principal processes: thermal (or Jeans) escape, and charge exchange with hot protons in the extended ionosphere. At sunspot minimum, the latter process dominates, with the thermal process taking over when the exospheric temperature becomes hot. The sum remains constant, and is regulated by the ability of hydrogen to diffuse through the lower thermosphere. Helium escapes mainly by nonthermal processes involving ions, and is in steady state with degassing, with an atmospheric lifetime around 1 Myr.

C. Solar Signatures in Geologic Records

In spite of numerous efforts to investigate the coupling between the Sun and the Earth's surface environment, only a few of the claimed correlations have stood the test of serious statistical analysis and have offered a reasonable physical link between solar and terrestrial effects. One of the major reasons for this situation is the very short timespan of direct climatic and aeronomic observations. Geologic records offer a potential tool to extend studies in the past, in principle as far as the oldest sedimentary rocks.

Sedimentary late Precambrian rocks of the Elatina Formation (650 Gyr) found in the Adelaide geosyncline in South Australia were claimed by Williams (1981) and Williams and Sonett (1985) to exhibit a climatic response to the solar cycle and longer-period solar variations (see the chapter by Anderson). The 10-m thick drilled-core sections of siltstones and fine-grained sandstones show periodicites in the thickness of the individual laminae forming 1580 consecutive "cycles" flagged by dark bands. Each cycle contains 8 to 16 laminae with an average of 12 laminae/cycle. In their first interpretation, Williams and Sonett (1985) and Sonett and Williams (1987) described the individual laminae as annual increments or varves produced by the superposition of silts and fine sands in spring and summer and clay suspended in quiet waters in winter. Such structures are frequently observed today in lake sediments deposited in periglacial climates in Northern Europe, Canada and Alaska. In this scenario, the thickest laminae would correspond to warm summers producing large runoff of snow and ice melt carrying sediments into the lake. Conversely, during cooler summers with little runoff, thinner more clay-like sediments would be deposited. The positive skewness of the cycles and the numerous numerical analogies between the Elatina records and the time-series of sunspot numbers observed since the last century led Sonett and Williams (1987) to propose that the 12, 22–25 and 103 varve-year variations found in the varves reflected similar cyclicities in solar activity. An increase in solar activity would generate a larger total solar irradiance or more intense solar ultraviolet flux which could induce warmer summers and greater runoff of meltwater and thicker varves. A similar correlation with the solar cycle was found in recent varves deposited in the periglacial Skilak Lake in Alaska (Sonett and Williams 1985). A weak solar influence was found in this 236-yr varve sequence spanning the period 1700 to 1930. A good correlation between the varve thickness and the sunspot index was found with common periods of 11 and 22 yr. By contrast, analysis of three varve time-series from the Eocene Green River Formation (50 Myr) (Crowley et al. 1986) showed only a weak 11-yr signal during a limited portion of the $\sim$ 7500-yr core.

   The hypothesis of a link between solar activity and climate in the Elatina deposits was partly challenged by Zahnle and Walker (1987) who suggested that the periodicities reflect a solar period of 10.8 yr and its beating with a presumed lunar nodal cycle of 20.3 yr. The physical causal relationship between solar activity and climate, fairly elusive today, is even more speculative for periods as remote as the Precambrian era. As mentioned in Sec. IV, Gérard and François (1987) examined the amplitude of the global surface-temperature response to solar variability in an atmosphere poorer in oxygen. They found that changes in total or ultraviolet solar irradiance similar to those observed in the modern solar cycle fail to produce significant temperature effects. New observations from two other Precambrian sites of South Australia exhibit structures and periodicities qualitatively similar to those of Elatina but with 14 to 15 and 15 to 24 (average 22.4) laminae/cycle, respectively. These measurements prompted Williams (1988,1989) to propose an alternative explanation to the observed periodicities and to assign a different time scale to individual laminae. In this view, the "Elatina cycle" and not individual laminae are taken as yearly climatic signals. The rythmites are then interpreted as deposits of marine ebb-tidal deltas recording variabilities in the velocity and range of the paleo-ebb tides. In this interpretation, the Elatina laminae are considered as diurnal increments and the lamina cycles as truncated fortnightly groupings through nondeposition at neap tides. From these data, Williams (1989) and Sonett et al. (1988) estimated orbital parameters of the late Precambrian Earth-Moon system and suggested little change in the Earth-Moon distance during the Phanerozoic.

   Anderson (1961) and Anderson et al. (1972) analyzed ancient evaporite varves and claimed the presence of 80 to 90 yr and $\sim$ 200-yr periodicities, two periods represented in the spectrum of the sunspot index and other proxy data. Anderson (see his chapter) suggests that production of sediments can be thermally forced more directly than clastic sediment. Changes in the evaporation rate are produced by small variations in the seasonal distribution of insolation which accompany the precession rate of the planet as demonstrated by the cycles observed in the Permian Castile Formation. Consequently, minerals produced under the strong control of evaporation are probably the best candidates for detecting geologic evidence of solar-climate coupling.

The analysis of the thermoluminescence of a recent sedimentary core spanning a period of 18 centuries shows the presence of 138, 59, 12.06 and 10.8 year periodicities (Castagnoli et al. 1988*b*). The sum of the two low-frequency signals may be considered as an 82.6-yr component modulated by a second wave with a period of 206 yr. The location of the last two maxima corresponds with periods of past solar diameter maxima, suggesting a possible link between the two sets of observations.

LONG-TERM HISTORY OF CLIMATE ICE AGES AND MILANKOVITCH PERIODICITY

A. BERGER
Université Catholique de Louvain

During the first half of the 20th century, Köppen, Spitaler and Milankovitch considered high winter and low summer insolation as favoring glaciation. After Köppen and Wegener related the Milankovitch new radiation curve to Penck and Brückner's subdivision of the Quaternary, there was a long-lasting debate as to whether or not such changes in the insolation can explain the Quaternary glacial-interglacial cycles. In the 1970s, with improvements in radiometric dating and in acquiring and interpreting the long continuous geologic records, with the advent of computers and with the development of more sophisticated astronomical and climate models, the astronomical theory and in particular its Milankovitch version were revived. Now, most geologic, astronomical and climatological obstacles have been overcome and the causal role of the orbital variations in long-term variation of climate over the last Myr is taken almost universally for granted.

I. INTRODUCTION

During the past two decades there has been ever increasing interest in the Earth's climate and in its variations. This concern stems partly from the recognition that mankind is becoming a significant factor in the climatic balance of the Earth and partly from man's increasing vulnerability to climatic fluctuations, whether they are natural or man-induced (Schneider 1989).

The urgency for climatic predictions has led to the development of numerous mathematical models of climate (Schlesinger 1986; Tricot and Berger 1987; Schlesinger and Mitchell 1987; Mitchell 1989; Fichefet et al 1989; Berger et al. 1989*a*, *b*). The analysis of environmental data found in natural archives has revealed a wealth of useful information on both the natural and perturbed behavior of the coupled Earth system (Oeschger and Eddy 1988). These quantitative data on global changes of the past can be used to put in broader context observed trends in contemporary data. In particular, the identification of the Milankovitch cycles and higher-frequency climatic oscillations in both the ocean and continental archives can help to evaluate climate models and identify unknown and often important interconnections between physical, chemical and biological processes.

It is not the purpose of this chapter to discuss in detail the reliability of the proxy records used to reconstruct past climates, nor the accuracy of the time scale (particularly for Pre-Quaternary times) on which these reconstructions are based. The impact of these uncertainties on the periodicities has been critically reviewed in Imbrie et al. (1984, 1989). Sensitivity analyses of the spectrum of geologic time series to some processes like bioturbation and accumulation rate (Pestiaux and Berger 1984; Morley and Shackleton 1984; Dalfes et al. 1984) and to the accuracy of the time scale (Pestiaux et al. 1988; Berger 1989*b*) have been made to avoid the classical criticism of circular reasoning.

The astronomical theory of ice ages has received substantial support from detailed analysis of deep-sea sedimentary cores, but the question as to what processes in the climate system are responsible for the translation of the insolation changes into large variations in continental ice volume remains unsettled (Rind et al. 1989). However, a series of snapshot experiments using general circulation models (Prell and Kutzbach 1987) and transient simulations using a coupled climate system model (Berger et al. 1989*a, b*) have laid the grounds on which the physical reality of the Milankovitch theory can be tested (see the chapter by Ghil). This chapter will review the important steps since 1975 which have significantly contributed to confirm the astronomical theory.

II. MILANKOVITCH ERA AND DEBATE

In the 19th century, Croll (1875) stressed the importance of severe winters as a cause of Quaternary ice ages. It is only during the first decades of the 20th century that Spitaler (1921) rejected Croll's theory that the conjunction of a long cold winter and a short hot summer provides the most favorable conditions for glaciation. He adopted the opposite view, first put forward by Murphy as early as 1876, that a long cool summer and short mild winter are the most favorable. The diminution of heat income during the summer half-year has also been claimed by Brückner et al. (1925) as the decisive factor in glaciation. Milankovitch (1920,1941), however, was the first to present a complete astronomical theory of Pleistocene ice ages by computing the changes of orbital elements over time and linking the changes in insolation to climate (Imbrie and Imbrie 1979; Berger 1988). Between 1915 and 1940,

Milankovitch put the astronomical theory of the Pleistocene ice ages on a firm mathematical basis. He calculated how the intensity of radiation striking the top of the atmosphere during the caloric summer and winter half-years varied as a function of latitude and orbital parameters e, ϵ, $e \sin \bar{\omega}$, then emphasized the importance of summertime insolation at 65°N as a controlling factor of northern hemisphere glaciations, with its dominate obliquity-driven periodicity of 41,000 yr. He then estimated the magnitude of ice-age departures from the present-day air-surface temperatures, estimating the radiation balance at the Earth's surface.

The essential product of the Milankovitch theory is his curve showing how the intensity of summer sunlight varied over the past 600 kyr. He identified certain low points with 4 European ice ages (reconstructed 15 yr earlier by Penck and Brückner [1909]) and concluded that these geologic data constituted a verification of his theory.

Milankovitch's success was only partial because the Quaternary has exhibited many more glacial periods than were claimed during the first part of the 20th century (see, e.g., Shackleton and Opdyke 1973, 1976; Morley and Hays 1981), and because the ice volume record is dominated by a 100-kyr rather than by a 41-kyr cycle. Moreover, during the late 1960s, detailed studies of Alpine terraces showed that the climatic reconstruction of Penck and Brückner was wrong: the time scale was grossly in error, as terraces are tectonic rather than climatic features (Kukla 1975a). Although it is clear that the empirical argument used by Milankovitch to support his theory was misinterpreted, modern evidence now strongly supports its essential concept that the orbital variations exert a significant influence on climate.

From approximately 1950 to 1970, the Milankovitch theory was largely based on fragmentary geologic records supported by incomplete and frequently incorrect radiometric data. The accuracy of the astronomical parameters and of the related insolation fields were not known, and climate was considered too resilient to react to "such small changes" as observed in the summer half-year caloric insolation of Milankovitch. Finally, the adjustment of the boundary conditions of energy-balance models (EBM) showed that the magnitude of the calculated response was indeed very small (see, e.g., Budyko 1969; Sellers 1970). If these early numerical experiments are viewed narrowly as a test of the astronomical theory, they are open to question because the models used are incomplete and contain untested parameterization of important physical processes.

III. MILANKOVITCH REVIVAL

In the late 1960s, radioactive dating and paleomagnetic techniques gradually clarified the Pleistocene time scale (Broecker et al. 1968). Better instrumental methods came on the scene by application of $\delta^{18}O$ measurements to ice-age foraminifera relics (Emiliani 1966; Shackleton and Opdyke 1973; Duplessy 1978); ecological methods of core interpretation were perfected (Imbrie and Kipp 1971); global climates in the past were reconstructed (CLIMAP 1976,1981); and atmospheric general circulation models (Smagorinsky 1963) and climate models became available (Alyea 1972). With these improvements, and with the advent of computers, and the development of astronomical and climate models, a more critical and deeper investigation became necessary of all four main steps of any astronomical theory of paleoclimates, namely:

1. Computation of the astronomical elements (Berger 1976a);
2. Computation of the appropriate insolation parameters (Berger 1978a,b);
3. Development of suitable climate models (Kutzbach 1985; Crowley 1988);
4. Analysis of geologic data in the time and frequency domains designed to investigate the physical mechanisms, and calibrate and validate the climate models (Berger et al. 1984).

It is this systematic approach with modern powerful techniques which has brought, since 1975, the following major discoveries supporting progressively the astronomical theory.

A. Bipartition of the Precessional Period

In 1976, Hays, Imbrie and Shackleton demonstrated from spectral analysis of the climate sensitive indicators in selected deep-sea records that the astronomical orbital periods (100, 41, 23 and 19 kyr) are superimposed upon a general red noise spectrum. The geologic observation of bipartition of the 21-kyr precessional peak into 23 and 19 kyr (confirmed in astronomical computations made independently by Berger, [1977]) was one of the first and most delicate and impressive tests for Milankovitch's theory.

B. Monthly Insolations

Long-term deviations from today's values of the caloric semi-annual insolation used by various workers (Milankovitch 1941; Bernard 1962; Kukla 1972; Sharaf and Budnikova 1969; Vernekar 1972; Berger 1978b) range up to 3 to 4% at maximum. However, if the monthly insolation values (Berger 1978a) are used, important fluctuations masked by the half-year averaging method become recognizable. For example, both 10 and 125 kyr ago, during the Holocene and Eemian interglacials, all latitudes were overinsolated in July with respect to the present, particularly in the northern polar regions, where the positive anomaly reached 12%. This is especially significant when a delay of some thousands of years, required for the ice sheets to melt, is taken into account in climate modeling.

A detailed treatment of the seasonal cycle is required to explain climate variations (Kukla 1975b) in realistic climate models (Berger 1976b). The well-known high sea-level stands of the Barbados III (~124 kyr ago) and II

(~103 kyr ago) and marine terraces (Broecker et al. 1968) clearly correspond in time to high summer insolation anomaly which amounts to some 10% of current values, particularly in the high latitudes. An abortive glaciation 115 kyr ago which separated these two warm intervals was successfully simulated by Royer et al. (1983) but not by Rind et al. (1989) using the associated insolation minimum as the only external forcing. The main glacial transition between stages 5 and 4, 72 kyr ago, was underinsolated and, more important, this drop in insolation was not compensated by any significant increase during the whole Würm glaciation phase. Another important decrease 25 kyr ago augers the 18 kyr ago maximum extent of ice in the northern hemisphere. Between 83 kyr and 18 kyr ago, there was an overall solar energy deficiency of 2.5×10^{25} calories north of 45°N, sufficient to compensate for the latent heat liberated in the atmosphere during the formation of snow required by the buildup of the huge 18-kyr-ago ice sheets (Mason 1976).

C. Astronomical Frequencies Documented in Diverse Geologic Records

Spectral analysis of climatic records of the past 800 kyr provides substantial evidence that, at least near the frequencies of variation in obliquity and precession, a considerable fraction of the climatic variance is driven in some way by insolation changes accompanying the perturbations of the Earth's orbit (Imbrie and Imbrie 1980; Berger 1989b). Berger and Pestiaux (1987) have shown that using long deep-sea cores the following significant peaks are present at: 103 ± 24 kyr; 42 ± 8 kyr; 23 ± 4 kyr.

However, the interpretation is uncertain. The 100-kyr cycle, so dominant a feature of the late Pleistocene record, does not exhibit a constant amplitude over the past 2 to 3 Myr (Pestiaux and Berger 1984). This periodicity disappeared 1 Myr before present, at a time when the ice sheets were much less developed over the Earth, reinforcing the idea that the growth of the major ice sheets may have played a role in the modulation of the 100-kyr cycle.

The spectrum of different cores varies with location depending on the nature of the climatic parameter analyzed (Hays et al. 1976). For example, the 41-kyr cycle is not seen in core V30–97 whereas the 23-kyr cycle is dominant in Atlantic summer sea-surface temperatures of the last 250 kyr (Ruddiman and McIntyre 1981; Imbrie 1982). This is not too surprising as these spectra depend on the way the climate system reacts to the insolation forcing and on which type of insolation it is sensitive to. Indeed, contrary to the Milankovitch model with high polar latitudes recording the obliquity signal (as shown in the Vostok core, e.g.; Jouzel et al. 1987) and low latitudes recording only the precessional signal, the latitudinal dependence of the insolation parameters is more complex. The mid-month high-latitude summer insolation displays a stronger signal in the precession band than in the obliquity band (Berger and Pestiaux 1984).

D. Phase Coherency

The variance components centered near a 100-kyr cycle which dominates most climatic records (Hays et al. 1976), seem to be in phase with the eccentricity cycle (high eccentricity at low ice volume). The exceptional strength of this cycle suggests stochastic (Hasselman 1976; Kominz and Pisias 1979) amplification of the insolation forcing, or nonlinear amplification through the deep-ocean circulation, involving: carbon dioxide; ice-sheet-related mechanisms and feedbacks; the isostatic rebound of the elastic lithosphere and of the viscous mantle; and the ocean-ice interactions. The 100-kyr climatic cycle can indeed be explained both,

(a) from the eccentricity signal directly, provided an amplification mechanism can be found (as in the double-potential theory of Nicolis [1980,1982] and Benzi et al. [1982]) and/or,

(b) by a beat between the two main precessional components as shown by Wigley (1976) from nonlinear climate theory.

It must be stressed here that the same arises for the 100-kyr eccentricity cycle in celestial mechanics: the frequency corresponding to the second period of the eccentricity (94945 yr in Table 3 of Berger [1977]) is obtained from precession frequencies number 3 and 1 in Berger's (1977) Table 2 (1/18976 – 1/23716). The other frequencies (number 1 and 3 to 6) come, respectively, from the following combinations: 2–1, 3–2, 4–1, 4–2 and 3–4 (see Berger and Loutre [1987] for more details).

Milankovitch theory requires a high eccentricity in order for an ice age to occur, which is just the reverse of the correlation claimed at the beginning of this chapter. Although Milankovitch included the effect of eccentricity through the precessional parameter alone, if a higher degree of accuracy is used in insolation computations, $(1 - e^2)^{-3/2}$ appears as a factor in all insolation parameters at all latitudinal bands, reflecting the full equation of the elliptical motion and the variation of the Earth-Sun mean distance in terms of the invariable semimajor axis of the Earth's orbit (Berger 1978b). Although its absolute effect is relatively small (1% at the most), this term increases the insolation at times of high eccentricity and decreases it during low eccentricity, a result seemingly coherent with recent findings; more details are given in Berger (1989a). It should, therefore, not be ignored, as it reinforces the impact on climate in the (a) variant of the 100-kyr cycle explanation above.

However, Imbrie and his collaborators (1984) show that the coherency of orbital and climatic variables in the 100-kyr band is enhanced significantly when the geologic record is tuned precisely to the obliquity and precession bands. Using the so-called orbitally tuned SPECMAP time scale refined by Martinson et al. (1987) for the last 300 kyr, they found a coherency in the astronomical bands significant at more than 99% level of confidence. This supports the second (b) mechanism (beat) rather than the first.

Such a rather coherent phase relationship was also reasonably well defined between insolation and ice volume by Kominz and Pisias (1979) where obliquity consistently led the δ ^{18}O record by about 10 kyr, whereas precession seemed to be in phase with the 23-kyr geologic signal. However, the recent results obtained by SPECMAP (Imbrie et al. 1989) show that these phase leads and lags are more complicated.

Indeed, CLIMAP (1976) and more recently SPECMAP (Imbrie et al. 1984) teams have shown that the phase lags in the climate response to orbital forcing depend on the nature of the climatic parameters themselves and on their geographical location. For example, in their data the sea-surface temperature of the southern oceans seems to lead the response of the northern hemisphere ice sheets by roughly 3 kyr.

E. Other Astronomical Frequencies

Ruddiman et al. (1986) found the 54-kyr astronomical period that was predicted in 1976 (Berger 1977). A similar period of 58-kyr was found in a 400-kyr record of the paleomagnetic field from Summer Lake in south-central Oregon (Negrini et al. 1988).

The 412-kyr period in the eccentricity, was predicted by Berger (1976). Briskin and Harrell (1980) using 2-Myr sediment cores from both the Atlantic and the Pacific found it in the oxygen isotopic record of planktonic foraminifers, in coarse sediment fraction and in the magnetic inclination (which led them to propose that a relationship may exist between the eccentricity and the Earth's core modulation of the magnetic field). Kominz et al. (1979) also reported it from two spectra covering 730-kyr-long records.

The investigation by Moore et al. (1982) of calcium carbonate concentrations in equatorial Pacific sediments (core RC11–209), showed that the Pacific carbonate spectrum has been dominated for the past 2 Myr by variance in the 400-kyr band, with more modest contributions in the 100-kyr and 41-kyr bands (matching the variations of the eccentricity and obliquity, respectively).

F. Data from the Pliocene and Late Miocene

Moore et al. (1982) discovered that the 400-kyr eccentricity and sedimentary cycles were in phase over the last 8 Myr, but the 100-kyr cycle that dominated climatic variability during the Pleistocene ice ages (29% of the total variability) had only a minor effect 5 to 8.5 Myr ago, when it accounted for 6 times less variability than today at the deep sea drilling project (DSDP) site 158. Prell (1982) also failed to find evidence for a strong 100-kyr cycle in pre-Pleistocene sediments at the DSDP site 502 in the western Caribbean and at the site 503 in the eastern equatorial Pacific. He did find evidence of the 41-kyr cycle in a 7 Myr sediment record, as well as evidence of a 250-kyr cycle.

Spectral analysis of DSDP Hole 552A reveals a dominant quasi periodicity associated with obliquity-induced temperature variations in surface water and weaker peaks at the eccentricity and precession periods (Backman et al. 1986). In the Mediterranean Pliocene, rhythmic lithological variations in the Trubi and Narbone Formations of Sicily and Calabria show cycles that could be related to precession and eccentricity (Hilgen 1987). In particular, the precession cycle corresponds well with the mean duration of the deposition of the basic rhythmites, the eccentricity cycle of about 400 kyr would match the average duration of the carbonate units and the 100-kyr cycle, the arrangement of the sapropelitic intercalations.

The upper Pliocene is at the limit of validity of the astronomical calculations for the time domain. However, there is still high confidence in the values of the astronomical frequencies, so that conclusions from comparison of geologic and astronomical data may still be made. Moreover, a new astronomical computation, valid over the last 10 Myr is underway (Berger and Loutre 1988).

G. Pre-Cenozoic Evidence of an Astronomical Signal

Evidence exists that orbital variations have been linked to climate at periods shorter than 100-kyr during the past few 100 Myr, although dating such remote past raises questions about the length of such periodicities. This appeared at times when major ice masses were probably absent. Walsh power spectra of the Blue Lias Formation (basal Jurassic) show two cycles of <93 kyr duration which may record changes in orbital precession and obliquity (Weedon 1985,1986). Carbonate production in pelagic mid-Cretaceous sediments, quantified by calcium carbonate and optical densitometry time series, reflects the orbital eccentricity and precessional cycles (Herbert and Fisher 1986). Fourier analysis of long sections of the Late Triassic Lockatong and Passaic formation of the Newark Basin show periods in thickness corresponding roughly to the astronomical periodicities (Olsen 1986). All these interesting results encourage research of the stability of the solar system in order to determine to what extent the changing Earth-Moon distance, for example, influenced the length of the main astronomical periods. Recent computations (Berger et al. 1989c; Loutre and Berger 1989) (Table I) show that the precession and obliquity cycles should indeed be reduced drastically prior to 2 Gyr ago with the obliquity cycle starting to approach the precessional ones even if we take into account this varying Earth-Moon distance together with the changes in the rotation and dynamical ellipticity of the Earth. Already 270 Myr ago, changes were not negligible: the main precessional periods were 17.6 kyr and 21.0 kyr and the obliquity was 35.1 kyr. The astronomical theory of paleoclimates may eventually provide astronomers with periodicities with which they can test the stability of the theories of the planetary system over much of the Earth's history.

TABLE I
Changes in the Periods of Precession and Obliquity Due to Changes in the Earth-Moon Distance, and in the Rotation and Dynamical Ellipticity of the Earth[a]

Epoch	Periods (kyr)		
(Gyr ago)	Precession		Obliquity
0	19	23	41
0.5	16	19	30
1	15	17	25
2	13	14	20
2.5	11	13	17

[a]Data from Berger et al. (1989c) and Loutre and Berger (1989).

H. Combination Tones

Pestiaux et al. (1988) used cores with a high sedimentation rate covering the last glacial-interglacial cycle to resolve the higher-frequency part of the spectrum. Besides the 19-kyr precessional peak, three other periods were detected at significant levels: 10.3 ± 2.2 kyr; 4.7 ± 0.8 kyr; 2.5 ± 0.5 kyr.

These preferential frequency bands of climatic variability outside the direct orbital forcing band are still too broad to allow for a definite physical explanation. A tentative interpretation, though, may be given in terms of the climatic system's nonlinear response to variations in the insolation available at the top of the atmosphere. The 10.3, 4.7 and 2.5 kyr near-periodicities are indeed rough combinations tones of the 41, 23 and 19 kyr peaks found in the main insolation perturbations. Moreover, Pestiaux et al. (1988) succeeded in predicting these shorter periods by using the Ghil - Le Treut - Kallen nonlinear oscillator-climate model (Ghil and Le Treut 1981; Le Treut and Ghil 1983; see the chapter by Ghil). On the other hand, a tendency to obtain free oscillations at periods up to several tens of kyr in complex climate models has also been mentioned by Sergin (1979), Kallen et al. (1979) and Ghil (1980).

I. Ice Sheets Modeling and the 100-kyr Periodicity

Hays et al. (1976) were among the first to suggest that the enhancement of the 100-kyr cycle may be due to nonlinearities in the climate response, while Wigley (1976) suggested that this period may be a beat effect between the two main precession periodicities. Imbrie and Imbrie (1980) have developed a simplified glacial-dynamics model designed especially for the explicit purpose of reproducing the Pleistocene ice volume record from orbital forcing. The rate of climatic change is made inversely proportional to a time constant that assumes one of two specified values, depending on whether the climate is warming or cooling. Such a model tuned over the last 150 kyr, is forced with orbital input corresponding to an irradiation curve for July 65 N, with a mean time constant of 17 kyr and a 4 to 1 difference between the time constants of glacial growth and melting. The model's simulation of the isotopic record of ice volume over the past 150 kyr is reasonably good but results for earlier times are mixed and parametric adjustments do little or nothing to improve the matter.

Models based on a beat or on a simple form of nonlinearity (such as the asymmetrical response of the models of Imbrie and Imbrie [1980], in which the time constant governing ice decay is smaller than the time constant governing the ice growth), may hide a second problem. It is difficult to introduce substantial 100-kyr power into the climate response without also introducing power reflecting the 413-kyr eccentricity cycle in amounts that are much greater than have been detected in most Late Pleistocene climatic records (Kukla et al. 1981); however, the fit is much better in the results obtained by Moore et al. (1982) where the whole Pleistocene is considered (see, e.g., Sec. III. F).

The role of ice sheets in determining the long-period climate response in the 100-kyr range, may be clarified with realistic parametric modeling of the ice-sheet dynamics. Following Birchfield and Weertman (1978), solar radiation variations seem to be large enough to account for the ice-age cycles when glacier mechanics are included. By adding calving by proglacial lakes in his ice-sheet model, Pollard (1982,1984) created a sharp 100-kyr cycle that stayed roughly in step with the geologic record of ice volume as far back as 600 kyr ago. He also found that the eccentricity cycle was impressed only indirectly through its accentuation of the precession cycle and that a sharp termination resulted when an eccentricity-strengthened precession cycle coincided with an existing large ice sheet.

Another mechanism for amplifying the effects of orbital variations, and specifically of the eccentricity peak, is the interaction between the ice sheet and the underlying bedrock through isostatic rebound (Oerlemans 1980; Birchfield et al. 1981). High-latitude topography (Birchfield et al. 1982) and ice albedo-temperature feedback (Birchfield and Weertman 1982) are also included in the models.

Recent ice-sheet models show that the 100-kyr cycle can be simulated with (Ghil and Le Treut 1981; Saltzman et al. 1984) or without (Lindzen 1986) internal free oscillations related to resonances when astronomically forced. It is significantly re-inforced when isostatic rebound (Hyde and Peltier 1985) and iceberg calving are taken into account (Pollard 1982).

J. Equilibrium Three-Dimensional Climate Models Astronomically Controlled

Geography may help explain the climate's sensitivity (North et al. 1983; Mitchell et al. 1988). For example, when orbital variations are used that favor increasingly cooler summers (as at the transition between 125 kyr and 115 kyr ago and at the last glacial maximum 18 kyr ago), models with a realistic distribution of continents and oceans generate the largest ice cap over northern Canada and Scandinavia. These are obviously the most sensitive regions of the climate system to orbital influences. However, it must be mentioned that the GISS model failed to maintain snow cover through the summer at location of suspected initiation of the major ice sheets, despite the astronomically reduced summer and fall insolation (Rind et al. 1989). This result is, in some way, in opposition to the Royer et al. (1983) experiment which showed that 115 kyr ago, an annual mean cooling of more than 2°C over Canada with an increased precipitation could have acted as a trigger for the initiation of the Laurentide ice-sheet's growth.

Although glacial ice has generally received the most attention, evidence exists that orbital variations also influence the behavior of the North Atlantic deep ocean water and atmospheric features such as the intensity of the westerly winds and of the Indian monsoon. For example, changing the orbital configurations to that of 9 kyr ago, when insolation seasonality was 14% higher than today, leads to an intensified southwest monsoon (Kutzbach 1981).

A simulation over the last 150 kyr (Prell and Kutzbach 1987) has shown that under glacial conditions, the simulated monsoon is weakened in southern Asia but precipitation is increased in the equatorial west Indian Ocean and equatorial North Africa. Moreover, the monsoon is strongly tied to the precession parameter (their maxima coincide) as is also the case for the variations in tropical (Bernard 1962; Rossignol-Strick 1983; Short and Mengel 1986) and equatorial climate (Pokras and Mix 1987).

K. Transient Response of the Climate System to the Orbital Forcing and Change in the Seasonal Pattern of Insolation

In addition to the calculation of the Earth's climate in equilibrium with a particular insolation pattern and other boundary conditions such as the ice sheets, the simulation of the transient response of a realistic climate system to orbital variations provides additional understanding of the physical mechanisms linking astronomical forcing with climate. Berger (1979) suggested earlier that the long-term astronomical variation of the latitudinal distribution of the seasonal pattern of insolation is the key factor driving the climate system, and the complex interactions among its different parts. A 2.5-dimension time-dependent physical climate model, which takes into account the feedbacks between the atmosphere, the upper ocean, the sea ice, the ice sheets and the lithosphere, thought to be the most important for the astronomical time scales, strongly supports this hypothesis (Berger et al. 1988; Fichefet et al. 1989; Berger et al. 1990a, b). In this model, the simulated long-term variations of the global ice volume over the past 125 kyr agree remarkably well with the reconstructed sea-level curves of Chappell and Shackleton (1986) and Labeyrie et al. (1987).

Using astronomically forced climate models of varying complexity, it has been shown that the dynamic behavior of the climate over the last 400 kyr is reproduced fairly well (Imbrie and Imbrie 1980; Berger 1980; Berger et al. 1988). Assuming sufficient predictability (Nicolis and Nicolis 1986) and no human interference at the astronomical scale, orbital forcing predicts that the general cooling that began 6 kyr ago will continue with the first moderate cold peak around 5 kyr from now, a major cooling about 23 kyr from now and full ice-age conditions 60 kyr from now (Berger 1980,1988).

IV. CONCLUSIONS

Recent evidence suggests that major orbital variations can force climatic changes such as the Pleistocene glacials and interglacials. Among the number of competing theories to explain the advance and retreat of the Pleistocene ice sheets and similar climatic variations of the past, only the astronomical theory (of which Milankovitch theory is a particular version) is supported so far by substantial physical evidence. But other factors, such as topography and plate tectonics, have played an important role in establishing the general framework in which the orbital changes have been able to act efficiently. Moreover, processes such as the reflectance of solar energy by snow and ice fields and the infrared absorption by greenhouse gases, have created feedbacks that amplify orbital signals.

The astronomical theory of climates provides an absolute clock with which to date Quaternary sediments with a precision several times greater than is otherwise possible (Martinson et al. 1987). It also provides the boundary conditions necessary for a better understanding of the climatic system and the interactions between the atmosphere, hydrosphere, cryosphere, biosphere and lithosphere, which, at astronomical time scales, all play a role. Lastly, it allows a better understanding of the seasonal cycle and it can be used to test the performance of climate models over a broad spectrum of climatic regimes.

Various advantages of the astronomical theory of climates may be listed as follows: (a) it allows a better understanding of the other forcings, in particular the CO_2 cycle (Pisias and Shackleton 1984; Barnola et al. 1987), by extracting the astronomical signal from climate variability; (b) it predicts gross natural climate changes to be expected on the geologic time scale in the next 100 kyr, an approximate decay period of radioactive wastes; (c) it allows a better understanding of the sensitivity of our present-day interglacial climate and of the possible superinterglacial that could be generated by human activities within the next 50 yr or so; (d) it may provide data for astronomers with which to test the stability of the planetary system in pre-Quaternary times; (e) it enables accurate computation of the insolation changes at the decadal time scale due to change in orbital elements, in relation to the satellite measurements of the solar constant and its variations; (f) it allows a better understanding of the planetary system and the climatic variations of the planets (Ward 1974; Ward et al. 1979; Pollack 1979); and finally (g) it allows us to transfer theoretical knowledge (spectral analysis, numerical schemes, etc.) and technologies (deep-sea drilling, satellites, supercomputers, etc.) to society at large.

Acknowledgement.   The author wishes to thank G. Kukla from Lamont Doherty Geological Observatory, Columbia University, New York, for his careful reading of the manuscript and for fruitful discussions, and N. Materne-Depoorter for preparing the manuscript.

QUATERNARY GLACIATIONS: THEORY AND OBSERVATIONS

MICHAEL GHIL

University of California, Los Angeles

Ice ages represent but a small fraction of the Earth's geologic history. Still, much has been learned about the causes of climatic variability on all time scales by studying glaciation cycles of the Quaternary. This chapter reviews the geochemical and micropaleontological records of Quaternary glaciations, and the history of glacial theories. It discusses a very simple model of self-sustained climatic oscillations, coupling the radiation balance of the atmosphere and oceans with the mass balance of ice sheets and with the isostatic adjustment of bedrock. The model's spectral response to small, quasi-periodic insolation variations, due to secular changes in the Earth's orbital parameters is described. This response exhibits the known peaks at 100 kyr, 40 kyr and 20 kyr and predicts additional peaks at 15 kyr, 13 kyr and 10 kyr. Observational evidence for these latter peaks is presented; this comes from three marine cores with high sedimentation rates in the Indian Ocean, and from the Antarctic ice core at Vostok. Other models of Quaternary glaciations are reviewed and confronted with the data. Open questions are mentioned, and the relevance of this problem, and of the methods used to solve it, are discussed for the solar-variability community. In particular, the net, globally and annually averaged change in solar radiation at the top of the atmosphere is very small (0.1%) for orbital changes with very long periods, as well as for solar irradiance changes with periods of one or two decades. This leaves a substantial role to internal climate variability on either time scale.

I. INTRODUCTION AND MOTIVATION

As dwellers on the Earth, we are particularly sensitive to the effects that changes in the Sun might have on our climate. Ancient Mesoamericans were sufficiently concerned by the possibility of the Sun's not rising on the next morning that they offered human sacrifices by way of maintaining its regular habits. Our concern with the Sun's changing behavior and its effects on climate is less extreme, but still quite substantial, as attested to by a huge literature on solar-terrestrial relationships in general (see, e.g., Castagnoli 1988), and Sun-climate relationships in particular (see, e.g., Siscoe 1978; Van Loon and Labitzke 1990).

To understand better the nature of time-varying Sun-climate relationships, if any, it is necessary to consider, more generally, the issue of large climatic variations in the presence of small external forcing. Climatic records exist on various time scales, from instrumental records on the time scale of months to hundreds of years, through historical documents and archaeological evidence, to geologic proxy records on the time scale of thousands to millions of years. These records indicate that climate varies on all time scales in an irregular fashion.

A summary of climatic variability on all time scales appears in Fig. 1. The most striking feature is the presence of sharp peaks superimposed on a continuous background. The relative power in the peaks is poorly known; it depends of course on the climatic variable whose power spectrum is plotted, which is left undefined in the figure. Furthermore, phenomena of small spatial extent will contribute mostly to the high-frequency end of the spectrum,

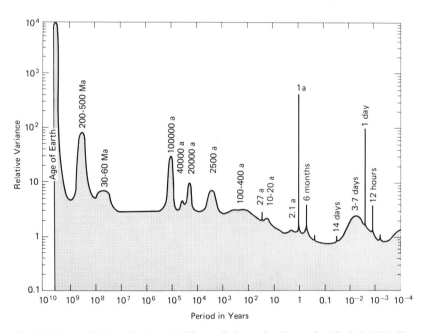

Fig. 1. Estimate of global climatic variability on all time scales (figure after Mitchell 1976). The position of the peaks indicated is relatively well known, but not their relative height and width. Units of time are abbreviated as 1 year = 1a and 10^6 years = 1Ma.

while large spatial scales play an increasing role towards the low-frequency end.

Many phenomena are believed to contribute to changes in climate. Anomalies in atmospheric flow patterns affect climate on the time scale of months and seasons. On the time scale of tens of Myr, plate tectonics and continental drift play an important role. Variations in the chemical composition of the atmosphere and oceans are essential on the time scale of Gyr and significant on time scales as short as decades.

The appropriate definition of the climatic system itself depends on the phenomena one is interested in, which determine the components of the system active on the corresponding time scale. No single model could encompass all temporal and spatial scales, include all the components, mechanisms and processes, and thus explain all the climatic phenomena at once.

The goal in this chapter is much more limited, and guided indirectly by the issue of Sun-climate relationships; it concentrates on the most striking phenomena to occur during the last 2 Myr of the Earth's climatic history, the Quaternary period, namely on glaciation cycles. The time scale of these phenomena ranges from 10^3 to 10^5 yr.

Glaciation cycles have an amplitude in global temperature of $2°$ to $5°C$ (CLIMAP 1976), while interdecadal climate oscillations documented in the instrumental record have an amplitude of 0.1 to 0.2 °C (Ghil and Vautard 1990). Clearly the latter have rates of temperature change 10-100 times larger than the former. The changes in globally and annually averaged insolation received at the top of the atmosphere are of $\sim 0.1\%$ both due to orbital changes on time scales of 10^4 to 10^5 yr (Berger 1978a) and due to solar luminosity changes on the time scale of 10 to 10^2 yr (chapter by Fröhlich); quasi-equilibrium sensitivity of global temperature to slow changes in insolation is ~ 1 °C in the former given 1% change in the latter (Ghil and Childress 1987,p.320).

Changes in temperature on the Quaternary time scale are thus too large to be explained solely by orbital changes, whereas interdecadal temperature changes are too rapid, given the thermal inertia of the upper ocean, to be explained entirely by solar-variability changes. Any explanation of the latter must rely, therefore, as much on the interaction between forcing and internal climate variability as the explanation of the former, which is undertaken in this chapter.

This chapter describes and models, in one of the simplest ways possible, the components of the climatic system active on Quaternary time scales—atmosphere, ocean, continental ice sheets, the Earth's upper strata—and their nonlinear interactions. In Sec. II, the discovery of geologic evidence for past glaciations is sketched, geochemical methods for the study of deep-sea cores are reviewed and the phenomenology of glaciation cycles is described as deduced from these cores. A near-periodicity of roughly 100,000 yr dominates continuous records of isotopic proxy data for ice volume, with smaller spectral peaks near 40,000 and 20,000 yr, as suggested in Fig. 1. The records themselves are rather irregular and much of the spectral power resides in a continuous background.

In Sec. III, a brief introduction to the physical processes of climate evolution is given. Atmospheric radiation balance, as well as the mass balance and viscoplastic flow of large ice sheets are reviewed. For the sake of clarity in exposition to a nonspecialist readership, one set of self-consistent, simple climate models is fully explored, while others are mentioned for comparison purposes. A broad review of paleoclimatic modeling is given by Saltzman (1985).

The equations derived and analyzed in Sec. III for radiation balance and ice flow are coupled in Sec. IV.A with an equation for bedrock response to yield a system of differential equations that govern stable, self-sustained, periodic oscillations. Changes in the orbital parameters of the Earth on the Quaternary time scales provide small changes in insolation. These quasi-periodic changes in the system's forcing produce forced oscillations of a quasi-periodic or aperiodic character. The power spectra of these oscillations show the above mentioned peaks with periodicities near 100 kyr, 40 kyr and 20 kyr, as well as the continuous background apparent in Fig. 1. In addition, they exhibit peaks at 15 kyr, 13 kyr, 10 kyr and even higher frequencies, up to and including 2.4 kyr (Sec.IV.A). These high frequencies are verified in recent marine and ice-core records with sufficiently high resolution (Sec.IV.B).

A few novel methods for the study of nonlinear, quasi-periodic and aperiodic phenomena have been developed in the context of theoretical climate dynamics. These methods are discussed in Secs. III and IV, hoping that they may serve the solar-variability community. A comparison between the results of different paleoclimatic models and their ability to reproduce certain features of existing paleoclimatic records is given in Sec. V. In Sec. V, it is also pointed out that no direct evidence of solar-variability effects on Quaternary glaciations is available. However, the presence of the 2.4 kyr peak in both climatic and ^{14}C records is intriguing enough to be worth mentioning.

II. PALEOCLIMATOLOGICAL EVIDENCE

A. History of the Problem

Changes in the extent of Alpine glaciers have been known to Swiss mountaineers for many generations. Huge erratic boulders lying in the low, currently unglaciated valleys, and parallel striations of permanently exposed, flat rock surface were easily associated with similar boulders carried on the surface of glaciers flowing in the higher valleys, and with the aspect of present-day glacier beds temporarily exposed by unusual summer melting or deep cracks. The boulders and striations spoke clearly to the unbiased eye of times when the glaciers extended further down into the valleys, and filled them to the crest of their dividing ranges.

Until the early 19th century, however, the nascent science of geology had another explanation for these and related observations: the Biblical flood. It was only with difficulty that a number of courageous scientists, led by Louis Agassiz, faced the facts and gradually convinced their colleagues of the past extent of glaciations and their alternation in time with warmer periods, such as the one in which we now live (Imbrie and Imbrie 1986).

Eventually, the geology of the Quaternary period became intimately linked with the study of the changing extent of continental ice sheets. In the northern hemisphere, the Laurentide ice sheet at various times covered most of Canada and New England, and much of the Middle West, Great Plains and Rockies. The Fenno-Scandian ice sheet extended into Eastern and Central Europe, without quite linking up with the smaller Alpine ice cap, mentioned above.

In the southern hemisphere, the Antarctic ice sheet was well established and reached its present volume before the Quaternary. The Antarctic continent being surrounded by oceans, this ice sheet maintained a nearly constant extent and volume throughout the Quaternary (Mercer 1983). Smaller ice sheets, however, developed over parts of Australia and New Zealand, and extended out from the Andes. In between its extreme advances, the ice in both hemispheres retreated to areas comparable to those it occupies today: Antarctica, Greenland, the Canadian Archipelago and small mountain glaciers.

The Quaternary in its entirety has to be considered an ice age when compared with the mean temperature of much longer, completely ice-free periods of the Earth's history, such as the Mesozoic era, the age of Dinosaurs, which lasted for 160 Myr. Other ice ages occurred in the Earth's past, 2.3 Gyr, 1 Gyr, 700, 450 and 270 Myr ago. All the episodes for which the presence of ice is recorded and temporally resolved in the geologic record seem to show higher climatic variability than the entirely ice-free episodes. The total duration of the ice-free episodes in the geologic past seems to be much longer than that of episodes when ice of variable extent was present (Crowley 1983).

Plate tectonics, continental drift and orographic changes certainly play a role in creating geographic situations in which land ice can appear and develop. We shall not concern ourselves, however, with the time scales of tens of Myr on which these phenomena are important. Ice will be assumed to be present in the system and the variations in its extent will be studied, along with the variations in temperature with which they interact.

Geochemical Proxy Data. Variations in continental ice extent leave clear traces in the geologic record. Besides the previously mentioned erratic boulders and striations, classical stratigraphic methods record glacial till, the coarse, unstratified debris left behind when glaciers melt, alternating with thin, stratified, interglacial deposits. The uplift of the Earth's crust as ice sheets melt is recorded in shifting shore lines along the Baltic Sea and the East Coast of North America. The pollen of temperate-climate plants alternates in stratigraphic sequence with that deposited during cold climates.

But the most important, detailed evidence of glaciations started to accumulate with the advent of geochemical, isotopic methods in the 1950s. Long piston cores raised from the bottom of the sea by oceanographic research vessels contain fossil shells of micro-organisms living in the ocean, among which foraminifera are particularly wide-spread and well-studied. The calcium carbonate in the shells of bottom dwelling, or benthic, foraminifera contains oxygen whose *isotopic abundance ratio* $R = {}^{18}O/{}^{16}O$ reflects to a large extent the same ratio in the water from which they precipitated. The isotopic abundance ratio of a sample R_{sample}, is commonly represented by

$$\delta^{18}O = (R_{sample}/R_{std} - 1) \times 1000 \qquad (1)$$

where R_{std} is a reference standard. The standard for water is called Standard Mean Ocean Water (SMOW); a slightly different standard (Pee Dee Belemnivella, or PDB) is used for the carbonate. The normal abundance of the heavy ${}^{18}O$ isotope in water is only 0.2% approximately and it varies little, hence the use of the factor 1000 in formula (1) for $\delta^{18}O$. Currently available mass spectrometers determine $\delta^{18}O$ in calcium carbonate with a nominal precision of $\pm$ 0.05 per mil (Duplessy 1978).

As part of the hydrologic cycle, water molecules which evaporate from the surface of the oceans preferentially contain the light isotope ${}^{16}O$. If these molecules precipitate into high-latitude ice sheets and are fixed there, the ${}^{16}O$ cannot return to the ocean as it does during warm episodes, with runoff and river water. Hence the ocean becomes impoverished in the light, abundant isotope ${}^{16}O$ when large ice sheets are present, so that $\delta^{18}O$ of ocean water, and of benthic foraminiferal shells deposited in it, are higher during glacial episodes.

The mixing time of the world ocean is of the order of 1000 yr, and bottom waters never have an entirely uniform $\delta^{18}O$. Moreover, the isotopic ratio of the shells differs from one species to another, is influenced by the temperature of the ambient water, as well as by its $\delta^{18}O$, and is not quite in thermochemical equilibrium with the water. Still, the $\delta^{18}O$ of the microfossils in deep-sea cores appears to be positively correlated with the ice volume and hence constitutes a (relatively) reliable proxy indicator of global ice volume on the time scales of kyr (Shackleton and Opdyke 1973).

The Nature of Glaciation Cycles.   What is the evidence from deep-sea cores as to climatic change during the Quaternary? Figure 2 shows the $\delta^{18}O$ record of a deep-sea core from the western equatorial part of the Pacific Ocean.

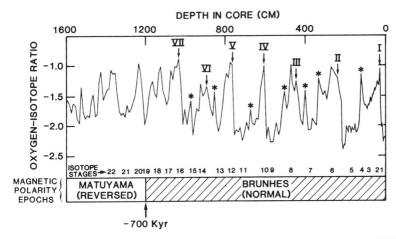

Fig. 2. Oxygen isotope record of deep-sea core V28-V238 (figure after Imbrie and Imbrie 1986).

The scale on top is simply the depth in the core. To translate this into a time scale, absolute dates for points along the core are necessary. Other isotopic methods, in particular the study of the potassium-argon ratio in lava flows, helped establish in the 1960s the dates of *polarity reversals* in the Earth's magnetic field. The current polarity is called normal, the opposite one is reversed. Reversals involve changes in the amplitude, as well as in the direction of the field and occur over a period of about 10 to 20 kyr (Ghil and Childress 1987, Sec.7.1). The first reversal can be dated to 730 kyr ago, to within 20 kyr. Two additional, short episodes of normal polarity have occurred during the Quaternary. They are called the Jaramillo and Olduvai normal events. The beginning of the Olduvai event, 1.8 Myr ago, is now agreed upon as the beginning of the Quaternary period.

Aside from the Brunhes/Matuyama polarity reversal a little over 700 kyr ago, no absolute dates were available to Shackleton and Opdyke (1973) for the analysis of the core in Fig. 2, so a uniform sedimentation rate is assumed in the time scale at the bottom of the figure. The isotopic stages marked along this scale represent episodes of algebraically higher $\delta^{18}O$ ratios: even numbers; or lower ratios: odd numbers.

The record in Fig. 2 shows an irregular evolution of $\delta^{18}O$, and hence ice volume, over its entire length. Sharp drops in ice volume approximately every 100 kyr are prominent; they are called *terminations* by Broecker and van Donk (1970) in their study of similar records. The growth of ice volume in between terminations seems more gradual in certain records, particularly in continental stratigraphic sequences (Kukla 1969), leading to the idea of a roughly sawtooth-like shape of *glaciation cycles*.

The "terminations" are not equally spaced, nor do the segments of record between terminations look alike: irregular variability on time scales

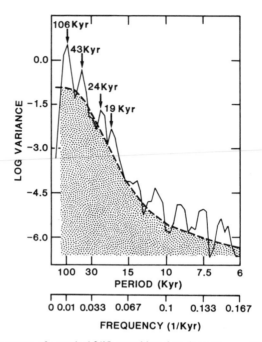

Fig. 3. Power spectrum of a patched δ¹⁸O record based on deep-sea cores RC11–120 and E49–18 (figure after Imbrie and Imbrie 1986).

shorter than 100 kyr is evident in Fig. 2. A few sharp spikes as high as the terminations, and many smaller ones, occur repeatedly, but aperiodically. Long plateaus with very little high-frequency variability appear in the $\delta^{18}O$ curve during some of the major cycles, but not others. These plateaus are mostly near the mean value of the record, $\delta^{18}O = -1.5$ per mil.

Figure 3 shows the power spectrum, due to Hays et al. (1976), of a combined $\delta^{18}O$ record taken from two cores in the southern part of the Indian Ocean, one of which goes back 450 kyr. The most prominent peak is at 106 kyr, with smaller peaks at 43 kyr, 24 kyr and 19 kyr. These peaks are superimposed on a red-noise-like continuous background.

Other deep-sea cores and other proxy indicators, both isotopic and micropaleontological, all seem to show irregular variations like Fig. 2, with a major near periodicity close to 100 kyr, terminations, spikes and plateaus. Their power spectra show peaks near 100 kyr, 40 kyr and 20 kyr, as well as occasional additional peaks. These are superimposed on a continuous background which decreases from low to high frequencies.

B. Current Status of Paleoclimatic Records

Information Content of Quaternary Records.   For the purposes of modeling, understanding and explaining glaciation cycles, we need to take a closer look at their information content. The δ¹⁸O record of four marine cores is shown in Fig. 4. Two of these, RC11–120 (Hays et al. 1976) and V19-240 (Shackleton 1977) cover much of the late Pleistocene and have moderate sedimentation rates on average. The other two, V28–238 and V28–239 (Shackleton and Opdyke 1973, 1976) cover most or all of the Quaternary, having much smaller sedimentation rates. The corresponding resolution is between a few kyr to tens of kyr. The trade-off between resolution and total length of such a time series has been imposed by the limit in physical length of piston cores (Crowley 1983). This trade-off implies that each core provides a band-limited view of climatic variability.

All four records show irregular changes occurring over the shortest, as well as the longest time scales resolved. The range of δ¹⁸O variations is nor-

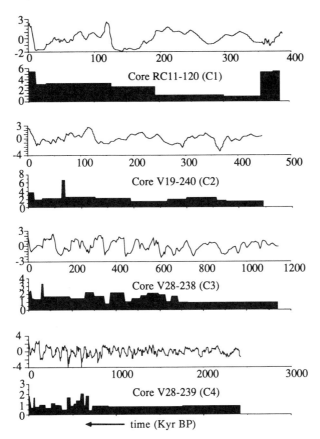

Fig. 4. Four marine records of δ¹⁸O, a proxy indicator of global ice volume. The ordinate is centered on the mean and normalized by the standard deviation of each record. Estimated sedimentation rates are given as solid bars underneath each core, in cm kyr⁻¹ (figure after Pestiaux 1984).

malized in the figure by the variance of the record in each panel. This $\delta^{18}O$ range translates into a change of global ice volume by a factor of 2 to 3 over the Quaternary. Microfaunal counts in marine sediments and $\delta^{18}O$ in ice cores indicate a corresponding change of global temperature by a few K (Crowley 1983; Lorius et al. 1985).

An approach to determining the *significant* number of degrees of freedom in time series such as those of Fig. 4 was proposed by Broomhead and King (1986), based on statistical concepts from signal processing (Pike et al. 1984), and, independently, by Fraedrich (1986). Similar concepts have been applied to the nonlinear dynamics of short-term climatic variability by Ghil (1987). Vautard and Ghil (1989) have developed and refined this approach, and applied it to the four cores in Fig. 4.

Their Singular Spectrum Analysis (SSA) estimates the *statistical dimension S*, rather than attempting to determine the *dynamical*, or correlation, dimension D (Grassberger and Procaccia 1983a; Mayer-Kress 1986). Values as different as $D = 3.1$ (Nicolis and Nicolis 1984) and $D \geq 14$ (Grassberger 1986) were obtained for D from core V28–238. This discrepancy is basically due to difficulties with limited sample size and measurement noise encountered by the Grassberger-Proccacia algorithm and related ones for computing D, whereas the algorithms work well for much longer and cleaner time series (Farmer et al. 1983; Eckmann and Ruelle 1985; Mayer-Kress 1986). SSA deals explicitly with the problems of measurement noise, and with the strong nonstationarity due, at least in part, to the large changes in sedimentation rate. The values of S given by SSA range between 5, for core RC11–120, and 10, for V28–239, with confidence intervals of $\pm$ 1. The difference in S between the two agrees well with visual inspection of the time series, which indicates a greater range of time scales in the variability of the latter than the former core.

It follows that a deterministic nonlinear model with a few lumped, or global, variables, interacting through a comparable number of physical mechanisms, should be able in principle to reproduce the dynamical information available in Quaternary proxy records, to within the accuracy of these records. Much greater spatial detail is available for selected times in the past (CLIMAP 1976), such as Last Glacial Maximum, at 18 kyr ago, the preceding glacial maximum at $\sim$ 125 kyr ago, and a few times during the Holocene. But the additional degrees of freedom associated with this spatial detail do not seem to be crucial for the long-term dynamics.

Spectral Information: the 100 kyr Dilemma.   The four cores in Fig. 4, as well as most other such cores, reveal by visual inspection a dominant near-periodicity of 100 kyr (see, e.g., Broecker and van Donk 1970). The visual result is confirmed by spectral analysis, which in addition yields smaller peaks near 40 kyr and 20 kyr (Hays et al. 1976), and still smaller ones between 15 kyr and 10 kyr (Pestiaux 1984; Pestiaux et al. 1988). These peaks are superimposed on a continuous background, which contains most of the variance. The spectral continuum has increasing power with decreasing frequency, as do stochastic processes of the type called red noise (Mitchell 1976).

The two types of forcing considered important in explaining this spectrum are orbital changes, and white-noise forcing by shorter-lived climatic phenomena (Hasselmann 1976). White noise has a flat spectrum; it produces a red-noise spectrum when passing through a linear, stable system, that acts as a low-pass filter. Orbital changes are quasi-periodic (Ghil and Childress 1987, Sec. 12.3), with an obliquity peak at 41 kyr, precessional peaks at 23 kyr and 19 kyr, and eccentricity peaks at 413, 95, 124, 100 and 131 kyr, in decreasing order of amplitude (Berger 1978a).

The idealized spectra of Quaternary records and of the relevant forcing are shown in Fig. 5. The climatic peaks at 40 kyr and 20 kyr are most likely related to the obliquity and precessional peaks, respectively. Part of the red-

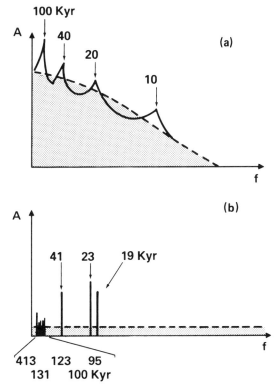

Fig. 5. The 100 kyr dilemma: (a) a composite power spectrum of marine $\delta^{18}O$ vs (b) the composite spectrum of orbitally caused insolation changes. The units on the ordinate are arbitrary (A); the continuous part of each spectrum is stippled (figure from Ghil 1989).

noise climatic background, but far from all can be explained by the white-noise forcing. This leaves the so-called "100 kyr dilemma" of explaining the peak near 100 kyr, which is dominant at least during the late Pleistocene (the last 700 to 900 kyr; see, e.g., Pisias and Moore 1981). It is this peak which is associated with temperature changes of a few degrees, while those at 40 kyr and 20 kyr correspond to much smaller changes in temperature. The 100 kyr peak cannot be produced by direct eccentricity forcing of a quasi-equilibrium model, neither in its amplitude (see Ghil 1989, and references therein), nor in its location, due to the widely distributed nature of the eccentricity variations, with only the fourth-largest peak at 100 kyr.

Many modeling efforts have focused on this 100 kyr dilemma. A complete explanation of the Quaternary record, however, requires modeling most of the variance in the red-noise continuum, as well as the high-frequency peaks to the right of the precessional ones in Fig. 5a. The latter peaks are the ones related to the relative suddenness of "terminations" (Broecker and van Donk 1970), as well as to large-amplitude, rapid "spikes" in the $\delta^{18}O$ records, such as the Alleröd-Younger Dryas event between 13 kyr and 9 kyr (Duplessy et al. 1981; Jouzel et al. 1987). The ingredients of such a complete explanation will be sketched in the following two sections, with references to complementary approaches. Sec. V includes a comparison of results from various models.

III. PHYSICAL PROCESSES OF CLIMATE EVOLUTION

A. Schematic Models of Climatic Variability

A popular explanation of Quaternary glaciations attributes them to changes in solar radiation received at the top of the atmosphere. These changes can have two causes: (1) variations in the Sun's energy output, and (2) variations in the mean distance between Sun and Earth, due to planetary perturbations of the Earth's orbit (Berger 1988; Milankovitch 1941). Insolation changes due to the first cause are not as well known as those due to the second, and unlikely to have been much larger, over the last few Myr at least (chapter by Kasting and Grinspoon). Moreover, orbital changes in globally and annually averaged insolation are themselves too small by about an order of magnitude to produce, directly and exclusively, the observed temperature changes of a few K (Ghil and Childress 1987, Secs. 10.4, 10.5, and references therein). This, combined with the 100 kyr dilemma and related mysteries, as illustrated in Fig. 5, suggests a search for additional causes of variability, internal to the climatic system, i.e., independent of any external insolation changes.

The simplest type of internal variability can be modeled as purely periodic, self-sustained oscillations. Two examples are given, and one of them is developed in greater detail in Sec. III. B. The first one may be described as follows.

Ice-albedo feedback, which plays a key role in energy-balance models (EBMs: Budyko 1969; Sellers 1969; Held and Suarez 1974; North 1975; Ghil 1976), states that the radiation balance becomes more negative, and hence temperature T goes down, when the planetary reflectivity, or albedo, α goes up,

$$\dot{T} \cong - \alpha. \qquad (2a)$$

Albedo in turn increases with the ice volume V,

$$\alpha \cong V. \qquad (2b)$$

Eliminating α between Eqs. (2a) and (2b) yields

$$\dot{T} \cong - V. \qquad (2c)$$

The schematic symbol "$\cong$" above should be read as "positively correlated with." It only indicates variations in the same sense, not direct, linear proportionality. This apparently vague concept of "co-variation" has been given an exact meaning within the mathematical framework of Boolean delay equations (BDEs), developed by Ghil and Mullhaupt (1985) and Ghil et al. (1987) for the rigorous analysis of so-called conceptual models of climatic change (see also Wright et al. 1990).

In addition to Eq. (2c) for $\dot{T}$, we need an equation for the evolution of V in order to close the system. Ice-sheet models are governed by an ice-mass balance equation, which can be schematized in this short-hand notation as

$$\dot{V} \cong p \qquad (3a)$$

where p is net solid precipitation over the entire ice sheet. This in turn is given by the difference

$$p \cong p_{ac} - p_{ab} \qquad (3b)$$

where the two terms are net accumulation and net ablation. Net accumulation rates on ice sheets are larger during warm climatic episodes (Lorius et al. 1985; Jouzel et al. 1987),

$$p_{ac} \cong T. \qquad (3c)$$

This has been called the *precipitation-temperature feedback* by Le Treut and Ghil (1983). Let us assume for the moment that the effect of T on p_{ac}, Eq. (3c), is stronger than its obviously opposite effect on p_{ab}, so that

$$p \cong T. \qquad (3c')$$

Eliminating p between Eqs. (3a) and (3c') leads to

$$V \cong T. \tag{3d}$$

Equations (2c) and (3d) together form an oscillatory system

$$\dot{T} \cong - V \tag{4a}$$

$$\dot{V} \cong T. \tag{4b}$$

The two mechanisms in the symbolic system (Eqs. 4a,b) are at the basis of the model of two nonlinear ordinary differential equations formulated by Källén et al. (1979), which will be presented in Sec. III.B. The results of this model show indeed a self-sustained oscillation, as suggested by Eqs. (4a,b), and provide supporting evidence for the assumption made here in replacing Eqs. (3b,c) by Eq. (3c').

The second example of coupling between two physical mechanisms which may yield an oscillation concerns ice-sheet models (Birchfield et al. 1981; Oerlemans 1982; Pollard 1984). The first equation is that for ice-mass balance, Eq. (3a). The second has been called *load-accumulation feedback* by Ghil (1985). It states that the net solid precipitation rate goes down as the ice load increases, because the ice sheet sinks into the bedrock, causing the snow line to move poleward,

$$\dot{p} \cong - V. \tag{5a}$$

Rewriting Eq. (3a) for convenience here as

$$\dot{V} \cong p \tag{5b}$$

we see that the system of Eqs. (5a,b) also represents an elementary oscillation.

More precisely, the physical mechanisms represented schematically in the system of Eqs. (4a,b) or (5a,b) give rise, through their coupling, to the *possibility* of an oscillation. Whether an oscillation does indeed occur and is stable depends on the numerical value of parameters in the respective system of nonlinear ordinary differential equations. The oscillation schematized by Eqs. (5a,b) has been modeled in a spatially one-dimensional ice-sheet model with bedrock dynamics by Birchfield and Grumbine (1985) and by Hide and Peltier (1985).

Substantial differences between the latter two models exist, which are not reflected by the simple conceptual Eqs. (5a,b). The parameter values

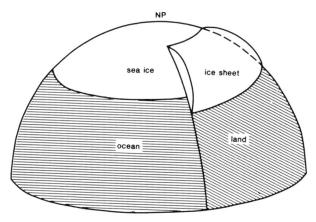

Fig. 6. Surface features of nonlinear oscillatory climate model, Eq. (7a) (figure from Ghil and Childress 1987).

used, especially for the visco-elastic properties of the bedrock, are also different. Consequently, the Birchfield and Grumbine model exhibits an oscillation with a period of 50 to 70 kyr, for a reasonable range of parameter values. The preliminary simple-forcing experiment of Peltier and Hyde (1984) indicates the presence of an oscillation with a period of 100 kyr, although no complete analysis is provided. Various other oscillatory mechanisms for paleoclimatic variability have been considered by Sergin (1979), Saltzman, Sutera and associates (see Saltzman [1985] for a detailed review) and by Ghil et al. (1987).

B. Radiation Balance and Ice-Sheet Dynamics

Here we examine how the potential for an oscillation contained in the interactive mechanisms (Eqs. 4a,b) is realized in a system of two coupled ordinary differential equations, formulated by Källén et al. (1979) and further analyzed by Ghil and Tavantzis (1983), Nicolis (1984) and Harvey and Schneider (1984). The model couples the radiation balance of EBMs, with their ice-albedo feedback (Eq. 4a), to the ice-mass balance of ice-sheet models, with the temperature-precipitation feedback, Eq. (4b). It is globally averaged, but distinguishes between open ocean, marine ice, bare land and continental ice sheet (Fig. 6).

The ice sheet, following Weertman (1976), is axially symmetric, and its meridional cross section is symmetric about a vertical axis half way between the Arctic coastline and its southern tip (see Fig. 7). Due to plastic flow, the profile of the ice sheet is parabolic, which is an excellent approximation for present-day ice sheets, caps and domes of widely varying sizes (Paterson 1972; Ghil and Childress 1987, Sec.11.2). The mass balance of the ice sheet is given by net accumulation with rate a, north of the snowline, and net ablation with rate a', south of it,

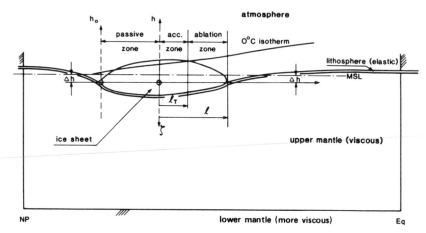

Fig. 7. Meridional cross section through the Earth's upper strata, Eq. (7b) (figure after Ghil and Childress 1987).

$$\dot{V} = aA - a'A' \qquad (6)$$

where A and A' are the accumulation and ablation areas, respectively. Equation (6) corresponds, in physical variables, to the conceptual Eqs. (3a,b). Empirically, the ratio $\varepsilon = a/a'$ is close to 1/3. The precipitation-temperature feedback of Eq. (3c) is expressed by making $\varepsilon = \varepsilon(T)$ a ramp function with a portion where it increases with T, as shown in Fig. 8. The snowline is the intersection of the 0°C isotherm with the surface of the ice sheet (see Fig. 7), at a meridional distance ℓ_T from the vertical symmetry axis. The position ℓ_T of the snowline, separating area A from A', is determined within the model by assuming that the 0°C isotherm rises in altitude as global temperature T rises.

The radiation balance associated with the surface features in Fig. 6 is given by

$$c_T\dot{T} = Q[1 - \gamma(\alpha_0 + \alpha_1\ell) - (1 - \gamma)\alpha_{oc}(T)] - \kappa(T - T_\kappa). \qquad (7a)$$

The first term on the right-hand side of this equation is the solar radiation absorbed by the system, the second is outgoing infrared radiation. The heat capacity is c_T, γ is the fraction of the Earth's surface occupied by land, 2ℓ is the meridional extent of the ice sheet, and T_κ a reference temperature. The linear approximation of infrared radiation is well justified by both the data (Budyko 1969) and model results (Källén et al. 1979). The value of c_T/κ is based on ocean overturning times of 1 kyr and larger (see Ghil et al. 1987).

Of the insolation Q at the top of the atmosphere, only an amount

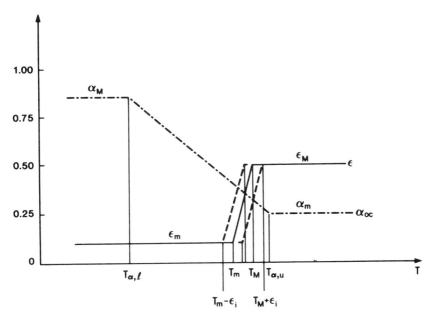

Fig. 8. The accumulation-to-ablation ratio $\varepsilon(T)$ (solid line) and the oceanic albedo α_{oc} (T) (dash-dotted line) as a function of globally and annually averaged temperature T (figure from Le Treut and Ghil 1983). The two parallel dashed lines are explained in Sec. IV.B, Eqs. (10a,b,c).

$Q(1 - \alpha)$ is absorbed. The albedo contains contributions from land and sea, $\alpha = \gamma\alpha_{land} + (1 - \gamma)\alpha_{oc}$. The land contribution is simply linear in the ice extent ℓ. The ocean contribution is a ramp function of temperature, also shown in Fig. 8 (cf. Sellers 1969). For fixed ℓ, Eq. (7a) is essentially an EBM; it exhibits multiple equilibria, arising from back-to-back saddle-node bifurcations, as discussed by Ghil (1989); see also Guckenheimer and Holmes (1983) for saddle-node bifurcations in general.

Taking into account the geometry of the ice sheet, as shown in Fig. 7, the mass-balance Eq. (6) can be written as:

$$c_L\ell = \ell^{-1/2} \{[1 + \varepsilon(T)]\ell_T(T,\ell) - \ell\} \qquad (7b)$$

where c_L is a relaxation constant for the ice sheet. At fixed T, Eq. (7b) yields two stable equilibria: a vanishingly small ice cap, and a sizable ice sheet. These are separated by an unstable equilibrium, and result from 2 saddle-node bifurcations, just like in EBMs (Ghil 1984; Ghil and Childress 1987). The coupled system (Eqs. 7a,b) also has multiple equilibria. Of these, an equilibrium (T_s,ℓ_s) close to the present climate is the only one to be stable for any parameter values at all (Ghil and Tavantzis 1983).

IV. CLIMATIC OSCILLATORS: SIMULATION AND PREDICTION

A. Coupled Oscillations of the Climatic System

Free Oscillations and Hopf Bifurcation.   The system given by Eqs.
(7a,b) is the detailed quantitative expression of the coupled conceptual mechanisms described by Eqs. (4a,b) and of additional, secondary mechanisms,
not subsumed by the minimal conceptual system (4a,b). The model (7a,b)
does in fact exhibit stable oscillatory solutions, as suggested by Eqs. (4a,b)
and shown in Fig. 9. When T is near its maximum (point A' in the figure),
the hydrologic cycle is most active, with large evaporation over the low-to-
midlatitude oceans, and larger snow accumulation rates, making ℓ grow
(point A''). As ℓ grows, albedo increases and temperature starts to decrease,
gradually cutting back evaporation and precipitation ($B' = B$). The ice sheet
is still expanding, as the snow mass accumulated along its surface flows
through the interior of the sheet to its periphery ($B'' = B$). The temperature
keeps sinking (C'), while the ice reaches its maximum extent (C''), further
lowering temperature (D').
     Due to small snow accumulation, continued ablation causes the ice sheet
to start shrinking (D''). This makes the temperature rise again ($E' = E$), leading to a renewal of snow accumulation. The increased accumulation competes
with the continued shrinking, due to the delay in the plastic flow of the ice
($E'' = E$). As the ice sheet starts to expand again (F''), the temperature continues to rise due to the thermal delay in the system (F'). This in turn accel-
erates the hydrologic activity and the continued expansion of the ice sheets,
completing the cycle (G', G'').
     The physics of this oscillation, as discussed in the preceding two para-

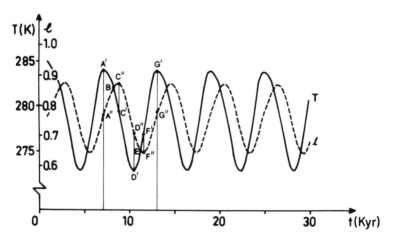

Fig. 9. Self-sustained oscillation of a coupled energy balance model-ice sheet model. Solid line
    is temperature T, ice extent ℓ is dashed (figure from Ghil and Childress 1987).

graphs, contains the basic mechanisms (2c) and (3a,b,c). Aside from these, it also contains the physics of the change of the 0 °C isotherm position, affecting the ratio A/A'. Together, these mechanisms yield an oscillation with T leading V by about one quarter period. Thus $\dot{T}$ and V are negatively correlated, while $\dot{V}$ and T are positively correlated, just as indicated by Eqs. (4a,b). This justifies *a posteriori* our replacing Eq. (3c) by Eq. (3c′).

The presence or absence of the oscillation in Fig. 9 depends most strongly on the parameter $\mu = c_T/\kappa c_L$. Large μ corresponds physically to a very small, possibly pre-Quaternary ice cap (see Ghil 1984; Saltzman and Sutera 1987). At these values, no oscillation exists and the stationary solution (T_s, ℓ_s) is stable. For Quaternary-size ice sheets, c_L is comparable in value to c_T. As μ decreases to O(1), i.e., as we proceed from very small to more substantial ice sheets, the equilibrium (T_s, ℓ_s) transfers its stability to a branch of periodic solutions, by a bifurcation mechanism. In contradistinction to EBMs (Ghil 1984,1989), which involve only (multiple) branches of stationary solutions, this transfer of stability is associated with a *Hopf bifurcation*.

The normal form of such a bifurcation from stationary to periodic solutions is written out and stability analysis performed in Ghil and Childress (1987, Sec. 12.2); see also Guckenheimer and Holmes (1983) for Hopf bifurcations in general. Here we only show the simplest bifurcation diagram for it (Fig. 10). Hopf bifurcation occurs in most physical situations where a periodic solution arises from a stationary one, as an oscillatory mode becomes unstable. Such a bifurcation represents essentially the growth to finite amplitude, and nonlinear equilibration, of an oscillatory instability in a system of arbitrarily many degrees of freedom.

The oscillation in Fig. 9 has about the right amplitude in temperature T and ice volume V, compared to paleoclimatic records. But its period is about 10 kyr, while the largest fluctuations during the Quaternary, especially during the last Myr (see Sec.III.A), are associated with the 100 kyr peak. Furthermore, Figs. 4 and 5 show the records to be irregular, rather than purely periodic, with additional peaks and with a red-noise background. These additional features will obtain by adding one mechanism, bedrock dynamics, and orbital forcing to the model, Eqs. (7a,b).

Orbitally-Forced Oscillations: Nonlinear Resonance. The augmented system of three coupled ordinary differential equations considered hereafter includes all three mechanisms of Eqs. (4a,b) and (5a,b): radiation balance, with ice-albedo feedback; ice-mass balance, with precipitation-temperature feedback; and isostatic rebound, with load-accumulation feedback. The third independent variable in the nonlinear ordinary differential equation is bedrock deflection ζ; its rate of change is due to the elasticity of the lithosphere and the viscous flow of the upper mantle (see Fig. 7). The complete system is coupled through the ice-mass balance equation, which is modified from the earlier form, Eq. (7b), while the energy balance stays the same:

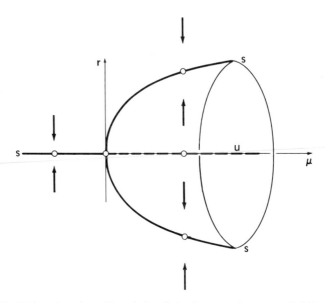

Fig. 10. Hopf bifurcation of a stable periodic solution from a stationary one. Solid branches are stable (s), dashed branches are unstable (u). The mean radius r of the periodic orbit (shown in false perspective as light solid) behaves parabolically, $r = \sqrt{\mu}$, near the bifurcation point. The open circles indicate: the bifurcation point ($r = 0$, $\mu = 0$), a unique stable equilibrium ($r = 0$, $\mu = \mu_- < 0$), an unstable equilibrium ($r = 0$, $\mu = \mu_+ > 0$), and a stable periodic solution ($r = \sqrt{\mu_+} > 0$). The arrows visualize attraction by the stable solutions, stationary or periodic, and repulsion by the unstable equilibrium (figure after Ghil and Childress 1987).

$$\dot{T} = Q(t)[1 - \gamma\,(\alpha_0 + \alpha_1 \ell) - (1 - \gamma)\alpha_{oc}(T)] - k(T - T_k) \quad (8a)$$

$$\ell = \{1 + (2/3)C\,\zeta\,\ell^{-1/2}\}^{-1}\,\{(3/2)\,\mu\ell^{-1/2}[(1 + \varepsilon(T,t))\,\ell_T(T,\ell,\zeta) - \ell] \\ - (2/3)\ell^{-1/2}(A\zeta\ell^{-2} + B\ell^{-3/2})\} \quad (8b)$$

$$\zeta = A\zeta\ell^{-2} + B\ell^{-3/2}. \quad (8c)$$

A, B and C are constants, and the relaxation constants c_T and c_L of Eq. (7a,b) have been absorbed into the parameter μ and time t by a suitable scaling.

For fixed insolation Q and accumulation-to-ablation ratio ε, the equilibria of system (8a,b) are in one-to-one correspondence with those of Eq. (7a,b), and the self-sustained oscillation of Eqs. (8a,b) also differs only slightly from that shown in Fig. 9 (Ghil and LeTreut 1981). Time-dependent external forcing, however, combines with this simple internal variability to yield more complex and realistic model behavior.

The orbitally caused insolation forcing is of two types: eccentricity forcing, which has a globally and annually averaged net effect, and obliquity and precessional forcing, which produces only larger or smaller differences in heating rates between hemispheres and seasons. The former is included by allowing Q in Eq. (8a) to change with the two representative periods $\tau_4 = 100$ kyr and $\tau_5 = 400$ kyr,

$$Q(t) = Q_o \{1 + \delta_4 \sin (2\pi t/\tau_4) + \delta_5 \sin [(2\pi t/\tau_5) + \phi_5]\} \quad (9)$$

where ϕ_5 is an arbitrary phase. The amplitudes δ_4 and δ_5 are proportional to $(\Delta e)^2/2$, where Δe is the total change in eccentricity e. As $\Delta e \leq 0.04$ over the Quaternary, δ_4 and δ_5 should not exceed 0.001. Using the complete spectrum of eccentricity forcing (see Fig. 5b above; cf. Berger 1978a), rather than the two representative frequencies $1/\tau_4$ and $1/\tau_5$ only, does not produce significantly different results (see Figs. 11 and 12 below).

For $\delta_4 = 0.001$, $\delta_5 = 0$, and a value of μ large enough for the system not to oscillate internally, the solution approaches in the limit of large times a small-amplitude forced oscillation with period 100 kyr and amplitude of roughly 0.1 K. This is consistent with the sensitivity results for EBMs of ~ 1 K change in mean global temperature T for 1% change in annual-mean insolation Q (Ghil 1985, 1989; Ghil and Childress 1987). For the same forcing, however, and a smaller μ value, at which internal oscillations would be present for $\delta_4 = 0$, the forced solution shows oscillations with an amplitude comparable to that in Fig. 9 and a period near 10 kyr, but modulated now at 100 kyr (LeTreut and Ghil 1983).

This represents a *nonlinear resonance* phenomenon, in which part of the internal variability is transferred to the forcing frequency. Both linear and nonlinear resonance involve an increase in the amplitude of the internal oscillation. But linear resonance is limited to a small difference between the internal and external frequencies, while nonlinear resonance also occurs for large differences between the two. The amplitude of the modulation at 100 kyr, when the internal, 10-kyr oscillation is present, is much larger than the amplitude of the purely forced oscillation when the internal one is absent (compare panels a and b of Fig. 13 in Ghil [1985]). This constitutes the *transfer of variability* from the internal to the external frequency.

Entrainment, Combination Tones and Chaotic Solutions. Forcing associated with obliquity and precessional changes has no net effect on the globally and annually averaged $Q(t)$. The usual approach, dating back to Milankovitch (1941), has been to calculate the insolation at a particular latitude and season as representative for their effect on the waxing and waning of the ice sheets. It seems simpler and less arbitrary to incorporate these orbital-forcing components through their net effect on the hydrologic cycle. In model (8a,b,c), this effect can be expressed by allowing the effective hydrologic temperatures T_m and T_M that determine the ramp portion of ε (T) in Fig. 8 to vary with the respective orbital changes. Thus in Eq. (8b)

$$\varepsilon(T,t) = \varepsilon[T(t); T_m (t), T_M(t)] \quad (10a)$$

while

$$[T_m(t),\, T_M(t)] = (T^o_m,\, T^o_M)\,\{1 + \varepsilon_j \sin[(2\pi t/\tau_j) + \phi_j]\}.\qquad (10b,c)$$

The periods $\tau_1 = 19$ kyr and $\tau_2 = 23$ kyr are associated with precession, while $\tau_3 = 41$ kyr is associated with obliquity (see also Fig. 5b). The phases ϕ_1, ϕ_2 and ϕ_3 in Eqs. (10a,b,c), like ϕ_5 in Eq. (9), are arbitrary. This modeling of orbital effects does not change the maximum or minimum accumulation or ablation rate. It only changes the range over which $\varepsilon = a/a'$ is sensitive to temperature variations. Values of ε_j used in the model are typically between 0.001 and 0.005.

We now apply hydrologic forcing of Eqs. (10a,b,c) to the model at a single frequency $f_j = 1/\tau_j$, $j = 1,2$ or 3. When this is done in the model's oscillating regime, increasing ε_j gradually, one observes partial and then total transfer of variability from the internal frequency $f_0 \cong 0.1$ kyr^{-1} to the forcing frequency f_j or to an integer multiple thereof, kf_j, $k = 1,2,3, \ldots$ This phenomenon is called *entrainment*, or frequency locking, and it was first observed by C. Huygens to occur between two clocks with different pendulum periods when brought close enough together. It is also found in the biological realm, where many animals and insects have internal clocks with periods close to, but not equal to, 24 hr, so-called circadian rhythms, which are entrained by the diurnal light cycle (Gleick 1987; Winfree 1980).

Further increase of ε_j leads to loss of entrainment, or detrainment: the solutions no longer have period f_0 or kf_j, but become irregular and their power spectrum becomes continuous. Such irregularity, produced by a few degrees of freedom interacting deterministically, is called *chaos*, rather than randomness. This concept has attracted considerable attention recently in the physical and biological, as well as atmospheric sciences (Ghil et al. 1991). The estimate given in Sec. II.B for the number S of statistically significant degrees of freedom, $5 \leq S \leq 10$, is compatible with deterministic chaos explaining much of the spectral continuum in Fig. 5a.

When precessional forcing with both ε_1 and ε_2 nonzero is applied to the model's internal oscillations, the transfer of variability from f_0 no longer occurs just to one frequency, nor to f_1 and f_2 only. *Combination tones* $k_1 f_1 + k_2 f_2$, with arbitrary integer k_1 and k_2, are excited. If both k_1 and k_2 are positive, we have a sum tone; if $k_1 k_2 < 0$, we have a difference tone.

In music, difference tones were discovered by the German organist Sorge and the Italian violinist Tartini in the middle of the 18th century. Sum tones were discovered by Helmholtz (1885), who developed a rigorous theory for their auditive perception. Difference tones produced in the inner ear by sound propagation from musical instruments are deeper and more easily perceived than sum tones, which explains the earlier discovery of the former.

Subjecting the model to both hydrologic and insolation forcing, with appropriate values of ε_1, ε_2, ε_3, δ_4 and δ_5, and arbitrary phases ϕ_1, ϕ_2, ϕ_3 and

ϕ_5 (Le Treut et al. 1988), one obtains the spectrum shown in Fig. 11a for a simulated marine core, and in Fig. 11b for a simulated ice core. The first panel uses log-log coordinates, showing clearly the continuous background with a negative slope, and the dominant peak near 100 kyr. The second panel uses log-linear coordinates to emphasize the range of frequencies for which most Quaternary records provide information.

The climatic variables T, V and ζ (not shown here) are given by the same forced solution of model Eqs. (8a,b,c), with the same parameter values in both panels. The isotopic variables in either panel are derived from the climatic ones by simple diagnostic relationships, similar to those used in interpreting actual isotopic data (Le Treut et al. 1988). The peaks in both panels are exactly at the same frequencies, and both panels use a logarithmic scale for the ordinate. The differences between the two panels in relative amplitude of the peaks are thus due entirely to the fact that marine $\delta^{18}O$ is dominated by ice volume V, while ice-core $\delta^{18}O$ is dominated by temperature T. Given the model's nonlinearity, these differences are not surprising, and are consistent with such differences between spectra of distinct paleoclimatic indicators which are known to occur even in the same marine core (Hays et al. 1976; Ruddiman and McIntyre 1981).

The dominant peak in the marine record (Fig. 11a) corresponds to a difference tone, 109 kyr $= 1/(f_1 - f_2)$. The directly forced orbital peaks f_1, f_2 and f_3 are smaller, as in the paleoclimatic records. Finally, sum tones 14.7 kyr $= 1/(f_2 + f_3)$, 13 kyr $= 1/(f_1 + f_3)$ and 10.4 kyr $= 1/(f_1 + f_2)$, as well as harmonics 11.5 kyr $= 1/2f_2$ and 9.5 kyr $= 1/2f_1$, are in evidence in the high-frequency band to the right of the orbital peaks. The dominance of

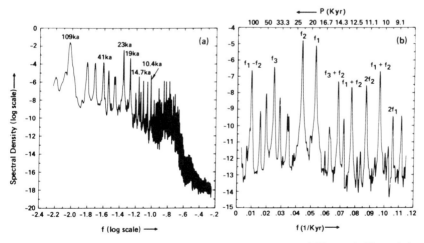

Fig. 11. Power spectrum of simulated (a) marine $\delta^{18}O$ and (b) ice-core $\delta^{18}O$ records. The periods associated with the labeled peaks in panel (b) are, from left to right, 109 kyr, 41 kyr, 23 kyr, 19 kyr, 14.7 kyr, 13 kyr, 11.5 kyr, 10.4 kyr and 9.5 kyr (figure from Le Treut et al. 1988).

the difference and sum tones $f_1 \pm f_2$ is probably due to the proximity of the two precessional frequencies, relative to the obliquity frequency f_3 (Le Treut and Ghil 1983). The magnitude of the peak near 100 kyr is affected but little by changing the amplitude of direct eccentricity forcing.

B. Observational Verification in Recent Isotopic Records

Marine Cores.   Sum tones and harmonics of the orbital frequencies are harder to perceive in marine sedimentary records, because they require higher sedimentation rates in order to be resolved, and because they are more affected by bioturbation and other post-depositional processes (Dalfes et al. 1984; Pestiaux 1984). They have not been reported so far in oscillatory models with internal frequencies to the left of obliquity (Birchfield and Grumbine 1985; Peltier and Hyde 1984; Saltzman and Sutera 1984). The presence of the high-frequency tones $\sum_{j=1}^{3} k_j f_j$, with substantial amplitudes, is thus a specific prediction of the model used for illustration purposes in Sec. IV.A (Ghil and Le Treut 1981; Le Treut and Ghil 1983; Ghil 1985).

Table I shows the observational verification of these high-frequency peaks by careful spectral analysis of three cores with high sedimentation rates from the Indian Ocean (Pestiaux et al. 1988). Six out of the eight peaks $f_i^{(d)}$ with periods between 5 kyr and 13 kyr, resolved in these cores, are very well approximated by combination tones with low-order K:

$$ K = \sum_{j=1}^{3} \left| k_j \right| \leq 3, \quad r_i \equiv \left| f_i^{(d)} - \sum_{j=1}^{3} k_j^{(i)} f_j \right| \leq 10^{-2}. \qquad (11a,b) $$

TABLE I
High-Frequency Peaks $f^{(d)}$ in Marine Cores[a] and in Theory $\Sigma k_j f_j$ [b]

$1 / f^{(d)}$ Data	k_1	k_2	k_3	K	$1 / \Sigma k_j f_j$ Theory	Good Agreement[c]
13.0	1	0	1	2	12.98	*
12.3	2	0	-1	3	12.37	*
10.2	1	1	0	2	10.41	*
9.5	2	0	0	2	9.51	*
8.8	0	2	1	3	8.99	*
7.4	1	2	0	3	7.17	*
5.6	0	4	0	4	5.76	
5.5	1	3	0	4	5.47	

[a] Pestiaux 1984; Pestiaux et al. 1988.
[b] Le Treut and Ghil 1983; Ghil 1985; Le Treut et al. 1988.
[c] Agreement according to Eqs. (11a, b) is indicated by an asterisk in the last column.

The constraint (11a) of low order is necessary because of Kronecker's lemma, which states that any given $f^{(d)}$ can be approximated arbitrarily well by a sum tone $kf' + \ell f''$, provided f'/f'' is irrational and $|k| + |\ell|$ is large enough.

The spectral information in Fig. 11 and Table I is also reflected in the time domain. The model-simulated time series of $\delta^{18}O$ in a marine core, whose power spectrum appears in Fig. 11a, is shown in Fig. 12. In contradistinction to Pollard (1984) or Saltzman (1987), no effort whatsoever was made to "tune" the solution in Fig. 12 to any particular record. In the author's view, such tuning should not be done, for two reasons. First, it is true that the four records in Fig. 4 (and other Quaternary records, not shown here) exhibit a number of features agreed to be common to all such records; still, considerable discrepancies in timing and magnitude of these features persist (Duplessy 1978; Jouzel et al. 1987; Winograd et al. 1988). Furthermore, any efforts to forcibly reconcile these differences by orbital tuning (Imbrie et al. 1984; Gribbin 1978) only reduces the precious information content of the records (Grassberger 1986; Vautard and Ghil 1989).

Second, deterministically chaotic time series, while more predictable than stochastically random ones, still have severely limited predictability, and hence reproducibility (Ghil and Childress 1987, Sec.12.6; Le Treut and Ghil 1983): if the paleoclimatic model were perfect, and our knowledge of the state of the climatic system at the beginning of the Quaternary arbitrarily

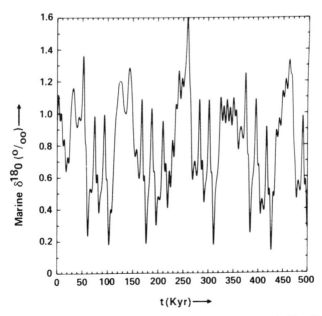

Fig. 12. Model-simulated marine $\delta^{18}O$ record; its power spectrum appears in Fig. 11a (figure from Le Treut et al. 1988).

good, model solution and paleoclimatic records would still diverge over a few hundreds of kyr. This is even true of models of planetary motion, which are quasi-periodic, rather than aperiodic (i.e., chaotic; see Ghil and Childress 1987, Secs.12.3, 12.7, and references therein). The difficulty with time-domain reproducibility is illustrated, for instance, by Pollard's (1984) failure, in spite of skillful multi-parameter tuning, to achieve a good fit between his model solution and the $\delta^{18}O$ record he used, over more than a few 100 kyr. A fortiori, it is remarkable that the qualitative aspect of Fig. 12 matches most of the important features in Figs. 2 and 4 for the late Quaternary (recall that time advances to the left in Fig. 4, and to the right in Figs. 2 and 12).

The rapidity of terminations and the sharp "spikes" are associated with the high-frequency peaks in the spectrum. The mean interval between terminations is near 100 kyr: there are 5 over 500 kyr in the model (Fig. 12), as compared to 7 since the well-dated Brunhes-Matuyama magnetic reversal near 700 kyr in the data (Figs. 2 and 4). The irregularity in the distance between terminations, and the varying number of spikes between them, are associated with the continuous background and the limited predictability that it entails.

Among the short events, or spikes, in Fig. 12, some have almost full glacial-to-interglacial amplitude, and follow the terminations, especially at 490 kyr, near the "present." The latter event has remarkable similarity with the Younger Dryas, during the last deglaciation. A few authors (e.g., Duplessy et al. 1981) have studied only the features of this event in and around the North Atlantic, while others have claimed that the event was in fact restricted to this region of the northern hemisphere (see, e.g., Broecker et al. 1985). But recent evidence indicates that the event affected the entire northern hemisphere (Kallel et al. 1988), and even the whole globe (Jouzel et al. 1987). This makes an explanation of the Younger Dryas along the lines of global mechanisms, as in Fig. 12, more plausible.

Ice Cores.   The Vostok ice core from East Antarctica provides a high-resolution record of $\delta^{18}O$ (Lorius et al. 1985), as well as of δD, the isotopic abundance ratio of D/H ($^2H/^1H$: Jouzel et al. 1987) over the last 160 kyr. δD and $\delta^{18}O$ in an ice core reflect both the local temperature at the time snow precipitated onto the surface of the ice sheet (Le Treut et al. 1988). The δD record has a resolution of about 0.1 kyr, much better than that in the marine cores used by Pestiaux et al. (1988), which was 0.6 kyr at best.

To analyze the spectrum of the δD-derived temperature record at Vostok, Yiou et al. (1989) used a multi-taper method (MTM) due to Thomson (1982). Besides its high spectral resolution, the MTM provides a statistical F-test for the validity of the location and amplitude of each peak obtained. This permits one to detect peaks with high statistical confidence, even when their amplitude is low. By contrast, the maximum-entropy method provides high resolution but no statistical confidence estimates, while the Blackman-Tukey au-

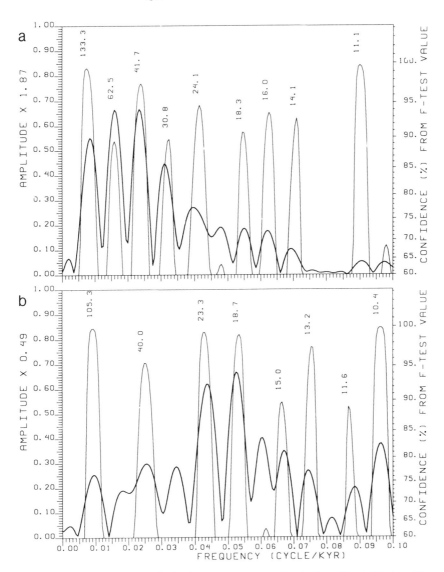

Fig. 13. Multi-taper method analysis of temperature time series: (a) from Vostok δD data; (b) from the Le Treut and Ghil (1983) model. The thick line represents amplitude (left axis); the thin line gives confidence value in percent (right axis) (figure from Yiou et al. 1989).

tocorrelation method gives confidence in the largest peaks only, and very limited resolution.

The MTM analysis of the Vostok temperature record is shown in Fig. 13a. For comparison purposes, the MTM spectrum of a temperature time series of length equal to that of the Vostok records from the Le Treut and Ghil

(1983) model is given in Fig. 13b. The peaks in the two spectra are in good correspondence with each other, taking into account the small record length. The peak with largest confidence in both records is the suborbital one, at 133 kyr in Vostok and 105 kyr in the model. Next come the orbital peaks, near 41 kyr, 23 kyr and 19 kyr. After this are the superorbital ones, associated with sum tones or second harmonics in the model; the peak at 10.4 kyr in the model and 11.1 kyr in the data almost equals in confidence the dominant peak near 100 kyr.

Peaks with even higher frequencies are present in the MTM analysis of the Vostok temperatures (not shown here). Prominent among these is a peak near 2.4 kyr, already identified in other ice cores, from Greenland (Dansgaard et al. 1982) and the Antarctica (Benoist et al. 1982). Given the uncertainties in the depth-vs-time profile at Vostok (Jouzel et al. 1987; Lorius et al. 1985), Yiou et al. (1989) examined the stability of the peaks detected by varying the chronology of the core and examining MTM spectra of every 40-kyr-long window in the record. The results of these two tests are shown in Table II.

From Table II, it is clear that all stable high-frequency peaks in the Vostok core, including the one at 2.4 kyr, can be explained as combination tones of the three orbital frequencies. This leaves only the peaks at 62.5 kyr and 30.8 kyr in Fig. 13a unexplained. They appear to be related to the internal periodicity of the ice-mass/bedrock oscillator governed by Eqs. (5a,b), and to its second harmonic (Yiou et al. 1989); this periodicity was found by Birchfield and Grumbine (1985) to be about 50-70 kyr, depending on exact parameter values.

V. CONCLUDING REMARKS

An examination of paleoclimatic records for the last 2 Myr, the Quaternary period, strongly suggests that the recorded variability cannot be attrib-

TABLE II
Dynamically Stable Frequencies in the Vostok Temperature Record[a]

Period (kyr)	Frequency (cycles/kyr)	Combination Tone	Relative Error[b]	Order of Combination
38.7	0.026	f_3	6.10^{-2}	1
20.0	0.050	$2f_3$	2.10^{-2}	2
11.1	0.090	$2f_2$	3.10^{-2}	2
5.98	0.160	$f_1 + 2f_2 + f_3$	2.10^{-2}	4
4.37	0.228	$f_1 + 4f_2$	9.10^{-3}	5
3.46	0.289	$3f_1 + 3f_2$	2.10^{-3}	6
2.66	0.376	$3f_1 + 5f_2$	2.10^{-3}	8
2.42	0.413	$7f_1 + f_2$	2.10^{-3}	8

[a]Table from Yiou et al. (1989).
[b]The relative error is $r_i/f_i^{(d)}$, where the absolute error r_i is defined in Eq. (11b) and $f_i^{(d)}$ is the observed frequency.

uted entirely to external factors, such as variations of solar luminosity or of the Earth's orbit around the Sun. The variations of insolation received at the top of the atmosphere due to either factor are too small in amplitude and too limited in spectral features to cause most of the observed variability in the records. The isotopic and micropaleontological records contain both periodic and aperiodic components, which are likely manifestations of deterministic order and chaos.

Simple nonlinear models are amazingly versatile tools in exploring the mechanisms of irregular climatic variability. The nonlinearity manifests itself in multiple stationary solutions, stable periodic, and chaotic ones. Deterministic chaos is not restricted to the simplest, spatially zero-dimensional models: it is pervasive in models with varying spatial detail and hierarchic complexity (Témam 1988), as well as in the climatic system itself (Ghil et al. 1991). Dynamical systems and bifurcation theory provide a common thread for the study of more and more complex behavior in models with increasing detail. This thread can be followed in moving back and forth between the simple and the detailed, to gain ideas and develop intuition from the simple ones, and confidence and realism from the detailed ones. It is the thread which can lead not only through the labyrinthine corridors of modeling, but through the bewildering labyrinth of nature itself.

Tables I and II, and the comparison of Figs. 11, 12 and 13b with Figs. 4 and 13a, respectively, leave us with both surprise and satisfaction that such a simple model as one given by Eqs. (8a,b,c) can explain so many features of Quaternary records. Obviously the model is semi-empirical, and many parameters are but best guesses. To obviate the need for this empiricism, much more detailed models would be necessary. Embedding the same mechanisms into a much more complex, spatially detailed and realistic model, can still produce the same results (Ghil and Childress 1987, p.320). This is supported by the similar behavior of zero and one-dimensional energy-balance and ice-sheet models, as well as by the limited information content of the data (Sec. II.B). Therefore, it is only the computer power and human perseverance which have to be larger in order to arrive at the same, or very similar, results when using less empirical, but more cumbersome models.

A systematic investigation of the information content of Quaternary records indicates that 5 to 10 active mechanisms should suffice to model the records' main, statistically significant features. This number is comparable to that of mechanisms proposed to explain Quaternary glaciation cycles, not in any individual model, but in all of them combined. A quick comparison of some of these models against the records is given in Table III.

It is clear that all models reproduce the relatively small obliquity and precession peaks (PCs #2, 3, 4 or 9). Each of models B, G and S reproduces some lower-frequency peak (PCs #1, 2, or 3), and only G yields the high-frequency peaks. An interesting feature emphasized by Vautard and Ghil's Singular Spectrum Analysis in the marine cores is the peak at 65 kyr, given only by the Birchfield and Grumbine (1985) model, and present also in the Vostok ice core. Thus no model by itself reproduces all the features of the data. This is not too surprising, when considering the fact that each model only uses two or three mechanisms, rather than five or ten, as suggested by the data.

The feedback mechanisms discussed so far, and represented in various pairings in Table III, are the ice-albedo feedback and greenhouse effect incorporated in Eq. (7a) (Budyko 1969; Sellers 1969), hydrologic effects (Eq. (7b), Fig. 8: Källen et al. 1979; Ghil and Le Treut 1981), ice-sheet/bedrock dynamics (Eqs. (7b) and (8b,c): Birchfield et al. 1981; Ghil and Le Treut 1981; Pollard 1984) and marine-ice sheet/thermohaline circulation interactions (Saltzman and Sutera 1984,1987). Other mechanisms likely to help generate internal climatic variability on the time scales of interest are ice-sheet thermodynamics (Oerlemans 1982), thermohaline circulation changes of the ocean (Ghil et al. 1987; Saltzman et al. 1984), and changes in the carbon cycle (see, e.g., Broecker et al. 1985). It is an open question of Quaternary glaciation theory which of these mechanisms are the most important ones.

The comparison of spectral features in Tables I and III suggests that ice-sheet/bedrock dynamics accounts for the peak near 65 kyr and its harmonic(s), while the coupling between ice-albedo and precipitation-temperature feedback accounts for the combination tones between 5 kyr and 15 kyr. Multiple theories account for the presence of the orbital peaks at 41 kyr, 23 kyr and 19 kyr, and for the suborbital one near 100 kyr (see also Oerlemans [1982], Peltier and Hyde [1984], and Pollard [1984], not mentioned in Table III for lack of space). To disentangle the various possibilities, spectral phase information, in addition to the spectral density information reviewed here, is necessary. The way to use such phase information for ascertaining the mechanisms most responsible for the observed power is described by Le Treut et al. (1988). Information of this type is now being systematically collected by Imbrie et al. (1989).

It is time to conclude with a wild speculation. The peak at 2.4 kyr revealed by MTM analysis of the Vostok temperature record with 99% confidence (Yiou et al. 1989) is tantalizingly close to the 2.3 kyr peak in ^{14}C records (Sonett and Finney 1989; Thomson 1990). The latter peak might well be due to changes in the global carbon cycle (Siegenthaler et al. 1980) with a similar period. These changes in turn might be related to an oscillation of the oceans' thermohaline circulation (Ghil et al. 1987) with a mean period of roughly 2.5 kyr.

As pointed out by Sonett and Finney (1989), the 2.3 kyr cycle seems to play a crucial role in modulating or participating in combination tones with two other long cycles present in the radiocarbon record. The shortest one of the 9 long periods identified by these authors is 208 yr. Castagnoli et al. (1988) find evidence in a thermoluminescence profile from a marine core spanning the last 2 millenia for modulation of both the Schwabe 11.4 yr cycle

TABLE III
Comparison of Spectral Features in the Four Cores, C1 Through C4 of Fig. 4, with Model Results[a]

	Low Frequencies			Orbital Frequencies (A,B,G,S)		High Frequencies
Period	109 kyr[b]	100 kyr[c]	50-70 kyr[c]	41 kyr	19-23 kyr	14.7 kyr, 13 kyr, 10.4 kyr
Prediction	G	S	B	Obliquity	Precession	G
C1		130kyr (PC#1, irregular)		40kyr (PC#2, variable)	20-25 kyr (PC#3)	15-20 kyr (PC#4) 10kyr (PC#5)
C2		100kyr (PC#1)		30-40kyr (PC#2)		10-15kyr (PC#4)
C3		103kyr (PC#1,#2)		40-50kyr (PC#3)	20-30kyr (PC#4)	
C4	100kyr (PC#1,#2)		65kyr (PC#3, also 300-500kyr)	40-50kyr (PC#9,10)		

The periods in the table heading correspond to predictions of the descriptive astronomical theory (A), of the ice sheet model of Birchfield and Grumbine (1985:B), of the model reviewed here (G), and of the marine-ice-sheet oscillatory models of Saltzman, Sutera and associates (S). The entries refer to principal components (PCs) of SSA: the higher the order of the PC, the lower its variance (table from Vautard and Ghil 1989).

[b] Difference tone of the precessional frequencies.

[c] Self-sustained oscillation.

and the Gleissberg 82.6 yr cycle (see also Attolini et al. [1988*b*] for evidence of the stability of the 11.4 yr cycle in ^{10}Be from a Greenland ice core over the last millenium) with a period of 206 yr. Overtones of a climatic oscillation at 2.4 kyr may thus contribute to the splitting of the 11.4 yr cycle into 12.1 yr and 10.8 yr cycles of geochemical indicators of cosmic rays, solar flux and geomagnetic field.

Acknowledgments.   It is a pleasure to thank colleagues too numerous to mention, except in the references, for many fruitful discussions, preprints and reprints. G. Cini Castagnoli, G. E. Morfill and C. P. Sonnett were among the conference participants who provided stimulating feedback. Comments from two anonymous referees helped improve the presentation. The work reported in a few sections of this review, and the writing, were supported by grants from NSF's Climate Dynamics Program. B. Gola and C. Monroe turned the magic trick of processing the chapter almost before I finished writing it.

SOLAR VARIABILITY CAPTURED IN CLIMATIC AND HIGH-RESOLUTION PALEOCLIMATIC RECORDS: A GEOLOGIC PERSPECTIVE

ROGER Y. ANDERSON
University of New Mexico

Natural filters in the ocean-atmosphere system and in geologic processes mask expression of the 11-yr cycle of solar variability in climatic and paleoclimatic records. However, paleoclimatic records from ~170 Myr and 250 Myr old varved evaporite sediments, believed to have been regulated by thermal climatic forcing, are cyclic at Gleissberg-, Maunder- and Sporer-cycle frequencies. Also long-term changes in the El Nino/Southern Oscillation and/or Quasi-biennial Oscillation climate system may be modulated by solar variability. If the cyclicity recorded in these climatic and paleoclimatic records can be linked to solar forcing through a physical model, some rare ancient sediments will provide information about solar variability over a broad range of frequencies.

I. INTRODUCTION

Many decades of research have failed to provide incontrovertible evidence that changes in solar activity have a measurable effect on the Earth's climate. Also, the search for a history of the Sun in geologic strata has yielded only a few climatic and paleoclimatic records that contain possible evidence of solar-induced cyclicity. One could argue that a strong solar/climate relationship would have been recognized long ago and therefore an association, if found, would be insignificant in terrestrial history, not relevant to future climatic change and unlikely to produce any geologic records of solar cyclicity.

On the other hand, paleoclimatic research over the past three decades suggests that the climate system is balanced between quasi-stable states and

that solar effects, even though weak and indirect, could lead to disproportion-
ate changes in climate. Geologists once believed, for example, that glacial
deposits resting on sediment that contained tropical vegetation implied that
the whole Earth had plunged from one climatic extreme to another. Today,
this same association is explained simply by repositioning a continent. After
this apparent mystery of extreme climatic change was removed, paleoclimatic
research focused on the Ice Ages and on small differences in the receipt of
solar radiation (Milankovitch forcing) as a source of climatic variability. It
was then recognized that subtle changes in the seasonal distribution of solar
radiation, with virtually no change in net radiation, was sufficient to trigger
the great climatic cycles of the Pleistocene (Hays et al. 1976). These advances
in paleoclimatology show that, indeed, the Earth's climatic system is deli-
cately balanced, and they improve the prospect that the Sun plays a subtle
but important role in the generation of climatic variability.

   This chapter briefly examines historical and geologic evidence that solar
variability can affect climate. It is not necessarily a review of evidence that
has accumulated since a previous survey (Anderson 1961). Instead, a direct
approach is taken of identifying those systems and geologic environments
that have a potential for collecting information about solar-generated climatic
change. In doing so, it is convenient to think of the atmosphere, the ocean-
atmosphere system and geologic processes as natural filters of the incoming
solar signal (Fig. 1). The signal, in this case, is considered to be irradiant
energy that is oscillating, more or less regularly, at recognized solar frequencies. Frequencies of interest include the $\sim$11-yr sunspot and irradiance cycle,

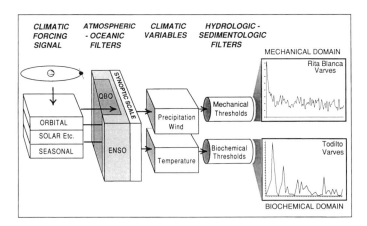

Fig. 1. Natural filtering along the path between solar radiation and varve deposition. The upper
   and middle atmosphere and interannual variability in the ocean-atmosphere system filter solar
   variability but may allow longer cycles to pass. In the geologic realm, mechanical thresholds
   are additional filters that remove a potential solar signal carried by wind and precipitation.
   Thermal forcing may pass through geochemical thresholds and regulate thickness of varves.

the ~22-yr Hale cycle, the ~70–90-yr Gleissberg cycle, and longer solar cycles for which there is less information about periodicity but that are recognized as aspects of solar variability. Maunder and Sporer digressions or oscillations have periods of ~180 and ~220 yr, respectively. Harmonic analysis of the 9600-yr ^{14}C proxy record of solar variability (Stuiver and Braziunas 1989) indicates that the Sporer oscillation of ~220 yr is a strong harmonic of a longer cycle at 420 yr. A weaker harmonic was identified at ~150 yr, as was the Gleissberg cycle and several weaker cycles at higher frequencies. If variability in the 9600-yr ^{14}C record is assumed to reflect variability on the Sun's surface, the temporal pattern appears to be far more complex than the regularity implicit in the 11-yr solar cycle and is dominated by repeated or alternating digressions of Sporer- and Maunder-like cycles. A still-longer cycle in ^{14}C production, that has a period of about ~2400 yr (Damon and Linick 1986; Damon 1988) is also recognized in the 9600-yr record.

The upper and especially the middle atmosphere effectively absorb and filter all variability generated by the Sun at wavelengths < 300 nm, and radiation at longer wavelengths that do reach the Earth's surface is essentially invariant (see the chapter by Hunten). Solar cycles do affect processes in the middle atmosphere but interannual variability in the ocean-atmosphere system serves as a subsequent filter for any weak solar-generated signal that might be transferred through a surface coupling between the stratosphere and troposphere. Finally, certain geologic processes act as filters placed between a solar/climatic signal and those geologic processes that preserve information about climatic change. Only under rare conditions might one expect changes in solar irradiance to pass indirectly through all three filters and be captured in the stratigraphic record.

Even with compelling arguments for believing that a causal connection between solar variability and tropospheric climate is extremely unlikely, examples of solar/climate associations continue to be uncovered and critically examined. In this chapter we look at some of these examples and also consider suggestions that a solar signal may enter the climate system through the Quasi-biennial Oscillation (QBO) and the El Nino/Southern Oscillation (ENSO). We then examine geologic filters and evidence for solar variability in high-resolution (varved) paleoclimatic records.

II. OCEANIC AND ATMOSPHERIC FILTERS AND THE HISTORICAL RECORD

Long instrumental records of climatic variables typically display quasi-periodic, almost chaotic, patterns of interannual to centennial climatic change. Variance in this mid-range frequency of oscillation is low relative to variance at higher (diurnal and annual cycle) and lower (Milankovitch) frequencies (Mitchell 1976; Kutzbach 1976). Within this middle range, interannual variability generated by the coupled ocean-atmosphere system would appear to preclude expression of the extremely weak signal associated with the solar cycle that passes through the tropopause (1/1400; chapter by Hunten). Even the estimate of a 0.14% change in irradiance for the Maunder digression, suggested in an article by Kerr (1987), is miniscule in comparison to energy exchanged during interannual heating and cooling of the ocean-atmosphere system. Interactive ocean-atmosphere climate models, such as the model developed by Zebiak and Cane (1987) suggest that such tiny changes in irradiance are insufficient to produce a model response (M. A. Cane, personal communication).

In spite of the effectiveness of atmospheric filters, some evidence exists in historical and climatic proxy records that the effects of solar variability do reach into the lower atmosphere and even into the ocean's surface layer. Stuiver and Braziunus (1989), for example, have re-examined an earlier conclusion, based on a shorter ^{14}C record (Stuiver and Quay 1980), that there is no convincing evidence for a Sun-climate relationship. Changes in ^{14}C in tree wood, linked to changes in solar activity (Stuiver and Quay 1980), are accompanied by changes in tree growth (Sonett and Suess 1984; chapter by Damon and Sonett) that suggest a climatic response to solar forcing. A 22-yr drought cycle in the western United States has been linked to the Hale solar cycle (Mitchell et al. 1979). Changes in nitrogen trapped in polar snow and ice are correlatable with solar activity and are accompanied by changes in sodium concentration (Parker and Zeller 1980; Zeller and Parker 1981). In this case, an association with the solar cycle suggests that sodium was picked up in droplets of sea water and advected to the sampling site by strong Antarctic winds (Zeller, personal communication), an explanation that would appear to require a solar/ocean-atmosphere connection. The most commonly cited solar/climate association is between prolonged minima and maxima in solar activity and global temperature, as found during the Maunder Minimum (Eddy 1976a; chapter by Ruben).

Evidence for an association in most of the above examples is strongest for solar cycles that are longer than 11-yr, suggesting that atmospheric filters may allow stronger and longer solar signals to pass (Eddy 1976b). It is not proven that changes in solar irradiance are directly or indirectly responsible for any of the associations but the associations do leave open the possibility that solar forcing, by some unknown pathway, affects climatic variability. One such pathway, suggested recently, may be through the ENSO/QBO system.

III. SOLAR CYCLE MODULATION OF ENSO/QBO?

The El Nino/Southern Oscillation and the closely related Quasi-biennial Oscillation are important contributors to interannual climatic variability, especially at middle-to-lower latitudes (Ropelewski and Halpert 1987). ENSO is a natural, generally longitudinal oscillation in the temperature and thickness of the mixed layer in the tropical and subtropical Pacific ocean that is accompanied by changes in wind stress from tropical easterly (trade) winds (Graham and White 1988). The average period of oscillation is slightly less than 4 yr, but El Nino events display a highly variable pattern of recurrence, with strong to very strong El Nino events commonly spaced a decade or more apart (Quinn et al. 1987; Enfield 1989; Michaelsen 1989). The QBO is related to ENSO and is a reversal in equatorial stratospheric winds with an average period of about 26 months.

Variability generated by ENSO and the QBO helps mask any solar signal (Fig. 1). At times, however, the ENSO/QBO system itself appears to be modulated by the solar cycle and if so, ENSO/QBO would be part of a pathway for solar forcing of the ocean-atmosphere system. For example, the QBO in stratospheric winds is correlated with the 11-yr solar cycle (van Loon and Labitzke 1988). Also, temperature anomalies in the lower stratosphere and upper troposphere correlate with both the 11-yr solar cycle and ENSO (Sellers 1990). The solar cycle/QBO association is still considered a mysterious effect and research on its expression in the troposphere is considered too new and active to assess (chapter by Hunten). Evidence presented by van Loon and Labitzke (1988) for a solar cycle/QBO association and by Tinsley (1988), who linked the same association to the position Atlantic storm tracks, however, suggests that the effects of solar variability reach the troposphere.

Indications that the ENSO/QBO system is modulated by solar activity and that the effects may reach the lower troposphere and even into the ocean's mixed layer are preliminary and controversial. The observed solar/ENSO/ QBO associations, however, are sufficiently interesting to report. Quinn et al. (1987) compiled a >400-yr record of the recurrence of El Nino events. When this historical El Nino record is compared to the record of sunspot numbers, El Nino events are two to three times more frequent when sunspot activity and solar irradiance are low, as, for example, during the Maunder and Dalton Minima (Fig. 2; Anderson 1990). Fewer El Ninos occur at times of increased solar activity. Although Michaelsen (1989) did not relate long-term associations between El Nino and tree rings to the solar cycle, his analysis of El Nino variability defines a similar relationship.

Barnett (1990), in a related unpublished study, using methods similar to those of van Loon and Labitzke (1988), compared changes in global and tropical sea-surface temperature fields with amplitude variations in the QBO frequency band. Barnett found an association between high solar flux and low sea-surface temperature for the relatively strong 11-yr solar cycles that occurred after 1950.

The solar-El Nino/QBO associations established in the younger part of the record, when the 11-yr solar cycle is strong, shift out of phase in the older part of the same records. For example, although the cycle period persists, the strong solar/QBO association found prior to 1950 is less clear before 1950 (Barnett 1990), during an interval when solar flux was generally low. A sim-

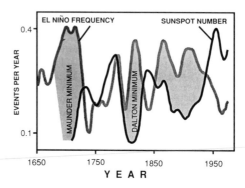

Fig. 2. Comparison of sunspot number and recurrence of El Nino events since 1650. About twice as many El Nino events occur during Maunder and Dalton sunspot minima when sunspot number is low and the 11-yr solar cycle is weakly developed. Stippled pattern emphasizes decadal intervals when solar cycle is weak and El Ninos are more frequent (figure modified from Anderson 1990).

ilar change in phase and a less-clear association occurs in the sunspot/El Nino record before 1850 (Anderson 1990). Inaccuracy of older data may contribute to the loss of a clear association in the older part of the two records or solar cyclicity may not modulate ENSO/QBO when solar activity is generally weak, as during the older intervals (Anderson 1990).

   The historical and instrumental records that support a solar/ENSO/QBO association are short and the associations might be coincidental. Disproof of the hypothesis, or finding a mechanism, may require decades of observation. If proof comes, the resulting effects may be climatologically important but would not necessarily improve chances for capturing a history of the ~11-yr solar cycle in ancient geologic records. This is because ENSO is a phenomenon of the Pacific Basin and owes its natural frequency of oscillation to the present geometry of the Pacific, a geometry that has existed for ≤ 100 Myr. Even if an analog to ENSO existed in the past, the effects of ENSO/QBO in near-decadal cycles might be expressed largely as changes in precipitation and runoff. For reasons discussed in the next section, an 11-yr solar signal is unlikely to be captured by this type of paleohydrologic processes.

IV. EVIDENCE FOR SOLAR CYCLICITY IN THE GEOLOGIC RECORD

   Most attempts to identify or characterize past solar variability in the geologic record have focused on climatic change as the means for transferring the solar signal to accumulations of sediment. Recently, it has been suggested that changes in the solar irradiance spectrum are transmitted directly to accumulating sediments through the effects of thermoluminescence (Castagnoli et al. 1988b, see also their chapter). These studies have reported finding evidence for Wolf, Sporer, Maunder and Dalton sunspot minima in young, recently accumulated sediments in which the age relations were established by the ^{210}Pb method. Application of this methodology is relatively new and unevaluated. Here we examine other, indirect mechanisms for which climatic forcing of biological, geochemical, and sedimentational processes would be the means for recording a solar signal.

The flux of sediments into geologic basins, where sediments are ultimately preserved, can be the result of climatically controlled processes. Therefore, the sedimentary/stratigraphic record is, to some extent, also a paleoclimatic record. If the stratigraphic record has sufficient temporal resolution (e.g., annual), and if accumulation is closely regulated by climatic factors, there exists a potential for recording solar-induced climatic variability as sedimentary cycles. As was the case for processes in the ocean-atmosphere, geologic processes also act as filters to further obscure and remove the signal of solar variability. Therefore, prospects for identifying a solar signal in ancient geologic deposits are more remote than in climatic records.

Geologic Filters of the Solar-Climatic Signal

Mechanical Filters. Varved sediments accumulate as a result of two distinctly different geologic or sedimentational processes. One class of processes is referred to as mechanical and sediments of this type are called clastics (e.g., sand, silt, clay). The other class of sediments is derived from dissolved materials and are precipitated by chemical/biologic processes (e.g., carbonate, evaporite minerals). The potential for recording solar-related climatic changes in mechanical sediments is poor because the flux of sediments is regulated by several physical thresholds that resist the movement of water or sediment (Fig. 1; Anderson 1986). For example, runoff after rainfall or snowmelt only occurs after an event of a certain magnitude and transport of sediment in runoff only occurs above a certain mechanical or shear threshold. Such thresholds act as low-pass filters that allow only large and infrequent events to be recorded. Fourier spectra for such time series tend to be "noisy" and have peaks in spectral density that are distributed over a broad frequency band (see Rita Blanca spectrum in Fig. 1). Much of the interannual climate signal is lost in such records and there is little chance of recording events that are regularly spaced in time.

A case in point is the geologically ancient Precambrian Elatina Formation. These mechanical, sand-to-silt-size sediments are laminated on a scale of millimeters and resemble varves. Each couplet of coarse and fine sediment was believed to represent the annual cycle. The incredible degree of regularity found in these laminations, in cycles of about 12 laminae couplets, was originally interpreted as evidence for strong solar forcing. It was believed that only the melting of glaciers by a variable Sun could account for such regularity (Williams 1981,1985; Williams and Sonett 1985; Sonett and Williams

1987). Further study showed that deposition of Elatina sediments by regular tidal currents was a more likely explanation for the clastic laminations (Sonett et al. 1988; Williams 1988; Williams 1989). Re-interpretation of the Elatina laminations helps confirm that records of solar forcing of climatic change are less likely to be recorded in mechanically generated sediments. The 11-yr solar cycle has been reported in many varved sequences, some of which have geochemical components that identify 11-yr cyclicity (see Crowley et al. 1986; Roberts 1988). For sediments comprised mainly of clastics, examination of many records suggests that spectral density near 11 yr is no stronger and no more consistent than for other interannual frequencies. The Hale cycle, however, has been reported in several varve time-series that contain clastics (Anderson 1961; Anderson and Koopmans 1963; Anderson and Koopmans 1969; Sonett and Williams 1985) and these studies have cited possible solar/climate associations.

Chemical/Biological Filters.   Thresholds, such as loss of solubility at chemical saturation or supersaturation, also occur in chemical/biological systems. However, such thresholds are relatively low compared to the strength of seasonal climatic forcing which can induce biological growth or precipitation each year. Because accumulation is usually triggered each year, the complete spectrum of seasonal, interannual, centennial and even millennial variability can more readily be recorded in such chemical/biological sediments than in clastic sediments. Production of biological sediments, mostly as carbonate and organic matter, is strongly affected by factors such as nutrient supply, turbidity and circulation. Because these factors are highly variable and influenced by runoff from the drainage basin, biological sediments are less likely to capture a solar-climate signal than chemical sediments. Evaporite varves, comprised of minerals such as calcite (calcium carbonate), anhydrite (calcium sulfate), and halite (sodium chloride) tend to record a more direct and less complicated response to seasonal climatic forcing and therefore offer the best hope for studying the history of the Sun in geologically ancient sediments.

    Critics of efforts to identify solar effects in laminated sediments point out, correctly, that there is no proof that ancient laminations were deposited annually. Proof of seasonal and annual accumulation, however, is available for many modern deposits (O'Sullivan 1983). Decades of direct observation of sedimentation has identified the solstice cycle as a dominant control on sediment flux in a wide range of geologic environments (Anderson and Dean 1988). In fact, almost all sediments with high accumulation rates would be seasonally laminated if left undisturbed after deposition. Undisturbed laminations are preserved only because they were deposited in low-oxygen or highly saline environments.

    The Castile evaporites, discussed more fully below, were deposited in a saline environment and its varved laminations contain evidence that Milan-kovitch forcing at precession and eccentricity frequencies regulated evaporite deposition (Anderson 1982,1984). Although the argument is circular, orbital cycles would not be recognizable in Castile laminations if the layers were not annual. In the case of the Castile, and most other laminated marine evapo-rites, the thickness of halite laminae is approximately equivalent to a year of evaporation of sea water. Although there is no proof positive, a search for solar cyclicity in climatically sensitive laminated evaporites can be carried out with confidence that evaporite minerals were deposited seasonally and that pairs or groupings of seasonal laminations comprise varves.

V. PALEOCLIMATIC EVIDENCE FOR
SOLAR VARIABILITY

Precambrian Varves

Microlaminated iron formations are common in ancient Precambrian de-posits and extend back in time beyond 2 Gyr. Garrels (1987) developed a geochemical model that suggests that the iron-rich laminations that accumu-lated in these ancient geologic basins are analogs for younger, Paleozoic and Mesozoic subaqueous varved evaporites, such as found in the Permian Castile Formation (Anderson et al. 1972). Trendall (1972) believes that the varved laminations in Australian iron formations can be correlated for 80 km, imply-ing that the processes that produced laminations responded seasonally over a wide area, a condition that indicates that deposition was regulated by external climatic forcing. The processes that generated laminations were biological/chemical, rather than mechanical, and the iron-rich varves reflect a favorable environment for capturing the effects of solar/climate variability.

Trendall and Blockley (1970) noted that the microlaminations in Austra-lian banded-iron formations were commonly organized into cycles of about 25 yr. Trendall (1973) later determined the average period to be 23.5 yr. An analysis by Sonett (personal communication) identified ∼22-yr and ∼11-yr periodicity in varve time series from certain well-preserved sections from the iron formations, with the ∼22-yr cycle more prominent than the ∼11-yr cycle. A re-interpretation of the origin of cyclicity in the Precambrian Elatina laminations does not affect results from the microlaminated iron deposits be-cause environments of deposition are markedly different. Hence, what ap-pears to be a dominant Hale cycle in varves from the Australian banded-iron formations may be the oldest reported evidence for solar-cycles.

The laminations in the Precambrian banded-iron formations discussed above are also organized into larger, mesoscale sedimentary cycles. Assum-ing that the laminae are varves, these larger cycles represent centennial to millennial climatic variability. However, the time resolution in these long

sequences of iron-rich laminations is obscured by diagenesis and metamorphism, making it difficult to investigate the periodicity of long climatic
cycles.

Post-Precambrian Varves

Favorable conditions for the study of periodicity in long varve sequences
are found in certain Paleozoic and Mesozoic varved evaporite deposits, such
as the Permian Castile Formation and Jurassic Todilto Formation. Neither a
11-yr nor a 22-yr cycle has been consistently recognized in many spectral
analyses from the geologically younger Permian Castile evaporites (~250
Myr). An 11-yr cycle was reported from the Jurassic Todilto (~170 Myr)
evaporites (Anderson and Kirkland 1960) but the evidence is equivocal (Anderson and Koopmans 1963).

A lack of clear evidence for 11-yr or 22-yr cyclicity in younger evaporites is not surprising, given the meagre evidence for an historical solar-
climate association. As for the ocean-atmosphere system, variability in physical and geochemical processes within a sedimentary basin also acts as a
low-pass filter of the climate signal. Therefore, evaporite varves, even without evidence for 11-yr solar forcing, still have the potential for recording
~70–90-yr and ~200-yr solar cycles, and possibly longer solar cycles.
Moreover, long solar cycles can be expected to be accompanied by greater
changes in solar irradiance, geomagnetic or other Earth-based solar effects.
Hence, a search of varved, chemical sediments can be carried out with some
assurance that finding strong and persistent climatic cycles at the longer solar-
cycle frequencies means a possible solar association.

Todilto Formation.   The Upper Jurassic Todilto limestone in New Mexico was deposited in a broad, shallow arid basin with a large evaporative
surface area as compared to volume and with its waters approaching saturation with respect to $CaSO_4$ (Anderson and Kirkland 1960). Todilto varves
consist of mm-scale alternations of sharply defined organic laminae and
nearly pure calcium carbonate (Fig. 3). Although individual laminae of calcium carbonate have a variable thickness over short distances that is related
to diagenesis and preservation (Fig. 3), groups of laminae can be correlated
with a high degree of confidence for distances of several kilometers in the
basin (Anderson and Kirkland 1966). The high degree of lateral continuity
indicates that changes in geochemistry and deposition occurred simultaneously over a wide area, and probably over the entire basin, in response to
seasonal climatic forcing. Clear definition of carbonate laminae increases the
reliability of the internal varve chronology and helps make the Todilto a
highly favorable sequence for recording changes in climate at solar-cycle periods. A graphic plot of raw thickness for the 900-yr Todilto varve time-series
(Fig. 4A) reveals a well-defined and persistent ~200-yr climatic cycle. Four
of the 5 smoothed cycles have an intermediate cycle between larger ~200-yr

Fig. 3. Etched slab of calcium carbonate (light) and organic (dark) laminations in Jurassic Todilto limestone. Average thickness of a couplet is 0.2 to 0.3 mm. Individual couplets have been traced for several km.

maxima. A power spectrum of the time series places the two strong-cycle periods at 82 and 180 yr (Fig. 4B).

Castile Formation. The Castile evaporite varves accumulated in a deep geologic basin in what is now a part of western Texas and New Mexico. In the Permian geologic period, 250 Myr ago, the basin was located near the paleoequator, along the arid western margin of the Permian supercontinent, and was almost completely isolated from the ocean. Varved laminations accumulated in the basin as the waters in the basin evaporated, reached saturation and seasonally precipitated calcium carbonate (calcite, $CaCO_3$), calcium sulfate (anhydrite, $CaSO_4$) and sodium chloride (halite, NaCl). A closed physiographic relief of ~600 m in the basin assured continuous accumulation of varves for more than a quarter-million years and resulted in the unparalleled Castile record of annual to millennial climatic change (Anderson 1982).

Most of the Castile varves are comprised of mm-scale couplets of calcite and anhydrite laminae that occur as beds of laminated anhydrite tens of m thick (e.g., Anhydrite I, II, III; Fig. 5). Other varves are comprised of cm-scale couplets of anhydrite and halite that occur as equally thick beds of

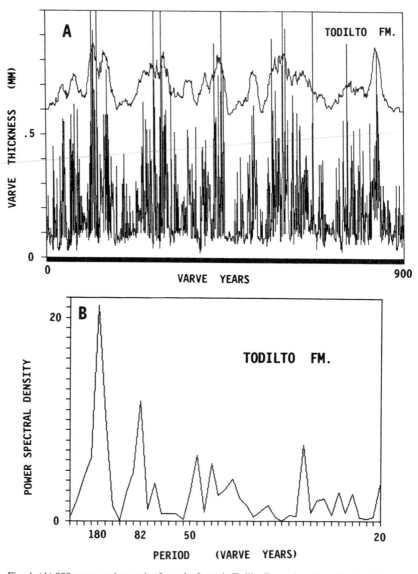

Fig. 4. (A) 900-yr varve time series from the Jurassic Todilto Formation, New Mexico. Note that a regular cycle with a period near 200 yr, as well as a lesser cycle between maxima, is apparent in both unsmoothed and smoothed data. (B) Power-spectral density in the time series above. Note that the weaker cycle has a period near 80 yr and the period of the stronger cycle is 180 yr.

laminated halite (e.g., Halite I, II). Individual laminae within these major units maintain a remarkably constant lateral thickness over more than 100 km, or almost the entire 14,000 km² basin (Anderson and Kirkland 1966; Anderson et al. 1972). As for the Todilto varves, changes in salinity are believed to have occurred seasonally and over a wide area in response to the annual climatic cycle.

Annual time series were reconstructed for lamination couplets (varves) in the major anhydrite and halite beds (e.g., Anhydrite I, Halite I). Within these major units, geochemical systems and evaporite minerals of greatest interest in searching for possible solar-cycle associations (see earlier discussion of natural filters) are anhydrite and halite. The solubility of minerals in the anhydrite and halite systems are largely regulated by evaporation. Of the two minerals, the controls on NaCl solubility and precipitation are less complex than for $CaSO_4$, a mineral whose solubility can be altered by common ion and other geochemical effects. For these reasons, the halite system, in which solubility is most directly related to temperature through evaporation and concentration, is the most likely system and environment to have a straightforward relationship with thermal forcing. Even for the halite system, however, some freshening by direct rainfall or marine inflow is a countervailing factor that may have helped regulate NaCl concentration.

The fidelity of the Castile recording differs during the history of the basin and for different frequencies of climatic change. The entire Castile time series shows a climatic response to the orbital cycles of precession and eccentricity (Anderson 1982,1984) indicating that the small changes in caloric half-year insolation that accompany the orbital cycles (Kutzbach and Gallimore 1988) were sufficient to produce systematic changes in varve thickness. Some parts of the time series for calcium sulfate laminae exhibit little variance in laminae thickness. In other segments, however, the sulfate laminae have a variable thickness and can be examined for evidence of solar-cycle periodicity. These segments occur above and below the major stratigraphic units of laminated halite (Halite I and Halite II; Fig. 5).

The beds of laminated halite that are bounded by sulfate laminae with variable thickness were deposited during episodes of strong evaporation which alternated with episodes of freshening (Dean and Anderson 1978). These profound cycles in salinity have an average period of ~2500 yr (Fig. 5; Anderson 1982). The volume of water in the basin was drawn down rapidly during the evaporative stage of strong salinity cycles, often resulting in the precipitation of laminated halite. Independent evidence for evaporative drawdown occurs as gypsum growth structures that may reflect a crystal growth under relatively shallow-water conditions (A. C. Kendall, personal communication, 1989).

Conditions that prevailed during drawdown include (1) a dominance of thermal forcing and evaporation, with a minimal effect of freshening; (2) high

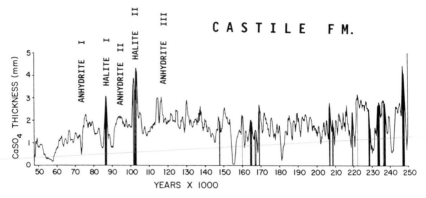

Fig. 5. Complete time series of Permian Castile varve thickness showing position of thick stratigraphic units of Anhydrite I, II and III and position of Halite I and II. Note that about 4 to 5
oscillations in calcium sulfate thickness occur in each 10,000-yr unit of the time series and
that this cycle maintains the same approximate frequency as the basin fills with sediment,
becomes shallower, and is more strongly affected by climatic forcing.

variance in sulfate laminae thickness; and (3) seasonal and annual precipitation of halite from a relatively simple geochemical system. These conditions
favored the recording of any solar/climate cycles that may have been reflected
in regional or global changes in temperature and expressed as changes in
evaporation. Accordingly, we look closely at cyclicity preserved in the Castile
varve time series before, during and after the drawdown stage for Halite I and
Halite II.

Two 1000-yr sequences of anhydrite varves (couplets of calcite/anhy-
drite) from immediately above and below Halite I, were reconstructed and
plotted contiguously with the ~1000 halite varves (anhydrite/halite couplets)
within Halite I (Fig. 6.A). Halite I is comprised of halite laminae ~5-cm
thick that occur in a continuous, uninterrupted sequence. A plot of halite
laminae thickness for part of Halite I (Fig. 7A), as well as a plot of power-
spectral density (Fig. 7B) shows that the strongest period within Halite I is
~190 yr. Examination of the entire 3000-yr drawdown sequence shows that
more sulfate was precipitated and varve thickness increased during the later
stage of evaporative drawdown, commencing about 600 yr before precipitation of Halite I (Fig. 6A). The cycles immediately before and after Halite I,
when brines were more concentrated, increased markedly in amplitude and
display less-regular oscillations in thickness, but have about the same
average-cycle period as in Halite I.

A 3800-yr segment of anhydrite and halite varves was plotted for the
interval before, during and after deposition of Halite II. The accumulation of
halite within Halite II was interrupted several times and as a result Halite II
is divided into alternating thick beds of laminated halite separated by thinner

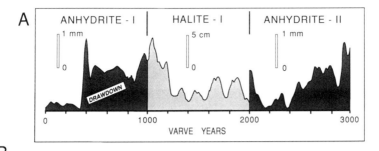

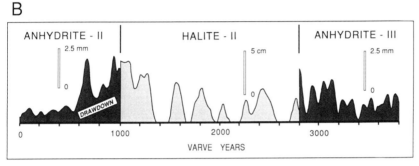

Fig. 6. Segments of the varve time series below, within, and above (A) Halite I and (B) Halite II salt beds (see Fig. 5). Note that a ~200-yr cycle is established before, during and after halite precipitation in both sequences. Increased amplitude in the ~200-yr cycle develops during a stage of evaporative drawdown when thermal forcing and evaporation are the dominant control on precipitation of both anhydrite ($CaSO_4$) and halite ($NaCl$).

beds of laminated anhydrite (Figs. 5, 6B), with a total elapsed time of ~1800 varve years for Halite II. A sharp increase in the rate of anhydrite deposition commenced about 400 varve years before precipitation of Halite II, as opposed to 600 yr before precipitation of Halite I. Within Halite II, the climatic cycle responsible for interrupting the deposition of halite has an average period of ~220 yr. Some varves within the intervening anhydrite beds lack definition and are difficult to count and measure, and the resulting inaccuracy may account for a less regular expression of a ~200-yr cycle in Halite II than in Halite I. A more complex oscillation with average periods of ~90 and ~200 yr occurs in the contiguous segment of Anhydrite III that lies above Halite II (Fig. 6B).

Solar Cyclicity in Castile and Todilto Evaporites

The varve time series from Todilto and Castile evaporites, in which varve thickness is believed to reflect mainly thermal forcing, recorded climatic cycles at the periods of Gleissberg, Maunder and Sporer deviations. In the Todilto Formation, the spectrum shows one cycle period (82 yr) near that of the Gleissberg cycle, with a second, stronger period (~180 yr) near that of

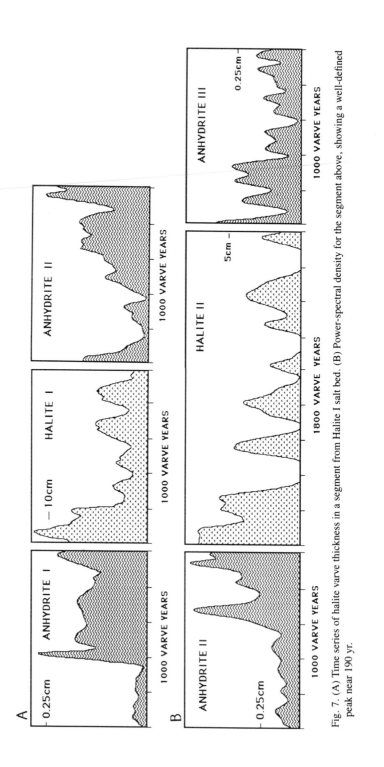

Fig. 7. (A) Time series of halite varve thickness in a segment from Halite I salt bed. (B) Power-spectral density for the segment above, showing a well-defined peak near 190 yr.

the Maunder deviation (Fig. 4B). In the segment from Halite I, the dominant period is ~190 yr (Fig. 7B). These results correspond, rather closely, to Cole's (1973) calculation of 84 and 190 yr for the long solar periods. In Halite II, where the cycles are interrupted and less regular, the cycle appears to be closer to the 220-yr Sporer deviation. A ~200-yr oscillation is also recognizable in Castile anhydrite varves elsewhere in the Castile sequence (Anderson 1982).

Given the fact that the historical record contains only a few long solar cycles, the apparent agreement in period between the historical record of solar variability and the climatic records from the two ancient evaporites is extraordinary. If the association is real it implies that the historical digressions in solar activity represents a typical pattern of past solar variability. The reality of a ~200-yr climatic cycle in the Castile is supported by its expression before, during and after halite deposition, with little change in period, and in two independent geochemical systems. Confidence that ~200-yr climate cyclicity is related to solar effects is improved by finding a ~200-yr periodicity in the long ^{14}C proxy record of solar activity (Stuiver and Braziunus 1989).

In this chapter an attempt has been made to show that the solar-period cycles recorded in these particular varve sequences were likely to have been produced by thermally induced changes in evaporation. Although the pathway for solar forcing of changes in temperature at the Earth's surface are unknown, regional and global changes in temperature are believed to accompany the Maunder digression (Eddy 1976a). The Todilto and Castile examples, because climatic cyclicity is expressed at the long solar periods and is clearly related to thermal forcing, may be ancient analogs for changes in climate that are suspected during major digressions in solar activity (e.g., Wolf, Sporer, Maunder and Dalton minima). Ancient analogs cannot substitute, as a form of evidence, for modern observations that are linked to an accepted physical model. Although the Todilto and Castile examples are reasonably convincing, given the assumptions involved, confirmation of a Sun-climate relationship will depend upon other lines of investigation.

VI. A MILLENNIAL SOLAR CYCLE?

The thick beds of halite in the Castile Formation (e.g., Halite I and Halite II) that contain ~200-yr climatic oscillations were themselves formed as part of a longer and stronger climatic cycle. These longer and stronger cycles were responsible for episodes of evaporative drawdown and freshening and for changes in salinity that accompanied the precipitation of major halite beds (Fig. 5; Anderson et al. 1972; Dean and Anderson 1978). The average period of oscillation for the strong salinity cycles is ~2500 varve yr. Increased variance associated with this millennial cycle is distributed over a band of the spectrum corresponding to periodicities between 1800 to 3000 yr, with the maximum spectral density somewhere between 2500 and 2700 varve years (Anderson 1982).

The ~2500-yr cycle can be seen in a plot of the entire Castile time series (Fig. 5), where it is resolved by a moving-average filter with a window of 900 yr. The 1800–3000-yr oscillation observable in this plot exhibits a 3- to 5-fold increase in amplitude between the start and the end of evaporite deposition. The systematic increase in amplitude is attributed to the gradual filling of the basin with sediment, a correspondingly shallower basin and smaller volume of water, and a greater impact of episodes of freshening and evaporation on salinity. Because the period remained essentially unchanged as the amplitude of the cycle increased through time, changes in salinity must have been "externally" forced upon the basin. External climatic forcing is also indicated by correlation of seasonally produced laminations throughout the basin and by varve cyclicity in the Milankovitch frequency band. The ~2500-yr cycle, occurring, as it does, between the annual and Milankovitch cycles, is part of a continuum of unexplained climatic variability between the two deterministic mechanisms. The "strength" of the ~2500-yr climatic cycle within this continuum, judging from its impact upon Castile geochemistry, is greater than either ~200-yr or ~20,000-yr climatic cycles.

Elsewhere, a ~2500-yr climate cycle has been depicted in a summary spectrum of climatic variability compiled by Mitchell (1976), and is documented in several different recording media and environmental settings. Evidence for a strong millennial climatic cycle is found in the advance and retreat of glaciers (Denton and Karlen 1973), in changes in isotopic ratio and CO_2 concentration of glacial ice (Oeschger et al. 1985), in changes in sea level (Fairbanks 1989), in changes in marine isotopic records from Foraminifera in the Atlantic (Broecker et al. 1988) and in the western Pacific (Linsley 1989), and in cycles of ocean upwelling and productivity in the eastern Pacific (Pisias 1978; Juillet-leclerc and Schrader 1987). A strong millennial climatic cycle has also been observed in late Pleistocene varved marine sediments off California. In this last example, the millennial climatic cycle is related to ENSO (Anderson et al. 1990).

Speculation about the origin of a ~2500-yr oscillation in climate has been limited by a recent appreciation of its ubiquity, persistence and importance. Volcanism and geomagnetic changes appear to be implicated, and the cycle period corresponds approximately to the time frame of deep ocean circulation. A solar role is also conceivable and cannot be discounted. For example, a 2400-yr cycle has been identified in ^{14}C in tree rings, where it is attributed to changes in the dipole moment of Earth's magnetic field (Damon and Linick 1986; Damon 1988; chapter by Damon and Sonett; see review in Bradley 1985). If subsequent studies should establish a connection between magnetic field changes and climate, prospects would improve for linking ~2500-yr climatic cycles to solar variability.

VII. CONCLUSIONS

Two ancient evaporite varve sequences, the Jurassic Todilto Formation and the Permian Castile Formation, believed to have been responsive to thermal climatic forcing, contain evidence for cyclicity at periods of solar variability (~70–90 yr, ~200 yr and ~2500 yr). The Todilto and Castile may be ancient analogs for a modern association between Maunder-like solar activity deviations and coincident changes in global temperature.

Long climatic cycles in ancient evaporites and in the modern ENSO/QBO system reflect climatic processes related to solar heating at equatorial and off-equatorial latitudes but no pathways are known to connect cycles in these systems to changes in the Sun. The absence of an obvious pathway, however, has not dulled the search to find solar/climate associations. If this search ultimately identifies a mechanism to explain the associations, ancient climatic records, such as those of the Todilto or the 200,000 yr varve time series from the Castile, will be a rich source of information about the long-term behavior of the Sun (see patterns of variability in Anderson [1982]).

Acknowledgments. I thank D. W. Kirkland and W. E. Dean for help in collecting data for Rita Blanca, Todilto and Castile time series. National Science Foundation sponsored early work on the Castile Formation and a grant from the National Science Foundation supported studies of ENSO.

SOLAR-TERRESTRIAL RELATIONSHIPS IN RECENT SEA SEDIMENTS

G. CINI CASTAGNOLI, G. BONINO and A. PROVENZALE
Istituto di Cosmogeofisica del CNR

The calcium carbonate ($CaCO_3$) and the bulk thermoluminescence (TL) profiles of two recent Ionian Sea cores that span the last four millennia are analyzed. The cores have been dated with a precision of 1% over the last 20 centuries and the constancy of the sedimentation rate has been verified. The high dating precision has been obtained by radiometric measurements which have been tested and extended by a careful analysis of the horizons produced by well-known historical volcanic events in the Campanian area (Pompei, Pollena, Ischia). The study of the $CaCO_3$ and the TL time series obtained in this way reveals the existence of several outstanding features and well-defined cyclic variations. The high precision of the time scale allows for carefully determining the periodicities present in the signal and for exploring their possible solar origin. The analysis of the carbonate profile reveals the presence of a century-scale wave with period of about 206 yr, this wave is accompanied by two sidebands suggesting that the 206-yr wave is modulated by a long component with period of ~2000 yr, as observed in the analysis of tree-ring radiocarbon. In addition, a significant spectral line at about 141 yr is found. The analysis of the TL profile shows the presence of well-defined spectral lines at 137.7 yr, 59 yr, 28.5 yr, 22 yr, 12.06 yr and 10.8 yr. The beats of the two decennial TL components produce a modulated wave train with an 11.4-yr period carrier (the eleven-year solar cycle?) and a 206-yr period amplitude modulation. The nodes of the modulated wave train correspond to the minimal sunspot number cycles in AD 1810 and AD 1913. The TL amplitude modulation coincides, both in period and phase, with the 206-yr wave detected in the carbonate profile. We finally compare the properties of the TL profile and of the $CaCO_3$ signal with those of the radiocarbon ($\Delta^{14}C$) record in tree rings. An overall similarity is observed between the carbonate profile and the radiocarbon record, at least up to about 3.5×10^3

yr BP. The 206-yr wave found in the carbonate profile leads by 90° the well-known 206-yr component detected in the radiocarbon record. Some of the possible implications of these results are discussed.

1. INTRODUCTION

Experimental time series obtained from ice and sediment cores and from tree rings are a typical source of information on the past solar-terrestrial relationships. These time series usually refer to changes in the global climate and in the Sun-Earth system which happen on long temporal scales. The time series from ice cores (i.e., ^{10}Be, CO_2, oxygen, deuterium and carbon isotopes) extend 15 to 20 × 10^4 yr BP (before present) with a typical time resolution of ~2000 yr (Hammer et al. 1978; Raisbeck et al. 1981b 1987; Neftel et al. 1982; Yiou et al. 1985; Barnola et al. 1987; Genthon et al. 1987; Jouzel et al. 1987; Beer 1988; Oeschger 1988; Yiou et al. 1989); only for the last few centuries, a time resolution of 3 to 5 yr has been obtained (Beer et al. 1988b; Neftel et al. 1985). The time series from sediment cores (carbonates, oxygen and carbon isotopes, ^{10}Be, etc.) usually refer to time intervals of geologic interest (Shackleton and Opdyke 1973; Hays et al. 1976; Kominz et al. 1979; Imbrie and Imbrie 1980; Kent 1982; Pisias et al. 1984; Paterne et al. 1986; Martinson et al. 1987; Southon et al. 1987). The measurements in ice and sediment cores typically allow for determining some of the climatic effects of the astronomical periodicities of the Sun-Earth system.

On the other hand, tree-ring time series (i.e, ^{14}C and ring width) contain information extending to 8 or 9 × 10^3 yr, with a time resolution of about 10 to 20 yr (Stuiver 1961; Ferguson 1971; Stuiver 1980; Stuiver and Quay 1980; Suess 1980a,b; Neftel et al. 1981; Pearson et al. 1983; Pearson and Baillie 1983; Sonett 1984; Sonett and Suess 1984; Damon and Linick 1986; Stuiver and Becker 1986; Suess 1986; Beer et al. 1988b; Sonett 1988; Stuiver and Braziunas 1988, 1989; Sonett and Finney 1990). The solar imprint on these time series seems to be almost surely assessed. In addition, the analysis of the ^{10}Be content (which has the same cosmogenic origin as the ^{14}C) in a recent Greenland ice core (Beer et al. 1990), spanning about 2 centuries and sampled on a yearly basis, has revealed the presence of the well-known 11-yr cycle of sunspot activity in this record. A cyclogram analysis of the ^{10}Be record in the Milcent ice core (spanning about 8 centuries) has also shown the presence of the 11-yr cycle (Beer et al. 1985; Attolini et al. 1988b). (Isotope records are also discussed in the chapters by Damon and Sonett and Beer et al. and varves in Anderson's chapter.)

By contrast, very few studies have been conducted on recent sediments in order to gain information on the Sun-Earth relationships in the last few

564 G. C. CASTAGNOLI ET AL.

millennia. This fact was mainly due to the difficulty of providing a precise dating of the cores spanning these time scales, as well as of obtaining signals with a time resolution fine enough to resolve the 11-yr solar cycle. To fill, at least partially, this lack of information, some years ago we started a research program devoted to the precise dating of an ensemble of recent sea sediment cores, in order to obtain several time series of the relevant properties of the sediment having a typical time resolution of a few (2 to 4) yr (Cini Castagnoli 1988). In particular, we analyzed the total calcium carbonate (CaCO$_3$) and the thermoluminescence (TL) profiles of the cores. The dating of the cores was obtained by testing the radiometric determinations of the sedimentation rate with the markers of historical volcanic events within the marine sediments. We stress that it is this high dating precision of the cores which allows us to draw some definite conclusions on the presence of 11-yr and 22-yr cycles in the sediment VTL records and consequently to study the possible imprint of the 11-yr solar cycle on this type of terrestrial reservoirs. In addition, the contemporary presence of the 11-yr cycle and of longer (century-scale) cycles allows us to study the possible solar origin of the centennial oscillations as well (Cini Castagnoli et al. 1988a,b, 1989, 1990a,b,c,d). The Gleissberg cycle has, in fact, now been revealed in a total carbonate time series.

II. THE CORES

The cores considered in this chapter, named GT14 and GT89/3, were taken by means of a gravity corer in the Ionian Sea, on the Apulian continental shelf. The coordinates are latitude 39° 45′55″N, longitude 17° 53′30″E for the first core and latitude 39° 45′43″ N, longitude 17° 53′55″ E for the second core. The extraction site is at about 18 km from the coast, in the Gulf of Taranto in front of Gallipoli, in water depths of 166 m and 178 m, respectively. The GT14 core is 1.17 m long and the GT89/3 core has a length of 2.81 m. The distance between the extraction sites of the two cores is about 1 km. The cores are fossiliferous calcareous mud with traces of terrigenous clastics. No signs of sedimentary stratification are visible (even under the microscope) through the cores. The only marker of stratification surfaces is given by preferred alignment of mica flakes and/or by the long axes of bioclastics. The cores are thus composed by a very monotonous sediment whose texture and microfossil associations are uniform. Figure 1 shows a map of the Ionian Sea continental shelf; the approximate site where the two sediment cores were extracted is indicated.

In order to study the existence of possible variations in the composition of the sediment and in the sedimentation rate itself, we have analyzed the profile of the total carbonate content in the two cores (see Cini Castagnoli et al. 1990b,c). Both the GT14 and the GT89/3 cores have been divided into samples of equal thickness $\Delta d = 0.25$ cm. The sediment samples are pro-

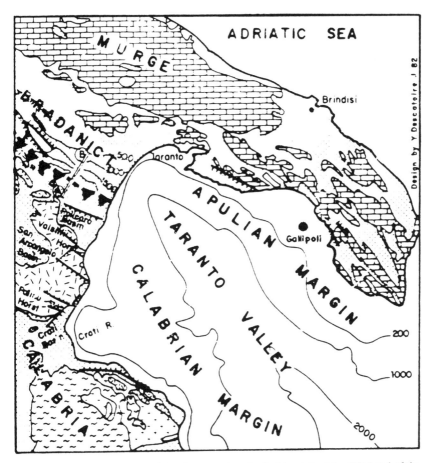

Fig. 1. Map of the Ionian Sea giving the extraction site (solid point) of the GT14 and of the GT89/3 cores considered in this study (figure from Rossi et al. 1983).

gressively numbered in the same way: the top of sample number 1 is at AD 1979 for both cores. For the present work we have analyzed the entire GT14 core for a total of 467 samples; the GT89/3 core has been analyzed from sample 61 to sample 948. The integrated $CaCO_3$ content of each sample was determined by titration with EDTA using the standard procedure given by Barnes (1959). The material of each sample was dried at 110 °C overnight and the $CaCO_3$ from the powder (typically 100 mg) was leached by boiling 2 or 3 times with 50 ml of 2% acetic acid. The solution was made to a known volume and, after addition of a buffer solution, it was titrated with a 0.01 M solution of sodium salt of EDTA. Titration was performed on three aliquots of the solution and frequent titration of standard solutions was carried out to ensure reproducibility. The triplicate analyses agreed within 0.3%. The

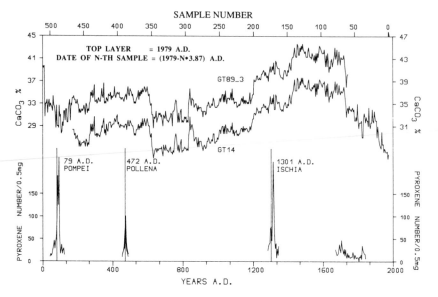

Fig. 2. Carbonate profiles of the GT14 and GT89/3 cores. The sample number is indicated above the figure and the corresponding time in years AD, as obtained by the dating procedure discussed in the text, is indicated at the bottom of the figure. We have indicated the pyroxene content of the samples where guide-horizons are expected.

$CaCO_3$ profiles of the two cores are shown in Fig. 2 for the upper 500 samples. The carbonate content in the core GT14 varies between 26% and 38%, with mean value of 31% and standard deviation of 3.9%; it varies between 27% and 47% with mean value 37% and standard deviation 4% in the much longer GT89/3 core. The carbonate content of the sediment varies significantly along the cores, potentially providing relevant information on the past climatic fluctuations. In addition, the fluctuations of the $CaCO_3$ content in the two cores display an astonishing layer-to-layer correspondence within a fraction of a percent, indicating both the almost perfect reproducibility of the carbonate measurement and the equality of the sedimentation rate in the two cores.

III. CORE DATING

To transform the depth profiles into the corresponding time series, it is necessary to determine the sedimentation rate in each core accurately, as well as to investigate whether this remained constant for the entire time interval spanned by the sediments. The determination of the sedimentation rate is conducted by considering: (a) the measurement of the activity of radioisotopes that are not in equilibrium in natural radioactive chains; and (b) the measurement, in the upper layers of the sediment, of the activity of the radioisotopes produced by nuclear bomb testing in the atmosphere. A check of the correct determination of the sedimentation rate (and of its constancy along the core) is then obtained by searching for the presence of event-markers within the marine sediments. These are provided by the ashes of the volcanic eruptions in the Campania volcanic area. In particular, we count under the microscope the number of pyroxene grains present in 0.5 mg of sediment for each sample. Whenever a pyroxene peak is found, we test for the signature of the corresponding volcanic explosion. When historical eruptions are recognized, a precise dating of the corresponding sediment layer is obtained.

For the Ionian Sea cores considered here the following procedure (see the next 3 subsections) was adopted:

Natural Radioisotopes.   We have measured the ^{210}Pb ($T_{1/2}$ = 22.3 yr) activity in excess with respect to the ^{226}Ra isotope ($T_{1/2}$ = 1600 yr). The presence of the ^{210}Pb isotope in marine sediments is due to atmospheric fallout; this isotope is a decay product of the ^{222}Rn isotope ($T_{1/2}$ = 3.82 days), emanating into the atmosphere mainly from the lithosphere. The measurement of the amount of ^{210}Pb present in a sediment as a function of depth allows a direct dating of the core until approximately 150 yr BP (Krishnaswami et al. 1971). The least-square fit of the ^{210}Pb *excess* activity with respect to an exponential decay function has a correlation coefficient $r^2 \geq 0.98$. This indicates the uniformity of the sedimentation rate in the last 2 centuries and allows its determination to be made with an error of about 5%.

Radioisotopes from Bomb-Testing.   A maximum concentration of ^{137}Cs has been attained during years AD 1963–1964, as a result of the intense activity of nuclear weapon testing in the atmosphere. We checked the presence of a peak of this radioisotope in the upper layers of the cores. This gives a validation of the results obtained by ^{210}Pb method; it also guarantees that the top of the core has not been disturbed (e.g, during the extraction operations).

Tephroanalysis.   Relevant volcanic events detectable in these sediments are the Pollena (AD 472) and Pompei (AD 79) plinian eruptions and the Ischia (AD 1301) event. These events are associated with high concentration peaks of pyroxene in the sediment layers deposited at the time of the eruptions and with the contemporary presence of typical tracer grains. Figure 2 reports, below the two $CaCO_3$ profiles, the pyroxene content of some of the sediment samples which have been analyzed. The progressive number n_i of the sample containing the ashes of a recognized historical eruption fixes the time interval Δt corresponding to the sampling thickness Δd = 2.5 mm, averaged on the sediment above the n_i-th sample, by means of the relationship Δt = (AD 1979 − historical date) / n_i. The results of the tephroanalysis indicate that the sedimentation rate is constant in both cores, at least for the both cores, at least for the last 2 millennia, and allow for the accurate evaluation of the time interval $\Delta t = 3.87 \pm 0.04$ yr. The sedimentation rate is $s = 0.0646 \pm 0.0007$ cm yr^{-1}. We show in Fig. 3 the details of the pyroxene content of the samples corresponding to the AD 472 Pollena eruption. An exact coincidence of the pyroxene content of the two cores is evident; this provides a further indication that the sedimentation rate is the same in both cores. Moreover, the narrow width of the pyroxene peak indicates that bioturbation in the core is negligible within our sampling accuracy. The analysis of the pyroxene content of the GT89/3 core samples, which are expected to contain the volcanic ashes of the Pompei eruption, gives the evaluated date of AD 78.8 (layer 491) for the well-known plinian eruption of AD 79. We note that also the modern, less violent Vesuvian eruptions may be recognized in the pyroxene profile. As an example, at the bottom right corner of Fig. 2, we show a part of the pyroxene profile containing the Vesuvian eruption of AD 1822. As a final result of the precise core dating, the depth profiles can thus be transformed into well-defined time series.

IV. THE CARBONATE PROFILES OF THE GT14 AND GT89/3 CORES

Because the carbonate profiles of the GT14 and of the GT89/3 cores are almost identical, here we discuss mainly the $CaCO_3$ profile of the longer core. The profile of the carbonate signal is reported in the inset of Fig. 4 together with its quartic trend. The most prominent feature of the carbonate profiles is the presence of large amplitude, long-term fluctuations in the total $CaCO_3$ content of the sediment. A long-term fluctuation with apparent periodicity of about 3500 yr seems to dominate the variations of the carbonates. A regular rising phase is observed from about AD 700 to 1700, followed by a steeper descent down to the present values with a concentration change from maximum to minimum of about 12% (see also Fig. 2). We note that the present-day values are similar to those observed during the 7th and 8th centuries (Lamb 1972). High $CaCO_3$ concentrations occurred during the 16th and 17th century, and also at the beginning of the record during the second millennium BC.

It is interesting to note that two basic facts resulting from the above analysis are that (a) the sedimentation rate in the two cores is constant, i.e, the same amount of sediment is deposited during equal time intervals, and (b) the total carbonate content of the sediment shows significant variations which are perfectly consistent in both cores. As the main components of the sediment are carbonates and silicates, the above results imply the existence of a balancing mechanism in the deposition of these two constituents.

Over the long-term trend, several short-term fluctuations of the carbonate content are superimposed. The power spectral density of the carbonate profile of the GT89/3 core is reported in Fig. 4; the $CaCO_3$ data have been detrended by removal of the quartic trend before the spectral analysis. Be-

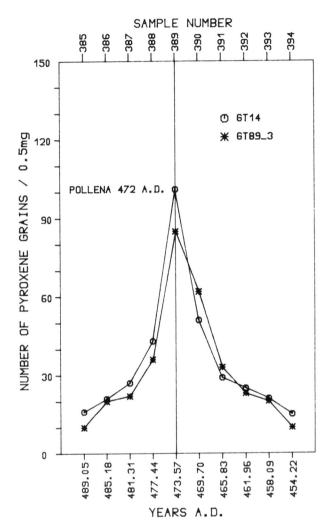

Fig. 3. Pyroxene content of the samples of the GT14 and of the GT89/3 cores corresponding to the Pollena eruption in AD 472. The correspondence between the dating obtained by the ^{210}Pb method and the dating given by the tephroanalysis is an important result which indicates the constancy of the sedimentation rate in these cores. The narrowness of the pyroxene peak indicates that bioturbation is negligible for the sampling interval $\Delta d = 0.25$ cm considered here.

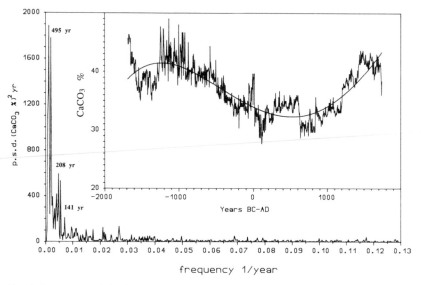

Fig. 4. Power spectral density of the detrended CaCO₃ series obtained from the GT89/3 core. The time series of the % CaCO₃ content of the GT89/3 core is shown in the inset. The long-term trend is given by the least-square fit of the data to a quartic polynomial.

cause there are three independent measurements for each data point in the carbonate profile, the significativity in the spectrum is $2/5P(f)$ $\leq \pi(f) \leq 5P(f)$, where $P(f)$ is our spectral estimate and $\pi(f)$ is the "true" spectrum. Dominant components in the carbonate spectrum are large-amplitude peaks at periods of approximately 495 yr, 248 yr, 208 yr, 180 yr and 141 yr. We note that no significant component is present at higher (de-cennial) frequencies. On the contrary, the analysis of the thermoluminescence profile of the GT14 core (see Sec. V) reveals the presence of well-defined decennial periodicities. As the measurements of the TL and CaCO₃ profiles of the GT14 core have been obtained from the same sediment layers, the absence of high-frequency periodicities in the CaCO₃ series suggests that the carbonate reservoir may have a slow reaction time that presumably damps out the fast (decennial) oscillations.

As carbonates are an important part of the carbon cycle, it may be inter-esting to compare the CaCO₃ data with the atmospheric $\Delta^{14}C$ record in tree rings, which represents the only data with high precision and comparable time resolution. Here we consider the time series of high-precision, decadal atmospheric tree-ring $\Delta^{14}C$ abundance given by Stuiver and Becker (1986) and we compare it with the CaCO₃ profile of the GT14 and GT89/3 cores. The radiocarbon time series for the last 2 millennia is shown in Fig. 5 together with the GT14 carbonate profile. A resemblance between the CaCO₃ series

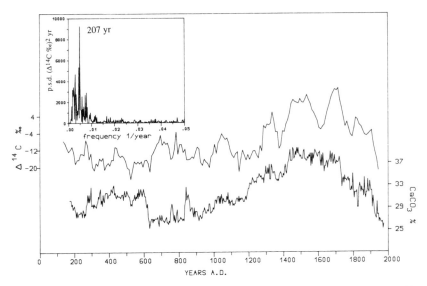

Fig. 5. Time series of the ‰ tree-ring $\Delta^{14}C$ as given by Stuiver and Becker (1986), together with the $CaCO_3$ series. The power spectral density of the radiocarbon record as detrended by Stuiver and Braziunas (1988) is shown in the inset.

and the $\Delta^{14}C$ data seems to be present, indicating that approximately the same long-term period is present in both records. This behavior is found also for the GT89/3 core, over a longer period of time extending at least up to 37 centuries BP. A significant cross correlation of about 0.8 between the $CaCO_3$ and the $\Delta^{14}C$ profiles is obtained, the carbonate record leading the tree-ring radiocarbon by a time lag between 0 and 60 yr. The power spectrum of the $\Delta^{14}C$ signal, as detrended by Stuiver and Braziunas (1988), is reported in the inset of Fig. 5. The 95% significance is $0.4\ P(f) < \pi(f) < 5\ P(f)$, where $\pi(f)$ is the true spectrum and $P(f)$ is our spectral estimate. Energetic spectral peaks at approximately 206 yr, 228 yr and 138 yr are present, indicating the presence of centennial waves with these average periodicities in the detrended tree-ring radiocarbon as well as in the carbonates. See Sonett and Finney (1990) for a detailed discussion of the radiocarbon spectrum.

A further step in the analysis is to search for close correlations between the century-scale fluctuations observed both in the carbonate profile and in the radiocarbon series. To this end, we have performed a detailed analysis of the amplitudes and phases of the Suess wiggles with a period of about 200 yr. This has been obtained by carefully applying the method of superposition of epochs (this is also called "stacking"; see, e.g, Worthing and Geffner [1962] for an introduction to this method). In this approach, a time series is partitioned into consecutive segments of fixed length T that are then superposed and averaged. The components with periods different from T/n are (approximately) filtered out and only the wave with period T and its harmonics survive in the superposed data. In our case, the carbonate profile of the GT89/3 core (starting in 1690 BC) and the tree-ring ^{14}C data (truncated at the same starting date as the carbonates) have been folded and superposed. By least-square fitting a sinusoid with period T to the superposed data, we have determined the phase and amplitude of the waves, in steps of 1 yr, for T varying from $T = 170$ yr to $T = 240$ yr. The correlation coefficients r^2 of the sinusoids to the folded experimental data for the radiocarbon series and for the carbonate profile of the GT89/3 core vs the folding period T are shown in Fig. 6. The dependence of r^2 on T for the carbonate (stars) and the radiocarbon (circles) series shows a striking similarity over the entire range of T between 170 and 240 yr. The correlation coefficients for the carbonate series have 3 well-defined, separate maxima at periods of about 179 yr, 205 yr and 226 yr. The correlation coefficients of the radiocarbon data show 3 distinct maxima at about 187 yr, 206 yr and 228 yr, in agreement with the results of Sonett and Finney (1990). The superposed data and the least-square sinusoids for the period $T = 206$ yr are shown in Fig. 7a for the $CaCO_3$ series and in Fig. 7b for the radiocarbon record. To better visualize the wave, we have repeated twice both the superposed experimental data and the least-square-fit sinusoid. The reference year (i.e., the year zero in the graph of the superposed data, to which the phase of the wave is referred) is AD 247. The phase of the radiocarbon wave is $\phi_{206}(^{14}C) = 1.05$ rad and the phase of the carbonate wave is $\phi_{206}(CaCO_3) = 2.73$ rad. Note that there is a phase shift of about 90° (corresponding to a time lag of about 50 yr) between the two waves,

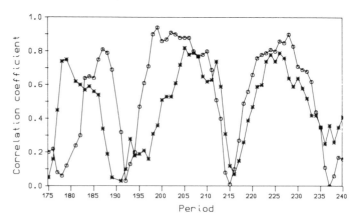

Fig. 6. Correlation coefficient between superposed data and least-square-fit sinusoids as obtained by the method of the superposition of epochs ("stacking"). Circles are for the radiocarbon record and stars for the $CaCO_3$ series. The folding period T varies between 170 yr and 240 yr.

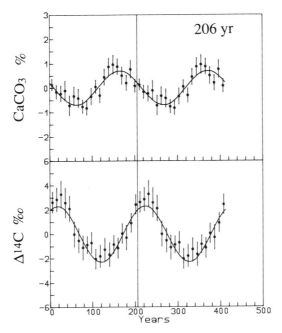

Fig. 7. Results of the method of the superposition of epochs for the folding period $T = 206$ yr. Panel (a) refers to the $CaCO_3$ series and panel (b) refers to the $\Delta^{14}C$ record. The solid line in each panel represents the least-square-fit sinusoid to the superposed data. To better visualize the wave, both the superposed data and the least-square-fit sinusoid have been repeated twice. The reference year for the superposition of epochs has been taken to be AD 247.

indicating that the carbonate wave leads the radiocarbon oscillations. Figures 8a and 8b report the results of the superposition of epochs for $T = 225$ yr for the two series, respectively. For this periodicity, the carbonates lead the radiocarbon by a lag of about 15 yr. These results suggest the existence of Suess wiggles with periodicity of about 200 yr in the carbonate data; they also confirm the fact that this spectral component seems to be split into at least 3 separate bands whose average frequencies are approximately the same for both the radiocarbon and the carbonate data.

The above results indicate a close relationship between the carbonate profiles and the tree-ring radiocarbon record. As the $CaCO_3$ content of the sea sediments is governed by biological, physical and chemical processes in surface sea waters, which in turn are strongly influenced by climatic conditions, and as the ~200-yr radiocarbon wave is known to be governed by the solar modulation of galactic cosmic rays through the action of interplanetary magnetic fields, the results presented here suggest a strong Sun-climate relationship. These results are consistent with the findings of Sonett and Suess

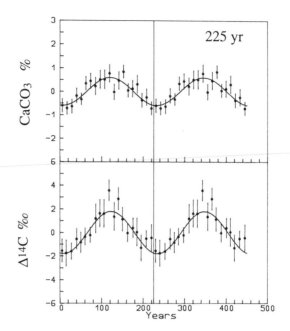

Fig. 8. Results of the method of the superposition of epochs for the folding period T = 225 yr. Same details as in Fig. 7 apply.

(1984), who found a good correlation between the spectra of the $\Delta^{14}C$ series in tree rings and that of the ring widths of bristlecone pine wood from Campito Mountain, California, and made a strong case for a Sun-climate relationship. In the words of Sonett and Finney (1990), "the two records are inversely correlated in the sense that thicker tree rings (more rapid growth), if associated with higher temperature and greater solar activity also infers greater interplanetary modulation and thus lessened production of radiocarbon." These conclusions obviously "depend upon the idea that solar activity is positively correlated with temperature." Our results may also indicate the importance of a climatic control of the atmospheric $\Delta^{14}C$ via p_{CO_2} variations; long-term changes in the radiocarbon record could be due to such effects.

We finally note that the three-line structure of the $\sim$200 yr oscillations recalls a situation in which a carrier wave (with period of about 206 yr) is modulated by a longer wave that generates two sidebands at lower and higher frequencies (Sonett 1984). The splitting frequency between the two sidebands at $\sim$185 yr and $\sim$225 yr provide some indication on the possible presence of a long modulation with a period of $\sim$2000 yr acting on the $\sim$200-yr oscillations present both in the carbonate profile of the GT89/3 core and in the tree-ring radiocarbon. We recall that a modulating wave with a period of about 2400 yr has been shown to be present in the long-running Vostok record

(Yiou et al. 1989), and a component with a period of approximately 2.3×10^3 yr has been detected in the ^{14}C record (Sonett and Finney 1990), where the strong 206-yr component was shown to be modulated by a longer cycle with a period of ~ 2000 yr. This latter component may well be due to changes in the global carbon cycle (Siegenthaler et al. 1980) with a similar period. As Ghil (see his chapter) pointed out, these changes might be related to an oscillation of the oceans' thermohaline circulation (Ghil et al. 1987) with a mean period of roughly 2.5×10^3 yr. Other authors (Hood and Jirikowic 1990) have inferred an essentially solar origin of the 2.3×10^3-yr cycle detected in the radiocarbon record.

V. THE THERMOLUMINESCENCE PROFILE OF THE GT14 CORE

The analysis of the $CaCO_3$ profiles of the two Ionian Sea cores discussed above provides information on the variations of the chemical composition of the sediment, which is mainly affected by climatic factors. It is now important to study other physical properties of the sediment that are linked with both the composition of the material of the core and with the possible solar and/or climatic irradiation conditions affecting the polymineral grains at the time of deposition. In this regard, the thermoluminescence (TL) of the sediment is an optimal property that is very sensitive to both these effects (Cini Castagnoli and Bonino 1988). In this section, we review the principal results which we have obtained in the analysis of the TL profile of the GT14 core.

The TL Signal

The TL signal which is analyzed here comes from the *bulk* TL level of the mixture of the polymineralic crystals forming the carbonate mud of the core, which have aeolic, precipitated *in situ,* and detrital origin. The advancing fronts of dust which are typically monitored in the atmosphere after a Saharian storm event (see, e.g., Prodi and Fea 1979) provide a small fraction of the thermoluminescent components (quartz and feldspars) of the fine materials that form the mud of the core. The two main components of the sediment are the carbonates and the silicates precipitated *in situ,* which are partly of biological origin. When crystalline minerals are exposed to ionizing radiation in their natural environment, free electrons and holes are produced. A small fraction of the former become trapped at defects in the crystals and it is these trapped charges that are released and are able to recombine giving rise to the TL signal in a radiative process during the heating of the sample. The recombination comes also as a consequence of the exposure of the material to light. This process is important for the formation of the TL signal in the grains precipitated from the atmosphere because it is in competition with the process due to the ionizing radiation. The TL variations in the core may thus reflect the variations in the environmental conditions at the time of deposition of the sediment, which may either affect the specific TL intensity of a given amount of thermoluminescent material and/or the composition of the sediment (in particular, the relative amount of carbonates). The TL signal (photons detected by a photomultiplier as a function of the heating temperature of the sample) is proportional to the number of trapped electrons. It is important to recall that the predepositional TL is augmented by the TL acquired after deposition from the local radioactivity concentration in the core. In very recent sediments (such as the GT14 core) the predepositional TL is predominant over the TL acquired after sedimentation.

In order to measure the TL profile of the GT14 core, a few grams of *bulk* material were taken from each layer. The preparation of the samples was done in red light, treating the still wet material of the core by successive washing in NaOH, water and acetone, and using a centrifuge in such a way as to preserve the original composition of the polyminerals. This is also the reason for which we do not leach the samples with HCl, as usually done in the standard TL procedure used for dating purposes. After drying in an oven at 40°C overnight, the powder was gently sieved and the fraction <44 μm was weighted in samples of 15 mg for the TL measurements. Glow curve measurements were made by using a TL analyzer described by Miono and Ohta (1969). The sample was spread on a 1-cm diameter platinum disk and was heated from room temperature to 400°C at a heating rate of 5°C s⁻¹, in an atmosphere of nitrogen gas. The TL signal from the photomultiplier, typically of the order of 5×10^4 photons s⁻¹, was registered by a XY recorder as a function of temperature. An example of the glow curves obtained from 4 different samples labeled *a, b, c,* and *d* is given in Fig. 9. Curves *a* and *a'* refer to two replicas from the same sample. Several replicas were prepared from the mud of each sample. Four of these provide four "natural" glow

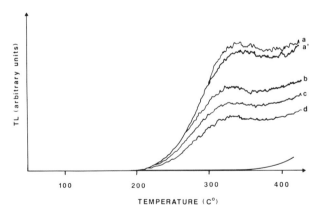

Fig. 9. Four example glow curves (labeled *a, b, c* and *d*) measured in four different samples of the core. Curves *a* and *a'* refer to two different replicas from the same sample.

curves for each sample, while a fifth replica was exposed to sunlamp and provides a reference, bleached glow curve. The bleaching procedure erases the information recorded in the TL profile of the sediment and provides a reference noise signal to be contrasted with the natural TL profile.

The TL level of a given sample is obtained by considering the value of the TL glow curve at an appropriate temperature. In Cini Castagnoli et al. (1988a,b, 1989, 1990a), we considered the TL level measured in the glow curve plateau at approximately 340°C, and in Cini Castagnoli et al. (1990d) we considered the TL signals provided by measuring the TL intensity of the glow curves at five different temperatures in the plateau, namely at 300°C, 320°C, 340°C, 360°C and 380°C. A total of 20 profiles of the natural TL intensity as a function of depth, plus 5 bleached TL profiles were thus obtained and analyzed (note that we consider 4 natural replicas and 1 bleached replica for each sample of the core). Even if the profiles measured at the different temperatures are not independent, the use of different reading temperatures allows us to reduce the random noise errors introduced by both the measurement and reading procedures. The time series obtained from the 4 different replicas for each sample have a very strong correlation among each other. For example, in the case of the profiles measured at 340°C, the correlation coefficient between any 2 natural TL profiles (i.e., profiles obtained from unbleached samples) is larger than 0.90. By contrast, all these natural profiles are very poorly correlated with the bleached TL profile obtained at the same temperature, the correlation coefficient between bleached and unbleached profiles being less than 0.26. As the TL profiles obtained from the 4 unbleached replicas for each sample are closely correlated, taking their average further reduces the effects of random noise. We are thus left with 5 average TL time series corresponding to the 5 reading temperatures selected. These 5 profiles have been studied in detail by Cini Castagnoli et al. (1990d). Here we consider only the TL signal obtained by a further average over these 5 reading temperatures. The resulting TL signal is shown in the inset of Fig. 10.

Spectral Properties of the TL Profile

Figure 10 reports the power spectral density of the average TL signal. Because the independence of the TL profiles obtained at the different reading temperatures is not assured, the 95% significance of the spectrum in Fig. 10 is conservatively maintained as that given by considering just 4 independent measurements (i.e., the 4 independent replicas per sample at a single reading temperature). The 95% significance in the spectrum is consequently given by $\frac{1}{2} P(f) \leq \pi(f) \leq 4P(f)$, where $\pi(f)$ is the "true" spectrum and $P(f)$ is our spectral estimate. The average "bleached" TL signal is reported in the inset of Fig. 11. This figure reports the power spectral density of the bleached signal. No significant peaks are present in the spectrum of the bleached signal, confirming the fact that the relevant information on the cycles contained

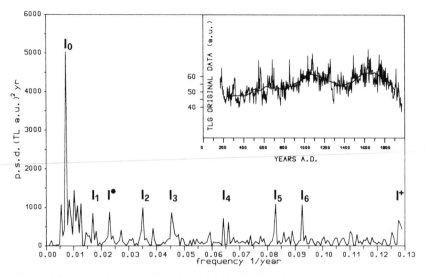

Fig. 10. Power spectral density of the detrended TL signal obtained by averaging the 5 natural (unbleached) readings at 300 °C, 320 °C, 340 °C, 360 °C and 380 °C. A 160-yr running average has been removed from the signal before the spectral analysis. Units on the ordinates are (TL arbitrary units)^2 × yr. The frequency step is $\Delta f = 0.605 \times 10^{-3}$ yr^{-1}. The symbols I_o, .., I_6, $I*$, I^+ indicate the main spectral peaks discussed in the text. The average TL signal is shown in the inset.

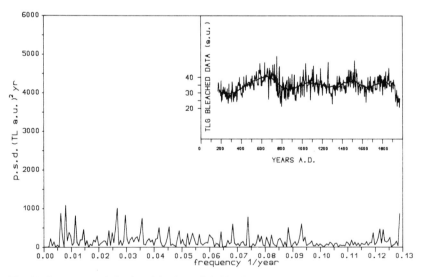

Fig. 11. Power spectral density of the detrended, bleached TL signal obtained by averaging the 5 readings at 300 °C, 320 °C, 340 °C, 360 °C and 380 °C. Same details as in Fig. 5 apply. The average bleached TL time series is shown in the inset.

TABLE I
Periods and Frequencies of the Main Significant Spectral Lines
Present in the TL Spectrum

	l_0	l_1	l^*	l_2	l_3	l_4	l_5	l_6
T (yr)	137.7	59.0	43.5	28.5	22.03	15.6	12.06	10.8
$v = 1/T$ (yr^{-1})	0.00726	0.0169	0.023	0.0351	0.0454	0.0641	0.0829	0.0926

in the TL profile of the sediment is destroyed by the bleaching procedure. Note that the significance of the 2 spectra reported, respectively, in Figs. 10 and 11 is vastly different as there is only 1 replica for the bleached signal, while there are 4 independent replicas for the natural TL profile.

The spectrum shown in Fig. 10 shows a well-defined line structure. The lines denoted by l_0, .., l_6 and l^* are all significant at the 95% level. Table I reports the periods and the amplitudes of these spectral components. A very strong peak (l_0) is visible at a period of 137.7 yr. This spectral component is discussed further below. We note the existence of two peaks (l_3 and l_2) at the period of the Hale cycle of solar magnetic activity (22 yr) and at a period of about 28.5 yr. We recall that a periodicity of 28.5 yr has already been found in the spectral analysis of the sunspot data at a significance level not lower than that of the 22-yr cycle (Otaola and Zenteno 1983). Two other strong peaks (l_5 and l_6) are visible at periods of 12.06 yr and 10.8 yr, respectively.

It is interesting to note that no spectral peaks at ~200 yr are present in the original (nondetrended) TL signal, nor in the detrended one. Conversely, the splitting of the decennial and ventennial TL spectral lines may be generated by a century-scale modulation with a period of approximately 200 yr. In particular, the beat of the 12.06 and 10.8 components produce a modulated wave train with a carrier wave with a period of 11.4 yr and an amplitude modulation with a period of 206 yr (see Cini Castagnoli et al. (1988b, 1989). This situation corresponds to a "suppressed carrier modulation." It is extremely interesting to note that this TL modulation has the period of the strong 206-yr component discussed above for both the tree-ring radiocarbon and for the carbonate profiles of the GT14 and GT89/3 cores. By applying the method of the superposition of epochs to the TL data, it is possible to obtain the amplitude and the phase of the two components l_5 and l_6 with periods of 12.06 and 10.8 yr. These are, respectively, $A_5 = 1.2$ TL AU, $\phi_5 = 4.84$ and $A_6 = 1.0$ TL AU, $\phi_6 = 0.94$ rad. Figure 12 shows the modulated wave train generated by the sum of the 2 decennial TL waves, together with the 206-yr wave detected in the analysis of the carbonate profile of the GT89/3 core (see Fig. 7a). The astonishing coincidence of the 206-yr carbonate wave with the century-scale modulation of the 11.4-yr TL cycle must be noted. We may attribute the splitting of the decennial lines in the TL record

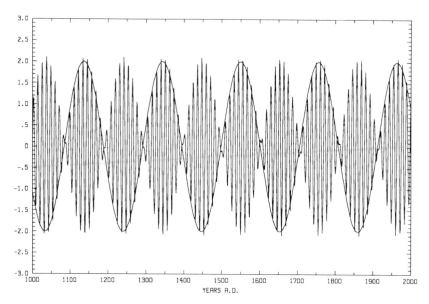

Fig. 12. The modulated wave train generated by the beat of the 12.06-yr and 10.8-yr TL components together with the 206-yr wave detected in the carbonate series. Note the perfect coincidence of the 206-yr carbonate wave with the century-scale TL modulation.

as due to the variations of the relative amount of carbonates in the sediment, which is modulated by the 206-yr cycle.

In addition, the squared amplitude TL modulation of 206 yr is in good agreement with the overall shape of the centennial variation of the sunspot number series R_z. Figure 13 reports the R_z series together with the squared amplitude modulation given by the beat of the 2 decennial periodicities. The nodes of the modulation (spaced by 103 yr) fall in AD 1810 and AD 1913, and closely correspond to the century-scale minima of the sunspot number series R_z (Cini Castagnoli et al. 1988b). This is possibly the only evidence for connecting the ubiquitous 206-yr wave found in terrestrial reservoirs to a solar origin.

We also recall that the 206-yr wave present in the $\Delta^{14}C$ record lags by approximately 90° the 206-yr carbonate wave (see Fig. 7), such that maxima or minima of the atmospheric radiocarbon data correspond to the nodes of the TL modulation. The origin of the 90° phase shift between the radiocarbon wave and the carbonate wave is unknown. However, we note that if the tree-ring radiocarbon can be considered as representative of the inventory of the ^{14}C in the atmosphere, it would then be the integral of its production rate dN/dt. As the atmospheric radiocarbon is produced by the interaction of galactic cosmic rays with the atmosphere, the ^{14}C production rate dN/dt is inversely related to the magnetic activity level of the Sun, which prevents the galactic cosmic rays from reaching the Earth. A sinusoidal variability of the inventory

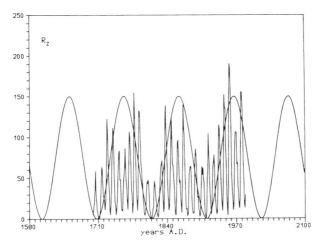

Fig. 13. Time series of the yearly sunspot number in the last 3 centuries. Superposed to this we show the squared amplitude modulation of the wave train generated by the beat of the 12.06-yr and 10.8-yr TL waves. Notice how this modulation describes the general shape of the sunspot secular variations.

$N(t) = A\sin(2\pi t/T + \phi)$ would thus imply a variability of the production rate $dN/dt = A\sin(2\pi t/T + \phi + \pi/2)$, and a corresponding variation of the solar magnetic output $S(t) = dN/dt$. From our results, the carbonate content of the sediment shows a century-scale sinusoidal variation $A\sin(2\pi t/T + \phi + \pi/2)$, which turns out to be the same sinusoidal variation (in period and phase) as that of dN/dt that has been inferred from the tree-ring radiocarbon data. On the other hand, the carbonate variations have most likely a climatic origin. Thus, if the mechanism responsible for the carbonate fluctuations is at the same time responsible also for the variation of the ^{14}C production rate, then our result may be a strong point in favor of a direct correlation between climate and solar variability. In this context, solar variability should thus be intended as a century-scale variation of both the solar magnetic and brightness outputs. Clearly, these remarks are nothing more than euristic speculations which have been stimulated by the close agreement in period and phase experimentally found for the century-scale variations detected in so different terrestrial archives like the tree-ring radiocarbon, the carbonate content of a sea sediment and the TL level of the sediment. These speculations find a physical motivation in the recent discovery that the solar "constant" is a variable quantity which depends upon "the predominance of the irradiance contribution of bright faculae over that of dark spots" (Foukal 1990). As a final point of caution, however, we recall that the 206-yr carbonate wave is maximum or minimum in correspondence of the centennial maxima of the sunspot number R_z. Consequently, the above interpretation would imply that centennial sunspot maxima would be associated one time with high magnetic solar

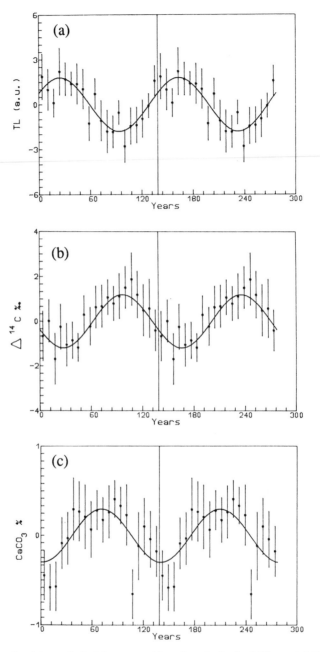

Fig. 14. Results of the method of the superposition of epochs for the folding period $T = 138.5$ yr. Panel (a) refers to the TL data, panel (b) refers to the $\Delta^{14}C$ record, and panel (c) refers to the $CaCO_3$ series. The solid line in each panel represents the least-square-fit sinusoid to the superposed data.

level, the next time with low magnetic solar level. Further experimental and theoretical work is clearly required on these intriguing issues.

We finally note that two significant spectral lines (I_o and I_1) at periods of 137.7 yr and 59 yr are present in the TL spectrum. These two lines are again at the same distance as I_5 and I_6. The 137.7-yr line is, by far, the strongest component in the TL spectrum, as it is about 5 times larger than all the other spectral lines, which have approximately the same amplitude. We also note that the period of this wave is the two thirds of the period of the 206 yr amplitude modulation.

A wave with a periodicity of $\sim$140 yr has also been detected in the carbonate profiles of the GT14 and of the GT89/3 cores and in the tree-ring radiocarbon series. To better explore the properties of the centennial cycle with this approximate periodicity we have applied the method of the super-position of epochs to the 3 records, for folding periods T varying from $T = 135$ yr to $T = 143$ yr. A maximum correlation coefficient between the superposed data and the least-square-fit sinusoid is found at a period of 138.5 yr ($r^2 = 0.8$) for the radiocarbon series. At this value of T, the correlation coefficients between the superposed data and the least-square-fit sinusoids are quite large for the TL and $CaCO_3$ records as well (being, respectively, 0.75 and 0.5). Therefore, we have chosen this common period to compare the phases of the waves appearing in the 3 records. Figures 14a, b and c report the results of the superposition of epochs for the 3 data sets for $T = 138.5$ yr. A well-defined oscillation with this periodicity is present in all 3 records. An interesting observation is that the 138.5-yr wave in the radiocarbon series is about opposite (i.e., phase shifted by 180°) with respect to the 138.5-yr wave in the TL signal. Again, the 138.5-yr wave in the radiocarbon record lags the carbonate wave by $\sim$90°.

Comparison between the TL Profile and the Radiocarbon Series

To further explore the possible similarities between the tree-ring radio-carbon record and the TL profile, we show in Fig. 15 the $\Delta^{14}C$ signal, plotted on an inverted scale (curve a), and the TL signal (curve b) for the period AD 1100 to 1900. The 206-yr wave obtained from the $\Delta^{14}C$ series is superposed to the data. We note that this 206-yr wave correctly reproduces, at least for the period considered, the principal characteristics of the $\Delta^{14}C$ data. The Wolf, the Sporer and the Maunder minima of solar activity are clearly visible as maxima in the original radiocarbon data as well as maxima of the 206-yr oscillation. Curve c and d of Fig. 15 represent, respectively, the sum of 137.7-yr and 59-yr waves and of the 12.06-yr and 10.8-yr waves obtained from the analysis of the TL data. The comparison between the radiocarbon series and the TL data indicate that the TL data display the expected Maunder minimum during the period AD 1645 to 1715 and the expected Wolf mini-mum around AD 1300. At these times, the sum of the 137.7-yr and 59-yr

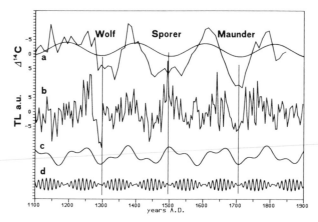

Fig. 15. (a) The decadal $\Delta^{14}C$ record as given by Stuiver and Braziunas (1988) together with the 206-yr radiocarbon wave; (b) the detrended TL signal from the GT14 core; (c) the sum of the two TL waves with periods of 137.7 and 59 yr; and (d) the sum of the two TL waves with periods of 12.06 and 10.8 yr.

waves displays well-defined minima. In the TL data, the Sporer minimum starts at the end of the 14th century, in agreement with the radiocarbon indications. From the TL data, however, this event seems to have recovered around AD 1500, showing a double feature not present in the $\Delta^{14}C$ record. In this regard, we note that the sum of the 137.7-yr and 59-yr TL waves displays a maximum around AD 1500. The end of the Sporer event is approximately at the same time for both TL and radiocarbon series. Note also that the Maunder, Sporer and Wolf secular minima of solar activity closely correspond to nodes of the beat of the 12.06-yr and 10.8-yr waves as shown in curve d of Fig. 15.

VI. DISCUSSION AND CONCLUSIONS

In this chapter, we have discussed the analysis of the experimental time series of carbonates and thermoluminescence which we have obtained from measurements in two recent Ionian sea cores. The possibility of carefully assessing the presence and the properties of the main periodicities dominating the records is given by the precise dating of the cores and by the fact that the sedimentation rate is constant. The preliminary dating of the cores has been pursued through the ^{210}Pb method; the analysis of the ^{137}Cs content of the upper layers of the sediment has indicated that the cores are undisturbed. A careful tephroanalysis of the known volcanic events of Pompei (AD 79), Pollena (AD 472) and Ischia (AD 1301) has confirmed the constancy of the sedimentation rate determined by the ^{210}Pb method. This has furnished a precise evaluation of the sedimentation rate, which is $s = 0.0646 \pm 0.0007$ cm yr.$^{-1}$. The TL and $CaCO_3$ depth profiles have been transformed into TL and $CaCO_3$ time series; the main features and periodicities present in the signals have been analyzed. The TL and carbonate profiles have then been compared with the radiocarbon record in tree rings and with the series of the mean annual sunspot number R_z. The main results of this analysis can be summarized as follows:

(1) The carbonate profiles of the GT14 and of the GT89/3 cores are practically coincident, indicating the reproducibility of carbonate measurements in recent sea sediments.

(2) A strong century-scale wave with period of 206 yr is present in the $CaCO_3$ profile of the GT14 and of the GT89/3 cores, consistently with the results found for the ^{14}C record in tree rings (Suess wiggles), suggesting a strong Sun-climate relationship. The 206-yr cycle detected in the radiocarbon record lags by a phase shift of 90° the 206-wave detected in the carbonate profile. The 206-yr cycle detected in both the carbonate profile and in the radiocarbon record seems to be splitted into 3 components with periods of about 187 yr, 206 yr and 228 yr. This result suggests that the 206-yr cycle may be modulated by a longer cycle with a period of $\sim$2000 yr.

(3) In the analysis of the TL profile of the GT14 core, high-frequency (decennial) periodicities have been precisely identified. The spectrum of the TL signal has a well-defined line structure with few dominant components. Among these, the 22-yr (Hale) cycle has been detected. The beat of two decennial TL waves with periods of 12.06 yr and 10.8 yr produces a modulated wave train with a 11.4-yr carrier and a 206-yr amplitude modulation. This 206-yr squared amplitude modulation seems to describe at least some aspects of the century-scale variations in the sunspot number series R_z.

(4) The 206-yr wave detected in the carbonate record coincides, both in period and phase, with the amplitude modulation of the two decennial TL waves. If only the 11-yr cycle is felt by the TL recording mechanism, and the amount of the thermoluminescent material (related to the composition of the core) is modulated with a period of 206 yr, then the splitting of the 11-yr cycle into the 2 observed TL decennial components would be naturally explained.

(5) A strong $\sim$138-yr wave dominates the spectrum of the TL time series. A wave with the same period is present also in the radiocarbon record and in the $CaCO_3$ profile.

As a general conclusion we stress the fact that different terrestrial reservoirs are apparently driven by similar frequencies, suggesting a common solar origin for the variations of such widely different systems. In particular, a strong Sun-climate relationship is suggested by several results. We also emphasize the interest of studying the properties of recent sea sediments and the importance of obtaining a precise dating of the cores. This allows for one to explore carefully, in a phenomenological way, some of the properties of the solar variability and of the recent variations in the Sun-Earth system.

Acknowledgments.   We acknowledge helpful discussion with C. Castagnoli, E. Callegari, Zhu Guang-mei and N. Bhandari. We are grateful to M. Serio for numerical assistance and to F. Caprioglio and A. Romero for technical help.

PART V

The Early Sun

MAGNETIC HISTORY OF THE SUN

E. H. LEVY
University of Arizona

A. A. RUZMAIKIN
Institute of Terrestrial Magnetism, Ionosphere and Radio Wave Propagation

and

T. V. RUZMAIKINA
O. Yu Schmidt Institute of Physics of the Earth

Magnetic fields play important roles in the behavior and evolution of many astrophysical objects. For the solar system, magnetic fields apparently had a critical role in at least some aspects of the formation and early evolution of the system and continue today to play a central role in controlling much of the Sun's activity and dynamical behavior. This review gives a brief overview of the major magnetic effects and behavior that characterized and influenced aspects of the gross history of the Sun. The chapter surveys the behavior of magnetic fields in the Sun from the early effect of the interstellar field on protosolar collapse through to the modern solar magnetic cycle. Emphasis is given to the evolution of the Sun's magnetic field, its memory of the past magnetic states, and its loss of memory, as these depend on physical episodes, conditions and phenomena during the Sun's development.

I. INTRODUCTION: THE INTERSTELLAR PRECURSOR

In this chapter, we consider the gross aspects of magnetic field behavior in the Sun. We emphasize the variety of physical processes that dominate the interaction between the magnetic field and the solar fluid during the several epochs of solar history. For the purpose of this discussion, the Sun's development can be thought of as encompassing three epochs: (1) the formation of the protosolar nebula from the collapse of interstellar matter; (2) the evolution of the early Sun and the Sun's interaction with the protoplanetary nebula; and (3) the modern Sun. This chapter follows the main stages of Sun's magnetic history. The present section describes the precursor material from which the Sun is thought to have formed and introduces some principal concepts needed to understand the Sun's magnetic history. In Sec. II, collapse of the protosolar cloud and contraction through an accretion phase is discussed. Sections III and IV cover the formation and early evolution of the Sun and its magnetic field. Section V discusses the gross behavior of the modern solar magnetic field. Sec. VI looks briefly ahead to the fate of the solar magnetic field during subsequent evolution of the Sun. We will see that one critical challenge that confronts our understanding of the present-day Sun and the Sun's evolution is to learn how much the Sun, at each stage of its evolution, remembers of its past. By necessity, this chapter focuses as much on questions as on answers— this focus on outstanding questions reflects the incomplete state of our current understanding, both about solar magnetic fields and their behavior, and about the Sun's history.

To a first approximation, the Sun is a static, spherically symmetric, hot ball of gas. However, many of the Sun's most interesting and puzzling behaviors are highly erratic and asymmetrical: flares, the solar wind, the early protosolar disk, energetic particles, etc. These latter behaviors result from the fact that the Sun rotates and is magnetized. The Sun's rotation and magnetization are closely related phenomena. Magnetic fields constrain rotation by effectively rearranging the distribution of angular momentum; and the organizing influence of rotation on internal fluid motions is essential to the dynamo process that efficiently produces large-scale magnetic fields in cosmical objects.

The magnetic history of the Sun begins before the Sun itself and continues to the present day. Presumably, the Sun formed as a typical star from the collapse of an interstellar cloud of gas and dust. The collapse of such a cloud to form a star is one possible resolution of the battle between the confining force of gravity and the opposing, internal, expansive forces that arise from magnetic fields, rotation, internal heat and fluid turbulence. The formation of a star like the Sun requires the action of mechanisms that remove internal energy: for example, energy associated with the fluid turbulence is transformed into heat through dissipative processes; internal heat is lost from an interstellar cloud through radiation. In addition, other processes, such as mechanical loss of energy through radiated Alfvén waves, may also be important. Supermagnetosonic turbulent motions are thought to be rapidly dissipated in internal shock waves; on this basis, it is believed that the turbulent velocities cannot persistently exceed the magnetosonic speed ($V_{MS} = \sqrt{B^2/4\pi\rho + \gamma p/\rho}$). If this is correct, then the turbulent contribution to the overall expansive stress should not substantially exceed that due to the combination of thermal and magnetic pressures. The high luminosity of protostellar objects is the external manifestation of these dissipative processes and the consequent radiation.

The removal of angular momentum and magnetic flux from protostellar material poses a set of questions that are still not fully answered. Naively, under most astrophysical conditions, both magnetic flux and rotation seem to be locked into gas, and removed only slowly under the action of classical molecular processes. The transport time for either the fluid's differential rotation or its embedded magnetic field is governed by a diffusion relationship, with the time scale given simply as

$$\tau \sim l^2/[\nu, \eta] \tag{1}$$

where ν is the kinematic viscosity of the fluid or η is the magnetic diffusivity, l is the system's spatial scale. For example, in a tenuous neutral gas, the viscosity is $\nu = 1/3\nu\lambda$, with thermal speed v and particle mean free path λ. Consider, for example, the situation with 1 $M_{\odot}$ of matter spread throughout a volume of space at, say 10^6 particles cm^{-3}, with a linear scale of about 10^{17} cm. If the temperature were as high even as 100 K, the kinematic viscosity would be $\nu \sim 5 \times 10^{13}$ cm^{2}s^{-1}. Then, from Eq. (1), the time scale for the rotation to dissipate, through classical viscous processes, would be some 10^{12} yr, substantially longer than age of the universe; yet stars form over time scales of some 10^6 to 10^7 yr. Under most circumstances, the time that would be needed for magnetic field lines to leak out of interstellar material as a result of classical electrical resistivity is similarly long. These simple examples indicate the extent to which more complex collective phenomena must play critical roles in star formation (see the chapter by Stauffer and Soderblom).

A. Precursor States

Now compare the Sun with its interstellar precursor material. Although the detailed history of protostellar gas and dust is not at all understood, consider that, if spread throughout a region of space at a density of one hydrogen atom per cubic centimeter, the Sun's matter would have occupied a region of some 3 parsec diameter, with a volume of 10^{57} cm^3—involving a linear scale of about 10^{19} cm. Leaving aside the possibility of any peculiar rotation, galactic rotation alone would imply more than 10^{55} g cm^2 s^{-1} of angular momentum in such a diffuse mass. However, the present angular momentum of the solar system is only several times 10^{50} g cm^2 s^{-1}, and the Sun itself has a hundred times again less angular momentum. Thus, less than one part in 10^5 of the original angular momentum of this hypothetical precursor matter still resides in the solar system, and only one part in 10^7 still resides in the Sun, where most of the solar system's matter is located. A similar point can be made for the magnetic field. Under the conditions just described, taking

3×10^{-6} gauss as the typical interstellar magnetic field intensity, the total magnetic flux threading the solar precursor gas would have been about 10^{32} G cm². If this magnetic flux were still threading the Sun, then the magnetic intensity at the solar surface would be some 10^{10} gauss, except in the unlikely event that the field could have become completely buried beneath the Sun's surface (leaving less than a part in 10^9 exposed) and remain buried against the effects of magnetic buoyancy (Parker 1979a). As we shall see, the Sun is unlikely to retain primordial magnetic flux in excess of a few $\times 10^{22}$ gauss cm². Of course the detailed history of presolar matter was not so simple as just described. However, the point is clear: the Sun has apparently forgotten much of its earlier life. The formation of the Sun must have involved the shedding of large amounts of both angular momentum and magnetic field.

Actually, we know something about the Sun's likely more proximate precursor states: stars apparently form from the dense cores of molecular clouds. Information about the magnetic fields, rotation, turbulence and densities in interstellar clouds is obtained through a variety of means. Information about the strengths and structures of cloud magnetic fields is obtained from Zeeman splitting of molecular lines in radio emission and from the polarization of optical and infrared emissions. Information about cloud internal turbulence and rotation is obtained by measuring the doppler-induced line widths and frequency-shift profiles across clouds. And information about physical conditions within clouds (mass density, molecular composition, etc.) is obtained from models of emission and absorption spectra observed in interstellar clouds and star-forming regions.

In order to set the initial ideas for the more detailed discussion that follows, suppose that a molecular cloud has a number density, rotation and magnetic field in the ranges

$$
\begin{aligned}
n_{\text{cloud}} &\sim 10^3 - 10^4 \text{ cm}^{-3} \\
\Omega_{\text{cloud}} &\sim 10^{-13} - 10^{-15} \text{ s}^{-1} \\
B_{\text{cloud}} &\sim 10^{-4} \text{ G}
\end{aligned}
\tag{2}
$$

and that the respective values for a typical cloud core may fall in the ranges

$$
\begin{aligned}
n_{\text{core}} &\sim 10^5 - 10^6 \text{ cm}^{-3} \\
\Omega_{\text{core}} &\sim 10^{-13} - 10^{-14} \text{ s}^{-1} \\
B_{\text{core}} &\sim 10^{-3} \text{ G}
\end{aligned}
\tag{3}
$$

(Bodenheimer 1981; Myers and Benson 1983; Goldsmith and Arquilla 1985). Then, for a cloud, the angular momentum and entrained magnetic flux are

$$
\begin{aligned}
J_{\text{cloud}} &\sim 10^{53} - 10^{55} \text{ g cm}^2 \text{ s}^{-1} \\
\Phi_{\text{cloud}} &\sim 10^{32} \text{ G cm}^2
\end{aligned}
\tag{4}
$$

and for a cloud core,

$$J_{core} \sim 10^{53} - 10^{54} \text{ g cm}^2 \text{ s}^{-1}$$
$$\Phi_{core} \sim 10^{31} \text{ G cm}^2. \tag{5}$$

However, it should be noted that the lower rotation bound given in Eq. (3) is constrained by observational limitations; actual cloud rotation rates could be substantially lower in many cases. It is reasonable to expect that the physical lower limit to rotation, in most cases, will be given by the rotation of the Galaxy as a whole, $\sim 10^{-15}$ s $^{-1}$, near the galactocentric distance of the Sun. Using that as a typical lower bound on cloud rotation rates would diminish the lower bound on angular momentum by an order of magnitude in Eqs. (5).

In the early stages of the Sun's formation, magnetic fields were likely to have provided the dominant mechanisms through which angular momentum was shed from the collapsing gas so that the Sun could form. Thus, the sequence in which magnetic field and angular momentum were shed from the collapsing gas would have been critical in determining the initial angular momentum of the solar system. At later stages of the Sun's life, magnetic fields may have continued to play an important role in the transport of angular momentum, and thus in controlling the Sun's state of rotation (although other effects, such as dissipative tidal torques, may also have played important roles at times during the evolution of the solar system). Additionally, in trying to understand the connection between the initial magnetic field carried by presolar matter and the subsequent magnetic fields of the Sun, it is important to recognize that the evolution of the magnetic field depends in detail on the subsequent state of motion and the electrical conductivity of the fluid in which it is embedded. Thus, depending on the physical conditions, through some episodes of its history the Sun may retain a persistent memory of its past magnetic state, while through other episodes, the Sun may suffer a loss of its magnetic memory (owing to electrical or turbulent dissipation) and emerge with a new magnetic field as a result of dynamo generation processes acting in contemporaneous fluid motions.

B. Evolution

If we make the most reasonable assumption that molecular clouds and their dense cores represent sequential stages in the evolution of interstellar matter toward stars, then from the general numbers given above we see that this phase of evolution tends more closely to preserve rotation than to preserve angular momentum. In addition, during this early phase of interstellar dynamics leading to star formation, the magnetic flux is approximately conserved.

The near conservation of magnetic flux threading the collapsing gas (*viz.*, approximately 10^{32} to 10^{31} G cm^2 per solar mass) is easy to understand. To the extent that the prestellar material contracts more or less spherically symmetrically, entraining the magnetic field, the magnetic flux per solar mass will remain constant. However, inasmuch as the cloud material is partially supported by the magnetic field in directions perpendicular to the field lines, the gas can contract more easily by sliding along the magnetic field lines than by moving perpendicular to the lines of force and carrying the field with it. Thus, although the flux per localized solar-mass sphere can diminish somewhat with increasing mass density, the magnetic field and interstellar matter seem to be effectively coupled during this stage of evolution.

The preservation of rotation *rate,* rather than angular momentum, is also easy to understand in general terms. The collapsing core is connected to the surrounding gas by the magnetic field. Because the field is relatively efficient at removing angular momentum from slowly contracting gas, the rotation of the inner cloud is kept close to that of the surrounding gas—angular momentum transferred outward by the magnetic torque.

From the preceding discussion, it appears that the precursor of the Sun probably contained some 3 to 4 orders of magnitude more angular momentum than does the present Sun (and some 2 orders of magnitude more angular momentum than the Sun and planetary system together). In addition, the solar precursor probably contained some 8 to 10 orders of magnitude more magnetic flux than does the present Sun.

Consider the time scale for subsequent evolution of the proto-Sun. In the case that the cloud core should collapse in gravitational free fall, the collapse time is of the order $1/\sqrt{G\rho}$, where G is the gravitational constant and ρ is the mass density. For atomic number density $n = 10^5$ H cm^{-3}, the collapse time is of the order 5×10^5 yr. Now compare this with the time scale for the removal of angular momentum from the core.

As the cloud core rotates, it radiates torsional Alfvén waves into the surrounding gas. These waves carry away angular momentum. We can estimate simply the time scale of angular momentum loss by integrating the resultant magnetic torque over the surface of the cloud core. Consider a magnetic field line entering the cloud core parallel to the rotation axis, and at a distance R from it: crudely, the pitch angle at which the field line enters the core is $\vartheta = 2\pi R/V_A P$, where V_A is the Alfvén speed in the surrounding cloud medium and P is the cloud rotation period. This expression is strictly applicable only for small ϑ, but it suffices for the purpose of making a general estimate. Consequently, the magnetic torque acting on the cloud core is $\mathcal{T} \sim \sqrt{4\pi\rho}B\Omega\langle R^4\rangle$, where B is the magnetic field in the surrounding cloud, Ω is the rate of core rotation, ρ is the mass density in the cloud, and the angular brackets denote an average over the cloud-core surface, which for the purpose of this order-of-magnitude estimate need not be defined precisely. (More detailed expressions for angular momentum radiation in Alfvén waves can be found in Ebert et al. [1960], Mouschovias [1977,1978] and Mouschovias and Paleolugou [1980].) The spin-down time for the core, T, is then

$$T = \frac{J_{core}}{\mathcal{J}} \sim \frac{MR^2}{\sqrt{4\pi\rho B \langle R^4 \rangle}}. \qquad (6)$$

Here M is cloud-core mass, say 1 $M_\odot$ and ρ is the mass density in the surrounding cloud. For $B \sim 10^{-4}$G and $\rho \sim 10^{-21}$g cm^{-3}, the spin-down time is of the order of 5 $\times$ 10^5 yr.

II. PROTOSOLAR COLLAPSE AND CONTRACTION

Altogether, the simple analysis given in the preceding section suggests that the basic (gravitational) collapse time of a cloud core and the angular momentum loss time, due to magnetic stress, can be comparable in magnitude. The similarity of these two time scales complicates a simple understanding of the actual collapse of real clouds. Moreover, the influence of other physical processes is likely to have further complicated the final contraction of protosolar material to form the solar system. For example, neutral matter in the cloud core may have slowly settled through the ionized component in the process of ambipolar diffusion as gravity pulled neutral particles slowly through the ionized gas that was supported by the magnetic field. Eventually, we expect that the increasing capability of computers will facilitate detailed numerical simulations of cloud collapse and star formation, from initial interstellar configurations to fully developed protostellar nebulae, and including the influence of all important physical effects (rotation, magnetic fields, etc.). Also, advanced observational techniques at infrared and submillimeter wavelengths will improve our empirical understanding of these still veiled stages of star formation. For the present, however, we are limited to trying to estimate our way through the gap in understanding that separates interstellar clouds and cloud cores from the protostars and developed stars, and to trying to understand what conditions and processes may have controlled the formation of the Sun.

A. Coupling of Magnetic Field and Matter

If magnetic flux were strictly frozen into protostellar matter, so that motion perpendicular to the field carried the lines of magnetic force with it while matter could flow freely along field lines, then the only physical problem in understanding the early magnetic history of the Sun would be to determine the geometry of cloud collapse. However, the degree to which the matter and the magnetic field are coupled depends sensitively on the physical conditions, especially the abundance of free electrical charge as well as the relevant dynamical time scales. The equilibrium ionization fraction of interstellar gas results from a balance between the rate of ionization and the rate of recombination. At low temperature, thermal ionization is small, and ionization re-

sults principally from ionizing stellar photons, from extrinsic ionizing radiation—*viz.*, the cosmic rays—and from intrinsic ionizing radiation in the
natural radioactivities of certain nuclides.

Free electric charge describes circular motion in the interstellar magnetic
field. Although the circular trajectories of charged particles may slide freely
along the magnetic field lines, the particles cannot move large distances perpendicular to the field: the charged particles and the magnetic field lines are
thus constrained to move together. The neutral gas, on the other hand, is free
to move independently of the magnetic field. The charged particles and the
neutrals are coupled by momentum-exchanging collisions; only to this extent
is the system of magnetic field, charged particles and neutral matter constrained to move together as a composite fluid.

In the tenuous, diffuse interstellar medium the density is so low that
extrinsic radiation easily penetrates through the gas; moreover, the low gas
density results in a very low rate of recombination. As a result, the ionization
level remains high enough to couple the gas to the fluid over the length scales
and time scales characteristic of the large-scale interstellar dynamics. However, as the gas contracts, other things remaining constant, the level of ionization will fall (Nakano and Tademaru 1972). At higher densities, electrons
are lost more rapidly from the gas, both to molecular recombination and by
absorption onto dust particles; in addition, dense interstellar matter within
clouds becomes optically thick and prevents the penetration of extrinsic ionizing radiation. Because of these two effects, the level of ionization falls
within dense clouds. Ultimately, the residual ionization is that resulting from
decay of the radioactive nuclides carried by interstellar matter, competing
against the high rate of recombination in the dense and dusty gas. As interstellar matter contracts to form a protostar, and as the ionization level falls,
three effects contribute to allowing the magnetic field to escape from the matter: finite electrical conductivity, ambipolar diffusion and turbulence.

Finite Electrical Conductivity: Ohmic Diffusion.   Because the ions and
electrons collide with the gas molecules, the electrical current is reduced by
the familiar process of joule dissipation, associated with finite electrical conductivity σ_e which, for simplicity, we take to be a scalar. Consequently, the
magnetic field diffuses through the ionized gas with a resistive or Ohmic time
scale τ_o which in terms of the resistive diffusion coefficient η is given by

$$\tau_o \sim \frac{l_B^2}{\eta} \quad \text{with} \quad \eta = \frac{c^2}{4\pi\sigma_e} \tag{7}$$

where l_B is the characteristic length scale of the magnetic field.

Ambipolar Diffusion.   Because the magnetic field is tied directly to the
charged particles, but is tied to the neutral component only indirectly by

collisions between the charges and the neutral particles, under the confining stress of gravity and the expansive stress of the magnetic field, the neutral matter may settle through the charged particles, as the latter escape with the magnetic field. The time scale τ_a for this process of so-called ambipolar diffusion is given by

$$\tau_a = \frac{8\pi l_B^2 \chi n^2 \mu_{ni} \langle \sigma_{ni} v \rangle}{B^2} \tag{8}$$

(Spitzer 1968; Parker 1963b), where χ is the degree of ionization, μ_{ni} is the neutral-ion reduced mass, l_B the spatial scale of the magnetic field, and σ_{ni} and v are, respectively, the neutral-ion collision cross section and speed.

Turbulence. In a turbulent fluid the transport of vector magnetic field through the *average* fluid is enhanced in a manner analogous to the way that mixing of a scalar (say, sugar) is enhanced by stirring (coffee). Crudely, a turbulent diffusion time τ_t can be defined, in terms of a turbulent diffusion coefficient n_t by

$$\tau_t = \frac{l_B^2}{\eta_t} \quad \text{with} \quad \eta_t \sim \frac{1}{3} v_t l_t \tag{9}$$

where v_t and l_t are the turbulent velocity and the length scale of the large turbulent eddies.

In the foregoing, we have described the evolution of the proto-Sun in only the most general terms, and defined the basic processes. In the following, we trace the Sun's hypothetical evolution in more detail.

III. FORMATION OF THE PROTO-SUN

Today the Sun is a slowly rotating star, of nearly spherical shape. The strength of the centrifugal force at the Sun's equator is a small fraction of the gravitational force:

$$\frac{\Omega_\odot^2 R_\odot^3}{GM_\odot} \sim 2 \times 10^{-5}. \tag{10}$$

However, the presence of the planetary system indicates the great importance of rotation during the Sun's formation: the planets are entirely supported in their orbits by centrifugal force. In this section, we discuss the formation of the Sun and solar nebula, focusing attention primarily on the effects of rotation and magnetic field during this initial phase of the Sun's life.

The solar system displays numerous characteristics that imply that it formed within a thin, extended, dissipative, pre-planetary disk surrounding the young Sun. The planetary orbits are closely confined to a single plane, are nearly circular and corevolve. Moreover, the bulk of the solar system's angular momentum is concentrated in the orbital motion of the planets (>95%), whose total mass is small in comparison with that of the Sun—the planetary masses total about $1/743$ $M_\odot$. These features indicate that the protoplanetary nebula had relaxed dissipatively and lost the memory of its previous collapse before the planets formed. At the same time, the similarity of elemental and isotopic abundances of the main nonvolatile elements in the Sun, Earth and meteorites and a similarity of chemical compositions of the Sun and Jupiter suggest that the Sun and the protoplanetary disk have formed from the same clump of interstellar matter, usually called the presolar nebula. The near coincidence of the solar equatorial plane and the average plane of the planetary orbits further supports the idea of a common origin of the Sun and planets from a single disk-shaped nebula.

Infrared and radio surveys of the nearby molecular clouds Ophiuchus and Taurus reveal such clouds to be the sites of present-day formation of T Tauri stars. These stars apparently are young (10^5 to 10^6 yr) and have low masses, in the range 0.5 to 2.5 $M_\odot$. Similarity of the masses of these stars and the Sun—mass being the most important parameter determining the class and evolution of a star—suggests the idea that T Tauri stars with masses near 1 $M_\odot$ can be identified generically as "Proto-Suns."

Observations suggest that T Tauri stars are formed as a result of the collapse of small dense cores of molecular clouds. The Ophiuchus and Taurus molecular clouds contain small (<0.1 pc), dense (> 10^{-20} to 10^{-18} g cm^{-3}) and cool ($\sim$ 10 K) cores, visible in ammonia lines (Myers and Benson 1983). Also, in other infrared surveys of specific morphological classes, some steep-spectrum infrared sources in the Taurus molecular cloud are identified with protostars. These seem to lie near the centers of small ammonia cores (Myers et al. 1987). It seems plausible that these cores, as well as infrared sources unidentified in optical emission, and embedded T Tauri stars, represent an evolutionary sequence leading to formation of T Tauri stars (Adams et al. 1987; see the chapter by Bertout et al.).

Elongated structures, presumably associated with disk-like distributions of material, have been detected in the past several years around 4 young stars HL Tau, DG Tau, Beta Pic and IR source L 1551 IRS 5 (Beckwith et al. 1985; Smith and Terrile 1984; Strom et al. 1985a; Sargent 1988). Gaseous disk-like envelopes associated with HL Tau, DG Tau and L 1551 IRS 5 have typical radii (1 to 2) $\times$ 10^3 AU, as measured in the $J = 1$–0 transitions of CO and ^{13}CO and in the continuum at 2.2 mm; their masses typically are not less than about 0.01$M_\odot$ (Sargent and Beckwith 1987). Analysis of the radial velocity distribution within the disk around HL Tau indicates rotation, with the gas velocity clearly decreasing with distance from the star, implying that the disk is gravitationally bound to the star and that stellar gravity dominates the gravitational potential. The broad infrared spectra (at $\lambda > 10\mu$m) that characterize at least 60% of T Tauri stars arise from heated dust located between about 0.01 and more than 100 AU from the star (Rucinski 1985). In order to account for the fact that these stars are visible at optical wavelengths, the simplest interpretation comes from assuming that the observed far-infrared radiation arises in an optically thick, but physically thin, circumstellar envelope, which implies a disk-like structure (Adams et al. 1987). These far-infrared excesses require a mass of emitting dust in the range 10^{-3} to 10^{-4} $M_\odot$; and, hence, a total disk mass in the range 0.01 to 0.1 $M_\odot$, provided that the ratio of dust to gas in the disks is similar to that in the interstellar medium. On the basis of existing observations, it is plausible to expect that all T Tauri stars that have infrared excess are surrounded by circumstellar disks (Bertout et al. 1988; chapter by Bertout et al.).

Altogether, the observations suggest a common origin of T Tauri stars and the disks surrounding them. Combined with the isotopic homogeneity of the solar system mentioned above, these observations suggest, by inference, that the Sun and the preplanetary disk also originated together during collapse of the presolar nebula. On this basis, the presolar nebula can provisionally be identified as evolving from a molecular cloud core with a mass in the range of about one to a few $M_\odot$. Much remains to be learned about conditions and physical processes associated with the formation of 1 $M_\odot$ stars.

A. Influence of Angular Momentum on Development of the Sun and Disk

Molecular cloud cores generally rotate. Although the minimal average cloud rotation is set by the overall rotation of the Galaxy, higher rates of rotation are frequently observed, resulting from turbulent motions in the interstellar medium. As noted previously, observations show interstellar molecular clouds to have angular velocities ranging from $\Omega \sim 10^{-13}$ rad s^{-1} to probably $< 10^{-15}$ rad s^{-1} (Bodenheimer 1981). Rotation of small cores ($\lesssim$ 0.1 pc) of molecular clouds has, so far, been measured only for angular velocities $\gtrsim 2 \times 10^{-14}$ rad s^{-1}. Such a rotation rate was not revealed in approximately 30% of studied cores, suggesting that a substantial number of cloud cores may have angular velocities below this observational limit (Myers and Benson 1983; Ungerechts et al. 1982).

The value of the angular momentum J and its internal distribution $J(r)$ are important characteristics of a collapsing cloud core. In the absence of torques, these would be conserved during contraction, resulting in a significant increase in the ratio of the rotational energy to gravitational energy and leading to the possibility that rotation could have a significant influence on the subsequent evolution. Exploration of the internal structures of cores is still in a rudimentary state and further observations of cloud cores at high angular resolution are needed. Crude estimates of the mean radial density profiles give $\rho \sim r^{-p}$, with $1 < p < 3$, in the radial range from 0.1 to 0.5 pc.

More detailed information on density and angular velocity distributions within small cloud cores is still lacking (Myers 1985). For the purpose of making estimates in the following discussion, we will follow convention and use two limiting analytic descriptions of protostellar cloud cores:

$$\Omega = \text{const} \quad \text{for } \rho = \text{const} \tag{11a}$$

$$\Omega = \text{const} \quad \text{for } \rho \propto r^{-2}. \tag{11b}$$

The first case, Eq. (11a), describes the angular momentum distribution corresponding to a sphere of uniform density and homogeneous rotation. The latter relation, Eq. (11b), corresponds to a uniformly rotating, singular, isothermal sphere (Shu 1977) with $\rho = c_s^2/(2\pi G r^2)$, where c_s is the sound speed and G is the gravitational constant. The distribution in Eq. (11b) could result from redistribution of angular momentum in an isothermal core with gravity balanced by the pressure gradient. The entrained interstellar magnetic field and turbulence could, in principle, produce such a redistribution within the time scale of core evolution; however, this question requires further investigation.

It is clear that rotation does play an important role in star formation generally. Stellar statistics show that 60% or more of solar-type stars may belong to binary or multiple systems (Abt 1983). Typically, binary systems have angular momenta in excess of 10^{52} g cm^2 s^{-1}. A careful analysis of the angular momentum distributions of spectroscopic binaries reveals a deficiency of very close unevolved binary stars in which the component separation is below 10 $R_\odot$ and, therefore, having angular momenta lower than 10^{52} g cm^2 s^{-1} (Kraicheva et al. 1978). Note that the maximum possible angular momentum of a 1 $M_\odot$ star is about 10^{51} g cm^2 s^{-1} (Ruzmaikina 1981), approximately the angular momentum of the Sun with its present-day structure, but rotating at the centrifugal breakup velocity at its equator. Most T Tauri stars rotate significantly more slowly than this breakup angular velocity (Vogel and Kuhi 1981). Hence, during the formation of a binary star pair, a substantial fraction of the total angular momentum winds up in the orbital motion of the stars. The formation of a disk around a single star may be an alternative way for a star to shed excess angular momentum (chapter by Stauffer and Soderblom).

Reconstruction of the solar nebula by restoring the planets with H and He to the point of recovering the solar chemical composition results in a total mass ranging from 0.01 to 0.1 $M_\odot$, a total angular momentum in the range of 3×10^{51} to 2×10^{52} g cm^2 s^{-1}, and a projected surface density distribution proportional to $R^{-3/2}$, where R is the radial distance from the Sun (Weidenschilling 1977). It should also be noted that gravitational perturbations produced by the giant planets in the process of their accumulation could result in ejection of a substantial mass of solid material from the solar system, perhaps comparable with the total mass of the outer planets. Altogether then, one can estimate the minimum mass and minimum angular momentum of the solar nebula, respectively, to be $M \sim 0.02\ M_\odot$ and $J \sim 5 \times 10^{51}$ g cm^2 s^{-1}. These estimates pertain to a gaseous disk of radius $R_d = 10$ AU, i.e. about the radius of Saturn's orbit. Assuming that the projected surface density is indeed proportional to $R^{-3/2}$, then M and J increase as $R_d^{1/2}$ and R_d, respectively, where again R_d is disk radius. In any case, the relation $J_d/M_d >> J_\odot/M_\odot$, where the subscript d denotes the disk, must have applied to the protoplanetary nebula. The partition of angular momentum between the Sun and disk is substantially more inhomogeneous than the distributions given by Eqs. (11a,b). Therefore, either the protosolar nebula started with a very highly inhomogeneous specific angular momentum distribution (with specific angular momentum increasing outward even faster than for a rigidly rotating sphere) or major angular momentum redistribution has occurred between the Sun and the surrounding nebula.

For a system of two bodies with fixed total mass and angular momentum, the energy of the system decreases when mass is transported from the body of lower mass to the body of larger mass and the angular momentum is transported into the orbit of the lower mass body (Lynden-Bell and Pringle 1974). Hence, the dissipation of rotational energy during formation of a star from a diffuse rotating cloud naturally involves the transfer of angular momentum outward from the star to the surrounding disk. The efficiency of angular momentum transport during collapse of the presolar nebula is a key problem in understanding the formation of the Sun and protoplanetary disk. The existence of long-period stellar binaries and the dynamics of rotating cloud contraction show that angular momentum redistribution is not always effective enough to allow the formation of a single star and disk.

The early stage of protostellar collapse has been studied in a number of numerical simulations carried out for nonrotating clouds (see, for example, Larson 1969; Stahler et al. 1980). The initial collapse is characterized by freefall, isothermal contraction and a density increase in the cloud center. The gas in the center becomes optically thick and heated at density $\rho \sim 10^{-13}$ g cm^{-3}. (The outer region, with density $\rho \lesssim 10^{-13}$ g cm^{-3}, remains optically thin and collapses isothermally.) The increase of pressure associated with heating results in a temporary slowing of the contraction in the central region until the density reaches $\sim 10^{-7}$ g cm^{-3}. After this, a new stage of dynamical contraction occurs, caused by dissociation of the hydrogen molecules. This stage ends with the formation of a star-like core (the solar embryo, for the purpose of this discussion) in the inner part of the cloud. The embryo mass and central density fall in the range $M \sim 10^{-3}$ to 10^{-2} $M_\odot$ and $\rho \sim 10^{-2}$ g

cm^{-3}. (Subsequent calculations suggest the possibility of oscillations of the stellar embryo [Tscharnuter 1987]. Such oscillations could be important for the early evolution of the core; however, more complete analysis is needed to decide this question.) The core is surrounded by an envelope of infalling material. The time scale over which the core accumulates to about 1 $M_\odot$ ranges from 10^5 to 10^6 yr for an initial density of the presolar nebula close to the Jeans limit (10^{-19} g cm^{-3} for $T = 10$ K).

One can expect that rotating clouds contract in a fashion similar to non-rotating clouds, so long as the centrifugal force remains small in comparison with both self-gravity and the gradient of internal pressure. The relative importance of rotation is measured by the ratio of rotational and gravitational energies β. If the angular momentum is conserved locally during contraction, then β grows with density as $\rho^{1/3}$.

Two-and three-dimensional simulations of rotating clouds have been performed only for relatively early stages of collapse, prior to dissociation of molecular hydrogen. The simulations indicate that a cloud becomes unstable at some critical value of β (Bodenheimer 1981; Boss 1987b); instability apparently sets in when $\beta > 0.27$. It is likely that the further fate of a cloud depends on the stage at which the instability develops. If such an instability becomes important early during the first stage of hydrodynamic collapse, a cloud may evolve into a binary or multiple system. In order for an opaque region to form before fragmentation, the angular momentum of the cloud must be $\leq 10^{53}$ g cm^2 s^{-1} for the distribution given in Eq. (11a) (Safronov and Ruzmaikina 1978). Boss (1985) pointed out the possibility of efficient outward transport of the angular momentum from the center of the quasi-hydrostatic opaque region resulting from gravitational torque. Such torque could be produced by a spiral density wave generated in the envelope when the axial symmetry of the density distribution in the envelope is broken. Such angular momentum transport could suppress fission of the quasi-hydrostatically contracting region.

It is questionable, however, whether outward transfer of angular momentum within (or from) the opaque core can suppress fragmentation of the central part of the cloud during its further contraction. This ambiguity arises from the fact that, in the case of spherically symmetric contraction, dissociation of molecular hydrogen in the central part of the cloud is accompanied by its hydrodynamical collapse. During this phase of collapse, the cloud density increases by several orders of magnitude (5 orders of magnitude in the case when rotation is unimportant) and β can approach criticality, even if it were as small as a few times 10^{-3} at the beginning of the dissociational stage. If the central part of a cloud breaks into two or more separate fragments, the collapse may result in formation of a binary or multiple star system unless angular momentum is rapidly shed to the surrounding medium. Continuing accretion of the remaining envelope can only add angular momentum to this embryonic multiple star system, and tidal interaction of components can transport the rotational angular momentum into orbital motion.

Formation of a single star, such as the Sun, surrounded by a disk is plausible when the prestellar nebula (more exactly, its central part) rotates so slowly that a hydrostatically balanced, star-like core forms in the cloud center before β reaches the critical value (Ruzmaikina 1980, 1981). In that case, the initial mass, radius and central density of the core are likely to be close to those found for a nonrotating cloud, i.e., $M_c \sim 10^{-2}\,M_\odot$, $R_c \sim 3 \times 10^{11}$ cm, and $\rho \sim 10^{-2}$ g cm^{-3}, respectively (Larson 1969). Such a core is expected to contract slowly, with a time scale that ranges between 10^2 and 10^5 yr for initial core masses ranging from $10^{-2}M_\odot$ to 1 $M_\odot$. This time scale is significantly longer than the free-fall time. In addition, such a core forms in a state of differential rotation, with angular velocity increasing inward, and it is likely to possess a magnetic field of the strength 10 to 10^3 gauss, resulting from intensification of the interstellar magnetic field (Ruzmaikina 1985; also see the next section). The outward transport of angular momentum by magnetic torque can prevent a stellar core from breaking apart, producing instead an outflow of gas from the core equator to form an embryonic disk (Ruzmaikina 1980, 1985). (Note that this flow into the disk seems to occur when the core mass exceeds a certain value, that is larger the lower the angular momentum of the cloud. Outflow into the disk can barely occur if the cloud angular momentum per solar mass is less than about 5×10^{50} g cm^2 s^{-1} [Ruzmaikina 1981].) Another mechanism that might inhibit core fragmentation at this stage is generation of a spiral density wave in the solar nebula by a nonaxisymmetric core (Yuan and Cassen 1985). The spiral wave transports angular momentum out of the core on a time scale that is not longer than the time of core evolution; this inhibits growth of β in the core and, consequently, limits the amplitude of deviations from axial symmetry.

The formation of a single stellar core during the initial collapse imposes an upper limit on J. If the initial distribution of angular momentum within the cloud follows Eq. (11a), then the maximum angular momentum J_{cr} that allows formation of a single core is not well determined, but can be given, for total mass m, within the range

$$4 \times 10^{51} \left(\frac{M}{M_\odot}\right)^{5/3} < J_{cr} < 2 \times 10^{52} \left(\frac{M}{M_\odot}\right)^{5/3} \qquad (12a)$$

(Ruzmaikina 1981). For the distribution given by Eq. (11b) the value of J_{cr} exceeds that for the distribution in Eq. (11a) by a factor of approximately $(M_c/M)^{-4/3} \sim 500$, when $M_c \sim 10^{-2}\,M$, i.e.,

$$J_{cr} \gtrsim 2 \times 10^{54} \left(\frac{M}{M_\odot}\right)^3 \text{ g cm}^2 \text{ s}^{-1}. \qquad (12b)$$

In the case considered here for illustration, the rotation has no remarkable influence on the collapse of the central part of the cloud before core
formation. Further evolution of the collapsing cloud and the characteristics
of the resultant disk are strongly dependent on the total angular momentum
of the cloud and the efficiency of the angular momentum redistribution inside
the core and disk.

Material accreting from a sufficiently slowly rotating cloud onto a core
of mass M_c crosses the equatorial plane inside the centrifugal radius

$$R_k(t) = \frac{j^2(t)}{GM_c(t)} \qquad (13)$$

where j is the specific angular momentum of the accreting material about the
average rotational axis of the cloud (see, e.g., Cassen and Moosman 1981).
If the angular momentum is not redistributed, a disk begins to form when R_k
becomes greater than the equatorial radius of the radiative star-like core. (For
simplicity, we will identify the radius of the star-like core with that resulting
from collapse of a nonrotating cloud, $R_c \sim 3 \times 10^{11}$ cm [Stahler et al. 1980].
In fact, R_c would be somewhat larger owing to the effects of rotation.) At
time t, the angular momentum of a sphere containing mass $M = Mt$ (in the
case of constant accretion rate) depends on the distribution of angular momentum inside the cloud; thus

$$j(t) = \frac{5}{2}\frac{J}{M} \times \left[\frac{M_c(t)}{M}\right]^{2/3} \qquad (14a)$$

or

$$j(t) = \frac{9}{2}\frac{J}{M} \times \left[\frac{M_c(t)}{M}\right]^{2} \qquad (14b)$$

respectively, in the cases corresponding to the examples given in Eqs. (11a,
b) above, where $\rho = $ const or $\rho \propto r^{-2}$ (and $\Omega = $ const). Cassen and Moosman
(1981) described the evolution of disk surface mass density σ resulting directly from accretion of cloud material under the gravitational influence of
only the central star. They assumed a distribution of angular momentum in
the cloud corresponding to $\Omega = $ const and $\rho \propto r^{-2}$; here we use the same
approach to give the surface density evolution in the case when $\Omega = $ const
and $\rho = $ const:

$$\sigma(u,t) = \frac{3}{5\sqrt{10}\pi}R_k^{-3/4}(0)\left(\frac{M_c(t)}{M}\right)u^{-5/4}\int_u^1 S_1(x)x^{1/4}dx \qquad (15)$$

where

$$S_1(x) = \frac{(1-x^{1/2})(3-2x)}{4x(1-x)^{3/2}} \quad \text{and} \quad u = \frac{R}{R_k(t)}. \tag{16}$$

Taking Eqs. (13) and (14), and assuming that a disk begins to form when R_k > R_c, we find that a disk begins to form when the core mass attains a value of about

$$M_c(0) \geq \left(k^2 \frac{R_c}{R_0}\right)^q M \tag{17}$$

where $R_0 = J^2/GM^3$, $k = 2/3$, and $q = 3$ when $\rho = $ const, and $k = 2/9$ and $q = 1/3$ when $\rho \propto r^{-2}$. (Again, homogeneous rotation is assumed in both cases.) Prior to the time when R_k exceeds R_c in this calculation, all mass is assumed to fall onto the core.

In order to see the implications of Eq. (17), take $M = 1$ $M_\odot$ and assume that the angular momentum J falls between the minimal value for the presolar nebula $J_{min} \sim 10^{51}$ g cm^2 s^{-1} and the critical value for instability J_{cr} as given in Eqs. (12 a, b). In the case (a), with $\rho = $ const, we find that $R_k < R_c$ for all $M_c(0) < M$, if $J = J_{min}$ and $M_c(0) \sim 0.017$ M if J is about 10^{52} g cm^2 s^{-1}. In the case (b), we find that $R_k < R_c$ when $M_c(t) \sim 0.96 - 0.046$ M if J is in the range between 3×10^{51} to 10^{54} g cm^2 s^{-1}. To the end of accretion, when $\dot{M}t = M$, the bulk of the cloud mass remains in the core if $J \sim J_{min}$ and, hence, $R_c > R_0$. In the opposite case, $R_c << R_k(0)$, the final mass of the core is

$$M_c = 0.36 \, M \left(\frac{R_c}{R_0}\right)^{1/3} \tag{18a}$$

in the case when $\rho \propto r^{-2}$ (Cassen and Summers 1983), and

$$M_c = 0.5 \, M \left(\frac{R_c}{R_0}\right)^3 \tag{18b}$$

in the case when $\rho = $ const. The mass of the disk is given by $M_d = M - M_c$.

The subsequent redistribution of angular momentum in the core and disk can have significant effects on their physical states and on their dynamical evolution. The energetic profitability of disk formation implies that rotational energy is dissipated, resulting in the net flow of matter inward and the net flow of angular momentum outward. Three mechanisms of redistribution of the angular momentum (and mass) within disks are thought to be the main candidates to drive this dynamical evolution: turbulent viscosity, magnetic forces and gravitational torques.

Several reasons have been advanced for believing that the protosolar disk was turbulent. Cameron (1962) suggested that shear motion between the disk and accreting envelope (with fluid Reynolds number $\mathcal{R}_e > 10^{10}$) during the accretional stage induced turbulence. Lin and Papaloizou (1980) developed the possibility that thermal convective instability in the local vertical direction would be induced by high opacity from dust grains after clearing of the envelope. But the theory of turbulence that is needed for quantitative estimation of turbulent viscosity in disks is still under investigation. In the absence of an adequate theory of turbulent viscosity, v_t is usually taken in a parameterized form suggested by Shakura and Sunyaev (1973)

$$v_t = \alpha_t c_s h = \alpha_t c_s^2 \, \Omega^{-1} \tag{19}$$

where $\alpha_t < 1$ is a dimensionless constant, c_s is the sound speed, h is the scale height of the disk, and Ω is the angular velocity of Keplerian motion. A value $\alpha_t \sim 10^{-2}$ seems a reasonable assumption during the accretional stage, although this assumption is admittedly arbitrary. Experiments on supersonic shear flow reveal that a few percent of the kinetic energy of the flow goes into excitation of turbulence (Morkovin 1972). This amount of energy is likely to be sufficient to support turbulence with $\alpha_t \sim 10^{-2}$. Such turbulence would result in a time scale of redistribution of the angular momentum of about 10^5 yr over a length scale comparable to the present-day radius of the solar system (Ruzmaikina 1980). It is not clear, however, whether shear flow associated with accretion can really make the disk turbulent throughout its whole thickness or only in a subsurface layer and at the inner and outer edges of the disk. The most carefully developed analysis of turbulent viscosity and related disk models is based on a theory of convective instability (Lin and Papaloizou 1980) and indicates a somewhat less effective transport of angular momentum, with a substantially longer time scale of the order of 10^6 yr (Cabot et al. 1987a,b).

The outer regions of disks may be susceptible to a Rayleigh instability because the specific angular momentum of material accreting there would have been less than the local specific angular momentum of the disk's Keplerian motion (once a disk has expanded beyond the centrifugal radius of the accreting material). It has also been suggested that, near the inner edge of a disk, turbulence could be generated in an intense boundary-shear layer; however, the dynamical characteristics of such a layer are not understood, and it is not clear whether intense shear actually occurs.

Gravitational torque may also be important in redistributing angular momentum in the outer part of a massive disk; when $M_d > M_c$, the disk can become gravitationally unstable (Cameron 1978; Cassen et al. 1980). A disk of moderate mass $M_d \sim M_\odot$ will be unstable to the growth of nonaxisymmetric distortions having growth rates comparable to the orbital frequency near the outer edge of the disk (Adams et al. 1989). Such instabilities can generate transient, spiral, shear waves that can redistribute angular momentum in the disk. Lin and Pringle (1987) derived a time scale for such angular momentum redistribution in the vicinity of $10 \, \Omega^{-1}$. However, it is not clear whether angular momentum is transported over large distances in the radial direction. Vertical temperature stratification in the disk can refract such waves to high altitudes, where they may steepen and dissipate (Lin et al. 1989). Adams et al. (1989) suggest that spiral waves with azimuthal wave number $m = 1$ may be most effective at transporting angular momentum over large radial distances. These are eccentric modes, in which the star and the disturbance orbit about their common center of mass. The unstable modes can encompass the entire disk and be driven nonlinearly by these eccentric motions.

As its mass decreases, a disk becomes less susceptible to gravitational instabilities. Whether low-mass disks exhibit spiral-wave gravitational instabilities at a dynamically important level is not yet clear. However, the variety of theoretical possibilities is very large. For example, Adams et al. (1989) suggest the possibility that, in a disk where the temperature profile results from the instability itself, nonlinear coupling could drive unstable modes even in a disk with mass much less than that of the central star. Sekiya and Miyama (1989) found that shear instability could occur in a nearly Keplerian disk. Growth times for such instability have been estimated to fall in the range 10^3 to 10^4 rotation periods. The effectiveness of these modes for transporting angular momentum remains to be determined.

Magnetic fields amplified during collapse of a cloud or generated by a hydromagnetic dynamo can play a role in the evolution of the protosolar disk (Lüst and Schlüter 1955; Hoyle 1960; Alfvén and Arrhenius 1976; Levy 1978; Levy and Sonett 1978; Ruzmaikina 1980, 1985; Hayashi 1981; Stepinski and Levy 1988). This is surely so for the central region of the protosolar nebula, where the temperature is high enough for the evaporation of dust particles and thermal ionization of the alkaline metals (≥ 1600 K). According to some models, the radius of this region changes during star formation and early evolution of the protosolar nebula from a few tenths to about 1 AU (Ruzmaikina and Maeva 1986). At larger distances, it is generally thought that the degree of ionization near the central plane would have been much lower, because the thermal ionization would have been negligible and the disk opaque to external sources of ionizing radiation, although during some episodes of rapid evolution, indicated by the sporadic FU Orionis phenomenon, high temperatures and thermal ionization could extend to much larger distances from the central star. With respect to internal ionization from the decay of radioactive nuclides, the abundance of ^{40}K was too low to create a high ionization fraction in the dusty gas. The only sufficient sources of ionization likely to have been present in this part of the nebula are the short-lived radioactive nuclides such as ^{26}Al. However, it is not clear just how high an ionization would have been produced. As a consequence, it is not clear to what extent the magnetic field and gas were decoupled at these larger distances from the center of the nebula; the role of magnetic field in angular momentum transport in these regions remains a question.

Meteorite Magnetization.   The discovery of remanent magnetization of meteorites suggests the existence of a protosolar nebular magnetic field (Levy 1978; Levy and Sonett 1978). The typical intensities of the inferred magnetizing fields range over some 2 orders of magnitude. For chondrules in the Allende meteorite (a type 3 carbonaceous chondrite), the inferred magnetizing fields range from 3 to 7 Gauss. Magnetization of achondrite meteorites typically indicate magnetizing field intensities of 0.1 Gauss or less. Remanent magnetization of ordinary chondrites seems to have been acquired in fields of about 1 Gauss (Cisowski 1987; chapter by Cisowski and Hood). Randomly directed magnetization revealed in the Olivenz chondrite on scales of ~ 1 mm or less are most easily interpreted as resulting from the accretion of previously magnetized pebbles (Collinson 1985); but the possibility cannot be excluded that randomization resulted from post-accretional mixing in a regolith-like layer.

     It is not clear at this time whether the large dispersion of inferred magnetizing field intensities is a result of the evolution of the magnetizing field or a result of other aspects of the mechanisms through which the meteorite material became magnetized. Generally speaking, objects magnetized in the nebula can be expected to have been rotating with respect to the ambient magnetic field. If the magnetization was acquired over a period of time long in comparison with the rotation period, then only the projected DC component of the magnetic field would have been effective in producing residual magnetization in the matter. Therefore, even in the simplest case, with a homogeneous magnetic field, a randomly rotating ensemble of objects would be expected to display a *distribution* of apparent magnetizing field intensities. In the real nebula, however, the field would not even have been homogeneous; therefore the actual distribution of inferred magnetizing intensities is expected to be complicated. Moreover, it is likely that the magnetizing mechanisms differ among the different meteorite classes, including, not only thermal and chemical mechanisms, but also mechanical magnetizing processes that work during parent-body accretion in the presence of a magnetic field. Because of these complexities, a detailed interpretation of the wide distribution of inferred meteorite magnetizing fields, both among different varieties of meteorite material and even among different samples of the same type of material, has not yet been accomplished.

     Despite these ambiguities in interpreting the variation of inferred magnetizing field intensities, it remains attractive to believe that meteorites and their parent bodies were magnetized in the protosolar nebula. It is important to keep in mind that the complex geometrical and physical processes involved in the acquisition of magnetization by a rotating body would act to diminish the apparent intensity of the magnetizing field, distributing the apparent intensities downward from the ambient field. Therefore, it is likely that the maximum inferred magnetizing intensities within each meteorite class are the ones with the most important physical significance for understanding nebula processes, and it is most important that these be verified and understood.

Meteorites are believed to originate in the asteroid belt, between 2 and 4.5 AU from the Sun. An outstanding question is whether all meteoritic material actually formed in that region, as is generally assumed, or whether some material was processed closer to or farther from the Sun (Morfill et al. 1985; Ruzmaikina et al. 1989), and later transported to the present asteroid belt. This question is, of course, of great importance for interpreting the meteorite magnetization record in terms of nebular and early solar dynamical processes.

Electrical conductivity of the gas is a critical issue in understanding the behavior of magnetic fields in a disk; significant magnetic effects require that the electrical conductivity be high enough to couple the field and gas. Restoration of the early minimal-mass solar nebula, obtained by adding to the planets the H and He that presumably was lost from the nebula, and distributing this material smoothly, results in a surface density $\sigma \gtrsim 10^3$ g cm^{-2} in the region $R \lesssim 1$ AU and, possibly, in the region of Mars and in the asteroid belt. Such a dense layer of gas shields its interior from extrinsic ionization, such as that caused by galactic cosmic rays and stellar photons. Leaving aside for now the possibility of ionization from intrinsic radioactivities such as ^{26}Al, the gas and magnetic field could have been closely coupled in the surface layers of the disk, containing a fraction of the projected mass density, and in the far outer regions of the disk, where the density was low enough that ionizing cosmic radiation could penetrate.

Take the characteristic dynamical time of fluid motion to be the Kepler rotation period, and the characteristic length scale of the magnetic field to be l_B. If the magnetic field and gas are to be closely coupled, then the magnetic Reynolds number must exceed unity. Writing this condition in terms of the ionization fraction χ, we obtain

$$\chi > \frac{m_e c^2}{4\pi e^2} \frac{\langle \sigma_{en} v_e \rangle}{\sqrt{GMR}} \frac{R}{l_B} \qquad (20)$$

for the condition that the field and gas will be closely coupled. The degree of ionization is determined by a balance between the rate of ionization ζ by cosmic rays ($\zeta \sim 10^{-17}$ s^{-1} in present-day molecular clouds; Nakano and Umebayashi 1980) or by intrinsic radioactive sources, and electron losses to molecular recombination and accretion on dust grains. The minimum rate of ionization in the presolar nebula is the ionization produced by ^{40}K, with $\zeta \sim$

2×10^{-21} s^{-1} (Cameron 1962). Assuming that electron/ion loss is controlled by dust, the ionization equilibrium becomes, very crudely,

$$\chi \sim \frac{n_i}{2n} = \frac{\zeta}{2n_d\sigma_d v_i} \qquad (21)$$

where n_d and σ_d are the number density and cross section of the dust particles, and v_i is the thermal velocity of ions.

It follows from Eqs. (20) and (21) that the magnetic field and gas are closely coupled in regions where the density of the gas is less than a critical density ρ', where

$$\rho' \sim \frac{4\pi e^2}{m_e c^2} \frac{\sqrt{GMR}}{\langle \sigma_{ne} v_e \rangle} \frac{\mu \zeta m_H}{(n_d/n)v_i\sigma_d} \frac{l_B}{R}. \qquad (22)$$

Now, assuming that the galactic cosmic-ray flux can reach the gas, so that $\zeta \sim 10^{-17}$s^{-1}, and taking $T \sim 300$ K, $\mu = 2$, $l_B \sim R \sim 1$ AU, $n_d/n \sim 10^{-13}$, $\sigma_d \sim 10^{-9}$ cm^2, $\langle \sigma_{ne} v_e \rangle \sim 4 \times 10^{-9}$ cm^3 s^{-1}, $M \sim 1$ M$_\odot$, and $v_i \sim c_s \sim 10^5$ cm s^{-1}, we obtain $\rho' \sim 10^{-8}$ g cm^{-3}. (It is worth noting that, in the minimal-mass solar nebula, the central-plane density is smaller than ρ' throughout.)

Altogether, the preceding discussion suggests that all parts of the proto-solar disk to which the galactic cosmic-ray flux could penetrate could have had an electrical conductivity high enough to couple the field and gas (except in the case of a very massive disk). But, as we already have noted, with some 10^3 g cm^{-2} of disk material intervening, the nebular midplane would have been shielded from galactic cosmic rays, which can penetrate only of the order 10^2 g cm^{-2}. Now consider that the density of nebular gas falls rapidly away from the midplane, as exp $(-z^2/\Lambda^2)$, where Λ is the local scale height. Thus, assuming that the cosmic rays are not otherwise shielded by surrounding material, the energetic particles could penetrate to within 1 to 1.5 scale heights of the midplane. At such altitudes the local mass density would fall to between 1/3 and 1/10 of the midplane value. This region of the disk (outward of 1 to 1.5 scale heights from the midplane) defines a region likely to be magnetohydrodynamically active, even under the conservative assumption that cosmic rays provide the only ionization source.

Consider the possibility that Keplerian differential rotation of this electrically conducting region could stretch a solar magnetic field to generate an azimuthal field with the amplitude inferred from meteorite remanence. The maximum intensity of azimuthal magnetic field that can be generated by differential rotation is limited by Ohmic dissipation or by ambipolar diffusion, whichever gives the smaller limit. In the presence of an axially symmetric poloidal magnetic field B_p, and differential rotation characterized by magnetic Reynolds number $\mathcal{R}_m$, Ohmic dissipation limits the azimuthal field B_ϕ according to

$$B_\phi \sim \mathcal{R}_m B_p \approx \frac{\rho'}{\rho} B_p \qquad (23)$$

where the final form of the relationship follows from the definition of ρ'. However, this limit can only be attained if the time scale of amplification by the Keplerian shear, $\sim (R\, d\Omega/dR)^{-1}$, is shorter than the time scale of ambipolar diffusion given in Eq. (8). For the situation discussed here, ambipolar diffusion limits the field strength according to:

$$B^2 \lesssim \frac{8\mu_{ni}\rho\sqrt{GMR}}{3\mu m_H} \frac{\zeta}{(n_d/n)} \frac{\langle \sigma_n v_i \rangle}{\langle \sigma_d v_i \rangle} \frac{l_B}{R} \lesssim 10^{10}\, \rho\, G^2 \qquad (24)$$

where the numerical estimation uses the same parameter values used earlier to estimate ρ'. Plausible values for ρ in the high-electrical-conductivity surface layers could be of the order of 10^{-10} g cm^{-3}, so that the ambipolar limit on magnetic field with intensity could be of the order of 1 gauss (Levy 1978).

Now suppose that the nebular magnetic field was generated by the winding up of a weak, ambient, poloidal field rooted in the Sun. The field from an axially aligned solar dipole has no component in the direction of Keplerian shear; therefore, such a field cannot effectively provide a seed from which a strong azimuthal field could be generated. If the dipole were tilted, then the radial component of dipole field in the plane of the disk would lie in the direction of the shear; but this field would reverse in each half rotation, and thus also would not result in generation of a strong azimuthal field. If the Sun had a quadrupole magnetic field, however, this would produce a radial magnetic field in the nebula that could, in principle, be amplified by differential rotation. The poloidal component of such a field would vary with distance from the Sun as

$$B_p = B_{p\odot} \left(\frac{R_{p\odot}}{R} \right)^4 \qquad (25)$$

where $B_{p\odot}$ is the quadrupole magnetic field intensity at the surface of the proto-Sun and $R_{p\odot} = 3 \times 10^{11}$ cm is its radius. For numerical estimation, take $B_{p\odot}$ even as high as 10^4 gauss. Then at a distance of 3 AU, $B_p \sim 10^{-5}$ gauss. Because $\mathcal{R}_m$ is unlikely to exceed a few tens, differential rotation of the disk is unlikely to have amplified such a solar magnetic field by the factor of 10^4 or more that is needed to account for the remanent magnetization of meteorites.

Now consider, very roughly, the possibility of a dynamo acting in the outer layers of the nebula. In the case when the solar nebula is turbulent, the generation of a magnetic field by a hydromagnetic dynamo is possible in the outer layers, where $\mathcal{R}_m = \rho'/\rho \gg 1$. Crudely, for a dynamo mechanism to operate, the dynamo number $D = \alpha G_\Omega R^3/\eta_t^2$ must exceed some critical value, of the order of 10 (where, in this context, $\alpha \sim v_t$ is the average helicity of the flow and $G_\Omega = -R\partial\Omega/\partial R$ measures the differential rotation).

In the case of Keplerian rotation, and taking the scale height for the vertical density distribution to be $h \approx c_s\Omega^{-1} \ll R$, the dynamo number D can be written the form

$$D \approx \frac{27}{2} \frac{\Omega R}{v_t} \left(\frac{R}{l_t}\right)^2 \approx \frac{27}{2}\alpha_t^{-1} \left(\frac{R}{h}\right)^3 \left(\frac{h}{l_t}\right) \approx 10^5 \, \alpha_t^{-1} \qquad (26)$$

We have seen earlier that in the surface layers of the nebula, where $\rho \lesssim 10^{-10}$ g cm^{-3}, the magnetic Reynolds number is high enough that the field and gas are well coupled. Inasmuch as $\alpha_t < 1$ by definition, Eq. (26) indicates that under such circumstances the dynamo number can be very large. Even making the conservative assumption that there are no significant ionization sources in the dense midplane regions of the nebula (cf. Consolmagno and Jokipii 1978), it is likely that some layers of the nebula—with local mass density $\sim 10^{-10}$ g cm^{-3}—could have generated a dynamo magnetic field. In Eq. (24), we estimated one dynamical limitation on the strength of such a field, of the order of 1 Gauss, imposed by ambipolar diffusion. Another limitation is likely to be imposed by a balance between the Lorentz and Coriolis forces, the latter of which motivates the crucial helicity of the turbulence. Thus

$$B_\phi B_p \lesssim \rho\Omega h v_t \qquad (27)$$

where the scale height h is assumed to be the characteristic scale of variation of the magnetic field, and v_t is the characteristic velocity of turbulent motions. With $\rho \sim 10^{-10}$ g cm^{-3}, and making use of Eq. (23), we find again that the magnetic field strength is limited to of the order of 1 Gauss.

Altogether, while many uncertainties remain, these crude estimates suggest that a magnetic field with intensity as high as 1 Gauss in the nebula could plausibly have been generated at distances from the Sun comparable to the location of the present-day asteroid belt (Levy 1978); such a field could account for the typical magnetizing fields inferred from measurements of meteorite remanence. However, much work remains to decide whether a fully consistent picture can be constructed of such a nebular magnetic field. A principle and critical question involves the electrical conductivity of the gas, especially its distribution with respect to altitude above the nebular midplane. If the only adequate source of ionization were cosmic rays, then only the "surface" layers of the disk would have had high enough ionization to generate a field. In that case, the toroidal component of the field would have been limited to the region above a scale height or two from the disk midplane, and could not have penetrated into a lower-electrical-conductivity region near the midplane. However, the poloidal field generated in such a dynamo would have penetrated to the midplane in much the same way that the Earth's magnetic field protrudes out of the planet's electrically conducting core.

It is important to realize that, while the typical magnetization of meteorites seems at least plausibly accountable on the basis of a dynamo field generated in a canonical protoplanetary-nebula model, the strongest of the inferred magnetic fields (ranging to the order of 10 Gauss) are less easily accounted for. One must consider at least two possibilities: on the one hand, these singularly high remanence measurements made on some small components of meteorites may be flawed; on the other hand, it is possible that the high fields occurred in some kind episodic phases during the nebula's evolution.

The maximum magnetic field intensities that we have discussed here would exert a magnetic pressure that can be substantial in comparison with the gas pressure (Levy 1989). If the magnetic pressure were to grow to a significant fraction of the total gas pressure, the disk would become unstable to the buoyant loss of magnetic field (Parker 1979a). In this case, loops of magnetic field could create and fill a corona above the disk. Such a corona could be the site of energetic flares and could produce wind-like mass loss from the disk (Levy and Araki 1989).

Summary. Whatever the original state of the protosolar cloud, redistribution of angular momentum in the forming proto-Sun and disk was likely to have played an important role in adjusting the structure of the early solar system, resulting in the present-day configuration of the Sun and planets. If the Sun formed from a relatively low-mass, low-angular-momentum cloud, then the outward flow of angular momentum and mass would have initiated the formation of an embryo disk from the outer equatorial region of the proto-Sun, followed by spreading of the disk to the present-day extent of the solar system, in excess of forty astronomical units. This could have occurred even in a low-angular-momentum nebula ($J < 10^{52}$ g cm^2 s^{-1}), for which R_k is very much less than the radius of the solar system (Ruzmaikina 1980; Cassen and Moosman 1981; Cassen and Summers 1983; Ruzmaikina and Maeva 1986). The ultimate dissipation of the disk may also have been influenced by similar angular momentum transport and rotational energy dissipation (Lynden-Bell and Pringle 1974; Ruden and Lin 1986). If, on the other hand, the Sun formed from a higher-mass and higher-angular-momentum precursor, then angular momentum transport would have been essential for the formation of the Sun as a central mass condensation. Magnetic fields provide one effective mechanism for transporting angular momentum. Astrophyical observations have indicated that magnetic fields are common in many such systems. The meteorite evidence from our own solar system suggests that dynamically significant magnetic fields were present in the protoplanetary nebula, which could have had an important effect on the nebula's evolution (Levy 1978, 1989).

A more detailed consideration of the joint formation of the Sun and solar nebular is beyond the scope of this chapter. However, we close this section with two observations that may be pertinent to the history of the solar magnetic field. First, if the angular momentum of the presolar cloud core was as low as indicated in the previous discussion, then, for the purpose of understanding the behavior of the magnetic field in the forming Sun, it suffices to consider the collapse of an approximately spherical, nonrotating system— this is the approach that we take in the next section, where we discuss gross characteristics of magnetic field behavior in a collapsing nebula with 10^{52} g cm^2 s^{-1} of angular momentum.

If disk material accreted back onto the Sun from the disk during the disk's late evolution, then the inner part of the Sun and the outer layers may have had significantly different evolutionary histories. In principle, this could result in burying a magnetic field in the Sun's interior, isolating it from the outer layers. Such a field configuration has been invoked in attempts to reconcile solar differential rotation measurements with the presence of a magnetic field.

IV. MAGNETIC FIELD OF THE PROTO-SUN

As discussed in the first section, a magnetic field in the core of a molecular cloud is partially decoupled from the gas because of the low degree of ionization. Thus, the total magnetic flux through the equatorial plane of a cloud core decreases during contraction. Moreover, the magnetic field stresses do not grow rapidly enough to prevent hydrodynamical collapse (Mestel 1965; Mestel and Paris 1979). Ambipolar diffusion (Mestel and Spitzer 1956) and ohmic dissipation are the main mechanisms of decay of the magnetic flux in such a collapsing cloud (Black and Scott 1982; Ruzmaikina 1980, 1985; Nakano and Umebayashi 1986). The behavior of the magnetic field is determined by the relation between these time scales and the contraction time scale τ_d

$$\tau_d = \frac{l}{V} \tag{28}$$

where l is the spatial scale and V is velocity of the flow of the gas. τ_d depends on the time scale and geometry of contraction of the presolar nebula, and hence on its precollapse history and the angular momentum. In the case considered in the previous section (with rotation assumed unimportant until formation of a star-like core), the initial stages of the spherically symmetric collapse provide a reasonable approximation on which to base investigation of the evolution of the magnetic field of a weakly magnetized initially presolar nebula.

The evolution of magnetic field intensity at the center of a spherically symmetric collapsing nonturbulent and nonrotating cloud is shown in Fig. 1. In the early stages of collapse, the behavior of the magnetic field is dominated by the drift of ions through the neutrals, i.e., ambipolar diffusion. Assuming that the strength of the magnetic field is limited by ambipolar diffusion, we can use Eq. (8) with Eq. (28) to obtain

$$B^2 \approx 8\pi\mu_{ni}n^2 \langle\sigma_{ni}v\rangle\chi Vl_B. \tag{29}$$

If the collapse starts from a homogeneous density, then during the initial stages of collapse the density distribution evolves toward $\rho \sim r^{-2}$. In the first approximation, the magnetic field can be taken to be coupled to the ionized component of the gas, i.e., $Bl_B^2 = $ const, where l_B is the length scale of the magnetic field. Under these circumstances the intensity of the magnetic field increases as

$$B \sim \rho^{0.4} \tag{30}$$

(Ruzmaikina 1985). A similar relation ($B \sim \rho^{0.44}$) was obtained by Black and Scott (1982) in two-dimensional calculations of collapse of magnetized clouds with ambipolar diffusion.

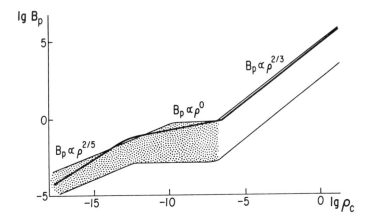

Fig. 1. Evolution of magnetic field intensity at the center of a spherically symmetric, collapsing cloud in the absence of rotation and turbulence. The shaded region follows the magnetic field evolution for ionization rates ξ between 10^{-21} and 10^{-17} s^{-1} (Ruzmaikina 1985). The dark line shows the field evolution for $\xi \sim 6 \times 10^{-18}$ s^{-1} (Dudurov et al. 1989).

When the density increases to

$$\rho = \frac{\mu e}{c} \left(\frac{m_H}{m_e}\right) \frac{B}{\sqrt{\langle\sigma v\rangle_{en}\langle\sigma v\rangle_{in}}} = 10^{-10}\, B\ \mathrm{g\ cm^{-3}} \tag{31}$$

(where B is measured in gauss), ohmic dissipation begins to dominate the behavior of the magnetic field. The time scale of ohmic dissipation is proportional to l_B^2 and is independent of the magnetic field strength. Therefore, after this stage, the length scale of the magnetic field in the inner part increases, and the field tends toward uniformity, with an intensity determined by the intensity at the inner edge of the region where the density satisfies Eq. (31).

At a density of 10^{-7} g cm^{-3}, the temperature reaches ~ 1600 K, the dust evaporates, alkaline metals are thermally ionized, and the magnetic field becomes coupled to the contracting gas. At this stage of contraction, the magnetic field intensifies by a factor of 10^3 to 10^4. This evolutionary sequence results in a star-like core with a central magnetic field intensity of 10 to 10^3 Gauss (Ruzmaikina 1985; Dudurov et al. 1989).

In the absence of turbulence, the poloidal magnetic field would be frozen into the core material during the time of accretion, and compressed by a factor of about 500, as the density increased from 10^{-2} to 10^2 g cm^{-3}. Figure 2 shows the central poloidal magnetic field intensity as a function of the radius of a nonturbulent proto-Sun approaching the main sequence (Dudurov et al. 1989).

Amplification of the toroidal magnetic field would be produced by differential rotation in the forming proto-Sun. The proto-Sun could form in a state of differential rotation because of the strong concentration of density to the center (much stronger than in the precursor molecular cloud core) and because of the tendency to conserve angular momentum during collapse. For example, assuming that angular momentum was carried in with the collapsing material from an initial state of uniform density and uniform angular velocity, a star-like core would have formed rotating with an angular velocity of 10^{-4} s^{-1} in the center (where $\rho_c \sim 10^{-2}$ g cm^{-3}) and 10^{-6} s^{-1} near the surface (where $\rho \sim 10^{-5}$ g cm^{-3}), provided that the initial presolar nebula had a total angular momentum of 10^{52} g cm^2 s^{-1}. The maximum intensity of a toroidal magnetic field generated by such nonuniform rotation is given by

$$B_\phi = \mathcal{R}_m^q\, B_p \tag{32}$$

where $q = 1$ if the poloidal magnetic field is aligned with the axis of rotation and $q = 1/3$ if it is perpendicular (Parker 1979a). $\mathcal{R}_m$ is the magnetic Reynolds number associated with the differential rotation

$$\mathcal{R}_m = \frac{4\pi\sigma_e}{c^2} l_B^2 \Omega. \tag{33}$$

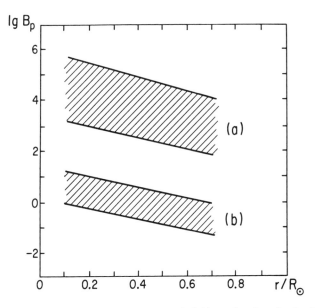

Fig. 2. Intensity of the solar internal poloidal magnetic field as a function of radius in the early Sun: (a) assuming no turbulence until the Sun has arrived on the main sequence; and (b) assuming evolution through a Hayashi turbulent stage according to the description given in the text.

For fully ionized gas in the core $\sigma_e = 10^7 T^{3/2}$ s^{-1}. Taking, for the core, $l_B \sim R \sim 10^{12}$ cm, $\Omega \sim 10^{-6}$ s^{-1} and $T \sim 10^4$K, we obtain $\mathcal{R}_m \sim 10^{11}$. The amplification of the magnetic field to its maximal value takes $\mathcal{R}_m$ and $\mathcal{R}_m^{1/3}$ rotational periods, which amounts to 10^{10} yr and about 10^3 yr for magnetic field parallel and perpendicular, respectively, to the rotation axis. Thus, differential rotation, if sustained in some way in a nonturbulent core, could increase B_ϕ throughout the entire time of accretion, resulting in the possibility of strong toroidal magnetic fields. In fact, however, transport of angular momentum and/or mixing of the inner regions of the proto-Sun can stop the amplification of the field much earlier. The time scale of redistribution of the angular momentum is

$$\tau_B \sim \frac{R}{V_A} \tag{34}$$

where V_A is the Alfvén velocity determined by poloidal component of the magnetic field, which is some 10^2 to 10^4 Gauss in the newly formed core, if the rate of ionization in the presolar nebular was between 10^{-17} and 10^{-21} s^{-1}.

The outer convection zone of the Sun has a short magnetic field retention time, as will be discussed presently. However, the radiative core, below 0.7 $R_\odot$, could retain a primordial magnetic field, resulting from enhancement of the interstellar magnetic field if the Sun did not evolve through a turbulent stage. Theoretical studies of the evolution of 1 $M_\odot$ stars approaching the main sequence indicate the occurrence of a fully convectively mixed (Hayashi) stage. The convection is likely to have been turbulent because of the large Reynolds number associated with the fluid motions. Turbulence tangles the magnetic field, reduces l_B and hastens the field's dissipation. However, in the presence of differential rotation left over from an earlier epoch, or supported by convection, and density gradients, the turbulence can also generate a magnetic field through dynamo action if $\Re_m \gg 1$. The rate of turbulent energy dissipation is approximately equal to the luminosity of the proto-Sun because the turbulent convection transports the heat from the central part to the surface. Thus

$$L_{p\odot} \sim \frac{\rho v_t^3 \mathcal{V}_t}{l_t} \tag{35}$$

where l_t is the main scale of turbulence and $\mathcal{V}_t$ is the volume of the turbulent region. It follows from Eq. (35) that

$$v_t \sim \left(\frac{L_{p\odot} l_t}{M_t}\right)^{1/3} \tag{36}$$

where $M_t = \mathcal{V}_t \rho$ is the mass of turbulent region.

As was noted earlier, B_ϕ and B_p for the turbulent $\alpha\omega$-dynamo can be estimated from Eq. (27), yielding

$$B_p = \left[\frac{8\pi}{3} h\rho\right]^{1/2} \left[\frac{l_t}{R}\right]^{1/2} \left[\frac{\Omega}{R}\right]^{1/4} v_t^{3/4} \tag{37a}$$

$$B_\phi = [24\pi h\rho]^{1/2} \left[\frac{R}{l_t}\right]^{1/2} [\Omega^3 R]^{1/4} v_t^{1/4}. \tag{37b}$$

In Eq. (37) we used $B_\phi \sim D^{1/2} B_p$. For turbulent velocities determined by Eq. (36) and parameters corresponding to a protostar of about 1 $M_\odot$ at the Hayashi stage: $L \sim 10 L_\odot$, $R = 3$ to $5 R_\odot$, $M_t = M_\odot$, and $\Omega \sim 3 \times 10^{-6}$ to 10^{-4} s^{-1}, which are typical for T Tauri stars, one obtains $B_p \sim 10^4$ to 10^5 Gauss and $B_\phi \sim 10^5$ to 10^6 Gauss in the core region, where $\rho \sim 10$ g cm^{-3}, and $B_p \sim 3 \times 10^3$ to 3×10^4 Gauss and $B_\phi \sim 3 \times 10^4$ to 3×10^5 Gauss under the bottom of the convective zone if $l_t \sim h \sim 0.1 R_\odot$ where $\rho \sim 10^{-1}$ g cm^{-3}. In the foregoing, we have assumed that the magnetic field intensities are determined by local kinematical and dynamical conditions. These results are not strongly dependent on the assumed ratio l_t/h. Mestel and Weiss (1987) arrived at a similar conclusion from a phenomenological argument. They noted that a magnetic field of about 500 Gauss (observed in one T Tauri star) would be compressed to a value of 5×10^3 Gauss if magnetic flux were conserved during the star's evolution to the main sequence, in the absence of a turbulent convective stage.

The mass density at the Sun's present photosphere is about $10^{-6.5}$ g cm^{-3}. Assuming that a T Tauri photosphere is similar to the present solar photosphere, and assuming that we can apply local dynamical balance in the form given above, it follows from Eq. (37), that the intensity of the poloidal magnetic field would be approximately 1/300 that at the base of the convection zone. This photospheric magnetic field would control the magnetic field in the star's corona and wind. Observations reveal the activity of T Tauri stars, which presumably is associated with a strong magnetic field. Flare emissions are apparently produced by thermal bremsstrahlung in hot plasma associated with solar-like flares (chapter by Feigelson et al.). Some T Tauri stars show the presence of large star spots covering more than 20% of the surfaces (Bouvier 1987). Also, magnetic braking is assumed to be responsible for rapid rotational deceleration of newly formed stars.

It has been suggested that a wind from such a magnetized star could magnetize protometeoritic material. First assume that the wind-born magnetic field falls off with distance from the Sun as r^{-2}. If such a magnetic field had an intensity of 0.1 to 1 Gauss in the asteroid belt, then that would imply a magnetic field intensity of 10^3 to 10^4 Gauss at a T Tauri photosphere, with radius taken to be about 3×10^{11} cm, a not unreasonable magnetic intensity. In fact, the magnetic field entrained in such a wind flowing from a rotating star would decrease with distance from the Sun even less rapidly than r^{-2}, because the field winds to a tight spiral at large distances from the Sun. It is possible, therefore, that at least some meteorite magnetization may have resulted from such an extended solar magnetic field. Such a magnetic field has also been suggested to provide a source of energy for driving asteroid evolution through magnetic induction processes (Sonett et al. 1970). See also the chapter by Herbert et al.

In order for a wind-extended solar magnetic field to cause meteorite magnetization, it would have been necessary for most of the disk mass to have dispersed before the magnetization occurred; otherwise, the disk would have prevented the material from having been exposed to the wind and its entrained magnetic field. This could place an important constraint on the evolutionary sequence in the protoplanetary system. Moreover, if the solar magnetic field had been carried by the wind, it would have been necessary that the solar-wind kinetic-energy density exceeded the entrained magnetic-energy density. It is instructive to see what this implies. For a magnetic field in the range of 0.1 to 1 Gauss, the magnetic energy density, $B^2/8\pi$, is 4 × 10^{-4} to 4 × 10^{-2} erg cm^{-3}. Suppose that the mass-loss rate from a vigorous T Tauri wind is $\dot{M}$ (M$_\odot$ yr^{-1}). Then, at a distance R astronomical units from the central star, a wind of speed v cm s^{-1} has a kinetic energy density ε given by ε ~0.01 $\dot{M}_v R^{-2}$ erg cm^{-3}. Taking v ~ 3 × 10^{-7} cm s^{-1} and $\dot{M}$ ranging from a nominal 10^{-8} to as high as 10^{-6} M$_\odot$ yr^{-1}, we find that ε ~ 1.5 × 10^{-4} to 1.5 × 10^{-2} erg cm^{-3}, at a distance of 3 AU. Thus, at 3 AU, a magnetic field approaching 0.1 Gauss could marginally be entrained by a 10^{-8} M$_\odot$ yr^{-1} wind flow, while a field approaching 1 Gauss could marginally be entrained by a 10^{-6} M$_\odot$ yr^{-1} wind flow (Levy and Sonett 1978). However, the energy density of such a wind-entrained magnetic field would increase toward the Sun at a rate between $1/R^2$ and $1/R^4$ (depending on how tightly wound the structure was), while the wind kinetic-energy density would increase only as $1/R^2$ toward the Sun. To determine the maximum magnitude of the field that could be carried by such a wind requires more detailed calculations, which are beyond the scope of this chapter.

Finally, the present-day solar wind magnetic field alternates its sign: in the equatorial plane, the sign of the average magnetic field changes typically 2 to 4 times per month (actually per solar rotation); at higher latitudes, the period of alteration is expected to correspond to the solar magnetic cycle, i.e., once in each 11 yr. However, this alternating-sign behavior is not fundamental to the system. The early Sun could have produced a steady, or only very slowly varying, magnetic field in the wind (if, for example, the wind carried a solar quadrupole-symmetry field, rather than the dipole-symmetry field of the contemporary Sun). Ultimately, the magnetization of meteorite matter needs to be understood in terms of a realistic picture of the magnetic field in the nebula. That question is beyond the purview of this chapter; but it is worth noting that the small-scale components of meteorites, which exhibit the strongest magnetization, were probably magnetized very rapidly, so that even an inhomogenous and varying magnetic field could account for that component of the magnetization.

The magnetic field in the convective zone of the present-day Sun is renewed in the time of the turbulent dissipation, of the order of a few years (Zel'dovich et al. 1983). However, in the radiative core (below about 0.7 R$_\odot$), the dissipation time of a magnetic field, having its main scale l_B ~ R$_\odot$, exceeds the age of the Sun, because of the high electrical conductivity of the solar interior and the absence of turbulent mixing. The question arises whether the magnetic field in the solar radiative core retains any memory of its earlier evolutionary stages. If so, then we would have a tool for studying some aspects of solar evolution. The answer depends on how the fully convective Sun evolved to have a radiative core. The internal field could have dissipated completely if the turbulence decayed slowly, so that turbulent dissipation continued after dynamo action had ceased. However, consider the possibility that the dynamo number remained approximately constant during the decay of convective turbulence in the inner part of the Sun; this is a possibility because both the dynamo α and η_t depend on v_t in the same way. Assuming this to be the case, and assuming that the intensity of the generated magnetic field decreases with v_t in accordance with the Eq. (37), then the intensity of the magnetic field would have been set by the conditions existing when the time scale of the turbulent dissipation of the magnetic field became larger than the time scale τ of the decay of the turbulence, i.e., when $\tau \geq R^2/\eta_t$. The time scale for decay of turbulence in the Sun's interior is likely to have been of the order of the Kelvin-Helmholtz time, τ_{KH}, which also governs the rate of the Sun's evolution to the main sequence; $\tau_{KH} \sim 10$ Myr. The condition $\tau \sim \tau_{KH}$ results in $v_t = 3R^2/l_B\tau_{KH}$; substitution into Eq. (37) gives a value for the remanent intensity of the magnetic field in the radiative core that could have persisted after the proto-Sun came through the Hayashi stage, and to the present day

$$ B_p \sim \left[\frac{8\pi hR\rho}{\sqrt{3}}\right]^{1/2} \left[\frac{R}{l_t}\right]^{1/4} \left[\frac{\Omega}{\tau_{KH}^3}\right]^{1/4} \tag{38a} $$

$$ B_\phi \sim (8\pi hR\rho)^{1/2} \left[\frac{R}{l_t}\right]^{3/4} \left[\frac{\Omega^3}{\tau_{KH}}\right]^{1/4} \tag{38b} $$

Assuming, again, that the magnetic field intensity is governed by kinematical and dynamical conditions locally, these equations result in $B_p \sim 1$ to 10 Gauss in the core and 0.1 to 1 Gauss just under the bottom of the convective zone, and $B_\phi \sim 10^5$ to 10^6 Gauss and 10^3 to 10^5 Gauss in the same regions, for the numerical values mentioned above. This argument suggests the possibility that the Sun could have made it to the main sequence with a relatively weak poloidal magnetic field in its radiative core. However, it is clear that other evolutionary sequences could be envisioned, differing in detail from what is described here; thus this argument must be considered only plausible, not compelling. Other cases have been considered elsewhere, including the possibility of an oscillating dynamo (Schüssler 1975; Parker 1981; Mestel and Weiss 1987). At our present stage of understanding, we must rely on observations to try to pin down these aspects of the evolutionary history; this is taken up in the next section.

In the presence of differential rotation, the toroidal magnetic field could subsequently be amplified to a value limited by magnetic buoyancy; this would limit B_ϕ to about 10^7 Gauss in the core. However, redistribution of angular momentum by magnetic torques is likely to drive the radiative interior toward homogeneous rotation on a time scale given by Eq. (34), and stop the continuing increase of B_ϕ soon after the decay of convection. This point is also discussed further in the next section.

V. THE MODERN SUN

The equilibrium state of the modern Sun is determined almost entirely by a balance between the gravitational force and the pressure gradient. In this approximation, the Sun is spherically symmetric. Internal pressure is provided by heat from thermonuclear burning in the center. This energy is transported by radiation up to about $0.7 R_\odot$ and mainly by turbulent convection in the shell between 0.7 and $1 R_\odot$. In the absence of a magnetic field, the Sun would be a quiescent, luminous ball, at least over the time scales of human observation. The presence of a magnetic field creates life in this otherwise quiet body.

As discussed earlier, a magnetic field is effective at redistributing angular momentum. One observes, today, a weak surface rotation of the Sun, 3×10^{-6} s^{-1}. It is believed that the Sun's rotation has slowed through loss of the angular momentum carried away by the solar wind. However, it is unclear to what extent the deeper layers of the Sun have been slowed by this loss of angular momentum from the surface. Some solar seismological measurements suggest that the Sun's interior is rotating as much as twice (or more) as fast as the surface. It is unclear whether such steep rotation gradients can be reconciled with the existence of a solar magnetic field producing strong stresses opposing the nonuniform rotation.

The main visible part of the Sun's magnetic field is believed to be generated in the convection zone through the action of a hydromagnetic dynamo. This field, interacting with the convective motion and rotation, produces all phenomena: the sunspots, prominences, flares, etc. As a component of the solar wind, the magnetic field plays an important role in the interplanetary medium and in the solar wind's interaction with planets. The generation of the solar magnetic field, the field's behavior, and its interaction with the solar rotation are discussed below.

A. Internal Rotation of the Sun

Information on the distribution of rotation inside the Sun is obtained from helioseismological investigations. The periods of the Sun's oscillatory modes cover a range from a few minutes to several hours. The frequencies and corresponding spatial structures of the modes carry evidence about the structure of regions of the Sun where the modes are concentrated. The amplitudes of solar oscillations are sufficiently small that they can be considered small deviations about the spherical equilibrium state. In this spherically symmetric approximation, the eigenfrequencies v_{nl} depend on the radial n and latitude l mode numbers and do not depend on the azimuth number m. Rotation breaks this symmetry and removes the degeneracy of the frequencies. First-order perturbation theory for slow rotation, with angular velocity depending only on radius, gives

$$\nu_{nlm} = \nu_{nl} + m \int_0^{R_\odot} K_{nl}(r)\Omega(r)dr \qquad (39)$$

where K_{nl} is the kernel defined by the eigenmodes of the spherically symmetric approximation (Gough 1981a). Thus, Eq. (39) provides a measure of the
angular velocity if the eigenfrequencies and eigenmodes are determined from
observations and theory.

Up to the present, the solar oscillations that are well studied observationally and theoretically are the acoustic modes with periods in the range
from 3 min to 1 hr. The observations from which Ω has been estimated are
of mostly sectoral modes ($m = l$, $1 \le l \le 130$) concentrated near the equator
(Duvall and Harvey 1984; Leibacher 1984; see, however, Brown 1986). The
region where $\Omega(r)$ is most reliably determined by the mode-splitting data is
located near the equatorial plane at radii between 0.3 and 0.9 $R_\odot$. One possible form of the distribution of rotation inferred from the data is presented
in Fig. 3 (Duvall et al. 1984). (Other model reductions, which differ in the
manner of data analysis, give results that are qualitatively similar.) The most
impressive feature in Fig. 3 is the radial gradient of angular velocity. Even in
the region 0.4 to 1 $R_\odot$, the rotation gradient stands out with a magnitude
about 2 times greater than the latitudinal gradient of angular velocity observed
on the solar surface. Other determinations of $\Omega(r)$ have suggested even
steeper rotation gradients (Hill et al. 1982).

Morrow (1988) has analyzed frequency splitting of modes with $l = 15$
to 99, and discovered that the surface latitudinal differential rotation pervades
the convective zone and perhaps even to deeper layers. He finds little or no

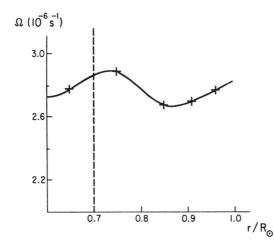

Fig. 3. Solar angular velocity $\Omega(r)$ in the equatorial region (figure from Duvall et al. 1984).

radial gradient of angular velocity inside the convective zone. This picture implies a significant radial gradient in a transitional layer between the convective zone and the radiative interior (see also Dziembowski et al. 1989).

B. Fossil Magnetic Field in the Radiative Interior

The solar magnetic field is variable on all observed spatial and temporal scales. A magnetic field fixed into the high-electrical-conductivity radiative core could only vary or decay over time scales too long to be compatible with these observations (Cowling 1957). This fact (and especially the regular and frequent polarity reversals at the largest scale of the field) gives evidence that at least most of the externally observable magnetic field is generated in the convective zone. The oscillating magnetic field can only penetrate a short distance into the static radiative core; a field oscillating with the observed 22-yr period has a skin depth (of the order of $\sqrt{\eta t}$) of only some tens of kilometers. Therefore, the magnetic field can permeate the deep radiative interior only as a remanent of the Sun's earlier evolution. The discussion of cosmogonical conditions and processes given in the previous section suggests the likely existence of magnetic fields in this part of the modern Sun. However, as we have pointed out, the considerable uncertainty about details of the Sun's evolution leave considerable uncertainty about the magnitude of such a field.

The possible presence of both nonuniform rotation and a magnetic field in the Sun's interior cannot be considered in isolation from one another; a dynamically consistent picture of the solar interior must take account of the inherent coupling of these two aspects of the system.

Differential rotation can produce, even from a weak poloidal magnetic field B_p, a much stronger toroidal field B_ϕ, and a magnetic torque ($\propto B_p B_\phi$), which is able to force the radiative interior toward uniform rotation. The azimuthal component of the induction equation is

$$\frac{\partial B_\phi}{\partial t} = r(\mathbf{B}_p \cdot \nabla\Omega) + \mathcal{R}_m^{-1} \Delta B_\phi \qquad (40)$$

where the magnetic Reynolds number $\mathcal{R}_m$ is typically very large in the Sun (possibly 10^{12} in the radiative core, depending on the amplitude of the nonuniform rotation there, and 10^{10} near the boundary with the convective zone). At first glance the situation seems simple: as long as $|B_\phi|$ is not large, the dissipation term is negligible, and Eq. (40) gives a linear growth for B_ϕ, whatever the structure of the poloidal field $\mathbf{B}_p$. However, the ultimate evolution of the magnetic field depends essentially on the structure of B_p. If the poloidal field is axisymmetric and $(\mathbf{B}_p \cdot \nabla\Omega) \neq 0$ then the azimuthal field evolves first in the manner just described and ultimately, after a characteristic time $\mathcal{R}_m\Omega^{-1}$ ($\sim 10^{10}$ yr at $\mathcal{R}_m = 10^{12}$), approaches a steady state $B_\phi \sim \mathcal{R}_m B_p$.

The magnetic torque associated with $(\mathbf{B}_p, B_\phi)$ reduces the angular velocity gradient over an Alfvénic time $(L\sqrt{4\pi\rho}/B_p \sim 10^4$ yr for $B_p = 1$ G, $L = 0.2$ $R_\odot)$, even if B_p is very small (Mestel and Weiss 1987; Dudorov et al. 1989). Thus, we are pressed by the question of how both solar internal rotation *and* even a weak internal magnetic field can co-exist.

For a nonaxisymmetric poloidal field, the development is different. Consider, as illustrative example, a uniform poloidal field B_0 perpendicular to the axis of rotation (Parker 1979a). Take the characteristic length scale across the differentially rotating region to be L, so that $\mathcal{R}_m = L^2\Omega/\eta$. Then the differential rotation produces an azimuthal magnetic field in the form of a nested double spiral, in which the neighboring lines of force are in opposite directions. Because this now-tangled field has small spatial scale, it will dissipate rapidly in the shearing region. After a characteristic time of the order of $\mathcal{R}_m^{1/3}\Omega^{-1}$ (10^4 yr for $\mathcal{R}_m = 10^{12}$) the magnetic field will be excluded from the nonuniformly rotating region. An extension of this analysis is given by Brodsky et al. (1988) who show that the magnetic field is less quickly expelled from the rotating region if the shear is concentrated in a boundary layer rather than being distributed (i.e., when the functional form of Ω approximates a step function). In this case, modes localized inside the rotating region decay with a rate that is not so much accelerated by the differential rotation, but instead is of the order L^2/η inside; however, the field in the shear layer itself decays rapidly, effectively reducing the magnetic torque transmitted across the boundary. Altogether, nonaxisymmetric magnetic field modes tend to be excluded from the rotating region (penetrating only to the extent of a skin depth) and have a much diminished long-term effect on the differential rotation.

Another possibility, that would allow co-existence of persistent differential rotation and an axisymmetric magnetic field is a field structure that satisfies Ferraro's isorotation condition $\mathbf{B}_p \cdot \nabla\Omega = 0$, i.e., the magnetic lines lie on surfaces $\Omega = $ const. (The axisymmetric condition $\partial\Omega/\partial\phi = 0$, can be generalized to $\mathbf{B} \cdot \nabla\Omega = 0$) To illustrate that the torque can vanish, note that a force-free magnetic field provides a sufficient condition for the vanishing of the torque: $\nabla \times \mathbf{B} = k\mathbf{B}$, where k is a pseudoscalar proportional to Ω. (Upon taking the divergence of this relation, one obtains the Ferraro condition.) It is simple to illustrate that such a force-free, isorotating field structure can occur in at least a part of the Sun. Take Ω to be constant in the sphere $r < R_c$, and to be a different constant, say 0, in the region $r > R_c$. Equation (41) below shows a simple force-free magnetic field in the region $r < R_c$, surrounded by a magnetic field in $r > R_c$, which tends to a uniform field in the z direction, $\mathbf{B}_p = B_0\mathbf{e}_z$, as $r \to \infty$. In this case, the differential rotation is localized at $r = R_c$. The magnetic field has the form

$$(B_r, B_\theta, B_\phi) = (-2\cos\theta, \sin\theta, kr\cos\theta)f(r) \qquad (41)$$

where

$$f(r) = \begin{cases} Ar^{-3/2}J_{3/2}(kr), & \text{if } r < R_c \\ -B_o\,(1 - R_c^3/r^3), & \text{if } r > R_c \end{cases} \tag{42}$$

and

$$A = 3B_o\frac{d}{dr}\left[J_{3/2}(kR_c)R_c^{-1/2}\right]^{-1} \tag{43}$$

(see Chandrasekhar 1956; and Moffatt 1978). The corresponding lines of force lie on a family of nested toroidal surfaces in the region $r < R_c$ and they are identical with the streamlines of an irrotational flow past this sphere. Of course, this force-free magnetic field does not disturb the angular velocity distribution; it is force free inside and no torque acts across the surface $r = R_c$, because the normal component of the magnetic field vanishes on that surface. In some sense, this solution is a realization of the idea of a magnetically decoupled core suggested by Rosner and Weiss (1985), and illustrates some of the possibilities that exist, at least in principal, for co-existence of a magnetic field and differential rotation. However, it is problematical to what extent this kind of situation can apply to the real Sun. There are at least two problems: first it is, of course, questionable whether such a field structure could be realized as a result of solar evolutionary processes; second, the structure given by Eq. (41) is not globally applicable to the Sun. The external field is not confined to a finite region of space and contains, in a manner of speaking, field sources at infinity. No magnetic field can be globally force free; it must satisfy an overall virial theorem for dynamical consistency. In the mathematical illustration given here, the apparent force-free character of the field is sustained only because an arbitrarily large magnetic stress is transmitted to infinity. Further work is necessary to demonstrate a self-consistent magnetic field structure that is compatible with persistent differential rotation while satisfying both evolutionary and dynamical requirements.

Now consider what happens if a field structure deviates from isorotation in the presence of differential rotation. A disturbance of the isorotation state by Ω' produces an additional azimuthal field $B'\phi$, leading to

$$\frac{\partial B'_\phi}{\partial t} = r(\mathbf{B}_p\,\nabla\Omega') \tag{44}$$

and an accompanying magnetic stress that acts back on the differential rotation as

$$r^2\frac{\partial\Omega'}{\partial t} = (\mathbf{B}_p\nabla)rB'_\phi. \tag{45}$$

If the scale of variation of Ω' along $\mathbf{B}_p$ is small in comparison with r and with the scale of variation of B_p itself, then these equations reduce to

$$\frac{\partial^2 \Omega'}{\partial t^2} = V_A^2 \frac{\partial^2 \Omega'}{\partial s^2} \qquad (46)$$

where V_A (s) = $B_p/\sqrt{(4\pi\rho)}$, the local Alfvén speed, is measured along $\mathbf{B}_p$ (see, for instance, Mestel and Weiss 1987). Thus, the disturbance propagates as the Alfvénic wave. For example, with B_p = 1 G and ρ = 100 g cm^{-3} at the center of the Sun, this wave traverses the distance L = 0.1 R$_\odot$ in < 10^4 yr. The density near the bottom of the convective zone is only 0.1 g cm^{-3}, in which case the disturbance propagation time decreases to $\sim$ 300 yr. Altogether then, one self-consistent state for a differentially rotating solar core and magnetic field is that of an oscillator with characteristic period in the range of some 10^3 to 10^4 yr. It is problematical whether such a state would persist for very long.

This discussion considered the possibility of a cosmogonic poloidal magnetic field in the radiative interior. Hinata (1986) has suggested the possibility that a dynamo magnetic field might be generated in the radiative interior through a nonlinear dynamo driven by random phase circularly polarized Alfvén waves created in the convective zone and propagating into the core. He suggested the possibility of an α^2-dynamo without differential rotation. If such a dynamo were able to operate, it is conceivable that it could produce a force-free magnetic field (at least in some limited volume).

C. Dynamo Magnetic Fields in the Convective Zone

The time scales characterizing evolution of the magnetic field in the solar convection zone are very short in comparison with the Sun's evolution time. The turbulent diffusion time over the convection zone, $(0.3R_\odot)^2/\eta_t$, is of the order of a few years ($\eta_t \approx 5 \times 10^{12}$ cm^2 s^{-1}). This is also of the order of the Sun's 22 yr magnetic cycle. However, solar magnetic variations are also observed over substantially longer periods of time (chapter by Damon and Sonett): secular modulation and epochs of the Maunder-like minima. Thus, questions arise about from where these longer periodicities arise and what controls them.

Magnetic field transport in the solar convective zone is determined by convection and differential rotation. The Rayleigh number is so large ($\sim 10^{12}$) that the convection must be turbulent. (The apparent order in the form of granulation and supergranulation cells can be explained as a synergetic phenomenon.) Convection of electrically conducting fluid, in a rotating body, with a large magnetic Reynolds number $\mathscr{R}_m$ = lv/η, where l and v are the characteristic values of the spatial scale and the velocity, and η is the magnetic diffusivity, acts as a hydromagnetic dynamo. In view of the prolonged and large-scale phenomena under consideration in this chapter, three mean measures of the convection are relevant: the turbulent diffusivity of the mean magnetic field, determined by (v^2); the mean helicity determined by $\langle v \cdot \nabla \times v \rangle$; and the differential rotation arising from the interaction between convection and rotation. The differential rotation is probably accompanied by a meridional motion.

The most straightforward and widely used method of treating of magnetic field evolution in a turbulent fluid is to apply averaged MHD equations in which the kinematic parameters (turbulent diffusion, mean helicity and differential rotation) are taken to be given functions (for reviews, see Moffatt 1978; Parker 1979a; Krause and Rädler 1980; Zel'dovich et al. 1983). This approximation resulted in the explanation and development of the important fact that the mean solar magnetic field is not stationary but rather has the form of waves propagating nearly along surfaces of constant angular velocity. By adjusting the kinematic parameters, the period of the waves can be made equal to 22 yr, the observed solar magnetic-cycle period.

Dynamo waves migrate in directions that coincide closely with surfaces of isorotation (Yoshimura 1975a). Dynamo theory gives a direction of migration determined by the sign of the product of the mean helicity and angular velocity gradient in each hemisphere. According to theoretical estimates, the mean helicity in a given hemisphere is of the same sign over almost the entire convective zone and can change sign only near the bottom of the zone. Meanwhile, helioseismological data suggests that the gradient of angular velocity changes sign at least twice in the convective zone. Altogether, the picture that emerges includes several dynamo wave modes of different amplitudes on the solar surface. Thus, there is the possibility to use the behavior of the magnetic field to probe the solar rotation in the convective zone (Ruzmaikin and Starchenko 1987). An attempt to construct a realistic picture of these waves has been made by Makarov et al. (1987; see also Brandenburg and Tuominen 1988). It has been shown that current helioseismological data for differential rotation suggest the existence of 2 (or 3) dynamo waves in each hemisphere (Fig. 4). One wave, generated near the surface, appears at the latitude 60° and propagates toward the pole. A second, apparently more intense, wave of magnetic field is excited in the middle of the convective zone and migrates toward the equator, manifesting itself in the form of the well-known butterfly diagram.

Further development of this approach in order to describe such phenomena as the Maunder minimum and the other so-called Grand Minima run into serious mathematical difficulties involved in the analysis of a continuous system of nonlinear MHD equations. Some light has been shed on this subject by a new approach in which we forego attempts to know detailed information about the solar dynamical system and study only crude properties of its attractors (Ruzmaikin 1981).

The phase space of the solar dynamical system is formed by magnetic field and fluid velocity variables. Magnetohydrodynamic behavior can be de-latitude

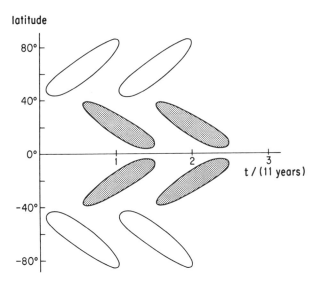

Fig. 4. A schematic depiction of magnetic dynamo waves on the solar surface. There are two waves in each hemisphere: one propagates toward the poles (the polar wave); the other propagates toward the equator (the Maunder butterfly wave).

scribed as a path in this phase space. It is natural to focus on the attractors rather than on the detailed phase trajectories in order to describe general aspects of a system's behavior; attractors are simply manifolds to which many or all of the phase-space paths are attracted. The simplest property of an attractor is its dimension d. A point attractor has dimension $d = 0$, a limiting Póincare cycle in the form of a closed curve has $d = 1$, etc. On attractors of integer dimension, phase trajectories are regular curves. However, some dynamical systems have attractors of fractal dimension on which the paths behave in an irregular manner. For example, an attractor of dimension $d = 3 + \varepsilon$, where $\varepsilon \ll 1$, can be pictured as a ball with an infinite number of small holes so that a path traveling over this manifold becomes completely tangled. Such behavior is called chaotic, to match its appearance.

The dimension of the solar attractor can, in principle, be found from observational data with the help of a method by Grassberger and Procaccia (1983a), as well other methods. The essence of the method is to consider a sequence of N points of one variable in equal time intervals of duration τ. It is assumed that the sequence corresponds to a path crossing many times the points of an attractor. The dimension of the attractor appears as an exponent of the correlation integral, $w(r) \sim r^d$, at small distances r between the points of the attractor (see the chapter by Morfill et al.).

As a possible sequence, one can take the Wolf number of sunspots: $W(t)$, $W(t + \tau), \ldots ,W(t + [N - 1]\tau)$. Preliminary estimates of this correlation dimension give d close to 2 (Kürths 1987; chapter by Morfill et al.; Makarenko and Ajmanova 1989). Unfortunately, the sequence of existing data is too short to encompass such phenomena as the Maunder-like minima. In this respect, a better estimate can be obtained from radiocarbon data, which is available over a time interval of about 9000 yr. (The intervals between the Maunder-like minima are less than 300 yr.) Gizzatulina et al. (1988), by analyzing the radiocarbon data for a 6000-yr period, have found $d \approx 3$, corresponding to a 3-torus ($d = 3$), that is unstable and transforms into a strange attractor, or directly to attractor of fractal dimension. In any case, the analysis suggests that the Sun is dynamically stochastic. It is important that the analysis of both $W(t)$ and the radiocarbon sequence show small dimensionality. To look at the matter another way, this indicates that the solar dynamical system is simple and can be described by approximately 4 or a slightly larger number of effective variables (or modes). The 3 quasi-periodic variations of solar activity seem to be the most probable candidates to form that quasi 3-torus: the 22-yr dynamo wave or corresponding 11-yr sunspot cycle, the secular variation of the solar cycle, and the Maunder-like minima epochs (Ruzmaikin 1981, 1985). Theoretical models consisting of several ordinary differential equations, and yielding attractors of this kind were constructed by Weiss et al. (1984) and Malinetsky et al. (1986). However, the values of parameters used in these models are still far from realistic for the Sun. The underlying merit of these studies is that they indicate the possibility of constructing low-dimension solar models with chaos. However, to construct a realistic model remains a challenge to future investigations.

D. Influence of a Fossil Field on the Solar Cycle

At least part of any magnetic field trapped in the Sun's radiative interior is likely to penetrate into the convective zone from below, both the poloidal and toroidal parts of the field. The boundary conditions on the surface separating the convective and radiative zones require that the normal component of induction B_r and tangential components of $\mathbf{H} = \mu^{-1}\mathbf{B}$, i.e., $\mu^{-1}B_\theta$ and $\mu^{-1}B_\phi$ are continuous (where here μ is the magnetic permeability). To the extent that the magnetic permeability is unity in the radiative interior and dominated by turbulence in the convective zone (where we can take $\mu = \mu_t$ $\ll 1$), the tangential component of the large-scale field (induction) is discontinuous at the boundary. According to the estimate by Ivanova and Ruzmaikin (1976), $\mu_t \approx 3 \times 10^{-5}$ in the Sun's convective turbulence. Thus, turbulent diamagnetism can strongly reduce the toroidal part of the fossil magnetic field inside the convective zone. However, it does not affect the radial component of the magnetic field.

Unless, as discussed earlier, there is some mechanism for completely burying a primordial magnetic field in the radiative interior in some dynamically self-consistent way, it is inevitable that a substantial part of any such field will penetrate into the convection zone from below. Levy and Boyer (1982; see also Pudovkin and Benevolenskaya 1982; Boyer and Levy 1984) have pointed out that a quasi-stationary magnetic field penetrating to the convective zone will induce polarity bias or asymmetry in the solar cycle, with alternate 11-yr halves of the full 22-yr cycle having different amplitudes (see also Sonett 1982, 1983). In a crude sense, this result becomes evident if one adds a constant bias field to the periodic solar magnetic oscillation. However, the properties of dynamos in the presence of an imposed nondynamo quasi-constant magnetic field presents richer and more difficult mathematical problems (see Levy [1989] for additional discussion). It is also important to have additional observational evidence for the existence and properties of such solar polarity asymmetry.

In 1948, Gnevyshev and Ohl' found, from analysis of the 22-yr cycle, that even 11-yr halves of the cycle are less active (as measured by sunspot number) than are the odd ones (see also, Sonett 1983). Later Chernovsky (1966) came to the same conclusion in a study of geomagnetic activity during 80 yr from 1884 to 1964. To explain this asymmetry, it is sufficient to introduce an approximately 0.5 gauss quasi-constant field to the solar dynamo (Pudovkin and Benevolenskaya 1982). The remanent intensity of the poloidal magnetic field in the solar core, given in Eq. (33), coincides with the internal magnetic field of the Sun estimated by Boyer and Levy (1982), who suggested that this primordial fossil field could be responsible for the observed polarity asymmetry of the 22-yr solar magnetic cycle. Such a field with a quadrupole component could also manifest itself in the form of asynchronous reversals of the Sun's north and south polarities, which effect was observed by Babcock in 1959 and Howard in 1974 (Boyer and Levy 1984).

Altogether, these results suggest the idea that the Sun evolved through a fully convective stage during its early evolution, followed by decay of convection in the interior and the formation of a radiative core containing a weak remanent magnetic field.

VI. THE SUN'S FUTURE

A naive, though not uncommon, picture of the evolution of stellar magnetic fields is that the magnetic flux threading a star is approximately conserved throughout the star's life, leading to a simple relationship between the star's radius and the magnetic field strength: $|\mathbf{B}| \propto R^{-2}$. One of the principal points that we have tried to make in this chapter is that, in fact, there exists *no* such simple relationship. Indeed, in real cosmical bodies, simple magnetic flux conservation seems to be the exception, rather than the rule. Even for stars, in which the classical electrical conductivity is nearly always high enough to enforce flux conservation, episodes of turbulent mixing cause rapid memory loss of previous magnetic fields and, from time to time, generate new magnetic fields through dynamo action. During stellar evolution then, it is the episodes of turbulent mixing that dominate the evolution of magnetic field (Levy and Rose 1974). Moreover, during such turbulent episodes, the magnetic memory loss is very rapid, with a memory span $\sim R^2/\eta_t$, as given earlier. The solar convection zone is the best known example of this short turbulent memory. The present magnetic memory of the Sun's convection zone lasts only of the order of a few oscillation times, *viz.*, a few decades. The presently observed solar magnetic field is therefore entirely a product of contemporary generation, except insofar as it may be modified by a weak internal remanent magnetic field.

After approximately another 5 Gyr as a main sequence star, the Sun is expected to evolve through red-giant phases and ultimately to fade into senescence as a white dwarf. At least two effects ensure that the present solar magnetic field will be forgotten by the aging Sun. First, as a red-giant star, the Sun will be convective throughout much of its interior; the associated turbulent mixing will destroy and regenerate magnetic fields on time scales very short in comparison with the Sun's evolution time. Second, the outer layers of the Sun are likely to be lost during red-giant and planetary nebula stages of evolution. Less than 5% of the Sun's mass is in the present convection zone. Thus, although it is uncertain just how much mass the Sun will lose before the end of its active life, even a relatively small mass loss would dissipate the entire region in which the present solar magnetic field is generated. In the future, as in the past, the magnetic evolution of the Sun will not be written by the simple equation of magnetic flux conservation. Instead, the Sun will destroy magnetic fields, and reconstruct them anew, with the details depending on the details of evolutionary processes.

PRE- AND MAIN-SEQUENCE EVOLUTION OF SOLAR ACTIVITY

FREDERICK M. WALTER
State University of New York, Stony Brook

and

DON C. BARRY
University of Southern California

The magnetic activity on single solar-like stars declines with stellar age. This has important consequences for the influence of the Sun on the early solar system. We review what is meant by stellar activity, and how it is measured. We discuss stellar activity in the pre-main-sequence phase of evolution, and conclude that the classical T Tauri stars do not exhibit solar-like activity, while the naked T Tauri stars do. The emission surface fluxes of the naked T Tauri stars are similar to those of the youngest main-sequence G stars. We discuss the activity-decay relations, and conclude that the best representation for solar-like stars is a decay proportional to exp(A × t^{0.5}), where A is a function of line excitation temperature. From these decay laws one can determine the interdependences of the activity, age and rotation periods. We discuss the fluxes of ionizing photons at the Earth early in its history, and show that there was sufficient fluence to account for the observed isotopic ratios of the noble gases. Finally, we discuss some of the observational tasks remaining before we can claim to understand fully the causal relations affecting stellar activity.

I. INTRODUCTION

Stellar activity is ubiquitous among the solar-like stars. Observable as photospheric spots and chromospheric and coronal emission, this activity is the manifestation of the nonradiative heating processes that occur in the outer atmospheres of the Sun and solar-like stars. The ultimate origin of this activity is probably the stellar magnetic field, generated in a dynamo by the interaction of the differential rotation with the stellar convection. Although this paradigm is generally accepted, the details of the dynamo process and the subsequent concentration of magnetic flux into flux tubes in the Sun are poorly understood (see, e.g., Parker 1986; Weiss 1981; Gilman and DeLuca 1986). The existence of flux tubes and loop-like structures in the solar atmosphere is well documented (see, e.g., Eddy 1979), but the details of how the gas constrained within the magnetic flux tubes is heated are again poorly understood (see, e.g., Chiuderi 1981; Cram 1987).

The Sun is a typical G dwarf. We assume that solar-like stars, if viewed in sufficient detail, will manifest the same physics and the same atmospheric structures, albeit with great latitude in scale. Indeed, observation discloses other stars that exhibit gross solar-like characteristics. There is evidence of starspots on RS CVn and BY Dra stars (Hall 1976), main-sequence stars (Radick et al. 1987; see also the chapter by Radick), and among pre-main-sequence stars (Rydgren and Vrba 1983). Strong magnetic-fields have been detected in solar-like stars (Saar 1987b; chapter by Saar; Hartmann 1987). Most, if not all, late-type dwarfs have chromospheres (Linsky 1980; Jordan and Linsky 1987) with surface fluxes significantly above the basal flux levels (Schrijver 1987). Most, if not all, solar-like stars also possess coronae (Rosner et al. 1985). It is important, however, to bear in mind that among the solar-like stars, the Sun is remarkably nondescript.

We examine here the behavior of the Sun over time. Our time machine is a sample of solar-like stars of diverse ages. It is implicit in our arguments that the Sun is indeed a solar-like star, and that we can extrapolate the mean behavior of the Sun in time from the observations of the solar-like stars. There are at least two reasons for wanting to know the long-term history of solar activity. The first is that by studying the activity levels as a function of macroscopic stellar parameters (i.e., mass, age, rotation rate, chemical composition) one can hope to deduce the functional dependences of and the physics behind the stellar dynamo. Observationally, the primary indicator of the mean activity level seems to be stellar rotation, not age; but the two are intimately related among the single stars as the rotation periods increase monotonically with time, presumably due to magnetic braking (see, e.g., Endal and Sofia 1981). Hence a time history maps into a rotation history of the activity. More appropriately for this work, we are interested in the history of the Sun because the solar activity levels have played a major role in shaping the present-day solar system (cf., Pepin et al. 1980; Feigelson 1982). The ionizing flux from the Sun, much higher in the past than now, may have played a major role in determining the chemistry of the atmospheres of the terrestrial planets (Pepin 1989). Solar flares, that probably correlate with mean activity levels, and their accompanying energetic particles, may also have had a major role in determining the chemistry of the early solar system.

II. MEASURING STELLAR ACTIVITY AND ITS CORRELATIONS

A. The Solar Atmosphere

The atmospheric temperature of a solar-like star reaches a minimum just above the photosphere. Were the solar atmosphere in pure radiative equilibrium, its temperature would continue to decrease outwards as the radiation field dilutes as r^{-2}. Yet in the Sun and the late-type stars, the temperature of the atmosphere rises as one climbs outwards from the photosphere. This is a consequence of nonradiative heating.

The chromosphere is that region of the solar atmosphere visible in emission lines during solar eclipse. Here nonradiative heating dominates the energy balance and the temperature gradients are generally small over many pressure scale heights (Linsky 1980). The temperature gradients are kept small because of the efficiency of the atmospheric cooling agents, among them the resonance lines of H I Lyman α, Ca II H and K, and Mg II h and k. At temperatures above about 30,000 K, H I Lyman α becomes optically thin and, due to the lack of efficient cooling agents, the temperature increases to about 10^6 K in less than one pressure scale height. In the Sun, the thickness of this transition region is a few thousand km (Mariska 1986). Among the strongest emission lines arising in the transition region are the C IV 1548,1551 Å, Si IV 1394,1403 Å and N V 1238,1242 Å doublets (Jordan and Linsky 1987). At higher temperatures, the cooling function increases and the temperature gradients again become small. Mean solar coronal temperatures run from about 10^6 K in the quiet corona (the large-scale structures) to a few times 10^6 K in active-region loops (Vaiana and Rosner 1978). At these temperatures, most of the emission occurs at X-ray wavelengths. The most active late-type stars seem to have higher coronal temperatures, up to a few times 10^7 K (see, e.g., Schrijver et al. 1984).

We note that one need not go to the stellar chromosphere or above to detect stellar activity. Starspots are visible (by their darkness) in the photospheric light. In addition, Simon et al. (1985) show that the ultraviolet continuum below 2000 Å is enhanced in active stars, suggesting that mechanical heating in the upper photosphere accompanies the activity in the outer atmosphere.

B. The Correlations of Activity

The strengths of the chromospheric and coronal emission features in solar-like stars are well correlated with each other. Ayres et al. (1981) showed that $R_l \propto R_{hk}^\alpha$, where R is the line flux normalized to the bolometric flux, and R_{hk} is the normalized flux ratio for the Mg II h and k lines 2795, 2802 Å. The power-law exponent α, a function of line formation temperature, ranges from unity for lower chromospheric lines to about 1.5 for transition-region lines and 3 for coronal X-ray emission. These power-law fits hold over about 2 orders of magnitude in R_{hk}, and for G and K dwarfs and giants. Further analyses (see, e.g., Basri et al. 1985; Basri 1987a; Bennett 1987) support this result.

A number of authors (White and Livingston 1981; Oranje 1983; Fisher et al. 1984; Skumanich et al. 1984) have discussed the disk-integrated solar spectrum, and how the Sun would appear as an unresolved star. When observed as a star, the Sun lies on these flux-flux correlations, although at the less-active end. Bennett et al. (1984) and Bennett (1987) show that the flux-flux relations observed for the disk-integrated Sun have power-law slopes similar to those observed in solar-like stars, despite covering a much smaller range in activity level.

The tight spatial and temporal correlations between the strength of chromospheric Ca II emission and magnetic flux observed across the face of the Sun (Skumanich et al. 1975; Schrijver et al. 1989b) suggests that the chromospheric heating mechanism is magnetically dominated. Schrijver and Cote (1987) find essentially the same relation between "excess" Ca emission and magnetic flux density $<fB>$, where f is the magnetic filling factor, on the resolved solar surface as do Saar and Schrijver (1987) for unresolved solar-like stars. Linsky and Saar (1987) discuss the constraints on dynamo models arising from stellar magnetic field observations.

C. Activity Parameters and Diagnostics

Lacking theoretical guidance, the determination of the best parameters and diagnostics for describing stellar activity is very much an empirical art. Activity is measured as excess emission above that of a purely radiative atmosphere. There are three simple observational parameters: the luminosity L (erg s^{-1}), the surface flux F (erg cm^{-2} s^{-1}) and relative flux R. These three measures are related: L is the surface flux F integrated over the stellar surface; R is the ratio of F to the bolometric flux σT^4. In physical terms, R may measure the efficiency of the dynamo while F measures the output of the dynamo per unit area. One needs to be wary in using L to compare stars of different radii, or R to compare stars of different T_{eff}. When considering only solar-like stars, as we do here, the three measures are equivalent. [For Ca II, one often uses R'_{HK} or $F'_{HK} \cdot F_{HK}$ and R_{HK}, as defined by Noyes et al. (1984a), represent not the actual flux in the emission core, but rather the flux within the 1.09 Å HK photometer passband. F'_{HK} and R'_{HK} refer to the values corrected for the photospheric emission outside the K_1 and H_1 minima. Linsky and Ayres (1978) and Linsky et al. (1979) define the primed quantities to refer to that flux between the K_1 minima which is chromospheric in origin, a more physical, if less easily determined, quantity. Here we use the Noyes et al. definition of R' because it is an observable quantity.]

In dealing with coronal lines (integrated into one X-ray flux) or transition-region lines, the measurement of line fluxes and conversion into either F or R is straightforward because little, if any, background correction

TABLE I
HD 115043 R'_{HK} Calibration

log F'_{HK}	log R'_{HK}	cal[a]	Source
6.59	−4.17	—	this work (spectral subtraction)
6.57	−4.15	R	Barry et al. (1987)
6.33	−4.43	M	Soderblom (1985)
6.60	−4.16	R	Soderblom (1985), with Rutten calibration applied

[a]R: Rutten (1984) calibration; M: Middelkoop (1982) calibration.

is necessary for cool stars at far-ultraviolet and X-ray wavelengths. However, most measurements of late-type stellar activity have been obtained at Ca II H and K, using the Mount Wilson HK photometer (Vaughan et al. 1978). This photometer produces a flux ratio S of the counts in a 1.09 Å triangular bandpass centered on the H and K emission cores to the counts in the nearby continuum. Middelkoop (1982) derived a calibration of S in terms of surface flux F. Using solar data from Oranje (1983), Rutten (1984) obtained an absolute calibration a factor of 1.7 larger. The difference lies in the fact that, as Oranje (1983) had illustrated, 45% of the response of the HK photometer to the solar K line profile comes from photons outside the K_1 minima. Middelkoop (1982) had argued that the contribution from outside the minima was less than 10%. In calibrating the background photospheric contribution R_{phot}, the flux arising outside the K_1 and H_1 minima which is subtracted from R_{HK} to yield R'_{HK}, Hartmann et al. (1984) find that for the Sun 45% of R_{HK} is accounted for by R_{phot}. [The absolute calibration depends on the shape of the line profile. This factor of 1.7 merely brings the two published calibrations into agreement. Oranje's (1983) solar calibration is for the quiet Sun; Schrijver et al. (1989b) argue that the absolute calibration for the plage K line profile is 20% larger. This is a consequence of the broader emission profile relative to the FWHM of the triangular photometer response.]

We attempted an independent verification of the calibration using moderate resolution stellar spectra. In Table I, we present various measurements of the Ca II H and K flux in the young active G1 dwarf HD 115043, a member of the UMa cluster. We observed this star (as a spectral standard) with the KPNO 2.1 m GOLDCAM spectrograph in December 1987 and December 1988. Emission is visible in the line core (resolution is 1.5 Å FWHM). Relative flux calibration was determined in the usual manner using spectrophotometric standard stars; absolute calibration was obtained from the relation between the 3925—3975 Å surface flux, the $(V\text{-}R)_J$ color (Linsky et al. 1979), and the Barnes-Evans relation (Barnes and Evans 1976). K and H line fluxes were determined by subtracting the spectrum of a star of similar spectral type but with less activity (HD 88725). The residual emission (see Fig. 1) is a lower limit to the chromospheric emission. Presumably all the photospheric components cancel, but HD 88725 is fairly active, and the chromo-

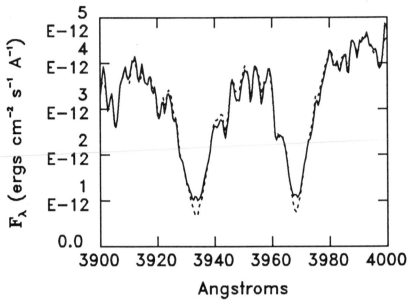

Fig. 1. A spectrum of the chromospherically active G1V star HD 115043, a member of the UMa cluster (solid line), with the spectrum of HD 88725, a less active G1V, overplotted (dashes). The spectra are fairly well matched, with the largest deviations at the bottom of the Ca II H and K lines. The residual emission corresponds to that estimated from Rutten's (1984) calibration of the S index.

spheric emission of HD 115043 may be partially oversubtracted. The resultant surface flux and the derived R'_{HK} are given in the first line of Table I. Note that these values are independent of the calibration of R'_{HK} and S. The second line shows the values determined by Barry et al. (1987) using Rutten's calibration. This determination is in good agreement with our measurement. Soderblom's (1985) value of R'_{HK}, using the Mount Wilson HK photometer, and the implied F'_{HK}, are given below. Use of Rutten's calibration (line 4) brings the values in line with our spectroscopic determination. If, however, Schrijver et al.'s plage profile calibration is applicable, then our spectral difference determination may underestimate the flux by 20% with this choice of "standard" star. In any event, Rutten's calibration appears preferable to that of Middelkoop. We conclude that those published values of R'_{HK} which are based on the Middelkoop (1982) calibration are miscalibrated, and should be corrected by multiplying by a factor of 1.7. This is a simple scaling and does not in itself affect the functional form of the activity decay laws.

R'_{HK}, and analogously F'_{HK}, as defined by Noyes et al. (1984b) include all the flux between the K_1 minima. Noyes et al. argue that all this radiation is chromospheric, i.e., the photosphere is black in the cores of the K and H lines. Linsky and Ayres (1978) show that a model in radiative equilibrium does not have black cores, and suggest that 40% of the radiation between the solar K_1 minima is photospheric. We show below (Sec. IV) that our adopted Ca II decay relation allows for some 20% of R'_{HK} of the solar Ca II emission to be nonmagnetic in origin. Schrijver (1987) and Rutten (1987a) suggest that there is a basal, or minimal, Ca II flux among the dwarfs which depends only on the color, or spectral type, of the star. The basal flux is most evident among the hottest of the solar-like stars (Walter et al. 1988a), and may be due to acoustic heating by a thin convection zone. Schrijver shows that one obtains tighter flux-flux correlations upon subtraction of the empirically determined basal flux and suggests that "excess" flux above the basal level represents the magnetically heated portion of the stellar activity. Examination of Schrijver's (1987) data suggests that $\approx$ 12% of the solar Ca II flux may be basal in origin.

Basri (1987a) argues that the existence of the flux-flux correlations implies that there is only one stellar diagnostic (i.e., one can use R_{hk} to predict R_l for any other line), and measurements of many atmospheric diagnostics provide only redundant information (see Sec. IV). That this is so is probably a consequence of the basic physics of stellar atmospheres, with the cooling rate (i.e., the integrated emission flux or the activity level) determined solely by the heating rate.

Zwaan (1986) concludes that the magnetic components of the chromospheric, transition region and coronal emission are determined by just one "activity parameter." In a sense, the choice of "best" parameter depends on what one chooses to use as the independent variable. The choices for the abscissa seem to come down to rotation—either the rotation period P (or its inverse Ω), the Rossby number Ro ($=P/\tau_c$, where τ_c is the convective overturn time), or v sin i. Many authors, including Pallavicini et al. (1981), Walter and Bowyer (1981) and Walter (1981,1982) have shown that X-ray activity correlates with stellar rotation. Basri (1987a) argues that the surface flux F is the optimum activity diagnostic, because use of R introduces a temperature dependence into the activity relations. Basri finds that F vs Ro leads to an excellent relation for both single stars and rapidly rotating stars in binaries, over a wide range in spectral types. Basri also argues that the Rossby number is the best parameter to unify the activity relations for single stars and stars in close binaries, but suggests that it is premature to select Rossby number over period solely on an observational basis. On the other hand, Rutten (1987b) shows that the activity relations of giants and dwarfs can be unified by using F vs P. We note that P is a directly measurable quantity, while τ_c (and Ro) is in essence still an adjustable parameter. Skumanich and Mac-Gregor (1986) also discuss the choice of variables to minimize temperature dependences of the activity relations.

In what follows we will use the two activity diagnostics, R and F, as convenient. The two are interchangeable among stars of the same T_{eff}. In order to integrate the pre-main-sequence stars into the picture, we will use surface fluxes rather than try to normalize the temperature-dependent R values to the solar T_{eff}.

Because we are interested in the time history of the chromospheric and coronal fluxes of a 1 solar-mass star, we will take liberties in the derivation of intermediate results. We will make extensive use of relations involving the Rossby number Ro because it has been used extensively in the literature, and is a convenient parameterization of the rotation. As we will be dealing only with dwarfs of a particular mass, Ro in our case is merely a slowly time-varying parameter times the stellar rotation period (see Sec. IV.C). Furthermore, details of the basal fluxes are of only passing importance. The effect of an ionizing photon on the chemistry of the terrestrial atmosphere is the same whether it was generated acoustically or by magnetically related processes, so the decay relations that we discuss include no basal flux corrections.

We will ultimately use stellar age as the independent variable. This is allowable, even though the stellar age has no direct influence on the activity. The prime driver of the activity is presumably differential rotation, which we expect to scale with the stellar rotation, which in turn, in single stars, is a function of the stellar age.

III. PRE-MAIN-SEQUENCE ACTIVITY

The earliest point at which one can measure the activity of a solar-like star is when it first becomes visible on its pre-main-sequence (PMS) evolutionary track. Until recently the T Tauri stars were considered synonymous with the low-mass PMS stars. However, Walter et al. (1988b) have argued that the classical T Tauri stars (CTTS; Appenzeller and Mundt 1989) are only the most spectacular of the low-mass PMS stars. Most low-mass PMS stars are naked T Tauri stars (NTTS). The NTTS occupy the same region of space in the HR diagram as the CTTS but lack the strong emission lines and the continuum excesses of the CTTS (Walter 1986,1987a; Walter et al. 1987). It is only by using the NTTS that it is possible to discuss in a meaningful manner the mean stellar activity levels of solar-like stars in their PMS phase of evolution. Feigelson et al., in their chapter, review the evidence for magnetic activity in low-mass PMS stars.

The CTTS are strong sources of chromospheric and transition-region emission lines (see, e.g., Brown et al. 1981), with typical surface fluxes orders of magnitude larger than observed in even the most active solar-like dwarfs. Simon et al. (1985) showed that the chromospheric and transition-region fluxes scattered by orders of magnitude within the CTTS stars, while for stars on the main sequence the normalized fluxes R scattered little at a given age. The fluxes of the CTTS, however, are not germane to the problem at hand, because they do not arise in a solar-like chromosphere. It has been shown (see, e.g., Calvet et al. 1984) that the deep chromospheric hypothesis for the CTTS emission lines cannot explain the flux of the Hα emission line, generally the strongest observed line. The presence of P Cygni line profiles and forbidden emission lines indicates the existence of an extended atmosphere (Edwards and Strom 1987). The near- and far-infrared excesses probably arise in an extended, flattened, disk-like structure (Bertout et al. 1988). The ultraviolet excesses are very likely the manifestation of the boundary layer between the stellar photosphere and the accretion disk (Kenyon 1987). It is not clear exactly where the "chromospheric" emission in the CTTS arises, but there is strong circumstantial evidence that it comes from an extended region, not a compact solar-like chromosphere (see the chapter by Basri and Bertout).

We therefore dismiss the CTTS on the grounds that the observed activity is not solar-like. Because the CTTS have massive (typically 10^{-2} to 10^{-3} $M_\odot$) circumstellar disks, it is likely that they have not (yet) formed planets; hence the radiation field of a CTTS will not directly affect the chemistry of its planets.

On the other hand, the NTTS show every indication of possessing solar-like atmospheres without any circumstellar contributions. [The CTTS very likely do have solar-like atmospheres; the emission is just swamped by the circumstellar emission. There is evidence that the CTTS possess normal stellar coronae (Walter et al. 1988b).] In Fig. 2 we show the distribution of Hα surface fluxes (from Walter et al. 1988b) with color for the low-mass PMS stars in Tau-Aur. The NTTS lie in a narrow region, with surface fluxes comparable to those observed in the most active late-type stars (dMe and RS CVn stars), whereas the CTTS scatter widely. Similarly, the Ca II, Mg II and C IV surface fluxes are comparable to those of the most active dwarfs. Walter et al. (1987) derived the emission measure distribution for HDE 283572, a PMS 2 $M_\odot$ G subgiant, and showed that the emission measure distribution was indeed solar-like. The NTTS lie on the flux-flux correlations for cool stars.

Ages of the NTTS are determined from their positions in the HR diagram, using the isochrones of Cohen and Kuhi (1979) and of VandenBerg (personal communication). Uncertainties in the stellar extinctions can be significant. Relative ages can be deduced quite well; absolute ages depend on the inherent accuracy of the theoretical isochrones. Masses are also determined from location in the HR diagram, using theoretical evolutionary tracks of the same authors. Masses implied by the two sets of tracks differ by over 25% in places. In general, the limits on masses deduced from spectroscopic binaries (Mundt et al. 1983; Mathieu et al. 1989) are consistent with the masses predicted by theory.

The NTTS are plagued by spots. Rydgren and Vrba (1983) and Rydgren et al. (1984) determined rotation periods for 5 NTTS from their photometric modulation. The color variations are consistent with the existence of dark spots (unlike the CTTS, where modulations seem to be due to hot spots, perhaps associated with the boundary layer [Vrba et al. 1989]), and the am-

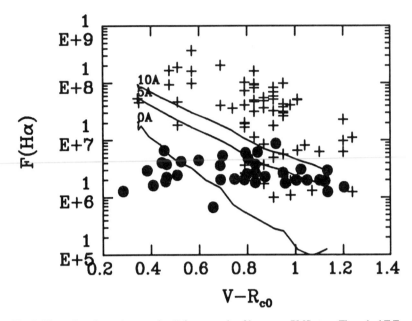

Fig. 2. Hα surface fluxes (erg cm⁻² s⁻¹) for a sample of low-mass PMS stars. The naked T Tauri stars (filled circles) occupy a limited range of surface flux, independent of $V - R$ color (unreddened, on the Cousins system). Catalogued T Tauri stars (+) scatter widely: their Hα emission does not arise in a solar-like chromosphere. Lines of constant equivalent width are overplotted (0Å corresponds to a completely filled in Hα line, with neither emission nor absorption relative to the continuum). Many of the T Tauri stars with Hα fluxes comparable to those of the NTTS may well be NTTS.

plitudes indicate filling factors of 10 to 30%. In addition, the optical and near-infrared photometry (Walter et al. 1989) is consistent with the existence of spots about 1000 K cooler than the immaculate photosphere covering up to 40% of the stellar surface. Among the restricted sample of solar-mass NTTS, spot filling factors are typically 5 to 10%. This is significantly larger than the few percent filling factors implied in the Hyades from the continuum photometric variations (Radick et al. 1987). At its most active phase, only about 1% of the solar surface is covered by sunspot groups (Bray and Loughhead 1964).

Although rotating faster than the Sun, the NTTS (and the CTTS) are not the extremely fast rotators PMS stars were once thought to be. Typical rotation periods are between 1 and 10 days, based on their photometric variability (Walter et al. 1988b; chapter by Stauffer and Soderblom).

Walter et al. (1988b) discuss the variation of chromospheric (Hα and Ca II H and K) and coronal (0.1-4.5 keV) X-ray surface fluxes with color or mass for the NTTS in Tau-Aur. Here we rederive F'_{HK} for the NTTS by subtracting a standard star of the same spectral type. For the hotter stars in our sample, this leads to surface fluxes significantly larger (up to a factor of 2) than reported by Walter et al. (1988b). Chromospheric and coronal surface fluxes for the solar mass ($0.8 \leq M/M_\odot \leq 1.2$) NTTS are given in Table II. Over their range in ages ($6 \leq \log t(\mathrm{yr}) \leq 7.7$) there is no significant variation in any of the diagnostics with stellar age. F'_{HK} and F_{hk} do show a dependence on mass, with $F'_{HK} \propto M$, the stellar mass, for $0.4 \leq M/M_\odot \leq 1.1$. Above this mass, the emission may saturate (Vilhu and Walter 1987). Currently there is no evidence that the surface fluxes correlate with stellar rotation. The ratios of F'_{HK} to F_{hk} and $F_{H\alpha}$ are similar to those observed in solar-like stars.

It is reasonable to assume that the Sun, some 4.6 Gyr ago, appeared at the stellar birthline (Stahler 1983) as a classical T Tauri star. At some point, perhaps when the inner protoplanetary disk collapsed into planetesimals, the Sun took on the appearance of a naked T Tauri star. At this point the plane of the protosolar system became optically thin, and a solar-like atmosphere became the dominant source of ionizing radiation.

IV. SOLAR ACTIVITY ON THE MAIN SEQUENCE

A. The R'_{HK}-Age Relation in Clusters

The most obvious chromospheric diagnostic in the visible spectrum of solar-like stars is the flux emitted in the central reversals of the Ca II H and K lines (Wilson [1966a] gives a prescient (p)review of this subject). The most reliable stellar ages are obtained by converting the observed color, magnitude diagrams of open star clusters into effective temperature, absolute luminosity diagrams and matching them with isochrones obtained from theoretically produced stellar evolutionary tracks (see, e.g., VandenBerg 1985).

Wilson (1963) compared histograms of eye-estimated Ca II emission strengths for stars in several clusters of known age with histograms for field stars. He showed that the strongest chromospheric emission originates in the youngest cluster (the Pleiades), the emission is weaker in the somewhat older

TABLE II
Solar-Mass Naked T Tauri Star Surface Fluxes

Line	$\log F$ (erg cm^{-2} s^{-1})
Ca II H and K	6.7 ± 0.3
Mg II h and k	6.6 ± 0.3
Hα	6.4 ± 0.2
C IV	6.2 ± 0.2
X-ray	7.1 ± 0.3

Hyades and Praesepe clusters, and is weakest for our (old) Sun and among the presumably mostly old field stars.

This analysis was not extended to clusters containing stars of solar age and older because all of the old clusters are distant, and their solar-like stars are faint. Consequently, several subsequent studies incorporated ages obtained from less-direct photometric methods (Wilson and Skumanich 1964) or statistical methods involving galactic orbital parameters or velocities which in turn are believed to be correlated with age (Skumanich 1965; Wilson and Woolley 1970).

Skumanich (1972) plotted photoelectric measures of Ca II H and K emission vs age for the Sun, and members of the Hyades and Ursa Major clusters along with indirect estimates for Pleiades members based on Kraft and Greenstein's (1969) equivalent-width measurements (Skumanich and Eddy 1981; Hartmann et al. 1984). Also, he plotted average equatorial velocities for solar-like stars in the Pleiades and Hyades, and for the Sun. Skumanich concluded that both Ca II emission and stellar rotation decay with the inverse square root of age. Skumanich and Eddy (1981), Soderblom and Clements (1987), and Soderblom (1988) largely confirm this conclusion.

However, most other recent chromospheric emission vs age studies (see, e.g., Catalano and Marilli 1983; Cabestany and Vazquez 1983; Herbig 1985; Barry et al. 1984; Simon et al. 1985; Barry et al. 1987; Barry 1988) do not support the conclusion that emission varies with the inverse square root of age. Neither do they support a linear relation between emission and rotation. Rather, they support an exponential relation between emission and rotation and between emission and age. Also, Noyes et al. (1984b) and Gilliland (1985) found exponential relations between emission and Rossby number which should be nearly the same as a relation between emission and rotation period at a given mass during the star's main-sequence life. Stepién (1987) found an exponential relation between the total emergent stellar magnetic flux and the Rossby number.

All of the studies are consistent with a power-law relation between rotation period and age of the form $P_{rot} \sim t^n$ with $0.36 < n < 1$. Rotational (v sin i) studies support a value of $n = 0.5$ although Smith (1987) stated that observational agreement with this relation has been demonstrated only in a general statistical sense. Rengarajan (1984) found $n = 0.46 \pm 0.1$ from a study of directly observed rotation periods.

Differences among the chromospheric decay laws derived in the above investigations arise from the use of different quantities supposedly representing chromospheric activity and from ages obtained by different methods. The question of which quantity (F, R, L) is most representative of the stellar activity is discussed by Hartmann and Noyes (1987) and Basri (1987a) (see also Sec. II). The question of which ages are best is settled: the cluster ages t_{cl} are best, but solar-like stars in distant clusters are difficult to observe. Observations of chromospheric emissions from clusters representing a wide range of ages have become available only recently (Barry et al. 1987). While these observations should be repeated at higher resolution and be extended to more stars, they appear to be of sufficient precision to clarify the potpourri of emission-age, emission-rotation and rotation-age dependences described above. For this reason, we describe them in some detail.

Figure 3 is a plot of each cluster's mean F'_{HK} (at effective $B - V = 0.60$) vs cluster age. The diagram is adapted from Barry et al. (1987), with the Rutten calibration applied. The horizontal lines through each data point indicate the extreme range of ages reported for that cluster. The vertical lines indicate the mean error in the mean value of F'_{HK} for the stars in that cluster.

The R'_{HK} values of the solar-like members of the clusters with $t_{cl} < 0.3$ Gyr have a greater intrinsic dispersion which causes a greater uncertainty in their mean R'_{HK} values. Consequently, for $t_{cl} < 0.3$ Gyr the curve represents the stars in a statistical sense. Individual stars can deviate substantially from the cluster mean. This may well be a consequence of enhanced flaring among the youngest and most active stars. The intrinsic dispersion in R'_{HK} within a cluster decreases with increasing age such that the Coma, Hyades and NGC

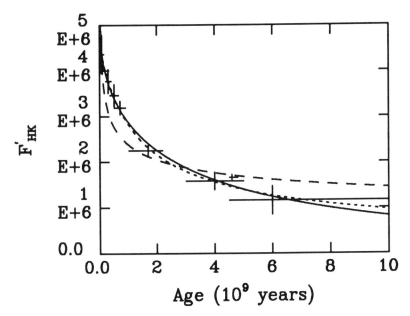

Fig. 3. F'_{HK} vs age for a sample of open clusters and the Sun, from Barry et al. (1987). The age of NGC 188 has been set at 6 Gyr. The error bars represent the range in estimated cluster ages and the mean error in the mean value of the F'_{HK} measures, normalized to $B - V = 0.60$. The solid line is the best-fit exponential decay in $t^{0.5}$. The dashed line represents the best power-law fit. The dotted line is a power-law fit with index -0.6, but with the time axis offset by 0.8 Gyr.

752 mean R'_{HK} values are defined better and are more representative of the individual member stars. Except for binaries with enhanced chromospheric emission (likely a consequence of tidally induced spinup), individual stars do not deviate widely. The uncertainty in the age of NGC 188 and the fact that only three stars have been measured, diminishes its value for deriving the age dependence of chromospheric decay.

The solid curve in Fig. 3 is an exponential decay in $t^{0.5}$; the exact relation is

$$F'_{HK} = 5.11 \times 10^6 \exp(-0.52 t_{cl,9}^{0.5}) \qquad (1)$$

or alternatively

$$R'_{HK} = 7.89 \times 10^{-5} \exp(-0.58 t_{cl,9}^{0.5}) \qquad (2)$$

where $t_{cl,9}$ is the cluster age in units of 1 Gyr. The dashed line represents the best-fit power law (in a least-squares sense) with $F'_{HK} \propto t^{-0.2}$. Taking the range in estimated ages as a 3σ estimate of the age uncertainty, and the *rms* scatter of the cluster F'_{HK} measurements as 1σ errors, the exponential fit cannot be rejected at 99% confidence, while the best-fit power law is easily rejected ($\chi_7^2 = 76$). The power laws show systematic deviations from the trend of the data. Subtraction of a constant flux equal to between 20% and 50% of the solar F'_{HK} value, to account for photospheric contamination and the basal flux and to yield a decay curve for the magnetic activity, has no significant effect on the fits. Subtraction of 50% of the solar F'_{HK} from each point increases the exponent of the best power-law fit to -0.3, but with no improvement in the χ^2 of the fit.

In Fig. 3 we overplot, in dots, a power-law decay with exponent -0.6 (see Feigelson and Kriss 1989), but with the time axis offset by $t_9 = 0.8$. This offset power-law decay is indistinguishable from the exponential decay, and there is no physical justification for choosing one over the other. However, the exponential decay is mathematically simpler, with one fewer degree of freedom. On the basis of what are arguably the best activity vs age data, we adopt an exponential in $t^{0.5}$ as the form of the activity decay law. Inverting this decay relation, chromospheric ages t_{Ca} may be estimated for stars having measured R'_{HK}.

B. Decay of Surface Li Abundance in Solar-Mass Stars

Ages of solar-like stars outside of clusters have been deduced from the surface lithium abundance, as the Li abundance declines with age (Herbig 1965; Bodenheimer 1965; Zappalà 1972; Duncan 1981). Duncan (1981) defined his Li vs age calibration by fitting the Li surface abundances for 3 clusters and the Sun with an exp(t) age dependence. Catalano and Marilli

(1983) showed that both Duncan's (1981) and Soderblom's (1983) data are better represented by $\exp(t^{0.5})$ Li decay rate.

Because both are ultimately based on cluster ages, lithium ages t_{Li} and chromospheric ages t_{Ca} should be similar. Comparison of the t_{Li} determined by Duncan (1981) with the chromospheric ages t_{Ca} for 20 of the Simon et al. (1985) stars having available Ca emission data yields $t_{Li} = 1.1(t_{Ca})^{(0.44 \pm 0.09)}$. The power-law dependence would be removed if the Li data had been calibrated with an $\exp(t^{0.5})$ age dependence. Barry (in preparation) also demonstrates that the galactic orbital peculiar velocities of solar-like stars, which should be a crude indication of age, correlate more tightly with Ca ages than with Li ages using the $\exp(t)$ calibration. We believe the chromospheric ages t_{Ca} are on a stronger observational footing than are the lithium ages, and argue that the preponderance of the data suggests an $\exp(t^{0.5})$ decay of Li in solar-like stars.

The choice of Li-age calibration does have a major impact upon the form of the derived decay law. Simon et al. (1985) present an analysis of the age dependences of five chromospheric, five transition region, and one coronal diagnostic, using Duncan's (1981) Li-age calibration for most of their age estimates. They fitted their data with an $\exp(t)$ decay, that differs from the $\exp(t^{0.5})$ decay deduced from the cluster R'_{HK}, t_{cl} data. The discrepancy would be removed if the lithium decayed as $\exp(t^{0.5})$. Consequently, $\exp(t)$ decay relations appear to be erroneous.

C. Rotation, Rossby Number, Activity and Age

Noyes et al. (1984b) found that their correlation of R'_{HK} with rotation for stars representing a range of masses was tightened considerably when the observed rotation periods are replaced by the Rossby number ($Ro = P_{rot}/\tau_c^{(2)}$ where $\tau_c^{(2)}$ is a convective overturn time). [The use of the Rossby number here was not theoretically prescribed; rather the color-dependent correction term used to minimize the scatter had the functional form of the theoretical convective turnover time. $\tau_c^{(2)}$ is a free parameter constrained by the data.] They report an alternative fit of the form $R'_{HK} \propto \exp(-0.9\,Ro)$. Using Rutten's calibration, we find

$$R'_{HK} = 1.05 \times 10^{-4}\exp(-0.88\,Ro). \qquad (3)$$

Elimination of R'_{HK} between this and the $R'_{HK}(t)$ relation gives the age dependence of the Rossby number: $Ro = 0.39 + 0.61\,t_9^{0.5}$, effective at $B - V = 0.60$ for solar-composition stars. This increase of the Rossby number with age accounts for the decrease of outer atmospheric emission with age. In this representation, the dynamo number n_D (inversely proportional to the square of the Rossby number) varies inversely with age.

Using this decay of Ro with time, mindful of the fact that Ro is a useful parameterization of the stellar rotation and not necessarily physically mean-ingful, one can predict the rotation history of a solar-mass star for comparison with observations. Barry (1988) compared Soderblom's (1985) calculated an-gular rotational velocities for 115 solar-like stars with their calculated chro-mospheric ages t_{Ca}, and found that the data can be fitted by relations of the form $(\Omega_*)/(\Omega_\odot) = Kt^{-n}$ where K depends on mass and $0.36 < n < 0.40$. A value of $n = 0.5$ is not consistent with the data.

The Rucinski and VandenBerg (1986) theoretical models predict a de-crease in $\tau(0.5)$, the convective turnover time at a particular position in the convection zone, which may be approximated by $\tau(0.5) = 15.3 - 0.29\,t_9$ during the main-sequence life of a 1 solar-mass, solar-composition star. Barry (1988) found only a 7% difference between Ro at $B-V = 0.60$ (at which $Ro(t)$ derived above is applicable) and at $B-V = 0.70$ appropriate for a young 1 $M_\odot$ star. Ignoring this small difference, the product of the linear representation of $\tau(0.5)$ and $Ro(t)$ derived above gives P_{rot} as a function of age for a 1 $M_\odot$ star because, by definition, $P_{rot} = Ro \times \tau_c$. The predicted P_{rot} for the Sun is 23.7 days, or 7% less than observed; this discrepancy may be due to the color dependences of Ro and τ_c. This polynomial relation is dom-inated by the $\sqrt{t_9}$ term for typical stellar ages, and approximates a Skumanich-type $\sqrt{t}$ relation, beginning with an initial rotation period of 6 days (the relation is also well described by a $t^{0.25}$ power law). This is consist-ent with the $P_{rot} \propto t^{0.5}$ relations found by Skumanich (1972), Soderblom (1983) and Stepién (1988b).

Barry (1988) found $\log_e(\Omega_*/\Omega_\odot) = A - 0.37\log_e(t)$ represents stellar spin-down where A is a function of mass. For the Sun's 4.6 Gyr age and 25.4 day rotation period, this leads to $P_{rot} = 14.44\,t_9^{0.37}$ (see Fig. 4). From $\tau_c(B-V)$ Noyes et al. (1984b) and the time dependence of R'_{HK}, one can derive a time history of the stellar rotation period. The Noyes et al. P_{rot} rela-tion for $B-V = 0.67$ agrees with the $t^{0.37}$ relation to better than 5% for ages between 0.2 and 10 Gyr. Given the uncertainty in the proper definition of τ, the color dependences, other modeling uncertainties, and an uncertainty of about 0.1 in the exponent of t derived by Barry et al. (1987), the reasonable agreement between these $P_{rot}(t)$ relations argues that the difference in the ex-ponents between the $P_{rot} \propto t^{0.37}$ and $Ro \propto t^{0.5}$ relations is not a discrepancy, but is to be expected if the convective overturn time τ_c varies slowly during a star's main-sequence lifetime.

D. The Activity-Decay Laws

R'_{HK} data exist for 20 of the Simon et al. (1985) stars, permitting an estimation of chromospheric ages t_{Ca}. Their fluxes for the other 4 lower-chromosphere diagnostics, their 5 transition-region diagnostics, and their co-ronal X-ray flux have been converted into flux ratios R_l analogous to R'_{HK}. In Table III, we present the recalculated decay rate A ($R_l \propto \exp[A \times t_9^{0.5}]$) for

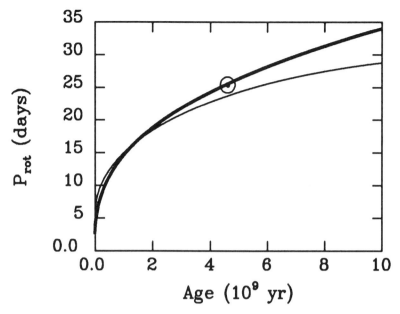

Fig. 4. The predicted increase of rotation period with age. The solid line is the power-law decay derived by Barry (1988), $P_{rot} = 14.44 \times t^{0.37}$. The thin line is the product of the Rossby number $Ro(t)$, derived in the text, and the theoretical convective turnover time τ_c, from Rucinski and VandenBerg. The former is normalized to the solar period and age (overplotted). The latter approximates a $t^{0.25}$ power law.

each diagnostic. The X-ray decay rate is calculated using the Simon et al. stars, the mean fluxes from the Pleiades and Hyades (Micela et al. 1985), and the NTTS fluxes. Note that these are observed decay relations for the total line strength. We have not attempted to separate the emission into its magnetic and basal components. As the basal flux should be roughly constant with time, the decay of the magnetic activity is steeper than the relations quoted here.

Within each atmospheric region, the decay rates are similar. Note that the Ca II decay is marginally slower than that of the other lower-chromospheric diagnostics. If real, this may be due to the presence of significant photospheric (not basal) flux between the K_1 and H_1 minima. We can force the rate of the Ca II decay to equal the mean of the other lower-chromospheric diagnostics, including Mg II that should behave in a manner very similar to Ca II, by subtracting a constant flux equal to 20% of the solar R'_{HK} value from each of the measurements. This appears to corroborate the Linsky and Ayres (1978) result that a purely radiative model solar atmosphere produces about 40% of the radiation observed within the solar K_1 minima. This constant fraction is the radiative fraction only; the basal, or acoustic,

TABLE III

Exponential Activity-Decay Laws

Line	A^a
Lower Chromosphere	
Ca II	-0.54 ± 0.02
Mg II	-0.60 ± 0.18
C I	-0.58 ± 0.35
O I	-0.64 ± 0.37
Si II	-0.63 ± 0.19
Transition Region	
C II	-0.85 ± 0.26
C IV	-0.98 ± 0.24
Si IV	-0.93 ± 0.23
He II	-0.96 ± 0.37
N V	-1.00 ± 0.51
Corona	
X-ray	-2.20 ± 0.22

^{a}Decay of the form $R \propto \exp(A \times t_9^{0.5})$.

fluxes are probably similar among all the lower-chromospheric lines, and not determinable by this kind of differential measurement.

The weighted mean of the slopes for the diagnostics representing each atmospheric level gives $\exp(-0.60 \pm 0.11 t_9^{0.5})$, $\exp(-0.94 \pm 0.15 t_9^{0.5})$ and $\exp(-2.20 \pm 0.21 t_9^{0.5})$ for the age dependences of emission from the lower chromosphere, transition region and corona, respectively. Elimination of the $t^{0.5}$ age dependence between $Ro(t)$ derived above and the decay laws presented in Table III leads to the prediction that the respective diagnostics will decay roughly as $\exp(-1.0\,Ro)$, $\exp(-1.5\,Ro)$ and $\exp(-3.7\,Ro)$. The ratio of the conversion rate of magnetic energy to luminous flux in the transition region to that in the lower chromosphere varies as $\exp(-0.5\,Ro)$ while the ratio in the corona to the lower chromosphere varies as $\exp(-2.7\,Ro)$. Using these relations, one can reproduce the flux-flux relations $R_l \propto R'^{\alpha}_{HK}$, with $\alpha =$ 1, 1.6 ± 0.4 and 3.7 ± 0.8 for the lower chromosphere, transition region and corona, respectively. For comparison, Ayres et al. (1981) found $\alpha = 1$, 1.5 and 3 and Schrijver (1987), correcting for the basal contribution to the flux, found $\alpha = 1$, 1 and 1.7, respectively.

The decay of X-ray surface flux is illustrated in Fig. 5. The solid line is an exponential decay in $t^{0.5}$, based on the fit to cluster Ca II data discussed in Sec. IV.A. Feigelson and Kriss (1989) have suggested that the true form of the X-ray decay is a Skumanich-like power law, with $L_X \propto t^{-0.6}$. Because solar-mass stars, with the exception of the PMS stars, have nearly identical radii, the L_X and F_X decay laws will be identical for stars on the main sequence. The larger radii of the PMS stars enhance their luminosities relative

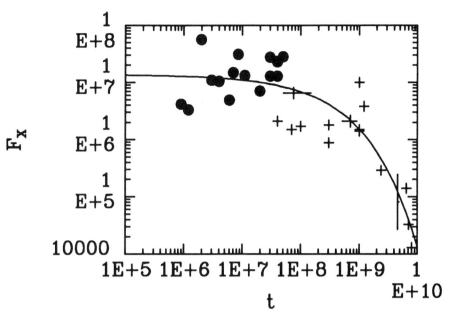

Fig. 5. The decay of X-ray surface flux with time. Filled circles are solar mass (0.8< $M/M_\odot$ <1.3) NTTS. Error bars represent the mean X-ray surface fluxes of the solar-like stars in the Pleiades and Hyades, and the Sun. Pluses are stars from Simon et al. (1985), plotted at their Ca II ages. The solid curve is the best-fit exponential in $t^{0.5}$.

to their surface fluxes, resulting in a luminosity which declines much more rapidly than an exponential at early times. However, Caillault and Helfand (1985) showed that the decline in L_x from the Pleiades through the Hyades to the Sun is exponential, and inconsistent with a Skumanich relation. Also, Stepien (1988a) finds that the X-ray filling factor f_x (proportional to L_x for solar-radius stars) decays as exp(Ro), or as exp($t^{0.5}$), after substituting the time dependence of Ro. We conclude that the power-law decay is an artifact of the choice of L_x as the activity parameter for stars with different radii, and that the balance of the evidence favors an exponential decay.

V. IMPLICATIONS FOR THE SOLAR SYSTEM

A. Solar Luminosity Variations

The Sun exerts significant influence on the planets. We are most interested in the history of solar luminosity (in specific emission lines) with time. For the purposes of this discussion, we assume that the solar gravity has not changed with time, either as a result of significant mass loss (Willson et al. 1987) or of a variable gravitational constant, and that the Earth has remained at a distance of 1 AU.

The bulk of the solar luminosity is emitted at visible wavelengths. The initial solar luminosity (based on evolutionary tracks by VandenBerg) was as large as a few tens of $L_\odot$ at the stellar birthline when the Sun emerged from its natal cloud. The total luminosity dropped rapidly as the Sun descended the Hayashi track, reaching a minimum of about 0.5 $L_\odot$ at the base of the Hayashi track at an age of 10 Myr. According to the standard model, the Sun arrived on the zero-age main sequence after 40 Myr with about 70% of its present luminosity. On the main sequence, the solar luminosity and temperature have increased to the present values (see the chapter by Dearborn). From the point of view of the solar influence on the planets, the increase in luminosity over geologic time is probably the most important effect (see the chapter by Kasting and Grinspoon). Simulations show that the terrestrial planets accumulated on time scales of about 10 Myr, and that the Earth and Venus reached their present masses only after about 100 Myr (Wetherill 1989). Therefore, discussions of the effects of solar activity upon the Earth and its atmosphere during the solar T Tauri ($\lesssim$ 10 Myr) phase are academic.

While the solar luminosity as a whole has increased with time, the solar activity has decreased significantly, as shown in the previous section. When the Sun was very young, the flux of ionizing photons and of protons from solar flares was significantly larger than observed today.

B. An Ionizing Flux History

Our interest in examining the Sun in time is to determine the ionizing flux rate at the Earth as a function of time. This number has major consequences for the evolution of the terrestrial atmosphere (see the chapter by Hunten et al.). The number of ionizing photons incident per unit area at one astronomical unit is given by

$$N_i = \int \frac{L_i(t)}{h\nu} \frac{1}{4\pi(1\mathrm{AU})^2} \, dt \qquad (4)$$

where L_i is the luminosity of the Sun at ionizing wavelengths ($<$912 Å).

Given the decay laws in Sec. IV, we can determine the ionizing flux at the Earth during its history. In Table IV, we present the time history of the fluence N_i of coronal X rays, transition region, and lower-chromospheric emission lines. Note that, as pointed out by Simon et al. (1985) among others, the decay is fastest at the highest energies. The critical photons for their effects on the terrestrial atmosphere are those with X-ray and extreme ultraviolet (EUV) wavelengths. In the absence of observations of EUV fluxes of cool stars, we assume that the EUV flux decays at the rate of the coronal X rays. The rate must lie between this and the decay rate of the transition-region lines; a slower decay will produce a greater fluence of ionizing radiation.

For X rays in the keV band (see Fig. 5), the cumulative number of photons incident upon the Earth from time 0 to the present is of order 10^{27}

TABLE IV
Solar Ionizing Flux History

Age Interval	Fractional Fluence ($F \times t$)		
1 Gyr	X ray	TR[a]	LC[a]
< 0.0001	0.01	0.004	0.003
0.0001 – 0.001	0.03	0.01	0.007
0.001 – 0.01	0.09	0.03	0.012
0.01 – 0.03	0.06	0.02	0.011
0.03 – 0.1	0.09	0.05	0.03
0.1 – 0.4	0.24	0.13	0.11
0.4 – 0.7	0.13	0.10	0.09
0.7 – 1.0	0.09	0.09	0.08
1.0 – 2.0	0.15	0.23	0.22
2.0 – 3.0	0.06	0.16	0.18
3.0 – 4.0	0.03	0.12	0.15
4.0 – 4.6	0.01	0.06	0.08

[a]TR = transition region; LC = lower chromosphere.

cm^{-2}, with half of these arriving in the first 0.4 Gyr (half the age of the Hyades). At all wavelengths less than 912 Å, the total flux is about 10 times that in the ≈0.1 to 4.5 keV soft X-ray band (Woods and Rottman 1990). The fluence per logarithmic time interval peaks at about 0.3 Gyr. Note that the integral distribution is largely insensitive to the contribution at PMS ages because of the short absolute times involved. Therefore, uncertainties due to the changing radii of the PMS stars, in the extinction on the midplane of the early solar system, and in the exact time when the terrestrial atmosphere first appeared are not important. The ionizing fluxes are most relevant for altering atmospheric chemistry once the planetary accretion episode is largely over and planetary atmospheres cease being disrupted by major impacts. The time scale for this is probably of order 100 Myr, so again the early history of the fluence may not be directly relevant. There appears to be sufficient ionizing radiation from the Sun during this period to account for the isotopic ratios of the noble gases using an exospheric escape model (Pepin 1989).

C. Stellar Flares and Proton Fluxes

Evidence for a probable second direct effect of the early solar activity upon the solar system is found in proton irradiation deduced from elemental compositions of inclusions in meteorites (see the chapters by Wood and Pellas and by Caffee et al.). If the irradiation did not arise from galactic sources, then the proton fluence inferred from the isotopic ratios of the noble gases appears to require either an exposure to an active Sun (assuming that flare proton fluxes scale in some way with activity levels) for a long period prior to accumulation of these primitive bodies into the meteorites, or to much higher levels, perhaps in T Tauri flaring episodes, for a shorter time. The meteoritic evidence appears to constrain only the proton fluence, and not the flux or the exposure times.

Unfortunately, we know little about either flare frequencies or flare intensities among young solar-like stars (see the chapter by Feigelson et al.) or the scaling of proton fluxes with X-ray fluxes in very large flares, and so we cannot estimate the proton irradiances to which the primitive material in the solar nebula may have been subject. However, it is more likely that continued exposure to flaring after, rather than during, the T Tauri phase is responsible for the spallation nucleosynthesis of noble gases in the oldest meteoritic inclusions. This conclusion comes from the observation that the T Tauri stars are surrounded by massive disks of material. Formation of solid bodies out of the dust probably occurs at the midplane of the disk, where the densities are highest. Such particles, by virtue of their mass and drag do not stray far from the midplane. In the midplane the opacities are large, and it is unclear whether flare-produced protons could penetrate the protoplanetary disk. Additionally, once solid bodies begin to accumulate, they proceed to grow and may clear the disk on very short ($\approx$ 0.1 Myr) time scales, so that as substantial solid bodies begin to form, the star will not remain a classical dusty T Tauri star for long, and becomes a naked T Tauri star.

VI. CAVEATS AND CONCLUSIONS

It would be a serious mistake to conclude that the decay of stellar activity with time is well understood. It is firmly established that the activity does decrease with time, and that the higher excitation lines decay more rapidly. The decay of the activity is an empirically described phenomenon, and therefore can be defined solely on an observational basis. However, as we have discussed, one can perform correlations and obtain decay relations that may or may not be physically meaningful. There is little in the way of predictions from dynamo theory on the time decay of surface fields or of stellar spindown to guide the observers.

The question of which parameter (L, F or R) is the better activity diagnostic, although moot for the solar-mass stars, is of fundamental importance. Equally important is the treatment of nonmagnetic, including basal, contributions to the fluxes. Decay of the total line fluxes is observable, and is relevant to our question of the time history of solar irradiation of the Earth. But the true decay of the magnetically sensitive portion of the emission, that part relevant to modelers of stellar activity and dynamos, is not observable, except perhaps at X-ray wavelengths. One must account for the basal flux before constructing the true activity—activity-decay relations (see Schrijver 1987). [If the basal fluxes are indeed due to acoustic waves generated in the convective zone, then there is no a priori reason to expect that the basal fluxes will be constant over the main-sequence lifetime of a solar-like star, because the depth of the convective zone, and the convective velocities, change with age.] The decay laws quoted herein, with the exception of that for coronal X rays, are probably shallower than the true decay laws for the magnetically active component of the solar-like stellar atmospheres.

A. Rotation Period Discrepancies

There are discrepancies between the observed and predicted rotational periods for solar-mass stars. Radick et al. (1987) compared directly observed rotation periods along the Hyades main sequence with calculated rotation periods after Noyes et al. (1984b). They found the predicted rotation periods are about 2 days (25%) shorter than the measured periods for the early G stars near $B - V = 0.60$ and that they are about 4 days (45%) longer near $B - V = 0.90$. They recalculated the rotation periods with a different formulation for the Rossby number and convective overturn time (Gilliland 1985) and found a small improvement in the agreement between the calculated and the observed rotation periods. Their results suggest that: (1) the Noyes et al. procedure should be improved upon to achieve better agreement between predicted and observed rotation periods among stars of the same age but different mass; and (2) this improvement might come from an improved theoretical understanding of the parameters which describe the dynamo mechanism as a function of mass and age. In particular, any time history of the stellar rotation period must take into account the changes in luminosity and color with time on the main sequence. The polynomial formulation of $P_{rot}(t)$ yields longer periods at the age of the Hyades than do either the Noyes et al. (1984b) or the Barry (1988) $t^{0.37}$ formulations, but the latter do better at predicting the solar rotation. One might start by looking for time and color (t)-dependent modifications to the formulations that would bring agreement between the predicted and observed periods.

B. Observational Tasks

As in most cases, one would dearly like more and better observational data in order to define the decay rates better and search for subtle effects in the relations that may be significant in terms of our theoretical understanding of the problem.

Although it is a difficult observational task, we suggest that studies of stellar activity concentrate on open clusters because the ages of cluster members can be estimated with far more precision than they can for the (brighter and more easily observed) field stars. It is crucial to increase the number of solar-mass stars observed in each cluster, to improve statistics and determine the variance as a function of age. The decline of the dispersion in the relation with age may indicate the time scale for the stellar braking mechanism to erase all memory of the initial rotation rates. Note that the time decay of the rotation period, together with a model of the angular momentum transfer mechanism, will yield the mass-loss rate as a function of stellar age. It is also important to observe cluster members over a range in masses. Activity levels in single stars depend on both rotation and mass. Knowledge of both parameters are necessary to provide constraints to dynamo theories. In addition, among the solar-like stars there are variations in metallicity [Fe/H]. The effect of metallicity upon activity levels is not known. We suggest that a study of the dependence of R'_{HK} on [Fe/H] at a given T_{eff} (or mass) is warranted, to determine whether any systematic effects exist.

X-ray observations of T associations and young clusters should be planned so that information on stellar coronal variability is obtained. This will permit estimation of the frequencies and intensities of stellar flares, which is an important starting point for estimation of proton fluences in the early solar system.

Ancillary to observations of activity in cluster stars, it is important as well to determine rotational periods for these stars. By definition, $P_{rot} = Ro \times \tau_c$. If $P_{rot} \propto t^n$, $Ro \propto t^m$ and $\tau_c \propto t^{-q}$, then $m = q + n$. Direct observations of P_{rot} as a function of age would improve our knowledge of n, but these are very difficult to obtain for old, solar-like stars. Both Ro and τ_c are unobservable. A problem with τ_c is that its definition depends on an arbitrary choice of position within the convection zone. However, if one defines τ_c to minimize the scatter in the R'_{HK} vs P relation, one can deduce a value for m. Helioseismology (and perhaps astroseismology) will shed light on the structure and dynamics of the convection zone, and may provide some physical basis for estimates of τ_c. But, without further insight into the dynamo mechanism, or into whether or not the Rossby number is a proper parameterization to use, clarification of these relations is little more than an algebraic exercise.

C. Summary

In single stars (and binaries where tidal coupling is unimportant), stellar activity decays with increasing age. The decay, at all levels in the stellar atmosphere, can be described by

$$F_l \propto \exp(A \times t^{0.5}) \tag{5}$$

where t is the stellar age and the coefficient A is a function of excitation temperature. This particular form of the decay law is based on observations of stars in clusters, which can be accurately dated. From the observed decay laws and stellar rotation periods, and estimates of the convective overturn time τ_c, one can derive a self-consistent set of equations describing the time decay of the Rossby number Ro and the stellar rotation period. For now, it seems that the Rossby number (or equivalent) is the best predictor of stellar activity levels.

One can integrate the decay relations to determine the ionizing radiation history from the Sun at the terrestrial orbit. The decay is rapid, and the fluence per logarithmic time interval peaks between 0.1 and 1 Gyr. The ionizing radiation from the Sun during this period appears sufficient to account for the isotopic ratios of the noble gases. The fluence of ionizing radiation at ages less than about 0.1 Gyr is of little practical consequence because the terrestrial planets as we know them had probably not fully formed much before that time.

MAGNETIC ACTIVITY IN PRE-MAIN-SEQUENCE STARS

ERIC D. FEIGELSON
Pennsylvania State University

MARK S. GIAMPAPA
National Optical Astronomy Observatories

and

FREDERICK J. VRBA
U.S. Naval Observatory

The extensive new evidence for extremely high levels of magnetic activity on the surface of pre-main-sequence stars is reviewed and analyzed in the context of solar evolution. Most of these observations concern "weak" T Tauri stars, in which the surface properties of the young stars are not dominated by circumstellar matter as in "classical" T Tauri stars. Optical photometry shows that weak T Tauri stars are moderately rapid rotators with cool starspots covering 5 to 40% of the surface. Optical and ultraviolet spectroscopy indicates that their chromospheres are enhanced ~50 times over contemporary solar levels. X-ray and radio intensities are of order 10^3 to 10^5 times greater than peak solar levels, and they vary on time scales of days and (in a few cases) hours or minutes. The variability can only be explained in terms of some extremely powerful magnetic-flare model. The large numbers of weak T Tauri stars, probably larger than the population of classical T Tauri stars, strongly suggests that the Sun was a weak T Tauri star for at least part of its pre-main-sequence evolutionary phase. We summarize these findings in a suggested portrait of the Sun as a 1 Myr old weak T Tauri star. These astronomical studies support models for enhanced solar flaring developed by researchers of the ancient solar system to explain certain isotopic and heating anomalies in carbonaceous chondrites.

I. INTRODUCTION

Our understanding of the Sun during its first 10 Myr of existence is based primarily on observations and models of pre-main-sequence (PMS) stars. Interstellar clouds of molecular gas and dust several hundred lightyears away are forming stars today, often referred to as T Tauri stars, that exhibit unusual properties presumably shared by the Sun immediately after its formation 4.6 Gyr ago. Since their discovery in the 1940s, astronomers' interpretation of T Tauri stars have developed around two principal themes. First, their broad optical emission lines were widely interpreted as signs of gas outflow from the surface, with mass-loss rates several orders of magnitude greater than those seen in main-sequence stars. This is the T Tauri wind mentioned in standard textbooks, and often considered to be the mechanism by which the solar nebula was dissipated during formation of the Sun's planetary system. Astronomers have also found collimated outflows on spatial scales much larger than the stellar surface in some PMS stars, as revealed by bipolar molecular flows, Herbig-Haro objects and emission-line jets. Recently, these findings have been incorporated into an elegant model involving an accretion disk, boundary layer and polar outflow. This new model, reviewed in the chapter by Bertout, et al, successfully accounts for many of the peculiarities of classical T Tauri stars (CTT): the broad permitted optical emission lines, asymmetrical forbidden-line profiles, bipolar flows, ultraviolet continuum and emission lines, and infrared excesses.

The second theme woven through studies of PMS stars involves high levels of magnetic activity, manifested in various enhanced forms of solar magnetic activity including sunspots, chromospheric emission and flares. Joy (1945) was the first to suggest that T Tauri emission lines "must have their origin under similar physical circumstances" as the solar chromosphere. This analogy led to the "deep chromosphere" model suggested by Herbig (1970). Also, researchers in Mexico (Haro and Chavira 1966) and the Soviet Union (Ambartsumian and Mirzoyan 1975) discovered hundreds of flash variable stars in star-forming regions, and argued for a close analogy between T Tauri stars and dwarf M flare stars. However, the PMS flash variables are much less massive than the Sun, and it was not clear that similar behavior occurs in more massive PMS stars. Herbig and Rao (1972), in their catalog of PMS emission-line stars, stated that "Orion-type variables without line emission . . . are actually more numerous than the T Tauri stars" and they may constitute a link between the strong emission stars and young main-sequence stars with high chromospheric activity. But few of these stars were individually identified and little research was directed towards them during the 1960–70s.

This situation changed dramatically in the 1980s, when images of nearby star-forming regions were obtained with astronomy's first major imaging X-ray instrument, the Imaging Proportional Counter on board the Einstein (HEAO-2) satellite launched in 1978. The detection of dozens of highly variable PMS stars with X-ray luminosities $\approx 10^3$ times solar levels was unexpected. Many of the X-ray-selected stars are missing the characteristic broad emission lines and unusual colors of the CTT stars, and thus were not prominent in traditional Hα and infrared surveys. These 'weak' T Tauri (WTT) stars are also found in optical surveys for Ca II H and K emission stars, weak Hα emission stars, and proper-motion members of young star associations. The X-ray, optical and radio properties of these stars, which are generally close to one solar mass, are best explained by extremely high levels of magnetic activity, including individual powerful flares. Although the X-ray surveys are often incomplete and these stars are difficult to distinguish from ordinary stars at other bands, it now appears that the population of WTT stars at least equals, and perhaps greatly exceeds, the population of CTT stars.

The exact relationship between CTT and WTT stars is still somewhat uncertain. Are some stars born as CTT and others as WTT? Or are all stars born CTT and evolve through a WTT phase as they descend to the main sequence? Are the underlying stellar surfaces in CTT stars identical to those of WTT stars, or are there differences in addition to those associated with the circumstellar material? In all of these cases, there is now considerable evidence that solar-type stars, and therefore the Sun, exhibited magnetic activity orders of magnitude stronger than is seen today during its 1 to 10 Myr PMS evolutionary phase.

This chapter summarizes the properties of WTT and CTT stars associated with magnetic activity during the PMS phase. We attempt to draw a portrait of the Sun as a weak T Tauri star and inquire into the implication of PMS magnetic activity for solar system studies. Due to the rapid growth of research on this topic, only a fraction of the relevant papers can be mentioned here. The reader is referred to more detailed reviews on the X-ray (Feigelson 1984), ultraviolet (Imhoff and Appenzeller 1987), optical (Walter 1987A) and radio (Feigelson 1987; André 1987) studies of WTT stars. Reviews on PMS magnetic activity (Giampapa and Imhoff 1985; Gahm 1986; Giampapa 1986; Montmerle and André 1988), and reviews of T Tauri stars (see, e.g., Basri 1987b; Appenzeller and Mundt 1989; Bertout 1989; see also the chapter by Bertout et al.) are also valuable.

II. BULK PROPERTIES OF WEAK T TAURI STARS

The basic stellar parameters (surface temperature, luminosity, mass, rotation rate, etc.) are often difficult to obtain for CTT stars, because the presence of circumstellar material often both masks the underlying photospheric absorption lines and extincts the stellar light. This problem is alleviated for WTT stars; several dozen have now been characterized with some degree of certainty.

Mass

Direct mass determinations of WTT stars can only be accomplished via visual orbit determinations or for stars in eclipsing double-lined spectroscopic binary systems. Currently there are only 10 PMS spectroscopic binary systems known (Mathieu 1989) of which 8 contain WTT stars. The spectroscopic binary frequency of WTT stars appears to be similar to that of main-sequence, near-solar-mass field stars (Mathieu, Walter and Myers, in preparation). In the two systems with the best mass contraints, one (V826 Tau) consists of two coeval $0.6M_\odot$ stars and the other (162814–2427) consists of $>1.02M_\odot$ and $>0.94M_\odot$ stars. Otherwise, the mass is estimated from positions in the Hertzsprung-Russell (HR) diagram combined with theoretical evolutionary tracks.

Luminosity and Surface Temperature

In WTT stars, the absence of active or even self-luminous accretion disks not only allows direct determinations of the pure stellar luminosities but also affords straightforward determinations of stellar reddening and effective temperatures. Once placed on a HR diagram, the stellar positions can be compared with theoretical evolutionary tracks providing estimates of masses and ages. Figure 1 shows the HR diagram of the best-studied star-forming region, the Taurus-Auriga cloud complex which contains about 36 CTT stars and 46 known WTT stars. More WTT stars are undoubtedly present, since the X-ray surveys are incomplete. Most have photospheric temperatures around 3500 to 4000 K and bolometric luminosities between 0.2 and $2L_\odot$. The convective evolutionary tracks indicate that the Sun would have had a constant temperature over an age range of 0.3 to 1.5 Myr of 4200 K, corresponding to a spectral type around K5.

Age

The Taurus-Auriga PMS stars in Fig. 1 show a spread of ages ranging from 0.3 to ≥ 20 Myr (Walter et al. 1988b). These ages are estimated from the theoretical isochrones of Cohen and Kuhi (1979) and may be too high by 0.1 to 0.2 Myr (Stahler 1983). WTT stars in the II Sco and Ori OB1c (Walter 1987B) associations show mass and age ranges similar to those seen in Taurus-Auriga WTT stars. Most of these stars lie in the mass range 0.3 to 0.8 $M_\odot$, with a tail of higher-mass stars ranging somewhat higher than 2 $M_\odot$. The CTT stars appear to be on average younger than WTT stars, although their mass distributions are nearly identical. K. Strom et al. (1989) find that the CTT stars are more common for ages $\lesssim$ 3 Myr, and WTT stars dominate at older ages. This is consistent with the view that all stars are born with CTT characteristics and evolve into WTT stars on this time scale. However, it does not exclude the possibility that some stars are born and remain WTT stars. The spatial distribution of PMS stars in X-ray images, for instance, show that

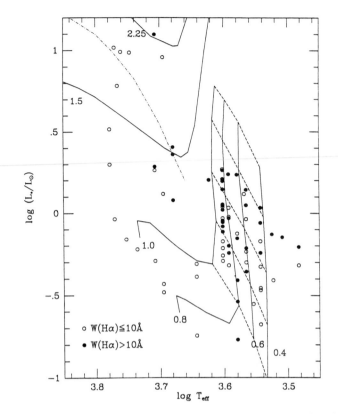

Fig. 1. The HR diagram for all known classical T Tauri (CTT:filled circles) and weak T Tauri (WTT:open circles) stars in the Taurus-Auriga cloud complex. CTT stars are defined here to be those with Hα equivalent width >1 nm. Photospheric luminosities without circumstellar contributions are plotted. Solid lines show theoretical evolutionary tracks, with mass in $M_\odot$ labeled. Dashed lines show isochrones for 0.1, 0.3, 1, 3 and 10 Myr for lower-mass stars, and the dot-dash line shows the isochrone for 3 Myr for higher-mass stars (figure from K. Strom et al. 1989).

many WTT stars lie as close to the molecular cloud cores as CTT stars (see, e.g., Montmerle et al. 1983; Feigelson and Kriss 1989) suggesting a common age distribution.

Radius

Using the Barnes and Evans (1976) relation, an empirical relation between color, luminosity and radius in cool stars, we have calculated radii for $0.5 \leq M_\odot \leq 1.5$ WTT stars in Taurus-Auriga using the data of Walter et al. (1988b) and Vrba et al. (1989). Taking a subset of these stars with masses 0.6 to 1.2 $M_\odot$, correcting their Barnes-Evans radii to 1.0 $M_\odot$ using evolutionary tracks from Cohen and Kuhi (1979) and the relation $L = 4\pi R^2 \sigma T_e^4$, we have plotted stellar radius as a function of stellar age in Fig. 2. The stars

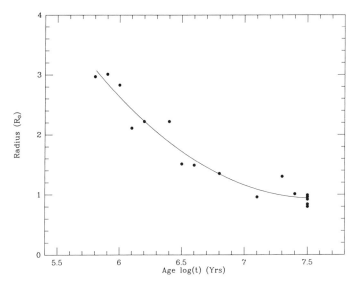

Fig. 2. Plot of photospheric radii for 17 WTT stars in the Taurus-Auriga cloud as a function of age, derived from the Barnes-Evans relation and the HR diagram isochrones as described in the text. The curve is a least-squares second-order polynomial fit to the data.

plotted at $\log(t) = 7.5$ yr have ages which are more difficult to determine because they are well out of the convective Hayashi phase of evolution and are nearing the main sequence along radiative tracks.

Several Taurus-Auriga WTT stars have rotation periods determined from photometric monitoring of light variations due to rotational modulation by cool star spots (see Sec. IV below), and have measured spectroscopic projected equatorial rotation velocities, V (sin i). These parameters can be combined via $V = 2\pi R/P$ and $V = V_{meas}/(\sin i)$ to determine lower limits to the stellar radii in the form of $R(\sin i)$. These values are roughly consistent with the Barnes-Evans radii computed above. These radii are also consistent with those inferred from the theoretical evolutionary tracks of Iben (1965), indicating that the main-sequence Barnes-Evans relation can be applied to PMS stars self-consistently. Figure 2 is thus likely to be a reasonably reliable semiempirical measure of how solar-mass, stellar radii change with age, and is the first such measure not based strictly on theoretical considerations. At an age of 1 Myr, the Sun likely had a radius of near 2.5 $R_\odot$, and contracted to its present radius at an age of 10 to 20 Myr.

Rotation

The projected surface rotation velocities have been measured spectroscopically in several dozen WTT stars (see, e.g., Bouvier et al. 1986; Hartmann et al. 1987; Walter et al. 1988b). These studies show that WTT stars generally have $V(\sin i) \leq 10$ to 15 km s^{-1}, a few at 20 to 30 km s^{-1}, and a few higher-mass and/or more highly evolved stars with $V(\sin i)$ as high as ≈ 75 km s^{-1}. It is of note that WTT stars of ≤ 1.0 M$_\odot$ which are still on their convective Hayashi tracks (ages ≤ 6 Myr) show no evidence of spinning-up as a function of age. Studies of older clusters with significant numbers of ≤ 1.0 M$_\odot$ stars nearing the ZAMS, however, show typically much higher rotation velocities (cf. Stauffer et al. 1985). The few WTT stars with photometrically measured rotation periods generally have rotation periods of 3 to 7 days, except for more massive (≈ 2 M$_\odot$) stars with periods of 1 to 2 days that are nearing the main-sequence along radiative tracks. To put these results in the perspective of the Sun, at age 1 Myr, the Sun probably had a radius 2.5 R$_\odot$, a typical V_{rot} around 20 km s^{-1}, and thus a rotation period of ≈ 6 days. These properties are summarized in Sec VIII below.

It is also interesting to compare the WTT star rotation rates with their CTT brethren at the same location in the HR diagram, but interacting with accretion disks. V ($\sin i$) and photometric monitoring studies find the same rotational velocity and period properties as for the WTT and CTT stars in the Taurus-Auriga clouds (Hartmann et al. 1987; Walter et al. 1988b). This similarity indicates either that WTT and CTT stars have similar age distributions, or that the CTT circumstellar disks and winds do not transport significant amounts of angular momentum into or out of the star.

III. WEAK T TAURI CHROMOSPHERES

As mentioned in Sec. I, the resemblance between CTT optical emission-line spectra and that of the solar chromosphere has been noted since the first T Tauri spectra were obtained. The connection is particularly striking in the ultraviolet spectrum, where the Mg II resonance line is strong, lines of Si I, C I, O I and Fe II indicate plasma around 10^4 K, and lines of C IV, Si IV, N V and He II indicate plasma around 10^5 K (Imhoff and Appenzeller 1987). The Mg II h and k surface fluxes are typically 10^7 to 10^8 erg cm^{-2} s^{-1}, approximately 50 times solar levels (Giampapa et al. 1981). The Mg II, Ca II, far ultraviolet emission and X-ray emission together can comprise 1% or more of the bolometric luminosity in T Tauri stars, a level of chromospheric nonradiative heating higher than seen in any other type of late-type star. While this general approach has been successful in explaining some essential aspects of T Tauri line spectra, the line strengths and profiles of many CTT spectra require a geometrically extended and high-velocity atmosphere in addition to compact chromosphere and coronal regions. In particular, it is not possible to explain simultaneously the strong Hα intensities and the large Balmer decrements seen in CTT stars without invoking an emitting region much larger than a compact stellar chromosphere.

One notable spectral characteristic of CTT stars is that the Ca II near-infrared triplet lines are always saturated (that is, the three lines have equal strengths, as they do in solar plages and flares), but their intensities vary greatly from star to star. Herbig and Soderblom (1980) interpret this to mean that T Tauri chromospheres are inhomogeneous, with differing fractions of hot and cool regions. This may be due to plage associated with the large cool starspots indicated by photometric studies (Sec. IV below).

While most optical and ultraviolet spectral studies have involved CTT stars, studies of WTT and weak-lined CTT stars have begun to reveal the same similarity to solar spectra. Finkenzeller and Basri (1987) show that most of the weak absorption lines in WTT stars are identical to those of matched main-sequence stars. The residual emission lines, particularly in the G2,WTT star Sz 19, are similar to those seen in solar chromospheric spectra obtained during eclipse. Another X-ray-emitting WTT, HDE 283572, has been studied at optical and ultraviolet wavelengths in detail (Walter et al. 1987). The total radiative losses between temperatures of 10,000 K and 200,000 K are $\approx 10^3$ times that of the quiet Sun. This star, however, has a mass of ≈ 2 $M_\odot$ and is a rather rapid rotator, so it may not be a good analog of the early Sun.

Theoretical model chromospheres of increasing sophistication were developed to account for the CTT spectra, following Herbig's (1970) suggestion that the atmospheric temperature minimum is located at high mass column density, resulting in higher chromospheric-line opacities and enhanced line intensities. Cram (1979) constructed such "deep chromosphere" models, giving predictions for the ultraviolet continuum, Balmer, calcium, sodium and iron line ratios and profiles. The model provided semi-quantitative explanation for a blue/ultraviolet continuum, narrow emission core and broad absorption wings of Ca II and Na D lines, the saturation of the Ca II infrared triplet lines, and other features. However, significant discrepancies with observed T Tauri spectra were also found; in particular, the broad hydrogen Balmer lines could not plausibly be due to a strong chromosphere but must be circumstellar in origin. More detailed models with improved radiative transfer were developed by Calvet et al. (1984), which predicted additional effects such as differential veiling of different absorption lines.

The deep chromosphere models, combined with a normal photosphere, thus appear to explain reasonably well the detailed optical and ultraviolet spectral characteristics of WTT stars. As a star exhibits more of the CTT properties, however, it is not always clear exactly which features are due to a strong chromosphere and which to an ionized wind or star-disk boundary layer. It is clear that a "classical chromosphere" alone can not account for all of the properties of the emission-line spectra in CTT stars. An extended region must be invoked.

IV. WEAK T TAURI STARSPOTS

Several photometric monitoring campaigns have been carried out in order to deduce rotation periods of WTT stars by rotational modulation of the stars' light by surface features (see, e.g., Rydgren and Vrba 1983; Rydgren et al. 1984; Bouvier et al. 1986; Bouvier and Bertout 1989). When the observations are carried out at several wavelengths it is, in principle, possible to model the starspot parameters of temperature and hemisphere filling factors. Details of the actual spot modeling vary from one group of investigators to another, and the photometric data are often insufficient to constrain completely the models, especially for low-amplitude stars. Nevertheless, a clear general picture of the nature of WTT starspots has emerged.

Table I summarizes these results for WTT stars where ΔV is the full range of amplitude of the lightcurves in the V filter during the epochs observed, f is the full range of the fraction of a stellar hemisphere covered by the spot, and ΔT is the full range of the difference in temperature between the star and the spot. The light variations are always due to spots cooler than the stellar photospheres by typically 500 to 1000 K and with filling factors 0.1 to 0.4. These spots cover much larger areas of the stellar surfaces than contemporary sunspots, or those on young main-sequence stars like the Hyades or Pleiades. The only late-type stars with comparably large cool starspots are the RS CVn and BY Dra-type binary stars, and even here filling factors >0.1 are rare (Rodono et al. 1986; Bouvier and Bertout 1989). Spots hotter than the normal stellar photospheres are ruled out since such spots would show a much steeper increase in amplitude of light variation toward shorter wavelength than is actually observed.

The data of Table I for stars with multiple epochs of observation indicate that the starspots can last for years, but typically evolve from one observational epoch to the next. Only V410 Tau currently has enough collected observational data to examine spot evolution in any detail. Herbst (1989) reports

TABLE I
WTT Stars With Modeled Starspots

Name	Epoch	ΔV (mag)	f (frac)	ΔT K
HP Tau/G2	1984[a]	0.10	.20–.35	300–500
HD 283447	1981[b]	0.10	.17	730
V410 Tau	1981[b]	0.21	.25	600
	1984–86[c]	0.26–0.56	.08–.17	780–1265
	1981–87[d]	0.21–0.60	.00–.42	1050–1400
FK2 (P3)	1981[b]	0.21	.27	510
	1984–86[c]	0.26–0.41	.09–.11	535–1060
WK2 (P4)	1984–86[c]	0.12–0.26	.04–.17	685–810

[a] Vrba et al. 1989.
[b] Rydgren and Vrba 1983.
[c] Bouvier and Bertout 1989.
[d] Vrba et al. 1988.

that two spots were present during 1983–84, while only a single larger spot was present in 1986–87. During this time span the secondary spot drifted in phase with respect to the primary and gradually disappeared.

Similar photometric monitoring campaigns aimed at finding rotation rates have been successful in determining spot parameters for CTT stars (Bouvier and Bertout 1989; Vrba et al. 1989, and references therein). In the cases of weak-lined, marginal CTT stars, dark spots similar to those on WTT stars are observed. However, the lightcurves of the stronger emission-line CTT stars are generally dominated by smaller spots ($f \leq 0.01$), several 1000 deg hotter than the stellar photospheres, and with lifetimes of at most a few weeks. Such hot spots are thought not to be photospheric in origin but to be due to accretion directed along magnetic field lines at the boundary layer between the accretion disk and stellar photosphere (Bertout et al. 1988). Cool spots may also be common in stronger-lined CTT photospheres but are generally difficult to detect due to masking effects from the hot spots. There is at least one clear case of co-existing hot and cool spots on a CTT star, DN Tau (Vrba et al. 1986).

V. VARIABILITY ON TIME SCALES OF DAYS TO YEARS

In the optical band, aperiodic variability with amplitudes up to several times the total photospheric luminosity is a common characteristic of CTT stars. A common mode for such variations is for the star to become fainter over a broad range of optical wavelengths for a few weeks or months. While long-time-scale variability can in principle be attributed to a variety of causes (e.g., changes in the photosphere or chromosphere, circumstellar disk or wind, or line-of-sight obscuration), the large amplitude and color characteristics of the changes are not easily explained. Variations in the energy output from the stellar interior might be involved. Appenzeller and Dearborn (1984), however, have suggested that CTT variability is due to the suppression of photospheric emission by large-scale magnetic fields near the stellar surface. If true, this would imply that starspots as large or larger than those seen on WTT stars (Sec. IV) are present on CTT stars.

Among WTT stars, long-term variability can only be assessed for those stars whose rotationally modulated lightcurves have been determined for more than one epoch. Such data are available to date for a few Taurus-Auriga WTT stars, and most reveal no evidence for systematic changes in the mean light levels of these stars beyond 0.05 mag in the V filter. The WTT star FK2 is an exception, and exhibitied ~ 0.20 mag decrease betwen 1981 and 1984 (Bouvier and Bertout 1989). While more work is needed before definitive statements can be made, WTT stars are clearly in stark contrast to the CTT stars whose mean light levels may vary by as much as 2 mag outside of their periodic variability.

The X-ray emission of WTT and CTT stars is virtually always variable by large factors, although accurate time scales and amplitudes are difficult to establish because the measurements are limited to the few dozen images of nearby star-forming regions obtained with the Einstein satellite between 1978 and 1981. The best data are from the ρ Ophiuchi cloud, for which two observations where made in 1979 and four within a single week in 1981 (Montmerle et al. 1983a, b). Different combinations of X-ray emitting stars are seen at different epochs, giving the impression of an "X-ray Christmas Tree". The variations are typically factors of 2 to 20 in amplitude and are statistically similar to the distribution of peak X-ray emission in solar flares, as shown in Fig. 3. The X-ray luminosities, however, are 10^{29} to 10^{31} erg s^{-1} in the 0.5 to 4 keV soft X-ray band, or $\approx 10^3$ times more powerful than even a Class 3 solar flare.

The newest band of the spectrum to show variability in pre-main-sequence stars is the radio continuum. Radio emission had been detected in a dozen or more CTT stars during the early 1980s, in efforts to measure free-free thermal bremsstrahlung from the T Tauri winds or collimated outflows. In a number of cases, the spectral or morphological characteristics of the emission agrees well with the thermal outflow model (see the chapter by Bertout et al.). However, radio continuum surveys of star-formation regions and pointings at known WTT stars showed a few examples of highly variable sources. V410 Tau, the rapidly rotating WTT with giant starspots (Sec. IV),

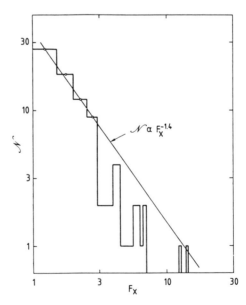

Fig. 3. Distribution of X-ray variability ranges for all ρ Ophiuchi sources, exhibiting a relation similar to M$_\odot$ flares (figure from Montmerle et al. 1983a).

was seen at the Very Large Array (VLA) in 1981–82 with a very strong ascending spectrum ranging from 4 mJy at 1.4 GHz in 1983 to 58 mJy (Cohen and Bieging 1986). It dropped a factor of $\approx$ 60 to 1 mJy at 15 GHz in 1983, and has remained at low but detectable levels in more recent observations. A number of other WTT stars have been found to be radio sources with flux densities variable by factor of 2 to 20 on time scales of days to years. They include two other spotted Taurus-Auriga WTT stars, the G2 star HP Tau and HD 283447, listed in Table I (Cohen and Bieging 1986; Kutner et al. 1986; O'Neal et al. 1990), the X-ray discovered star in ρ Ophiuchi ROX 31 (Stine et al. 1988), and at least two stars in the Orion nebula (Garay et al. 1987).

It should be noted that, whereas a large fraction of pre-main-sequence stars in nearby star-forming regions are detected X-ray sources at levels of 10^{29} to 10^{31} erg s^{-1}, only about 10% are demonstrated variable radio sources at levels of 10^{17} erg s^{-1} Hz^{-1} or brighter, although the fraction detected appears to rise at lower levels (Leous et al. 1991). This low detection rate is probably due to the comparatively poor sensitivity of the radio telescopes to magnetic flare emission. The existing X-ray sensitivity limit corresponds to $\approx 10^3$ times the emission of a Class 3 solar flare at a distance of 150 pc, while the radio limit corresponds to $\approx 10^5$ times the emission of such a flare.

VI. VARIABILITY ON SHORT TIME SCALES AND FLARE MODELS

Very rapid variations can only be due to localized phenomena involving dense gas at or very close to the stellar surface. Two types of short-time-scale optical variations have been reported, both principally in the U photometric band. First, low-amplitude (up to a few percent) flickering is inferred to be present from the power spectrum of photometric data from several CTT stars (Worden et al. 1981). The power law index of the fluctuations is consistent with a superposition of solar-flare-like events, requiring several flares to be active on the surface at all times. In light of the recent CTT stellar models (chapter by Bertout et al.), it seems possible that this flickering occurs in the boundary layer of the accretion disk and stellar surface, rather than in magnetic loops.

Second, analysis of photometric monitoring observations of WTT stars designed to study the rotationally modulated starspots (Sec. IV) have also revealed U-band variations on time scales of hours. Since these stars show little evidence of boundary layers and strong evidence for large starspots, the variations are probably magnetic in origin. The effect ranges from stars which generally display only 0.05 mag flickering (e.g., V410 Tau, HD 283447) to stars which routinely flare at the $\geq$0.50 mag level (e.g., FK2, WK2). Vrba et al. (1988b) show that for the best-studied of the weak-flaring stars V410 Tau, its flaring tends to occur more often at minimum light, indicating an associa-

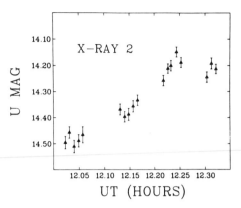

Fig. 4. U-band photometry showing a powerful flare in the Taurus-Auriga WTT star FK 2 = V826 Tau. The U magnitude without a flare should have been 14.74 at this rotational phase (figure from Rydgren and Vrba 1983).

tion of flaring with active cool starspot regions. We have looked at the data for the most active flaring stars, FK2 and WK2, and find no obvious correlation between flaring and light phase. Fortunately, one WTT flare was observed with high time resolution. Figure 4 shows a flare in FK 2 caught in progress, although the observations were terminated due to dawn. The rise time is about 40 min and amplitude about 0.5 U mag. No U brightening was observed during monitoring 4 hr earlier or 21.4 hr later.

In the X-ray band, all CTT and WTT stars show variability on long time scales between Einstein observations (Sec. V) and a handful show flares within a single observation. Two such cases are shown in Figs. 5 and 6. One strong-line CTT star, DG Tau, exhibited the most rapid known event, rising

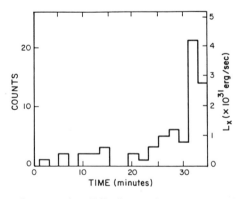

Fig. 5. A rapid X-ray flare event in a T Tauri pre-main-sequence star, the CTT star DG Tau, showing Einstein IPC counts between 0.5 and 4.5 keV. (figure from Feigelson and DeCampli 1981).

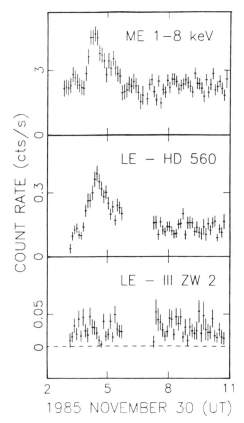

Fig. 6. A rapid *X*-ray flare event in a T Tauri pre-main-sequence star, the young visual binary system HD 560, showing EXOSAT medium-energy counts, low-energy counts from the stars, and low-energy counts from a background galaxy. The flare most likely arises from the G5 post T Tauri component HD 560B (figure from Tagliaferri et al. 1988).

a factor of >6 in about 4 min to a luminosity of $\approx 4 \times 10^{31}$ erg s^{-1} (Feigelson and Decampli 1981). A complete *X*-ray event, with rise and fall within 4 hr and a peak luminosity of $\approx 1.5 \times 10^{31}$ erg s^{-1}, was seen in the CTT star AS 205 (Walter and Kuhi 1984). A similar event, though with peak luminosities around several times 10^{30} erg s^{-1}, has been seen with the EXOSAT *X*-ray satellite from a G5 probable post-T Tauri star HD 560B (Tagliaferri et al. 1988).

A single rapid, and extraordinarily luminous, event has been seen in a WTT star at radio wavelengths (Feigelson and Montmerle 1985). The reddened $\approx$G0 star DoAr 21, with bolometric luminosity around 25 L$_\odot$, lies close to the core of the ρ Ophiuchi cloud and is one of the most *X*-ray luminous stars in the cloud. With strong variable Hα line emission in the 1940s–60s and no emission lines in the 1960s-80s, it may be the only known case of a CTT star changing into an WTT star in historical times. During a 1983 VLA observation, it was seen to rise to 48 mJy at 5 GHz within 5 hr, with a spectral index of $+1.2$ and peak luminosity of 1.5×10^{18} erg s^{-1} Hz^{-1} (Fig. 7). At other epochs, DoAr 21 was detected at various flux levels between 1 and 5 mJy and with a spectral index around 0.0 (Stine et al. 1988). DoAr 21 and several other WTT stars have recently been detected and resolved using Very Long Baseline Interferometry techniques, demonstrating that their emission is produced on scales around a few stellar radii (Phillips et al. 1991).

The principal interpretative framework adopted to explain these luminous and rapid variations is a scaled-up solar flare. Due to the paucity of the data, only isothermal, homogeneous, single magnetic-loop models have been used. This is clearly an unrealistic simplification, given the complexity of the geometry and physical processes known to be present in solar flares. The model inputs for the X-ray events seen in DG Tau, AS 205 and ROX 20 are typically a peak soft X-ray luminosity of 10^{31} erg s^{-1} ($\approx 10^3$ times that of a Class 3 solar flare) and a cooling time of 10^3 to 10^4 s (≈ 10 times that of a

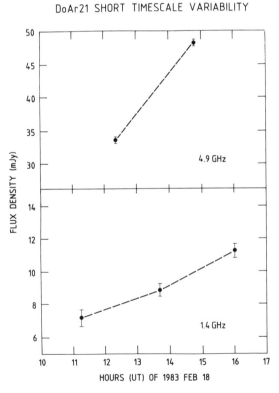

Fig. 7. Powerful radio-continuum flare in the WTT DoAr 21 in February 1983 (figure from Feigelson and Montmerle 1985).

solar flare). The model outputs under simple assumptions are plasma temperatures around 10^7 K, thermal electron densities around 10^{11} cm^{-3} and loop lengths around 10^{11} cm. The temperatures and densities are comparable to those inferred for solar flares, but the loop size is hundreds of times larger. The CTT magnetic loops are larger than the stars themselves in this model.

Similar results are obtained for single-loop models of the DoAr 21 radio event (André 1987). Assuming a dipole magnetic loop with a power law nonthermal electron population emitting gyrosynchrotron radiation, and including self-absorption, a $\approx 10^{11}$ cm loop with field strengths of ~ 100 G at the base and $\approx 10^{-2}$ G at the top is derived. The fields are not unusually strong, but again the loop size is enormous compared to solar flares, and continuous particle injection for hours is needed.

Mullan (1985) has proposed an alternative to single-loop models to explain the long-lived radio events seen on RS CVn stars, which might also apply to the events seen on T Tauri stars. He suggests that the outbursts are due instead to the efficient conversion of mechanical energy from the convection zone into electric fields. The conversion occurs when a small magnetic loop rises from the photosphere, reaching a density where the electrodynamic coupling with the turbulent environment becomes high. Magnetic-field amplification and reconnection, runaway electron acceleration, nonthermal radio emission and rapid coronal heating ensue. The emission is thus produced over a large surface area near many reconnection sites, rather than in a single large magnetic loop. This model is not consistent with the recent VLBS findings (Phillips et al. 1991).

VII. THE MAGNETIC-ACTIVITY INTERPRETATION OF WEAK T TAURI STARS

The studies described above provide considerable support for extremely high levels of magnetic surface activity in weak T Tauri stars: cool starspots covering 5% to 40% of the stellar surface; X-ray and radio continuum emission several orders of magnitude higher than main-sequence levels; and, in a few stars, the detection of rapid, powerful events that may be scaled-up versions of solar flares. However, some doubts and alternative explanations for the data can be raised. Direct measurement of magnetic field strengths and filling factors through Zeeman splitting of absorption lines have not been successful in CTT stars, with upper limits of several hundred gauss (Johnstone and Penston 1986), though fields at this level are indirectly inferred to be present from infrared polarization data (Gnedin et al. 1988). Measurements have not yet been reported for naked T Tauri stars, though with the largest telescopes, such a measurement may soon be feasible (see the chapter by Saar). The deep chromosphere interpretation satisfactorily explains the optical and ultraviolet spectrum of WTT stars, but is open to question. Kenyon and Hartmann (1987) suggest the spectral features could arise in a boundary layer between an accretion disk and the star, as a weak version of the model successful in explaining CTT spectra. Early models for the X-ray emission in PMS stars were generally not based on a magnetic interpretation. Various researchers predicted a high-temperature corona analogous to the contemporary solar corona, a hot component of the T Tauri wind, a high-temperature shock as the T Tauri wind hits the surrounding interstellar cloud, or a high-temperature shock as infalling material hits the stellar surface (see Feigelson 1984, for references). However, these models fail to explain the rapid variability, hard spectrum and lack of correlation with Hα emission in the observed X-ray-emitting stars. A thermal explanation for the radio continuum emission in WTT stars, which is probably valid in some CTT stars with ionized winds, similarly cannot explain the rapid variability, small size and lack of correlation with Hα emission in WTT stars (Feigelson 1987).

There are two arguments in favor of the magnetic activity model for WTT stars in addition to those outlined above. First, in the absence of direct measures of magnetic fields, astronomers have adopted a variety of radiative proxies for the magnetic field in the Sun and other late-type stars. Magnetic flux is thought to be produced by a regenerative dynamo in the stellar interior, emerging on the stellar surface where it couples with the atmosphere through poorly understood nonradiative heating mechanisms. The result is chromosphere, coronae, active regions, flares and other manifestations of solar-type activity. The intensity of these radiative emissions is believed to scale with the amount of magnetic flux on the stellar surface. Recent studies indicate that, in at least some of these proxy measures, WTT stars follow the patterns seen in other late-type stars. Using the surface flux of chromospheric emission lines, e.g., the indicator R_{HK} (where R is the chromospheric flux in the calcium H and K lines normalized by the stellar flux), surface flux of X-rays, and total X-ray emission, WTT stars extend the relationships between activity and stellar youthfulness already established for main sequence stars of different ages (Walter et al. 1988b; Feigelson and Kriss 1989; chapter by Walter and Barry). While CTT stars dominated by circumstellar emission do not follow some of these trends, WTT stars generally do. Similarly, Bouvier (1989) has recently shown that WTT and CTT X-ray emission increases with higher surface rotational velocities, consistent with the empirical correlation known in main-sequence stars (see e.g., Pallavicini et al. 1981). This is shown in Fig. 8.

Second, a number of researchers have noted empirical similarities between WTT stars and the class of magnetically active binary stars known as RS CVn- and BY Dra-type binaries. These are post-main-sequence or main-sequence stars rather than pre-main-sequence stars, but share many characteristics with WTT stars. Both classes consist of approximately solar-mass stars that are temporarily located in the red-subgiant region of the HR diagram, although their internal energy generation and transport differ. Both classes have rapid surface rotation rates compared to the Sun, WTT stars

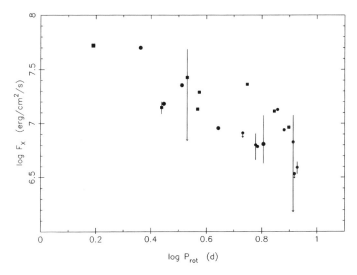

Fig. 8. X-ray surface flux rotation period for WTT (filled squares) and CTT (filled circles) stars (figure from Bouvier 1989).

(because of the angular momentum inherited from gravitational collapse) and RS CVn stars (because of tidal orbit-spin coupling with its close binary companion). Both classes have large, long-lived cool starspots that cause rotationally modulated photometric variations of $\simeq 10\%$ of the visual light. Both classes show variable and spectrally hard X-ray emission around 10^{30} erg s^{-1}, although WTT X-ray spectral data are as yet insufficient to detect the multi-temperature thermal-emission lines seen in some RS CVn systems. The classes exhibit similar behavior in the radio band: continuous variable emission around 10^{16} to 10^{17} erg s^{-1} Hz^{-1} with occasional outbursts lasting hours to days with 10^{18} erg s^{-1} Hz^{-1}. To date, however, the circular polarization often seen in RS CVn radio emission has not been detected in WTT stars, possibly because of polarity cancellations in a complex magnetic field geometry. Such a range of similar behavior cannot be coincidental, and suggests that the physical causes of these manifestations of stellar activity are similar in both pre- and post-main-sequence evolutionary phases.

A final point to be made is that the WTT phase is not restricted to a small class of stars as, for example, the RS CVn phenomenon is restricted to close tidally locked binary systems. This is established by the high number of WTT compared to CTT stars. The relative size of the WTT to the CTT populations is difficult to measure accurately: it depends on the chosen definition of the classes; it may differ from one star-forming region to another; and it is likely to be underestimated because X-ray surveys for WTT stars are often spatially incomplete and flux limited. In the Taurus-Auriga cloud complex, the number of WTT stars already outnumbers the CTT stars by $\approx 50\%$

(Herbig and Bell 1988), and is likely to increase when X-ray data improve. In the Chameleon cloud, the WTT and CTT populations appear to be approximately equal, and in the Ophiuchus cloud, the WTT population may again dominate. These population studies imply either that the WTT phase is more long lived than the CTT phase, and/or that some pre-main-sequence stars are born weak. In either case, a given star is equally or more likely to pass through a WTT phase than a CTT phase. Our Sun, with a solar system that clearly points to a disk during the pre-main-sequence phase, is likely to have passed through both a CTT and WTT phase.

VIII. PORTRAIT OF THE SUN AS A WEAK T TAURI STAR

From the characteristics of WTT stars reviewed in Secs. II-VI, together with the arguments in Sec VII concerning the likelihood that the Sun shared these properties for at least some of its pre-main-sequence phase, we can now attempt to paint the portrait of the pre-main-sequence Sun. The results are given in Table II and are illustrated in Fig. 9. We hope this brief summary will prove useful to solar and solar system scientists who do not wish to become experts in T Tauri stellar astronomy. We list properties averaged over the ensemble of well-studied WTT stars, and adjusted where possible to a mass of 1 $M_\odot$ and an age of 1 Myr. This age represents the youngest stars for which reliable WTT observations are available. Clearly, the Sun will spend more time with properties intermediate between the 1 Myr WTT and the contemporary Sun than with the extreme properties shown. The figure represents our attempts to illustrate the large starspots and large magnetic flare loops inferred to exist on WTT stars.

TABLE II
Properties of the Sun Today and as a Weak T Tauri Star

	Today[a] Age = 5 Gyr	Weak T Tauri Age = 1 Myr
Mass	$\equiv 1M_\odot$	$\approx 1M_\odot$
Radius	$\equiv 1R_\odot$	$\approx 2.5R_\odot$
Luminosity	$\equiv 1L_\odot$	$\approx 1.7L_\odot$
T_{eff}	$= 5770$ K	≈ 4200 K
Spectral Type	$= $ G2V	$\approx$ K5IV
V_{rot}	≈ 2.0 km s^{-1}	≈ 25 km s^{-1}
Rot Period	≈ 25 days	≈ 5 days
f_{spots}	≈ 0.002 (spot max)	≈ 0.25
T_{spots}	≈ 4200 K	≈ 3300 K
$L_{x\text{-ray}}$	$10^{27} - 10^{28}$ erg s^{-1}	10^{30} erg s^{-1}

[a] 1.0 $M_\odot$ = 1.99 × 10^{33}g; 1.0 $R_\odot$ = 6.96 × 10^{10} cm; 1.0 $L_\odot$ = 3.90 × 10^{33} erg s^{-1}.

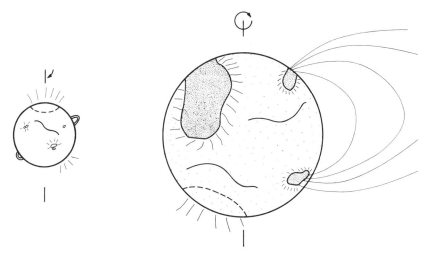

Fig. 9. Sketches comparing the appearance of the Sun today (left) with its appearance as a 1 Myr old weak T Tauri (WTT) star (right). Active regions, plage, filaments, coronal magnetic loops, coronal holes, and rotation rate are all indicated. Dot density indicates decreasing temperature. Details are provided in the text and Table II.

The physical properties listed in Table II for the WTT Sun do not all have the same accuracy. The mass is probably accurate to a few percent or better, as classical T Tauri mass-loss rates are probably no more than 10^{-8} $M_\odot yr^{-1}$ and most of this mass may come from the circumstellar disk. The radius is probably accurate to $\pm 10\%$ (see Fig. 2). The bolometric photospheric luminosity could be uncertain by a factor of ± 2, since we have no independent measure of stellar age to determine the dispersion of T Tauri luminosities. The effective temperature and spectral type may be off by perhaps ± 500 K and ± 2 subclasses. The listed rotational velocity and period are accurate to about $\pm 50\%$; the wider distribution of V (sin i) is presumably due to the range of inclination angles. The starspot fractional coverage and temperature are uncertain by perhaps a factor of 2 and ± 500 K, respectively (see Table I). The X-ray luminosity represents an average of an ensemble of pre-main-sequence stars (Feigelson and Kriss 1989). Due to the high level of variability, a given star can range over a factor of 5 or more, and different stars are also likely to have different mean levels. The X-ray luminosity of the contemporary Sun also varies several fold over an 11-yr activity cycle, and more during flare events.

Finally, we mention again (see Sec. II)—but do not resolve—the important issue concerning the relation between the classical and weak T Tauri phase as applied to the Sun. It is conceptually simplest to identify the CTT phase with the young Sun surrounded by the solar nebula, which then evolved into the WTT phase as planetisimals formed and the gaseous nebula dissipated. In the context of pre-main-sequence stellar studies, Shu et al. (1987)

call these Phases 4 and 5 in their proposed outline of the star-formation process. K. Strom et al. (1989), noting that Taurus-Auriga WTT stars are generally closer to the main sequence than CTT stars, have estimated that this transformation takes ≈3 Myr. While we have no strong objection to this appealing model, we note that other lines of evidence suggest that many WTT stars are coeval with CTT stars. Their spatial and rotational distributions are virtually indistinguishable, and visual CTT/WTT binaries (presumably born together) are known. Half of the WTT stars with masses ≤1 $M_\odot$ have estimated ages <3 Myr (Fig. 1), clearly contemporaneous with CTT stars. Many of the WTT have infrared and millimeter excesses indicating that dusty disks are present (K. Strom et al. 1989; Beckwith et al. 1990; André et al. 1990), although not touching the stellar surface. It is thus possible that some stars are born with circumstellar matter closely interacting with the protostar, and others are not.

Whatever the resolution of this nature vs nuture debate, it is likely that the Sun was for at least part of its pre-main-sequence phase a magnetically active star. Even during a classical T Tauri phase, the magnetic activity is probably present, as evidenced by the X-ray flares of CTT stars (Secs. V-VI and Figs. 5 and 6). The stellar surface of a CTT star within the complex environment of disk, boundary layer and wind may be very similar to that more readily recognized in WTT stars.

IX. IMPLICATIONS FOR THE EARLY SOLAR SYSTEM

Solar system scientists studying meteorites dating back to the formation of the Sun have repeatedly considered the possibility that the young Sun was very magnetically active to explain certain unusual phenomena. Three problems in particular have motivated such models. First, the presence of isotopic abundance anomalies in small fractions of ancient meteorites might be due to spallation of normal abundance material in the solar nebula by energetic protons produced in powerful flares. Originally suggested by Fowler et al. (1962), this idea was quantitatively developed by Clayton et al. (1977) and others. The general result was that proton fluences orders of magnitude higher than produced by the contemporary Sun are needed to explain the ^{26}Al anomaly, and that such particle fluxes would produce other unobserved anomalies. For these reasons, and the recognition of other possible sources of galactic ^{26}Al (supernovae, novae, asymptotic red giants and Wolf-Rayet stars), this "local proton irradiation" model for isotopic anomalies was not widely accepted. However, the solar-flare spallation model has recently been reexamined by Wasserburg and Arnould (1987) in light of improved spallation cross-section data and a range of energetic proton spectra. The spallation model has an advantage over various thermonuclear generation models in that it predicts that the ^{26}Al anomaly will be accompanied by a ^{53}Mn anomaly; such an effect has recently been reported to be present in certain Allende inclusions (Birck and Allègre 1985). They find that, for different flare-proton spectra, fluences of order 10^{18} to 10^{22} cm^{-2} are needed to produce observed ^{26}Al and ^{53}Mn excesses. Some of these models are several orders of magnitude less demanding than the 10^{21} to 10^{25} cm^{-2} fluences estimated in earlier calculations.

Second, chondrules are small silicate structures present in most meteorites that were once partially or fully molten. Radiometric dating indicates that most cannot predate the Sun, and were thus probably melted in the solar nebula. Deducing the mechanism by which they were melted is one of the most difficult problems in solar system studies (see, e.g., Grossman 1988; Levy 1988). Among the many models are impacts between asteroids or planetisimals in the early nebula, volcanic production on planetisimals, condensation in a hot solar nebula, and transient heating of dust in a cold solar nebula. The latter class of models seems most acceptable, although it is not clear which of several sources of heating (e.g., lightning, magnetic reconnection, chemical energy, radiation or particles from solar flares) caused the melting. It seems possible that some manifestation of WTT magnetic activity—radiational heating by flares, nebular shocks produced by flares, injection of strong magnetic fields into the nebula by giant solar-flare loops—may be responsible. We estimate that WTT stars convert of order 10^{-3} to 10^{-2} of their bolometric energy into some form of nonphotospheric emission (chromosphere, X-rays, particles, transient phenomena, etc.), compared to the 10^{-5} to 10^{-4} for the contemporary Sun.

Third, detailed examination of the irradiation histories of certain meteoritic components has shown evidence for enhanced solar flares in the Sun during the PMS phase (Caffee et al. 1988; chapter by Caffee et al.). While it is generally difficult to determine the duration of exposures to solar particles, and hence to estimate the particle fluence, Caffee et al. (1987) have carefully measured the amounts of ^{21}Ne nuclides produced by solar proton spallation in individual meteoritic grains with high and low solar-flare VH ion track densities. Large excesses of spallation-produced ^{21}Ne are found in the grains with high track densities. They argue this is best explained by a ≈ 1000-fold increase in proton fluence and an extension of the proton spectrum to higher energies than seen in the contemporary Sun. This and related studies of carbonaceous chondrites are discussed in detail in the chapters by Caffee et al. and by Cisowski and Hood.

The studies of WTT stars discussed in this chapter have a clear contribution to these debates concerning meteoritic properties: *invoking high levels of energetic particle or radiation fluxes associated with magnetic activity in the early Sun is no longer ad hoc.* Studies of young stars provide a wealth of evidence independent of the meteoritic data that strong radio, ultraviolet and X-ray emission along with giant flares and their associated particle fluxes and shock waves should have been present in the pre-main-sequence Sun. This point was made by Worden et al. (1981) and Feigelson (1982) based on early indications of T Tauri flaring. The evidence accumulated more recently for WTT starspots and flares supports this conclusion. Of course, the probability that the early Sun was very magnetically active does not by itself prove that the isotopic anomalies, chondrule formation and early grain irradiation were flare induced. The quantitative relationships between proton fluences relevant to the meteoritic studies and the electromagnetic flare output observed in PMS stars is also very uncertain. We suggest, however, that these models deserve more attention and other effects of early solar activity should also be sought in ancient solar system bodies.

X. DIRECTIONS FOR FURTHER RESEARCH

The study of magnetic activity on pre-main-sequence stars has become a field of very active observational research during the 1980s. The research will undoubtedly continue along the directions described above, but there are also prospects for observations using new telescopes during the next few years and a serious need for detailed theoretical investigations of the phenomena uncovered. Some of the new lines of research that we envision follow.

(1) The discovery of new WTT stars and further investigation of their flaring should progress through the 1990s with new X-ray astronomical satellite missions. The German ROSAT mission to be launched in early 1990 will provide an all-sky soft X-ray survey with 10^5 X-ray sources, several hundred of which are likely to be pre-main-sequence stars. Pointed ROSAT observations should also give data several times more sensitive than the best obtained with the Einstein satellite, though the soft energy response will preclude penetrating deep into star-forming clouds. The Japanese Astro-D satellite to be launched in early 1994 will provide much higher sensitivity and high-energy response than Einstein or ROSAT (Roentgen Satellite), though with only moderate spatial resolution. It should provide unique X-ray spectral and timing data on individual T Tauri stars. The U.S. Advanced X-ray Astrophysics Facility (AXAF) is planned to be a long-lived space observatory to be launched around 1997. Its X-ray CCD camera in particular will have an ideal combination of high sensitivity, good high-energy response, and high spatial and spectral resolution to provide a complete census of pre-mainsequence stars in all nearby star-forming regions. A single deep AXAF image of the Orion nebula core, for example, may show as many WTT stars as are known today throughout the entire sky. Finally, analysis of the Einstein Observatory archives has not yet been completed. It may, for example, establish the presence of pre-main-sequence stars in small molecular clouds where observations at other bands had not found them.

(2) Progress is likely in detailed optical and ultraviolet spectral studies of WTT surfaces. High spectral resolution studies with the 8-m class telescopes currently under construction should permit direct detection of magnetic fields, and discrimination between flare, prominence, plage and cool-

spot activity as the stars rotate. Detailed modeling of the inhomogeneous surfaces should then be possible.

(3) Theoretical astrophysical research linking the surface magnetic activity of pre-main-sequence stars to the processes in their interiors is needed. For example, it is uncertain how much magnetic field from the parent molecular cloud is incorporated into the protostar. If present, a "fossil" magnetic field from the birth process may have a significant effect on the convection and magnetic dynamo along the Hayashi track (Tayler 1987). The creation, eruption and dissipation of magnetic energy may differ in rapidly vs slowly rotating PMS stars (Shu and Terebey 1984), and in CTT vs WTT stars due to the presence or absence of a magnetically driven wind (Shu et al. 1988). The new data on WTT stars may constrain some of the complex relationships between angular momentum, magnetic dynamos and surface activity during the T Tauri phase. Levy, Ruzmaikin and Ruzmaikina (see their chapter) give an excellent analysis of our current theoretical understanding of the magnetic field expected to be present in pre-main-sequence stars. Insights into the physics of stellar magnetic fields may also emerge from a thoughtful comparison of WTT and RS CVn stars.

(4) The research community devoted to the study of the early solar system should become more familiar with these findings and renew consideration of models that involve a magnetically active T Tauri Sun. Early irradiation models are clearly boosted by the stellar findings. An important line of research might seek to quantify the effects of giant flares on solar nebula material. Under what circumstances do the energetic protons penetrate to the radii at which meteoritic grains were formed? Can the shock waves associated with giant flares melt ices or disrupt loose aggregates in the inner solar nebula? Can flare radiation or shocks melt rocky particles and thus be involved in chondrule formation? These questions are among the many that need answers.

Acknowledgments.   We thank the referees, N. Calvet and an anonymous scientist, for thoughtful comments that significantly improved this chapter. The authors would also like to thank their colleagues T. Montmerle, W. Herbst and L. Cram for many stimulating discussions. E. D. F.'s work on pre-main-sequence stars is supported by a Presidential Young Investigator Award cofunded by the NSF, TRW Inc. and Sun Microsystems Inc., and by NASA grants under its Astrophysics Data Program. The NOAO is operated by AURA, Inc., under cooperative agreement with the National Science Foundation.

THE CLASSICAL T TAURI STARS: FUTURE SOLAR SYSTEMS?

CLAUDE BERTOUT
Institut d'Astrophysique de Paris

GIBOR BASRI
University of California, Berkeley

and

SYLVIE CABRIT
Service d'Astrophysique, CEN-Saclay

Observed properties of low-mass young stellar objects are reviewed with emphasis on their circumstellar (possibly) protoplanetary disks and on their powerful winds. We discuss proposed theoretical wind mechanisms in the light of observational constraints, and conclude that although none of the existing mass-loss theories is entirely convincing, they demonstrate that both the disk and the magnetic field must play key roles in protostellar and T Tauri mass ejection.

I. INTRODUCTION

Since the realization that the Hα emission stars (first studied by Joy [1945]) are actually solar-type pre-main-sequence stars, we have stars to observe with conditions that may have prevailed at the beginning of our own solar system. The advent of infrared and radio astronomy have further enabled us to study these stars as they are forming out of original molecular cloud cores, while they are still embedded in the surrounding dust and optically invisible. Even though theories of star formation are still being developed, now at least we have substantial observational information about the early conditions around young stars.

Adams et al. (1987) used infrared energy distributions to define three classes of young stellar objects representing different physical stages. The first (Class I) includes mostly embedded sources with spectra that rise to the far infrared (i.e., d log λF_λ/ d log $\lambda \gtrsim 0$). These objects have high amounts of extinction caused by circumstellar dust (which contributes to their spectral shape), as well as emission from cool dust; they are probably the youngest class, i.e., true protostars whose luminosity is derived primarily or entirely from accretion. The second (Class II stars) has infrared spectra which fall beyond a few μm (or rarely are flat), but with shallower slopes than can be explained by a Rayleigh-Jeans blackbody tail. These are classical T Tauri stars, with optical spectra of late-type stars, but also with spectral anomalies that include various emission lines (e.g., Hα), as well as "veiling" of the photospheric lines by a blue continuum excess (Herbig 1962). The third class (III) of young stars has continuum distributions which are basically stellar in shape, and their ultraviolet and optical spectra show only excess emission that can be ascribed to strong magnetic activity on the star. These are the "weak-lined" or (sometimes called "naked") T Tauri stars.

Particularly exciting in the context of this book is the rapidly developing consensus that classical T Tauri stars are different from weak-lined T Tauris primarily due to the presence of a circumstellar accretion disk of gas and dust, i.e., a "solar nebula." This means that we can observe conditions analogous to those in the early solar nebula directly, instead of being restricted to the historical approach on solar system bodies. It also seems increasingly clear that young solar systems undergo protracted phases of massive outflows simultaneous with accretion through the disk, starting very early in the star formation and continuing in some cases for a few Myr. Continuum distributions have been shown to be consistent with disks (Adams, et al. 1987) for Class I stars, and similarly for Class II stars (see, e.g., Bertout, et al. 1988). Outflows are seen for Class I stars as bipolar CO outflows (cf. Lada 1985) and occasional radio jets (Snell and Bally 1986), and as optical jets and through the presence of associated Herbig-Haro (HH) objects (cf. Mundt 1988). Except for the occasional exception, Class II stars have no radio signatures of outflow, but display outflow indicators in the optical range, particularly blue-shifted absorption features in permitted lines and blue-displaced emission in forbidden lines.

After reviewing the observational evidence for accretion disks, we discuss the various observational signatures of outflow: high-velocity molecular gas and radio-continuum emission, jets and HH objects, and finally optical-emission lines in classical T Tauri stars. The final section summarizes the constraints we can put on mass loss and its driving mechanism as well as current theoretical ideas about the outflow origin.

II. FLOW FIELDS AROUND YOUNG STARS

A. Circumstellar Disks

Because the subject of disks has been amply reviewed in the last few years (Basri 1987b; Kenyon 1987; Hartmann and Kenyon 1988; Bertout 1989), we only briefly summarize the important points here with a few references to the literature. Lynden-Bell and Pringle's (1974) suggestion that classical T Tauri stars might be accretion-disk systems did not catch on at the time, probably because the infrared and ultraviolet evidence that now makes it compelling was not yet available. The main reason for thinking disks are present is the shape of the infrared continuum distribution (log λ F_λ vs log λ), which is too flat to come from a single-temperature blackbody. Its shape suggests the presence of dust at a variety of distances from the star; the nearby warm dust contributes near-infrared light and the more distant cooler dust emits at increasingly longer wavelengths. This dust has to be distributed in a disk rather than a sphere around the star because the optical extinction that would arise from the spherical distribution of dust required to explain the infrared excess is much too high to be consistent with the observed optical spectrum (Myers et al. 1987). The slope of the infrared excess predicted by a simple flat accretion-disk model is consistent with at least some of the observations, although a number of classical T Tauri stars have slopes that are too shallow for this simple model to explain.

Other, more indirect, lines of evidence are also available. The shape of optical forbidden-line profiles is typically asymmetric with the red wing either diminished or missing. This has been interpreted by Appenzeller et al. (1984) and Edwards et al. (1987) as a sign of mass loss from the star in which our view of the receding flow is cut off by the presence of a disk. The size of the disk implied by the emission measure in the forbidden lines is on the order of 100 AU. Another line of evidence for disks comes from the analysis of FU Orionis objects (Hartmann and Kenyon 1985,1987a,b). These objects have recently undergone a large ($\sim$ 5 mag) increase in brightness, and are thought to have been classical T Tauri stars prior to this outburst. The luminosity increase is suggested as due to a large increase in accretion rate within the disk that now contributes virtually all the light observed. The overall continuum distribution is consistent with a disk, while the average absorption-line profile in these systems shows some evidence of double-peaked disk profiles. Perhaps even more convincing is that the average line width is larger in the optical lines than in the near-infrared lines, an effect expected if the optical lines are formed nearer the star at higher temperatures and faster orbital velocities than are the infrared lines.

Another indirect piece of evidence that circumstellar material is responsible for both infrared excess and emission-line activity in classical T Tauri stars can be found by looking at the correlation between the two. The strongest competing explanation for the emission lines has been strong magnetic activity on the star (see, e.g., Calvet et al. 1984). If this were a major contributor to the emission lines, we should expect to find no correlation between them and the infrared excess, which is not believed to have anything to do with stellar magnetic activity. On the other hand, if the emission lines are produced in conjunction with the accretion disk, either in a boundary layer or in mass-loss activity caused by the disk, then there should be such a correlation. In Fig. 1, we show the relation between emission-line activity and infrared excess. The ordinate is the logarithm of the ratio of line luminosity to total luminosity. The lines included in the luminosity estimate are Hα and Hβ as well as the Ca II and Mg II resonance lines. The abscissa is the excess color index $E(V-L)$, defined with respect to main-sequence stars of the same spectral type. This number increases for stars with greater infrared excess. It is clear that there is a definite correlation between the amount of energy radiated in the emission lines and in the near infrared.

Finally, there is the matter of the ultraviolet and blue excesses that veil the optical absorption lines and present their own energy problem. Ultraviolet and blue excesses were originally suggested as symptomatic of strong mag-

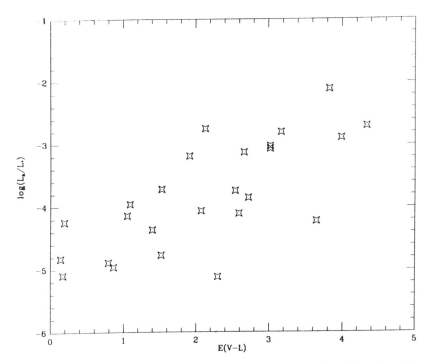

Fig. 1. Emission-line activity in T Tauri stars vs infrared excess. The ordinate is the ratio of the flux in several strong lines to the bolometric luminosity. The abscissa is the excess in the (V-L) color of the stars compared to the (V-L) color of main-sequence stars. Thus, it is a measure of the near-infrared excess in units that become larger for larger excesses.

netic activity on the surface of the star—chromospheric and flare heating. Calvet et al. (1984) showed that the continuum would begin to pick up a contribution from chromospheric layers if the chromosphere begins at a sufficiently high density in the stellar atmosphere. The primary difficulty with such an explanation comes when one realizes that the total luminosity in the ultraviolet excess is often a large percentage of total stellar luminosity, even exceeding it in the most extreme cases. It hardly seems possible for the star to arrange matters so that a major part of its energy is not radiated from the photosphere, but instead goes into nonradiative heating of the outer layers (cf. Basri 1988).

The disk hypothesis solves this problem in a self-consistent way. If the star is rotating slowly compared to breakup, as observations confirm, then the disk material loses as much kinetic energy in slowing onto the stellar surface as it has radiated gravitational potential energy in spiraling toward the star (Lynden-Bell and Pringle 1974). Because this energy emerges not from a large-disk area but rather from the small boundary layer between star and disk, its average temperature must be rather high. Models predict an optically thick boundary layer to be 8000 to 10,000 K, which means the emission comes out primarily in the ultraviolet and visible passbands. Actual observations show that classical T Tauri stars as a rule show Balmer continuum emission jumps, which implies that the boundary layer is partially optically thin in the Paschen continuum (Basri and Bertout 1989). At any rate, the energy available in the accretion disk provides a natural source for the observed excess energy.

How many young stars have disks and how long the disks last are questions of great relevance for discussions of the early solar system. If there is a substantial near-infrared excess (say $\Delta K \gtrsim 0.1$ mag), Balmer continuum emission, and strong Hα emission (say W_λ (Hα) $\gtrsim 5\text{Å}$), then one could have a classical T Tauri system in which the young star interacts with an accretion disk. It is not certain what fraction of visible young stars fall into this category because searches for weak-lined T Tauri stars are still incomplete to an unknown degree. It seems likely, however, that a substantial fraction of young solar-type stars (perhaps less than one-half and more than one-tenth) appear as classical T Tauri stars for some length of time. The typical time a disk lasts in these systems is also somewhat uncertain at present. The current technique is to try to position the star itself in the HR diagram and then use quasi-static pre-main-sequence convective-radiative evolutionary tracks to infer an age. The presence of a disk renders this procedure uncertain from both observational and theoretical points of view. The observational effects of the disk luminosity and its occultation of the star must be accounted for; this requires a detailed modeling of the system, in which the difficult-to-determine inclination plays a crucial role. Furthermore, if accretion of disk matter onto the star proceeds at rates $\gtrsim 10^{-7}$ $M_\odot \text{yr}^{-1}$, it may seriously alter the theoretical evolution of the star in the HR diagram (Hartmann and Kenyon 1990). An initial attempt to assess disk ages (Strom et al. 1989) using isochrones derived from classical quasi-static pre-main-sequence models leads to the conclusion that many disks last for a few Myr. The hypothesis that a typical T Tauri star is a pre-main-sequence star with an accreting circumstellar gas and dust disk around it has interesting implications for scenarios about the young Sun. First, it means that classical T Tauri stars are likely relevant to assess the solar nebula, and that solar nebulae may be a relatively common occurrence in the population of young solar-type stars. Second, the fact that the T Tauri phenomenon may be in large part due to the star/disk interaction means that once formed, and after the disk is cleared away, planets will not be subjected to the strong radiation fields characteristic of the classical T Tauri environment. This in turn implies that large ultraviolet flux and strong stellar winds were probably not a factor for the Earth at the time life may have started. It is less clear whether these conditions are relevant to the problem of planetesimal melting (see the chapters by Wood and Pellas and by Herbert et al.); this depends on how rapidly the formation of planetesimals clears out the disk. This is not to imply that the Sun looked as it does today during the first Gyr of the solar system (chapter by Kasting and Grinspoon). As emphasized in the chapter by Feigelson et al., it was still vastly more magnetically active than it is today, with substantially stronger ultraviolet, X-ray, and particle fluxes. Clearly, one must look at weak-lined T Tauri stars to assess the early Sun.

B. Protostellar Outflows

Molecular line observations in the millimeter range have revealed energetic collimated outflows from many newly formed stars still embedded in their parental molecular cloud (see Lada [1985] and Snell [1987] for reviews).

High-velocity molecular gas was first observed toward the core of the Orion A molecular cloud, where CO line profiles in the immediate vicinity of the luminous infrared cluster were found to possess broad wings extending out to $V \sim \pm 90$ km s^{-1} from cloud velocity (Zuckerman et al. 1976; Kwan and Scoville 1976). Extensive CO surveys have since then detected cold, supersonic molecular gas ($T \sim 10$ to 20 K; $V > 5$ km s^{-1}) toward ~ 100 embedded young stellar objects with luminosities ranging from 10^4 to 10^5 L$_\odot$ (see, e.g., Bally and Lada 1983) down to a few L$_\odot$ (see, e.g., Myers et al. 1988; Terebey et al. 1989). In several cases, low-level CO wings are found which extend out to ± 70 to 150 km s^{-1} (Koo 1989; Margulis and Snell 1989).

Several facts indicate that the high-velocity CO gas traces a phase of strong, collimated mass loss from the embedded object: (1) mass inside the high-velocity emission region is too small for velocities to be due to gravitationally bound rotation or collapse; (2) in most cases, the high-velocity emission region is bipolar with blue-shifted and red-shifted gas symmetrically distributed in two lobes about the central source, which rules out turbulence or shear between colliding clouds as the origin of the supersonic motions; (3) in several cases, direct evidence for outflow is provided by the presence within the high-velocity lobes of Herbig-Haro objects or of water masers with proper-motion vectors pointing away from the stellar source (see, e.g., Snell et al. 1980; Genzel et al. 1981; see Sec. II.C below).

Although CO outflows have been found around a few very active T Tauri stars (T Tauri itself, AS 353a, HL Tau, RNO 91), searches around less extreme stars have so far proved unsuccessful (Edwards and Snell 1982; Levreault 1988). While this negative result could have been caused by the insufficient sensitivity and resolution of early surveys, preliminary observations using the (IRAM) 30-m antenna confirm the lack of high-velocity molecular gas around most classical T Tauri stars (P. André, personal communication). This can be interpreted either (a) as lack of molecular material in the vicinity of T Tauri stars, or (b) as evidence for modest mass-loss rates during the T Tauri phase. We note the related absence of H_2O masers in the vicinity of classical T Tauri stars (Thum et al. 1981), which also led to the suggestion that there is not enough molecular material around these stars for collisionally exciting the maser transition. While this view is consistent with the moderate visual extinction of most T Tauri stars, there has been some dispute over the last decade about the magnitude of mass loss in these stars (see, e.g., DeCampli 1981; also see Sec. II.D). It is therefore difficult at this point to choose between possibilities (a) and (b).

An example of bipolar molecular flow from a low-luminosity, embedded young stellar object is shown in Fig. 2. The flow source, shown as a filled triangle, is a cold infrared source ($T \sim 15$ K) of total luminosity 3 $L_\odot$ (assuming a distance of 250 pc) embedded in the dark globule B 335. Solid and dashed lines indicate contours of constant integrated-intensity in the blue-shifted and red-shifted wings of the $J = 1 \rightarrow 0$ line of CO. The high-velocity gas is confined to two conical lobes with opening angle $\sim 30°$ and size ~ 0.3 to 0.9 pc. In this particular outflow, each lobe shows both blue-shifted and red-shifted emission, because the flow is oriented nearly perpendicular to the line of sight and has an expansion component perpendicular to its axis; therefore, the front of the expanding lobes is approaching us while the rear is receding from us.

Observations of bipolar flows have also been made in millimeter-range lines of less abundant molecules like OH, CS, HCO^+, H_2CO, NH_3, SO, SiO and in the infrared vibrational transitions of H_2. All these lines provide useful information about high-density regions and shocks within the high-velocity gas (cf. Lada 1985; Snell 1987). However, the rotational lines of CO remain the most widely used means of tracing the large-scale properties of molecular flows because of their ubiquity and strength (antenna temperatures ~ 0.1 to 10 K). CO data are especially useful because they allow one to derive the

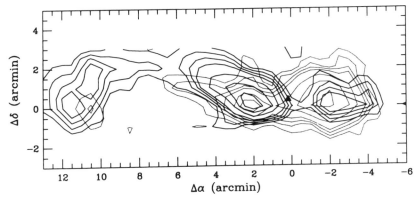

Fig. 2. A CO map of the low-luminosity bipolar outflow source B 335. Solid and dashed lines indicate contours of constant blue-shifted and red-shifted integrated emission, respectively. They are superposed on each other because the flow is nearly perpendicular to the line of sight (figure adapted from Cabrit et al. 1988).

mass, velocity, spatial extent and dynamical age of the flow, and hence indirectly to measure the mass-loss rate from the central object averaged over the flow lifetime.

The total H_2 mass in the flow is estimated by assuming optically thin ^{12}CO ($J = 1 \rightarrow 0$) emission in thermodynamic equilibrium at the cloud temperature and a standard CO/H_2 abundance ratio. This procedure yields a mass of $0.06\ M_\odot$ for the B 335 flow in Fig. 1. Similar values are obtained for other molecular flows around low-luminosity young stellar objects (Myers et al. 1988; Levreault 1988). However, this value is only a lower limit because the $^{12}CO/^{13}CO$ intensity ratio over the line wings is often less than the assumed isotopic ratio $\sim 60/90$, indicating that the high-velocity ^{12}CO emission is optically thick. The ^{12}CO ($J = 1 \rightarrow 0$) optical depth $\tau \sim 4$ inferred for the B 335 flow (Goldsmith et al. 1984) suggests that the total flow mass is more likely $\sim 0.25\ M_\odot$. Since the mass of high-velocity molecular gas is a large fraction of the mass of the central object, the flow is believed to consist mostly of ambient cloud material entrained by the energetic wind from the young stellar source, although the very high-velocity, low-level CO wings seen in some flows may be associated with the wind itself (see, e.g., Lizano et al. 1988).

Because CO line profiles only offer information about the line-of-sight velocity of the outflowing molecular gas, estimated values of the mean flow velocity depend on assumptions about projection affects and velocity gradients in the flow and range from 2.5 to 10 km s^{-1} in B 335 and other low-luminosity young stellar objects (Cabrit et al. 1988; Myers et al. 1988; Levreault 1988). High-velocity CO emission typically extends out to ~ 0.1 to 1 pc from the central source, so the dynamical time scale of the flow is $\sim 10^4$

to 10^5 yr. Dividing the total momentum and kinetic energy in the outflowing gas by the dynamical time scale then yields an estimate of the force and mechanical luminosity of the flow. Though uncertain by 1 to 2 orders of magnitude (cf. Cabrit and Bertout 1990), these values are large and imply that mass ejection from the young stellar object is very energetic (see Sec. III). The force in the B 335 flow is estimated to lie between 8×10^{-5} and 2×10^{-5} $M_\odot$ km s^{-1} yr^{-1} while the flow mechanical luminosity lies between 2×10^{-4} and 2×10^{-2} $L_\odot$. Wind mass-loss rates calculated assuming momentum conservation at the wind/flow interface and a wind velocity of 200 km s^{-1} then range from 4×10^{-9} to 10^{-7} $M_\odot$ yr^{-1} (Cabrit et al. 1988).

Luminous molecular outflow sources ($L_{bol} > 10^3$ $L_\odot$) also often possess detectable radio centimetric continuum emission (Snell and Bally 1986) attributed to thermal free-free radiation from an ionized envelope with temperature $\sim 10^4$ K. The spectral index $\alpha = $ d log S_ν/d logν of the radio flux S_ν is usually close to the 0.6 value expected for a constant-velocity, fully ionized, spherical wind (cf. Panagia and Felli 1975), although sources with moderate luminosity tend to show flatter spectra ($-0.1 \lesssim \alpha \lesssim 0.5$) that suggest optically thin emission. The density law in the envelope should then drop less steeply than r^{-2}. This could be achieved in an infalling envelope or in a collimated wind whose opening angle increases with distance from the star (cf. André 1987). Direct evidence for collimated mass loss is found toward the molecular flow source L1551-IRS5, where the 6-cm radio continuum emission is resolved into a narrow jet of length ~ 500 AU and width < 50 AU (Bieging and Cohen 1985). Free-free emission was also detected toward a few classical T Tauri stars: DG Tau, T Tau, HL Tau and NX Mon (Bertout and Thum 1982; Cohen and Bieging 1986; Brown et al. 1985; Evans et al. 1987). When available, the spectral index is again close to 0.6, but the star often appears surrounded by a marginally resolved halo that cannot originate in a spherical stellar wind and therefore makes the observed flux difficult to interpret.

If the wind is spherical and fully ionized and if its velocity is known, the mass-loss rate in the ionized wind can be directly deduced from the radio flux (Panagia and Felli 1975). Values obtained for a wind velocity of 200 km s^{-1} range from 10^{-6} $M_\odot$ yr^{-1} in luminous sources down to 10^{-8} to 10^{-7} $M_\odot$ yr^{-1} in detectable classical T Tauri stars. That other T Tauri stars surveyed in Taurus-Auriga and ρ Ophiuchi are not detected at 6 cm suggests that their ionized mass-loss rate is $<2 \times 10^{-8}$ and 7×10^{-8} $M_\odot$ yr^{-1}, respectively (Bieging et al. 1984; André et al. 1987). However, these values are probably uncertain within at least 1 order of magnitude due to uncertainties in the envelope geometry, ionization and velocity field.

C. Optical Jets and Herbig-Haro Objects

Optical jets and Herbig-Haro (HH) objects are the most obvious and spectacular manifestation of strong outflows from young stars. Recent summaries of this field, with a number of excellent illustrations, can be found in Mundt (1988) and Reipurth (1989b). This review will therefore be very brief and only mention the main points. Jets are highly collimated structures that can be seen in emission lines such as Hα and [S II] $\lambda\lambda$6716,6731. They can extend out to 0.5 pc from the star, and typically have length-to-width ratios of 5/20. They trace outflows that were often first studied through their associated HH objects. These are small nebular emission knots that are found in star-forming regions, and that primarily display emission lines normally thought to arise in shocked interstellar material. They are sometimes found in linear chains, that point back to a known pre-main-sequence star or embedded infrared object. Proper motions can be found for a number of HH objects and their space motions show that they are moving at several 100 km s^{-1} away from the exciting star. There are examples in which both sides of a bipolar outflow are seen (e.g., HH 1–2; cf. Herbig and Jones 1981), though often the receding HH objects are probably buried too deeply in the surrounding molecular cloud to be seen.

It is not clear yet whether the Herbig-Haro objects are "bullets" that were shot from the exciting star and are now colliding with the interstellar medium, or whether they are shocked interstellar gas that result from the collision of a fast wind with a stationary cloudlet. In both cases, one envisions a bow shock developing, and indeed bow-shock models can explain a number of features of the observed line profiles (Hartigan 1988). In particular, the existence of motions spanning a range of several 100 km s^{-1} emanating from a region with typical size of 1000 AU or smaller can be understood with this model. It also provides an explanation for the co-spatial existence of highly ionized species and neutral species. At the moment, it is still unclear which scenario is more applicable; both are likely needed in order to describe a clumpy jet working its way into a clumpy surrounding medium. Examples can be found in which the HH objects seem to be merely the brightest knots in a jet emanating from the star (HH 7–11) or in which they look like the working surface of a jet (HH 34), or in which it is not so clear what they are (HH 1–2). In general, the Herbig-Haro objects tend to be associated only with partially or completely embedded exciting stars; they are relatively rare among classical T Tauri stars. This implies either that they tend to occur in an early stage of star formation, or that they are more easily seen in sources with strong mass loss. Recent high-resolution studies of forbidden lines seem to indicate that unresolved HH objects or jets are in fact present in the vicinity of many T Tauri stars (S. Edwards, personal communication).

Perhaps the best example of a collimated young stellar object wind is HH 34. Figure 3 shows the surroundings of this object in the light of [S II]. Here a straight jet that is partially broken into a series of knots extends for more than 30″ out of a point source. The jet then disappears, but directly along its extension another knot or two can be seen followed by a large conical nebula (HH 34S) with its apex pointing away from the star. HH 34S looks

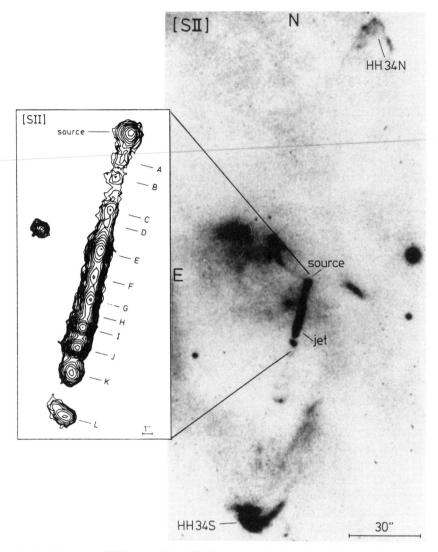

Fig. 3. Right panel: [S II] image of the HH 34 region. A knotty jet is pointing away from the source, toward HH 34 S. Both this HH object and the fainter one of the opposite side of the source (HH 34 N) are shaped like bow shocks. HH 34 S and the jet are blue shifted while HH 34 N is red shifted. Left panel: Iso-intensity contour plot of the jet in [S II]. Twelve knots are seen, with a typical separation of 2″. (Figure supplied by courtesy of R. Mundt.)

like the result of the jet boring its way into the interstellar medium, and apparently marks the end of the flow. A similar but fainter nebula with the symmetry reversed (HH 34N) can be seen emanating in the opposite direction from the star.

Mundt et al. (1987) summarize observed properties of jets, that typically have a velocity of hundreds of km s^{-1}, with an internal velocity dispersion of 30 km s^{-1} or less (a few have of order 100 km s^{-1}), and electron densities on the order of 10^3 cm^{-3}. The exciting sources appear typically to have properties expected from solar-type young stars. Even though these jets are unambiguously high-velocity collimated outflows, their physics and driving mechanisms are not well understood. It is possible that they only show optical emission along part of their length because internal shocks and fluid instabilities are required to excite the emission; when the flow is too smooth the jet cannot be seen. If the conical HH objects are indeed the working surface of the jets, the observed deceleration of material can be explained; but the expected cocoon of return flow has yet to be observed. How the flow is accelerated and collimated is still a matter of controversy; the reservoir of energy and momentum afforded by an accretion disk is very likely involved (see Sec. III). In any case, the apparent youth of most of the stars driving jets and HH objects seems to imply that this phase of collimated outflow only lasts < 0.1 Myr or so, with rates of mass loss in the jet estimated to span the range 10^{-10} to 10^{-8} M$_\odot$ yr^{-1}, i.e., sizeably smaller than mass-loss rates derived from the molecular lines. It is therefore unlikely that jets are directly driving their associated molecular outflows, and the presence of an atomic wind component has therefore been hypothesized and detected in a few cases (Bally and Stark 1983; Lizano et al. 1988). High-sensitivity observations of atomic hydrogen can be made using the Arecibo dish, and surveys of star-forming regions are now underway.

D. T Tauri Winds

Classical T Tauri stars were first identified by their strong Hα emission line, the properties of which suggest that it is formed in a wind emanating from the young star. Normal main-sequence stars of similar spectral types have a fairly strong Hα absorption line that is increasingly weakened in those stars known to be chromospherically active as the line core fills in. In a few unusually active main-sequence stars (e.g., the dMe stars), there is actually weak Hα emission, that is also characteristic of the magnetically active naked T Tauri stars. It should be noted that it is easier to see an emission line carrying a specified flux against a fainter photosphere, so that line emission will be more apparent in an M star than in a G star at any given flux level. The chromospheric emission has an equivalent width of at most 5 Å or so and is usually not very wide: with FWHM (full width at half maximum) perhaps 50 km s^{-1} at most, because broadening is caused by pressure rather than by velocity gradients. By contrast, classical T Tauri stars have Hα equivalent widths $\gtrsim$ 10 Å, ranging up to over 100 Å. Additionally, these emission lines are typically quite broad, with FWHM $\sim$ 250 to 300 km s^{-1}. Although Hα is qualitatively quite different in classical T Tauri stars and in main-sequence stars with comparable spectral types, most other photospheric lines look relatively normal (apart from possible veiling by the boundary-layer continuum and some chromospheric filling in of strong line cores).

High-resolution profiles of emission and absorption lines provide evidence for complex gaseous flows in classical T Tauri envelopes. Several studies of Hα show a variety of line profiles (see, e.g., Kuhi 1964; Mundt 1984). According to Kuhi (1978) a majority of T Tauri stars displays Type III P Cygni profiles (in Beals' classification) at Hα; in these profiles, the emission is broad and symmetric, and a blue-displaced absorption feature at typically 80 km s^{-1} is superimposed on the emission. While symmetric emission profiles are also common, only a few T Tauri stars display Type I P Cygni profiles with blue-displaced absorption components reaching below the photospheric continuum. Type I profiles unambiguously indicate that mass loss is taking place, in contrast to Type III profiles (see, e.g., Bertout 1984). Inverse P Cygni profiles indicative of mass accretion are not observed at Hα, although a subclass of classical T Tauri stars named YY Orionis stars after their prototype (Walker 1972) displays such profiles at the higher members of the Balmer series. We must caution the reader that almost any statement about emission lines in classical T Tauri stars has many exceptions, since the lines in a given star are always variable. Sometimes, they alter their shape drastically in a few days or weeks, and sometimes they are relatively stable for several years.

In spite of these problems, there were several bold attempts to derive mass-loss rates from the Hα line profiles of T Tauri stars (see, e.g., Kuhi 1964; Hartmann et al. 1982). Inferred values span the range 10^{-9} to 10^{-7} M$_\odot$ yr^{-1}. All the complexities discussed above as well as some of the recent data presented in the following paragraphs make it clear that these values might well be completely wrong. It is therefore intriguing that analysis of optical [S II] forbidden lines, which is less complicated because they are optically thin, yields similar wind mass-loss rates (cf. Edwards et al. 1987). However, the possible contribution of unresolved shock regions to the forbidden-line emission in T Tauri stars makes these values uncertain as well (see e.g., Cabrit et al. 1990). In any case, it is puzzling that the large ionized regions required for the forbidden-line emission in these stars have managed to escape detection in the radio continuum at 6 cm (cf. Sec. II.B). One way out of this problem would be that the number ratio of [S II] ions to electrons in the emitting region is larger than the assumed cosmic S/H abundance (e.g., due to incomplete ionization of hydrogen at temperatures $T \sim 10^4$ K), so that the observed [S II] line fluxes could be produced with lower wind-emission measures and hence lower radio-continuum fluxes. Quantitative investigation of this possibility remain to be undertaken.

The faintness of classical T Tauri stars in the optical range was a problem until recently for high-resolution spectroscopic observations, but the study of T Tauri line profiles is undergoing a radical improvement with the recent advent of good echelle spectrographs on several large telescopes. They provide for the first time high-quality simultaneous observations of several of the most important optical wind diagnostics. By way of example of the new data just becoming available, we present a few lines here from three stars which span the range of line strengths found among the T Tauri stars: DR Tau, DF Tau and DN Tau. (The overall spectral-energy distributions as well as moderate-dispersion optical spectra of these three objects are given by Bertout [1989].) DN Tau is the least active of the three, meaning that the Hα equivalent width and continuum excesses are smallest, and there is the least indication of mass-loss activity. DF Tau is an excellent example of a star likely to have a disk, and its Hα line is rather variable in strength but fairly consistent in shape. DR Tau is an "extreme" T Tauri star, having experienced a brightness increase of several magnitudes over the last few decades. It has strong emission lines and continuum excesses and displays evidence of high outflow and/or inflow velocities in both emission and absorption. The first four Balmer lines are shown for each star Fig. 4, on common continuum scales. The Balmer decrement stands out the most; the Hα line is much stronger than the higher-series lines. This cannot happen if the line is formed on the stellar surface because either low densities or a large emitting volume are needed to produce both the observed Hα flux and Balmer decrement. At the same time, if the higher Balmer lines are scaled up so they have the same peak heights, then they often look remarkably similar to Hα (although there are notable exceptions). The main difference is usually that the absorption component is less blue shifted and often weaker as one goes up the series. It is also more common for the absorption component to go below the continuum in the higher lines.

One puzzling point about these emission lines is the relatively high symmetry of the line wings. If one ignores the absorption component, the lines tend to extend to about the same velocity redward and blueward of the stellar-rest velocity, and often even have very similar shapes. This is surprising for a line formed in an expanding wind, and doubly surprising if the wind is occulted at the midplane by a disk. Indeed, DeCampli (1981) notes that such lines need not imply any wind at all; the broadening could all be due to some combination of turbulence and orbital motions. In this regard, it is interesting that the typical velocity widths are also approximately those expected from Keplerian orbits just off the star.

A physical explanation proposed for line broadening is Alfvén waves; one needs turbulence of < 100 km s^{-1} to produce the observed widths. Re-

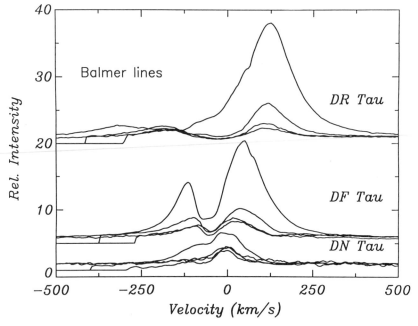

Fig. 4. The first four Balmer lines from three representative T Tauri stars. The lines for each star are shown on a common continuum scale. Velocity is measured in the stellar rest frame, as determined from photospheric absorption lines.

cently, Hartmann et al. have studied in great detail the line formation problem in a spherical Alfvén wave-driven wind. Although these authors do not actually think that is an appropriate description of a T Tauri wind, their work is important in defining limitations of spherical wind models. Not surprisingly, the computed profiles tend to have a P Cygni appearance, and a number of adjustments have to be made to the simplest models before they begin to resemble typical T Tauri profiles. In particular, the absorption component is generally broader and deeper than observed, and the line peak is more redward-displaced than observed. Furthermore, the Ca II infrared triplet is hard to drive into emission in these models, while it is almost always in emission in the classical T Tauri stars. Figure 5 shows one of the Ca II triplet lines for the three stars discussed above. In DN Tau, the line could arise purely from a strong chromosphere, but there is high-velocity material in the other cases. We note that on different occasions the line profiles can appear rather different; in one instance the DF Tau profile shows no emission at all. The DR Tau profile is typical of the extreme stars, and unshifted triangular profiles such as this are fairly common among T Tauri profiles.

The three basic types of profiles generally seen in classical T Tauri stars

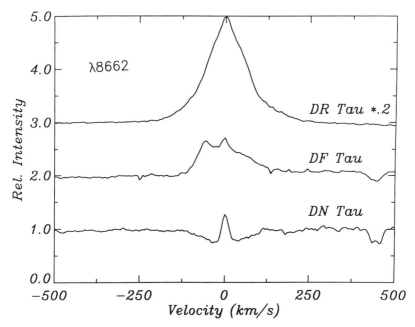

Fig. 5. One of the Ca II infrared triplet lines from each of the stars in Fig. 4. These profiles are from the same nights as in Fig. 4, but the profiles look rather different on different nights.

are shown in Fig. 6. The most common one is the Type III P Cygni profile (bottom profile in Fig. 6) with a peak near rest, a blue-shifted absorption feature, and relatively symmetric far emission wings. There is a smaller class of Type III profiles in which the absorption feature is essentially unshifted and the profile almost symmetric. The second, rarer class of profiles has flat tops that are often cut by a blue-shifted absorption feature or occasionally by several absorption features. The third class of profiles (upper profile Fig. 6) is relatively symmetric about a peak at zero velocity, with no obvious absorption features and triangular or symmetric wings.

The interpretation of these profiles is actually not clear at this time. As discussed by Bertout (1984), Type III P Cygni profiles can be interpreted in several different ways, depending on assumptions on the origin of the apparent absorption feature. Flat-topped profiles can be formed either in an optically thin wind or in an accelerating optically thick wind (cf. Bertout and Magnan 1987). Triangular profiles may impose interesting constraints on models since they seem to require comparable strengths of "turbulent" and organized velocity components (Bertout, in preparation). Obviously, a successful physical model for the formation of T Tauri emission lines should allow for all three profile classes, since the profiles of Fig. 6 are all the same line from the same star in three different months. They are Ca II λ8662 pro-

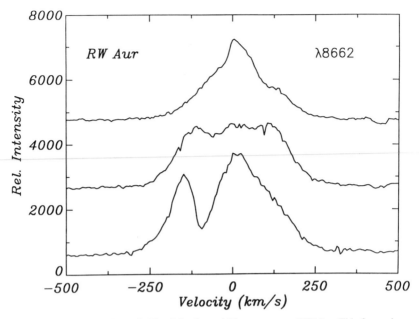

Fig. 6. The same Ca II line as in Fig. 5 for the variable extreme star RW Aur. This figure shows the appearance of the line on 3 different nights. These line profiles cover the three main types of line profiles seen in T Tauri stars.

files from the rather extreme variable star RW Aur, but they are also indicative of profiles seen in Hα and other lines from other stars. Because they are all present in this one star, they may imply that the variety of profiles observed in various stars are all manifestations of one underlying set of phenomena. Obviously, then, the optical lines contain a rich lode of information on the flow fields around young stars, and the mining operations have only just begun.

E. Summary

Observational evidence for strong outflows from young stars is now completely convincing, and evidence is building that classical T Tauri stars are also surrounded by accretion disks. Neither of these characteristics is likely to hold for weak-lined T Tauris, so that they define the difference between the two classes. The underlying stars, which are probably rather similar in both cases, exhibit much stronger magnetic activity than today's Sun. The accretion disks might well be examples of solar nebulae that hold potential to allow observations relevant to the planetary formation process. Their presence around a fair fraction of low-mass young stellar objects means that they are quite possibly the agents responsible for the continuum excesses in both the ultraviolet and infrared, for the strong, broad emission lines, and for the strong outflows from these stars. The outflows tend to be strong and collimated and begin very early in the life of a star. While the star remains embedded in the cloud (as an infrared but not optical source), the outflows can be seen as bipolar radio sources and occasionally as radio continuum sources. Optical evidence of the outflow comes with the appearance of Herbig-Haro objects and sometimes of optical jets. These manifestations seem to fade as the star clears away its surroundings, but so long as the accretion disk is present there is still evidence for substantial outflows in the optical lines from T Tauri stars.

III. THEORIES OF MASS LOSS AND COMPARISON WITH OBSERVATIONS

One can in principle derive mass-loss rates from the mass-loss diagnostics reviewed in the previous sections, albeit with considerable uncertainty. Two of these diagnostics, molecular outflows and optical forbidden lines, appear most promising today for quantitative work because they provide relatively precise information about the momentum rate and mechanical luminosity necessary to drive outflows from both embedded and optically visible young stellar objects.

A. Observational Constraints

As seen in Sec. II.B, molecular outflows are found mostly in connection with embedded sources of various luminosities. There are also a few CO flows associated with classical T Tauri stars but they are rare and rather inconspicuous. The lack of detectable flows around optically selected young stellar objects has been attributed to the absence of significant amounts of molecular gas around them, a reasoning which assumes that CO flows result from the interaction of a protostellar wind with the circumstellar medium. Whether the wind from young stars is made up primarily of molecular, atomic or ionized material remains a subject of controversy; but in any case, CO flows, that are unaffected by extinction, are quite useful for deriving global wind properties during the protostellar phase.

A major result of observation is that the CO momentum rate F_{CO} and the mechanical luminosity of the CO flow L_{CO} are correlated with the bolometric luminosity L_{bol} of the protostellar source driving the flow. This correlation extends over 5 orders of magnitude, from low-luminosity T Tauri stars to OB stars. Typically one finds

$$F_{CO} \approx 100 \, L_{bol}/c, \qquad L_{CO} \approx 0.001 \, L_{bol} \qquad (1)$$

but for individual sources the dispersion about these average numbers is considerable (see, e.g., Lada 1985). This result is usually interpreted as evidence for a mass ejection mechanism that is common to all these objects and somehow related to the central object's radiative energy source. A major problem with this interpretation arises once one assumes that the radiative energy is mainly of stellar origin. High-mass stars, that have presumably reached the main sequence, have luminosities high enough to radiatively drive massive outflows (see below). Low-luminosity stars, on the other hand, do not possess enough radiative energy to do this, so the various mechanisms reviewed below try to remedy this problem by using other energy sources. But, if we have to invoke different mass-loss mechanisms for driving the outflows of low-luminosity and high-luminosity objects, why does the correlation between L_{CO} and L_{bol} extend over the entire luminosity range? One way out of this dilemma is to assume that neither the source of the flow nor the bolometric luminosity are primarily of stellar origin. It comes to mind then that a circumstellar self-luminous (accretion) disk could be instrumental in driving a wind (see, e.g., Edwards and Strom 1988), although the exact mechanism by which this may occur is still a matter of debate. The observed correlation between the infrared excess of T Tauri stars (which is presumably related to the disk accretion rate), and their flux in the [O I] $\lambda 6300$ line (which measures the wind's mass-loss rate) provides a first line of evidence supporting this suggestion (Cabrit et al. 1990; Cohen et al. 1989).

If molecular flows result mostly from the interaction of the wind with the circumstellar material (i.e., if the amount of molecular material ejected from the central source is negligible compared to accelerated material), F_{wind} and L_{wind} can be derived from the CO data. There exist two possible types of interaction between disk and wind (cf. Dyson 1984). If the shocked wind cools down on a time scale shorter than the CO flow's dynamical time scale, then the CO flow is accelerated by the wind's momentum and one can write

$$F_{wind} \approx F_{CO}, \qquad L_{wind} \approx (V_{wind}/V_{CO})L_{CO}. \tag{2}$$

If radiative losses are low in the shocked-wind region, the wind is pushed instead by the wind's thermal pressure; then

$$F_{wind} \approx (V_{CO}/V_{wind})F_{CO}, \qquad \text{and } L_{wind} \approx L_{CO} \tag{3}$$

The CO is expected to be energy driven if the wind velocity is larger than a critical velocity V_c, which itself depends on the mass-loss rate and the density distribution of circumstellar matter (cf. Dyson 1984). For a constant density $\approx 10^4 \text{cm}^{-3}$ and mass-loss rate $M_{wind} \lesssim 10^{-6} M_\odot \text{ yr}^{-1}$, one finds $V_c \approx 200$ to 300 km s^{-1}, a value comparable to those derived from most mass-loss diagnostics in the optical range (see above). It is therefore unclear at this time whether the CO is driven by the wind's momentum or its pressure. This introduces an additional uncertainty in the wind parameters derived from CO data, so that "average" values are

$$F_{wind} \approx 10 \text{ to } 100 \, L_{bol}/c \tag{4}$$

and

$$L_{wind} \approx 0.001 \text{ to } 0.01 L_{bol}. \tag{5}$$

One can derive similar relations for optical T Tauri stars by using the range of mass-loss rate values estimated from the forbidden lines or from the Hα line

$$10^{-9} \lesssim \dot{M}_{wind} \lesssim 10^{-7} \text{ M}_\odot \text{ yr}^{-1}. \tag{6}$$

One then finds

$$F_{wind} = \dot{M}_{wind} V_{wind} \approx 10 \text{ to } 10^3 \, L_{bol}/c \tag{7}$$

and

$$L_{wind} = \frac{1}{2}\dot{M}_{wind} V^2_{wind} \approx 0.003 \text{ to } 0.3 \, L_{bol}. \tag{8}$$

While constraints on the efficiency of the wind mechanism thus appear even stronger in optical young stellar objects than in the embedded ones, one must recall that mass-loss rate values derived from optical lines are quite uncertain (see Sec. II.D). We now proceed to discuss various plausible mass-loss mechanisms in the light of these observational constraints.

B. Mass-Loss Mechanisms

In pre-main-sequence stars, three energy sources that can be used to drive a stellar wind are:

1. Photons, that themselves result from gravitational contraction;
2. Rotation;
3. Magnetic field.

These three energy sources are probably also present in the circumstellar disk associated with the young stellar objects. Regarding the first, if energy dissipation occurs within the disk because of viscous or some other torques, then gravitational energy will be transformed into photons and the produced luminosity will be comparable to the accretion luminosity. As for the second, in optical T Tauri stars that have low rotational velocities, the rotational energy is presumably much larger in the Keplerian disk than in the central star. Concerning the third, whether typical protostellar disks are magnetized is unknown at present, but it is not unlikely if, for example, the protostellar dynamical collapse is magnetically controlled. The following mechanisms taking advantage of these energy sources have been proposed to drive outflows from both the stellar object and its circumstellar disk.

Radiation Pressure.   Mass loss in luminous OB stars is likely to be caused by radiation pressure, and one might expect that this mechanism is also effective in driving outflows of luminous protostellar sources. Two conditions must be fulfilled for this to occur. The first states that the luminosity of the young stellar object must be larger than the wind mechanical luminosity

$$L_{bol} \gtrsim 4\pi GM^*c/\kappa \qquad (9)$$

where κ is the effective continuous opacity. If graphite grains are the main opacity source, one finds $L_{bol} \gtrsim 10^3 L_\odot$ (Königl 1986). If ice is the main opacity source, then $L_{bol} \gtrsim 10^2 L_\odot$ (Bally 1986). The second condition states that the wind momentum rate cannot be greater than the stellar momentum output rate, i.e.,

$$\tau \approx cF_w/L_{bol} \qquad (10)$$

where τ is the wind's optical depth. This ratio ranges typically from 1 to 10 for high-luminosity protostellar objects, which implies that winds have moderate optical thickness in the wavelength range where the above effective opacity is defined.

Thus, while radiation pressure could well be driving the winds of high-mass protostars (e.g., V645 Cyg; cf. Levreault 1988), another mechanism must be driving the winds of solar-type young stellar objects.

Thermal Expansion.   The solar wind is driven by thermal expansion. In that case, however, radiative losses are proportional to $\dot{M}^2_{wind}$, and the observed X-ray flux of T Tauri stars indicates that thermal coronal expansion cannot be responsible for mass-loss rates $\gtrsim 10^{-9} M_\odot$ yr^{-1} in these objects (DeCampli 1981).

More recently, several thermal-expansion mechanisms have been proposed which use the disk as a starting point. In Torbett's (1984) model, for example, matter is ejected from the boundary layer between disk and star because the radiative diffusion time there is larger than the dynamical accretion time, which then results in large pressure gradients. Numerical models, however, indicate that the efficiency of this process in converting mass accretion into mass outflow, defined by $\beta = \dot{M}_{wind}/\dot{M}_{acc}$, is $< 10^{-3}$. Assuming that the bolometric luminosity of young stellar objects is larger than the accretion luminosity one then finds that

$$\frac{cF_{wind}}{L_{bol}} \lesssim \frac{\beta c}{V_{wind}} \lesssim 1 \qquad (11)$$

and

$$\frac{L_{wind}}{L_{bol}} \lesssim \frac{\beta}{2} \lesssim 5 \times 10^{-4}. \tag{12}$$

The efficiency of this mechanism is thus too low by at least 1 order of magnitude to drive protostellar or T Tauri star outflows. Pudritz (1985) proposed instead that the disk's surface is heated up and ionized by photons from the central regions. Thermal pressure then results in an expansion of disk matter at radii $< 10^{15}$ cm from the disk's center along poloidal magnetic field lines. In that model, the mass accretion rate is $10^{-3} M_\odot$ yr^{-1} and the mass-loss rate is $10^{-6} M_\odot$ yr^{-1}; the mechanism's efficiency is comparable to that of Torbett's model, and the same conclusions thus apply.

Alfvén Wave Propagation. A T Tauri wind model in which matter is accelerated by the magnetic pressure resulting from the propagation of Alfvén waves in the stellar envelope was worked out by DeCampli (1981), Hartmann et al. (1982) and Lago (1979,1984). Hydromagnetic waves are supposed to have their origin in magnetic-field perturbations caused by turbulent motions in the outer convective layers of the star. For this mechanism to work, the waves must stay undamped up to the Alfvén radius; its efficiency in transforming "Alfvén-wave luminosity" into wind-mechanical luminosity thus depends on the damping mechanisms, as well as on the stellar magnetic-field strength. For undamped waves, models show that a mass-loss rate of $10^{-8} M_\odot$ yr^{-1} can be produced if the average stellar field is a few 100 gauss and that about 10% of the Alfvén-wave luminosity is transformed into mechanical wind luminosity. The above relation between L_{wind} and L_{bol} in T Tauri stars then implies that more than 1% of the bolometric luminosity of these stars must be converted into Alfvén-wave luminosity. Note that none of the above assumptions can be independently supported, nor is the damping of the waves understood.

Centrifugal Ejection. Suppose now that a stellar object rotating at close to breakup velocity possesses a magnetic field strong enough to drag along a corotating envelope. The stellar rotational energy is then transferred to the envelope where the centrifugal force can accelerate a wind (Mestel 1968). This mechanism is clearly not directly at work in T Tauri stars, which rotate at a fraction of their breakup velocity (Vogel and Kuhi 1981; Hartmann et al. 1986; Bouvier et al. 1986). The rotation velocities of younger, embedded objects are, however, unknown, and it is conceivable that they are rotating faster than more-evolved sources. Hartmann and McGregor (1982) have applied this mechanism to the case of a massive protostar; they found that a ratio $cF_{wind}/L_{bol} \sim 50$ similar to that inferred for the Orion A outflow could be obtained for rotational velocities $> 90\%$ of the breakup speed and magnetic fields stronger than 10 gauss; the resulting strong wind led to a large angular momentum loss, with the result that the central object was slowed down on a time scale of a few 10^3 yr.

A modification of this mechanism was proposed by Shu et al. (1988) for a protostellar object rotating at breakup velocity and surrounded by an accretion disk. For accretion to occur onto the central object, most of the angular momentum of the incoming matter must be blown away in a wind originating near the equator where effective gravity is lowest. The gas is first pushed out by local thermal pressure, set in corotation by the magnetic field, and accelerated up to the Alfvén radius by the centrifugal force. Using parameters typical of a low-mass protostellar object, Shu et al. find that an average magnetic field of 150 gauss is necessary to produce a mass loss of mostly atomic gas at a rate of about $3 \times 10^{-6} M_\odot$ yr^{-1}. They suggest that the onset of deuterium burning triggers convection throughout the star and that convection in turn leads to dynamo processes and strong magnetic-field amplification. An advantage of the Shu et al. model is that it makes precise predictions concerning the ratio of mass accretion to mass loss. Assuming steady-state rotation and ejection, one finds

$$\frac{M_{\text{wind}}}{M_{\text{acc}}} = f = \frac{1 - 2b}{J - 2b} \qquad (13)$$

where J is the ratio of the specific angular momentum of ejected gas l_{wind} to the specific angular momentum of gas orbiting the star in the equatorial plane $l_{\text{eq}} = \Omega R_{eq}^2$ and where b is the ratio of the specific stellar angular momentum to l_{eq}. For an entirely convective, uniformly rotating star, one finds $b = 0.14$. Furthermore, Shu et al. show that

$$J = \left(\frac{V_{\text{wind}}}{2V_{\text{eq}}}\right)^2 + \frac{3}{2}$$

where V_{wind} is the wind's terminal velocity. The wind parameters and the disk accretion parameters are thus related according to the following relations

$$\frac{\dot{M}_{\text{wind}}}{\dot{M}_{\text{acc}}} = f = \frac{2 - 4b}{\left(\dfrac{V_{\text{wind}}}{V_{\text{eq}}}\right)^2 - 2b + 3} \approx 0.3 \qquad (14)$$

$$\frac{L_{\text{wind}}}{L_{\text{acc}}} = f \cdot \frac{1}{2} \left(\frac{V_{\text{wind}}}{V_{\text{eq}}}\right)^2 \approx 0.3 \qquad (15)$$

$$\frac{cF_{\text{wind}}}{L_{\text{acc}}} = 10^3 f \left(\frac{V_{\text{wind}}}{V_{\text{eq}}}\right)^2 \left(\frac{300 \text{ km} \cdot \text{s}^{-1}}{V_{\text{wind}}}\right) \approx 900. \qquad (16)$$

Numerical values given above are for a deuterium-burning low-mass protostar with $V_{eq} = 140$ km s^{-1} and $V_{wind} = 200$ km s^{-1}. This model thus meets, and even exceeds, the energetic and dynamic requirements of observed protostellar winds (Eqs. 4 to 8), and can even account for the strongest mass-loss rates observed in bipolar outflows. However, the efficiency of this mechanism drops drastically if the protostar does not rotate close to breakup. In particular, it does not work for the slowly rotating optical classical T Tauri stars, which is vexing since the energetic properties of embedded and optical young stellar objects seem so similar.

In a different but related approach, Pudritz and Norman (1983,1986) worked out an analytical description of centrifugal ejection from a magnetized disk, which assumes that the field lines are globally parallel to the system's rotation axis but are pinched toward the center in the disk plane because of accretion in the disk. If the angle between field lines and rotation axis is larger than $\sim 30°$, the radiatively heated disk envelope is accelerated along magnetic field lines by the centrifugal force up to the Alfvén radius r_A (Blandford and Payne 1982). The ejected gas thus reaches an asymptotic speed V_{wind} given by

$$V_{wind} = \Omega r_A = (r_A/r_D)V_D \tag{17}$$

where V_D is the rotational velocity of disk matter at its starting radius r_D and where $\Omega = V_D \neq r_D$. Pudritz and Norman assumed $r_A \neq r_D = 10$, so that the specific angular momentum of the ejected matter is 100 times larger than its original specific angular momentum, resulting in mass accretion within the disk with rate $\dot{M}_{acc}$. In steady-state, mass-accretion and mass-loss rates are related by

$$\frac{\dot{M}_{wind}}{\dot{M}_{acc}} = \left(\frac{r_D}{r_A}\right)^2 \approx 0.01. \tag{18}$$

An interesting feature of Pudritz and Norman's model is its ability to produce the different wind components observed; the cool outer parts of the disks eject a massive molecular wind component that reaches end velocities in the range 10 to 50 km s^{-1} at $r_A \approx 0.1$ pc, while the innermost part of the disk, that are photoionized by ultraviolet radiation from the accretion zone, eject an ionized wind with mass-loss rate equal to only 2% of the molecular mass-loss rate but with velocity in the range 200 to 600 km s^{-1}. In intermediate regions, atomic disk matter is ejected and reaches intermediate velocities.

Given, the escape velocity from the central object, $V_{esc} = (GM_*/R_*)^{1/2}$, and assuming again that $r_A = 10r_D$, the following relations are derived from Eqs. (17) and (18)

$$\frac{L_{wind}}{L_{acc}} = \frac{1}{2}V_{wind}^2 \left(\frac{r_D}{r_A}\right)^2 \left(\frac{R_*}{GM_*}\right) = \frac{1}{2}\left(\frac{V_D}{V_{esc}}\right)^2 \approx 0.001 \left(\frac{V_{wind}}{V_{esc}}\right)^2 \qquad (19)$$

$$\frac{cF_{wind}}{L_{acc}} = \left(\frac{r_D}{r_A}\right)\left(\frac{V_D}{V_{esc}}\right)\left(\frac{c}{V_{esc}}\right) \approx 10 \left(\frac{300 \text{ km} \cdot \text{s}^{-1}}{V_{wind}}\right)\left(\frac{V_{wind}}{V_{esc}}\right)^2. \qquad (20)$$

Therefore, this model is able to account for the average correlations observed in young stellar objects between L_{bol} and the wind parameters (cf. Eqs. 4–7), provided (a) $L_{bol} \approx L_{acc}$ and (b) $V_{wind} \approx 1$ to $5 \cdot V_{esc}$, i.e., $V_D \approx 0.1$ to $0.5V_{esc}$. For classical T Tauri stars with $V_{wind} \sim V_{esc}$ and Keplerian disks, the wind must originate from the inner parts of the disk. One problem that has been raised concerning this model is that the required accretion rates for objects with $L_{bol} \geq 10^4 L_\odot$ become unreasonably large ($\sim 10^{-3}$ to 10^{-2} $M_\odot$ yr^{-1}); another problem is whether or not the poloidal field lines remain attached to their footpoints in the disk long enough for the centrifugal acceleration to take place. Pringle (1989) argues that the disk field is a dynamo field with too short a characteristic time scale for this requirement to be met. For this, and indeed for all models involving magnetic fields, the field strengths and geometries currently remain *ad hoc* theoretical constructs (but more detailed treatments are now under way, see e.g., König] [1989]).

Magnetodynamical Ejection.   While using the same tools (i.e., rotation at breakup velocity and magnetic field) for driving an outflow, Draine's (1983) mechanism differs from the preceding one in that the magnetic field connects the stellar object to the circumstellar medium in which it is frozen. Field-line twisting efficiently converts stellar rotational energy into magnetic energy, that accumulates in the stellar surroundings. Propagation of this perturbation then results in a shock wave moving outwards. There, magnetic energy is converted into kinetic energy of expansion while a dense expanding shell is created at the boundary of this "magnetic bubble." For this transient phenomenon to work, a strong stellar magnetic field ($\approx 10^3$G) is needed, and analytical computations predict magnetic-bubble properties comparable to observed molecular flows around massive objects.

The wind-driving mechanism proposed by Uchida and Shibata (1985) for low-mass protostars is quite similar, but it converts gravitational rather than rotational energy into magnetic energy. It assumes transient ejection from a nonequilibrium circumstellar disk rotating more slowly than at Keplerian velocities. The magnetic field is initially parallel to the rotation axis, and field-line twisting occurs in the central, free-falling parts of the disk. The Lorentz force created by the accumulated toroidal field component ejects the disk's envelope toward the poles, while the denser parts keep collapsing toward the central object. The resulting molecular outflow is therefore confined to the walls of a cylinder, in contrast to Pudritz and Norman's model. Outflowing disk matter is then gradually accelerated along the helicoidal field lines as the perturbation propagates outwards. In order to explain observed small-scale collimated jets, Uchida and Shibata add an additional twist to this story. They hypothesize that disk matter penetrates the stellar magnetic field when reconnection occurs between the dipolar stellar field and the poloidal disk field. Disk matter then free falls onto the stellar magnetic poles and an accretion shock wave propagates outwards, thereby dragging away ionized gas along open field lines. Ejection is therefore mainly caused by the thermal pressure of shocked matter. Again, the low X-ray luminosity of young stellar objects limits the applicability of this mechanism to mass-loss rates smaller than $10^{-9} M_\odot$ yr^{-1}.

Pringle (1989) recently outlined another possible mechanism of magnetodynamical ejection which rests on the transformation of shear energy into magnetic energy within the boundary layer between disk and star. Pringle argues that the magnetic field of disk material flowing into the boundary layer is amplified by the strong shear forces there and finds that a large fraction of the boundary-layer luminosity emerges from the boundary layer in the form of a strongly toroidal field. The magnetic flux could then be driven away in a magnetic wind. If the wind speed is comparable to the maximum disk rotational velocity $v_\phi = (GM_*/R_*)^{1/2}$, then mass-loss and mass-accretion rates are found to be related by

$$\frac{M_{wind}}{M_{acc}} \lesssim \frac{H}{R} \qquad (21)$$

where H is the disk scale height and R is essentially the stellar radius. Equality in Eq. (21) would occur in the case of complete efficiency of the mechanism converting magnetic energy into the wind's kinetic energy. Using the maximum temperature reached in the disk to compute H, Pringle uses Eq. (21) to find

$$\frac{L_{wind}}{L_{acc}} \lesssim 0.03 \left(\frac{M_*}{0.5\ M_\odot}\right)^{-3/8} \left(\frac{R_*}{3\ R_\odot}\right)^{1/4} \left(\frac{\dot{M}}{10^{-5} M_\odot yr^{-1}}\right)^{1/8} \qquad (22)$$

and one can then calculate

$$\frac{cF_{wind}}{L_{acc}} \lesssim \left(\frac{c}{v_\phi}\right)\left(\frac{H}{R}\right) = 100 \left(\frac{M_*}{0.5\ M_\odot}\right)^{-7/8} \left(\frac{R_*}{3\ R_\odot}\right)^{3/4} \left(\frac{\dot{M}}{10^{-5} M_\odot yr^{-1}}\right)^{1/8}. \qquad (23)$$

These upper limits demonstrate that the expected efficiency of this possible boundary-layer wind-driving mechanism, while perhaps not quite large enough to drive the most energetic observed winds and flows, make it worthwhile to investigate the hydromagnetic processes taking place in the boundary layer in detail. Only then will one be able to determine both the feasibility of the mechanism and its efficiency.

C. Collimation Mechanisms

Observed flow anisotropy is another observational constraint that must be dealt with by models. Collimated jets are observed in the optical and radio range on scales ranging from 10^{15} to 10^{16} cm; Herbig-Haro objects can be approximately aligned on distances up to about 10^{17} cm; and bipolar molecular flows extend on even larger scales, up to 10^{18} cm. When several of these outflow manifestations are observed around the same source, as in L1551 IRS5, they are approximately aligned along the same axis. Several mechanisms have been proposed to explain the observed collimation of protostellar outflows and are discussed below.

Magnetic Collimation. Flow collimation occurs naturally in those mass-loss models that use a magnetized rotating central object or circumstellar disk in order to drive the outflow (Hartmann and McGregor 1982; Draine 1983; Pudritz and Norman 1986; Uchida and Shibata 1985; Shu et al. 1988; Pringle 1989). This occurs because the toroidal field component caused by rotation is larger at the equator than near the rotation axis, so that the resulting magnetic pressure forces the field lines and the ejected matter to curve toward the system's poles. A numerical simulation of this effect starting from an initial radial field configuration was made by Sakurai (1985,1987). Once a wind is collimated close to the star, it will remain so at greater distances if it is pressure confined by the circumstellar medium.

Pressure Collimation. In ejection models that do not use the magnetic field, mass loss is likely to be isotropic, so that collimation must be enforced by the surrounding medium. This occurs in the strongly anisotropic circumstellar medium with strong density gradients that can be expected when the protostellar cloud flattens either along the rotation axis of the initial condensation (Boss 1987a) or along the magnetic field direction (Mouschovias 1976). The structure and evolution of a stellar wind bubble expanding in an anisotropic medium depends on the type of interaction between wind and circumstellar medium. Königl (1982) shows that the bubble resulting from an energy-driven interaction expanding in a density gradient $\propto r^{-2}$ could form a bipolar Laval nozzle in which the gas is accelerated to supersonic values. Barral and Cantó (1981) studied the collimation by an isothermal and hydrostatic circumstellar disk of an initially isotropic stellar wind in the case of a momentum-driven interaction with a constant density circumstellar medium. They find a steady-state configuration in which the ram pressure of the wind matches the external pressure exactly. The resulting cavity is bipolar and the wind is refocused on the system's axis at a distance $R \approx 10\,R_o$, where R_o is the cavity radius in the disk plane, which is given by

$$\frac{R_o}{10^{15}\text{cm}} \approx \left[7\left(\frac{\dot{M}_{\text{wind}}}{10^{-8}M_\odot\text{yr}^{-1}}\right)\left(\frac{V_{\text{wind}}}{200\text{ km}^{-1}}\right)\left(\frac{n_o}{10^8\text{cm}^{-3}}\right)^{-1}\left(\frac{T_o}{10\text{ K}}\right)^{-1}\right]^{1/2} \quad (24)$$

and where n_o and T_o are the density and temperature in the disk. An oblique shock is formed at the refocusing point and the wind is accelerated in a collimated jet propagating along the flow axis; computed jet properties are comparable to observations if $\dot{M}_{\text{wind}} \approx 10^{-8}$ to $10^{-7}M_\odot$ yr^{-1} and $V_{\text{wind}} \approx 100$ to 400 km s^{-1} (Cantó et al. 1988). From Eq. (24), one then finds that the disk density must be $> 10^{10}$ cm^{-3} at $R_o \approx 10^{14}$cm for the jet to be formed at a distance from the star $< 10^{15}$cm, as observed. Whether typical disk densities are this high so far away from the star is unknown. This mechanism may also be invoked to explain the large-scale (0.1 to 1 pc) collimation of bipolar molecular flows; Sakashita and Hanami (1986) computed the evolution of the molecular lobes swept up by the wind in the two regimes of a momentum-driven and energy-driven interaction using the same density as Barral and Cantó, and find that collimation degree, acceleration and speed of the outflow are closer to observed values when an energy-driven flow is assumed.

D. Conclusions

Among the several wind mechanisms discussed above, radiation pressure and thermal expansion appear unable to drive mass loss from low-luminosity young stellar objects. Other mechanisms, all of which depend in some way or another on magnetic fields, are more plausible, although there are problems with every one of them. We were able to gather enough quantitative information from published work on three models (those of Shu et al. [1988], Pudritz and Norman [1986] and Pringle [1989]) to derive predicted mechanical luminosities and forces from the wind models, and all three appear compatible with the typically observed values of these physical quantities. Nonetheless, each one has limitations and uncharted areas which make its applicability to the whole class of embedded and optical young stellar objects questionable. It is therefore fair to conclude at this point that current work is convincing on one aspect, namely, the crucial role of magnetic fields in driving protostellar winds. Those who had hoped to go through an entire astrophysical career without looking at the equations of magnetohydrodynamics—and many prefer this rather pleasant but possibly foolish idea—will probably have to stay away from theories of mass loss from young solar-type stars.

PROTOPLANETARY THERMAL METAMORPHISM: THE HYPOTHESIS OF ELECTROMAGNETIC INDUCTION IN THE PROTOSOLAR WIND

FLOYD HERBERT, CHARLES P. SONETT
University of Arizona

and

MICHAEL J. GAFFEY
Rensselaer Polytechnic Institute

Observed taxonomic types of asteroids are identified with thermal histories deduced from asteroidal reflection spectra and meteoritic compositional trends. These thermal histories are interpreted assuming electromagnetic-induction heating arising from an active pre-main-sequence Sun. The immediate cause of the hypothetical induction is a strong electric field present in a high-density solar wind. This wind should have been at least partly ionized and therefore magnetized and would have driven electrical currents in protoplanets that were sufficiently electrically conductive. Many present-day meteoritic materials and postulated protoplanetary material candidates satisfy this condition. We also review some aspects of heating from extinct short-lived radionuclides. The fossil evidence of primordial asteroidal heating is suggestive of the pattern expected from electrical induction. The types of asteroids that predominantly occur in the inner belt (i.e., at small heliocentric distance) appear to be igneous on the basis of telescopic observations. The middle-belt types likewise appear to be the result of thermal metamorphosis in the presence of liquid water, indicating the attainment of intermediate temperatures. By contrast, the outer-belt types appear to contain considerable water but lack hydrated minerals, indicating a perpetually cold existence. There is also some statistical evidence that the major igneous types are preponderant at certain sizes, a trend which could be a *relict of a primordial distribution. The existence of an asteroidal heat source dependent on a particular evolutionary phenomenon of the Sun such as a dense plasma outflow can be used to deduce or constrain aspects of solar history if the fossil evidence is strong enough. In this chapter we argue that the fossil evidence is compelling enough to attempt to deduce from it some basic features of the plasma outflow such as its duration (possibly 1 Myr) and time-integrated mass loss (of order 10^{-2} to 10^{-1} $M_\odot$). One possible model that yields the dense equatorial ionized solar wind needed by the induction model is the "X-celerator" mechanism of Shu et al.*

I. INTRODUCTION

It has long been known from meteorite studies that one or more episodes of protoplanetary heating occurred in the early solar system (see, e.g., Fish et al. 1960; Wasserburg et al. 1965). Meteorites exhibit the effects of early heating, ranging from negligible to complete melting of the metal and silicate phases. Since meteorites are generally presumed to arise from larger parent objects, probably some subset of the present asteroids was involved in this range of thermal events (Wetherill and Chapman 1988). Although the heliocentric dependence of early solar system heating events inferred from meteoritic evidence is vague unless correlated with telescopic studies of asteroidal mineralogy, meteorites do provide excellent chronologies of those events. From radiometric age studies, the heating event(s) are inferred to have taken place very early in the evolution of the solar system, probably within the first few 10 Myr (see, e.g., the chapter by Wood and Pellas).

Larger bodies of terrestrial class, such as Mars, Venus and the Earth have small surface-to-volume ratios and sufficient gravitational potential or accretional energy to have favored complex and heterogeneous thermal histories. The smaller terrestrial planets, such as Mercury (Siegfried and Solomon 1974) and the Moon, may require a primordial energy source in addition to gravitational self-energy in order to produce the inferred early melting and igneous activity. Mercury appears to have undergone nearly total melting very early in its history (Murray et al. 1974; Solomon 1976; Siegfried and Solomon 1974) and its present-day magnetic field suggests that a molten interior may have survived to the present day (Ness et al. 1974; Schubert et al. 1988). The Moon also seems to have undergone widespread, though not necessarily complete melting (Wood et al. 1970; Smith 1970; Walker et al. 1975; see also the review by Binder 1986, and references therein).

Although acceptable heat source(s) responsible for these thermal episodes are varied in the case of the larger planets just mentioned, nearly all asteroids are too small to have been heated by any but two known mechanisms. For these objects, the two viable heat sources appear to be electromagnetic induction (Sonett et al. 1968,1970) and heating from decay of now extinct short-lived radioisotopes (Brown 1947; Urey 1955; Fish et al. 1960). In this chapter we primarily review the evidence for primordial electromagnetic heating. Planetary thermal events arising via this mechanism call attention to the central role that the Sun must have had in establishing the requisite interplanetary electric fields and associated strong solar-plasma-flow field. The primary critical requirement for the occurrence of induction heating is the presence of a massive primordial ionized solar wind, which we discuss in Sec. VII. The thermal histories of the asteroids would in that case imply the occurrence of a massive primordial ionized solar outflow. A number of igneous meteorites, including many of the basaltic achondrites, have ages very close (< 20 Myr) to the age of the solar system derived from the study of chondritic meteorites. Their petrologies and mineralogies are the result of the differentiation of solid and liquid components by density-controlled gravitational segregation. The antiquity of such differentiated meteorites requires the presence of planetesimals of significant size (10s to 100s of km) and fairly high temperatures at the time of the heating event in order to provide gravitational forces sufficient to segregate the phases with different densities (Taylor 1988a). Since such bodies have cooling times to the closure of the radiogenic systems of a few Myr or more, depending upon size, the time scale of the early heating event and of planetesimal growth must be very short. Only a very limited set of heating mechanisms can plausibly be invoked to produce the early thermal episodes recorded in the meteorites.

Several classes of asteroids (types S, M, E—asteroidal types are conventionally denoted by letter; see the summary in Table I) show strong evidence in their spectra of igneous processes, and therefore are conventionally interpreted as having undergone strong heating (Gaffey 1984; Bell 1986b; Gaffey 1986; Gaffey and Ostro 1987; Gaffey 1988; Bell et al. 1989; Gaffey et al. 1989) but see Wetherill and Chapman (1988) for a summary of an opposing view based primarily on the statistics of meteorite types. These putatively igneous asteroid types occur primarily in the inner belt (i.e., at smaller heliocentric distance). Since these small bodies have large surface-to-volume ratios and little accretional energy release, exotic heat sources are generally required for any significant heating. The asteroid types (primarily C but also including the rarer types B, F, G) occurring most frequently in the middle belt are believed to be the result of thermal metamorphosis in the presence of liquid water, indicating the attainment of intermediate temperatures (see the sources cited above and Lebofsky [1978], Lebofsky et al. [1989], Jones [1988] and Vilas and Gaffey [1989]). By contrast, the dominant outer-belt types (P, D) appear to contain considerable water but no hydrated minerals, indicating a perpetually cold existence (Hartmann et al. 1985; Lebofsky et al. 1988; Jones 1988; Vilas and Gaffey 1989, and references therein).

The relative frequencies of the major igneous types are somewhat greater at certain sizes (Gradie et al. 1989), which may be a relict of a primordial distribution, the result of collisional evolution or a statistical artifact of the observations. These possibilities are discussed more fully in the next section, but the induction-heating model predicts a preferred radius for maximum heating and any evidence for such behavior in the primordial asteroid population would have important consequences for the induction-heating model.

II. COMPOSITION AND DISTRIBUTION OF ASTEROID TAXONOMIC TYPES

A. Interpretation of the Taxonomic Types

It has been known for some time that there are systematic trends in the spatial distribution of asteroid taxonomic types defined by their reflection spectra (Chapman et al. 1975; Zellner and Bowell 1977; Bowell et al. 1978). The distribution of the major taxonomic types with respect to heliocentric distance corrected for observational bias effects (Gradie et al. 1989) is shown in Fig. 1. The nature and origin of the various taxonomic types of asteroids have been the subjects of considerable dispute. The two dominant scenarios are that the observed variations in the asteroid-belt composition were primarily determined by the original nebular condensation processes and alterna-

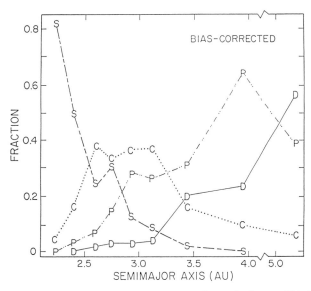

Fig. 1. Distribution of asteroid compositional types vs heliocentric distance. This figure shows preliminary bias-corrected estimates of the distribution with respect to heliocentric distance of the major asteroid taxonomic types in the main belt (Gradie et al. 1989). This outward succession of types is thought to represent a sequence of metamorphic types from igneously differentiated through aqueous altered to primitive. The different curves are labeled by the asteroid taxonomic types included. If this sequence is the result of primordial heating (as is generally believed), then the heat source required must fall off with heliocentric distance fairly rapidly in this region.

tively that there has been considerable post-condensation metamorphism in parts of the belt. The two principal viewpoints have been the result of primary reliance either on the reflection spectra alone or on inferences from the properties and statistical distributions of meteorites collected over the last few Myr by the Earth's surface. A full exposition of this controversy is beyond the scope of this chapter; more complete discussion is given by Wetherill and Chapman (1988), Bell et al. (1989), Gaffey et al. (1989) and Gradie et al. (1989). It seems most likely to the present authors that these trends in the taxonomic distributions of asteroids are reliably interpreted by their direct observation and thus they represent variations in the primordial thermal histories of the asteroids. Consequently, we adopt this point of view in interpreting the asteroidal compositional trends as representing important clues to aspects of solar system formation and early thermal evolution.

A number of attempts have been made to use the distribution of asteroid types to constrain the nature and processes of the late solar nebula and early solar system (see, e.g., Gradie and Tedesco 1982; Tholen 1984; Zellner et al. 1984; Bell 1986b; Gradie et al. 1989; Bell et al. 1989). The conclusions of such analyses are critically dependent upon the particular assumptions which were made concerning the compositional nature of each taxonomic type.

Expanded asteroid observational data bases combined with enhanced interpretive capabilities have produced considerable improvement in compositional analysis of asteroids over the last decade. This work has been reviewed in Gaffey et al. (1989). Although the more numerous asteroid classes are likely to include a diverse range of assemblages and to have experienced a variety of thermal histories (e.g., primitive C3-assemblages as subtype within the predominantly igneous S-type; Bell et al. 1987,1989; Bell 1988), useful generalizations can be made concerning the compositional nature of each asteroid class. The probable compositions and the meteoritic analogs for the taxonomic types are summarized in Table I (Gaffey et al. 1989; Bell et al. 1989).

Of these generalizations, the S-asteroid interpretation is the most critical since it will have the greatest effect on the models derived from a consideration of the heliocentric distribution of the asteroid types. The spectral data for S asteroids have been interpreted as indicating primitive assemblages analogous to the ordinary chondritic meteorites by some investigators, and as indicating igneous assemblages analogous to the stony-iron meteorites by others. Gaffey (1984), Wetherill and Chapman (1988) and Gaffey et al. (1989) have reviewed the history of this controversy. To summarize briefly, the ordinary chondrite interpretation of S asteroids is associated with the view that the distribution of taxonomic types is due to compositional variation of the primordial condensate (this hypothesis is variously discussed by Anders [1964], Larimer and Anders [1967], Grossman and Larimer [1974], Lewis [1974], Wasson and Wetherill [1979], Gradie and Tedesco [1982] and Wetherill and Chapman [1988]). In this scenario it is argued that the S asteroids

TABLE I
General Mineralogy and Meteoritic Analogs for the Asteroid Taxonomic Types

Asteroid Type[a]	Mineralogy[b]	Meteoritic Analog[b]
A [4]	Olivine or olivine-metal	Olivine achondrite or Pallasite (Igneous)
B, C, F, and G [112]	Hydrated silicates + carbon/organics/opaques	CI1-CM2 assemblages and assemblages produced by aqueous alteration and/or metamorphism of CI/CM precursor materials
P and D [49]	Carbon/organic-rich silicates (anhydrous?) assemblages	Organic-rich cosmic dust grains? CI1-CM2 plus organics? C3 plus organics? (primitive)
E [8]	Enstatite or possibly other iron-free silicates	Enstatite achondrites (igneous)
M [21]	Nickel-iron metal possibly with trace silicates	Iron meteorites (igneous) Enstatite chondrites? (primitive)
Q [1]	Olivine + pyroxene + metal	Ordinary chondrites (primitive)
R [1]	Pyroxene + olivine	Pyroxene-olivine achondrite (igneous)
S [144]	Metal $\geq$ olivine > pyroxene	Pallasites with accessory pyroxene; Olivine-dominated stony-irons; Ureilites & Primitive Achondrites (igneous); Small component of CV/CO Chondrites & possibly ordinary chondrites (primitive)
V [1]	Pyroxene + feldspar	Basaltic achondrites (igneous)
T [4]	—[c]	—[c]

[a] Asteroid classification from Tholen (1984). The number in brackets is the number of classified asteroids of that type.
[b] Mineralogy and meteoritic analogs from Gaffey et al. (1989, and references therein).
[c] Mineralogy & meteoritic analogues: Possibly similar to types P/D.

are undifferentiated chondritic assemblages and the change from an S-dominated inner belt to a C-dominated outer belt is interpreted as the fossil signature of the nebular compositional gradient, with increasing oxidation state and lower temperatures at greater heliocentric distances across this interval.

The alternative view of the igneous assemblage is that an energetic primordial heating episode has considerably modified the primordial mineralogical distribution by melting some asteroids and warming others sufficiently to mobilize liquid water.

For the few S asteroids which have been subjected to detailed spectral study (e.g., 8 Flora by Gaffey [1984]; 15 Eunomia by Gaffey and Ostro

[1987]), their igneous nature has been solidly established by their nonchondritic rotational spectral variations. The individual points on Fig. 2 show the wavelength position of the 2 μm absorption feature relative to the ratio of the areas (total absorbance) for the 1 and 2 μm absorption features for the S asteroid 15 Eunomia. The values obtained for sequential rotational aspects, where different portions of the surface are observed, are connected by the light lines. The heavy line represents the error-weighted, linear least-squares fit to this set of points. The stippled area is the permitted region for nebular anhydrous silicate assemblages which have not undergone melting and differentiation, such as the chondritic meteorites.

These spectral parameters are functions of mineral composition and mineral abundance. The band area ratio increases with increasing pyroxene abundance in olivine-orthopyroxene mixtures (see, e.g., Cloutis et al. 1986), while the 2 μm band center shifts toward longer wavelength with increasing iron content (see, e.g., Adams 1974). The compositions and relative abundances of these mafic silicate minerals in undifferentiated (chondritic) assemblages are strongly correlated as a consequence of their essentially constant bulk chemistry. The abundance of olivine increases by the conversion of pyroxene plus iron to olivine. Thus, within the chondritic suite, increasing olivine correlates with decreasing pyroxene and with increasing iron content in the mafic silicates. In Fig. 2, such a correlation produces a negative slope, such as is seen for the actual chondritic meteorites. By contrast, the slope of the Eunomia distribution is positive. While calibrational uncertainties, systematic errors in the observational data, and artifacts of interpretive methodologies may all cause inaccuracies in the spectral parameters, none of these can materially affect the slope of the distribution. The only plausible means of producing such a positive slope is by magmatic differentiation within the parent body of Eunomia, a process which decouples the mineral abundance from mineral composition.

Although detailed study of individual S asteroids is a continuing effort, it is a slow process. Application of several less-diagnostic criteria to the approximately 150 S asteroids, for which spectral data are available, consistently indicates differentiated assemblages. While the presence of a small component of undifferentiated objects within the S population is permitted by the present data, the bulk of the population must have undergone a strong post-accretionary heating episode.

Bell (1986b; see also Bell et al. 1989) defined three asteroid superclasses—primitive, metamorphic and igneous—based upon the processes required to produce the inferred asteroid surface assemblages. The igneous superclass (taxonomic types V,R,S,A,M,E) include those objects whose surface mineralogy require melting and magmatic differentiation within their parent planetesimal. The metamorphic superclass (types B,G,F,T) exhibit surface assemblages indicating subsolidus aqueous alteration and thermal metamorphism. The primitive superclass (types D,P,C,K,Q) are consistent with unal-

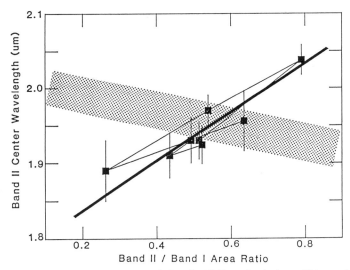

Fig. 2. Near-infrared spectral parameter correlations for 15 Eunomia, the largest S-type asteroid. The individual points were made in successive observations as Eunomia rotated and are connected by the narrow lines to show the observational sequence. The heavy line shows a least-squares fit to these data points, which is to be contrasted with the stippled region indicating the trend expected for nebular condensates that have not been melted.

tered nebular (chondritic) material (although the C class may be more representative of the metamorphic superclass). The absence of hydrated silicate minerals in the P and D types of the outer belt is interpreted as an indication that these objects never achieved post-accretionary temperatures sufficient to melt ice (Lebofsky et al. 1989; Jones 1988).

B. Distribution with Heliocentric Distance

Bell (1986*b*) and Bell et al. (1989) showed the relative abundance of his superclass objects binned into seven intervals of heliocentric distance between 1.9 and 5.2 AU. The relative abundance of the igneous superclass decreases in a nearly linear manner from approximately 100% at 1.9 AU to 0% at 3.4 AU. The relative abundance of the primitive superclass increases from approximately 0% to 100% over the same interval, reaching 100% at 4.0 AU. The metamorphic superclass shows a broad low distribution peaking (about 10%) near 3.0 AU. It was concluded that this pattern indicates a planetesimal heating mechanism whose efficiency decreased with increasing heliocentric distance.

Gaffey (1988) divided the asteroid types classified by Tholen (1984) into igneous (those which had surface assemblages resulted from at least partial melting) and nonigneous types based upon spectral analysis of a few members of each type. The igneous group includes asteroid types S, M, A, E, V and R. The relative proportions of these "heated" objects in eleven dynamically defined asteroid zones (Zellner et al. 1985) are shown vs heliocentric distance on the upper half of Fig. 3. The solid line is a linear least-squares fit to the points between 1.9 and 3.5 AU. For comparison, the lower portion of Fig. 3 shows a histogram of total asteroid number density vs heliocentric distance. From 1.9 to 3.5 AU, across the interval of the main belt, the proportion of objects which underwent strong (T> 1000° C) post-accretionary heating decreases linearly from 100% to 0%, with a slope of approximately 67% per AU. Based upon plausible heating mechanisms for such small bodies, upon the chronology of meteorite heating events, and upon the size requirements for density-controlled magmatic segregation by gravitation forces within the planetesimals, these temperatures seem to have been attained very early in solar system history (the first few Myr) but after the planetesimals had grown to their present sizes or somewhat larger.

The asteroid compositional variation with heliocentric distance appears to be primarily a function of the intensity of or susceptibility to a very early heating mechanism, perhaps modified by a radial gradient in the composition of the initial material preserved from the solar nebula. The simplest model would invoke a Sun-related heat source with a strong dependence on heliocentric distance. In the following section, we discuss the applicability of the electromagnetic-induction heating hypothesis, a candidate for such a Sun-centered energy source.

A non-Sun centered heat source such as [26]Al or other short-lived radioisotope can be made consistent with the observed pattern of asteroidal type distribution as well. Such consistency would, however, require some additional complexity such as an accretionary wave moving outward through the early solar system, so that the inner-belt bodies accreted early and incorporated a large amount of the heating element while the outer-belt bodies accreted several half-lives later and incorporated little or none. Although certainly possible, this hypothesis seems more *ad hoc* and thus perhaps less probable than a single stage process.

C. Distribution with Respect to Size

The induction-heat source is also size dependent, so one might expect to see some relict of this effect in the distribution of asteroidal types. Figure 4 (derived from Fig. 6 of Gradie et al. 1989) shows the distribution of groups of asteroidal compositional types with respect to size in the main belt (out to zone IIIa as defined by Zellner et al. [1985]). The two putatively igneous types S and M both become relatively more abundant than the low-temperature metamorphic types (C, B, F, G) at small asteroidal diameters. In fact, there appears to be a peak in the S-type distribution at about 50 km of about 50 %, superimposed on a general rising trend. In the original Gradie et al. figure, the other classes are not tabulated below the values shown here, but there is one additional S class point at 20 km which is consistent with isolation of the peak at 50 km; the significance of this peak is presently un-

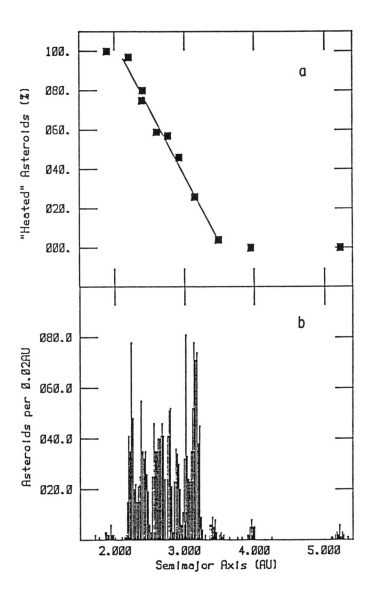

Fig. 3. (a) The relative proportions of apparently strongly heated asteroid taxonomic types S, M, A, E, V and R plotted vs heliocentric distance. (b) A histogram of total asteroid number density vs heliocentric distance.

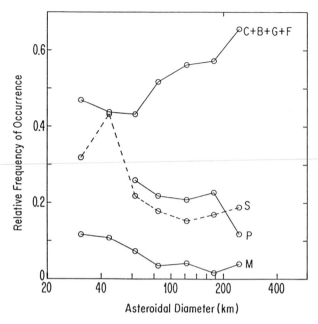

Fig. 4. Distribution of asteroid compositional types vs asteroid size. This figure replots in relative frequency from estimates of the bias-corrected distribution with respect to size of subsets of asteroid taxonomic types in the main belt out to zone IIIa (Gradie et al. 1989). The different curves are labeled by the asteroid taxonomic types included. The S and M types are believed to be igneous differentiates.

known. In what follows we shall interpret this peak as a relict of the primordial size distribution, but the reader should view the conclusions we draw from it with caution. However, the general preponderance of S and M types at small radius (or the related excess of C types at large radius) seems not in much doubt.

Even if the features noted in Fig. 4 are accepted as real, they must be interpreted in the light of later collisional evolution. Models of collisional evolution of asteroids can be very complex, but they typically involve the production of a distribution of smaller asteroids from larger ones (see, e.g., Chapman and Davis 1975; Davis et al. 1979,1989). Under the conditions usually assumed in such models, fragmentation typically prevails over accretion for all but the very largest asteroids (Davis et al. 1979). Thus, the expectation is that there will be some bodies remaining from the primordial accretion phase and a large distribution of smaller fragments (possibly dominating the general population of asteroids). The fragments would be expected to follow a power-law distribution except for two important effects noted by Chapman and Davis (1975): self-gravitation (important at large sizes) and material strength (important at smaller sizes and dependent on the type of material and its history of fragmentation and melting, if any). If the 50-km feature in the S distribution is primordial, it is probably the result of one or the other of these two effects or else is merely a residue of the primordial distribution.

Possibly the most important strength effect is the difference between the ease of fragmentation of stones of various types of iron. Chapman (1974) proposed that removal by impacts of most of the stony mantles of asteroids with iron cores would occur more frequently than complete disruption of the core and that the expected result of this event, a largely iron body with a considerable residue of surficial rock, would be a likely candidate for the typical S asteroid. Thus, this interpretation of S-type spectra is that these objects are complete (because of the strength of iron) asteroid cores that have had most of the associated mantle material removed and so have surfaces showing comparable amounts of iron and rock. This view implies that the S asteroids are cores that have not been disrupted (as opposed to M types which presumably have) and thus retain their original size distribution. Although there have been many subsequent proposals for the nature of S asteroids, the Chapman (1974) model is still one of the principal contenders (for further exposition of this picture, see Gaffey [1984], Gaffey et al. [1989] and Bell et al. [1989]).

An interesting corollary of this idea (first pointed out by Chapman [1974] in a different connection) is that the sizes of S asteroids would be substantially the same as that of the iron core of the original body. Therefore if the iron-to-rock ratios of the original bodies were similar, the present S asteroid sizes would be an approximately constant fraction (1/2 to 2/3, say) of the sizes of their original parent bodies. M asteroids would thus represent the result of the presumably rarer more energetic collisions which shattered the iron cores and exposed the relatively pure iron of their interiors to dominate the surface area and hence the spectra. If the original parent bodies of the S and M asteroids were concentrated at a particular size and broken up, the S objects would primarily have sizes at some fraction of the original parent-body peak size and the M asteroids would form a power-law distribution at smaller sizes. Figure 4 is consistent with a picture of this type if the original S and M parent bodies were most frequently about 70 to 100 km in diameter, and breaking up to 50-km S objects and smaller M objects.

It is also possible that the survival of the apparent peak in the size distribution of S asteroids is merely the result of an original population excess there which has not been completely combed out by the later collisional evolution. In either case, interpretation of the apparent excess points to the role of electromagnetic heating.

Thus a simplistic but plausible interpretation of Fig. 4 is that the primordial heat source that melted the S and M parent bodies might have been a strongly peaked function of asteroid size. This has been interpreted by Herbert (1989; see also the next section) as evidence for and a constraint on the electromagnetic-heating hypothesis. In this picture, the size- and distance-dependent induction heat source would have primarily melted parent bodies at the inner edge of the belt with diameters around 70 to 100 km, which were subsequently broken down into 50-km S asteroids (complete cores) and smaller M objects (core fragments). Those asteroids that escaped melting, the C and similar types, would be those whose sizes were significantly different from those of the S and M parent bodies or whose heliocentric distance was too great for melting, or both. Inner-belt C objects would be preferentially larger or smaller (but this latter possibility lies beyond the range of the data).

III. PROTOPLANETARY INDUCTION HEATING

Electromagnetic heating is forced by the electric field generated within an object immersed in the solar wind. In the generation of the interplanetary magnetic and electric fields, both the plasma outflow velocity and the spin rate of the Sun are involved. Furthermore, a high-density plasma flow from the Sun is required, and in some models also an elevated ambient temperature. Additionally, the object to be heated must obviously be at least partially formed by the time of the induction epoch. Theoretical justification for the very rapid formation of very small (1 to 10 km) protoplanets on time scales of a few heliocentric orbits has been given by Goldreich and Ward (1973). Even at this size, electromagnetic heating already becomes possible (Sonett et al. 1970; Herbert 1989). Asteroidal growth in size has been modeled by Greenberg et al. (1978) and Patterson and Spaute (1988), based upon damping of relative encounter velocities by lossy collisions with an abundant population of the smallest-planetesimal material, and found to be quite rapid over at least the next order of magnitude. They calculate that growth from 10 to 100 km requires only 10^5 to 10^6 yr. Thus it appears that there is no fundamental difficulty with the relative timing of events.

In the context of electromagnetic theory, both the transverse electric (TE) and transverse magnetic (TM) modes (see, e.g., Stratton 1941) can drive joule heating depending upon the properties of the solar wind. Both induction modes are known to occur in the present epoch of the solar system, though far reduced in intensity from that hypothesized for the primordial solar system. The extension of electromagnetic excitation from vacuum to the case of a supersonic plasma has not been widely studied. In particular, that of the TM mode to the case of the steady state has no vacuum counterpart. It corresponds to unipolar induction in the case of steady fields. The TE or eddy-current mode exists, of course, only for time-varying fields.

Discontinuities (contact surfaces) and other disturbances in the frame comoving with the supersonic plasma, are perceived as time dependent in frames attached to planetary objects and thus generate solenoidal electric fields yielding (TE-mode) eddy currents. But the TM mode is more complex. Here, because the TM electric field (and therefore current density) has a radial component at the planetary surface, the current system must close into the external plasma. The interaction between the external and internal current and field systems is a problem of significant complexity.

The TM mode was, in a somewhat limited context, originally identified in connection with attempts to explain orbital angular momentum loss in an artificial satellite that was thought to originate in the interaction of the ionosphere with the satellite (Drell et al. 1965). TM induction was also hypothesized in the interaction of Io with the Jovian magnetosphere (Goldreich and Lynden-Bell 1969), and as a test of TM induction, has been shown successfully to exist in a plasma machine experiment (Srnka 1973). The TE mode was readily identified by intense eddy-current induction in the Apollo magnetometer experiment (Sonett et al. 1971). Direct observations of induction in solar system bodies have also been made at (a) Io (Ness et al. 1979; Neubauer 1980; Belcher et al. 1981); (b) Titan (Ness et al. 1982; Hartle et al. 1982; Neubauer et al. 1984); (c) Venus and Mars (see review by Luhmann 1986); and (d) comets (Smith et al. 1986; Neubauer et al. 1986). For the early solar system, the application is a matter of parameterization of magnetic fields and plasmas, the physics having been amply demonstrated in the examples just cited.

The existence of joule heating in the presence of induced currents is inevitable; its significance arises from the extreme value of induction resulting from the hypothesized high solar plasma efflux and (for the TM mode) high solar spin rate. Joule heating in protoplanetary interiors is analogous to the greenhouse effect—the energy can be coupled to the deep interior where it cannot easily escape, thus raising temperatures considerably. Although the assumed solar-wind energy flux is small (comparable to that of present-day sunlight), its coupling to the deep interior would have generated large interior temperatures in order to develop the thermal gradient necessary to drive enough outwardly diffusing heat.

The maximum power level which can be extracted under these conditions is roughly comparable to the kinetic energy flux in the plasma stream; the actual power level depends on a kind of impedance match between the planetary resistance and the hydromagnetic waves coupling the planetary and plasma current systems (Sonett et al. 1970; Neubauer 1980). The impedance match is controlled by a number of planetary and plasma variables; it is a principal aspect of the thermal models discussed in this chapter. The plasma kinetic-energy flux is ultimately related to the solar mass loss; this relation is the reason for the importance of that parameter in earlier work. The electric field at the asteroid's surface is assumed to be diminished from the value of the motional field by the magnetic-flux deflection factor (Sonett et al. 1970). The solar-wind plasma density is assumed $= \Delta M_\odot/(4\pi v_{sw} T_\odot \ d^2)$, where d is heliocentric distance, $T_\odot$ is the induction epoch duration and v_{sw} is the solar-wind speed. The mass loss $\Delta M_\odot$ is assumed isotropic; since the relevant parameter is the solar-wind density, if the mass loss is predominantly equatorial, the required net mass loss is correspondingly reduced.

It is because of the current-closure requirement across the planetary surface in the case of TM induction that the net electrical resistance of the surface layers is important. Because most rock electrical conductivity is quite temperature dependent, the surface temperature has a strong influence on the heating. This surface-temperature dependence is one of the principal reasons for the strong heliocentric distance dependence of the electrical induction models. The TE and TM electromagnetic induction modes in the body interiors are governed by their respective wave equations

$$\nabla^2 \Omega^{te} + k^2 \Omega^{te} = 0 \tag{1}$$

and

$$\nabla^2 \Omega^{tm} - \frac{1}{k_1 r} \frac{dk_1}{dr} \frac{\partial}{\partial r}(r\Omega^{tm}) + k^2 \Omega^{tm} = 0 \tag{2}$$

where $k_1 = i\omega\varepsilon - \sigma$, $k_2 = i\omega\mu$, $k^2 = -k_1 k_2 = \omega^2\mu\varepsilon + i\sigma\mu\omega$, and Ω^{te} and Ω^{tm} are scalar potential functions of r, θ, ϕ and frequency ω (Wyatt 1962; see also Horning and Schubert 1975; Stern 1976). [Equations (1) and (2) given earlier in Sonett and Colburn (1974, p. 421) are incorrectly referred to there as TM (Eq. 1) and TE (Eq. 2), respectively, rather than as given here.] In the external plasma, the wave number $k = 2\pi/\lambda = \omega/v$, where ω is doppler shifted and v is the resultant of the wave propagation and plasma convection velocities. The electric and magnetic fields are computed from

$$\mathbf{E} = \frac{1}{k_1}\boldsymbol{\nabla} \times \boldsymbol{\nabla} \times \Omega^{tm}\mathbf{r} + \boldsymbol{\nabla} \times \Omega^{te}\mathbf{r} \tag{3}$$

and

$$\mathbf{H} = \frac{1}{k_2}\boldsymbol{\nabla} \times \boldsymbol{\nabla} \times \Omega^{te}\mathbf{r} - \boldsymbol{\nabla} \times \Omega^{tm}\mathbf{r} \tag{4}$$

with $\mathbf{B} = \mu\mathbf{H}$.

Within the body where $\sigma \neq 0$, Ω^{te} and Ω^{tm} are oscillatory and evanescent when $\omega \neq 0$. When $\omega \to 0$, the boundary conditions require $\Omega^{te} \to 0$ also but Ω^{tm} remains finite, being related to the electrostatic electric potential Φ by $\Phi = 1/\sigma(\partial r\Omega^{tm}/\partial r)$. In this case Eq. (2) reduces to the much simpler "unipolar" (time-independent TM) mode

$$\nabla^2\Phi = -\frac{1}{\sigma}\nabla\sigma \cdot \nabla\Phi \qquad (5)$$

and $E = -\nabla\Phi$.

The boundary conditions for waves in a plasma with sufficient ram and/or thermal pressure to overcome the pressure of the induced internal magnetic field are such as to compress the induced response entirely within the body. When this condition is violated, the situation is approximated in the heating models by a heuristic pressure-matching formalism described by Sonett et al. (1970) and related to the impedance matching effect described earlier.

Dependence of induction on fluctuations in the fields (such as the TE mode and skin-depth effects) are minimal in bodies as small as asteroids if most of the power lies below a few mHz in frequency and the asteroidal conductivity is typically $< 10^{-1}$ S m^{-1}. For this case, induction is primarily by the unipolar mode as described in Eq. (5). The unipolar equation is solved for a constant electric field at the boundary (long-wavelength limit) (Sonett et al. 1970). The motional electric field in the solar wind is given by $v_{sw}B_{\theta}$ where the wind speed v_{sw} is assumed independent of heliocentric distance d and B_{θ} the magnetic field transverse to the flow is assumed $\propto d^{-1}$; both assumptions are satisfied for the present-day solar wind.

The electric fields are used to compute the joule heating rate $\dot{Q} = \sigma(E_{te}^2 + E_{tm}^2)$ (the usual factor of 1/2 is cancelled by including negative ω) which in turn is used to drive the inhomogeneous heat transfer equation

$$\rho c_p \frac{\partial T}{\partial t} = -\nabla \cdot F_Q + \dot{Q} \qquad (6)$$

where ρ is mass density, c_p specific heat per unit mass at constant pressure, T temperature, F_Q heat flux, $\dot{Q}$ heating rate, and t time. In the present work, $\dot{Q}$ can be electrical and/or nuclear. Heat is assumed to be transported by conduction ($F_Q = -k\nabla T$ with $k =$ thermal conductivity) when the material is unmelted and by completely efficient convection

$$F_Q = \left(\frac{c_p\rho}{r^3\Delta t}r\right)\int_{r_o}^{r} max\{0,[T(\bar{\rho}) - T_s(\bar{\rho})]\}\bar{\rho}^2\,d\bar{\rho} \qquad (7)$$

where $T(r)$ is computed after a time step Δt (in a radially outward sequence following the prescription of Reynolds et al. [1966]) at temperatures above the solidus $T_s(r)$. Here r_o is the lowest level of the melt region associated with r.

The variation of heating with depth has been investigated by Sonett et

al. (1970) for the unipolar mode. An example from that work is shown in
Fig. 5a assuming a 300-km radius asteroid at a solar distance of 2.5 AU.
Additional parameters that define the heating rate are the solar field, the solar
spin rate, an e-fold time for solar despin, the ambient nebular temperature
and a large solar mass loss. This model assumed an olivine composition with
very high electrical resistivity, necessitating solar parameters much more pro-
ductive of induction than those used in more current asteroid heating models
described below. Here, the model temperature increases at all radii for the
first 1 Myr and flattens out subsequently for radii $\leq$ 290 km, the outer 10 km

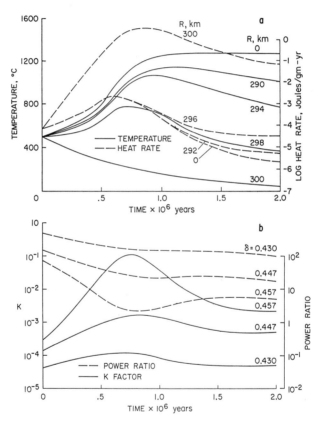

Fig. 5. (a) Temperature vs time in a sample of the model asteroid calculations by Sonett et al.
1970. Both heating rate and temperature are shown for different levels in the body, demonstrating both heat storage and loss from shells near the surface. (b) Loss factor, k defining the
fraction of the effective unipolar potential applied to the asteroid of Fig. 5a as a function of
time due to variations in the induced magnetic pressure. The effect of induced magnetic pressure is described in the text and more fully by Sonett et al. (1970). Also shown is the ratio of
solar-wind kinetic energy flux integrated over the asteroid's cross section to the total power
actually absorbed. Three models are shown, with the curves labeled δ = 0.457 corresponding
to Fig. 5a. δ is a.

being the most strongly affected by radiative heat loss. The deepest parts of
the body attain the highest temperatures. The variation of the effective impedance matching magnetic-field flux loss factor in this model is shown in
Fig. 5b.

In Fig. 6, we illustrate an interesting complexity of the heating model
calculated for the case of the Moon during that body's accretion phase (Herbert et al. 1977). Although the focus of this chapter is asteroid heating, the
lunar case is included because it shows behavior which may be relevant to
asteroid evolution. During the slow initial accretion, melting occurs while the
Moon is still small. Eventually, the rising interior pressure increases the (pres-
sure-dependent) interior solidus temperature at a rate that outstrips the interior
temperature and the Moon resolidifies. With continued growth, skin-depth
localization and accretional heating raise the temperature of the outer layer
(and with it the local electrical conductivity) to stimulate a current-concen-
trating instability (described below; see also by Herbert and Sonett 1979)

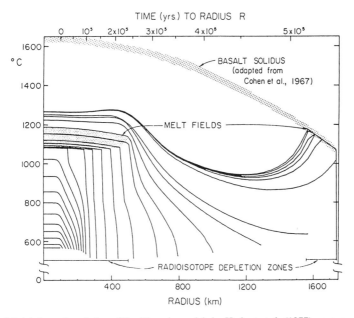

Fig. 6. Model thermal evolution of the Moon in models by Herbert et al. (1977), an example of
the thermal gradient heating instability and dependence on accretion time. The solid curves
show successive thermal profiles within the Moon as it accretes in a time of 5.3×10^5 yr. A
rather resistive rock type (model lunar pyroxenite; Ringwood and Essene 1970; Duba et al.
1973) is assumed, requiring a high environmental temperature of $500°$ C. The Moon reaches
the solidus when its radius reaches 250 km. Resolidification due to overburden pressure occurs
when the radius grows to 800 km. A second melt field is generated near the surface after
accretion is completed. The basalt solidus indicated is for the pressure variation within the
fully accreted Moon.

which concentrates the bulk of the available heating power in the outer layers that ultimately melt, creating a magma ocean. Here again, the solar parameters are more potent than those used in current asteroid models. Although a strong radial variation of the solidus temperature is not likely for small bodies such as asteroids, localization of heating may be important in the larger ones.

Shorter time scales preferentially favor heating in smaller bodies. The actual outflow time is not well known; some theoretical treatments of the source of the flow (see Sec. VII; Shu et al. 1987) suggest that the flow could be driven by inflowing protoplanetary disk material and thus might be episodic. In such a case, there could be significant energy input at a number of time scales. Recent work illustrating this time-scale dependence is summarized in Figs. 7 and 8, showing computations of asteroid heating vs size and solar distance. Each case considered is actually a grid of asteroid thermal-history models, with all parameters held constant except heliocentric distance and asteroid size. The results are summarized by a contour plot of the maximum interior temperature reached in each asteroid model plotted as a function of the size and distance of that asteroid. The parameter α in these figures is described in Sec. IV.

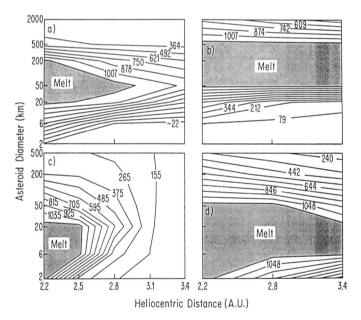

Fig. 7. Contour plots of maximum temperature attained by each asteroid within each of four suites of recently computed models (Herbert 1989). The effects of induction epoch duration and temperature dependence of electrical conductivity are investigated by variation of these parameters. Figs. 7a and 7b are for 10 Myr and figures 7c and 7d are for 10^4 yr. Figures 7a and 7c were generated using $\alpha = 0.025$ K^{-1} and Figs. 7b and 7d are for 0.003 K^{-1}.

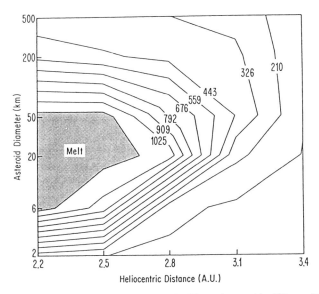

Fig. 8. Contour plots of maximum temperature attained by each asteroid within a suite of recent models (Herbert 1989). The conditions assumed are adjusted to yield an outcome similar to what has been inferred from the data in Figs. 1, 3 and 4. The outflow time is 10^5 yr and α = 0.025 K^{-1}.

The dependence on radius arises from the dependence of the magnetic deflection of plasma upon size, controlling the induction impedance match. Total current I at a given current density (and hence joule heating, holding electrical conductivity constant) through a body of radius R is $\propto R^2$ while the induced magnetic field is $\propto I/R$. Thus, for a given joule heating rate, the induced magnetic pressure is an increasing function of R. The saturation of joule heating with increase in radius limits heating for larger body sizes, while thermal conduction limits the temperature rise in smaller bodies. Consequently, there exists a body radius for which maximum temperature increase occurs (Sonett et al. 1970; Herbert and Sonett 1978,1979,1980; Herbert 1989).

This size and distance dependence of heating was originally suggested (Herbert and Sonett 1978) as being the cause of the difference between the apparently igneous surface of 4 Vesta (originally suggested as the eucrite parent body by McCord and Gaffey [1974] and Drake [1979], but recent observations of Earth-crossing asteroids by Cruikshank et al. [1989,1990] suggest that its relationship may be limited to mere resemblance) and the unmelted surfaces of 2 Pallas and 1 Ceres (Chapman et al. 1975; Lebofsky 1978; Larson and Veeder 1979). However, for reasons discussed later, this view has been somewhat modified. Model calculations for electrical heating show a pronounced falloff in heating with heliocentric distance, especially for materials with large temperature dependence of electrical conductivity. Models also show a heating maximum as a function of body size, that depends on the heating epoch duration. This can be made to match what is inferred from the distribution of asteroidal taxonomic types with seemingly reasonable parameterizations. Recent work such as that shown in Figs. 7 and 8 explores this suggestion further, in application to the now greatly expanded observational data base summarized in Figs. 1, 3 and 4. Thus the inferred maximum primordial heat input at intermediate asteroid sizes and at the inner edge of the belt is compared with the results of model calculations of induction heating.

The metamorphism pattern seen in the asteroid belt and described in Sec. II is suggestive of the heating pattern just described for electrical heating. For example, if the solar mass lost is small but confined to a short time interval, such as 10^4 yr, the size at which maximum heating occurs is around 10 km. If the time interval is very long, such as 10^7 yr, then the heating maximum occurs around 100 km. This contrasts with the 10^6-year interval case with large solar mass loss assumed by Herbert and Sonett (1978,1979,1980), a model that yielded a maximum temperature at the even larger asteroid diameter of about 200 km. Models that assume the assembly of larger asteroids from already-melted protoasteroids lead to another interesting suite of possibilities analogous to scenarios investigated by Wood (1979). In particular this possibility may be relevant to the case of the parent body of the basaltic achondrites where different melt regions occurring at widely separated times (up to 100 Myr apart) have been inferred by Mittlefehldt (1979).

An alternative model for producing melt in large bodies like Vesta was also proposed by Herbert and Sonett (1979). This still incompletely developed scenario depends on the fact that the current density distribution in large bodies composed of material with highly temperature-dependent electrical conductivity can be unstable. In particular, warmer regions tend to attract more than their share of the current, leading to positive feedback and local thermal runaway. Since this effect could lead to high temperatures in just a portion of a large asteroid, simulating the electromagnetic induction behavior of a smaller asteroid, it is conceivable that large asteroids could undergo melting in a subregion of size comparable to that predicted for maximum heating by the simplified model. However, at present this possibility remains speculative.

IV. BULK ELECTRICAL CONDUCTIVITY

Induction heating is strongly dependent on the electrical conductivity of the material to be heated. To understand the models, it is necessary first to provide details of bulk electrical conductivity insofar as it may be inferred for primordial matter. Many candidate materials' conductivities have been measured, but since the nature of the primordial mix of materials that originally accreted into the asteroids is unknown, its electrical characteristics are still mysterious. Whatever the protoasteroidal composition was, it is likely to have consisted of a large variety of compounds, a situation that favors conductivities greater than that typical for igneous minerals and thus favors the likelihood of significant electrical currents.

For the purposes of modeling a function of the form $min\{\sigma_o e^{\alpha T}$, 0.1 S m$^{-1}\}$ is sometimes used. At high temperatures, the electrical conductivity of real candidate materials often stops increasing with temperature, as though a conductive mineral were being volatized or destroyed by chemical reactions (cf. Fig. 9). Although a more familiar temperature dependence for the electrical conductivity of rocky semiconductors is $e^{-E/kT}$, with E the activation energy of the conduction mode, most real minerals have several conduction modes with differing activation energies which often produce an overall con-

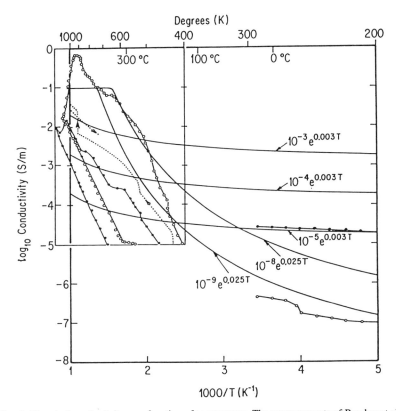

Fig. 9. Electrical conductivity as a function of temperature. The measurements of Brecher et al. (1975) at low temperature and Duba and Boland (1984) at high temperature of the conductivities of samples of the meteorites Allende and Murcheson are shown. The solid-dot curves are for Allende and the open-circle curves are for Murcheson. Since different samples were used, the different determinations are not necessarily consistent (e.g., see the curves for Allende). Also shown are some of the analytic conductivity functions used in the models of the present work labeled by their analytic form.

ductivity profile with $\partial \log \sigma/\partial T$ only weakly variable and thus approxima-
table by a constant ($\equiv \alpha$) as assumed here.

Examples of this behavior are shown in Fig. 9, in which measurements
of $\sigma(T)$ made by Brecher et al. (1975) and Duba and Boland (1984) are compared with the constant $\partial \log \sigma/\partial T$ form. Given that the nature of the true
primordial conductivity function is so poorly known, parameterization of the
model conductivity function by more than two variables seems unjustified at
the present level of approximation.

Candidate materials whose conductivities have been measured are carbonaceous chondrites (Schwerer et al. 1971; Brecher et al. 1975; Duba and
Boland 1984) and ordinary chondrites (Evernden and Verhoogen 1956;
Schwerer et al. 1971; Brecher 1973; Brecher et al. 1975). These possible
analogs of primordial material have room-temperature conductivities varying
over 10^{-10} to 1 S m^{-1}. Electrical conductivity is strongly dependent in most
cases on temperature, oxygen fugacity and the presence of impurities such as
water or carbon (Parkhomenko 1967; Duba and Boland 1984).

The presence of a deep regolith (with thickness more than a few percent
of body size) can further complicate the scenario; limitation of electrical current by the series resistance associated with porosity reduces heating but the
thermal blanketing effect can help retain heat.

V. HEATING BY RADIONUCLIDES

The hypothesis that one or more now extinct radionuclides were thermally active in the early solar system appears to have been first put forward
by Brown (1947) followed by Urey (1955). Urey subsequently abandoned the
^{26}Al hypothesis because, in the concentrations required to melt asteroidal objects, larger bodies such as the Moon in a homogeneous solar system would
have been "superheated." Nevertheless, this proposition was further examined
by Fish et al. (1960) and a search was begun, following a suggestion by
Sonett in the mid 1960s for vestigial ^{26}Mg. The search for evidence of primordial ^{26}Al (aluminum-correlated ^{26}Mg excesses) in meteoritic, lunar and
terrestrial feldspar samples failed to find such evidence at the level of about
^{26}Al/Al ≤ 0.01 to 3×10^{-6}, depending on the sample (Schramm et al. 1970).
These values represent ^{26}Al concentrations and heating rates as of the time of
solidification of the feldspars investigated, and would correspond variously
to temperature increases (in the case of negligible losses to the body surface,
establishing an upper bound) in the range of 30 to 260 K, making reasonable
assumptions about the ratio of Al to Si (*ibid.*).

Interest revived when Lee et al. (1977) and Hutcheon et al. (1978) found
inferred primordial ^{26}Al/Al values of about 50×10^{-6} in Ca/Al-rich inclusions in several chondrules from the C3 (i.e., unmelted) meteorite Allende.
This concentration, when present in bulk material, corresponds to an energy
equivalent to a temperature increase of over 5000 K, although the average

Allende [26]Al concentration is far less. The Ca/Al-rich inclusion concentration would have produced melting in bodies a few km in radius or after a delay (from the inclusion solidification epochs) of three half-lives (or about 2 Myr).

Evidence for primordial [26]Al/Al $= 8.4 \times 10^{-6}$ (temperature increase $\leq$ 750 K) has been found in a 70-μm hibonite clast in the H3 chondrite Dhajala (Hinton and Bischoff 1984). No evidence for [26]Al was found in a hibonite-rich inclusion in Allende (Lee et al. 1979; Hinton and Bischoff 1984), in the Al-rich chondrules of Dhajala, or in plagioclase grains in other H3 or LL3 chondrites (Hinton and Bischoff 1984). These results have been interpreted as an indication that either the different grains formed over an interval at least several half-lives of [26]Al in length, or that significant heterogeneities in [26]Al abundance were present in the solar nebula with significant [26]Al abundance being rather rare.

Recently, a value of [26]Al/Al $= (8 \pm 2) \times 10^{-6}$ has been inferred for the plagioclase component of a chondrule found in the LL ordinary chondrite Semarkona (Hutcheon and Hutchison 1989). The bulk composition, mineralogy, crystal textures and rare-earth element distribution patterns among the major minerals were interpreted as indicating that the parent material of this chondrule had undergone melting and magmatic differentiation prior to the formation of the chondrule—and indeed that [26]Al abundance would, if present, be likely to produce high temperatures. However, the rest of the meteorite had never been subjected to such high temperatures, so presumably the chondrule is a fragment of an earlier generation of rock. It is inferred that this earlier parent rock of the chondrule was formed on a body which had melted in the presence of (because of?) [26]Al. The relevance of this finding for asteroidal metamorphism in general is unclear, as this is the only such specimen ever found.

It is very difficult to estimate the primordial bulk [26]Al abundance in meteoritic material since the conditions for detection (an initially magnesium-free or very magnesium-poor mineral in which a present-day excess of [26]Mg can be attributed to the decay of [26]Al) do not apply to most meteoritic minerals. For example, a search for indications of primordial [26]Al in Vaca Muerta, a mesosiderite and thus derived from a parent body that melted, has been negative at the [26]Al/Al $\leq 0.4 \times 10^{-6}$ level (Papanastassiou et al. 1984). Indeed, no fossil traces of [26]Al have been found in any igneous meteorite (Wasserberg 1985; Podosek and Swindle 1988), although it is commonly accepted among meteoriticists that this short-lived radioisotope could have been the heat source for the parent bodies of such meteorites.

The absence of fossil evidence for [26]Al does not necessarily eliminate [26]Al as the heat source for that body, but rather constrains the cooling time. That is, it remains possible that [26]Al had originally been abundant but had fallen below the stated values by the time that the Mg-free plagioclase phase crystallized from the melt. If crystallization were delayed 5 Myr (seven half-lives), the present upper limit of the [26]Al abundance in this mesosiderite would be seen even if the initial ^{26}Al /^{27}Al was as high as that seen in the highest Allende Ca/Al inclusions. Cooling times of many Myr are expected for 100-km bodies, but the requirement becomes more stringent when one considers that cooling times become very fast near the surfaces of these bodies. This is particularly true in eruptive events such as might be expected for the production of basaltic meteorites. For example, the cooling time drops to the order of 10 yr for 10-m burial. Thus, although a deeply buried igneous rock that later is excavated to become a meteorite can cool slowly, the ^{26}Al heating hypothesis predicts that fossil evidence ought to be found in most eruptive minerals. To date, only the one small sample found in Semarkona (Hutcheon and Hutchison, 1989) corroborates this prediction. Nevertheless, except for rapid crystallization of the parental melt, the absence of ^{26}Al in igneous meteorites cannot be used to eliminate this potential heat source for many meteorite parent bodies.

If ^{26}Al actually was a significant heat source in early solar system bodies (see, e.g., Lee et al. 1977), its abundance varied widely between bodies at the time or times of their formation. If the larger terrestrial bodies accreted from a population of planetesimals more than a few Myr after the formation of those planetesimals were formed, then the "superheating" problem does not arise since essentially all of the ^{26}Al will have decayed by then. However, it is clear that within the planetesimal population, the heating intensity varied enormously. The bulk material of the ordinary and carbonaceous chondrites never approached the melting point. Apparently the P and D asteroids never even reached the melting point of water. Clearly then, the inferred ^{26}Al did not melt the source bodies of the chondritic meteorites either because of their very small size or the low ^{26}Al abundance, although it has been invoked to produce the metamorphism of the higher metamorphic types. Just as clearly, other meteorite parent bodies were undergoing very intense heating. However, without a knowledge of the spatial distribution of this heating and the sizes of the bodies involved, the ^{26}Al heat source cannot be tested (see the chapter by Wood and Pellas).

VI. OTHER HEAT SOURCES

Significant heating by long-lived radionuclides such as ^{40}K, ^{232}Th, ^{235}U and ^{238}U is likely only for the largest asteroids (Lazarewicz and Gaffey 1980; Scott et al. 1989) because these isotopes' heat is released on a time scale longer than the heat diffusion time scale of small bodies. Thus, only a very large (and unlikely) abundance would lead to melting.

Likewise, the gravitational self-energy released by accretion is clearly insufficient to melt any but the largest asteroids. The differential energy of crossing orbits, released by collisions, could however be significant. A useful comparison exists between the energy density required to melt protoasteroidal material (about half going to the 1000 to 1500 K temperature rise and half supplying the heat of fusion) and the velocity required for an equal kinetic-energy density. Because impacts put a large fraction of their energy into ejecta kinetic energy (Cintala et al. 1979), the velocity giving a kinetic-energy density equivalent to that required for melting (about 1.7 to 1.9 km s^{-1} for typical meteoritic materials) may usefully be compared to the asteroidal escape velocity (about 1.0 to 1.3 m s^{-1} for each km of radius). These velocities become equal at an asteroid radius of about 1500 km, or about 1000 km if only enough energy is to be supplied to reach the point where partial melting just begins.

The largest apparently partially melted asteroid, Vesta, has an escape velocity equivalent to about 20% of the energy required to completely melt a projectile, while the equivalent energy fraction shrinks rapidly with radius (7% for a 100 km asteroid, 0.7% for 10 km, and so forth). Thus, when there is anything like equipartition of energy between melting and kinetic energy, the latter will result in ejection of fragments and a net mass loss instead of accretion. This behavior is roughly what was found in models by Wasson et al. (1987). Thus, it is to be expected that most impact-generated melt will escape and cool very rapidly as small particles. A small portion of this material could conceivably be re-accreted by the original target asteroids (or others nearby in phase space) but only in the form of cool solid (glassy?) droplets. However, even if re-accreted, this material would not contribute to magma bodies or significant heating on the asteroid.

Other heat sources have been discussed by Hostetler and Drake (1980). Tidal heating has been discussed in the case of the Moon (Sonett and Reynolds 1979; Hostetler and Drake 1980) but seems problematic even there and is certainly insufficient for asteroids. Release of chemical energy (Kuskov and Khitarov 1979), adiabatic compression of nebular gasses (Hostetler and Drake 1980), a hypothetical superluminous phase of the Sun (Wasson 1974) and explosive magnetic reconnection events in the solar nebula (Sonett 1979) are speculations which have occasionally been mentioned but seem unlikely to have supplied enough heat to protoplanets to do more than perhaps melt some ices.

VII. THE ASTROPHYSICAL SCENARIO

Most work on primordial inductive heating has assumed a solar-wind plasma flux comparable to the large H fluxes (up to nearly 10^{-6} $M_\odot$ yr^{-1}) seen in T Tauri stars (Kuhi 1964; chapter by Bertout et al.). Even larger outflows occur at FU Orionis stars, which appear to be T Tauri stars in an outburst phase (Herbig 1977). Very massive outflows appear to occur frequently in what is believed to be the pre-T-Tauri stage of solar-mass stars, involving large electron densities (10^8 to 10^{11} cm^{-3}), mass losses in the range

10^{-7} to 10^{-6} $M_\odot$ yr^{-1}, and outflow velocities around 100 km s^{-1} (Lada 1985,1987). Edwards et al. (1987) have observed S$^+$ and mass losses of 10^{-9} to 10^{-7} $M_\odot$ yr^{-1} with velocities up to 200 km s^{-1} in spectra of T Tauri stars; they infer the presence of opaque protoplanetary accretion disks. We expect that the presumably even greater mass and angular momentum outflow rates characteristic of the protostellar stages preceding the T Tauri epoch are the best candidates for the induction heating scenario. If such fluxes continue over time scales of order 1 Myr, a large fraction of a solar mass can be lost. For the Sun to have lost more than about half of its original mass is unlikely and the limit may well be more stringent than that (Weidenschilling 1978), but intense solar-wind fluxes such as those mentioned above but of shorter duration are not ruled out. The virial theorem can be used to establish an upper limit for solar mass loss of around 0.5 $M_\odot$ (Weidenschilling 1978). However, $\Delta M \leq 0.1 M_\odot$ seems realistic.

Principally because of the highly diffuse nature of molecular clouds, believed to be the source of protostars, there is an enormous concentration of matter involved in the formation of a star and so conservation of angular momentum erects a formidable centrifugal barrier against star formation. Thus, the shedding of angular momentum is an essential part of star formation. One possibility in the later stages of collapse is magnetic coupling to stellar winds. Shu et al. (1987) suggest that this occurs at the onset of deuterium burning, when convection throughout most of the protostar coupled with its rapid rotation is likely to generate a strong magnetic field. They associate the existence of a "stellar birthline" in the HR diagram with deuterium burning onset and the consequent turning-on of a strong magnetic field which they suggest drives an accretion-terminating stellar wind.

The loss of considerable angular momentum, consistent with strong magnetic braking in the presence of intense mass loss, in the pre-T-Tauri stages of protostellar evolution has been inferred from observations of T Tauri stars by Vogel and Kuhi (1981) and Hartmann et al. (1986). A theory of accretion-driven mass loss with strong magnetic braking has been developed by Shu et al. (1988; see the chapter by Stauffer and Soderblom). In this picture, a thin accretion disk feeds mass inwards toward a rapidly rotating highly magnetic protostar through a sonic transition X-point. Ionization of part of the infalling matter couples it to the rapidly rotating magnetic field, accelerating it toward co-rotation velocity. Because the X-point is beyond the synchronous orbital radius, the acceleration of the newly created plasma causes it to move outwards. The plasma is continually accelerated towards the increasing co-rotation speed as it moves out until it passes the Alfvén transition radius. In this way, the angular momentum of the infalling accretion disk is acquired by the star and coupled to the outflowing plasma wind. The magnetic-field configuration is poloidal at the poles (where a neutral "ordinary" wind is blown outward along the rotational axis) and open at low latitudes (outwards of the X-point), allowing the ionized "extraordinary" wind to expand outward indefinitely.

Many astrophysical outflows are bipolar, i.e., perpendicular to the equatorial plane. The induction-heating model requires a plasma outflow in the equatorial plane. The relationship between polar and equatorial outflows is not well understood. Clearly, however, the interaction between an equatorial wind and a nebular disk can force the wind into a bipolar outflow, perhaps on a distance scale of 10 to 100 AU. Another possibility is the Shu et al. (1988) model, which couples angular momentum of matter flowing into the Sun from the disk inner edge to an outflow by MHD processes, with an additional bipolar wind as a by-product.

VIII. DISCUSSION AND CONCLUSIONS

The available meteoritic and asteroidal evidence indicates the presence of an intense, selective, and short-lived heat source in the very early solar system. The chronology of meteoritic heating events, the spatial distribution of asteroid types, and the possible size-sensitive nature of the heating mechanism permit a general evaluation of the two leading candidates.

The apparent strong dependence of asteroid thermal evolution on heliocentric distance would require special accretionary conditions for ^{26}Al to be the primary heat source. In particular, accretion to form the asteroid parent bodies would have to have been nearly complete at 2.0 AU when ^{26}Al was still available but have been uniformly delayed for several ^{26}Al half-lives at 3.5 AU in order that the outer asteroid belt not melt throughout. The relative completeness of accretion at any specific time during this interval must have been a nearly linear function of heliocentric distance. An accretionary wave propagating slowly (e.g., 0.3 to 0.5 AU Myr^{-1}) would satisfy the observations, but is an *ad hoc* solution. It is also occasionally suggested that the primordial condensation of a large fraction of water ice outside of a sharp boundary in the middle of the belt due to the decrease of temperature with heliocentric distance might have prevented rock melting in the outer belt but this would likely have left much more water visible in the present-day belt.

An additional difficulty for the ^{26}Al hypothesis is the simultaneous existence of large melted and unmelted asteroids at similar heliocentric distances. Vesta at 2.36 AU has a surface with igneous assemblages while Ceres and Pallas at 2.77 AU have surfaces of low temperature and/or aqueously altered assemblages. If the rate for the accretionary wave postulated in Sec. II were taken seriously, unmelted asteroids such as Ceres and Pallas should form about 1 Myr or 1 to 2 half-lives later than igneous ones like Vesta, despite their similar heliocentric distances. If larger planetesimals accrete at faster rates than smaller ones as models of runaway accretion suggest, then the problem becomes much more acute. Moreover, the presence of smaller differentiated bodies at similar (e.g., 2.64 AU: 15 Eunomia; Gaffey and Ostro 1987) and greater heliocentric distances (e.g., 2.92 AU: 349 Dembowska; Feierberg et al. 1980) would require a delayed accretion for Ceres and Pallas. While the ^{26}Al accretional wave model would imply a range (several half-lives) of planetesimal accretionary intervals at any heliocentric distance between 2.0 and 3.5 AU, the requirement that the two largest asteroids be late accreters is also *ad hoc*.

If the apparent particular preferential size discussed in Sec. II for the iron cores of fragmented asteroids is borne out by subsequent investigation, then this would also present a problem for the ^{26}Al hypothesis. Assuming that ^{26}Al abundance would maximize in bodies of intermediate size seems implausible at best. While none of the above problems with the ^{26}Al hypothesis are individually fatal, they do place severe and perhaps unreasonable constraints on the rate and timing of planetesimal accretion as a function of both heliocentric distance and, perhaps, body size. Whether such constraints may be fatal is beyond the scope of this chapter. By contrast, the natural pattern of induction heating-induced metamorphism is quite similar to that deduced from asteroidal taxonomic distributions. There is naturally a strong gradient in heating intensity with heliocentric distance, if typical primitive meteoritic materials are assumed as a starting point for the thermal evolution. There is a natural preferred size for asteroid melting, though the value of this size is a function of such model parameters as induction-epoch duration. In addition, the other model parameters required all tend to be relatively reasonable—no "forcing" is required.

Although the ^{26}Al and induction-heating scenarios are not mutually exclusive, the present evidence suggests that induction heating had the dominant role. If the induction-heating hypothesis is correct, one might extrapolate back toward an inference of the properties of the proto-Sun. Obviously, the induction-heating hypothesis implies a massive equatorial solar wind. If the wind was spherically symmetric, according to the work (Herbert 1989) summarized in Figs. 6 and 7, the total integrated mass loss would have been in the range of 0.1 $M_\odot$. This work also suggests a mass-loss epoch duration on the order of 1 Myr, which is somewhat longer than the estimates of the angular-momentum loss times by Vogel and Kuhi (1981) and Hartmann et al. (1986). It should be noted that the original conclusion of Herbert (1988) that the mass-loss duration was 10^5 yr should probably be modified; 1 Myr is probably a better estimate. A duration of 10 Myr, which would seem marginally consistent with Fig. 4, is ruled out from a detailed examination of the model outcomes.

A qualitative conclusion that can be drawn from the inferred occurrence of primordial induction heating is that the massive solar outflow must have included the equatorial plane or very close to it out to a distance of at least 4 AU so as to have impinged upon the asteroids. Thus, if the wind was associated with the polar outflows frequently seen in astrophysical objects (see, e.g., Shu et al. 1987), it must have occurred in addition to the polar outflow or must have been diverted toward the polar axis at a distance greater than 4 AU; either seems possible.

A mechanism for driving a dense solar wind of the type inferred in the foregoing would lend credibility to the induction-heating hypothesis. One attractive candidate is the so-called "X- celerater" model of Shu et al. (1988). This model drives a strong equatorial plasma outflow whose angular momentum, energy and mass flux is derived from a co-existing protoplanetary accretion disk.

One consequence of a dense primordial solar wind that has not been clearly observed in meteorites is evidence for enhanced regolith accretion of solar-wind gasses (chapter by Wood and Pellas). A ready rationalization of this fact is the possibility that the earliest protoplanets exposed to the solar wind may have undergone repeated generations of catastrophic disruption and reformation of breccias. However, this idea remains to be developed in greater specificity. Searches in the most ancient low-temperature meteoritic breccia clasts are thus desirable.

In summary, the hypothesis explored in this chapter is that joule heating occurred during the formation epoch of the terrestrial planets (including asteroids) and was in some cases a significant driver of thermal evolution. Heating from electromagnetic induction requires (1) that a large part of any mass loss from the protosolar system was centrally driven in contrast to mass shedding at the outer edge of the protoplanetary nebular disk and (2) that the net mass loss was a few percent of the solar mass. Other effects on protoplanetary objects (such as imbedded solar-wind gas, surface heating effects, etc.) are also expected.

Acknowledgments. Informative conversations with W. V. Boynton, C. R. Chapman, D. R. Davis, L. L. Hood and P. Pellas have materially improved this work.

Note added in proof: A possibility, discussed by Wood and Pellas (their chapter), is that the fraction of protoplanetary nebular water incorporated into asteroidal material increased rapidly with heliocentric distance. The large phase change enthalpy of the water could then absorb excess radionuclear heat that would otherwise raise the internal temperature above the water's melting (or boiling) point in those asteroids with large water fractions. However, the outermost part of the belt seems never to have been exposed to liquid water and those asteroids are not dominated by their water content today (Jones 1988; Lebofsky et al. 1988), in contradiction to the hypothesis.

One requirement of the Shu et al. and other protostellar scenarios that the energy and mass source driving the outflow originates in inflow from a protoplanetary disk, implying that electromagnetic heating occurred primarily during the disk epoch. The role of a disk near 2 to 4 AU depends on thickness, patchiness, and gas content. Current astrophysical data is inconclusive as to which stage of stellar evolution was most likely.

WHAT HEATED THE PARENT METEORITE PLANETS?*

JOHN A. WOOD
Harvard-Smithsonian Center for Astrophysics

and

PAUL PELLAS
Muséum National d'Histoire Naturelle

We examine the plausibility of the two most widely discussed mechanisms for heating the small planetesimals in which the meteorites formed: the decay of short-lived ^{26}*Al in the planetesimals, and joule heating by electric currents in the planetesimals induced by the motion past them of an early intense solar wind plasma. Both hypotheses are seen to have problems. Noncumulate eucrites (meteoritic fragments of lava erupted on the surface(s) of early planetesimals) did not contain enough* ^{26}*Al to leave detectable excess* ^{26}*Mg signatures; if* ^{26}*Al melted the lava, the latter had to wait underground 1.2 Myr or more while its* ^{26}*Al decayed before erupting, a requirement that may be difficult to satisfy mechanically. In the case of inductive heating, it appears increasingly unlikely that there was an epoch of solar system history in which an intense (i.e., T Tauri) solar wind streamed past young planetesimals. The wind appears to have blown (mostly in poleward directions) only while the solar nebula (accretion disk) was present and shielding any planetesimals that had formed. Solar wind inductive heating cannot be modeled accurately because the parameters most important to modeling, such as the early solar wind flux in the ecliptic plane*

---

*General-Editor's Note: The Space Science Series aims to produce textbooks at the most advanced level. That is where disagreement occurs, for that is, after all, what science is all about. We publish disagreement too, and identify where it occurs. There are serious discrepancies between this chapter and the one by Herbert et al. in this book. So, here is something for our readers to work on!

and the electrical conductivity function of the material in planetesimals, are very poorly known. No model has been published that makes conservative assumptions across the board and achieves melting in planetesimals.

I. INTRODUCTION

Virtually all meteorites experienced heating inside planetesimals of which they were once a part, soon after the solar system was formed. Some (*irons, achondrites*) are the products of melting that occurred in their parent planetesimals. Others (the so-called *ordinary chondrites*) were transformed (metamorphosed) by anhydrous solid-state reactions at temperatures up to 1200 K. Still others (*carbonaceous chondrites*) were metamorphosed at lower temperatures in watery systems.

Portions of the thermal histories of meteorites can be read from compositional gradients in the crystals of Fe,Ni alloy they contain, and from the pattern of preservation of nuclear fission tracks in various minerals having different threshold temperatures for track annealing. The relatively rapid rates at which the meteorites cooled (1 to 100 K Myr^{-1}) show that their parent bodies were small, asteroidal in dimension. The source of heat that wrought these effects in the meteorite parent bodies is not known. The decay of long-lived radioactivity (the principal source of heat in the Earth) would have been ineffective, because the time scale for release of energy by decay of K, U, Th isotopes (~ 1 Gyr) is long compared with the time scale for conductive loss of heat from asteroid-size bodies (~ 100 Myr). Transformation of the kinetic energy of accretion into heat (which is thought to have melted the early lunar magma ocean) also can be ruled out, because the growth of a small, low-gravity object could not proceed if accretional encounters occurred at high enough velocities (>1 km s^{-1}) to correspond to kinetic energies adequate to heat the growing object to melting or metamorphic temperatures.

Other sources of heat that have been suggested for the meteorite planetesimals include the protosolar accretion disk or nebula (this would require that the nebula was hot at the radial distance where the meteorite planetesimals formed, and that the latter accreted before the nebula cooled); radiant energy from an early Hayashi-phase Sun (Wasson 1974, p. 192); and exothermic chemical reactions between unstable presolar compounds after they joined the nebula and accreted into planetesimals (Clayton 1980). None of these potential heat sources has seemed very plausible to most workers in planetary science. Two possible heating mechanisms that *have* been taken seriously are (a) decay of ^{26}Al (a radionuclide with a very short half-life, 0.72 Myr, that is known to have been present in the early solar system); and (b) joule heating by electric currents induced in planetary matter by the movement through interplanetary space of an intense early solar wind and magnetic field. The efficacy of the latter process would, of course, depend upon the degree and pattern of activity in the early Sun, which is the subject of this book.

Each of the heating mechanisms named would have operated on a particular time scale, and would have deposited heat at particular positions in planetesimals. Meteorites contain evidence of the time scale of the heating cycle they experienced and the planetary depths at which they were heated. In this chapter we critically compare the meteoritic evidence with predictions of the ^{26}Al and joule heating models.

II. KINETICS OF HEATING AND COOLING OF METEORITE PARENT BODIES

We begin by reviewing the best radiometric ages that have been determined for meteoritic materials (Table I). These are the times that have elapsed since the samples cooled from an earlier high-temperature state to their *closure temperatures*, the limiting temperatures beneath which the relevant radioactive parents and daughters are not redistributed by igneous action or solid-state diffusion, and daughters can be expected to remain where their parents once resided. Closure temperatures, which are known only approximately, have been estimated in a variety of ways (see, e.g., Turner et al.'s [1978] discussion of the closure temperature for ^{40}Ar-^{39}Ar dating).

Table I includes ages obtained by four different dating methods, in columns ordered by decreasing estimated closure temperatures. U,Pb dating is based on the decay of ^{235}U to ^{207}Pb and ^{238}U to ^{206}Pb. The ^{147}Sm-^{143}Nd and ^{87}Rb-^{87}Sr methods rely on the decay of the first nuclide named to the second. The ^{40}Ar-^{39}Ar method is based on the decay of ^{40}K to ^{40}Ar (neutron irradiation of samples in a reactor produces ^{39}Ar from remaining ^{40}K, which makes the measurement of relative amounts of parent and daughter easier and more precise).

Application of a radiometric dating technique requires consideration of three quantities: the amounts of parent p and daughter d nuclides in a sample, and the amount of nuclide d that was already present in the material of the sample when it formed (and hence was not created by subsequent radioactive decay). When all three quantities can be derived formally from measurements made on the sample, an *absolute* age can be calculated. When one or more of the quantities is based on assumptions, however defensible, the resulting age is called a *model* age. (This generalization is not strictly applicable to the ^{40}Ar-^{39}Ar technique.) In the case of U,Pb dating, sometimes both parent and daughter nuclides are measured, yielding absolute ages. Results are especially reliable if ages are determined by both the ^{235}U-^{207}Pb and ^{238}U-^{206}Pb methods and found to be *concordant*, i.e., to agree closely. When this is the case, measurement of the ^{207}Pb/^{206}Pb ratio yields an absolute age that is highly precise, because of the large difference in the rates at which the two daughter nuclides accumulate. Often only the ratio of daughter nuclides, ^{207}Pb/^{206}Pb, is measured, without accompanying measurements of the parent U isotopes. Ages can be derived from these data if assumptions are made about the iso-

TABLE I
The Oldest Ages (in Gyr) Obtained by Four Radiometric Dating Techniques for Meteorites and Meteorite Components

Approximate Closure Temperature	U,Pb Ages 900 K	^{147}Sm-^{143}Nd Absolute Ages 900 K	^{87}Rb-^{87}Sr Absolute Ages (a) 570 K	^{40}Ar-^{39}Ar 500 K
CHONDRITES				
Allende				
Whole rock	4.553 ± 0.004 (b,d)			
Ca,Al-rich inclusions				
BR-9	4.570 ± 0.001 (b,e)			
G-2	4.568 ± 0.001 (b,e)			
EGG-6	4.565 ± 0.005 (b,f)			
WA-5	4.561 ± 0.006 (b,f)			
H chondrites (whole rock)	4.55 ± 0.12 (b,g,h)		4.52 ± 0.05 (i)	4.52 – 4.44 (j)
Tieschitz (H3)			4.53 ± 0.06 (i)	4.45 ± 0.05 (j)
Guareña (H6)				
whole rock			4.46 ± 0.08 (l)	4.44 ± 0.03 (j)
phosphates	4.504 ± 0.0005 (c,k)			
Kernouve (H6)				
whole rock				4.45 ± 0.03 (j)
phosphates	4.521 ± 0.0005 (c,k)			
LL chondrites	4.536 ± 0.008 (b,g,h)		4.48 ± 0.04 (m)	
Chainpur (whole rock)			4.517 ± 0.056 (m)	
St. Séverin (LL6)				
whole rock		4.55 ± 0.33 (n)	4.51 ± 0.33 (n)	
phosphates	4.552 ± 0.004 (c,f)			
ACHONDRITES				
Noncumulate eucrites				
Ibitira (whole rock)	4.556 ± 0.006 (b,r)			4.42-4.38 (± 0.03) (p)
	4.560 ± 0.003 (b,e)			4.35-4.34 (± 0.03) (q)

TABLE I
Continued

Approximate Closure Temperature	U,Pb Ages 900 K	^{147}Sm-^{143}Nd Absolute Ages 900 K	^{87}Rb-^{87}Sr Absolute Ages (a) 570 K	^{40}Ar-^{39}Ar 500 K
Y-75011				
whole rock				4.0-3.9 (t)
clast		4.55 ± 0.14 (s)		
matrix			4.50 ± 0.05 (s)	
			4.46 ± 0.06 (s)	
Juvinas (whole rock)	4.539 ± 0.004 (b,g,u)	4.50 ± 0.06 (v)	4.50 ± 0.07 (w)	
Cumulate eucrites				
Moama	4.43 ± 0.04 (b,x)	4.46 ± 0.03 (n)		
Serra de Magé	4.397 ± 0.007 (b,x)	4.41 ± 0.02 (y)		
Angra dos Reis				
whole rock	4.551 ± 0.004 (c,f)	4.564 ± 0.037 (n)		
phosphates	4.553 ± 0.008 (c,f)			
IRONS				
Silicate inclusions in IAB irons Mundrabilla, Woodbine				4.57 ± 0.03 (z)

a: Decay constant for ^{87} Rb used in this table = 1.42×10^{-11} yr^{-1};
b: ^{207}Pb-^{206}Pb model ages;
c: Absolute ages based on ^{235}U-^{207}Pb, ^{238}U-^{206}Pb decay;
d: Tatsumoto et al. (1976);
e: Manhès et al. (1987);
f: Chen and Wasserburg (1981);
g: Manhès (1982);
h: Averages of ages;
i: Minster and Allègre (1979);
j: Turner et al. (1978);
k: Göpel et al. (1989);
l: Wasserburg et al. (1969);

m: Minster and Allègre (1981);
n: Jacobsen and Wasserburg (1984);
p: Hohenberg et al. (1981);
q: Trieloff et al. (1989);
r: Chen and Wasserburg (1985);
s: Nyquist et al. (1986);
t: Nyquist et al. (1988);
u: Manhès et al. (1984);
v: Lugmair et al. (1975);
w: Allègre et al. (1975);
x: Tera et al. (1989);
y: Lugmair et al. (1977);
z: Niemeyer (1979).

topic composition of Pb already present in the systems when they crystallized, but these are *model* ages because the amounts of parent nuclides present have not been directly measured.

The oldest (and very precise) absolute ages in Table I are the concordant ages obtained by U-Pb plus $^{207}Pb/^{206}Pb$ analysis. Phosphate minerals (which are enriched in U) in the ordinary chondrites Guareña, Kernouve and St. Séverin give especially precise ages in the range 4.551 to 4.504 Gyr, in most cases with an error of less than 1 Myr (Chen and Wasserburg 1981; Göpel et al. 1989). (Ages obtained by Göpel et al. for two L5 chondrites, Homestead and Knyahinya, are less well founded because of the high level of shock these meteorites experienced during preterrestrial impacts [Göpel, personal communication].) The most accurate $^{206}Pb/^{207}Pb$ *model* ages in the U,Pb column are for samples in which $^{206}Pb/^{204}Pb > 100$, i.e., where the amount of parent U was large compared to the amount of primordial Pb that was present at the outset (nonradiogenic ^{204}Pb is a measure of the latter, radiogenic ^{206}Pb a measure of the former). Of the materials in Table I, $^{206}Pb/^{204}Pb > 100$ for the CAI's (Ca, Al-rich inclusions) and the eucrites.

Among these oldest ages, we seek evidence of when heat was applied to the meteorite planetesimals, and for how long. ^{26}Al heat presumably would have been applied as a spike immediately after the planetesimals accreted, and would have decayed on a time scale of ~ 1 Myr. Solar wind heating could have been applied appreciably (10 Myr?) after the planetesimals accreted, maybe even episodically (Fig. 1). The error bars on some U,Pb ages should permit 10 Myr time differences to be resolved.

The oldest objects in the table appear to be the Allende (CAI's), at 4.56 to 4.57 Gyr. CAI's are mm- or cm-size objects that were once dispersed in the solar nebula, after which they became incorporated in meteorite planetesimals when the latter accreted. Using the $^{207}Pb-^{206}Pb$ model ages of CAI's as a zero point for planetesimal history, we note that the other meteorite samples cooled through isotopic closure ~ 10 Myr or more later than CAI's (ages by all four radiometric techniques). This would appear to militate against ^{26}Al heating for the planetesimals, and support solar wind heating. However, in most cases the age difference can be ascribed to the finite time it would take a planetesimal to cool to isotopic closure, after it reached peak temperature; it may have little to do with how soon the planetesimal heated up after it accreted. Note that ages in Table I tend to decrease to the right, which (with the possible exception of the U,Pb vs $^{147}Sm-^{143}Nd$ columns) is the direction of decreasing closure temperatures.

Most significant are the age differences between CAI's and *non-cumulate eucrites,* the latter being basaltic igneous rocks that erupted to the surfaces of their parent planetesimals, where they cooled promptly. If these lavas erupted immediately after they were melted in planetesimal interiors, their ages should express the time scale of heating only; cooling should not be an issue. The eucrite Juvinas appears to have cooled ~ 27 Myr (many half-lives of ^{26}Al)

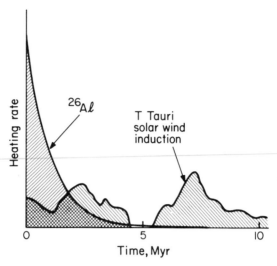

Fig. 1. Possible differences in the time scales of planetesimal heating by [26]Al decay vs T Tauri solar wind induction (schematic).

later than the time when the CAI's were formed (Manhès et al. 1984). Further, these authors conclude from the meteorite's Pb isotopic systematics that < 2 Myr elapsed between the time when differentiation (melting) produced the eucritic lava and the time when it crystallized. The [207]Pb-[206]Pb age for Ibitira is ~8 Myr younger than that of the CAI's, but in this case the age difference is close to the errors associated with the ages (Table I).

Table I shows that the ages of *cumulate eucrites* are as much as 150 Myr younger than those of noncumulate eucrites. Cumulate eucrites contain coarser mineral grains than the latter, and there is textural evidence that the minerals tended to sink and accumulate (hence "cumulate") before the rock solidified. This takes time, in the feeble gravitational field of a planetesimal, so the melt must have cooled slowly in a buried, insulated position inside its parent body instead of chilling quickly on its surface. These young ages do not necessarily testify to late heating, however, because the insulating effect of a thick surface layer of broken rubble can be great enough to delay the cooling of magma emplaced beneath the layer by >100 Myr (Haack et al. 1990).

The time intervals before cooling for noncumulate eucrites point to a problem with the [26]Al heating mechanism: eucritic lava may have been molten in the solar system too long after the origin of the solar system (when [26]Al presumably ceased to be added to it) for its melting to be attributable to heat generated by the decay of [26]Al. We will return to this point in Sec. IV. There is no serious conflict between the ages and the concept of solar-wind-induced heating, which can have occurred any time in the first 10 Myr, or even more, of solar system history.

III. SIZE AND THERMAL STRUCTURE OF PARENT
METEORITE BODIES

As noted, the fact that ordinary chondrites cooled to isotopic closure for the ^{87}Rb-^{87}Sr and ^{40}K-^{40}Ar systems later than they did for the ^{147}Sm-^{143}Sm and U-Pb systems is an expression of the finite rate at which their parent bodies cooled. The age differences correspond to mean cooling rates of 1 to 20 K Myr^{-1} through the temperature range above isotopic closure. Independent estimates of meteorite cooling rates made from metallographic and ^{244}Pu fission track data tend to confirm these values (Lipschutz et al. 1989).

The various types of meteorites reached different peak temperatures in their parent bodies. Melting ($t \gtrsim 1500$ K) was required to produce irons and achondrites. Ordinary chondrites have been metamorphosed to different degrees, reflecting the dissimilar peak temperatures and cooling times they experienced. The classification system used for ordinary chondrites includes a number (e.g., Guareña is an H6 chondrite) that expresses the severity of metamorphism, ranging from 3 (very mild metamorphism, peak temperature ~ 700 K) to 6 (highly metamorphosed, ~ 1200 K).

Model studies of the thermal histories of asteroid-size bodies (see, e.g., Fricker et al. 1970) allow these peak temperatures and cooling rates to be related to depths in and sizes of parent bodies. Among H4, H5 and H6 ordinary chondrites, there are correlations between cooling rate and peak metamorphic temperature consistent with residence in a body of 75 to 100 km radius (Pellas and Fieni 1988). Among LL4, LL5, and LL6 chondrites (which are chemically distinct from H chondrites), the data point to a different body of 125 to 150 km radius. In both cases, the most highly metamorphosed chondrites (highest peak temperatures) tend to have the youngest ^{40}Ar-^{39}Ar ages, so they cooled slowest and presumably were most deeply buried (Fig. 2). The least metamorphosed chondrites (type 4) cooled most rapidly. Thus type 6 chondrites can be thought of as once occupying the deepest zones of their parent bodies, enclosed by layers of type 5 and then type 4 chondritic material: this is the classic onion-shell model of chondrite metamorphism.

The onion-shell model is supported by the differences in metamorphic intervals shown in Table II. (A metamorphic interval Δt is the difference in ages obtained by two radiometric techniques, and represents the time needed to cool from the higher to the lower of the closure temperatures.) The second column of the table shows absolute metamorphic intervals based on the differences between U,Pb and ^{40}Ar-^{39}Ar ages. Guareña, Kernouve and St. Séverin are all highly metamorphosed chondrites, but Δt for St. Séverin is three times as great as it is for the other two specimens. This suggests that the LL parent body was larger, and cooled more slowly, than the H parent body. Also shown in the second column of Table II are approximate values of Δt for two LL5 (somewhat less metamorphosed than LL6) chondrites, Tuxtuac and Olivenza. These metamorphic intervals are based on the (conservative) assump-

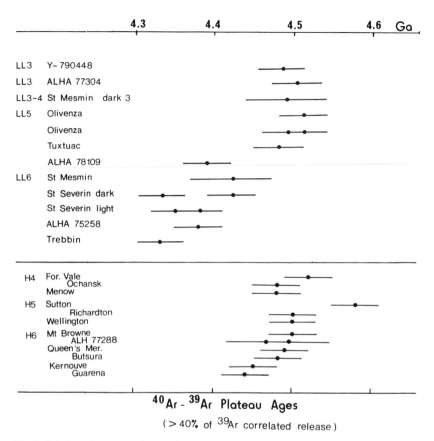

^{40}Ar - ^{39}Ar Plateau Ages

(> 40% of ^{39}Ar correlated release)

Fig. 2. Relationship between degree of metamorphism of chondritic meteorites (which increases from 3 to 6; i.e., LL6 is more metamorphosed than LL5, etc.) and times when they cooled through the closure temperature for the ^{40}Ar-^{39}Ar dating method (in Gyr ago). Included are all the measurements in the literature for which >40% of the ^{39}Ar released correlated with ^{40}Ar (a measure of the reliability of the age). The longer times taken by highly metamorphosed chondrites to cool through closure is an expression of their slower cooling rates, and therefore greater depths of burial in parent planetesimals (figure adapted from Lipschutz et al. 1989).

tion that their U,Pb ages are similar to that of St. Séverin phosphates (Table I). The fact that these two less-metamorphosed LL chondrites cooled more rapidly than St. Séverin is consistent with the onion-shell model. (Estimated closure temperatures for the U,Pb and ^{40}Ar-^{39}Ar systems have uncertainties of about ± 100 K [see, e.g., Turner et al. 1978].)

    The third column of Table II shows another type of absolute metamorphic interval: one defined by the difference between ^{244}Pu fission track densities displayed in merrillite (a phosphate mineral that was enriched in ^{244}Pu) and adjacent orthopyroxene minerals. The two minerals have different fission track retention temperatures (i.e., temperatures above which tracks tend to be

TABLE II
Metamorphic Intervals (Δt, Myr) for Five Ordinary Chondrites.

	Absolute Δt: U,Pb age − ^{40}Ar − ^{39}Ar age[a] Temperature Range, ~1100 K − ~550 K[c]	Absolute Δt from ^{244}Pu Fission Tracks[b] Temperature Range, ~500 K − ~350 K[d]
Guareña (H6)	64 ± 30[e]	87 ± 44
Kernouve (H6)	71 ± 30[e]	72 ± 47
St. Séverin (LL6)	180 ± 34[f, g]	179^{+25}_{-97}
Tuxtuac (LL5)	(82 ± 34)[h]	77^{+39}_{-50}
Olivenza (LL5)	(62 ± 34)[h]	133^{+55}_{-62}

a: ^{40}Ar-^{39}Ar ages from Turner et al. (1978), Hohenberg et al. (1981), Bernatowicz et al. (1988), Trieloff et al. (1989).
b: Time differences based on ^{244}Pu fission track densities in merrillite grains and adjacent orthopyroxenes (P. Pellas, unpublished data).
c: An educated guess at the temperature range of metamorphism, from Dodd (1969), Olsen and Bunch (1984), and Turner et al. (1978). U,Pb closure temperatures and argon retention temperatures are estimated to be uncertain by ± 100 K.
d: Temperature dependence of annealing rate of fission tracks in pyroxene is based on laboratory experiments, extrapolated to times of 10 Myr (Pellas and Storzer 1981). Data for merrillite from Mold et al. (1984).
e: Absolute ^{235}U-^{207}Pb and ^{238}U-^{206}Pb ages, from Göpel et al. (1989).
f: U,Pb age of 4.552 ± 0.004 Gyr, from Chen and Wasserburg (1981).
g: ^{40}Ar-^{39}Ar age of 4.37 ± 0.3 Gyr: averaged data of Hohenberg et al. (1981) and Trieloff et al. (1989).
h: U,Pb age not known; assumed equal to that of St. Séverin.

annealed out and lost), and this coupled with the known half-life of ^{244}Pu provides the basis for determining cooling rates. The track retention temperatures used (~500 K for orthopyroxene, ~350 K for merrillite) were based on extrapolations to 10 Myr of the results of laboratory annealing experiments (Pellas and Storzer 1981; Mold et al. 1984). The substantial uncertainties shown in Table II stem from large grain-to-grain variations in observed track densities.

The metamorphic intervals of columns two and three span different and approximately adjacent temperature ranges. By summing the two, we obtain an approximate duration of metamorphism in the temperature interval ~1100 K − ~350 K that is ~150 Myr for the two H6 chondrites Guareña and Kernouve, and is 160 to 200 Myr for the two LL5 meteorites (Tuxtuac and Olivenza) and ~360 Myr for the LL6 chondrite St. Séverin. Thus, again a longer cooling time is indicated for a more highly metamorphosed (deeply buried) LL6 chondrite than for LL5 chondrites; and the LL parent body appears to have been larger (and cooled more slowly) than the H chondrite parent.

Below we list and comment upon several objections that have been raised to the onion-shell model of chondrite metamorphism:

(a) There is a third type of ordinary chondrite, the L (as opposed to H and LL) group, among which cooling rates do not correlate well with degree of metamorphism (and hence, presumably, position in the planetesimal). Models for creating a parent body with a less regular structure (both mechanical and thermal) have been proposed to account for this observation: perhaps small planetesimals that had reached differing internal temperatures coalesced into a thermally heterogeneous mass which then cooled (Scott and Rajan 1981), or an onion-shell planetesimal may have been disrupted by a collision while still hot, after which much of the debris re-accreted into a thermally disorganized structure (Rubin et al. 1982).

(b) Taylor et al. (1987) also reject an onion-shell structure for the H chondrite parent body, citing a lack of correlation of cooling rates (derived only from metallographic data) with degree of metamorphism. However, Pellas and Fieni (1988) show that cooling rates based on absolute radiometric ages and closure temperatures (as discussed above) and on the preservation of ^{244}Pu fission tracks do support an onion-shell model.

(c) The least-metamorphosed ordinary chondrites (H3, L3, LL3) should have cooled fastest in an onion-shell planetesimal, but cooling rates obtained by the metallographic method appear to indicate just the opposite; their nominal cooling rates are very slow, 0.1 to 1 K Myr^{-1} (Wood 1967). The fine-grained, little-metamorphosed substance of these meteorites does not permit the estimation of cooling rates by the fission track method, and there is little basis for estimating their cooling rates from radiometric ages and closure temperatures. It is possible that the high minor-element (especially P) content of metal grains in little-metamorphosed chondrites invalidates the use of this technique in those meteorites.

IV. ALUMINUM-26 AS A HEAT SOURCE

^{26}Al would be a particularly effective heat source for planetesimals because the time scale on which it decays and generates heat (the half-life of the radionuclide, 0.72 Myr) is short compared with the time scale for conductive heat loss from even very small rocky objects. There is unequivocal evidence that live ^{26}Al existed in the early solar system. Excess ^{26}Mg created by the decay of ^{26}Al has been found in the CAI's of most carbonaceous chondrites and one ordinary chondrite (Lee et al. 1976; Hinton and Bischoff 1984), and in anorthite from a pyroxene-olivine clast of the Semarkona (LL3) ordinary chondrite (Hutcheon and Hutchison 1989). In some cases (e.g., the CAI's labeled EGG-6 and WA-5 in Table I), the amount of excess ^{26}Mg found corresponds to an initial abundance of ^{26}Al such that the ratio ^{26}Al/^{27}Al $\sim 5 \times 10^{-5}$. If all the Al in accreting material had this isotopic composition, decay of the ^{26}Al would have released $\sim 10^4$ J of energy per g of planetary material, sufficient to melt the interiors of planetesimals unless they were small enough (less than a few km in radius) to dissipate the energy as fast as it was deposited (Lee et al. 1976).

However, it is by no means clear that all Al in early solar system material did contain this much ^{26}Al. Some CAI's contain amounts of radiogenic ^{26}Mg corresponding to an initial ^{26}Al/^{27}Al $<5 \times 10^{-5}$, some contain no measurable radiogenic ^{26}Mg at all. Either (i) ^{26}Al was heterogeneously distributed in early solar system material, and the mean value of ^{26}Al/^{27}Al for solar system Al was less than 5×10^{-5}, perhaps much less; or (ii) CAI's were formed over a period of millions of years, and by the time the last ones appeared, ^{26}Al no longer existed in the system that gave rise to them. Since CAI's with and without initial ^{26}Al occur mixed together in chondrites, however, [ii] would mean that accretion did not occur until ^{26}Al had become extinct, in which case ^{26}Al decay could not have heated the planetesimals. A third possibility is that mild metamorphism in parent planetesimals may have redistributed Mg isotopes in some CAI's, blurring or erasing the evidence for ^{26}Al initially present (MacPherson et al. 1988). Mineralogical differences among CAI's could cause them to be differently affected by metamorphism.

More important, not all Al in chondrites occurs in CAI's. CAI's are very rare in most chondrite subtypes, or missing altogether. The Al in these chondrites occurs in their chondrules and matrix. *Chondrules* are mm-sized igneous spheroids, rich in Si, Mg and Fe, of unknown origin, which make up the bulk of the substance of chondrites. *Matrix* is μm-scale mineral matter, thought once to have been dust dispersed in the nebula, which now fills the interstices between chondrules and CAI's. Even in Allende, which has the highest observed content of Al-rich CAI's, most Al resides in the chondrules and matrix. CAI's are unique objects in many respects; it is possible that they formed in ways far removed from the processes which created chondrules and dust, and the latter did not have the opportunity that CAI's did to incorporate ^{26}Al.

A difficulty for the ^{26}Al-heating model is the fact that the noncumulate eucrites Juvinas and Pasamonte have been found to contain no radiogenic ^{26}Mg to within experimental errors (Schramm et al. 1970). If ^{26}Al was abundant enough to partially melt the interior of these eucrites' planetesimal(s), surely some of the live ^{26}Al would be carried aloft with the lava, and would be incorporated in the eucritic rock when it solidified, where it would leave a detectable isotopic signature when it decayed. Either (a) ^{26}Al was not in fact the heat source, or (b) the eucritic lava lingered in the interior of the planet long enough for its ^{26}Al to decay to indetectability before it erupted to the surface. Applying the same reasoning that Schramm et al. did for the cumulate eucrite Moore County, at least 1.2 Myr would have to have elapsed between the time when ^{26}Al was still abundant enough to cause melting, and the time when the Juvinas and Pasamonte lavas, no longer containing enough ^{26}Al to produce ^{26}Mg anomalies above the detection limit of Schramm et al.

erupted and solidified. This would be consistent with the millions of years that ^{207}Pb-^{206}Pb ages seem to say elapsed, between the formation of the solar system and the time when the noncumulate eucrites crystallized (Sec. II).

Many members of the meteoritics community, long vexed by the puzzle of how such small planetary bodies could have been heated, have been quick to accept rationalization (b) above. If the cumulate eucrite magma(s) were emplaced under an insulating megaregolith for >100 Myr, where they eventually cooled and crystallized (Sec. II), why could not other magma be held in such a site for only 1.2 Myr, then vented to the surface, where it cooled rapidly (as the textures of noncumulate eucrites require)? An impact might excavate a magma chamber, for example, releasing its contents onto the surface of the asteroid (Taylor 1988*b*). This scenario cannot be excluded, but it does not square well with the distribution of eucritic material on the asteroid 4 Vesta. Vesta is essentially covered with eucritic material (Gaffey 1983), which would appear to require the melting and widespread eruption of more lava than could be sequestered for 1 Myr in one or a few localized magma chambers. However, it is possible that the eucrites which fall to Earth do not derive from Vesta; perhaps eucritic rock on Vesta *does* contain radiogenic ^{26}Mg.

V. ELECTROMAGNETIC INDUCTION AS A HEAT SOURCE

This heating mechanism is based on the premise that early planetesimals were exposed to an intense flux of plasma with an embedded magnetic field, during a T Tauri phase of evolution of the pre-main sequence Sun (Sonett et al. 1968; Sonett 1969). When Sonett and co-workers first proposed the concept in the 1960s, the phenomenon of T Tauri outflow had been recently discovered (Kuhi 1964). Kuhi described mass loss rates as great as 5.8 × 10^{-7} $M_{\odot}$ yr^{-1}; estimates of total stellar mass loss during the T Tauri phase of pre-main-sequence evolution approached 50% of the initial mass (see, e.g., Ezer and Cameron 1971). It was widely assumed that the T Tauri outflow was isotropic, and that it continued after the solar nebula dissipated; indeed, the suggestion was made that the wind was responsible for "blowing the nebula away." Planetesimals that had begun to accrete in the nebula would have remained in Keplerian orbits when the nebular gases were dissipated, and thereafter they were bathed in the T Tauri flux.

There are two electromagnetic modes of interaction of an early intense solar wind with planetesimals which have the potential for generating heat (Sonett and Colburn 1968; Sonett et al. 1968). One is transverse magnetic inductive (joule) heating, which would have been caused by the steady motion of the solar wind and its embedded magnetic field past a planetesimal. The other is the transverse electric mode of heating, which would be promoted by fluctuations in the interplanetary magnetic field encountered by the planetesimal. Sonett and co-workers have argued that the latter mode of in-

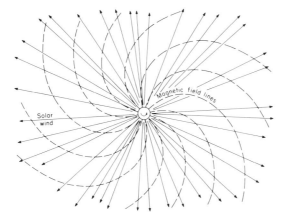

Fig. 3. Solar wind stream lines and magnetic field lines in the early solar system (schematic). As the magnetic field lines turn, each field line moves outward along the stream lines it crosses.

teraction would have been ineffective in heating asteroid-size bodies (though it may have heated the surface layers of larger objects, such as the Moon); therefore only transverse magnetic (TM) induction will be discussed in this chapter.

The early solar-wind plasma is pictured as streaming out radially from the Sun, sweeping past planetesimals at a velocity as high as several hundred km s^{-1}. The magnetic field of the Sun was embedded in the plasma. Near the ecliptic plane the field consisted (as it does now) of magnetic lines of force spiralling out from the rotating Sun (Fig. 3). As the solar wind swept past the planetesimals, the field lines cut through them at an angle to the direction of motion of the plasma. The planetesimals had some finite electrical conductivity, and motion of magnetic field lines through a conductor induces a current in it, in a direction perpendicular to the motion of the field lines. In the solar wind case, the polarity of the field lines was such that the current induced flowed from south to north through the planetesimals (Fig. 4). Upon leaving a planetesimal in the northern hemisphere, the current flowed through the solar wind plasma and completed the circuit in the southern hemisphere. Electromagnetic interaction between a planetesimal and the plasma/magnetic field system would retard the streaming motion of the latter. At the same time, the passage of current through the planetesimal would heat it to some extent; effectively, kinetic energy of the plasma was converted to heat in the conductive planetesimal.

Factors Affecting the Heating of Planetesimals

The physics underlying induction heating is fairly complex (see Herbert et al. 1990; Sonett and Colburn 1968; Sonett et al. 1968; Herbert and Sonett

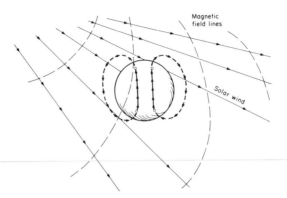

Fig. 4. Current flow (heavy lines) induced in a planetesimal by the passage of magnetic field lines embedded in a solar wind (schematic).

1978; Herbert 1989; chapter by Herbert et al.). However, it is straightforward to list the factors that would affect the amount of heat deposited in a planetesimal.

1. The intensity, duration and ionization state of the early intense solar wind. Clearly the more magnetic-field-bearing plasma sweeps past a planetesimal, the more kinetic energy will be converted to electrically induced heat in the body.
2. The solar wind velocity, as it determines the kinetic energy density of the plasma, and this kinetic energy is the ultimate source of the heat generated.
3. The electrical conductivity of planetesimal material, as the joule heating rate varies directly as σE^2, where σ is conductivity and E is the electric field at the planetesimal's surface.
4. The surface magnetic field and rotation rate of the T Tauri Sun, the latter because rapid rotation winds the magnetic field lines in the plasma more tightly, and the amplified magnetic field makes inductive heating more efficient (Sonett et al. 1970).
5. The size of the planetesimal. TM induction becomes self-defeating in objects larger than a critical size (which varies with the heating rate), because the current systems generated in large bodies create secondary planetary magnetic fields strong enough to deflect partially the approaching solar wind, diminishing its effect on the planetesimal. On the other hand, the planetesimal has to be large enough not to radiate away immediately the heat generated in it.

In the quarter-century since Kuhi's (1964) paper, T Tauri stars have been studied intensively (chapter by Bertout et al.). The advances in our understanding of these objects make it increasingly difficult to defend the proposition that a T Tauri solar wind inductively heated early planetesimals.

First, estimates of the T Tauri wind flux have gone down. Bertout et al. (see their chapter) cite current mass-loss estimates in the range 10^{-9} to 10^{-7} $M_\odot$ yr^{-1}, but stress the many uncertainties in making such an estimate. DeCampli (1981) has argued that the evidence for any significant mass loss at all is ambiguous in most T Tauri stars, and in the others, it is difficult to make a case for a mass-loss rate greater than several $\times$ 10^{-8} $M_\odot$ yr^{-1}. In the current view, it is unlikely that the Sun, in a T Tauri stage of its evolution, lost more than a few percent of its mass as solar wind.

Second, the T Tauri outflow is now seen to be nonisotropic. Modeling of the wind density and velocity fields needed to reproduce T Tauri spectral line profiles shows that most of the wind blows off at high stellar latitudes (Edwards et al. 1987), avoiding the equatorial plane where accreting planetesimals orbit.

Third, the evolutionary stage pictured earlier in which the solar nebula has dissipated but the T Tauri wind continues to flow past planetesimals does not seem to exist. T Tauri outflows appear to depend upon the existence of a disk (see, e.g., Edwards et al. 1989; Bertout and Bouvier 1989). While the disk is present it must deflect the wind flow (perhaps accounting for the observed nonisotropic outflow noted in the previous paragraph), and prevent it from interacting with planetesimals that have formed. Edwards et al. (1989) are explicit about this: "The effect of energetic TTS winds on forming planetary systems may be considerably smaller than is usually assumed. Not only would winds with $M_w \sim 10^{-8}$ $M_\odot$ yr^{-1} stop when the disks finish actively accreting, but when present, the energetic winds may be directed largely perpendicular to the disk plane, and thus have minimal interaction with forming planets."

These difficulties notwithstanding, Herbert (1989) has carried out a model study of the T Tauri wind heating effects that solar system planetesimals might experience. In some cases, his planetesimals get hot enough to melt internally. It is difficult to model the situation, because the values of so many of the parameters that control the process are extremely uncertain. This being the case, a positive result for the model study carries conviction only if conservative assumptions were made for these parameter values. The following paragraphs discuss the assumptions made or results obtained by Herbert relative to the five pertinent factors listed above.

Herbert's Model Assumptions/Results

1. Among the several models of Herbert (1989) that achieved melting, his model 4b is the most conservative in terms of the solar wind flux: 3×10^{-9} $M_\odot$ yr^{-1} for 10 Myr, for a total mass loss of 0.03 $M_\odot$. Many astrophysicists would consider this a defensible T Tauri outflow, though of course it

sidesteps the issue of poleward-collimated outflow and the question of whether there was ever a stage in solar system history when the Sun's T Tauri wind streamed directly past its planetesimals.

2. All of Herbert's (1989) models assume a wind velocity equal to that of the present solar wind, 400 km s^{-1}. This is not a conservative assumption: studies of T Tauri wind velocities point to lower values. Kuhi (1964), in his classic paper on T Tauri stars, obtained velocities of 225 to 325 km s^{-1}. A more recent study (Mundt 1984) attributes the Doppler shift of spectral lines of T Tauri stars to terminal wind velocities that range between 100 and 400 km s^{-1}. The kinetic energy density of the early wind, which is converted by electrical induction to heat energy in planetesimals, of course scales as the square of this velocity; the kinetic energy *flux* of the wind scales as the *cube* of the velocity.

3. The electrical conductivity of the rock in planetesimals subject to induction heating is a particularly crucial parameter, and one that is poorly known. Measured conductivities of various rock and meteorite types vary by orders of magnitude. Some simple substances, such as pure olivine [$(Mg,Fe)_2SiO_4$], are not conductive enough to permit any significant degree of electrical induction heating for plausible solar wind flux conditions (Duba et al. 1974). Other materials, such as carbonaceous chondrites, which are more complex and probably better analogs to primitive planetesimal material than pure olivine, have much higher conductivities (Brecher et al. 1975; Duba and Boland 1984).

All materials tend to increase very rapidly in conductivity with temperature. This, of course, has the potential for creating a thermal runaway if the rock is being heated electrically: the more conductive the rock gets, the more current it passes, and this makes it hotter and more conductive yet. Herbert's model 4b assumes the electrical conductivity function to be $\sigma = (10^{-5} S/m)e^{0.003T}$ (T = temperature, K), which yields $\sigma = 2.3 \times 10^{-5}S/m$ at room temperature and $2 \times 10^{-4}S/m$ at 1000 K. This is a conservative value for the high-temperature conductivity, but the low-temperature conductivity may be unrealistically high. The low-temperature conductivity assumed agrees with laboratory measurements by Brecher et al. (1975), but the data of Duba and Boland (1984) point to much lower values. Low-temperature conductivity is particularly important to the concept of induction heating of planetesimals, because presumably the planetesimals were relatively cool at the outset; if electrical conductivity was too low at this stage, inductive heating would be incapable of depositing enough heat to start a thermal runaway.

4. Herbert (1989) assumes a solar surface magnetic field of 100 G and a solar rotational velocity of 10^{-4} rad s^{-1} (a rotation period of 0.7 day) for all his models. This is a realistic value for the magnetic field, but T Tauri stars rotate more slowly than assumed (periods of 2 to 8 days; Bouvier et al. 1986).

5. Though other models tested by Herbert (1989) do not heat objects as

large as Vesta to the melting temperature, his model 4b does achieve melting in bodies between 50 and ~500 km diameter.

VI. COMMENTS ON EVIDENCE HELD TO FAVOR ELECTROMAGNETIC HEATING

Heating Effects in Planetesimals as a Function of Radial Distance in the Solar System

Gradie and Tedesco (1982) have shown that the asteroid belt is stratified, i.e., mean distances of the several compositional (spectrophotometric) types of asteroids are distributed differently. The asteroid types (E, R, S, M, F) with spectra indicative of high-temperature silicate minerals are concentrated between 1.8 and 3 AU, while dark carbon-rich types (C, P, D) orbit farther out (2 to >5 AU). Carbon-rich asteroids are presumably the source of carbonaceous chondrites, meteorite types that have experienced only slight heating—enough to melt water ice, but not enough to dehydrate minerals or pyrolize organic compounds. The orbital stratification is not perfect; there is extensive overlap of compositional types, but the tendency is clear for less-heated carbonaceous asteroids to occur farther out than more-heated noncarbonaceous asteroids.

Herbert and Sonett (1978) have cited this as evidence favoring solar wind heating of the asteroids. Clearly the effectiveness of solar wind heating would have been inversely related to distance from the Sun (though in point of fact, in Herbert's [1989] model 4b, discussed above, the dependency is so weak that melting is produced throughout the asteroid belt). The question is whether solar wind induction heating is the only mechanism that would produce a radial dependency of thermal processing.

It has become increasingly apparent that there was no sharp line of demarcation between the properties of early chondritic planetesimals and comet nuclei (Wood 1991). The C-rich parent bodies of the carbonaceous chondrites may have incorporated water ice when they accreted in the outer asteroid belt, which later melted, furnishing the liquid water that altered silicate minerals in these bodies into the assemblage of clay minerals which we observe in carbonaceous chondrites today (see, e.g., Grimm and McSween 1988). It is quite possible that the component of accreted ice increased with radial distance through the asteroid belt. In that case, the planetary heating mechanism, whatever it was, would have had its effects diluted to increasing degrees at greater radial distances by the need to melt ice and vaporize water. (Water ice is a potent thermal buffer: ~3 kJ g^{-1} of energy is needed to melt and vaporize ice, vs only ~1.6 kJ g^{-1} to heat rock from 200 K to 1400 K.) If heating occurred only in the rocky component of planetesimals (e.g., [26]Al heating), objects having a very high ice/rock ratio might not even have become warm enough to initiate melting of the ice.

In more general terms, all the processes that we can imagine happening in the early solar system would have operated more intensely and energetically at smaller radial distances from the Sun. Orbital velocities and velocity gradients; the pressure, temperature, shear and turbulent velocities of nebular gas; the infall velocity of interstellar material joining the nebula; radiant energy received from the young Sun; all these diminished with distance from the Sun. Whatever the ultimate energy source was for the unknown process that heated early planetesimals, the chances are good that it diminished with distance from the Sun.

Solar Effects in Gas-Rich Regolith Breccias.

Many meteorites are *breccias* (welded-together assemblages of angular rock fragments, *clasts*, of all sizes, including finely comminuted material between the fragments), obviously the product of shattering collisions between the meteorite parent bodies. Some of these were formed in the *regoliths* of their parent bodies. (A regolith is a relatively thin residual layer of debris and dust on the surface of an airless, or nearly so, body, which is stirred and comminuted on a continuing basis by sporadic meteoroid impacts [Housen et al. 1979; Housen and Wilkening 1982].) *Regolith breccias* are formed by impacts, in subsurface zones where momentary shock pressure has the effect of consolidating loose debris into solid rock. Regolith breccias were abundant among the lunar samples returned by the Apollo program.

As regoliths are exposed to the solar wind, and solar wind particles striking at velocities of several 100 km s^{-1} implant themselves $\sim$0.1 μm deep in the surfaces of mineral grains, it is to be expected that regolith breccias (both lunar and meteoritic) would contain high concentrations of solar wind gases; this is precisely the case. Meteoritic regolith breccias contain 10^2 to 10^5 cm^3 STP g^{-1} of excess ^{4}He, which is attributable to solar wind implantation. Regolith breccias also contain *cosmogenic* nuclides (produced by the spallation of target nuclei in the clasts by cosmic rays, at a time when the clasts were close enough to the surface of their parent planetesimal to be unshielded from cosmic rays).

If the regoliths preserved as meteorite breccias were formed at the same time when a hypothetical solar wind having 10^5 times the present solar wind flux was inductively heating the planetesimals, the breccias should preserve correspondingly elevated concentrations of solar wind noble gases and cosmogenic nuclides produced by solar cosmic rays. Can the abundance of these products in meteoritic regolith breccias be used to test whether a greatly enhanced solar wind bathed the meteorite planetesimals? The answer is probably not. Housen et al. (1979) cite studies of radiometric ages of regolith breccias that show they formed long after the hypothetical intense solar wind epoch, i.e., 1 to 3.9 Gyr ago. However, Caffee et al. (1987) have interpreted high levels of cosmogenic ^{21}Ne and ^{38}Ar which they found in clasts of the Kapoeta regolith breccia as evidence of an active early Sun. These nuclides can have been formed either by galactic cosmic rays, acting over a period of ~100 Myr, or by the current flux of solar cosmic rays, again over a period of ~100 Myr, or by an enhanced flux of solar cosmic rays, acting over a correspondingly shorter time interval. Caffee et al. argue from dynamical studies and petrographic properties of the Kapoeta breccia that the residence time of particles in the exposed zone of its regolith was only ~1 Myr (i.e., individual clasts and grains tended to be ejected to escape, or buried > 2 m deep, on that time scale), and on this basis they reject the first two possibilities. They are driven to conclude that the Kapoeta regolith was irradiated by cosmic rays from an early Sun ~10^3 times as active as the present Sun.

We point out that the ~1 Myr lifetime of planetesimal regoliths, on which this conclusion depends, has not been firmly established. In fact, there is evidence for much older planetesimal regoliths: Lorin and Pellas (1979) showed from a careful study of the cosmic-ray-produced nuclides in the Djermaia breccia that its regolith persisted, unshielded from cosmic rays, for ~15 Myr (~30 Myr, if the present-day production rate of ^{21}Ne is assumed). More recently Wieler et al. (1989a,b) have argued that enhanced levels of cosmogenic Ne and He in matrix and clasts of the regolith breccia Fayetteville can best be understood as resulting from galactic cosmic ray spallation in a regolith over a period of >20 Myr. If regoliths can survive this long, there is no need to invoke an early period of enhanced solar activity to account for their cosmic ray record.

Finally, we note that Papanastassiou et al. (1974) studied two igneous clasts in Kapoeta (the breccia on which Caffee et al. [1987] base their argument), and found them to have relatively young ^{87}Rb-^{87}Sr isochron ages (3.89 ± 0.05 and 3.63 ± 0.08 Gyr); they concluded that implanted solar wind gases and tracks in the breccia must record events extending over 1 Gyr after the formation of the solar system, not early solar system processes.

VII. CONCLUSIONS

We feel that both the ^{26}Al and the solar wind induction heating mechanisms have been accepted too uncritically (by different populations of workers, of course). Both mechanisms have problems. The principal embarrassment to ^{26}Al heating is the fact that eucritic lavas, melted by the mysterious heating mechanism in some early planetesimal, did not contain enough ^{26}Al to decay to radiogenic ^{26}Mg at levels detectable by Schramm et al. (1970) when they erupted to their planetesimal surface and cooled. It is necessary to postulate that the eucritic lavas lingered underground in their planetesimal(s) while their ^{26}Al decayed away.

The solar wind induction heating concept has a worse problem. Astrophysical evidence has made it seem increasingly unlikely that an intense solar wind flux, associated with the Sun's T Tauri stage of evolution, blew past planetesimals in the early solar system. Instead, it was probably collimated in the direction of the Sun's poles by the persistence, during the T Tauri epoch, of the solar nebula. In addition to this problem, it is difficult to assess the importance of early solar wind induction heating because the parameters important to the process, such as the early solar wind flux as a function of time and the electrical conductivity of the rocky substance of planetesimals as a function of temperature and time, are very poorly known. The most successful attempt to model the effect, Herbert's (1989) model 4b, employs several model parameters that cannot be considered conservative.

We think it quite possible that neither of these popular mechanisms was responsible for the heating of planetesimals in the early solar system. The true heating mechanism may have eluded us so far; this would not be surprising in view of our ignorance of the processes that operated in the early solar system.

Acknowledgments   We are grateful to A. Duba, S. J. Kenyon and C. P. Sonett for expert discussions, and to F. Herbert, I. D. Hutcheon and D. A. Kring for constructive reviews of our manuscript. Our research was supported by a grant from the National Aeronautics and Space Administration (J. A. W.) and the Institut National des Sciences de l'Univers, ATP Planètologie and the Centre National de la Recherche Scientifique (P. P.).

THE RELICT MAGNETISM OF METEORITES

S. M. CISOWSKI
University of California, Santa Barbara

and

L. L. HOOD
University of Arizona

The remanent magnetism recorded in chondritic and achondritic meteorites can potentially constrain the existence and properties of large-scale magnetic fields in the early solar system. Unlike terrestrial rocks, the main ferromagnetic carriers in meteorites are metallic Fe and NiFe alloys whose thermomagnetic properties are only partially understood. Nevertheless, paleointensity experiments involving repeated heatings of selected samples suggest that carbonaceous chondrites experienced magnetic fields of the order of one tenth to several gauss during low-temperature (~ 300° C) metamorphic events on their parent bodies occurring within a few tens of Myr of protosolar collapse. Chondrules, whose primary relict magnetism could have been acquired in the solar nebula at an even earlier epoch, have produced somewhat higher paleointensities but these are more uncertain because of difficulties in identifying the acquisition mechanism of the remanence and because of the possibility of terrestrial contamination. Paleointensities derived for ordinary and enstatite chondrites are even less reliable. The magnetic remanence recorded in achondrites (brecciated fragments of a parent body that experienced igneous differentiation) is better understood and available paleointensity estimates are generally < 0.1 gauss. A possible source of the magnetism recorded in carbonaceous chondrites and chondrules is a large-scale magnetic field in the protoplanetary nebula—either a compressed interstellar field or a field generated by a turbulent nebular dynamo. Limited directional evidence from oriented chondrules suggests that the field was oriented mainly in the ecliptic plane. An alternative possibility is a

*solar field embedded in a protosolar wind after dispersal of the nebular gas;
however, it is unclear whether the long cooling times of carbonaceous chondrite
parent bodies are compatible with this hypothesis. Finally, intrinsic parent-
body fields and fields due to local surface processes such as impacts may have
contributed to the magnetization of all meteorite classes but particularly to the
achondrites and ordinary chondrites.*

I. INTRODUCTION

Investigations of the remanent magnetism of meteorites offer the possibility of observationally constraining the magnetic field and plasma environment of interplanetary space during the earliest stages of solar system evolution. Since meteorite parent bodies may have accreted during the first $\sim 10^5$
yr when a protoplanetary nebula was still extant, the originally recorded remanence may be more representative of nebular magnetic fields (e.g., a compressed interstellar field) rather than direct solar or solar-wind magnetic
fields. However, many parent bodies experienced re-heating events at later
times during or after dissipation of the nebula when a protosolar wind field
could have been present in the equatorial plane. Thus, meteorites may also
afford an opportunity to characterize the early Sun when a full understanding
of their magnetic and metamorphic histories is obtained. Any information
obtained from meteorites about the magnetic environment of the early solar
system is of cross-disciplinary importance, as magnetic fields also play a
crucial role in the rotational evolution of solar-like stars (chapter by Saar).
    The magnetizations of chondrules could be expected to provide the most
primitive information about the early magnetic environment, while the magnetizations of their surrounding matrix would relate to a somewhat later time
when the bulk meteorite accreted and lithified. However, as all chondritic
meteorites have, as assembled units, suffered varying degrees of thermal and
shock metamorphism, the original magnetic signal of all component chondrules would have been modified and degraded in proportion to the severity
of these later events. The magnetizations associated with differentiated parent
bodies (i.e., the achondrites) would relate solely to this later phase of parent-
body evolution, and might provide information on the duration and decay
rate of any nascent magnetic fields, or alternately may offer evidence for the
generation of localized (i.e., nonsolar) fields on or within the achondritic
parent bodies.
    While measurements of the intensity of magnetic remanence for meteo-
rite specimens is a straightforward process, particularly with the ultrahigh
sensitivities of modern cryogenic sample magnetometers, conversion of this
raw data into usable information about the intensity of the magnetizing field
is generally a difficult and uncertain chore. Even in the case of terrestrial
samples, where greater availability allows for the selection of those with
simple mineralogies and histories, and where the magnetic properties of the
remanent carriers are generally better understood, paleointensity estimates

are often imprecise, difficult or entirely elusive. In some cases, terrestrial samples (e.g., lodestones) are known to exhibit remanence intensities far in excess of what can be expected to result from magnetizations acquired in the ambient terrestrial field. Sometimes these anomalous intensities may derive from strong internal microscopic fields relating to mineral intergrowths, or from intense but short-lived external fields of a local nature (i.e., lightning strikes).

Methods for paleointensity determinations on extraterrestrial samples generally derive from techniques which were developed for use with terrestrial materials, particularly archeological specimens for which paleomagnetic determinations are often a valuable age-dating tool. The most widely employed methods involve heating and cooling the samples in a precisely measured laboratory field. This represents an attempt to impart a thermal remanence (i.e., a magnetic remanence which is acquired by mineralogically stable magnetic carriers in cooling below their Curie temperatures) to the sample, in order to compare it to the intensity of remanence exhibited by the sample in the natural state. Other simpler paleointensity methods involve normalizing the measured remanence intensity of a collected sample to a variety of laboratory induced remanent magnetizations that are believed to be related to the total remanence-carrying ability of the sample. These latter methods are less precise than the heating methods, and are more useful in producing relative rather than absolute paleointensity estimates. Their principal advantage is that they are less time consuming than the thermal methods, and can be applied even to samples that are thermally unstable.

Most paleomagnetic investigations of meteorites have focused on paleointensity, as the direction of remanent magnetization within a meteorite is obviously of no significance without an external frame of reference relating it to its orientation within its parent body or to the nebular disk. However, the internal consistency of remanence directions between various meteorite components can be a powerful tool in determining either the relative timing of remanence acquisition or the accretionary and metamorphic history of the parent body. Thus, the careful determination of vector directions of remanence can often provide important insights and constraints which supplement the paleointensity data. In some cases, the degree of directional stability for individual chondrules might also provide information either on the temporal behavior of the magnetizing field, or alternately on the motion of the chondrule within a fixed magnetic field geometry.

The following discussion and evaluation of meteorite paleomagnetic studies begins with a short review of the magnetic mineralogies of meteorites, with special emphasis on the thermal instabilities of many of the extraterrestrial opaque phases which make both the experimental simulations of acquisition of thermal remanence and the interpretation of meteorite magnetizations so difficult. In the next section, possible sources of the magnetic fields recorded by meteorites are discussed. As these sources may relate either to large-scale processes pertaining to the early solar nebula or the proto-Sun, or to local or even internal causes entirely independent of solar evolution, a careful consideration of these potential mechanisms is critical in an evaluation of the paleomagnetic record of the meteorites and how it relates to the primitive solar environment. In Sec. IV, a detailed review of paleointensity results derived from the various classes of meteorites is given, and the likelihood that the results might relate to solar or nebula-wide magnetic fields is considered. This is followed finally by a discussion of the implications of meteorite magnetism for early solar system evolution (Sec. V).

II. MAGNETIC MINERALOGIES OF METEORITES

Metallic Fe and Ni-Fe are ubiquitous magnetic phases in meteorites. Because of their high-saturation magnetization values, they generally dominate the high field magnetic properties and induced magnetization of extraterrestrial samples. Other magnetic carriers in meteorites include the Fe-oxide magnetite, which occurs in ultrapure form in many carbonaceous chondrites (Kerridge et al. 1979), and various Fe-sulfides (e.g., pyrrhotite) and Fe-carbide (cohenite).

Figure 1 summarizes how the thermomagnetic properties of Ni-Fe varies as a function of Ni content. At 20° C, kamacite, the body-centered cubic crystal structure, or alpha phase, is the stable crystal form for metallic Fe with a low Ni content, whereas taenite, the face-centered cubic structure, or gamma phase, is preferred at higher Ni contents. At higher temperatures, kamacite transforms to taenite, but because the transition temperature is above the gamma-phase Curie temperature for lower Ni contents, remanent magnetization is lost at this point. The delayed regression of taenite to ka-

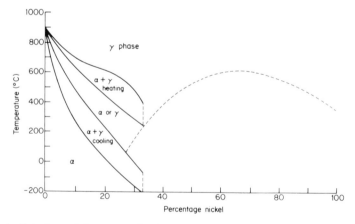

Fig. 1. Alpha (kamacite) to gamma (taenite) transition temperatures, and taenite Curie temperature curve (dashed line) as a function of Ni content, for NiFe (figure adapted from Pickles and Sucksmith 1940).

macite upon cooling produces a characteristic thermal hysteresis effect which allows an estimate of the relative contribution of the Ni-poor alpha phase to the magnetic signature of the meteorite. Since the classification of meteorites is in large part based on the Ni contents of their component magnetic phases (see Sec. IV.B), a classification of meteorites based solely on thermomagnetic properties is possible (Nagata 1980a).

The unique thermomagnetic properties of Ni-Fe alloys have important implications for paleointensity experiments on meteorite specimens. Because of the thermal hysteresis effect for kamacite, the magnetization acquired during cooling in the laboratory field will occur at a substantially lower temperature than the loss of NRM (natural remanent magnetization) during heating for metal containing even a few percent Ni (Fig. 1). However, since the TRM (thermoremanent magnetization) gained and the NRM lost for a particular temperature interval are generally derived from measurements made after cooling to room temperature (see Sec. IV.A), this effect should not influence the results of the standard paleointensity experiments, unless the remanence measurements are made at high temperature.

A potentially more troublesome problem comes from the demonstration by Wasilewski (1976) that shock can transform kamacite to metastable taenite. Since the nonmagnetic gamma phase will revert to the magnetic alpha phase with laboratory heating, new remanence carriers may be created in shocked samples that are subjected to thermal paleointensity experiments. This could lead to an underestimation of the paleofield, as the laboratory-induced thermal remanence would be exaggerated by the newly created alpha phase.

Tetrataenite is a highly ordered form of Fe-Ni (between 48% and 57% Ni) which has recently been recognized as an important remanence carrier in chondritic meteorites. Because it attains its atomic ordering below 320°C, or ~ 200° below its Curie point, with an accompanying dramatic change occurring in its magnetic properties, its original TRM may not be preserved. Also, laboratory heating above ~ 500°C will destroy the ordered structure, while subsequent laboratory cooling is unlikely to restore ordered tetrataenite (Wasilewski 1988). In combination, these thermomagnetic characteristics of tetrataenite make it a difficult mineral from which to obtain meaningful paleointensity information. Because the formation of tetrataenite requires exceptionally slow cooling rates, its common presence in rapidly cooled chondrules implies that the acquisition of chondrule remanent magnetizations often postdates their formation (Nagata 1988).

III. SOURCES OF THE MAGNETIC FIELDS RECORDED BY METEORITES

In this section, possible mechanisms for producing early solar system magnetic fields that could have been responsible for the magnetization of

some or all meteorite classes are explored. A comprehensive earlier review
of the probable origins of meteoritic paleomagnetism has been given by Levy
and Sonett (1978). Supplementary discussions based on summaries of laboratory paleomagnetic investigations have been presented by Nagata (1979),
Sugiura and Strangway (1981,1988), Hood and Cisowski (1983) and by Cisowski (1987).

A. Large-Scale Field Generation Mechanisms

A natural initial source of large-scale magnetic fields in the early solar
nebula is compression of a pre-existing interstellar magnetic field. Since stars
and their associated accretion disks originate via gravitational collapse of
weakly ionized interstellar cloud clumps, there is little doubt that some amplification of a pre-existing interstellar field will occur during the collapse
(Fig. 2). The question is how much amplification can occur and for how long.
In the interstellar medium, typical densities and magnetic field strengths are
$\sim$ 1 hydrogen atom per cm^3 and $\sim$ 2 $\times$ 10^{-6} gauss, respectively (see e.g.,
Spitzer 1968). Viscous accretion-disk models based on hydrodynamic col-
lapse calculations yield representative mass densities in the protoplanetary
nebula of the order of 10^{-11} g cm^{-3} at distances of $\sim$ 3 AU, although this
value depends on the assumed disk mass and protostellar mass accretion rate
(see, e.g., Morfill 1985). Under the extreme assumption that the gas is fully
ionized and that magnetic flux is fully conserved, the compressed field
strength B is related to the original field strength B_o and the ratio of compressed to original densities ρ/ρ_o by $B = B_o(\rho/\rho_o)^{2/3}$. Thus, a rough upper
limit for the compressed interstellar field strength in the protoplanetary nebula
where the meteorites would have formed is $\sim$ 700 gauss, more than sufficient
to explain meteorite paleointensities (Sec. IV). In practice, however, the gas
is only partially ionized; flux is therefore not conserved due to slippage of the
magnetic field and charged plasma component through the "sea" of neutrals
(ambipolar diffusion). Final field amplitudes can be estimated at present only
through theoretical considerations.

According to most current protosolar accretion-disk models, temperatures are relatively low ($<$ 500 K at 3 AU) so that ionization production is
mainly by cosmic rays and radioactive elements in rocky and metallic materials (Umebayashi and Nakano 1988). These authors have calculated correspondingly low ionization fractions and effective gas electrical conductivities
σ in order to estimate the diffusion lifetime τ_d for dissipation of magnetic
fields in the nebula. For τ_d greater than typical transport times, the field is
advected with the gas. In particular, differential rotation will result in an amplification of the azimuthal field component if $\tau_k \ll \tau_d$, where τ_k is the Keplerian orbit period. (The azimuthal component of gas motion is essentially
Keplerian in a low-temperature nebula.) They find that τ_k is much greater
than τ_d in the region presently occupied by the terrestrial planets but becomes
less than τ_d at larger radial distances. Thus, in the inner region, the field is

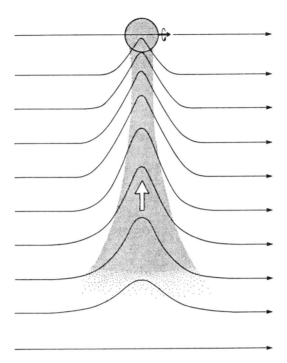

Fig. 2. Schematic illustration of the compression of an interstellar cloud magnetic field in a partially ionized protosolar accretion disk. Only the meridional component of the field orientation is shown. Near the equatorial plane, the gas velocity in the plane of the figure is nearly radially inward, as indicated. The field outside the disk is parallel to the rotation axis as is often observed for interstellar-cloud magnetic fields near young stellar objects. However, differential rotation in the disk amplifies the azimuthal field component so that the dominant orientation near the equatorial plane is azimuthal.

decoupled from the gas and little amplification occurs while, in the region occupied now by the Jovian planets, amplification by differential rotation could have been important. On the other hand, some solar-nebula models are characterized by much higher temperatures than that considered by Umebayashi and Nakano (1988; see also Consolmagno 1978) so that direct thermal ionization could become important throughout the terrestrial planet region. For example, nebular temperatures in the high-accretion-rate models investigated by Cameron (1985) exceed 10^3 K at distances out to ~ 4 AU. The latter models may be more appropriate for the early high-luminosity protosolar accretion phase occurring at times $< 10^5$ yr after collapse. During this period, much stronger gas-field coupling and field amplification via differential rotation are to be expected. Earlier theoretical treatments by Umebayashi and Nakano (1984) have already indicated that compressed interstellar fields of the order of several tenths of a gauss are likely to have been present in the region where meteorite parent bodies could have formed. This estimate combined with the likelihood of stronger gas-field coupling during earlier high-temperature nebular phases make compressed interstellar fields an attractive large-scale field source in the early solar system.

Because most current solar-nebula models are turbulent as well as differentially rotating (see, e.g., Wood and Morfill 1988), it is possible that hydromagnetic dynamo action may also assist in amplifying pre-existing nebular fields (Levy and Sonett 1978; Levy 1978). This is especially true during the early high-temperature phase referred to above when the ohmic diffusion lifetime may have been longer than the turbulent velocity time scale in the inner zone.

Subsequent to the dispersal of the solar nebula, it is inevitable that meteorite parent bodies would have been exposed to a solar magnetic field extended into interplanetary space by a protosolar wind. It has been suggested that this epoch may also have resulted in electromagnetic heating of the same parent bodies (Sonett et al. 1970; Herbert 1989; chapter by Herbert et al.), a process that could have left a magnetic record in meteorites. Although recent astrophysical evidence has solidified the idea of massive protostellar winds (see reviews by Cohen 1984; Lada 1985; Shu et al. 1987), this evidence has also shown that the wind structure is initially bipolar, consisting of oppositely directed jets perpendicular to the accretion disk plane. As time proceeds, the wind becomes more isotropic and the surrounding interstellar cloud dissipates revealing the protostellar object as a T Tauri star. Many or most T Tauri stars probably still have accretion disks (Bertout et al. 1988; see also the chapter by Bertout et al.), implying the persistence of nebular gas that would continue to shield meteorite parent bodies from direct exposure to the wind. Nevertheless, at some point (possibly > 10 Myr after cloud collapse), the wind will dissipate the nebula (Cameron 1973; Horedt 1978; Ohtsuki and Nakagawa 1988), allowing direct exposure of solid objects to wind magnetic fields. According to Cohen (1984), T Tauri stars of solar mass typically have rotational velocities < 25 km s^{-1} (period ~ 3 days for a radius of 10^{11} cm) and are characterized by inhomogeneous surfaces with strong flare activity indicative of the presence of strong magnetic fields. If, for simplicity, we assume that the field takes the form of an Archimedean spiral near the plane of the ecliptic, then the radial and azimuthal components are given by $B_r = B_s R_s^2/r^2$ and $B_\phi = B_s \Omega R_s^2/(V_\omega r)$, respectively, where B_s is the mean surface field, Ω is the protosolar angular rotation frequency, R_s is the protosolar radius, and V_ω is the wind radial velocity. For an outflow velocity of $V_\omega \sim 300$ km s^{-1}, a surface field strength of ~ 100 gauss, and a rotation period of 3 days, one obtains $B_r \approx 0.004$ gauss r^{-2} (AU) and $B_\phi = 0.05$ gauss r^{-1}(AU). Thus, even at a distance of several AU, the field amplitude would be > 0.01 gauss. Given current uncertainties in the properties of T Tauri stars (surface field strength, radii, wind velocity, etc.), such a field may have contributed to the magnetization of meteorites whose parent bodies experienced metamorphic heating while exposed to the protosolar wind (whether or not electromagnetic induction caused the heating). Although the estimated amplitudes are consistent only with the low paleointensity estimates for achondrites (Sec. IV.C), stronger fields will exist near the boundary between the stellar wind and the nebula by analogy with the interaction of the present solar wind with planetary ionospheres (see, e.g., Ip 1984).

B. Small-Scale Field Generation Mechanisms

In addition to large-scale fields pertaining to the macroscopic evolution of the Sun or nebular disk, a variety of mechanisms exist for producing local and/or temporally transient fields that may have contributed substantially to the observed magnetism of meteorites, particularly those that experienced significant shock and/or re-heating events after their initial formation.

Achondritic meteorites are petrologically similar to lunar igneous samples and appear to have originated on one or more differentiated parent objects. These meteorites are therefore most likely to exhibit magnetization produced locally within or on parent bodies. Although most achondrite parent bodies were probably at least as small as the Moon, these bodies may have possessed proprotionately larger metallic cores capable of dynamo field generation. The parent object of the SNC meteorites (probably Mars) clearly has a relatively large metallic core although no large-scale intrinsic field appears to exist at present (Russell 1980).

In addition to core dynamo fields, local surface processes must also be considered as a contributor to the magnetization of meteorites originating on parent bodies. A variety of local surface mechanisms may be envisaged (see, e.g., Daily and Dyal 1979). However, impacts are likely to play a dominant role as indicated from lunar studies and laboratory experiments. At incident velocities ≥ 10 km s^{-1}, silicate meteoroid impacts on planetary surfaces result in vaporization of a portion of both the target material and the meteoroid. The resulting partially ionized vapor cloud expands thermally outward from the impact point at a velocity exceeding that of even the fastest solid and molten ejecta (Melosh 1989; Hood and Vickery 1984). Laboratory experiments have shown that transient magnetic fields are generated by currents forced thermally within the plasma (Crawford and Schultz 1988). From lunar experience, it is doubtful that these intrinsic impact fields produce large-scale magnetization of the surface. (For example, there is no correlation between lunar crustal magnetic anomalies and impact crater locations.) Physically, this is because of the reduced thermal gradients in large-scale impacts which should theoretically produce weaker currents and fields. Nevertheless, these fields may have contributed substantially to *small-scale* magnetization such as would be preserved in highly shocked meteorites. In addition, the expanding plasma cloud interacts strongly with any ambient magnetic field producing a transient field amplification near the interface between the plasma and the external medium. An especially large compression of an ambient field occurs antipodal (diametrically opposite) to the impact point. This interaction may help to explain the concentrations of lunar crustal magnetization observed to occur antipodal to young large impact basins (Lin et al. 1988; Hood 1987a). By implication, impact processes may have played an important role in producing magnetization on sizeable meteorite parent bodies.

IV. SPECIFIC PALEOINTENSITY STUDIES ON METEORITES

Paleointensity experiments on natural samples represent an attempt to determine the strength of the ambient magnetic field, or paleofield, when the samples acquired their remanent magnetizations. Generally, only samples that possess a single component of remanent magnetization are suitable for paleointensity studies, unless the several components that sometimes comprise a sample's remanence are confined to discrete temperature intervals.

A number of further assumptions must be made before accepting paleointensity results derived from extraterrestrial samples. The first is that the natural remanent magnetization (NRM) carried by the sample is of extraterrestrial origin. This implies that the sample did not acquire a significant viscous remanent magnetization in the Earth's magnetic field since its rendezvous with the Earth (a potential problem for the meteorites collected from Antarctica, due to their long terrestrial residence), and was not exposed to stray DC or AC magnetic fields (e.g., from hand magnets or electric motors) since its collection (more of a potential problem for museum specimens).

The second assumption is that the extraterrestrial remanence residing in the sample is of a thermal origin, i.e., that the remanence was acquired during cooling through its blocking temperature (a temperature, always below the Curie temperature of the particular magnetic phase involved, at which the magnetization becomes stable in direction over time, or locked in due either to the fixing of domain walls or to the attainment of thermal stability in single-domain particles). This type of magnetization is called thermal remanent magnetization (TRM). One possible exception to this TRM criterion involves submicron magnetic grains which acquire their magnetization as they grow to a size in which their magnetization becomes stable through time; this will be discussed in more detail below.

The third assumption is that most of the TRM signal, which for many extraterrestrial samples was acquired several Gyr ago, has survived until today. The original NRM that a meteorite or lunar sample acquired has been subjected not only to viscous decay through time, but can also be diminished due to shock demagnetization or thermal cycling while in solar orbit.

While the long list of potential problems which can afflict meteorite magnetism may seem discouraging, the observation that the remanence properties within the various types of meteorites are generally similar (Sugiura 1977; Sonett 1978) suggests that most meteorites do contain an underlying magnetic signal of extraterrestrial origin.

A. Paleointensity Methods

Generally, the most reliable paleointensity experiments involve heating the sample twice, in a stepwise fashion to progressively higher temperatures, until the NRM is entirely erased. In the first heating the sample is allowed to cool in a precisely measured laboratory field. In the second heating and cooling cycle, the sample is either rotated 180°, or the sample is allowed to cool in a zero ambient field. In either case, both the NRM lost and the partial TRM gained for each heating step can be determined by either vector addition and subtraction or, in the zero field case, vector subtraction, once measurements are made at room temperature after each heating and cooling cycle. If rigorously applied, the paleofield, which is derived from the slope of the "NRM remaining vs partial TRM gained" curve (Fig. 3), times the ambient field strength, is reliable only if a linear curve is attained (nonlinearity can result from chemical alteration, among other things). Unfortunately, such linearity is rare for extraterrestrial samples, except within limited temperature intervals. This type of dual-heating paleointensity experiment is generally referred to as the KTT experiment, after the three early paleomagnetists (Koenigsberger 1938; Thellier and Thellier 1959) who independently devised this experiment. Representative dual-heating paleointensity estimates on the various classes of meteorites are summarized in Table I.

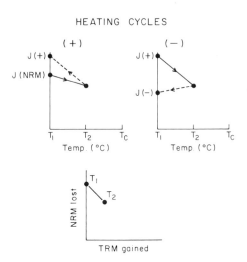

Fig. 3. Illustration of the KTT paleointensity experiment, where a sample carrying an NRM is heated and cooled from temperature T_2 first in the presence of an applied field in a positive direction, and second in an antipodal (negative) applied field. By adding the resultant vectors and dividing by 2, the NRM remaining after heating to T_2 is obtained; then by subtracting the vectors and dividing by 2, the TRM gained is obtained. The slope of the NRM vs TRM gained-curve times the applied-field intensity gives the paleofield intensity.

TABLE I
Representative Dual-Heating Paleointensity Results for Various Types of Meteorites

Meteorite	Type of Sample	Temp. (°C)	Field (gauss)	Reference[a]
Carbonaceous Chondrites				
Orgueil (C1)	bulk	20–120	0.67	1
Murchinson (C2)	bulk	20–90	0.18	1
Allende (C3)	bulk	20–130	1.09	1
Allende (C3)	bulk	20–330	0.79–3.3 (median–1.23)	2
Leoville (C4)	matrix	20–330	1.9–3.0	
	chondrules	20–330	0.7–1.3	
	bulk	20–350	0.97	3
Ordinary Chondrites				
Dagley Downs (L4)	bulk	20–470	0.68	3
Fukutomi (L5)	bulk	20–400	0.10	3
Seminole (H4)	bulk	20–350	0.39	3
Yonozu (H4 or 5)	bulk	20–470	0.18	3
Enstatite Chondrites				
Abee (E4)	clast	74–228	6.1	4
	matrix	56–216	7.0	
Indarch (E4)	bulk	20–384	17.8	5
Achondrites				
Yamato 7308	bulk	20–240	0.07	6
Yamato 791195	bulk	20–150	0.10	
		150–350	0.025–0.05	7
		350–550	0.01	
Shergoty	bulk	20–138	0.02	8
		362–598	< 0.01	

[a] 1: Banerjee and Hargrraves 1972; 2: Nagata and Funaki, 1983; 3: Nagata and Sugiura 1977; 4: Sugiura and Strangway 1981; 5: Sugiura and Strangway 1981; 5: Sugiura and Strangway 1982; 6: Nagata 1979; 7: Cisowski 1991; 8: Cisowski 1986.

Other paleointensity techniques that have been employed on extraterrestrial samples either require only a single heating, with alteration determined by other magnetic tests, or no heating, with paleointensity determined after normalization to some nonthermal type of remanence. However, these latter normalization methods do ultimately require a calibration of the nonthermal normalization parameters through laboratory heatings on selected samples, in order to produce absolute paleointensity results.

B. Paleointensity Estimates on Chondrites

Chondrites include all undifferentiated meteorites with a nonvolatile solar composition (see, e.g., Wasson 1985). These meteorites are apparently derived from parent bodies which, because of their size, greater distance from the Sun, or lesser concentrations of radioactive elements, have not experienced any significant partial melting. They usually contain small spherical forms (chondrules) which appear to have formed from molten droplets produced in their parent chondrites' prehistory. A wide variety of mechanisms have been proposed for the origin of chondrules, including meteorite impacts on the surface of the chondrite parent bodies, liquid condensates from hot nebular gases, and electrical discharge (lightning strikes) in the nebular environment.

Chondrites are generally classified according to their metal (i.e., NiFe) contents, with carbonaceous chondrites having the least native metal, while the LL, L, H and E (enstatite) chondrites contain progressively more Fe (albeit with lesser percentages of Ni). This progression reflects the oxidation state of the various types of chondrites, with Fe in the carbonaceous variety being in the highest oxidation states (including its presence as Fe^{+3} in magnetite), in contrast to the high abundance of native Fe in the enstatite chondrites.

Chondrites are also classified according to the degree of thermal metamorphism which they have experienced, with type 3 having been heated to 400–600°C, while type 6 experienced heating of up to 950°C. In general, the H, L and LL chondrites (referred to as ordinary chondrites) and enstatite chondrites represent higher petrologic types (i.e., they have been heated to higher temperatures) than the carbonaceous chondrites. Many carbonaceous chondrites have been subjected to substantial low-temperature hydrothermal alteration, with type 2 specimens having been flushed with water below 20°C, while type 1 meteorites may have been altered by water possibly as warm as 140°C. As many chondritic meteorites (perhaps the majority of ordinary chondrites; Scott et al. 1985) are actually breccias, the petrologic grade assigned to them often reflects the highest temperature seen by individual clasts within the meteorite, and does not necessarily reflect the temperature at which the meteoritic breccia was assembled. This distinction is important in considering the origin of the magnetization of a bulk meteorite sample, which, if it consists of a large number of clasts, will not exceed either the age or the blocking temperature of the assemblage event.

1. Carbonaceous Chondrites. Dual-heating paleointensity results are available for only one type 1 (Orgueil) and one type 2 (Murchison) carbonaceous chondrite. In both cases, the principal component of the NRM was confined to temperatures below 150°C, as might be expected from their low-temperature hydrothermal origin. Magnetite, along with NiFe and pyrrhotite, are the likely remanence carriers in both meteorites. In Orgueil, the magnetite occurs both as spherules up to 40 μm in diameter, probably formed as a product of hydrous alteration at low temperatures (Kerridge et al. 1979), and as μm-sized irregular grains, perhaps representing an early high-temperature phase (Bostrom and Fredriksson 1966). In Murchison, most of the magnetite grains are 2 to 5 μm and rounded to slightly angular (Lewis and Anders 1975). A paleofield of 0.18 gauss was derived for Murchison, while a 0.67

gauss paleofield resulted from the NRM of Orgueil (Banerjee and Hargraves 1972), comparable to the Earth's present field intensity. $^{129}I/^{127}I$ ratios on magnetic separates from these C1 chondrites indicate that their magnetite grains formed within 2×10^5 yr of each other, close to the condensation time of the solar nebula (Lewis and Anders 1975).

Numerous paleomagnetic investigations of the type 3 Allende carbonaceous chondrite have been undertaken since its fall in 1969. Paleointensities derived both from the matrix and chondrules of this massive meteorite have mostly fallen in the range of 1 to 3 gauss, as illustrated in Fig. 4, and as summarized in Table I. However, the accuracy of many of the paleointensity experiments on Allende subsamples has been degraded by heating-induced magnetochemical changes that begin at temperatures of $<100°C$ (Cisowski 1987, Fig. 6). Paleointensities as high as 7 gauss have been reported for preseparated chondrules from Allende (Lanoix et al. 1978a,b), although a later investigation suggested that these particular specimens may have suffered terrestrial contamination (Wasilewski 1981).

The directionally stable NRM's of Allende bulk samples and components are generally demagnetized by about 300°C, which, for this type 3 carbonaceous chondrite, may reflect the relatively low temperatures at which its remanence-carrying grains acquire their magnetizations. Heating experiments have demonstrated that only $\sim 20\%$ of the IRM_s (saturation remanence, or maximum remanence that a sample can carry, measured after exposure to a strong magnetic field) remains for the bulk Allende sample after cooling from 350°C in a zero field (Cisowski 1987, Fig. 6), although the thermomagnetic curve indicates the dominant magnetic phase (magnetite or taenite) has a Curie temperature (T_c) close to 600°C. This indicates that either the magnetite and taenite are not contributing substantially to the remanence carried by this meteorite, or that their grain sizes are either very small (near-superparamagnetic) or large (multidomain), such that they acquire their magnetization well below their Curie temperatures.

Alternatively, Nagata and Funaki (1983) interpreted the limited thermal stability of Allende to mean that the NRM resides primarily in pyrrhotite $(T_c = 320°C)$. Its existence is inferred primarily from the nonstoiciometric composition of a magnetic "troilite" component, which, if truly FeS, should be antiferromagnetic and hence incapable of significantly contributing to the remanence. Wasilewski and Saralker (1981) also concluded that the stable remanence in Allende was carried by a magnetic phase formed during a sulfidation event in its early history. However, these authors regarded the acquisition mechanism for Allende magnetization as being of the nature of chemical remanence (CRM) and so inappropriate for any thermal paleointensity technique. Sugiura and Strangway (1985) interpreted a weak directional coherency of the remanence vectors for the small oriented Allende chondrules as a reflection of a low-temperature thermochemical event experienced by the meteorite sometime after its consolidation.

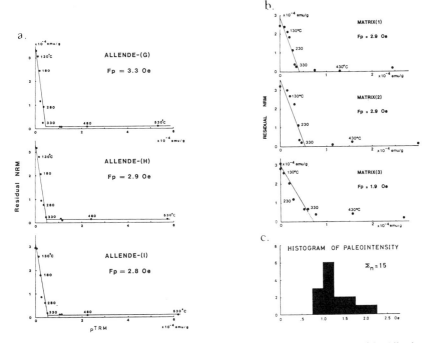

Fig. 4. Results of KTT experiments on (a) three bulk, and (b) three matrix samples of the Allende carbonaceous chondrite, and (c) histogram of KTT paleointensities for 15 small bulk specimens. The ambient laboratory field for all experiments was 0.44 gauss (oersteds). The indicated paleofields (F_p) are given in oersteds (figure from Nagata and Funaki 1983).

The whole-rock [207]Pb-[206]Pb age for Allende of 4.553 Gyr indicates that it was originally heated only shortly after condensation of the solar nebula, although its Th-He[4] age suggests a more recent reheating (or sulfidation) event perhaps at 3.4 Gyr (Fireman et al. 1970).

While any nonthermal remanence would seem to be not amenable to paleointensity experiments, an exception may occur if the CRM is acquired by a new, submicron magnetic phase growing through the critical threshold size of stable single-domain behavior, even if this occurs at a temperature well below its T_c (Kobayahi 1959). This concept may be best understood by considering the equation for the characteristic relaxation time of magnetization (Neel 1955)

$$\tau^{-1} = f \exp(-K_u V/kT) \qquad (1)$$

where f is a frequency factor, K_u is the uniaxial anisotropy, k is Boltzmann's constant, V is the particle volume and T is temperature. Remanence blocking (the setting in of a stable remanence) occurs rapidly with a small increase in the exponential term, when τ becomes larger than the observational time frame. When this occurs, the grain is said to go from superparamagnetic to stable single-domain behavior. Note that any increase in volume V has the same effect as a decrease in temperature T, meaning that as far as remanence blocking is concerned, the growing of a new magnetic phase is mathematically equivalent to the cooling of a nongrowing magnetic phase. Thus, application of the KTT experiment to samples which carry a CRM may not be entirely unwarranted, although experimental results indicate that remanence is less efficiently acquired via the grain growth process vis-à-vis true thermal blocking (Kobayashi 1962). This difference in magnetization efficiency, if a general case, would produce an underestimate of the paleofield strength for CRM-bearing samples. (The efficiency of magnetization for a particular sample can be defined as the intensity of remanence divided by the strength of the ambient field.)

A paleointensity estimate of 0.97 gauss has been derived from a dual-heating experiment (Fig. 5) on the carbonaceous chondrite Leoville (Nagata and Sugiura 1977). This result, combined with the ~1 gauss or greater paleointensity results associated with Allende and the other higher-grade carbonaceous chondrites, suggests that carbonaceous chondrites subjected to dry thermal metamorphism (types 3 and 4) were formed in stronger fields than those subjected to a high degree of hydrothermal metamorphism (types 1 and 2). Alternatively, this difference may reflect the abundance of secondary, low-temperature magnetite (Bostrom and Fredriksson 1966) in these later types, which, because it carries a CRM, may be less efficiently magnetized than high-temerpature magnetite that carries a TRM (Kobayashi 1962).

2. Paleointensities of Ordinary and Enstatite Chondrites. Ordinary (LL, L and H) and enstatite (E) chondrites most often fall into the highest petrologic grades (types 5 and 6). Thus the majority of their individual clasts were heated to temperatures in excess of the Curie temperature of their component magnetic phases. However, the frequent domination of their remanence by complex and metastable magnetic phases makes interpretation of their remanence, particularly in response to laboratory heating, difficult. The presence of Ni, along with decreasing grain size, tends to increase a metal grain's "magnetic hardness" or resistance to alternating-field demagnetization. Because of these two effects, LL meteorites, characterized by smaller, high-Ni metal grains, tend to possess NRM's that are the most stable to alternating-field demagnetization and H and E chondrites, that contain large, low-Ni metal grains, exhibit weak resistance to alternating-field demagnetization (Brecher and Arrhenius 1974).

Dual-heating paleointensity experiments by Nagata and Sugiura (1977) on a variety of ordinary chondrites gave paleofield estimates between 0.10 and 0.68 gauss (Fig. 5). Additional KTT experiments on a chondrule from the Bjurbole L4 chondrite, and on Antarctic chondrites Yamato 74160 and ALHA 77260, resulted in apparent paleofield estimates of about 0.05 gauss.

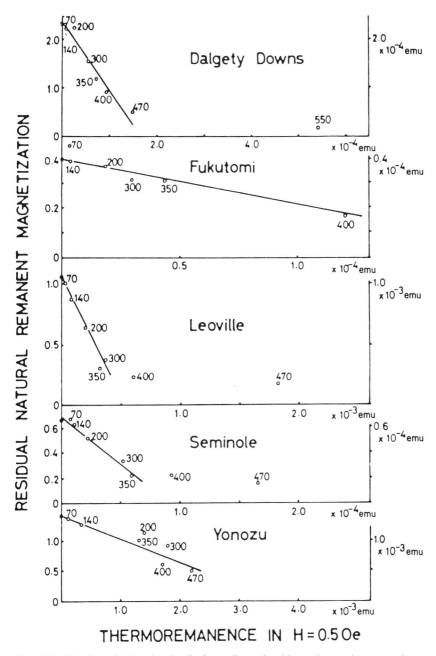

Fig. 5. Dual-heating paleointensity plots for four ordinary chondrites and one carbonaceous chondrite (Leoville). The paleofields, that range from 0.1 to 0.68 gauss for the ordinary chondrites, are derived by multiplying the slope of linear segments by the laboratory field of 0.5 gauss. (figure from Nagata and Sugiura 1977).

These later results must be considered with some reservation because of the presence of plessite and tetrataenite as remanence carriers in Bjurbole (Sugiura and Strangway 1982), and high-magnetic anisotropy in the latter two L and LL chondrites (Nagata and Funaki 1982). Tetrataenite, as discussed in Sec. II, presents particular problems to the interpretation of stable meteorite magnetism, as its high coercivity insures its contribution to the most resistant "primary" remanence, while its thermal instability and low-formational temperature ($<320°C$) complicate the interpretaton of any paleointensity experiment on meteorites which contain this phase (Nagata and Funaki 1982).

Thirteen oriented chondrules from Antarctic L6 chondrite ALHA 76009 displayed stable but randomly directed remanences, although matrix samples showed some evidence for a magnetic overprinting event. The chondrules may have undergone a slow rotational precession in a directionally stable magnetic field as they cooled, because the remanence vectors derived from step-wise thermal demagnetizations to progressively higher temperatures described a circular path consisting of several revolutions. Dual-heating paleointensity estimates of 0.47 to 0.97 gauss again cannot be unequivocally accepted because of uncertainty about the thermal nature of the remanence and evidence for interactions in the KTT "NRM vs partial TRM" curves (Nagata and Funaki 1981). Collinson (1987) also found evidence for randomly directed magnetization in the Olivenza (LL5) chondrite, on a scale of ~ 1 mm or less within the sample. This was interpreted as possibly resulting from the magnetization of individual grains which acquired their magnetizations during condensation in an early, strong solar system magnetic field, before their incorporation into the parent meteorite.

Dual-heating paleointensity experiments on seven additional H chondrites by Westphal and Whitechurch (1983) produced paleointensity estimates of up to 1.88 gauss. Unfortunately, these samples were not heated under vacuum conditions, and so the rapid loss of NRM and sluggish acquisition of partial TRM may reflect the destruction of fine metal carriers due to oxidation at high temperatures, rather than a high paleofield environment.

Single-heating paleointensity estimates have been made on a number of other H, L and LL chondrites by comparing the alternating field demagnetization behaviors of their NRM to their laboratory induced TRM's. These experiments, which must be considered much less reliable than the dual-heating experiments because irreversible magnetochemical changes are not always resolvable, generally gave (up to an order of magnitude) lower paleofield results than the dual-heating experiments (Brecher and Ranganayaki 1975; Brecher et al. 1977). In contrast, a similar single-heating type of paleointensity experiment on 12 other chondrites gave results more consistent with the preferred KTT results on ordinary chondrites listed in Table I (Nagata 1979, his Table 7).

Anomalously high paleofield determinations of 6 to 7 gauss and 17.8 gauss have been reported for clasts and a bulk sample of the enstatite chondrites Abee and Indarch (Sugiura and Strangway 1981,1982). The results from Abee contrast with a much lower paleofield of 0.33 gauss derived by a single-heating experiment on this same meteorite (Brecher and Ranganayaki 1975). Interpretation of the strong Abee NRM as a TRM is complicated by its unusual magnetic mineralogy (including cohenite and schreibersite), the intergrown textures of their magnetic grains (reflected in magnetic behavior characteristic of magnetostatic interactions), and noncoherent directions of magnetizations between matrix and clasts.

Although metamorphic ages of ordinary chondrites are generally within about 100 Myr of the ~4.53 Gyr age of the solar system, about 1/3 have undergone significant reheating since formation of their parent bodies, probably as a result of impact events. Gas-retention ages of the more heavily shocked samples indicate that many of these events occurred within the last 1 Gyr (Taylor and Heymann 1969). Thus the magnetizations of many chondrites, particularly those displaying mineralogic shock effects or brecciation, may postdate the consolidation of their parent bodies by a considerable time period.

In summary, the remanent magnetizations of chondrites present a mixed picture, clouded by uncertainties about the thermal nature of their remanence, the time frame of remanence acquisition, and their questionable credibility as magnetic recorders. Taken at face value, ordinary chondrites appear to have become magnetized in fields comparable to the present field of the Earth, while carbonaceous chondrites assembled in slightly stronger fields of about 1 gauss. It may be noted that this range of paleofield intensity is roughly consistent with that inferred by Sonett (1978) from a compilation of susceptibility/mass and magnetic moment/mass for a large sample of meteorites. The few enstatite chondrites studied magnetically to date indicate paleofields even more than an order of magnitude higher, although their remanence appears to be the most complex, and the least amenable to straightforward paleointensity techniques.

As the variable oxidation states of chondrites are sometimes attributed to their relative positions in the solar nebula during their formation (enstatite chondrites forming closest to the center, carbonaceous chondrites forming the farthest from the center), the lower paleointensities associated with the ordinary chondrites as compared to carbonaceous chondrites cannot be reconciled with a simple decrease of magnetic field strength with increasing distance from the proto-Sun. Such a decrease has been inferred by the apparent confinement of differentiated asteroids to the inner regions of the asteroid belt (Bell 1986a), and would be consistent with a magnetic field associated with either a nebular dynamo or an enhanced solar wind (see Sec. III.A).

C. Paleointensities of Achondrites

Achondrites represent impact ejected fragments of the outer layers of a large asteroidal body which underwent melting and differentiation early in its

history. A comparison of the spectral reflections of achondrites to larger asteroidal bodies suggests that 4 Vesta, the third largest asteroid (~550 km in
diameter), at a distance of 2.36 AU, is a potential source for differentiated
meteorites. If nearly all achondrites were derived from one parent body (as is
suggested by their oxygen isotope ratios), the eucrites would represent basaltic flows within the upper crust of the asteroid, while the diogenites would
represent ultramafic plutonic rocks formed at deeper levels. Howardites can
be accounted for as breccias composed of both extrusive and intrusive rocks,
brought together by large, deep-penetrating impacts. Nearly all achondritic
meteorites, with the exception of a few eucrites, appear to be breccias composed of numerous clasts. As with many chondrites, the composite remanent
magnetizations measured for most bulk achondritic samples must therefore
be regarded as dating only to the time of consolidation and lithification of
clasts.

Two dual-heating paleointensity estimates, both < 0.1 gauss, have been
derived from ordinary achondrites, including a brecciated howardite (Y7308)
and an unbrecciated eucrite (Y791195) (Table I). These more rigorous results
have been supplemented by paleofield estimates ranging from < 0.24 gauss
for an unbrecciated diogenite to more typical values of from 0.03 to 0.09
gauss for five other achondrites, all derived by comparing their NRM alternating-field demagnetization spectra to that of an anhysteretic remanence
(Nagata 1979,1980b). Anhysteretic remanence magnetization is imparted by
exposing the sample to an alternating field of high intensity while in the
presence of a nonvarying weaker ambient field. As anhysteretic remanence
magnetization is sometimes regarded as an isothermal analogue to the acquisition of remanence through cooling, this method offers a simpler alternative
to the KTT experiment.

Brecher and Fuhrman (1979) have interpreted the intense NRM's (more
than 10% of their corresponding saturation remanence, or IRM_s values) in a
pair of ureilites (Kenna and Goalpara) as evidence for a strong nebular magnetizing field of about 1 gauss. Ureilites are a unique type of highly shocked,
ultramafic achondrite with high carbon content. However, this assignment of
a high paleofield intensity to the environment in which the ureilites formed is
inconsistent with lower paleofield values of from 0.048 to 0.34 gauss reported
for other ureilites collected from Antarctica (Nagata 1981).

In order to derive further paleointensity information from this important
class of meteorites, NRM values for all measured low-metal achondrites (ureilites excluded) can be normalized to corresponding IRM_s values, measured
after exposing each sample to a strong, saturating field (Fig. 6). This normalization technique has been useful in discerning a period of apparently high
lunar field strength, between about 3.9 and 3.6 Gyr ago, possibly relating to
a short-lived but energetic lunar dynamo. Laboratory heatings on fine-
grained, metal-bearing lunar samples in a variety of applied magnetic fields
allow for the calibration of this paleointensity technique in terms of an abso-

ACHONDRITE PALEOINTENSITIES FROM Irms NORMALIZATION

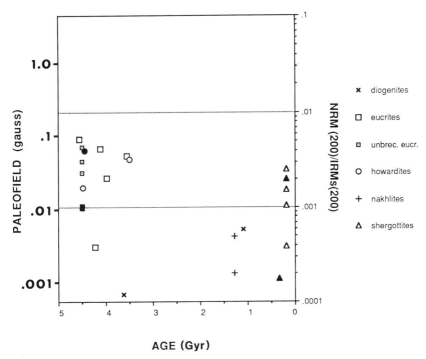

Fig. 6. Paleofield vs radiometric age for differentiated achondritic meteorites. The paleofield estimates are based on the ratio of NRM to IRM$_s$ (after alternating-field demagnetization to 200 gauss to remove spurious components of magnetization, and to eliminate the contribution of coarser-grained remanence carriers), which has been calibrated to absolute field values through heatings of fine-grained lunar samples in a variety of laboratory fields (Fuller and Cisowski 1987). KTT results on a howardite and a shergottite are shown as solid symbols. Magnetic data is from Brecher et al. (1979), Nagata (1979) and Cisowski (1986,1991).

lute paleofield estimate (Cisowski and Fuller 1986). Unlike many lunar mare basalts, the indicated paleointensities for the achondrites fall below 0.1 gauss.

Included in this diagram are the shergottites and nakhlites, whose relatively young radiometric ages (<1.5 Gyr) indicate their origin from a relatively large celestial body, most probably Mars. A KTT paleointensity experiment on a subsample of the Shergotty meteorite indicated a paleofield of about 0.02 gauss for an NRM component, most of which demagnetizes with zero field heatings below 200°C (Cisowski 1986). As a considerable portion of the remanence for the more highly shocked Antarctic shergottite EETA 79001 persisted to ~400°C, this paleofield might represent the strength of the Martian surface field when this class of meteorites experienced a strong shock event about 180 Myr ago (Collinson 1986). In this case, the Martian paleofield could possibly represent a fossil field derived from volcanic flows which erupted while the parent body possessed a strong internal dynamo field, as shergottites have about an order of magnitude more remanence-carrying potential than lunar samples (lunar surface fields of up to hundreds of gammas were detected by the Apollo astronauts).

D. Remanence of Iron Meteorites

Because TRM acquired by iron meteorites has been demonstrated to be not directly proportional to the strength of the ambient field (Brecher and Albright 1977), it is doubtful that meaningful paleointensity information can be derived from these specimens. The high thermal conductivity of metallic meteorites and their potential of acquiring viscous remanence also means that they are likely to have become remagnetized in the Earth's field, either during passage through the atmosphere or since their landing. As with many chondritic meteorites, tetrataenite appears to dominate the NRM behavior of at least some iron meteorites (see, e.g. Funaki et al. 1988).

V. DISCUSSION

The interpretation of the magnetic record of meteorites and its relationship to the early history of the solar system is highly dependent upon how selective one is in accepting published paleointensity estimates. If all results are unequivically accepted, including the paleofield estimates of several gauss or more, a case could be made for the systematic decay of a very strong solar system-wide field while the various meteoritic components and parent bodies evolved. Thus the very high fields (up to 7 gauss) recorded in some chondrules might represent the earliest recorded fields that originated from primitive solar and nebular processes. The existence of early, intense paleofields might also be supported by the strong magnetizations associated with enstatite chondrites and ureilites, which have been described as relics of primordial nebular material (Wasson and Wai 1970; Wiik 1972). The progressively diminishing fields associated with the higher thermal grade, chemically primitive carbonaceous chondrites (~1 to 3 gauss), the ordinary chondrites (several tenths of a gauss), and finally the achondrites (<0.1 gauss) would then reflect the dimunition of the early strong fields between the time of chondrule formation, accretion and heating of the various chondritic parent bodies, and lastly differentiation on the achondrite parent bodies.

However, this interpretation requires a certain amount of naiveté regarding the complexities of meteorite magnetism and paleointensity methods, and also ignores small-scale sources for magnetizing fields which do not directly relate to solar processes (see Sec. III.B). For example, since TRM can only be acquired during cooling, it is unlikely that substantial fields would be recorded by all but the surficial portions of the chondrite parent bodies, if induction heating is their primary energy source. This is particulary true if recent best estimates of $\sim 10^5$ yr for the duration of strong induction heating during the high solar mass-loss phase (Herbert 1989) are correct, since this time frame is orders of magnitude shorter than the calculated cooling rates for the interiors of parent bodies <10 km in radius and the meteoritic cooling rates as estimated from the compositional gradients in component metallic grains and preservation and annealing of nuclear fission tracks (Fricker et al. 1970; chapter by Wood and Pellas). Also high precision iodine-xenon ages indicate that the retention of the daughter isotope ^{129}Xe, corresponding to a temperature of $\sim 360°C$, differed by more than 10 Myr between various meteorites (Podosek 1970; Lewis and Anders 1975). As most meteorite paleointensity estimates are based on lower-temperature components (Table I; Figs. 4 and 5), a similar spread in the time of acquisition of magnetization, extending long after the cessation of the protosolar high-mass-loss phase, can be inferred. Perhaps the strong magnetizations associated with chondrites require other short-term heat sources (i.e., radioactive decay of ^{26}Al with a half-life of 0.72 Myr), which would allow for some cooling while the strong solar wind was still active.

The directional data obtained from the ALHA 76009 L6 chondrules (Nagata and Funaki 1981) suggests that their magnetizing field was nearly normal to the rotational axis of the chondrules. If the chondrules are assumed to have had a rotational axis approximately parallel to the nebular axis, this would imply that the field was predominantly in the ecliptic plane. Such an orientation would be consistent with a field carried by a partially ionized, differentially rotating solar nebula, but would be inconsistent with a dipole field of solar or nebular origin, that would be normal to the rotational axis at the ecliptic plane.

The general absence of strong magnetization associated with achondrites, particularly the unbrecciated varieties, presents several problems regarding the evolution of the early solar system. One is that the strong fields implied from the magnetizations of chondrites were apparently absent during differentiation of the achondrite parent body. This is suprising in that many achondrite radiometric ages are not appreciably different from those of chondrites (see Table I in the chapter by Wood and Pellas), while the chemistry of some achondrites indicates that upward migration and cooling of partial melts occurred even as the interior of the body was continuing to be heated, presumably in part by electrical induction (Mittlefehldt 1979).

The other extreme of meteorite paleointensity interpretation is to accept only the low-intensity paleofield results associated with those samples with the least complicated magnetic mineralogies and thermal histories, the achondrites. This produces the less satisfying conclusion that there is no strong evidence for the existence of intense, large-scale magnetic fields at the time of differentiation of the achondritic parent bodies. However, even this conservative approach leaves open the possibility that strong magnetic fields did exist at a somewhat earlier time, but that deciphering of an earlier magnetic record awaits a better understanding of the mechanisms of remanence acquisition of chondritic meteorites, and their thermal histories. One might even argue that a rapid spin rate of the achondrite parent body would preclude recording of a solar-wind field parallel to the ecliptic (Sonett 1978), even for rapidly cooled eucrites. Presumably only detailed thermal demagnetization experiments, as performed by Nagata and Funaki on the ALHA 76009 chondrules, could reveal a time-varying orientation of basaltic achondrites in an intense T Tauri solar-wind field.

Finally, it must be remembered that small-scale processes are also likely to have contributed to the magnetization of some meteorites, particularly the ordinary chondrites. In most or all cases, impact-generated magnetic fields are too transient to be preserved as a simple thermoremanence acquired by slow cooling through the Curie blocking spectrum. Thus, shock and/or rapid post-shock cooling must be considered as more probable remanence mechanisms. Ordinary chondrites have often experienced strong shock events that may have resulted in significant acquisition of magnetic remanence in the presence of a transient magnetic field. However, the thermal and shock histories of these meteorites are complex and the nature of their magnetic remanence carriers is only marginally understood. Only when both the physical histories and thermomagnetic properties are fully unraveled will it be possible to identify the magnetizing field as transient or steady.

Acknowledgments. The authors thank T. Nagata and P. J. Wasilewski for discussions which contributed to our interpretation of meteorite magnetic data and their significance to the early solar system magnetic environment.

PART VI
Solar-like Stars

THE LUMINOSITY VARIABILITY OF SOLAR-TYPE STARS

RICHARD R. RADICK
Solar Research Branch, Geophysics Directorate (AFSC)

High-precision photometric measurements of stellar variability offer insight into the past history, present state, and future course of solar brightness changes. Both dynamo theory and a wide range of observational evidence argue that valid stellar proxies for the Sun exist and may be discriminated by spectral type and rotation rate. Observations further reveal that the brightness variability of a solar-type star decreases with age, in step with its rotation rate. The brightness variations exhibited by the Sun during its youth were probably 10 to 100 times stronger than the present, relatively feeble fluctuations of 0.1% or so detected during the past decade by spacecraft instruments.

I. INTRODUCTION

The past decade was marked by the first unambiguous detections of solar irradiance variability by radiometers aboard the Solar Maximum Mission and Nimbus 7 spacecraft (chapter by Fröhlich et al.). Contemporary solar observations, paleoclimatic studies, and analyses of lunar and meteoritic samples, however, yield only limited understanding of solar variability and its evolution. Fortunately, we can augment this incomplete picture by studying stars similar to the Sun. Assuming that valid stellar proxies for the Sun exist, photometric observations of such stars can provide a unique probabalistic description of past, present and future solar variability.

As the term "luminosity" signifies the total radiative output of a star, a "luminosity variation" must, strictly speaking, involve some real change in that output. In contrast, the term "irradiance" denotes the radiative flux received at some specific place; in particular, it often designates the solar flux incident on Earth. In this sense, it is used as a substitute for the older term: "solar constant." Because irradiance variations can be induced by anisotropies in the emitted flux, such variations do not necessarily imply luminosity changes. For example, rotational modulation, which is a prominent constituent of solar irradiance variability, is a phenomenon that need not involve any fluctuation in intrinsic luminosity.

From a geophysical perspective, the distinction between luminosity and irradiance variability is often not very critical: what matters is simply that the amount of solar radiation arriving at the Earth is changing. The astrophysicist, who seeks to identify and study the physical mechanisms underlying the variability, worries more about the difference. Because it is often somewhat unclear, for the Sun as well as the stars, whether or not we are dealing with true luminosity variations, and because the term "irradiance" is now strongly associated with the Sun, the word "brightness" offers a convenient term for use whenever the specificity implied by the term "luminosity" is too restrictive. It will appear frequently.

This chapter focuses on solar and stellar brightness variations on time scales ranging from days to years, and the evolution of these variations during the main-sequence lifetime of a solar-type star. Transient effects caused by flare-like events will not be considered, nor will the long-term secular luminosity increase associated with interior evolution (chapter by Dearborn). After reviewing the physical causes and characteristics of solar irradiance variability (Sec. II), we will consider criteria for differentiating valid stellar proxies for the Sun (Sec. III). A discussion of modern, high-precision stellar photometry (Sec. IV) is then followed by a detailed description of the photometric variability of a solar-type star (and, by analogy, the Sun) at four points during its life: (1) shortly after its arrival on the main sequence, i.e., a Pleiades-age "infant" Sun, several times 10 Myr old; (2) a Hyades-age "young" Sun, several times 100 Myr old; (3) a "mature" star like the present Sun, several times 1 Gyr old; and (4) an "aging" Sun, approaching the end of its main-sequence evolution at ~10 Gyr (Sec. V). The concluding summary highlights the evolutionary trends implicit in the preceding snapshots and points out several areas where our current understanding remains especially incomplete (Sec. VI).

II. THE SUN AS A VARIABLE STAR

It is widely believed that solar brightness variations on time scales ranging from days to years are primarily caused by the presence and evolution of active regions on the solar surface. Young active regions typically contain dark sunspots as their most prominent features. Sunspots have long been known to contain intense magnetic fields (Hale 1908), and are dark because these fields interfere with convective energy transport beneath the solar surface, or photosphere. Surrounding the sunspots are extensive aggregations of bright, but very small, faculae. The faculae apparently mark where individual magnetic flux tubes penetrate the solar photosphere. These flux tubes are too small to inhibit convection seriously, so they are not dark. Instead, they appear bright, because they increase the local emissivity by exposing hotter gas underlying the surrounding photosphere (the physics of magnetic structures on the Sun is reviewed in the chapter by Spruit). Excess heating of the solar chromosphere above the facular regions creates emission plages, which are easily detected by their enhanced radiation in certain spectral lines, such as the H and K resonance lines of singly ionized calcium (Ca II H and K). As an active region ages, its spots disappear, and its faculae disperse, merge with the active network, and gradually also disappear. Large sunspots can live for several weeks, a time scale which accidentally corresponds fairly closely to the approximately 25-day (sidereal) solar rotation period, and the faculae of an active region may persist for a few months. The Sun characteristically shows azimuthal nonuniformities in the distribution of its active regions, which it maintains by its tendency to create successive active regions in relatively fixed locations, or activity complexes, that can last for several months or even years (Bumba and Howard 1965; Gaizauskas et al. 1983)—much longer than the lifetime of an individual active region.

Variations in total solar irradiance involve radiative flux deficits produced by sunspots, flux enhancements associated with active region faculae, and additional contributions from other bright magnetic network elements. The first positive detections of such variations were made by the Active Cavity Radiometer Irradiance Monitor (ACRIM) and the Earth Radiation Budget (ERB) radiometer experiments aboard the Solar Maximum Mission (SMM) and the Nimbus 7 spacecraft, respectively (Willson et al. 1981; Hickey et al. 1981). On the time scale of days to weeks, dark sunspots and bright faculae cause solar irradiance fluctuations amounting to 0.1% or so, with sunspots producing the most obvious effects. These variations arise mainly from the emergence and decay of individual solar active regions, and from solar rotation that carries these features into and out of view on the solar disk.

The discovery of solar irradiance variability by SMM/ACRIM and Nimbus 7/ERB was quickly followed by the elaboration of competing irradiance models, which disagreed, in particular, about the importance of the facular contribution, and whether or not there exists any fundamental physical connection between the sunspot deficit and the facular excess from any particular active region (see, e.g., Sofia et al. 1982; Hudson et al. 1982; Hoyt et al. 1983; see also the papers and, especially, the discussion transcripts in LaBonte et al. eds. [1983]). The latter dispute has not yet been fully settled, although it is now generally agreed that the facular excess roughly balances the sunspot deficit over the entire lifetime of a solar active region (Chapman et al. 1986; also Foukal and Lean 1986).

The relative prominence of the sunspot signature in the early SMM/ ACRIM radiometry helped create a widespread (but mistaken) expectation that the Sun would become *brighter* as its 11-yr activity cycle waned and sunspots grew scarce (see, e.g., Eddy 1983). In fact, sustained measurements from both SMM/ACRIM and Nimbus 7/ERB showed that the Sun became *fainter* by almost 0.1% between activity maximum (1980) and activity minimum (1986), and then started to recover as solar activity began its current rise (Willson and Hudson 1988). This behavior implies that solar irradiance variations on the time scale of the 11-yr solar activity cycle are dominated by excess radiation from faculae and bright magnetic network features, rather than the sunspot deficit (de Charentenay and Koutchmy 1985; Foukal and Lean 1988). Alternatively, it has also been suggested that the effective temperature of the activity belts varies slightly, in step with the activity cycle (Kuhn et al. 1988).

Because solar activity complexes long outlive the sunspots and plages of individual active regions, the rotational modulation signal found embedded in solar time-series measurements can remain relatively stable in frequency, phase and amplitude for many solar rotation periods (cf. the chapter by Morfill et al.). Of course, neither the presence of rotation modulation nor even a change in its waveform necessarily implies true luminosity variability: one could imagine the same permanent markings drifting about on the solar surface, coalescing at one time to accentuate the rotational modulation and dispersing at another to attenuate it, but always keeping their total number and contrast (and, therefore, the luminosity) fixed. We know this does not happen on the Sun, but only because we can resolve its surface features and watch them change. This is not easily done in stellar studies. The distinctive periodic signature of rotational modulation, however, does make it a powerful tool for measuring stellar rotation. This is particularly true for stars that rotate as slowly as the Sun ($\sim$ 2 km s^{-1}), which make difficult targets for spectroscopic measurements of rotational velocity.

Before focusing our attention on the photometric brightness variations of other stars, we should remind ourselves how pervasively the Sun influences our thinking and expectations. To start with, the Sun provides our vocabulary: spots, faculae, plages, active regions, chromospheres, coronae—such terms have all been borrowed from solar astronomy, and underlie the conceptual framework used to interpret a wide range of stellar observations. The Sun also suggests possible variability mechanisms, such as rotational modulation, active region evolution and activity cycles. We must beware, lest the power and familiarity of the solar example entice us into forcing every stellar observation to fit a preconceptual mold based on the Sun. Indeed, it seems prudent to seek examples of stellar behavior that deviate from the solar paradigm, in order to test its generality, establish its limits, and avoid traps of circular reasoning.

III. STELLAR PROXIES FOR THE SUN

The contemporary branch of astrophysics known as "the solar-stellar connection" has accumulated persuasive evidence arguing that magnetic activity following a predictable evolution is a normal feature of lower-main-sequence stars, including the Sun. This evidence may be consolidated under three propositions which, in turn, offer criteria for differentiating valid stellar proxies for the Sun. These propositions are: (A) the magnetic activity of a lower-main-sequence star is governed by its rotation, mass and chemical composition, (B) the surface rotation rate of a lower-main-sequence star quickly loses its memory of initial conditions, after which time it is specified by stellar mass and age alone, and (C) the Sun is an ordinary lower-main-sequence star, and has experienced the usual evolution of such a star. Discussion of each of these propositions follows.

A. The Influence of Rotation, Mass and Chemical Composition on Stellar Magnetic Activity

Although it was first proposed over seventy years ago that the surface magnetic fields of the Sun are produced by dynamo action (Larmor 1919), theoretical understanding of the process remains far from complete (for a recent review, see Weiss 1989). Current theory suggests that the solar dynamo is seated in or immediately beneath the Sun's convective envelope. The key ingredients of the dynamo are rotation and convection, which interact to produce both differential rotation and cyclonic eddies or helicity. Differential rotation readily converts poloidal magnetic flux into toroidal flux and amplifies it, while helicity twists toroidal flux, thereby regenerating reversed poloidal field. Something resembling this oscillatory, self-sustaining process is generally believed to underlie the 11-yr solar sunspot cycle.

On both theoretical and experimental grounds, it appears likely that the basic convective pattern in the outer envelope of a lower-main-sequence star changes as stellar rotation rate decreases, from something unlike anything observed on the solar surface, namely, a regular distribution of elongated "banana" cells, or rolls, aligned parallel to the stellar rotation axis, to a more irregular, cellular pattern at least vaguely reminiscent of the solar granulation and supergranulation patterns (Knobloch et al. 1981; Hart et al. 1986). This transition should affect both differential rotation and the behavior of the highly nonlinear stellar dynamo: the amplification of surface magnetic fields is greater for more rapidly rotating stars, but their dynamos may not be reversing. Thus, there is reason to suspect that the magnetic activity and the attendant brightness variability of young, rapidly rotating stars may not be simply more vigorous versions of what is now observed on the Sun. Instead, the young Sun may have shown strong, noncyclic activity with relatively complicated spatial and temporal behavior, and the infant Sun may have exhibited wildly irregular activity.

In any case, dynamo theory predicts that magnetic-field amplification should vary strongly and directly with stellar rotation rate, and observations agree that rapidly rotating stars show relatively more vigorous magnetic activity. Thus, the existence of a relation between magnetic activity and rotation among lower-main-sequence stars is widely accepted, even though the empirical details of the relation are still being debated (see, e.g., Noyes et al. 1984b; Basri 1987a; Radick and Baliunas 1987; Rutten and Schrijver 1987; Dobson and Radick 1989; chapter by Walter and Barry). In practice, the activity-rotation relation is indirectly expressed in terms of chromospheric or coronal emission rather than magnetic flux or field strength, because direct measurements of stellar magnetic fields are both difficult and scarce (chapter by Saar). Strictly speaking, commonly observed activity indicators, such as chromospheric Ca II H + K line-core emission or coronal soft X-ray emission, prove only that a star's outer atmosphere is being heated by nonradiative processes. Solar studies, however, have demonstrated that regions of enhanced chromospheric and coronal emission coincide with underlying concentrations of photospheric magnetic flux, and that emission strength scales directly with mean field strength in these regions (Skumanich et al. 1975; Poletto et al. 1975; Schrijver et al. 1989b). Thus, although temperature, surface gravity and chemical composition probably all effect the complex and incompletely understood connection between surface magnetic flux and its radiative indicators, these indicators are, nevertheless, widely considered to be reliable diagnostics of stellar magnetic activity.

Early investigations of stellar activity tended to focus on its relation to stellar age, and modern discussions still often emphasize this aspect. In fact, both mean activity and surface rotation rate decline with age in nearly identical ways among lower-main-sequence stars. The classic power-law dependence (Ca II H + K emission and rotation rate are both proportional to the inverse square root of stellar age for stars of similar mass: Skumanich 1972) remains a simple and useful rule for lower-main-sequence stars throughout much of the age range considered in this chapter, despite the fact that it fails for very young stars, including pre-main-sequence stars, as well as for some exotic binary star varieties. Therefore, it seems natural to ask whether stellar activity is more fundamentally linked to rotation rate or age. The clearest answer comes from a class of close binary stars, the RS CVn systems (Hall 1976). These stars maintain rapid rotation as they age by drawing on the angular momentum reservoir of their orbital motion through tidal interaction (DeCampli and Baliunas 1979). They also maintain high levels of activity, thereby proving that the causal determinant for stellar activity is rotation, rather than age.

On theoretical grounds, stellar mass should also strongly affect magnetic activity. For example, the dynamo number, which parameterizes the efficiency of the hydromagnetic dynamo in mean field theory (Durney and Latour 1978; Noyes et al. 1984b), depends as strongly on the time scale for convective turnover at the base of the stellar envelope as it does on rotation rate. This time scale in turn depends strongly on stellar mass, increasing monotonically with decreasing mass among lower-main-sequence stars and rising especially steeply between mid F- and mid G-type main-sequence stars (Gilliland 1985). Thus, lower-mass stars should be relatively more active, among otherwise comparable stars. Outside the realm of theory, however, the dependence of activity on mass has engendered protracted debate, with the dispute centering on whether activity is better measured in terms of flux (see, e.g., Rutten and Schrijver 1987), or as a nondimensional ratio relative to stellar bolometric flux (see, e.g., Noyes et al. 1984b). On the one hand, the empirical activity-rotation relation is found to show little mass dependence when activity is measured in terms of flux (cf. Noyes et al. 1984b, Fig. 4). On the other, mass influences stellar activity about as strongly as rotation when normalized ratios are used (Radick and Baliunas 1987). Accordingly, until the issue of how best to measure activity is settled, the importance of mass as a direct determinant for stellar magnetic activity will also remain uncertain.

Like stellar mass, chemical composition strongly affects convective zone structure, and should therefore influence magnetic activity. In brief, a metal-rich star will have a deeper convective zone and a longer convective turnover time scale than a lower-Z star of the same mass (see, e.g., Ruciński and VandenBerg 1986). In this respect, metal enhancement mimics lower mass, and a metal-rich star should be relatively more active than its lower-Z counterpart of comparable mass. In practice, however, stars of strictly equal mass are not easy to select, because stellar masses are generally estimated from intermediaries, such as photometric color indices, that are also sensitive to metallicity. Indeed, both models (Ruciński and VandenBerg 1990) and observations (Radick et al. 1990a) suggest that the various effects conspire in such a way that a more massive, metal-rich star and a less massive, lower-Z star with the same $B - V$ color index will show comparable activity and photometric variability.

Even if chemical composition had no intrinsic effect on stellar magnetic activity, it would enhance the apparent activity of metal-rich stars through its influence on radiative activity diagnostics such as chromospheric Ca II H + K emission. Calculations (Restaino 1990) suggest that Ca II H + K emission depends strongly and directly on metallicity. Emission indices, such as the Mount Wilson S-index, which measures the ratio of the combined fluxes in two 1-Å continuum windows centered on the H and K lines to the sum of the fluxes in two 20-Å continuum windows on either side of the doublet (Vaughan et al. 1978), could respond even more strongly because of enhanced line-blanketing effects outside the cores of the lines. Not surprisingly, the identification, separation and measurement of the real and apparent effects of composition on stellar magnetic activity also remains an unfinished task.

B. The Dependence of Surface Rotation Rate on Stellar Age and Mass

Whereas the magnetic activity and variability of ordinary lower-main-sequence stars depends strongly on rotation, rotation itself is governed by age and mass (Stauffer and Hartmann 1986; the chapter by Stauffer and Soderblom reviews the rotation of lower-main-sequence stars). Once such a star arrives on the main sequence, its surface rotation rate slows as the star's magnetically coupled wind carries away angular momentum (Schatzman 1962; Kawaler 1988; Pinsonneault et al. 1989). This magnetic braking is strongly dependent on rotation rate, so the deceleration is, initially, very rapid. Theory suggests that the rotational velocity of a solar mass star becomes independent (to a few percent) of initial angular momentum within 100 Myr (Kawaler 1989), and measured rotation periods for stars somewhat older than this in the Hyades and Coma Berenices clusters (Radick et al. 1987; Radick et al. 1990a), which show little scatter at any given color, support this conclusion. After the early spin-down episode is completed, the star's rotation continues to slow, albeit at a more leisurely rate. Observations of cluster stars imply that the Pleiades-age "infant Sun" rotated 5 to 10 times faster than the present Sun (Benz et al. 1984; Stauffer et al. 1984; Stauffer and Hartmann 1987), while the Hyades-age "young Sun" rotated only ~ 3 times faster than the present Sun (Radick et al. 1987). The activity-age relation suggests that the "old Sun" will still rotate at perhaps half its present rate (Soderblom 1985).

The surface rotation rate of lower-main-sequence Hyades stars shows some dependence on stellar mass. Among these stars, rotational velocity declines from ~ 7 km s^{-1} at $B - V = 0.6$ (G0V) to ~ 4 km s^{-1} at $B - V = 1.2$ (K5V), and shows a peak-to-peak scatter of only ~ 2 km s^{-1} at any given color within this range (Radick et al. 1987). This trend in mean rotation rate may simply be a relic of the initial spin-down episode, because the temporal evolution of rotation does not appear to be strongly mass dependent after an age of 100 Myr or so (Soderblom 1985; Kawaler 1989). Until direct measurements of rotation for stars in an old cluster such as M67 are made to complement the currently available observations of Hyades-age stars, however, this assertion will remain perched rather insecurely on age estimates for field stars. At the other extreme, elucidation of the mass dependence of rotational evolution for stars much younger than about 100 Myr is complicated by the breakdown of the coevality assumption for very young cluster stars: differences in the contraction time scale for stars of different masses make it difficult to assign accurate main-sequence lifetimes to very young stars. Because the rotation rates of stars of comparable mass in the Hyades and Coma Berenices clusters, which are believed to be about the same age, are essentially indistinguishable (Benz et al. 1984; Radick et al. 1990a), stellar rotation must not be sensitive to modest differences in chemical composition.

C. The Sun: an Ordinary Lower-Main-Sequence Star

In terms of its mass, age and chemical composition, the Sun appears to be a rather ordinary star, a conclusion that is reinforced by the fact that a single theory of stellar structure and evolution suffices for both the Sun and other stars. In other words, the Sun appears to obey the Russell-Vogt theorem, which states that the structure of any star of some specified age depends only on its initial mass and chemical composition.

Since the Russell-Vogt theorem excludes both rotation and magnetic field as primary factors affecting stellar structure and evolution, however, it also implies that magnetic activity is, in some sense, merely an incidental characteristic of stars. Thus, adherence to the Russell-Vogt theorem constitutes no certain proof that the Sun is normal in terms of its rotation or magnetic activity, the properties that most strongly influence its brightness variability. Indeed, it has been suggested that solar rotation and chromospheric Ca II H + K emission are both well below the mean values shown by nearby field stars of similar age and spectral type (Smith 1979; Blanco et al. 1974). More recent investigations have concluded that neither is the case (see, e.g., Soderblom 1983,1985); the Sun appears to obey the same activity-rotation-age relations that describe the behavior of other lower-main-sequence stars.

What is now clear is that chromospheric and coronal activity are characteristic of most, if not all, ordinary lower-main-sequence stars, and that the older stars among these, at least, also often exhibit activity cycles that are strongly reminiscent of the 11-yr solar cycle (chapter by Baliunas). Activity and variability observations of ordinary lower-main-sequence stars are almost always interpreted in terms of familiar solar phenomena, namely, magnetic active regions, containing both dark spots and bright emission features, distributed nonuniformly over the stellar surface. A notable exception is the "starpatch" model (Toner and Gray 1988), which was developed recently to reconcile peculiarities in the observed activity of a young, rapidly rotating star, HD131156A (ξ Bootis A, spectral type G8V). Although the proposed starpatch, like a sunspot, is dark, it is also distinguished from a sunspot by its high latitude, very large size, low contrast, long lifetime, apparent lack of strong magnetic fields, and its tendency to enhance, rather than suppress, surface velocity fields. Indeed, the proposed starpatch in several ways resembles more closely an activity complex, or even a coronal hole, rather than a single active region. The behavior of the velocity fields, in particular, is a puzzle.

A bolder challenge to the Sun's assumed normalcy is posed by the recent suggestion that the early Sun shed up to half its mass during the first few 100 Myr of its main-sequence existence (Willson et al. 1987; Guzik et al. 1987). Thus, this mass-loss hypothesis implies that some older, solar-mass stars (including the Sun) started out their lives as A-type stars, whereas others were always G-type stars. Under such circumstances, of course, it seems unlikely that a group of stars destined to become solar-type stars would all follow the same rotational evolution, as the magnetic braking experienced by stars of significantly dissimilar mass differs substantially. In other words, a group of Hyades-age solar-mass stars might be expected to exhibit a rather wide spread in rotation rate, with the mass-losing late arrivals rotating more rapidly. Likewise, one might expect the lithium depletion occurring among such a group of stars to proceed at different rates, with the late arrivals being relatively lithium-rich. Neither effect is observed. As mentioned previously, the dispersion in rotation among both Hyades and Coma stars is actually rather small, and, although the observed lithium abundance of Hyades and Coma stars shows interesting behavior as a function of spectral type (Boesgaard and Tripicco 1986; Boesgaard 1987), the dispersion at any given color is, again, rather small. Inconsistencies such as these suggest that the mass-loss hypothesis remains far from established, even though it does offer an interesting challenge to the current orthodoxy.

In summary, currently available evidence, both theoretical and observational, tends to support the expectation that valid stellar proxies for solar activity and its attendant brightness variability do exist. The prescription for selecting solar analogs is, in fact, quite simple: choose solar-type (i.e., solar-mass) stars rotating at the solar rate. Chemical composition, while important in principle, appears to be of less practical concern, perhaps because the composition of stars in the solar neighborhood is too homogeneous for its influence to be strongly evident in activity and variability measurements. Furthermore, good proxies for the early Sun should be available among young solar-type stars. The existence of a tight relation linking stellar age and surface rotation rate suggests that the ages of these solar proxies can be satisfactorily established from measurements of their rotation. Of course, the ages of solar-type cluster stars can also be estimated, independent of rotation, from evolutionary considerations.

IV. HIGH-PRECISION STELLAR PHOTOMETRY

Conventional stellar photometry typically satisfies itself with a measurement precision of 1% or so, which is orders of magnitude poorer than that achieved by the SMM/ACRIM and Nimbus 7/ERB radiometry, and also clearly inadequate for detecting signals as weak as those provided by the Sun—~ 0.1%. Furthermore, groundbased photometry is unlikely ever to match the precision attained by spaceborne radiometry, because even the best stellar photometry must contend with a host of deleterious effects, each capable of contributing 0.1% (0.001 mag) or more to the overall noise budget (see, e.g., Young 1974; Rufener 1985; Lockwood and Skiff 1988). Some of these, such as photon and scintillation noise, can be reduced to arbitrarily low levels by using larger telescopes or longer integration times. Others, such as extinction effects, are more difficult to compensate, at least in the context of sequential, single-star, photoelectric photometry. Instrumental effects, which can arise from a variety of thermal, electrical and mechanical instabilities, can be particularly bothersome. Perversely, attempts to cure one problem can sometimes exacerbate others. For example, use of small photometric apertures to limit sky light can accentuate noise from variable seeing, or from centering and tracking errors. Despite such handicaps, it appears that the total measurement imprecision of stellar photometry can be reduced and held to a few tenths of a percent when good equipment and conservative observing procedures are used in conjunction with proper calibration, quality control, and data reduction techniques. The Geneva Observatory photometric system, which has sustained a measurement precision of about 0.5% (0.005 mag) for over 20 yr, is a notable example of this discipline (see, e.g., Rufener and Nicolet 1988; Rufener 1989). The differential Strömgren b,y photometry from Lowell Observatory (described more fully below) has achieved comparable quality. Although precision can be improved somewhat further through the statistical leverage of numerous observations, measurement error remains an important limiting factor for current investigations of stellar brightness variability.

The dominant component in the error budget for Lowell Observatory photometry, for example, appears to be photon statistics, followed by centering and tracking effects (Lockwood and Skiff 1988). A total of at least 90,000 counts, compensated for sky, is demanded for all stars by increasing as necessary the integration times for fainter stars (Radick et al. 1987). Thus, the contribution to the nightly measurement error from photon statistics is 0.3% (0.003 mag) or less, an amount comparable to the total *rms* dispersion (0.2 to 0.3%) found to characterize observations of stars that show no sign of variability. The observing season for any given star typically spans several months, during which time 20 to 30 observations are acquired. Thus, the mean standard deviation of the seasonal average should approach 0.1%, assuming the presence of no additional systematic instrumental or stellar effects. This long-term photometric precision is, indeed, attained by the Lowell measurements.

High-precision stellar photometry is relative photometry: stars of interest are compared to other (presumably) nonvariable standard stars. Of course, the stability of these photometric standards is hardly assured, especially at the high-precision limit, and stellar photometrists are now beginning to discover just how common low-level variability really is among certain groups and spectral types of stars. For example, it now appears that virtually all young stars later than mid-F spectral type are variable at a level of 1 to 3% on time scales ranging from days to years (see, e.g., Radick et al. 1990b). Among lower-main-sequence field stars of all ages, the incidence of photometric variability at the level of 1% or greater appears to be about 33% (Lockwood and Skiff 1988). Unfortunately, the contaminating presence of a variable comparison star is often discovered well after the observations have begun. High-precision CCD photometry of ensembles of stars offers an effective way to bypass this problem, provided adequate long-term measurement stability can be achieved and sustained (Gilliland and Brown 1988; Frandsen et al. 1989). For photoelectric measurements, a simple way to guard against the hazard is to observe more comparison stars: at Lowell Observatory, for example, stars are typically observed in groups of four, with two comparison stars included in each quartet.

Four stars provide six different pairings, so the observations of a quartet form six separate (but not independent) differential time series [e.g., (star 1-star 2), (1–3), (1–4), (2–3), (2–4), and (3–4)]. Since the nightly instrumental error (0.2 to 0.3%) is known, a measured dispersion significantly larger than this amount is evidence that the pair in question includes at least one variable star. A more sophisticated diagnostic is to compute correlation coefficients for pairs of lightcurves involving a common star [e.g., for star 3, (1–3) vs (2–3), (1–3) vs (3–4), and (2–3) vs (3–4)], and formally assess the statistical significance of these coefficients. Periodic variability arising from rotational modulation is often easy to spot through periodogram analysis, and the inferred rotation can frequently be confirmed through spectroscopic measurements of Doppler line broadening.

Broadband photometric brightness measurements are certainly not the easiest way to monitor stellar activity. For example, variations in Ca II K-line emission from the solar chromosphere are much stronger than the minute continuum brightness fluctuations associated with sunspots and faculae. Observations have demonstrated that the rotational signal present in solar Ca II K-line time series can have an amplitude as large as 7% during activity maximum, and that the amplitude of the activity cycle itself approaches 20% (Lean et al. 1983; Keil et al. 1990). Since stellar Ca II H + K observations routinely achieve a precision of 1% (chapter by Baliunas), the stellar S/N ratios should lie in a comfortable working regime, if solar and stellar signals are comparable. In contrast, the S/N ratios available from broadband stellar photometry are likely to be substantially less favorable.

V. THE PHOTOMETRIC VARIABILITY OF SOLAR-TYPE STARS

A. Stars Several Times 10 Myr Old: The "Infant" Sun

Our present picture of the photometric variability of very young lower-main-sequence stars is based on three principal sources: (1) observations of G- and K-type stars in the α Persei cluster (age ≈ 50 Myr: Mermilliod 1981); (2) extensive, but heterogeneous, observations of numerous stars in the Pleiades cluster (age ≈ 80 Myr: Mermilliod 1981); and (3) intensive observations of HD129333, an early G-type member of the Pleiades moving group.

1. Stars in the α Persei Cluster.   Broadband photometric observations of six G- and K-type members of the α Persei cluster, obtained over a 12-day interval, show all six to be variable (Stauffer et al. 1985). The brightness variations arise primarily from rotational modulation. Five of the six stars, all with visual (*V*-band) peak-to-peak variability amplitudes ranging from ~10% to over 15% (0.10 to 0.15 mag), are rotating very rapidly, with periods of less than one day. The sixth, with a variability amplitude of about 7%, has a rotation period of ~ 5 days. The photometric periods are consistent with spectroscopic measurements of rotational velocity. Thus, this small sample of very young main-sequence stars shows one of the characteristic properties of stellar activity: brightness variability, like other diagnostics of magnetic activity, tends to correlate with rotation rate. This behavior is apparently established very early in the main-sequence lifetime of solar-type stars.

2. Stars in the Pleiades Cluster.   The existence of photometric brightness variations at levels of 5 to 15% on time scales of hours to days among Pleiades dwarfs in the spectral-type range K3V-M0V was first discovered over 15 yr ago (Robinson and Kraft 1974). More recently, another survey, "Cloudcroft," showed that photometric variability at levels of 3 to 15% (peak-to-peak) on time scales of days to weeks also occurs among Pleiades stars in the spectral-type range F8V-K2V (Radick et al. 1982; Radick et al. 1983*a*), with the variability amplitude increasing rather sharply toward the cooler end of this range. The Cloudcroft survey, which spanned 2 yr, also demonstrated that the variability is persistent, with episodes lasting several months or even longer. At about the same time, a third photometric study, "ESO," of three late G-type and 16 K-type Pleiades members completed this initial exploration of variability among lower-main-sequence Pleiades stars (Alphenaar and van Leeuwen 1981; van Leeuwen and Alphenaar 1982; van Leeuwen 1983). The ESO study sustained much denser temporal coverage than did the Cloudcroft survey. All 19 of the ESO program stars were found to vary on short time scales, with peak-to-peak *V*-band amplitudes ranging from ~2% to > 20%. In addition, two of seven ESO comparison stars (both F9V) were found to vary with amplitudes of ~ 2%. Four of the 19 ESO program stars happened to be in the Cloudcroft survey. Of these, three, all of which vary at low amplitude, were not judged to be variable from the Cloudcroft survey. The fourth, with an amplitude of ~ 20%, was found to be the most strongly variable star in both studies. Four of the seven ESO comparison stars, including the two variable ones, were also in the Cloudcroft survey, and one of these two was also judged to vary from that survey.

The ESO observations showed that rotational modulation is a prominent component of the photometric variability of Pleiades stars, just as it is among the somewhat younger stars of the α Persei cluster. Rotation periods ranging from less than a quarter day to slightly over one day were measured for 12 of the 19 late G- and K-type ESO program stars; one of the two variable F-type ESO comparison stars showed a period of ~1 day, also. Like the α Persei stars, Pleiades stars show a strong positive correlation between variability amplitude and rotation rate.

Five of the 19 ESO program stars were observed during two consecutive seasons. None of these showed any significant change in lightcurve period, although changes in lightcurve waveform occurred for two of them. Three showed changes of 2 to 3% in mean brightness. The Cloudcroft survey also uncovered evidence of year-to-year changes in mean brightness amounting to a few percent among F-type Pleiades stars. Thus, the surface markings on these very young stars are not completely static features.

Subsequent studies (Stauffer 1984; Stauffer et al. 1984: Stauffer and Hartmann 1987) have confirmed that the photometric periods obtained from the ESO study are consistent with spectroscopic rotational velocities, and have also revealed the photometric variability of several additional late G- and K-type Pleiades stars. Attention has been strongly focused on the rapidly rotating K-type stars, however. No further photometric or rotational studies of more strictly solar-type Pleiades stars have been made in recent years.

3. HD129333. Fortunately, this deficiency has been neatly covered by two recent, complementary, long-term studies of a relatively bright $V \approx 7.5$) G0V field star, HD129333 (Lockwood and Skiff 1988; Dorren and Guinan 1990). This star first attracted attention because it has the highest known chromospheric Ca II H + K emission level of any single, early G-type star (Soderblom 1985). Furthermore, its space motion suggests that it is a member of the Pleiades moving group (Soderblom and Clements 1987), which, if true, establishes its age at about 80 Myr. Its high lithium abundance and rapid rotation offer additional evidence of its youth.

Differential photometric observations of HD129333 relative to the same comparison star have been obtained from several observatories since 1983. Fortunately, this common comparison star appears to be exceptionally stable (Lockwood and Skiff 1988), so there is little doubt that the observed variations originate with HD129333. The short-term photometric variability of HD129333 typically has a peak-to-peak amplitude of ~ 5% in the visual (Strömgren y-band), and ranges from ~ 1% to > 7%. The typical amplitude in the ultraviolet (Strömgren u-band) is larger: ~ 8%. Major changes in the amplitude of variation can occur within a few days. Like other very young stars, much of the day-to-day variation of HD129333 arises from rotational modulation. The rotation period, which has been measured to a precision of 1%, has ranged between 2.70 and 2.80 days at various times between 1983 and 1988. This variation in period has been adduced as evidence for differential rotation (Dorren and Guinan 1990). There is evidence that both Ca II H + K and Hα emission becomes stronger as the photometric lightcurve approaches brightness minimum, which suggests that the stellar surface mark-

ings comprise bright emission features accompanied by dark spots, a configuration reminiscent of solar active regions.

Between 1983 and 1988, the mean brightness of HD129333 declined by
~ 5% in the visible continuum, with much of this change occurring during
the 1985–1987 interval. At the same time, its mean Ca II H + K emission was
was increasing, especially during 1985–1987. Thus, unlike the Sun, this very
young star becomes fainter as its mean level of chromospheric activity rises.
Between 1988 and 1989, the star's mean brightness recovered by almost 2%.
Although it has been suggested that these year-to-year variations mark an
activity cycle with a time scale of 8 yr or so (Dorren and Guinan 1990), the
fact that young, active stars generally do not show obvious, smooth, regular
activity cycles in Ca II H + K emission (chapter by Baliunas) dictates caution
in drawing this conclusion until several cyclic repetitions have actually been
observed.

In summary, very young solar-type main-sequence stars commonly display short-term brightness variations ranging from 2 to 15% in the visible
continuum. Rotational modulation is a prominent component of this short-
term variability, although effects caused by the evolution or, at least, the redistribution of surface features are also certainly present. These stars also
show longer-term changes of several percent in mean brightness. Such year-
to-year variations must reflect real temporal changes in overall coverage or
contrast arising from the evolution of the stellar surface markings, and represent true luminosity variations. The surface markings on these very young
stars bear at least some resemblance to solar active regions, insofar as they
appear to contain both dark spots and bright emission features.

B. Stars Several Times 100 Myr Old: The "Young" Sun

Our understanding of the photometric variability of somewhat older
solar-type main-sequence stars is grounded on: (1) observations of HD39587
(χ^1 Ori, spectral type G1V), a member of the Ursa Major group (age ~ 270
Myr: Giannuzzi 1979); (2) extensive and largely homogeneous observations
over the past decade of numerous stars in the Hyades cluster (age ~ 660 Myr:
Mermilliod 1981) and HD1835 (9 Cet, spectral type G2V), a member of the
Hyades group; (3) four years' observations of several members of the Coma
Berenices cluster, which is generally believed to be about Hyades-age, or a
bit younger; and (4) observations during the past decade of various field stars
whose comparative youth is implied by their relatively high activity.

1. HD39587. The young, solar-type star HD39587 shows short-term
brightness variations that have a typical peak-to-peak amplitude of ~ 3% in
the visual continuum (Strömgren *b*- and *y*-bands) (Lockwood and Skiff 1988;
Dorren and Guinan 1990). Although rotational modulation contributes to
these variations (the measured rotation period is ~ 5 days), its prominence
relative to other sources of short-term variation, both stellar and instrumental,

seems to be somewhat less than it is among the very young, Pleiades-age stars just considered. The star also shows year-to-year changes in mean brightness that span a total range of $\sim 2\%$ over 6 yr, but never amount to more than $\sim 0.5\%$ between two consecutive seasons (Lockwood and Skiff 1988).

2. *Stars in the Hyades Cluster.*   The most sustained of all photometric studies of lower-main sequence stars, now completing its tenth year, involves the Hyades, a nearby cluster $\sim 15\%$ the age of the Sun (Radick et al. 1982; Radick et al. 1983a; Lockwood et al. 1984; Radick et al. 1986,1987). A total of 62 cluster members, ranging in spectral type between F0V and M2V, have been observed for one or more seasons at the Cloudcroft and Lowell Observatories. It is now apparent that persistent brightness variability with peak-to-peak visual (Strömgren *b*- and *y*-band) amplitudes ranging between 1 and 4% on time scales of days to weeks is a ubiquitous characteristic of lower-main-sequence Hyades stars later than spectral type F8V. In contrast, stars earlier than spectral type F7V do not show dectectable photometric variations. The amplitude of this short-term variability shows little dependence on spectral type among the G- and K-type stars of the cluster (Radick et al. 1990a). As is true for other solar-type stars, rotational modulation plus evolving stellar surface features adequately explain the observed short-term photometric variations. Photometric rotation periods have been measured for 23 Hyades stars with spectral types between F8V and K8V. These periods range between 6 and 8 days for early G-type stars. The F-type stars are rotating somewhat more rapidly, whereas K-type stars (with one exception) tend to have rotation periods of 10 to 14 days. Parallel photometric and chromospheric Ca II H + K emission observations of several Hyades stars from the Lowell and Mount Wilson Observatories, respectively, show two distinctive short-term patterns: (a) rotational modulation signals present in the photometric and the Ca II H + K time series vary inversely with respect to each other; and (b) abrupt, secular changes in photometric brightness are frequently accompanied by opposite (albeit more gradual) changes in chromospheric Ca II H + K emission. These patterns reinforce the conclusion that the surface markings on young stars are active regions containing both dark spots and bright emission plages.

Hyades stars later than $\sim$ F8V also show clear evidence of longer-term, year-to-year brightness changes that rarely exceed 1% between consecutive years, but can amount to as much as 4% over several years. Like the short-term fluctuations, these longer-term brightness variations are accompanied by inverse changes in mean chromospheric activity. This is the same pattern shown by HD129333, the very young star discussed earlier, and is opposite to the Sun's behavior.

The variability characteristics of the Hyades stream star, HD1835, are very similar to those of Hyades cluster members (Lockwood and Skiff 1988).

Its rotation period is slightly less than 8 days. In brief, HD1835 is essentially indistinguishable from similar stars in the Hyades cluster itself.

3. Stars in the Coma Berenices Cluster.   Photometric observations of nine members of the Coma Berenices star cluster, spectral types F3V to K0V, were made at the Lowell Observatory between 1984 and 1987, inclusive (Radick et al. 1990*a*). The four G-type stars in this sample were all found to vary on both short-term and season-to-season time scales, and the single K-type star also showed hints of variability. The photometry shows that the *b-y* color of these stars increases slightly (i.e., they appear cooler) as they become fainter. Overall, the variations are indistinguishable from those of Hyades stars of similar spectral type. The four F-type stars in the sample, none later than F8V, were not detectably variable. Rotation periods for the four G-type stars were measured from modulation signals present in the time series, and were found to be closely similar to those of Hyades stars of comparable *B-V* color. The closest analog to the Sun among these four stars (according to *B-V* color) has a rotation period slightly longer than 8 days.

4. Young Solar-Type Field Stars.   Exploratory observations made about a decade ago suggested that the photometric variability of active lower-main-sequence field stars generally conform to the patterns shown by similar stars in young clusters such as the Hyades (Dorren and Guinan 1982; Radick et al. 1983*b*). These initial findings have since been solidly confirmed by a long-term, high-precision photometric study of 36 F-, G- and K-type field stars, which was initiated in 1984 at the Lowell Observatory (Lockwood and Skiff 1988). Although most of the 36 Lowell program stars are dwarfs, the sample also includes one subdwarf, six or seven subgiants, and one giant. Twenty-nine stars in the Lowell study were selected from Wilson's (1978) chromospheric activity survey. Four more were from the extension of that survey, the ongoing Mount Wilson HK Project, bringing to 33 the total number of stars for which regular, parallel photometric brightness and Ca II H + K emission observations exist (Radick et al. 1990*b*).

Field stars, of course, cannot usually be age dated on an individual basis, except through appeal to activity-rotation-age relations or, occasionally, lithium depletion. The Lowell survey does, however, contain a dozen or so stars whose mean chromospheric activity, as measured by Ca II H + K emission, is comparable to that from Hyades stars of similar spectral type. These stars are presumably Hyades age. All of these stars show both short-term and year-to-year brightness variations, typically amounting to a few percent. Furthermore, in all but one case, the brightness variations anticorrelate with chromospheric Ca II H + K emission changes on the year-to-year time scale, just as they do among other young stars.

In summary, young solar-type stars 10 to 20% the age of the Sun display the same short-term and year-to-year brightness variations shown by even younger stars of similar type, except at lesser amplitudes. Short-term variations range from 2 to 5% in the visible continuum. Year-to-year variations tend to have comparable amplitudes, or perhaps a bit less, over time spans of several years. Rotational modulation remains an important component of the short-term variability, but effects caused by the evolution of surface features, which presumably reflect true luminosity variations, become relatively more pronounced, in comparison with rotational modulation among the slightly older stars.

C. Stars Several Gyr Old: The "Mature" Sun

Although there are clusters that are comparable in age to the Sun (e.g., M67) or even older (e.g., NGC 188), none are nearby. Such clusters should make excellent targets for photometric observations using modern CCD detectors (see, e.g., Gilliland and Brown 1988) and tunable narrowband filters (Radick 1987), but long-term variability studies of lower-main-sequence stars in distant clusters are just beginning. Accordingly, our present knowledge of the photometric variability of solar-age lower-main-sequence stars still rests on photoelectric observations of relatively nearby field stars. Although ages can be assigned to such stars from their kinematic properties or through appeal to activity-age and rotation-age relations, the age discrimination is crude and uncertain, and really serves mainly to separate young stars from older stars.

The Cloudcroft survey (Radick et al. 1982; Radick et al. 1983a) included photometric observations of 42 stars in the Malmquist field near the north galactic pole. These stars are mainly old disk stars comparable in age to the Sun. The survey suggested that both the incidence and the amplitudes of short-term and year-to-year brightness variability among Malmquist field stars are lower than among younger stars in the Pleiades and Hyades clusters. It is, however, uncertain whether or not any intrinsic stellar variability was actually detected among the stars of the Malmquist field: in retrospect, it seems possible that stellar variations would have been swamped by the instrumental noise present in the Cloudcroft photometry.

Eighteen stars in the high-precision Lowell photometric program can be considered "mature" on the basis of their chromospheric Ca II H + K emission. This assessment is supported by measurements of rotation period, which are available for a few of the stars. For five G- and K-type stars among the 18, the measured rotation periods average slightly more than 30 days: i.e., these stars as a group are rotating just slightly more slowly than the Sun is. Of the 18 stars, only one exhibits statistically significant short-term brightness variations, and the photometry of that star is compromised by variability among its comparison stars. The infrequent detection of short-term variability among these older, less active stars is probably a consequence of the threshold imposed by the night-to-night measurement imprecision (0.2 to 0.3%), rather than evidence that such stars lack intrinsic variability. The Sun itself, if measured by Lowell photometry, would be judged quiescent, because its intrinsic short-term brightness variability rarely, if ever, exceeds the nightly instrumental precision of the Lowell observations.

In contrast to short-term variations, significant brightness variability is commonly found among the older, less active stars on the year-to-year time scale. Ten of 17 older stars for which a meaningful determination could be made exhibited such variability. The year-to-year variation of these stars rarely exceeds 2 or 3 times the solar level of $\sim 0.1\%$, which suggests that the amplitude of the year-to-year solar variation is fairly typical for older, less active stars. Analysis of annual means for the 18 stars reveals that continuum brightness variations accompany changes in stellar chromospheric activity on the time scale of years. In sharp contrast to the pattern shown by younger, more active stars, however, these older, less active stars tend to show solar-like behavior: in 15 of 18 cases, their brightness varies directly with changes in chromospheric activity, and in two of the three exceptions, the photometry is, once again, compromised by variable comparison stars.

In summary, mature, solar-age stars continue a now-familiar pattern: the amplitude of their photometric variability declines as their rotation rate slows. However, the correlation between mean brightness and chromospheric activity level on the year-to-year time scale switches from the inverse relation shown by young stars to the direct relation characteristic of the Sun.

D. Stars Approaching 10 Gyr in Age: The "Aging" Sun

Although it is possible to speculate about the photometric variability of aging solar-type stars, the observational basis for such speculation remains fragmentary. The solar neighborhood contains at least two bright stars, α Centauri A (G2V) and β Hydri (G2IV), very similar to the Sun, except older, that are obvious targets for some future photometric study. Today, however, the only source of observational information about the variability of aging lower-main-sequence stars is the Lowell Observatory study (Lockwood and Skiff 1988). One subdwarf, HD103095 (Groombridge 1830, spectral type G8VI) and three inactive F- and G-type subgiants are among the stars of the Lowell study. Of these stars, HD103095 stands out as one of the most quiescent stars in the entire study: the observed variation is probably all instrumental, on both short-term and year-to-year time scales. Likewise, none of the subgiants shows significant short-term brightness variability. On the other hand, two of the three exhibit low-level, year-to-year brightness variations that correlate directly with chromospheric activity changes. Thus, the little evidence available suggests that aging lower-main-sequence stars simply maintain the variability pattern established by the time they reach solar age, albeit at ever decreasing levels.

VI. SUMMARY AND PROSPECT

The low-level brightness variations exhibited by the Sun on time scales ranging from days to years are widely believed to originate from magnetic activity, in general, and from the presence and evolution of active regions on the solar surface, in particular. On time scales of days to months, these variations, which typically amount to 0.1% or so, arise from nonuniformities in the spatial distribution of the active regions about the solar surface (rotational modulation), and from the formation and decay of individual active regions. Variations caused by the latter are apt to be true luminosity effects. Likewise, variability on the time scales of months to years, which also amounts to 0.1% or so, and which is connected to the evolution of activity complexes and the 11-yr activity cycle, probably involves true luminosity changes.

Solar magnetic activity and its stellar analogs are currently believed to arise from the operation of a magnetohydrodynamic dynamo somewhere in the stellar interior. The behavior of this dynamo and, by extension, the characteristics of magnetic activity and its attendant brightness variations, are governed by stellar rotation, and by properties of the stellar convective zone which are, in turn, determined mainly by stellar mass and chemical composition. Furthermore, the rotation of a solar-type star quickly loses its memory of initial formation conditions, whereupon it is governed by stellar mass and age. Dynamo theory, the evolution of rotation among lower-main-sequence stars, and the empirical fact that such stars, including the Sun, seem to obey a single activity-rotation-age relation, together suggest that the Sun is a normal star in terms of its rotation, activity and brightness variability, and argue that valid stellar proxies for the Sun do exist, and may be discriminated by spectral type and rotation rate.

Most, if not all, of the brightness variability observed among ordinary lower-main-sequence stars can be interpreted in terms of magnetic active regions containing both dark and bright features (i.e., spots and faculae), which are distributed nonuniformly over the stellar surface. On young stars, these active regions are more numerous, or larger, or have higher contrast than on older stars such as the Sun, as the activity signatures observed on young stars are stronger and vary with greater amplitudes. The dark spots must also be more prominent relative to bright continuum features on young stars, because the spots dominate the photometric brightness variability on all time scales for such stars, rather than on just the short-term time scale, as for the Sun. Young stars apparently prefer to arrange their larger amount of surface magnetic flux into dark spots or spot groups, rather than into bright faculae and magnetic network features.

As a solar-type star ages, its rotation slows, and its overall level of magnetic activity declines. Its continuum brightness variability, which is controlled by the competing effects of dark spots and bright features (including both faculae and other bright magnetic network elements), also evolves. Initially, the dark spots dominate on all time scales: brightness variations are seen to anticorrelate with chromospheric activity changes on both short-term and year-to-year time scales. Gradually, however, the balance shifts toward the bright features on the year-to-year time scale, and must temporarily pass through equality. A star at this point will show little year-to-year brightness variability. The mean activity of such a star, however, will still be high, relative to that of the present Sun, and it will also continue to exhibit considerable short-term brightness variability arising from rotational modulation and active region evolution. A solar-type star apparently does not reach this stage until it is older than the stars of the Hyades cluster. By the age of the Sun, the bright-feature excess has become dominant on the time scale of the activity cycle, and year-to-year brightness variability has reappeared, albeit at a lesser amplitude than during the earlier, spot-dominated episode. These year-to-year variations, however, now correlate directly with changes in mean chromospheric activity. Short-term variability continues to show a prominent spot signature, and also continues to anticorrelate with chromospheric emission changes on the same time scale. These patterns, once established, apparently persist as the star approaches the end of its main-sequence evolution. This complex and contrasting evolution of brightness variability on the two time scales suggests that there is no strong physical coupling between spot deficits and facular excesses.

Although the stars do provide us with a probabilistic picture of solar brightness variability and its evolution, major parts of that picture remain incomplete. For example, somewhere in the long interval between Hyades-age stars and solar-age stars, the relation between brightness variability and chromospheric activity on the year-to-year time scale changes its sense. At the present time, this transformation may as well be accomplished through magic, because there are absolutely no observations available to illuminate it. More generally, our current knowledge of the temporal evolution of stellar variability for stars older than the Hyades is based almost exclusively on observations of field stars, for which accurate ages are rarely available. There is no fundamental reason why this needs to be so: clusters older than the Hyades (indeed, older than the Sun) do exist, and there is no reason, other than the nonavailability of instruments, telescopes and support, why the activity, variability and rotation of large, homogeneous samples of lower-main-sequence stars in such clusters cannot be studied. Indeed, even young clusters offer further opportunities. For example, the activity, variability and rotation of solar-type stars in the Pleiades cluster, itself, remain largely unknown, despite the impressive effort directed at the rapidly rotating K-type members of that cluster. Praesepe remains almost totally unexplored; the α Persei cluster has only been touched.

Even though further observations should much improve our picture of stellar variability, we will also need theoretical advances if we are to understand what we are seeing. For example, we need to understand better how magnetic flux is organized on the surface of stars. What determines its distribution between dark spots and bright features such as faculae? How does this distribution evolve as the star ages? More fundamentally, exactly how does the stellar dynamo, which we believe to be responsible for stellar magnetic activity and activity cycles, operate? How predictable is it? Clearly, much remains to be done before we are able to take full advantage of the stars' ability to tell us about the Sun in time.

THE PAST, PRESENT AND FUTURE OF SOLAR MAGNETISM: STELLAR MAGNETIC ACTIVITY

SALLIE BALIUNAS

Harvard-Smithsonian Center for Astrophysics

Many stars show magnetic activity, and of those that do, the majority appear to show variability that is cyclic, on time scales ranging roughly from as short as 2 yr to as long as several tens of yr. Studies of magnetic activity of other stars may yield understanding of the complex physics underlying solar magnetism. In this regard, two broad classes of observations of magnetic activity are useful: (1) studies of the dependence of the mean level of magnetic activity on stellar properties such as mass, age and rotation; and (2) studies of the variability with time of magnetic activity of individual stars. Results from the first class of studies are briefly described as the underpinning for discussion of the time variability of stellar magnetism. Solar-type stars display activity in one of two states: approximately two-thirds of the solar-type stars show magnetic variability that is cyclic, analogous to the 11-yr sunspot cycle. About one-third of the solar-type stars have lower mean levels of magnetic activity and little or no variability. Such prolonged states of low and constant magnetic activity may be similar to the Sun during the Maunder minimum. These tentative results suggest that solar-type stars undergo significant changes in magnetic activity on time scales of centuries. If correlated with magnetic variations, accompanying brightness changes for solar-type stars could be larger during entry to and exit from the Maunder minimum state than irradiance changes observed for the Sun during the 11-yr cycle.

I. INTRODUCTION

The origin, maintenance and variability of the Sun's magnetism are poorly understood. Models of solar magnetism rest upon knowledge of the solar interior and the physics of the interplay of subsurface motions with the internal magnetic field, both of which are uncertain. The currently favored model is that of a hydromagnetic dynamo, which requires subsurface convection and differential rotation to amplify and regenerate the internal solar magnetic field on a time scale of 11 yr. During the 11-yr cycle, a poloidal field is transformed to a toroidal one which replenishes the poloidal field and begins the cycle anew (Parker 1955,1979a; Babcock 1961; Steenbeck and Krause 1969; Leighton 1969; Yoshimura 1975b; Stix 1981; Belvedere 1985; Ruzmaikin 1985).

Another characteristic of solar magnetic activity is its universality; similar activity is evident on most stars. The dynamo, or other physical mechanism responsible for magnetism and its variability, must be operating inside a wide variety of stellar laboratories. Furthermore, studies of samples of stars with a range of masses and ages show a dependence of the mean (time-averaged) level of magnetism on these physical properties. Solar and stellar magnetic activity have three attributes which must be explained by any successful theory:

1. The mean level of magnetism (temporally and spatially averaged);
2. Its time variability, which may have periodic components;
3. Its universal occurrence on lower main-sequence stars.

The lesson from theoretical studies of solar magnetism is that its physical mechanism is complex and its understanding is incomplete. For example, one ingredient of the dynamo models is the dependence of the rate of rotation on depth. Analysis of the observed frequency modes of the p-mode (5-min) oscillations observed on the Sun's surface results in the finding that interior solar rotation may be constant with depth through much of the convective zone (Brown 1985; Brown and Morrow 1987). This key observation disagrees with the expected workings of the accepted dynamo models (see Gilman et al. 1989).

Progress in understanding solar and stellar magnetism can be expected if our view is broadened beyond the Sun. Attempting to get at the root cause of stellar magnetism by the study of just one star, the Sun, is like trying to learn about human nature from the study of one person. A full understanding of solar magnetism may only be gained from the study of magnetic phenomena on other stars, including those with properties similar to and dissimilar from the Sun's.

The study of stellar magnetic activity is an essential complement to solar physics. First, the impact of different physical properties on stellar magnetism can be explored. Second, time variability within a large sample of stars can be used to infer long-term changes of individual stars in a shorter interval of time, and finally, the study of stars of different ages provides the mechanism of unfolding the past, present and future history of solar magnetism.

II. MAGNETISM BY PROXY

The surfaces of dwarf stars other than the Sun cannot be resolved. This means that individual starspots cannot be directly counted, as in the case of sunspots on the Sun. Only the spatially averaged signal from a pinpoint of starlight can be sampled and the contribution from nonmagnetic areas of the stellar surface dilutes the signal from magnetic areas.

A promising technique of imaging stellar magnetic inhomogeneities is the search for tell-tale Zeeman broadening in magnetically sensitive lines (those with high Lande g-values). The observations and their analysis are very demanding, but are not impossible, and can contain much insight (see the chapter by Saar). However, the analysis of Zeeman broadening is inefficient for surveys of many stars.

One solution is to use an indirect but simple indicator of magnetic activity. On the surface of the Sun, the flux and configuration of magnetic fields can be inferred from the intensity of certain spectral lines that provide cooling for the input of nonradiative heating caused by the presence of magnetic fields. The ultraviolet and high-energy spectral regions contain many emission lines from the upper atmosphere of the Sun that brighten with increasing magnetic activity (Tousey 1963). Such lines are, however, observable only by space-borne instrumentation, which is not cost-effective for studies of many stars.

A. Calcium H and K Spectral Lines

One important exception is the singly ionized spectrum of calcium, and specifically, the Fraunhofer H ($\lambda3967$) and K ($\lambda3934$) doublet. Accessible to groundbased telescopes, the cores of the H and K lines are emission features that brighten with increasing magnetism (see Fig. 1). Direct measurements of the Sun show that areas of increased magnetic field flux are accompanied by bright H and K emission cores (Babcock and Babcock 1955; Skumanich et al. 1975; Schrijver et al. 1989b). Furthermore, changes in magnetic activity, for example, through the 11-yr sunspot cycle, produce relative changes in the disk-integrated fluxes of the H and K emission cores with amplitudes on the order of 20% (White and Livingston 1981; White et al. 1987), a variability that can be confidently measured.

The convenience and economy of groundbased observations has led to the creation of a rich library of H and K measurements from a variety of stars. Hence, the fluxes of the calcium H and K emission cores are the most useful and powerful proxy available for the description of stellar magnetism and its time behavior.

B. Cross-sectional and Time-serial Techniques

Two kinds of H and K studies are valuable in furthering our knowledge of solar and stellar magnetic activity: cross-sectional and time-serial. Cross-

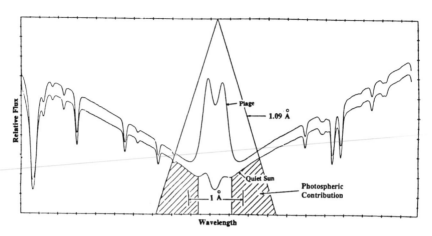

Fig. 1. The central portion of the calcium K spectrum line from two regions of the solar surface:
an area with little magnetic activity, the quiet sun, and one with strong magnetic activity, a
plage. The emission core of the calcium K line brightens with increased magnetic activity. The
passband of the HK photometer is superposed on the spectrum. An estimate of the nonmagnetic contribution to the measured flux in the passband is indicated (figure adapted from Hartmann et al. 1984).

sectional studies are conducted with snapshots of the fluxes of the calcium
emission cores in a variety of stars with different physical parameters, for
example, mass, age, rotation or chemical composition. The cross-sectional
analysis seeks to uncover the influence of physical properties on the mean
level of activity. One example is the study of stars of a given mass but different ages. Such studies have been conducted over four decades, and are discussed below (see also the chapters by Stauffer and Soderblom and by Walter
and Barry).

Temporal studies follow changes in stellar magnetism of individual stars.
Many temporal scales are important: for example, the counterpart of solar
flares can be seen over time scales of minutes, while starspot cycles must be
tracked for decades. In essence, we first place the Sun in the context of its
mean (time-averaged and spatially averaged) level of magnetic activity. Second, we compare the Sun's temporal behavior to that of other stars.

III. CROSS-SECTIONAL STUDIES OF STELLAR MAGNETISM

Early studies (Wilson 1963; Wilson and Skumanich 1964; Kraft 1967;
Skumanich 1972) of the calcium H and K emission cores of lower main-
sequence stars reveal a seemingly universal property of stellar magnetism:
Stars less massive than about 1.5 $M_\odot$ show magnetic activity that is analogous to solar activity. The extreme limit of the *lowest* mass that supports
magnetic activity is still unsettled; magnetic activity is certainly present in

stars as small as a few 0.1 $M_\odot$ (see Giampapa and Liebert 1986; Giampapa 1987; Fleming and Giampapa 1989). Only magnetic phenomena of other lower main-sequence (and some evolved, red stars) which are analogous to those occurring on the Sun are considered to be the subject of solar-stellar magnetic activity. Other stars may display plasma that is substantially hotter than their photospheric temperatures (for example, O- and B-type stars) or steady magnetic fields (such as the Ap stars), but their hot plasma or magnetism is different from the by-products thought to be created by the solar dynamo (see Rosner et al. 1985). In this connection, it is significant that the stellar angular momentum per unit mass suffers a sharp drop near 1.5 $M_\odot$; dwarf stars smaller than this have relatively slow rotation (Kraft 1970). For example, a star near 2 $M_\odot$ has an angular velocity of about 120 km s^{-1} and no magnetic activity; while the Sun, at 1 $M_\odot$ has a velocity of merely 2 km s^{-1} and definite magnetic activity.

Theoretical models predict that subsurface convective motions also first appear in stars with masses near 1.5 $M_\odot$. Stars below this limit have large subsurface radiative temperature gradients and thus convective zones; these convective zones are predicted to deepen with decreasing mass (see Cox and Giuli 1968; Gilman 1980). The subsurface convective motions generate acoustic waves which contribute to the nonradiative heating of the lower levels of a star's atmosphere (see Ulmschneider 1980) and to the emission seen in the calcium H and K emission cores.

The theoretical prediction of the existence of subsurface convective zones provides a clue to the cause of surface magnetism and its variability. The actions of subsurface shear and helicity, generated by the motions of convection and differential rotation, empower the recurrent nature of the magnetic dynamo process (see Parker 1955,1979a; Durney and Latour 1978). Those magnetic fields provide further nonradiative heating of the atmosphere, beyond that provided by the convective motions alone, and further contribute to emission in the H and K fluxes.

Three ingredients combine to create magnetic activity analogous to solar activity in lower main-sequence stars (Durney and Latour 1978):

1. Subsurface convection;
2. Rotation through differential rotation;
3. Seed internal magnetic fields.

Of these three components, convection is the crucial one that allows the magnetic dynamo to operate in lower main-sequence stars but prevents it in more massive main-sequence stars; significant convective zones are absent from more massive stars. Rotation (i.e., differential rotation) and seed, internal magnetic fields are also necessary for dynamo action, but may be present in all stars. It is the combination of all three dynamo ingredients (rotation, subsurface convection and internal magnetic fields) that explains the occurrence of magnetic activity on lower main-sequence stars.

Cross-sectional studies of magnetic activity reveal an additional aspect of stellar magnetic activity: its mean level decreases gradually with age and rotation during the lengthy main-sequence phase of a dwarf star. It is significant that rotation also slows with increasing main-sequence age (chapter by Stauffer and Soderblom). Initial studies of stars whose ages can be independently assessed show that stars younger than the Sun tend to rotate faster and also have stronger magnetic activity, as reflected in brighter H and K emissions fluxes (Skumanich 1972).

Figure 2 shows the roughly parallel decrease in magnetic activity and rotation as a function of age, for stars near 1 $M_\odot$. The youngest stars, such as those of the Pleiades, with ages ~0.01 $\tau_\odot$, show the strongest magnetic activity levels and the greatest rotational velocities. Stars of the Hyades cluster, with ages ~0.1 $\tau_\odot$, have intermediate levels of magnetic activity and rotation. The old stars, like the Sun, display the weakest magnetic activity levels and slowest rotation. The quantitative form of the relationship for the decay of rotation with age is approximately an inverse square root (Skumanich 1972; Soderblom 1983; Hartmann and Noyes 1987; Barry 1988; chapters by Stauffer and Soderblom and by Walter and Barry).

The explanation for the decay of rotation and magnetic activity with age relies upon the magnetic field, which is discussed in detail in the chapters by

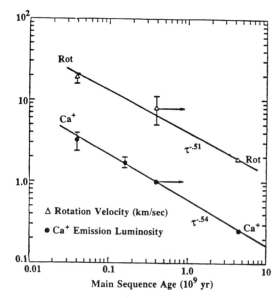

Fig.2. Rotation and magnetic activity strength (parameterized by the relative calcium emission flux) as a function of age for stars near 1 $M_\odot$ on the main sequence. The slopes of the fits describing the decay of rotation and activity with age τ are also shown (figure adapted from Skumanich 1972).

Saar, by Stauffer and Soderblom, and by Walter and Barry. Briefly, the magnetism permeating the outer solar atmosphere helps to support the hot (2×10^6 K) plasma of the corona which is evaporating into a wind. Carried with the wind, magnetic fields provide a long and effective moment arm which brakes rotation as angular momentum is lost over the lifetime of a star. The slowing of rotation also promotes the weakening of magnetic activity with age, perhaps as a consequence of decreasing efficiency of the dynamo, as mirrored by the fading of the H and K fluxes with age.

Pre-main-sequence stars and stars newly arrived on the main sequence show a broad distribution of angular momentum per unit mass (see Hartmann and Noyes 1987; chapters by Stauffer and Soderblom and by Walter and Barry). However, shortly after their arrival on the main sequence, stars rapidly shed any excess angular momentum so that the distribution of angular momentum per unit mass becomes a smooth and narrow function of mass. Thus, the dynamo process is not only universal, but also powerful in quickly erasing natal angular momentum and forcing a prescribed decay with age and mass.

A. Implications for the Sun's Magnetic History

In the early stages of the Sun's existence on the main sequence, it had fairly rapid rotation (tens of km s^{-1}) and intensified magnetic activity. The persistent angular momentum loss through the magnetic solar wind subsequently slowed the surface angular velocity. This slower rotation fed back into the efficiency of the magnetic dynamo, which weakened with time. The result was ever-lessening magnetic activity. As a consequence, the Sun's magnetism and rotation have eroded over the several Gyr it has spent on the main sequence. The future angular momentum of the Sun can be predicted from the observed relation of rotation to main-sequence age as shown in Fig 2 (see also Soderblom 1985; Barry 1988) and the measured strength of magnetic activity of stars in clusters older than the Sun (e.g., NGC 188; see Barry 1988). Over the next few Gyr, continued loss of angular momentum will probably slow the Sun's rotation and weaken its magnetic activity further; these decreases are likely to be small in absolute terms if the mass-loss rate remains small. However, it should be noted that accurate measurements of rotation and age do not exist for main-sequence stars that represent the Sun at a very great age.

B. Stellar Magnetic Activity

An ongoing program of cross-sectional and time-serial monitoring of magnetic activity, aimed at documenting the magnetic fluxes of single, solar and nonsolar type stars, has been underway for over two decades at Mount Wilson Observatory. This unique effort, the HK Project, began in 1966 with monthly measurements of the chromospheric fluxes of lower main-sequence stars. In 1980, the HK Project expanded to nightly observations. It has since measured the fluxes of over 1000 stars and has gathered over 260,000 measurements. Over the past decade the program has broadened to include not only lower main-sequence stars but post-main-sequence stars as well.

In 1977 the HK Project was transferred permanently to the Mount Wilson Observatory 60-inch reflector. The HK instrument is a spectrophotometer with good long-term stability, so that the long-term precision of its records is 1 to 2%. For each measurement of magnetic activity of a star, the fluxes in the H and K passbands (see Fig. 1), as well as two nearby reference passbands from the nonvariable stellar continuum, are recorded. The H and K passbands must be kept narrow (1-Å width) to isolate the narrow, chromospheric emission from the dilution of the nearby, photospheric flux. Passbands as narrow as this must be tunable to compensate for the radial velocity of the Earth's orbital motion and each star's motion relative to the Sun (Vaughan et al. 1978).

The measure of magnetic activity, labeled S, is the ratio of the fluxes in the H and K emission cores to that of the nearby continuum fluxes. [For a given star, variations in S are assumed to be caused predominantly by changes in the fluxes of the calcium H and K emission cores. The fluxes of the continuum passbands are assumed to remain approximately constant.] The flux ratio is a quantity that is not very sensitive to fluctuations of the intervening sky. Long-term stability of the flux ratio S is achieved through measurements of a standard lamp and non varying stars (Wilson 1978; Vaughan et al. 1978).

C. Magnetic Activity of Stars in the Solar Neighborhood

To place the Sun in the context of modern, quantitative measurements of magnetic activity, we turn first to a cross-sectional survey of the globally averaged strength of magnetic activity of stars in the solar neighborhood. Figure 3 shows the distribution of magnetic activity with lower main-sequence mass (which is parameterized in terms of an observed quantity, the B-V color index) of the stars in the local solar neighborhood, i.e., within 25 parsecs of the Sun (Woolley et al. 1970). The data have been described and analyzed by several researchers (Vaughan and Preston 1980; Hartmann et al. 1984; Soderblom 1985). The Sun is among those stars relatively weak in chromospheric emission.

In Fig. 3, the distribution of S values gradually rises with increasing B-V color index (or, equivalently, with decreasing stellar mass). This upward trend toward the right of the diagram is an artifact of the measurement of S, which is the ratio of the flux of the H and K emission to the nearby stellar continuum. The background continuum of stars fades with decreasing mass as their photospheric temperatures fall. Hence, the rise in S with increasing B-V is *not* a physical strengthening of magnetic activity toward lower-mass stars, but is a numerical consequence of a relatively decreasing flux in the reference bands, which are the denominator of S (see Noyes et al. 1984b).

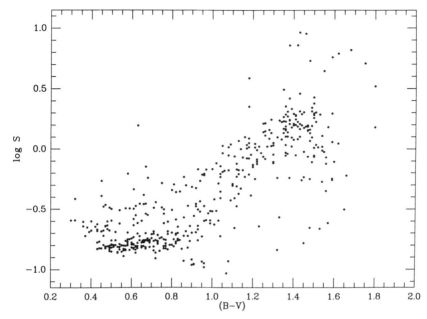

Fig. 3. The H and K relative fluxes, labeled S (common logarithm), of the stars in the Solar Neighborhood Survey are plotted as a function of the B-V color index, which is related to the mass of a lower main-sequence star. For the Sun, the mean value of $S \sim 0.171$ and B-$V \sim 0.66$ (data from Mount Wilson Observatory HK Project).

The results of the Solar Neighborhood Survey show a range of magnetic activity (represented by S), that translates roughly into a range of ages. This distribution of ages occurs at all masses represented in the sample. Several properties of the solar neighborhood, for example, the star-formation rate as a function of time, can also be extracted from this survey if the decay of activity with age is accurately known (see Hartmann et al. 1984; Barry 1988). Some problems with the quantitative details of the relation between activity and age are discussed below.

D. Magnetic Activity, Age and Rotation

Extensive cross-sectional surveys of magnetic activity can also be used to study the dependence of rotation with age of lower main-sequence stars. Here, another observed quantity, rotation, is added to the observing program.

As discussed above, the rotation of solar-type stars had been known to decay as a function of main-sequence age. Rotation in slowly spinning stars such as the Sun (~ 2 km s^{-1}) is notoriously difficult to measure with classical techniques that seek to distinguish such small levels of Doppler broadening from other line-broadening mechanisms (see, e.g., Smith and Gray 1976; Soderblom 1983).

A second and more powerful way to accurately determine slow rotation in lower main-sequence stars makes use of the time-serial records of *nightly* H and K measurements (Vaughan et al. 1981; Baliunas et al. 1983). By tracking H and K variations of stars over several months, we gather records that contain a signal modulated by rotation. Apparently, the inhomogeneity of the stellar surface causes the H and K flux to vary as surface features are carried into and out of the line of sight by axial rotation. Periodicities in the H and K records on time scales of days to months are almost certainly caused by rotation, and the measured rotation periods can be confirmed by repetition and by comparison with Doppler-broadening measurements of rotational velocities. Thus, rotation periods can be measured directly from the H and K records. From a knowledge of the radius of a star, the equatorial velocity can be inferred.

While it requires a substantial amount of telescope time, this technique of rotational modulation produces valuable results because modulation directly marks the rotation periods, independent of the angle of inclination of the stellar rotation axis to our line of sight. Doppler-broadening measurements supply the projected rotational velocity, or the component of rotation along our line of sight, giving only an upper limit of the rotation period. The same technique has been successfully used to measure rotation from photometric darkening caused by the presence of starspots (see the chapter by Radick).

Accurate measurements of rotation in lower main-sequence stars permit us to investigate the quantitative dependence of rotation upon age and magnetic activity (Noyes et al. 1984b; Simon et al. 1985; Rutten and Schrijver 1987; Dobson and Radick 1989; Basri 1987a; Schrijver et al. 1989a; Noyes et al. 1991). Such studies lead to an important conclusion: the mean level of magnetic activity depends mainly upon two parameters: the main-sequence age (or, equivalently, rotation) and the mass.

In their parameterization, Noyes et al. (1984b) express magnetic activity in terms of the Rossby number, the ratio of rotation period to the convective turnover time (which is believed to be a function of lower main-sequence mass). The Rossby number is an important linking between the observations to dynamo theory, which relies upon the motions of rotation and convection (see Durney and Latour 1978).

However, while the Noyes et al. parameterization has a physical appeal, it may not be correct or complete. The detailed form of the parameterization, and the question of the existence of a nonmagnetic, or basal, level of emission have been vigorously debated in the literature (Rutten and Schrijver 1987; Schrijver et al. 1989b; Dobson and Radick 1989). But, the evidence supports the broad, two-parameter characterization of stellar magnetism given above (Noyes et al. 1991): a main-sequence star of a particular mass will display a prescribed amount of magnetic activity at a given age; and the magnetic activity and rotation will both decay with time. Furthermore, this behavior is independent of the star's initial angular momentum at the time of its formation.

IV. SOLAR MAGNETIC VARIABILITY

Sunspot number is the longest, direct record of solar variability. [Some researchers consider the radiocarbon record a direct record of solar magnetic activity; in that case, the radiocarbon records, going back nearly 10^4 yr, would be the longest direct record. However, analyses of the radiocarbon records do not provide information about solar variability on short time scales, that is, from year to year, or about the length of sunspot cycles over time.] Sunspot number displays obvious periodicities, and its power spectrum contains a group of peaks that clump near 11 yr. That period is, however, an average; the time between sunspot number maxima has been as short as 8 yr and as long as 15 (Eddy 1976). Longer periods, such as the 90-yr Gleissberg cycle, or a 200-yr period, may also be present, which could help explain the modulation in frequency and amplitude of the higher frequencies (see the chapter by Damon and Sonett).

The extended epoch of very low sunspot number, which occurred during much of the 17th century, the Maunder minimum, displays an aspect of the Sun's behavior that is quite unlike its modern behavior. Lasting approximately 70 yr (ca. A.D. 1640 - 1715), the Maunder minimum was characterized by the near total absence of the 11-yr sunspot cycle (Maunder 1894; Eddy 1976,1977,1988). Supporting evidence for the existence of the Maunder minimum derives from radiocarbon abundances in tree rings (see the chapter by Damon and Sonett).

Evidence that the 11-yr solar magnetic periodicity is not the sole component of solar variability is provided by radiocarbon studies of tree rings. Sequences of the abundance of the isotope ^{14}C in tree rings chronicle the envelope of solar magnetic variability over long time scales. Analyses of those radiocarbon records suggest that the Maunder minimum, a time of depressed solar magnetic activity, is not a rare feature of solar variability. Indeed, the Sun may have spent as much as one-third of its recent history in such a magnetic lull (Damon 1977). Furthermore, magnetically low episodes have tended to recur in the past approximately every two centuries (chapter by Damon and Sonett).

Thus, the behavior of the Sun's magnetic periodicities is not that of a precise, 11-yr clock, but rather a clock with random perturbations, multiple periodicities, or a combination of both. There are several components to solar magnetic variability—a remarkably strong, average 11-yr period, deviations from that period, and its suppression or disappearance every few centuries. Magnetic variations appreciably longer than a few centuries also appear to exist, as suggested by an analysis of the radiocarbon data (chapter by Damon and Sonett).

Models of the solar dynamo must produce not only the nearly 11-yr periodicity, but also its suppression every few centuries during Maunder minima. Some chaos models do predict a lull in periodicities every few centuries (Weiss et al. 1984; Ruzmaikin 1985); the models, however, are still tentative. For the present, it appears that observations of magnetic activity may provide a very useful input to theoretical attempts at the formulation of a dynamo model. The remainder of this discussion will focus upon the observational information provided by the stellar magnetic variability records.

V. STELLAR MAGNETIC VARIABILITY

Time serial monitoring is a powerful technique that helps to unravel the complex behavior of stellar magnetic activity. Advances in technology and availability of telescope time have contributed to the program of monitoring magnetic activity of lower main-sequence stars.

A. The Wilson Program

In 1966, Wilson began an unprecedented observing program at the coudé focus of the venerable 100-inch Hooker telescope of Mount Wilson Observatory. For a dozen years, Wilson made monthly observations of a group of 91 dwarf stars, ranging from early F- to M-dwarf spectral types. Wilson's sample also included stars that are more and some less active than the Sun.

These pioneering efforts provided the first glimpse of long-term changes in magnetic activity of other single, dwarf stars (Wilson 1978). At every sampled interval of stellar mass, long-term fluctuations in magnetic activity were present. After Wilson's decade of monitoring, the H and K records could be broadly classified into three categories:

1. Some stars varied smoothly, with a cycle on the order of a decade, reminiscent of the 11-yr sunspot cycle;
2. Other stars varied erratically—that is, significant, long-term fluctuations occurred but not with a clear periodicity;
3. The remaining stars were constant with time—no discernible variations were present.

B. Properties of the Wilson Sample

The stars in the Wilson sample range in spectral type from early F dwarf (~ 1.5 M$_\odot$) to early M dwarf (several 0.1 M$_\odot$). There is only one M dwarf in the sample, because these cool stars have an extremely faint photospheric flux in the blue that makes them difficult to measure.

Magnetically weak and active stars in each mass interval were selected; as a result, the sample includes two broad categories of age: old stars, which means, in the case of stars the size of the Sun, an age of several Gyr; and young stars, which correspond roughly to a 1 $M_\odot$ star at an age <1 Gyr. Finally, the stars are single stars as far as can be judged; they lack significant radial velocity variations caused by orbital motions or other evidence that would suggest the presence of a companion star. Binary systems, especially those systems whose stars are in close proximity, further complicate the question of the mechanism for long-term magnetic activity. A close pair of binary stars is forced quickly to synchronous rotation by tidal forces that affect stellar interior conditions. As a consequence, angular momentum loss occurs differently in close binaries than in single stars. Such close binaries usually rotate more rapidly than their single counterparts. Rapid rotation of the binary star components produces greatly enhanced magnetic activity compared to the activity level of the Sun (Hall 1980; Marcy and Basri 1989).

C. Extension of Wilson's Records

The initial survey of stellar magnetic variations that Wilson had provided for the years 1966 - 1978 needed to be extended in order to determine periodicities with accuracy. Today, the H and K records of Wilson's 91 lower main-sequence stars span over two decades (Baliunas and Vaughan 1985; Baliunas 1988; Soderblom and Baliunas 1988; Noyes et al. 1991). Four examples are shown in Fig. 4, where the H and K flux is shown as a function of time, with monthly averages for the post-1980 data, which were sampled nightly. The observational program has a drawback that warrants comment: long intervals of observations, spanning years and preferably decades, are required. We are therefore endeavoring to operate the program several decades more.

D. Limitations of the Records

Although the magnetic records, in the form of chromospheric H and K relative fluxes, already span two decades, that time interval is relatively short from the viewpoint of solar activity; up to now, not quite two 11-yr solar cycles have been observed. Our experience with one star, the Sun, for which we have records reaching back about 3 centuries, is that its period averages 11 yr, but as noted before, the times between successive sunspot maxima have ranged between 8 and 15 yr. In addition, the sunspot cycle may have vanished, as it appears to have done during the Maunder minimum. In the case of the Sun and similar stars, lengthy intervals of observation are necessary to define the average periodicity and its deviations, and multiple periodicities. The defect of the brevity of the H and K records (only 20 yr) can be ameliorated by monitoring a large sample of stars; however, only 100 lower main-sequence stars have been monitored thus far, and many of these are not strictly identical to the Sun in their physical properties. On the contrary, some

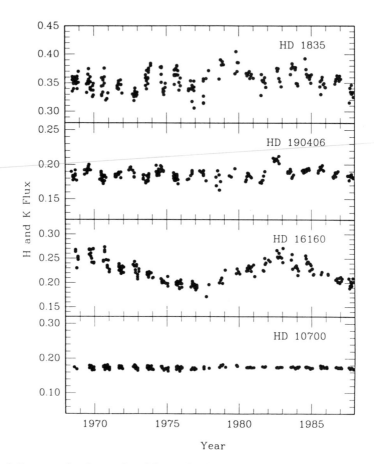

Fig. 4. Four examples of magnetic activity as a function of time. The top two records are typical
of young, solar-type stars; the bottom two panels show the types of records seen in solar-type
stars whose ages are similar to the Sun's. Beginning in 1980, monthly averages of nightly
observations are plotted (data from Mount Wilson Observatory HK Project).

program stars were chosen precisely because they are quite different than the
Sun, so that the dependence of magnetic periodicities upon physical properties can be explored.

E. Magnetic Variability of the Stars in the Wilson Sample

In this section, first the results of the entire sample of dwarf stars is
discussed; then stars that are similar to the Sun are considered (cf. Baliunas
and Vaughan 1985; Baliunas 1988; Noyes et al. 1991).

As noted above, Wilson (1978) discovered 3 types of long-term magnetic variability:

1. Stars whose long-term variations resembled the Sun's, i.e., appeared smooth and cyclic, often with periods on the order of a decade;
2. Stars whose variations were significant, but erratic, with no easily identified period;
3. Stars with no significant variation with time, i.e., stars whose magnetic flux is constant with time.

Each of these three types of magnetic activity records can be seen in Fig. 4.

In further refinement of the above classifications, Vaughan (1980) noted that variations resembling the Sun's (category 1) were generally found among the old stars of the sample, including the Sun and other slowly rotating stars. The stars with flat records (category 3) also tend to be those with weak magnetism and are therefore presumably old. In contrast, all young stars in the sample displayed substantial variability of magnetic activity. Moreover, the time scale of the variability in the young stars is often observed to be erratic (category 2) rather than repetitive. In other words, young stars are never "quiet" and are usually not cyclic (Baliunas and Vaughan 1985; Baliunas 1988).

F. Period Analysis

One goal of extending Wilson's records is to be able to determine periodicities quantitatively. The technique used for detecting the absence or presence of periodicities involved Fourier analysis modified for unevenly sampled time series, which are characteristic of stellar data. Unlike the Sun, most stars cannot be observed throughout the year. This fact causes substantial gaps between "seasons" of observations. In addition, nights of inclement weather or nonfunctioning equipment sometimes occur and produce additional short gaps in the records. Rather than bias the data by interpolating through gaps, or by binning (a procedure apt to lead to spurious results because of the months separating the windows of nighttime accessibility for each star), we used a modified Fourier analysis developed by Scargle (1982; see Horne and Baliunas 1987). The Scargle technique computes the periodogram only from the original time series, without binning or interpolation of data. An important development in the Scargle analysis is the concept of the "false-alarm probability," which evaluates the probability of a peak in the periodogram of a given height that could be produced by gaussian noise with the same variance as the data. The false-alarm probability provides an objective criterion for judging the significance of a peak in a periodogram computed from unevenly sampled date.

Based upon observations of many cycles on the Sun, we conclude that no technique for period search can accurately identify the mean cycle period or full range of periodicities within time intervals of only several decades. A possible, qualitative criterion for the longest period that can reasonably be detected in a record is that at least two cycles are observed, so that an assessment of the deviations in the length of two cycles can be made. By this requirement, the 20-yr span of H and K records can only reliably reveal mean periods on the order of 10 yr. The records suggest that some longer cycles exist, but a lengthier observing interval is necessary to quantify those periods. The shortness of the records may also be a source of systematic errors that bias the results of the search for the dependence of cycle periods upon stellar parameters.

VI. RESULTS OF THE CURRENT SURVEY

With those limitations in mind, the results to date for the extended survey can be summarized as follows, for the three categories of stars found by Wilson (see Baliunas and Vaughan 1985; Baliunas 1988; Noyes et al. 1991).

(1) About 60% of the sample has periodic, or apparently periodic, magnetic activity cycles (HD 16160; Fig. 4) which can range in period from 2.6 yr (HD 190406; Fig. 4) to longer than 20 yr, the span of the measurements. In the latter cases, the magnetic activity has been increasing or decreasing over two decades. Presumably, such trends must eventually reverse.

(2) About 15% are variable, but with no obvious periodicity, according to the periodogram analysis (HD 1835; Fig. 4). Multiple, shorter periods may be interfering and causing such time behavior; on the other hand, each "cycle" may be a unique event, unrelated to past and future events. This behavior is unlike the Sun's. Because the stars that vary erratically with time are generally young, rapidly rotating and magnetically active, other time-dependent phenomena may dominate the records and thereby bury a smooth, low-amplitude cycle (Gilliland and Baliunas 1987; Noyes et al. 1991). For example, large flares and active regions may introduce dominant signals on time scales of days, months or even years. These phenomena might be expected to occur randomly, and hence disguise periodic phenomena of a lower amplitude.

(3) About 10 to 15% of the stars are constant with time (HD 10700; Fig. 4). Stars with flat magnetic activity records may be in a Mauder-minimum-like state, perhaps as the Sun was during the 17th century. This possibility will be explored below, in the context of solar behavior.

A. Detailed Results for Stars with Periodicities

We first consider those stars (the majority of Wilson's sample) that are definitely or possibly periodic. Stars of nearly all masses and activity levels are included in this group. In this heterogeneous group of stars, cycle periods by and large do not depend in an obvious way on rotation, magnetic activity level, age, mass, or any combination of these parameters, such as the Rossby number (which predicts the mean level of magnetic activity extremely well; see above). Thus, predictions of the periods of activity cycles by the solar dynamo models, which generally indicate a dependence upon mass (through convective motions) or rotation (or on mean level of magnetic activity or age through rotation), or both, have not been confirmed. However, because of the brevity of the H and K records, the observed mean periods, especially the longer ones, may not yet be accurately defined to permit an unbiased comparison with models.

Upon close inspection of subsets of this group of stars with possible or obvious periodicities, we find some possible correlations between cycle period and physical properties. One subset of stars has periods shorter than 5 yr; such short periods have never been observed on the Sun. This group of stars includes HD 190406, a star whose cycle is 2.6 yr (Fig. 4). These nonsolar periods occur only in stars that are at least as massive (that is, as hot or hotter) than the Sun and presumably have shallower convective zones. More massive stars also rotate faster than the Sun. But while it is a necessary condition that a star be at least as massive as the Sun to show a short-cycle period, that is not a sufficient condition. Some of the more massive stars have long-cycle periods.

A second subset of stars appears to have an extremely interesting behavior, namely, cycle periods that are proportional to rotation periods (Noyes et al. 1984b,1990; Baliunas 1988). A star is included in this subset if it has slow rotation (slower than about 20 days) and a pronounced, large-amplitude cycle. All the stars so chosen have cycles lengths that are relatively long, on the order of 10 yr or more. The cycle periods within this subset tend to increase proportionately with rotation periods, in agreement with simple scalings from dynamo models (Noyes et al. 1984b). However, there is a strong selection bias in this sample because mean-cycle periods appreciably longer than about a decade are as yet difficult to detect with accuracy in the 20-yr observing interval. In addition, several stars clearly violate this relationship (Baliunas 1988; Noyes et al. 1991). One example is the star HD 81809 (Fig. 4), that is extremely close to the Sun in spectral type and in mean S. However, the rotation period of HD 81809 is 40 days (Baliunas and Jastrow 1990), much longer than the solar rotation period, but its cycle period is shorter, 8 yr. With longer records and larger samples, this potentially interesting result could be verified.

VII. RESULTS FOR SOLAR-TYPE STARS

We now turn to the stars that represent the Sun over its main-sequence lifetime. In order to build up statistics for the past, present and future Sun, we observe solar-type stars, that is, single stars with masses and ages close to that of the Sun. Stars similar in mass to the Sun can be chosen from a narrow range in the B-V, or color index (see Fig. 3). An approximately vertical cut near the Sun's mass, corresponding to B-$V \sim 0.66$, selects stars that are close the Sun's mass, but with differing levels of magnetic activity. These gradations of magnetic activity presumably represent the Sun at different ages of its main-sequence phase. It can be difficult to determine ages precisely and independently of the level of magnetic activity in single, dwarf stars in the field; see Baliunas and Jastrow (1989,1990) for a discussion of the sources of information for ages of the specific stars discussed here.

Table I compares the characteristics of the current Sun, solar-type stars similar in age to the Sun, and solar-type stars representing the young Sun (at an age of approximately 1 Gyr).

The young, solar-type stars have a mean $S \sim 0.31$, from which their mean age is estimated (Barry 1988) to be about 1 Gyr. The mean rotation period, measured from the modulation of S (Baliunas et al. 1983) is 9.1 days. The patterns of their long-term magnetic activity fall into two categories: apparently periodic or variable with no clear periodicity (labeled "erratic" in Table I). All the young, solar-type stars show magnetic variability, that is, their records with time are never constant, in contrast to the older stars.

The old, solar-like stars are similar in mass and age to the Sun at present, as inferred from their mean level of magnetic activity and rotation periods (which have been measured for only 5 of the stars in this group). Their long-term magnetic records are either apparently periodic or constant with time. The fact that their mean level of S is slightly lower than the Sun's is discussed in the next section.

The final entry in the table, concerning the correlation between brightness and magnetic activity changes, is discussed in Sec. IX, and in detail by Radick et al. (1990b) and the chapter by Radick.

A. Recent History of the Sun

Thirteen of the Wilson stars can be considered as similar to the Sun in mass and age (Baliunas and Jastrow 1989,1990). This sample produces a

TABLE I
Evolution of Solar Magnetic Activity

Argument	Young Sun-like Stars	Old Sun-like Stars	Sun
Number of stars	7	12	1
Main-sequence age	~1 Gyr	few Gyr	4.6 Gyr
Mean chromospheric flux ratio, S	0.314	0.165	0.171
Mean rotation period	9.1 d	27 d ($N=5$)	25 d
Cycle behavior	periodic or erratic; no minima	periodic; minima: ¼ time	11.1 yr (plus others); minimum: ⅓ time
Irradiance and magnetic activity changes	inverse correlation, <few%	positive correlation, ≤0.4%	positive correlation, ~0.1%

picture of the Sun's behavior over time scales of decades and perhaps centuries. The H and K records of these stars fall into two categories: (1) cyclic, with periods on the order of a decade; and (2) constant with time. Four stars in the sample are flat, suggesting that roughly one-third of solar-type stars show such constant magnetic behavior records. These stars may represent the Sun within a Maunder minimum, when magnetic variations with time are relatively flat. The occurrence of a Maunder minimum in solar-type stars appears to be related to the behavior of solar magnetism displayed in radiocarbon studies of tree rings. These studies indicate that the Sun appears to have spent about one-third of the last 8000 yr in a states of low activity resembling the Maunder minimum (Damon 1977). As in the case of the Sun, the Maunder-minima of other stars must also last longer than just 20 yr or so, otherwise some stars would have entered or exited this state during the time of the records. When the records are extended to a half-century, such entries and exits ought to occur on 1 or 2 stars (out of the sample of 13) if Maunder minima persist for about a century.

The usual excursion of the Sun in H and K fluxes of the solar-type stars yields more quantitative information (Baliunas and Jastrow 1989,1990). First, the sporadic observations of 61 solar-type stars in the Solar Neighborhood Survey (see Sec. III.C above) were combined with the measurements of the 13 Wilson stars whose H and K records span 20 yr. This merged sample includes 74 solar-type stars. In order to build up the number of independent measurements of magnetic fluxes of solar-type stars, averages (at 0.1-yr intervals) of the 20 yr of data of each of 13 stars were computed; each average was considered as an independent snapshot of long-term magnetic behavior. The sporadic observations were also averaged over 0.1-yr intervals.

The resulting frequency distribution of magnetic activity among solar-like stars yields a picture of magnetic variability over decades and centuries (Fig. 5). A preliminary analysis of the distribution contains some intriguing results.

The usual excursion of the Sun in H and K flux is about 0.164 to 0.178 in S, and its mean S value is 0.171 during the 11-yr cycle (Wilson 1978). Observations for about 75% of the stars in the sample fall within a broad distribution including the solar values of magnetic variability that occur during the sunspot cycle, but extending to somewhat higher and lower values. However, there is, in addition, a narrow distribution at low levels of magnetic activity, clearly separated from the broad distribution at high-flux levels. The center of this narrow peak is at $S \sim 0.145$. The level of H and K flux that corresponds to near-zero magnetic flux on the Sun is $S \sim 0.14$ (Schrijver et al. 1989b). S has contributions from nonmagnetic and instrumental sources that produce this offset in its relation to the quantitative amount of magnetic flux. Thus, the stars with H and K fluxes in this narrow distribution of low S values have magnetic fluxes that are significantly lower than at the minimum of the 11-yr sunspot cycle, and are close to zero magnetic flux as observed

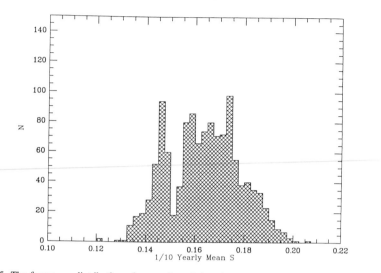

Fig. 5. The frequency distribution of magnetic activity of the solar-type stars with ages similar to that of the Sun. Two distinct groups are evident. One is a broad distribution centered nearly on the Sun's magnetic variability during the 11-yr cycle ($0.164 \leq S \leq 0.178$); the other, a narrow distribution, is centered much lower than the magnetic activity at the bottom of the sunspot cycle and is very close to the level of zero magnetic activity as inferred from the Sun. The stars with very low magnetic activity (those within the narrow distribution) tend to have flat H and K records (data from Mount Wilson Observatory HK Project).

on the Sun. The counterargument that the stars with low levels of magnetic activity namely, those in the narrow distribution are all extremely old is refuted by individual examples, discussed briefly below (see also Baliunas and Jastrow 1990).

Furthermore, the low levels of magnetic activity tend to occur in stars whose H and K records have been flat with time. If these magnetically flat, solar-type stars can be identified with the Maunder minimum, their level of magnetic activity indicates that the Maunder minimum phenomenon is accompanied by a significant drop in mean surface magnetism, to levels lower than those at the bottom of the sunspot cycle (Baliunas and Jastrow 1990). These tentative results for solar-type stars confirm the results from the radiocarbon records, namely that solar magnetism drops to exceedingly low levels during a Maunder minimum.

The H and K observations suggest that although the surface of the Sun is equally unspotted in a sunspot minimum and a Maunder minimum, its magnetic state may be quite different in these two phases. The number of points in the narrow distribution is about one-quarter of the total points in the histogram. This ratio is not far from the estimate of one-third given above for the fraction of solar-type stars with magnetically flat records, which are presumably in a Maunder minimum, and for the Sun's own behavior inferred from the radiocarbon records.

These results should be confirmed with a much larger sample of solar-type stars whose ages can be determined accurately. However, the method shows promise as a way to infer the statistics of long-term behavior of solar magnetism by observations over a few years rather than centuries.

VIII. HISTORY OF THE SOLAR CYCLE

The H and K records shown in Fig. 4 depict the history of solar magnetic variations on time scales of Gyr. The top two panels are representative of the behavior of solar-like stars at an age no greater than ~ 1 Gyr on the main sequence. The bottom two panels represent the magnetic behavior of stars whose age is similar to the Sun, namely, several Gyr.

At its current age, the magnetic variations of the Sun appear to occur in two states: the *cyclic* stage, with pronounced amplitudes of variability and periods on the order of a decade; or the *Maunder minimum* stage, with either very low-amplitude, or essentially no, magnetic variations and average levels lower (and close to zero magnetic flux in the disk average) than during the cyclic phase (Baliunas and Jastrow 1989,1990).

From these results, it can be inferred that the young Sun had a mean level of magnetic flux several times higher than today in absolute terms. The relative amplitude of the cycle was similar to that of the present Sun. However, the morphology of the variations was quite different. First, the young Sun never experienced a Maunder minimum. Second, the long-term magnetic variations could have been quite different than today. They may have been erratic, without a clear cycle period (for example, HD 1835; Fig 4), or periodic, but with a shorter period (as in HD 190406, period ~ 2.6 yr; Fig 4). The limited size of the sample of solar-type stars prevents a more quantitative analysis.

A. The Sun's Future

As for the future of solar magnetic variability, again, only a qualitative conclusion is possible. Dynamo models that depend upon rotation (see Durney and Latour 1978; Noyes et al. 1984*b*) would suggest that the Sun's magnetic activity will not change dramatically in the future if its future angular momentum decrease is limited. In this connection, observations of very old stars tend to agree with such dynamo predictions. Several solar-type stars in our group appear to be very old, perhaps on the order of 10 Gyr. One example is HD 103095, a metal-deficient subdwarf star. The cycle in this star, whose mean level of magnetic activity matches the Sun's, has an amplitude also equal to the Sun's (see Baliunas and Vaughan 1985). If the comparison to the

Sun is valid, it appears that the Sun's present cyclic magnetic behavior will continue for the remainder of its main-sequence lifetime. Whether the weakened magnetic states of the Maunder minimum will become more or less frequent as the Sun ages cannot be determined from the current sample.

IX. MAGNETIC AND BRIGHTNESS VARIATIONS

Recent satellite observations have had a significant impact upon our knowledge of solar magnetic variability and its possible terrestrial influence. The Solar Maximum Mission and Nimbus-7 satellites monitored the solar irradiance through much of the past sunspot cycle (cycle 21; Hickey et al. 1981; Willson et al. 1981). Those measurements show that over the course of the 11-yr sunspot cycle, the Sun's increased coverage by sunspots is accompanied by a brightening in its total irradiance. Models (Foukal and Lean 1988) suggest that the irradiance variations are governed by bright, magnetic regions such as faculae, that thereby produce a positive correlation between sunspots and irradiance changes.

This aspect of solar magnetic behavior has also been confirmed for solar-type stars (Radick et al. 1990b; chapter by Radick). At Lowell Observatory, extremely precise measurements of the photometric brightness changes of some of the Mount Wilson dwarf stars have been underway for several years (Lockwood and Skiff 1988). A cross analysis of the H and K observations from Mount Wilson and the photometric measurements from Lowell reveal behavior that is similar to the Sun's for the stars of solar age that occur in both samples, namely, that brightness and magnetic activity both increase (or decrease) together.

However, for stars representing the young Sun, brightness and magnetic changes are anticorrelated in the Lowell-Mount Wilson observations. In the young Sun, the presence of large, dark sunspots (which must have covered several percent of the area of the disk) would have dominated the bright emission from magnetic areas. Thus, in contrast to the current Sun, the irradiance variations of the young Sun would have been controlled by the dark sunspots. During the course of a sunspot cycle, as magnetic activity increased, the brightness of the young Sun ought to have faded, perhaps by as much as several percent (Radick et al. 1990b).

The Mount Wilson Observatory's HK Project and the Lowell Observatory stellar photometry program can provide the foundation for a theoretical and observational understanding of the Sun in time, from a unique perspective: the behavior of statistically meaningful samples of solar-like stars. Not only do such stars provide laboratories for the study of the different conditions under which magnetic activity occurs; they also provide a means of obtaining information on the past, present and future of solar magnetism without the need for monitoring the Sun for billions of years.

Acknowledgments.   I would like to thank the staff of Mount Wilson Observatory for their devoted efforts. The Observatory's former and current operators, the Carnegie Institution of Washington and Mount Wilson Institute, have been essential to the inception and continuation of our program. This research is supported by the Scholarly Studies Program, the Langley Abbot Program and other funds of the Smithsonian Institution, a grant from the National Science Foundation, the Center of Excellence in Information Systems of Tennessee State University, the Richard Lounsberry Foundation, the George C. Marshall Institute and the Mobil Foundation, Inc., as well as by several generous individuals.

THE EVOLUTION OF ANGULAR MOMENTUM IN
SOLAR MASS STARS

JOHN R. STAUFFER
University of California, Santa Cruz

and

DAVID R. SODERBLOM
Space Telescope Science Institute

At least some solar-mass stars go through a brief period of very rapid rotation at the beginning of their main-sequence career. Within less than 30 Myr after their arrival on the main sequence, the surface rotational velocities of these stars decrease from 100–200 km s^{-1} to < 30 kms s^{-1}. At least one third of the stars less massive than the Sun also arrive on the main sequence with large rotational velocities, with the duration of the rapid-rotation phase increasing with decreasing stellar mass. Subsequently, low-mass stars slowly continue to lose angular momentum via analogs of the solar wind. The spindown process includes considerable feedback, so that by the age of the Hyades, solar-type stars of a given mass all have nearly the same rotational velocity. The Sun has "normal" rotation for its mass and age, but because of the feedback in the spindown, whether or not the Sun arrived on the main sequence as a rapid rotator cannot be determined. We outline the observations leading to these conclusions and the implications they may have for understanding other properties of these stars as well as the possible connection between disks, planet formation and rotation.

I. INTRODUCTION

Until recently, the study of the rotation of solar-type stars was impossible. The only observational technique available was to determine line breadths in stars from photographic spectrograms, from which one inferred v sin i, the equatorial rotational velocity projected onto the line of sight. Herbig and Spaulding (1955) surveyed late-type stars, but in general could only determine conservative upper limits because their measurement limit of 17 km s^{-1} was larger than what we now know to be the average rotational velocity for low-mass stars. Kraft (1965,1967) pushed the photographic technique to its limits to measure v sin i near the 10 km s^{-1} level. For rotation much less than that, rotation is no longer the dominant line broadener (the intrinsic breadth of solar lines is about 7 km s^{-1}), and is therefore difficult to measure. The net result was that photographic spectroscopy limited the study of rotation mainly to the upper-main sequence ($M \gtrsim 1.5$ M$_\odot$).

Our knowledge of rotation among the massive stars has not changed much in recent years, but we do now know many critical and fascinating details for rotation among low-mass stars. Kraft's (1967) study (the culmination of photographic work) was a milestone for it showed two correlations that strongly influenced subsequent work: (1) a connection between rotation and age; and (2) a connection between rotation and chromospheric activity. These were quantified shortly afterward by Skumanich (1972), in his now-famous $t^{-1/2}$ laws.

Another milestone was provided by Gray in a series of papers (Gray 1973,1975; Smith and Gray 1976), that showed how to extract rotational broadening from a line profile even when it was far from the dominant broadener. At about the same time, the group at Mount Wilson who were surveying Ca II H and K emission in late-type dwarfs started to observe a subset of their sample often enough to be able to see rotational modulation. This discovery has been of fundamental importance, because rotational periods can be determined to much higher precision than v sin i, and are free of projection effects (except that a star viewed along the spin axis shows no modulation). Moreover, rotation periods have been determined for a few stars at rates that would be impossible to see as line broadening.

Angular momentum holds the key to understanding star formation, and is an important determinant of stellar evolution as well. That angular momentum is important for the evolution of the Sun was well demonstrated by Kraft (1970) (and understood before that). He showed how most high-mass stars (i.e., above 1.5 M$_\odot$) have specific angular momenta that fall on a relation $\langle J/M \rangle \propto M^{2/3}$. The Sun and similar stars fall below this relation by fully 2 orders of magnitude. However, if the angular momentum of our planetary system is included, the total falls right on the mean relation. This quite naturally has led to speculation that the young Sun dissipated its angular momentum into the surrounding planetary or protoplanetary material.

Low-mass stars may indeed form planets. However, the idea that planetary formation is the mechanism that explains the slow rotation of low-mass stars has been made less plausible by the recent discovery of near-solar-mass stars that rotate at up to 100 times the solar rate. This phase of extremely fast rotation as stars approach the zero-age main sequence (ZAMS) is surely the single most interesting part of the rotational evolution of our proto-Sun, and we will discuss these observations in some detail.

Other central problems in the study of rotation among solar-type stars include:

1. Is there radially differential rotation in the Sun and other stars? Can angular momentum "hide" in the core of a star until later in its life?
2. Can latitudinal differential rotation be detected on stars other than the Sun? This is important for studying magnetic dynamos.
3. Does rotation influence other observables? For example, lithium depletion occurs at or near the base of the convective zone of a solar-type star. In this region, a discontinuity in the rotation rate may occur, particularly during some stages of evolution, as the convective zone slows down more quickly than the radiative core. This leads to shear and turbulence, and so may aid Li depletion.
4. What happens to a solar-type star after it leaves the main sequence?
5. Are the low-metallicity stars of Population II different in their angular momentum than Population I?

Kraft's (1970) review provides an excellent summary of the state of knowledge of stellar rotation at that time. More recent reviews (see, e.g., Stauffer and Hartmann 1986; Hartmann and Noyes 1987) treat some of the above subjects. The present discussion will center on what information is available to delineate the past and future behavior of the Sun itself, since this book is a biography of the Sun. We will treat other related stars and phenomena as an aid to explaining the Sun.

II. INITIAL ANGULAR-MOMENTUM DISTRIBUTION FOR 1 $M_\odot$ STARS

In order to understand the evolution of angular-momentum of solar-mass stars, it is useful to know the initial condition from which they evolve. Quite simple considerations indicate that newly formed stars might be expected to have large rotational velocities. In fact, if stars did not exist, one would predict their absence based on angular-momentum considerations. That is, if one takes typical densities and sizes for a molecular-cloud core, and imparts to that core a net rotational velocity corresponding only to what one would derive based on the average shear in the rotation of the galactic disk, and then allows that cloud to shrink to stellar dimensions, the predicted surface rotational velocity exceeds the breakup velocity for the star by a large factor. This observation is typically referred to as the "angular-momentum problem." (Note, as we discuss below, that the random orientation of stellar rotation axes indicates that stellar angular momentum does not have its origin in the rotation of the Galaxy. However, any plausible scheme for estimating the characteristic angular momentum of protostellar clouds leads to the same angular-momentum problem.)

The seemingly most straightforward way to estimate the extent to which real stars face an angular-momentum problem is to measure the rotation of dense molecular cloud cores, assuming that they will eventually collapse and become single stars. Because the net rotation rate for such clouds is small, and the clouds themselves have small angular sizes, measuring such a rotation is difficult, and has only been accomplished for a small number of clouds (Goldsmith and Arquilla 1985; Adams 1989). When such rotation can be measured, angular rotation rates of order 3×10^{-14} are typical for clouds with size scales of order 1 parsec, which is still much too large for clouds to collapse to stellar dimensions without considerable angular momentum loss.

An alternative means to estimate the initial angular-momentum distribution for solar-mass stars is by extrapolation from the rotational velocities observed for high-mass stars. High-mass stars with spectral type earlier than about F0 do not have outer convective envelopes, and hence are thought not to have the dynamo-driven winds that inexorably carry angular momentum away from low-mass stars. Therefore, it is generally assumed that the rotational velocities for high-mass, main-sequence stars are indicative of their initial rotation rates. As noted earlier, Kraft (1970) showed that observations for high-mass stars are consistent with $\langle J/M \rangle \propto M^{2/3}$. A re-analysis of this question including more recent data (Kawaler 1987) largely confirms Kraft's result, though Kawaler derives a somewhat larger exponent for the relation. Either relation predicts rotation rates of order 100 km s^{-1} for 1 M$_\odot$ stars upon their arrival on the main sequence if there is no additional pre-main-sequence loss of angular momentum.

There is a small amount of observational data which relate to the orientation of the rotation axis for stars in addition to the rotation rates. For stars in open clusters, several indirect arguments suggest that the rotation axes are randomly oriented. The most compelling of these arguments derives from consideration of the rotational velocity distribution for high-mass stars in open clusters as a function of galactic latitude. The most likely alignment direction for cluster stars would be for the stellar rotation axes to be parallel to the galactic rotation axis. If this were the case, stars in clusters located at high galactic latitude would have systematically smaller projected rotational velocites. No such correlation between observed rotation rates and galactic latitude has been found, from which we conclude that stellar rotation axes are not preferentially aligned with the rotation axis of the Galaxy (Kraft 1965). Note also that the orientations of the orbital angular-momentum axes of binary stars appear to be random as well (Kraft 1965).

While open cluster stars apparently have randomly distributed rotation axes, there is evidence that the rotational axes of stars in associations may, in some cases, be nonrandom. Recent observations suggest that rotation axes of at least some young stars in associations are preferentially aligned parallel to the direction of the local magnetic field, and perpendicular to the long axis of filaments (or sheets) in the parent molecular cloud (Heyer et al. 1987; Vrba et al. 1988a; Strom et al. 1985b). Shu et al. (1987) ascribe this to fundamentally different star formation modes. Gravity dominates magnetic pressures for the open-cluster environment, and hence there is no alignment of axes (referred to as the "supercritical case" by Shu et al.). Star formation in molecular-cloud filaments begins in a much lower-density environment, where collapse parallel to the ambient magnetic field direction is favored. If this model is correct, the initial angular momentum for a solar-mass star may be a function of whether it was born in a cluster or an association. There is as yet insufficient observational data to confirm or refute this prediction. Also, other lines of evidence suggest that T Tauri stars (as a group) have randomly oriented rotation axes (Bertout et al. 1988; Hartmann and Stauffer 1989). This would not necessarily exclude the model described above, because the rotation axes could still have been aligned with the local magnetic field at formation, if the local magnetic-field orientation is sufficiently varied over the star-forming region.

III. ANGULAR MOMENTUM DURING EARLY PRE-MAIN-SEQUENCE EVOLUTION

As discussed in the preceding section, there is a general consensus that considerable angular momentum must be "lost" between the time when a dense molecular-cloud core begins to collapse and the time when a young T Tauri star first becomes visible. A broad variety of observational data for T Tauri stars (including spectral energy distributions, forbidden line-profile asymmetries, and direct imaging in the optical, infrared and mm-continuum) strongly indicate that at least an intermediate solution to this angular-momentum problem is the formation of a disk (Terebey et al. 1984; Strom et al. 1988; Adams et al. 1987). Typical sizes and masses estimated for these disks are 100 AU and of order 0.1 $M_\odot$, so the disks are clearly an important step in the star-formation process. What is not yet clear is the extent to which all stars form such disks, the time scale and mechanism by which the disks dissipate, and the extent to which mass and/or angular momentum is transferred between the disk and the central star.

Perhaps the most striking examples of the influence of disks on the early evolution of low-mass stars are the FU Ori objects. FU Ori objects are young stars that quickly increase in luminosity by a hundred-fold, then slowly decline in brightness over decades. Hartmann and Kenyon (1985) interpret the outburst as the onset of steady-state accretion in a pre-existing disk. Plausible estimates of the necessary accretion rates and the number of known FU Ori stars indicates that if all stars pass through the FU Ori stage, then at least 0.01 $M_\odot$ of their mass should be derived from disk accretion (Kenyon et al. 1988). Since this matter arrives on the star's surface with approximately the

Keplerian velocity for that radius, this accretion could have a considerable influence on the observed rotational velocities of young stars.

Indirect evidence for disk accretion exists for some stars even beyond the FU Ori stage in the form of peculiar spectral energy distributions for some T Tauri stars and from the apparent "veiling" of the absorption-line spectra of the more active T Tauri stars (Kenyon and Hartmann 1987; Bertout et al. 1988). Kenyon and Hartmann (1987) estimate that the most active T Tauri stars could accrete 5 to 10% of their mass through a disk, leading (in the absence of angular-momentum loss) to rotation at 1/4 to 1/2 the stellar breakup velocity. It is very likely that the rapid rotation induced by this accretion would in fact lead to the development of a strong wind and considerable angular-momentum loss. Shu et al. (1988) posit that this type of accretion does spin up the central star to nearly breakup rotation, and that further accretion initiates the development of a centrifugally driven wind which powers the bipolar outflows seen from very young stars. Mercer-Smith et al. (1984) have constructed models whereby low-mass stars form primarily by disk accretion. Since the time between the end of disk accretion and the appearance of the star as a visible object is presumably quite short, these models face the difficulty that even the apparently youngest T Tauri stars have rotational velocities that are quite low, as we shall discuss in the next section. For further discussion of T Tauri stars see the chapter by Bertout et al.

It is apparent from the preceding discussion that the formation and evolution of disks during the pre-main-sequence contraction of solar-mass stars intimately affect the rotational velocity that these stars will have upon arrival on the main sequence. Do all young, solar-mass stars have disks? A negative response to that question is suggested by Walter (1986), since there appears to be at least as many naked T Tauri stars as classical T Tauri stars, and the former do not show the emission-line activity or infrared excesses indicative of disks. If the naked T Tauri stars do really form without disks, one might predict two populations of stars on the main sequence with systematically different rotational properties. However, this conclusion is not yet warranted. Strom et al. (1989) claim that infrared properties of naked T Tauri stars overlap significantly with classical T Tauri stars, and that there is simply a range in initial disk masses; alternatively, the differences in the two populations may simply indicate an age difference, with the naked T Tauri stars being older on average, allowing their disks to have dissipated somewhat.

IV. ROTATION DURING PRE-MAIN-SEQUENCE EVOLUTION

Prior to the measurement of rotational velocities for T Tauri stars, it was generally assumed that those rates would be relatively high. This followed either from the angular-momentum considerations just mentioned, or from a straightforward extrapolation of the rotation-activity relation of normal stars to the realm of extreme activity seen among the T Tauri stars. Indeed, the first measurements of rotation in T Tauri stars suggested rapid rotation (Walker 1956; Herbig 1957). This belief was not dispelled until the application of Fourier-quotient techniques to photographic spectra of T Tauri stars by Vogel and Kuhi (1981), who showed that most T Tauri stars had rotational velocities below their detection limit of 30 km s^{-1}. Subsequent studies by Hartmann et al. (1986), Bouvier et al. (1986), and Hartmann and Stauffer (1989) have shown that the actual mean rotational velocity of T Tauri stars is only of order 15 km s^{-1}, and that there is little or no correlation of rotation with optical emission-line activity; in fact, there is a slight tendency for the least active stars, the naked T Tauri stars, to rotate faster than the active T Tauri stars. For approximately solar-mass T Tauri stars (those with masses in the range 0.5 to 1.0 M$_\odot$ as estimated from their location in an HR diagram), the range in rotation for the T Tauri stars observed to date is quite small, ~ 8 km s^{-1} to 35 km s^{-1}. The naked T Tauri stars have essentially the same rotational velocity distribution, but with a slight weighting towards higher velocities and with one notable rapid rotator (V410 Tau), with a v sin i of approximately 70 km s^{-1}.

The surprise comes when one looks at an open cluster whose age is such that solar-mass stars should have just arrived on the main sequence. The α Persei cluster, with an age of 50 Myr and a distance of only 150 parsecs is the best candidate for such a study (Stauffer et al. 1985; Stauffer et al. 1989). Figure 1 shows the rotational velocity distribution for low-mass stars in this cluster. Instead of the relatively slow rotational velocities observed for T Tauri stars, we now find rotational velocities for solar-mass stars that even exceed 150 km s^{-1}. These large rotational velocities and velocity range are found for all spectral types from late F to early M. (See the chapter by Herbert et al. on the importance of stellar rotation for electrical heating of planetesimals.)

The α Persei data are not, in fact, contradictory to the T Tauri rotational-velocity distribution, because the considerable contraction and increase in central concentration as stars evolve from the Hayashi track to the main sequence predict that there should be a sizeable increase in surface rotational velocities. However, matching the quite large rotational velocites observed for some α Persei stars requires that there be little angular-momentum loss during this period and that what angular-momentum loss there is, not be a strong function of rotation rate (Stauffer and Hartmann 1987). This may not be a problem, because other measures of the wind-forming region of young, relatively rapidly rotating stars (e.g., their optical and ultraviolet emission-line activity) also do not correlate strongly with rotation (Caillault and Helfand 1985; Hartmann and Stauffer 1989).

Additional information on the early spin-down of solar-mass stars can be obtained by examination of a slightly older open cluster. The best candidate for this step is the Pleiades, with an age only 20 Myr older than α Per (i.e., Pleiades $\simeq$ 70 Myr). Figure 1b shows the rotational velocity distribution

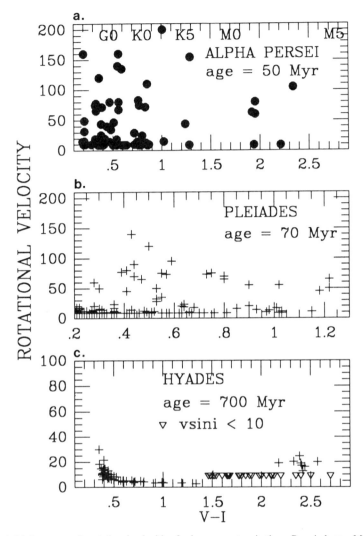

Fig. 1.(a) Spectroscopic rotational velocities for low-mass stars in the α Persei cluster. Many of the stars plotted at 10 km s^{-1} simply have measured upper limits of 10 km s^{-1}. Accuracies for the rotational velocity measurements are typically of order 10% of the velocity. (b) As for (a), but for the Pleiades cluster. (c) Rotational velocity data for the Hyades cluster. Rotational velocities blueward of $V-I = 1.4$ are primarily from photometric variability studies. The inverted triangles indicate rotational velocity upper limits of 10 km s^{-1}, while the small number of stars redward of $V-I = 1.5$ with spectroscopic rotational velocities > 10 km s^{-1} are indicated by + symbols.

for the Pleiades (Stauffer et al. 1984; Stauffer and Hartmann 1987). The rotational velocities for the K and M dwarfs in the Pleiades are not greatly different from those observed for α Perseus; the Pleiades G dwarfs, however, are nearly all slow rotators. The simplest explanation for this observation is that the G dwarfs which arrive on the main sequence as very rapid rotators spin-down to relatively low rotational velocities in a time period of order 20 Myr. Endal and Sofia (1981) actually predicted this brief period of rapid rotation upon arrival on the main sequence for solar-mass stars. The short spin-down time for G dwarfs in their model derives largely from the fact that the outer convective envelope was essentially uncoupled from the radiative core. With only about 10% of the angular momentum for a solar-mass star contained in the outer convective envelope, a plausible wind can provide enough angular-momentum loss to give the necessary rapid spin-down to the observed rotation rate. This model also suggests why the Pleiades K dwarfs still show an appreciable number of very rapid rotators. The explanation is that the fraction of a star's angular momentum contained in the outer convective envelope increases with decreasing effective temperature, so that early K dwarfs have 30 to 40% of their angular momentum in the outer convective envelope. Their spin-down time is longer simply because a larger fraction of the star's total angular momentum is involved. Figure 2 illustrates the variation in the moment of inertia of the outer convective envelope for main-sequence stars as a function of mass, as derived from models by D. Vandenberg (personal communication, 1988).

    If the above model is correct for the rapid spin-down of solar-mass stars, a simple prediction is that observations of an even older cluster would show slowly rotating G and K dwarfs, but some rapidly rotating M dwarfs. Rota-

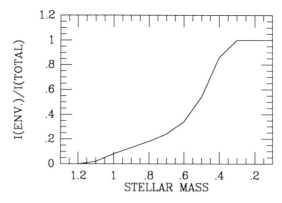

Fig. 2. Ratio of the moment of inertia of the convective envelope to the total moment of inertia for zero-age main sequence stars of different masses, as derived from unpublished models by D. VandenBerg.

tional velocities for low-mass stars in the Hyades (Fig. 1c) (age $\simeq$ 700 Myr) confirm this prediction (Radick et al. 1987; Stauffer et al. 1987).

The Endal and Sofia (1981) model predicted essentially no coupling between the outer convective envelope and the core, leading to a very rapidly rotating solar core even at an age of several billion years. This rapid core rotation is ruled out by the recent helioseismology results. Therefore, it is assumed that there must be some coupling between the core and envelope which acts on a relatively long time scale. Attempts to develop theoretical models for low-mass stars which match the observed rotational velocity data for young open clusters are in progress (Kawaler 1988; Pinsonneault et al. 1989).

Finally, we note that consideration of binary stars may provide useful insights into the resolution of the angular-momentum problem. Forming two stars in orbit around a common center-of-mass is an obvious means to avoid forming a single, very rapidly rotating star. In fact, the orbital angular momentum of a typical binary is much larger than the rotational angular momentum of even a very rapidly rotating star. A possible conclusion is that the protostars that suffer most from excessive angular momentum form binaries rather than single, rapidly rotating stars. Radial and rotational velocity surveys of complete samples of low-mass stars in young clusters and associations will provide a first step towards illuminating this issue.

V. ROTATIONAL EVOLUTION ON THE MAIN SEQUENCE

Upon its arrival on the zero-age main sequence, and following the stage of rapid rotation just described, a 1 $M_\odot$ star rotates at $\sim$ 10 times the solar rate, i.e., with a v sin i of about 20 km s^{-1}. This continues to decline, so that by the age of the Hyades (0.7 Gyr), that star rotates at 5 to 7 km s^{-1}. Solar lines have an intrinsic breadth of roughly 7 km s^{-1}, so that for stars older than the Hyades (which is to say virtually all in the field), rotation is not the dominant line-broadening mechanism. How is rotation then measured? Three techniques are available.

First, Smith and Gray (1976) have shown how a high-resolution line profile can be made to yield a precise v sin i with the aid of the profile's power spectrum. The basic idea is that the functional form of rotational broadening leads to strict geometric limits on the maximum displacement of specific intensity profiles from the stellar limb. These "corners" then produce sidelobes in the power spectrum. The other major broadening mechanism, macroturbulence, has a functional form like a gaussian. The power spectrum of a gaussian is another gaussian, which is smooth. This distinction makes it straightforward to distinguish these broadening agents which produce subtler differences in the profile itself. Using this method, it is possible to see the solar rotation (at v sin i = 1.7 km s^{-1}), and, in general, to determine v sin i to ± 0.5 km s^{-1} or so (Soderblom 1982).

Second, and more important, the surface inhomogeneity of a solar-type star leads to photometric variability as rotation carries those inhomogeneities across the stellar disk. This is especially evident in a chromospheric feature such as the Ca II K line because of better contrast. By monitoring a number of stars on a nightly basis, the Mount Wilson group has determined rotation periods. These are important for studying rotation among solar-type stars because of their high precision and because they are free of any aspect dependence. However, older solar-type stars (like the Sun itself) have very weak chromospheres, so weak that no detectable modulation of the Ca II H and K lines has been seen. It is to be hoped that future observations with higher sensitivity will detect modulation in older stars because the v sin i values are not very useful for them.

A third method is to note the tight correlation between rotation and chromospheric activity and to use the known rotation periods to estimate more. It is easy to survey the H and K emission of nearby solar-type stars (see Vaughan and Preston 1980), and from that to infer the distribution of rotation among the same stars (Soderblom 1985). Individual rotation periods estimated in this fashion appear to be good to about 20%, which is good enough for statistical studies.

The main observational data have been reported in Smith (1978), Soderblom (1983), Baliunas et al. (1983), and Benz et al. (1984). These data are consistent with the power-law relationship between rotation and age put forth by Skumanich (1972): $\Omega \propto t^{-1/2}$. Such a relationship follows from naive theory if one assumes that the magnetic-field strength is proportional to the angular velocity.

Like many power laws, this $t^{-1/2}$ relation is *consistent* with the observations, but one is not compelled toward it. The details for any one star may be much more complex. However, several factors suggest that the rotational evolution of a solar-type star is not accidental but inexorable:

1. Solar-type stars in the Hyades show little or no scatter in rotation rates about a mean relation (Radick et al. 1987). Thus, stars of the same age and composition appear to end up rotating at the same rate.
2. The distribution of rotation rates estimated from H and K emission is likewise consistent with the Skumanich relation if the star-formation rate has been uniform.

On this last point, the distribution of H and K emission (hence rotation) does not change over a mass range of at least 0.9 to 1.1 M$_\odot$ (Soderblom 1985), which means that other stars of similar mass share the same general evolution. Stars less massive than the Sun rotate more slowly at a given age, while higher-mass stars rotate faster.

We know little enough of the internal rotation of the Sun, and nothing about the role that that effect might play in other stars. This is clearly a critical issue if it is possible for a significant reserve of angular momentum to hide deep beneath the surface, where it can be tapped at a later date. Latitudinal differential rotation is also important, especially for understanding dynamos. Here some evidence does exist for stars. The most relevant observations are those of Baliunas et al. (1985), who found evidence for multiple periods in their monitoring of stellar H and K emission. Although differential rotation is not the only potential source of such an effect, it is the most plausible agent.

Very young, highly active stars and close binaries show pronounced photometric variability. As such they are not strictly representative of the Sun, but evidence for differential rotation has been seen (Vogt 1981; Vogt and Penrod 1983; Busso et al. 1985) which, reassuringly, is in the same sense as the solar differential rotation.

More recently, Baliunas and her co-workers (see her chapter) have seen evidence for solar-like differential rotation on other solar-type stars in the form of period changes over time or double periods within a season. Their data cannot distinguish the sense of differential rotation relative to latitude, nor is there evidence for any consistent correlation between change in P_{rot} and overall activity.

Finally, in the study of the rotational evolution of solar-type stars, the most basic question of all is: Is the Sun typical for its mass and age? Is there anything about the Sun that sets it off from other otherwise similar stars? If so, is the difference linked to planets? Soderblom (1982,1983) showed that there was no evidence that the Sun differed in any significant way in its rotation from other old stars of similar mass, despite Smith's (1978) assertion. This conclusion is reinforced by the proxy data of the H and K emission: again the Sun looks just like other old stars of the solar neighborhood (Soderblom 1985). Only very careful long-term studies that determine actual rotation periods and the degree of differential rotation will definitively settle this question.

VI. AFTER THE MAIN SEQUENCE

As noted, detecting rotation in the oldest solar-type stars is very difficult: v sin i values are small, and chromospheric emission is generally too weak to produce measurable photometric variability. Given the slow rotation of the Sun, and the apparent absence of any additional angular momentum in its core, one expects 1 $M_\odot$ stars to rotate at very slow rates indeed as they bloat upon leaving the main sequence, even if no additional angular momentum is lost.

This expectation appears to be fulfilled. For example, the chromospheric emission strengths of the oldest main-sequence stars are consistent with slower-than-solar rotation if the same rotation-activity relation is obeyed. The subgiant β Hydri, a slightly evolved star near 1 $M_\odot$ (Spectral type G2 IV), likewise fits this mold.

A possible contradiction is provided by Arcturus (spectral type K2IIIp). Its mass is not well known since it is a single star, but spectroscopic determinations of its gravity seem very consistent in predicting a mass less than the Sun's (cf. Bell et al 1985). The problem is that Arcturus has a measurable v sin i ($\sim$ 2.5 km s^{-1}), that it should not have if it left the main sequence with the Sun's current angular momentum. (Arcturus' radius is twenty times the Sun's, so that its main-sequence rotation rate would have to have been roughly 50 km s^{-1}.) The Sun does not appear to be hiding angular momentum deep within that it can use at a later date. Perhaps Arcturus has undergone considerable mass loss after leaving the main sequence, so that it used to have a mass consistent with an appreciable rotation rate. Perhaps the line broadening has been misinterpreted. A final alternative is that the mass estimates for Arcturus have been too low; a recent estimate (Edvardsson 1988) suggests a mass of 2 to 3 $M_\odot$, which would resolve the rotation discrepancy.

Another possible contradiction to the expectation arises from observations of the rotational velocities of horizontal-branch stars. Peterson (1983) has shown that up to 1/3 of the horizontal-branch stars in some globular clusters have rotational velocities in the range 15 to 30 km s^{-1}. Because the progenitors of these stars presumably had surface rotational velocities of $<$ 10 km s^{-1}, this may indicate that the cores of these stars were rotating considerably more rapidly than their exteriors. An alternative explanation would be that these Population II stars have different rotational velocity histories from the Population I stars with which we are comparing them.

VII. ANGULAR MOMENTUM AND PLANETARY FORMATION

Our solar system is testimony that one of the means for stars to solve their angular momentum problem is to leave behind bits of high angular-momentum material in the form of a disk. What is not yet known is whether the particular solution used by our Sun (i.e., the formation of a number of planets with masses much less than the solar mass) is the norm or an aberration. Attempts to detect planets around other stars have generally not been successful, though a few controversial claims of possible planetary companions have been made (Campbell et al. 1988; van de Kamp 1986). When the first large scale programs to measure rotational velocities of field and cluster stars showed a break at about spectral type F0, with high-mass stars being rapid rotators and low-mass stars primarily slow rotators, one explanation advanced was that the cause of the break was that low-mass stars formed planets, whereas high-mass stars did not (Huang 1965). An alternative explanation is that the outer convective envelope, found only in the stars later than F0, drives a wind from the late-type stars (analogous to the solar wind), and the wind slowly carries away angular momentum during their main-sequence lifetime. The rotational velocities derived by Wilson (1966b) and Kraft (1967) (which show a progressive decrease in rotation for older stars, rather than an abrupt drop in rotation very early in the life of low-mass stars and a steady rotation rate thereafter) suggest that the latter explanation is better. The first to propose magnetic braking as the cause for slow rotation for late-type stars was Schatzmann (1962).

The rotational-velocity data for young clusters discussed in Sec. III provides a new opportunity to consider the connection between observed stellar rotational velocities and planet formation. As noted by van Leeuwen and Alphenaar (1982) and Rucinski (1988), the Pleiades and α Persei rapid rotators have angular momenta that follow Kraft's power-law relation derived from high-mass stars. In the same mass range where the rapid rotators are found, however, an approximately equal number of stars in each cluster have rotational velocities at or below the detection limit ($\sim$10 km s^{-1}) for the surveys (see Hartmann and Noyes 1987, Fig. 1). Several explanations for this slowly rotating population have been proposed (Stauffer and Hartmann 1986). One speculative possibility is that the slow rotators formed planets (or binary companions), whereas the rapid rotators did not. An observation suggestive of this is that the Pleiades K dwarf rapid rotators are all located along the single star main sequence in an HR diagram, whereas the slow rotators are approximately evenly divided between photometric binaries and single stars. Some theoretical support for this speculation is provided by Durisen et al. (1989), who predict radically different disk characteristics above and below a critical angular rotation rate of protostars. The division between super-critical and sub-critical star-formation modes postulated by Shu et al. (1987) also suggests that stars forming in different environments might have different angular momenta and planetary-system characteristics. Similarly, if naked T Tauri stars are born with less massive disks than classical T Tauri stars (an interpretation of the data which is plausible but not compelling), then the planet-forming characteristics of naked and classical T Tauri stars may be significantly different.

VIII. CONCLUSIONS AND HOPES FOR THE FUTURE

At this point, we believe we understand the major phases of evolution of the angular momentum of a solar-type star, and that no dramatic surprises wait to be discovered (post-main-sequence evolution could be an exception). Most of the lifetime of a star like the Sun is spent rotating very slowly; after all, the Sun rotates at a surface velocity of $<$ 2 km s^{-1}, and it is only halfway through its life. At least some, and perhaps all, of such stars go through a phase of extraordinarily rapid rotation just as they arrive on the zero-age main sequence, with a short-lived phase of angular-momentum loss, before the star starts its long, slow decline of rotation with advancing age.

The fact that these stars spin up to such rapid rates at all probably indicates that that process takes place in an extremely short time, for if it did not, the stars would lose angular momentum as fast as the surface accelerated, and the surfaces would not acquire the very high rates that they do.

Nevertheless, there remain a number of key questions to address observationally. Observers will be kept hard at work until the end of this century answering questions already posed by existing star-formation theories. New observational tools which should become available within approximately the next decade (HST to image T Tauri disks; ROSAT (Roentgen Satellite) and AXAF (Advanced X-ray Astrophysics Facility) to allow a complete census of low-mass star formation in associations; ISO and SIRTF to detect β Pic type disks one thousand times fainter than possible with IRAS; cryogenic-echelle spectrographs to provide high-resolution spectra of heavily embedded protostars) may make it possible to answer many of these questions. A difficult but important program will be to determine the extent to which low-mass stars born in bound clusters have systematically different rotational or disk properties relative to their cousins born in associations. Similarly, what is the dependence of disk properties and stellar rotation properties on the presence (and orbital separation and mass ratio) of a binary companion?

In contrast to the lack of correlation between rotation and optical emission line activity for T Tauri and naked T Tauri stars, Bouvier (1987) has shown that rotation and X-ray activity are well correlated for these stars. Bouvier interprets these data (cf. Bertout 1989) as indicating that the primary energy source for the optical emission lines in the classical T Tauri stars is accretion from a circumstellar disk, whereas the X-ray emission primarily arises from a solar-type magnetic dynamo. This suggests that deep X-ray imaging of the Taurus dark-cloud region with either ROSAT or AXAF should succeed in finding more slowly rotating Taurus members, with little bias between naked or classical T Tauri stars.

The possibility of significant age spreads in open clusters has been used to explain various facets of the rotational velocity distribution, HR-diagram properties, and lithium abundance properties for these clusters. The observational tools exist now to place much better limits on the possible age spread for young clusters, which in turn will limit the effects of such an age spread on the rotational velocity distribution.

Binary stars also offer the hope of placing strong constraints on models of the pre-main-sequence evolution of low-mass stars. At least a plausible assumption for wide binaries is that they are coeval. Hence, the two components of a pre-main-sequence binary should both fit on a single isochrone. Similarly, one might expect the lithium abundances of the binary components to fall on a lithium isochrone. It is also of interest to determine the rotational velocities, infrared excesses, etc. for binary systems. Probably most important would be the derivation of masses for a few pre-main-sequence binary systems. The opportunity provided by binary systems is now becoming a reality as a variety of surveys are providing lists of binary stars in young stellar systems (see, e.g., Reipurth 1989b; Mathieu et al. 1989; Zinnecker 1989).

A task that may go beyond the capabilities of this century is the observational confirmation of the idea that the cores of many Pleiades-age G dwarfs should be rotating much more rapidly than their outer convective envelopes. Astroseismology of nearby stars and eventually Hyades solar-mass stars may help accomplish this task. Extending the number of horizontal-branch stars for which rotational-velocity estimates are available will provide another useful constraint on the internal rotation of old, solar-mass stars. Finally, attempts to understand better the details of angular-momentum loss in young stars should be pursued. Collier-Cameron and Robinson's (1989) spectra of HD 36705, a very young, field star analog to the open-cluster rapid rotators, suggest that considerable mass loss and presumably angular-momentum loss occurs by ejection of discrete, partially ionized clouds rather than as a steady wind. High-resolution, high signal-to-noise ratio spectra of similar stars should be obtained in order to determine if HD 36705 is a prototype or a peculiarity.

Theoreticians also have some important contributions to make. First, as far as we are aware, no one has published tabular data indicating the change in moment of inertia of solar-type stars as they approach the zero-age main sequence. The numbers would be readily available to anyone calculating stellar structure at these phases, and are essential for interpreting and understanding the observations of rotation in young stars. Second, more sophisticated models of angular-momentum loss are needed, especially models that incorporate improved treatment of coupling between the core and the convective envelope. The dependence of angular-momentum loss on mass also needs to be known. Finally, it is possible that the study of lithium depletion may reveal some aspects of processes related to angular-momentum loss. Butler et al. (1987) and Balachandran et al. (1988) have shown that the rapid rotators in young clusters have depleted much less lithium than slow rotators of the same effective temperature. One interpretation of this effect is that the rapid rotators are younger than the slow rotators, which would then also explain the lithium difference. However, there is no proof that the rapid rotators are significantly younger than the slow rotators, and it may well be that rapid rotation by itself somehow inhibits lithium depletion. Pre-main-sequence models that incorporate rotation and lithium burning are being constructed; it will be interesting to see if they can match the observations.

Note added in proof: Since this article was originally written, new field detections of several young stars, in particular HD 17925 K2V, $t \sim 0.07$ Gyr, $fB \sim 525$ G (Saar 1991) and the weak T Tauri Tap 35 (G8, $t \sim 0.008$ Gyr, $fB \sim 1000$ G (Basri and Marcy 1991) have been made. These new data strongly point towards an exponential (or log-normal) relationship between magnetic flux and stellar age. An exponential fit to the enlarged data set yields $fB \propto e^{-0.8t \text{ Gyr}}$ (Saar 1991b).

THE TIME EVOLUTION OF MAGNETIC FIELDS ON SOLAR-LIKE STARS

STEVEN H. SAAR
Harvard-Smithsonian Center for Astrophysics

Magnetic fields play a crucial role in governing the rotational evolution of solar-like stars. Techniques for measuring stellar magnetic fields are reviewed, and the most recent determinations of magnetic-field strengths and surface-area coverages on cool stars are analyzed for correlations with age. Tentative evidence for a decrease in magnetic flux with age is found, primarily caused by a reduction in the magnetic area filling factor with time.

I. INTRODUCTION

The idea that magnetic fields play an important role in the evolution of late-type stars has been slowly pieced together from a number of important discoveries made in this century. Very early, it was clear that magnetic fields profoundly affect the physical structure and energetics of the outer envelope and atmosphere of the Sun. Even before the discovery of magnetic fields in sunspots (Hale 1908), their presence was strongly suggested by the topology study of the solar corona. Later, it became evident that there was an intimate relationship between magnetic fields and surface activity, such as Ca II emission in plages (see, e.g., Babcock and Babcock 1955) and flares (see, e.g., Giovanelli 1947). Discovery of similar phenomena on cool stars indicated that they, too, possessed magnetic fields. Indeed, the strength of activity on some late-type stars implied that their magnetic fields were stronger and/or more widespread than on the Sun (see, e.g., Golub 1984; Linsky 1985).

Parallel with these discoveries, Kraft (1967) noted that cool stars with surface convective layers showed a distinct decline in rotational velocity with age. Viewed in the context of dynamo theory (Parker 1955), Kraft's result would suggest that magnetic-field generation, which is dependent on rotational shear in the convective zone, would also decline as a star evolved. Observational support for this idea came when Wilson (1963) found (through studies of open-cluster stars of known age) that the magnetic activity diagnostic, chromospheric Ca II emission, shows a similar drop with time. Schatzman (1962) placed all these observations in a theoretical perspective by demonstrating that magnetic fields, rooted on a rotating star and "frozen" into its nonrotating, expanding wind, could apply sufficient torque to slow down the star's rotation.

Thus, magnetic fields are directly or indirectly connected with not only the structure and energy balance in the outer atmospheres of cool stars, but the very rotational evolution of the convection zone itself. The detailed physics of all these phenomena, however, will not be fully understood until information on the stellar magnetic fields is available. Such measurements are just now becoming available (beginning with the work of Robinson et al. [1980]) and are the subject of this chapter.

This chapter concentrates on the time evolution of stellar magnetism from the observational point of view. Section II begins with a brief review of the techniques used in measuring the magnetic fields of cool solar-like stars. Difficulties with the methods of analysis and a summary of the main results are given. In Sec. III, correlations between derived magnetic parameters and stellar age are explored by using a critically selected subset of the measurements made to date. A discussion of the results in the context of other activity-age relations is included, as well as a brief discussion of the data for pre- and post-main-sequence stars. Finally, in Sec. IV, possible directions for future research are considered.

II. TECHNIQUES

The most direct way of measuring magnetic fields in astronomical sources is through the Zeeman effect in spectral lines. The presence of a magnetic field will lift the degeneracy of certain quantum levels in an atom (denoted by the quantum number m_J), separating them in energy by an amount proportional to the field strength itself. For transitions with $\Delta m_J = 0$, the resulting spectral line is linearly polarized, while for $\Delta m_J = \pm 1$, the line is right- or left-handed circularly polarized. The latter effect is generally larger and more easily detected, and hence has been the basis for extensive studies of solar magnetic fields over the years. Unfortunately, similar observations of circular polarization on late-type stars have yielded virtually no positive detections (see, e.g., Babcock 1958; Borra et al. 1984).

The null result can be understood by realizing that, while high-resolution magnetograms of the Sun can show strong magnetic signals, the inferred flux decreases with decreasing spatial resolution. Full-disk observations of the

Sun show a field of only a few gauss. Physically, as more magnetic flux regions of opposite polarity fill the field of view, the circular polarization signal cancels. Unresolved stars thus will rarely show any significant net magnetic flux in circularly polarized light. Broadband linear polarization has been observed on late-type stars (see, e.g., Huovelin et al. 1988), but the results are difficult to interpret uniquely in the absence of additional geometric information (Landi Degl'Innocenti 1982).

The solution arrived at by Robinson (1980) is to observe magnetic broadening in *unpolarized* light. However, even this method has difficulties. The relative size of the Zeeman effect can be quantified by comparing it with other broadening mechanisms. Consider the simplest case: the so-called *simple triplet*—a spectral line which, under the influence of a magnetic field, splits into two circularly polarized σ components split symmetrically about an unshifted, linearly polarized π component. In this case, the wavelength separation of the σ components from line center $\Delta\lambda_B$ due to a magnetic field of strength B (gauss) is given by $\Delta\lambda_B = 4.67 \times 10^{-13} g_{\text{eff}} \lambda^2 B$, where λ is in Å and g_{eff} is the *effective* Landé g value (an intensity-weighted mean of the magnetic sensitivities of the σ components; see Beckers 1969). If we convert this wavelength shift into a velocity shift, i.e., $v = (\Delta\lambda/\lambda)c$

$$\Delta v_B = \frac{\Delta\lambda_B}{\lambda}c = 1.4 \times 10^{-7} g_{\text{eff}} \lambda B \qquad (1)$$

with Δv_B in km s^{-1}. Note the dependence on wavelength. The Doppler width (in velocity units) of the line originating from an element with atomic mass m is given by

$$\Delta v_D = \left(\frac{2kT_{\text{eff}}}{m} + \xi^2 \right)^{1/2} \qquad (2)$$

where T_{eff} is the stellar effective temperature and ξ is the microturbulent velocity in the photosphere. Thus, for an iron line at $\lambda = 6000$ Å with $g_{\text{eff}} = 2.5$ (i.e., very magnetically sensitive), and taking solar values of $T_{\text{eff}} = 5800$ K, $B \approx 1000$ gauss, and $\xi = 1$ km s^{-1}, one obtains $\Delta v_B = 2.1$ km s^{-1}, which is on the order of $\Delta v_D = 1.7$ km s^{-1}. Doppler broadening, however, is only one component of the nonmagnetic line width. One must also consider the effects of rotation (v sin $i > 1$ to 2 km s^{-1}, typically) and convective motions (i.e., macroturbulence, generally > 2 km s^{-1}; Gray 1984). Indeed, solar lines at 6000 Å typically have half widths of ≈ 0.08 Å, corresponding to a velocity half width of 4 km s^{-1}. Clearly, magnetic broadening is small in the optical portion of the spectra of solar-like stars and will not dominate the width of stellar line profiles. Careful analysis techniques using high-resolution ($\lambda/\Delta\lambda \geq 80,000$ at 6000Å; i.e., $\Delta\lambda \approx 2\Delta\lambda_B$) high signal-to-noise

(S/N $>$ 100 to 200) spectral data are therefore required. Even then, magnetic fields are increasingly difficult to measure at high v sin i, and nearly impossible, in practice, for v sin $i >$ 10 km s^{-1} (see, e.g., Saar 1988a).

The situation is still more complicated by the presence of nonmagnetic regions on the stellar surface. These areas will contribute an unshifted component to the line profile with a strength proportional to the fraction of the stellar surface they occupy. The magnetic signal can therefore be considerably diluted. The Sun, for example, has a surface coverage of magnetic regions (i.e., active network, plage, filagree, pores, spots) of roughly 1 to 2%, undetectable by the methods described here. Line blending will further confuse matters if not accounted for in some way (Gondoin et al. 1985; Saar 1988a; Basri and Marcy 1988). Finally, since pores and spots are dark compared with the photosphere at optical wavelengths, stellar magnetic measurements in the optical region refer to brighter structures, perhaps analogous to solar network and plage.

Within the inherent restrictions listed above, astronomers could finally obtain the necessary high-precision observations with the advent of modern detectors such as CCDs and Reticons. Robinson (1980) devised a scheme whereby low and high g_{eff} lines of very similar properties (equivalent width, excitation potential) are ratioed in the Fourier frequency domain. Marcy (1982,1984) developed an analogous procedure for comparing lines directly. In essence, both methods model an observed magnetically sensitive (high g_{eff}) line profile (F_{obs}) as observed in a star with active regions covering a fraction f (the filling factor) of its visible surface using

$$F_{obs} = fF_{mag}(B) + (1 - f)F_{nonmag}(B = 0) \qquad (3)$$

where B is the mean field strength in the active regions, and F_{mag} and F_{nonmag} represent the line profiles in the magnetic and nonmagnetic regions, respectively. In both methods, a low g_{eff} line is used for F_{nonmag}, and F_{mag} is formed from a weighted sum of three low g-line profiles. The first large survey of f and B on cool stars using these techniques was made by Marcy (1984). By linearly adding lines to form a Zeeman triplet, however, these early methods ignored radiative-transfer effects and did not include the exact Zeeman patterns. Also, no compensation was made for the ubiquitous weak-line blends present in the spectra of cool stars. In retrospect, all of these effects are important (Hartmann 1987; Saar 1988a; Basri and Marcy 1988) and likely have contributed to some unusual results, such as all stars having the *same* magnetic flux (Gray 1985; but see Stepién 1987). Subsequent modeling efforts have included radiative-transfer effects in computing F_{mag} and F_{nonmag}, full Zeeman patterns, and explicit integration of the magnetic-intensity profiles over the stellar disk (see, e.g., Saar and Linsky 1985; Basri and Marcy 1988; Bopp et al. 1989; Marcy and Basri 1989). A typical example of magnetic broadening effects is shown in Fig. 1.

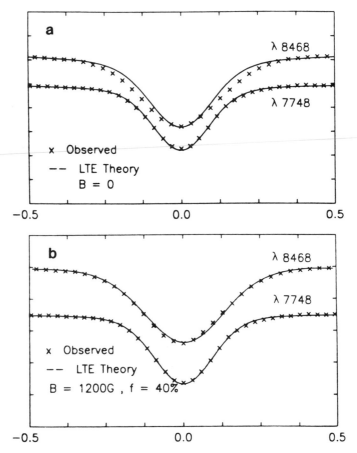

Fig. 1. Example of a modern analysis of stellar absorption lines for magnetic-field parameters (adapted from Basri and Marcy 1988). Data for a magnetically sensitive line (8468 Å; g_{eff} = 2.5) and a magnetically insensitive (7748 Å; g_{eff} = 1.1) line of the active G8 dwarf ξ Bootis A (age ≈ 3 × 10^8 yr) are shown (crosses) and compared with radiative-transfer models (solid lines). Plot (a) shows models for the two lines (with identical rotational and turbulent broadening parameters) for B = 0; clearly, the high-g_{eff} line is poorly fit in the wings of the line. Plot (b) shows models for the two lines with B = 1200 gauss and f = 40% rotational and turbulent broadening) with B = 0; both lines are fit well.

III. STELLAR MAGNETISM AND AGE

Evolution Near the Main Sequence

In order to explore the age dependence of stellar magnetism, I have compiled all measurements of magnetic-field strengths and filling factors made to date. These measurements, however, span some nine years and nearly as many analysis techniques. To construct a more homogeneous data set, and since earlier measurements have been shown to have systematic effects (Saar 1988a; Stepién 1987; see also above), the data set has been restricted to those f and B values which were derived by more modern methods (i.e., those which include magnetic radiative-transfer effects in the line modeling). Further, the analysis is confined to dwarf stars with reasonably well-determined ages. These restrictions leave data from the following sources: Saar and Linsky (1986a); Saar (1987a), and Marcy and Basri (1989). Most of the stars used have ages derived from cluster membership or Li I strength. In the cases of HD 22049 (ϵ Eri) and HD 131156A (ξ Boo A), measurements of both Marcy and Basri (1989) and Saar (1987a) are included in the analysis. Half weight was given to all age determinations that were not based on cluster membership, since these ages are generally less accurate. The final data set, summarized in Table I, is unfortunately quite small, consisting of only 13 stars.

Several functional forms have been proposed for the age-activity relationships in late-type stars. These include $X \propto t^{-0.5}$ (Skumanich 1972), $X \propto e^{-t}$ (Soderblom 1983) and log-normal ($X \propto e^{\alpha t^{0.5}}$; Simon et al. 1985; Barry 1988) forms where X denotes the flux or luminosity in some activity diagnostic. Several authors have noted correlations between derived magnetic param-

TABLE I
Stellar Magnetic Parameters, Spectral Types and Ages[a]

HD Number	Spec. Type	B (gauss)	f (%)	Ref.	Log Age (yr)	Method	Cluster/ Ref.[b]
39857	G0V	1000	60	1	8.5	C	UMa Stream
190406	G1V	1800	10	1	9.4	L	4
Sun	G2V	1500	2	1	9.6	L + A	4
1835	G2V	1400	32	1	8.8	C	Hyades
28099	G2V	1700	30	2	8.8	C	Hyades
20630	G5V	1500	35	1	9.0	L + A	4,5
131156A	G8V	1600	22	3	8.5	C	UMa Stream
131156A	G8V	1800	35	1	8.5	C	UMa Stream
152391	G8V	1700	18	1	8.5	A	5
22049	K2V	1000	30	3	9.0	A	6
22049	K2V	1900	13	1	9.0	A	6
45088	K3V	2400	50	2	8.5	C	UMa Stream
131156B	K4V	if 2600	≤20	1	8.5	C	UMa Stream
201091	K5V	if 1500	≤5	1	9.6	C	Wolf 630 group
201092	K7V	if 1500	≤10	1	9.6	C	Wolf 630 group

[a]Method of age determination is denoted by: C = cluster membership; L = Li I abundance; A = activity level.

[b]References: [1]Saar (1987a); [2]Saar and Linsky (1986b); [3]Marcy and Basri (1989); [4]Duncan (1981); [5]Barry (1988); [6]Soderblom and Däppen (1989).

eters and nonradiative emission from stellar chromospheres, transition regions and coronae (Marcy 1984; Saar and Schrijver 1987; Saar 1988*b*;
Stepień 1988*a*; Marcy and Basri 1989). Since such activity is also dependent
on stellar age (see, e.g., Hartmann and Noyes 1987), it is reasonable to expect similar relations between the magnetic parameters and age. Hence, I
have searched the magnetic data set defined above for correlations with these
functional forms.

Many theoretical models of magnetically driven rotational braking in
late-type stars assume that $B \propto \Omega^{\alpha}$, where α is a positive integer (see, e.g.,
Stix 1972; Kawaler 1988; Stepień 1988*b*; Weber and Davis 1967). If this
assumption is correct, the unambiguous dependence of v sin *i* on age then
implies that the magnetic-field strength should also decline with age. A plot
of the stars in the present sample, however (Fig. 2), shows no discernable
dependence of B on stellar age. This is consistent with the similar lack of
significant correlation between B and Ω (Linsky and Saar 1987). Instead,
there appears to be a significant negative correlation between B and T_{eff}. Saar
and Linsky (1986*b*) interpreted this to be a result of $B \propto B_{eq} \equiv (8\pi P_{gas})^{0.5}$, the
equipartition magnetic-field strength, where P_{gas} was computed at continuum
optical depth unity from atmospheric models. In this scenario, the external
photospheric gas pressure confines flux tubes and limits B. More recent analysis of slower rotating K dwarfs (Basri and Marcy 1988; Marcy and Basri
1989) suggests that rather than a strict proportionality, $B \leq B_{eq}$ may be more
appropriate (Galloway and Weiss 1981). The situation is somewhat confused,
however, by the use of different lines in the analyses, and the role of differing
depths of line formation must be sorted out to confirm the $B - B_{eq}$ relationship.

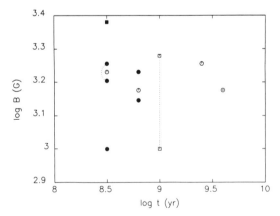

Fig. 2. Log B vs log age. Circles are G stars, squares are K stars, filled symbols are stars which
are members of clusters. The Sun is designated by a ⊙. Stars in common to different observers
are connected by dotted lines. No correlation is evident.

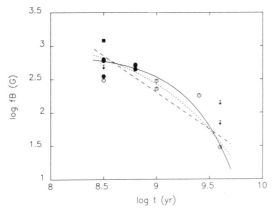

Fig. 3. Logarithm of the magnetic flux density fB vs log stellar age. Symbols are as in Fig. 1. Least-squares fits for $fB \propto t^{-1.0}$ (long dashed), $fB \propto e^{-\alpha t}$ (solid), and $fB \propto e^{-\alpha t^{0.5}}$ (short dashed) are shown.

The next obvious possibility is a correlation between age and magnetic flux. A search has been made for correlations with magnetic-flux density, i.e., the product $f \times B$, in order to reduce errors due to uncertainties in stellar radius. A reasonably clear correlation can be seen (Fig. 3), showing fB decaying with age. The data are insufficient to distinguish which decay law is best; the flatter part of the data at high flux levels, however, suggests that the exponential forms may be superior, with the simple exponential ($\propto e^{-t}$) form yielding marginally better fits. The reason for the decline of fB with time is the analogous decline in f (Fig. 4). Thus, it appears that changes in the surface-area coverage of magnetic regions are the main cause for the decline in magnetic activity and rotation rates with age on the main sequence. These results are consistent with the correlation between f and fB and Ω (Saar and Linsky 1986b; Linsky and Saar 1987; Marcy and Basri 1989) combined with the well-known Ω $-$ age relations (see, e.g., Soderblom 1983).

Due to the hint of a saturated state (or the approach to one) in the data, the dynamo theory developed by Skumanich and MacGregor (1986) is in better agreement with our data than that of Durney and Robinson (1982), as first suggested by Linsky and Saar (1987). This saturation is likely the result of the covering of the stellar surface with active regions (i.e., a filling factor approaching a maximum; see, e.g., Vilhu 1984). The saturation level in X-ray luminosity is proportional to R^2 (Fleming et al. 1989), also suggesting that saturation is primarily a surface-area effect (i.e., filling factor) and not strongly dependent on field strength. Thus, young main-sequence stars may have much of their surfaces covered with magnetic regions (see, e.g., Cayrel et al. 1983). Note that the filling factors derived by the magnetic-analysis techniques may be incorrect by a scaling factor related to the temperature

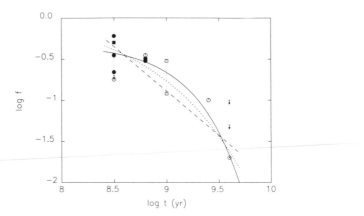

Fig. 4. Logarithm of the filling factor vs log stellar age. Symbols are as in Fig. 1. Least-squares fits for $f \propto t^{-1.0}$ (long dashed), $f \propto e^{-\alpha t}$ (solid), and $f \propto e^{-\alpha t^{0.5}}$ (short dashed) are shown. Comparing Figs. 3 and 4 shows that the decline of magnetic flux fB with age is due to a decrease in the surface coverage f of magnetic regions.

difference inside and outside of the magnetic regions; thus the "saturation" may not take place at $f \approx 1.0$.

Post- and Pre-Main-Sequence Evolution

The evolution of magnetic fields in solar-like post-main-sequence stars is still largely uncharted territory from the observational standpoint. Indeed, there have been no detections of magnetic fields on post-main-sequence stars of the types into which the Sun is expected to evolve (slowly rotating yellow and red giants). The only stars above the main sequence with measured magnetic parameters are RS CVn systems, and the results are sparse and somewhat contradictory. Giampapa et al. (1983), using a blended but highly Zeeman-sensitive line in the infrared, found a 1290 gauss field covering 48% of the surface on λ And. Marcy and Bruning (1984) surveyed 8 active giants, including λ And, at optical wavelengths and observed no significant magnetic broadening. They interpreted this negative result as due to lower field strengths in these lower-gravity stars. Gondoin et al. (1985) re-observed λ And, using less-blended, lower-g_{eff} infrared lines, this time making only a marginal detection at $B = 600$ gauss and $f = 0.20$. They interpret the varying results for this star in terms of varying coverage of starspots with larger magnetic-field strengths. Bopp et al. (1989) discovered a 2000 gauss field covering some 66% of the surface of the active subgiant HD 17433. The bulk of the evidence, however, suggests that magnetic-field strengths above the main sequence are somewhat smaller than those seen in dwarfs, consistent with the expectations of Zwaan and Cram (1989). No clear trends are visible in the filling factors measured to date.

Observationally, the pre-main-sequence evolution of magnetic fields in solar-like stars is even less studied than the post-main-sequence: there are no detections at all. This is largely due to the faint visual magnitudes and high v sin i of potential target stars: the brightest naked T Tauri stars currently known have $V \approx 9$ (Walter et al. 1988b), for example.

IV. SUMMARY AND FUTURE DIRECTIONS

In summary, the very limited data available suggest that the magnetic-field strength on stars depends largely on photospheric properties (gas pressure), and thus will vary from main-sequence values only in the pre-main-sequence T Tauri stage and in post-main-sequence evolution (see the chapters by Feigelson et al. and by Bertout and Basri). The magnetic flux, on the other hand, declines with stellar age, largely because of a similar decline in f. It is therefore likely that the decrease in stellar activity as a star ages is directly attributable to the decrease in f with time. I caution the reader, however, that these results are based on a very small data set.

In terms of future directions for stellar magnetic research, the suggestions of Saar (1987b) and Linsky (1989) are still valid. Investigations of cluster stars with relatively well-known ages should be undertaken. New larger-aperture telescopes will be helpful here. Much of the difficulty in measuring magnetic fields from optical spectra can be eliminated by observing in the middle (Saar et al. 1987) or far infrared (Deming et al. 1988). Since the ratio of the Zeeman splitting to the Doppler width is proportional to wavelength, the individual Zeeman components can often be resolved (see, e.g., Saar and Linsky 1985) in the infrared making the accurate determination of f and B considerably easier. The problem of blends is also reduced in the infrared due to the lower number density of lines. The new generation of infrared detectors and spectrographs under development should revolutionize stellar magnetic studies. Combined optical and infrared studies of magnetic fields, by exploiting the differing depths of line formation and differing continuum contributions in the two wavelength regions, may be able to separate the bright plage and cool spot components on stars (see, e.g., Sun et al. 1987). The relationships between magnetic flux and upper atmospheric heating and emission are best explored through simultaneous measurement of both magnetic fields and as many activity diagnostics as possible. The new echelle spectrographs coupled with large-format CCDs, capable of obtaining nearly the entire optical spectrum of a star in one or two exposures, will greatly aid this work. The possibility of magnetic imaging by using magnetic measurements in both unpolarized and linear polarized light (Saar et al. 1988) may bear fruit. The first successful measurements of weak broadband circular polarized light from cool stars (Kemp et al. 1987) should be followed up.

Improvements in the physics of the line modeling and advances in observational strategies will also be important. The effect of convection-related line asymmetries should be studied (Toner and Gray 1988). The effect on line profiles of differing thermodynamic properties in the magnetic and nonmagnetic regions should also be investigated in detail. Techniques using many lines (see, e.g., Mathys and Solanki 1989) should also be explored, since, once properly calibrated, they offer the possibility for deriving more accurate results, approximate information on the thermodynamic differences between magnetic and quiet regions, and some rough idea of actual magnetic geometry as well (see Solanki and Mathys 1988).

Acknowledgments.   This research is supported by a Smithsonian Institution postdoctoral research fellowship, a grant from NASA and an Interagency Transfer. I would like to extend special thanks to G. Marcy, G. Basri and D. Barry for sharing some results with me prior to publication, and to the referees for helpful comments.

ACKNOWLEDGMENTS

The editors acknowledge National Aeronautics and Space Administration Grant NASW-4357, National Science Foundation Grant AST-8815404 and The University of Arizona for support of the preparation of this book. They wish to thank J. E. Frecker, who volunteered as one of the proofreaders of this book. The following authors wish to acknowledge specific funds involved in supporting the preparation of their chapters.

Anderson, R. Y.: NSF Grant ATM-8707462
Baliunas, S. L.: NSF Grant AST-8616545
Becker, R. H.: NASA Grant NAG-9-60
Cisowski, S. M.: NASA Grant NAG-9-22414
Damon, P. E.: NSF Grant EAR-8309448
Ghil, M.: NSF Grants ATM-8615424 *and* ATM 90–13217
Jokipii, J. R.: NSF Grant ATM-8618260 *and* NASA Grant NSG-7101
Kasting, J. F.: NASA Joint Interchange NCA 2–369 *and* NSF Grant ATM-8901775
Kerridge, J. F.: NASA Grant NAG 9-27
Lal, D.: NSF Grant EAR 8904484
Lingenfelter, R. E.: NSF Grant ATM-8717676
Marti, K.: NASA Grant NAG 9-41
Pellas, P.: ATP Planètologie Contracts 87-67-02 and 88-37-07
Pepin, R. O.: NASA Grant NAG 9-60
Reedy, R. C.: NASA/JSC Order T-294M
Saar, S.: NASA Grant NAGW 112 *and* Interagency Transfer W15130
Signer, P.: Swiss National Science Foundation Grants 2.813.0.85 *and* 2000-025261
Walter, F. M.: NASA Grant NAG-8-508 *and* NSF Grant AST-8996308
Wieler, R.: Swiss National Science Foundation Grants 2.813.0.85 *and* 2000-025261
Wood, J. A.: NASA Grant NAG 9-28

BIBLIOGRAPHY

BIBLIOGRAPHY

Compiled by Mary Guerrieri

Abazov, A. I., Abdurashitov, D. N., Anosov, O. V., Avdeyev, A. V., Belouska, Yu. I., By-chuk, O. V., Danshin, S. N., Eroshkina, L. A., Faizov, E. L., Gavrin, V. N., Gayevsky, V. I., Girin, S. V., Grigoryev, A. M., Kalikhov, A. V., Kireyev, S. M., Knodel, T. V., Knyshenko, I. I., Kornoukhov, V. N., Mezentseva, S. A., Mirmov, I. N., Nemstsveridze, Sh. M., Ostrinsky, A. I., Petukhov, V. V., Pikhulya, O. E., Pshukov, A. M., Resvin, N. Ye., Sedredinov, R. K., Shikhin, A. A., Shilo, Yu. I., Slyusareva, Ye. D., Stepanyuk, M. V., Tikhonov, A. A., Timofeyev, P. V., Veretenkin, E. P., Vermul, V. M., Yants, V. E., Zakharov, Yu. I., Zatsepin, G. T., Zhandarov, V. L., Cleveland, B. T., Bowles, T. J., El-liot, S. R., O'Brien, H. A., Wark, D. L., Wilkerson, J. F., Davis, R., Lande, K., and Cherry, M. L. 1989. The Baksan neutrino observatory Soviet-American gallium solar neu-trino experiment. In *Neutrino '88: Proc. 13th Intl. Conf. on Neutrino Physics and Astrophys-ics*, eds. J. Schneps, T. Kafka, W. A. Mann and P. Nath (Singapore: World Scientific), pp. 317–330.

Abazov, A. I., Abdurashitov, D. N., Anosov, O. V., Eroshkina, L. A., Gavrin, V. N., Kali-khov, A. V., Knodel, T. V., Knyshenko, I. I., Kornoukhov, V. N., Mezentseva, S. A., Mir-mov, I. N., Ostrinsky, A. I., Petukhov, V. V., Pshukov, A. M., Resvin, N. Ye., Shikhin, A. A., Timofeyev, P. V., Veretenkin, E. P., Vermul, V. M., Zakharov, Yu., Zatsepin, G. T., Zhandarov, V. I., Bowles, T. J., Cleveland, B. T., Elliot, S. R., O'Brien, H. A., Wark, D. L., Wilkerson, J. F., Davis, R., Lande, K., Cherry, M. L., and Kouzes, R. T. 1990. The Soviet-American gallium experiment (SAGE) at the Baksan neutrino observatory. *Neutrino '90*, in press.

Abraham, Z., and Iben, I., Jr. 1970. The abundance of ^{3}He and ^{4}He in the initial Sun implied by the Kocharov and Starbunov assumption. *Astrophys. J.* 162:L125-L127.

Abt, H. A. 1983. Normal and abnormal binary stars. *Ann. Rev. Astron. Astrophys.* 21:343–372.

Adams, F. C. 1989. The infrared signature of star formation. In *Infrared Spectroscopy in Astron-omy*, ed. B. H. Kaldeich (Dordrecht: Kluwer), pp. 233–243.

Adams, F. C., Lada, C. J., and Shu, F. 1987. Spectral evolution of young stellar objects. *Astro-phys. J.* 312:788–806.

Adams, F. C., Ruden, S. P., and Shu, F. H. 1989. Eccentric gravitational instabilities in nearly Keplerian disks. Harvard-Smithsonian Center for Astrophysics Preprint No. 2829.

Adams, J. B. 1974. Visible and near-infrared diffuse reflectance spectra of pyroxenes as applied to remote sensing of solid objects in the solar system. *J. Geophys. Res.* 79:4829–4836.

Akaike, H. 1969a. Fitting autoregressive models for prediction. *Ann. Inst. Statist. Math.* 21:243–247.

Akaike, H. 1969b. Power spectrum estimation through autoregressive model fitting. *Ann. Inst. Statis. Math.* 21:407–419.

Alcock, C. R., Fuller, G. M., and Mathews, G. J. 1987. The quark hadron phase transition and primordial nucleosynthesis. *Astrophys. J.* 320:439–447.

Alfvén, H., and Arrhenius, G. 1976. *Evolution of the Solar System*, NASA SP-345.

Allègre, C. J., Birck, J. L., Fourcade, S., and Semet, M. P. 1975. Rubidium-87/strontium-87 age of Juvinas basaltic achondrite and early igneous activity in the solar system. *Science* 187:436–438.

Allen, C. W. 1976. *Astrophysical Quantities* (London: Athlone Press).

Aller, L. H. 1986. In *Spectroscopy of Astrophysical Plasmas*, eds. A. Dalgarno and D. Layzer (Cambridge: Cambridge Univ. Press), pp. 89–124.

Alphenaar, P., and van Leeuwen, F. 1981. Variable stars in the Pleiades cluster. *Info. Bull. Variable Stars* No. 1957, pp. 1–5.

Altschuler, M. D., Trotter, D. E., Newkirk, G., Jr., and Howard, R. 1969. Magnetic fields and the structure of the solar corona. I. Methods of calculating coronal fields. *Solar Phys.* 9:131–149.

Alvarez, L. W., Alvarez, W., Asaro, F., and Michel, H. V. 1980. Extraterrestrial cause for the Cretaceous-Tertiary extinction. *Science* 208:1095–1108.

Alyea, F. N. 1972. Numerical simulation of an ice-age paleoclimate. Colorado State Univ. Atmos. Sci. Paper 193.

Ambartsumian, V. A., and Mirzoyan, L. V. 1975. Flare stars in star clusters and associations. In *Variable Stars and Stellar Evolutions,* eds. V. E. Sherwood and L. Plaut (Dordrecht: D. Reidel), pp. 3–14.

Ambronn, L. 1905. Die Messungen des Sonnendurchmessers an dem Repsold'schen 6-zolligen Heliometer der Sternwarte zu Gottingen. *Astron. Mittheilungen Koniglichen Sternwarte Gottingen* III:1–125.

Anders, E. 1964. Origin, age, and composition of meteorites. *Space Sci. Rev.* 3:583–714.

Anders, E. 1975. Do stony meteorites come from comets? *Icarus* 24:363–371.

Anders, E. 1978. Most stony meteorites come from the asteroid belt. In *Asteroids: An Exploratory Assessment,* eds. D. Morrison and W. C. Wells, NASA CP-2053, pp. 57–76.

Anders, E. 1988. Circumstellar material in meteorites: Nobel gases, carbon and nitgrogen. In *Meteroites and the Early Solar System,* eds. J. F. Kerridge and M. S. Matthews (Tucson: Univ. of Arizona Press), pp. 927–955.

Anders, E., and Ebihara, M. 1982. Solar-system abundances of the elements. *Geochim. Cosmochim. Acta* 46:2363–2380.

Anders, E., and Grevesse, N. 1989. Abundances of the elements: Meteoritic and solar. *Geochim. Cosmochim. Acta* 53:197–214.

Anderson, L. S., and Athay, R. G. 1989a. Chromospheric and coronal heating. *Astrophys. J.* 336:1089–1091.

Anderson, L. S., and Athay, R. G. 1989b. Model solar chromosphere with prescribed heating. *Astrophys. J.* 346:1010–1018.

Anderson, R. Y. 1961. Solar-terrestrial climatic patterns in varved sediments. In *Solar Variations, Climatic Change and Related Geophysical Problems,* ed. R. W. Fairbridge. Annals New York Acad. Sci., vol. 95, pp. 424–439.

Anderson, R. Y. 1982. A long geoclimatic record from the Permian. *J. Geophys. Res.* 87:7285–7294.

Anderson, R. Y. 1984. Orbital forcing of evaporite sedimentation. In *Milankovitch and Climate: Understanding the Response to Orbital Forcing,* eds. A. L. Berger, J. Imbrie, J. Hays, G. Kukla and B. Saltzman (Dordrecht: D. Reidel), pp. 147–162.

Anderson, R. Y. 1986. The varve microcosm: Propagator of cyclic bedding. *Paleoceanography* 1(4):373–382.

Anderson, R. Y. 1990. Solar-cycle modulation of ENSO: A mechanism for Pacific and global climate change. In *Sixth Pacific Climate (PACLIM) Workshop,* eds. J. Betancourt and A. M. MacKay (Asilomar, Calif.: Calif. State Dept. of Water Resources), pp. 77–81.

Anderson, R. Y., and Dean, W. E. 1988. Lacustrine varve formation through time. *Paleogeogr. Paleoclimatol. Paleoecol.* 62:215–235.

Anderson, R. Y., and Kirkland, D. W. 1960. Origin, varves, and cycles in the Jurassic Todilito Formation. *Amer. Assoc. Petrol. Geol. Bull.* 44:37–52.

Anderson, R. Y., and Kirkland, W. 1966. Intrabasin varve correlation. *Geol. Soc. Amer. Bull.* 77:241–256.

Anderson, R. Y., and Koopmans, L. H. 1963. Harmonic analysis of varve time series. *J. Geophys. Res.* 68:877–893.

Anderson, R. Y., and Koopmans, L. H. 1969. Statistical analysis of the Rita Blanca varve timeseries. In *Paleoecology of an Early Pleistocene Lake on the High Plains of Texas,* eds. R. Y. Anderson and D. W. Kirkland, Geol. Soc. Amer. Memoir 113, pp. 59–75.

Anderson, R. Y., Dean, W. E., Kirkland, D. W., and Snider, H. I. 1972. Permian Castile varved evaporite sequence, west Texas and New Mexico. *Geol. Soc. Amer. Bull.* 83:59–86.

Anderson, R. Y., Linsley, B. K., and Gardner, J. V. 1990. Expression of seasonal and ENSO forcing in climatic variability at lower than ENSO frequencies: Evidence from Pleistocene marine varves off California. In *Paleoclimates: The Record from Lakes, Ocean, and Land*, eds. L. V. Benson and P. A. Myers. *Paleogeogr. Paleoclimatol. Paleoecol.* vol. 78.

André, P. 1987. Radio emission from young stellar objects. In *Protostars and Molecular Clouds*, eds. T. Montmerle and C. Bertout (Saclay: CEA/Dox), pp. 143–187.

André, P., Montmerle, T., and Feigelson, E. D. 1987. A VLA survey of radio-emitting young stars in the ρ Ophiuchi dark cloud. *Astron. J.* 93:1182–1198.

André, P., Montmerle, T., Feigelson, E. D., and Steppe, H. 1990. Cold dust around young stellar objects in the ρ ophiuchi cloud core. *Astron. Astrophys.* 240:321–331.

Angell, J. K., and Korshover, J. 1973. Quasi-biennial and long-term fluctuations in total ozone. *Mon. Weather Rev.* 101:426–443.

Angell, J. K., and Korshover, J. 1976. Global analysis of recent total ozone fluctuations. *Mon. Weather Rev.* 104:63–75.

Angione, R. 1981. Review of ground-based measurements. In *Variations of the Solar Constant*, ed. S. Sofia, NASA CP-2191, pp. 11–30

Antonucci, E., Gabriel, A. H., Acton, L. W., Culhane, J. L., Doyle, J. G., Leibacher, J. W., Machado, M. E., Orwig, L. E., and Rapley, C. G. 1982. Impulsive phase of flares in soft X-ray emission. *Solar Phys.* 78:107–123.

Antonucci, E., Gabriel, A. H., and Dennis, B. R. 1984. The energetics of chromospheric evaporation in solar flares. *Astrophys. J.* 287:917–925.

Appenzeller, I., and Dearborn, D. S. 1984. Brightness variations caused by surface magnetic fields in pre-main sequence stars. *Astrophys. J.* 278:689–694.

Appenzeller, I., and Mundt, R. 1989. T Tauri stars. *Astron. Astrophys. Rev.* 1:291–334.

Appenzeller, I., Jankovics, I., and Oestriecher, R. 1985. Forbidden line profiles of T Tauri stars. *Astron. Astrophys.* 141:108–115.

Applegate, J. H., and Hogan, C. J. 1985. Relicts of cosmic quark condensation. *Phys. Rev.* D31:3037–3045.

Applegate, J. H., Hogan, C. J., and Scherrer, R. J. 1987. Cosmological baryon diffusion and nucleosynthesis. *Phys. Rev.* D35:1151–1160.

Armstrong, T. W., and Alsmiller, R. G., Jr. 1971*a*. Calculation of cosmogenic radionuclides in the Moon and comparison with Apollo measurements. In *Proc. Lunar Sci. Conf.* 2:1729–1745.

Armstrong, T. W., and Alsmiller, R. G., Jr. 1971*b*. Calculation of cosmogenic radionuclides in the Moon and comparison with Apollo measurements. ORNL TN-3267 (Oak Ridge, Tenn.: Oak Ridge National Lab).

Armstrong, T. W., Chandler, K. C., and Barish, J. 1973. Calculations of neutron flux spectra induced in the Earth's atmosphere by galactic cosmic rays. *J. Geophys. Res.* 78:2715–2726.

Arnoldy, R. L. 1971. Signature in the interplanetary medium for substorms. *J. Geophys. Res.* 76:5189–5201.

Arter, W. F. 1983. Magnetic-flux transport by a conveting layer–topological, geometrical and compressible phenomena. *J. Fluid Mech.* 132:25–48.

Atkinson, R. d'E., and Houtermans, F. G. 1929. Transmutation of the lighter elements in stars. 1929. *Nature* 123:567–568 (abstract).

Atmanspacher, H., Scheingraber, H., and Voges, W. 1988. Global scaling properties of a chaotic attractor reconstructed from experimental data. *Phys. Rev.* A37:1314–1322.

Attolini, M. R., Galli, M., and Nanni, T. 1988*a*. Long and short cycles in solar activity during the last millenium. In *Secular, Solar, and Geomagnetic Variations in the Last 10,000 Years*, eds. F. R. Stephenson and A. W. Wolfendale (Dordrecht: Kluwer), pp. 49–68.

Attolini, M. R., Cecchini, S., Cini Castagnoli, G., Galli, M., and Nanni, T. 1988*b*. On the existence of the 11-year cycle in solar activity before the Maunder minimum. *J. Geophys. Res.* 93:12729–12734.

Audouze, J., Epherre, M., and Reeves, H. 1967. Survey of experimental cross sections for proton-induced spallation reactions in He^4, C^{12}, N^{14}, and O^{16}. In *High-Energy Nuclear Reactions in Astrophysics*, ed. B. S. P. Shen (New York: Benjamin), pp. 255–271.

Audouze, J., Bibring, J. P., Dran, J. C., Maurette, M., and Walker, R. M. 1976. Heavily irradiated grains and neon isotope anomalies in carbonaceous chrondrites. *Astrophys. J.* 206:L185-L189.

Auzout, A. 1729. Du micromètre ou manière exacte. *Mem. Acad. Sci. Paris* 7:118–130.

Axford, W. I. 1972. The interaction of the solar wind with the interstellar medium. In *Solar*

Wind: Proc. Second Intl. Conf., eds. C. P. Sonnett, P. J. Coleman, Jr. and J. M. Wilcox, NASA SP-308, p. 609.

Ayres, T. R., Marstad, N. C., and Linsky, J. L. 1981. Outer atmospheres of cool stars. IX. A survey of ultraviolet emission from F-K dwarfs and giants with IUE. *Astrophys. J.* 247:545–559.

Babcock, H. W. 1958. A catalog of magnetic stars. *Astrophys. J. Suppl.* 3:141–210.

Babcock, H. W. 1961. The topology of the Sun's magnetic field and the 22 year cycle. *Astrophys. J.* 133:572–587.

Babcock, H. W., and Babcock, H. D. 1955. The Sun's magnetic field: 1952–1954. *Astrophys. J.* 121:349–366.

Backman, J., Pestiaux, P., Zimmerman, H., and Hermelin, O. 1986. Paleoclimatic and paleoceanographic development in the Pliocene North Atlantic: Discoaster accumulation and coarse fraction data. In *North Atlantic Palaeoceanography,* eds. C. P. Summerhayes and N. J. Shackleton, Geol. Soc. of London SP-21 (Oxford: Blackwell Scientific), pp. 231–242.

Bahcall, J. N. 1978. Solar neutrino experiments. *Rev. Mod. Phys.* 50:881–903.

Bahcall, J. N. 1985. Solar neutrino experiments: Theory. In *Solar Neutrinos and Neutrino Astronomy,* eds. M. L. Cherry, W. A. Fowler and K. Lande (New York: American Inst. of Physics), pp. 60–68.

Bahcall, J. N. 1989. *Neutrino Astrophysics* (Cambridge: Cambridge Univ. Press).

Bahcall, J. N., and Ulrich, R. K. 1971. Solar neutrinos. III. Composition and magnetic-field effects and related inferences. *Astrophys. J.* 170:593–603.

Bahcall, J. N., and Ulrich, R. K. 1988. Solar models, neutrino experiments, and helioseismology. *Rev. Mod. Phys.* 60:297–372.

Bahcall, J. N., Huebner, W. F., Lubow, S. F., Parker, P. D., and Ulrich, R. K. 1982. Standard solar models and the uncertainties in predicted capture rates of solar neutrinos. *Rev. Mod. Phys.* 54:767–799.

Bahcall, J. N., Cleveland, R. T., Davis, R., and Rowley, J. K. 1985. Chlorine and gallium solar neutrino experiments. *Astrophys. J.* 292:L79–L83.

Bahcall, J. N., Baldo-Ceolin, M., Cline, D. B., and Rubbia, C. 1986. Predictions for the liquid argon detector. *Phys. Lett.* B178:324–328.

Bahcall, J. N., Field, G. B., and Press, W. H. 1987. Is solar neutrino capture rate correlated with sunspot number? *Astrophys. J.* 320:L69-L73.

Bahcall, J. N., Davis, R., and Wolfenstein, L. 1988. Solar neutrinos: A field in transition. *Nature* 334:487–493.

Bai, T. 1986. Two classes of gamma-ray/proton flares: Impulsive and gradual. *Astrophys. J.* 308:912–928.

Bai, T. 1987. Distribution of flares on the Sun: Superactive regions and active zones of 1980–1985. *Astrophys. J.* 314:795–807.

Balachandran, S., Lambert, D., and Stauffer, J. R. 1988. Lithium in lower main sequence stars of the Alpha Persei cluster. *Astrophys. J.* 333:267–276.

Baldini, U., and Coyne, G. V. 1984. In *The Louvain Lectures of Bellarmine and the Autograph Copy of his 1616 Declaration to Galileo.* Studi Galileiani, vol. I (Vatican: Vatican Obs.), pp. 1–48.

Baliunas, S. L. 1988. Long-term variations of stellar magnetic activity in lower main sequence stars. In *Formation and Evolution of Low-Mass Stars,* eds. A. K. Dupree and M. T. V. T. Lago (Boston: D. Reidel), pp. 319–330.

Baliunas, S. L., and Jastrow, R. 1989. Long-term changes in surface activity of solar-type stars. *Bull. Amer. Astron. Soc.* 21:115 (abstract).

Baliunas, S. L., and Jastrow, R. 1990. Evidence for long-term brightness changes of solar-type stars. *Nature* 348:520.

Baliunas, S. L., and Vaughan, A. H. 1985. Stellar activity cycles. *Ann. Rev. Astron. Astrophys.* 23:379–412.

Baliunas, S. L. 1983. Stellar rotation in lower main-sequence stars measured from time variations in H and K emission-line fluxes. II. Detailed analysis of the 1980 observing season data. *Astrophys. J.* 275:752–772.

Baliunas, S. L., Horne, J. H., Porter, A., Duncan, D. K., Frazer, J., Lanning, H., Misch, A.,

Mueller, J., Noyes, R. W., Soyumer, D., Vaughan, A. H., and Woodard, L. 1985. Time-series measurements of the CaII H and K emission in cool stars and the search for differential rotation. *Astrophys. J.* 294:310–325.

Bally, J. 1986. Massive bipolar outflows around young stars. *Irish Astron. J.* 17:270–279.

Bally, J., and Lada, C. J. 1983. The high-velocity molecular flows near young stellar objects. *Astrophys. J.* 265:824–847.

Bally, J., and Stark, A. A. 1983. Atomic hydrogen associated with the high-velocity flow in NGC 2071. *Astrophys. J.* 266:L61-L64.

Bame, S. J., Asbridge, J. R., Feldman, W. C., Montgomery, M. D., and Kearney, P. D. 1975. Solar wind heavy ion abundances. *Solar Phys.* 43:463–473.

Bame, S. J., Asbridge, J. R., Feldman, W. C., and Gosling, J. T. 1977. Evidence for structure-free state at high solar wind speeds. *J. Geophys. Res.* 82:1487–1492.

Bandeen, W. E., and Maran, S. P., eds. 1975. *Possible Relationships Between Solar Activity and Meteorological Phenomena,* NASA SP-366.

Bander, C. M., and Orszag, S. A. 1978. Introduction to boundary-layer theory. In *Advanced Mathematical Methods for Science and Engineering* (New York: McGraw-Hill), pp. 419–426.

Banerjee, S. K., and Hargraves, R. B. 1972. Natural remanent magnetizations of carbonaceous chrondrites and the magnetic field in the early solar system. *Earth Planet. Sci. Lett.* 17:110–119.

Barabanov, I. R., Veretenkin, E. P., Gavrin, V. N., Danshin, S. N., Eroshkina, L. A., Zatse-pin, G. T., Zakharov, Yu. I., Klimova, S. A., Klimov, Yu. B., Knodel, T. V., Kopylov, A. V., Orekhov, I. V., Tikhonov, A. A., and Churmaeva, M. I. 1985. Pilot installation of the gallium-germanium solar neutrino telescope. In *Solar Neutrinos and Neutrino Astronomy,* eds. M. L. Cherry, K. Lande and W. A. Fowler (New York: American Inst. of Physics), pp. 175–184.

Barashenkov, V. S., Gudima, K. K., and Toneev, V. D. 1968. Energy dependence of nuclear cross sections for nucleons above 50 MeV. Dubna Rept. P2–4183 (Moscow: Joint Inst. for Nuclear Research). In Russian.

Barbetti, M. 1976. The Lake Mungo geomagnetic excursion. *Phil. Trans. Roy. Soc. London* A281:515–542.

Barnes, H. A. 1959. *Apparatus and Methods of Oceonography* (London: Allen and Unwin).

Barnes, H. A., Sargent, H. H., III, and Tryon, P. V. 1980. Sunspot cycle simulation using ran-dom noise. In *The Ancient Sun: Fossil Record in the Earth, Moon and Meteorites,* eds. R. O. Pepin, J. A. Eddy and R. B. Merrill (New York: Pergamon Press), pp. 159–163.

Barnes, T. G., and Evans, D. S. 1976. Stellar angular diameters and visual surface brightness. I. Late spectral types. *Mon. Not. Roy. Astron. Soc.* 174:489–502.

Barnett, T. 1990. A solar-ocean relation: Fact or fiction? *Geophys. Res. Lett.* 16:803–806.

Barnola, J. M., Raynaud, D., Korotkevitch, Y. S., and Lorius, C. 1987. Vostok ice core: A 160,000 year record of atmospheric CO_2. *Nature* 329:408–414.

Barral, J. F., and Cantó, J. 1981. A stellar wind model for bipolar nebulae. *Rev. Mexicana Astron. Astrof.* 5:101–108.

Barry, D. C. 1988. The chromospheric age dependence of the birthrate, composition, motions, and rotation of late F and G dwarfs within 25 parsecs of the Sun. *Astrophys. J.* 334:436–448.

Barry, D. C., Hege, K., and Cromwell, R. H. 1984. The time dependence of chromospheric decay for solar type stars. *Astrophys. J.* 277:L65–68.

Barry, D. C., Cromwell, R. H., and Hege, E. K. 1987. Chromospheric activity and ages of solar-type stars. *Astrophys. J.* 315:264–272.

Barth, C. A., Tobiska, W. K., Siskind, D. E., and Cleary, D. D. 1987. Solar-terrestrial cou-pling: Low-latitude thermospheric nitric oxide. *Geophys. Res. Lett.* 15:92–94.

Basri, G. S. 1987a. Stellar activity in synchronized binaries. II. A correlation analysis with single stars. *Astrophys. J.* 316:377–388.

Basri, G. S. 1987b. The T Tauri stars. In *Cool Stars, Stellar Systems and the Sun: Proc. Fifth Cambridge Workshop,* eds. J. L. Linsky and R. E. Stencel (New York: Springer-Verlag), pp. 411–421.

Basri, G. S. 1988. Emission activity in T Tauri stars. In *Formation and Evolution of Low Mass Stars,* eds. A. K. Dupree and M. T. V. T. Lago (Dordrecht: Kluwer), pp. 247–255.

Basri, G. S., and Bertout, C. 1989. Accretion disks around T Tauri stars. II. Balmer emission. *Astrophys. J.* 341:340–358.

Basri, G. S., and Marcy, G. W. 1988. Physical realism in the analysis of stellar magnetic fields. *Astrophys. J.* 330:273–285.

Basri, G. S., and Marcy, G. W. 1991. In *The Sun and Cool Stars: Proc IAU Coll. 130*, eds. I. Tuominen et al. (Dordrecht: Kluwer), in press.

Basri, G. S., Laurent, R., and Walter, F. M. 1985. Stellar activity in synchronized binaries. I. Dependence on rotation. *Astrophys. J.* 298:761–771.

Baxter, M. S., and Farmer, J. G. 1973. Radiocarbon: Short-term variations. *Earth Planet. Sci. Lett.* 20:295–299.

Baxter, M. S., and Walton, A. 1971. Fluctuations of atmospheric carbon-14 concentrations during the past century. *Proc. Roy. Soc. London* A321:105–127.

Baxter, M. S., Farmer, J. G., and Walton, A. 1973. Comments on "On the magnitude of the 11-year radiocarbon cycle" by P. E. Damon, A. Long, and E. I. Walick. *Earth Planet. Sci. Lett.* 20:307–310.

Bazilevskaya, G., Stozhkov, Yu. I., and Charakhch'yan, T. N. 1982. Cosmic rays, solar activity and neutrino flux from the Sun. *J. Experim. Theoret. Phys. Lett.* 35:341–343.

Beardsley, B. J. 1987. The Visual Shape and Multipole Moments of the Sun. Ph.D. Thesis, Univ. of Arizona.

Beardsley, B. J., Czarnowski, R. J., Kroll, R. J., Cornuelle, C. S., Oglesby, P. H., Yi, L., and Hill, H. A. 1989. Variations in solar luminosity since 1957 using differential radius observations. *Bull. Amer. Astron. Soc.* 20:1011 (abstract).

Becker, R. H. 1989. Solar wind gases in a metal separate from lunar soil 68501. *Lunar Planet. Sci.* XX:54–55 (abstract).

Becker, R. H., and Clayton, R. N. 1977. Nitrogen isotopes in lunar soils as a measure of cosmic-ray exposure and regolith history. *Proc. Lunar Sci. Conf.* 8:3685–3704.

Becker, R. H., and Kramer, F. E. 1986. Extension of the holocene dendrochronology by the preboreal pine series, 8800–10,100 BP. *Radiocarbon* 28:961–967.

Becker, R. H., and Pepin, R. O. 1989. Long-term changes in solar wind elemental and isotopic ratios: A comparison of two lunar ilmenites of different antiquities. *Geochim. Cosmochim. Acta* 53:1135–1146.

Beckers, J. M. 1969. *A Table of Zeeman Multiplets*. Air Force Cambridge Research Lab Report 471.

Beckwith, S. V. W., Zuckerman, B., Skrutskie, M. F., and Dyck, H. M. 1984. Discovery of solar system-size halos around young stars. *Astrophys. J.* 278:793–800.

Beckwith, S. V. W., Sargent, A. I., Chini, R. S., and Güsten, R. 1990. A survey or circumstellar disks around young stellar objects. *Astron. J.* 99:924–945.

Beer, J. 1988. Radioisotopes in natural archives: Information about the history of the solar-terrestrial system. In *Solar-Terrestrial Relationships and the Earth Environment in the Last Millennia*, ed. G. Cini Castagnoli (Amsterdam: North Holland Press), pp. 183–198.

Beer, J., Siegenthaler, U., Oeschger, H., Andrée, M., Bonani, G., Hofmann, H., Nessi, M., Suter, M., Wölfli, W., Finkel, R., and Langeway, C. C. 1983. Temporal ^{10}Be variations. In *Proc. 18th Intl. Cosmic Ray Conf.*, vol. 9, eds. N. Durgaprasad, S. Ramadurai, P. V. Raman Murthy, M. V. S. Rao and K. Sivaprasad (Bombay: Tata Inst. of Fundamental Research), pp. 317–320.

Beer, J., Oeschger, H., Andrée, M., Bonani, B., Suter, M., Wölfli, W., and Langway, C. C. 1984. Temporal variations in the ^{10}Be concentration levels found in the Dye 3 ice core, Greenland. *Annals Glaciol.* 5:16–18.

Beer, J., Oeschger, H., Finkel, R. C., Cini Castagnoli, G., Bonino, G., Attolini, M. R., and Galli, M. 1985. Accelerator measurements of ^{10}Be: The 11-yr solar cycle from 1180–1800 AD. *Nucl. Instrum. Meth.* B10–11:415–418.

Beer, J., Siegenthaler, U., and Blinov, A. 1988a. Temporal ^{10}Be variations in ice: Information on solar activity and geomagnetic field intensity. In *Secular, Solar and Geomagnetic Variations in the Last 10,000 Years*, eds. F. R. Stephenson and W. Wolfendale (Dordrecht: Kluwer), pp. 297–313.

Beer, J., Siegenthaler, U., Bonani, G., Finkel, R. C., Oeschger, H., Suter, M., and Wölfli, W. 1988b. Information on past solar activity and geomagnetism from ^{10}Be in the Camp Century Ice Core. *Nature* 331:675–679.

Beer, J., Blinov, A., Bonani, G., Finkel, R. C., Hofmann, H. J., Lehmann, B., Oeschger, H., Sigg, A., Schwander, J., Staffelbach, T., Stauffer, B., Suter, M., and Wölfli, W. 1990. Use of ¹⁰Be in polar ice to trace the 11-year cycle of solar activity: Information on cosmic ray history. *Nature* 347:164–166.

Begemann, F., Born, W., Palme, H., Vilcsek, E., and Wänke, H. 1972. Cosmic-ray produced radioisotopes in Apollo 12 and Apollo 14 samples. In *Proc. Lunar Sci. Conf.* 3:1693–1702.

Belcher, J. W., and Davis, L., Jr. 1971. Large amplitude Alfvén waves in the interplanetary medium. 2. *J. Geophys. Res.* 76:3534–3563.

Belcher, J. W., Goertz, C. K., Sullivan, J. D., and Acuña, M. H. 1981. Plasma observations of the Alfvén wave generated by Io. *J. Geophys. Res.* 86:8508–8512.

Bell, J. F. 1986a. Mineralogical evidence of meteorite parent bodies. *Lunar Planet. Sci.* XVII:985–986 (abstract).

Bell, J. F. 1986b. Mineralogical evolution of the asteroid belt. *Meteoritics* 21:333–334 (abstract).

Bell, J. F. 1988. A probable asteroidal parent body for the CV or CO chondrites. *Meteoritics* 23:256–257 (abstract).

Bell, J. F., Hawke, B. R., and Owensby, P. D. 1987. Carbonaceous chondrites from S-type asteroids. *Bull. Amer. Astron. Soc.* 19:841 (abstract).

Bell, J. F., Davis, D. R., Hartmann, W. K., and Gaffey, M. J. 1989. Asteroids: The big picture. In *Asteroids II*, eds. R. P. Binzel, T. Gehrels and M. S. Matthews (Tucson: Univ. of Arizona Press), pp. 921–945.

Bell, R. A., Edvardsson, B., and Gustafsson, B. 1985. The surface gravity of Arcturus from MgH lines, strong metal lines and the ionization equilibrium of iron. *Mon. Not. Roy. Astron. Soc.* 212:497–515.

Belvedere, G. 1985. Solar and stellar activity: The theoretical approach. *Solar Phys.* 100:363–383.

Bender, C. M., and Orszag, S. A. 1978. *Advanced Mathematical Methods for Science and Engineering* (New York: McGraw-Hill).

Benioff, P. A. 1956. Cosmic-ray production rate and mean removal time of beryllium-7 from the atmosphere. *Phys. Rev.* 104:1122–1130.

Benkert, J.-P. 1989. Solare Edelgase bei stufenweisem Aetzen von Ilmeniten und Pyroxenen aus dem Mondregolith. Ph.D. Thesis No. 8812, ETH Zürich.

Benkert, J.-P., Baur, H., Pedroni, A., Wieler, R., and Signer, P. 1988. Solar He, Ne, and Ar in regolith minerals: All are mixtures of two components. *Lunar Planet. Sci.* XIX:59–60 (abstract).

Bennett, J. O. 1987. A Comprehensive Survey of Late-Type Stars in the Far-Ultraviolet. Ph.D. Thesis, Univ. of Colorado.

Bennett, J. O., Ayres, T. R., and Rottman, G. J. 1984. The solar stellar connection in the far ultraviolet. In *Future of Ultraviolet Astronomy Based on Six Years of IUE Research*, NASA CP-2349, pp. 437–440.

Benoist, J. P., Glangeaud, F., Martin, N., Lacoume, J. L., Lorius, C., and Oulahman, A. 1982. Study of climatic time series by time-frequency analysis. In *Proc. Intl. Conf. Acousics, Speech and Signal Processing* (New York: Inst. of Electrical and Electronics Engineers Press), pp. 1902–1905.

Benz, W., Mayor, M., and Mermilliod, J. 1984. Rotational studies of lower main sequence stars in nearby open galactic clusters. I. Velocity distributions and age dependence. *Astron. Astrophys.* 138:93–100.

Benzi, R., Parisi, G., Sutera, A., and Vulpiani, A. 1982. Stochastic resonance in climatic change. *Tellus* 34:10–16.

Berger, A. 1976a. Obliquity and precession for the last 5,000,000 years. *Astron. Astrophys.* 51:127–135.

Berger, A. 1976b. Long-term variations of daily and monthly insolation during the last Ice Age. *Eos: Trans. AGU* 57:254.

Berger, A. 1977. Support for the astronomical theory of climatic change. *Nature* 268:44–45.

Berger, A. 1978a. Long-term variations of daily insolation and Quaternary climatic changes. *J. Atmos. Sci.* 35:2362–2367.

Berger, A. 1978b. Long-term variations of caloric insolation resulting from the Earth's orbital elements. *Quat. Res.* 9:139–167.

Berger, A. 1979. Insolation signatures of Quaternary climatic changes. *Nuovo Cimento* 2C(1):63–87.

Berger, A. 1980. Milankovitch astronomical theory of paleoclimates: A modern review. *Vistas in Astron.* 24(2):103–122.

Berger, A. 1988. Milankovitch theory and climate. *Rev. Geophys.* 26:624–657.

Berger, A. 1989a. The spectral characteristics of pre-Quaternary climatic records, an example of the relationship between the astronomical theory and geo-sciences. In *Climate and Geo-Sciences: A Challenge for Sciences and Society in the 21st Century*, eds. A. Berger, S. Schneider and J.-C. Duplessy (Dordrecht: Kluwer), pp. 47–76.

Berger, A. 1989b. Pleistocene climatic variability at astronomical frequencies. *Quat. Intl.* 2:1–14.

Berger, A., and Andjelic, T. P. 1988. Milutin Milankovitch, père de la théorie astronomique des paléoclimats. *Histoire et Mesure,* vol. III-3 (Paris: Editions du CNRS), pp. 385–402.

Berger, A., and Loutre, M. F. 1987. *Origine des fréquences des éléments astronomiques intervenant dans le calcul de l'insolation.* Scientific Report 1987/1.3 (Lovain-la-Neuve: Inst. Astron. Geophys. Georges Lemaître).

Berger, A., and Loutre, M. F. 1988. *Insolation Values for the Climate of the Last 10 million Years.* Scientific Rept. 1988/13 (Louvain-la-Neuve: Inst. Astron. Geophys. Georges Lemaître).

Berger, A., and Pestiaux, P. 1984. Accuracy and stability of the Quaternary terrestrial insolation. In *Milankovitch and Climate,* eds. A. Berger, J. Imbrie, J. Hays, G. Kukla and B. Saltzman (Dordrecht: D. Reidel), pp. 83–112.

Berger, A., and Pestiaux, P. 1987. Astronomical frequencies in paleoclimatic data. In *The Climate of China and Global Climate,* eds. Y. Duzheng, F. Congbin, C. Jiping and M. Yoshino (Berlin: Springer-Verlag), pp. 106–114.

Berger, A., Imbrie, J., Hays, J., Kukla, G., and Saltzman, B., eds. 1984. *Milankovitch and Climate: Understanding the Response to Orbital Forcing,* vols. 1 and 2 (Dordrecht: D. Reidel).

Berger, A., Gallée, H., Fichefet, Th., Marsiat, I., and Tricot, Ch. 1988. Transient response of the climate system to the astronomical forcing. In *Global Change IGBP,* eds. A. H. Cottenie and A. Teller (Scope, Belgium: Académie Royale des Sciences, des Lettres et des Beaux-Arts de Belgique), pp. 10–24.

Berger, A., Loutre, M. F., and Dehant, V. 1989. Influence of the changing lunar orbit on the astronomical frequencies of pre-Quaternary insolation patterns. *Paleoceanog.* 4(5):555–564.

Berger, A., Gallée, H., Fichefet, Th., Marsiat, I., and Tricot, Ch. 1990a. Testing the astronomical theory with a coupled climate-ice sheet model. *Global and Planetary Change* 3:125–142.

Berger, A., Fichefet, Th., Gallée, H., Marsiat, I., Tricot, Ch., and van Ypersele de Strihou, J. P. 1990b. Physical interactions within a coupled climate model over the last glacial-interglacial cycle. *Trans. Roy. Soc. Edinburgh: Earth Sciences* 82:357–369.

Bernard, E. A. 1962. Théorie astronomique des pluviaux et interpluviaux du quaternaire africain. *Mem. Acad. Roy. Sci, Outre-Mer, Brussels, Classe Sci. Tech.,* tome 12(1).

Bernatowicz, T. J., Hohenberg, C. M., and Podosek, F. A. 1979. Xenon component organization in 14301. *Proc. Lunar Planet. Sci. Conf.* 10:1587–1616.

Bernatowicz, T. J., Kramer, F. E., Podosek, F. A., and Honda, M. 1982. Adsorption and excess fission Xe: Adsorption of Xe on vacuum crushed minerals. *Proc. Lunar Planet. Sci. Conf.* 13:A465-A476.

Bernatowicz, T. J., Podosek, F. A., Swindle, T. D., and Honda, M. 1988. I-Xe systematics in LL chrondites. *Geochim. Cosmochim. Acta* 52:1113–1122.

Berner, R. A., Lasaga, A. C., and Garrels, R. M. 1983. The carbonate-silicate geochemical cycle and its effect on atmospheric carbon dioxide over the past 100 million years. *Amer. J. Sci.* 283:641–683.

Berry, P. A. M. 1987. Periodicities in sunspot cycle. *Vistas in Astron.* 30:97–108.

Berthomieu, G., Provost, J., and Schatzman, E. 1984. Solar gravity modes as a test of turbulent diffusion mixing. *Nature* 308:254–256.

Bertout, C. 1984. T Tauri stars: An overview. *Rept. Prog. Phys.* 47:111–174.

Bertout, C. 1989. T Tauri stars: Wild as dust. *Ann. Rev. Astron. Astrophys.* 27:351–395.

Bertout, C., and Bouvier, J. 1989. T Tauri disk models. In *ESO Workshop on Low Mass Star*

Formation and Pre-Main Sequence Objects, ed. B. Reipurth (Garching: European Southern Obs.), pp. 215–232.

Bertout, C., and Magnan, C. 1987. Line profiles from moving spherical shells. *Astron. Astrophys.* 183:319–323.

Bertout, C., and Thum, C. 1982. Radio observations of pre-main-sequence stars: Results and interpretation. *Astron. Astrophys.* 107:368–375.

Bertout, C., Basri, G., and Bouvier, J. 1988. Accretion disks around T Tauri stars. *Astrophys. J.* 330:350–373.

Bertsch, D. L., Biswas, S., and Reams, D. V. 1974. Solar cosmic ray composition above 10 MeV/nucleon and its energy dependence in the 4 August 1972 event. *Solar Phys.* 39:489–491.

Bethe, H. A. 1939. Energy production in stars. *Phys. Rev.* 55:434–456.

Bethe, H. A. 1986. Possible explanation of the solar-neutrino puzzle. *Phys. Rev. Lett.* 56:1305–1308.

Bethe, H. A., and Critchfield, C. L. 1938. The formation of deuterons by proton combination. *Phys. Rev.* 54:248–254.

Bhandari, N. 1981. Records of ancient cosmic radiation in extraterrestrial rocks. *Proc. Indian Acad. Sci. (Earth Planet. Sci.)* 90:359–382.

Bhandari, N., Goswami, J. N., Lal, D., and Tamhane, A. S. 1973a. Long-term fluxes of heavy cosmic ray nuclei based on observations in meteorites and lunar samples. *Astrophys. J.* 185:975–983.

Bhandari, N., Goswami, J. N., and Lal, D. 1973b. Kinetic energy spectra of 5–1500 McV/n VH nuclei: Long-term averaged fluxes at 1–3 AU. In *Proc. 13th Intl. Cosmic Ray Conf.,* vol. 1 (Denver: Univ. of Denver Press), pp. 281–286.

Bhandari, N., Bhattacharya, S. K., and Padia, J. T. 1976. Solar proton fluxes during the last million years. *Proc. Lunar Sci. Conf.* 7:513–523.

Bhattacharya, S. K., Goswami, J. N., Gupta, S. K., and Lal, D. 1973. Cosmic ray effects induced in a rock exposed on the Moon or in free space: Contrast in patterns for track and isotopes. *Moon* 8:253–286.

Bibring, J. P., Borg, J., Burlingame, A. L., Langevin, Y., Maurette, M., and Vassent, B. 1975. Solar-wind and solar-flare maturation of the lunar regolith. *Proc. Lunar Sci. Conf.* 5:3471–3493.

Beiging, J. H., and Cohen, M. 1985. Multifrequency radio images of L1551-IRS5. *Astrophys. J.* 289:L5-L8.

Bieging, J. H., Cohen, M., and Schwartz, P. R. 1984. VLA observations of T Tauri stars. II. A luminosity-limited survey of Taurus-Auriga. *Astrophys. J.* 282:699–708.

Biermann, L. 1941. *Vierteljahrschr. Astr. Ges.* 76:194.

Binder, A. B. 1986. The initial thermal state of the Moon. In *Origin of the Moon,* eds. W. K. Hartmann, R. J. Phillips and G. J. Taylor (Houston: Lunar and Planetary Inst.), pp. 425–433.

Birchfield, G. E., and Grumbine, R. W. 1985. "Slow" physics of large continental ice sheets, the underlying bedrock and the Pleistocene ice ages. *J. Geophys. Res.* 90:11294–11302.

Birchfield, G. E., and Weertman, J. 1978. A note on the spectral response of a model continental ice sheet. *J. Geophys. Res.* 83:4123–4125.

Birchfield, G. E., and Weertman, J. 1982. A model study of the role of variable ice albedo in the climate response of the Earth to orbital variations. *Icarus* 50:462–472.

Birchfield, G. E., Weertman, J., and Lunde, A. T. 1981. A paleoclimate model of northern hemisphere ice sheets. *Quat. Res.* 15:126–142.

Birchfield, G. E., Weertman, J., and Lunde, A. T. 1982. A model study of the role of high-latitude topography in the climatic response to orbital insolation anomalies. *J. Atmos. Sci.* 39:71–87.

Birk, J.-L., and Allègre, P. 1985. Evidence for the presence of Mn-53 in the early solar system. *Geophys. Res. Lett.* 12:745–748.

Black, D. C. 1972. On the origins of trapped helium, neon and argon isotopic variations in meteorites. I. Gas-rich meteorites, lunar soil and breccias. *Geochim. Cosmochim. Acta* 36:347–375.

Black, J. H., and Scott, E. H. 1982. A numerical study of the effect of ambipolar diffusion on the collapse of magnetic gas clouds. *Astrophys. J.* 293:696–773.

Blanco, C., Catalano, S., Marilli, E., and Rodono, M. 1974. Absolute fluxes of K chromospheric emission in main sequence stars. *Astron. Astrophys.* 33:257–264.

Bland, C. J., and Cioni, G. 1968. Geomagnetic cutoff rigidities in non-vertical directions. *Earth Planet Sci. Lett.* 4:339–405.

Blandford, R. D., and Payne, D. G. 1982. Hydromagnetic flows from accretion discs and the production of radio jets. *Mon. Not. Roy. Astron. Soc.* 199:883–903.

Blanford, G. E., Fruland, R. M., and Morrison, D. A. 1975. Longterm differential energy spectrum for solar flare iron group particles. *Proc. Lunar Sci. Conf.* 6:3557–3576.

Bochsler, P. 1984. Helium and Oxygen in the Solar Wind: Dynamic Properties and Abundances of Elements and Helium Isotopes as Observed with the ISEE-3 Plasma Composition Experiment. Habilitation Thesis, Univ. of Bern.

Bochsler, P. 1987. Solar wind ion composition. *Phys. Scripta* T18:55–60.

Bochsler, P. 1989. Velocity and abundances of silicon ions in the solar wind. *J. Geophys. Res.* 94:2365–2373.

Bochsler, P., and Geiss, J. 1973. Solar abundances of light nuclei and mixing of the Sun. *Solar Phys.* 32:3–11.

Bochsler, P., and Geiss, J. 1989. Composition of the solar wind. In *Solar System Plasma Physics,* eds. J. H. Waite Jr., J. L. Burch and R. L. Moore. AGU Geophysical Mono. 54 (Washington, D.C.: American Geophysical Union), pp. 133–141.

Bochsler, P., Geiss, J., and Kunz, S. 1986. Abundances of carbon, oxygen, and neon in the solar wind during the period from August 1978 to June 1982. *Solar Phys.* 103:177–201.

Bochsler, P., Geiss, J., and Maeder, A. 1990. The abundance of ^{3}He in the solar wind—A constraint for models of solar evolution. *Solar Phys.* 128:203–215.

Bodenheimer, P. 1965. Studies in stellar evolution. II. Lithium depletion during pre-main-sequence contraction. *Astrophys. J.* 142:451–461.

Bodenheimer, P. 1981. The effect of rotation during star formation. In *Fundamental Problems in the Theory of Stellar Evolution,* eds. D. Sugimoto, D. Q. Lamb and D. N. Schramm (Dordrecht: D. Reidel), pp. 5–26.

Boeckl, R. S. 1972. A depth profile of ^{14}C in the lunar rock 12002. *Earth Planet. Sci. Lett.* 16:269–272.

Boesgaard, A. M. 1987. Lithium in the Coma star cluster. *Astrophys. J.* 321:967–974.

Boesgaard, A. M., and Tripicco, M. J. 1986. Lithium in the Hyades cluster. *Astrophys. J.* 302:L49-L53.

Boesgaard, A. M., Budge, K. G., and Burck, F. F. 1988. Lithium and the metallicity of the Ursa Major group. *Astrophys. J.* 325:749–758.

Bogart, R. S., and Bai, T. 1985. Confirmation of a 152 day periodicity in the occurrence of solar flares inferred from microwave data. *Astrophys. J.* 299:L57-L61.

Böhm-Vitense, E. 1958. Über die Wasserstoffonvektionszone in Sternen verschiedener Effektiv-temperaturen und Leuchtkräfte. *Z. Astrophys.* 46:108–143.

Booth, N. E. 1987. The indium solar neutrino projects. *Sci. Prog. Oxford* 71:563–592.

Bopp, B. W., Saar, S. H., Ambruster, C., Feldman, P., Dempsey, R., Allen, M., and Barden, S. P. 1989. The active chromosphere binary HD 17433 (VY Ari). *Astrophys. J.* 339:1059–1072.

Borg, J., Chaumont, J., Jouret, C., Langevin, Y., and Maurette, M. 1980. Solar wind radiation damage in lunar dust grains and the charactreistics of the ancient solar wind. In *The Ancient Sun: Fossil Record in the Earth, Moon and Meteorites,* eds. R. O. Pepin, J. A. Eddy and R. B. Merrill (New York: Pergamon Press), pp. 431–461.

Borra, E. F., Edwards, G., and Mayor, M. 1984. The magnetic field of late-type stars. *Astrophys. J.* 284:211–222.

Borrini, G., Gosling, J. T., Bame, S. J., Feldman, W. C., and Wilcox, J. M. 1981. Solar wind helium and hydrogen structure near the heliospheric current sheet: A signal of coronal streamers at 1 AU. *J. Geophys. Res.* 86:4565–4573.

Bos, A. J., and Hill, H. A. 1982. Detection of individual normal modes of oscillation of the Sun in the period range from 2 hr to 10 min in solar diameter studies. *Solar Phys.* 82:89–102.

Boss, A. P. 1985. Angular momentum transfer by gravitational torques and the evolution of binary protostars. *Mon. Not. Roy. Astron. Soc.* 209:543–567.

Boss, A. P. 1987a. Bipolar flows, molecular gas disks, and the collapse and accretion of rotating interstellar clouds. *Astrophys. J.* 316:721–732.

Boss, A. P. 1987*b*. Protostellar formation in rotating interstellar clouds. IV. Nonuniform initial conditions. *Astrophys. J.* 319:149–161.

Bostrom, K., and Frederiksson, K. 1966. Surface conditions of the Orgueil meteorite parent body as indicated by mineral associations. *Smithsonian Misc. Collection* 151(3):4–5.

Bostrom, C. O., Williams, D. J., Arens, J. F., and Kohl, J. W. 1967–1973. Solar proton monitor experiment. *Solar Geophys. Data*, vols. 282–341, monthly summaries.

Bouvier, J. 1989. Rotation in T Tauri stars. II. Clues for magnetic activity. In preparation.

Bouvier, J., and Bertout, C. 1989. Spots on T Tauri stars. *Astron. Astrophys.* 211:99–114.

Bouvier, J. 1987. Pre-main sequence stellar activity. In *Protostars and Molecular Clouds*, eds. T. Montmerle and C. Bertout (Toulouse: CEN-Saclay), pp. 189–210.

Bouvier, J., Bertout, C., Benz, W., and Mayor, M. 1986. Rotation in T Tauri stars. I. Observations and immediate analysis. *Astron. Astrophys.* 165:110–119.

Bowell, E., Chapman, C. R., Gradie, J. C., Morrison, D., and Zellner, B. 1978. Taxonomy of asteroids. *Icarus* 35:313–335.

Boyer, D., and Levy, E. H. 1984. Oscillating dynamo magnetic field in the presence of external nondynamo field. *Astrophys. J.* 277:848–861.

Bracewell, R. N. 1978. *The Fourier Transform and Its Applications*, 2nd ed. (New York: McGraw-Hill).

Bradley, J. P., Brownlee, D. E., and Fraundorf, P. 1984. Discovery of nuclear tracks in interplanetary dust. *Science* 226:1432–1434.

Bradley, R. S. 1985. *Quaternary Paleoclimatology* (Boston: Allen and Unwin).

Brandenburg, A., and Tuominen, I. 1988. Variation of magnetic fields and flows during the solar cycle. COSPAR paper N12.4.6 (Espoo, Finland).

Brandt, J. C. 1970. *Introduction to the Solar Wind* (San Francisco: Freeman).

Brasseur, G., De Rudder, A., Keating, G. M., and Pitts, M. C. 1987. Response of middle atmosphere to short-term solar ultraviolet variations. 2. Theory. *J. Geophys. Res.* 92:903–914.

Bray, R. J., and Loughhead, R. E. 1964. *Sunspots* (New York: Dover).

Brecher, A. 1973. The electrical conductivity of carbonaceous meteorites. *Meteoritics* 8:330 (abstract).

Brecher, A., and Albright, L. 1977. The thermomagnetic hypothesis and the origin of magnetization in iron meteorites. *J. Geomag. Geoelect.* 29:379–400.

Brecher, A., and Arrhenius, G. 1974. The paleomagnetic record in carbonaceous chondrites: Natural remanence and magnetic properties. *J. Geophys. Res.* 79:2081–2106.

Brecher, A., and Fuhrman, M. 1979. The magnetic effects of brecciation and shock in meteorites. II. The ureilites and evidence for strong nebular magnetic fields. *Moon and Planets* 20:251–263.

Brecher, A., and Ranganayaki, R. P. 1975. Paleomagnetic systematics of ordinary chrondrites. *Earth Planet. Sci. Lett.* 25:57–67.

Brecher, A., Briggs, P. L., and Simmons, G. 1975. The low-temperature electrical properties of carbonaceous meteorites. *Earth Planet Sci. Lett.* 28:37–45.

Brecher, A., Fuhrman, M., and Stein, J. 1979. The magnetic effect of brecciation and shock in meteorites. III. The achondrites. *Moon and Planets* 20:265–279.

Brenemann, H. H., and Stone, E. C. 1985. Solar coronal and photospheric abundances from solar energetic particle measurements. *Astrophys. J.* 299:L57-L61.

Bretthorst, L. G. 1988. Bayesian Spectrum Analysis and Parameter Estimation. *Lecture Notes in Statistics*, vol. 48 (Berlin: Springer-Verlag).

Briskin, M., and Harrell, J. 1980. Time series analysis of the Pleistocene deep-sea paleoclimatic record using periodic regression. *Marine Geol.* 36:1–22.

Brodsky, Yu. A., Krasheninnikova, Yu. S., and Ruzmaikin, A. A. 1988. MHD action of differential rotation. *Magnitnaya Gidrodinamika* 3:13–18.

Broecker, W. S., and van Donk, J. 1970. Insolation changes, ice volumes, and the O^{18} record in deep-sea cores. *Rev. Geophys. Space Phys.* 8:169–197.

Broecker, W. S., Thurber, D. L., Goddard, J., Ku, T., Matthews, R. K., and Mesolella, K. J. 1968. Milankovitch hypothesis supported by precise dating of coral reefs and deep sea sediments. *Science* 159:297–300.

Broecker, W. S., Peteet, D. M., and Rind, D. 1985. Does the ocean-atmosphere system have more than one stable mode of operation? *Nature* 315:21–25.

Broecker, W. S., Andrée, M., Bonani, G., Wölfli, W., Oeschger, H., and Klas, M. 1988. Can the Greenland climate jumps be identified in records from ocean and land? *Quat. Res.* 30:1–6.

Broomhead, D. S., and King, G. P. 1986. Extracting qualitative dynamics from experimental data. *Physics* 20D:217–236.

Brown, A., Ferraz, M. C. de M., and Jordan, C. 1981. The chromosphere and corona of T Tauri. *Mon. Not. Roy. Astron. Soc.* 207:831–859.

Brown, A., Mundt, R., and Drake, S. A. 1985. Radio continuum emission from pre-main sequence stars and associated structures. In *Radio Stars*, eds. R. M. Hellming and D. M. Gibson (Dordrecht: D. Reidel), pp. 105–110.

Brown, H. 1947. An experimental method for the estimation of the age of the elements. *Phys. Rev.* 72:348.

Brown, H. 1949. Rare gases and the formation of the Earth's atmosphere. In *The Atmospheres of the Earth and Planets*, ed. G. P. Kuiper (Chicago: Univ. of Chicago Press), pp. 258–266.

Brown, J. C. 1971. The deduction of energy spectra of non-thermal electrons in flares from the observed dynamic spectra of hard X-ray bursts. *Solar Phys.* 18:489–502.

Brown, T. M. 1982. Seeing-independent definitions of the solar limb position. *Astron Astrophys.* 116:260–264.

Brown, T. M. 1985. Solar rotation as a function of depth and latitude. *Nature* 317:591–594.

Brown, T. M. 1986. Measuring the Sun's internal rotation using solar *p*-mode oscillations. In *Seismology of the Sun and the Distant Stars*, ed. D. O. Gough (Dordrecht: D. Reidel), pp. 199–214.

Brown, T. M. 1987. Grown-based observations relating to the global state of the Sun. In *Solar Radiative Output Radiation*, ed. P. Foukal (Boulder: Natl. Center for Atmosphehric Research), pp. 176–188.

Brown, T. M. 1988*a*. Ground-based observations relating to the global state of the Sun. In *Solar Radiative Output Variations*, ed. P. V. Foukal (Cambridge, Mass.: Cambridge Research and Instrumentation), pp. 176–188.

Brown, T. M. 1988*b*. Solar diameter measurements. *Bull. Amer. Astron. Soc.* 20:376 (abstract).

Brown, T. M., and Morrow, C. A. 1987. Depth and latitude dependence of solar rotation. *Astrophys. J.* 314:L21-L26.

Brown, T. M., Elmore, D. F., Lacey, L., and Hull, H. 1982. Solar diameter monitor: An instrument to measure long-term changes. *Appl. Opt.* 21:3588–3596.

Browne, E., Firestone, R. B., and Shirley, V. S. 1986. *Table of Radioactive Isotopes* (New York: Wiley).

Brownlee, D. E., and Rajan, R. S. 1973. Micrometeorite craters discovered on chondrule-like objects from Kapoeta meteorite. *Science* 182:1341–1344.

Brückner, E., Köppen, W., and Wegener, A. 1925. Uber die Klimate der geologischen Vorzeit. *Zeit. Gletscherkunde* 14.

Bucha, V. 1969. Changes of the Earth's magnetic moment and radiocarbon dating. *Nature* 224:681–683.

Bucha, V. 1970. Influence of the Earth's magnetic field on radiocarbon dating. In *Radiocarbon Variations and Absolute Chronology*, ed. I. U. Olsson (New York: Wiley), pp. 501–511.

Budyko, M. I. 1969. Effect of solar radiation variations on the climate of Earth. *Tellus* 21:611–620.

Bumba, V., and Howard, R. 1965. Large-scale distribution of solar magnetic fields. *Astrophys. J.* 141:1502–1512.

Bunch, T. E., and Rajan, R. S. 1988. Meteorite regolith breccias. In *Meteorites and the Early Solar System*, eds. J. F. Kerridge and M. S. Matthews (Tucson: Univ. of Arizona Press), pp. 144–164.

Burchuladze, A. A., Pagava, S. V., Povinec, P., Togonidze, G. J., and Usacev, S. 1980. Radiocarbon variations with the 11 year solar cycle during the last century. *Nature* 287:320–322.

Burg, J. P. 1975. Maximum Entropy Spectral Analysis. Ph.D. Thesis, Stanford Univ.

Bürgi, A., and Geiss, J. 1986. Helium and minor ions in the corona and solar wind: Dynamics and charge states. *Solar Phys.* 103:347–383.

Burlaga, L. F., McDonald, F. B., Ness, N. F., Schwenn, R., Lazarus, A. J., and Mariani, F. 1984. Interplanetary flow systems associated with cosmic ray modulation in 1977–1980. *J. Geophys. Res.* 89:6579–6587.

Burnett, D. S., and Woolum, D. S. 1977. Exposure ages and erosion rates for lunar rocks. *Phys. Chem. Earth* 10:63–101.

Busse, F., and Whitehead, J. A. 1974. Oscillatory and collective instabilities in large Prandtl number convection. *J. Fluid Mech.* 66:67–79.

Busso, M., Scaltriti, F., and Cellino, A. 1985. Differential rotation and activity cycles in RS CVn binaries. *Astron. Astrophys.* 148:29–34.

Butler, R. P., Cohen, R. D., Duncan, D. K., and Marcy, G. W. 1987. The Pleiades rapid rotators: Evidence for an evolutionary sequence. *Astrophys. J.* 319:L19-L22.

Cabestany, J., and Vazquez, M. 1983. The age decay of stellar magnetic activity in MS field solar-type stars. *Astrophys. Space Sci.* 97:151–156.

Cabot, W., Canuto, V. M., Kubickyj, O., and Pollack, J. B. 1987*a*. The role of turbulent convection in the primitive solar nebula. I. Theory. *Icarus* 69:387–422.

Cabot, W., Canuto, V. M., Kubickyj, O., and Pollack, J. B. 1987*b*. The role of turbulent convection in the primitive solar nebula. II. Results. *Icarus* 69:423–457.

Cabrit, S., and Bertout, C. 1990. CO line formation in bipolar flows. II. Decelerated outflow case and summary of results. *Astrophys. J.* 348:530–541.

Cabrit, S., Goldsmith, P. F., and Snell, R. L. 1988. Identification of RNO 43 and B335 as two highly collimated bipolar flows oriented nearly in the plane of the sky. *Astrophys. J.* 334:196–208.

Cabrit, S., Edwards, S., Strom, S. E., and Strom, K. M. 1990. Forbidden line emission and infrared excesses in T Tauri stars: Evidence for accretion-driven mass loss? *Astrophys. J.* 354:687–700.

Caffee, M. W., and Macdougall, J. D. 1988. Compaction ages. In *Meteorites and the Early Solar System*, eds. J. F. Kerridge and M. S. Matthews (Tucson: Univ. of Arizona Press), pp. 289–298.

Caffee, M. W., Goswami, J. N., Hohenberg, C. M., and Swindle, T. D. 1983. Cosmogenic neon from precompaction irradiation of Kapoeta and Murchison. In *Proc. Lunar Planet. Sci. Conf.* 14, *J. Geophys. Res. Suppl.* 88:B267-B273.

Caffee, M. W., Goswami, J. N., Hohenberg, C. M., and Swindle, T. D. 1986. Pre-compaction irradiation of meteorite grains. *Lunar Planet. Sci.* XVII:99–100 (abstract).

Caffee, M. W., Hohenberg, C. M., Swindle, T. D., and Goswami, J. N. 1987. Evidence in meteorites for an active early Sun. *Astrophys. J.* 313:L31-L35.

Caffee, M. W., Goswami, J. N., Hohenberg, C. M., Marti, K., and Reedy, R. C. 1988. Irradiation records in meteorites. In *Meteorites and the Early Solar System*, eds. J. F. Kerridge and M. S. Matthews (Tucson: Univ. of Arizona Press), pp. 205–245.

Caillault, J.-P., and Helfand, D. J. 1985. The EINSTEIN soft X-ray survey of the Pleiades. *Astrophys. J.* 289:279–299.

Callis, L. B., Alpert, J. C., and Geller, M. A. 1985. An assessment of thermal, wind, and planetary wave changes in the middle and lower atmosphere due to 11-year UV flux variations. *J. Geophys. Res.* 90:2273–2282.

Calvet, N., Basri, G., and Kuhi, L. V. 1984. The chromospheric hypothesis for the T Tauri phenomenon. *Astrophys. J.* 277:725–737.

Cameron, A. G. W. 1962. The formation of the Sun and planets. *Icarus* 1:13–69.

Cameron, A. G. W. 1973. Accumulation processes in the primitive solar nebula. *Icarus* 18:407–450.

Cameron, A. G. W. 1978. Physics of primitive solar nebula and of giant gaseous protoplanets. In *Protostars and Planets*, ed. T. Gehrels (Tucson: Univ. of Arizona Press), pp. 453–487.

Cameron, A. G. W. 1982. Elemental and nuclidic abundances in the solar system In *Essays in Nuclear Astrophysics*, eds. C. A. Barnes, D. D. Clayton and D. N. Schramm (Cambridge: Cambridge Univ. Press), pp. 23–43.

Cameron, A. G. W. 1985. Formation and evolution of the primitive solar nebula. In *Protostars and Planets II*, eds. D. C. Black and M. S. Matthews (Tucson: Univ. of Arizona Press), pp. 1073–1099.

Cameron, A. G. W. 1986. Magnetic studies on Shergotty and other SNC meteorites. *Geochim. Cosmochim. Acta* 50:1043–1048.

Campbell, B., Walker, G. A. H., and Yang, S. 1988. A search for substellar companions to solar-type stars. *Astrophys. J.* 331:902–921.

Cane, H. V., McGuire, R. E., and von Rosenvinge, T. T. 1986. Two classes of solar energetic particle events associated with impulsive and long duration soft X-ray flares. *Astrophys. J.* 301:448–459.

Cantó, J., Tenorio-Tagle, G., and Rozyczka, M. 1988. The formation of interstellar jets by the convergence of supersonic conical flows. *Astron. Astrophys.* 192:287–294.

Canuto, V. M., Levine, J. S., Augustsson, T. R., and Imhoff, C. L. 1982. UV radiations from the young Sun and oxygen and ozone levels in the prebiological paleoatmosphere. *Nature* 296:816–820.

Canuto, V. M., Levine, J. S., Augustsson, T. R., and Imhoff, C. L. 1983a. Oxygen and ozone in the early Earth's atmosphere. *Precambrian Res.* 20:109–120.

Canuto, V. M., Levine, J. S., Augustsson, T. R., Imhoff, C. L., and Giampapa, M. S. 1983b. The young Sun and the atmosphere and photochemistry of the early Earth. *Nature* 305:281–286.

Cassé, M., and Goret, P. 1978. Ionization models of cosmic ray sources. *Astrophys. J.* 221:703–712.

Cassen, P. M., and Moosman, A. 1981. On the formation of protostellar disks. *Icarus* 48:353–376.

Cassen, P. M., and Summers, A. L. 1983. Models of the formation of the solar nebula. *Icarus* 53:26–40.

Cassen, P. M., Smith, B. F., Miller, R. H., and Reynolds, R. T. 1980. Numerical experiments on the stability of preplanetary disks. *Icarus* 48:377–392.

Catalano, S., and Marilli, E. 1983. Ca II chromospheric emission and rotation of main sequence stars. *Astron. Astrophys.* 121:190–197.

Caudell, T. P., and Hill, H. A. 1980. On the statistical significance of the reported phase coherence for solar oscillations. *Mon. Not. Roy. Astron. Soc.* 193:381–383.

Cayrel, R., Cayrel de Strobel, G., Campbell, B., Mein, N., Nein, P., and Dumont, S. 1983. Evidence of high chromospheric activity in Hyades dwarfs from spectroscopic observations. *Astron. Astrophys.* 123:89–92.

Chamberlain, J. W. 1960. Interplanetary gas. II. Expansion of a model solar corona. *Astrophys. J.* 131:47–56.

Chamberlain, J. W., and Hunten, D. M. 1987. *Theory of Planetary Atmospheres* (Orlando: Academic Press).

Chandrasekhar, S. 1956. On free-force magnetic fields. *Proc. Natl. Acad. Sci.* 42:1–5.

Chapman, A., 1987. Gauging angles. *Sky & Telescope* 73:362–364.

Chapman, C. R., and Davis, D. R. 1975. Asteroid collisional evolution: Evidence for a much larger early population. *Science* 190:553–556.

Chapman, C. R., Morrison, D., and Zellner, B. 1975. The surface properties of asteroids: A synthesis of polarimetry, radiometry, and spectrophotometry. *Icarus* 25:104–130.

Chapman, G. A. 1984. On the energy balance of solar active regions. *Nature* 308:252–254.

Chapman, G. A. 1987. Variations of solar irradiance due to magnetic activity. *Ann. Rev. Astron. Astrophys.* 25:633–667.

Chapman, G. A., Herzog, A. D., Lawrence, J. K., and Shelton, J. C. 1984. Solar luminosity fluctuations and active region photometry. *Astrophys. J.* 282:L99-L101.

Chapman, G. A., Herzog, A. D., and Lawrence, J. K. 1986. Time-integrated energy budget of a solar activity complex. *Nature* 319:654–655.

Chapman, S. 1957. Notes on the solar corona and the terrestrial ionosphere. *Smithsonian Contrib. Astrophys.* 2:1–11.

Chappell, J., and Shackleton, N. J. 1986. Oxygen isotopes and sea level. *Nature* 324:137–140.

Chen, H. H. 1985. Direct approach to resolve the solar neutrino problem. *Phys. Rev. Lett.* 55:1534–1536.

Chen, J. H., and Wasserburg, G. J. 1981. The isotopic composition of uranium and lead in Allende inclusions and meteoritic phosphates. *Earth Planet. Sci. Lett.* 52:1–15.

Chen, J. H., and Wasserburg, G. J. 1985. U-Th-Pb isotopic studies on meteorites ALH 81005 and Ibitira. *Lunar Planet. Sci.* XVI:119–120 (abstract).

Chernovsky, E. J. 1966. Double sunspot cycle variation in terrestrial magnetic activity. *J. Geophys. Res.* 71:965–974.

Cherry, M. L., Lande, K., and Fowler, W. A., eds. 1985. *Solar Neutrinos and Neutrino Astronomy* (New York: American Inst. of Physics).

Chiang, W. H., and Foukal, P. V. 1985. The influence of faculae on sunspot heat blocking. *Solar Phys.* 97:9–20.

Christensen-Dalsgaard, J., Duvall, T. L., Jr., Gough, D. O., Harvey, J. W., and Rhodes, E. J., Jr. 1985*a*. Speed of sound in the solar interior. *Nature* 315:378–382.

Christensen-Dalsgaard, J., Gough, D., and Toomre, J. 1985*b*. Seismology of the Sun. *Science* 229:923–931.

Christensen-Dalsgaard, J., Gough, D. O., and Thompson, M. J. 1988. Determination of the solar internal sound speed by means of a differential asymptotic inversion. In *Seismology of the Sun and Sun-Like Stars,* ed. E. J. Rolfe, ESA SP-286, pp. 493–497.

Chuideri, C. 1981. Magnetic heating in the Sun. In *Solar Phenomena in Stars and Stellar Systems,* eds. R. M. Bonnet and A. K. Dupree (Dordrecht: D. Reidel), pp. 269–288.

Chupp, E. L. 1984. High-energy neutral radiations from the Sun. *Ann. Rev. Astron. Astrophys.* 22:359–387.

Chupp, E. L. 1988*a*. High-energy particle production in solar flares (SEP, gamma-ray, and neutron emissions). *Physica Scripta* T18:5–19.

Chupp, E. L. 1988*b*. Solar neutron observations and their relation to solar flare acceleration problems. *Solar Phys.* 118:137–154.

Chupp, E. L., Forrest, D. J., Higbie, P. R., Suri, A. N., Tsai, C., and Dunphy, P. P. 1973. Solar gamma ray lines observed during the solar activity of August 2 to August 11, 1972. *Nature* 241:333–335.

Cimino, M. 1946. Una conferma del ciclo solare ventiduennale dedotta dalle oscillazioni dei diametri solari. *Contrib. Osserv. Astron. Roy.* 1(8):614–618.

Cini Castagnoli, G. 1988. Introduction to the Course. In *Solar-Terrestrial Relationships and the Earth Environment in the Last Millennia,* ed. G. Cini Castagnoli (Amsterdam: North-Holland Press).

Cini Castagnoli, G., and Bonino, G. 1988. Solar imprint in sea sediments. The thermoluminescence profile: A new proxy record. In *Secular, Solar and Geomagnetic Variations in the Last 10,000 Years.* eds. F. R. Stephenson and A. W. Wolfendale (Dordrecht: Kluwer), pp. 341–348.

Cini Castagnoli, G., and Lal, D. 1980. Solar modulation effects in terrestrial production of carbon-14. *Radiocarbon* 22:133–158.

Cini Castagnoli, G., Bonino, G., Attolini, M. R., Galli, M., and Beer, J. 1984. Solar cycles in the last centuries in ^{10}Be and σ^{18} in polar ice and in thermoluminescence signals of a sea sediment. *Nuovo Cimento* 7C:235–244.

Cini Castagnoli, G., Bonino, G., and Provenzale, A. 1988*a*. On the thermoluminescence profile of an Ionian Sea sediment: Evidence of 137, 118, 12.1 and 10.8 y cycles in the last two millennia. *Nuovo Cimento* 11C:1–12.

Cini Castagnoli, G., Bonino, G., and Provenzale, A. 1988*b*. The thermoluminescence profile of a recent sea sediment core and the solar variability. *Solar Phys.* 117:187–197.

Cini Castagnoli, G., Bonino, G., and Provenzale, A. 1989. The 206-year cycle in tree ring radiocarbon data and in the thermoluminescence profile of a recent sea sediment. *J. Geophys. Res.* 94:11971–11976.

Cini Castagnoli, G., Bonino, G., Provenzale, A., and Serio, M. 1990*a*. On the presence of regular periodicities in the thermoluminescence profile of a recent sea sediment core. *Phil. Trans. Roy. Soc. London* A330:481.

Cini Castagnoli, G., Bonino, G., Caprioglio, F., Serio, M., Provenzale, A., and Bhandari, N. 1990*b*. The $CaCO_3$ profile in a recent Ionian Sea core and the tree ring radiocarbon record over the last two millennia. *Geophys. Res. Lett.* 17:1545–1548.

Cini Castagnoli, G., Bonino, G., Caprioglio, F., Provenzale, A., Serio, M., and Zhu, G.-M. 1990*c*. The carbonate profile of two recent Ionian Sea cores: Evidence that the sedimentation rate is constant over the last millennium. *Geophys. Res. Lett.* 17:1937–1940.

Cini Castagnoli, G., Bonino, G., Provenzale, A., and Serio, M. 1990*d*. On the solar origin of the thermoluminescence profile of the GT14 core. *Solar Phys.* 127:357–377.

Cintala, M. J., Head, J. W., Wilson, L. 1979. The nature and effects of impact cratering on small bodies. In *Asteroids,* ed. T. Gehrels (Tucson: Univ. of Arizona Press), pp. 579–600.

Cisowski, S. M. 1986. Magnetic studies on Shergotty and other SNC meteorites. *Geochim. Cosmochim. Acta* 50:1043–1048.

Cisowski, S. M. 1987. Magnetism of meteorites. In *Geomagnetism,* ed. J. A. Jacobs, vol. 2 (New York: Academic Press), pp. 525–560.

Cisowski, S. M. 1988. Paleomagnetism of unbrecciated eucrites. *Lunar Planet. Sci.* XIX:190–191 (abstract).

Cisowski, S. M. 1991. A remanent magnetic property of unbrecciated eucrites. *Earth Planet. Sci Lett.,* in press.

Cisowski, S. M., and Fuller, M. 1986. Lunar paleointensities via the IRMs normalization method and the early magnetic history of the Moon. In *Origin of the Moon,* eds. W. K. Hartmann, R. J. Phillips and G. J. Taylor (Houston: Lunar and Planetary Inst.), pp. 411–424.

Clancy, R. T., and Rusch, D. W. 1989. Climatology and trends of mesospheric (58–90 km) temperatures based upon 1982–1986 SME limb scattering profiles. *J. Geophys. Res.* 94:3377–3393.

Clark, D. H., and Stephenson, F. R. 1978. An interpretation of pre-telescope sunspot records from the Orient. *Q. J. Roy. Astron. Soc.* 19:3878–410.

Clark, M. J., and Smith, F. B. 1988. Wet and dry deposition of Chernobyl releases. *Nature* 332:245–249.

Clark, R. M. 1975. A calibration curve for radiocarbon dates. *Antiquity* 49:251–266.

Clayton, D. D. 1974. Maxwellian relative energies and solar neutrinos. *Nature* 249:131–134.

Clayton, D. D. 1980. Internal chemical energy: Heating of parent bodies and oxygen isotopic anomalies. *Nukleonika* 25:1477–1490.

Clayton, D. D., Dwek, E., and Woosley, S. E. 1977. Isotopic anomalies and proton irradiation in the early solar system. *Astrophys. J.* 214:300–315.

Clayton, R. N., and Thiemens, M. H. 1980. Lunar nitrogen: Evidence for secular change in the solar wind. In *The Ancient Sun: Fossil Record in the Earth, Moon and Meteorites,* eds. R. O. Pepin, J. A. Eddy and R. B. Merrill (New York: Pergamon Press), pp. 463–473.

Cleghorn, T. F., Freier, P. S., and Waddington, C. J. 1971. On the modulation and energy spectrum of highly charged cosmic rays. *Astrophys. Space Sci.* 14:422–430.

Cleveland, R. T., Davis, R., and Rowley, J. K. 1984. The chlorine and bromine solar neutrino experiments. In *Resonance Ionization Spectroscopy and Its Applications* (Bristol: A. Hilger), pp. 241–250.

CLIMAP Project Members. 1976. The surface of the Ice-Age Earth. *Science* 191:1131–1137.

CLIMAP Project Members. 1981. Seasonal reconstruction of the Earth's surface at the last glacial maximum. Geol. Soc. America Map and Chart Ser. MC-36, eds. A. McIntyre and R. Cline, pp. 1–18.

Cliver, E. W., Kahler, S. W., Shea, M. A., and Smart, D. F. 1982. Injection onsets of ~2 GeV protons, ~1 MeV electrons, and ~100 keV electrons in solar cosmic ray flares. *Astrophys. J.* 260:362–370.

Cliver, E. W., Forrest, D. J., Cane, H. V., Reames, D. V., McGuire, R. E., von Rosenvinge, T. T., Kane, S. R., and MacDowall, R. J. 1989, Solar flare nuclear gamma rays and interplanetary proton events. *Astrophys. J.* 343:953–970.

Cloutis, E., Gaffey, M. J., Jackowski, T. L., and Reed, K. L. 1986. Calibrations of phase abundance, composition, and particle size distribution for olivine-orthopyroxene mixtures from reflectance spectra. *J. Geophys. Res.* 91:11641–11653.

Cogley, J. G., and Henderson-Sellers, A. 1984. The origin and earliest state of the Earth's hydrosphere. *Rev. Geophys. Space Phys.* 22:131.

Cohen, L. H., Keisuke, I., and Kennedy, G. C. 1967. Melting and phase relations in an anhydrous basalt to 40 kilobars. *Amer. J. Sci.* 265:475–518.

Cohen, M. 1984. The T Tauri stars. *Phys. Rept.* 116:173–249.

Cohen, M., and Bieging, J. H. 1986. Radio variability and structure of T Tauri stars. *Astrophys. J.* 92:1396–1402.

Cohen, M., and Kuhi, L. V. 1979. Observational studies of pre-main-sequence evolution. *Astrophys. J. Suppl.* 41:743–843.

Cohen, M., Emerson, J. P., and Beichmann, C. A. 1989. A reexamination of the luminosity sources in T Tauri stars. I. Taurus-Auriga. *Astrophys. J.* 339:445–473.

Cohen, T. J., and Lintz, P. R. 1974. Long term periodicities in the sunspot cycle. *Nature* 250:398–400.

Colburn, D. S., and Sonett, C. P. 1966. Discontinuities in the solar wind. *Space Sci. Rev.* 5:439–506.

Cole, T. W. 1973. Periodicities in solar activity. *Solar Phys.* 30:103–110.

Collier-Cameron, A. C., and Robinson, R. D. 1989. Fast H alpha variations on a rapidly rotating cool main sequence star. I. Circumstellar clouds. *Mon. Not. Roy. Astron. Soc.* 236:57–87.

Collinson, D. W. 1985. Magnetic properties of the Olivenza chondrite—a cautionary tale. *Meteoritics* 20:629 (abstract).

Collinson, D. W. 1986. Magnetic properties of antarctic shergottite meteorites EETA 79001 and ALHA 77005: Possible relevance to a Martian magnetic field. *Earth Planet. Sci. Lett.* 77:159–164.

Collinson, D. W. 1987. Magnetic properties of the Olivenza meteorite—possible implications for its evolution and an early solar system magnetic field. *Earth Planet. Sci. Lett.* 84:369–380.

Consolmagno, G. J. 1978. Electromagnetic Process in the Evolution of the Solar Nebula. Ph.D. Thesis, Univ. of Arizona.

Coplan, M. A., Ogilvie, K. W., Bochsler, P., and Geiss, J. 1984. Interpretation of ^{3}He abundance variations in the solar wind. *Solar Phys.* 93:415–434.

Cornell, M. E., Hurford, G. J., Kiplinger, A. L., and Dennis, B. R. 1984. The relative timing of microwaves and hard X-rays in solar flares. *Astrophys. J.* 279:875–881.

Costa, J. E. R., Kaufmann, P., and Takakura, T. 1984. Timing analysis of hard X-ray emission and 22 GHz flux and polarization in a solar burst. *Solar Phys.* 94:369–374.

Cowan, G. A., and Haxton, W. C. 1982a. Solar neutrino production of technetium-97 and technetium-98. *Science* 216:51–54.

Cowan, G. A., and Haxton, W. C. 1982b. Solar variability, glacial epochs, and solar neutrinos. *Los Alamos Sci.* 3(2):46–56.

Cox, A. N. 1965. Stellar absorbtion coefficients and opacities. In *Stars and Stellar Systems*, eds. I. H. Aller and D. R. McLaughlin, vol. 8 (Chicago: Univ. of Chicago Press), pp. 195–267.

Cox, A. N. 1969. Geomagnetic reversals. *Science* 163:237–245.

Cox, J. P., and Giuli, R. T. 1968. Stability of the radiative gradient. In *Principles of Stellar Structure*, vol. 1 (New York: Gordon and Breach), pp. 262–280.

Cram, L. E. 1979. Atmospheres of T Tauri stars: The photosphere and low chromosphere. *Astrophys. J.* 234:949–957.

Cram, L. E. 1987. Heating of chromospheres and coronae: Present status of theory. In *Cool Stars, Stellar Systems and the Sun: Proc. Fifth Cambridge Workshop*, eds. J. L. Linsky and R. E. Stencel (New York: Springer-Verlag), pp. 123–134.

Cravens, T. E., and Stewart, A. I. 1978. Global morphology of nitric oxide in the lower E region. *J. Geophys. Res.* 83:2446–2456.

Cravens, T. E., Gérard, J. C., Lecompte, M., Stewart, A. I., and Rusch, D. W. 1985. The global distribution of nitric oxide in the thermosphere as determined by the Atmosphere Explorer D satellite. *J. Geophys. Res.* 90:9862–9870.

Crawford, D. A., and Schultz, P. H. 1988. Laboratory observations of impact-generated magnetic fields. *Nature* 336:50–52.

Crawford, H. J., Price, P. B., Cartwright, B. G., and Sullivan, J. D. 1975. Solar flare particles: Energy dependent composition and relationship to solar composition. *Astrophys. J.* 195:213–221.

Creer, K. M. 1988. Geomagnetic field and radiocarbon activity through Holocene time. In *Secular, Solar and Geomagnetic Variations in the Last 10,000 Years*, eds. F. R. Stephenson and A. W. Wolfendale (Dordrecht: Kluwer), pp. 381–397.

Croll, J. 1875. *Climate and Time in Their Geological Relations* (New York: Appleton).

Crowley, K. D., Duchon, C. E., and Rhi, J. 1986. Climate record in varved sediments of the Eocene Green River Formation. *J. Geophys. Res.* 91:8637–8647.

Crowley, T. J. 1983. The geologic record of climatic change. *Rev. Geophys. Space Phys.* 21:828–877.

Crowley, T. J. 1988. Paleoclimate modelling. In *Physically-Based Modelling and Simulation of Climate and Climatic Change*, ed. M. Schlesinger (Dordrecht: Kluwer), pp. 883–949.

Crozaz, G. 1980. Solar flare and cosmic ray particle tracks in lunar samples and meteorites: What they tell us about the ancient Sun. In *The Ancient Sun: Fossil Record in the Earth,*

Moon and Meteorites, eds. R. O. Pepin, J. A. Eddy and R. B. Merrill (New York: Pergamon Press), pp. 331–346.

Crozaz, G., Taylor, G. J., Walker, R. M., and Seitz, M. G. 1974. Early active Sun?: Radiation history of distinct components in fines. *Proc. Lunar Sci. Conf.* 5:2591–2596.

Cruikshank, D. P., Hartmann, W. K., Tholen, D. J., Bell, J. F., and Brown, R. H. 1989. Basaltic achondrites: Discovery of source asteroids. *Meteoritics* 24:260 (abstract).

Cruikshank, D. P., Tholen, D. J., Hartmann, W. K., Bell, J. F., and Brown, R. H. 1990. Three basaltic Earth-crossing asteroids and the source of the basaltic meteorites. *Icarus* 89:1–13.

Crutzen, P. J., Isaksen, I. S. A., and Reid, G. C. 1975. Solar proton events: Stratospheric sources of nitric oxide. *Science* 189:457–459.

Cummings, A. C., Stone, E. C., and Webber, W. R. 1987. Latitudinal and radial gradients of anomalous and galactic cosmic rays in the outer heliosphere. *Geophys. Res. Lett.* 14:174–177.

Currie, R. G. 1974. Solar cycle signal in surface air temperature. *J. Geophys. Res.* 79:5657–5660.

Dai, K. M., and Fan, C. Y. 1986. Bomb produced ^{14}C content in tree rings grown at different latitudes. In *Proc. 12th Intl. Radiocarbon Conf.,* eds. M. Stuiver and R. Kra. *Radiocarbon* 28(2A):346–349.

Daily, W. D., and Dyal, P. 1979. Theories for the origin of lunar magnetism. *Phys. Earth Planet. Int.* 20:255–270.

Dalfes, H. N., Schneider, S. H., and Thompson, S. L. 1984. Effects of bioturbation on climatic spectra inferred from deep sea cores. In *Milankovitch and Climate: Understanding Orbital Forcing,* eds. A. Berger, J. Imbrie, J. Hays, G. Kukla and B. Saltzman (Dordrecht: D. Reidel), pp. 481–492.

Damon, P. E. 1968. Radiocarbon and climate. *Meteorol. Monographs* 8:151–154.

Damon, P. E. 1970. Climatic vs. magnetic perturbations of the atmospheric C-14 reservoir. In *Radiocarbon Variations and Absolute Chronology,* ed. I. U. Olsson (New York: Wiley), pp. 571–593.

Damon, P. E. 1977. Solar induced variations of energetic particles at one AU. In *The Solar Output and Its Variation,* ed. O. R. White (Boulder: Colorado Assoc. Univ. Press), pp. 429–448.

Damon, P. E. 1988. Production and decay of radiocarbon and its modulation by geomagnetic field-solar activity changes with possible implications for global environment. In *Secular, Solar and Geomagnetic Variations in the Last 10,000 Years,* eds. F. R. Stephenson and A. W. Wolfendale (Dordrecht: Kluwer), pp. 267–285.

Damon, P. E., and Kunen, S. M. 1976. Global cooling? *Science* 193:447–453.

Damon, P. E., and Linick, T. W. 1986. Geomagnetic-heliomagnetic modulation of atmospheric radiocarbon production. In *Proc. 12th Intl. Radiocarbon Conf.,* eds. M. Stuiver and R. Kra. *Radiocarbon* 28(2A):266–278.

Damon, P. E., and Sternberg, R. S. 1990. Global production and decay of radiocarbon. In *Proc. 13th Intl. Radiocarbon Conf,* eds. A. Long and R. Kra. *Radiocarbon* 31(3):695–703.

Damon, P. E., Long, A., and Wallick, E. I. 1973*a.* On the magnitude of the 11-year radiocarbon cycle. *Earth Planet. Sci. Lett.* 20:300–306.

Damon, P. E., Long, A., and Wallick, E. I. 1973*b.* Comments on "Radiocarbon: Short-term variations" by M. S. Baxter and J. G. Farmer. *Earth Planet Sci. Lett.* 20:311–314.

Damon, P. E., Lerman, J. C., and Long, A. 1978. Temporal fluctuations of atmospheric ^{14}C: Causal factors and implications. *Ann. Rev. Earth Planet. Sci.* 6:457–494.

Damon, P. E., Sternberg, R. S., and Radnell, C. J. 1983. Modelling of atmospheric radiocarbon fluctuations for the past three centuries. In *Proc. 11th Intl. Radiocarbon Conf.,* eds. M. Stuiver and R. Kra. *Radiocarbon* 25(2):249–258.

Damon, P. E., Cheng, S., and Linick, T. W. 1990. Fine and hyperfine structure in the spectrum of secular variations of atmospheric ^{14}C. In *Proc. 13th Intl. Radiocarbon Conf.,* eds. A. Long and R. Kra. *Radiocarbon* 31(3):704–718.

Danjon, A., and Couderc, A. 1979. Lunettes et téléscopes. In *Librairie Scientifique et Technique Blanchard,* pp. 414–418.

Dansgaard, W., Johnsen, S. J., Clausen, H. B., and Langway, C. C. 1980. Climatic record revealed by the Camp Century ice core. In *The Late-Glacial Ages,* ed. K. Turekian (New Haven, Conn.: Yale Univ. Press), pp. 37–46.

Dansgaard, W., Clausen, H. B., Gundestrup, N., Hammer, C. U., Johnson, S. F., Kristins-dothy, P. M., and Rech, N. 1989. A new Greenland deep ice core. *Science* 218:1273–1277.

Dansgaard, W., Johnsen, S. J., Clausen, H. B., Dahl-Jensen, D., Gunderstrup, N., Hammer, C., and Oeschger, H. 1984. North Atlantic climate oscillations revealed by deep Greenland ice cores. In *Climate Processes and Climate Sensitivity*, eds. J. E. Hansen and T. Takahashi, AGU Geophysical Mono. 29 (Washington, D.C.: American Geophysical Union), pp. 288–298.

Däppen, W. 1983. Hydrostatic reaction of the Sun to local disturbances. *Astron. Astrophys.* 124:11–22.

Dartyge, E., Duraud, J. P., Langevin, Y., and Maurette, M. 1978. A new method for investigating the past activity of ancient solar flare cosmic rays over a time scale of a few billion years. *Proc. Lunar Planet. Sci. Conf.* 9:2375–2398.

Davis, D. R., Chapman, C. R., Greenberg, R., Weidenschilling, S. J., and Harris, A. W. 1979. Collisional evolution of asteroids: Populations, rotations and velocities. In *Asteroids*, ed. T. Gehrels (Tucson: Univ. of Arizona Press), pp. 528–557.

Davis, D. R., Weidenschilling, S. J., Farinella, P., Paolicchi, P., and Binzel, R. P. 1989. Asteroid collisional history: Effects on sizes and spins. In *Asteroids II*, eds. R. P. Binzel, T. Gehrels and M. S. Matthews (Tucson: Univ. of Arizona Press), pp. 805–826.

Davis, L., Jr. 1966. Models of the interplanetary fields and plasma flow. In *The Solar Wind*, eds. R. J. Mackin, Jr. and M. Neugebauer (Pasadena: Jet Propulsion Lab), pp. 147–164.

Davis, R. 1987. In *Proc. 13th Texas Symp. Relativistic Astrophysics* (Philadelphia: World Science).

Davis, R. 1988. In *Proc. Seventh Workshop on Grand Unification. ICOBAN '66* (Toyama, Japan).

Davis, R., Jr., Harmer, D. S., and Hoffman, K. C. 1968. Search for neutrinos from the Sun. *Phys. Rev. Lett.* 20:1205–1209.

Davis, R., Jr., Lande, K., Cleveland, B. T., Ullman, J., and Rowley, J. K. 1988. Report on the chlorine solar neutrino experiment. In *Neutrino '88: Proc. 13th Intl. Conf. on Neutrino Physics and Astrophysics*, eds. J. Schneps, T. Kafka, W. A. Mann and P. Nath (Singapore: World Scentific), pp. 518–525.

Dean, W. E., and Anderson, R. Y. 1978. Salinity cycles—evidence for subaqueous deposition in the Castile Formation and lower part of the Salado Formation, Delaware Basin, Texas and New Mexico. *New Mexico Bur. Mines and Mineral Resources Circular* 159:47–52.

Dearborn, D. S. P., and Blake, J. R. 1980a. Is the Sun constant? *Astrophys. J.* 237:616–619.

Dearborn, D. S. P., and Blake, J. R. 1980b. Magnetic fields and the solar constant. *Nature* 287:365–366.

Dearborn, D. S. P., and Blake, J. R. 1982. Surface magnetic fields and the solar luminosity. *Astrophys. J.* 257:896–900.

Dearborn, D. S. P., and Eggleton, P. P. 1976. CNO abundances in red giant envelopes. *Quart. J. Roy. Astron. Soc.* 17:448–456.

Dearborn, D. S. P., and Eggleton, P. P. 1977. Life of a mixed-up model. *Astrophys. J.* 213:177–182.

Dearborn, D. S. P., and Fuller, G. 1989. Neutrino oscillations and uncertainties in the solar model. *Phys. Rev.* D39:3542–3548.

Dearborn, D. S. P., and Newman, M. 1978. Efficiency of convection and time variation of the solar constant. *Science* 201:150–151.

Dearborn, D. S. P., Schramm, D. N., and Steigman, G. 1986a. Astrophysical constraints on the couplings of Axions, Majorons, and Familons. *Phys. Rev. Lett.* 56:26–29.

Dearborn, D. S. P., Schramm, D. N., and Steigman, G. 1986b. Destruction of ^{3}He in stars. *Astrophys. J.* 307:35–38.

Dearborn, D. S. P., Marx, G., and Ruff, I. 1987. A classical solution for the solar neutrino puzzle. *Prog. Theor. Phys.* 77:12–15.

de Bellefon, A., Espigat, P., and Waysand, G. 1985. Indium solar neutrino experiment using superconducting grains. In *Solar Neutrinos and Neutrino Astronomy*, eds. M. L. Cherry, K. Lande and W. A. Fowler (New York: American Inst. of Physics), pp. 227–236.

Debrunner, H., Fluckiger, E., Chupp, E. L., and Forrest, D. J. 1983. The solar cosmic ray neutron event on June 3, 1982. In *Proc. 18th Intl. Cosmic Ray Conf.*, vol. 4, eds. N. Durgaprasad, S. Ramadurai, P. V. Raman Murthy, M. V. S. Rao and K. Sivaprasad (Bombay: Tata Inst. of Fundamental Research), pp. 75–78.

DeCampli, W. M. 1981. T Tauri winds. *Astrophys. J.* 244:124–146.

DeCampli, W. M., and Baliunas, S. L. 1979. What tides and flares do to RS Canum Venaticorum binaries. *Astrophys. J.* 230:815–821.

de Charentenay, C., and Koutchmy, S. 1985. Sur la variabilité de la constante solaire. *Compt. Rend.* 301:151–156.

Deinzer, W., Hensler, G., Schüssler, M., and Weisshaar, E. 1984. Model calculations of magnetic flux tubes. *Astron. Astrophys.* 139:426–434.

de Jong, A. F. M., and Mook, W. G. 1980. Medium-term atmospheric ^{14}C variations. *Radiocarbon* 22:267–272.

Delache, Ph. 1988. Variability of the solar diameter. *Adv. Space Res.* 8:119–128.

Delache, Ph., and Scherrer, P. H. 1983. Detection of solar gravity mode oscillations. *Nature* 306:651–653.

Delache, Ph., Laclare, F., and Sadsaoud, H. 1985. Long period oscillations in solar diameter measurements. *Nature* 317:416–418.

de la Zerda Lerner, A., and O'Brien, K. 1987. Atmospheric radioactivity and variations in the solar neutrino flux. *Nature* 330:353–354.

Deliyannis, C. P., Demarque, P., Kawaler, S. D., Krauss, I. M., and Romanelli, P. 1989. Primordial lithium and the standard models. Yale Preprint YCTP-P13–88, pp. 1–17.

Deming, D., Boyle, R. J., Jennings, D. E., and Wiedemann, G. 1988. Solar magnetic field studies using the 12 micron emission lines. I. Quiet Sun time series and sunspot slices. *Astrophys. J.* 333:978–995.

Dennis, B. R. 1985. Solar hard X-ray bursts. *Solar Phys.* 100:465–490.

Denton, G. H., and Karlen, W. 1973. Holocene climatic variations—their pattern and possible cause. *Quat. Res.* 3:155–205.

DesMarais, D. J., Hayes, J. M., and Meinschien, W. G. 1974. Retention of solar wind-implanted elements in lunar soils. *Lunar Sci. Conf.* V:168–170 (abstract).

Deubner, F. I., Ulrich, R. K., and Rhodes, F. J. 1979. Solar *p* modal oscillations as a tracer of differential rotation. *Astron. Astrophys.* 72:177–185.

de Vries, H. 1958. Variations in concentration of radiocarbon with time and location on Earth. *Proc. K. Ned. Akad. Wet.* B61:94–102.

de Vries, H. 1959. Measurement and use of natural radiocarbon. In *Researches in Geochemistry*, ed. P. H. Abelson (New York: Wiley), pp. 169–189.

Dicke, R. H., and Goldenberg, H. M. 1967. Solar oblateness and general relativity. *Phys. Rev. Lett.* 18:313–315.

Dicke, R. H., Kuhn, J. R., and Libbrecht, K. G. 1984. The variable oblateness of the Sun: Measurement of 1984. *Astrophys. J.* 311:1025–1030.

Dicke, R. H., Kuhn, J. R., and Libbrecht, K. G. 1985. Oblateness of the Sun in 1983 and relativity. *Nature* 316:687–690.

Diethorn, W. S., and Stocko, W. L. 1972. The dose to man from atmospheric ^{85}Kr. *Health Phys.* 23:653–662.

Dilke, F. W. W., and Gough, D. O. 1972. Solar spoon. *Nature* 240:262–264.

Djurovic, D., and Paquet, P. 1988. The solar origin of the 50-day fluctuation of the Earth rotation and atmospheric circulation. *Astron. Astrophys.* 204:306–312.

Dobson, A. K., and Radick, R. R. 1989. Coronal activity-rotation relations for lower main-sequence stars. *Astrophys. J.* 344:907–914.

Dodd, R. T. 1969. Metamorphism of the ordinary chondrites: A review. *Geochim. Cosmochim. Acta* 33:161–203.

Dodge, J. C. 1975. Source regions of Type II radio bursts. *Solar Phys.* 42:445–459.

Dodson, H. W., and Hedeman, E. R. 1970. Time variations in solar activity. In *Solar Terrestrial Physics Part I*, eds. E. Dyer and C. de Jager (Dordrecht: D. Reidel), pp. 151–172.

Dohnanyi, J. S. 1978. Particle dynamics. In *Cosmic Dust*, ed. J. A. M. McDonnell (New York: Wiley), pp. 527–605.

Donnelly, R. F. 1988. *Ann. Geophys.* 6:417–424.

Donnelly, R. F., Heath, D. F., Lean, J. L., and Rottman, G. J. 1983. Differences in the temporal variations of solar UV flux, 10.7-cm solar radio flux, sunspot number, and Ca K plage data caused by solar rotation and active region evolution. *J. Geophys. Res.* 88:9883–9888.

Dorren, J. D., and Guinan, E. F. 1982. Evidence for starspots on single solar-type stars. *Astrophys. J.* 87:1546–1557.

Dorren, J. D., and Guinan, E. F. 1990. HD 129333—the Sun in its infancy. *Astrophys. J.*, submitted.

Draine, B. T. 1983. Magnetic bubbles and high-velocity outflows in molecular clouds. *Astrophys. J.* 270:519–536.

Drake, M. J. 1979. Geochemical evolution of the eucrite parent body: Possible nature and evolution of asteroid 4 Vesta. In *Asteroids*, ed. T. Gehrels (Tucson: Univ. of Arizona Press), pp. 765–782.

Drake, S. 1970. Sunspots, Sizzi, and Scheiner. In *Galileo Studies* (Ann Arbor: Univ. of Michigan Press), pp. 177–199.

Dran, J. C., Durrieu, L., Jouret, C., and Maurette, M. 1970. Habit and texture studies of lunar and meteoritic materials with a 1 MeV electron microscope. *Earth Planet. Sci. Lett.* 9:391–400.

Drell, S. D., Foley, H. M., and Ruderman, M. A. 1965. Drag and propulsion of large satellites in the ionosphere: An Alfvén propulsion engine in space. *J. Geophys. Res.* 70:3131–3146.

Dreschhoff, G., and Zeller, E. J. 1989. Evidence of individual solar proton events in Antarctic snow: Reconstruction of a past solar activity record. The Sun in Time, 6–10 March, Tucson, Arizona, Abstract Booklet, p. 14.

Drobyshevski, E. M., and Yuferev, V. S. 1974. Topological pumping of magnetic flux by three-dimensional convection. *J. Fluid Mech.* 65:33–44.

Drozd, R. J., Hohenberg, C. M., Morgan, C. J., and Ralston, C. E. 1974. Cosmic-ray exposure history at the Apollo 16 and other lunar sites: Lunar surface dynamics. *Geochim. Cosmochim. Acta* 38:1625–1642.

Drozd, R., Hohenberg, C., and Morgan, C. 1975. Krypton and xenon in Apollo 14 samples: Fission and neutron capture effects in gas-rich samples. *Proc. Lunar Sci. Conf.* 6:1857–1877.

Drozd, R. J., Kennedy, B. M., Morgan, C. J., Podosek, F. A., and Taylor, G. J. 1976. The excess fission xenon problem in lunar samples. *Proc. Lunar Sci. Conf.* 7:599–623.

Duba, A., and Boland, J. N. 1984. High temperature electrical conductivity of the carbonaceous chondrites Allende and Murchison. *Lunar Planet. Sci.* XV:232–233 (abstract).

Duba, A., Boland, J. N., and Ringwood, A. E. 1973. The electrical conductivity of pyroxene. *J. Geology* 81:727–735.

Duba, A., Heard, H. C., and Schock, R. N. 1974. Electrical conductivity of olivine at high pressure and under controlled oxygen fugacity. *J. Geophys. Res.* 79:1667–1673.

Dudorov, A. E., Krivodubsky, V. N., Ruzmaikina, T. V., and Ruzmaikin, A. A. 1989. The internal large scale magnetic field of the Sun. *Astron. J. (USSR)*, in press.

Duijveman, A., Hoyng, P., and Machado, M. E. 1982. X-ray imaging of 3 flares during the impulsive phase. *Solar Phys.* 81:137–157.

Duncan, D. K. 1981. Lithium abundances, K line emission and ages of nearby solar type stars. *Astrophys. J.* 248:651–669.

Dungey, J. W. 1961. Interplanetary magnetic field and the auroral zones. *Phys. Rev. Lett.* 6:47–48.

Dunham, D. W., Sofia, S., Fiala, A. D., and Muller, P. M. 1980. Observations of a change in the solar radius between 1717 and 1719. *Science* 210:1243–1244.

Duplessy, J.-C. 1978. Isotope studies. In *Climatic Change*, ed. J. Gribbin (Cambridge: Cambridge Univ. Press), pp. 46–68.

Duplessy, J.-C., Delibrias, G., Turon, J. L., Pujol, C., and Duprat, J. 1981. Deglacial warming of the Northeastern Atlantic Ocean: Correlation with the paleoclimatic evolution of the European continent. *Paleogeogr. Paleoclimatol. Paleoecol.* 35:121–144.

Durisen, R. H., Yang, S., Cassen, P., and Stahler, S. W. 1989. Numerical models of rotating protostars. *Astrophys. J.* 345:959–971.

Durney, B. R., and Latour, J. 1978. On the angular momentum loss of late-type stars. *Geophys. Astrophys. Fluid Dyn.* 9:241–255.

Durney, B. R., and Robinson, R. D. 1982. On an estimate of the dynamo-generated magnetic field in late-type stars. *Astrophys. J.* 253:290–297.

Duvall, T. L., and Harvey, J. W. 1984. Rotational frequency splitting of solar oscillations. *Nature* 310:19–21.

Duvall, T. L., Dziembowski, W. A., Goode, P. R., Gough, D. O., Harvey, J. W., and Lei-bacher, J. W. 1984. Internal rotation of the Sun. *Nature* 10:22–25.

Duvall, T. L., Jr., Harvey, J. W., and Pomerantz, M. A. 1986. Latitude and depth variation of solar rotation. *Nature* 321:500–501.

Duvall, T. L., Jr., Harvey, J. W., Libbrecht, K. G., Popp, R. D., and Pomerantz, M. A. 1988. Frequencies of solar *p* modal oscillations. *Astrophys. J.* 324:1158–1171.

Dyson, J. E. 1984. The interpretation of flows in molecular clouds. *Astrophys. Space Sci.* 106:181–197.

Dziembowski, W. A., Goode, P. R., and Libbrecht, K. G. 1989. The radial gradient in the Sun's rotation. *Astrophys. J.* 337:L53-L57.

Earle, E. D., Davidson, W. F., and Ewan, G. T. 1988. Observing the Sun from two kilometres underground: The Sudbury Neutrino Observatory (SNO). *Phys. Canada* 44(2):49–55.

Eberhardt, P., Geiss, J., and Grögler, N. 1965. Further evidence on the origin of trapped gases in the meteorite Khor Temiki. *J. Geophys. Res.* 70:4375–4378.

Eberhardt, P., Geiss, J., Graf, H., Grögler, N., Krähenbühl, U., Schwaller, H., Scharzmüller, J., and Stettler, A. 1970. Trapped solar wind noble gases, exposure ages and K/Ar ages in Apollo 11 lunar fine material. In Proc. Apollo 11 Lunar Sci. Conf. *Geochim. Cosmochim. Acta Suppl. 4,* 2:1037–1070.

Eberhardt, P., Geiss, J., Graf, H., Grögler, N., Mendia, M. D., Mörgeli, M., Schwaller, H., Stettler, A., Krähenbühl, U., and von Gunten, H. R. 1972. Trapped solar wind noble gases in Apollo 12 lunar fines 12001 and Apollo 11 breccia 10046. *Proc. Third Lunar Sci. Conf., Geochim. Cosmochim. Acta Suppl. 3,* 12:1821–1856.

Ebert, R., von Hoerner, S., and Temesváry, S. 1960. *Die Entsehung von Sternen durch Konden-sation diffuser Materie* (Berlin: Springer-Verlag).

Eckmann, J.-P., and Ruelle, D. 1985. Ergodic theory of chaos and strange attractors. *Rev. Mod. Phys.* 57:617–656.

Eddington, A. S. 1920. Internal constitution of stars. *Observatory* 43:341–358.

Eddy, J. A. 1976*a*. The Maunder minimum. *Science* 192:1189–1202.

Eddy, J. A. 1976*b*. The Sun since the Bronze age. In *Physics of Solar Planetary Environments,* ed. D. J. Williams, vol. 2 (Washington, D.C.: American Geophysical Union), pp. 958–972.

Eddy, J. A. 1977. Historical evidence for the existence of the solar cycle. In *The Solar Output and Its Variation,* ed. O. R. White (Boulder: Colorado Assoc. Univ. Press), pp. 51–71.

Eddy, J. A. 1979. *A New Sun: The Solar Results from Skylab,* NASA SP-402.

Eddy, J. A. 1980. The historical record of solar activity. In *The Ancient Sun: Fossil Record in the Earth, Moon and Meteorites,* eds. R. O. Pepin, J. A. Eddy and R. B. Merrill (New York: Pergamon Press), pp. 119–134.

Eddy, J. A. 1983. Keynote address: An historical review of solar variability, weather, and cli-mate. In *Weather and Climate Responses to Solar Variations,* ed. B. M. McCormac (Boul-der: Colorado Assoc. Univ. Press), pp. 1–15.

Eddy, J. A. 1988. Variability of the present and ancient Sun: A test of solar uniformitarianism. In *Secular Solar and Geomagnetic Variations in the Last 10,000 Years,* eds. F. R. Stephenson and A. W. Wolfendale (Dordrecht: Kluwer), pp. 1–23.

Eddy, J. A., and Boornazian, A. A. 1979. Secular decrease in the solar diameter, 1863–1953. *Bull Amer. Astron. Soc.* 11:437 (abstract).

Eddy, J. A., Stephenson, F. R., and Yau, K. K. C. 1989. On pre-telescopic sunspot records. *Quart. J. Roy. Astron. Soc.* 30:60–73.

Edvardsson, B. 1988. Spectroscopic surface gravities and chemical compositions for 8 nearby single subgiants. *Astron. Astrophys.* 190:148–166.

Edwards, S., and Snell, R. 1982. A search for high-velocity molecular gas around T Tauri stars. *Astrophys. J.* 261:151–160.

Edwards, S., and Strom, S. E. 1987. Energetic winds from low mass young stellar objects. In *Cool Stars, Stellar Systems and the Sun: Proc. Fifth Cambridge Workshop,* eds. J. L. Linsky and R. E. Stencel (New York: Springer-Verlag), pp. 443–454.

Edwards, S., Cabrit, S., Strom, S. E., Heyer, I., Strom, K. M., and Anderson, E. 1987. For-bidden line and Hα profiles in T Tauri star spectra: A probe of anisotropic mass outflows and circumstellar disks. *Astrophys. J.* 321:473–495.

Edwards, S., Cabrit, S., Ghandour, L. O., and Strom, S. E. 1989. Forbidden lines in T Tauri star spectra: A clue to the origin of T Tauri winds? In *ESO Workshop on Low Mass Star Formation and Pre-Main Sequence Objects*, ed. B. Reipurth (Garching: European Southern Obs.), pp. 385–397.

Eggen, O. J., and Sandage, A. R. 1964. New photoelectric observations of stars in the old galactic cluster M67. *Astrophys. J.* 140:130–143.

Ehmert, A. 1959. On the modulation of the primary spectrum of cosmic rays from solar activity. In *Proc. Moscow Cosmic Ray Conf.*, vol. 4, ed. L. I. Dorman (Amsterdam: North-Holland Press), pp. 142–148. In Russian.

Ellison, D. C., and Ramaty, R. 1985. Shock acceleration of electrons and ions in solar flares. *Astrophys. J.* 298:400–408.

Elmore, D., and Phillips, F. M. 1987. Accelerator mass spectrometry for measurement of long-lived radionuclides. *Science* 236:543–550.

Elsasser, W., Ney, E. P., and Winckler, J. R. 1956. Cosmic ray intensity and geomagnetism. *Nature* 178:1226–1227.

Emiliani, C. 1966. Isotopic paleotemperatures. *Science* 154:851–857.

Endal, A S., and Schatten, K. H. 1962. The faint young Sun-climate paradox. Continental influences. *J. Geophys. Res.* 87:7295–7302.

Endal, A. S., and Sofia, S. 1981. The evolution of rotating stars. I. Method and exploratory calculations for a 7 M$_\odot$ star. *Astrophys. J.* 210:184–198.

Endal, A. S., and Twigg, L. W. 1982. The effect of perturbation of convective energy transport on the luminosity and radius of the Sun. *Astrophys. J.* 260:342–352.

Endal, A. S., Sofia, S., and Twigg, L. W. 1985. Changes of solar luminosity and radius following secular perturbations in the convective envelope. *Astrophys. J.* 290:748–757.

Endfield, D. 1989. Is El Niño becoming more common? *Oceanography* 1(2):23–27.

Etique, P. 1981. L'utilisation des plagioclases du régolithe lunaire comme détecteurs des gaz rares provenant des rayonnements corpusculaires solaires. Ph.D. Thesis, ETH Zürich.

Etique, P., Signer, P., and Wieler, R. 1981. An in-depth study of neon and argon in lunar soil plagioclases, revisited: Implanted solar flare noble gases. *Lunar Planet. Sci.* XII:265–267 (abstract).

Eugster, O. 1985. Multistage exposure history of the 74261 soil constituents. *Proc. Lunar Planet. Sci. Conf.* 16, *J. Geophys. Res. Suppl.* 90:D95-D102.

Eugster, O., Grögler, N., Eberhardt, P., and Geiss, J. 1979. Double drive tube 74001/2: History of the black and orange glass: Determination of a pre-exposure 3.7 AE ago by ^{136}Xe/^{235}U dating. *Proc. Lunar Planet. Sci. Conf.* 10:1351–1379.

Eugster, O., Grögler, N., Eberhardt, P., and Geiss, J. 1980. Double drive tube 74001/2: Composition of noble gases trapped 3.7 AE ago. *Proc. Lunar Planet. Sci. Conf.* 11:1565–1592.

Eugster, O., Geiss, J., and Grögler, N. 1983*a*. Dating of early regolith exposure and the evolution of trapped ^{40}Ar/^{36}Ar with time. *Lunar Planet. Sci.* XIV:177–178 (abstract).

Eugster, O., Eberhardt, P., Geiss, J., and Grögler, N. 1983*b*. Neutron-induced fission of uranium: A dating method for lunar surface material. *Science* 219:170–172.

Evans, J. C., Reeves, J. H., and Reedy, R. C. 1987. Solar cosmic ray produced radionuclides in the Salem meteorite. In *Lunar Planet. Sci.* XVIII:271–272 (abstract).

Evans, N. J., Levreault, R. M., Beckwith, S., and Skrutskie, M. 1987. Observations of infrared emission lines and radio continuum emission from pre-main-sequence objects. *Astrophys. J.* 320:364–375.

Evenson, P., Meyer, P., and Pyle, K. R. 1983. Protons from the decay of solar flare neutrons. *Astrophys. J.* 274:875–882.

Evenson, P., Meyer, P., Yanagita, S., and Forrest, D. J. 1984. Electron rich particle events and the production of gamma rays by solar flares. *Astrophys. J.* 283:439–449.

Evernden, J. F., and Verhoogen, J. 1956. Electrical resistivity of meteorites. *Nature* 178:106–107.

Ewan, G. T., Evans, H. C., Lee, H. W., Leslie, J. R., Mak, H.-B., McLatchie, W., Robertson, B. C., Skiensved, P., Allen, R. C., Bühler, G., Chen, H. H., Doe, P. J., Sinclair, D., Tanner, N. W., Anglin, J. D., Bercovitch, M., Davidson, W. F., Hargrove, C. K., Mes, H., Storey, R. S., Earle, E. D., Milton, G. M., Jagam, P., Simpson, J. J., McDonald, A. B., Hallman, E. D., Carter, A. L., and Kessler, D. 1987. *Sudbury Neutrino Observatory Proposal* (Ottawa: SNO-87-12).

Ezer, D., and Cameron, A. G. W. 1968. Solar spin-down and neutrino fluxes. *Astrophys. Lett.* 1:177–179.

Ezer, D., and Cameron, A. G. W. 1971. Pre-main sequence stellar evolution with mass loss. *Astrophys. Space Sci.* 10:52–70.

Fabian, P., Pyle, J. A., and Wells, R. J. 1979. The August 72 solar proton event and the atmospheric ozone layer. *Nature* 277:458–460.

Fabricius, J. 1611. Ioh. Fabricii Phrysii de Maculis in Sole Observatis et apparente earum cum Sole conversion Narratio. (Wittenberg).

Fairbanks, R. G. 1989. A 17,000-year glacio-eustatic sea level record: Influence of glacial melting rates on the Younger Dryas event and deep-ocean circulation. *Nature* 342:637–642.

Fan, C. Y., Chen, T. M., Jun, S. X., and Dai, K. M. 1986. Radiocarbon activity variations in dated tree rings grown in the MacKenzie delta. In *Proc. 12th Intl. Radiocarbon Conf.*, eds. M. Stuiver and R. Kra. *Radiocarbon* 28(2A): 300–305.

Farmer, J. D., Ott, E., and Yorke, J. A. 1983. The dimension of chaotic attractors. *Physica* 7D:153–180.

Faulkner, J., and Gilliland, R. L. 1985. Weakly interacting massive particles and the solar neutrino flux. *Astrophys. J.* 299:994–1000.

Faulkner, J., and Swenson, F. J. 1988. Main sequence evolution with efficient central energy transport. *Astrophys. J.* 329:L47-L50.

Feeley, H. W., Larsen, R. J., and Sanderson, C. G. 1988. *Annual Rept. of the Surface Air Sampling Program*, U.S. Dept. of Energy Rept. EML-497.

Feierberg, M. A., Larson, H. P., Fink, U., and Smith, H. A. 1980. Spectroscopic evidence for two achondrite parent bodies: Asteroids 349 Dembowska and 4 Vesta. *Geochim. Cosmochim. Acta* 44:513–524.

Feigelson, E. D. 1982. X-ray emission from young stars and implications for the early solar system. *Icarus* 51:155–163.

Feigelson, E. D. 1984. X-ray emission from pre-main sequence stars. In *Cool Stars, Stellar Systems and the Sun: Proc. Third Cambridge Workshop*, eds. S. L. Baliunas and L. Hartmann (New York: Springer-Verlag), pp. 27–42.

Feigelson, E. D. 1987. Microwave observations of nonthermal phenomena in pre-main sequence stars. In *Cool Stars, Stellar Systems and the Sun: Proc. Fifth Cambridge Workshop*, eds. J. L. Linsky and R. E. Stencel (New York: Springer-Verlag), pp. 455–465.

Feigelson, E. D., and DeCampli, W. M. 1981. Observations of X-ray emission from T Tauri stars. *Astrophys. J.* 243:L89-L93.

Feigelson, E. D., and Kriss, G. A. 1989. Soft X-ray observations of pre-main sequence stars in the chamaeleon dark cloud. *Astrophys. J.* 338:262–276.

Feigelson, E. D., and Montmerle, T. 1985. An extremely variable radio star in the Rho Ophiuchi clouds. *Astrophys. J.* 289:L19-L23.

Ferguson, C. W. 1971. Dendrochronology of bristle cone pine, Pinus aristata: Establishment of a 7484-yr chronology in the White Mountains of Eastern-central California. In *Radiocarbon Variations and Absolute Chronology*, ed. I. U. Olsson (New York: Wiley), p. 237.

Ferguson, C. W., and Graybill, D. A. 1983. Dendrochronology of bristlecone pine: A progress report. In *Proc. 11th Intl. Radiocarbon Conf.*, eds. M. Stuiver and R. Kra. *Radiocarbon* 25(2):287–290.

Fernandez, J. A., and Ip, W.-H. 1989. Statistical and evolutionary aspects of cometary orbits. In *Comets in the Post-Halley Era*, vol. 1, ed. R. L. Newburn, Jr. (Dordrecht: Kluwer), pp. 487–535.

Feynman, J., and Fougere, P. 1985. Eighty year cycle in solar terrestrial phenomena confirmed. *J. Geophys. Res.* 89:3023–3027.

Feynman, J., Armstrong, T. P., Dao-Gibner, L., and Silverman, S. 1990. A new interplanetary proton fluence model. *J. Spacecraft Rockets* 27:403–410.

Feyssel, M., and Nitschelm, N. 1985. Time-dependent aspects of the atmospheric driven fluctuations in the duration of the day. *Ann. Geof.* 3:181–186.

Fichefet, Th., Tricot, Ch., Berger, A., Gallée, H., and Marsiat, I. 1989. Climate studies with a coupled atmosphere-upper ocean-ice sheet model. *Phil. Trans. Roy. Soc. London.* A329:249–261.

Fink, D., Paul, M., Hollos, G., Theis, S., Vogt, S., Stueck, R., Englert, P., and Michel, R.

1987. Measurements of ^{41}Ca spallation cross sections and ^{41} Ca concentrations in the Grant meteorite by accelerator mass spectrometry. *Nucl. Instr. Meth.* B29:275–280.

Finkel, R. C., Arnold, J. R., Imamura, M., Reedy, R. C., Fruchter, J. S., Loosli, H. H., Evans, J. C., Delany, A. C., and Shedlovsky, J. P. 1971. Depth variation of cosmogenic nuclides in a lunar surface rock and lunar soil. In *Proc. Lunar Sci. Conf.* 2:1773–1789.

Finkenzeller, U., and Basri, G. 1987. The atmospheres of T Tauri stars. I. High-resolution calibrated observations of moderately active stars. *Astrophys. J.* 318:832–843.

Finney, S. A. 1988. The Spectrum of Radiocarbon. M. S. Thesis, Univ. of Arizona.

Fireman, E. L. 1980. Solar activity during the past 10^4 years from radionuclides in lunar samples. In *The Ancient Sun: Fossil Record in the Earth, Moon and Meteorites*, eds. R. O. Pepin, J. A. Eddy and R. B. Merrill (New York: Pergamon Press), pp. 365–386.

Fireman, E. L., and Beukens, R. P. 1989. Carbon-14 production by 155-MeV protons in meteorites. *Lunar Planet. Sci.* XX:291–292 (abstract).

Fireman, E. L., DeFilce, J., and Norton, E. 1970. Ages of the Allende meteorite. *Geochim. Cosmochim. Acta* 34:873–881.

Fireman, E. L., Cleveland, B. T., Davis, R., and Rowley, J. K. 1985. Cosmic-ray depth studies at the Homestake mine with $^{39}K \rightarrow ^{37}Ar$ detectors. In *Solar Neutrinos and Neutrino Astronomy*, eds. M. L. Cherry, K. Lande and W. A. Fowler (New York: American Inst. of Physics), pp. 22–31.

Fish, R. A., Goles, G. G., and Anders, E. 1960. The record in the meteorites. III. On the development of meteorites in asteroidal bodies. *Astrophys. J.* 132:243–258.

Fisher, D. A. 1982. Carbon-14 production compared to oxygen isotope records from Camp Century, Greenland and Devon Island, Canada. *Climate Change* 4:419–426.

Fisher, R., McCabe, M., Mickey, D., Seagraves, P., and Sime, D. G. 1984. The Sun as a star: 1982 June 14–August 13. *Astrophys J.* 280:873–878.

Fisk, L. A. 1978. ^{3}He-rich flares: A possible explanation. *Astrophys. J.* 224:1048–1055.

Fleischer, R. L., Price, P. B., and Walker, R. M. 1975. *Nuclear Tracks in Solids* (San Francisco: Univ. of California Press).

Fleischer, R. L., Price, P. B., Walker, R. M., and Maurette, M. 1967a. Origin of fossil charged particle tracks in meteorites. *J. Geophys. Res.* 72:333–353.

Fleischer, R. L., Price, P. B., Walker, R. M., Maurette, M., and Morgan, G. 1967b. Tracks of heavy cosmic ray primaries in meteorites. *J. Geophys. Res.* 72:355–366.

Fleming, T. A., and Giampapa, M. S. 1989. A search for chromospheres at faint magnitudes. *Astrophys. J.* 346:299–302.

Fleming, T. A., Gioia, I. M., and Maccacaro, T. 1989. The relation between X-ray emission and rotation in late-type stars from the perspective of X-ray selection. *Astrophys. J.* 340:1011–1032.

Forman, M. A., and Schaeffer, O. A. 1980. ^{37}Ar and ^{39}Ar in meteorites and solar modulation of cosmic rays during the last thousand years. In *The Ancient Sun: Fossil Record in the Earth, Moon and Meteorites*, eds. R. O. Pepin, J. A. Eddy and R. B. Merrill (New York: Pergamon Press), pp. 279–292.

Forman, M. A., Ramaty, R., and Zweibel, E. G. 1986. The acceleration and propagation of solar flare energetic particles. In *The Physics of the Sun*, ed. P. A. Sturrock, vol. 2 (Dordrecht: D. Reidel), pp. 249–289.

Formisano, V., and Moreno, G. 1971. Helium and heavy ions in the solar wind. *Nuovo Cimento* 1:365–422.

Forrest, D. J. 1983. Solar γ-ray lines, positron electron pairs. In *Astrophysics*, eds. M. L. Burns, A. K. Harding and R. Ramaty (New York: American Inst. of Physics), pp. 3–14.

Forrest, D. J., Vestrand, W. T., Chupp, E. L., Rieger, E., Cooper, J., and Share, G. 1985. Neutral pion production in solar flares. In *Proc. 19th Intl. Cosmic Ray Conf.*, vol. 4 (Washington, D.C.: NASA), pp. 146–169.

Fossat, E., Gelly, B., Grec, G., and Pomerantz, M. 1987. Search for solar p-mode frequency changes between 1980 and 1985. *Astron. Astrophys.* 177:L47-L48.

Foukal, P. V. 1990. Solar luminosity variations over timescales of days to the past few solar cycles. *Phil. Trans. Roy. Soc. London.* A330:591–599.

Foukal, P. V., and Fowler, L. A. 1984. A photometric study of heat flow at the solar photosphere. *Astrophys. J.* 281:442–454. Erratum 286:377.

Foukal, P. V., and Lean, J. 1986. The influence of faculae on total solar irradiance and luminosity. *Astrophys. J.* 302:826–835.

Foukal, P. V., and Lean, J. 1988. Magnetic modulation of solar luminosity by photospheric activity. *Astrophys. J.* 328:347–357.

Foukal, P. V., and Vernazza, J. 1979. The effect of magnetic fields on solar luminosity. *Astrophys. J.* 234:707–715.

Foukal, P. V., Fowler, L. A., and Livshits, M. 1983. A thermal model of sunspot influence on solar luminosity. *Astrophys. J.* 267:863–871.

Fowler, L. A., Foukal, P. V., and Duvall, T. L., Jr. 1983. Sunspot bright rings and the thermal diffusivity of solar convection. *Solar Phys.* 84:33–44.

Fowler, W. A. 1972. What cooks with solar neutrinos? *Nature* 238:24–26.

Fowler, W. A. 1987. Age of the observable universe. *Quart. J. Roy. Astron. Soc.* 78:87–108.

Fowler, W. A., Greenstein, J. H., and Hoyle, F. 1962. Nucleosynthesis during the early history of the solar system. *Geophys. J. Roy. Astron. Soc.* 6:148–220.

Fowler, W. A., Caughlan, G., and Zimmerman, R. 1967. Thermonuclear reaction rates. *Ann. Rev. Astron. Astrophys.* 5:525–570.

Fowler, W. A., Caughlan, G., and Zimmerman, R. 1975. Thermonuclear reaction rates. II. *Ann. Rev. Astron. Astrophys.* 13:69–112.

Fraedrich, K. 1986. Estimating the dimensions of weather and climate attractors. *J. Atmos. Sci.* 43:419–432.

Frakes, L. A. 1979. *Climates Throughout Geologic Time* (New York: Elsevier).

François, L. M. 1987. Reducing power of ferrous iron in the Archean ocean. 2. Role of $FEOH^+$ photoxidation. *Paleocean.* 2:395–408.

François, L. M., and Gérard, J. C. 1986. Reducing power of ferrous iron in the Archaen ocean. 1. Contribution of photosynthetic oxygen. *Paleocean.* 1:355–368.

François, L. M., and Gérard, J. C. 1988. Ozone, climate and biospheric environment in the ancient oxygen-poor atmosphere. *Planet. Space Sci.* 361:1391–1414.

Frandsen, S., Dreyer, P., and Kjeldsen, H. 1989. Stellar photometric stability. I. The open clusters Melotte 105, NGC 2660, and NGC 4755. *Astron. Astrophys.* 215:287–304.

Fredrikson, K., and Keil, K. 1963. The light-dark structure in the Pantar and Kapoeta stone meteorites. *Geochim. Cosmochim. Acta* 27:717–739.

Freedman, M. S. 1978. ^{205}Tl as a low energy neutrino detector. In *Proc. Informal Conf. on Status and Future of Solar Neutrino Research,* ed. G. Friedlander, BNL-50879 (Upton, N.Y.: Brookhaven National Lab), pp. 313–360.

Freedman, M. S. 1988. Measurement of low energy neutrino absorption probability in thallium 205. *Nucl. Instr. Meth.* A271:267–276.

Freedman, M. S., Stevens, C. M., Horwitz, E. P., Fuchs, L. H., Lerner, J. L., Goodman, L. S., Childs, W. J., and Hessler, J. 1976. Solar neutrinos: Proposal for a test. *Science* 193:1117–1119.

Freier, P. S., and Waddington, C. J. 1965. Modulation of cosmic rays by an electric field. In *Proc. 9th Intl. Cosmic Ray Conf.,* vol. 1 (Amsterdam: North-Holland Press), pp. 176–179.

Freier, P. S., and Waddington, C. J. 1966. The helium nuclei of the primary cosmic radiation as studied over a solar cycle of activity interpreted in terms of electric field modulation. *Space Sci. Rev.* 4:313–372.

Frick, U., Becker, R. H., and Pepin, R. O. 1988. Solar wind record in the lunar regolith: Nitrogen and noble gases. *Proc. Lunar Planet. Sci. Conf.* 18:87–120.

Fricke, K. J., and Kippenhahn, R. 1972. Evolution of rotating stars. *Ann. Rev. Astron. Astrophys.* 10:45–72.

Fricker, P. E., Goldstein, J. I., and Summers, A. L. 1970. Cooling rates and thermal histories of iron and stony-iron meteorites. *Geochim. Cosmochim. Acta* 34:475–491.

Friedlander, G., ed. 1978. *Proc. Informal Conf. on Status and Future of Solar Neutrino Research,* BNL-50879 (Upton, N.Y.: Brookhaven National Lab).

Friedlander, G., and Wenesser, J. 1987. Solar neutrinos: Experimental approaches. *Science* 235:760–765.

Fröhlich, C. 1977. Contemporary measures of the solar constant. In *The Solar Output and Its Variation,* ed. O. R. White (Boulder: Colorado Assoc. Univ. Press), pp. 93–109.

Fröhlich, C. 1987a. Variability of the solar "constant" on time scales of minutes to years. *J. Geophys. Res.* 92:796–800.

Fröhlich, C. 1987*b*. Solar oscillations and helioseismology from ACRIM/SMM irradiance data. In *New and Exotic Phenomena*, eds. O. Fackler and J. T. T. Vân (Gif-sûr-Yvette: Editions Frontières), pp. 395–402.

Fröhlich, C., and Pap, J. 1989. Multi-spectral analysis of total solar irradiance variations. *Astron. Astrophys.* 220:272–280.

Fruchter, J. S., Evans, J. C., Reeves, J. H., and Perkins, R. W. 1982. Measurement of ²⁶Al in Apollo 15 core 15008 and ²²Na in Apollo 17 rock 74275. In *Lunar Planet. Sci.* XIII:243–244 (abstract).

Funaki, M., Taguchi, I., Damon, J., Nagata, T., and Kondo, T. 1988. Magnetic and metallographic studies of the bocaiuva iron meteorite. *Proc. NIPR Symp. Antarctic Meteorites* 1:231–246.

Funk, H., Podosek, F., and Rowe, M. W. 1967. Spallation yields of krypton and xenon from irradiation of strontium and barium with 730 MeV protons. *Earth Planet. Sci. Lett.* 3:193–196.

Gaffey, M. J. 1983. The asteroid (4) Vestra: Rotational spectral variations, surface material heterogeneity, and implications for the origin of the basaltic achondrites. *Lunar Planet Sci.* XIV:231–232 (abstract).

Gaffey, M. J. 1984. Rotational spectral variations of asteroid (8) Flora: Implications for the nature of the S-type asteroids and for the parent bodies of the ordinary chondrites. *Icarus* 60:83–114.

Gaffey, M. J. 1986. The spectral and physical properties of metal in meteorite assemblages: Implications for asteroid surface materials. *Icarus* 66:468–486.

Gaffey, M. J. 1988. Thermal history of the asteroid belt: Implications for accretion of the terrestrial planets. *Lunar Planet Sci.* XIX:369–370 (abstract).

Gaffey, M. J., and Ostro, S. J. 1987. Surface lithologic heterogeneity and body shape for asteroid (15) Eunomia: Evidence from rotational spectral variations and multi-color lightcurve inversions. *Lunar Planet Sci.* XVIII:310–311 (abstract).

Gaffey, M. J., Bell, J. F., and Cruikshank, D. P. 1989. Reflectance spectroscopy and asteroid surface mineralogy. In *Asteroids II*, eds. R. P. Binzel, T. Gehrels and M. S. Matthews (Tucson: Univ. of Arizona Press), pp. 98–127.

Gahm, G. F. 1986. Flarelike variability in T Tauri stars. In *Flares: Solar and Stellar*, ed. P. M. Gondhalekar (Chilton, Didcot, England: Rutherford Appleton Lab), pp. 124–139.

Gahm, G. F. 1989. Some aspects of T-Tauri variability. In *Formation and Evolution of Low Mass Stars*, eds. A. K. Dupree and M. T. V. T. Lago (Dordrecht: Kluwer), pp. 295–304.

Gaisser, T. K., and Stanev, T. 1985. Atmospheric neutrino fluxes at low energy. In *Solar Neutrinos and Neutrino Astronomy*, eds. M. L. Cherry, W. A. Fowler and K. Lande (New York: American Inst. of Physics), pp. 276–281.

Gaizauskas, V., Harvey, K. L., Harvey, J. W., and Zwaan, C. 1983. Large-scale patterns formed by solar active regions during the ascending phase of cycle 21. *Astrophys. J.* 265:1056–1062.

Galileo, G. 1613. Istoria e dimostrazioni intorno alle macchie solari e loro accidenti. In *Le Opere di Galileo Galilei*, ed. A. Favaro, vol. 5 (Florence), p. 95. Trans. by S. Drake. *Discoveries and Opinions of Galileo Galilei* (Garden City: Doubleday, 1957), pp. 89–143.

Galileo, G. 1623. Il saggiatore. In *Le Opere di Galileo Galilei*, ed. A. Favaro (Florence). Trans. by S. Drake. *Discoveries and Opinions of Galileo Galilei* (Garden City: Doubleday, 1957).

Galloway, D. J., and Weiss, N. O. 1981. Convection and magnetic fields in stars. *Astrophys. J.* 243:945–953.

Galvin, A. B., Ipavich, F. M., Gloeckler, G., Hovestadt, D., Klecker, B., and Scholer, M. 1984. Solar wind ionization temperatures inferred from the charge state composition of diffuse particle events. *J. Geophys. Res.* 89:2655–2671.

Galzauskas, V., Harvey, K. L., Harvey, J. W., and Zwaan, C. 1983. Large-scale patterns formed by solar active regions during the ascending phase of cycle 21. *Astrophys. J.* 265:1056–1065.

Gamow, G. 1938. Nuclear energy sources and stellar evolution. *Phys. Rev.* 53:595–604.

Garay, G., Moran, J. M., and Reid, M. J. 1987. Compact continuum radio sources in the Orion nebula. *Astrophys. J.* 314:535–550.

Garcia, R. R., Solomon, S., Roble, R. G., and Rusch, D. W. 1984. A numerical response of the middle atmosphere to the 11-year solar cycle. *Planet. Space Sci.* 32:411–423.

Garcia-Muñoz, M., Mason, G. M., and Simpson, J. A. 1975. The anomalous ⁴He component in the cosmic-ray spectrum at < 50 MeV per nucleon during 1972–1974. *Astrophys. J.* 202:265–275.

Garcia-Muñoz, M., Meyer, P., Pyle, K. R., Simpson, J. A., and Evenson, P. 1986. The dependence of solar modulation on the sign of the cosmic ray particle charge. *J. Geophys. Res.* 91:2858–2866.

Garrels, R. M. 1987. A model for the deposition of the microbanded Precambrian iron formations. *Amer. J. Sci.* 287:81–106.

Gault, D. E., Hörz, F., Brownlee, D. E., and Hartung, J. B. 1974. Mixing of the lunar regolith. *Proc. Lunar Sci. Conf.* 5:2365–2386.

Gaustad, J. E., and Rogerson, J. B. 1961. The solar limb intensity profile. *Astrophys. J.* 134:323–330.

Gavrin, V. N., and the SAGE collaboration. 1990. The Baksan gallium solar neutrino experiment. In *Inside the Sun: Proc. IAU Coll. 121*, eds. G. Berthomieu and M. Cribier (Dordrecht: Kluwer), pp. 201–212.

Geiss, J. 1973. Solar wind composition and implications about the history of the solar system. *Proc. 13th Intl. Cosmic Ray Conf.*, vol. 5 (Denver: Univ. of Denver), pp. 3375–3398.

Geiss, J. 1982. Processes affecting abundances in the solar wind. *Space Sci. Rev.* 33:201–217.

Geiss, J., and Bochsler, P. 1981. On the abundances of rare ions in the solar wind. In *Solar Wind: Proc. Fourth Intl. Conf.*, ed. H. Rosenbauer, Rept. No. MPAE-W-100-81-31, pp. 403–413.

Geiss, J., and Bochsler, P. 1982. Nitrogen isotopes in the solar system. *Geochim. Cosmochim. Acta* 46:529–548.

Geiss, J., and Bochsler, P. 1985. Ion composition in the solar wind in relation to solar abundances. In *Isotopic Ratios in the Solar System*, ed. D. Gautier (Toulouse: Cepadues-Editions), pp. 213–228.

Geiss, J., and Bochsler, P. 1986. Solar wind composition and what we expect to learn from out-of-ecliptic measurements. In *The Sun and the Heliosphere in Three Dimensions*, ed. R. G. Marsden (Dordrecht: D. Reidel), pp. 173–186.

Geiss, J., and Reeves, H. 1972. Cosmic and solar system abundances of deuterium and helium-3. *Astron. Astrophys.* 18:126–132.

Geiss, J., Eberhardt, P., Bühler, F., Meister, J., and Signer, P. 1970a. Apollo 11 and 12 solar wind composition experiments: Fluxes of He and Ne isotopes. *J. Geophys. Res.* 75:5972–5979.

Geiss, J., Hirt, P., and Leutwyler, H. 1970b. On acceleration and motion of ions in corona and solar wind. *Solar Phys.* 12:458–483.

Geiss, J., Bühler, F., Cerutti,H., Eberhardt, P., and Filleux, Ch. 1972. Solar wind composition experiments. In *Apollo 16 Prelim. Sci. Rept.*, NASA SP-315, pp. 14.1–14.10.

Geiss, J., Lind, D. L., and Stettler, W. 1975. Magnetospheric particle composition experiment. In *The Skylab Results*, eds. W. C. Schneider and T. E. Hanes. *Advances in the Astronautical Sciences*, vol. 31 (Washington, D.C.: American Astronautical Soc.), pp. 849–867.

Gelly, B., Fossat, E., and Grec, G. 1988. Solar *p*-mode frequency variations between 1980 and 1986. In *Seismology of the Sun and Sun-Like Stars*, ed. E. J. Rolfe, ESA SP-286, pp. 275–278.

Gensho, R., Nitoh, O., Makino, T., and Honda, M. 1979. Some long-lived and stable nuclides produced by nuclear reactions. *Phys. Chem. Earth* 11:11–18.

Genthon, C., Barnola, J. M., Raynaud, D., Lorius, C., Jouzel, J., Barkov, N. I., Korotkevich, Y. S., and Kotlyakov, V. M. 1987. Vostok ice core: Climatic response to CO_2 and orbital forcing changes over the last climatic cycle. *Nature* 329:414–418.

Genzel, R., Reid, M. J., Moran, J. M., and Downes, D. 1981. Proper motions and distances of H_2O maser sources. I. The outflow in Orion-KL. *Astrophys. J.* 244:884–902.

Gérard, J. C. 1990. Modelling the climate response to solar variability. *Phil. Trans. Roy. Soc. London* A, in press.

Gérard, J. C., and Barth, C. A. 1977. High-latitude nitric oxide in the lower thermosphere. *J. Geophys. Res.* 82:674–680.

Gérard, J. C., and François, L. M. 1987. A model of solar-cycle effects of paleoclimate and its implications for the Elatina Foundation. *Nature* 326:577–580.

Gerling, E. K., and Levskii, L. K. 1956. On the origin of the rare gases in stony meteorites. *Dokl. Akad. Nauk. SSR* 110:750–755.

Ghil, M. 1976. Climate stability for a Sellers-type model. *J. Atmos. Sci.* 33:3–20.

Ghil, M. 1980. Internal climatic mechanisms participating in glaciation cycles. In *Climatic Variations and Variability: Facts and Theories*, ed. A. Berger (Dordrecht: D. Reidel), pp. 539–557.

Ghil, M. 1984. Climate sensitivity, energy balance models, and oscillatory climate models. *J. Geophys. Res.* 89:1280–1284.

Ghil, M. 1985. Theoretical climate dynamics: An introduction. In *Turbulence and Predictability in Geophysical Fluid Dynamics*, eds. M. Ghil, R. Benzi and G. Parisi (Amsterdam: North-Holland Press), pp. 347–402.

Ghil, M. 1987. Dynamics, statistics and predictability of planetary flow regimes. In *Irreversible Phenomena and Dynamical Systems Analysis in the Geosciences*, eds. C. Nicolis and G. Nicolis (Dordrecht: D. Reidel), pp. 241–283.

Ghil, M. 1989. Deceptively-simple models of climatic change. In *Climate and Geosciences*, eds. A. Berger, J.-C. Duplessy and S. H. Schneider (Dordrecht: D. Reidel), pp. 211–240.

Ghil, M., and Childress, S. 1987. *Topics in Geophysical Fluid Dynamics: Atmospheric Dynamics, Dynamo Theory and Climate Dynamics* (New York: Springer-Verlag).

Ghil, M., and LeTreut, H. 1981. A climate model with cryodynamics and geodynamics. *J. Geophys. Res.* 86:5262–5270.

Ghil, M., and Mullhaupt, A. 1985. Boolean delay equations. II. Periodic and aperiodic solutions. *J. Stat. Phys.* 42:125–173.

Ghil, M., and Tavantzis, J. 1983. Global Hopf bifurcation in a simple climate model. *SIAM J. Appl. Math.* 43:1019–1041.

Ghil, M., and Vautard, R. 1990. Interdecadal oscillations and the warming trend in global temperatures. *Nature*, submitted.

Ghil, M., Mullaupt, A., and Pestiaux, P. 1987. Deep water formation and Quaternary glaciations. *Climate Dyn.* 2:1–10.

Ghil, M., Kimoto, M., and Neelin, J. D. 1991. Chaos and predictability in the atmospheric sciences. *Rev. Geophys.*, in press.

Gialanella, L. 1943. Le variazioni del diametro solare nel sessantennio 1874–1937 secondo le osservazioni eseguite nell'Osservatorio del Campidoglio. *Comment. Pontificia Acad. Sci.* VI:1139–1195.

Giampapa, M. S. 1984. Photometric variations of solar type stars: Results of the cloudcroft survey. In *Solar Irradiance Variations and Active Region Time Scales*, eds. B. J. LaBonte, G. A. Chapman, H. S. Hudson and R. C. Wilson, NASA CP-2310, pp. 173–184.

Giampapa, M. S. 1986. T Tauri stars: Flare characteristics and flare models. In *Flares: Solar and Stellar*, ed. P. M. Gondhalekar, Rutherford Appleton Lab Rept., pp. 232–240.

Giampapa, M. S. 1987. Atmospheres of stars in the limit of thin and thick convection zones: The M dwarf stars. In *Cool Stars, Stellar Systems and The Sun: Proc. Fifth Cambridge Workshop*, eds. J. L. Linsky and R. E. Stencel (New York: Springer-Verlag), pp. 236–249.

Giampapa, M. S., and Imhoff, C. L. 1985. The ambient radiation fields of young solar systems: Ultraviolet and X-ray emission from T Tauri stars. In *Protostars and Planets II*, eds. D. C. Black and M. S. Matthews (Tucson: Univ. of Arizona Press), pp. 386–404.

Giampapa, M. S., and Liebert, J. 1986. Hα observations of M dwarf stars: Implications for stellar dynamo models and stellar binematic properties at faint magnitudes. *Astrophys. J.* 305:784–794.

Giampapa, M. S., Calvet, N., Imhoff, C. L., and Kuhi, L. V. 1981. IUE observations of pre-main sequence stars. I. MgII and Ca II resonance line fluxes for T Tauri stars. *Astrophys. J.* 251:113–125.

Giampapa, M. S., Golub, L., and Worden, S. P. 1983. The magnetic field of the RS Canus Venatacorum star Lambda Andromedae. *Astrophys. J.* 268:L121-L124.

Giannuzzi, M. A. 1979. On the eclipsing binaries of the Ursa Major stream. *Astron. Astrophys.* 77:214–244.

Gibson, E. G. 1973. *The Quiet Sun*, NASA SP-303.

Gilliland, R. L. 1980. Solar luminosity variations. *Nature* 286:838–839.

Gilliland, R. L. 1981. Solar radius variations over the past 265 years. *Astrophys. J.* 248:1144–1155.

Gilliland, R. L. 1982. Modeling solar variability. *Astrophys. J.* 253:399–405.

Gilliland, R. L. 1985. The relation of chromospheric activity to convection, rotation, and evolution off the main sequence. *Astrophys. J.* 299:286–294.

Gilliland, R. L. 1988. Mechanisms of slow change in solar luminosity. In *Solar Radiative Output Variations*, ed. P. V. Foukal (Cambridge, Mass.: Cambridge Research and Instrumentation), pp. 289–300.

Gilliland, R. L. 1989. Solar evolution. *Palaeogeogr. Palaeoclimat. Palaeoecol.* 75:35–55.

Gilliland, R. L., and Baliunas, S. L. 1987. Objective characterization of stellar activity cycles. I. Methods and solar cycle analyses. *Astrophys. J.* 314:766–781.

Gilliland, R. L., and Brown, T. M. 1988. Time-resolved CCD photometry of an ensemble of stars. *Publ. Astron. Soc. Pacific* 100:754–765.

Gilliland, R. L., and Däppen, W. 1988. Oscillations in solar models with weakly interacting massive particles. *Astrophys. J.* 324:1153–1157.

Gilliland, R. L., Faulkner, J., Press, W. H., and Spergel, D. N. 1986. Solar models with energy transport by weakly interacting particles. *Astrophys. J.* 306:703–709.

Gilman, P. A. 1980. Differential rotation in stars with convective zones. In *Stellar Turbulence*, eds. D. F. Gray and J. L. Linsky (New York: Springer-Verlag), pp. 19–38.

Gilman, P. A. 1985. The solar dynamo: Observations and theory of solar convection, global circulation, and magnetic fields. In *Physics of the Sun*, eds. P. A. Sturrock, T. E. Holzer, D. Mihalas and R. K. Ulrich (Dordrecht: D. Reidel), pp. 95–153.

Gilman, P. A., and DeLuca, E. E. 1986. Dynamo theory for the Sun and stars. In *Cool Stars, Stellar Systems and the Sun: Proc. Fourth Cambridge Workshop*, eds. M. Zeilik and D. M. Gibson (New York: Springer-Verlag), pp. 163–172.

Gilman, P. A., and Miller, J. 1986. Nonlinear convection of a compressible fluid in a rotating spherical shell. *Astrophys. J. Suppl.* 61:585–608.

Gilman, P. A., Morrow, C. A., and DeLuca, E. E. 1989. Angular momentum transport and dynamo action in the Sun: Implications of recent oscillation measurements. *Astrophys. J.* 338:508–537.

Giovanelli, R. G. 1947. Magnetic and electric phenomena in the Sun's atmosphere associated with sunspots. *Mon. Not. Roy. Astron. Soc.* 107:338–355.

Gizzatulina, S. M., Rukavishnikov, V. D., Ruzmaikin, A. A., and Tavastsherna, K. S. 1988. Radiocarbon evidence of global stochasticity of the solar activity. Preprint N40(794), IZMIRAN, Moscow. Submitted to *Solar Phys.*

Glatzmaier, G. 1985a. Numerical simulations of convective dynamos. Part II. Field propagation in the convective zone. *Astrophys. J.* 291:300–307.

Glatzmaier, G. A. 1985b. Numerical simulations of stellar convetive dynamos. III. At the base of the convective zone. *Geophys. Astrophys. Fluid Dyn.* 31:137–150.

Gleeson, L. J., and Axford, W. I. 1967. Cosmic rays in the interplanetary medium. *Astrophys. J.* 149:L115–L118.

Gleeson, L. J., and Axford, W. I. 1968. Solar modulation of galactic cosmic rays. *Astrophys. J.* 154:1011–1019.

Gleick, J. 1987. *Chaos: Making a New Science* (New York: Viking).

Gleissberg, W. 1944. A table of secular variations of the solar cycle. *Terrest. Magnet. Atmos. Elect.* 49:243–244.

Gleissberg, W. 1966. Ascent and descent in the eighty-year cycles of solar activity. *J. British Astron. Assoc.* 76:265–270.

Gloeckler, G., and Geiss, J. 1989. The abundances of elements and isotopes in the solar wind. In *Cosmic Abundances of Matter*, ed. C. J. Waddington (New York: American Inst. of Physics), pp. 49–71.

Gloeckler, G., Ipavich, F. M., Hamilton, D. C., Wilken, B., Studemann, W., Kremser, G., and Hovestadt, D. 1986. Solar wind carbon, nitrogen, and oxygen abundances measured in the Earth-magnetosheath with AMPTE/CCE. *Geophys. Res. Lett.* 13:793–796.

Gloeckler, G., Ipavich, F. M., Hamilton, D. C., Wilken, B., and Kremser, G. 1989. Heavy ion abundances in coronal hole solar wind flows. *Eos: Trans. AGU* 70:424.

Gnedin, Yu.N., Red'kina, N. P., and Tarasov, K. V. 1988. Influence of the magnetic field on the spectrum of linear polarization of the emission of T Tauri stars. *Soviet Astron.* 32:186–193.

Gold, T. 1964. Magnetic energy shedding in the solar atmosphere. In *The Physics of Solar Flares*, ed. W. Hess, NASA SP-50, pp. 389–395.

Goldreich, P., and Lynden-Bell, D. 1969. A Jovian unipolar inductor. *Astrophys. J.* 156:56.

Goldreich, P., and Ward, W. R. 1973. The formation of planetesimals. *Astrophys. J.* 183:1051–1061.

Goldsmith, P. F., and Arquilla, F. 1985. Rotation in dark clouds. In *Protostars and Planets II*, eds. D. C. Black and M. S. Matthews (Tucson: Univ. of Arizona Press), pp. 137–149.

Goldsmith, P. F., Snell, R. L., Hemeon-Heyer, M., and Langer, W. D. 1984. Bipolar outflows in dark clouds. *Astrophys. J.* 286:599–608.

Golub, L. 1984. Quiescent coronae of active chromosphere stars. In *Cool Stars, Stellar Systems and the Sun: Proc. Third Cambridge Workshop*, eds. P. Byrne and M. Rodono (Dordrecht: D. Reidel), pp. 83–108.

Gondoin, P., Giampapa, M. S., and Bookbinder, J. A. 1985. Stellar magnetic field measurements utilizing IR spectral lines. *Astrophys. J.* 297:710–718.

Göpel, C., Manhès, G., and Allègre, C. J. 1989. U-Pb study of phosphates in chondrites. *Meteoritics* 24:270–271 (abstract).

Gordon, B. E., and Hollweg, J. V. 1983. Collisional damping of surface waves in the solar corona. *Astrophys. J.* 266:373–382.

Goswami, J. N., and Lal, D. 1975. Compositional studies of solar-iron and heavier nuclei of kinetic energy below 10 MeV/nucleon. *Proc. Lunar Sci. Conf.* 6:3531–3555.

Goswami, J. N., and Lal, D. 1979. Formation of the parent bodies of the carbonaceous chondrites. *Icarus* 10:510–521.

Goswami, J. N., Hutcheon, I. D., and Macdougall, J. D. 1976. Microcraters and solar flare tracks in crystals from carbonaceous chondrites and lunar breccias. *Proc. Lunar Sci. Conf.* 7:543–562.

Goswami, J. N., Lal, D., and Macdougall, J. D. 1979. Composition and energy spectra of solar flare heavy nuclei during the early history of the solar system. In *Proc. 16th Intl. Cosmic Ray Conf.*, vol. 5 (Kyoto: Univ. of Tokyo), pp. 116–121.

Goswami, J. N., Lal, D., and Macdougall, J. D. 1980. Charge composition and energy spectra of ancient solar flare heavy nuclei. In *The Ancient Sun: Fossil Record in the Earth, Moon and Meteorites*, eds. R. O. Pepin, J. A. Eddy and R. B. Merrill (New York: Pergamon Press), pp. 347–364.

Goswami, J. N., Lal, D., and Wilkening, L. L. 1984. Gas-rich meteorites: Probes for particle environment and dynamical processes in the inner solar system. *Space Sci. Rev.* 37:111–159.

Goswami, J. N., McGuire, R. E., Reedy, R. C., Lal, D., and Jha, R. 1988. Solar flare protons and alpha particles during the last three solar cycles. *J. Geophys. Res.* 93:7195–7205.

Gough, D. O. 1981a. A new measure of solar rotation. *Mon. Not. Roy. Astron. Soc.* 196:731–745.

Gough, D. O. 1981b. On the seat of the solar cycle. In *Variations of the Solar Constant*, ed. S. Sofia, NASA CP-2191, pp. 185–206.

Gough, D. O. 1981c. Solar interior structure and luminosity variations. *Solar Phys.* 74:21–34.

Gough, D. O. 1988. Theory of solar variation. In *Solar-Terrestrial Relationships and the Earth Environment in the Last Millennia*, ed. G. C. Castagnoli (Amsterdam: North-Holland Press), pp. 90–132.

Gough, D. O. 1989. Deep roots of solar cycles. *Nature* 336:618–619.

Gradie, J. C., and Tedesco, E. 1982. Compositional structure of the asteroid belt. *Science* 216:1405–1407.

Gradie, J. C., Chapman, C. R., and Tedesco, E. F. 1989. Distribution of taxonomic classes and the compositional structure of the asteroid belt. In *Asteroids II*, eds. R. P. Binzel, T. Gehrels and M. S. Matthews (Tucson: Univ. of Arizona Press), pp. 316–335.

Graham, N. E., and White, W. B. 1988. The El Niño cycle: A natural oscillator of the Pacific ocean-atmosphere system. *Science* 240:1293–1302.

Grassberger, P. 1986. Do climatic attractors exist? *Nature* 323:609–612.

Grassberger, P. 1983. Generalized dimensions for strange attractors. *Phys. Lett.* A97:227–230.

Grassberger, P. 1988. Finite sample corrections to entropy and dimension estimates. *Phys. Lett.* A128:369–350.

Grassberger, P., and Procaccia, I. 1983a. Characterization of strange attractors. *Phys. Rev. Lett.* 50:346–349.

Grassberger, P., and Procaccia, I. 1983b. Measuring the strangeness of strange attractors. *Physica* 9D:189–197.

Gray, D. F. 1973. On the existence of classical microturbulence. *Astrophys. J.* 184:461–468.

Gray, D. F. 1975. Atmospheric turbulence measures in stars above the main sequence. *Astrophys. J.* 202:148–157.

Gray, D. F. 1984. Measurements of Zeeman broadening in F, G, and K dwarfs. *Astrophys. J.* 277:640–647.

Gray, D. F. 1985. An apparent universal magnetic constant for cool stars. *Publ. Astron. Soc. Pacific* 97:719–724.

Greenberg, R., Wacker, F., Hartmann, W. K., and Chapman, C. R. 1978. Planetesimals to planets: Numerical simulation of collisional evolution. *Icarus* 35:1–26.

Grevesse, N. 1984. Accurate atomic data and solar photospheric spectroscopy. *Phys. Scripta* T8:49–58.

Grey, D. 1972. Fluctuations of Atmospheric Radiocarbon. Ph.D. Thesis, Univ. of Arizona.

Gribov, V., and Pontecorvo, B. 1969. Neutrino astronomy and lepton charge. *Phys. Lett.* 28B:493–495.

Grimm, R. E., and McSween, H. Y., Jr. 1988. Water and the thermal history of the CM carbonaceous chondrite parent body. *Lunar Planet. Sci.* XIX:427–428 (abstract).

Grinspoon, D. H. 1988. Large Impact Events and Atmospheric Evolution on the Terrestrial Planets. Ph.D. Thesis, Univ. of Arizona.

Grinspoon, D. H., and Sagan, C. 1987. Was the early Earth shrouded in impact-generated dust? *Bull. Amer. Astron. Soc.* 19:872 (abstract).

Grinspoon, D. H., and Sagan, C. 1991. Impact dust and climate on primordial Earth. In preparation.

Grossman, J. N. 1988. Formation of chondrules. In *Meteorites and the Early Solar System*, eds. J. F. Kerridge and M. S. Matthews (Tucson: Univ. of Arizona Press), pp. 680–696.

Grossman, L., and Larimer, J. W. 1974. Early chemical history of the solar system. *Rev. Geophys. Space Phys.* 12:71–101.

Grossmann-Doerth, U., Knölker, M., Schüssler, M., and Weisshaar, E. 1989. Models of magnetic flux sheets. In *Solar and Stellar Granulation*, eds. R. Rutten and G. Severino (Dordrecht: D. Reidel), pp. 481–490.

Grün, E., Zook, H. A., Fechtig, H., and Giese, R. H. 1985. Collisional balance of the meteoritic complex. *Icarus* 62:244–272.

Guckenheimer, J., and Holmes, P. 1983. *Nonlinear Oscillation, Dynamical Systems, and Bifurcations of Vector Fields* (New York: Springer-Verlag).

Guenther, D. B., and Sarajedini, A. 1988. Insensitivities of the solar p-mode frequencies to changes in the helium abundance. *Astrophys. J.* 327:993–997.

Guenther, D. B., Jaffe, A., and DeMarque, P. 1989. The standard solar model: Composition, opacities, and seismology. *Astrophys. J.* 345:1022–1033.

Guzik, J. A., Wilson, L. A. and Brunish, W. M. 1987. A comparison between mass-losing and standard solar models. *Astrophys. J.* 319:957–965.

Haack, H., Rasmussen, K. L., and Warren, P. H. 1990. Effects of regolith/megaregolith insulation on the cooling histories of differentiated asteroids. *J. Geophys. Res.* 95:5111–5124.

Hale, G. E. 1908. On the probable existence of a magnetic field in sunspots. *Astrophys. J.* 28:315–343.

Hall, D. N. B. 1975. Spectroscopic detection of solar ^{3}He. *Astrophys. J.* 197:509–512.

Hall, D. S. 1976. The RS CVn binaries and binaries with similar properties. In *Multiple Periodic Variable Stars*, ed. W. S. Fitch (Dordrecht: D. Reidel), pp. 287–345.

Hall, D. S. 1980. The RS canum Venaticorum binaries. In *Solar Phenomena in Stars and Stellar Systems*, eds. R. M. Bonnet and A. K. Dupree (Dordrecht: D. Reidel), pp. 431–448.

Halsey, T. C., Jensen, M. H., Kadanoff, L. P., Procaccia, I., and Shraiman, B. I. 1986. Fractal measures and their singularities: The characterization of strange sets. *Phys. Rev.* A33:1141–1151.

Hammer, C. U. 1980. Acidity of polar ice cores in relation to absolute dating, past volcanism, and radio-echos. *J. Glaciol.* 25:359–372.

Hammer, C. U., Clausen, H. B., Dansgaard, W., Gundestrup, N., Johnsen, S. J., and Reeh, N. 1978. Dating of Greenland ice cores by flow models, isotopes, volcanic debris, and continental dust. *J. Glaciol.* 20:3–26.

Hampel, W. 1980. Low-energy neutrinos in astrophysics. In *Neutrino Physics and Astrophysics*, ed. E. Fiorini (New York: Plenum), pp. 61–79.

Hampel, W. 1985. The gallium solar neutrino detector. In *Solar Neutrinos and Neutrino Astronomy*, eds. M. L. Cherry, K. Lande and W. A. Fowler (New York: American Inst. of Physics), pp. 162–174.

Hansen, J., and Lebedeff, S. 1988. Global surface air temperatures: Update through 1987. *Geophys. Res. Lett.* 15:323–326.

Hardy, D. A., Gussenhoven, M. S., Raistrick, R., and McNeil, W. J. 1987. Statistical and functional representations of the pattern of auroral energy flux, number flux and conductivity. *J. Geophys. Res.* 92:12275–12294.

Haro, G., and Chavira, E. 1966. Flare stars in stellar aggregates of different ages. *Vistas in Astron.* 8:89–107.

Hart, J. E. Toomre, J., Deane, A. E., Hurlburt, N. E., Glatzmaier, G. A., Fichtl, G. H., Leslie, F., Fowlis, W. W., and Gilman, P. A. 1986. Laboratory experiments on planetary and stellar convection performed on Spacelab 3. *Science* 234:61–64.

Hart, M. H. 1978. The evolution of the atmosphere of the Earth. *Icarus* 33:23–39.

Hartigan, P. 1988. Large line widths in HH objects: The bow shock model. In *Formation and Evolution of Low Mass Stars*, eds. A. K. Dupree and M. T. V. T. Lago (Dordrecht: Kluwer), pp. 285–294.

Hartle, R. E., Sittler, E. C., Jr., Ogilvie, K. W., Scudder, J. D. Lazarus, A. J., and Atreya, S. K. 1982. Titan's ion exosphere observed from Voyager 1. *J. Geophys. Res.* 87:1383–1394.

Hartmann, L. W. 1987. Stellar magnetic fields: Optical observations and analysis. In *Cool Stars, Stellar Systems and the Sun: Proc. Fifth Cambridge Workshop*, eds. J. L. Linsky and R. E. Stencel (New York: Springer-Verlag), pp. 1–9.

Hartmann, L. W., and Kenyon, S. J. 1985. On the nature of FU Orionis objects. *Astrophys. J.* 299:462–478.

Hartmann, L. W., and Kenyon, S. J. 1987a. Further evidence for disk accretion in FU Orionis objects. *Astrophys. J.* 312:243–253.

Hartmann, L. W., and Kenyon, S. J. 1987b. High spectral resolution infrared observations of V 1057 Cygni. *Astrophys. J.* 322:393–398.

Hartmann, L. W., and Kenyon, S. J. 1988. Accretion disks around young stars. In *Formation and Evolution of Low Mass Stars*, eds. A. K. Dupree and M. T. V. T. Lago (Dordrecht: Kluwer), pp. 163–179.

Hartmann, L. W., and Kenyon, S. J. 1990. Optical veiling, disk accretion, and the evolution of T Tauri stars. *Astrophys. J.* 349:190–196.

Hartmann, L. W., and MacGregor, K. B. 1982. Protostellar mass and angular momentum loss. *Astrophys. J.* 259:180–192.

Hartmann, L. W., and Noyes, R. W. 1987. Rotation and magnetic activity in main-sequence stars. *Ann Rev. Astron. Astrophys.* 25:271–301.

Hartmann, L. W., and Stauffer, J. 1989. Additional rotational velocities of T Tauri stars. *Astron J.* 97:873–880.

Hartmann, L. W., Edwards, S., and Avrett, A. 1982. Wave-driven wind models from cool stars. II. Models for T Tauri stars. *Astrophys. J.* 261:279–292.

Hartmann, L. W., Soderblom, D., Noyes, R., Burnham, N., and Vaughan, A. 1984. An analysis of the Vaughan-Preston Survey of chromospheric emission. *Astrophys. J.* 276:254–265.

Hartmann, L. W., Hewett, R., Stahler, S., and Mathieu, R. D. 1986. Rotational and radial velocities of T Tauri stars. *Astrophys. J.* 309:275–293.

Hartmann, L. W., Soderblom, D. R., and Stauffer, J. R. 1987. Rotation and kinematics of the pre-main sequence stars in Taurus-Auriga with Ca II emission. *Astron J.* 93:907–912.

Hartmann, W. K. 1980. Dropping stones in magma oceans: Effects of early lunar cratering. In *Proc. Conf. Lunar Highlands Crust* (New York: Pergamon Press), pp. 155–173.

Harvey, J. 1984. Helium 10830Å irradiance: 1975–1983. In *Solar Irradiance Variations on Active Region Time Scales*, eds. B. J. LaBonte, G. A. Chapman, H. S. Hudson and R. C. Willson, NASA CP-2310, pp. 197–212.

Harvey, L. D. D., and Schneider, S. H. 1984. Sensitivity of internally-generated climate oscillations to ocean model formulation. In *Milankovitch and Climate: Understanding Orbital Forcing*, eds. A. Berger, J. Imbrie, J. Hays, G. Kukla and B. Saltzman (Dordrecht: D. Reidel), pp. 653–667.

Hasselman, K. 1976. Stochastic climate models, part I. *Tellus* 28:473–485.

Hauglustaine, D., and Gérard, J. C. 1989. Possible composition and climatic changes due to past energetic particle precipitation. *Ann. Geophys.*, in press.

Haxton, W. C. 1987. Analytical treatments of matter-enhanced solar-neutrino oscillations. *Phys. Rev.* D35:2352–2364.

Haxton, W. C. 1988. Radiochemical neutrino detection via $^{127}I(v_e e^-)^{127}Xe$. *Phys. Rev. Lett.* 60:768–771.

Haxton, W. C., and Cowan, G. A. 1980. Solar neutrino production of long-lived isotopes and secular variations in the Sun. *Science* 210:897–899.

Haxton, W. C., and Johnson, C. W. 1988. Geochemical integrations of the neutrino flux from stellar collapses. *Nature* 333:325–329.

Hays, J. D. 1971. Faunal extinctions and reversals of the Earth's magnetic field. *Geol. Soc. Amer. Bull.* 82:2433–2447.

Hays, J. D., Imbrie, J., and Shackleton, N. J. 1976. Variations in the Earth's orbit: Pacemaker of the Ice Ages. *Science* 194:1121–1132.

Heaps, M. G. 1978. The effects of a solar proton event on the monor neutral constituents of the summer polar mesosphere. U.S. Army Atmospheric Science Laboratory Report ASL-TR-0012.

Heath, D. F. 1980. A review of observational evidence for short and long term ultraviolet flux variability of the Sun. In *Sun and Climate* (Toulouse: CNES), pp. 447–471.

Heath, D. F., and Schlesinger, B. M. 1984. Temporal variability of UV solar spectral irradiance from 160–400 nm over periods of the evolution and rotation regions from maximum to minimum phases of the sunspot cycle. In *Current Problems in Atmospheric Radiation,* ed. G. Fiocco (Huntington, Va.: A. Deepak), pp. 315–319.

Heath, D. F., and Schlesinger, B. M. 1986. The Mg 280-nm doublet as a monitor of changes in solar ultraviolet irradiance. *J. Geophys. Res.* 91:8672–8682.

Heath, D. F., and Thekaekara, M. P. 1977. The solar spectrum between 1200 and 1300 Å. In *The Solar Output and Its Variation,* ed. O. R. White (Boulder: Colorado Assoc. Univ. Press), pp. 193–212.

Heath, D. F., Krueger, A. J., and Crutzen, P. J. 1977. Solar proton events: Influence on stratospheric ozone. *Science* 197:888–889.

Hedin, A. E. 1983. A revised thermospheric model based on mass spectrometer and incoherent scatter data: MSIS-83. *J. Geophys. Res.* 88:10170–10188.

Hedin, A. E. 1987. MSIS-86 thermospheric model. *J. Geophys. Res.* 92:4649–4662.

Held, I. M., and Suarez, M. J. 1974. Simple albedo feedback models of the icecaps. *Tellus* 36:613–628.

Helmholtz, H. L. F. 1885. *On the Sensations of Tone as a Physiological Basis for the Theory of Music,* 2nd English ed., reprinted by Dover, New York, 1954.

Henderson-Sellers, A. 1979. Clouds and the long term stability of the Earth's atmosphere and climate. *Nature* 279:786–788.

Henning, W., Kutschera, W., Ernst, H., Kubic, P., Mayer, W., Moringa, H., Nolte, E., Ratzinger, U., Müller, M., and Schüll, D. 1985. The ^{205}Tl experiments. In *Solar Neutrinos and Neutrino Astronomy,* eds. M. L. Cherry, K. Lande and W. A. Fowler (New York: American Inst. of Physics), pp. 203–211.

Herbert, F. 1989. Primordial electrical induction heating of asteroids. *Icarus* 78:402–410.

Herbert, F., and Sonett, C. P. 1978. Primordial metamorphism of asteroids via electric induction in a T Tauri-like solar wind. *Astrophys. Space Sci.* 55:227–239.

Herbert, F., and Sonett, C. P. 1979. Electromagnetic heating of minor planets in the early solar system. *Icarus* 40:484–496.

Herbert, F., and Sonett, C. P. 1980. Electromagnetic inductive heating of the asteroids and Moon as evidence bearing on the primordial solar wind. In *The Ancient Sun: Fossil Record in the Earth, Moon and Meteorites,* eds. R. O. Pepin, J. A. Eddy and R. B. Merrill (New York: Pergamon Press), pp. 563–576.

Herbert, F., and Sonett, C. P., and Wiskerchen, M. J. 1977. Model "zero-age" lunar thermal profiles resulting from electrical induction. *J. Geophys. Res.* 82:2054–2060.

Herbert, T. D., and Fisher, A. G. 1986. Milankovitch climatic origin of mid-Cretaceous black shale rhythms, central Italy. *Nature* 321:739–743.

Herbig, G. H. 1957. The widths of absorption lines in T Tauri-like stars. *Astrophys. J.* 125:612–613.

Herbig, G. H. 1962. The properties and problems of T Tauri stars and related objects. *Adv. Astron. Astrophys.* 1:47–103.

Herbig, G. H. 1965. Lithium abundances in F5-G8 dwarfs. *Astrophys. J.* 135:588–609.

Herbig, G. H. 1970. Pre-main sequence evolution: Introductory remarks. *Mem. Roy. Soc. Sci. Liège, Ser. 5* 19:13–25.

Herbig, G. H. 1977. Eruptive phenomena in early stellar evolution. *Astrophys. J.* 217:693–715.

Herbig, G. H. 1985. Chromospheric II-alpha emission in F8-G3 dwarfs, and its connection with the T Tauri stars. *Astrophys. J.* 289:269–278.

Herbig, G. H., and Bell, K. R. 1988. Third catalog of emission-line stars of the Orion population. *Lick Obs.Bull.* 1111:1–90.

Herbig, G. H., and James, B. F. 1981. Lange proper motions of the Herbig-Haro objects HH1 and HH2. *Astron. J.* 86:1232–1244

Herbig, G. H., and Rao, N. K. 1972. Second catalog of emission-line stars of the Orion population. *Astrophys. J.* 174:401–423.

Herbig, G. H., and Soderblom, D. R. 1980. Observations and interpretation of the near-infrared line spectra of T Tauri stars. *Astrophys. J.* 242:628–637.

Herbig, G. H., and Spaulding, J. 1955. Axial rotation and line broadening in stars of spectral type F0-K5. *Astrophys. J.* 121:118–143.

Herbold, G., Ulmschneider, P., Spruit, H. C., and Rosner, R. 1985. Propagation of nonlinear, radiatively damped longitudinal waves along magnetic flux tubes in the solar atmosphere. *Astron. Astrophys.* 145:157–169.

Herbst, W. 1989. Spots on the weak T Tauri star V410: The Sun at one million years? *Astron. J.* 98:2268–2274.

Herr, R. 1980. Solar rotation in 1612: Galileo versus Harriott. *Bull. Amer. Astron. Soc.* 12:504 (abstract).

Herzog, G. F. 1972. Relation between solar and planetary neon in carbonaceous chondrites. *J. Geophys. Res.* 77:6219–6225.

Hess, W. N., Canfield, E. H., and Lingenfelter, R. E. 1961. Cosmic ray demography. *J. Geophys. Res.* 66:665–677.

Heyer, M. H., Vrba, F. J., Snell, R. L., Scholerb, F. P., Strom, S. E., Goldsmith, P. F., and Strom, K. M. 1987. The magnetic evolution of the Taurus molecular clouds. I. Large-scale properties. *Astrophys. J.* 321:855–876.

Heymann, D., and Dziczkaniec, M. 1976. Early irradiation of matter in the solar system: Magnesium (proton, neutron) scheme. *Science* 191:79–81.

Heymann, D., and Yaniv, A. 1970. Ar^{40} anomaly in lunar samples from Apollo 11. *Proc. Lunar Sci. Conf.* 2:1261–1267.

Heymann, D., Yaniv, A., Adams, J. A. S., and Fryer, G. E. 1970. Inert gases in lunar samples. *Science* 167:555–558.

Heyvaerts, J., and Priest, E. R. 1983. Coronal heating by phase-mixed shear Alfvén waves. *Astron. Astrophys.* 117:220–234.

Hickey, J. R., Frieden, R. G., Griffin, F. J., Cone, A. S., Maschhoff, R. H., and Gnaidy, J. 1977. The self-calibrating sensor of the eclectic satellite pyrheliometer ESP program. In *Proc. Annual Meeting of the Amer. Solar Energy Soc.* 15:1–4.

Hickey, J. R., Alton, B. M., Griffin, F. J., Jacobowitz, H., Pellegrino, P., Smith, E. A., Vonder Haar, T. H., and Maschoff, R. H. 1981. Solar variability indications from Nimbus 7 satellite data. In *Variations of the Solar Constant*, ed. S. Sofia, NASA CP-2191, pp. 59–72.

Hickey, J. R., Alton, B. M., Kyle, H. L., and Major, E. R. 1988. Observation of total solar irradiance variability from Nimbus satellites. *Adv. Space Res* 8(7):5–10.

Hickey, J. R., Alton, B. M., Kyle, H. L., and Hoyt, D. V. 1989. Total solar irradiance measurements by ERB/NIMBUS 7: A review of nine years. *Space Sci. Rev.* 48:321–342.

Hilgen, F. J. 1987. Sedimentary rhythms and high resolution chronostratigraphic correlations in the Mediterranean Pliocene. *Newslett. Stratig.* 17(2):109–127.

Hill, H. A., and Caudell, T. P. 1979. Global oscillations of the Sun: Observed as oscillations in the solar limb darkening function. *Mon. Not. Roy. Astron. Soc.* 186:327–342.

Hill, H. A., and Stebbins, R. T. 1975. The intrinsic visual oblateness of the Sun. *Astrophys. J.* 200:471–483.

Hill, H. A., Stebbins, R. T., and Oleson, J. R. 1975a. The finite Fourier transform definition of an edge on the solar disk. *Astrophys. J.* 200:484–498.

Hill, H. A., Stebbins, R. T., and Brown, T. M. 1975*b*. Recent solar oblateness observations: Data, interpretation and significance for experimental relativity. In *Atomic Masses and Fundamental Constants*, vol. 5, eds. J. H. Sanders and H. Wapstra (New York: Plenum), pp. 622–630.

Hill, H. A., Bos, R. J., and Goode, P. R. 1982. Preliminary determination of the Sun's gravitational quadrupole moment from rotational splitting of global oscillations and its relevance to tests of general relativity. *Phys. Rev. Lett.* 49:1794–1797.

Hinata, S. 1986. Is it possible to maintain magnetic fields in the radiative cores of rotating stars with convective envelopes? *Astron. Astrophys.* 169:161–163.

Hines, C. O. 1974. A possible mechanism for the production of Sun-weather correlations. *J. Atmos. Sci.* 31:589–591.

Hintenberger, H., Vilcsek, E., and Wänke, H. 1965. Über die Isotopenzusammensetzung und über den Sitz der leichten Uredelgase in Steinmeteoriten. *Z. Naturforsch.* 20a:939–945.

Hintenberger, H., Weber, H. W., and Takaoka, N. 1971. Concentrations and isotopic abundances of the rare gases in lunar matter. *Proc. Lunar Sci. Conf.* 2:1607–1625.

Hintenberger, H., Weber, H. W., and Schultz, L. 1974. Solar, spallogenic and radiogenic rare gases in Apollo 17 soils and breccias. *Proc. Lunar Sci. Conf.* 5:2005–2022.

Hinteregger, H. E., Dedo, D. E., Manson, J. E., and Skillman, D. 1977. EUV flux variations with solar rotation observed during 1974–1976 from the AE-C Satellite. *Space Res.* 17:533–544.

Hinton, R. W., and Bischoff, A. 1984. Ion microprobe magnesium isotope analysis of plagioclase and hibonite from ordinary chondrites. *Nature* 308:169–172.

Hirata, K. S., Kajita, T., Koshiba, M., Nakahata, M., Oyama, Y., Sato, N., Suzuki, A., Takta, M., Totsuka, Y., Kifune, T., Suda T., Takahoshi, K., Tanimori, T., Miyano, K., Yamada, M., Beier, E. W., Feldscher, L. R., Kim, S. B., Mann, A. K., Newcomer, F. M., Van Berg, R., and Zhang, W. 1987. Observation of a neutrino burst from supernova SN1987A. *Phys. Rev. Lett.* 58:1490–1493.

Hirata, K. S., Kajita, T., Kifune, T., Kihara, K., Nakamura, K., Ohara, S., Oyama, Y., Sato, N., Takita, M., Totsuka, Y., Yaginuma, Y., Mori, M., Suzuki, A., Takahashi, K., Tanimori, T., Yamada M., Koshiba, M., Suda, T., Miyano, K., Miyata, H., Takei, H., Kaneyuki, K., Nagashima, Y., Suzuki, Y., Beier, E. W., Feldscher, L. R., Frank, E. D., Frati, W., Kim S. B., Mann, A. K., Newcommer, F. M., Van Berg, R., and Zhang, W. 1988. Observation in the Kamiokande-II detector of the neutrino burst from supernova SN1987A. *Phys. Rev.* D38:448–458.

Hirayama, T., Okamoto, T., and Hudson, H. S. 1984. Facular limb darkening function for irradiance modelling. In *Solar Irradiance Variations on Active Region Time Scales*, eds. B. J. LaBonte, G. A. Chapman, H. S. Hudson and R. C. Willson, NASA CP-2310, pp. 59–72.

Hirshberg, J., Alksne, A., Colburn, D. S., Bame, S. J., and Hundhausen, A. J. 1970. Observation of a solar flare induced interplanetary shock and helium-enriched driver gas. *J. Geophys. Res.* 75:1–15.

Hobbs, L. M., and Pilachowski, C. 1988. Lithium in old open clusters: NGC 188. *Astrophys. J.* 334:734–745.

Hoeksema, J. T., and Scherrer, P. H. 1982. The solar magnetic field - 1976 through 1985. Rept. 87, 10, 331.

Hofmann, H. J., Beer, J., Bonani, G., von Gunten, H. R., Raman, S., Suter, M., Walker, R. L., Wölfli, W., and Zimmermann, D. 1987. ^{10}Be half-life and AMS-standards. *Nucl. Instr. Meth.* B29:32–36.

Hohenberg, C. M., Marti, K., Podosek, F. A., Reedy, R. C., and Shirck, J. R. 1978. Comparisons between observed and predicted cosmogenic noble gases in lunar samples. In *Proc. Lunar Planet. Sci. Conf.* 9:2311–2344.

Hohenberg, C. M., Hudson, B., Kennedy, B. M., and Podosek, F. A. 1981. Noble gas retention chronologies for the St. Séverin meteorite. *Geochim. Cosmochim. Acta* 45:535–546.

Hohenberg, C. M., Nichols, R. H., Jr., Olinger, C. T., and Goswami, J. N. 1990. Cosmogenic neon from individual grains of CM meteorites: Extremely long pre-compaction exposure histories or an enhanced early particle flux. *Geochim. Cosmochim. Acta* 54:2133–2140.

Holland, H. D. 1978. *The Chemistry of the Atmosphere and Oceans* (New York: Wiley).

Holland, H. D. 1984. *The Chemical Evolution of the Atmosphere and Oceans* (Princeton, N.J.: Princeton Univ. Press).

Hollweg, J. V. 1981. Alfvén waves in the solar atmosphere. *Solar Phys.* 70:25–66.

Hollweg, J. V. 1984. Resonances of coronal loops. *Astrophys. J.* 277:392–403.

Hollweg, J. V., and Sterling, A. C. 1984. Resonant heating: An interpretation of coronal loop data. *Astrophys. J.* 282:L31-L33.

Holt, S. S., and Ramaty, R. 1969. Microwave and hard X-ray bursts from solar flares. *Solar Phys.* 8:119–141.

Holzer, T. E. 1989. Interaction between the solar wind and the interstellar medium. *Ann. Rev. Astron. Astrophys.* 27:199–234.

Hones, E. W., Jr., ed 1984a. *Magnetic Reconnection in Space and Laboratory Plasmas,* AGU Geophysical Mono. 30 (Washington, D.C.: American Geophysical Union).

Hones, E. W., Jr. 1984b. Plasma sheet behavior during substorms. In *Magnetic Reconnection and Laboratory Plasmas,* ed. E. W. Hones, Jr., AGU Geophysical Mono. 30 (Washington, D.C.: American Geophysical Union), pp.178–184.

Hood, L. L. 1986. Coupled stratospheric ozone and temperature responses to short-term changes in solar ultraviolet flux: An analysis of Nimbus 7 BUV and SAMS data. *J. Geophys. Res.* 91:5264–5276.

Hood, L. L. 1987a. Magnetic field and remanent magnetization effects of basin-forming impacts on the Moon. *Geophys. Res. Lett.* 14:844–847.

Hood, L. L. 1987b. Solar ultraviolet radiation induced variations in the stratosphere and mesosphere. *J. Geophys. Res.* 92:876–888.

Hood, L. L., and Cantrell, S. 1988. Stratospheric ozone and temperature responses to short-term solar ultraviolet variations: Reproducibility of low-latitude response measurements. *Ann. Geophys.* 6:525–530.

Hood, L. L., and Cisowski, S. M. 1983. Paleomagnetism of the Moon and meteorites. *Rev. Geophys. Space Phys.* 21:676–684.

Hood, L. L., and Douglass, A. R. 1989. Stratospheric responses to solar ultraviolet variations: Comparisons with radiative-photochemical models. In *Ozone in the Atmosphere: Proc. Quadrennial Ozone Symp.,* eds. R. D. Bojkov and P. Fabian (Hampton, Va.: A. Deepak), pp. 380–383.

Hood, L. L., and Jirikowic, J. L. 1990. Recurring variations of probable solar origin in the atmospheric Δ^{14}C time record. *Geophys. Res. Lett.* 17:85.

Hood, L. L., and Vickery, A. 1984. Magnetic field amplification and generation hypervelocity meteoroid impacts with application to lunar paleomagnetism. *J. Geophys. Res.* 89:211–223.

Horedt, G. P. 1978. Blow-off of the protoplanetary cloud by a T Tauri like solar wind. *Astron. Astrophys.* 64:173–178.

Horne, J. H., and Baliunas, S. L. 1987. A prescription for period analysis of unevenly sampled time series. *Astrophys. J.* 302:757–764.

Horning, B. L., and Schubert, G. 1975. Lunar electromagnetic scattering. 3. Propagation at arbitrary angles to the cavity axis. *J. Geophys. Res.* 80:4215–4229.

Hostetler, C. J., and Drake, M. J. 1980. On the early global melting of the terrestrial planets. *Proc. Lunar Planet. Sci. Conf.* 11:1915–1929.

Housen, K. R., and Wilkening, L. L. 1982. Regoliths on small bodies in the solar system. *Ann. Rev. Earth Planet. Sci.* 10:355–376.

Housen, K. R., Wilkening, L. L., Chapman, C. R., and Greenberg, R. 1979. Asteroidal regoliths. *Icarus* 39:317–351.

Housley, R. M. 1980. Toward a model of grain surface exposure in planetary regoliths. In *The Ancient Sun: Fossil Record in the Earth, Moon and Meteorites,* eds. R. O. Pepin, J. A. Eddy and R. B. Merrill (New York: Pergamon Press), pp. 401–410.

Houtermans, J. C. 1971. Geophysical Interpretation of Bristlecone Pine Radiocarbon Measurements Using a Method of Fourier Analysis of Unequally Spaced Data. Ph.D. Thesis, Univ. of Bern.

Houtermans, J. C., Suess, H. E., and Oeschger, H. 1973. Reservoir models and production rate variations of natural radiocarbon. *J. Geophys. Res.* 78:1897–1908.

Hovestadt, D., Vollmer, O., Gloeckler, G., and Fan, C. Y. 1973a. Differential energy spectra of low-energy (< 8.5 MeV per nucleon) heavy cosmic rays during solar quiet times. *Phys. Rev. Lett.* 31:650–653.

Hovestadt, D., Vollmer, O., Gloeckler, G., and Fan, C. Y. 1973b. Measurement of elemental abundance of very low energy solar cosmic rays. In *Proc. 13th Intl. Cosmic Ray Conf.*, vol. 2 (Denver: Univ. of Denver Press), pp. 1498–1503.

Howard, R. A., Sheeley, N. R., Jr., Michels, D. J., and Koomen, M. J. 1986. The solar cycle dependence of coronal mass ejections. In *The Sun and the Heliosphere in Three Dimensions*, ed. R. G. Marsden (Dordrecht: D. Reidel), pp. 107–111.

Hoyle, F. 1960. On the origin of the solar system. *Quart. J. Roy. Astron. Soc.* 1:28–55.

Hoyng, P., Duijveman, A., Machado, M. E., Rust, D. M., Svestka, Z., Boelee, A., de Jager, C., Frost, K. J., LaFleur, H., Simnett, G. M., van Beek, H. F., and Woodgate, B. E. 1981. The site of the hard X-ray emission in solar flares. *Astrophys. J.* 246:L155–L159.

Hoyt, D. V. 1979. The Smithsonian Astrophysical Observatory Solar Constant Program. *Rev. Geophys. Space Res.* 17:427–458.

Hoyt, D. V., and Eddy, J. A. 1982. *An Atlas of Variations in the Solar Constant Caused by Sunspot Blocking and Facular Emissions from 1874 to 1981*, NCAR TN194 + STR (Boulder: Natl. Center for Atmospheric Research).

Hoyt, D. V., Eddy, J. A., and Hudson, H. S. 1983. Sunspot areas and solar irradiance variations during 1980. *Astrophys. J.* 275:878–888.

Hua, X.-M., and Lingenfelter, R. E. 1987a. Solar flare neutron production and the angular distribution of the capture gamma ray emission. *Solar Phys.* 107:351–383.

Hua, X.-M., and Lingenfelter, R. E. 1987b. A determination of the ^{3}He/H ratio in the solar photosphere from solar gamma ray line observations. *Astrophys. J.* 319:555–566.

Huang, S.-S. 1965. Rotational behavior of the main sequence stars and its plausible consequences concerning formation of planetary systems. *Astrophys. J.* 141:985–992.

Hubbard, F., and Dearborn, D. S. P. 1980. Magnetic mixing and the Arcturus problem. *Astrophys. J.* 239:248–252.

Hubbard, F., and Dearborn, D. S. P. 1981. Can radiative core stars hide in the H-R diagram? *Astrophys. J.* 247:236–238.

Hudson, B., Flynn, G. J., Fraundorf, P. Hohenberg, C. M., and Shirck, J. 1981. Noble gases in stratospheric dust particles: Confirmation of extraterrestrial origin. *Science* 211:383–386.

Hudson, H. S. 1988a. Observed variability of the solar luminosity. *Ann. Rev. Astron. Astrophys.* 26:473–508.

Hudson, H. S. 1988b. Modelling of total solar irradiance variability: An overview. *Adv. Space Res.* 8(7):15–20.

Hudson, H. S. and Willson, R. C. 1981. SMM experiment: Initial observations by the active cavity radiometer. *Adv. Space Res.* 1(13):285–288.

Hudson, H. S., and Willson, R. C. 1982. Sunspots and solar variability. In *Physics of Sunspots*, eds. L. Cram and J. Thomas (Sunspot, N.M.: Sacramento Peak Obs.), pp. 434–445.

Hudson, H. S., and Woodard, M. 1983. Solar surface granulation and variations of total solar irradiance. *Bull. Amer. Astron. Soc.* 15:715 (abstract).

Hudson, H. S., Silva, S., Woodard, M., and Willson, R. C. 1982. The effects of sunspots on solar irradiance. *Solar Phys.* 76:211–219.

Huebner, W. F., Merts, A. L., Magee, N. H., and Argo, M. F. 1977. Astrophysical Opacity Library. Los Alamos Natl. Lab Sci. Rept. LA-6760-m.

Hummer, D. G., and Mihalas, D. 1988. The equation of state for stellar envelopes. I. An occupation probability formalism for the truncation of internal partition functions. *Astrophys. J.* 331:794–818.

Hundhausen, A. J. 1972. *Coronal Expansion and the Solar Wind* (New York: Springer-Verlag).

Hundhausen, A. J. 1977. An interplanetary view of coronal holes. In *Coronal Holes and High Speed Wind Streams*, ed. J. B. Zirker (Boulder: Colorado Assoc. Univ. Press), pp. 225–329.

Huneke, J. C., Smith, S. P., Rajan, S. R., Papanastassiou, D. A., and Wasserburg, G. J. 1977. Comparison of the chronology of the Kapoeta parent planet and the Moon. *Lunar Sci.* VIII:484–486 (abstract).

Hunten, D. M. 1990. Escape of atmospheres, ancient and modern. *Icarus* 85:1–20.

Hunten, D. M., Pepin, R. O., and Walker, J. C. G. 1987. Mass fractionation in hydrodynamic escape. *Icarus* 69:532–549.

Hunten, D. M., Pepin, R. O., and Owen, T. C. 1988. Planetary atmospheres. In *Meteorites and the Early Solar System*, eds. J. F. Kerridge and M. S. Matthews (Tucson: Univ. of Arizona Press), pp. 565–591.

Hunten, D. M., Donahue, T. M., Walker, J. C. G., and Kasting, J. F. 1989. Escape of atmospheres and loss of water. In *Origin and Evolution of Planetary and Satellite Atmospheres,* eds. S. K. Atreya, J. B. Pollack and M. S. Matthews (Tucson: Univ. of Arizona Press), pp. 386–422.

Huovelin, J., Saar, S. H., and Tuominen, I. 1988. Relations between broad-band linear polarization and Ca II H and K emission in late-type stars. *Astrophys. J.* 329:882–893.

Hurst, G. S., Chen, C. H., Kramer, S. D., Cleveland, B. T., Davis, R., Rowley, R. K., Gabbard, F., and Schima, F. J. 1984. Feasibility of the ^{81}Br (v, e^-) experiment. *Phys. Rev. Lett.* 53:1116–1119.

Hurst, G. S., Chen, C. H., Kramer, S. D., and Allman, S. L. 1985. A proposed solar neutrino experiment using ^{81}Br$(v, e^-)^{81}$Kr. In *Solar Neutrinos and Neutrino Astronomy,* eds. M. L. Cherry, K. Lande and W. A. Fowler (New York: American Inst. of Physics), pp. 152–161.

Hut, P., Alvarez, W., Elder, P. W., Hansen, T., Kauffman, E. G., Keller, G., Shoemaker, E. M., and Weissman, P. R. 1987. Comet showers as a cause of mass extinctions. *Nature* 329:116–126.

Hutcheon, I. D., and Hutchison, R. 1989. Evidence from the Semarkona ordinary chondrite for ^{26}Al heating of small planets. *Nature* 337:238–241.

Hutcheon, I. D., Macdougall, J. D., and Price, P. B. 1974. Improved determination of the long-term Fe spectrum from 1 to 460 MeV/amu. *Proc. Lunar Sci. Conf.* 5:2561–2576.

Hutcheon, I. D., Steele, I. M., Smith, J. V., and Clayton, R. N. 1978. Ion microprobe, electron microprobe and cathodoluminescence data for Allende inclusions with emphasis on plagioclase chemistry. *Proc. Lunar Planet. Sci. Conf.* 9:1345–1368.

Hyde, W. T., and Peltier, W. R. 1985. Sensitivity experiments with a model of the Ice Age cycle. The response to harmonic forcing. *J. Atmos. Sci.* 42:2170–2188.

Iben, I., Jr. 1965. Stellar evolution. I. The approach to the main sequence. *Astrophys. J.* 141:993–1018.

Iben, I., Jr. 1967. Stellar evolution. VI. Evolution from the main sequence to the giant branch for stars of mass 1.0 $M_\odot$, 1.25 $M_\odot$, and 1.5 $M_\odot$. *Astrophys. J.* 147:624–649.

Iben, I., Jr., and Mahaffy, J. 1976. On the Sun's accoustical spectrum. *Astrophys. J.* 209:L39–L43.

Ichimoto, K., Kubota, J., Suzuki, M., Thomura, I., and Kurokawa, H. 1985. Periodic behaviour of solar flare activity. *Nature* 316:422–424.

Iglesias, C. A., Rogers, F. J., and Wilson, R. G. 1987. Reexamination of the metal contribution to astrophysical opacity. *Astrophys. J.* 322:L45–L48.

Imbrie, J. 1982. Astronomical theory of the Pleistocene Ice Ages: A brief historical. *Icarus* 50:408–422.

Imbrie, J., and Imbrie, K. P. 1979. *Ice Ages: Solving the Mystery* (Short Hills, N.J.: Enslow).

Imbrie, J., and Imbrie, J. Z. 1980. Modelling the climatic response to orbital variations. *Science* 207:943–953.

Imbrie, J., and Imbrie, K. P. 1986. *Ice Ages: Solving the Mystery,* 2nd ed. (Cambridge: Harvard Univ. Press).

Imbrie, J., and Kipp, N. G. 1971. New micropaleontological method for quantitative paleoclimatology: Application to a late Pleistocene Caribbean core. In *Late Cenezoic Glacial Ages,* ed. K. K. Turekian (New Haven, Conn.: Yale Univ. Press), pp. 71–181.

Imbrie, J., Hays, J. D., Martinson, D. G., McIntyre, A., Mix, A. C., Morley, J. J., Pisias, N. G., Prell, W. L., and Shackleton, N. J. 1984. The orbital theory of Pleistocene climate: Support from a revised chronology of the marine δ^{18}O record. In *Milankovitch and Climate: Understanding Orbital Forcing,* eds. A. Berger, J. Imbrie, J. Hays, G. Kukla and B. Saltzman (Dordrecht: D. Reidel), pp. 269–305.

Imbrie, J., McIntyre, A., and Mix, A. 1989. Oceanic response to orbital forcing in the late Quaternary: Observational and experimental strategies. In *Climate and Geosciences,* eds. A. Berger, J.-C. Duplessy and S. Schneider (Dordrecht: D. Reidel), pp. 121–164.

Imhoff, C. L., and Appenzeller, I. 1987. Pre-main sequence stars. In *Exploring the Universe with the IUE Satellite,* eds. Y. Kondo and W. Wamsteker (Dordrecht: D. Reidel), pp. 295–319.

Ionson, J. A. 1984. Unified theory of electrodynamic coupling in coronal magnetic loops: The coronal heating problem. *Astrophys. J.* 276:357–368.

Ip, W.-H. 1984. Magnetic field amplification in the solar nebula through interaction with the T Tauri wind. *Nature* 312:625–626.

Ipavich, F. M., Galvin, A. B., Gloeckler, G., Hovestadt, D., Bame, S. J., Klecker, B., Scholer, M., Fisk, L. A., and Fan, C. Y. 1986. Solar wind Fe and CNO measurements in high-speed flows. *J. Geophys. Res.* 91:4133–4141.

Isaak, G. R. 1981. Is the Sun an oblique magnetic rotator? *Nature* 296:130–131.

Ivanova, T. S., and Ruzmaikin, A. A. 1976. Magnetohydrodynamic dynamo model of the solar cycle. *Astron. J. (USSR)* 53:398–410.

Jackman, C. H., Frederick, J. E., and Stolarski, R. S. 1980. Production of odd nitrogen in the stratosphere and mesosphere: An intercomparison of source strengths. *J. Geophys. Res.* 85:7495–7505.

Jacobsen, S. B., and Wasserburg, G. J. 1984. Sm-Nd evolution of chondrites and achondrites. II. *Earth Planet. Sci. Lett.* 67:137–150.

Janes, K., and Demarque, P. 1983. The ages and compositions of the oldest clusters. *Astrophys. J.* 264:206–214.

Jefferies, S. M., Pomerantz, M. A., Duvall, T. L., Jr., Harvey, J. W., and Jaksha, D. B. 1988. Helioseismology from the South Pole: Comparison of 1987 and 1981 results. In *Seismology of the Sun and Sun-Like Stars*, ed. E. J. Rolfe, ESA SP-286, pp. 279–284.

Jeffery, P. M., and Reynolds, J. H. 1961. Origin of excess Xe [129] in stone meteorites. *J. Geophys. Res.* 66:2582–2583.

Johnsen, S. J., Dansgaard, W., Clausen, H. B., and Langway, C. C. 1970. Climatic oscillations 1200–2000 AD. *Nature* 227:482–483.

Johnstone, R. M., and Penston, M. V. 1986. A search for magnetic fields in the T Tauri stars GW Ori, CoD-34 7151, and RU Lupi. *Mon. Not. Roy. Astron. Soc.* 219:927–941.

Jokipii, J. R. 1971. Propagation of cosmic rays in the solar wind. *Rev. Geophys. Space Phys.* 9:27–87.

Jokipii, J. R. 1985. Effects of three-dimensional heliospheric structures on cosmic-ray modulation. In *The Sun and Heliosphere in Three Dimensions*, ed. R. G. Marsden (Dordrecht: D. Reidel), pp. 375–387.

Jokipii, J. R., and Davila, J. 1981. Effects of drift on the transport of cosmic rays. IV. More realistic diffusion coefficients. *Astrophys. J.* 248:1156–1163.

Jokipii, J. R., and Kota, J. 1988. The polar heliospheric magnetic field. *Geophys. Res. Lett.* 16:1–4.

Jokipii, J. R., and Lee, L. C. 1974. An effect of cosmic rays on the distant solar wind. In *Solar Wind: Proc. Third Intl. Conf*, ed. C. T. Russell (Los Angeles: Univ. of California Press), pp. 224–229.

Jokipii, J. R., and Thomas, B. 1981. Effects of drift on the transport of cosmic rays. IV. Modulation by a wavy interplanetary current sheet. *Astrophys. J.* 243:1115–1122.

Jokipii, J. R., Levy, E. H., and Hubbard, W. B. 1977. Effects of particle drift on cosmic ray transport. 1. General properties, applications to solar modulation. *Astrophys. J.* 213:861–868.

Jones, P. D., Wigley, T. M. L., and Wright, P. B. 1986. Global temperature variations between 1861 and 1984. *Nature* 322:430–434.

Jones, T. D. 1988. An Infrared Reflectance Study of Water in Outer Belt Asteroids: Clues to Composition and Origin. Ph.D. Thesis, Univ. of Arizona.

Jordan, C., and Linsky, J. L. 1987. Chromospheres and transition regions. In *Exploring the Universe with the IUE Satellite*, ed. Y. Kondo (Boston: D. Reidel), pp. 259–293.

Jouzel, J., Lorius, C., Petit, J. R., Genthon, C., Barkov, N. I., Kotlyakov, V. M., and Petrov, V. M. 1987. Vostok ice core: A continuous isotope temperature record over the last climatic cycle (160,000 years). *Nature* 329:403–408.

Joy, A. H. 1945. T Tauri variable stars. *Astrophys. J.* 102:168–195.

Juillet-Leclerc, A., and Schrader, H. 1987. Variations of upwelling intensity recorded in varved sediment from the Gulf of California during the past 3000 years. *Nature* 329:146–149.

Kaiser, T., and Wasserburg, G. J. 1983. The isotopic composition and concentration of Ag in iron meteorites and the origin of exotic silver. *Geochim. Cosmochim. Acta* 47:43–58.

Kallel, N., Labeyrie, L. D., Arnold, M., Okada, H., Dudley, W. C., and Duplessy, J.-C. 1988. Evidence of cooling during the Younger Dryas in the western North Pacific. *Oceanol. Acta* 11:369–375.

Källén, E., Crafoord, C., and Ghil, M. 1979. Free oscillations in a coupled atmosphere-hydrosphere-cryosphere system. *J. Atmos. Sci.* 36:2292–2303.

Kallmann-Bijl, R., Boyd, L. F., Lagow, H., Poloskov, S. M., and Priester, W. 1961. *COSPAR Intl. Reference Atmosphere* (Amsterdam: North-Holland Press).

Kane, S. R., Evenson, P., and Meyer, P. 1985. Acceleration of interplanetary solar electrons in the 1982 August 14 solar flare. *Astrophys. J.* 299:L107–L110.

Karhu, J., and Epstein, S. 1986. The implication of the oxygen isotope records in coexisting cherts and phosphates. *Geochim. Cosmochim. Acta* 50:1745–1756.

Karpen, J. T. Doschek, G. A., and Seeley, J. F. 1986. High resolution X-ray spectra of solar flares. VIII. Mass upflows in the large flare of 1980 November 7. *Astrophys. J.* 306:327–339.

Kashkarov, L. L., Genaeva, L. I., Korotkova, N. N., and Lavrukhina, A. K. 1977. Energy spectrum of the cosmic ray iron group nuclei in the region 1–100 MeV nucleon^{-1} by fossil tracks. In *Proc. 15th Intl. Cosmic Ray Conf.*, vol. 5 (Sofia: Bulgarian Academy of Sciences), pp. 60–63.

Kasting, J. F. 1982. Stability of ammonia in the primitive terrestrial atmosphere. *J. Geophys. Res.* 87:3091–3098.

Kasting, J. F. 1985. Photochemical consequences of enhanced CO_2 levels in Earth's early atmosphere. In *The Carbon Cycle and Atmospheric CO_2: Natural Variations Archean to Present*, eds. E. T. Sundquist and W. S. Broecker, AGU Geophysical Mono. 32 (Washington, D.C.: American Geophysical Union), pp. 612–622.

Kasting, J. F. 1987. Theoretical constraints on oxygen and carbon dioxide concentrations in the Precambrian atmosphere. *Precambrian Res.* 34:205–229.

Kasting, J. F. 1988. Runaway and moist greenhouse atmospheres and the evolution of Earth and Venus. *Icarus* 74:472–494.

Kasting, J. F. 1989. Long-term stability of the Earth's climate. *Palaeogeog. Paleaeoclimat. Palaeoecol.* 75:83–95.

Kasting, J. F. 1990a. Bolide impacts and the oxidation state of carbon in the Earth's early atmosphere. *Origins of Life* 20:199–231.

Kasting, J. F. 1990b. Evolution of the Earth's atmosphere and hydrosphere: Hadean to recent. In *Organic Geochemistry*, eds. M. H. Engel and S. A. Macko, in press.

Kasting, J. F., and Ackerman, T. P. 1986. Climatic consequences of very high CO_2 levels in Earth's early atmosphere. *Science* 234:1383–1385.

Kasting, J. F., and Donahue, T. M. 1980. The evolution of atmospheric ozone. *J. Geophys. Res.* 85:3255–3263.

Kasting, J. F., and Walker, J. C. G. 1981. Limits on oxygen concentration in the prebiological atmosphere and the rate of abiotic fixation of nitrogen. *J. Geophys. Res.* 86:1147–1158.

Kasting, J. F., Liu, S. C., and Donahue, T. M. 1979. Oxygen levels in the prebiological atmosphere. *J. Geophys. Res.* 84:3097–3107.

Kasting, J. F., Pollack, J. B., and Crisp, D. 1984. Effects of high CO_2 levels on surface temperature and atmospheric oxidation state on the early Earth. *J. Atmos. Chem.* 1:403–428.

Kasting, J. F., Holland, H. D., and Pinto, J. P. 1985. Oxidant abundances in rainwater and the evolution of atmospheric oxygen. *J. Geophys. Res.* 90:10497–10510.

Kasting, J. F., Zahnle, K. J., Pinto, J. P., and Young, A. T. 1989. Sulfur, ultraviolet radiation, and the early evolution of life. *Origins of Life* 19:95–108.

Kawaler, S. D. 1987. Angular momentum in stars: The Kraft curve revisited. *Publ. Astron. Soc. Pacific* 99:1322–1328.

Kawaler, S. D. 1988. Angular momentum loss in low-mass stars. *Astrophys. J.* 333:236–247.

Kawaler, S. D. 1989. Rotational dating of middle-aged stars. *Astrophys. J.* 343:L65–L68.

Keating, G. M., Pitts, M. C., Brasseur, G., and De Rudder, A. 1987. Response of middle atmosphere to short-term solar ultraviolet variations. 1. Observations. *J. Geophys. Res.* 92:889–902.

Keil, S. L., Fleck, B., and Sarajadini, A. 1990. Variations of the solar Ca K line 1976–1989: Correlation with other activity indicators and rotational modulation with the solar cycle. Preprint.

Kemp, J. C., Henson, G. D., Kraus, D. J., Dunaway, M. H., Hall, D. S., Boyd, L. J., Genet, R. M., Guinan, E. F., Wacker, S. W., and McCook, G. P. 1987. Discovery of optical circular polarization in Lambda Andromedae. *Astrophys. J.* 317:L29–L32.

Kent, D. V. 1982. Apparent correlation of paleomagnetic intensity and climatic records in deep-sea sediments. *Nature* 299:538–539.

Kenyon, S. J. 1987. Accretion as an energy source for pre-main sequence stars. In *Cool Stars, Stellar Systems and the Sun: Proc. Fifth Cambridge Workshop*, eds. J. L. Linsky and R. E. Stencel (New York: Springer-Verlag), pp. 431–442.

Kenyon, S. J., and Hartmann, L. W. 1987. Spectral energy distributions of T Tauri stars: Disk flaring and limits on accretion. *Astrophys. J.* 323:714–733.

Kenyon, S. J., Hartmann, L. W., and Hewett, R. 1988. Accretion disk models for FU Orionis and V1057 Cygni: Detailed comparisons between observations and theory. *Astrophys. J.* 325:231–251.

Kerr, R. A. 1985. Periodic extinctions and impacts challenged. *Science* 227:1451–1453.

Kerr, R. A. 1987. How the Sun faded even as the sunspots did. *Science* 236:1624–1625.

Kerr, R. A. 1988. Sunspot-weather link holding up. *Science* 242:124–125.

Kerridge, J. F. 1975. Solar nitrogen: Evidence for a secular increase in the ratio of nitrogen-15 to nitrogen-14. *Science* 188:162–164.

Kerridge, J. F. 1980. Secular variations in composition of the solar wind: Evidence and causes. In *The Ancient Sun: Fossil Record in the Earth, Moon and Meteorites*, eds. R. O. Pepin, J. A. Eddy and R. B. Merrill (New York: Pergamon Press), pp. 475–489.

Kerridge, J. F. 1989. What is causing the secular increase in solar nitrogen-15? *Science*, submited.

Kerridge, J. F., and Matthews, M. S., eds. 1989. *Meteorites and the Early Solar System* (Tucson: Univ. of Arizona Press).

Kerridge, J. F., Kaplan, I. R., Lingenfelter, R. E., and Boynton, W. V. 1977. Solar wind nitrogen: Mechanisms for isotopic evolution. *Proc. Lunar Sci. Conf.* 8:3773–3789.

Kerridge, J. F., Mackay, A. L., and Boynton, W. F. 1979. Magnetite in C1 carbonaceous meteorites: Origin by aqueous activity on a planetesimal surface. *Science* 205:395–397.

Kiehl, J. T., and Dickinson, R. E. 1987. A study of the radiative effects of enhanced atmospheric CO_2 and CH_4 on early Earth surface temperatures. *J. Geophys. Res.* 92:2991–2998.

Killeen, T. L. 1987. Energetics and dynamics of the Earth's thermosphere. *Rev. Geophys.* 25:433–454.

Killeen, T. L., and Roble, R. G. 1988. Thermosphere dynamics: Contributions from the first 5 years of the Dynamics Explorer program. *Rev. Geophys.* 26:329–367.

Kim, S. B. 1989. Real Time, Directional Measurement of ^8B Solar Neutrinos in the Kamiokande-II Detector and Search for Short-Time Variation. Ph.D. Thesis, Univ. of Pennsylvania.

King, J. H. 1974. Solar proton fluences for 1977–1983 space missions. *J. Spacecrafts and Rockets* 11:401–408.

Kirschner, S. 1988. Chimie de la Neige ($NaCl$, NO_3, SO_4) à la station Pole Sud. Ph.D. Thesis, Univ. of Grenoble.

Kirsten, T. 1990. The Gallex project. In *Inside the Sun: Proc. IAU Coll. 121*, eds. G. Berthomieu and M. Cribier (Dordrecht: Kluwer), pp. 187–199.

Kitazawa, K. 1970. Intensity of the geomagnetic field in Japan for the past 10,000 years. *J. Geophys. Res.* 75:7403–7411.

Klein, J., Lerman, J. C., Damon, P. E., and Linick, T. 1980. Radiocarbon concentrations in the atmosphere: 8000 year record of variations in tree rings. *Radiocarbon* 22:950–961.

Klein, J., Middleton, R., Fink, D., Dietrich, J. W., Aylmer, D., and Herzog, G. F. 1988. Beryllium-10 and aluminum-26 contents of lunar rock 74275. In *Lunar Planet. Sci.* XIX:607–608 (abstract).

Klein, J., Fink, D., Herzog, G. F., Pierazzo, E., Middleton, R., and Vogt, S. 1989. SCR produced ^{41}Ca in lunar basalt 74275. *Meteoritics* 24:286 (abstract).

Knobloch, E., and Rosner, R. 1981. On the spectrum of turbulent magnetic fields. *Astrophys. J.* 247:300–311.

Knobloch, E., Rosner, R., and Weiss, N. O. 1981. Magnetic fields in late-type stars. *Mon. Not. Roy. Astron. Soc.* 197:45P–49P.

Knölker, M., Schüssler, M., and Weisshaar, E. 1988. Model calculations of magnetic flux tubes. *Astron. Astrophys.* 194:257–267.

Kobayashi, K. 1959. Chemical remanent magnetization of ferromagnetic minerals and its application to rock magnetisms. *J. Geomagnet. Geoelec.* 10:99–117.

Kobayashi, K. 1962. Magnetization-blocking process by volume development of ferromagnetic particles. *J. Phys. Soc. Japan* 17:695–698.

Kocharov, G. E. 1973. Once more on solar neutrinos. In *Proc. 5th Intl. Seminar on Solar Cosmic Rays and Their Penetration into the Earth's Magnetosphere*, ed. G. E. Dergacheva, pp. 193–204 (in Russian). *Bull. Acad. Sci. USSR Phys. Series (USA)* 41:129–152.

Kocharov, G. E. 1977. Neutrino astrophysics. *Izv. Akad. Nauk SSSR, Seriya Fiz.* 41:1916–1948.

Kocharov, G. E. 1980. *Nuclear Astrophysics of the Sun* (Munich: Thiemig).

Kocharov, G. E. 1984. Plasma and nuclear processes in solar matter. In *Plasma Astrophysics*, eds. T. D. Guyenne and I. M. Zeleny, ESA SP-207, pp. 65–81.

Kocharov, G. E. 1986. Cosmic ray archeology, solar activity and supernova explosions. In *Plasma Astrophysics*, eds. T. D. Guyenne and I. M. Zeleny, ESA SP-251, pp. 259–270.

Kocharov, G. E. 1987. Nuclear processes in the solar atmosphere and the particle acceleration problem. *Astrophys. Space Sci.* 6:155–202.

Kocharov, G. E., and Kocharov, L. G. 1978. Modern state of experimental and theoretical investigations of solar events enriched by helium-3. In *Proc. 10th Leningrad Seminar on Space Physics*, pp. 37–72.

Kocharov, G. E., and Starbunov, Yu. N. 1969. Concerning thermonuclear reaction in the interior of the Sun and solar neutrinos. Ioffe Physical-Technical Institute, USSR Academy of Sciences Preprint No. 240; Upton: BNL-tr-340 (Eng. trans. 1970).

Kocharov, G. E., and Starbunov. Yu. N. 1970a. Concerning thermonuclear reaction in the interior of the Sun and solar neutrinos. *JETP Lett.* 11:81–83.

Kocharov, G. E., and Starbunov, Yu. N. 1970b. Point sources of neutrino radiation. In *Proc. 11th Intl. Cosmic Ray Conf., Acta Phys. Acad. Sci. Hungary* 29 (Suppl. 4), pp. 353–379.

Kocharov, G. E., and Starbunov, Yu. N. 1971. Solar models with increased concentration of ³He and neutrino fluxes. In *Proc. 12th Intl. Cosmic Ray Conf.*, vol. 7, pp. 2865–2869.

Kocharov, L. G., and Kocharov, G. E. 1984. ³He-rich solar flares. *Space Sci. Rev.* 38:89–141.

Koeningsberger, J. G. 1938. Natural remanent magnetism of eruptive rocks, Parts 1 and 2. *Terrest. Magnet. Atmos. Phys.* 43:119–127, 299–320.

Kohl, C. P., Murrell, M. T., Russ, G. P., III, and Arnold, J. R. 1978. Evidence for the constancy of the solar cosmic ray flux over the past ten million years: ^{53}Mn and ^{26}Al measurements. In *Proc. Lunar Planet. Sci. Conf.* 9:2299–2310.

Kominz, M. A., and Pisias, N. G. 1979. Pleistocene climate: Deterministic or stochastic? *Science* 204:171–173.

Kominz, M. A., Heath, G. R., Ku, T. L., and Pisias, N. G. 1979. Brunhes time scales and the interpretation of climatic changes. *Earth Planet. Sci. Lett.* 45:394–410.

Königl, A. 1982. On the nature of bipolar sources in dense molecular clouds. *Astrophys. J.* 261:115–134.

Königl, A. 1986. Stellar and galactic jets: Theoretical issues. *Canadian J. Phys.* 64:362–368.

Königl, A. 1989. Self-similar models of magnetized accretion disks. *Astrophys. J.* 342:208–223.

Konstantinov, and Kocharov, G. 1965. Astrophysical phenomena and radiocarbon. *Soviet Academy Rept. DAN SSSR* 165:63–64.

Koo, B.-C. 1989. Extremely high-velocity molecular flows in young stellar objects. *Astrophys. J.* 337:318–331.

Korzennik, S. G., and Ulrich, R. K. 1989. Seismic analysis of the solar interior. I. Can opacity changes improve the theoretical frequencies? *Astrophys. J.* 339:1144–1155.

Kota, J., and Jokipii, J. R. 1983. Effects of drift on the transport of cosmic rays. VI. A three-dimensional model, including diffusion. *Astrophys. J.* 265:573–581.

Kraft, R. P. 1965. Rotational velocities of stars in the Pleiades and Hyades clusters. *Astrophys. J.* 142:681–702.

Kraft, R. P. 1967. Studies of stellar rotation. V. The dependence of rotation on age among solar-type stars. *Astrophys. J.* 150:551–569.

Kraft, R. P. 1970. Stellar rotation. In *Spectroscopic Astrophysics*, ed. G. H. Herbig (Berkeley: Univ. of California Press), pp. 385–422.

Kraft, R. P., and Greenstein, J. L. 1969. A new method for finding faint members of the Pleiades. In *Low Luminosity Stars*, ed. S. S. Kumar (New York: Gordon and Breach), pp. 65–82.

Krajcheva, Z. T., Popova, E. T., Tutukov, A. V., and Jungelson, L. E. 1978. Some properties of spectroscopic binary stars. *Astron. J. (USSR)* 55:1175–1178.

Krause, F., and Rädler, K.-H. 1980. *Mean-Field Magnetohydrodynamics and Dynamo Theory* (Berlin: Akademie-Verlag).

Krishnamurthy, R. V., Bhattacharya, S. K., and Kusumgar, S. 1986. Paleoclimate changes deduced from $^{13}C/^{12}C$ and C/N ratios of Karewa lake sediments, India. *Nature* 323:150–152.

Krishnaswamy, S., Lal, D., Martin, J. M., and Meybeck, M. 1971. Geochronology of lake sediments. *Earth Planet. Sci. Lett.* 11:407–414.

Kroner, B., Rhein, M., Bruns, M., Schoch-Fischer, H., Munnich, K. O., Stuiver, M., and Becker, B. 1986. Radiocarbon calibration data for the 6th to the 8th millennia BC. In *Proc. 12th Intl. Radiocarbon Conf.*, eds. M. Stuiver and R. Kra. *Radiocarbon* 28 (2B):954–960.

Krummenacher, D., Merrihue, C. M., Pepin, R. O., and Reynolds, J. H. 1962. Meteoritic krypton and barium versus the general isotopic anomalies in xenon. *Geochim. Cosmochim. Acta* 26:231–249.

Ku, T. L. 1979. ^{10}Be in the marine environment. In *The Natural Radioactive Isotopes in Beryllium in the Environment*, eds. K. K. Turekian and H. L. Volchok. Rept. and abstracts of a meeting sponsored by Yale Univ. and the U.S. Dept. of Energy; unpaged.

Kuhi, L. V. 1964. Mass loss from T Tauri stars. *Astrophys. J.* 140:1409–1433.

Kuhi, L. V. 1978. Spectral characteristics of T Tauri stars. In *Protostars and Planets*, ed. T. Gehrels (Tucson: Univ. of Arizona Press), pp. 708–717.

Kuhn, J. R. 1988. Helioseismological splitting measurements and the non-spherical solar temperature structure. *Astrophys. J.* 331:L131–L134.

Kuhn, J. R. 1989. Helioseismic observations of the solar cycle. *Astrophys. J.* 339:L45–L47.

Kuhn, J. R., Libbrecht, K. G., and Libbrecht, K. G. 1985. Observations of a solar latitude dependent limb brightness variation. *Astrophys. J.* 290:758–764.

Kuhn, J. R., Libbrecht, K. G., and Dicke, R. H. 1988. The surface temperature of the Sun and changes in the solar constant. *Science* 242:908–911.

Kuhn, W. R., and Atreya, S. K. 1979. Ammonia photolysis and the greenhouse effect in the primordial atmosphere of the Earth. *Icarus* 37:207–213.

Kuhn, W. R., and Kasting, J. F. 1983. The effects of increased CO_2 concentrations of surface temperature of the early Earth. *Nature* 301:53–55.

Kukla, G. 1972. Insolation and glacials. *Boreas* 1:63–96.

Kukla, G. 1975a. Loess stratigraphy of central Europe. In *After the Australopithecines*, eds. K. W. Butzer and G. U. Isaac (The Hague: Monton), pp. 99–188.

Kukla, G. 1975b. Missing link between Milankovitch and climate. *Nature* 253:600–603.

Kukla, G., Berger, A., Lotti, R., and Brown, J. 1981. Orbital signature of interglacials. *Nature* 290:295–300.

Kukla, J. 1969. The cause of the Holocene climate change. *Geol. Mijnbouw* 48(3):307–334.

Kundu, M. R. 1985. High spatial resolution microwave observations of the Sun. *Solar Phys.* 100:491–514.

Kuperus, M., Ionson, J. A., and Spicer, D. S. 1981. On the theory of coronal heating mechanisms. *Ann. Rev. Astron. Astrophys.* 19:7–40.

Kürths, J. 1987. Estimating parameters of attractors in some astrophysical time series. In *Proc. Intl. Conf. on Nonlinear Oscillations*, ed. M. Farkas (Budapest), pp. 664–667.

Kusakabe, M., Ku, T. L., Southon, J. R., Vogel, J. S., Nelson, D. E., Measures, C. I., and Nozaki, Y. 1987. Distribution of ^{10}Be and ^9Be in the Pacific Ocean. *Earth Planet. Sci. Lett.* 82:231–240.

Kuskov, O. L., and Khitarov, N. I. 1979. Thermal effects of chemical reactions in the undifferentiated Earth. *Phys. Earth Planet. Int.* 18:20–26.

Kutner, M. L., Rydgren, A. E., and Vrba, F. J. 1986. Detection of 6 cm radio emission from late-type PMS stars with weak chromospheric emission. *Astron. J.* 92:895–897.

Kutzbach, J. E. 1976. The nature of climate and climate variation. *Quat. Res.* 6:471–480.

Kutzbach, J. E. 1981. Monsoon climate of the early Holocene: Climate experiment with the Earth's orbital parameters for 9,000 years ago. *Science* 214:59–61.

Kutzbach, J. E. 1985. Modeling of paleoclimates. *Adv. Geophys.* 28:159–196.

Kutzbach, J. E., and Gallimore, R. G. 1988. Sensitivity of a coupled atmosphere/mixed layer ocean model to changes in orbital forcing at 9000 years B. P. *J. Geophys. Res.* 93:803–821.

Kuzminoff, V. V., and Pomansky, A. A. 1983. ^{81}Kr production rate in the atmosphere. In *Proc. 18th Intl. Cosmic Ray Conf.*, vol. 2, eds N. Durgaprasad, S. Ramadurai, P. V. Raman Murthy, M. V. S. Rao and K. Sivaprasad (Bombay: Tata Inst. of Fundamental Research), pp. 356–360.

Kuzminov, V. V., Pomansky, A. A., and Chailhladze, V. L. 1988. New possibilities for the geochemical $^{81}Br^{-81}$ solar neutrino experiment. *Nucl. Instr. Meth.* A271:257–258.

Kwan, J., and Scoville, N. 1976. The nature of the broad molecular line emission at the Kleinmann-Low nebula. *Astrophys. J.* 210:L39-L43.

Labeyrie, L. D., Duplessy, J.-C., and Blanc, P. L. 1987. Deep water formation and temperature variation over the last 125,000 years. *Tellus* 327:477–482.

Labitzke, K. 1987. Sunspots, the QBO, and the stratospheric temperature in the north pole region. *Geophys. Res. Lett.* 14:535–537.

Labitzke, K., and van Loon, H. 1988. Associations between the 11-year cycle, the QBO and the atmosphere. Part I: The troposphere and stratosphere of the northern hemisphere in winter. *J. Atmos. Terrest. Phys.* 50:197–206.

LaBonte, B. J., Chapman, G. A., Hudson, H. S., and Willson, R. C. eds. 1983. *Solar Irradiance Variations on Active Region Time Scales,* NASA CP-2310.

Laclare, F. 1983. Mesures du diamètre solaire à l'astrolabe. *Astron. Astrophys.* 125:200–203.

Laclare, F. 1987. Sur les variations du diamètre du soleil observées à l'astrolabe solaire du CERGA. *Compt. Rend. II* 305:451–454.

Laclare, F., Demarcq, J., and Cholet, F. 1980. Sur la réalisation d'un astrolabe solaire au Centre d'Etudes et de Recherches Géodynamiques et Astronomiques. *Compt. Rend.* B291:189–192.

Lada, C. J. 1985. Cold outflows, energetic winds, and enigmatic jets around young stellar objects. *Ann. Rev. Astron. Astrophys.* 23:267–317.

Lada, C. J. 1987. On the importance of outflows for molecular clouds and star formation. In *Galactic and Extragalactic Star Formation,* eds. R. E. Pudritz and M. Fich (Dordrecht: D. Reidel).

Lagache, M. 1965. Contribution à l'étude de l'altération des feldspaths, dans l'eau, entre 100 et 200°C, sous diverses pressions de CO_2, et application à la synthèse des minéraux argileux. *Bull. Soc. France Minerol. Crist.* 88:223–253.

Lagache, M. 1976. New data on the kinetics of the dissolution of alkali feldspars at 200°C in CO_2 charged water. *Geochim. Cosmochim. Acta* 40:157–161.

Lago, M. T. V. T. 1979. Ph.D. Thesis, Univ. of Sussex.

Lago, M. T. V. T. 1984. A new investigation of the T Tauri star RU Lupi. II. The wind model. *Mon. Not. Roy. Astron. Soc.* 210:323–340.

La Hire, Ph. 1683–1718. Manuscripts D2.1–10. Archives Obs. Paris.

Laird, C. M., Zeller, E. J., and Armstrong, T. P. 1982. Solar activity and nitrate deposition in South Pole snow. *Geophys. Res. Lett.* 9:1195–1198.

Lake, J. A. 1988. Origin of the eukaryotic nucleus determined by rate-invariant analysis of rRNA sequences. *Nature* 331:184–186.

Lal, D. 1972. Hard rock cosmic ray archaeology. *Space Sci. Rev.* 14:3–102.

Lal, D. 1985. Carbon cycle variations during the past 50,000 years: Atmospheric $^{14}C/^{12}C$ ratios as an isotopic indicator. In *The Carbon Cycle and Atmospheric CO_2: Natural Variations Archean to Present,* eds. E. T. Sundquist and W. S. Broecker, AGU Geophysical Mono. 32 (Washington, D.C.: American Geophysical Union), pp. 221–233.

Lal, D. 1987. ^{10}Be in polar ice: Data reflect changes in cosmic ray flux on polar meteorology. *Geophys. Res. Lett.* 14:785–788.

Lal, D. 1988a. Temporal variations of low-energy cosmic-ray protons on decadal and million year time-scales: Implications to their origin. *Astrophys. Space Phys.* 144:337–346.

Lal, D. 1988b. Theoretically expected variations in the terrestrial cosmic-ray production rates of isotopes. In *Solar-Terrestrial Relationships and the Earth Environment in the Last Millennia,* ed. G. C. Castagnoli (Amsterdam: North-Holland Press), pp. 216–233.

Lal, D. 1990. The influence of climatic and oceanic processes on the atmospheric change ^{14}C record. In preparation.

Lal, D., and Marti, K. 1977. On the flux of low-energy particles in the solar system: The record in St. Séverin meteorite. *Nucl. Track. Det.* 1:127–130.

Lal, D., and Peters, B. 1962. Cosmic ray produced isotopes and their application to problems in geophysics. In *Progress in Elementary Particle and Cosmic Ray Physics,* eds. J. G. Wilson and S. A. Wouthuysen (Amsterdam: North-Holland Press), pp. 1–74.

Lal, D., and Peters, B. 1967. Cosmic ray produced radioactivity on the Earth. In *Handbuch der Physik*, ed. S. Flügge 46(2):551–612.

Lal, D., and Rajan, R. S. 1969. Observations on space irradiation of individual crystals of gas-rich meteorites. *Nature* 233:269–271.

Lal, D., Laffoon, M. A., and Erickson, D. J., III. 1990. Inferences about climatic changes based on δ^{14}, $\sigma^{13}C$, $CaCo_3$, pCO_2 and ^{10}Be paleodata. In *Interaction of the Global Carbon and Climate Systems*, eds. R. S. Keir and W. H. Berger. Proc. EPRI Conf., Lake Arrowhead, Calif., Oct. 24–28, 1988, in press.

Lalande, J. 1771. *L'Astronomie*, 2nd ed. (Paris: Editions Desaimp).

Lamb, H. H. 1972. *Climate, Present, Past and Future: Fundamentals and Climate Now* (London: Methuen).

Lamb, H. H. 1977. *Climate: Present, Past and Future: Climate History and the Future* (London: Methuen).

Landi degl'Innocenti, E. 1982. Expected broadband linear polarization from cool stars with magnetic structures. *Astron. Astrophys.* 110:25–29.

Landini, M., Monsignori-Rossi, B. C., Pallavicini, R., and Piro, L. 1986. EXOSAT detection of an X-ray flare from a solar type star π^1 $UM_\odot$, 157, 212–222.

Langevin, Y., and Maurette, M. 1980. A model for small body regolith evolution: The critical parameters. *Lunar Planet. Sci.* XI:602–604 (abstract).

Langevin, Y., Arnold, J. R., and Nishiizumi, K. 1982. Lunar surface gardening processes: Comparison of model calculations with radionuclide data. *J. Geophys. Res.* 87:6681–6691.

Lanoix, M., Strangway, D. W., and Pearce, G. W. 1978a. The primordial magnetic field preserved in chondrules of the Allende meteorite. *Geophys. Res. Lett.* 5:73–76.

Lanoix, M., Strangway, D. W., and Pearce, G. W. 1978b. Paleointensity determinations from Allende chondrules. *Lunar Planet. Sci.* IX:630–632 (abstract).

Lanzerotti, L. J., Reedy, R. C., and Arnold, J. R. 1973. Alpha particles in solar cosmic rays over the last 80,000 years. *Science* 179:1232–1234.

Larimer, J. W., and Anders, E. 1967. Chemical fractionations in meteorites. Abundance patterns and their interpretation. *Geochim. Cosmochim. Acta* 31:1239–1270.

Larmor, J. 1919. How could a rotating body such as the Sun become a magnet? *Engl. Mech.* 110:113–114.

Larson, H. P., and Veeder, G. J. 1979. Infrared spectral reflectances of asteroid surfaces. In *Asteroids*, ed. T. Gehrels (Tucson: Univ. of Arizona Press), pp. 724–744.

Larson, R. B. 1969. Numerical calculations of the dynamics of a collapsing protostar. *Mon. Not. Roy. Astron. Soc.* 168:271–295.

Lavielle, B. 1987. Etude de la production par rayonnement cosmique du krypton et du xénon dans la météorite de Saint-Séverin. Ph.D. Thesis, Univ. of Bordeaux.

Lavielle, B., and Regnier, S. 1984. Cross-section measurements of krypton 78 to 86 isotopes in the spallation of palladium and silver. A new semi-empirical spallation formula in the Y to Ag target mass region. *J. Phys.* 45:981–988.

Lawrence, J. K., Chapman, G. A., Herzog, A. D., and Shelton, J. C. 1985. Solar luminosity fluctuations during the disk transit of an active region. *Astrophys. J.* 292:297–308.

Lazarewicz, A. R., and Gaffey, M. J. 1980. Thermal, stress and mineralogic evolution of small planetary objects. *Lunar Sci. Conf.* XI:616–620 (abstract).

Lazear, G., Damon, P. E., and Sternberg, R. S. 1980. The concept of DC gain in modelling secular variations in atmospheric ^{14}C. In *Proc. 10th Intl. Radiocarbon Conf.*, eds. M. Stuiver and R. Kra. *Radiocarbon* 22(2):318–327.

Lean, J. L. 1987. Solar UV irradiance variation: A review. *J. Geophys. Res.* 92:839–868.

Lean, J. L., and Brueckner, G. E. 1989. Intermediate-term solar periodicities: 100 to 500 days. *Astrophys. J.* 337:568–578.

Lean, J. L., and Foukal, P. 1988. A model of solar luminosity modulation by magnetic activity between 1954 and 1984. *Science* 240:906–908.

Lean, J. L., Livingston, W. C., White, O. R., and Skumanich, A. 1983. Modelling solar spectral irradiance variations at ultraviolet wavelengths. In *Solar Irradiance Variations on Active Region Time Scales*, eds. B. J. LaBonte, G. A. Chapman, H. S. Hudson and R. C. Willson, NASA CP-2310, pp. 253–288.

Lebofsky, L. A. 1978. Asteroid 1 Ceres: Evidence for water of hydration. *Mon. Not. Roy. Astron. Soc.* 182:17P–21P.

Lebofsky, L. A., Jones, T. D., and Herbert, F. 1989. Asteroid volatile inventories. In *Origin and Evolution of Planetary and Satellite Atmospheres*, eds. S. K. Atreya, J. B. Pollack and M. S. Matthews (Tucson: Univ. of Arizona Press), pp. 192–229.

Lebreton, Y., and Maeder, A. 1987. Stellar evolution with turbulent diffusion mixing. VI. The solar model, surface ^{7}Li and ^{3}He abundances, solar neutrinos and oscillations. *Astron. Astrophys.* 175:99–112.

Lederer, C. M., Hollander, J. M., and Perlman, I. *Table of Isotopes*, 6th ed. (New York: Wiley).

Lee, M. A. 1988. The solar wind termination shock and the heliosphere beyond. In *Solar Wind: Proc. Sixth Intl. Conf.*, eds. V. J. Pizzo, T. E. Holzer and D. G. Sime, NCAR TN-306 (Boulder: Natl. Center for Atmospheric Research), pp. 635–650.

Lee, R. B., III, Barkstrom, B. R., and Cess, R. D. 1987. Characteristics of the Earth Radiation Budget Experiment solar monitors. *Appl. Opt.* 26:3090–3096.

Lee, T. 1978. A local proton irradiation model for isotopic anomalies in the solar system. *Astrophys. J.* 224:217–226.

Lee, T., Papanastassiou, D. A., and Wasserburg, G. J. 1976. Demonstration of ^{26}Mg excess in Allende and evidence for ^{26}Al. *Geophys. Res. Lett.* 3:109–112.

Lee, T., Papanastassiou, D. A., and Wasserburg, G. J. 1977. Aluminum-26 in the early solar system: Fossil or fuel? *Astrophys. J.* 211:L107–L110.

Lee, T., Russell, W. A., and Wasserburg, G. J. 1979. Calcium isotopic anomalies and the lack of aluminum-26 in an unusual Allende inclusion. *Astrophys. J.* 228:L93–L98.

Legrand, J. P., and Legoff, J. 1987. Louis Morin et les observations météorologiques sous Louis XVI. *Vie des Sci.* 4:251–281.

Leibacher, J. 1984. Helioseismological determination of solar internal rotation. In *Theoretical Problems in Stellar Stability and Oscillations* (Liège: Univ. of Liège), pp. 298–301.

Leighton, R. B. 1961. Preview on granulation, observations, studies. *Nuovo Cimento Suppl.* 22:321–325.

Leighton, R. B. 1964. Transport of magnetic fields on the Sun. *Astrophys. J.* 140:1547–1562.

Leighton, R. B. 1969. A magneto-kinematic model of the solar cycle. *Astrophys. J.* 156:1–26.

Leister, N. V. 1989*a*. Orientacao do Sistema fundamental de Referencia. Observacoes do Sol com o Astrolabio de Valinhos. Ph.D. Thesis, IAG-USP.

Leister, N. V. 1989*b*. Melidas do diametro solar. *Bol. Soc. Astron. Brasileira* 10:36.

Lemaire, J. 1989. Meteoritic ions in the corona and solar wind. *IAGA Bull.* 53:267.

Leos, J. A., Feigelson, E. D., André, P., and Montmerle, T. 1991. A rich cluster of radio stars in the ρ Ophiuchi cloud cores. *Bull. Amer. Astron. Soc.*, in press.

Lerman, J. C., Mook, W. G., and Vogel, J. C. 1970. C-14 in tree rings from different localities. In *Radiocarbon Variations and Absolute Chronology*, ed. I. U. Olsson (New York: Wiley), pp. 275–301.

Leroy-Ladurie, E. 1967. *Histoire du climat depuis l'an mil* (Paris: Editions Flammarion).

Letaw, J. R., Silberberg, R., and Tsao, C. H. 1987. Radiation hazards on space missions. *Nature* 330:709–710.

LeTreut, H., and Ghil, M. 1983. Orbital forcing, climatic interactions, and glaciation cycles. *J. Geophys. Res.* 88:5167–5190.

LeTreut, H., Portes, J., Jouzel, J., and Ghil, M. 1988. Isotopic modeling of climatic oscillations: Implications for a comparative study of marine and ice-core records. *J. Geophys. Res.* 93:9365–9383.

Levine, J. S., and Boughner, R. E. 1979. The effect of paleoatmospheric ozone on surface temperature. *Icarus* 39:310–314.

Levine, J. S., Hays, P. B., and Walker, J. C. G. 1979. The evolution and variability of atmospheric ozone over geological time. *Icarus* 39:295–309.

Levine, J. S., Augustsson, T. R., Boughner, R. E., Natarajan, M., and Sacks, L. J. 1981. Comets and the photochemistry of the paleoatmosphere. In *Comets and the Origin of Life*, ed. C. Ponamperuma (Dordrecht: Reidel), pp. 161–190.

Levine, J. S., Augustsson, T. R., and Natarajan, M. 1982. The prebiological paleoatmosphere: Stability and composition. *Origins of Life* 12:245–259.

Levreault, R. M. 1988. Molecular flows and mass-loss in pre-main-sequence stars. *Astrophys. J.* 330:897–910.

Levy, E. H. 1978. Magnetic field in the primitive nebula. *Nature* 276:481.

Levy, E. H. 1988. Energetics of chondrule formation. In *Meteorites and the Early Solar System*, eds. J. F. Kerridge and M. S. Matthews (Tucson: Univ. of Arizona Press), pp. 697–711.

Levy, E. H., and Araki, S. 1989. Magnetic reconnection flares in the protoplanetary nebula and the possible origin of meteorite chondrules. *Icarus* 81:74–91.

Levy, E. H., and Boyer, D. 1982. Oscillating dynamo in the presence of a fossil magnetic field. The solar cycle. *Astrophys. J.* 254:L19–L22.

Levy, E. H., and Rose, W. K. 1974. Production of magnetic fields in the interiors of stars and several effects on stellar evolution. *Astrophys. J.* 193:37–48.

Levy, E. H., and Sonett, C. P. 1978. Meteorite magnetism and early solar system magnetic fields. In *Protostars and Planets*, ed. T. Gehrels (Tucson: Univ. of Arizona Press), pp. 516–532.

Levy, R. H., Petshek, H. E., and Siscoe, G. L. 1963. Aerodynamic aspects of magnetospheric flow. *AAIA J.* 2:2065–2076.

Lewis, J. S. 1974. The temperature gradient in the solar nebula. *Science* 186:440–443.

Lewis, R. S., and Anders, E. 1975. Condensation time of the solar nebula from extinct ^{129}I in primitive meteorites. *Proc. Natl. Acad. Sci.* 172:268–273.

Libbrecht, K. G. 1988. Solar p-mode frequency splittings. *Seismology of the Sun and Sun-Like Stars*, ed. E. J. Rolfe, ESA SP-286, pp. 3–10.

Libby, W. 1952. *Radiocarbon Dating* (Chicago: Univ. of Chicago Press).

Light, E. S., Merker, M., Verschell, S. J., Mendell, R. B., and Korff, S. A. 1973. Time dependent worldwide distribution of atmospheric neutrons and of their products. 2. Calculation. *J. Geophys. Res.* 78:2741–2762.

Lightner, B. D., and Marti, K. 1974. Lunar trapped xenon. In *Proc. Lunar Sci. Conf.* 5:2023–2031.

Lin, D. N. C., and Papaloizou, J. 1980. On the structure and evolution of the primordial solar nebula. *Mon. Not. Roy. Astron. Soc.* 191:37–48.

Lin, D. N. C., and Pringle, J. E. 1987. A viscosity prescription for a self-gravitating accretion disc. *Mon. Not. Roy. Astron. Soc.* 225:607–613.

Lin, D. N. C., Papaloizou, J. C. B., and Savonije, G. J. 1989. Propagation of tidal disturbances in gaseous accretion disks. In preparation.

Lin, R. P. 1985. Energetic solar electrons in the interplanetary medium. *Solar Phys.* 100:537–562.

Lin, R. P. 1987. Solar particle acceleration and propagation. *Rev. Geophys.* 25:676–684.

Lin, R. P., and Hudson, H. S. 1976. Non-thermal processes in large solar flares. *Solar Phys.* 50:153–178.

Lin, R. P., Schwartz, R. A., Pelling, R. M., and Hurley, K. C. 1981. A new component of hard X-rays in solar flares. *Astrophys. J.* 251:L109–L114.

Lin, R. P., Anderson, K. A., and Hood, L. L. 1988. Lunar surface magnetic field concentrations antipodal to young large impact basins. *Icarus* 74:529–541.

Lind, D. L., Geiss, J., and Stettler, W. 1979. Solar and terrestrial noble gases in magnetospheric precipitation. *J. Geophys. Res.* 84:6435–6442.

Lindzen, R. S. 1986. A simple model for 100K-year oscillations in glaciation. *J. Atmos. Sci.* 43:986–996.

Lingenfelter, R. E. 1963. Production of carbon-14 by cosmic-ray neutrons. *Rev. Geophys.* 1:35–55.

Lingenfelter, R. E., and Hudson, H. S. 1980. Solar particle fluxes and the ancient Sun. In *The Ancient Sun: Fossil Record in the Earth, Moon and Meteorites*, eds. R. O. Pepin, J. A. Eddy and R. B. Merrill (New York: Pergamon Press), pp. 69–79.

Lingenfelter, R. E., and Ramaty, R. 1970. Astrophysical and geophysical variations in ^{14}C production. In *Radiocarbon Variations and Absolute Chronology*, ed. I. U. Olsson (New York: Wiley), pp. 513–537.

Linick, T. W., Suess, H. E., and Becker, B. 1985. La Jolla measurements of radiocarbon in South German Oak tree-ring chronologies. *Radiocarbon* 27:20–32.

Linick, T. W., Long, A., Damon, P. E., and Ferguson, C. W. 1986. High precision radiocarbon dating of bristlecone pine from 6554 to 5350 BC. In *Proc. 12th Intl. Conf. on Radiocarbon*, eds. M. Stuiver and R. Kra. *Radiocarbon* 28(2B):943–953.

Link, F. 1977. Sur l'activité solaire au 17ème siècle. *Astron. Astrophys.* 54:857–861.

Linsky, J. L. 1980. Stellar chromospheres. *Ann. Rev. Astron. Astrophys.* 18:439–88.

Linsky, J. L. 1985. Nonradiative activity across the H-R diagram: Which stars are solar-like? *Solar Phys.* 100:333–362.

Linsky, J. L. 1989. Solar and stellar magnetic fields and structures: Observations. *Solar Phys.* 121:187–196.

Linsky, J. L., and Ayres, T. R. 1978. Stellar model chromospheres. VI. Empirical estimates of the chromospheric radiative losses of late-type stars. *Astrophys. J.* 220:619–628.

Linsky, J. L., and Saar, S. 1987. Measurements of stellar magnetic fields: Empirical constraints on stellar dynamo and rotational evolution theories. In *Cool Stars, Stellar Systems and The Sun: Proc. Fifth Cambridge Workshop*, eds. J. L. Linsky and R. E. Stencel (New York: Springer-Verlag), pp. 44–46.

Linsky, J. L., Worden, S. P., McClintock, W., and Robertson, R. M. 1979. Stellar model chromospheres. X. High-resolution, absolute flux profiles of the Ca II H and K lines in stars of spectral types F0–M2. *Astrophys. J. Suppl.* 41:47–74.

Linsley, B. K. 1989. Carbonate sedimentation in the Sulu Sea linked to the onset of northern hemisphere glaciation. *EOS: Trans. AGU* 70:1135.

Lipschutz, M. E., Gaffey, M. J., and Pellas, P. 1989. Meteoritic parent bodies: Nature, number, size, and relation to present-day asteroids. In *Asteroids II*, eds. R. P. Binzel, T. Gehrels and M. S. Matthews (Tucson: Univ. of Arizona Press), pp. 740–777.

Livingston, W., Wallace, L., and White, O. R. 1988. Spectrum line intensity as a surrogate for solar irradiance variations. *Science* 240:1765–1767.

Lizano, S., Heiles, C., Rodríguez, L. F., Koo, B.-C., Shu, F. H., Hasegawa, T., Hayashi, S., and Mirabel, I. F. 1988. Neutral stellar winds that drive bipolar outflows in low mass protostars. *Astrophys. J.* 328:763–776.

Lockwood, G. W., and Skiff, B. A. 1988. *Luminosity Variations of Stars Similar to the Sun.* Air Force Geophys. Lab. Tech. Rept. 88–0221.

Lockwood, G. W., Thompson, D. T., Radick, R. R., Osborn, W. H., Baggett, W. E., Duncan, D. K., and Hartmann, L. W. 1984. The photometric variability of solar-type stars. IV. Detection of rotational modulation among Hyades stars. *Publ. Astron. Soc. Pacific* 96:714–722.

Lomb, N. B., and Anderson, A. P. 1980. The analysis and forecasting of the Wolf sunspot numbers. *Mon. Not. Roy. Astron. Soc.* 190:723–732.

Lorenz, E. N. 1963. Deterministic nonperiodic flow. *J. Atmos. Sci.* 20:130–141.

Lorin, J. C., and Pellas, P. 1979. Pre-irradiation history of Djermaia (H) chondritic breccia. *Icarus* 40:502–509.

Lorius, C., Jouzel, J., Ritz, C., Merlivat, L., Barkov, N. I., Korotkevitch, Y. S., and Kotlyakov, V. M. 1985. A 150,000-year climatic record from Antarctic ice. *Nature* 316:591–596.

Loutre, M. F., and Berger, A. 1989. *Pre-Quaternary Amplitudes in the Expansion of the Obliquity and Climatic Precession*, Scientific Report 1989/4 (Louvain-la-Neuve: Inst. Astron. Geophys. Georges Lemaître).

Lovelock, J. E. 1979. *Gaia: A New Look at Life on Earth* (Oxford: Oxford Univ. Press).

Lovelock, J. E. 1988. *The Ages of Gaia* (New York: Norton).

Lovelock, J. E., and Whitfield, M. 1982. Life span of the biosphere. *Nature* 296:561–563.

Lugmair, G. W., Scheinin, N. B., and Marti, K. 1975. Sm-Nd age and history of Apollo 17 basalt 75075: Evidence for early differentiation of the lunar exterior. *Proc. Lunar Sci. Conf.* 6:1419–1429.

Lugmair, G. W., Scheinin, N. B., and Carlson, R. W. 1977. Sm-Nd systematics of the Serra de Mage eucrite. *Meteoritics* 12:300–301 (abstract).

Luhmann, J. G. 1986. The solar wind interaction with Venus. *Space Sci. Rev.* 44:241–306.

Luhn, A., Klecker, B., Hovestadt, D., and Möbius, E. 1987. The mean ionic charge of silicon in ^{3}He-rich events. *Astrophys. J.* 317:951–955.

Lund, S. P., and Banerjee, S. K. 1985. Late quaternary paleomagnetic field secular variation from two Minnesota lakes. *J. Geophys. Res.* 90:803–825.

Lüst, R., and Schlüter, A. 1955. Angular momentum transport by magnetic fields and the braking of rotating stars. *Z. Astrophys.* 38:190.

Lynden-Bell, D., and Pringle, J. E. 1974. The evolution of viscous disks and the origin of nebular variables. *Mon. Not. Roy. Astron. Soc.* 168:603–637.

Macdougall, J. D., and Kothari, B. K. 1976. Formation chronology for C2 meteorites. *Earth Planet. Sci. Lett.* 33:36–44.

Macdougall, J. D., and Lugmair, G. W. 1989. Chronology of chemical change in the Orgueil CI chondrite based on Sr isotopic systematics. *Meteoritics* 24:297 (abstract).

Macdougall, J. D., and Phinney, D. 1977. Olivine separates from Murchison and Cold-Bokkeveld: Particle tracks and noble gases. *Proc. Lunar Sci. Conf.* 8:293–311.

Macdougall, J. D., Lugmair, G. W., and Kerridge, J. F. 1984. Early solar system aqueous activity: Sr isotope evidence from the Orgueil CI meteorite. *Nature* 307:249–251.

MacElroy, J. M. D., and Manuel, O. K. 1986. Can intrasolar diffusion contribute to isotope anomalies in the solar wind? *Proc. Lunar Planet. Sci. Conf.* 16, *J. Geophys. Res. Suppl.* 91:D473–D482.

MacPherson, G. J., Fahey, A. J., Lundberg, L. L., and Zinner, E. 1988. Al-Mg isotopic systematics and metamorphism in five coarse-grained Allende CAIs *Meteoritics* 23:286 (abstract).

Magee, N. H., Merts, A. L., and Huebner, W. F. 1984. Is the metal contribution to astrophysical plasmas correct? *Astrophys. J.* 283:264–272.

Maher, K. A., and Stevenson, D. J. 1988. Impact frustration of the origin of life. *Nature* 331:612–614.

Mahoney, W. A., Ling, J. D., Wheaton, W. A., and Jacobson, A. S. 1984. HEAO3 discovery of ^{26}Al in the interstellar medium. *Astrophys. J.* 286:578–585.

Makarenko, N. G., and Ajmanova, G. K. 1989. K-entropy and dimension of solar attractor. *Pisma Astron. J. (USSR)*, submitted.

Makarov, V. I., Ruzmaikin, A. A., and Starchenko, S. V. 1987. Magnetic waves of solar activity. *Solar Phys.* 111:267–277.

Malinetsky, G. G., Ruzmaikin, A. A., and Samarsky, A. A. 1986. A model for prolong variations of the solar activity. Keldysh Inst. of Applied Math., Moscow, Preprint N 170.

Manabe, S., and Wetherald, R. T. 1980. On the distribution of climate change resulting from an increase in CO_2-content of the atmosphere. *J. Atmos. Sci.* 37:99–118.

Manhès, G. 1982. Développement de l'ensemble chronométrique U-Th-Pb. Contribution à la chronologie initiale du système solaire. Thesis, Univ. of Paris VII.

Manhès, G., Allègre, C. J., and Provost, A. 1984. U-Th-Pb systematics of the eucrite "Juvinas:" Precise age determination and evidence for exotic lead. *Geochim. Cosmochim. Acta* 48:2247–2264.

Manhès, G., Göpel, C., and Allègre, C. J. 1987. High resolution chronology of the early solar system based on lead isotopes. *Meteoritics* 22:453–454 (abstract).

Manka, R. H., and Michel, F. C. 1971. Lunar atmosphere as a source of lunar surface elements. *Proc. Lunar Sci. Conf.* 2:1717–1728.

Mann, A. K. 1988. Present status of the solar neutrino puzzle. Univ. of Penn. UPR-0160E. In *Neutrino '88: Proc. 13th Intl. Conf. on Neutrino Physics and Astrophysics*, eds. J. Schneps, T. Kafka, W. A. Mann and P. Nath (Singapore: World Scientific), pp. 105–122.

Mann, C. J. 172. Faunal extinctions and reversal of the Earth's magnetic field: Discussion. *Geol. Soc. Amer. Bull.* 83:2211–2214.

Marcy, G. W. 1982. A technique for measuring magnetic fields on solar-type stars. *Publ. Astron. Soc. Pacific* 94:989–996.

Marcy, G. W. 1984. Observations of magnetic fields on solar-type stars. *Astrophys. J.* 276:286–304.

Marcy, G. W., and Basri, G. S. 1989. Physical realism in the analysis of stellar magnetic fields. II. K dwarfs. *Astrophys. J.* 345:480–488.

Marcy, G. W., and Bruning, D. H. 1984. Magnetic field observations of evolved stars. *Astrophys. J.* 281:286–291.

Margulis, L., and Lovelock, J. E. 1974. Biological modulation of the Earth's atmosphere. *Icarus* 21:471–489.

Margulis, M., and Snell, R. L. 1989. Very high velocity emission from molecular outflows. *Astrophys. J.* 343:779–784.

Mariska, J. T. 1986. The quiet solar transition region. *Ann. Rev. Astron. Astrophys.* 24:23–48.

Marshall, H. G., Walker, J. C. G., and Kuhn, W. R. 1988. Long-term climate change and the geochemical cycle of carbon. *J. Geophys. Res.* 93:791–801.

Marti, K. 1967. Mass-spectrometric detection of cosmic-ray-produced ^{81}Kr in meteorites and the possibility of Kr-Kr dating. *Phys. Rev. Lett.* 18:264–266.

Marti, K. 1969. Solar-type xenon: A new isotopic composition of xenon in the Pesyanoe meteorite. *Science* 166:1263–1265.

Marti, K. 1982. Krypton-81-krypton dating by mass spectrometry. In *Nuclear and Chemical Dating Techniques: Interpreting the Environmental Record,* ed. L. A. Currie, ACS Symp. Series No. 176 (Washington, D.C.: American Chemical Soc.), pp. 129–137.

Marti, K., and Lugmair, G. W. 1971. Kr^{81}-Kr and K-Ar^{40} ages, cosmic-ray spallation products, and neutron effects in lunar samples from Oceanus Procellarum. In *Lunar Sci. Conf.* 2:1591–1605.

Marti, K., Lightner, B. D. and Lugmair, G. W. 1973. On ^{244} Pu in lunar rocks from Fra Mauro and implications regarding their origin. *Moon* 8:241–250.

Martinson, D. G., Pisias, N. G., Hays, J. D., Imbrie, J., Moore, T. C., and Shackleton, N. J. 1987. Age dating and the orbital theory of the ice ages: Development of a high-resolution 0 to 300,000-year chronostratigraphy. *Quat. Res.* 27(1):1–29.

Mason, B. J. 1976. Towards the understanding and prediction of climatic variations. *Quart. J. Roy. Meteorol. Soc.* 102:473–499.

Mathieu, R. D. 1989. Spectroscopic binaries among low-mass pre-main sequence stars. In *Highlights in Astronomy,* ed. D. McNally (Dordrecht: Kluwer), pp. 111–115.

Mathieu, R. D., Walter, F. M., and Myers, P. C. 1989. The discovery of six pre-main sequence spectroscopic binaries. *Astron. J.* 98:987–1001.

Mathys, G., and Solanki, S. K. 1989. Magnetic fields in late-type dwarfs: Preliminary results of a multiline approach. *Astron. Astrophy.* 208:189–197.

Maunder, E. W. 1890. *Mon. Not. Roy. Astron. Soc.* 1:251.

Maunder, E. W. 1894. A prolonged sunspot minimum. *Knowledge* 17:173–176.

Maunder, E. W. 1922. The prolonged sunspot minimum of 1645–1715. *J. Brit. Astron. Assoc.* 32:140–145.

Mayer-Kress, G. 1986. *Dimensions and Entropies in Chaotic Systems* (Berlin: Springer-Verlag).

Mayfield, E. B., and Lawrence, J. K. 1985. The correlation of solar flare production with magnetic energy in active regions. *Solar Phys.* 96:293–305.

McCord, T. B., and Gaffey, M. J. 1974. Asteroids: Surface composition from reflectance spectroscopy. *Science* 185:352–355.

McCormac, B. M., and Seliga, T. A., eds. 1979. *Solar-Terrestrial Influences on Weather and Climate* (Dordrecht: D. Reidel).

McDonald, F. B., ed. 1963. *Solar Proton Manual,* NASA TR-R-169.

McDonald, F. B., and Lal, N. 1986. Variations of galactic cosmic rays with helio-latitude in the outer heliosphere. *Geophys. Res. Lett.* 13:781–784.

McDonald, F. B., Fichtel, C. E., and Fisk, L. A. 1974. Solar particles (observations, relationships to the Sun, acceleration, interplanetary medium). In *High Energy Particles and Quanta in Astrophysics,* eds. F. B. McDonald and C. E. Fichtel (Cambridge: Massachusetts Inst. of Technology), pp. 212–272.

McDonald, F. B., Lal, N., Trainor, J. B., Van Hollebeke, M. A. I., and Webber, W. R. 1981. The solar modulation of galactic cosmic rays in the outer heliosphere. *Astrophys. J.* 249:L71-L75.

McElhinny, M. W., and Senanayake, W. E. 1982. Variations in the geomagnetic dipole. 1. The past 50,000 years. *J. Geomag. Geoelect.* 34:39–51.

McElroy, M. B., and Donahue, T. M. 1972. Stability of the Martian atmosphere. *Science* 177:986–988.

McGuire, R. E., and von Rosenvinge, T. T. 1984. The energy spectra of solar energetic particles. *Adv. Space Res.* 4:117–125.

McGuire, R. E., von Rosenvinge, T. T., and McDonald, F. B. 1986. The composition of solar energetic particles. *Astrophys. J.* 301:938–961.

McKay, D. S., Bogard, D. D., Morris, R. V., Korotev, R. L., Johnson, P., and Wentworth, S. J. 1986. Apollo 16 regolith breccias: Characterization and evidence for early formation in the mega-regolith. *Proc. Lunar Planet. Sci. Conf.* 16, *J. Geophys. Res. Suppl.* 91:D277–D303.

McKibben, R. B. 1988. Cosmic ray modulation. In *Solar Wind: Proc. Sixth Intl. Conf.,* eds. V. J. Pizzo, T. E. Holzer and D. G. Sime, NCAR TN-306 (Boulder: Natl. Center for Atmospheric Res.), pp. 615–633.

McKibben, R. B., Pyle, K. R., and Simpson, J. A. 1982. The galactic cosmic ray radial intensity gradient and large scale modulation in the heliosphere. *Astrophys. J.* 254:L23–L27.

Mecherikunnel, A. T., Lee, R. B., Kyle, H. L., and Major, E. R. 1988. Intercomparison of solar total irradiance from recent spacecraft measurements. *J. Geophys. Res.* 93:9503–9509.

Melosh, H. J. 1989. *Impact Cratering: A Geologic Process* (New York: Oxford Univ. Press).

Mercer, M. J. 1983. Cenozoic glaciation in the southern hemisphere. *Ann. Rev. Earth Planet. Sci.* 11:99–132.

Mercer-Smith, J. A., Cameron, A. G. W., and Epstein, R. I. 1984. On the formation of stars from disk accretion. *Astrophys. J.* 279:363–366.

Merker, M. 1970. Solar Cycle Modulation of Fast Neutrons in the Atmosphere. Ph.D. Thesis, New York Univ.

Mermilliod, J. C. 1981. Comparative studies of young open clusters. III. Empirical isochronous curves and the zero age main sequence. *Astron. Astrophys.* 97:235–244.

Merrill, R. T., and McElhinny, M. W. 1983. *The Earth's Magnetic Field, Its History, Origin, and Planetary Perspective* (London: Academic Press).

Mestel, L. 1966. The magnetic field of contracting gas cloud. I. Strict flux-freezing. *Mon. Not. Roy. Astron. Soc.* 133:265–284.

Mestel, L. 1968. Magnetic braking by a stellar wind. *Mon. Not. Roy. Astron. Soc.* 138:359–391.

Mestel, L., and Paris, R. B. 1979. Magnetic braking during star formation. III. *Mon. Not. Roy. Astron. Soc.* 187:337–356.

Mestel, L., and Spitzer, L., Jr. 1956. Star formation in magnetic dust clouds. *Mon. Not. Roy. Astron. Soc.* 116:505–514.

Mestel, L., and Weiss, N. O. 1987. Magnetic fields and non-uniform rotation in stellar radiative zones. *Mon. Not. Roy. Astron. Soc.* 226:123–135.

Mewaldt, R. A. 1980. Spacecraft measurements of the elemental and isotopic composition of solar energetic particles. In *The Ancient Sun: Fossil Record in the Earth, Moon and Meteorites,* eds. R. O. Pepin, J. A. Eddy and R. B. Merrill (New York: Pergamon Press), pp. 81–101.

Mewaldt, R. A., Spalding, J. D., and Stone, E. C. 1984. A high-resolution study of the isotopes of solar flare nuclei. *Astrophys. J.* 280:892–901.

Meyer, J. P. 1981. A tentative ordering of all available solar energetic particles abundance observations. II. Discussion and comparison with coronal abundances. In *Proc. 17th Intl. Cosmic Ray Conf.,* vol. 3, pp. 149–152.

Meyer, J. P. 1985. The baseline composition of solar energetic particles. *Astrophys. J. Suppl.* 57:151–171.

Micela, G., Sciortino, S., Serio, S., Vaiana, G. S., Bookbinder, J., Golub, L., Harnden, F. R., Jr., and Rosner, R. 1985. EINSTEIN X-ray survey of the Pleiades: The dependence of X-ray emissions on stellar age. *Astrophys. J.* 292:172–180.

Michaelsen, J. 1989. Long-period fluctuations in El Niño amplitude and frequency reconstructed from tree rings. In *Aspects of Climate Variability in the Pacific and Western Americas,* ed. D. H. Peterson. AGU Mono. 55 (Washington, D.C.: American Geophysical Union), pp. 69–74.

Michel, R., and Brinkmann, G. 1980. On the depth-dependent production of radionuclides ($44 \leq A \leq 59$) by solar protons in extraterrestrial matter. *J. Radioanal. Chem.* 59:467–510.

Michel, R., Brinkmann, G., Weigel, H., and Herr, W. 1979. Measurement and hybrid-model analysis of proton-induced reactions with V, Fe, and Co. *Nucl. Phys.* A322:40–60.

Michel, R., Brinkmann, G., and Stück, R. 1982. Solar cosmic-ray-produced radionuclides in meteorites. *Earth Planet. Sci. Lett.* 59:33–48. Erratum 64:174.

Michel, R., Peifer,F., Thesis, S., Begemann, F., Weber, H., Signer, P., Weiler, R., Cloth, P., Dragovitsch, P., Filges, D., and Englert, P. 1989. Production of stable and radioactive nuclides in thick stony targets (R = 15 and 25 cm) isotropically irradiated with 600 MeV protons and simulation of the production of cosmogenic nuclides in meteorites. *Nucl. Instr. Meth. Phys. Res.* 42:76–100.

Middlekoop, F. 1982. Magnetic structures in cool stars. IV. Rotation and Ca II emission of main-sequence stars. *Astron. Astrophys.* 107:31–35.

Mihalas, D. 1988. Atomic transfer and radiative transfer in stars. In *Atomic Processes in Plasmas,* ed. A. Hauer (New York: American Inst. of Physics), pp. 42–50.

Mikheyev, S. P., and Smirnov, A. Yu. 1985. Resonance enhancement of oscillations in matter and solar neutrino spectroscopy. *Soviet Nucl. Phys.* 42:913–917.

Mikheyev, S. P., and Smirnov, A. Yu. 1986. Resonance amplification of ν oscillations in matter and solar neutrinos spectroscopy. *Nuovo Cimento* 9:17–26.
Mikheyev, S. P., and Smirnov, A. Yu. 1987. Resonant oscillations of solar neutrinos: Present status. In *Underground Physics 87* (Moscow: Nauka Press), pp. 34–39.
Milankovitch, M. M. 1920. Théorie mathématique des phénomènes thermiques produits par la radiation solaire. *Académie Yugoslave des Sciences et des Arts de Zagreb* (Paris: Gauthier-Villars).
Milankovitch, M. M. 1941. *Kanon der Erdbestrahlung*. Royal Serbian Academy Spec. Publ. 132, Sec. of Math. and Natural Sci. vol. 33. English trans.: *Canon of Insolation and the Ice Age Problem*. 1969. (Jerusalem: Israel Program for Scientific Trans.).
Miller, J. A., and Ramaty, R. 1987. Ion and relativistic electron acceleration by Alfvén and whistler turbulence in solar flares. *Solar Phys.* 113:195–201.
Miller, J. A., and Ramaty, R. 1989. Relativistic electron transport and Bremsstrahlung production in solar flares. *Astrophys. J.* 344:973–990.
Minster, J. F., and Allègre, C. J. 1979. ^{87}Rb-^{87}Sr chronology of H chondrites: Constraint and speculations on the early evolution of their parent body. *Earth Planet. Sci. Lett.* 43:333–347.
Minster, J. F., and Allègre, C. J. 1981. ^{87}Rb-^{87}Sr dating of LL chondrites. *Earth Planet. Sci. Lett.* 56:89–106.
Miono, S., and Ohta, M. 1969. The possible variation of ^{14}C predicted from thermoluminescence dating of the Japanese ancient pottery. In *Proc. 16th Intl. Cosmic Ray Conf.*, vol. 2, pp. 263–268.
Mitalas, R. 1972. Reduction of the solar neutrino flux by primordial ^{3}He in the Sun. *Astrophys. Lett.* 12:35–36.
Mitchell, D. G., Roelof, E. C., and Bame, S. J. 1983. Solar wind iron abundance variations at speeds > 600 km s^{-1}, 1972–1976. *J. Geophys. Res.* 88:9059–9068.
Mitchell, J. F. B. 1989. The "greenhouse" effect and climate change. *Rev. Geophys.* 27:115–140.
Mitchell, J. F. B., Gahame, N. S., and Needham, K. J. 1988. Climate simulations for 9,000 years before present: Seasonal variations and the effect of the Laurentide ice sheet. *J. Geophys. Res.* 93:8283–8304.
Mitchell, J. M., Jr. 1972. The natural breakdown of the present interglacial and its possible intervention by human activities. *Quat. Research* 2:436–445.
Mitchell, J. M., Jr. 1976. An overview of climatic variability and its causal mechanisms. *Quat. Res.* 6:481–493.
Mitchell, J. M., Jr., Stockton, C. W., and Meko, O. M. 1979. Evidence of 22-year rhythm of drought in the western United States related to the Hale solar cycle since the 17th century. In *Solar-Terrestrial Influences on Weather and Climate*, eds. B. M. McCormac and T. A. Seliga (Dordrecht: D. Reidel).
Mittlefehldt, D. W. 1979. The nature of asteroidal differentiation processes: Implications for primordial heat sources. *Proc. Lunar Planet. Sci. Conf.* 10:1975–1993.
Moffatt, H. K. 1978. *Magnetic Field Generation in Electrically Conducting Fluids* (Cambridge: Cambridge Univ. Press).
Mold, P., Bull, R. K., and Durrani, S. A. 1984. Fission-track annealing characteristics of meteoritic phosphates. *Nucl. Tracks* 9:119–128.
Monaghan, M. C. 1987. Greenland ice ^{10}Be concentrations and average precipitation rates north of 40°N to 45°N. *Earth Planet. Sci. Lett.* 84:197–203.
Montmerle, T., and André, P. 1988. X-rays, radio emission, and magnetism in low-mass young stars. In *Formation and Evolution of Low-Mass Stars*, eds. A. Dupree and T. Lago (Dordrecht: D. Reidel), pp. 225–246.
Montmerle, T., Koch-Miramond, L., Falgaronne, E., and Grindlay, J. E. 1983a. Einstein observations of the Rho Ophiuchi dark cloud: An X-ray Christmas tree. *Astrophys. J.* 269:182–201.
Montmerle, T., Koch-Miramond, L., Falgarone, E., and Grindlay, J. E. 1983b. X-ray variability of pre-main sequence objects associated with the Rho Oph dark cloud. *Phys. Scripta* T7:59–61.
Moore, T. C., Pisias, N. G., and Dunn, D. A. 1982. Carbonate time series of the Quaternary

and late Miocene sediments in the Pacific ocean: A spectral comparison. *Marine Geol.* 46:217–233.

Moraal, H., Jokipii, J. R., and Mewaldt, R. 1991. Heliospheric effects on cosmic-ray electrons. *Astrophys. J.* 367:191–199.

Moreno-Insertis, F. 1983. Rise times of horizontal magnetic flux tubes in the convection zone of the Sun. *Astron. Astrophys.* 122:241–250.

Moreno-Insertis, F. 1986. Nonlinear time evolution of kink-unstable magnetic flux tubes in the convective zone of the Sun. *Astron. Astrophys.* 166:291–305.

Morfill, G. 1985. Physics and chemistry in the primitive solar nebula. In *Birth and Infancy of Stars,* eds. R. Lucas, A. Omont and R. Stora (New York: North-Holland Press), pp. 693–794.

Morfill, G. E., Tscharnuter, W., and Völk, H. J. 1985. Dynamical and chemical evolution of the protoplanetary nebula. In *Protostars and Planets II* (Tucson: Univ. of Arizona Press), pp. 493–533.

Morfill, G. E., Atmanspacher, H., Demmel, V., Scheingraber, H., and Voges, W. 1989. Deterministic chaos in accreting neutron star systems. In *Timing Neutron Stars,* eds. H. Ögelman and E. P. J. van den Heuvel (Dordrecht: Kluwer), pp. 71–93.

Morin, L. 1665–1713. Manuscripts. Bibliothèque de l'Institut de France, M 1488.

Morinaga, H., ed. 1986. *Workshop of the Feasibility of the Solar Neutrino Detection with* ^{205}Pb *by Geochemical and Accelerator Mass Spectroscopical Measurements* (Darmstadt: GSI-86–9).

Morkovin, M. V. 1972. Effects of compressibility on turbulent flows. In *Mécanique de la Turbulence, No. 108: Colloques Internationaux du Centre National de la Recherche Scientifique,* Paris.

Morley, J. J., and Hays, J. D. 1981. Towards a high-resolution, global, deep-sea chronology for the last 750,000 years. *Earth Planet. Sci. Lett.* 53:279–295.

Morley, J. J., and Shackleton, N. J. 1984. The effect of accumulation rate on the spectrum of geologic time series: Evidence from two South Atlantic sediment cores. In *Milankovitch and Climate,* eds. A. Berger, J. Imbrie, J. Hays, G. Kukla and B. Saltzman (Dordrecht: D. Reidel), pp. 467–480.

Morrison, D. A., and Zinner, E. 1977. 12054 and 76215: New measurements of interplanetary dust and solar flare fluxes. *Proc. Lunar Sci. Conf.* 8:841–863.

Morrison, L. V., and Ward, C. G. 1975. An analysis of the transits of Mercury: 1677–1973. *Mon. Not. Roy. Astron. Soc.* 173:183–206.

Morrow, C. A. 1988. A New Picture for the Internal Rotation of the Sun. Ph.D. Thesis, Univ. of Colorado.

Moses, D., Dröge, W., Meyer, P., and Evenson, P. 1989. Characteristics of energetic solar flare electron spectra. *Astrophys. J.* 346:523–530.

Moss, D. 1987. The internal magnetic field of the Sun. *Mon. Not. Roy. Astron. Soc.* 224:1019–1029.

Mount, G. H., and Rottman, G. J. 1981. The solar spectral irradiance between 150 and 200 nm. *J. Geophys. Res.* 86:9193–9198.

Mount, G. H., Rottman, G. J., and Timothy, J. G. 1980. The solar spectral irradiance between 1200 and 2550 Å at solar maximum. *J. Geophys. Res.* 85:4271–4274.

Mouschovias, T. Ch. 1976. Nonhomologous contraction and equilibria of self-gravitating, magnetic interstellar clouds embedded in an intercloud medium: Star formation. II. Results. *Astrophys. J.* 207:141–158.

Mouschovias, T. Ch. 1977. A connection between the rate of rotation of interstellar clouds, magnetic fields, ambipolar diffusion, and the periods of binary stars. *Astrophys. J.* 211:147–151.

Mouschovias, T. Ch. 1978. Formation of stars and planetary systems in magnetic interstellar clouds. In *Protostars and Planets,* ed. T. Gehrels (Tucson: Univ. of Arizona Press), pp. 209–242.

Mouschovias, T. Ch., and Paleologou, E. V. 1980. Magnetic braking of an aligned rotator during star formation. *Astrophys. J.* 173:877–902.

Mullan, D. J. 1983. Isotopic anomalies among solar energetic particles: Contribution of preacceleration in collapsing magnetic neutral sheets. *Astrophys. J.* 268:385–395.

Mullan, D. J. 1985. Radio outbursts in RS Canum Venaticorum stars: Coronal heating and electron runaway. *Astrophys. J.* 295:628–633.

Mundt, R. 1984. Mass-loss in T Tauri stars: Observational studies of the cool parts of the stellar winds and expanding shells. *Astrophys. J.* 280:749–770.

Mundt, R. 1988. Flows and jets from young stars. In *Formation and Evolution of Low Mass Stars*, eds. A. K. Dupree and M. T. V. T. Lago (Dordrecht: Kluwer), pp. 257–279.

Mundt, R., Walter, F. M., Feigelson, E. D., Finkenzeller, U., Herbig, G. H., and Odell, A. P. 1983. Observations of suspected low-mass post-T Tauri stars and their evolutionary status. *Astrophys. J.* 269:229–238.

Mundt, R., Brugel, E. W., and Bührke, T. 1987. Jets from young stars: CCD imaging, long-slit spectroscopy, and interpretation of existing data. *Astrophys. J.* 319:275–303.

Murphy, J. J. 1876. The glacial climate and the polar ice-cap. *Quart. J. Roy. Geol. Soc.* 32:400–406.

Murphy, R. J., and Ramaty, R. 1984. Solar flare neutrons and gamma rays. *Adv. Space Res.* 4:127–136.

Murphy, R. J., Ramaty, R., Forrest, D. J., and Kozlovsky, B. 1985. Abundances from gamma-ray line spectroscopy. In *Proc. 19th Intl. Conf. on Cosmic Rays*, vol. 4, pp. 249–252.

Murphy, R. J., Dermer, C. D., and Ramaty, R. 1987. High energy processes in solar flares. *Astrophys. J. Suppl.* 63:721–748.

Murphy, R. J., Share, G. H., Forrest, D. J., and Vestrand, W. T. 1990. Nuclear line spectroscopy using SMM/GRS gamma-ray data from solar flares. In *Proc. 21st Intl. Cosmic Ray Conf.*, vol. 5, ed. R. J. Protheroe (Adelaide: Univ. of Adelaide), pp. 48–51.

Murray, B. C., Belton, M. J. S., Danielson, G. E., Davies, M. E., Gault, D., Hapke, B., O'Leary, B., Strom, R. G., Suomi, V., and Trask, N. 1974. Mariner 10 pictures of Mercury: First results. *Science* 184:459–461.

Murty, S. V. S. 1990. Cosmogenic nitrogen in Kapoeta: An attempt to isolate solar cosmic ray produced component. *Lunar Planet Sci.* XXI:829–830 (abstract).

Myers, P. S. 1985. Molecular cloud cores. In *Protostars and Planets*, eds. D. C. Black and M. S. Matthews (Tucson: Univ. of Arizona Press), pp. 87–103.

Myers, P. S., and Benson, P. J. 1983. Dense cores in dark clouds. II. NH_3 observations and star formation. *Astrophys. J.* 266:309–320.

Myers, P. S., Fuller, G. A., Mathieu, R. D., Beichman, C. A., Benson, P. J., Schild, R. E., and Emerson, J. P. 1987. Near-infrared and optical observations of IRAS sources in and near dense cores. *Astrophys. J.* 319:340–357.

Myers, P. S., Heyer, M., Snell, R. L., and Goldsmith, P. F. 1988. Dense cores in dark clouds. V. CO outflow. *Astrophys. J.* 324:907–919.

Nagata, T. 1979. Meteorite magnetism and the early solar system magnetic field. *Phys. Earth Planet. Int.* 20:324–341.

Nagata, T. 1980a. Magnetic classification of Antaractic meteorites. *Proc. Lunar Planet. Sci. Conf.* 11:1789–1799.

Nagata, T. 1980b. Paleomagnetism of antarctic achondrites. *Mem. Natl. Inst. Polar Res. Special Issue* 17:233–242.

Nagata, T. 1981. Paleointensity of antarctic achondrites. *Sixth Symp. Antarctic Meteorites*, pp. 78–79 (abstract).

Nagata, T. 1988. Magnetic analysis of Antarctic chondrites on the basis of a magnetic binary model system. *Proc. NIPR Symp. on Antarctic Meteorites* 1:247–260.

Nagata, T., and Funaki, M. 1981. The comparison of natural remanent magnetization of an antarctic chondrite, ALHA 76009 (L6). *Proc. Lunar Planet. Sci. Conf.* 12:B1229–B1241.

Nagata, T., and Funaki, M. 1982. Magnetic properties of tetrataenite-rich stony meteorites. *Mem. Natl. Inst. Polar Res. Special Issue* 25:222–250.

Nagata, T., and Funaki, M. 1983. Paleointensity of the Allende carbonaceous chondrite. *Mem. Natl. Inst. Polar Res. Special Issue* 30:403–434.

Nagata, T., and Sugiura, N. 1977. Paleomagnetic field intensity derived from meteorite magnetism. *Phys. Earth Planet. Int.* 13:373–379.

Nakahata, M. 1990. The Kamiokande solar neutrino experiment. In *Inside the Sun: Proc. IAU Coll. 121*, eds. G. Berthomieu and M. Cribier (Dordrecht: Kluwer), pp. 179–186.

Nakano, T., and Tademaru, E. 1972. Decoupling of magnetic fields in dense clouds with angular momentum. *Astrophys. J.* 173:87–101.

Nakano, T., and Umebayashi, T. 1980. Behavior of grains in the drift of plasma and magnetic fields in dense interstellar clouds. *Bull. Astron. Soc. Japan* 32:613–621.

Nakano, T., and Umebayashi, T. 1986. Dissipation of magnetic fields in very dense interstellar clouds. I. Formulation and conditions for efficient dissipation. *Mon. Not. Roy. Astron. Soc.* 218:663–684.

Naujokat, B. 1986. An update of the observed quasi-biennial oscillation of the stratospheric winds over the tropics. *J. Atmos.Sci.* 43:1873–1877.

Nautiyal, C. M., Padia, J. T., Rao, M. N., and Venkatesan, T. R. 1986. Solar flare neon composition and solar cosmic-ray exposure ages based on lunar mineral separates. *Astrophys. J.* 301:465–470.

Neel, T. 1955. Some theoretical aspects of rock magnetism. *Adv. in Phys.* 4:191–243.

Neftel, A., Oeschger, H., and Suess, H. E. 1981. Secular non-random variations of cosmogenic carbon-14 in the terrestrial atmosphere. *Earth Planet. Sci. Lett.* 56:127–147.

Neftel, A., Oeschger, H., Schwander, J., Stauffer, B., and Zumbrunn, R. 1982. Ice core sample measurements give atmospheric CO_2 content during the past 40,000 yr. *Nature* 295:220–223.

Neftel, A., Moor, E., Oeschger, H., and Stauffer, B. 1985. Evidence from polar ice cores for the increase in atmospheric CO_2 in the past two centuries. *Nature* 315:45–47.

Negrini, R. M., Verosub, K. L., and Davis, J. O. 1988. The middle to late Pleistocene geomagnetic field recorded in fine-grained sediments from Summer Lake, Oregon and Double Hot Springs, Nevada, USA. *Earth Planet. Sci. Lett.* 87:173–192.

Ness, N. F., Behannon, K. W., Lepping, R. P., Whang, Y. C., and Schatten, H. H. 1974. Magnetic field observations near Mercury: Preliminary results from Mariner 10. *Science* 185:151–160.

Ness, N. F., Acuña, M. H., Behannon, K. W., and Neubauer, F. M. 1979. Magnetic field studies at Jupiter by Voyager 1: Preliminary results. *Science* 204:982–987.

Ness, N. F., Acuña, M. H., Behannon, K. W., and Neubauer. 1982. The induced magnetosphere of Titan. *J. Geophys. Res.* 87:1369–1381.

Neubauer, F. M. 1980. Nonlinear standing Alfvén wave current system at Io: Theory. *J. Geophys. Res.* 85:1171–1178.

Neubauer, F. M., Gurnett, D. A., Scudder, J. D., and Hartle, R. E. 1984. Titan's magnetospheric interaction. In *Saturn*, eds. T. Gehrels and M. S. Matthews (Tucson: Univ. of Arizona Press), pp. 760–787.

Neubauer, F. M., Glassmeier, K. H., Pohl, M., Raeder, J., Acuña, M. H., Burlaga, L. F., Ness, N. F., Musmann, G., Mariani, F., Wallis, M. K., Ungstrup, E., and Schmidt, H. U. 1986. First results from the Giotto magnetometer experiment at comet Halley. *Nature* 321:352–355.

Neugebauer, M. 1981. Observations of solar-wind helium. *Fundam. Cosmic Phys.* 7:131–199.

Newkirk, G., Jr. 1980. Solar variability on time scales of 105 years to $10^{9.6}$ years. In *The Ancient Sun: Fossil Record in the Earth, Moon and Meteorites*, eds. R. O. Pepin, J. A. Eddy and R. B. Merrill (New York: Pergamon Press), pp. 290–320.

Newkirk, G., Jr., Hundhausen, A. J., and Pizzo, V. 1981. Solar cycle modulation of galactic cosmic rays: Speculation on the role of coronal transient. *J. Geophys. Res.* 86:5387–5396.

Newkirk, L. L. 1963. Calculations of low-energy neutron flux in the atmosphere by the Sn method. *J. Geophys. Res.* 68:1825–1833.

Newman, M. J., and Rood, R. T. 1977. Implications of solar evolution for the Earth's early atmosphere. *Science* 198:1035–1037.

Nicolis, C. 1980. Response of the Earth-atmosphere system to a fluctuating solar input. In *Sun and Climate* (Toulouse: CNES), pp. 385–396.

Nicolis, C. 1982. Stochastic aspects of climatic transitions-response to a periodic forcing. *Tellus* 34:1–9.

Nicolis, C. 1984. Self-oscillations, external forcings, and climate predictability. In *Milankovitch and Climate: Understanding Orbital Forcing*, eds. A. Berger, J. Imbrie, J. Hays, G. Kukla and B. Saltzman (Dordrecht: D. Reidel), pp. 637–652.

Nicolis, C., and Nicolis, G. 1984. Is there a climatic attractor? *Nature* 311:529–532.

Nicolis, C., and Nicolis, G. 1986. Reconstruction of the dynamics of the climate system from time series data. *Proc. Natl. Acad. Sci. Geophys.* 83:536–540.

Niemeyer, S. 1979. ^{40}Ar-^{39}Ar dating of inclusions from IAB iron meteorites. *Geochim. Cosmochim. Acta* 43:1829–1840.

Nier, A. O., and Schlutter, D. J. 1990. Helium and neon isotopes in individual stratospheric particles—A further study. *Lunar Planet. Sci.* XXI:883–884 (abstract).

Nishiizumi, K., Gensho, R., and Honda, M. 1981. Half-life of ^{59}Ni. *Radiochim. Acta* 29:113–116.

Nishiizumi, K., Klein, J., Middleton, R., Kubik, P. W., Elmore, D., and Arnold, J. R. 1986. Exposure history of shergottites. *Geochim. Cosmochim. Acta* 50:1017–1021.

Nishiizumi, K., Imamura, M., Kohl, C. P., Nagai, H., Kobayashi, K., Yoshida, K., Yamashita, H., Reedy, R. C., Honda, M., and Arnold, J. R. 1987. ^{10}Be profiles in lunar surface rock 68815. In *Proc. Lunar Planet. Sci. Conf.* 18:79–85.

Nishiizumi, K., Kubik, P. W., Elmore, D., Reedy, R. C., and Arnold, J. R. 1988. Cosmogenic ^{36}Cl productoin rates in meteorites and the lunar surface. In *Proc. Lunar Planet. Sci. Conf.* 19:305–312.

NOAA Natl. Geographic Data Center. 1988. *Zürich Sunspot Number Series* (Boulder: NOAA).

Nolte, E., and Pavicevic, M. K., eds. 1988. Solar neutrino detection. *Nucl. Instr. Meth.* A271:237–341.

Norris, J. P., and Matilsky, T. A. 1989. Is Hercules X-1 a strange attractor? *Astrophys. J.* 346:912–918.

Norris, T. L., Gancarz, A. J., Rokop, D. J., and Thomas, K. W. 1983. Half-life of ^{26}Al. *Proc. Lunar Planet. Sci. Conf. 14, J. Geophys. Res. Suppl.* 88:B331-B333.

North, G. R. 1975. Analytical solution to a simple climate model with diffusive heat transport. *J. Atmos. Sci.* 32:1301–1307.

North, G. R., Mengel, J. G., and Short, D. A. 1983. Simple energy balance model resolving the seasons and the continents. Application to the astronomical theory of the Ice Ages. *J. Geophys. Res.* 88:6576–6586.

Noyes, R. W., Weiss, N. O., and Vaughan, A. H. 1984a. The relation between stellar rotation rate and activity cycle periods. *Astrophys. J.* 287:769–773.

Noyes, R. W., Hartmann, L. W., Baliunas, S. L., Duncan, D. K., and Vaughan, A. H. 1984b. Rotation, convection, and magnetic activity in lower main-sequence stars. *Astrophys. J.* 279:763–777.

Noyes, R. W., Baliunas, S. L., and Guinan, E. F. 1991. What can other stars tell us about the Sun? In *The Solar Interior and Atmosphere* (Tucson: Univ. of Arizona Press), in press.

Nyquist, L. E., Takeda, H., Bansal, B. M., Shih, C. Y., Wiesmann, H., and Wooden, J. L. 1986. Rb-Sr and Sm-Nd internal isochron ages of a subophitic basalt clast and a matrix sample from the Y75011 eucrite. *J. Geophys. Res.* 91:8137–8150.

Nyquist, L. E., Bogard, D., Takeda, H., Bansal, B., Johnson, P., Shih, C. Y., and Wiesmann, H. 1988. Comparative chronologies of basaltic clasts in Antarctic eucrites Y-75011 and Y-792510. *Meteoritics* 23:295 (abstract).

O'Brien, K. 1971. Cosmic-ray propagation in the atmosphere. *Nuovo Cimento* A3:521–547.

O'Brien, K. 1972. Propagation of muons underground and the primary cosmic-ray spectrum below 40 TeV. *Phys. Rev.* D5:597–605.

O'Brien, K. 1975a. *The Cosmic Ray Field at Ground Level. Natural Radiation Environment II.* USERDA Rept. CONF-72085, part 1 (Springfield, Va.: U.S. Dept. of Commerce), pp. 15–54.

O'Brien, K. 1975b. Calculated cosmic-ray pion and proton fluxes at sea level. *J. Phys.* A8:1530–1534.

O'Brien, K. 1979. Secular variations in the production of cosmogenic isotopes in the Earth's atmosphere. *J. Geophys. Res.* 84:423–431.

O'Brien, K. 1983. The Production of Krypton-81 and Krypton-85 by Cosmic Rays. U.S. Dept. of Energy Rept. EML-420.

O'Brien, K., and Burke, G. de P. 1973. Calculated cosmic ray neutron monitor response to solar modulation of galactic cosmic rays. *J. Geophys. Res.* 78:3013–3019.

O'Dell, C. R., and Van Helden, H. 1988a. On the accuracy of the seventeenth century micrometer measurements. *Nature* 330:629–631.

O'Dell, C. R., and Van Helden, H. 1988b. Size of the Sun in the seventeenth century. *Nature* 332:690.

Oerlemans, J. 1980. Model experiments on the 100,000-yr glacial cycle. *Nature* 287:430–432.

Oerlemans, J. 1982. Glacial cycles and ice sheet modelling. *Cliamtic Change* 4:353–374.

Oeschger, H. 1988. The global carbon cycle, a part of the Earth system. In *Solar-Terrestrial Relationships and the Earth Environment in the Last Millennia*, ed. G. C. Castagnoli (Amsterdam: North-Holland Press), pp. 329–352.

Oeschger, H., and Eddy, J. A. 1988. *Global changes of the past. The International Geosphere-Biosphere Programme: A study of global change of the International Council of Scientific Unions*. IGBP Global Change Rept. 6 (Stockholm, Sweden).

Oeschger, H., Houtermans, J., Loosli, H., and Wahlen, M. 1970. The constancy of cosmic radiations from isotope studies in meteorites and on the Earth. In *Radiocarbon Variations and Absolute Chronology*, ed. I. U. Olsson (New York: Wiley Interscience), pp. 471–500.

Oeschger, H., Siegenthaler, U., Schotterer, U., and Gugelmann, A. 1975. A box diffusion model to study the carbon dioxide exchange in nature. *Tellus* 27:168–192.

Oeschger, H., Beer, J., Siegenthaler, U., Stauffer, B., Dansgaard, W., and Langway, C. C. 1984. Late glacial climate history from ice cores. In *Climate Processes and Climate Sensitivity*, eds. S. E. Hansen and T. Takahashi, AGU Geophysical Mono. 29 (Washington, D.C.: American Geophysical Union), pp. 299–306.

Oeschger, H., Stauffer, B., Finkel, R., and Langway, C. C., Jr. 1985. Variations of the CO_2 concentration of occluded air and of anions and dust in polar ice cores. In *The Carbon Cycle and Atmospheric CO_2: Natural Variations Archean to Present*, eds. E. T. Sundquist and W. S. Broecker, AGU Geophysical Mono. 32 (Washington, D.C.: American Geophysical Union), pp. 132–142.

Ogilvie, K. W., and Wilkerson, T. D. 1969. Helium abundance in the solar wind. *Solar Phys.* 8:435–449.

Ogilvie, K. W., Coplan, M. A., Bochsler, P., and Geiss, J. 1980. Abundance ratios $^4He^{++}/^3He^{++}$ in the solar wind. *J. Geophys. Res.* 85:6021–6024.

Ogilvie, K. W., Coplan, M. A., Bochsler, P., and Geiss, J. 1989. Solar wind observations with the Ion Composition Instrument aboard the ISEE-3/ICE spacecraft. *Solar Phys.* 124:167–183.

Ohtsuki, K., and Nakagawa, Y. 1988. Dissipation of the solar nebula. *Prog. Theor. Phys. Suppl.* 96:161–166.

Olinger, C. T., Garrison, D. H., and Hohenberg, C. M. 1983. Laser extraction of cosmogenic neon from microgram-size grains. *Lunar Planet. Sci.* XIX:889 (abstract).

Olinger, C. T., Hohenberg, C. M., Nichols, R. H., Jr., and Goswami, J. N. 1989. Precompaction irradiation of individual grains from the Murchison meteorite: An active early Sun? The Sun in Time, 6–10 March, Tucson, Arizona, Abstract Booklet, p. 36.

Olsen, E. J., and Bunch, T. E. 1984. Equilibrium temperatures of the ordinary chondrites: A new evaluation. *Geochim. Cosmochim. Acta* 48:1363–1365.

Olsen, P. E. 1986. A 40-million-year lake record of early Mesozoic orbital climatic forcing. *Science* 234:842–848.

Olson, S. 1980. Solar tracks in the snow. *Science News* 118:313–314.

O'Neal, D. B., Feigelson, E. D., Myers, P. C., and Mathieu, R. D. 1990. A VLA survey of weak T Tauri stars in Tauris-Auriga. *Bull. Amer. Astron. Soc.*, in press.

Oranje, B. J. 1983. The Ca II emission from the Sun as a star. II. The plage emission profile. *Astron. Astrophys.* 124:43–49.

Oster, L., Sofia, S., and Schatten, K. 1982. Solar irradiance variations due to active regions. *Astrophys. J.* 256:768–773.

O'Sullivan, P. E. 1983. Annually-laminated lake sediments and the study of Quaternary environmental changes—a review. *Quat. Sci. Rev.* 1:245–313.

Otaola, J. A., and Zenteno, G. 1983. On the existence of long-term periodicities in solar activity. *Solar Phys.* 89:209–213.

Owen, T., Cess, R. D., and Ramanathan, V. 1979. Early Earth: An enhanced carbon dioxide greenhouse to compensate for reduced solar luminosity. *Nature* 277:640–642.

Pace, N. R., Olsen, G. J., and Woese, C. R. 1986. Ribosomal RNA phylogeny and the primary lines of evolutionary descent. *Cell* 45:325–326.

Padia, J. T., and Rao, M. N. 1989. Neon isotope studies of Fayetteville and Kapoeta meteorites and clues to ancient solar activity. *Geochim. Cosmochim. Acta* 53:1461–1467.

Pagel, B. E. J., Terlevich, R. J., and Melnick, J. 1986. XII. New measurements of helium in H_{II} galaxies. *Publ. Astron. Soc. Pacific* 98:1005–1008.

Pal, Y. 1967. Cosmic rays and their interactions. In *Handbook of Physics*, 2nd ed., eds. E. U. Condon and H. Odishaw (New York: McGraw-Hill), pp. 9/272–328.

Pallavicini, R., Serio, S., and Vaiana, G. S. 1977. A survey of soft X-ray limb flare images: The relation between their structure in the corona and other physical parameters. *Astrophys. J.* 216:108–122.

Pallavicini, R., Golub, L., Rosner, R., Vaiana, G. S., Ayres, T., and Linsky, J. L. 1981. Relations among stellar X-ray emission observed from EINSTEIN, stellar rotation, and bolometric luminosity. *Astrophys. J.* 248:279–290.

Pallé, P. L., Perez, J. C., Regulo, C., Roca, Cortès, T., Isaak, G. R., Mcleod, C. P., and Van der Raay, H. B. 1986. The global oscillation spectrum of the Sun. *Astron. Astrophys.* 170:114–119.

Pallé, P. L., Régulo, C., and Roca Cortès, T. 1988. In *Seismology of the Sun and Sun-Like Stars*, ed. E. J. Rolfe, ESA SP-286, pp. 285–289.

Pallé, P. L., Régulo, C., and Roca Cortès, T. 1988. Rotational splitting of low l solar p-mode. In *Seismology of the Sun and Sun-Like Stars*, ed. E. J. Rolfe, ESA SP-286, pp. 125–130.

Panagia, N., and Felli, M. 1975. The spectrum of the free-free radiation from extended envelopes. *Astron. Astrophys.* 39:1–5.

Papanastassiou, D. A., Rajan, R. S., Hunecke, J. C., and Wasserburg, G. J. 1974. Rb-Sr ages and lunar analogs in a basaltic achondrite: Implications for early solar system chronology. *Lunar Sci.* V:583–585 (abstract).

Papanastassiou, D. A., Wasserburg, G. J., and Marvin, U. B. 1984. Absence of excess ^{26}Mg in anorthite from the Vaca Muetra mesosiderite. *Meteoritics* 19:289–290 (abstract).

Parker, B. C., and Zeller, E. J. 1980. Nitrogenous chemical composition of Antarctic ice and snow. *Antarctic J. U.S.* 15:79–81.

Parker, E. N. 1955. Hydrodynamic dynamo models. *Astrophys. J.* 122:293–314.

Parker, E. N. 1955a. The formation of sunspots from the solar toroidal magnetic field. *Astrophys. J.* 121:491–507.

Parker, E. N. 1955b. Hydromagnetic dynamo models. *Astrophys. J.* 122:293–314.

Parker, E. N. 1975. The generation of magnetic fields in astrophysical bodies. X. Magnetic bouyancy and the solar dynamo. *Astrophys. J.* 198:205–210.

Parker, E. N. 1963a. *Interplanetary Dynamical Processes* (New York: Interscience).

Parker, E. N. 1963b. The solar flare phenomenon and the theory of reconnection and annihilation of magnetic fields. *Astrophys. J. Suppl.* 8:177–224.

Parker, E. N. 1965. The passage of energetic charged particles through interplanetary space. *Planet. Space Sci.* 13:9–49.

Parker, E. N. 1974. The dynamical properties of twisted ropes of magnetic field and the vigor of new active regions on the Sun. *Astrophys. J.* 191:245–254.

Parker, E. N. 1975. The generation of magnetic fields in astrophysical bodies. X. Magnetic buoyancy and the solar dynamo. *Astrophys. J.* 198:205–209.

Parker, E. N. 1979a. *Cosmical Magnetic Fields* (Oxford: Clarendon Press).

Parker, E. N. 1979b. Sunspots and the physics of magnetic flux tubes. I. The general nature of the sunspot. *Astrophys. J.* 230:905–913.

Parker, E. N. 1981. Residual fields from extinct dynamos. *Geophys. Astrophys. Fluid Dyn.* 18:175–195.

Parker, E. N. 1983a. Magnetic neutral sheets in evolving fields. I. General theory. *Astrophys. J.* 264:635–641.

Parker, E. N. 1983b. Magnetic neutral sheets in evolving fields. II. Formation of the solar corona. *Astrophys. J.* 264:642–647.

Parker, E. N. 1986. The magnetic structure of solar and stellar atmospheres. In *Cool Stars, Stellar Systems and the Sun: Proc. Fourth Cambridge Workshop*, eds. M. Zelik and D. M. Gibson (New York: Springer-Verlag), pp. 341–362.

Parker, E. N. 1985a. Thermal shadows and forced convection. *Astrophys. J.* 286:666–676.

Parker, E. N. 1985b. Magnetic fields in the radiative interior of stars. II. Forced convection and the ^{7}Li abundance. *Astrophys. J.* 286:667–680.

Parkhomenko, E. I. 1967. *Electrical Properties of Rocks* (New York: Plenum Press).

Parkinson, J. H. 1983. New measurements of the solar diameter. *Nature* 304:518–520.

Parkinson, J. H., Morrison, L. V., and Stephenson, F. V. 1980. The constancy of the solar diameter over the past 250 years. *Nature* 288:548–551.

Passow, C. 1962. Phenomenologische Theorie zur Berechnung einer Kaskade aus schweren Teilchen (Nukleonenkaskade) in der Materid. Notiz A 285, Deut. Elektron. Synchroton, Hamburg.

Paterne, M., Guichard, F., Labeyrie, J., Gillot, P. Y., and Duplessy, J. C. 1986. Thyrrenian Sea tephrochronology of the oxygen isotope record for the past 60,000 years. *Marine Geol.* 72:259–285.

Paterson, W. S. B. 1972. Laurentide ice sheet: Estimated volumes during Late Wisconsin. *Rev. Geophys. Space Phys.* 10:885–917.

Patterson, C., and Spaute, D. 1988. Self-termination of runaway growth in the asteroid belt. Asteroids II, March 8–11, Tucson, Arizona, Abstract Booklet.

Pawelzik, K., and Schuster, H. G. 1987. Generalized dimensions and entropies from measured time series. *Phys. Rev.* A35:481–484.

Pearson, G. W., and Baillie, M. G. L. 1983. High-precision ^{14}C measurement of Irish oaks to show the natural atmospheric ^{14}C variations of the AD time period. In *Proc. 11th Intl. Radiocarbon Conf.*, eds. M. Stuiver and R. Kra. *Radiocarbon* 25(2): 187–196.

Pearson, G. W., and Stuiver, M. 1986. High precision calibration of the radiocarbon time scale, AD500–2500 BC. In *Proc. 12th Intl. Radiocarbon Conf.*, eds. M. Stuiver and R. Kra. *Radiocarbon* 28(2B):839–862.

Pearson, G. W., Pilcher, J. R., Baillie, M. G. L., Corbett, D. M., and Qua, F. 1986. High-precision ^{14}C measurements of Irish oaks to show the natural ^{14}C variations from AD 1840–5120 BC. In *Proc. 12th Intl. Radiocarbon Conf.*, eds. M. Stuiver and R. Kra. *Radiocarbon* 28(2B):911–934.

Pedroni, A. 1989. Die korpuskulare Bestrahlung der Oberflächen von Asteroiden; eine Studie der Edelgase in den Meteoriten Kapoeta und Fayetteville. Ph.D. Thesis, ETH Zürich.

Pedroni, A., Baur, H., Wieler, R., and Signer, P. 1988. T-Tauri irradiation of Kapoeta grains? *Lunar Planet. Sci.* XIX:913–914 (abstract).

Pellas, P., and Fieni, C. 1988. Thermal histories of ordinary chondrite parent asteroids. *Lunar Planet. Sci.* XIX:915–916 (abstract).

Pellas, P., and Storzer, D. 1981. ^{244}Pu fission track thermometry and its application to stony meteorites. *Proc. Roy. Soc. London* A374:253–270.

Pellas, P., Poupeau, G., Lorin, J. C., Reeves, H. and Audouze, J. 1969. Primitive low-energy particle irradiation of meteoritic crystals. *Nature* 223:272–274.

Peltier, R., and Hyde, W. 1984. A model of the ice age cycle. In *Milankovitch and Climate: Understanding Orbital Forcing*, eds. A. Berger, J. Imbrie, J. Hayes, G. Kukla and B. Saltzman (Dordrecht: D. Reidel), pp. 565–580.

Penck, A., and Brückner, A. 1909. *Die Alpen im Eiszeeitalter* (Leipzig: Tauchnitz).

Pepin, R. O. 1980. Rare gases in the past and present solar wind. In *The Ancient Sun: Fossil Record in the Earth, Moon and Meteorites*, eds. R. O. Pepin, J. A. Eddy and R. B. Merrill (New York: Pergamon Press), pp. 411–421.

Pepin, R. O. 1989. On the relationship between early solar activity and the evolution of the terrestrial planet atmospheres. In *The Formation and Evolution of Planetary Systems*, eds. H. Weaver, F. Paresce and L. Danly (New York: Cambridge Univ. Press), pp. 55–70.

Pepin, R. O. 1991. On the origin and evolution of terrestrial planet atmospheres and meteoritic volatiles. *Icarus*, in press.

Pepin, R. O., Eddy, J. A., and Merrill, R. B., eds. 1980. *The Ancient Sun: Fossil Record in the Earth, Moon and Meteorites* (New York: Pergamon Press).

Pestiaux, P. 1984. Approche spectrale en modélisation paléoclimatique. Ph.D. Thesis, Univ. Catholique de Louvain.

Pestiaux, P., and Berger, A. 1984. An optimal approach to the spectral characteristics of deep-sea climatic records. In *Milankovitch and Climate: Understanding Orbital Forcing*, eds. A. Berger, J. Imbrie, J. Hays, G. Kukla and B. Saltzman (Dordrecht: D. Reidel), pp. 417–446.

Pestiaux, P., Duplessy, J. C., and Berger, A. 1987. Paleoclimate variability at frequencies from 10^{-4} cycle per year to 10^{-3} cycle per year—evidence for nonlinear behavior of the climate system. In *Climate: History, Periodicity, and Predictability*, eds. M. R. Rampino, J. E. Sanders, W. S. Newman and L. K. Königsson (New York: Van Nostrand Reinhold), pp. 285–299.

Pestiaux, P., Duplessy, J. C., van der Mersch, I., and Berger, A. 1988. Paleoclimatic variability at frequencies ranging from 1 cycle per 10,000 years to 1 cycle per 1,000 years: Evidence for nonlinear behavior of the climate system. *Climatic Change* 12(1):9–37.

Peterson, L. E., and Winckler, J. R. 1959. Gamma-ray burst from a solar flare. *J. Geophys. Res.* 64:697–707.

Peterson, R. C. 1983. The rotation of horizontal-branch stars. II. Members of the globular clusters M3, M5, and M13. *Astrophys. J.* 275:737–751.

Phillips, R. B., Lonsdale, C. J., and Feigelson, E. D. 1991. Milliarcsecond radio structure of weak-lined T Tauri stars. *Astrophys. J.,* in press.

Picard, J. 1662–1682. Journaux des observations. *Arch. Obs. Paris* D1–14, D1–15, D1–16.

Pickles, A. T., and Sucksmith, W. 1940. A magnetic study of the two-phase iron-nickel alloys. *Proc. Roy. Soc. London* A175:331–344.

Pike, E. R., McWhirter, J. G., Bertero, M., and de Mol, C. 1984. Generalized information theory for inverse problems in signal processing. *Proc. IEEE* 131:660–667.

Pinsonneault, M., Kawaler, S., Sofia, S., and Demarque, P. 1989. Evolutionary models of the rotating Sun. *Astrophys, J.* 338:424–452.

Pisias, N. G. 1978. Paleoceanography of the Santa Barbara Basin during the last 8000 years. *Quat. Res.* 10:366–384.

Pisias, N. G., and Moore, T. C., Jr. 1981. The evolution of the Pleistocene climate: A time series approach. *Earth Planet, Sci. Lett.* 52:450–458.

Pisias, N. G., and Shackleton, N. J. 1984. Modelling the global climate response to orbital forcing and atmospheric carbon dioxide changes. *Nature* 310:757–759.

Pisias, N. G., Martinson, D. G., Moore, T. C., Shackleton, N. J., Prell, W., Hays, J., and Boden, G. 1984. High resolution stratigraphic correlations of benthic oxygen isotopic records spanning the last 300,000 years. *Marine Geol.* 56:119–136.

Pittock, A. B. 1978. A critical look at long-term Sun-weather relationships. *Rev. Geophys. Space Phys.* 16:400–420.

Pittock, A. B. 1979. Solar cycles and the weather: Successful experiments in autosuggestion? In *Solar-Terrestrial Influences on Weather and Climate*, eds. B. M. McCormac and T. A. Seliga (Dordrecht: D. Reidel), pp. 181–191.

Pittock, A. B. 1983. Solar variability, weather, and climate: An update. *Quart. J. Roy. Meterol. Soc.* 109:23–55.

Pneuman, G. W., and Kopp, R. A. 1971. Gas-magnetic field interactions in the solar corona. *Solar Phys.* 18:258–270.

Podosek, F. A. 1970. Dating of meteorites by the high-temperature release of iodine-correlated Xe^{129}. *Geochim. Cosmochim. Acta* 34:341–365.

Podosek, F. A., and Swindle, T. D. 1988. Extinct radionuclides. In *Meteorites and the Early Solar System*, eds. J. F. Kerridge and M. S. Matthews (Tucson: Univ. of Arizona Press), pp. 1093–1113.

Pokras, E. M., and Mix, A. C. 1987. Earth's precession cycle and Quaternary climatic change in equatorial Africa: Tropical Africa. *Nature* 326:486–487.

Poletto, G., Vaiana, G. S., Zombeck, M. V., Krieger, A. S., and Timothy, A. F. 1975. A comparison of coronal X-ray structures of active regions with magnetic fields computed from photospheric observations. *Solar Phys.* 44:83–99.

Pollack, J. B. 1979. Climatic change on the terrestrial planets. *Icarus* 37:479–553.

Pollard, D. 1982. A simple ice sheet model yields realistic 100 kyr glacial cycles. *Nature* 296:334–338.

Pollard, D. 1984. Some ice-age aspects of a calving ice-sheet model. In *Milankovitch and Climate: Understanding Orbital Forcing*, eds. A. Berger, J. Imbrie, J. Hays, G. Kukla and B. Saltzman (Dordrecht: D. Reidel), pp. 541–564.

Pomerantz, M. A., and Duggal, S. P. 1974. The Sun and cosmic rays. *Rev. Geophys. Space Phys.* 12:343–361.

Pontecorvo, B. 1946. *Inverse β Decay Processes*, PD-205 (Chalk River, Ont.: Atomic Energy of Canada).

Pontecorvo, B. 1958. Inverse β-processes and nonconservation of lepton charge. *Z. Eksp. Teor. Fiz.* 34:247–249.

Pontecorvo, B., and Bilenkij, S. 1987. Neutrino today. In *Annuario Enciclopedia Scienca e Technica*, Mondadore, Italy, in press; Joint Inst. for Nucl. Res. Dubna Rept. JINR-E1, 2–87–567 (Russian).

Poor, C. L. 1908. An investigation of the figure of the Sun and of possible variations in its size and shape. *Ann. New York Acad. Sci.* 18:385–424.

Porter, H. S., Jackman, C. H., and Green, A. E. S. 1976. Efficiency for production of atomic nitrogen and oxygen by relativistic proton impact in air. *J. Chem. Phys.* 65:154–167.

Potgieter, M. S., and Moraal, H. 1985. A drift model for the modulation of galactic cosmic rays. *Astrophys. J.* 294:425–435.

Poupeau, G., and Berdot, J. L. 1972. Irradiations anciennes et récentes des aubrites. *Earth Planet. Sci. Lett.* 14:381–396.

Poupeau, G., Chetrit, G. C., Berdot, J. L., and Pellas, P. 1973. Etude par la méthode des traces nucléaires du sol de la Mer da la Fécondite (Luna-16). *Geochim. Cosmochim. Acta* 37:2005–2016.

Poupeau, G., Walker, R. M., Zinner, E. and Morrison, D. A. 1975. Surface exposure history of individual crystals in the lunar regolith. *Proc. Lunar Sci. Conf.* 6:3433–3448.

Pouquet, A., Lesieur, M., Andre, J. C., and Basedevant, C. 1976. Evolution of high Reynolds number two-dimensional turbulence. *J. Fluid Mech.* 72:305–319.

Prell, W. L. 1982. Oxygen and carbon isotope stratigraphy for the Quarternary of Hole 502B: Evidence for two modes of isotopic variability. In *Initial Reports of the Deep Sea Drilling Project.* vol. LXVIII (Washington, D.C.: National Science Foundation), pp. 455–464.

Prell, W. L., and Kutzbach, J. E. 1987. Monsoon variability over the past 150,000 years. *J. Geophys. Res.* 92:8411–8425.

Press, W. H., and Spergel, D. N. 1985. Capture by the Sun of a galactic population of weakly interacting massive particles. *Astrophys. J.* 296:679–684.

Price, P. B., Hutcheon, I. D., Cowsik, R., and Barber, D. J. 1971. Enhanced emission of iron nuclei in solar flares. *Phys. Rev. Lett.* 26:916–919.

Price, P. B., Lal, D., Tamhance, A. S., and Perelygin, V. P. 1973. Characteristics of tracks of ions with $14 \leq Z \leq 36$ in common rock silicates. *Earth Planet. Sci. Lett.* 19:377–395.

Price, P. B., Hutcheon, I. D., Braddy, D., and Macdougall, J. D. 1975. Track studies bearing on solar-system regolith. *Proc. Lunar Sci. Conf.* 6:3449–3469.

Priest, E. R. 1984. *Solar Magneto-Hydrodynamics* (Dordrecht: D. Reidel).

Pringle, J. E. 1989. A boundary-layer origin for bipolar flows. *Mon. Not. Roy. Astron. Soc.* 236:107–115.

Prinn, R. G., and Fegley, B., Jr. 1989. Solar nebula chemistry: Origin of planetary, satellite, and cometary volatiles. In *Origin and Evolution of Planetary and Satellite Atmospheres,* eds. S. K. Atreya, J. B. Pollack and M. S. Matthews (Tucson: Univ. of Arizona Press), pp. 78–136.

Proctor, M. R. E., and Weiss, N. O. 1982. Magnetoconvection. *Rept. Prog. Phys.* 45:1317–1379.

Prodi, F., and Fea, G. 1979. A case of transport and deposition of Saharan dust over the Italian peninsula and southern Europe. *J. Geophys. Res.* 84:6951–6960.

Pudovkin, M. I., and Benevolenskaya, E. E. 1982. The quasisteady primordial magnetic field of the Sun, and the intensity variations of the solar cycle. *Soviet Astron. Lett.* 8:273–275.

Pudritz, R. E. 1985. Star formation in rotating, magnetized molecular disks. *Astrophys. J.* 293:216–229.

Pudritz, R. E., and Norman, C. A. 1983. Centrifugally driven winds from contracting rotating molecular disks. *Astrophys. J.* 274:677–697.

Pudritz, R. E., and Norman, C. A. 1986. Bipolar hydromagnetic winds from disks around protostellar objects. *Astrophys. J.* 301:571–586.

Quinn, W. H., Neal, V. T., and Autuñez de Mayolo, S. E. 1987. El Niño occurrences over the past four and a half centuries. *J. Geophys. Res.* 92:1449–14461.

Radick, R. R. 1987. The solar-stellar connection at low spectral resolution. In *The SHIRSOG Workshop,* ed. M. S. Giampapa (Tucson: Natl. Solar Obs.), pp. 133–135.

Radick, R. R., and Baliunas, S. L. 1987. Stellar activity and the rotation of Hyades stars. In *Cool Stars, Stellar Systems and the Sun: Proc. Fifth Cambridge Workshop,* eds. J. L. Linsky and R. E. Stencel (New York: Springer-Verlag), pp. 217–219.

Radick, R. R., Hartmann, L., Mihalas, D., Worden, S. P. Africano, J. L., Klimke, A., and Tyson, E. T. 1982. The photometric variability of solar-type stars. I. Preliminary results for the Pleiades, Hyades, and the Malmquist field. *Publ. Astron. Soc. Pacific* 94:934–944.

Radick, R. R., Lockwood, G. W., Thompson, D. T., Warnock, A., Hartmann, L. W., Mihalas,

D., Worden, S. P., Henry, G. W., and Sherlin, J. M. 1983a. The photometric variability of solar-type stars. III. Results from 1981–82, including parallel observations of thirty-six Hyades stars. *Publ. Astron. Soc. Pacific* 95:621–634.

Radick, R. R., Wilkerson, M. S., Worden, S. P., Africano, J. L., Klimke, A., Ruden, S., Rogers, W., Armandroff, T. E., and Giampapa, M. S. 1983b. The photometric variability of solar-type stars. II. Stars selected from Wilson's chromospheric activity survey. *Publ. Astron. Soc. Pacific* 95:300–310.

Radick, R. R., Lockwood, G. W., and Thompson, D. T. 1986. Variability characteristics of lower main-sequence Hyades stars. In *Cool Stars, Stellar Systems and the Sun: Proc. Fourth Cambridge Workshop,* eds. M. Zeilik and D. M. Gibson (New York: Springer-Verlag), pp. 209–211.

Radick, R. R., Thompson, D. T., Lockwood, G. W., Duncan, D. K., and Baggett, W. E. 1987. The activity, variability, and rotation of lower main-sequence Hyades stars. *Astrophys. J.* 321:459–472.

Radick, R. R., Skiff, B. A., and Lockwood, G. W. 1990a. The activity, variability, and rotation of lower main-sequence members of the Coma star cluster. *Astrophys. J.* 353:524–532.

Radick, R. R., Lockwood, G. W., and Baliunas, S. L. 1990b. Stellar activity and brightness variations: A glimpse at the Sun's history. *Science* 247:39–44.

Raghavan, R. S. 1976. Inverse β decay of ^{115}In$\rightarrow^{115}$Sn*: A new possibility for detecting solar neutrinos from the proton-proton reaction. *Phys. Rev. Lett.* 37:259–262.

Raghavan, R. S. 1986. Inverse beta decay of argon-40. *Phys. Rev.* D34:2088–2091.

Raghavan, R. S., Pakvasa, S., and Brown, B. A. 1986. New tools for solving the solar-neutrino problem. *Phys. Rev. Lett.* 57:1801–1804.

Raisbeck, G. M., and Yiou, F. 1985. ^{10}Be in polar ice and atmospheres. *Annals Glaciol.* 7:138–140.

Raisbeck, G. M., and Yiou, F. 1988. ^{10}Be as a proxy indicator of variations in solar activity and geomagnetic field intensity during the last 10,000 years. In *Secular, Solar and Geomagnetic Variations in the Last 10,000 Years,* eds. F. R. Stephenson and W. Wolfendale (Dordrecht: Kluwer), pp. 287–296.

Raisbeck, G. M., Yiou, F., Fruneau, M. Loiseux, J. M., Lieuvin, M., and Ravel, J. C. 1981a. Cosmogenic ^{10}Be/^{7}Be as a probe of atmospheric transport processes. *Geophys. Res. Lett.* 8:1015–1018.

Raisbeck, G. M., Yiou, F., Fruneau, M., Loiseaux, J. M., Lieuvin, M., Ravel, J. C., and Lorius, C. 1981b. Cosmogenic ^{10}Be concentrations in Antarctic ice during the past 30,000 years. *Nature* 292:825–826.

Raisbeck, G. M., Yiou, F., Bourles, D., and Kent, D. V. 1985. Evidence for an increase in cosmogenic ^{10}Be during a geomagnetic reversal. *Nature* 315:315–317.

Raisbeck, G. M., Yiou, F., Bourles, D., Lorius, C., Jouzel, J., and Barkov, N. I. 1987. Evidence for two intervals of enhanced ^{10}Be deposition in Antarctic ice during the last glacial period. *Nature* 326:273–277.

Rajan, R. S. 1974. On the irradiation history and origin of gas-rich meteorites. *Geochim. Cosmochim. Acta* 38:777–788.

Rajan, R. S., and Lugmair, G. W. 1988. Solar flare tracks and neutron capture effects in gas-rich meteorites. *Lunar Planet. Sci.* XIX:964–965 (abstract).

Rajan, R. S., Brownlee, D. E., Tomandl, D., Hodge, P. W., Farrar, H., and Britten, R. A. 1977. Detection of ^{4}He in stratospheric particles gives evidence of extraterrestrial origin. *Nature* 267:133–134.

Ramaty, R. 1979. Energetic particles in solar flares. In *Particle Acceleration Mechanisms in Astrophysics,* eds. J. Arons, C. McKee and C. Max (New York: American Inst. of Physics), pp. 135–154.

Ramaty, R. 1986. Nuclear processes in solar flares. In *Physics of the Sun,* ed. P. A. Sturrock, vol. 2 (Dordrecht: D. Reidel), pp. 291–323.

Ramaty, R., and Lingenfelter, R. E. 1971. Astrophysical and geophysical variations in carbon 14 production. In *Radiocarbon Variation and Absolute Chronology,* ed. I. U. Olsson (Uppsala: Almquist and Wiksells), pp. 513–537.

Ramaty, R., and Murphy, R. J. 1987. Nuclear processes and accelerated particles in solar flares. *Space Sci. Rev.* 45:213–268.

Ramaty, R., Colgate, S. A., Dulk, G. A., Hoyng, P., Knight, J. W., Lin, R. P., Melrose, D. B.,

Orrall, F., Paizis, C., Shapiro, P. R., Smith, D. F., and Van Hollebeke, M. 1980. Energetic particles in solar flares. In *Solar Flares,* ed. P. A. Sturrock (Boulder: Colorado Assoc. Univ. Press), pp. 117–185.

Ramaty, R., Murphy R. J., and Miller, J. A. 1990. Solar accelerated particles: Comparisons of abundances and energy spectra from particle and gamma-ray observations. In *Particle Astrophysics,* eds. W. V. Jones, F. J. Kerr and J. F. Ormes (New York: American Inst. of Physics), pp. 143–154.

Rancitelli, L. A., Perkins, R. W., Felix, W. D., and Wogman, N. A. 1972. Lunar surface processes and cosmic ray characterization from Apollo 12–15 lunar sample analyses. In *Proc. Lunar Sci. Conf.* 3:1681–1691.

Rao, M. N., and Padia, J. T. 1989. Ancient solar activity: Clues from neon in meteorites. The Sun in Time, 6–10 March, Tucson, Arizona, Abstract Booklet, p. 40.

Rao, U. R. 1972. Solar modulation of galactic cosmic radiation. *Space Sci. Rev.* 12:719–809.

Rapaport, J., Welch, P., Bahcall, J., Sugarbaker, E., Taddeucci, T. N., Goodman, C. D., Foster, C. F., Horen, D., Gaarde, C., Larsen, J., and Masterson, T. 1985. Solar-neutrino detection: Experimental determination of Gamow-Teller strengths via ^{98}Mo and ^{115}In (p,n) reactions. *Phys. Rev. Lett.* 54:2325–2328.

Raychaudhuri, P. 1984. Solar neutrino data and the 11-year solar activity cycle. *Solar Phys.* 93:397–401.

Razdan, H., Colburn, D. S., and Sonett, C. P. 1965. Recurrent $SI^+ - SI^-$ impulse pairs and shock structure in M region beams. *Planet. Space Sci.* 13:1111–1123.

Reagan, J. B., Meyerott, R. E., Nightingale, R. W., Gunton, R. C., Johnson, R. G., Evan, J. E., Imhof, W. L., Hezath, D. F., and Krueger, A. J. 1981. Effects of the August 1972 solar particle events on stratospheric ozone. *J. Geophys. Res.* 86:1473–1494.

Reames, D. V. 1990. Energetic particles from impulsive solar flares. *Astrophys. J. Suppl.* 73:235–251.

Reames, D. V., von Rosenvinge, T. T., and Lin, R. P. 1985. Solar ^{3}He-rich events and nonrelativistic electron events: A new association. *Astrophys. J.* 292:716–724.

Reames, D. V., Ramaty, R., and von Rosenvinge, T. T. 1988. Solar Ne abundances from gamma-ray spectroscopy and ^{3}He-rich particle events. *Astrophys. J.* 332:L87-L91.

Reedy, R. C. 1977. Solar proton fluxes since 1956. In *Proc. Lunar Sci. Conf.* 8:825–839.

Reedy, R. C. 1980. Lunar radionuclide records of average solar-cosmic-ray fluxes over the last ten million years. In *The Ancient Sun: Fossil Record in the Earth, Moon and Meteorites,* eds. R. O. Pepin, J. A. Eddy and R. B. Merrill (New York: Pergamon Press), pp. 365–386.

Reedy, R. C. 1987a. Predicting the production rates of cosmogenic nuclides in extraterrestrial matter. *Nucl. Instr. Meth.* B29:251–261.

Reedy, R. C. 1987b. Solar-proton-produced nuclides in meteorites. In *Lunar Planet. Sci.* XVIII:822–823 (abstract).

Reedy, R. C., and Arnold, J. R. 1972. Interaction of solar and galactic cosmic-ray particles with the Moon. *J. Geophys. Res.* 77:537–555.

Reedy, R. C., Arnold, J. R., and Lal, D. 1983. Cosmic-ray record in solar system matter. *Science* 219:127–135.

Rees, M. H., Lummerzheim, D., Roble, R. G., Winningham, J. D., Craven, J. D., and Frank, L. A. 1988. Auroral energy deposition rate, characteristic electron energy, and ionospheric parameters derived from Dynamics Explorer 1 images, *J. Geophys. Res.* 93:12841–12860.

Regnier, S., Hohenberg, C. M., Marti, K., and Reedy, R. C. 1979. Predicted versus observed cosmic-ray-produced noble gases in lunar samples: Improved Kr production ratios. In *Proc. Lunar Planet. Sci. Conf.* 10:1565–1586.

Regnier, S., Lavielle, B., Simonoff, M., and Simonoff, G. N. 1982. Nuclear reactions in Rb, Sr, Y, and Zr targets. *Phys. Rev.* C26:931–943.

Reid, G. C., Isaksen, I. S. A., Holger, T. E., and Crutzen, P. J. 1976. Influence of ancient solar-proton events on the evolution of life. *Nature* 259:177–179.

Reid, G. C., McAfer, J. F., and Crutzen, P. J. 1978. Effects of intense stratospheric ionisation events. *Nature* 275:489–492.

Reipurth, B. 1989a. Herbig-Haro objects: New observations of jets and bow shocks. In *Low Mass Star Formation and Pre-Main Sequence Objects,* ed. B. Reipurth (Garching: European Southern Obs.), pp. 247–279.

Reipurth, B. 1989*b*. Pre-main sequence binaries. In *Formation and Evolution of Low Mass Stars,* eds. A. K. Dupree and M. T. V. T. Lago (Dordrecht: Kluwer), pp. 305–319.

Rengarajan, T. N. 1984. Age-rotation relationship for late-type main-sequence stars. *Astrophys. J.* 283:L63-L65.

Restaino, S. R. 1990. On the variation of the Ca II K line profile with abundance. *Solar Phys.,* submitted.

Reynolds, J. H. 1967. Isotopic abundance anomalies in the solar system. *Ann. Rev. Nucl. Sci.* 17:254–316.

Reynolds, R. T., Fricker, P. E., and Summers, A. C. 1966. Effects of melting upon thermal models of the Earth. *J. Geophys. Res.* 71:573–582.

Rhodes, F. J., Ulrich, R. K., and Simon, G. W. 1977. Observations of nonradial *p* mode oscillations in the Sun. *Astrophys. J.* 218:901–919.

Ribes, E. 1989. Astronomical determinations of the solar variability. *Phil. Trans. Roy. Soc. London* 330:487–497.

Ribes, E., Ribes, J. C., and Barthalot, R. 1987. Evidence for a larger Sun with a slower rotation during the seventeenth century. *Nature* 326:52–55.

Ribes, E., Ribes, J. C., and Barthalot, R. 1988. Size of the Sun in the seventeenth century. *Nature* 332:689–690.

Ribes, E., Ribes, J. C., Vince, I., and Merlin, Ph. 1988. Sur l'oscillation du diamètre solaire. *Compt. Rend. II* 307:1195–1201.

Ribes, E., Merlin, Ph., Ribes, J. C., and Barthalot, R. 1989*a*. Absolute periodicities in the solar diameter, derived from historical and modern observations. *Ann. Geophys.* 7:321–329.

Ribes, E., Ribes, J. C., and Merlin, Ph. 1989*b*. Solar variability and its possible connection with the Earth's weather. In *Annales de l'Institut National d'Agronomie de Tunisie, Colloque Programme International de Corrélations Géologiques,* no. 152, Colloque de Djerba, October 1989, in press.

Rieger, E. 1989. Solar flares: High energy radiation and particles. *Solar Phys.* 121:323–345.

Rieger, E., Share, G. H., Forrest, D. J., Kanbach, G., Reppin, C., and Chupp, E. L. 1984. A 154-day periodicity in the occurrence of hard solar flares? *Nature* 312:623–625.

Rind, D., Peteet, D., and Kukla, G. 1989. Can Milankovitch orbital variations initiate the growth of ice sheets in a general circulation model? *J. Geophys. Res.* 12851–12871.

Ringwood, A. E., and Essene, E. 1970. Petrogenesis of Apollo 11 basalts, internal constitution and origin of the Moon. *Proc. Apollo 11 Lunar Sci. Conf.* 1:769–799.

Risbo, T. C., Lausen, H. B., and Rasmussen, K. L. 1981. Supernovae and nitrate in the Greenland ice sheet. *Nature* 294:637–639.

Roberts, L. T. 1988. Solar Cyclicity in the Lacustrine Green River Formation (Eocene, Wyoming). M.S. Thesis, Univ. of Southern California.

Robinson, E. L., and Kraft, R. P. 1974. On the variability of dwarf K- and M- type stars in the Pleiades and Hyades. *Astron J.* 79:698–701.

Robinson, R. D. 1980. Magnetic field measurements on stellar sources: A new method. *Astrophys. J.* 239:961–967.

Robinson, R. D., Worden, S. P., and Harvey, J. W. 1980. Observations of magnetic fields on two late-type dwarf stars. *Astrophys. J.* 236:L155-L158.

Rodono, M., Cutispoto, G., Pazzani, V., Catalano, S., Byrne, P. B., Doyle, J. D., Butler, C. J., Andrews, A. D., Blanco, C., Marilli, E., Linsky, J. L., Scaltriti, F., Busso, M., Cellino, A., Hopkins, J. L., Okazaki, A., Hayashi, S. S., Zeilik, M., Helston, R., Henson, G., Smith, P., and Simon, T. 1986. Rotational modulation and flares on RS CVn and BY Dra-type stars. *Astron. Astrophys.* 165:135–156.

Rogers, F. J. 1981. Equation of state of dense, partially degenerate, reacting plasmas. *Phys. Rev.* A74:1531–1543.

Rogers, F. J. 1983. Integral-equation method for partially ionized plasmas. *Phys. Rev.* A79:868–879.

Rogers, F. J. 1986. Occupation numbers for reacting plasmas: The role of the Planck Larkin Partition Function. *Astrophys. J.* 310:723–728.

Rogerson, J. B. 1959. The solar limb intensity profile. *Astrophys. J.* 130:985–990.

Rokop, D. J., Schroeder, N., and Wolfsberg, K. 1990. Mass spectrometry of technetium at the sub-picogram level. *Analyt. Chem.* 62:1271–1274.

Ronov, A. B., and Yaroshevsky, A. A. 1967. Chemical structure of the Earth's crust. *Geochem.* 11:1041–1066.

Rood, R. T., Steigman, G., and Tinsley, R. T. 1976. Stellar production as a source of ³He in the interstellar medium. *Astrophys. J.* 207:L29-L60.

Ropelewski, C. F., and Halpert, M. S. 1987. Global and regional scale precipitation patterns associated with the El Niño/Southern oscillation. *Mon. Weather Rev.* 115:1606–1626.

Rosch, J., and Yerle, R. 1983. Solar diameters. *Solar Phys.* 82:139–150.

Rosen, S. P. 1991. Theory of neutrino oscillations and the Sun. In *Solar Interior and Atmosphere,* eds. A. N. Cox, W. C. Livingston and M. S. Matthews (Tucson: Univ. of Arizona Press), in press.

Rosen, S. P., and Gelb, J. M. 1986. Mikheyev-Smirnov-Wolfenstein oscillations as a possible solution to the solar-neutrino problem. *Phys. Rev.* D34:969–979.

Rosner, R., and Weiss, N. O. 1985. Differential rotation and magnetic torques in the interior of the Sun. *Nature* 317:790–792.

Rosner, R., Golub, L., Coppi, B., and Vaiana, G. S. 1978. Heating of coronal plasma by anomalous current dissipation. *Astrophys. J.* 222:317–332.

Rosner, R., Golub, L., and Vaiana, G. S. 1985. On stellar X-ray emission. *Ann. Rev. Astron. Astrophys.* 23:413–452.

Ross, J. E., and Aller, L. H. 1976. The chemical composition of the Sun. *Science* 191:1223–1229.

Rossi, S., Auroux, C., and Mascle, J. 1983. The Gulf of Taranto (southern Italy): Seismic stratigraphy and shallow structure. *Marine Geol.* 51:327–346.

Rossignol-Strick, M. 1983. African monsoons, an immediate climate response to orbital insolation. *Nature* 303:46–49.

Rossow, W. B., Henderson-Sellers, A., and Weinrich, S. K. 1982. Cloud feedback: A stabilizing effect for the early Earth? *Science* 217:1245–1247.

Rottman, G. J. 1981. Rocket measurements of the solar spectral irradiance during solar minimum. *J. Geophys. Res.* 86:6697–6705.

Rottman, G. J. 1985. Solar ultraviolet irradiance 1982 and 1983. In *Atmospheric Ozone: Proc. Quadrennial Ozone Symp.* eds. C. S. Zerefos and A. Ghazi (Dordrecht: D. Reidel), pp. 656–660.

Rottman, G. J. 1987. Results from space measurements of solar UV and EUV flux. In *Solar Radiation Output Variations,* ed. P. V. Foukal (Cambridge, Mass.: Cambridge Research and Instrumentation), pp. 71–86.

Rottman, G. J., and London, J. 1984. Solar UV irradiance observations of 27-day variation and evidence for solar cycle variation. In *Current Problems in Atmospheric Radiation,* ed. G. Fiocco (Huntington, Va.: A. Deepak), pp. 320–324.

Rowley, J. K., Cleveland, B. T., Davis, R., Hampel, W., and Kirsten, T. 1980. The present and past luminosity of the Sun. In *The Ancient Sun: Fossil Record in the Earth, Moon and Meteorites,* eds. R. O. Pepin, J. A. Eddy and R. B. Merril (New York: Permagon Press), pp. 45–62.

Rowley, J. K., Cleveland, R. T., and Davis, R. 1985. The chlorine solar neutrino experiment. In *Solar Neutrinos and Neutrino Astronomy,* eds. M. L. Cherry, W. A. Fowler and K. Lande (New York: American Inst. of Physics), pp. 1–21

Royer, J. F., Deque, M., and Pestiaux, P. 1983. Orbital forcing of the inception of the Laurentide ice sheet. *Nature* 304:43–46.

Rubin, A. E., Taylor, G. J., Scott, E. R. D., and Keil, K. 1982. Petrologic insights into the fragmentation history of asteroids. In *Workshop on Lunar Breccias and Soils and their Meteoritic Analogs,* eds. G. J. Taylor and L. L. Wilkening, LPI Tech. Rept. 82–02 (Houston: Lunar and Planetary Inst.), pp. 107–110.

Rucinski, S. M. 1985. IRAS observations of T Tauri and Post-T Tauri stars. *Astron J.* 90:2321–2330.

Rucinski, S. M. 1988. Rotational properties of composite polytrope models. *Astron J.* 95:1895–1907.

Rucinski, S. M., and VandenBerg, D. A. 1986. Activity-related characteristics of the convective envelopes in evolving low-mass stars. *Publ. Astron. Soc. Pacific* 98:669–684.

Rucinski, S. M., and VandenBerg, D. A. 1990. Dependence of the Rossby number on helium and metal abundances. *Astron. J.* 99:1279–1283.

Ruden, S. P., and Lin, D. N. C. 1986. The global evolution of the primordial solar nebula. *Astrophys. J.* 308:831–901.

Ruddiman, W. F., and McIntyre, A. 1981. Oceanic mechanisms for amplification of the 23,000-year ice-volume cycle. *Science* 212:617–627.

Ruddiman, W. E., Shackleton, N. J., and McIntyre, A. 1986. North Atlantic sea-surface temperatures for the last 1.1 million years. In *North Atlantic Palaeoceanography*, eds. C. P. Summerhayes and N. J. Shackleton, Geol. Soc. of London SP-21 (Oxford: Blackwell Scientific), pp. 155–173.

Ruderman, M. A., and Chamberlain, J. W. 1975. Origin of the sunspot modulation of ozone: Its implications for stratospheric NO injection. *Planet. Space Sci.* 23:247–268.

Rudstam, G. 1966. Systematics of spallation yields. *Z. Naturforsch.* A21:1027–1041.

Ruelle, D. 1990. Deterministic chaos: The science and the fiction. *Proc. Roy. Soc. London* A427:241–248.

Rufener, F. 1985. Approaches to photometric calibrations. In *Calibration of Fundamental Stellar Quantities*, ed. D. S. Hayes (Dordrecht: D. Reidel), pp. 253–269.

Rufener, F. 1989. Catalogue of stars measured in the Geneva Observatory photometric system (fourth edition). *Astron. Astrophys. Suppl.* 78:469–481.

Rufener, F., and Nicolet, B. 1988. A new determination of the Geneva photometric passbands and their absolute calibration. *Astron. Astrophys.* 206:357–374.

Runcorn, S. K. 1983. Lunar magnetism, polar displacements and primeval satellites in the Earth-Moon system. *Nature* 304:589–596.

Rusch, D. W., and Barth, C. A. 1975. Satellite measurements of nitric oxide in the polar region. *J. Geophys. Res.* 809:3719–3721.

Rusch, D. W., and Clancy, R. T. 1989. A study of the time and spatial dependence of ozone near 1.0 mbar with emphasis on the springtime. In *Ozone in the Atmosphere: Proc. Quadrennial Ozone Symp.*, eds. R. D. Bojkov and P. Fabian (Hampton, Va.: A. Deepak), pp. 218–222.

Rusch, D. W., Gérard, J. C., Solomon, S., Crutzen, P. J., and Reid, G. C. 1981. The effect of particle precipitation events on the neutral and ion chemistry of the middle atmosphere I. Odd nitrogen. *Planet. Space Sci.* 29:767–774.

Russ, G. P., III, and Emerson, M. T. 1980. [53]Mn and [26]Al evidence for solar cosmic ray constancy—an improved model for interpretation. In *The Ancient Sun: Fossil Record in the Earth, Moon and Meteorites*, eds. R. O. Pepin, J. A. Eddy and R. B. Merrill (New York: Pergamon Press), pp. 387–399.

Rust, D. M. 1982. Solar flares, proton showers, and the space shuttle. *Science* 216:939–946.

Rutten, R. G. M. 1984. Magnetic structure in cool stars. VII. Absolute surface flux in Ca II H and K line cores. *Astron. Astrophys.* 130:353–360.

Rutten, R. G. M. 1987a. Magnetic Activity of Cool Stars and Its Dependence on Rotation and Evolution. Ph.D. Thesis, Utrecht.

Rutten, R. G. M. 1987b. Magnetic structures in cool stars. XII. Chromospheric activity and rotation in giants and dwarfs. *Astron. Astrophys.* 177:131–142.

Rutten, R. G. M., and Schrijver, C. J. 1987. Magnetic structure in cool stars. XIII. Appropriate units for the rotation-activity relation. *Astron. Astrophys.* 177:155–162.

Ruzmaikin, A. A. 1981. Solar cycle as strange attractor. *Comments on Astrophys.* 9:85–93.

Ruzmaikin, A. A. 1985. The solar dynamo. *Solar Phys.* 100:125–140.

Ruzmaikin, A. A., and Starchenko, S. V. 1987. Magnetic manifestations of solar rotation. *Astron. J. (USSR)* 64:1057–1065.

Ruzmaikina, T. V. 1980. On the role of magnetic fields and turbulence in the evolution of the presolar nebula. *Adv. Space Res.* 1:49–53.

Ruzmaikina, T. V. 1981. The angular momentum of protostars creating the protoplanetary disks. *Pisma Astron J. (USSR)* 7:188–192.

Ruzmaikina, T. V. 1985. Magnetic fields in the collapsing presolar nebula. *Astron. Vestnik (USSR)* 18:101–112.

Ruzmaikina, T. V., and Maeva, S. V. 1986. Investigation of the process of the solar nebula formation. *Astron. Vestnik (USSR)* 20:212–227.

Ruzmaikina, T. V., Safronov, V. S., and Weidenschilling, S. J. 1989. Radial mixing of material in the asteroidal zone. In *Asteroids II*, eds. R. P. Binzel, T. Gehrels and M. S. Matthews (Tucson: Univ. of Arizona Press), pp. 681–700.

Rydgren, A. E., and Vrba, F. J. 1983. Periodic light variations in four pre-main-sequence K stars. *Astrophys. J.* 267:191–198.

Rydgren, A. E., Zak, D. S., Vrba, F. J., Chugainov, P. F., and Zajtseva, G. V. 1984. UB-VRI monitoring of five late-type pre-main-sequence stars. *Astron. J.* 89:1015–1021.

Saar, S. H. 1987*a*. Observations and Analysis of Photospheric Magnetic Fields on G, K, and M Dwarf Stars. Ph.D. Thesis, Univ. of Colorado.

Saar, S. H. 1987*b*. The photospheric magnetic fields of cool stars: Recent results of survey and time-variability programs. In *Cool Stars, Stellar Systems and the Sun: Proc. Fifth Cambridge Workshop*, eds. J. L. Linsky and R. E. Stencel (New York: Springer-Verlag), pp. 10–20.

Saar, S. H. 1988*a*. Improved methods for the measurement and analysis of stellar magnetic fields. *Astrophys. J.* 324:441–465.

Saar, S. H. 1988*b*. The magnetic fields on cool stars and their correlation with chromospheric and coronal emission. In *Hot Thin Plasmas in Astrophysics*, ed. R. Pallavicini (Dordrecht: Kluwer), pp. 139–146.

Saar, S. H. 1991*a*. In *Mechanisms of Chromospheric Heating*, eds. P. Ulmschneider, E. Priest and R. Rosner (New York: Springer-Verlag), pp. 273–278.

Saar, S. H. 1991*b*. In *The Sun and Cool Stars: Proc IAU Coll. 130*, eds. I Tuominen et al. (Dordrecht: Kluwer), in press.

Saar, S. H., and Linsky, J. L. 1985. The photospheric magnetic field of the dM3.5e flare star AD Leonis. *Astrophys. J.* 299:L47-L50.

Saar, S. H., and Linsky, J. L. 1986*a*. Further observations of magnetic fields on active dwarf stars. 1986. *Astrophys. J.* 299:278–280.

Saar, S. H., and Linsky, J. L. 1986*b*. New measurements of stellar magnetic fields and emerging trends. *Adv. in Space Phys.* 6:(8):235–238.

Saar, S. H., and Schrijver, C. J. 1987. Empirical relations between magnetic fluxes and atmospheric radiative-losses for cool dwarf stars. In *Cool Stars, Stellar Systems and the Sun*, eds. J. L. Linsky and R. E. Stencel (New York: Springer-Verlag), pp. 38–40.

Saar, S. H., Linsky, J. L., and Giampapa, M. S. 1987. Four meter FTS observations of photospheric magnetic fields on M dwarfs. In *Proc. 27th Intl. Astrophys. Coll. on Observational Astrophysics with High Precision Data*, eds. L. Delbouille and A. Monfils (Liège: Univ. de Liège), pp. 103–108.

Saar, S. H., Huovelin, J., Giampapa, M. S., Linsky, J. L., and Jordan, C. 1988. Multiwavelength observations of magnetic fields and related activity on Xi Bootis A. In *Activity in Cool Star Envelopes*, eds. O. Havnes, B. R. Pettersen, J. H. M. M. Schmitt and J. E. Solheim (Dordrecht: Kluwer), pp. 45–48.

Safronov, V. S., and Ruzmaikina, T. V. 1978. On the angular momentum transfer and accumulation of solid bodies in the solar nebula. In *Protostars and Planets*, ed. T. Gehrels (Tucson: Univ. of Arizona Press), pp. 545–564.

Sagan, C., and Mullen, G. 1972. Earth and Mars: Evolution of atmospheres and surface temperatures. *Science* 177:52–56.

Sakashita, S., and Hanami, H. 1986. Formation of bipolar flow within a plane-parallel gas layer by stellar wind. *Publ. Astron. Soc. Japan* 38:879–894.

Sakurai, T. 1985. Magnetic stellar winds: A 2-D generalization of the Weber-Davis model. *Astron. Astrophys.* 152:121–129.

Sakurai, T. 1987. Magnetically collimated winds from accretion disks. *Publ. Astron. Soc. Japan* 39:821–835.

Saltzman, B. 1985. Paleoclimatic modeling. In *Paleoclimate Analysis and Modeling*, ed. A. D. Hecht (New York: Wiley), pp. 341–396.

Saltzman, B. 1987. Modeling the δ^{18}O derived record of Quaternary climatic change with low-order dynamical systems. In *Irreversible Phenomena and Dynamical Systems Analysis in Geosciences*, eds. C. Nicolis and G. Nicolis (Dordrecht: D. Reidel), pp. 355–380.

Saltzman, B., and Sutera, A. 1984. A model of the internal feedback system involved in late Quaternary climatic variations. *J. Atmos. Sci.* 41:736–745.

Saltzman, B., and Sutera, A. 1987. The mid-Quaternary climatic transition as the free response of a three-variable dynamical model. *J. Atmos. Sci.* 44:236–241.

Saltzman, B., Hansen, A. R., and Maasch, K. A. 1984. The late Quaternary glaciations as the response of a three-component feedback system to Earth-orbital-forcing. *J. Atmos. Sci.* 41:3380–3389.

Sanderson, T. R., Reinhard, R., van Nes, P., and Wenzel, K.-P. 1985. Observations of 35- to 1600-keV protons and low-frequency waves upstream of interplanetary shocks. *J. Geophys. Res.* 90:3973–3980.

Sanford, S. A. 1986. Solar flare track densities in interplanetary dust particles: The determination of an asteroidal versus cometary sources of the zodiacal dust cloud. *Icarus* 68:377–394.

Sargent, A. I. 1988. Molecular disks and their link to planetary systems. In *Formation and Evolution of Planetary Systems,* eds. H. A. Weaver, F. Paresce and L. Danly (New York: Cambridge Univ. Press), pp. 111–127.

Sargent, A. I., and Beckwith, S. 1987. Kinematics of the circumstellar gas of HL Tau and R Monocerotis. *Astrophys. J.* 323:294–305.

Sarton, G. 1947. Early observations of sunspots. *Isis* 37:69.

Sasaki, S., and Nakazawa, K. 1988. Origin of isotopic fractionation of terrestrial Xe: Hydrodynamic fractionation during escape of the primordial H_2-He atmosphere. *Earth Planet. Sci. Lett.* 89:323–334.

Scargle, J. D. 1982. Studies in astronomical time-series analysis. II. Statistical aspects of spectral analysis of unevenly spaced data. *Astrophys. J.* 263:835–853.

Schatten, K. H. 1988. A model for solar constant secular changes. *Geophys. Res. Lett.* 15:121–124.

Schatten, K. H., and Mayr, H. G. 1985. On the maintenance of sunspots: An ion hurricane mechanism. *Astrophys. J.* 299:1051–1062.

Schatten, K. H., Wilcox, J. M., and Ness, N. F. 1969. A model of interplanetary and coronal magnetic fields. *Solar Phys.* 6:442–455.

Schatten, K. H., Miller, N., Sofia, S., Endal, A., and Chapman, G. A. 1985. The importance of improved facular observations in understanding solar constant variations. *Astrophys. J.* 294:689–696.

Schatzman, E. 1962. A theory of the role of magnetic activity during star formation. *Ann. Astrophys.* 25:18–29.

Schatzman, E., and Maeder, A. 1981. Stellar evolution with turbulent diffusion mixing. III. The solar model and the neutrino problem. *Astron. Astrophys.* 96:1–16.

Scheiner, C. 1612*a*. *Tres Epistolae de Maculis Solaribus* (Augsburg).

Scheiner, C. 1612*b*. *Accuratior Disquisitio* (Augsburg).

Scheiner, C. 1630. *Rosa Ursina* (Bracciano).

Schidlowski, M., Eichmann, R., and Junge, C. E. 1975. Precambrian sedimentary carbonates: Carbon and oxygen isotope geochemistry and implications for terrestrial oxygen budget. *Precam. Res.* 2:1–69.

Schiøtt, H. E. 1970. Approximations and interpolation rules for ranges and range stragglings. *Radiation Effects* 6:107–113.

Schlesinger, M. E. 1986. Equilibrium and transient climatic warming induced by increased atmospheric CO_2. *Climate Dyn.* 1:35–52.

Schlesinger, M. E., and Mitchell, J. F. B. 1987. Climate model simulations of the equilibrium climatic response to increased carbon dioxide. *Rev. Geophys.* 25(4):760–798.

Schmid, J., Bochsler, P., and Geiss, J. 1988. Abundance of iron ions in the solar wind. *Astrophys. J.* 329:956–966.

Schmidt, B., and Bruhle, W. 1988. Klima, radiokohlenstoffgehalt und dendrochronologie. *Naturwissen. Rundschau* 5:177–182.

Schmidtke, G. 1981. Solar irradiance below 120 nm and its variations. *Solar Phys.* 74:251–263.

Schmidtke, G. 1984. Modelling the solar extreme ultraviolet irradiance for aeronomic applications. In *Handbuch der Physik,* XLIX-7, Geophys. III (Berlin: Springer-Verlag), pp. 1–55.

Schmidtke, G., Rawer, K., Botzek, H., Norbert, D., and Holzer, K. 1977. Solar EUV photon fluxes measured aboard Aeros A. *J. Geophys. Res.* 82:2423–2427.

Schneider, S. H. 1989. The greenhouse effect: Science and policy. *Science* 243:771–781.

Schonberg, M., and Chandrasekhar, S. 1942. On the evolution of the main sequence stars. *Astrophys. J.* 96:161–173.

Schove, D. J. 1955. The sunspot cycle 649 BC to 2000 AD. *J. Geophys. Res.* 60:127–145.

Schove, D. J. 1983. *Sunspot Cycles* (Stroudsburg, P. A.: Hutchinson Ross).

Schramm, D. N., Tera, F., and Wasserburg, G. J. 1970. The isotopic abundance of ^{26}Mg and limits on ^{26}Al in the early solar system. *Earth Planet. Sci. Lett.* 10:44–59.

Schrijver, C. J. 1987. Heating of stellar chromospheres and coronae: Evidence for non-magnetic heating. In *Cool Stars, Stellar Systems and the Sun: Proc. Fifth Cambridge Workshop*, eds. J. L. Linsky and R. E. Stencel (New York: Springer-Verlag), pp. 135–145.

Schrijver, C. J., and Coté, J. 1987. The relation between the Ca II K line-core flux density and the magnetic flux density on the Sun. In *Cool Stars, Stellar Systems and the Sun: Proc. Fifth Cambridge Workshop*, eds. J. L. Linsky and R. E. Stencel (New York: Springer-Verlag), pp. 51–53.

Schrijver, C. J., Mewe, R., and Walter, F. M. 1984. Coronal activity in F-, G-, and K-type stars. II. Coronal structure and rotation. *Astron. Astrophys.* 138:258–266.

Schrijver, C. J., Dobson, A. K., and Radick, R. R. 1989*a*. The magnetic, basal and radiative-equilibrium components in Mount Wilson Ca II H-K fluxes. *Astrophys. J.* 341:1035–1044.

Schrijver, C. J., Coté, J., Zwaan, C. and Saar, S. H. 1989*b*. Relations between the photospheric magnetic field and the emission from the outer atmospheres of cool stars. I. The solar Ca II K line core emission. *Astrophys. J.* 337:964–976.

Schroder, W. 1988. Aurorae during the Maunder minimum. *Meteorol. Atmos. Phys.* 38:246–251.

Schubert, G., Ross, M. N., Stevenson, D. J. and Spohn, T. 1988. Mercury's thermal history and the generation of its magnetic field. In *Mercury*, eds. F. Vilas, C. R. Chapman and M. S. Matthews (Tucson: Univ. of Arizona Press), pp. 429–460.

Schultz, L., and Signer, P. 1977. Noble gases in the St. Mesmin chondrite: Implications to the irradiation history of a brecciated meteorite. *Earth Planet. Sci. Lett.* 36:363–371.

Schultz, L., Signer, P., Lorin, J. C., and Pellas, P. 1972. Complex irradiation history of the Weston chondrite. *Earth Planet. Sci. Lett.* 15:403–410.

Schüssler, M. 1975. Axisymmetric α^2-dynamos in the Hayashi phase. *Astron. Astrophys.* 323:294–305.

Schüssler, M. 1979. Magnetic buoyancy revisited: Analytical and numerical results for rising flux tubes. *Astron. Astrophys.* 71:79–91.

Schwabe, H. 1843. Solar observations during 1843. *Astron Nach.* 20, no. 495. Quoted in *Early Solar Physics*, ed. A. J. Meadows (New York: Pergamon Press), 1970.

Schwartzman, D. W., and Volk, T. 1989. Biotic enhancement of weathering and the habitability of Earth. *Nature* 340:457–460.

Schwarzbach, M. 1985. *Wegener, le père de la dérive des continents* (Paris: Bélin).

Schwarzschild, M. 1958. *Structure and Evolution of the Stars* (Princeton, N.J.: Princeton Univ. Press), pp. 30–96.

Schwenn, R. 1983*a*. The "average" solar wind in the inner heliosphere: Structures and slow variations. In *Solar Wind: Proc. Fifth Intl. Conf.*, ed. M. Neugebauer, NASA CP-2280, pp. 489–507.

Schwenn, R. 1983*b*. Direct correlations between coronal transients and interplanetary disturbances. *Space Sci. Rev.* 34:85–99.

Schwerer, F. C., Nagata, T., and Fisher, R. M. 1971. Electrical conductivity of lunar surface rocks and chondritic meteorites. *Moon* 2:408–422.

Scott, E. R. D., and Rajan, R. S. 1981. Metallic minerals, thermal histories and parent bodies of some xenolithic, ordinary chondrite meteorites. *Geochim. Cosmochim. Acta* 45:53–67.

Scott, E. R. D., Lusby, D. and Keil, K. 1985. Ubiquitous brecciation after metamorphism in equilibrated ordinary chondrites. *Proc. Lunar Planet. Sci. Conf. 16, J. Geophys. Res. Suppl.* 90:D137-D148.

Scott, E. R. D., Taylor, G. J., Newsom, H. E., Herbert, F., Zolensky, M., and Kerridge, J. F. 1989. Chemical, thermal and impact processing of asteroids. In *Asteroids II*, eds. R. P. Binzel, T. Gehrels and M. S. Matthews (Tucson: Univ. of Arizona Press), pp. 701–739.

Scott, R. D. 1976. Possibility of using [81]Kr to detect solar neutrinos. *Nature* 264:729–730.

Secchi, P. 1872. Observation des variations de diamètre solaire. *Compt. Rend.* 75:606–613.

Sekiya, M., Nakazawa, K., and Hayashi, C. 1980. Dissipation of the rare gases contained in the primordial Earth's atmosphere. *Earth Planet. Sci. Lett.* 50:197–201.

Sekiya, M., Hayashi, C., and Nakazawa, K. 1981. Dissipation of the primordial terrestrial atmosphere due to irradiation of the solar far-UV during T Tauri stage. *Prog. Theor. Phys.* 66:1301–1316.

Sellers, W. D. 1969. A climate model based on the energy balance of the Earth-atmosphere system. *J. Appl. Meteorol.* 8:392–400.

Sellers, W. D. 1970. The effect of changes in the Earth's obliquity on the distribution of mean annual sea-level temperatures. *J. Appl. Meteorol.* 9:960–961.

Sellers, W. D. 1990. Climate signals in the lower stratosphere and upper troposphere. In *Proc. Sixth Annual Pacific Climate (PACLIM) Workshop,* eds. J. Betancourt and A. M. MacKay (Asilomar, Calif.: Calif. State Dept. of Water Resources), pp. 33–35.

Sepkoski, J. J., and Raup, D. M. 1986. Periodicity in marine extinction events. In *Dynamics of Extinction,* ed. D. K. Elliot (New York: Wiley).

Sergin, V. Ya. 1979. Numerical modeling of the glaciers-ocean-atmosphere global system. *J. Geophys. Res.* 84:3191–3204.

Severny, A., Kotov, V. A., and Tsap, T. T. 1984. Power spectrum of long-period solar oscillations and 160 minute pulsations during 1974–1982. *Nature* 307:247–249.

Shackleton, N. J. 1977. The oxygen isotope stratigraphic record of the late Pleistocene. *Phil. Trans. Roy. Soc. London* B280:169–182.

Shackleton, N. J., and Opdyke, N. D. 1973. Oxygen isotope and paleomagnetic stratigraphy of equatorial Pacific core V-28-238: Oxygen isotope temperatures and ice volumes on a 10^5 and 10^6 year scale. *Quat. Res.* 3:39–55.

Shackleton, N. J., and Opdyke, N. D. 1976. Oxygen isotope and paleomagnetic stratigraphy of Pacific core V28–239 Late Pliocene to Latest Pleistocene. In *Investigation of Late Quarternary Paleoceanography and Paleoclimatology,* eds. R. M. Cline and J. O. Hays, Geol. Soc. America Memoir 145 (Boulder: Geol. Soc. of America), pp. 449–464.

Shakura, N. I., and Sunyaev, R. A. 1973. Black holes in binary systems. Observational appearance. *Astron. Astrophys.* 24:337–355.

Shapiro, I. I. 1980. Is the Sun shrinking? *Science* 208:51–53.

Sharaf, G. S., and Budnikova, N. A. 1969. Secular perturbations in the elements of the Earth's orbit and the astronomical theory of climate variations. *Tr. Inst. Teor. Astron.* 14:48–85.

Shea, M. A., and Smart, D. F. 1975. A five by fifteen degree world grid of calculated cosmic-ray vertical cut-off rigidities for 1965 and 1975. In *Proc. 14th Intl. Cosmic Ray Conf.,* vol. 4 (Munich: MPI für Extraterrestrische Physik), p. 415.

Shea, M. A., and Smart, D. F. Preliminary search for cosmic radiation and solar-terrestrial parameters correlated with the reversal of the solar magnetic field. *Adv. Space Res.* 1(3):147–150.

Shea, M. A., and Smart, D. F. 1983. A world grid of calculated cosmic ray vertical cutoff rigidities for 1980. In *Proc. 18th Intl. Conf. on Cosmic Rays,* vol. 3, eds. N. Durgaprasad, S. Ramadurai, P. V. Raman Murthy, M. V. S. Rao and K. Sivaprasad (Bombay: Tata Inst. of Fundamental Research), pp. 415–418.

Shea, M. A., and Smart, D. F. 1990. A summary of major solar proton events. *Solar Phys.* 127:297–320.

Shea, M. A., Smart, D. F., and McCall, J. R. 1968. A five degree by fifteen degree world grid of trajectory-determined vertical cutoff rigidities. *Canadian J. Phys.* 46:1098–1101.

Shedlovsky, J. P., Honda, M., Reedy, R. C., Evans, J. C., Jr., Lal, D., Lindstrom, R. M., Delany, A. C., Arnold, J. R., Loosli, H. H., Fruchter, J. S., and Finkel, R. C. 1970. Pattern of bombardment-produced radionuclides in rock 10017 and in lunar soil. *Science* 167:574–576; *Proc. Apollo 11 Lunar Sci. Conf.,* pp. 1503–1532.

Sheeley, N. R., Stewart, R. T., Robinson, R. D., Howard, R. A., Koomen, M. J., and Michels, D. J. 1984. Associations between coronal mass ejections and metric type II bursts. *Astrophys. J.* 279:839–847.

Shirk, E. K. 1974. Observation of transiron solar flare nuclei in an Apollo 16 command-module window. *Astrophys. J.* 190:695–702.

Shoemaker, E. M. 1982. The collision of solid bodies. In *The New Solar System,* eds. J. K. Beatty, B. O'Leary and A. Chaikin (Cambridge: Cambridge Univ. Press), pp. 33–44.

Short, D. A., and Mengel, J. G. 1986. Tropical climate phase lags and Earth's precession cycle. *Nature* 322:48–50.

Shu, F. H., and Terebey, S. 1984. Formation of cool stars from cloud cores. In *Cool Stars, Stellar Systems and the Sun: Proc. Third Cambridge Workshop,* eds. S. L. Baliunas and L. Hartmann (New York: Springer-Verlag), pp. 78–89.

Shu, F. H., Adams, F. C., and Lizano, S. 1987. Star formation in molecular clouds: Observation and theory. *Ann. Rev. Astron. Astrophys.* 25:23–81.

Shu, F. H., Lizano, S., Ruden, S. P., and Najita, J. 1988. Mass loss from rapidly-rotating magnetic protostars. *Astrophys. J.* 328:L19-L23.

Siegenthaler, U., Heimann, M., and Oeschger, H. 1980. ^{14}C variations caused by changes in the global carbon cycle. *Radiocarbon* 22:177-191.

Siegfried, R. W., and Solomon, S. C. 1974. Mercury: Internal structure and thermal evolution. *Icarus* 23:192-205.

Sienkiewicz, R., Paczyński, R., and Ratcliff, S. J. 1988. Neutrino emission from solar models with a metal-depleted core. *Astrophys. J.* 326:392-394.

Sienkiewicz, R., Bahcall, J. N., and Paczynski, B. 1990. Mixing and the solar neutrino problem. *Astrophys. J.* 349:641-646.

Signer, P. 1964. Primordial rare gases in meteorites. In *The Origin and Evolution of Atmospheres and Oceans*, eds. P. J. Brancazio and A. G. W. Cameron (New York: Wiley), pp. 183-190.

Signer, P., Baur, H., Derksen, U., Etique, Ph., Funk, H., Horn, P., and Wieler, R. 1977. Helium, neon and argon records of lunar soil evolution. *Proc. Lunar Sci. Conf.* 8:3657-3683.

Signer, P., Baur, H., Etique, P., and Wieler, R. 1986. Nitrogen and noble gases in the 71501 bulk soil and ilmenites as records of the solar wind exposure: Which is correct? In *Workshop on Past and Present Solar Radiation*, LPI Tech. Rept. 86-02 (Houston: Lunar and Planetary Inst.), pp. 34-35.

Silberberg, R., and Tsao, C. H. 1973a. Partial cross-sections in high-energy nuclear reactions and astrophysical applications. I. Targets with Z to 28. *Astrophys. J. Suppl.* 25:315-333.

Silberberg, R., and Tsao, C. H. 1973b. Partial cross-sections in high-energy nuclear reactions and astrophysical applications. II. Targets heavier than nickel. *Astrophys. J. Suppl.* 25:335-367.

Silverman, S. M., and Blanchard, D. C. 1983. Wilson Bentley's auroral observations. *Planet. Space Sci.* 31:1131-1135.

Simnett, G. M. 1971. The release of energetic particles from the Sun. *Solar Phys.* 20:448-461.

Simnett, G. M. 1974. Relativistic electron events in interplanetary space. *Space Sci. Rev.* 16:257-323.

Simnett, G. M. 1983. Flares on the Sun: Selected results from SMM. In *Activity in Red Dwarf Stars: Proc. IAU Coll. 71*, eds. M. Rodono and P. B. Byrne (Dordrecht: D. Reidel), pp. 289-305.

Simnett, G. M. 1986a. Interplanetary phenomena and solar radio bursts. *Solar Phys.* 104:67-91.

Simnett, G. M. 1986b. A dominant role for protons at the onset of solar flares. *Solar Phys.* 106:165-183.

Simnett, G. M., and Haines, M. G. 1990. A new approach to hard X-ray production in solar flares. In *Proc. 21st Intl. Cosmic Ray Conf.*, vol. 5, ed. R. J. Protheroe (Adelaide: Univ. of Australia), pp. 44-47.

Simnett, G. M., and Strong, K. T. 1984. The impulsive phase of a solar limb flare. *Astrophys. J.* 284:839-847.

Simon, P. C. 1981. Solar irradiance between 120 and 400 nm and its variations. *Solar Phys.* 74:273-291.

Simon, P. C., and Brasseur, G. 1983. Photodissociation effect of solar UV radiation. *Planet. Space Sci.* 31:987-999.

Simon, P. C., Rottman, G. J., White, O. R., and Knapp, B. G. 1988. Short term variability between 120 and 300 nm from SME observations. In *Solar Radiation Output Variations*, ed. P. V. Foukal (Cambridge, Mass.: Cambridge Research and Instrumentation), pp. 125-134.

Simon, T., Herbig, G., and Boesgaard, A. M. 1985. The evolution of chromospheric activity and the spin-down of solar-type stars. *Astrophys. J.* 293:551-574.

Simpson, J. A. 1983. Elemental and isotopic composition of the galactic cosmic rays. *Ann. Rev. Nucl. Part. Sci.* 33:323-381.

Simpson, J. A., and Garcia-Muñoz, M. 1988. Cosmic-ray lifetime in the galaxy: Experimental results and models. *Space Sci. Rev.* 46:205-224.

Sinclair, D., Carter, A. L., Kessler, D., Earle, E. D., Jagam, P., Simpson, J. J., Allen, R. C., Chen, H. H., Doe, P. J., Hallman, E. D., Davidson, W. F., Storey, R. S., McDonald, A. B., Ewan, G. T., Mak, H.-B., and Robertson, B. C. 1986. Proposal to build a neutrino observatory in Sudbury, Canada. *Nuovo Cimento* 9C:308-317.

Siscoe, G. L. 1972. Structure and orientation of solar wind interaction fronts: Pioneer 6. *J. Geophys. Res.* 77:27–34.

Siscoe, G. L. 1980. Evidence in the auroral record for secular solar variations. *Rev. Geophys. Space Phys.* 18:647–658.

Sizzi, F. 1613. Sizzi to Orazio Morandi, 10 April 1613. In *Le Opere di Galileo Galilei*, ed. A. Favaro (Florence), vol. XI, pp. 491–492.

Skumanich, A. 1965. Population differences among bright G dwarfs and Ca II emission reversals. *Astron. J.* 70:692.

Skumanich, A. 1972. Time scales for Ca II emission decay, rotational braking, and lithium depletion. *Astrophys. J.* 171:565–567.

Skumanich, A., and Eddy, J. A. 1981. Aspects of long-term variability in Sun and stars. In *Solar Phenomena in Stars and Stellar Systems*, eds. R. M. Bonnet and A. K. Dupree (Dordrecht: D. Reidel), pp. 349–397.

Skumanich, A., and MacGregor, K. 1986. Evolution of activity signatures during the main sequence phase. *Adv. in Space Res.* 6(8):151–159.

Skumanich, A., Smythe, C., and Frazier, E. N. 1975. On the statistical description of inhomogeneities in the quiet solar atmosphere. I. Linear regression analysis and absolute calibration of multichannel observations of the Ca + emission network. *Astrophys. J.* 200:747–764.

Skumanich, A., Lean, J. L., White, O. R., and Livingston, W. C. 1984. The Sun as a star: Three-component analysis of chromospheric variability in the calcium K line. *Astrophys. J.* 282:776–783.

Smagorinsky, J. 1963. General circulation experiments with the primitive equations. I. The basic experiments. *Mon. Weather Rev.* 91:99–165.

Smith, B. A., and Terrile, R. J. 1984. A circumstellar disk around β Pictoris. *Science* 226:1421–1424.

Smith, E. J. 1985. Interplanetary shock phenomena beyond 1 AU. In *Collisionless Shocks in the Heliosphere: Reviews of Current Research*, eds. B. T. Tsurutani and R. G. Stone, vol. 35 (Washington, D.C.: American Geophysical Union), pp. 69–83.

Smith, E. J., and Thomas, B. T. 1986. Latitudinal extent of the heliospheric current sheet and modulation of galactic cosmic rays. *J. Geophys. Res.* 91:2933–2942.

Smith, E. J., and Wolfe, J. H. 1977. Pioneer 10, 11 observations of evolving solar wind streams and shocks beyond 1 AU. In *Study of Travelling Interplanetary Phenomena*, eds. M. A. Shea, D. F. Smart and S. T. Wu (Dordrecht: D. Reidel), pp. 227–257.

Smith, E. J., Tsurutani, B. T., and Rosenberg, R. L. 1978. Observations of the interplanetary sector structure to heliographic latitudes of 16°: Pioneer 11. *J. Geophys. Res.* 83:717–724.

Smith, E. J., Tsurutani, B. T., Slavin, J. A., Jones, D. E., Siscoe, G. L., and Mendis, D. A. 1986. International Cometary Explorer encounter with Giacobini-Zinner: Magnetic field observations. *Science* 232:382–385.

Smith, L. A. 1988. Intrinsic limits on dimension calculations. *Phys. Lett.* A133:283–288.

Smith, M. A. 1978. An anticorrelation between macroturbulence and age in G stars near the main sequence. *Astrophys. J.* 224:584–594.

Smith, M. A. 1979. Rotational studies of lower main-sequence stars. *Publ. Astron. Soc. Pacific* 91:737–745.

Smith, M. A. 1987. Post-zero-age main sequence rotation among late-type stars. In *Cool Stars, Stellar Systems and the Sun: Proc. Fifth Cambridge Workshop*, eds. J. L. Linsky and R. E. Stencel (New York: Springer-Verlag), pp. 192–204.

Smith, M. A., and Gray, D. F. 1976. Fourier analysis of spectral line profiles: A new test for an old art. *Publ. Astron. Soc. Pacific* 88:809–823.

Snell, R. L. 1987. Bipolar outflows and stellar jets. In *Star Forming Regions*, eds. M. Peimbert and J. Jugaku (Dordrecht: D. Reidel), pp. 213–237.

Snell, R. L. and Bally, J. 1986. Compact radio sources associated with molecular outflows. *Astrophys. J.* 303:683–701.

Snell, R. L., Loren, R. B., and Plambeck, R. L. 1980. Observations of CO in L1551-IRS5: Evidence for stellar wind driven shocks. *Astrophys. J.* 239:L17-L22.

Soderblom, D. R. 1982. Rotational studies of late-type stars. I. Rotational velocities of solar-type stars. *Astrophys. J.* 263:239–251.

Soderblom, D. R. 1983. Rotational studies of late-type stars. II. Ages of solar-type stars and the rotational history of the Sun. *Astrophys. J. Suppl.* 53:1–15.

Soderblom, D. R. 1985. A survey of chromospheric emission and rotation among solar-type stars in the solar neighborhood. In *Astron J.* 90:2103–2115.

Soderblom, D. R. 1988. The main sequence evolution of rotation and chromospheric activity in solar-type stars. In *Formation and Evolution of Low Mass Stars*, eds. A. K. Dupree and M. T. V. T. Lago (Dordrecht: Kluwer), pp. 389–398.

Soderblom, D. R., and Baliunas, S. L. 1988. The Sun among the stars: What stars indicate about solar variability. In *Secular, Solar and Geomagnetic Variations in the Last 10,000 Years*, eds. F. R. Stephenson and A. W. Wolfendale (Dordrecht: Kluwer), pp. 25–48.

Soderblom, D. R., and Clements, S. D. 1987. Chromospheric and transition region emission from young solar-type stars in clusters, kinematic groups, and the field. *Astron. J.* 93:920–937.

Soderblom, D. R., and Däppen, W. 1989. Modeling Epsilon Eridani and its oscillations. *Astrophys. J.* 342:945–950.

Sofia, S., and Chan, K. L. 1982. Quick matching technique to study the relationship between solar radius and luminosity variations. *Solar Phys.* 76:145–153.

Sofia, S., O'Keefe, J., Lesh, J., and Endal, A. 1979. Solar constant - Constraints on possible variations derived from solar diameter variations. *Science* 204:1306–1308.

Sofia, S., Oster, L., and Schatten, K. 1982. Solar irradiance modulation by active regions during 1980. *Solar Phys.* 80:87–98.

Solanki, S. K., and Mathys, G. 1988. A new technique for the measurement of stellar magnetic fields: First results. In *Activity in Cool Star Envelopes*, eds. O. Havnes, B. R. Pettersen, J. H. M. M. Schmitt and J. E. Solheim (Dordrecht: Kluwer), pp. 39–43.

Solomon, S. C. 1976. Some aspects of core formation in Mercury. *Icarus* 28:509–521.

Solomon, S. C., Reid, G. C., Rusch, D. W., and Thomas, R. J. 1983. SPE - Mesospheric ozone depletion during the solar proton event of July 13, 1982 - Part 2: Comparison between theory and measurements. *Geophys. Res. Lett.* 10:257–260.

Sonett, C. P. 1969. Fractionation of iron: A cosmogonic sleuthing tool. II. Heating by electrical induction. *Comments on Astrophys. Space Phys.* 1:41–48.

Sonett, C. P. 1978. Evidence for a primordial magnetic field during the meteorite parent body era. *Geophys. Res. Lett.* 5:151–154.

Sonett, C. P. 1979. On the origin of chondrules. *Geophys. Res. Lett.* 6:677–680.

Sonett, C. P. 1982. Sunspot time series: Spectrum from square law modulation of the Hale cycle. *Geophys. Res. Let.* 9:1313–1316.

Sonett, C. P. 1983. Sunspot index infers a small relict magnetic field in the Sun's core. *Nature* 306:670–673.

Sonett, C. P. 1984. Very long solar periods and the radiocarbon record. *Rev. Geophys. Space Phys.* 22(3):239–254.

Sonett, C. P. 1985. Suess "wiggles" - A comparison between radiocarbon records. *Meteoritics* 20:383–394.

Sonett, C. P. 1988. Time series, radiocarbon and varves. In *Solar-Terrestrial Relationships and the Earth Environment in the Last Millennia*, ed. G. Cini Castagnoli (Amsterdam: North-Holland Press), pp. 274–311.

Sonett, C. P., and Colburn, D. S. 1968. The principle of solar wind induced planetary dynamos. *Phys. Earth Planet. Int.* 1:326–346.

Sonett, C. P., and Colburn, D. S. 1974. Mie scattering of the interplanetary magnetic field by the whole Moon. In *Planets, Stars and Nebulae*, ed. T. Gehrels (Tucson: Univ. of Arizona Press), pp. 419–433.

Sonett, C. P., and Finney, S. A. 1990. The spectrum of radiocarbon. *Phil. Trans. Roy. Soc. London* A330:413–426.

Sonett, C. P., and Mihalov, J. D. 1972. Lunar fossil magnetism and perturbations of the solar wind. *J. Geophys. Res.* 77:588–603.

Sonett, C. P., and Reynolds, R. T. 1979. Primordial heating of asteroidal parent bodies. In *Asteroids*, ed. T. Gehrels (Tucson: Univ. of Arizona Press), pp. 822–848.

Sonett, C. P., and Suess, H. E. 1984. Correlation of bristlecone pine ring widths with atmospheric ^{14}C variations: A climate-Sun relation. *Nature* 307:141–143.

Sonett, C. P., and Williams, G. E. 1985. Solar periodicities expressed in varves from glacial Skilak Lake, Southern Alaska. *J. Geophys. Res.* 90:12019–12026.

Sonett, C. P., and Williams, G. E. 1987. Frequency modulation and stochastic variability of the Elatina varve record: A proxy for solar cyclicity? *Solar Phys.* 100:397–410.

Sonett, C. P., Colburn, D. S., and Schwartz, K. 1968. Electrical heating of meteorite parent bodies and planets by dynamo induction from a pre-main sequence T Tauri "solar wind." *Nature* 219:924–926.

Sonett, C. P., Colburn, D. S., Schwartz, K., and Keil, K. 1970. The melting of asteroidal-sized bodies by unipolar dynamo induction from a primordial T Tauri Sun. *Astrophys. Space Sci.* 7:446–488.

Sonett, C. P., Dyal, P., Parkin, C. W., Colburn, D. S., Mihalov, J. D., and Smith, B. F. 1971. Whole body response of the Moon to electromagnetic induction by the solar wind. *Science* 172:256–258.

Sonett, C. P., Morfill, G. E., and Jokipii, J. R. 1987. Interstellar shock waves and ^{10}Be from ice cores. *Nature* 330:458–460.

Sonett, C. P., Finney, S. A., and Williams, G. E. 1988. The lunar orbit in the late Precambrian and the Elatina sandstone laminae. *Nature* 335:806–808.

Southon, J. R., Ku, T. L., Nelson, D. E., Reyss, J. L., Duplessy, J. C., and Vogel, J. S. 1987. ^{10}Be in a deep-sea core: Implications regarding ^{10}Be production changes over the past 420 ka. *Earth Planet. Sci. Lett.* 85:356–364.

Spiegel, E. A., and Weiss, N. O. 1980. Magnetic activity and variations in solar luminosity. *Nature* 287:616–617.

Spiro, R. W., Reiff, Ph. H., and Maker, L. J. 1982. Precipitating electron energy flux and auroral zone conductance: An empirical model. *J. Geophys. Res.* 87:8215–8227.

Spitaler, R. 1921. Das Klima des Eiszeitalters, Prag (lithographed).

Spitzer, L., Jr. 1968. *Diffuse Matter in Space* (New York: Wiley-Interscience).

Spörer, F. W. G. 1889. Über die Periodictat der Sonnenflecken seit Jahre 1618. *Roy. Caroline Acad.* (quoted by E. W. Maunder).

Spruit, H. C. 1976. Pressure equilibrium and energy balance of small photospheric flux tubes. *Solar Phys.* 50:269–295.

Spruit, H. C. 1977. Heat flow near obstacles in the solar convection zone. *Solar Phys.* 55:3–14.

Spruit, H. C. 1980. Motion of magnetic flux tubes in the solar convection zone and chromosphere. *Astron. Astrophys.* 98:155.

Spruit, H. C. 1982a. Effect of spots on a star's radius and luminosity. *Astron. Astrophys.* 108:348–355.

Spruit, H. C. 1982b. The flow of heat near a starspot. *Astron. Astrophys.* 108:356–360.

Spruit, H. C. 1984. Interaction of flux tubes with convection. In *Small Scale Dynamical Phenomena in Quiet Stellar Atmospheres,* ed. S. L. Keil (Sunspot, N.M.: Sacramento Peak Obs.), pp. 249–259.

Spruit, H. C. 1988. Influence of magnetic activity on the solar luminosity and radius. In *Solar Radiative Output Variations,* ed. P. V. Foukal (Cambridge, Mass.: Cambridge Research and Instrumentation), pp. 254–288.

Spruit, H. C., and Weiss, A. 1986. Colors and luminosities of stars with spots. *Astron. Astrophys.* 166:167–176.

Spruit, H. C., and Zwaan, C. 1981. The size dependence of contrasts and numbers of small magnetic flux tubes in an active region. *Solar Phys.* 70:207–228.

Srnka, L. J. 1973. Steady Induction in the Solar Wind. Ph.D. Thesis, Univ. of Newcastle.

Stacey, F. D., and Banerjee, S. K. 1974. *The Physical Principles of Rock Magnetism* (New York: Elsevier).

Stahler, S. W. 1983. The birthline for low mass stars. *Astrophys. J.* 274:822–829.

Stahler, S. W., Shu, F. H., and Taam, R. E. 1980. The evolution of protostars. I. Global formation and results. *Astrophys. J.* 241:637–654.

Stauffer, J. R. 1984. Optical and infrared photometry of late-type stars in the Pleiades. *Astrophys. J.* 280:189–201.

Stauffer, J. R., and Hartmann, L. W. 1986. The rotational velocities of low mass stars. *Publ. Astron. Soc. Pacific* 98:1233–1251.

Stauffer, J. R., and Hartmann, L. W. 1987. The distribution of rotational velocities of low mass stars in the Pleiades. *Astrophys. J.* 318:337–355.

Stauffer, J. R., Hartmann, D., Soderblom, D., and Burnham, J. N. 1984. Rotational velocities of low mass stars in the Pleiades. *Astrophys. J.* 280:202–212.

Stauffer, J. R., Hartmann, L. W., Burnham, J. N., and Jones, B. F. 1985. Evolution of low-mass stars in the Alpha Persei Cluster. *Astrophys. J.* 289:247–261.

Stauffer, J. R., Hartmann, L. W., and Latham, D. W. 1987. Rotational velocities of low-mass stars in the Hyades. *Astrophys. J.* 320:L51-L55.

Stauffer, J. R., Hartmann, L. W., and Jones, B. F. 1989. Rotational velocities of newly discovered low mass members of the Alpha Persei cluster. *Astrophys. J.* 346:160–167.

Steenbeck, M., and Krause, F. 1969. Zur Dynamotheorie stellarer und planetarer Magnetfelder. I. Berechnung sonnenähnlicher wechselfeldgeneratoren. *Astron. Nachr.* 291:49–84.

Steinitz, R., and Eyni, M. 1980. Global properties of the solar wind. I. The invariance of the momentum flux density. *Astrophys. J.* 241:417–424.

Steljes, J. F. 1973. *Cosmic Ray NM-64 Neutron Monitor Data*, Repts. XVI-XXIII, May 1970 to Dec. 1972 (Chalk River, Ont.: Atomic Energy of Canada).

Stepién, K. 1987. On the nonexistence of the universal magnetic constant for cool stars. *Publ. Astron. Soc. Pacific.* 99:1292–1297.

Stepién, K. 1988a. Coronal activity in late-type main sequence stars. *Astrophys. J.* 335:892–913.

Stepién, K. 1988b. Spin-down of cool stars in their main sequence life. *Astrophys. J.* 335:907–913.

Stepinski, T. F., and Levy, E. H. 1988. Generation of dynamo magnetic fields in protoplanetary and other astrophysical accretion disks. *Astrophys. J.* 331:916–934.

Stern, D. P. 1976. Representation of magnetic fields in space. *Rev. Geophys. Space Phys.* 14:199–214.

Stern, R. A. 1983. Einstein observations of cool stars. *Adv. Space Res.* 2(9):39–50.

Stern, R. A., Underwood, J. H., and Antiochos, S. K. 1983. A giant X-ray flare in the Hyades. *Astrophys. J.* 264:L55-L59.

Stetson, P. B., and Harris, W. E. 1988. CCD photometry of the globular cluster M92. *Astron. J.* 96:909–975.

Stevenson, D. J. 1983a. The nature of the Earth prior to the oldest known rock record: The Hadean Earth. In *Earth's Earliest Biosphere: Its Origin and Evolution*, ed. J. W. Schopf (Princeton, N.J.: Princeton Univ. Press), pp. 32–40.

Stevenson, D. J. 1983b. Planetary magnetic fields. *Rept. Prog. Phys.* 46:555–620.

Stine, P. C., Feigelson, E. D., André, P., and Montmerle, T. 1988. Multi-epoch radio continuum surveys of the ρ Ophiuchi Dark Cloud. *Astron. J.* 96:1394–1406.

Stix, M. 1972. Non-linear dynamo waves. *Astron. Astrophys.* 20:9–12.

Stix, M. 1979. In *Proc. Freiburg Coll.: Small Scale Motions on the Sun*, Mitt. Kiepenheuer-Inst. No. 179, p. 87.

Stix, M. 1981. Theory of the solar cycle. *Solar Phys.* 74:79–101.

Stöffler, D., Gault, D. W., Wedekind, J., and Polkowski, G. 1975. Experimental hypervelocity impact into quartz sand: Distribution and shock metamorphism of ejecta. *J. Geophys. Res.* 80:4062–4077.

Stratton, J. A. 1941. *Electromagnetic Theory* (New York: McGraw-Hill).

Stringfellow, G. S., Swenson, F. J., and Faulkner, J. 1987. Is there a classical Hyades lithium problem? *Bull. Amer. Astron. Soc.* 19:1020 (abstract).

Strom, K. M., Strom, S. E., Edwards, S., Cabrit, S., and Skrutskie, M. F. 1989. Circumstellar material associated with solar-type pre-main sequence stars: A possible constraint on the time-scale for planet building. *Astron. J.* 97:1451–1470.

Strom, S. E., Strom, K. M., Grasdalen, G. L., Capps, R. W., and Thompson, D. A. 1985a. High spatial resolution studies of young stellar objects. II. A thick disk surrounding Lynds 1551, IRS 5. *Astrophys. J.* 90:2575–2580.

Strom, S. E., Strom, K. M., Grasdalen, G. L., Sellgren, K., Wolff, S., Morgan, J., Stocke, J., and Mundt, R. 1985b. An optical and infrared study of the region surrounding Herbig-Haro objects 1 and 2. *Astron. J.* 90:2281–2290.

Strom, S. E., Strom, K. M., and Edwards, S. 1988. Energetic winds and circumstellar disks associated with low mass young stellar objects. In *Galactic and Extragalactic Star Formation*, eds. R. Pudritz and M. Fich (Dordrecht: D. Reidel), pp. 53–88.

Strom, S. E., Strom, K. M., Edwards, S., Cabrit, S., and Skrutskie, M. 1989. Circumstellar material associated with solar-type pre-main sequence stars: A possible constraint on the timescale for planet-building. *Astron. J.* 97:1451–1470.

Stuiver, M. 1961. Variations in radiocarbon concentration and sunspot activity. *J. Geophys. Res.* 66:273–276.

Stuiver, M. 1965. Carbon-14 content of 18th and 19th century wood, variations correlated with sunspot activity. *Science* 149:533–535.

Stuiver, M. 1980. Solar variability and climate change during the current millennium. *Nature* 286:868–871.

Stuiver, M., and Becker, B. 1986. High-precision decadal calibration of the radiocarbon time scale 1950–2500 BC. In *Proc. 12th Intl. Radiocarbon Conf.*, eds. M. Struiver and R. Kra. *Radiocarbon* 28(2B):863–910.

Stuiver, M., and Braziunas, T. F. 1988. The solar component of the atmospheric ^{14}C record. In *Secular, Solar and Geomagnetic Variations in the Last 10,000 Years*, eds. F. R. Stephenson and A. W. Wolfendale (Dordrecht: Kluwer), pp. 245–266.

Stuiver, M., and Braziunas, T. F. 1989. Atmospheric ^{14}C and century-scale solar oscillations. *Nature* 338:405–408.

Stuiver, M., and Kra, R., eds. 1986. *Proc. 12th Intl. Radiocarbon Conf. Radiocarbon* 28(2B).

Stuiver, M., and Pearson, G. W. 1986. High precision calibration of the radiocarbon time scale. In *Proc. 12th Intl. Radiocarbon Conf.*, eds. M. Stuiver and R. Kra. *Radiocarbon* 28(2B):805–838.

Stuiver, M., and Quay, P. D. 1980. Changes in atmospheric carbon-14 attributed to a variable Sun. *Science* 207:11–19.

Stuiver, M., and Quay, P. D. 1981. Atmospheric ^{14}C changes resulting from fossil fuel CO_2 release and cosmic ray variability. *Earth Planet. Sci. Lett.* 53:349–362.

Stuiver, M., Kromer, B., Becker, B., and Ferguson, C. W. 1986a. Radiocarbon calibration back to 13,300 year BP and the ^{14}C age matching of the German oak and U.S. bristelcone pine chronologies. In *Proc. 12th Intl. Radiocarbon Conf.*, eds. M. Stuiver and R. Kra. *Radiocarbon* 28(2B):969–979.

Stuiver, M., Pearson, G. W., and Braziunas, T. 1986b. Radiocarbon age calibration of marine samples back to 9000 CAL YR BP. In *Proc. 12th Intl. Radiocarbon Conf.*, eds. M. Stuiver and R. Kra. *Radiocarbon* 28(2B):980–1021.

Suess, H. E. 1949. Die Haufigkeit der edelgase auf der Erde in Kosmos. *J. Geol.* 57:600–607.

Suess, H. E. 1955. Radiocarbon concentration in modern wood. *Science* 122:415–417.

Suess, H. E. 1965. Secular variations of the cosmic-ray produced carbon 14 in the atmosphere and their interpretations. *J. Geophys. Res.* 70:5937–5952.

Suess, H. E. 1968. Climate changes, solar activity, and cosmic-ray production rate of natural radiocarbon. *Meteorol. Mono.* 8:146–150.

Suess, H. E. 1980a. The radiocarbon record in tree rings of the last 8000 years. In *Proc. 10th Intl. Radiocarbon Conf.*, eds. M. Stuiver and R. Kra. *Radiocarbon* 22(2):200–209.

Suess, H. E. 1980b. Radiocarbon geophysics. *Endeavour* 4:113–117.

Suess, H. E. 1986. Secular variations of cosmogenic ^{14}C on Earth: Their discovery and interpretation. In *Proc. 12th Intl. Radiocarbon Conf.*, eds. M. Stuiver and R. Kra. *Radiocarbon* 28(2B):259–265.

Suess, H. E., Wänke, H., and Wlotzka, F. 1964. On the origin of gas-rich meteorites. *Geochim. Cosmochim. Acta* 28:595–607.

Suess, S. T. 1979. The solar wind during the Maunder minimum. *Planet. Space Sci.* 27:1001–1013.

Sugiura, N., and Strangway, D. W. 1981. The magnetic properties of the Albee meteorite: Evidence for a strong magnetic field in the early solar system. *Proc. Lunar Planet. Sci. Conf.* 12:1243–1256.

Sugiura, N., and Strangway, D. W. 1982. Magnetic properties of low-petrologic grade noncarbonaceous chondrites. *Mem. Natl. Inst. Polar Res. Special Issue* 25:260–280.

Sugiura, N., and Strangway, D. W. 1985. NRM directions around a centimeter-sized dark inclusion in Allende. *Proc. Lunar Planet. Sci. Conf. 15, J. Geophys. Res. Suppl.* 90:C729-C738.

Sugiura, N., and Strangway, D. W. 1988. Magnetic studies of meteorites: In *Meteorites and the Early Solar System*, eds. J. F. Kerridge and M. S. Matthews (Tucson: Univ. of Arizona Press), pp. 595–618.

Sun, W.-H., Giampapa, M. S., and Worden, S. P. 1987. Magnetic field measurements on the Sun and implications for stellar magnetic field observations. *Astrophys. J.* 312:930–942.

Svalgaard, L., and Wilcox, J. M. 1974. The spiral interplanetary magnetic field: A polarity and sunspot cycle variation. *Science* 186:51–53.

Svenonius, B., and Olausson, E. 1979. Cycles in solar activity, especially of long periods and certain terrestrial relationships. *Palaeogeo. Palaeoclim. Palaeoecol.* 26:89–97.

Sweet, P. A. 1955. The structure of sunspots. *Vistas in Astron.* 1:675–685.

Swindle, T. D., Caffee, M. W., Hohenberg, C. M., Hudson, G. B., Laul, J. C., Simon, S. B., and Papike, J. J. 1985. Noble gas component organization in Apollo 14 breccia 14318: [129]I and [244]Pu regolith chronology. *Proc. Lunar Planet. Sci. Conf.* 15, *J. Geophys. Res. Suppl.* 90:C517-C539.

Swindle, T. D., Caffee, M. W., Hohenberg, C. M., and Taylor, S. R. 1986. I-Pu-Xe dating and the relative ages of the Earth and Moon. In *Origin of the Moon,* eds. W. K. Hartmann, R. J. Phillips and G. J. Taylor (Houston: Lunar and Planetary Inst.), pp. 331–358.

Tagliaferri, G., Giommi, P., Angelini, L., Osborne, J. P., and Pallavicini, R. 1988. An X-ray flare from A B9 + Post-T Tauri star system in the field of the Seyfert Galaxy III Zw 2. *Astrophys. J.* 331:L113–116.

Takahashi, K., and Boyd, R. W. 1988. The possible consequences of strange quark matter on hydrogen burning in the Sun. *Astrophys. J.* 327:1009–1019.

Takens, F. 1981. Detecting strange attractors in turbulence. In *Dynamical Systems and Turbulence,* eds. D. A. Rand and L. S. Young (Berlin: Springer-Verlag), pp. 366–381.

Tamers, M. A., and Delibrias, G. 1961. Sections efficaces de l'oxygène 16 pour la production de carbone 14 des protons de hautes énergies. *Compt. Rend.* 253:1202–1203.

Tanaka, S., Sakamoto, K., and Komura, K. 1972. Aluminum 26 and manganese 53 produced by solar-flare particles in lunar rock and cosmic dust. *J. Geophys. Res.* 77:4281–4288.

Tandon, J. N., and Das, M. K. 1982. The effect of a magnetic field on solar luminosity. *Astrophys. J.* 260:338–341.

Tassoul, M., and Tassoul, J. 1984. Meridional circulation in rotating stars. VII. The effects of chemical inhomogeneities. *Astrophys. J.* 270:384–389.

Tatsumoto, M., Unruh, M., and Desborough, G. A. 1976. U-Th-Pb and Rb-Sr systematics of Allende and U-Th-Pb systematics of Orgueil. *Geochim. Cosmochim. Acta* 40:617–634.

Tayler, R. J. 1987. Magnetic activity in pre-main-sequence stars. *Mon. Not. Roy. Astron. Soc.* 227:553–561.

Taylor, G. J. 1988a. Metal segregation in asteroids. *Lunar Planet. Sci.* XX:1109–1110 (abstract).

Taylor, G. J. 1988b. The role of impacts in asteroidal differentiation. *Meteoritics* 23:304–305 (abstract).

Taylor, G. J., and Heymann, D. 1969. Shock, reheating, and gas retention ages of chondrites. *Earth Planet. Sci. Lett.* 7:151–161.

Taylor, G. J., Maggiore, P., Scott, E. R. D., Rubin, A. E., and Keil, K. 1987. Original structures and fragmentation and reassembly histories of asteroids: Evidence from meteorites. *Icarus* 69:1–13.

Témam, R. 1988. *Infinite-Dimensional Dynamical Systems in Mechanics and Physics* (New York: Springer-Verlag).

Tera, F., Carlson, R. W., and Boctor, N. Z. 1989. Contrasting Pb-Pb ages of the cumulate and non-cumulate eucrites. *Lunar Planet. Sci.* XX:1111–1112 (abstract).

Terebey, S., Shu, F. H., and Cassen, P. 1984. The collapse of the cores of slowly rotating isothermal clouds. *Astrophys. J.* 286:529–551.

Terebey, S., Vogel, S. N., and Myers, P. C. 1989. High-resolution CO observations of young low-mass stars. *Astrophys. J.* 340:472–478.

Thellier, E., and Thellier, O. 1959. Sur l'intensité du champ magnétique terrestre dans le passé historique et géologique. *Ann. Geophys.* 15:285–376.

Thiemens, M. H., and Clayton, R. N. 1980. Ancient solar wind in lunar microbreccias. *Earth Planet. Sci. Lett.* 47:34–42.

Thiemens, M. H., and Heidenreich, J. E., III. 1983. The mass-independent fractionation of oxygen: A novel isotope effect and its possible cosmochemical implications. *Science* 219:1073–1075.

Tholen, D. J. 1984. Asteroid Taxonomy from Cluster Analysis of Photometry. Ph.D. Thesis, Univ. of Arizona.

Thomas, B. T., a.id Smith, E. J. 1980. The Parker spiral configuration of the interplanetary magnetic field between 1 and 8.5 AU. *J. Geophys. Res.* 85:6861–6867.

Thomas B. T., and Smith, E. J. 1981. The structure and dynamics of the heliospheric current sheet. *J. Geophys. Res.* 86:11105–11110.

Thomas, J. H. 1979. Variations in the Sun's radius and temperature due to magnetic buoyancy. *Nature* 280:662–663.

Thomson, D. J. 1982. Spectrum estimation and harmonic analysis. *Proc. IEEE* 70:1055–1096.

Thomson, D. J. 1990. Time series analysis of Holocene climate data. *Phil. Trans. Roy. Soc. London* 330A:601–616.

Thorne, R. M. 1977. Energetic radiation belt electron precipitation: A natural depletion mechanism for stratospheric ozone. *Science* 195:287–289.

Thorne, R. M. 1980. The importance of energetic particle precipitation on the chemical composition of the middle atmosphere. *Pure Appl. Geophys.* 118:128–147.

Thum, C., Bertout, C., and Downes, D. 1981. A search for H_2O emission from Orion population stars. *Astron. Astrophys.* 94:80–84.

Tinsley, B. A. 1988. The solar cycle and the QBO influences on the latitude of storm tracks in the North Atlantic. *Geophys. Res. Lett.* 15:409–412.

Toner, C. G., and Gray, D. F. 1988. The starpatch on the G8 dwarf ξ Bootis A. *Astrophys. J.* 334:1008–1020.

Toon, O. B., Pollack, J. B., Ackerman, T. P., Turco, R. P., McKay, C. P. and Liu, M. S. 1982. *Evolution of an Impact-Generated Dust Cloud and Its Effects on the Atmosphere.* Geol. Soc. America SP-190, pp. 187–200.

Toptygin, I. N. 1985. *Cosmic Rays in Interplanetary Magnetic Fields* (Dordrecht: D. Reidel).

Torbett, M. 1984. Hydrodynamic ejection of bipolar flows from objects undergoing disk accretion: T Tauri stars, massive pre-main-sequence objects, and cataclysmic variables. *Astrophys. J.* 278:318–325.

Torr, M. R., and Torr, D. G. 1985. Ionization frequencies for solar cycle 21: Revised. *J. Geophys. Res.* 90:6675–6678.

Torr, M. R., Torr, D. G., and Hinteregger, H. E. 1980. Solar flux variability in the Schumann-Runge continuum as a function of solar cycle 21. *J. Geophys. Res.* 85:6063–6068.

Totsuka, Y. 1988. Solar neutrinos. ICR-181–88–27, Tanashi, Tokyo.

Tousey, R. 1963. The extreme ultraviolet spectrum of the Sun. *Space Sci. Rev.* 2:3–69.

Trendall, A. F. 1972. Revolution in Earth history. *J. Geol. Soc. Australia* 19(3):287–311.

Trendall, A. F. 1973. Varve cycles in the Weeli Wolli Formation of the Precambrian Hamersley Group, western Australia. *Econ. Geology* 68:1089–1097.

Trendall, A. F., and Blockley, G. B. 1970. The iron formations of the Precambrian Hammersley Group, western Australia. *Geol. Surv. Western Australia Bull.* No. 119.

Tricot, Ch., and Berger, A. 1987. Modelling the equilibrium and transient responses of temperature to past and future trace gases concentrations. *Climate Dyn.* 2:39–61.

Trieloff, M., Jessberger, E. K., and Oehm, J. 1989. Ar-Ar ages of LL chondrites. *Meteoritics* 24:332 (abstract).

Tscharnuter, W. M. 1987. A collapse model of the turbulent presolar nebula. *Astron. Astrophys.* 188:55–73.

Tsurutani, B. T., and Meng, C.-I. 1972. Interplanetary magnetic field variations and substorm activity. *J. Geophys. Res.* 77:2964–2970.

Tuniz, C., Smith, C. M., Moniot, R. K., Kruse, T. H., Savin, W., Pal, D. K., Herzog, G. F., and Reedy, R. C. 1984. Beryllium-10 contents of core samples from the St. Severin meteorite. *Geochim. Cosmochim. Acta* 48:1867–1872.

Turck-Chièze, S., Cahen, S., Casse, M., and Doom, C. 1988. Revisiting the solar model. *Astrophys. J.* 335:415–424.

Turner, G., Enright, M. C., and Cadogan, P. H. 1978. The early history of chondrite parent bodies inferred from ^{40}Ar-^{39}Ar ages. *Proc. Lunar Planet. Sci. Conf.* 9:989–1025.

Uchida, Y., and Shibata, K. 1985. Magnetodynamical acceleration of CO and optical bipolar flows from the region of star formation. *Publ. Astron. Soc. Japan* 37:515–535.

Ulmschneider, P. 1980. Theories of heating solar and stellar chromospheres. In *Solar Phenomena* in Stars and Stellar Systems, eds. R. M. Bonnet and A. K. Dupree (Dordrecht: D. Reidel), pp. 239–263.

Ulmschneider, P., and Stein, R. F. 1982. Heating of stellar chromospheres when magnetic fields are present. Astron. Astrophys. 106:9–13.

Ulrich, R. K. 1970. The five minute oscillations on the solar surface. Astrophys. J. 162:993–1002.

Ulrich, R. K. 1971a. Evidence for ³He in young open clusters. Astrophys. J. 165:L95-L99.

Ulrich, R. K. 1971b. Evolution of stars containing ³He. Astrophys. J. 168:57–70.

Ulrich, R. K. 1975. Solar neutrinos and variations in the solar luminosity. Science 190:619–624.

Ulrich, R. K. 1982. The influence of partial ionization and scattering states on the solar interior structure. Astrophys. J. 258:404–413.

Ulrich, R. K. 1986. Interpretation of global solar oscillation frequencies - WIMPS vs initial composition inhomogeneities. Bull. Amer. Astron. Soc. 53:989 (abstract).

Ulrich, R. K., and Rhodes, E. J. 1977. The sensitivity of nonradial p mode eigenfrequencies to the solar envelope. Astrophys. J. 218:521–529.

Umebayashi, T., and Nakano, T. 1984. Origin and dissipation of magnetic fields in protostars and in the primitive solar nebula. In Proc. 16th ISAS Lunar Planet. Symp., pp. 118–121.

Umebayashi, T., and Nakano, T. 1988. Ionization state and magnetic fields in the solar nebula. Prog. Theor. Phys. Suppl. 96:151–160.

Underhill, A. B., and Fahey, R. P. 1984. Do bipolar magnetic regions exist on the surfaces of early-type stars? Astrophys. J. 280:712–728.

Ungerechts, H., Walmsley, C. M., and Winneswisser, G. 1982. Ammonia observations of cold cloud cores. Astron. Astrophys. 111:339–345.

UNSCEAR. 1972. Ionizing Radiation Levels and Effects, vol. 1 (New York: United Nations).

Urata, K. 1986. Time series analysis of the Boussinesq convection. In Hydrodynamic and Magnetohydrodynamic Problems in the Sun and Stars, ed. Y. Osaki (Tokyo: Univ. of Tokyo), pp. 109–115.

Urey, H. C. 1952. The Planets: Their Origin and Development (New Haven, Conn.: Yale Univ. Press).

Urey, H. C. 1955. Proc. Natl. Acad. Sci. U.S. 41:127.

U.S. Standard Atmosphere Supplements. 1976. (Washington, D.C.: U.S. Govt. Printing Office).

Vaiana, G. S. 1981. Low luminosity galactic X-ray sources. Space Sci. Rev. 30:151–179.

Vaiana, G. S., and Rosner, R. 1978. Recent advances in coronal physics. Ann. Rev. Astron. Astrophys. 16:393–428.

Van Allen, J. A. 1980. Galactic cosmic ray intensity to a heliocentric distance of 18 AU. Astrophys. J. 238:763–767.

van de Kamp, P. 1986. Dark companions of stars. Space Sci. Rev. 43:211–327.

VandenBerg, D. A. 1983. Star cluster and stellar evolution. I. Improved synthetic color-magnitude diagrams for the oldest clusters. Astrophys. J. Suppl. 51:29–66.

VandenBerg, D. A. 1985. Evolution of 0.7 - 3.0 $M_\odot$ stars having - 1.0 < [Fe/H] < 0.0. Astrophys. J. Suppl. 58:711–769.

Van Hollebecke, M. A. I., Ma Sung, L. S., and McDonald, F. B. 1975. The time variation of solar proton energy spectra and size distribution with heliolongitude Solar Phys. 41:189–223.

van Leeuwen, F. 1983. The Pleiades. An Astrometric and Photometric Study of an Open Cluster. Ph.D. Thesis, Univ. of Leiden.

van Leeuwen, F., and Alphenaar, P. 1982. Variable K type stars in the Pleiades. ESO Messenger 28:15–19.

van Loon, H., and Labitzke, K. 1988. Association between the 11-year solar cycle, the QBO, and the atmosphere. Part II. Surface and 700 mb on the northern hemisphere in winter. J. Climate 1:905–920.

van Loon, H., and Labitzke, K. 1990. Association between the 11-year solar cycle and the atmosphere. Part IV. The stratosphere, not grouped by the phase of the QBO. J. Climate 3:827–837.

Van Nes, P., Reinhard, R., Sanderson, T. R., Wentzel, K.-P., and Zwickl, R. D. 1984. The energy spectrum of 35–1600 keV protons associated with interplanetary shocks. J. Geophys. Res. 89:2122–2132.

Vasiliev, S. S., and Kocharov, G. E. 1977. The function of distribution in the interior of the Sun and neutrino deficit. In Proc. 15th Intl. Conf. on Cosmic Rays 5:211–214.

Vasyliunas, V. M. 1975. Theoretical models of magnetic field line merging. *Rev. Geophys. Space Phys.* 13:303–336.

Vauclair, S., and Meyer, J. P. 1985. Diffusion in the chromosphere, and the composition of the solar corona and energetic particles. In *Proc. 19th Intl. Cosmic Ray Conf.*, vol. 4, pp. 233–236.

Vaughan, A. H. 1980. Comparison of activity cycles in old and young main-sequence stars. *Publ. Astron. Soc. Pacific* 92:392–396.

Vaughan, A. H., and Preston, G. W. 1980. A survey of chromospheric Ca II H and K emission in field stars of the solar neighborhood. *Publ. Astron. Soc. Pacific* 92:385–391.

Vaughan, A. H., Preston, G. W., and Wilson, O. C. 1978. Flux measurements of Ca II H and K emission. *Publ. Astron. Soc. Pacific* 90:267–274.

Vaughan, A. H., Baliunas, S. L., Middelkoop, F., Hartmann, L. W., Mihalas, D., Noyes, R. W., and Preston, G. W. 1981. Stellar rotation in lower main sequence stars measured from time variations in H and K emission-line fluxes. I. Initial results. *Astrophys. J.* 250:276–283.

Vautard, R., and Ghil, M. 1989. Singular spectrum analysis in nonlinear dynamics, with applications to paleoclimatic time series. *Physica D* 35:395–424.

Veck, N. J., and Parkinson, J. H. 1981. Solar abundances from X-ray flare observations. *Mon. Not. Roy. Astron. Soc.* 197:41–45.

Venkatesan, D., Decker, R. B., and Krimigis, S. M. 1984. Radial gradient of cosmic ray intensity from a comparative study of data from Voyager 1 and 2 and IMP8. *J. Geophys. Res.* 89:3735–3746.

Venkatesan, D., Decker, R. B., and Krimigis, S. M. 1987. Cosmic-ray intensity gradients during 1984–86. In *Proc. 20th Intl. Cosmic Ray Conf.*, vol. 3, pp. 385–388.

Vernazza, J. E., Avrett, E. H., and Loeser, R. 1981. Structure of the solar chromosphere. III. Models of the EUV brightness components of the quiet Sun. *Astrophys. J. Suppl.* 45:635–725

Vernekar, A. D. 1972. Long-period global variations of incoming solar radiation. *Meteorol. Mon.* 12(34).

Veselov, A. I., Vysotskii, M. I., and Yurov, V. P. 1987. Half-year variations in the solar neutrino flux given by the Davis data of 1979–1982. *Soviet J. Nucl. Phys.* 45:865–871.

Vilas, F., and Gaffey, M. J. 1989. Phyllosilicate absorption features in main-belt and outer-belt asteroid reflectance spectra. *Science* 246:790–792.

Vilhu, O. 1984. The nature of magnetic activity in lower main sequence stars. *Astron. Astrophys.* 133:117–136.

Vilhu, O., and Walter, F. M. 1987. Chromospheric-coronal activity at saturated levels. *Astrophys. J.* 321:958–966.

Vogel, S. N., and Kuhi, L. V. 1981. Rotational velocities of pre-main-sequence stars. *Astrophys. J.* 245:960–976.

Vogt, S. S. 1981. A method for unambiguous determination of starspot temperatures and areas: Application to II Pegasi, BY Dracons, and HD 209813. *Astrophys. J.* 250:327–340.

Vogt, S. S., and Penrod, G. D. 1983. Doppler imaging of spotted stars: Application to the RS Canum Venaticorum Star HR 1099. *Publ. Astron. Soc. Pacific* 95:565–576.

Völk, H. J. 1975. Cosmic ray propagation in interplanetary space. *Rev. Geophys. Space Phys.* 13:547–566.

Volk, T. 1987. Feedbacks between weathering and atmospheric CO_2 over the last 100 million years. *Amer. J. Sci.* 287:763–779.

von Steiger, R., and Geiss, J. 1989. Supply of fractionated gases to the corona. *Astron. Astrophys.* 225:222–238.

von Weizsäcker, C. F. 1937. Über Elementumwandlungen im Innern der Sterne I. *Physik. Z.* 38:176–191.

Vrba, F. J., Rydgren, A. E. Chugainov, P. F., Shakovskaaya, N. I., and Zak, D. S. 1986. Further evidence for rotational modulation of the light curves from T Tauri stars. *Astrophys. J.* 306:199–214.

Vrba, F. J., Strom, S. E., and Strom, K. M. 1988*a*. Optical polarization measurements in the vicinity of nearby star-forming complexes. I. The Lynds 1641 giant molecular cloud. *Astron. J.* 96:680–694.

Vrba, F. J., Herbst, W., and Booth, J. F. 1988*b*. Spot evolution on the T Tauri star V410 Tau. *Astron J.* 96:1032–1039.

Vrba, F. J., Rydgren, A. E. Chugainov, P. F., Shakovskaya, N. I., and Weaver, B. W. 1989. Additional UBVRI photometric searches for periodic light variability from T Tauri stars. *Astron. J.* 97:483–498.

Wagoner, R. V., Fowler, W. A., and Hoyle, F. 1967. On the synthesis of elements at very high temperatures. *Astrophys. J.* 148:3–49.

Wahlen, M. 1969. Aktivierung von kosmischem Material im interplanetaren Raum, Messtechnische und experimentelle Beiträge, sowie Modell-Rechnungen. Ph.D. Thesis, Univ. of Bern.

Waldmeier, M. 1961. *The Sunspot Activity in the Years 1610–1960* (Zürich: Schulthess).

Walker, D., Longhi, J., and Hays, J. F. 1975. Differentiation of a very thick magma body and implications for the source regions of mare basalts. *Proc. Lunar Sci. Conf.* 6:1103–1120.

Walker, J. C. G. 1977. *Evolution of the Atmosphere* (New York: Macmillan).

Walker, J. C. G. 1982. The earliest atmosphere of the Earth. *Precam. Res.* 17:147–171.

Walker, J. C. G. 1985. Carbon dioxide on the early Earth. *Origins of Life* 16:117–127.

Walker, J. C. G., Hays, P. B., and Kasting, J. F. 1981. A negative feedback mechanism for the long-term stabilization of Earth's surface temperature. *J. Geophys. Res.* 86:9776–9782.

Walker, J. C. G., Klein, C., Schidlowski, M., Schopf, J. W., Stevenson, D. J., and Walter, M. R. 1983. Environmental evolution of the Archean-Early Proterozoic Earth. In *Earth's Earliest Biosphere: Its Origin and Evolution,* ed. J. W. Schopf (Princeton, N.J.: Princeton Univ. Press), pp. 260–290.

Walker, M. F. 1956. Studies of extremely young clusters. I. NGC 2264. *Astrophys. J. Suppl.* 2:365–387.

Walker, M. F. 1972. Studies of extremely young clusters. VI. Spectroscopic observations of the ultraviolet-excess stars in the Orion nebula cluster and NGC 2264. *Astrophys. J.* 175:89–116.

Walker, R. M. 1975. Interaction of energetic nuclear particles in space with the lunar surface. *Ann. Rev. Earth Planet. Sci.* 3:99–128.

Walker, R. M. 1980. Nature of the fossil evidence: Moon and meteorites. In *The Ancient Sun: Fossil Record in the Earth, Moon and Meteorites,* eds. R. O. Pepin, J. A. Eddy and R. B. Merrill (New York: Pergamon Press), pp. 11–28.

Walker, R. M., and Yuhas, D. 1973. Cosmic ray track production rates in lunar materials. In *Proc. Lunar Sci. Conf.* 4:2379–2389.

Wallerstein, G. 1988. Mixing in stars. *Science* 240:1743–1750.

Wallerstein, G., and Dominy, J. F. 1988. Technitium and niobium abundances in four stars of MS and S type. *Astrophys. J.* 330:937–941.

Walter, F. M. 1981. On the coronae of rapidly rotating stars. II. A period-activity relation in G stars. *Astrophys. J.* 245:677–681.

Walter, F. M. 1982. On the coronae of rapidly rotating stars. III. An improved coronal rotation-activity relation in late-type dwarfs. *Astrophys. J.* 253:745–751.

Walter, F. M. 1986. X-ray sources in regions of star formation. I. The naked T Tauri stars. *Astrophys. J.* 306:573–586.

Walter, F. M. 1987a. The naked T Tauri stars: The low-mass pre-main-sequence unveiled. *Publ. Astron. Soc. Pacific* 99:31–37.

Walter, F. M. 1987b. Naked T Tauri stars in II Sco and Ori OBIc. In *Cool Stars, Stellar Systems and the Sun: Proc. Fifth Cambridge Workshop,* eds. J. L. Linsky and R. E. Stencel (New York: Springer-Verlag), pp. 422–430.

Walter, F. M., and Bowyer, S. 1981. On the coronae of rapidly rotating stars. I. The relation between rotation and coronal activity in RS CVn systems. *Astrophys. J.* 245:671–676.

Walter, F. M., and Kuhi, L. V. 1984. X-ray photometry and spectroscopy of T Tauri stars. *Astrophys. J.* 284:194–201.

Walter, F. M., Brown, A., Linsky, J. L., Rydgren, A. E., Vrba, F., Roth, M., Carrasco, L., Chugainov, P. F., Shakovskaya, N. I., and Imhoff, C. L. 1987. X-ray sources in regions of star formation. II. The pre-main sequence G star HDE 283572. *Astrophys. J.* 314:297–307.

Walter, F. M., Schrijver, C. J., and Boyd, W. 1988a. Transition region fluxes in A-F dwarfs: Basal fluxes and dynamo activity. In *A Decade of UV Astronomy with the IUE Satellie,* eds. N. Longston and E. J. Rolfe, ESA SP-281, vol. 1, pp. 323–326.

Walter, F. M., Brown, A., Mathieu, R. D., Myers, P. C., and Vrba, F. J. 1988b. X-ray sources in regions of star formation. III. Naked T Tauri stars associated with the Taurus-Auriga complex. *Astrophys. J.* 96:297–325.

Walter, F. M., Brown, A., Vrba, F. J., and Mathieu, R. D. 1989. How naked are the naked T Tauri stars? *Bull. Amer. Astron. Soc.* 20:1092 (abstract).

Wang, H. T., and Ramaty, R. 1974. Neutron propagation and 2.2 MeV gamma-ray line production in the solar atmosphere. *Solar Phys.* 36:129–137.

Wänke, H. 1965. Der Sonnenwind als Quelle der Uredelgase in Steinmeteoriten. *Z. Naturf.* B20a:946–949.

Ward, W. R. 1974. Climatic variations on Mars. 1. Astronomical theory of insolation. *J. Geophys. Res.* 79:3375–3387.

Ward, W. R., Burns, J. A., and Toon, O. B. 1979. Past obliquity oscillations of Mars: The role of the Tharsis uplift. *J. Geophys. Res.* 84:243–259.

Warren, P. H. 1985. The magma ocean concept and lunar evolution. *Ann. Rev. Earth Planet. Sci.* 13:201–240.

Wasilewski, P. J. 1976. Shock-loaded meteoritic bcc metal above the pressure transition: Remnant-magnetization stability and microstructure. *Phys. Earth Planet. Int.* 11:5–11.

Wasilewski, P. J. 1981. New magnetic results from Allende C3(V). *Phys. Earth Planet. Int.* 26:134–148.

Wasilewski, P. J. 1988. Magnetic characterization of the new magnetic mineral tetrataenite and its contrast with isochemical taenite. *Phys. Earth Planet. Int.* 52:150–158.

Wasilewski, P. J., and Saralker, C. 1981. Stable NRM and mineralogy in Allende: Chondrules. *Proc. Lunar Planet. Sci. Conf.* 12:1217–1227.

Wasserburg, G. J. 1985. Short-lived nuclei in the early solar system. In *Protostars and Planets II*, eds. D. C. Black and M. S. Matthews (Tucson: Univ. of Arizona Press), pp. 703–737.

Wasserburg, G. J., and Arnould, M. 1987. A possible relationship between extinct ^{26}Al and ^{53}Mn in meteorites and early solar activity. In *Nuclear Astrophysics*, ed. W. Hillebrandt (Berlin: Springer-Verlag), pp. 262–276.

Wasserburg, G. J., Burnett, D. S., and Frondel, C. 1965. Strontium-rubidium age of an iron meteorite. *Science* 150:1814–1818.

Wasserburg, G. J., Papanastassiou, D. A. and Sanz, H. G. 1969. Initial strontium for a chondrite and the determination of a metamorphism or formation interval. *Earth Planet. Sci. Lett.* 7:33–43.

Wasserburg, G. J., Tera, F., Papanastassiou, D. A., and Huneke, J. C. 1977. Isotopic and chemical investigations on Angra Dos Reis. *Earth Planet. Sci. Lett.* 35:294–316.

Wasserburg, G. J., Papanstassiou, D. A., and Lee, T. 1980. Isotopic heterogenities in the early solar system. In *Early Solar System Processes and the Present Solar System*, ed. D. Lal (Bologna: Soc. Italiana di Fisica), pp. 144–191.

Wasson, J. T. 1974. *Meteorites: Classification and Properties* (New York: Springer-Verlag).

Wasson, J. T. 1985. *Meteorites: Their Record of Early Solar-System History* (New York: W. H. Freeman).

Wasson, J. T., and Wai, C. H. 1970. Composition of the metal, schreibersite and perryite of enstatite achondrites and the origin of enstatite chondrites and achondrites. *Geochim. Cosmochim. Acta* 34:169–184.

Wasson, J. T., and Wetherill, G. W. 1979. Dynamical, chemical and isotopic evidence regarding the formation locations of asteroids and meteorites. In *Asteroids*, ed. T. Gehrels (Tucson: Univ. of Arizona Press), pp. 926–974.

Wasson, J. T., Benz, W., and Rubin, A. E. 1987. Heating of primitive, asteroid-size bodies by large impacts. In *Papers Presented at the 5th Annual Meteoritical Soc. Meeting*, p. 185 (abstract).

Watson, A. J., Donahue, T. M., and Walker, J. C. G. 1981. The dynamics of a rapidly escaping atmosphere: Applications to the evolution of Earth and Venus. *Icarus* 48:150–166.

Webber, W. R., and Lockwood, J. A. 1981. A study of long term variations and radial gradient of cosmic rays out to 23 AU. *J. Geophys. Res.* 86:11458–11462.

Webber, W. R. Benbrook, J. R. Thomas, J. R. Hunting, A., and Duincan, R. 1963. An evaluation of the radiation hazard due to solar-particle events. Boeing Company Rept. D2–90469, NTIS number AD-805442 (Seattle: Boeing).

Weber, E. J., and Davis, L. 1967. The angular momentum of the solar wind. *Astrophys. J.* 148:217–227.

Weedon, G. P. 1985/86. Hemipelagic shelf sedimentation and climatic cycles: The basal Jurassic (Blue Lias) of South Britain. *Earth Planet. Sci. Lett.* 76:321–335.

Weeks, L. H. 1967. Lyman-alpha emission from the Sun near solar minimum. *Astrophys. J.* 147:1203–1205.

Weeks, L. H., Cuikay, R. S., and Corbin, J. R. 1972. Ozone measurements in the mesosphere during the solar proton event of 2 November 1969. *J. Atmos. Sci.* 29:1138–1142.

Weertman, J. 1976. Milankovitch solar radiation variations and ice age ice sheet sizes. *Nature* 261:17–20.

Weertman, J., and Peel, D. A. 1981. On Antarctic glaciology: Ice sheets and ice cores. *Nature* 294:210–212.

Weidenschilling, S. J. 1977. The distribution of mass in the planetary system and solar nebula. *Astrophys. Space Sci.* 51:153–158.

Weidenschilling, S. J. 1978. A constraint on pre-main-sequence mass loss. *Moon and Planets* 19:279–287.

Weiss, N. O. 1981. Stellar magnetic structure and activity. In *Solar Phenomena in Stars and Stellar Systems,* eds. R. M. Bonnet and A. K. Dupree (Dordrecht: D. Reidel), pp. 449–462.

Weiss, N. O. 1989. Dynamo processes in stars. In *Accretion Disks and Magnetic Fields in Astrophysics,* ed. G. Belvedere (Dordrecht: Kluwer), pp. 11–29.

Weiss, N. O., Cattaneo, F., and Jones, C. A. 1984. Periodic and aperiodic dynamo waves. *Geophys. Astrophys. Fluid Dyn.* 30:305–341.

Wentzel, D. G. 1987. Solar oscillations: Generation of a g mode by two p models. *Astrophys. J.* 319:966–970.

Wenzel, K.-P., Reinhard, R. Sanderson, T. R., and Sarris, E. T. 1985. Characteristics of energetic particle events associated with interplanetary shocks. *J. Geophys. Res.* 90:12–18.

Westphal, M., and Whitechurch, H. 1983. Magnetic properties and paleointensity determination of seven H-group chondrites. *Phys. Earth Planet Int.* 31:1–9.

Wetherill, G. W. 1989. The formation of the solar system: Concensus, alternatives, and missing factors. In *The Formation and Evolution of Planetary Systems,* eds. H. A. Weaver, F. Paresce and L. Danly (New York: Cambridge Univ. Press), pp. 1–24.

Wetherill, G. W., and Chapman, C. R. 1988. Asteroids and meteorites. In *Meteorites and the Early Solar System,* eds. J. F. Kerridge and M. S. Matthews (Tucson: Univ. of Arizona Press), pp. 35–67.

White, O. R., and Livingston, W. C. 1981. Solar luminosity variation. III. Calcium K variation from solar minimum to maximum in cycle 21. *Astrophys. J.* 249:798–816.

White, O. R., Livingston, W. C., and Wallace, L. 1987. Variability of chromospheric and photospheric lines in solar cycle 21. *J. Geophys. Res.* 92:823–827.

Wieler, R., Etique, P., Poupeau, G., and Signer, P. 1983. Decrease of the solar flare/solar wind flux ratio in the past several aeons deduced from solar neon and tracks in lunar soil plagioclases. *Proc. Lunar Planet. Sci. Conf.* 13, *J. Geophys. Res. Suppl.* 88:A713-A724.

Wieler, R., Baur, H., and Signer, P. 1985. The isotopic composition of neon emitted in solar flares and trapped in lunar soil minerals. In *Isotopic Ratios in the Solar System* (Toulouse: Cepadues-Editions), pp. 137–146.

Wieler, R., Baur, H., and Signer, P. 1986. Noble gases from solar energetic particles revealed by closed system stepwise etching of lunar soil minerals *Geochim. Cosmochim. Acta* 50:1997–2017.

Wieler, R., Baur, H., Benkert, J.-P., Pedroni, A., and Signer, P. 1987. Noble gases in the meteorite Fayetteville and in lunar ilmenite originating from solar energetic particles. *Lunar Planet. Sci.* XVIII:1080–1081 (abstract).

Wieler, R., Baur, H., Pedroni, A., Pellas, P., and Signer, P. 1989a. Exposure history of the regolithic chondrite Fayetteville. I. Solar-gas-rich matrix samples. *Geochim. Cosmochim. Acta* 53:1441–1448.

Wieler, R., Graf, Th., Pedroni, A., Signer, P., Pellas, P., Fieni, C., Suter, M., Vogt, S., Clayton, R. N., and Laul, J. C. 1989b. Exposure history of the regolithic chondrite Fayetteville. II. Solar-gas-free light inclusions. *Geochim. Cosmochim. Acta* 53:1449–1459.

Wigley, T. M. L. 1976. Spectral analysis: Astronomical theory of climatic change. *Nature* 264:629–631.

White, O. R., ed. 1977. *The Solar Output and Its Variation* (Boulder: Colorado Assoc. Univ. Press).

Wiik, H. B. 1972. The chemical composition of the Havero meteorite and the genesis of the ureilites. *Meteoritics* 7:553–557.

Wild, J. P., Smerd, S. F., and Weiss, A. A. 1963. Solar bursts. *Ann. Rev. Astron. Astrophys.* 1:291–366.

Wilkening, L. L. 1971. *Particle Track Studies and the Origin of Gas-Rich Meteorites.* Center for Meteorite Studies Publ. No. 11 (Tempe, Ariz.: Center for Meteorite Studies).

Wilkening, L. L. 1973. Foreign inclusions in stony meteorites. I. Carbonaceous chondritic xenoliths in the Kapoeta Howardite. *Geochim. Cosmochim. Acta* 37:1985–1989.

Wilkening, L. L. 1977. Meteorites in meteorites: Evidence for mixing among the asteroids. In *Comets, Asteroids, and Meteorites: Interrelations, Evolution and Origins* ed. A. H. Delsemme (Toledo: Univ. of Toledo Press).

Williams, G. E. 1981. Sunspot periods in the late Precambrian glacial climate and solar-planetary relations. *Nature* 291:624–628.

Williams, G. E. 1985. Solar affinity of sedimentary cycles in the late Precambrian Elatina formation. *Australian J. Phys.* 38:1027–1043.

Williams, G. E. 1988. Cyclicity in the late Precambrian Elatina Formation, South Australia: Solar or tidal signature? *Climat. Change* 13:117–128.

Williams, G. E. 1989. Late Precambrian tidal rythmites in South Australia and the history of the Earth's rotation. *J. Geol. Soc. London* 146:97–111.

Williams, G. E., and Sonett, C. P. 1985. Solar signature in sedimentary cycles from the late Precambrian Elatina Formation, Australia. *Nature* 318:523–527.

Willson, L. A., Bowen, G. H., and Struck-Marcell, C. 1987. Mass loss on the main sequence. *Comments on Astrophys.* 12:17–34.

Willson, R. C. 1979. Active cavity radiometer type IV. *Appl. Opt.* 18:179–188.

Willson, R. C. 1982. Solar irradiance variations and solar activity. *J. Geophys. Res.* 87:4319–4324.

Willson, R. C. 1984. Measurements of solar total irradiance and its variability. *Space Sci. Rev.* 38:203–242.

Willson, R. C., and Hudson, H. S. 1988. Solar luminosity variations in solar cycle 21. *Nature* 332:810–812.

Willson, R. C., Gulkis, S., Janssen, M., Hudson, H. S., and Chapman, G. A. 1981. Observations of solar irradiance variability. *Science* 211:700–702.

Willson, R. C., Hudson, H. S., Fröhlich, C., and Brusa, R. W. 1986. Long-term downward trend in total solar irradiance. *Science* 234:1114–1117.

Wilson, M. D., and Bercovitch, M. 1975. *Cosmic Ray NM-64 Neutron Monitor Data,* Repts. XXIV–XXIX, Jan. 1973–Dec. 1974 (Chalk River, Ont.: Atomic Energy of Canada).

Wilson, O. C. 1963. A probable correlation between chromospheric activity and age in main-sequence stars. *Astrophys. J.* 138:832–848.

Wilson, O. C. 1966a. Stellar chromospheres. *Science* 151:1487–1498.

Wilson, O. C. 1966b. Stellar convection zones, chromospheres, and rotation. *Astrophys. J.* 144:695–708.

Wilson, O. C. 1978. Chromospheric variations in main-sequence stars. *Astrophys. J.* 226:379–396.

Wilson, O. C., and Skumanich, A. 1964. Dependence of chromospheric activity upon age in main-sequence stars: Additional evidence. *Astrophys. J.* 140:1401–1408.

Wilson, O. C., and Woolley, R. 1970. Calcium emission intensities as indicators of stellar age. *Mon. Not. Roy. Astron. Soc.* 148:463–475.

Wilson, R. C. 1984. Solar total irradiance variability measurements by the SMM/ACRIM I experiment. In *Solar Irradiance Variations on Active Region Time Scales,* eds. B. J. LaBonte, G. A. Chapman, H. S. Hudson and R. C. Willson, NASA CP-2310, pp. 1–43.

Wilson, R. M. 1988. A prediction for the maximum phase and duration of sunspot cycle 22. *J. Geophys. Res.* 93:10011–10015.

Winfree, A. T. 1980. *The Geometry of Biological Time* (New York: Springer-Verlag).

Winograd, I. J., Coplen, T. B., Szabo, B. J., and Riggs, A. C. 1988. A 250,000-year climatic record from Great Basin vein calcite: Implications for Milankovitch theory. *Science* 242:1275–1280.

Wittman, A. D. 1973. The solar diameter at 5000 Å and H-alpha from photoelectric drift scans. *Solar Phys.* 29:333–340.

Wittman, A. D. 1979. Tobias Mayer's Meridian-Sonnendurchmessers. *Mitt. Astron. Ges.* 48:25–28.

Wittman, A. D. 1980. Tobias Mayer's observations of the Sun: Evidence against a secular decrease of the solar diameter. *Solar Phys.* 66:223–229.

Wittman, A. D., and Debarbat, S. 1990. Der Sonnendurchmesser und Seine Variabiliat. *Sterne und Weltraum* 7–8:420–426.

Wittman, A. D., and Xu, Z. T. 1987. A catalogue of sunspot observations from 165 BC to AD 1684. *Astron. Astrophys. Suppl.* 70:83–94.

Wittman, A. D., Bonet Navarro, J. A., and Wohl, H. 1981. On the variability of the solar diameter. In *Physics of Sunspots*, eds. L. Cram and J. H. Thomas (Sunspot, N.M.: National Solar Obs.), pp. 422–433.

Wlotzka, F. 1963. Über die Hell-Dunkel-Struktur der urgashaltigen Chondrite Breitscheid und Pantar. *Geochim. Cosmochim. Acta* 27:419–429.

Wolfenstein, L. 1978. Neutrino oscillations in matter. *Phys. Rev.* D17:2369–2374.

Wolfenstein, L. 1979. Neutrino oscillations and stellar collapse. *Phys. Rev.* D20:2634–2635.

Wolff, C. 1984. Solar irradiance changes caused by g-modes and large scale convection. *Solar Phys.* 93:1–13.

Wolff, C., and Hickey, J. 1987. Multiperiodic irradiance changes caused by r-modes and g-modes. *Solar Phys.* 109:1–18.

Wolfsberg, K., Cowan, G. A., Bryant, E. A., Daniels, K. S., Downey, S. W., Haton, W. C., Niesen, V. G., Nogar, N. S., Miller, C. M., and Rokop, D. J. 1985. The molybdenum solar neutrino experiment. In *Solar Neutrinos and Neutrino Astronomy*, eds. M. L. Cherry, K. Lande and W. A. Fowler (New York: American Inst. of Physics), pp. 196–202.

Woolley, R., Epps, E. A., Penston, M. J., and Pocock, S. B. 1970. Catalogue of stars within twenty-five parsecs of the Sun. *Roy. Obs. Ann* No. 5.

Wollin, G., Ericson, D. B., and Ryan, W. B. 1971. Variations in magnetic intensity and climatic changes. *Nature* 232:549–551.

Wood, J. A. 1967. Chondrites: Their metallic minerals, thermal histories, and parent planets. *Icarus* 6:1–49.

Wood, J. A. 1979. Review of the metallographic cooling rates of meteorites and a new model for the planetesimals in which they formed. In *Asteroids*, ed. T. Gehrels (Tucson: Univ. of Arizona Press), pp. 849–891.

Wood, J. A. 1991. Refractory solids in chondrites and comets: How similar? In *Analysis of Returned Comet Nucleus Samples*, ed. S. Chang, NASA SP, in press.

Wood, J. A., and Morfill, G. E. 1988. A review of solar nebula models. In *Meteorites and the Early Solar System*, eds. J. F. Kerridge and M. S. Matthews (Tucson: Univ. of Arizona press), pp. 329–347.

Wood, J. A., Dickey, J. S., Jr., Marvin, U. B., and Powell, B. N. 1970. Lunar anorthosites and a geophysical model of the Moon. *Proc. Apollo 11 Lunar Sci. Conf.*, pp. 965–988.

Woodard, M. F., and Hudson, H. S. 1983. Frequencies, amplitudes and linewidths of solar oscillations from total solar irradiance observations. *Nature* 305:589–593.

Woodard, M. F., and Noyes, R. C. 1985. Change of the solar oscillation eigenfrequencies with the solar cycle. *Nature* 318:449–450.

Woods, T. N., and Rottman, G. J. 1990. Solar EUV irradiance derived from a sounding rocket experiment on 10 November 1988. *J. Geophys. Res.* 95:6227–6236.

Worden, S. P., Schneeberger, T. J., Kuhn, J. R., and Africano, J. L. 1981. Flare activity on T Tauri stars. *Astrophys. J.* 244:520–527.

Worthing, A. G., and Geffner, J. 1962. *Treatment of Experimental Data* (New York: Wiley).

Wright, C. S. 1980. On the longitudinal distribution of solar type II bursts. *Proc. Astron. Soc. Australia* 4:59–61.

Wright, D. G., Stocker, T. F., and Mysak, L. A. 1990. A note on Quaternary climate modelling using Boolean delay equations. *Climate Dyn.* 4:263–267.

Wyatt, P. J. 1962. Scattering of electromagnetic plane waves from inhomogeneous spherically symmetric objects. *Phys. Rev.* 127:1837–1843.

Xu, Z.-T., and Jiang, Y.-T. 1982. The solar activity in the seventeenth century re-assessed in the light of sunspot records in the local gazettes of China. *Chinese Astron. Astrophys.* 2:84–90.

Yanagita, S., Yamakoshi, K., and Gensho, R. 1978. Production of ^{59}Ni from α bombardment of iron. *Nucl. Phys.* A303:254–258.

Yang, J., Turner, M. S., Steigman, G., Schramm, D. N., and Olive, K. A. 1984. Primordial nucleosynthesis: A critical comparison of theory and observations. *Astrophys. J.* 281:493–511.

Yaniv, A., and Marti, K. 1981. Detection of stopped solar flare helium in lunar rock 68815. *Astrophys. J.* 247:L143-L146.

Yaniv, A., Marti, K., and Reedy, R. C. 1980. The solar cosmic-ray flux during the last two million years. In *Lunar Planet. Sci.* XI:1291–1293 (abstract).

Yasyulenis, R.Yu., Luyanas, V.Yu., and Styro, B. I. 1974. Argon spallation yields of cosmogenic radioisotopes. In *Proc. of the All-Union Conf. on Nuclear Meteorology*, trans. A. Baruch and N. Olaru (Springfield, Va.: U.S. Dept. of Commerce).

Yau, K. K. K. C., and Stephenson, F. R. 1988. A revised catalogue of far eastern observations of sunspots (165 BC to AD 1918). *Quart. J. Roy. Astron. Soc.* 29:175–197.

Yerle, R. 1981. Oscillations in the solar atmosphere. *Astron. Astrophys.* 100:L23–L25.

Yiou, F., Raisbeck, G. M., Bourles, D., Lorius, C., and Barkov, N. I. 1985. ^{10}Be in ice at Vostok Antarctica during the last climatic cycle. *Nature* 316:616–617.

Yiou, P., Genthon, C., Jouzel, J., Ghil, M., Le Treut, H., Barnola, J.-M., Lorius, C., and Korotkevitch, Y. N. 1989. High-frequency paleo-variability in climate and in CO_2 levels from Vostok ice-core records. In *Interaction of the Global Carbon and Climate Systems*, ed. R. Keir (Palo Alto, Calif.: Electric Power Research Inst.), pp. 28.1–28.22.

Yokoyama, Y., Auger, R., Bibron, R., Chesselset, R., Guichard, F., Leger, C., Mabuchi, H., Reyss, J. L., and Sato, J. 1972. Cosmonuclides in lunar rocks. In *Proc. Lunar Sci. Conf.* 3:1733–1746.

Yoshimori, M., Okudaira, K., Harasima, Y., and Kondo, I. 1983. Gamma-ray observations from Hinotori. *Solar Phys.* 86:375–382.

Yoshimura, H. 1975a. Dynamo wave propagation. *Astrophys. J.* 201:740–748.

Yoshimura, H. 1975b. A model of the solar cycle driven by the dynamo action of the global convection in the solar convection zone. *Astrophys. J. Suppl.* 29:467–494.

Young, A. T. 1974. Photomultipliers: Their cause and cure. Other components in photometric systems. Observational techniques and data reduction. In *Methods of Experimental Physics, Vol. 12: Astrophysics, Part A: Optical and Infrared*, ed. N. Carleton (New York: Academic Press), pp. 1–192.

Young, J. A., Thomas, C. W., and Wogman, N. A. 1970. Cosmogenic radionuclide production rates in argon in the stratosphere. *Nature* 227:160–161.

Yuan, C., and Cassen, P. 1985. Protostellar angular momentum transport by spiral density waves. *Icarus* 64:435–447.

Zahnle, K. 1988. Escape from Planet X. *Lunar Planet. Sci.* XIX:1313–1314 (abstract).

Zahnle, K. J., and Kasting, J. F. 1986. Mass fractionation during transonic escape and implications for loss of water from Venus and Mars. *Icarus* 68:462–480.

Zahnle, K. J., and Walker, J. C. G. 1982. The evolution of solar ultraviolet luminosity. *Rev. Geophys. Space Phys.* 20:280–292.

Zahnle, K. J., and Walker, J. C. G. 1987. Climatic oscillations during the Precambrian era. *Climat. Change* 10:269–284.

Zappala, R. R. 1972. Lithium abundances of stars in open clusters. *Astrophys. J.* 172:57–74.

Zbinden, H., Andrée, M., Oeschger, H., Ammann, B., Lotter, A., Bonani, G., and Wölfli, W. 1990. Atmospheric radiocarbon at the end of the last glacial: An estimate based on AMS radiocarbon dates on terrestrial macrofossils from lake sediments. In *Proc. 13th Intl. Radiocarbon Conf.*, eds. A. Long and R. Kra. *Radiocarbon* 31(3):793–802.

Zebiak, S. E., and Cane, M. A. 1987. A model El Niño-Southern oscillation. *Mon. Weather Rev.* 115:2262–2278.

Zel'dovich, Ya. B., and Raizer, Yu. P. 1966. Rarefied ionized gases. In *Physics of Shock Waves and High-Temperature Phenomena*, vol. 1, eds. W. D. Hays and R. F. Probstein (New York: Academic Press), pp. 216–218.

Zel'dovich, Ya. B., and Ruzmaikin, A. A. 1980. Dynamo problems in astrophysics. *Soviet Sci. Rev.* 333–383.

Zel'dovich, Ya. B., Ruzmaikin, A. A., and Sokoloff, D. D. 1983. *Magnetic Field in Astrophysics* (New York: Gordon and Breach).

Zeller, E. J., and Parker, B. C. 1981. Nitrate ion in Antarctic firn as a marker for solar activity. *Geophys. Res. Lett.* 8:895–898.

Zellner, B., and Bowell, E. 1977. Asteroid compositional types and their distribution. In *Comets, Asteroids, Meteorites—Interrelations, Evolution and Origins,* ed. A. H. Delsemme (Toledo: Univ. of Toledo Press), pp. 185–197.

Zellner, B., Tholen, D. J., and Tedesco, E. F. 1985. The eight-color asteroid survey: Results for 589 minor planets. *Icarus* 61:355–416.

Zhang, Z. 1983. The Korean auroral records of the period AD 1507–1747 and the ser arcs. Inst. for Plasma Res., Stanford Univ. Rept. no. 955.

Zinnecker, H. 1989. Pre-main sequence binaries. In *ESO Workshop on Low Mass Star Formation and Pre-Main Sequence Objects,* ed. B. Reipurth (Garching: European Southern Obs.), pp. 447–469.

Zinner, E. 1980. On the constancy of solar particle fluxes from track, thermoluminescence and solar wind measurements in lunar rocks. In *The Ancient Sun: Fossil Record in the Earth, Moon and Meteorites,* eds. R. O. Pepin, J. A. Eddy and R. B. Merrill (New York: Pergamon Press), pp. 201–226.

Zirker, J. B. 1977. Coronal holes—an overview. In *Coronal Holes and High Speed Wind Streams,* ed. J. B. Zirker (Boulder: Colorado Assoc. Univ. Press), pp. 1–26.

Zook, H. A. 1980. On lunar evidence for a possible large increase in solar flare activity $\sim 2 \times 10^4$ years ago. In *The Ancient Sun: Fossil Record in the Earth, Moon and Meteorites,* eds. R. O. Pepin, J. A. Eddy and R. B. Merrill (New York: Pergamon Press), pp. 245–266.

Zuckerman, B., Kuiper, T. B. H., and Rodriguez-Kuiper, E. N. 1976. High-velocity gas in the Orion infrared nebula. *Astrophys. J.* 209:L137–L142.

Zwaan, C. 1978. On the appearance of magnetic flux in the solar photosphere. *Solar Phys.* 60:213–240.

Zwaan, C. 1986. Relations between magnetic activity and steller properties. In *Cool Stars, Stellar Systems and the Sun: Proc. Fourth Cambridge Workshop,* eds. M. Zeilik and D. M. Gibson (New York: Springer-Verlag), pp. 19–29.

Zwaan, C., and Cram, L. E. 1989. Theory of magnetic fields in cool dwarfs. In *FGK Stars and T Tauri Stars,* eds. L. E. Cram and L. V. Kuhi, NASA SP-502, pp. 215–251.

GLOSSARY

GLOSSARY*

Compiled by Mary Guerrieri

ablation	usually thermal erosion of material, e.g., by high speed passage through the atmosphere.
accelerator mass spec- trometry (AMS)	measurement technique for getting extremely small ratios (to $\sim 10^{-15}$) of isotopes by accelerating ions to high energies and using nuclear techniques to identify and count the atoms of the rare isotope. Used commonly since about 1980 for cosmogenic radionuclides like ^{10}Be and ^{14}C.
accretion disk	the flattened disk of gas and dust in the equatorial plane around a young star, in which accretion of disk material around the stellar surface occurs as a result of viscous dissipation.
achondrite	a differentiated stony meteorite devoid of chondrules.
ACRIM	Active Cavity Radiometer Irradiance Monitor; experiment on the Solar Maximum Mission satellite designed to measure the total radiative output of the Sun.

---

*We have used some definitions from *Glossary of Astronomy and Astrophysics* by J. Hopkins (by permission of the University of Chicago Press, copyright 1980 by the University of Chicago), from *Astrophysical Quantities* by C. W. Allen (London: Athlone Press, 1973), from *Glossary of Geology,* edited by M. Gary, R. McAfee, and C. L Wolff (Washington, D.C.: American Geological Institute, 1972), from *Glossary of Geology,* 3rd ed., edited by R. L. Bates and J. A. Jackson (Alexandria, Va.: American Geological Institute, 1987), and from *The Planetary System* by David Morrison and Tobias Owen (Reading, Mass.: Addison-Wesley Publishing Co., 1988). We also acknowledge definitions and helpful comments from various chapter authors.

adiabat

the locus in pressure-temperature space followed by a parcel of matter which undergoes changes in volume without exchanging heat with its surroundings.

agglutinate

a lithology characteristic of planetary surfaces, consisting of regolith particles bonded together with impact-generated glass.

Alfvén waves

a generic term referring to waves in a magnetized plasma.

alias

refers to the folding of a spectrum around the Nyquist frequency and the replication of a spectrum at multiples of a sampling frequency.

almucantar

(arabic; french: almicantarat): circle on the celestial local sphere, parallel to the horizon.

alpha particle

the nucleus of a ^{4}He atom.

anorthosite

an igneous rock made up almost entirely of plagioclase feldspar.

Archean

an era of Earth's history ending about 2.5×199 years ago and followed by the Proterozoic.

archeomag-
netic

relating to the intensity and direction of the geomagnetic field in past time.

astrolabe

an ancient astronomical instrument for determining the orbital parameters of the Earth-Sun system, i.e., eccentricity, obliquity, inclination of the orbit, position of equinox and longitude of the perihelion.

attractor

that subspace in phase space which is occupied by a system in the limit $t \rightarrow 1$, starting with different initial conditions (Ci). The whole set of Ci's is called a "basin of attraction."

AU

astronomical unit. The mean distance of the Earth from the Sun, equal to 1.5×10^{13} cm.

AXAF

Advanced X-ray Astrophysics Facility.

Balmer continuum	radiation associated with the recombination of free electrons to the $n = 2$ bound level of the hydrogen atom.
barn	a measure of interaction cross section in high energy nuclear particle interactions (1 barn $= 20^{-24}$ cm^2).
basal flux	the minimum radiative flux associated with spectral lines arising in the chromosphere and transition region in late type stars.
basaltic achondrite	collective name for eucrites and howardites which superficially resemble terrestrial basalts or their fragmentation products.
Beal's classification	see P Cygni star.
biosphere	the part of the Earth containing biological organisms.
bolometric	total radiative output of a star integrated over all wavelengths.
Boltzmann's constant (k)	the constant of proportionality between energy per degree of freedom and temperature; $k = 1.38 \times 10^{-16}$ ergs per degree Kelvin.
Boltzmann's equation	the continuity equation in phase space describing density of position and momentum of an ensemble of particles.
Bond albedo	fraction of the total incident light reflected by a spherical body. Bolometric Bond albedo refers to reflectivity over all wavelengths.
bound-bound transitions	transitions between energy levels of an electron bound to a nucleus (the electron is bound both before and after the transition).
Boussinesq approximation	convection in a system where buoyancy depends only on temperature.
Boussinesq convection	convection where the Boussinesq approximation is made.

BP   before present.

breccia   rock composed of fragments derived from previous gen-
                            erations of rocks, cemented together to form a new li-
                            thology.

bremsstrah-   radiation emitted when a charged particle is accelerated.
lung   (Also called a free-free transition because the particle
                            executes an unbound transition.) The inverse of brems-
                            strahlung is free-free absorption.

brightness   measure of the luminosity of a body in a given spectral
                            region.

Ca II H and K   emission in the resonance lines of singly ionized calcium
emission   which observationally identifies emission in late-type
                            stars.

carbonaceous   a primitive meteorite, generally containing more than
chondrite   about 0.2% C. Most such chondrites are highly oxidized
                            and have nearly solar composition of all but the most
                            volatile elements.

CCD   charge-coupled device. A solid state detector used for
                            low-light level imaging.

Cerenkov ra-   visible (and also more energetic) radiation caused by an
diation   electromagnetic shock wave arising from charged par-
                            ticles moving with superluminal velocities (greater than
                            the speed of light in a medium).

charged-   path of radiation damage in certain solid dielectric media
particle   caused by the passage of heavily ionizing radiation. In
tracks   extraterrestrial material, tracks made by heavy ($Z \gtrsim 20$)
                            cosmic-ray nuclei can be observed in certain minerals.

chondrite   a stony meteorite characterized by chondrules embedded
                            in a finely crystalline matrix consisting of orthopyrox-
                            ene, olivine, and nickel-iron with or without glass.
                            Chondrites constitute over 80% of meteorite falls and are
                            usually classified according to the predominant pyrox-
                            ene: e.g., "enstatite chondrite," "bronzite chondrite,"
                            and "hypersthene chondrite." From *AGI Glossary*
                            (1987).

chromosphere   the portion of the solar atmosphere above the photosphere with temperatures between about 6000 and 20,000 K.

Class 3 solar flare   a very energetic flare seen on the Sun.

color excess   the difference between the observed and the intrinsic color indices, usually denoted, for example, by $E(V\text{-}L)$, where V and L are the magnitudes in the V and L bandpasses, respectively.

compaction age   time at which the constituents of a breccia were cemented together in their final configuration.

complex history   case where an object was exposed to significant fluxes of cosmic rays in two (or more) different geometries, such as a lunar rock being irradiated both while slightly buried and then while on the lunar surface.

convective overturn time   the time scale overwhich the stellar convective zone completes 1 overturn cycle.

corona   that portion of the solar atmosphere with temperatures above about 250,000 K. Coronal emission is generally seen in the EUV and X-ray portions of the spectrum.

coronal hole   a large-scale, low-density, low-temperature region in the solar corona where the EUV and X-ray coronal emission is abnormally low or absent; a coronal region apparently associated with diverging magnetic fields.

cosmic-ray exposure age   the period of time during which a meteorite was exposed to cosmic radiation, commonly the time between its final reduction in size because of collisions and its arrival on Earth. More generally, it is the time spent screened by not more than a few meters of overburden from the space environment. Nuclear reactions between the radiation and nuclides in the meteorite produce new nuclides, or associated phenomena such as tracks, whose abundances can be used to estimate the exposure age; also used with respect to exposure time on the lunar surface.

cosmic rays	energetic ($E \gtrsim 1$ MeV) ionized nuclei (mainly protons) and electrons in space.
cosmic-ray spectrum	usually refers to the number density of energetic (cosmic-ray) particles vs energy or rigidity.
cosmogenic	isotopes produced by interaction with cosmic radiation, e.g., ^{21}Ne is a cosmogenic radionuclide made by spallation reactions.
Coulomb drag	electrostatic interactions resulting in net transfer of momentum between charged particles, normally in a plasma.
Cretaceous	geologic time period lasting from about 145 Myr ago until 65 Myr ago; noted for its climatic warmth.
Curie	the quantity of radioactive material decaying at the rate of 3.7×10^{-10} disintegrations per second, corresponding to about one gram of radium.
Curie temperature	the temperature of transition above which the phenomena of ferromagnetism transforms into paramagnetism; also the ferroelectric phase and the paraelectric phase.
cutoff rigidity	the momentum vs geomagnetic latitude below which a primary cosmic-ray particle cannot penetrate the geomagnetic field and enter the Earth's atmosphere.
P Cygni star	a type of star named after the 5th magnitude peculiar variable Ble star P Cygni (HD 193272), about 1200 pc distant, whose spectrum shows strong emission lines, like those of the Be and Wolf-Rayet stars, with blue-shifted absorption components and redshifted emission components which are presumed to arise in an expanding shell of low-density matter. Beal's classification: Type 1 profiles have a broad emission feature with a violet-displaced absorption core which dips below the level of the continuum; Type II line profiles involve a violet-displaced dip in an emission line; in Type III, additional emission appears blueward of the blueshifted absorption feature.
Debye-Huckle model	the standard plasma model of an ionized classical gas.

diagenesis	a complex of sediment transformations, e.g., small to large crystals in benthic calcium carbonate.
differentiation (in a planet)	a process whereby the primordial substances are separated via density differences, e.g., metal sinking towards the center of a planet to form a core, displacing lighter silicates which form a crust and mantle.
(Henry) Draper system (HD system)	a classification of stellar spectra into the sequence O, B, A, F, G, K, M in order of decreasing temperature.
dwarf-M flare stars	a subset of the M-dwarf stars that have been observed to flare. Often, flares more powerful than those seen on the Sun occur in the dwarf-M flare stars.
eddy diffusion coefficient	the coefficient in the diffusion equation when diffusive transport is governed by eddies rather than molecular interactions.
effective Landé g factor	a measure of the sensitivity of a spectral line to magnetic fields, the effective Landé g value is the intensity-weighted average of the magnetic splitting of individual Zeeman components in an atomic transition.
eV	electron volt; unit of energy for ionizing radiation, with $1 \text{ eV} = 1.62 \times 10^{-19}$ joules. Nuclei in the cosmic rays typically have energies from about a mega-electron-volt (MeV, 10^6 eV) to many giga-electron-volts (GeV, sometimes BeV, 10^9 eV) per nucleon.
ENSO	El Niño Southern Oscillation. A 2–4 year quasi-periodicity in sea level pressure, surface temperature and meteorological parameters observed over a large part of the Pacific Ocean and adjacent land areas south of the equator.
erosion	slow but steady removal of the surfaces of an object. For a lunar rock, erosion is mainly caused by micrometeoroid "sandblasting."
exobase	base of an exosphere.
EXOSAT	X-ray Satellite.

exosphere

outermost region of an atmosphere, where atoms and molecules are on ballistic or escaping trajectories and collisions can usually be neglected.

exposure age

see cosmic-ray exposure age.

extinction

attenuation of starlight due to absorption and scattering by Earth's atmosphere, or by interstellar dust.

extinct nu-
clides

radioactive nuclides with short half-lives (compared with the age of the solar system) which were present when a meteorite or meteoritic component formed but which have now decayed below detection limits. Their one-time presence in the material is indicated by their decay products.

faculae

bright regions of the photosphere seen in white light, visible only near the limb of the Sun.

F-dwarf

main-sequence star somewhat hotter and more massive than the Sun (approximately 1.5 solar masses).

Fermi pres-
sure

pressure due to fermion degeneracy, usually but not necessarily electrons.

filling factor

fraction of the stellar surface covered with magnetic regions. Measurements to date have been of the bright (in continuum light) magnetic regions, probably analogous to solar plage and network. The filling factor appears to decrease with the age of the star as the star loses angular momentum through magnetic braking and its generation of magnetic flux via the dynamo mechanism becomes less efficient.

fission

a radioactive decay process in which a heavy nucleus fragments into two or more pieces, but of approximately unequal mass. The process can be spontaneous or induced by particle bombardment.

flavors

quantum mechanical descriptors. Lepton flavors include electrons, muons and tauons. Corresponding neutrino flavors are electron neutrinos, muon neutrinos and tauon neutrinos.

flux (f)	integral of the flux density f_λ over a passband; units are erg cm^{-2} s^{-1}.
flux density	integral of the specific intensity I_λ over angle. Units are erg cm^{-2} s^{-1} Å^{-1} (f_λ) or erg cm^{-2} s^{-1} υ^{-1} ($f\upsilon$).
F0 star	star of spectral class F, subtype 0. A main sequence star for this spectral type has a mass of about 1.6 times the Sun's.
Fu Ori star	a type of pre-main-sequence star, defined originally by the characteristics of the prototype FU Orionis. FU Ori stars are apparently low-mass T Tauri stars that undergo an outburst that causes them to brighten by a factor of 100 in a period less than one year, and then to stay bright for decades. Instabilities in an accretion disk are a plausible source for the outburst.
FWHM	full width at half maximum (minimum), e.g., the full width of a spectral line at half-maximum intensity.
Gaia	the hypothesis that the Earth's atmospheric composition and climate are actively regulated by the biota.
galactic cluster	*see* open cluster.
galactic cosmic rays	highly energetic (usually ~ 0.1–10 GeV) particles (mainly protons) incident on the solar system.
gardening	vertical or horizontal mixing of a regolith by micrometeoroid bombardment.
gas-rich meteorite	a meteorite that has resided in the regolith of its parent body, thereby acquiring a population of light noble gases (He, Ne, Ar) of solar-wind origin. They are present among different meteorite classes.
G-dwarf	The spectral class of the Sun and stars similar to the Sun. The Sun is classified as a G2-dwarf.
geomagnetic field	the Earth's magnetic field.

Gleissberg period	A period of approximately 80–90 years found in aurorae, the sunspot index, the radiocarbon variability, and possibly other data. It is attributed to variation in solar activity.
globular cluster	a tightly packed, symmetrical group (mass range 10^4 - 10^6 $M_\odot$) of very old (pure Population II) stars. The stellar density is so great in the center that the nucleus is usually unresolved. Their spectra indicate low abundances of heavy elements. Globular clusters are probably the oldest stellar formations in the Galaxy. They are generally found in the halo (as far as 50 kpc from the galactic center) and are "high-velocity" objects with very elongated orbits around the galactic center. Many are X-ray sources.
g-modes	solar acoustic waves as modified by gravitation. Buoyancy forces serve as the restoring forces that can lead to this mode of oscillation.
greenhouse effect	the warming of a planet's surface caused by radiation trapped by the infrared opacity of the atmosphere.
Gyr	10^{-9} years.
Hadean Era	the first 700 million years of Earth's history (4.5 to 3.8 Gyr) which preceded the beginning of the sedimentary rock record.
Hale period	22 year period of the solar magnetic field.
half-life	the length of time over which half of an initial population of radionuclide has decayed.
Hallstattzeit	a period of low average solar activity from 800 to 300 B.C., named after a contemporaneous cool and wet period in central Europe that preceeded the Little Ice Age by about 2100 years and had a similar duration of about 500 years.
HAO	High Altitude Observatory.
Hayashi track	a nearly vertical evolutionary track in the HR diagram toward the main sequence during phases when the star is largely or completely in convective equilibrium. The lu-

minosity, originally very high, decreases rapidly with contraction, but the surface temperature remains almost constant.

heliometer   a refracting telescope whose objective lens has been split into two halves. Each half objective forms a solar image. The contact of the two images is ensured by a micrometer screw. The apparent solar diameter is obtained by contact of the two images.

heliosphere   the solar wind "bubble" containing the solar system, terminating (at the heliopause) at perhaps 100 AU due to pressure balance with interstellar gas.

hep neutrinos   neutrinos resulting from the ^{3}He $+ p$ reaction.

Herbig-Haro objects (HH objects)   small, faint patches of nebulosity seen on the surfaces of many dark clouds, believed to be a very early phase of stellar evolution. All known Herbig-Haro objects have been found within the boundaries of dark clouds. They are strong infrared sources and are characterized by mass loss.

HH objects   *see* Herbig-Haro objects.

high Q   usually referring to a low dissipation, sharply tuned oscillator.

Holocene   pertaining to or designating the Recent epoch, beginning at about 11,000 B.C., and the development of man from the neolithic stage onward. The Holocene period began at the end of the last Ice Age.

homosphere   the well-mixed part of the atmosphere; its upper boundary (homopause) is at 100 km.

Hopf bifurcation   in nonlinear dynamics, mechanism for the generation of periodic solitons. As a physical parameter changes, a branch of steady-state solitons loses its stability, and the associated oscillatory instability grows to finite amplitude.

horizontal branch star   a stage during the post-main-sequence evolution following the first ascent of the giant branch, characterized by core helium burning. One particular subset of the horizontal branch population are the RR Lyrae stars.

HR diagram	Hertzsprung-Russell diagram. A plot of luminosity versus spectral type. More loosely, a plot of magnitude versus color. In present usage, a plot of bolometric magnitude against effective temperature for a population of stars. Related plots are the color-magnitude plot (absolute or apparent visual magnitude against color index) and the spectrum-magnitude plot (visual magnitude versus spectral type, the original form of the HR diagram).
HST	Hubble Space Telescope.
IDP	interplanetary dust particle, also known as a micrometeoroid or, after entry into the Earth's atmosphere, a micrometeorite.
ilmenite	an iron-titanium oxide with the formula $FeTiO_3$.
IMF	interplanetary magnetic field.
ion microprobe	an analytical instrument in which a finely focused beam of ions ionizes atoms in the sample and ejects them into a mass spectrometer where they may be analyzed.
IRAS	Infrared Astronomical Satellite.
irradiance	the distribution by wavelength of the Sun's photon output. The solar constant is the integral of the irradiance over all wavelengths.
ISEE-3	International Sun Earth Explorer. A spacecraft launched in 1978 to study particles and magnetic fields predominantly from the Sun and the magnetosphere. In 1983 ISEE-3 was moved from its near-Earth orbit to interplanetary space in order to study particles and fields associated with comet Giacobini-Zinner. In this latter role, ISEE-3 was renamed ICE, International Cometary Explorer.
ISM	Interstellar Medium.
ISO	Infrared Space Observatory, a European space telescope to be launched in (approximately) 1993.
jets	geyserlike eruptions on the Sun, causing prominences that rise from 25,000 to 50,000 miles or higher.

joule heating	the dissipation of an electric current as heat in overcoming circuit resistance.
K-dwarfs	main sequence stars that are intermediate in size, mass and temperature between the solar-like G-dwarfs and the cool M-dwarfs.
Keplerian orbit	an orbit within a gravitational field that is described by Kepler's laws; planetary orbits are generally described as Keplerian orbits.
Keplerian velocity	the instantaneous velocity of an object in a Keplerian orbit.
KeV	energy in units of thousands of electron volts.
kinematic viscosity (v)	the ratio of the absolute viscosity coefficient to the mass density.
Landé g value	the constant of proportionality relating the separations of lines of successive pairs of adjacent components of the levels of a spectral multiplet to the larger of the two J-values for the respective pairs. The interval between two successive components J and $J + 1$ is proportional to $J + 1$.
LTE	Local Thermodynamic Equilibrium. The assumption that all the distribution functions characterizing the material and its interaction with the radiation field (but not that of the radiation itself) at a point in the star are given by thermodynamic equilibrium relations at local values of the temperature and density.
luminosity	the integral of the surface flux F over the stellar surface; $4\pi R^2 \sigma T^4$, where R is the radius of a star, T is its effective temperature and σ is the Stefan-Boltzmann constant.
mafic	term used to describe the silicate mineral whose cations are predominantly Mg and/or Fe. It is also used for rocks made up principally of such minerals.
magmatic	a term associated with molten silicate material.

magnetic dy-namo	a mechanism for the cyclic generation of magnetic fields through the interaction of the convection of conducting matter with rotation and rotational shear.
main sequence	the principal sequence of stars on the HR diagram, containing more than 90% of the stars we observe, that runs diagonally from the upper left (high temperature, high luminosity) to the lower right (low temperature, low luminosity). Most of the lifetime of a star is spent on the main sequence.
mantle	the interior zone of a planet or satellite below the crust and above the core, which in the case of the Earth is divided into the upper mantel and the lower mantle with a transition zone in between.
mass spec-trometer	an analytical instrument in which the sample is converted into a beam of ions which can be separated from each other on the basis of their mass-to-charge ratio, generally by a magnetic or electrostatic field, so that the relative proportions of entities of different mass, commonly isotopes, can be determined.
Maunder min-imum	the period roughly between 1645 and 1715, identified with an apparent lack of sunspots and aurorae; named after E. W. Maunder (1851–1928).
M-dwarf	the spectral type designation of the most common type of star in the Galaxy. These dwarfs are smaller, less massive and cooler than the Sun.
mean-life	the mean (average) length of time that a radionuclide in an initial population will take to decrease to $1/e$ of the original amount, i.e., the half-life divided by the natural logarithm of 2.
mesosiderite	a stony-iron meteorite in which the silicates are mainly pyroxene (usually orthopyroxene) and calcic plagioclase. Mesosiderites often appear to be breccias made up of fragments of widely different chemical and mineralogical composition, cemented together by a nickel-iron matrix. Olivine is sometimes present, generally as separately enclosed crystals of fairly large size. From *AGI Glossary* (1987).

mesosphere	the atmospheric region lying above the stratosphere, with a negative temperature gradient. It extends typically from 10^{-3} to 10^{-6} bar for planetary atmospheres.
metamor- phism	solid-state modification of a rock, e.g., recrystallization, caused by elevated temperature (and possibly pressure).
meteorite	a natural object of extraterrestrial origin that survives passage through the atmosphere.
meteoroid	a small object in space that, after surviving passage through the Earth's atmosphere, is called a meteorite.
MHD shocks	magnetohydrodynamic shocks. Shock waves mediated by the magnetohydrodynamic properties of a plasma rather than by molecular collisions alone, as for a shock in a neutral gas. As do all shocks, MHD shocks form when disturbances move through the gas faster than the characteristic wave speed. In many cases (for MHD shocks) the appropriate wave velocity is the Alfvén velocity, which is the speed of one of the most commonly found wave modes in magnetized and ionized gases.
microme- teorite	a small extraterrestrial particle that has survived entry into the Earth's atmosphere. The actual size is not rigorously constrained but is operationally defined by the collection procedure because small particles are more abundant than large ones. In practice, the micrometeorites being studied in the laboratory after collection in the stratosphere are rarely as large as 50 μm.
microme- teoroid	very small meteoroid (radii as small as km), present in fairly large fluxes in space.
Milankovitch theory	the idea that Earth's climate is affected by orbital variations (precession, obliquity and eccentricity).
mixing-length theory	a semi-empirical theory used to describe convection phenomena in stars. It is based on the concept that a convective element of fluid transports its excess energy a characteristic distance (the mixing length) and then dissipates and deposits its energy into the ambient medium.
MWE	mean water equivalent.

naked T Tauri stars	pre-main-sequence stars of approximately the same age and mass as classical T Tauri stars, but with much weaker optical emission lines (and no "blue veiling").
noble gases	the elements He, Ne, Ar, Kr, Xe, and Rn, which are almost always monoatomic and rarely undergo chemical reactions. They are also called rare gases or inert gases.
nuclear stars	inelastic collisions of high-energy particles with nuclei, so-called because of their appearance in a photographic plate.
nuclear track	radiation damage trials in dielectric solids caused by passage of energetic charged particles.
nuclide	a species of atomic nucleus, analogous to the word "isotope" for a species of atom. The word is also used to distinguish between atomic nuclei that are in different energy states.
Nyquist frequency	one half the sampling frequency in an evenly sampled sequence. It determines the frequency of spectral repetition and the frequencies about which spectra appear to fold backwards upon themselves.
oersted (Oe)	unit of magnetic field intensity (mks).
the Older and the Younger Dryas	the exceptionally cold periods during the last glacial epoch (Wisconsin or Würm) that were followed by distinct warmer periods.
olivine	an iron-magnesium silicate with the formula $(Mg,Fe)^2SiO_4$. Olivine is the most abundant mineral in chondritic meteorites.
open cluster	a comparatively loose grouping (mass range 10_2 - 10_3 $M_\odot$) of Population I stars, strongly concentrated in the spiral arms or the disk of the Galaxy (in fact, open clusters give a good indication of where the spiral arms are). Unlike associations, open clusters are dynamically stable. Depending on their age, stars in open clusters "peel off" from the main sequence at different points (the higher the turnoff point, the younger the cluster). Melotte 66 is the oldest known open cluster.

paleointensity	the strength of the ancient magnetic field which was present when a sample acquired its remanent magnetization in nature.
parsec (pc)	parallax second; the distance at which one astronomical unit subtends as angle of 1 second of arc. 1 pc = 206,265 AU = 2.086×10^{13} lightyear.
Paschen continuum	radiation associated with the recombination of free electrons to the $n = 3$ bound level of the hydrogen atom.
Phanerozoic	the era of Earth's history following the Proterozoic.
plage	a bright region near a center of activity (e.g., a sunspot) as observed in emission in monochromatic light of some spectral line (Hα or Ca II). It is a chromospheric phenomenon associated with and often confused with a facula.
plagioclase	an alumino-silicate containing calcium and/or sodium with a formula lying between end-member compositions given by $NaAlSi_3O_8$ and $CaAl_2Si_2O_8$.
plasmoid	a quasi-stable toroidal aggregate of magnetized plasma.
p-mode	a normal mode of oscillation of the Sun in which pressure is the restoring force.
Poisson noise	"white noise" which is distributed according to Poisson statistics, i.e., the distribution function is given by $\omega(x) = (\lambda^x)/x!)\exp(-\lambda)$, where λ is the variance and $\sigma = \sqrt{\lambda}$ is the standard deviation.
population I and II	rough classification system for stars in galaxies. Population I stars are young stars, formed primarily in the disk of spiral galaxies. Population II stars are stars generally formed in the spheroid of spirals or ellipticals, primarily at the time the galaxy formed (hence, these stars are old and generally have low metallicity).
Prandtl number	ratio of the product of the viscosity coefficient and the specific heat at constant pressure to the thermal conductivity.

Precambrian	the period of geological prehistory prior to about 650 million years ago identified with the beginning of the significant fossil record.
pre-main-se-quence stars	young stars still undergoing gravitational contraction prior to their arrival on the main sequence.
primitive	in planetary science and meteoritics, material that is little changed chemically since its formation, hence representative of early solar system conditions.
Proterozoic	an era of Earth's history following the Archean and succeeded by the Phanerozoic; the period from 0.6 to 2.5 billion years after the formation of Earth.
proton	the hydrogen (^1H) nucleus.
pyroxenes	a group of common rock-forming silicates which have ratios of metal oxides (MgO, FeO or CaO) to SiO_2 of 1:1. These are called metasilicates. Pure members of this group are $MgSiO_3$ (enstatite), $FeSiO_3$ (ferrosilite). Pure $CaSiO_3$ does not crystallize with the pyroxene structure. Ca does substitute for up to 50% of the Mg and Fe in the pyroxene structure.
QBO	quasi-biennial oscillation.
R	the ratio of the surface flux in a spectral line F to the bolometric flux σT^4.
radiogenic	that which is made by radioactive decay.
radionuclide	an atomic nucleus that is unstable to radioactive decay, such as ^{26}Al, which decays with a half-life of 7.05 × 10^5 years to ^{26}Mg.
reaction cross section	proportionally constant that relates to the rate that a nucleon will react with a target nucleus to produce a specific product by a nuclear reaction.
Recent	post (last) glacial.
regolith	a fragmented layer found on the surface of many planetary or subplanetary objects. It is created from the local

competent lithologies by meteoroid impact and subsequently comminuted and turned over by such impacts.

remanent magnetism
: the magnetization that a sample retains through time, as measured in a zero magnetic field environment.

Reynold's number
: a dimensionless number ($R = Lv\nu$), where L is a typical dimension of the system, v is a measure of the velocities that prevail, and ν is the kinematic viscosity that governs the conditions for the occurrence of turbulence in fluids.

rigidity (R)
: momentum of a particle per unit charge, pc/ze. The flux of SCR (solar cosmic ray) particles as a function of rigidity is fairly well described as $dF/dR = $ const. $\exp(R/R_o)$, where R_o is a constant describing the spectral shape. For SCR, the units used for R and R_o are megavolts.

ROSAT
: Rötgen Satellite and X-ray Observatory.

Rossby number (R_o)
: the product of the rotation period and the convective overturn time. R_o is inversely proportional to the square root of the dynamo number n_D.

Rosseland mean opacities
: a coefficient of opacity which is a weighted inverse mean of the opacity over all frequencies. It is applied when the optical depth is very large and radiative transport reduces to a diffusion process (e.g., in stellar interiors).

Russell-Vogt theorem
: if the pressure, the opacity, and the energy generation rate are functions of the local values of density, temperature, and the chemical composition only, then the structure of a star is uniquely determined by the mass and the chemical composition of the star. (When isothermal cores occur in the interiors of stars, then multiple-valued solutions become possible.)

Saha equations
: equations determining the number of atoms of a given species in various stages of ionization that exist in an assembly of atoms and electrons in thermal equilibrium.

Schumann-Runge continuum
: a region of strong absorption of O_2, approximately 1300 to 1700 Å.

Schwabe cycle	the 11 year sunspot cycle.
SCLERA	Santa Catalina Laboratory for Experimental Relativity by Astrometry.
SDM	Solar Diameter Monitor; a fully automated photoelectric transit telescope. The horizonal semi-diameter measurement is obtained by timing the duration of the Sun's meridian transit.
Silurian period	the period from 430 to 395 Myr ago.
simple history	case where an object received essentially all of its cosmogenic products in one location and exposure geometry, such as a lunar rock ejected from a great depth and then not moved by another impact event.
SNC meteorites	a group of uncommon but apparently genetically related meteorite types which are highly differentiated (the shergottites, nahklites, and Chassigny). They may originate from Mars.
SNU	*see* solar neutrino unit.
solar cosmic rays	energetic (usually $\gtrsim 1$ MeV but seldom > 100 MeV) particles, mainly protons and electrons, accelerated as a result of very energetic processes on the Sun; also called solar energetic particles or solar-flare-associated particles.
solar flare	an explosive outburst of material on the solar photosphere that is observationally related to local magnetic fields on the Sun.
solar modulation	the process by which the solar wind reduces the intensity of the primary cosmic-ray flux within the heliosphere.
solar neutrino unit (SNU)	1 SNU = 10^{-36} solar-neutrino captures per second per target atom.
solar wind	expansion of the solar corona to form supersonic plasma streaming away from the Sun.

solidus   the lowest temperature at which a solid solution begins
                        to melt.

sonic X-point   a changeover in the physics of a fluid at a point where
                        bulk flow speed equals one of the relevant wave speeds.

spallation   type of nuclear reaction whereby a high-energy hadrons
                        such as a cosmic-ray proton, removes one or more nu-
                        clear fragments from an atomic nucleus, such as in the
                        production of ^{10}Be from oxygen. The residual nucleus of
                        such a reaction is called a spallogenic nuclide.

spallogenic   that which is produced by the violent partial disintegra-
                        tion of an atomic nucleus; in a meteoritic context, this is
                        generally due to cosmic-ray bombardment.

spectral fold-   the folding of a spectrum about the Nyquist frequency.
ing

spectral type   classification of a star according to its luminosity and
                        temperature; see (Henry) Draper system.

Spörer mini-   a period of low sunspot activity in the 15th century
mum   (about A.D. 1420–1570). (Compare Maunder mini-
                        mum.)

standard solar   a theoretical model of the evolution and properties of the
model   Sun, based on physical principles and constrained by as-
(SSM)   sumed initial composition, age and observed parame-
                        ters. Theoretical neutrino fluxes are calculated from the
                        model.

Stefan-   the constant of proportionality relating the radiant flux
Boltzmann   per unit area from a blackbody to the fourth power of its
constant (σ)   absolute temperature: $\sigma = 5.67 \times 10^{-8}$ W m^{-2} K^{-4}.

stellar activity   emission in excess of that expected from a purely radia-
                        tive atmosphere, and indicative of additional heating
                        sources in the stellar atmosphere.

STP   standard temperature (273 K) and pressure (760 mm).

stratopause   the upper boundary of the stratosphere (see Fig. 1 in the
                        chapter by Hunten et al.).

stratosphere	the atmospheric region lying above the troposphere, with a positive temperature gradient. It extends typically from 200 to 1 millibar in planetary atmospheres.
surface flux (F)	usually referring to emittance from the surface of a star in units of erg $cm^{-2}\,s^{-1}$.
Surveyor glass	a glass lens filter from the Surveyor III camera that was returned to Earth by the Apollo 12 astronauts after it had experienced 2.5 years of exposure to the space environment while on the lunar surface.
$t^{-\frac{1}{2}}$	empirically describes the temporal decay of chromospheric emission as a star ages.
tellurian	pertaining to the planet Earth, including its seas and atmosphere; distinguished from "terrestrial," which refers only to the solid part of the planet.
thermopause	the upper boundary of the thermosphere.
thermosphere	region of temperature rise due to ionospheric heating.
tracks	see charged-particle tracks.
trapped gas	gas contained within a solid substance, such as a rock, that was not produced by *in situ* processes such as radioactive decay or cosmic-ray interactions.
T Tauri phase	a period early in the Sun's history when it radiated much more strongly in the far ultraviolet (see Chapter by Walter and Barry).
T Tauri stars	young, late-type stars that are precursors to solar-mass stars. T Tauri stars are characterized by emission line spectra, infrared excesses and irregular variability. The prototype for this class of stars is T Tau. Classical T Tauri (CTT), naked T Tauri (NTT) and weak T Tauri (WTT) are commonly made distinctions of the T Tauri class.
tropopause	upper boundary of the troposphere (about 15 km), where the temperature gradient goes to zero.

troposphere	lowest level of Earth's atmosphere, from zero altitude to about 15 km above the surface. This is the region where most weather occurs. Its temperature decreases from about 290 K to 240 K.
Type II radio emission burst	radio emission produced in the solar atmosphere by MHD shocks.
Type III radio emission burst	radio emission produced in the solar atmosphere by fast electron streams.
Type IV radio emission burst	radio emission produced in the solar atmosphere by trapped relativistic electrons.
veiling (or blue veiling)	term used to describe the anomalously weak absorption lines, particularly in the blue portion of the spectrum, for very active T Tauri stars.
virial theorem	for a bound gravitational system the long-term average of the kinetic energy is one-half of the potential energy.
$v \sin i$	the rotational velocity of a star projected onto the line of sight; v = equatorial rotational velocity of a star and i = inclination of its rotational access with respect to the observer's line of sight.
Vogt-Russel theorem	*see* Russell-Vogt theorem.
Wilson depression	the depth difference between the surface of the sunspot's umbra and that of the surrounding photosphere at the same geometrical depth.
WIMPs	(hypothetical); weakly interacting massive particles.
Wisconsin (Wisconsinan; Würm) glacial	pertaining to the classical fourth glacial stage (and the last definitely ascertained, although there appear to be others) of the Pleistocene epoch in North America, following the Sangamonian interglacial stage and preceding the Holocene. From *AGI Glossary* (1987).

ZAMS   Zero-Age Main Sequence.

Zeeman com-   collectively, the linearly polarized (π) and circularly po-
ponents   larized (σ) components which comprise a line split in the
                       presence of a magnetic field.

Zeeman effect   line broadening due to the influence of magnetic fields.
                       A multiplet of lines is produced, with distinct polariza-
                       tion characteristics.

Zeeman split-   the splitting of atomic spectral lines in a magnetic field.
ting   In the simplest case, denoted as normal Zeeman split-
                       ting, a line splits into three components. One component
                       (the π component) is not displaced, and is linearly po-
                       larized. Two components (the σ components) are shifted
                       by equal amounts to lower and higher frequencies, the
                       magnitude of the shift being proportional to magnetic
                       field strength. In the general case, the σ components are
                       both circularly and linearly polarized.

zenith angle   an angle measured with respect to the vertical direction
                       above the Earth's surface.

zenith dis-   the zenith distance of the Sun (or star) is the angular
tance   distance between the zenith point and the Sun (or the
                       star).

INDEX

INDEX*

Aa index, 216, 335
ACRIM satellite, 12, 13, 14F, 15, 33, 95, 96, 789
Abscissa, 639, 685
Accelerated particles in solar flares, 232
  ancient Sun and, 256–59
  chemical and isotopic abundances, 254–56
  gamma-ray and neutron observations, 236–37, 238–39Fs
  hard X-rays, 239–40, 241–42Fs, 243
  models, 246–54
  radio observations, 243, 244F, 245–46
  space observations, 233–34, 235–36Fs
Accelerator mass spectrometry, 272, 286
Accretion, 683, 720, 734–35, 741
Accretion-disk model, 684, 686, 766, 768
Achondrites, 608, 741, 762, 769
  paleointensities, 779–80, 781F, 782
Activity-decay relations, 633, 648–49, 650T, 651F
Activity-rotation relation, 792
Adiabatic cooling, 205, 210, 318
Adiabatic expansion, 137
Adiabatic transport of energy, 168
Aerosols, 345
Age of Sun, 107–08, 161, 798, 801–05
  See also History of Sun; Magnetic fields, solar; Magnetic history of Sun; Solar wind
Ages of stars, 792, 794, 806–07, 814, 815, 817–18, 846, 847
Agglutinates, 390, 391
Alfvén waves, 150, 590, 594, 695–96, 703
Allende meteorite, 608, 733, 734, 745, 774, 775
Alpha particles, 261, 262, 264, 277–78, 309
  in cosmic rays, 318–19
Aluminum-26, 281–84, 678, 741
  hypothesis regarding, 732–34, 737–38
  planetesimal heating and, 745, 746, 750–52, 760–61
Ambipolar diffusion, 596–97, 610–11, 614
Ammonia, 459, 462

Amplitude modulation, 371–72
Angular momentum, 591, 594, 595, 613–14, 736
  See also Magnetic history of Sun; Solar-mass stars
Angular momentum loss, 815, 821, 829
Anhydrite system, 555–57
Antarctic meteorites, 776, 778
Archimedian spiral, 177
Arcturus, 844
Argon, 280, 292F, 293T, 294
Argon-40, 307, 391, 393, 395, 404, 405
Asteroid belt, 730, 757
Asteroids, 679, 711–12
  electromagnetic heating in
    size, 719, 720F, 721–22
    taxonomic types, 713F, 714, 715T, 716, 717F
  heliocentric distance distribution, 718, 719F
Astrolabe, 68, 69F, 72F
Atmosphere and solar irradiation, 463
  ancient atmosphere response, 483–87, 488–89Fs, 490–91, 492F, 493
  penetration depths of wavelengths, 465
  present atmosphere response, 476–82
  radiation-atmosphere interactions, 466–67
  solar source variability, 494–97
    energetic particles, 472–74
    PMS phase, 474–75
    radiative component, 468–69
    ultraviolet radiation, 469–70, 471F, 472
  traits of solar irradiation, 464–65
Atmospheric cosmic-ray propagation, 321–23
Atmospheric filter. See Climate of Earth
"Attractor," 35–37, 629
Aurorae, 175, 176, 190, 191, 193, 194, 196, 199, 386, 465
Auroral zone, 473, 474
Autoregressive order (AR), 369–370
Azimuthal diffusion, 253, 254
Azimuthal magnetic field, 610

_____
* See also Table of Contents

[979]